Lecture Notes in Computer Science 4403

Commenced Publication in 1973
Founding and Former Series Editors:
Gerhard Goos, Juris Hartmanis, and Jan van Leeuwen

Shigeru Obayashi   Kalyanmoy Deb
Carlo Poloni   Tomoyuki Hiroyasu
Tadahiko Murata (Eds.)

Evolutionary Multi-Criterion Optimization

4th International Conference, EMO 2007
Matsushima, Japan, March 5-8, 2007
Proceedings

Springer

Volume Editors

Shigeru Obayashi
Tohoku University
Sendai 980-8577, Japan
E-mail: obayashi@ieee.org

Kalyanmoy Deb
Indian Institute of Technology
Kanpur, PIN 208 016, India
E-mail: deb@iitk.ac.in

Carlo Poloni
University of Trieste
34142 Trieste, Italy
E-mail: poloni@units.it

Tomoyuki Hiroyasu
Doshisha University
Kyoto 610-0321, Japan
E-mail: tomo@is.doshisha.ac.jp

Tadahiko Murata
Kansai University, Osaka 569-1095, Japan
E-mail: murata@res.kutc.kansai-u.ac.jp

Library of Congress Control Number: 2007921125

CR Subject Classification (1998): F.2, G.1.6, G.1.2, I.2.8

LNCS Sublibrary: SL 1 – Theoretical Computer Science and General Issues

ISSN 0302-9743
ISBN-10 3-540-70927-4 Springer Berlin Heidelberg New York
ISBN-13 978-3-540-70927-5 Springer Berlin Heidelberg New York

Springer is a part of Springer Science+Business Media

springer.com

Typesetting: Camera-ready by author, data conversion by Scientific Publishing Services, Chennai, India
Printed on acid-free paper SPIN: 12021017 06/3142 5 4 3 2 1 0

Preface

Multicriterion optimization refers to problems with two or more objectives (normally in conflict with each other) which must be simultaneously satisfied. Evolutionary algorithms have been used for solving multicriterion optimization problems for over two decades, gaining an increasing attention from industry.

The 4th International Conference on Evolutionary Multi-criterion Optimization (EMO2007) was held during March 5–8, 2007, in Matsushima/Sendai, Japan. This was the fourth international conference dedicated entirely to this important topic, following the successful EMO 2001, EMO 2003 and EMO 2005 conferences, which were held in Zürich, Switzerland in March 2001, in Faro, Portugal in April 2003, and in Guanajuato, México in March 2005. EMO2007 was hosted by the Institute of Fluid Science, Tohoku University. EMO2007 was co-hosted by the Graduate School of Information Sciences, Tohoku University, the Japan Aerospace Exploration Agency (JAXA), and the Policy Grid Computing Laboratory, Kansai University.

The EMO2007 scientific program included four keynote speakers: Hirotaka Nakayama on aspiration level methods, Kay Chen Tan on large and computationally intensive real-world MO optimization problems, Carlos Fonseca on decision making, and Gary B. Lamont on design of large-scale network centric systems.

In response to the call for papers, 124 papers from 30 countries were submitted, each of which was independently reviewed by at least three members of the Program Committee. This volume contains the 65 papers that were accepted for presentation at the conference, together with contributions based on the invited talks. It is worth noting that the number of submissions to the EMO conference has steadily increased over the years. For EMO 2001, 87 papers were submitted (from which 45 were accepted). For EMO 2003, 100 papers were submitted (from which 56 were accepted). For EMO 2005, 115 papers were submitted (from which 59 were accepted). This is a clear indication of the growing interest in this research field.

We would like to express our appreciation to the keynote speakers for accepting our invitation. We thank all the authors who submitted their work to EMO 2007, and the members of the Program Committee for their thorough reviews. We wish to thank the Air Force Office of Scientific Research, Asian Office of Aerospace Research and Development for their contribution to the success of this conference. The organizers are particularly thankful to industrial sponsors, CD-adapco JAPAN Co., Ltd., Engineous Japan, Inc. and Honda Research Institute Japan Co., Ltd. for Dinner Sponsorship, Itochu Techno-Solutions Corporation and Sumisho Computer Systems Corporation for Lunch Sponsorship, BestSystems Co., Ltd. Fujitsu Limited, Hitachi, Ltd., Mitsubishi Heavy Industries, Ltd., SGI Japan, Ltd., for Refreshment Sponsorship, and Honda Research Institute Europe GmbH., Platform Computing Inc. and Microsoft Co., Ltd. for Student Support Sponsorship.

We also thank Alfred Hofmann and Ronan Nugent of Springer for their continued support in publishing EMO proceedings.

March 2007

Shigeru Obayashi
Carlo Poloni
Kalyanmoy Deb
Tomoyuki Hiroyasu
Tadahiko Murata

Organization

EMO2007 was co-hosted by the Graduate School of Information Sciences, Tohoku University, the Japan Aerospace Exploration Agency (JAXA), and the Policy Grid Computing Laboratory, Kansai University.

General Chairs

Shigeru Obayashi	IFS, Tohoku University, Japan
Carlo Poloni	University of Trieste, Italy
Kalyanmoy Deb	IIT Kanpur, India

International Program Committee

Chair

Tadahiko Murata	Kansai University, Japan

Co-chairs

Hernan Aguirre	Shinshu University, Japan
Hisao Ishibuchi	Osaka Prefecture University, Japan
Tatsuya Okabe	Honda Research Institute, ATR, Japan

Members

Hussein Abbass	University of New South Wales, Australia
Hernan Eduardo Aguirre Duran	Shinshu University, Japan
Robert T. F. Ah King	University of Mauritius, Mauritius
Enrique Alba	University of Malaga, Spain
Shapour Azarm	University of Maryland, College Park, USA
Luigi Barone	The University of Western Australia, Australia
Matthieu Basseur	University of Sciences and Technologies of Lille, France
Stefan Bleuler	ETH Zurich, Switzerland
Juergen Branke	University of Karlsruhe, Germany
Carlos Alberto Brizuela	CICESE, Mexico
Dirk Bueche	University of Applied Sciences Northwestern Switzerland, Switzerland
Carlos A. Coello Coello	CINVESTAV-IPN, Mexico
David W. Corne	Heriot-Watt University, UK
Lino Costa	University of Minho, Portugal
Kalyanmoy Deb	Indian Inst. of Technology Kanpur, India
Rolf Drechsler	University of Bremen, Germany

Matthias Ehrgott	The University of Auckland, New Zealand
Daryl Essam	Australian Defence Force Academy, Australia
Richard Everson	University of Exeter, UK
Jonathan Fieldsend	University of Exeter, UK
Carlos M. Fonseca	Universidade do Algarve, Portugal
Xavier Gandibleux	University of Nantes, France
António Gaspar-Cunha	IPC-University of Minho, Portugal
Christian Grimme	Robotics Research Institute Dortmund, Germany
Thomas Hanne	Fraunhofer Institute for Industrial Mathematics (ITWM), Germany
Martina Hasenjaeger	Honda Research Institute Europe, Germany
Christian Haubelt	University of Erlangen-Nuremberg, Germany
Arturo Hernandez-Aguirre	Centre for Research in Mathematics (CIMAT), Mexico
Malcolm I. Heywood	Dalhousie University, Canada
Philip Hingston	Edith Cowan University, Australia
Tomoyuki Hiroyasu	Doshisha University, Japan
Evan J Hughes	Cranfield University, UK
Christian Igel	Ruhr University Bochum, Germany
Hisao Ishibuchi	Osaka Prefecture University, Japan
Andrzej Jaszkiewicz	Poznan University of Technology, Poland
Shinkyu Jeong	Tohoku University, Japan
Yaochu Jin	Honda Research Institute Europe, Germany
Hajime Kita	Academic Center for Computing and Media Studies, Kyoto University, Japan
Joshua Knowles	MIB, University of Manchester, UK
Rajeev Kumar	Indian Institute of Technology Kharagpur, India
Mary E. Kurz	Clemson University, USA
Gary B. Lamon	Air Force Institute of Technology, USA
Dario Landa Silva	University of Nottingham, UK
Marco Laumanns	ETH Zurich, Switzerland
Xiaodong Li	RMIT University, Australia
José Antonio Lozano	University of the Basque Country, Spain
Carlos Mariano	Mexican Institute for Water Technology, Mexico
Efren Mezura-Montes	National Laboratory of Advanced Informatics (LANIA), Mexico
Sanaz Mostaghim	University of Karlsruhe, Germany
Tadahiko Murata	Kansai University, Japan
Kaname Narukawa	Honda R&D, Japan
Antonio J. Nebro	University of Malaga, Spain

Yusuke Nojima	Osaka Prefecture University, Japan
Shigeru Obayashi	Tohoku University, Japan
Tatsuya Okabe	Honda Research Institute Japan and Advanced Telecommunications Research Institute International, Japan
Johan Olvander	Linköping University, Sweden
Isao Ono	Tokyo Institute of Technology, Japan
Andrzej Osyczka	AGH University of Science and Technology, Poland
Akira Oyama	Japan Aerospace Exploration Agency, Japan
Luis Paquete	University of Algarve, Portugal
Geofferey Parks	University of Cambridge, UK
Valentino Pediroda	Department of Mechanical Engineering, University of Trieste, Italy
Silvia Poles	Esteco, Italy
Robin Purshouse	PA Consulting Group, UK
Ranji S. Ranjithan	North Carolina State University, USA
Patrick Reed	The Pennsylvania State University, USA
Margarita Reyes-Sierra	CINVESTAV-IPN, México
Peter Rockett	University of Sheffield, UK
Katya Rodríguez-Vázquez	IIMAS-UNAM, México
Jun Sakuma	Tokyo Institute of Technology, Japan
Dragan Savic	University of Exeter, UK
Mark Savill	Cranfield University and Girtoon College, Cambridge, UK
J. David Schaffer	Philips Research, USA
Hartmut Schmeck	Universitaet Karlsruhe, Germany
Detlef Seese	Universitaet Karlsruhe, Germany
Bernhard Sendhoff	Honda Research Institute Europe, Germany
Patrick Siarry	University Paris 12 (LiSSi), France
Thomas Stützle	Université Libre de Bruxelles, Belgium
El-Ghazali Talbi	University of Lille, France
Kay Chen Tan	National University of Singapore, Singapore
Kiyoshi Tanaka	Shinshu University, Japan
Jürgen Teich	University of Erlangen-Nuremberg, Germany
Jose Antonio Tenreiro Machado	Instituto Superior de Engenharia do Porto, Portugal
Lothar Thiele	ETH Zürich, Switzerland
Dirk Thierens	Utrecht University, The Netherlands
Ashutosh Tiwari	North Carolina State University, USA
Andrea Toffolo	University of Padova, Italy
Holgeer Ulmer	Robert Bosch GmbH, Germany
Godfrey Walters	University of Exeter, UK
Shinya Watanabe	Ritsumeikan University, Japan

Lyndon While	The University of Western Australia, Australia
Gary G. Yen	Oklahoma State University, USA
Yeboon Yun	Kagawa University, Japan
Andreas Zell	University of Tuebingen, Germany
Eckart Zitzler	ETH Zurich, Switzerland

Local Organizing Committee

Chair

Shigeru Obayashi	IFS, Tohoku University, Japan

Secretariat

Tomoyuki Hiroyasu	Doshisha University, Japan

Members

Hernán Aguirre	Shinshu University, Japan
Tetsushi Higashimura	Cd Adapco Japan, Japan
Hisao Ishibuchi	Osaka Prefecture University, Japan
Shinkyu Jeong	IFS, Tohoku University, Japan
Keiji Kudo	Engineous Japan, Japan
Tadahiko Murata	Kansai University, Japan
Kaname Narukawa	Honda R&D, Japan
Tatsuya Okabe	HRI-JP, ATR, Japan
Akira Oyama	JAXA-ISAS, Japan
Jun Sakuma	Tokyo Institute of Technology, Japan
Koji Shimoyama	IFS, Tohoku University, Japan
Tetsuya Uchimoto	IFS, Tohoku University, Japan
Shinya Watanabe	Ritsumeikan University, Japan
Kazuomi Yamamoto	JAXA-ISTA, Japan

EMO Steering Committee

David Corne	Heriot-Watt University, UK
Kalyanmoy Deb	IIT Kanpur, India
Peter J. Fleming	University of Sheffield, UK
Carlos Fonseca	Universidade do Algarve, Portugal
J. David Schaffer	Philips Research, USA
Lothar Thiele	ETH Zürich, Switzerland
Eckart Zitzler	ETH Zürich, Switzerland

Acknowledgements

Keynote Speakers

We thank the keynote speakers for their talks given at the conference.

Hirotaka Nakayama	Konan University, Japan
Kay Chen Tan	National University of Singapore, Singapore
Carlos Fonseca	Universidade do Algarve, Portugal
Gary B. Lamont	Air Force Institute of Technology, USA

Local Sponsors

Support by the following organizations and companies is gratefully acknowledged.

Sponsors

Institute of Fluid Science, Tohoku University
Graduate School of Information Sciences, Tohoku University
Japan Aerospace Exploration Agency (JAXA)
Policy Grid Computing Laboratory, Kansai University

Co-sponsors

Air Force Office of Scientific Research, Asian Office of Aerospace Research and Development (AFOSR/AOARD)+
Inoue Foundation for Science
The Kajima Foundation

Cooperation Companies

BESTSYSTEMS Co., Ltd.
Cd-Adapco Japan Co., Ltd.
Cray Japan Inc.
Engineous Japan, Inc.
Fujitsu Limited
Hitachi, Ltd.
Honda Research Institute Japan Co., Ltd.
Honda Research Institute Europe GmbH.
ITOCHU Techno-Solutions Corporation
Microsoft Co., Ltd.

+ AFOSR/AOARD support is not intended to express or imply endorsement by the U.S. Federal Government.

Mitsubishi Heavy Industries, Ltd.
Platform Computing Inc.
SGI Japan, Ltd.
Sumisho Computer Systems Corporation

Table of Contents

Invited Talks

Algorithm Design

Algorithm Improvements

Alternative Methods

Applications

Engineering Design

Many Objectives

Objective Handling

Performance Assessments

Aspiration Level Methods in Interactive Multi-objective Programming and Their Engineering Applications
(Abstract of Invited Talk)

Hirotaka Nakayama

Konan University
8-9-1, Okamoto, Higashinada-ku, Kobe Hyogo, 658-8501, Japan
nakayama@konan-u.ac.jp

Abstract. One of the most important tasks in multi-objective optimization is "trade-off analysis" which aims to make the total balance among objective functions. The trade-off relation among alternatives can be shown as Pareto frontier. In cases with two or three objective functions, the set of Pareto optimal solutions in the objective function space (i.e., Pareto frontier) can be depicted relatively easily. Seeing Pareto frontiers, we can grasp the trade-off relation among objectives totally. Therefore, it would be the best way to depict Pareto frontiers in cases with two or three objectives. (It might be difficult to read the trade-off relation among objectives with three dimension, though). In cases with more than three objectives, however, it is impossible to depict Pareto frontier. There are some cases with a large number (e.g., a few hundreds) of objective functions in engineering applications such as erection management of cable stayed bridges and optical lens design. Under this circumstance, interactive methods can help decision makers (DMs) to make local trade-off analysis through interaction of DMs and computers by showing a Pareto solution nearest to their desire. Along this line, aspiration level methods were developed, and have been observed to be effective in many practical problems in various fields. Satisficing Trade-off Method proposed by the author is one of aspiration level methods, and has several devices for making trade-off analysis easily, i.e., automatic trade-off and exact trade-off. This paper discusses those methods for multi-objective optimization, in particular, from a viewpoint of engineering application.

S. Obayashi et al. (Eds.): EMO 2007, LNCS 4403, p. 1, 2007.

Improving the Efficacy of Multi-objective Evolutionary Algorithms for Real-World Applications
(Abstract of Invited Talk)

Kay Chen Tan

National University of Singapore
4 Engineering Drive 3, Singapore 117576
eletankc@nus.edu.sg

Abstract. Multi-objective evolutionary algorithms (MOEAs) are a class of stochastic optimization techniques that simulate biological evolution to solve problems with multiple objectives. Multi-objective (MO) optimization is a challenging research topic because it involves the simultaneous optimization of several (and normally conflicting) objectives in the Pareto optimal sense. It requires researchers to address many issues that are unique to MO problems, such as fitness assignment, diversity preservation, balance between exploration and exploitation, elitism and archiving. In this talk, a few advanced features for handling large and computationally intensive real-world MO optimization problems will be presented. These include a distributed cooperative coevolutionary approach to handle large-scale problems via a divide-and-conquer strategy by harnessing technological advancements in parallel and distributed systems and a hybridization scheme with local search heuristics for combinatorial optimization with domain knowledge. The talk will also discuss the application of these techniques to various engineering problems including scheduling and system design, which often involve different competing specifications in a large and highly constrained search space.

S. Obayashi et al. (Eds.): EMO 2007, LNCS 4403, p. 2, 2007.

Decision Making in Evolutionary Optimization
(Abstract of Invited Talk)

Carlos M. Fonseca

Universidade do Algarve
Campus de Gambelas, 8005-139 Faro, Portugal
cmfonsec@ualg.pt

Abstract. Current evolutionary multiobjective optimization (EMO) approaches tend to emphasize the approximation of the Pareto-optimal front as a whole, thereby dissociating the optimization process from the selection of the final compromise solution by a decision maker. This has the advantage of removing subjective preference information from the optimization problem formulation, but it also makes the resulting problem computationally more demanding. In order to concentrate the search effort on the regions of potential interest to the decision maker, techniques for the progressive articulation of preferences in EMO have been proposed, casting EMO as the interaction between an evolutionary search mechanism and a decision maker. It is worth noting that even the promotion of diversity across the Pareto-optimal front, which is generally regarded as an optimizer design issue, may be successfully addressed by the decision maker within this framework, as it has been proposed recently by others. Regarding the evolutionary search mechanism, the main question at each iteration consists of determining the next candidate solution(s) to be evaluated, given the information acquired since the beginning of the run. This may be seen as another decision-making problem, but one with (very) incomplete attribute information, since objective values are generally not known for most potential alternatives. Alternatively, it may be seen as a control problem, where actions (new solutions) are to be selected based on the feedback provided by the decision maker. Either way, some model, however weak, of the underlying optimization problem must be assumed. In this talk, both the evaluation of current solutions and the generation of new candidate solutions in EMO will be discussed from a decision making perspective. From the discussion, opportunities for incorporating more explicit decision making in EMO will be identified.

S. Obayashi et al. (Eds.): EMO 2007, LNCS 4403, p. 3, 2007.

MOEAs in the Design of Network Centric Systems
(Abstract of Invited Talk)

Gary B. Lamont

Air Force Institute of Technology
2950 Hobson Way, Wright-Patterson AFB, Dayton, OH 45433-7765, USA
Gary.Lamont@afit.edu

Abstract. Advances in information and communications technology are changing network design techniques quantitatively and qualitatively. This technology is supporting the design of large scale network centric systems which are required in many contemporary real-world situations. These high-level robust centric systems by definition must provide improved information sharing and collaboration between network entities. Such systems enhance the quality of information awareness, improving sustainability, and mission effectiveness and efficiency. The hierarchical development of network centric systems includes all dynamic information elements and is applied so as to maximize the desired decision and action impact. Associated network information flow problems can have as objectives costs, delays, robustness, vulnerability, and reliability with related constraints of network flow capacities, rates, and quantities of information. The optimization of coupled complex capacitated network flow problems is therefore an integral and basic element of network centric systems design. Thus, the focus of the discussion is on the efficacy of multiobjective evolutionary algorithms (MOEAs) to solve effectively and efficiency variations of associated network flow problems, given sophisticated mathematical models. Also to be addressed are dynamic network environments where various information channels become non-available, change their characteristics, or information priorities are modified. Discrimination between possible MOEA operators (recombination, mutation, selection) and associated MOEA parameter values is discussed as related to solving effectively variations of multiobjective network centric information flow problems including real-time behavior. Example network flow applications provide insight to choosing appropriate MOEA characteristics. Included is a discussion of opportunities for future MOEA research in this arena.

S. Obayashi et al. (Eds.): EMO 2007, LNCS 4403, p. 4, 2007.

Controlling Dominance Area of Solutions and Its Impact on the Performance of MOEAs

Hiroyuki Sato, Hernán E. Aguirre, and Kiyoshi Tanaka

Shinshu University, Faculty of Engineering
4-17-1 Wakasato, Nagano, 380-8553 JAPAN
{sato@iplab., ahernan@, ktanaka@}shinshu-u.ac.jp

Abstract. This work proposes a method to control the dominance area of solutions in order to induce appropriate ranking of solutions for the problem at hand, enhance selection, and improve the performance of MOEAs on combinatorial optimization problems. The proposed method can control the degree of expansion or contraction of the dominance area of solutions using a user-defined parameter S. Modifying the dominance area of solutions changes their dominance relation inducing a ranking of solutions that is different to conventional dominance. In this work we use 0/1 multiobjective knapsack problems to analyze the effects on solutions ranking caused by contracting and expanding the dominance area of solutions and its impact on the search performance of a multi-objective optimizer when the number of objectives, the size of the search space, and the complexity of the problems vary. We show that either convergence or diversity can be emphasized by contracting or expanding the dominance area. Also, we show that the optimal value of the area of dominance depends strongly on all factors analyzed here: number of objectives, size of the search space, and complexity of the problems.

1 Introduction

Multiobjective evolutionary algorithms (MOEAs) [1,2] are being increasingly investigated for solving multiobjective optimization problems. MOEAs are particularly suitable for this task because they evolve simultaneously a population of potential solutions to the problem in hand, which allows us to search a set of Pareto non-dominated solutions in a single run of the algorithm.

Some important features of the latest generation MOEAs are that selection incorporates elitism and it is biased by Pareto dominance and a diversity preserving strategy in objective space. Pareto dominance based selection is thought to be effective for problems with convex and non-convex fronts and has been successfully applied, especially in two and three objective problems. However, some current research reveals that ranking by Pareto dominance on problems with an increased number of objectives might not longer be effective [3,4,5]. It has been shown that the characteristics of multiobjective landscapes viewed in terms of non-dominated fronts (that are found in the process of non-domination sorting) can change drastically as the number of objectives increases, i.e. the

S. Obayashi et al. (Eds.): EMO 2007, LNCS 4403, pp. 5–20, 2007.

number of fronts reduces substantially and become denser (more solutions per front) just by increasing the number of objectives [5]. In this case, most sampled solutions at a given time turn to be non-dominated. That is, most solutions are assigned the same rank of non-dominance and Pareto selection weakens since it has to discriminate mostly based on diversity of solutions. Another factor that affects the density of the fronts is the complexity of the individual single objective landscapes. It has been shown that the top non-dominated fronts become denser as the complexity of the landscapes reduces, and vice-versa [5]. This has been observed for multiple and many objectives landscapes and affects the behavior and effectiveness of Pareto selection in two ways. First, although the effect of the landscapes complexity on front density is not as strong as the effect of increasing the number of objectives, in practice the increased density of the top non-dominated fronts combined with elitism could make the instantaneous elite-population to be mostly composed of individuals with the same non-domination rank since early generations. In this case, again, selection has to rely mostly on diversity rather than on Pareto dominance ranking. Second, on problems of increased complexity could happen that there are too many but sparse fronts, in which case Pareto selection could become too strong increasing the likelihood that the algorithm gets trapped in local fronts. These studies suggest that for selection to be effective a more careful analysis of Pareto dominance relation is required when dealing with problems that have more than three objectives. In addition, for any number of objectives, the dominance relation should be appropriately revised according to the characteristics of the multi-objective landscape.

There are a few works on relaxed forms of Pareto dominance, such as ϵ-dominance [6] and α-domination [7]. ϵ-dominance acts as an archiving strategy and was proposed as a way of regulating convergence of a MOEA. The algorithm maintains a finite-size archive of non-dominated solutions, in which new points are only accepted if they are not ϵ-dominated by any other point of the current archive. ϵ-dominance strengthens selection during the archiving process. On the other hand, α-domination permits a solution $\boldsymbol{x}$ to dominate a solution $\boldsymbol{y}$ if $\boldsymbol{x}$ is slightly inferior to $\boldsymbol{y}$ in an objective but largely superior to $\boldsymbol{y}$ in some other objectives. α-domination was tried on an ad hoc continuous problem created specifically to illustrate a potential problem that Pareto selection could face. In addition, α-domination only introduces a method to strengthen selection and its effects have not been explained nor tested on standard test suit problems.

In this work, we propose a method to control the dominance area of solutions in order to induce appropriate ranking of solutions for the problem at hand, enhance selection, and improve the performance of MOEAs on combinatorial optimization problems. The proposed method can control the degree of expansion or contraction of the dominance area of solutions using a user-defined parameter S. Modifying the dominance area of solutions changes their dominance relation inducing a ranking of solutions that is different to conventional dominance. Contrary to ϵ-dominance and α-domination, the proposed method can strengthen or weaken selection by expanding or contracting the area of dominance and conceptually can be considered as a generalization of Pareto dominance. In addition, the motivation and method itself of the proposed approach is different to ϵ-dominance and α-domination. See **3** and **4** for a detailed explanation about ϵ-dominance, α-domination, and the proposed method.

In this work we analyze the effects on solutions ranking caused by contracting and expanding the dominance area of solutions and its impact on the search performance of a multi-objective optimizer when the number of objectives, the size of the search space, and the complexity of the problems vary. We chose NSGA-II as a representative elitist algorithm that uses dominance [8] and compare its performance with NSGA-II enhanced by the proposed method. We conduct our study on 0/1 multiobjective knapsack problems with $m = \{2, 3, 4, 5\}$ objectives varying the number of items n (size of search space is given by 2^n) and the feasibility ratio ϕ of the search space, which is a good indicator of the complexity of the landscapes in this kind of problems. This work clearly shows that either convergence or diversity can be emphasized by contracting or expanding the dominance area. Also, this work shows that the optimal value of S^* that controls the area of dominance depends strongly on all factors analyzed here: number of objectives, size of the search space, and complexity of the problems.

2 Multiobjective Optimization Concepts and Definitions

A multiobjective optimization problem including m kinds of objective functions is defined as follows:

$$\begin{cases} \text{Maximize } \boldsymbol{f}(\boldsymbol{x}) = (f_1(\boldsymbol{x}), f_2(\boldsymbol{x}), \ldots, f_m(\boldsymbol{x})) \\ \text{subject to } \boldsymbol{x} \in \mathcal{F} \end{cases} \tag{1}$$

where, $\boldsymbol{x} \in \mathcal{F}$ is a feasible solution vector in the solution space $\mathcal{S}(\mathcal{F} \subseteq \mathcal{S})$, and $f_i (i = 1, 2, \cdots, m)$ are the m objectives to be maximized. That is, we try to find a feasible solution vector $\boldsymbol{x} \in \mathcal{F}$ in the solution space maximizing each objective function $f_i (i = 1, 2, \ldots, m)$ in a vector fitness function $\boldsymbol{f}$. Important concepts used in determining a set of solutions for multiobjective optimization problems are dominance, Pareto optimality, Pareto set and Pareto front. Next we define *dominance* between solutions $\boldsymbol{x}, \boldsymbol{y} \in \mathcal{F}$ as follows: If

$$\begin{aligned} &\forall i \in \{1, 2, \ldots, m\} : f_i(\boldsymbol{x}) \geq f_i(\boldsymbol{y}) \ \wedge \\ &\exists i \in \{1, 2, \ldots, m\} : f_i(\boldsymbol{x}) > f_i(\boldsymbol{y}). \end{aligned} \tag{2}$$

are satisfied, $\boldsymbol{x}$ dominates $\boldsymbol{y}$. In the following, $\boldsymbol{x}$ dominates $\boldsymbol{y}$ is denoted by $\boldsymbol{f}(\boldsymbol{x}) \succeq \boldsymbol{f}(\boldsymbol{y})$. A solution vector $\boldsymbol{x}$ is said to be *Pareto optimal* with respect to $\mathcal{F}$ if it is not dominated by other solution vectors in $\mathcal{F}$. The presence of multiple objective functions, usually conflicting among them, gives rise to a set of optimal solutions. The set of Pareto optimal solutions (POS) is defined as

$$\mathcal{POS} = \{\boldsymbol{x} \in \mathcal{F} \mid \neg\exists \boldsymbol{y} \in \mathcal{F} : \boldsymbol{f}(\boldsymbol{x}) \succeq \boldsymbol{f}(\boldsymbol{y})\}, \tag{3}$$

and the Pareto front is defined as

$$\mathcal{F}ront = \{\boldsymbol{f}(\boldsymbol{x}) \mid \boldsymbol{x} \in \mathcal{POS}\}. \tag{4}$$

A convenient method to assign rank to solutions is by classifying them into non-dominated fronts [8]. Let us denote $\mathcal{Z}$ the set of solution we want to classify. The first front $\mathcal{F}ront_1$ is obtained from $\mathcal{Z}$ and corresponds to the set of POS in $\mathcal{Z}$. Let us denote this set as $\mathcal{POS}_1$. The subsequent fronts $\mathcal{F}ront_j; j > 1$, contain lower level non-dominated solutions and are obtained by disregarding solutions corresponding to the previous higher non-dominated fronts, i.e. $\mathcal{F}ront_j; j > 1$, is obtained from the set $\mathcal{Z} - \bigcup_{k=1}^{j-1} \mathcal{POS}_k$.

3 Related Works

Recently, some researchers have proposed the use of relaxed forms of Pareto dominance as a way of regulating convergence of a MOEA. Laummans et al. [6] proposed a relaxed form of dominance for MOEAs called ϵ-dominance seeking to ensure both properties of convergence towards the Pareto-optimal set and properties of diversity among the solutions found. A solution $\boldsymbol{x}$ ϵ-dominates a solution $\boldsymbol{y}$ for some $\epsilon > 0$, assuming maximization in all objectives, if

$$\forall i \in \{1, 2, \ldots, m\} : (1 + \epsilon) \cdot f_i(\boldsymbol{x}) \geq f_i(\boldsymbol{y}). \tag{5}$$

ϵ-dominance acts as an archiving strategy, where new points are only accepted if they are not ϵ-dominated by any other point of the current archive. Thus, it strengthens Pareto selection during the archiving process. In addition, ϵ-dominance uses a set of boxes to cover the Pareto front, where the size of such boxes is set by the user-defined parameter ϵ. Within each box only one non-dominated solution is retained. Thus, by using a larger value of ϵ the user can accelerate convergence, while sacrificing the quality (preciseness) of the Pareto front obtained. In contrast, if a high quality of the front is required, a small value of ϵ must be adopted. The definition of ϵ is very important. However, it is not simple to find the most appropriate value of ϵ, especially if nothing is known in advance about the shape of the Pareto front. Also, to correlate the number of desired solutions with the value of ϵ chosen is not easy. In addition, ϵ-dominance eliminates the extreme points of the Pareto front, which may be undesirable in some cases.

Another strategy that relaxes Pareto dominance is α-domination proposed by Ikeda et al. [7] to strengthen selection. The fundamental idea of α-domination is setting upper/lower bounds of trade-offs rates between two objectives. α-domination permits a solution $\boldsymbol{x}$ to dominate a solution $\boldsymbol{y}$ if $\boldsymbol{x}$ is slightly inferior to $\boldsymbol{y}$ in an objective but largely superior to $\boldsymbol{y}$ in some other objectives. To calculate α-dominance, first a relative fitness vector $\boldsymbol{g}(\boldsymbol{x}, \boldsymbol{y})$ between two solutions must be established. The i-th component of $\boldsymbol{g}(\boldsymbol{x}, \boldsymbol{y})$ is calculated by

$$g_i(\boldsymbol{x}, \boldsymbol{y}) = f_i(\boldsymbol{x}) - f_i(\boldsymbol{y}) + \sum_{j \neq i}^{m} \alpha_{ij}(f_i(\boldsymbol{x}) - f_i(\boldsymbol{y})) \tag{6}$$

where $f_i(\boldsymbol{x})$ is the fitness value of solution $\boldsymbol{x}$ on the i-th objective, and α_{ij} is the trade-off rate between the i-th and j-th objectives.

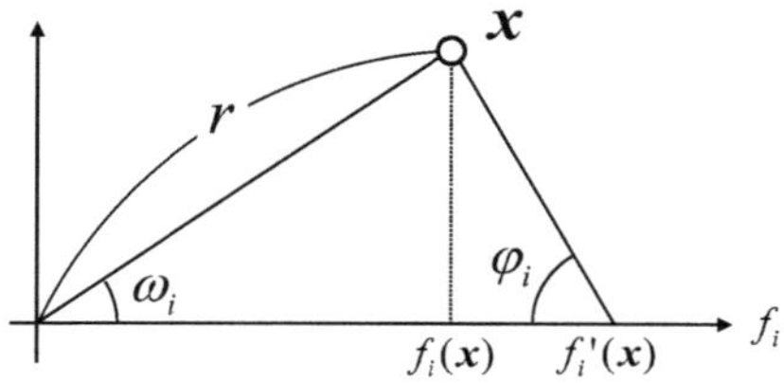

Fig. 1. Fitness modification to change the covered area of dominance

A solution $\boldsymbol{x}$ α-dominates a solution $\boldsymbol{y}$, assuming maximization in all objectives, if

$$\begin{aligned} &\forall i \in \{1, 2, \ldots, m\} : g_i(\boldsymbol{x}, \boldsymbol{y}) \geq 0 \ \wedge \\ &\exists i \in \{1, 2, \ldots, m\} : g_i(\boldsymbol{x}, \boldsymbol{y}) > 0. \end{aligned} \tag{7}$$

To calculate α-domination, α_{ij} trade-off rates must be properly set for each pair of objectives. Assessing the appropriate trade-offs between objectives could be a difficult problem, especially if nothing is known in advance about the landscape and shape of Pareto front. In addition, note that α-domination strengthens selection only.

4 Proposed Method

4.1 Contraction and Expansion of Dominance Area

In this work, we try to control the covered area of dominance. Normally, the dominance area is uniquely determined with a fitness vector $\boldsymbol{f}(\boldsymbol{x}) = (f_1(\boldsymbol{x}), f_2(\boldsymbol{x}), \cdots, f_m(\boldsymbol{x}))$ in the objective space when a solution $\boldsymbol{x}$ is given. To contract and expand the dominance area of solutions, we modify fitness value for each objective function by changing the user defined parameter S_i in the following equation

$$f_i'(x) = \frac{r \cdot \sin(\omega_i + S_i \cdot \pi)}{\sin(S_i \cdot \pi)} \qquad (i = 1, 2, \cdots, m) \tag{8}$$

where $\varphi_i = S_i \cdot \pi$. This equation is derived from the Sine theorem. We illustrate the fitness modification in **Fig. 1**, where r is the norm of $\boldsymbol{f}(\boldsymbol{x})$, $f_i(\boldsymbol{x})$ is the fitness value in the i-th objective, and ω_i is the declination angle between $\boldsymbol{f}(\boldsymbol{x})$ and $f_i(\boldsymbol{x})$. In this example, the i-th fitness value $f_i(\boldsymbol{x})$ is increased to $f_i'(\boldsymbol{x}) > f_i(\boldsymbol{x})$ by using $\varphi_i < \pi/2$ ($S_i < 0.5$). In case of $\varphi_i = \pi/2$ ($S_i = 0.5$), $f_i(\boldsymbol{x})$ does not change and $f_i'(\boldsymbol{x}) = f_i(\boldsymbol{x})$. Thus, this case is equivalent to the conventional dominance. On the other hand, in case of $\varphi_i > \pi/2$ ($S_i > 0.5$), $f_i(\boldsymbol{x})$ is decreased so $f_i'(\boldsymbol{x}) < f_i(\boldsymbol{x})$. Such fitness modification changes the dominance area of solutions. We show an example in **Fig. 2** (a)-(c), where three solutions $\boldsymbol{a}$, $\boldsymbol{b}$ and $\boldsymbol{c}$ are distributed in 2-dimensional objective space. In **Fig. 2** (a), $\boldsymbol{a}$ dominates $\boldsymbol{c}$, but

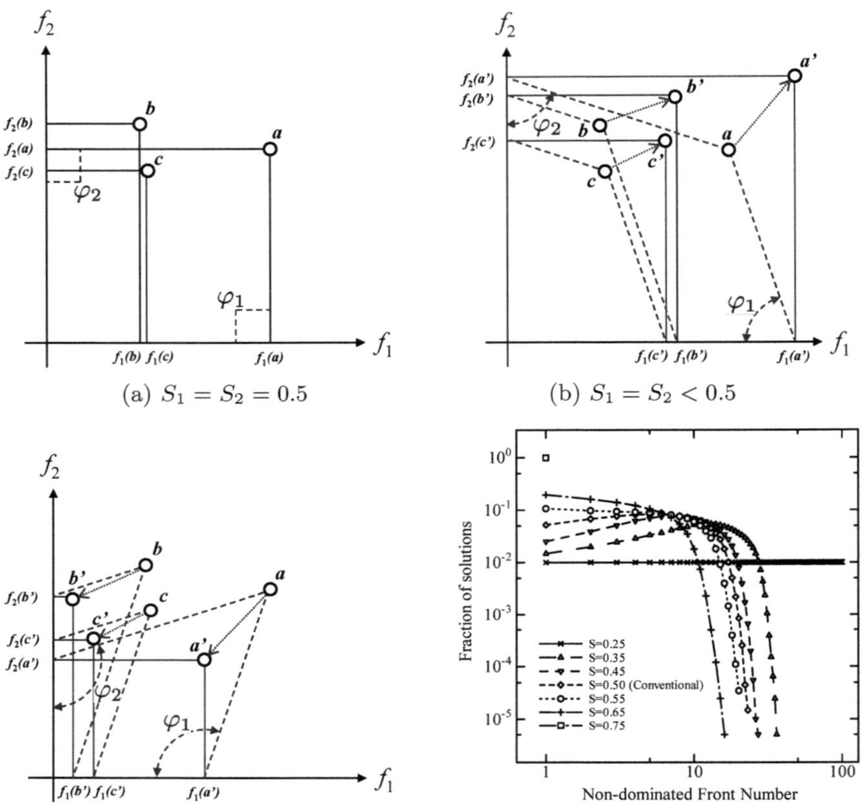

(a) $S_1 = S_2 = 0.5$ (b) $S_1 = S_2 < 0.5$

(c) $S_1 = S_2 > 0.5$ (d) Solutions per front

Fig. 2. Conventional dominance (a), examples of expanding (b) and contracting (c) the dominance area of solutions, and solutions per front varying the parameter S (d)

$\boldsymbol{a}$ and $\boldsymbol{b}$, and $\boldsymbol{b}$ and $\boldsymbol{c}$ do not dominate each other. However, if we modify fitness values for each solution by using **Eq. (8)**, the location of each solution moves in the objective space, and consequently the dominance relationship among solutions changes. For example, if we use $S_1 = S_2 < 0.5$ as shown in **Fig. 2** (b), the dominance area of solutions $\boldsymbol{a}'$, $\boldsymbol{b}'$ and $\boldsymbol{c}'$ is expanded from the original one of $\boldsymbol{a}$, $\boldsymbol{b}$ and $\boldsymbol{c}$. This causes that $\boldsymbol{a}'$ dominates $\boldsymbol{b}'$ and $\boldsymbol{c}'$, and $\boldsymbol{b}'$ dominates $\boldsymbol{c}'$. That is, expansion of dominance area by smaller $S_i(< 0.5)$ works to produce a more fine grained ranking of solutions and would strengthen selection. On the other hand, if we use $S_1 = S_2 > 0.5$ as shown in **Fig. 2** (c), the dominance area of solutions $\boldsymbol{a}'$, $\boldsymbol{b}'$ and $\boldsymbol{c}'$ is contracted from the original one of $\boldsymbol{a}$, $\boldsymbol{b}$ and $\boldsymbol{c}$. This causes that $\boldsymbol{a}'$, $\boldsymbol{b}'$ and $\boldsymbol{c}'$ do not dominate each other. That is, contracting the area of dominance by larger $S_i(> 0.5)$ works to produce a coarser ranking of solutions and would weaken selection.

4.2 Effects of Controlling Dominance Area

As indicated above, expanding or contracting the dominance area of solutions change the dominance relation of some solutions and therefore modify the distribution of the fronts (number of fronts and solutions per front). Since front distribution significantly relates to selection, we verify and illustrate the effect of expanding or contracting the dominance area on the distribution of the fronts changing the parameter S_i in **Eq. (8)**. Here, we randomly generate 100 solutions in the 2-dimensional objective space of $[0,1]^2$, calculate dominance among them after recalculating fitness with **Eq. (8)**, and perform a non-domination sorting to obtain the fronts. We repeat the above steps a 1000 times and calculate the average number of fronts and solutions per front, for each value of S_i. In this work, we use a common parameter $S = S_i(i = 1, 2, \cdots, m)$ for all objective functions, because we assume that all objective functions are normalized. **Fig. 2** (d) shows the fraction of number of solutions per front varying S in the range $[0.25, 0.75]$ in intervals of 0.1 along with results for conventional dominance ($S = 0.5$).

From this figure, note that if we gradually expand the area of dominance by decreasing S below 0.5, the number of fronts increases and the ranking of solutions by non-dominance can be fine grained. Note that for maximum expansion of the dominance area $S = 0.25$ there is one solution per front. On the other hand, if we gradually contract the area of dominance by increasing S above 0.5, the number of fronts decreases and ranking of solutions by non-dominance becomes coarser. Note that for maximum contraction of the dominance area $S = 0.75$ there is only one front that contains all solutions. Since different rankings can be produced, we can expect that the optimum parameter S^* that yields maximum search performance exists for a given kind of problem.

5 Benchmark Problems, Metrics, and Parameters

In this paper we use multiobjective 0/1 knapsack problems [9] as benchmark problems to study and compare the effects on search performance of ranking solutions by expanding or contracting their dominance area. The problem (KPn-m) is formulated to maximize the function

$$f_j(\boldsymbol{x}) = \sum_{i=1}^{n} x_i \cdot p_{i,j} \tag{9}$$

subject to

$$g_j(\boldsymbol{x}) = \sum_{i=1}^{n} x_i \cdot w_{i,j} \leq W_j \tag{10}$$

where $x_i \in \{0, 1\}$ $(i = 1, 2, \cdots, n)$ are elements of solution vector $\boldsymbol{x} = (x_1, x_2, \cdots, x_n)$, which gives a combination of items. Thus, we use binary representation in this work. Note that here we are interested in finding a set of non-dominated Pareto solutions. Also, $p_{i,j}$ and $w_{i,j}$ $(j = 1, 2, \cdots, m)$ denote profit and weight of item i according to knapsack (objective) j. W_j is the capacity of knapsack j, and solutions not satisfying this condition are considered as infeasible solutions $\bar{\mathcal{F}} = (\mathcal{S} - \mathcal{F})$. In this paper, we use benchmark problems with $m = \{2, 3, 4, 5\}$ objectives, $n = \{100, 250, 500, 750\}$ items and feasibility ratio $\phi = \{0.75, 0.5, 0.25\}$ downloaded from [10], for which we know the true Pareto non-dominated set only in case of two objectives $m = 2$. In these particular problems, we use a constant S for all objectives because the scale of each objective function is similar.

The hypervolume is used as a metric to evaluate sets of non-dominated solutions obtained by MOEAs. The hypervolume measures the m-dimensional volume of the region in objective space enclosed by the obtained non-dominated solutions and a dominated reference point [11]. Here we use $(f_1, f_2, \cdots, f_m)=(0, 0, \cdots, 0)$ as the reference point to calculate the hypervolume. A set of non-dominated solutions showing higher value of hypervolume can be considered as a better set of solutions from both convergence and diversity viewpoints. The hypervolume metric is a reliable metric and it is among the few recommended metrics to compare non-dominated sets [12]. To provide additional information separately on convergence and diversity of the obtained solutions in this work we also use Inverse Generational Distance (IGD) [13] and Spread (SP) [1], respectively. IGD takes the average distance for all members in the true Pareto front to their nearest solutions in the obtained set of non-dominated solutions (exactly the inverse process followed by Generational Distance GD [14]).

In our study we compare the performance of a conventional NSGA-II [8] with NSGA-II enhanced by the proposed method. We adopt two-point crossover with a crossover rate $p_c = 1.0$ for recombination, and apply bit-flipping mutation with a mutation rate $p_m = 1/n$. In the following experiments, we show the average performance with 30 runs, each of which spent 2,000 generations. Population size is set to $|P| = 200$ and the parent and offspring population sizes $|Q|$ and $|R|$ are set to half the population size $|P|$, i.e. $|Q| = |R| = 100$.

6 Experimental Results and Discussion

6.1 Performance Varying the Number of Objectives

In the following sections we observe the effects of varying the parameter S that controls the area of dominance of the solutions on the performance of the algorithm measured by the hypervolume. Recall that $S = 0.5$ indicates conventional dominance, values of $S > 0.5$ indicate contraction of the dominance area of the solutions, and values of $S < 0.5$ indicate expansion of the dominance area of the solutions.

First, we observe the effect of varying S on problems with different number of objectives. **Fig. 3** shows the values of the hypervolume achieved varying S in the range $[0.25, 0.75]$ in intervals of 0.05 on problems with $m = \{2, 3, 4, 5\}$ objectives, $n = 500$ items, and feasibility ratio $\phi = 0.50$. From this figure important observations are as follow. First, there is an optimum value S^* for each number of objectives that maximizes the hypervolume. Note however that the maximum value of hypervolume is not achieved by conventional dominance ($S = 0.5$) for any number of objectives. Second, to achieve the maximum value

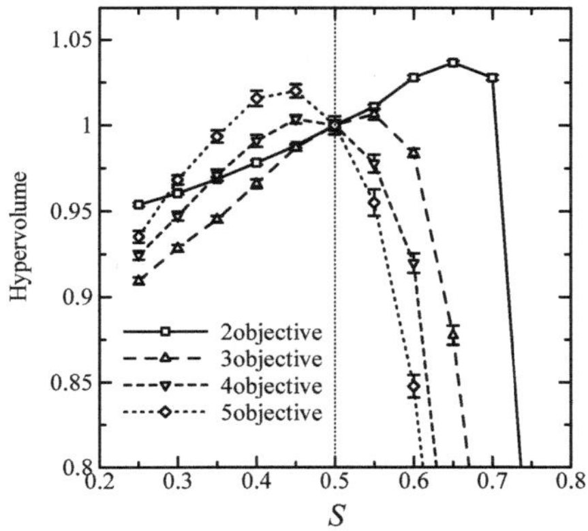

Fig. 3. Hypervolume as we increase the number of objectives m for problems with $n = 500$ items and $\phi = 0.5$ feasibility ratio

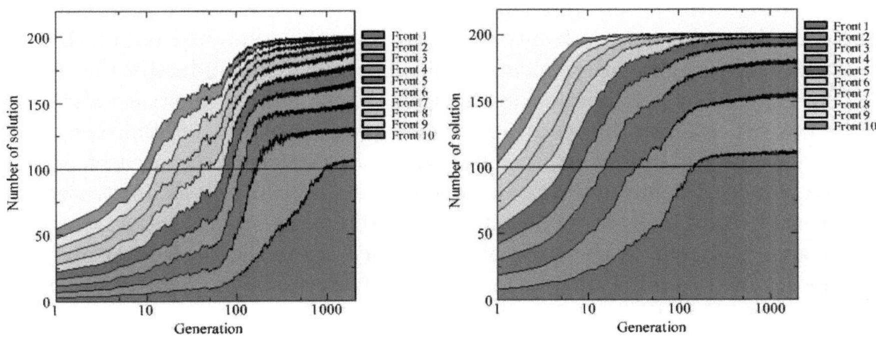

(a) $S = 0.5$, conventional dominance (b) $S^* = 0.65$, best by contracting dominance

Fig. 4. Front distribution over generation $m = 2$ objectives, $n = 500$ items, and $\phi = 0.5$ feasibility ratio

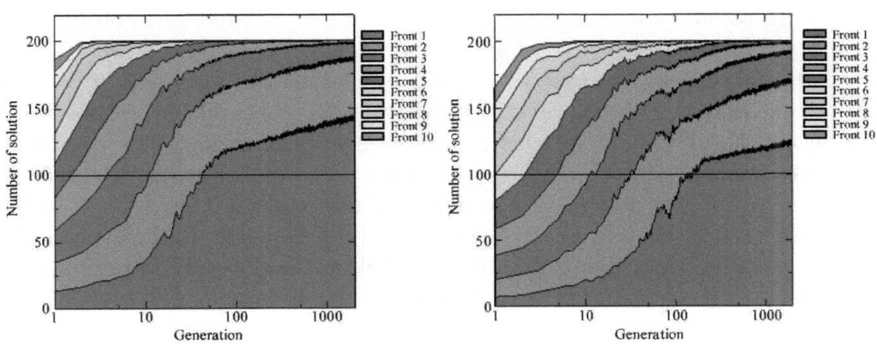

(a) $S = 0.5$, conventional dominance (b) $S^* = 0.45$, best by expanding dominance

Fig. 5. Front distribution over generation $m = 4$ objectives, $n = 500$ items, and $\phi = 0.5$ feasibility ratio of hypervolume, the degree of expansion or contraction of dominance area of solutions should be adjusted accordingly to the number of objectives. Note that maximum values of the hypervolume are achieved for two and three objectives by contracting the dominance area of the solutions ($S > 0.5$), whereas for four and five objectives the maximum hypervolume values are achieved by expanding the dominance area of the solutions ($S < 0.5$). Third, as a general trend in problems with $n = 500$ items and feasibility ratio $\phi = 0.50$, we observe that the optimum value S^* reduces as we increase the number of objectives. That is, increasing the number of objectives the area of dominance should be expanded by using smaller values of S^* to achieve maximum hypervolume.

Fig. 4 (a) and (b) show the front distribution over generation by conventional dominance ($S = 0.5$) and by contracting dominance with the optimum parameter ($S^* = 0.65$), respectively, on $m = 2$ objectives, $n = 500$ items and feasibility ratio $\phi = 0.5$. Similarly, **Fig. 5** (a) and (b) show on $m = 4$ objectives the front distributions by conventional dominance ($S = 0.5$) and by expanding dominance with the optimum parameter ($S^* = 0.45$), respectively. Results are presented for the ten top fronts obtained from the combined population of parents and offspring before truncation. The horizontal line indicates the truncation point after front non-domination sorting. These figures illustrate and corroborate our expectation that contraction or expansion of area of dominance changes the ranking of solutions. Remember that contraction of the area of dominance weakens selection and induces a coarse ranking of solution, as illustrated in **Fig. 4** (a) and (b), which works better for two and three objectives. Also, remember that an expansion of the area of dominance strengthen selection and induces a fine grained ranking of solutions, as illustrated in **Fig. 5** (a) and (b), which works better for four and five objectives.

6.2 Performance Varying the Size of the Search Space

Second, we observe the effects of varying S on problems with different number of items n. Note that the size of the search space is given by 2^n. **Fig. 6** shows the hypervolume varying S on problems with $n = \{100, 250, 500, 750\}$ items and feasibility ratio $\phi = 0.5$ for $m = \{2, 3, 4, 5\}$ objectives. From **Fig. 6** (a) we can see that in the case of $m = 2$ objectives the optimum S^* is similar for all n, around 0.65. However, from **Fig. 6** (b),(c), and (d) we observe that increasing the number of items n produces a clear shift of the optimum S^* towards smaller values (greater expansion of area of dominance), especially in the case of $m = 4$ and $m = 5$ objectives. For example, for $m = 4$, note the optimal $S^* = \{0.55, 0.5, 0.45, 0.45\}$ on $n = \{100, 250, 500, 750\}$, respectively. In the previous section, fixing the number of items to $n = 500$, results suggested that the degree of expansion or contraction of dominance area of solutions should be adjusted according to the number of objectives. The results presented in this section suggest that the degree of expansion or contraction of dominance area of solutions should also be adjusted according to the size of the search space, especially for an increased number of objectives.

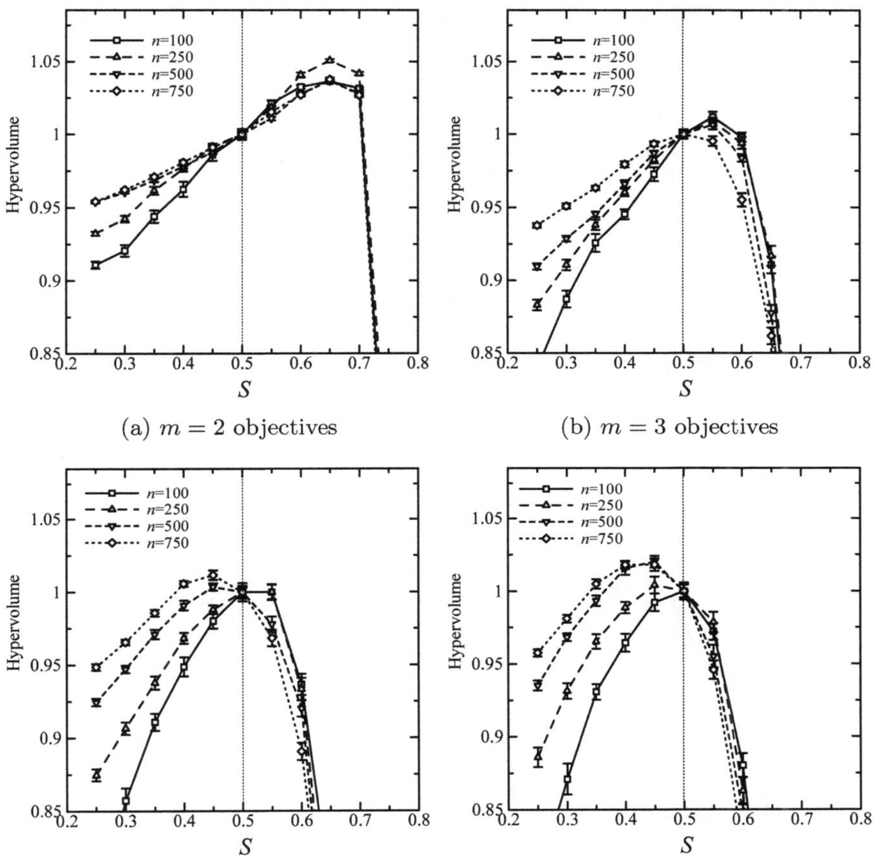

(a) $m = 2$ objectives

(b) $m = 3$ objectives

(c) $m = 4$ objectives

(d) $m = 5$ objectives

Fig. 6. Hypervolume as we increase the number of items n for problems with $m = \{2, 3, 4, 5\}$ objectives and $\phi = 0.5$ feasibility ratio

6.3 Performance Varying the Search Space Feasibility Ratio ϕ

Third, we observe the effects of varying S on problems with different feasibility ratio ϕ. **Fig. 7** shows the hypervolume varying S on problems with feasibility ratio $\phi = \{0.75, 0.5, 0.25\}$ and $n = 500$ items for $m = \{2, 3, 4, 5\}$ objectives. From **Fig. 7** (a)-(d) note that the effects on problems with different feasibility ratio ϕ resemble those observed on problems with different number of items. That is, in $m = 2$ objectives the optimum S^* is the same for all ϕ. However, reducing the feasibility ratio ϕ from 0.75 to 0.25, there is a shift of the optimum S^* towards smaller values, which becomes more notorious for $m = 4$ and $m = 5$ objectives. For example, for $m = 4$, note the optimal $S^* = \{0.55, 0.45, 0.4\}$ on $\phi = \{0.75, 0.5, 0.25\}$, respectively. These results suggest that the optimum degree of expansion or contraction of dominance area of solutions also depends on the

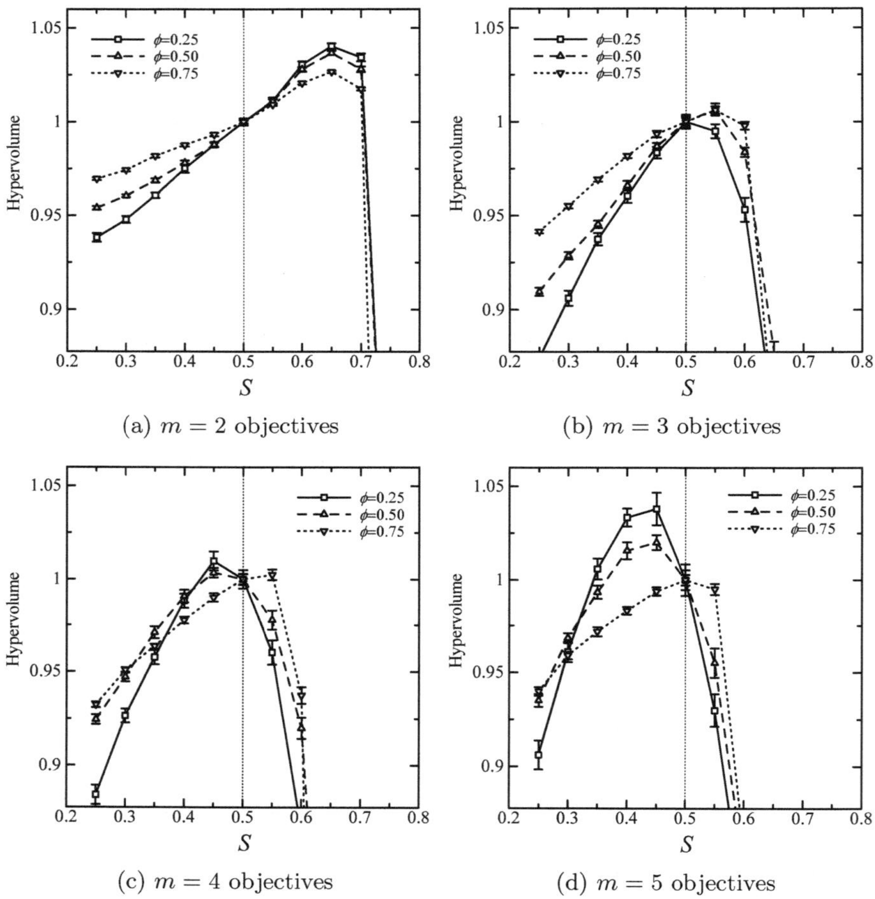

Fig. 7. Hypervolume as we decrease feasibility ratio for problems with $m = \{2, 3, 4, 5\}$ objectives and $n = 500$ items

feasibility ratio of the search space (complexity of the landscapes), especially for an increased number of objectives.

Summarizing, the optimum degree of expansion or contraction of the dominance area depends on the three aspects investigated in this work; that is, number of objectives, size of the search space, and feasibility ratio of the search space. For most real world combinatorial problems we can know in advance the number of objectives and size of the search space. Based on these information, we can use the results presented here as a good initial guidelines to properly set the degree of expansion or contraction of the area of dominance in order to achieve higher performance. However, the feasibility ratio (or complexity of the single objective landscapes) is usually unknown. It would be interesting to find adaptive ways to fine tune the parameter S for problems of different complexity.

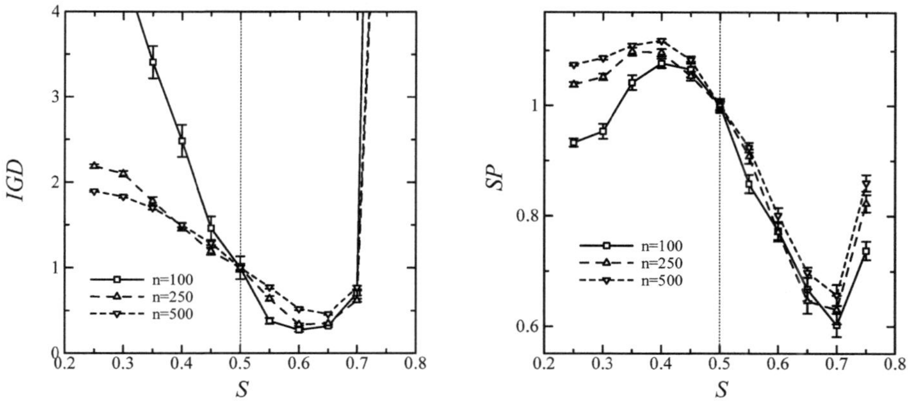

(a) IGD (b) SP

Fig. 8. Inverse generational distance IGD and Spread SP varying the number of items n on problems with $m = 2$ objectives and $\phi = 0.5$ feasibility ratio

6.4 Results on Complementary Metrics and Obtained Solutions

Fig. 8 (a) and (b) show the Inverse Generational Distance IGD and Spread SP, respectively, varying the number of items on problems with $m = 2$ objectives and feasibility ratio $\phi = 0.5$. From these figures note that optimum IGD and SP (smaller values) are achieved when the dominance area of solutions is contracted ($S > 0.5$) rather than by conventional dominance ($S = 0.5$), similar to the results shown in **Fig. 6** (a). The values S achieving minimum IGD are almost coincident with $S^* = 0.65$ achieving maximum hypervolume. However, in case of SP the values S are slightly shifted towards larger values. Also, note that the graph of SP shows a maximum peak in the area of $S < 0.5$ and a minimum peak in the area of $S > 0.5$.

To analyze the above observations further, **Fig. 9** illustrates the obtained solutions in the final generation for all 30 simulations by conventional dominance $S = 0.5$, contracting dominance $S^* = 0.65$, and expanding dominance $S = 0.4$ for $m = 2$ objectives, $n = 500$ items, and $\phi = 0.5$ feasibility ratio. Note that solutions obtained by conventional dominance are close to the true Pareto front but are clustered in a limited region of objective space. By contracting dominance with the optimum parameter $S^* = 0.65$, we can spread the obtained solutions showing the maximum hypervolume, although convergence of some of them seems to deteriorate. On the other hand, by expanding dominance with $S = 0.4$ showing the maximum SP (worst spread), we can further enhance convergence of the solutions within a narrower region of objective space.

7 Conclusions

In this work we have proposed a method that can control dominance area of solutions by a user defined parameter S. We showed that contracting or expanding

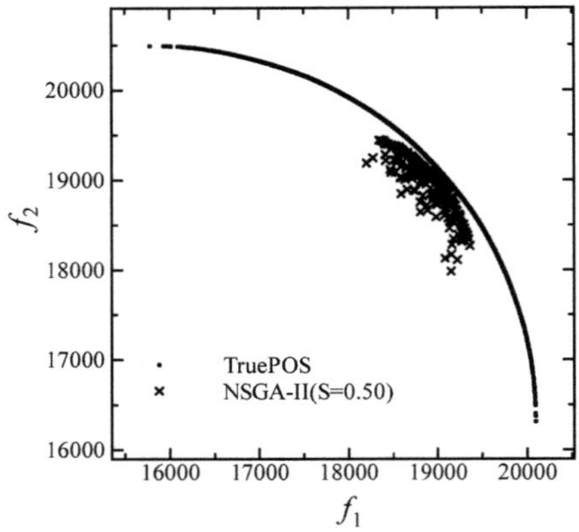

(a) Conventional dominance ($S = 0.5$)

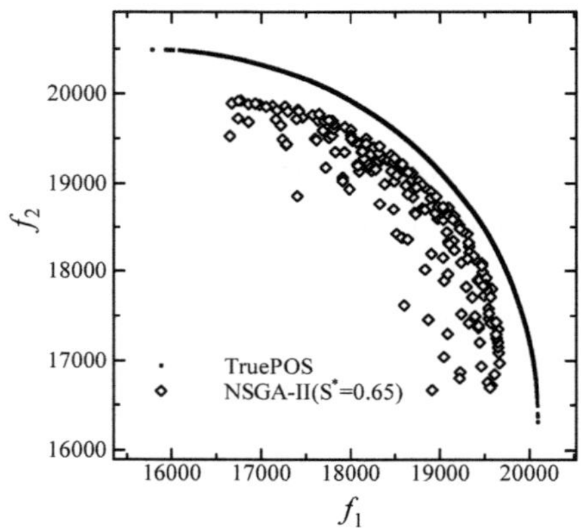

(b) Contracting dominance ($S^* = 0.65$)

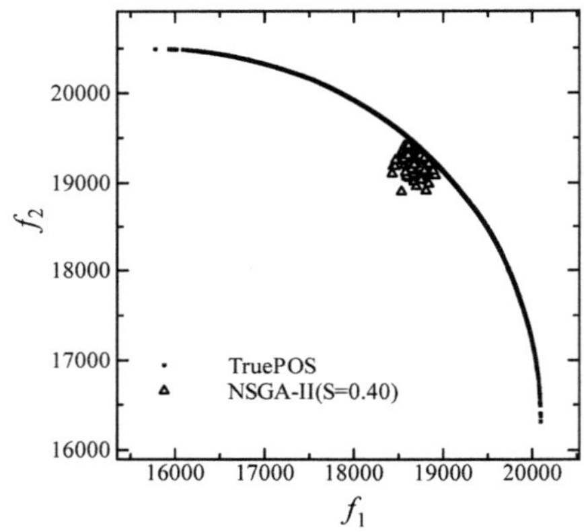

(c) Expanding dominance ($S = 0.40$)

Fig. 9. Obtained solutions by conventional dominance $S = 0.5$, contracting dominance $S^* = 0.65$, and expanding dominance $S = 0.4$ for $m = 2$ objectives, $n = 500$ items, and $\phi = 0.5$ feasibility ratio

the dominance area of solutions changes their dominance relation, modifying the distribution of solutions (number of fronts and number of solutions per front) in the multiobjective landscape. Since front distribution significantly relates to selection, we analyzed the effects on solutions ranking caused by contracting and expanding the dominance area of solutions and its impact on the search performance of a multi-objective optimizer. We used 0/1 multiobjective knapsack problems as benchmark problems and showed that the optimum value of S^* depends strongly on number of objectives, size of the search space, and feasibility ratio of the search space (complexity). In addition, we showed that significantly better performance can be achieved either on convergence or diversity of obtained solutions by contracting or expanding the dominance area rather than by using conventional dominance.

In this work, we have assumed a constant parameter $S_i = S$ on all objectives ($i = 1, 2, \cdots, m$) to control the expansion or contraction of dominance area. It would be interesting in the future to investigate the effect of varying S_i for each objective and control S adaptively, especially for problems of unknown characteristics. In addition, it could be valuable to combine this approach with the inclusion of preferences to guide the search towards a particular region of objective space. With the proposed method, we can improve either convergence or diversity of solutions but not simultaneously both. Therefore, we would like to combine the proposed method with other selection methods to achieve higher convergence while covering the whole true Pareto front. Furthermore, we should try our method on other kind of problems with more objectives and compare our method with other approaches that aim to solve many objective optimization problems.

References

1. K. Deb, *Multi-Objective Optimization using Evolutionary Algorithms*, John Wiley & Sons, 2001.
2. C. A. C. Coello, D. A. Van Veldhuizen, and G. B. Lamont, *Evolutionary Algorithms for Solving Multi-Objective Problems*, Kluwer Academic Publishers, 2002.
3. R. C. Purshouse and P. J. Fleming, "Conflict, Harmony, and Independence: Relationships in Evolutionary Multi-criterion Optimisation", *Second Intl. Conf. on Evolutionary Multi-Criterion Optimization*, Lecture Notes in Computer Science, Springer, Vol. 2632, pp.16-30, April 2003.
4. E. J. Hughes, "Evolutionary Many-Objective Optimisation: Many Once or One Many?", *Proc. 2005 IEEE Congress on Evolutionary Computation*, Vol.1, pp.222-227, September 2005.
5. H. Aguirre and K. Tanaka, "Working Principles, Behavior, and Performance of MOEAs on MNK-Landscapes", *European Journal of Operational Research, Special Issue on Evolutionary Multi-Objective Optimization*, Sep. 2006. (in press)
6. M. Laumanns, L. Thiele, K. Deb and E. Zitzler, "Combining Convergence and Diversity in Evolutionary Multi-objective Optimization", *Evolutionary Computation*, Vol.10, No.3, pp.263-282, Fall 2002.
7. K. Ikeda, H. Kita and S. Kobayashi, "Failure of Pareto-based MOEAs: Does Non-dominated Really Mean Near to Optimal?", *Proc. 2001 IEEE Congress on Evolutionary Computation*, Vol. 2, pp.957-962, May 2001.
8. K. Deb, S. Agrawal, A. Pratap and T. Meyarivan, "A Fast Elitist Non-Dominated Sorting Genetic Algorithm for Multi-Objective Optimization: NSGA-II", *KanGAL report 200001*, 2000.
9. E. Zitzler and L. Thiele, "Multiobjective optimization using evolutionary algorithms – a comparative case study", *Proc. of 5th Intl. Conf. on Parallel Problem Solving from Nature (PPSN-V)*, pp.292-301, 1998.
10. http://www.tik.ee.ethz.ch/~zitzler/testdata.html
11. E. Zitzler, *Evolutionary Algorithms for Multiobjective Optimization: Methods and Applications*, PhD thesis, Swiss Federal Institute of Technology, Zurich, 1999.
12. J. Knowles and D. Corne, "On Metrics for Comparing Non-dominated Sets", *Proc. 2002 IEEE Congress on Evolutionary Computation*, pp.711–716, 2002.

13. H. Sato, H. Aguirre, and K. Tanaka, "Local Dominance Using Polar Coordinates to Enhance Multi-objective Evolutionary Algorithms", *Proc. 2004 IEEE Congress on Evolutionary Computation*, Vol. 1, pp. 188-195, 2004.
14. D. A. V. Veldhuizen and G. B. Lamont, "On Measuring Multiobjective Evolutionary Algorithm Performance", *Proc. IEEE 2000 Congress on Evolutionary Computation*, Vol.1, pp.204–211, 2000.

Designing Multi-objective Variation Operators Using a Predator-Prey Approach

Christian Grimme and Joachim Lepping

Robotics Research Institute - Section Information Technology, Dortmund University, 44221 Dortmund, Germany
{christian.grimme,joachim.lepping}@udo.edu

Abstract. In this paper, we propose a new conceptual method for the design, investigation, and evaluation of multi-objective variation operators for evolutionary multi-objective algorithms. To this end, we apply a modified predator-prey model that allows an independent analysis of different operators. Using this model problem specific operators can be combined to more complex operators. Additionally, we review the simplex recombination, a new rotation-independent recombination scheme, and examine its impact concerning our design method. We show exemplarily as a first attempt the advantageous combination of several standard variation operators that lead to better results for selected test functions.

Keywords: Predator-Prey Model, Multi-Objective Operators, Evolutionary Multi-Objective Algorithm, Operator Design.

1 Introduction

In the last years of research several multi-objective evolutionary algorithms have been proposed for the simultaneous optimization of multiple and competing objectives. The manifold experiences with different algorithms and multi-objective optimization problems reveal that only an adroit combination and problem specific adjustment of the different evolutionary operators decide on the system's success or failure. All the more, this insight is ubiquitous in the case of single-objective optimization.

It is therefore even more astonishing that within the multi-objective optimization, the conceptual approaches are still mainly concerned with the selection operator. Instead of adapting all evolutionary operators, like in the single-objective algorithms, external or internal archives [18], metrics as new selection criteria [11] or other even more complicated and time consuming procedures are developed. Consequently, the selection operators are in most cases not well concerted with the rest of the applied operators. Büche [3] shows for some state-of-the-art evolutionary multi-objective algorithms that the approximation of the set of efficient solutions cannot be done with an arbitrary precision. The distance between the true Pareto front and the approximated set can be reduced only by a noticeable rise of the size of the archives - otherwise stagnation occurs.

S. Obayashi et al. (Eds.): EMO 2007, LNCS 4403, pp. 21–35, 2007.

However, research focusing on the field of variation operators or representations remains rare, so only a few approaches can be found in literature: Kursawe [8] examined for example the use of diploid representations for two objective test functions while Rudolph [13] and Hanne [7] are concerned with the problem of finding an appropriate controlling mechanism for the mutation strength in the multi-objective case. Only Rudenko and Schoenauer [12] design a special recombination operator for real value coded multi-objective problems.

One fact that is emerging from those studies is that traditional single objective operators are not suitable for the multi-objective case. It is therefore quite conceivable that interaction between the evolutionary operators cannot be taken over from the single-objective case and the exclusive change of the selection operator is not sufficient to meet the requirements of multi-objective optimization. Those requirements can be formulated as the simultaneous ability of diversity preservation and a good convergence to the optimum. In contrast to existing selection schemes we try to avoid complex and time consuming computations for the variation operators.

The scope of this paper is the analysis and development of variation operators for multi-objective optimization. To this end, we provide a model for the analysis of single-objective operators in a multi-objective problem context. Our model is based on the predator-prey model of Laumanns et al. [9] but includes many modifications. With this analysis environment it is possible to identify potential advantageous properties of single-objective operators for multi-objective problems. Therefore, this model can be applied for the design of a multi-objective variation operator by the tunable combination of several operators.

The remainder of the paper is organized as follows. First we describe our predator-prey model that consists of many modifications compared to Laumans' original model. Those changes are described and explained in Section 3. Further, in Section 4 we motivate and sketch a recombination scheme that has already been proposed by Grimme and Schmitt [6]. Afterwards, in Section 5 we show how existing variation operators can be combined to a problem specific adapted multi-objective operator. The paper ends with a brief conclusion and a motivation for future work.

2 Background

This section provides a relatively short introduction into the predator-prey model of Laumanns et al. [9]. Additionally, some existing extensions are reviewed and problems of this model are outlined which motivate modifications done to the model later on.

2.1 Laumanns' Predator-Prey Model

The idea that individuals interact in *time* and *space* within their own species as well as with other species forms the basis of Laumanns' asynchronous spatial structured predator-prey model, as shown in Figure 1 (a). The "prey" are the usual individuals of the evolutionary multi-objective algorithm representing the possible solutions of the multi-objective optimization problem. These prey are placed at vertices of a two-dimensional toroidal grid as the spatial population structure. One advantage of the toroidal structure is that by a random walk in this structure all places can be reached with equal probability [9]. Further advantages of spatial structured populations are broadly discussed by Tomassini [17] and will be omitted here. Due to this population structure the neighborhood of a particular grid point v is defined in terms of the number of steps taken from that grid point. The number of steps will be further on called radius r of neighborhood $N(v,r)$ which is given generally by the recursive Equation (1).

$$\begin{aligned} N(v, r = 0) &= v \\ N(v, r) &= \bigcup_{(v,\nu) \in E} N(\nu, r-1) \end{aligned} \tag{1}$$

This holds if the torus is considered to be represented by a graph $G(V, E)$ with V as set of vertices and E as set of edges. The amount of neighbors enclosed by $N(v,r)$ is given by Equation (2).

$$c_{neighbors} = (r+1)^2 + r^2 \tag{2}$$

The selection mechanism is realized by the predators where each one represents a single objective of the multi-objective problem. Thus, if there are m objectives for a multi-objective problem there have to be exactly m predator species.

The behavior of the predators is illustrated in Figure 1 (a). They move across the spatial structure according to a random walk which is retained as a uniformly distributed movement in the direct neighborhood of the position of a predator. The predator chases the prey only within its current neighborhood and according to the objective assigned to it. In Figure 1, $N(v, 1)$ as a neighborhood of $r = 1$ is depicted. The *worst* prey within this neighborhood is "eaten". As soon as the grid point for the prey becomes free, a reproduction neighborhood is spanned around the empty vertex. The actual reproduction refills the empty place. The new prey is created using standard recombination of $(c_{neighbors} - 1)$ enclosed prey. In addition, a mutation operator is applied to the created offspring.

It is important to annotate that the recombination neighborhood (RN) and the selection neighborhood (SN) do not contain necessarily the same set of prey since the neighborhoods are constructed from the predator grid point and the freed grid point respectively.

Because there are several predators for different objectives those prey which perform best with respect to all objectives are able to survive and represent the approximated Pareto set after a while.

2.2 Extensions to the Original Model

In the original study two major problems were observed: loss of diversity and stagnation of the process of convergence to the true Pareto front. The results of the preliminary study made obvious that suitable search operators were missing

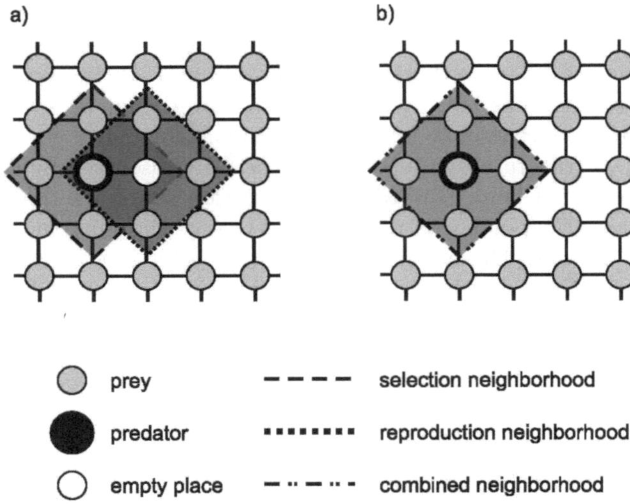

Fig. 1. Schematic illustration of the original predator-prey model by Laumanns (a) and of the modified predator-prey model which is be used in this paper (b)

to develop a simple evolutionary multi-objective algorithm. Nevertheless, only few extensions were proposed.

Deb [4], for example, macerates the strict mapping of one predator to one objective with an individual weighted vector in each predator. This individual selection allows each predator to steer prey to a specific region on the Pareto front. Based on both approaches, Li proposed a real-coded predator-prey (RCPPGA) [10] model. In his approach, he uses a genetic algorithm as the underlying search heuristic. Additionally, he applies a dynamic population, where predators as well as prey are able to move within the structured environment. Due to this behavior recombination only takes place if two prey individuals are in the same neighborhood. If a prey has no neighbors, no duplication is allowed. Consequently, Li has to define a special migration for both species to prevent extermination of them. Another model was developed by Schmitt [14]. Here the steady-state approach of the predator-prey model was replaced by the well-known controlling mechanism of the self-adapting evolution strategy [15]. In addition the weighted intermediate recombination operator proposed by Schwefel and Rudolph [16] was used.

Viewed together, all existing extensions so far mainly concentrate on the selection mechanism. Unfortunately, only little attention has been payed to the design of suitable variation operators.

3 A Model for Variation Operator Design

The lack of adequate variation operators for evolutionary multi-objective algorithms may arise from several difficulties in designing such complex methods.

Obviously, considering the large amount of variation operators this claim does not hold for the single-objective case. In this context, there is only a need for converging as close as possible to *one* optimum solution. However, multi-objective problems always have a whole set of optimal solutions so that a degree of diversity is desired for the optimization result. As a first step towards a design concept for multi-objective variation operators we construct an environment for the analysis of single-objective variation operators in a multi-objective problem context. To this end, we first modify the previously introduced predator-prey model. Second, we propose a methodology for the development of composed multi-objective operators from the analyzed single-objective operators.

3.1 Adaptation of the Predator-Prey Model

In order to make the predator-prey model suitable for our purpose we have to introduce some changes which are detailed next.

Combining Selection and Reproduction Neighborhoods: Since the reproduction mechanism of Laumanns' model works with two different neighborhoods, evolutionary operators may fail due to the often rather different sets of individuals in these neighborhoods. The prey removed by the predator is chosen from the selection neighborhood SN while the new prey is bred from individuals out of the reproduction neighborhood RN. This is shown in Figure 2 for $r = 1$ (left) and $r = 2$ (right). Hence, an individual is placed at the boundary of SN. Then, it is replaced by almost only unknown individuals from RN i.e. by individuals which did not take part in the selection process. Consequently, if $SN \neq RN$, individuals from $RN \backslash \{SN \cap RN\}$ (those individuals that are labeled with a question mark in Figure 2) must be looked upon as being of uncertain nature. Therefore, they are possibly worse than the removed individual. To come along with this drawback we consider an identical recombination neighborhood ($SN = RN$) for our model, see Figure 1 (b).

Separation of Operators by Speciation: Our modified model provides the possibility to realize various evolutionary operators independently of each other. Therefore, it becomes easier to investigate the effect of different variation operators on the computation of solutions. However, it may be necessary to switch off other operators to margin or eliminate their influence in the evolutionary search process.

In order to achieve this property for the predator-prey model the selection mechanism represented by the predators is split up in several species. That is a predator triggers the application of only one evolutionary operator. Although the operators are still strongly associated with the selection mechanism, they now perform independently. Thus, it becomes possible to apply even an arbitrary ratio of mutation and recombination predators to the model by chasing the prey with a different number of specific predators. Such a flexible ratio is considered to reach better solutions with respect to convergence and diversity. This ratio serves as the basic idea for the operator design concept proposed here.

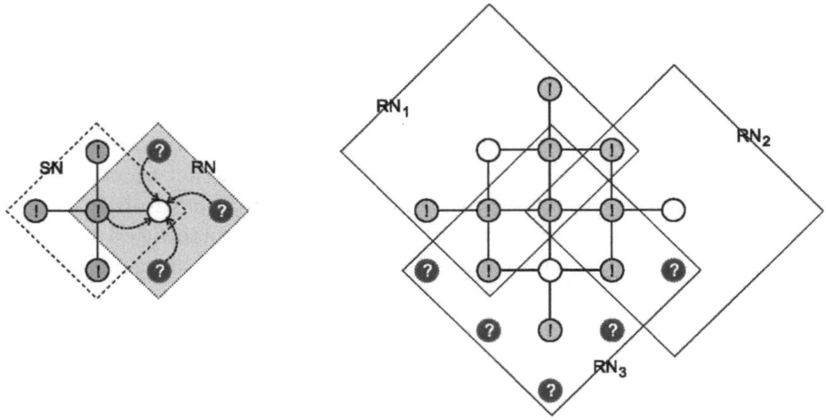

Fig. 2. Illustration of Laumanns' selection and reproduction neighborhood for $r = 1$ (left) and $r = 2$ (right). Individuals labeled with exclamation tags are known due to the selection process while those with question marks are of uncertain quality. Symmetric cases are obtained for clarity.

3.2 A Building Block Approach for Variation Operator Design

With our decoupled predator-prey model we are able to associate the single-objective selection with several different variation operators, that is each predator is specifically sensitive for one objective and causes exclusively one variation in the population. Naturally, different predators may trigger a variety of variation operators for the same objective, as schematically depicted in Figure 3. Predators that are configured in the described fashion perform their random walk independently on the torus. Thereby, we are able to investigate several effects on the population. First, in an experiment a single variation operator is applied to observe its impact on different objectives. In this way it is possible to identify advantageous properties of an operator for the multi-objective case. Second, we can apply the gained knowledge for the combination of multiple operators to possibly benefit from the positive effects of every single operator. Finally, the impact of the single operators on the composition is flexibly tunable. This is realized by introducing more identically configured predators in the system and adjusting their ratio which makes the whole configuration scalable.

4 Standard Recombination Operators on Multi-objective Problems

As mentioned in the previous section our aim is the composition of several standard variation operators. For the recombination of parent individuals a large variety of operators is available. Therefore, we have to make a first selection in order to give a proof of our concept. Note that of course the selection is not

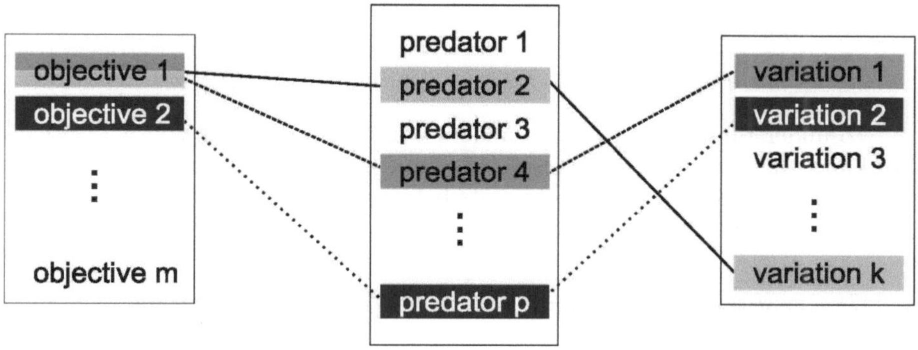

Fig. 3. Schematic depiction of exemplary predator configurations. Every predator is assigned to an objective as well as a variation operator.

restricted to the operators presented here. Any other set of variations may be possible as well.

The discrete recombination [16] is the first selected operator as it is considered to preserve good diversity within the population. Concurrently, we have learned from experience that a lack of convergence comes along with the application of this variation scheme. Furthermore, we decide to apply the simplex recombination [6] as this scheme combines the properties of intermediate and weighted intermediate [16] recombination for high dimensional problems. Moreover, simplex recombination solves the drawback of intermediate recombination that arises by choosing only one point in the center of gravity. Similar to intermediate recombination, simplex recombination is insensible to rotations within decision space. Furthermore, it is a quite new operator that has not been investigated comprehensively so far. We briefly introduce simplex recombination in the following paragraph to use it for experiments in the predator-prey environment later on.

Simplex Recombination

In order to tackle the problem of equal distributed selection of an offspring and to coordinate rotation in an n-dimensional space we want to focus on utilizing a geometric shape to restrict the reproduction within the search space similar to weighted intermediate recombination. To this end, we consider a triangular or in general an n-simplex which is defined as an n-dimensional polytop with $n+1$ points. However, it is difficult to choose a descendant uniformly distributed out of the n-dimensional simplex. The general technique to generate a random point in an n-simplex is presented here.

Definition 1 (Barycentric Coordinates). *Hence* $(n+1)$ *vectors* $p_1, \ldots, p_{n+1}$ *in the* n*-dimensional space. If* $p_2 - p_1 \ldots p_{n+1} - p_1$ *are linear independent every point* q *may be represented as a* $(n+1)$*-tupel of barycentric coordinates*

$(\beta_1, \ldots, \beta_{n+1})$ qualifying q by

$$q = \beta_1 p_1 + \cdots + \beta_{n+1} p_{n+1} \text{ and } \beta_1 + \cdots + \beta_{n+1} = 1$$

To understand how to pick a random point from an n-simplex it is essential to know the representation of a point P in a 1-simplex which is a line between two points A and B. The set of points between A and B is given by

$$[A, B] := \{aA + bB | a, b \in \mathbb{R}^+, a + b = 1\}.$$

If we choose $a = (1 - \lambda)$ and $b = \lambda$ then $P \in [A, B]$ results from

$$P = (1 - \lambda)A + \lambda B \tag{3}$$

The tupel $((1 - \lambda), \lambda)$ yields the barycentric coordinates of P. This construction procedure can be used in n-simplizia as well. As depicted in Figure 4 for a triangle, one creates, starting from a point A, two points P' and P'' with the same barycentric coordinates. Finally, point P is created on the line between these new points in an analogous way. For a tetrahedron a third point P''' is created to form a triangle. Then P is generated as described before.

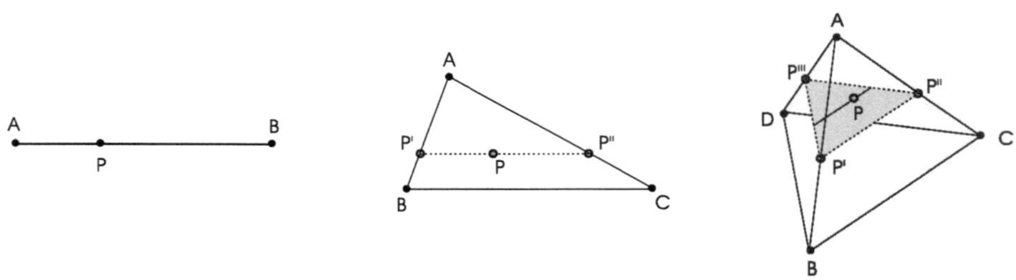

Fig. 4. Schematic procedure to generate a random point in a simplex depicted for a 1-, 2-, and 3-simplex.

The described recursive procedure can easily be transformed to an iterative one. Generally, the point P_{nS} in an n-simplex is yielded by:

$$P_{nS} = \sum_{i=1}^{n+1} \left((1 - \lambda_i) \prod_{j=0}^{i-1} \lambda_j \right) A_i \qquad \text{with } \lambda_0 := 1 \text{ and } \lambda_{n+1} := 0 \tag{4}$$

Using Equation (4) the new recombined offspring can be computed rather efficient as it has a complexity of $\mathcal{O}(n^2)$.

The vectors A_i with $i = \{1, \ldots, n+1\}$ span the actual simplex. To choose P_{nS} uniformly distributed for every $\lambda_j \in \{\lambda_1, \ldots, \lambda_n\}$ a random number $z_j \sim \mathcal{U}(0, 1)$ is generated and applied in:

$$\lambda_j = \sqrt[k]{z_j} \quad \text{with} \quad k = (n + 1) - j. \tag{5}$$

The square root till n-th root is taken to weight all portions of the simplex equally. In a two-dimensional space the area ratio $v^2 = V_{\Delta_1}/V_{\Delta_2}$ of two similar triangles $\Delta_1(g_1, h_1)$ and $\Delta_2(g_2, h_2)$ is equal to the ratio of the corresponding triangle sides $v = \frac{h_1}{h_2} = \frac{g_1}{g_2}$. This is followed from the theorem of intersecting lines and can be extended to hypervolumes in an n-dimensional space.

5 An Operator Design Case Study

Within this section we show exemplarily the application of our proposed design concept for new multi-objective variation operators. To this end, we first define two simple multi-objective problems with two and three objectives respectively. Both of them are based on the multisphere problem which is defined in Equation (6). While n is the dimension of the decision space, m denotes the number of objectives.

$$\mathcal{F}_m : \mathbb{R}^n \rightarrow \mathbb{R}^m \text{ with } \boldsymbol{x} \in \mathbb{R}^n, n, m \in \mathbb{N}$$

$$\mathcal{F}_m(\boldsymbol{x}) = \begin{pmatrix} f_1(\boldsymbol{x}) \\ \vdots \\ f_m(\boldsymbol{x}) \end{pmatrix} = \begin{pmatrix} (\boldsymbol{x} - \boldsymbol{c}_1)^2 \\ \vdots \\ (\boldsymbol{x} - \boldsymbol{c}_m)^2 \end{pmatrix} \tag{6}$$

The $\boldsymbol{c}_1, \ldots, \boldsymbol{c}_m \in \mathbb{R}^n$ are constant values. The special problem $\mathcal{F}_2(\boldsymbol{x})$, as defined in Equation (7), has constants $\boldsymbol{c}_1 = (0,0)^T$ and $\boldsymbol{c}_2 = (2,0)^T$ while $\mathcal{F}_3(\boldsymbol{x})$, see Equation (8), has constants $\boldsymbol{c}_1 = (0,0,0)^T$, $\boldsymbol{c}_2 = (2,0,0)^T$, and $\boldsymbol{c}_3 = (0,0,2)^T$. Both problems are convex and the corresponding Pareto sets are defined by a single line given as $0 \leq x_1 \leq 2$ and $x_2 = 0$ as well as a triangular shaped plain spanned between (0,0,0), (2,0,0), and (0,0,2) respectively.

$$\mathcal{F}_2(\boldsymbol{x}) = \begin{pmatrix} x_1^2 + x_2^2 \\ (x_1 - 2)^2 + x_2^2 \end{pmatrix} \quad \text{with } \boldsymbol{x} \in [-10, 10]^2 \tag{7}$$

$$\mathcal{F}_3(\boldsymbol{x}) = \begin{pmatrix} x_1^2 + x_2^2 + x_3^2 \\ (x_1 - 2)^2 + x_2^2 + x_3^2 \\ x_1^2 + x_2^2 + (x_3 - 2)^2 \end{pmatrix} \quad \text{with } \boldsymbol{x} \in [-10, 10]^3 \tag{8}$$

For investigation purpose we construct our new complex variation operator from simple standard operators. The applied operators are discrete recombination and standard mutation as described by Schwefel and Rudolph [16]. Further, we incorporate our previously described simplex recombination. The predator-prey set up is characterized by a torus size of 40 × 40 and we performed 50,000 function evaluations per experiment. As performance metric we compute the average euclidean distance from the true Pareto set. Note that we do not aim to outperform other approaches for multi-objective optimization. This metric is only used to compare the performances of different examined operators and their corresponding combination. First, we show the design method for the two objectives problem $\mathcal{F}_2(\boldsymbol{x})$.

5.1 Operator Design for Two Objectives

In order to design a complex operator for a two objective problem we first analyze the impact of the simple mutation as well as the simplex recombination. The optimization results are shown in Figure 5. Obviously, the mutation operator favors the extrema solutions of each objective, see Figure 5(a). By contrast, the simplex recombination crowds the population in the center of gravity of the initial solution space. This is due to the definition of the simplex recombination scheme as a new offspring is only located within the convex hull of the participating parental solutions. This property is desirable to achieve a good convergence to the optimum. In order to preserve all properties of both operators a weighted combination is used for our predator-prey model. A suitable ratio of mutation and simplex recombination is achieved by the different number of predators, as described in Section 3.1. Figure 5(f) shows the results for the combined operator of mutation and simplex recombination. Apparently, all advantageous properties can be combined to achieve both convergence and diversity.

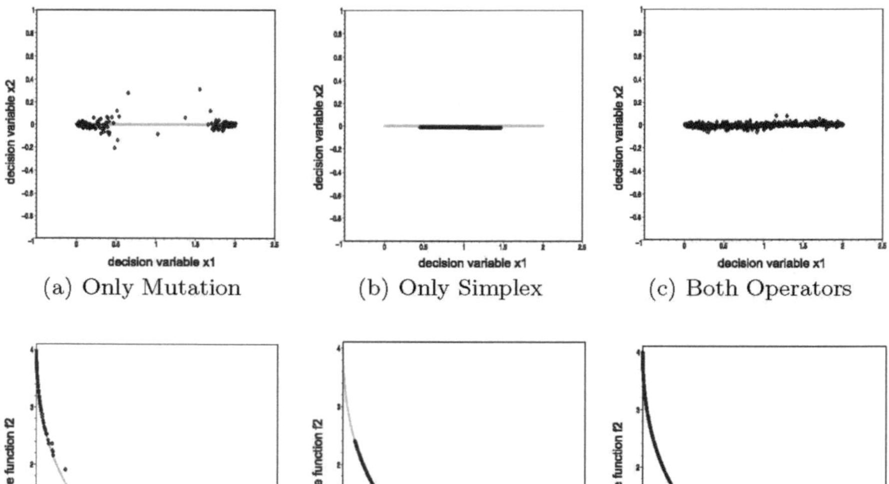

(a) Only Mutation (b) Only Simplex (c) Both Operators

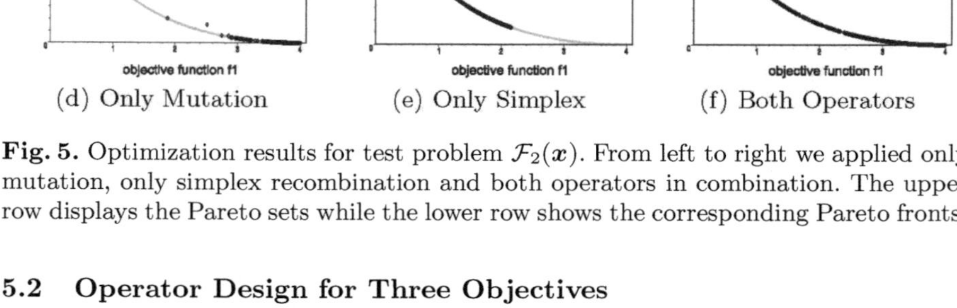

(d) Only Mutation (e) Only Simplex (f) Both Operators

Fig. 5. Optimization results for test problem $\mathcal{F}_2(\boldsymbol{x})$. From left to right we applied only mutation, only simplex recombination and both operators in combination. The upper row displays the Pareto sets while the lower row shows the corresponding Pareto fronts.

5.2 Operator Design for Three Objectives

To confirm the results from the last section for a slightly more complex problem we conduct an analogous experiment for test problem $\mathcal{F}_3(\boldsymbol{x})$. Here, we

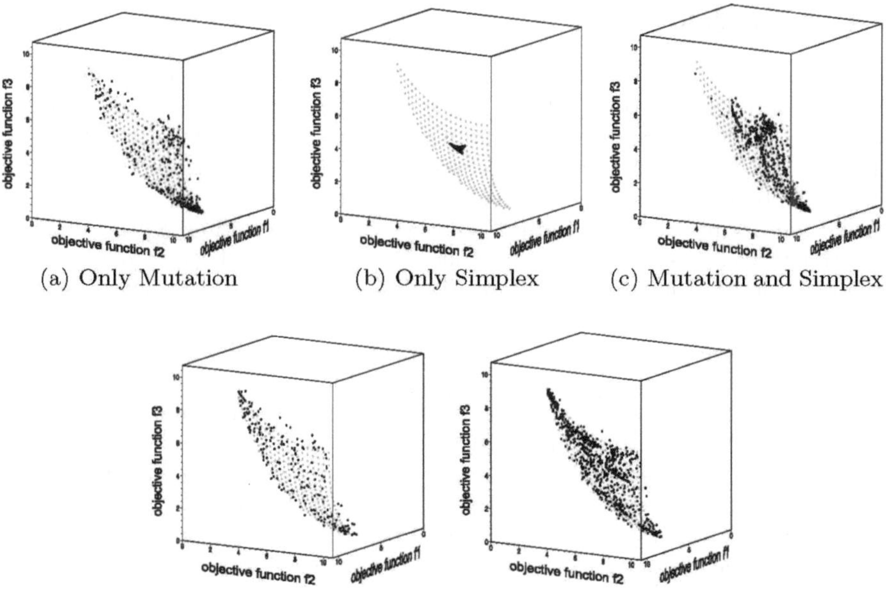

(a) Only Mutation (b) Only Simplex (c) Mutation and Simplex

(d) Mutation and Discrete Recombination (e) Combined Operator

Fig. 6. Optimized Pareto front results for test problem $\mathcal{F}_3(\boldsymbol{x})$. From left to right we applied only mutation, only simplex recombination, and mutation combined with simplex recombination in the upper row. In the lower row the combination of discrete recombination and mutation is shown as well as the finally composed operator.

consider the discrete recombination operator as well. As shown in Figure 6(a) mutation alone favors convergence to an extrema solution of one objective. Again, the simplex recombination is strongly collapsing the population while the solutions are very close to the true Pareto front in this area, see Figure 6(b). The combination of both operators yields the result shown in Figure 6(c). The influence of mutation is dominant, so that most solutions are still tending to one objective which leads to a loss of diversity. Nevertheless, the convergence behavior is quite better which can be attributed to the simplex recombination's influence. To overcome the diversity problem an alternative combination of mutation and discrete recombination is reviewed so that the diversity preserving properties of discrete recombination can be examined, see Figure 6(d). Finally, the combination of all three operators composes the previously observed properties of all examined simple operators as well. It is shown in Figure 6(e) that the resulting Pareto front favors no objective anymore, preserves diversity, and has a good convergence behavior, see also Table 1.

5.3 Discussion

In the presented figures the quality of diversity can be observed for the different operators and their combination. To be more precise with respect to the convergence to the optimum we provide the average distance from the Pareto set in Table 1. Noteworthy, we have not applied single discrete recombination as it is not possible to converge to the optimum. For discrete recombination additional mutation is always required as otherwise only an interchange of parental solutions takes place. Furthermore, the examination of discrete recombination has been omitted for problem $\mathcal{F}_2(\boldsymbol{x})$. It should be apparent from the presented results that we are able to find a good combination of the different operators for these simple test functions. Although the test problems are rather simple they suffice as a first proof of concept. For completeness, Table 2 shows the applied configurations of predators for the different experiments. Note that all these parameters are tuned manually.

As already mentioned in the introduction, Rudenko and Schoenauer [12] proposed a recombination operator for the multi-objective case which is mainly based on the BLX-α operator [5] for real coded genetic algorithm. For ($\alpha > 0$) BLX-α can be decomposed into a recombination and mutation part as this scheme allows also the discovery of new points within the search space. Therefore, also BLX-α fits our approach conceptional as it also combines two standard variations into one complex operator.

However, the main problem arises in finding a good ratio of those operators in general. As mentioned above, the predator-prey model allows an arbitrary ratio of different operators but it does not support any external or self-adaptation of this ratio yet. Thus, it is necessary to tune not only each single operator by the number of predators for each objective but also their interaction. It cannot be taken for granted that a configuration that supports the desired properties of a single operator also performs well for the combination. Apparently, there are many side effects that may originate from the superposition of different variation operator characteristics. One possible way to adjust the different optimal

Table 1. Average euclidean distance of the solutions from the Pareto set for the two test problems, the different operators, and their combinations

	Avg. Dist. to Pareto set	
Problem	$\mathcal{F}_2(\boldsymbol{x})$	$\mathcal{F}_3(\boldsymbol{x})$
Only Mutation	0.01098	0.09062
Only Simplex Recombination	0.01286	0.04243
Only Discrete Recombination	-	-
Mutation + Simplex Recombination	0.00705	0.06669
Mutation + Discrete Recombination	-	0.10924
Mutation + Simplex Recomb. + Discrete Recomb.	-	0.06064

Table 2. The amount of predators for the conducted experiments of this paper. Tuples (f_1, f_2) and (f_1, f_2, f_3) denote the number of predators chasing for objective f_1 and f_2 or f_1, f_2, and f_3 respectively. The predator's type is given by the column heading.

	P_{Mut}	$P_{Simplex}$	$P_{Discrete}$
$\mathcal{F}_2(x)$	(f_1, f_2)	(f_1, f_2)	(f_1, f_2)
Fig. 5(a)	(1,1)	-	-
Fig. 5(b)	-	(3,5)	-
Fig. 5(c)	(2,2)	(2,3)	-
$\mathcal{F}_3(x)$	(f_1, f_2, f_3)	(f_1, f_2, f_3)	(f_1, f_2, f_3)
Fig. 6(a)	(1,1,3)	-	-
Fig. 6(b)	-	(2,2,2)	-
Fig. 6(c)	(1,1,3)	(2,2,2)	-
Fig. 6(d)	(1,1,3)	-	(1,1,1)
Fig. 6(e)	(1,1,2)	(2,2,1)	(1,1,2)

ratios is the application of parameter optimization techniques. Besides all well known evolutionary methods [1] also statistical procedures like SPO [2] might be sufficient to tune those ratios.

6 Conclusion and Future Work

So far we have proposed a predator-prey model based design concept for multi-objective variation operators. Our approach allows to analyze single-objective variation operators for the multi-objective case and easily combine them to more complex but powerful operators. Thereby, every arbitrary ratio of various recombination and mutation schemes can be applied to the population. Additionally, as every possible variation scheme is represented by one predator all compositions can be realized. We have shown exemplarily that it is possible to combine properties of different single-objective mutation and recombination operators for the optimization of a test problem with two and three objectives.

Looking ahead, the model could be applied to more difficult test problems to analyze different variation operators' behavior. With these tools at hand one could identify reasonable combinations for different problems and perform a step towards a more general design pattern for multi-objective variation operators. To this end, the problem of finding good ratios of variations must be tackled, which leads to an underlying optimization problem of steering the search operators' collaboration. This may be solved by external optimization technique as well as self-adaptation strategies.

For future research it is possible to apply our model to genetic algorithms, evolution strategies, and related operators in general. After all, it seems to be promising to design specialized operators for solving more complex multi-objective real world problems. This is due to the capability to integrate specific knowledge into a larger problem context in a modular fashion.

Acknowledgement

Joachim Lepping is member of the Collaborative Research Center 531, "Computational Intelligence", at the Dortmund University with financial support of the Deutsche Forschungsgemeinschaft (DFG).

References

1. T. Bäck and H.-P. Schwefel. An Overview of Evolutionary Algorithms for Parameter Optimization. *Evolutionary Computation*, 1(1):1–23, 1993.
2. T. Bartz-Beielstein, C. Lasarczyk, and M. Preuss. Sequential parameter optimization. In B. McKay et al., editor, *Proceedings of 2005 Congress on Evolutionary Computation*, volume 1, pages 773–780, Piscataway, New Jersey, 2005. IEEE Press.
3. D. Büche, S. Müller, and P. Koumoutsakos. Self-Adaptation for Multi-objective Evolutionray Algorithms. In C. M. Fonseca, P. J. Fleming, E. Zitzler, K. Deb, and L. Thiele, editors, *Evolutionary Multi-Criterion Optimization Conference*, number 2632 in Lecture Notes in Computer Science (LNCS), pages 267–281, Berlin, 2003. Springer.
4. K. Deb. *Multi-Objective Optimization using Evolutionary Algorithms*. Wiley-Interscience Series in Systems and Optimization. Wiley, Chichester, 2001.
5. L. J. Eshelman and J. D. Schaffer. Real-coded genetic algorithms and interval-schemata. In *Second Workshop on Foundations of Genetic Algorithms*, pages 187–202. Morgan Kaufmann, 1992.
6. C. Grimme and K. Schmitt. Inside a Predator-Prey Model for Multi-Objective Optimization: A Second Study. In H.-G. Beyer et al., editors, *Genetic and Evolutionary Computation Conference*, pages 707–714, New York, 2006. ACM Press.
7. T. Hanne. On the convergence of multiobjective evolutionary algorithms. *European Journal of Operational Research*, 117:553–564, 1999.
8. F. Kursawe. A Variant of Evolution Strategies for Vector Optimization. In H.-P. Schwefel and R. Männer, editors, *Parallel Problem Solving from Nature*, volume 496 of *Lecture Notes in Computer Science (LNCS)*, pages 193–197, Berlin, 1991. Springer.
9. M. Laumanns, G. Rudolph, and H.-P. Schwefel. A Spatial Predator-Prey Approach to Multi-Objective Optimization: A Preliminary Study. In Th. Bäck, A. E. Eiben, M. Schoenauer, and H.-P. Schwefel, editors, *Parallel Problem Solving From Nature V*, pages 241–249. Springer, Berlin, 1998.
10. X. Li. A Real-Coded Predator-Prey Genetic Algorithm for Multiobjective Optimization. In *Evolutionary Multi-Criterion Optimization Conference*, pages 207–221, 2003.
11. B. Naujoks, N. Beume, and M. Emmerich. Metamodel-assisted SMS-EMOA applied to airfoil optimization tasks. In R. Schilling, W. Haase, J. Périaux, and H. Baier, editors, *EUROGEN 2005* (CD-ROM). FLM, TU München, 2005.
12. O. Rudenko and M. Schoenauer. Dominance based crossover operator for evolutionary multi-objective algorithms. In X. Yao et al., editors, *Parallel Problem Solving from Nature VIII*, volume 3242 of *Lecture Notes in Computer Science (LNCS)*, pages 812–821. Springer, 2004.
13. G. Rudolph. On a Multi-Objective Evolutionary Algorithm and Its Convergence to the Pareto Set. In *Proc. of the 5th IEEE CEC*, pages 511–516, Piscataway, New Jersey, 1998. IEEE Press.

14. K. Schmitt, J. Mehnen, and T. Michelitsch. Using predators and preys in evolution strategies. In H.-G. Beyer et al., editors, *Genetic and Evolutionary Computation Conference*, volume 1, pages 827–828, New York, 2005. ACM Press.
15. H.-P. Schwefel. *Evolution and Optimum Seeking.* Wiley, New York, 1995.
16. H.-P. Schwefel and G. Rudolph. Contemporary evolution strategies. In F. Morán, A. Moreno, J. J. Merelo, and P. Chacón, editors, *Advances in Artificial Life – Third European Conference on Artificial Life*, pages 893–907, Berlin, 1995. Springer.
17. M. Tomassini. *Spatially Structured Evolutionary Algorithms: Artificial Evolution in Space and Time.* Natural Computing Series. Springer, 2005.
18. E. Zitzler, M. Laumanns, and L. Thiele. SPEA2: Improving the Strength Pareto Evolutionary Algorithm. Technical Report 103, Computer Engineering and Communication Networks Lab (TIK), Swiss Federal Institute of Technology (ETH), Zürich, 2001.

Capabilities of EMOA to Detect and Preserve Equivalent Pareto Subsets

Günter Rudolph, Boris Naujoks, and Mike Preuss

Universität Dortmund, Lehrstuhl für Algorithm Engineering
44221 Dortmund, Germany
{mike.preuss,boris.naujoks,guenter.rudolph}@uni-dortmund.de
http://ls11-www.cs.uni-dortmund.de

Abstract. Recent works in evolutionary multiobjective optimization suggest to shift the focus from solely evaluating optimization success in the objective space to also taking the decision space into account. They indicate that this may be a) necessary to express the users requirements of obtaining distinct solutions (distinct Pareto set parts or subsets) of similar quality (comparable locations on the Pareto front) in real-world applications, and b) a demanding task for the currently most commonly used algorithms. We investigate if standard EMOA are able to detect and preserve equivalent Pareto subsets and develop an own special purpose EMOA that meets these requirements reliably.

1 Introduction

Almost all publications about evolutionary multiobjective algorithms (EMOA) put their emphasis on approximating the Pareto front in the objective space whereas the relevance of an appropriate approximation of the Pareto set is widely neglected. The knowledge about the Pareto front is important for the product designer. But as soon as a solution in objective space has been selected it is important to know for the product engineer if there are alternative solutions in the decision space that lead to the same objective vector. Such Pareto-optimal solutions in decision space exist if there are symmetries in the objective function. This phenomenon occurs for example in the test problems considered by Chan and Ray [1] or Preuss et al. [2]. Basically, the Pareto set could be partitioned into subsets where the images of each subset are identical, i.e., each Pareto subset of this partition represents the entire Pareto front. Figure 1 illustrates and distinguishes different cases that may occur in multiobjective problems.

Apart from artificial test problems, there are of course real-world problems that exhibit such symmetries. For example, consider the problem of designing a proper diet for people with special needs. Besides taking into account nutrient and non-nutrient requirements, there are also aesthetic standards regarding shape, colors and others (cf. Seljak [3]). Of course, there are numerous ways to compile alternative but equally valuable meals that differ only in the exchange of some vegetables.

S. Obayashi et al. (Eds.): EMO 2007, LNCS 4403, pp. 36–50, 2007.

Here, we are interested in the capabilities of standard EMOA of detecting and/or preserving Pareto subsets of equivalent quality. A more detailed view of our aims and methods is given in section 2. For our analysis, we construct an artificial problem class that exploits symmetries in the objective function in an extreme manner along with various geometric transformations. The same blue print can be used to construct further test classes in future. This approach is presented in section 3, which is enriched with an experimental investigation of the problem hardness via *design of experiment* (DOE) methods. Section 4 evaluates standard EMOA and a special purpose EMOA on this problem class which leads to the observation that standard EMOA and even the special purpose EMOA do not provide fully satisfying results. Therefore, we develop a new EMOA approach that is based on the multistart technique along with several scalarization methods. We can show empirically that this approach delivers a reliable and accurate approximation of all Pareto subsets with equivalent quality. We finish with our conclusions in section 6.

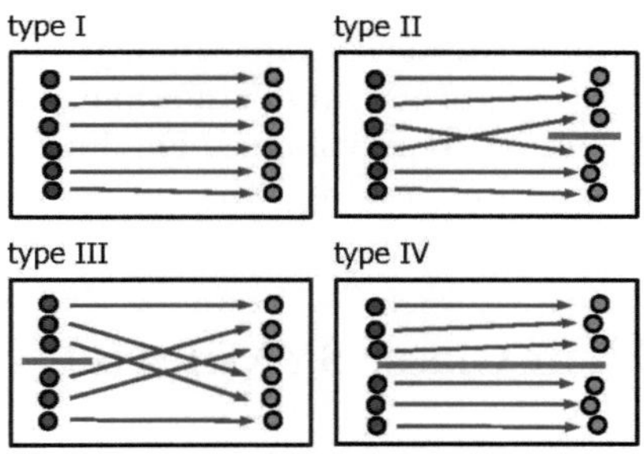

Fig. 1. Different Pareto set and Pareto front type combinations: One Pareto set and one Pareto front (type I), one Pareto set and multiple Pareto front parts (type II), multiple Pareto subsets and one Pareto front (type III), and multiple Pareto subsets and Pareto front parts (type IV). Type III problems are rarely investigated, although they potentially provide multiple preimages for every objective vector of interest.

2 Aims and Methods

To investigate the behavior of EMOA and their operators in presence of multiple Pareto set parts (type III problems), we concentrate on three main questions:

- Which properties make these problems especially hard or simple for standard EMOA?
- What are the mechanisms in EMOA that lead to better or worse performance in terms of Pareto set preservation and Pareto front approximation?
- How can Pareto set preservation in EMOA be improved?

Obviously, standard performance measures for multiobjective optimization algorithms disregard how Pareto sets are dealt with; they only refer to population distributions in the objective space. We therefore define two simple new measures which require knowledge about Pareto subset numbers and locations and are thus not applicable to real-world application problems.

Covered sets (cs): The number of covered Pareto subsets (which comprise at least one individual in their vicinity).

$$\mathrm{cs}(P, S) := |\{\mathrm{set} \in S : \exists \mathrm{ind} \in P, \mathrm{near}(\mathrm{ind}, \mathrm{set})\}| \tag{1}$$

Set population spread (sps): The standard deviation of the Pareto subset population counts (the number of individuals found on a Pareto subset).

$$\mathrm{sps}(P, S) := \sqrt{\mathrm{VAR}(\{\forall \mathrm{set} \in S : \ |\mathrm{ind} \in P, \mathrm{near}(\mathrm{ind}, \mathrm{set})|\})} \tag{2}$$

The formal definitions refer to a population P of points (ind) in decision space and a set S of Pareto subsets (set). The boolean function `near`(ind,set) becomes true if the tested individual reaches the vicinity of the tested set. For determining when exactly this is the case, the concrete problem must be taken into account. VAR stands for the sample variance s^2, determined to $s^2 = \frac{1}{n-1} \sum_{i=1}^{n} (x_i - \overline{x})^2$.

For measuring the Pareto front approximation quality of a population, we utilize the common S-metric (hypervolume). Furthermore, standard experiment layout and visualization techniques from the *design of experiments* (DOE) field (see Montgomery [4]) are employed.

3 A Test-Problem Class: SYM-PART

In a previous work [2], a configurable type III test problem with two distinct Pareto sets, overlapping only in the decision space origin, has been investigated. These distinct Pareto sets were caused by the point symmetry of the bi-modal objective function. It is easy to see that such property entails loss of surjectiveness by creating two or more preimages of the optima and search points in their vicinity. As soon as at least the global optimum of one objective function (which is by definition part of the Pareto front of the resulting multiobjective function) is affected, multiple, possibly connected Pareto subsets emerge. In the following, we use this reasoning to construct SYM-PART (symmetrical parts) test problems with a controllable number of Pareto subsets, heavily relying on symmetry properties of the underlying singleobjective functions.

3.1 Construction of the Test Problems

Starting point is a very simple and well known test problem with two objectives and two-dimensional search space, namely,

$$f(x_1, x_2) = \begin{pmatrix} (x_1 + a)^2 + x_2^2 \\ (x_1 - a)^2 + x_2^2 \end{pmatrix} \tag{3}$$

for some $a > 0$. The Pareto set $\mathcal{X}^* = \{x \in \mathbb{R}^2 : x = (x_1, 0)' \text{ with } x_1 \in [-a, a]\}$ maps to the Pareto front $\mathcal{F}^* = f(\mathcal{X}^*) = \{z \in \mathbb{R}^2 : z = (4\,a^2\,\nu^2, 4\,a^2$

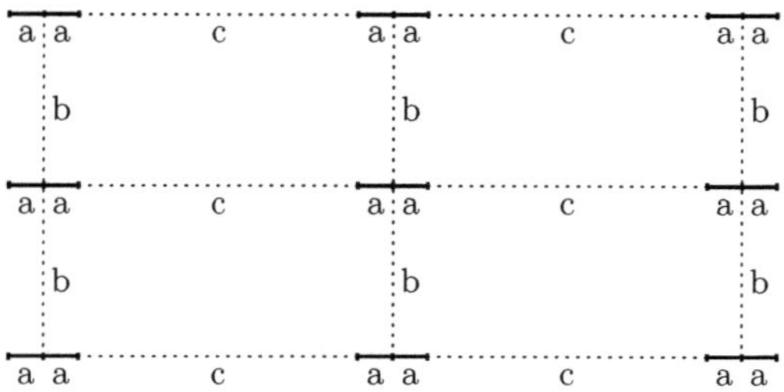

Fig. 2. Blue print of the initial test problem: Each subset of the Pareto set is a line of length $2a$. Parameter b specifies the vertical distance between neighboring Pareto subsets, whereas parameter c specifies the distance to the next Pareto subset on the horizontal line. Each Pareto subset maps to the same Pareto front.

$(1-\nu)^2)'$ with $\nu \in (0,1)\}$. Our idea is to translate the problem above to different regions in search space (see Fig. 2), such that each of these Pareto subsets are of equivalent quality since each Pareto subset maps to the same Pareto front.

For this purpose we define test problem (3) only in a certain neighborhood. Such a neighborhood will be called *tile* hereinafter (see Fig. 3).

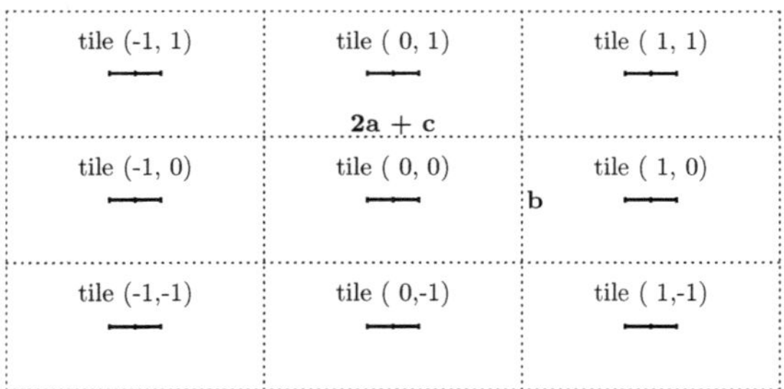

Fig. 3. Tile pattern for function (3) translated to tiles (i, j) that are defined by a rectangular region with width $2a+c$ and height b. Here, (i, j) denotes the tile identifier.

The tile identifiers are determined via

$$\hat{t}_1 = \operatorname{sgn}(x_1) \times \left\lceil \frac{|x_1| - (a + \frac{c}{2})}{2a + c} \right\rceil \tag{4}$$

$$\hat{t}_2 = \operatorname{sgn}(x_2) \times \left\lceil \frac{|x_2| - \frac{b}{2}}{b} \right\rceil \tag{5}$$

where a, b and c are the parameters for specifying the tile pattern. We restrict the problem to 9 tiles, i.e., the tile identifiers t_i only attain values in $\{-1, 0, 1\}$ by using the relation $t_i = \operatorname{sgn}(\hat{t}_i) \times \min\{|\,\hat{t}_i|, 1\}$. Now we are in the position to define the first test problem instance:

$$f^{(1)}(x_1, x_2) = f(x_1 - t_1\,(c + 2\,a), x_2 - t_2\,b)$$

The second test problem instance requires that x is rotated by $\omega = 45°$ via

$$r(x) = \begin{pmatrix} \cos\omega & -\sin\omega \\ \sin\omega & \cos\omega \end{pmatrix} x$$

before calculating the tile identifiers t_1, t_2 in (4) and (5). This leads to (see Fig. 4, left)

$$f^{(2)}(x_1, x_2) = f^{(1)}(r_1(x), r_2(x))\,.$$

Finally, we add a transformation that distorts the shape of the Pareto subsets:

$$d(x_1, x_2) = x_1 \times \left(\frac{x_2 - L + \varepsilon}{U - L}\right)^{-1}$$

for some small $\varepsilon > 0$ and where U and L denote the upper and lower bound of the search space, respectively. When transforming x_1 prior to calculating the tile identifiers, the third test problem instance is defined by (see Fig. 4, right)

$$f^{(3)}(x_1, x_2) = f^{(2)}(d(x_1, x_2), x_2)\,.$$

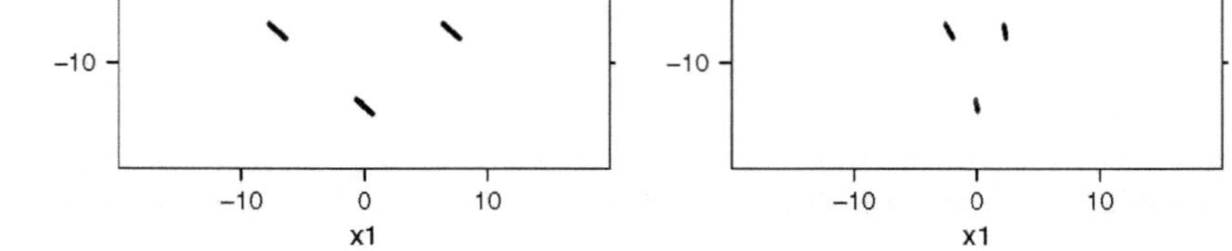

Fig. 4. Empirically detected (randomly enumerated) Pareto sets of SYM-PART test problems instances 2 and 3 (instance 1 refers to the original problem depicted in Fig. 3). Instance 2: 45° rotation, no transformation (left), instance 3: 45° rotation with transformation (right). Note that Pareto subset sizes differ here.

Needless to say, we are aware of the weaknesses of these test instances since they exploit only one type of symmetry and since they are defined only for two dimensions in search and objective space. But as can be seen shortly, these simple test problems can be used to demonstrate interesting phenomena occurring in standard EMOA and some special purpose EMOA presented here.

3.2 Experimental Investigation of Problem Hardness

In the following sections, several EMOA are tested for their ability to reach and preserve many or all existing Pareto subsets. It is therefore necessary to establish differently difficult problem instances of the SYM-PART problem class. In particular, the three problem instances developed in §3.1 shall be assessed. Apart from the fact that we do not employ any evolutionary algorithm but simple, deterministic, grid-based search methods, exploring the effect of problem modifications onto optimization methods is related to the approach of Langdon and Poli [5].

Experiment 1 relies on the utilization of *design of experiments* (DOE) techniques as first introduced by Fisher [6]. The controllable input variables or factors—in this case problem properties—are varied systematically in discrete levels. Observing the resulting performance changes then enables estimating the impact of single properties (main effects) and combined properties (interaction effects). An experimental layout that requires to actually test all possible factor level combinations is called a fully factorial design. For larger numbers of factors, one often uses fractional factorial designs. These reduce the number of runs by ignoring certain factor level combinations at the expense of explanatory power regarding higher-order interaction effects. For a more thorough introduction into DOE methods we refer to standard textbooks (e.g. Montgomery [4]).

Experiment 1: Problem hardness of different SYM-PART configurations.

Pre-experimental planning: First experiments revealed that a standard operator/value NSGA2 (see Tab. 3) performs reasonably well in preserving Pareto sets over a long time (30,000 evaluations). Replacing search operators or parameter values seems to weaken this ability. The NSGA2 is therefore chosen as constant base algorithm when modifying the treated problem.

Task: Detect which SYM-PART modifications have a large impact on the ability of an EMOA to discover and preserve as many Pareto sets as possible. Recommend few considerably different SYM-PART instances for further use.

Setup: We apply a full factorial design: NSGA2 is run with 30 repeats on each factor level combination (16). Low and high factor levels are given in Tab. 1. Bounds refers to the rectangular search space bounds, shift stands for translation of the whole tile structure relative to the origin, rotation and transformation are as stated in §3.1.

Results/Visualization: The mean number of covered sets (cs) and the set population spread (sps) are used to compute main and interaction effects. These

Table 1. SYM-PART problem designs, made of combinations of 4 factors, each of which has a low (left) and high (right) level. Chosing all 4 low levels results in the original problem as described in §3.1.

parameter	**bounds** $(L{:}U)$	**shift** vs. origin	**rotation** angle	**transformation**
factor levels	$-50{:}50/-20{:}20$	$(0,0)/(2,2)$	$0°$ / $45°$	no / yes

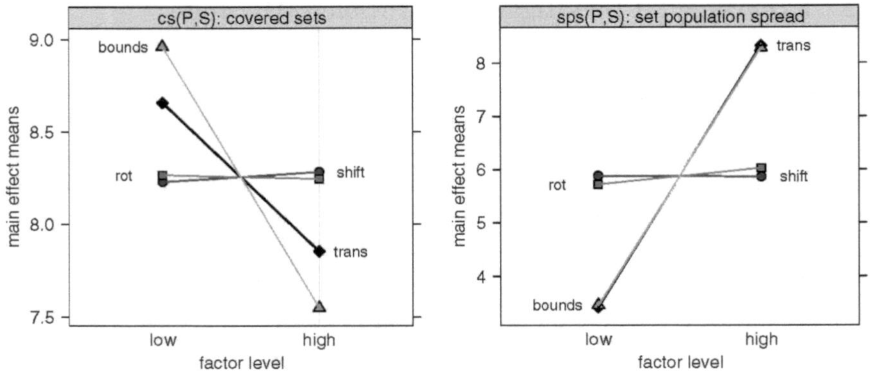

Fig. 5. DOE main effects, original mean values, without adjustment towards the average. As the standard deviations of the observed cs(P,S) values are almost 1 (up to ≈ 5 for the largest values of sps(P,S)), all but the largest two effects are insignificant.

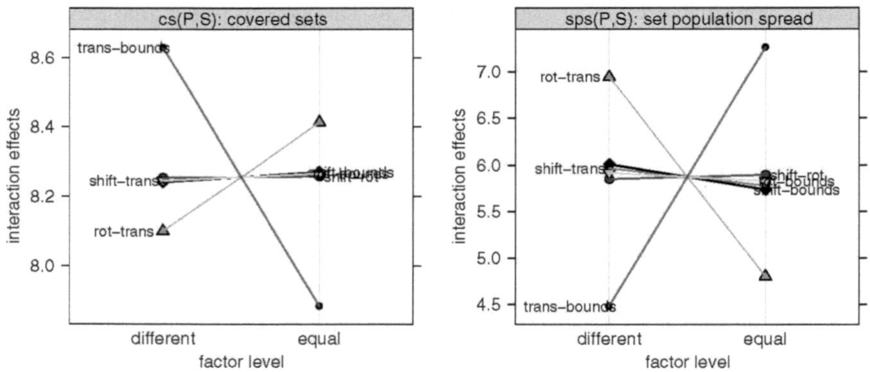

Fig. 6. DOE interaction effects, as Fig. 5 at the original location, but differently scaled, without adjustment. All except the trans-bound and rot-trans interactions are insignificant due to the high variance level.

are depicted in Fig. 5 and Fig. 6, respectively. Due to space limitations and to enhance comparability, all effects are plotted into one diagram, thereby deviating from standard DOE practice. Higher-order interaction effects (of more than two factors) are disregarded.

Observations: The strongest main effects are caused by the transformation and the extent of decision space bounds (trans and bounds in Fig. 5). Measures cs(P,S) and sps(P,S) return consistent values: For smaller decision spaces, less Pareto subsets are kept, and the spread of set populations increases. The transformation has a similar effect and obviously makes the problem harder if switched on. Shift and rotation apparently do not affect problem hardness. The interaction effect plots Fig. 6 document that only two interactions need to be considered: Trans-bound and rot-trans. Both interaction effects are much weaker than the important main effects. Whereas trans-bound signals harder problems if both factors are either set to their low or to their high levels, rot-trans points into the other direction. If only rotation or only transformation is switched on, the problem appears to be harder than if both are on or off.

Discussion: Surprisingly, changing decision space bounds has a large effect on performance in terms of $\mathrm{cs}(P, S)$ and $\mathrm{sps}(P, S)$. If the relative amount of search space that must be covered for placing individuals in all Pareto subsets approaches 1, the EMOA gets more and more difficulties. We attribute this behavior at least in part to the *polynomial mutation* (PM) operator which uses the upper and lower bounds for adjusting its step size distribution. We must however state that the PM operator works reasonably well even under very tight bounds around the Pareto subsets. As setting the bounds to the high factor level (-20/20) greatly increases problem difficulty, we consider only these in the following.

Dissecting the impact of the 4 possible combinations of rotation and transformation leads to an unexpected order of increasing hardness: $\neg$rot $\wedge$ $\neg$trans (mean/stddev(cs)=8.83/0.33) < rot $\wedge$ $\neg$trans (8.49/0.49) < rot $\wedge$ trans (8/0.60) < $\neg$rot $\wedge$ trans (7.71/0.76). To keep the number of problem instances for further testing as low as possible, we select only 3 of these, namely the simple one ($\neg$rot $\wedge$ $\neg$trans), the rotated one (rot $\wedge$ $\neg$trans), and the rotated and transformed one ($\neg$rot $\wedge$ $\neg$trans). Instead of the latter, one could also chose the not rotated but transformed instance. However, we refrained from doing so because the difference between these two is rather small, and it is currently not clear why the instance without rotation may be more difficult.

4 Evaluation of Standard EMOA on SYM-PART

Compared to §3.2, we now follow the opposite approach and test several common EMOA on the three previously selected SYM-PART problem instances.

Experiment 2: Investigate convergence/diversity tradeoff for different EMOA.

Pre-experimental planning: First results confirmed the expected behavior: Standard techniques do not perform well even on the simplest instance of the SYM-PART problem. The algorithms only kept a very limited number of tiles ($\mathrm{cs}(P,S)$).

Later, it was discovered that this unwanted behavior was seemingly caused by adaptive mutation featuring $n = 2$ step sizes [7]. After changing the variation operator to polynomial mutation [8], which became the standard mutation operator within EMOA in recent years, the quality of results increased significantly. This is indicated by the average number of tiles preserved by different EMOA, in turn using the two mentioned mutation operators. Mean values for $\mathrm{cs}(P,S)$ are given next to the corresponding standard deviations (in brackets) in Tab. 2. As

Table 2. Test of standard EMOA with different mutation operators, namely polynomial mutation (PM) and adaptive mutation with two step sizes (AM). The values give average cs(P,S) values of 18 runs with 10,000 evaluations each (standard deviations are given in brackets).

algorithm	AM	PM
NSGA-II	1.65 (0.745)	8.61 (0.608)
SPEA2	1.94 (0.873)	8.94 (0.236)

polynomial mutation performed much better, this operator was applied in the investigation of different EMOA on all instances of the SYM-PART problem.

Task: The performance of EMOA is to be tested on all instances of the SYM-PART problem. More detailedly, we look for drawbacks of the standard techniques in contrast to an algorithm that is explicitly developed to keep diversity in solution space as well as in decision space. Are the algorithms able to discover new tiles and can they keep the new tiles for the rest of the optimization run?

Setup: We invoke two standard techniques next to a new development within the field. The Pisa framework[1] is used to conduct the referred optimization runs with the standard techniques. Here, all specifications of the SYM-PART problem have been implemented as a variator, which can be optimized with respect to different objectives and multiple selectors. Among the set of available selectors, NSGA-II and SPEA2 are chosen, because these appear to be the currently most well-known and commonly used algorithms in the field [8,9]. Additionally, the more recent KP1 by Chan and Ray [1] is tested.

KP1 was designed for maintaining diversity in decision space as well as in objective space. Therefore, two criteria to measure the diversity of solutions in the corresponding spaces are defined and applied in each generation. These are dominated hypervolume of each individual for the objective space and a neighborhood counting approach for the decision space. Both are described in detail by Chan and Ray [1]. The OMNI-Optimizer by Deb et al. [10] considers only one of such measurements in the different space at a time and is not included in this study.

The parameters of the variation operators are set to standard values, i.e. SBX and PM with distribution indices $\eta_c = 15$ and $\eta_m = 20$, respectively. Crossover and mutation probability are set to one. Selection is performed using a (100+100) selection scheme for 300 generations in either cases, resulting in 30,000 fitness function evaluations per run (see Tab. 3).

The additional effort for a third algorithm in the study seems to be justified as the development aims of this algorithm directly address the difficulties of the chosen test problems.

Results/Visualization: Tab. 4 and 5 give average final results of the 30 runs performed for every algorithm on every instance of the SYM-PART problem.

[1] PISA - Platform and Programming Language Independent Interface for Search Algorithms, ETH Zurich, `www.tik.ee.ethz.ch/pisa/`

Table 3. Parameter setting for standard EMOA depicting mutation and crossover probabilities (mut.prop. and cross.prob.), distribution indices (η_m and η_c), and the selection scheme in use.

parameter	mut.prob.	η_m	cross.prob.	η_c	selection
value	1	20	1	15	$(100 + 100)$

Table 4. Test of different algorithms on all instances of the SYM-PART problem. The values give the average $cs(P, S)$ of 30 runs with 30,000 evaluations each (standard deviations are given in brackets).

algorithm	simple	rotated	rot.+trans.
NSGA-II	6.333 (1.446)	5.633 (1.450)	4.667 (1.124)
SPEA2	6.3 (1.022)	5.2 (1.157)	5 (1.364)
KP1	8.3 (1.290)	6.733 (1.818)	6.5 (0.9738)

Table 5. Test of different algorithms on all instances of the SYM-PART problem. The values give the average dominated hypervolume after 30 runs with 30,000 evaluations each (standard deviations are given in brackets).

algorithm	simple	rotated	rot.+trans.
NSGA-II	22.254 (0.00353)	22.255 (0.00305)	22.254 (0.00358)
SPEA2	22.257 (0.00237)	22.257 (0.00243)	22.255 (0.00278)
KP1	22.241 (0.00712)	22.231 (0.00689)	22.220 (0.00781)

Tab. 4 more detailedly depicts the average number of tiles preserved by the indicated algorithm after 30,000 evaluations. The mean hypervolume received after the corresponding runs is contained in Tab. 5. But, these averaged values of the final results do not give evidence for the behavior of the different algorithms during the runs. This aspect is tackled in Fig. 7, where all repetitions of runs have been averaged generation by generation. For example, the upper left diagram of Fig. 7 depicts three curves, one for each instance of the SYM-PART problem. Each curve is generated averaging the results achieved after the first generation, the second one, up to the 300th one. The same holds for all other curves within all diagrams in Fig. 7. The middle row holds SPEA2 results while the lower one displays the results of KP1 by Chan and Ray. The upper row is dedicated to NSGA2 and the left column to the generation-wise averaged number of tiles kept as can be seen from the example above.

The right column gives the generation-wise average values of the dominated hypervolume. Here, the displayed area is shortened to the starting phase of the runs up to generation 50. This is done to highlight the interesting developments during the runs and implies that no major changes in the behavior take place after the depicted interval of the run. The final results of the averaged runs can be taken from Tab. 4 and 5 as described above.

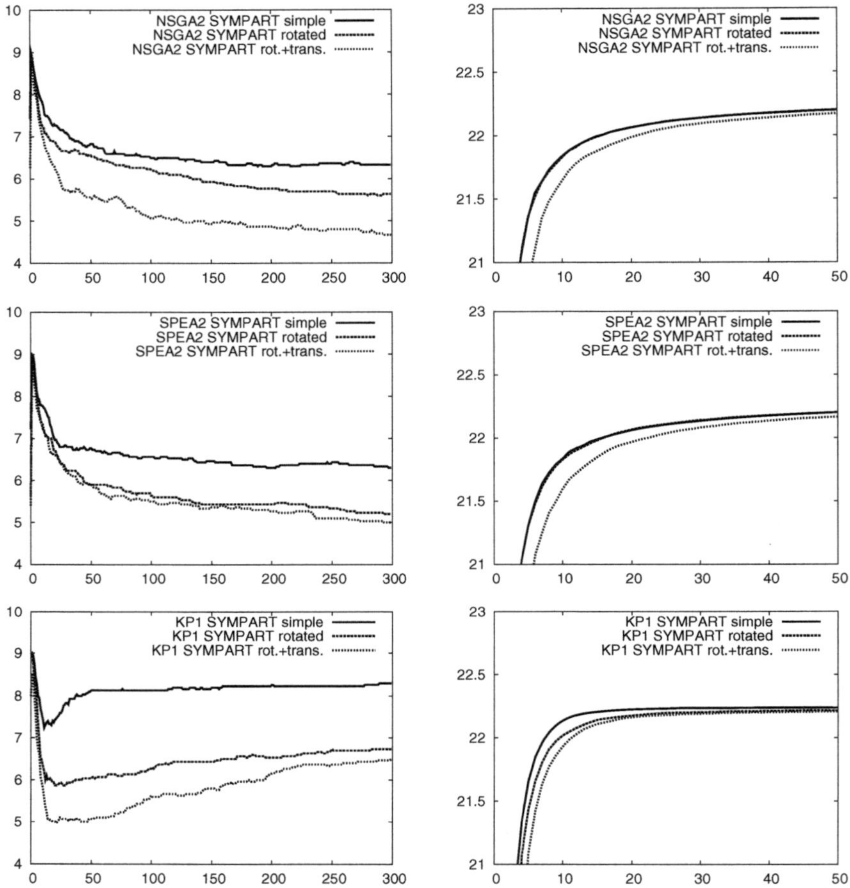

Fig. 7. Average runs of NSGA-II, SPEA2 and KP1 of Chan and Ray (labeled KP1) on all instances of the SYM-PART function. The left column presents the average $cs(P, S)$ values over the evaluations while the right one gives the average dominated hypervolume. The average runs have been received from 30 runs performed, 30,000 fitness function evaluations each. Only the first part up to 5,000 evaluations is presented in case of the hypervolume plot due to better observability of results.

Observations: With respect to the number of tiles kept, Tab. 4 shows the expected behavior of the algorithms within this study: The number of tiles kept decreasing with growing hardness of the considered instance of the SYM-PART problem. This means, most of the tiles are kept on SYM-PART 1 by all algorithms. Here, KP1 clearly outperforms the other algorithms keeping 8.3 of 9 tiles on average. This is the highest value achieved within all experiments. The lowest number of tiles is received for SYM-PART 3, the rotated and transformed instance and therefore the most difficult one. On this problem, NSGA2 receives the lowest value achieved within all experiments (4.667). For all algorithms, the values for SYM-PART 2 are greater than the ones for SYM-PART 3 and smaller than the ones for SYM-PART 1. KP1 performs better than the other algorithms on all instances. Interestingly, NSGA2 is better than SPEA2 on SYM-PART 1 and SYM-PART 2, while SPEA2 performs better on SYM-PART 3.

The behavior of the algorithms changes when taking the dominated hypervolume into account (see Tab. 5). SPEA2 receives the best results on all instances, followed shortly by NSGA2. KP1 clearly achieves the worst values of dominated hypervolume on all instances. Furthermore, the values from this algorithm decrease with problem complexity. This behavior can not be observed for NSGA2 and SPEA2. Here, the largest dominated hypervolume is obtained on the rotated instance of SYM-PART, while the lowest values are achieved on the rotated and transformed SYM-PART 3.

More dramatic differences in the behavior of KP1 in contrast to NSGA2 and SPEA2 can be observed in the diagrams of Fig. 7 considering the average $cs(P, S)$ values per generation. In contrast to the behavior of KP1, NSGA2 and SPEA2 loose tiles during the averaged optimisation runs. KP1 first looses tiles as well but turns its behavior after about 20 generation on all three instances. Starting here, KP1 almost constantly captures tiles back. Interestingly, steps can be observed even in the averaged runs. This is due to increasing as well as decreasing $cs(P, S)$ values within single runs. As a consequence, also KP1 is not able to keep all newly discovered tiles for the rest of the run. Some are lost again after only a few generations. But, in contrast to NSGA2 and SPEA2, this algorithm is able to keep more tiles than get lost. This leads to the over all increasing number of tiles on average.

The curves depicting the hypervolume do not yield such interesting results. The values here increase rapidly to almost optimal values for all algorithms. More detailedly, NSGA2 and SPEA2 act almost comparable on SYM-PART 1 and SYM-PART 2. The dominated hypervolume increases a bit more slightly on SYM-PART 3. This also holds for KP1, where a more distinct difference can be observed between SYM-PART 1 and SYM-PART 2. Over all, the results for KP1 seem to converge to the almost optimal values for the run a bit faster. But, as can be seen from Tab. 5, these values are worse than the ones for SPEA2 and NSGA2.

Discussion: With respect to the course of the tiles kept, an important difference in the behavior of the algorithms is observed. While this course decreases for NSGA2 and SPEA2, it increases for KP1. The final conclusion that all but one tile are lost after more generations of NSGA2 and SPEA2 while all tiles are captured back using KP1 is not shown, but is an self-evident assumption.

The values for the dominated hypervolume reveal that the more tiles are kept, the less hypervolume is achieved. This leads to the assumption that both criteria are conflicting. The fact that no hypervolume is lost with increasing number of tiles in the KP1 runs contradicts this assumption. Therefore, KP1 can be stated to be the best algorithm within this study, although not dominating all the hypervolume the other algorithms do. This is due to KP1 preserving diversity not only in the solution space, but also in the decision space. Considering both criteria, it would be better to stop the runs of the standard algorithms more early, i.e. after about 50 generations. At his point, they already dominate almost all possible hypervolume and occupy the highest number of tiles.

What is not tackled in this investigation is the distribution of individuals over tiles $sps(P, S)$. In the most comprehensive algorithms, the user would like the number of individuals to grow on newly occupied tiles. At the end of a run, a uniform distribution of individuals over all Pareto sets within tiles is aspired.

5 A Multistart Approach for Pareto Subset Detection

An alternative approach to detect and maintain several Pareto subsets of equivalent quality is provided by the multistart technique. The algorithm described here is still of experimental state but very promising. The main idea is as follows: We run a singleobjective optimizer for each objective function. Since the optimal solution of each objective function is Pareto-optimal we have a kind of anchor that can be used to approximate the associated Pareto subset successively by deploying some singleobjective optimizer repeatedly with different weights of the scalarized multiobjective function.

Let $f(x) = (\, f_1(x), \ldots, f_d(x)\,)$ be the objective function with $x \in \mathbb{R}^n$. At first, N runs with a standard $(1, \lambda)$-ES are made for each of the d objectives. The ES stops if the standard deviation σ of the mutation operator is below some threshold $\delta > 0$. Each solution x^* is stored and annotated with the index of the objective function used: $(x^*, k) \in \mathbb{R}^n \times \{1, \ldots, d\}$. Thus, we obtain $N \cdot d$ candidate solutions in this manner.

Suppose there are $s \in \mathbb{N}$ Pareto subsets with equivalent quality. If all Pareto subsets are hit by the multistart approach then we need only $s \cdot d$ anchor solutions as starting points of the singleobjective search with the scalarized multiobjective function to approximate all Pareto subsets. Since the number s of the equivalent Pareto subsets is unknown in general, we deploy an unsupervised clustering method to reduce the $N \cdot d$ candidate solutions to $s \cdot d$ anchor solutions required for the next step. Actually, it is possible to reduce the number of anchor solutions to s since we can apply the clustering method to the N solutions of each objective separately (recall that we have annotated each candidate solution with the index of the objective function used). Since the different objective functions may be of varying difficulty for the optimization, we can use the d outcomes of the clustering method as a consistency check. This idea, however, is currently not implemented. We simply cluster the candidate solutions of the objective function with index 1 and proceed with $\hat{s}$ estimated anchor solutions.

The scalarization used in the sequel is known as the *weighted Tchebycheff method* [11]: The multiobjective function $f : \mathbb{R}^n \to \mathbb{R}^d$ is scalarized via

$$f^{<s>}(x) = \max_{i=1,\ldots,d} \{\, w_i \, |f_i(x) - u_i^*| \,\}$$

where $u^* \in \mathbb{R}^d$ is the *utopian solution*. Since we have made N singleobjective optimizations of each objective $f_i : \mathbb{R}^n \to \mathbb{R}$ in the first phase of our algorithm, we have obtained an accurate estimator of the *ideal solution* z^* with $z_i^* = \min\{f_i(x) : x \in \mathbb{R}^n\}$ for $i = 1, \ldots, d$. As a consequence, we may set $u_i^* = z_i^* - 1$ to get a valid utopian solution required for the weighted Tchebycheff method (WTM). We have chosen WTM because of its ability to find also solutions whose images are on a *concave* Pareto front. Needless to say, here we tacitly assume that the Pareto subsets are connected.

The user may choose how many representatives of each Pareto subset are desired. Suppose we like to obtain k representatives. Then for each of the $\hat{s}$ anchor solutions x^* we start a standard $(1, \lambda)$-ES with initial $\sigma_0 = 10\,\delta$, seeding point x^*, and weights that cover all possible weight assignments with maximal uniformity. In case of $d = 2$ objectives the weights are given by $w_1 = j/(k-1)$ and $w_2 = 1 - w_1$ for $j = 1, \ldots, k-1$. Notice that the anchor solution x^* is used as initial parent of the ES for $j = 1$ only. The best solution found in this run serves as initial parent for $j = 2$. And so forth until $j = k - 1$. In this vein, we finally arrive at an approximation of all Pareto subsets that were detected in the first phase of the algorithm.

For an assessment of this approach, we made some experiments for the three test instances introduced previously. The parametrization was as follows: $\lambda = 5$, $N = 50$, $\delta = 10^{-5}$, $k = 10$. The initialization of the ES in the first phase used $\sigma_0 = 20/6$ and the starting point was sampled uniformly from the region $[-20, 20]^2$.

Each run out of 30 in total detected the 9 Pareto subsets reliably and approximated the Pareto subset with high accuracy. In the first phase each run of the $(1, \lambda)$-EA stops on average in less than 60 generations. Thus, we required less than $60 \times \lambda \times N \times d = 30,000$ function evaluations of the *single*-objective functions, which is equivalent to 15,000 function evaluations of the multiobjective function. The second phase (clustering) does not evaluate the objective function. The third phase required less than 5,000 function evaluations of the scalarized multiobjective function. Thus, this approach required less than the equivalent of 20,000 multiobjective function evaluations for a reliable and accurate approximation of all Pareto subsets for all test instances. Figure 8 shows typical results for the three test instances.

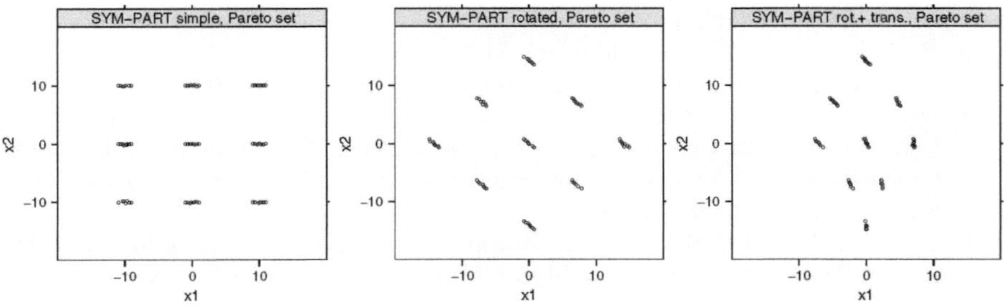

Fig. 8. Typical runs of the multistart approach on all instances of the SYM-PART problem (from left to right: instances 1, 2, and 3, as described in §3.1)

6 Conclusions and Future Work

We have shown that standard EMOA are not able to reliably detect and/or preserve all Pareto subsets of equivalent quality. This is not surprising as they have not been designed for this purpose. Moreover, this property is not required in some cases. But if we need this property we have to deploy special purpose EMOA. We have tested one such EMOA given in the literature and we have developed another EMOA that is based on a multistart approach which meets our requirements. It is imaginable that EMOA with niching can be successful in this case, too. But this analysis remains for future research, as well as the development of additional problem classes that exploit different types of symmetries and that are defined in higher-dimensional decision and objective spaces.

References

1. Chan, K.P., Ray, T.: An Evolutionary Algorithm to Maintain Diversity in the Parametric and the Objective Space. In: Proceedings of IEEE International Conference on Computational Robotics and Autonomous Systems (CIRAS 2005), Centre for Intelligent Control, National University of Singapore (2005) ISSN: 0219-6131.
2. Preuss, M., Naujoks, B., Rudolph, G.: Pareto Set and EMOA Behavior for Simple Multimodal Multiobjective Functions. In Runarsson, T.P., et al., eds.: Parallel Problem Solving from Nature (PPSN IX). Volume 4193 of Lecture Notes in Computer Science., Springer, Berlin (2006) 513–522
3. Seljak, B.K.: Dietary Menu Planning by Evolutionary Computation. In Filipic, B., Silc, J., eds.: Bioinspired Optimization Methods and their Applications (BIOMA 2006), Jozef Stefan Institute, Ljubljana, Slovenia (2006) 87–98
4. Montgomery, D.C.: Design and Analysis of Experiments. 5th edn. Wiley, New York (2001)
5. Langdon, W.B., Poli, R.: Evolving problems to learn about particle swarm and other optimisers. In McKay, B., et al., eds.: Congress on Evolutionary Computation (CEC'05). Volume 1., IEEE Press, Piscataway NJ (2005) 81–88
6. Fisher, R.A.: The Design of Experiments. Oliver and Boyd, Edinburgh (1935)
7. Beyer, H.G., Schwefel, H.P.: Evolution strategies – A comprehensive introduction. Natural Computing **1** (2002) 3–52
8. Deb, K.: Multi-Objective Optimization using Evolutionary Algorithms. Wiley, Chichester, UK (2001)
9. Coello Coello, C.A., Van Veldhuizen, D.A., Lamont, G.B.: Evolutionary Algorithms for Solving Multi-Objective Problems. Kluwer, New York (2002)
10. Deb, K., Tiwari, S.: Omni-optimizer: A Procedure for Single and Multi-objective Optimization. In Coello Coello, C.A., et al., eds.: Evolutionary Multi-Criterion Optimization (EMO 2005). Volume 3410 of Lecture Notes in Computer Science., Springer, Berlin (2005) 47–61
11. Steuer, R.E.: Multiple Criteria Optimization: Theory, Computation, and Application. Wiley series in probability and mathematical statistics. Wiley, New York (1986)

Optimization of Scalarizing Functions Through Evolutionary Multiobjective Optimization

Hisao Ishibuchi and Yusuke Nojima

Department of Computer Science and Intelligent Systems, Graduate School of Engineering, Osaka Prefecture University, 1-1 Gakuen-cho, Naka-ku, Sakai, Osaka 599-8531, Japan
hisaoi@cs.osakafu-u.ac.jp, nojima@cs.osakafu-u.ac.jp
http://www.ie.osakafu-u.ac.jp/~hisaoi/ci_lab_e

Abstract. This paper proposes an idea of using evolutionary multiobjective optimization (EMO) to optimize scalarizing functions. We assume that a scalarizing function to be optimized has already been generated from an original multiobjective problem. Our task is to optimize the given scalarizing function. In order to efficiently search for its optimal solution without getting stuck in local optima, we generate a new multiobjective problem to which an EMO algorithm is applied. The point is to specify multiple objectives, which are similar to but different from the scalarizing function, so that the location of the optimal solution is near the center of the Pareto front of the generated multiobjective problem. The use of EMO algorithms helps escape from local optima. It also helps find a number of alternative solutions around the optimal solution. Difficulties of Pareto ranking-based EMO algorithms in the handling of many objectives are avoided by the use of similar objectives. In this paper, we first demonstrate that the performance of EMO algorithms as single-objective optimizers of scalarizing functions highly depends on the choice of multiple objectives. Based on this observation, we propose a specification method of multiple objectives for the optimization of a weighted sum fitness function. Experimental results show that our approach works very well in the search for not only a single optimal solution but also a number of good alternative solutions around the optimal solution. Next we evaluate the performance of our approach in comparison with a hybrid EMO algorithm where a single-objective fitness evaluation scheme is probabilistically used in an EMO algorithm. Then we show that our approach can be also used to optimize other scalarizing functions (e.g., those based on constraint conditions and reference solutions). Finally we show that our approach is applicable not only to scalarizing functions but also other single-objective optimization problems.

1 Introduction

Evolutionary multiobjective optimization (EMO) is one of the most active research areas in the field of evolutionary computation. EMO algorithms have been successfully applied to various application areas involving multiple objectives [2]. In some cases, EMO algorithms can outperform single-objective evolutionary algorithms even when they are used to solve single-objective problems. It was reported in some studies on multiobjectivization [15], [18] that better results were obtained by

S. Obayashi et al. (Eds.): EMO 2007, LNCS 4403, pp. 51–65, 2007.

transforming single-objective problems into multiobjective ones (see [15] for multiobjectivization).

Motivated by these studies on multiobjectivization, we examined the use of EMO algorithms to optimize the sum of multiple objectives in our former studies [8], [10]. We obtained promising results when we used NSGA-II [3] to optimize the simple sum fitness function for a two-objective 500-item (i.e., 2-500) knapsack problem of Zitzler & Thiele [19]. That is, NSGA-II outperformed its single-objective version in finding the optimal solution of the sum of the two objectives. This is because the use of NSGA-II helps escape from local optima.

Usually EMO algorithms are very good at finding Pareto-optimal or near Pareto-optimal solutions around the center of the Pareto front of a two-objective problem. EMO algorithms, however, are not always good at finding good solutions near the edge of the Pareto front. This is illustrated in Fig. 1 where NSGA-II was applied to the 2-500 knapsack problem [19] using two different settings. In Fig. 1 (a), standard parameter values were used (i.e., 0.8 crossover probability and 1/500 mutation probability). In this case, we observe a good convergence of solutions to the Pareto front. Actually NSGA-II outperformed its single-objective version in finding the optimal solution of the simple sum fitness function: $fitness(\mathbf{x}) = f_1(\mathbf{x}) + f_2(\mathbf{x})$. On the other hand, lower crossover and higher mutation probabilities were used in Fig. 1 (b) in order to increase the diversity of solutions. The increase in the diversity of solutions in Fig. 1 (b) was achieved at the cost of the deterioration in the convergence to the Pareto front. Experimental results in Fig. 1 suggest that the direct use of EMO algorithms is not a good choice for finding the optimal solution of a weighted sum fitness function with very different weight values such as $fitness(\mathbf{x}) = 0.1 f_1(\mathbf{x}) + 0.9 f_2(\mathbf{x})$.

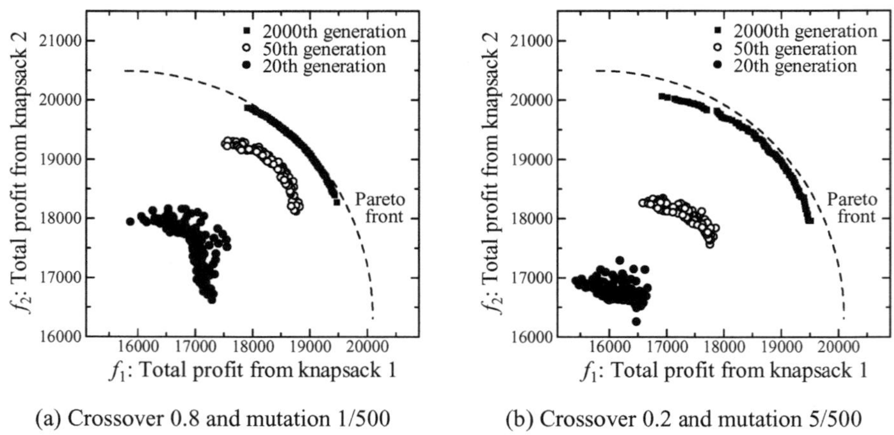

Fig. 1. Experimental results of NSGA-II on the 2-500 knapsack problem using two different settings of the crossover and mutation probabilities

Another weakness of EMO algorithms is the difficulty in the handling of many objectives. Most EMO algorithms are based on Pareto ranking to evaluate the fitness of each solution. Pareto ranking-based EMO algorithms, however, do not work well on many-objective problems (e.g., see [6], [7], [14], [17]). This is because solutions rarely dominate other solutions in the presence of many objectives. Hughes [7] showed that multiple runs of single-objective optimizers outperformed a single run of EMO algorithms in their applications to many-objective problems. Similar results were also reported in Jaszkiewicz [14]. These results in the literature suggest that the use of EMO algorithms is not a good choice for finding the optimal solution of a scalarizing function generated from many objectives such as the simple sum fitness function of four objectives: $fitness(\mathbf{x}) = f_1(\mathbf{x}) + f_2(\mathbf{x}) + f_3(\mathbf{x}) + f_4(\mathbf{x})$.

The above-mentioned experimental results can be summarized as follows:

(1) EMO algorithms work well for optimizing a scalarizing function if the location of its optimal solution is near the center of the Pareto front of a two-objective optimization problem. For example, EMO algorithms can easily find good solutions in the region A in Fig. 2 (a) as shown in Fig. 1 (a).
(2) EMO algorithms do not always work well for optimizing a scalarizing function if the location of its optimal solution is near the edge of the Pareto front of a two-objective optimization problem. For example, EMO algorithms do not always easily find good solutions in the region B or C in Fig. 2 (a) as shown in Fig. 1 (b).
(3) EMO algorithms are not likely to work well for optimizing a scalarizing function if they are applied to a many-objective problem.

In this paper, we propose an idea of using an EMO algorithm to efficiently optimize a scalarizing function even in the last two cases: (2) and (3). We generate a new multiobjective problem to which an EMO algorithm is applied. The point is to specify multiple objectives, which are similar to but different from the given scalarizing function, so that the location of the optimal solution is near the center of the Pareto front of the generated multiobjective problem. Our idea is illustrated in Fig. 2 (b) where we generate two objectives g_1 and g_2 in order to efficiently find good solutions in the region B. Slow convergence of EMO algorithms in the case of many objectives is remedied by the use of similar objectives as we will show later in this paper.

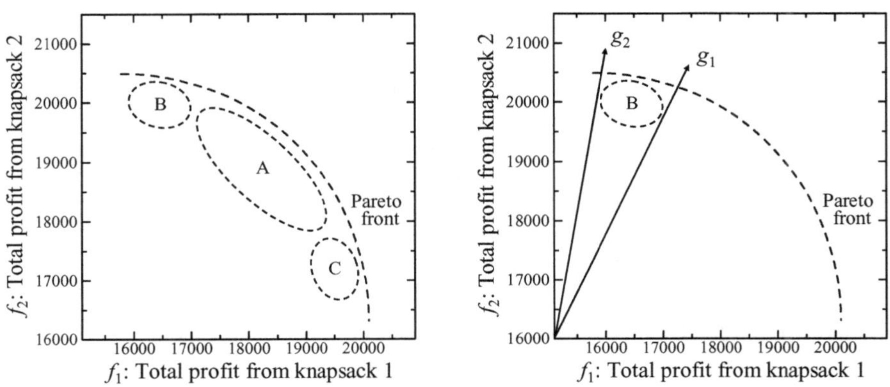

(a) Three regions in the objective space (b) Newly generated objectives: g_1 and g_2

Fig. 2. Illustration of the proposed idea

In this paper, we first demonstrate that the performance of EMO algorithms as single-objective optimizers of scalarizing functions highly depends on the choice of multiple objectives in Section 2. Based on this observation, we propose a specification method of multiple objectives for the optimization of a weighted sum fitness function in Section 3. Experimental results show that our approach works very well in the search for not only a single optimal solution but also a number of alternative solutions around the optimal solution. We also show that EMO algorithms work well as single-objective optimizers even in the case of many objectives. In Section 4, the effectiveness of our approach is compared with a hybrid EMO algorithm where a single-objective fitness evaluation scheme is probabilistically used in an EMO algorithm. Then we show that our approach is applicable not only to weighted sum fitness functions but also other scalarizing functions (e.g., those based on constraint conditions and reference solutions) and more general single-objective optimization problems in Section 5. Finally we conclude this paper in Section 6.

2 Optimization of Scalarizing Functions by EMO Algorithms

In this section, we examine the effectiveness of EMO algorithms as single-objective optimizers of scalarizing functions through computational experiments on multiobjective 0/1 knapsack problems in Zitzler & Thiele [19]. As a representative EMO algorithm, we use NSGA-II [3]. For comparison, we also use its single-objective version.

2.1 Scalarizing Functions

Let us consider the following k-objective maximization problem:

$$\text{Maximize } \mathbf{f}(\mathbf{x}) = (f_1(\mathbf{x}),\ f_2(\mathbf{x}),\ ...,\ f_k(\mathbf{x}))\,, \tag{1}$$

where $\mathbf{f}(\mathbf{x})$ is the k-dimensional objective vector, and $\mathbf{x}$ is the decision vector.

One of the frequently used scalarizing functions is the weighted sum fitness function with the non-negative weight vector $\mathbf{w} = (w_1, w_2, ..., w_k)$:

$$fitness(\mathbf{x}) = w_1 \cdot f_1(\mathbf{x}) + w_2 \cdot f_2(\mathbf{x}) + \ ...\ + w_k \cdot f_k(\mathbf{x})\,. \tag{2}$$

We assume that the weight vector $\mathbf{w}$ is normalized (i.e., the sum of the weight values is 1). The weight vector $\mathbf{w}$ in (2) is usually supposed to be given by human users.

The weighted sum fitness function with various weight vectors was successfully used to directly realize various search directions in multiobjective genetic local search (MOGLS) algorithms [9], [11], [12]. High performance of MOGLS of Jaszkiewicz [12] was reported [1], [13], [16]. The weighted sum fitness function was also used in hybrid or multi-stage EMO algorithms (e.g., see [8], [10], [16]).

When a reference vector $\mathbf{f}^* = (f_1^*, f_2^*, ..., f_k^*)$ is given as a desired point in the objective space, the distance from $\mathbf{f}^*$ can be used as a scalarizing function:

$$fitness(\mathbf{x}) = distance(\mathbf{f}^*, \mathbf{f}(\mathbf{x}))\,. \tag{3}$$

In this paper, we use the Euclidean distance. The incorporation of reference points into EMO algorithms was examined in Deb & Sundar [4].

Another scalarizing function is based on the transformation of some objectives into inequality conditions. Let us assume that the minimum requirement level for each of the first $(k-1)$ objectives is given as an inequality condition:

$$f_i(\mathbf{x}) \geq \varepsilon_i \text{ for } i = 1, 2, ..., k-1. \tag{4}$$

The following scalarizing fitness function is usually formulated from the maximization problem of $f_k(\mathbf{x})$ with the $(k-1)$ inequality conditions in (4):

$$fitness(\mathbf{x}) = f_k(\mathbf{x}) - \alpha \sum_{i=1}^{k-1} \max\{0, \varepsilon_i - f_i(\mathbf{x})\}, \tag{5}$$

where α is the unit penalty with respect to the violation of the inequality conditions in (4). In computational experiments of this paper, we specified α as $\alpha = 1$.

2.2 NSGA-II and Its Single-Objective Version

NSGA-II is an elitist EMO algorithm with the $(\mu + \lambda)$-ES generation update mechanism. The outline of NSGA-II can be written as follows:

[NSGA-II]
```
Step 1: P = Initialize(P)
Step 2: While the stopping condition is not satisfied, do
Step 3:                P' = Parent Selection(P)
Step 4:                P'' = Genetic Operations(P')
Step 5:                P = Generation Update(P∪P'')
Step 6: End while
Step 7: Return Non-dominated(P)
```

In NSGA-II, each solution in the current population P is evaluated using Pareto ranking and a crowding measure in the following manner for parent selection in Step 3. First the best rank is assigned to all the non-dominated solutions in the current population. Solutions with the best rank are tentatively removed from the current population. Next the second best rank is assigned to all the non-dominated solutions in the remaining population. In this manner, ranks are assigned to all solutions in the current population. The rank of each solution is used as the primary criterion in parent selection. A crowding measure is used to compare solutions with the same rank as the secondary criterion in parent selection (for details, see [2], [3]).

A prespecified number of pairs of parent solutions are selected from the current population by binary tournament selection to form a parent population P' in Step 3. An offspring solution is generated from each pair of parent solutions by crossover and mutation to form an offspring population P'' in Step 4. The current population P and the offspring population P'' are merged to form an enlarged population. Each solution in the enlarged population is evaluated by Pareto ranking and the crowding measure as in the parent selection phase. A prespecified number of the best solutions are chosen from the enlarged population as the next population P in Step 5. Usually the number of offspring solutions is the same as the population size (i.e., $\mu = \lambda$ in the $(\mu + \lambda)$-ES generation update mechanism).

We can easily implement a single-objective version (single-objective genetic algorithm: SOGA) of NSGA-II by using a scalarizing fitness function for parent selection and generation update. Such an SOGA has the $(\mu + \lambda)$-ES generation update mechanism with $\mu = \lambda$. We compare NSGA-II with SOGA through computational experiments on multiobjective 0/1 knapsack problems in Zitzler & Thiele [19].

2.3 Computational Experiments

As in Fig. 1 (a), we applied NSGA-II to the 2-objective 500-item (i.e., 2-500) knapsack problem [19] using the following parameter specifications:

Population size: 200 (i.e., $\mu = \lambda = 200$),
Crossover probability: 0.8 (uniform crossover),
Mutation probability: 1/500 (bit-flip mutation) where 500 is string length,
Termination condition: 2000 generations.

Average results over 50 runs of NSGA-II are summarized as the 50% attainment surface [5] in Fig. 3 (a) where average results of SOGA are also shown for comparison. The weight vector, the reference vector and the minimum requirement level were specified in SOGA as $\mathbf{w} = (0.5, 0.5)$, $\mathbf{f}^* = (19250, 19250)$ and $\varepsilon_1 = 18750$ in Fig. 3 (a), respectively. We can observe that NSGA-II outperformed SOGA in Fig. 3 (a).

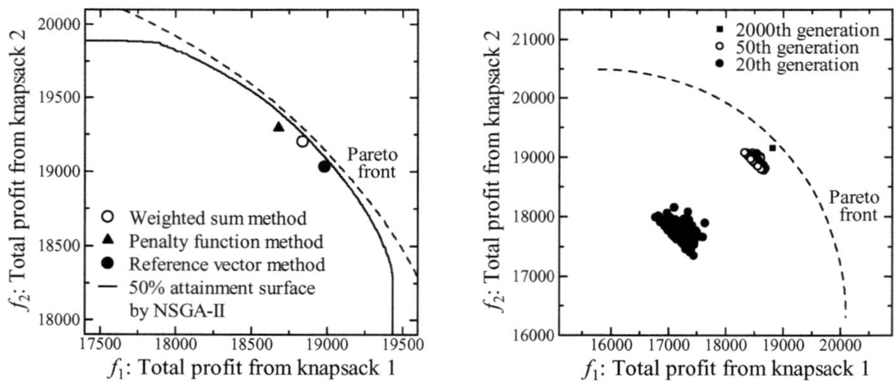

(a) Comparison between NSGA-II and SOGA (b) SOGA with the weighted sum fitness

Fig. 3. Experimental results of NSGA-II and SOGA on the 2-500 knapsack problem

In Fig. 3 (b), we show two intermediate and final populations during a single run of SOGA with the weighted sum fitness function with $\mathbf{w} = (0.5, 0.5)$. It should be noted that SOGA in Fig. 3 (b) was executed under the same parameter specifications as NSGA-II in Fig. 1 (a). From the comparison between these two figures, we can see that NSGA-II maintained a larger diversity of solutions. The decrease in the diversity of solutions during the execution of SOGA seems to be the main reason of the inferior performance of SOGA in Fig. 3 (a) in comparison with NSGA-II.

NSGA-II and SOGA are further compared with each other in Fig. 4 (a) for the 2-500 knapsack problem using the weighted sum fitness function with $\mathbf{w} = (0.5, 0.5)$. Fig. 4 (a) shows the distribution of obtained solutions from 50 runs of each algorithm. Whereas good solutions were almost always obtained by NSGA-II, the quality of the final solution by each run of SOGA seems to highly depend on the initial population.

In Fig. 3 (a) and Fig. 4 (a), NSGA-II outperformed SOGA when they were used to optimize the three scalarizing functions. The advantage of NSGA-II over SOGA, however, disappears as the increase in the number of objectives. In Fig. 4 (b), we show experimental results on the four-objective knapsack problem. The performance of NSGA-II as a single-objective optimizer was deteriorated in Fig. 4 by the increase in the number of objectives from two in Fig. 4 (a) to four in Fig. 4 (b). Pareto ranking-based EMO algorithms usually do not work well on many-objective problems.

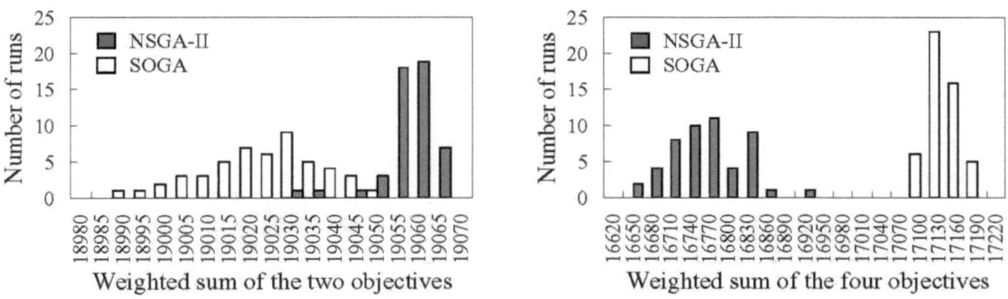

(a) Results on the two-objective problem (b) Results on the four-objective problem

Fig. 4. Comparison between NSGA-II and SOGA. The 2-500 knapsack problem with $\mathbf{w} = (0.5, 0.5)$ in (a) and the 4-500 knapsack problem with $\mathbf{w} = (0.25, 0.25, 0.25, 0.25)$ in (b).

The advantage of NSGA-II over SOGA also disappears when the location of the optimal solution of a scalarizing function is near the edge of the Pareto front as we have already explained using Fig. 1. For example, NSGA-II did not work well on the 2-500 knapsack problem as a single-objective optimizer of scalarizing functions in the following cases: $\mathbf{w} = (0.1, 0.9)$, $\mathbf{f}^* = (16750, 20500)$ and $\varepsilon_1 = 17000$. In each case, the location of the optimal solution is close to the top-left edge of the Pareto front (see Fig. 1 for the spatial relation between the obtained solutions by NSGA-II and the Pareto front). NSGA-II had difficulties in efficiently searching for good solutions near the edge of the Pareto front in comparison with SOGA in these cases.

In the above-mentioned three difficult cases, the performance of NSGA-II as a single-objective optimizer was improved when we used the following two objectives:

$$g_1(\mathbf{x}) = 0.5 f_1(\mathbf{x}) + 0.5 f_2(\mathbf{x}), \tag{6}$$

$$g_2(\mathbf{x}) = -0.3 f_1(\mathbf{x}) + 1.3 f_2(\mathbf{x}), \tag{7}$$

where $f_1(\mathbf{x})$ and $f_2(\mathbf{x})$ are the original two objectives of the 2-500 knapsack problem. Average results over 50 runs of NSGA-II and SOGA are summarized in Fig. 5 (a) where NSGA-II was applied to the two-objective problem in (6) and (7). In Fig. 5 (a),

NSGA-II outperformed SOGA in their applications to the optimization of the scalarizing functions in the above-mentioned three difficult cases. Fig. 5 (b) shows two intermediate and final populations during a single run of NSGA-II. The multiobjective search of NSGA-II was appropriately driven toward the desired region by the two objectives in (6) and (7) as we can see from the comparison of Fig. 5 (b) with Fig. 1 (a). As a result, NSGA-II found better solutions of the scalarizing functions than SOGA in Fig. 5 (a).

Experimental results in Fig. 5 suggest that the performance of NSGA-II as a single-objective optimizer highly depends on the specification of multiple objectives. In the next section, we propose a specification method of multiple objectives for the optimization of a weighted sum fitness function. The proposed idea can be used for other scalarizing functions as shown in Section 4.

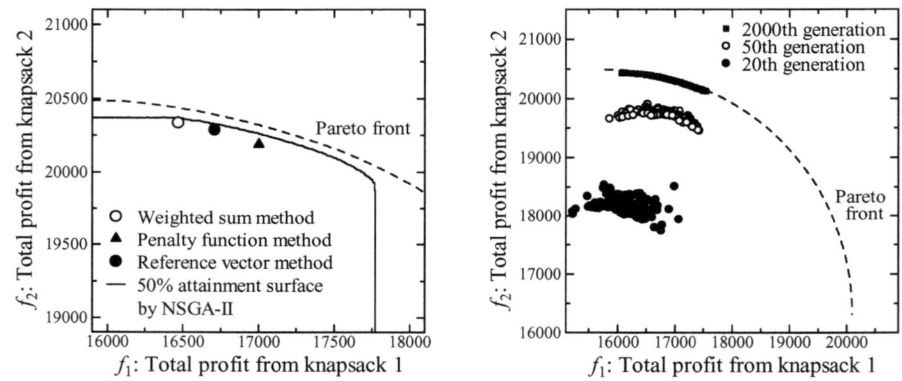

(a) Comparison between NSGA-II and SOGA (b) NSGA-II with the new two objectives

Fig. 5. Experimental results of NSGA-II and SOGA on the 2-500 knapsack problem

3 Handling of Weighted Sum Fitness Functions

When we use EMO algorithms to optimize a scalarizing function, it is essential to generate multiple objectives so that the location of the optimal solution is near the center of the Pareto front of the generated multiobjective problem. In this section, we show how we can generate such a multiobjective problem to optimize a weighted sum fitness function.

3.1 Weighted Sum Fitness Function of Two Objectives

Our task in this subsection is to optimize the weighted sum of two objectives:

$$fitness(\mathbf{x}) = w_1 \cdot f_1(\mathbf{x}) + w_2 \cdot f_2(\mathbf{x}) \,. \tag{8}$$

Two objectives can be newly generated by changing the weight vector as follows:

$$g_1(\mathbf{x}) = (w_1 + \beta) f_1(\mathbf{x}) + (w_2 - \beta) f_2(\mathbf{x}), \tag{9}$$

$$g_2(\mathbf{x}) = (w_1 - \beta) f_1(\mathbf{x}) + (w_2 + \beta) f_2(\mathbf{x}). \tag{10}$$

For example, the weight vectors of the newly generated two objectives are specified as (0.2, 0.8) and (0.4, 0.6) from the weight vector $\mathbf{w} = (0.3, 0.7)$ when $\beta = 0.1$.

Using various specifications of β, we applied NSGA-II to the two-objective problem in (9) and (10) generated from the weighted sum fitness function with $\mathbf{w} = (0.3, 0.7)$ for the 2-500 knapsack problem. Average results over 50 runs of NSGA-II are summarized in Fig. 6 (a). It should be noted that NSGA-II with $\beta = 0$ is the same as SOGA because $g_1(\mathbf{x})$ and $g_2(\mathbf{x})$ become the same as the original weighted sum fitness function. We also show experimental results for the case of $\mathbf{w} = (0.1, 0.9)$ in Fig. 6 (b). We can observe in Fig. 6 that the multiobjectivization by (9) and (10) clearly improved the quality of the obtained solutions. We can also observe that the performance of NSGA-II as a single-objective optimizer was not sensitive to the value of β.

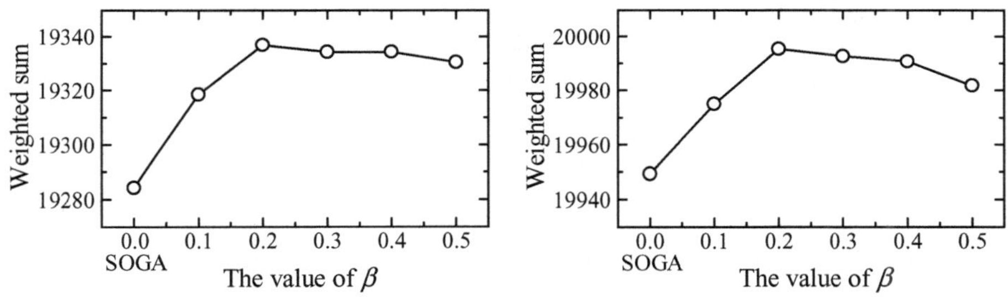

(a) Average results with $\mathbf{w} = (0.3, 0.7)$ (b) Average results with $\mathbf{w} = (0.1, 0.9)$

Fig. 6. Weighted sum optimization by NSGA-II for the 2-500 knapsack problem

The multiobjectivization by (9) and (10) is effective in the search not only for a single optimal solution but also for multiple good solutions around the optimal solution. In each plot of Fig. 7, we show two intermediate and final populations during a single run of NSGA-II for each of the two cases: $\beta = 0.3$ and $\beta = 0.5$. In both cases, the weight vector of the original weighted sum fitness function was specified as $\mathbf{w} = (0.3, 0.7)$. In Fig. 7, multiple solutions were obtained along the Pareto front of the original 2-500 knapsack problem. Moreover, the spread of the finally obtained solution set in each plot depended on the value of β. These observations suggest that the multiobjectivization by (9) and (10) can drive the population toward an appropriate search region and adjust its diversity. This means that the proposed idea has a potential usefulness as an approach to the focused search by EMO algorithms.

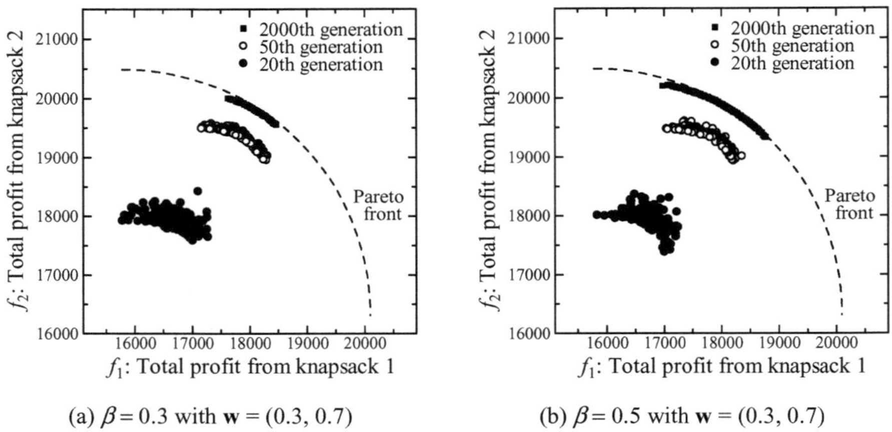

Fig. 7. Behavior of NSGA-II with the newly generated two objectives from the 2-500 problem

3.2 Weighted Sum Fitness Function of Many Objectives

The proposed idea in the previous subsection can be easily generalized to the case of more than two objectives. Let us consider the weighted sum of three objectives with the weight vector $\mathbf{w} = (w_1, w_2, w_3)$. Three objectives can be generated by changing the weight vector $\mathbf{w} = (w_1, w_2, w_3)$ toward the three directions: (1, 0, 0), (0, 1, 0) and (0, 0, 1). More specifically, the weight vectors of the three objectives are generated in the following manner:

$$\mathbf{w}_A = \mathbf{w} + \beta \cdot \frac{\mathbf{a}}{\|\mathbf{a}\|}, \quad \mathbf{w}_B = \mathbf{w} + \beta \cdot \frac{\mathbf{b}}{\|\mathbf{b}\|}, \text{ and } \mathbf{w}_C = \mathbf{w} + \beta \cdot \frac{\mathbf{c}}{\|\mathbf{c}\|}, \tag{11}$$

where $\|\mathbf{a}\|$ denotes the length of the vector, and **a**, **b**, **c** are specified as follows:

$$\mathbf{a} = (1, 0, 0) - \mathbf{w}, \quad \mathbf{b} = (0, 1, 0) - \mathbf{w}, \text{ and } \mathbf{c} = (0, 0, 1) - \mathbf{w}. \tag{12}$$

This method can be directly generalized to the case with more objectives. We applied NSGA-II to the weighted sum fitness function with $\mathbf{w} = (0.8, 0.1, 0.1)$ for the 3-500 knapsack problem. We also applied SOGA to the same problem. When we used NSGA-II, we generated a new three-objective problem using (11) and (12) with $\beta = 0.2$. The distribution of the obtained solutions by 50 runs of each algorithm is shown in Fig. 8 (a). We can see from Fig. 8 (a) that NSGA-II outperformed SOGA in their applications to the optimization of the weighted sum of the three objectives. It should be noted that NSGA-II did not work well for the same task as a single-objective optimizer when it was applied to the original 3-500 knapsack problem.

We also performed the same computational experiments on the weighted sum fitness function with $\mathbf{w} = (0.25, 0.25, 0.25, 0.25)$ for the 4-500 knapsack problem. Experimental results are shown in Fig. 8 (b). As in Fig. 8 (a), NSGA-II outperformed SOGA in Fig. 8 (b). The comparison between Fig. 4 (b) and Fig. 8 (b) clearly demonstrates the effect of the proposed idea on the performance of NSGA-II as a single-objective optimizer of the weighted sum fitness function. It should be noted that the difficulty of Pareto ranking-based EMO algorithms in the handling of many objectives was remedied by the use of similar objectives in our approach as shown in Fig. 8.

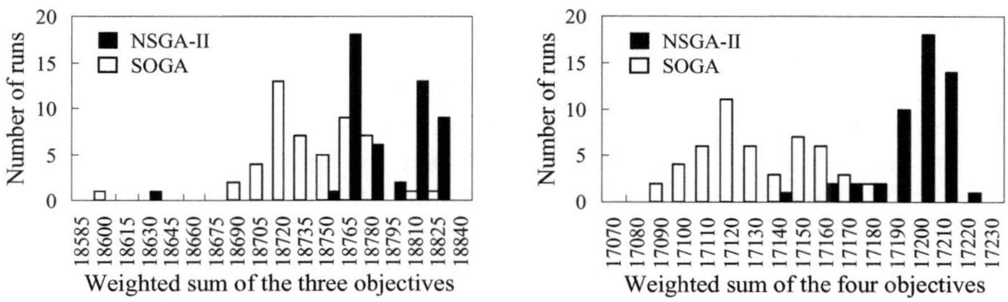

(a) 3-500 with **w** = (0.8, 0.1, 0.1) (b) 4-500 with **w** = (0.25, 0.25, 0.25, 0.25)

Fig. 8. Weighted sum optimization by NSGA-II using the proposed approach with $\beta = 0.2$

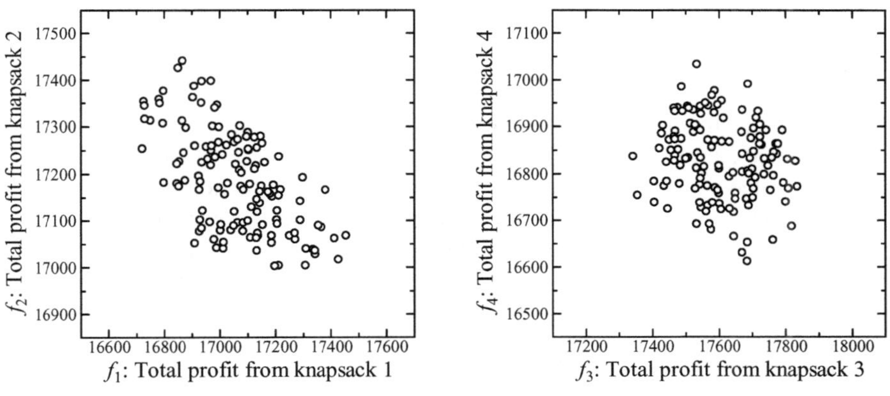

(a) Projection onto the f_1-f_2 space (b) Projection onto the f_3-f_4 space

Fig. 9. Obtained solutions by a single run of NSGA-II on the modified 4-500 problem

As shown in Fig. 8 (b), NSGA-II worked very well as a single-objective optimizer to search for the optimal solution of the weighted sum of the four objectives of the 4-500 knapsack problem. It also has the ability to find multiple non-dominated solutions as an EMO algorithm. In Fig. 9, we show the obtained non-dominated solution set by a single run of NSGA-II on the modified 4-500 knapsack problem, which was generated from the 4-500 knapsack problem with **w** = (0.25, 0.25, 0.25, 0.25) using (11) and (12) with $\beta = 0.2$. Each plot in Fig. 9 shows the projection of the obtained non-dominated solution set from the original four-dimensional objective space onto a two-dimensional space. From Fig. 9, we can see that a number of non-dominated solutions were obtained by a single run of NSGA-II.

4 Application of a Hybrid EMO Algorithm

In our former studies [8], [10], we proposed a hybrid EMO algorithm where a weighted sum fitness function was probabilistically used in NSGA-II. We introduced two probabilities P_{PS} and P_{GU}, which specified how often the weighted sum fitness function was used for parent selection and generation update in the hybrid EMO algorithm, respectively. In this section, we compare our approach (i.e., application of NSGA-II to modified multiobjective problems) with the hybrid EMO algorithm.

One extreme case of the hybrid EMO algorithm with $P_{PS} = P_{GU} = 0.0$ is exactly the same as NSGA-II since the weighted sum fitness function is never used. Another extreme case with $P_{PS} = P_{GU} = 1.0$ is the same as SOGA since the weighted sum fitness function is always used. The balance between single-objective and multiobjective search can be adjusted between the two extreme cases using the two probabilities.

In our computational experiments in this section, we examined the following 11x11 combinations of the two probabilities:

Probability P_{PS}: 0.0, 0.1, 0.2, 0.3, 0.4, 0.5, 0.6, 0.7, 0.8, 0.9, 1.0,
Probability P_{GU}: 0.0, 0.1, 0.2, 0.3, 0.4, 0.5, 0.6, 0.7, 0.8, 0.9, 1.0.

We applied the hybrid EMO algorithm to the 4-500 knapsack problem to optimize the weighted sum fitness function with the weight vector $\mathbf{w} = (0.25, 0.25, 0.25, 0.25)$. This weighted sum fitness function was used for parent selection with the probability P_{PS} and generation update with the probability P_{GU} in the hybrid EMO algorithm. When the weighted sum fitness function was not used, the multiobjective fitness evaluation scheme in NSGA-II was invoked to evaluate each solution based on the original four objectives in the 4-500 knapsack problem. Average results over 50 runs are summarized in Fig. 10 (a). The bottom-left bar with $P_{PS} = P_{GU} = 0.0$ shows the result of NSGA-II while the top-right bar with $P_{PS} = P_{GU} = 1.0$ shows the result of SOGA. Whereas the performance of NSGA-II was very poor in Fig. 10 (a), it was significantly improved by the probabilistic use of the weighted sum fitness function. Better results than SOGA were obtained by the hybrid EMO algorithm in the top-left corner with $P_{PS} = 0.0$ and $P_{GU} = 1.0$ in Fig. 10 (a). We also applied the hybrid EMO algorithm to the same problem after modifying the 4-500 knapsack problem using the proposed approach with $\beta = 0.2$. Experimental results were shown in Fig. 10 (b) where good results were obtained even when the weighted sum fitness function was not used (i.e., the bottom-left bar with $P_{PS} = P_{GU} = 0.0$). That is, the hybridization is not necessary in Fig. 10 (b) where we modified the 4-500 knapsack problem by the proposed approach. Moreover, we can observe that better results were obtained in Fig. 10 (b) after the modification of the 4-500 problem than Fig. 10 (a).

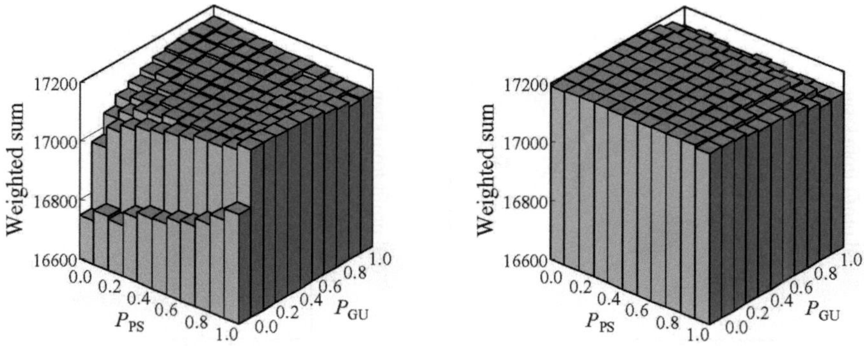

(a) Original 4-500 problem (b) Modified 4-500 problem

Fig. 10. Effect of our approach on the performance of the hybrid EMO algorithm

5 Handling of Other Scalarizing Fitness Functions

The basic idea of our approach is to generate multiple objectives, which are similar to but different from the given scalarizing function, so that the location of its optimal solution is near the center of the Pareto front of the generated multiobjective problem. This idea can be also implemented for other scalarizing functions.

For example, let us assume that we have a reference vector (19000, 20000) for the 2-500 knapsack problem. In this case, we can generate two objectives by specifying two reference vectors around (19000, 20000). Experimental results with newly generated two reference vectors (18000, 21000) and (20000, 19000) are shown in Fig. 11 (a). On the other hand, when we have a minimum requirement level (e.g., 18000) for the first objective of the 2-500 knapsack problem, we can generate two objectives by specifying two minimum requirement levels around 18000 (e.g., 17000 and 19000). Experimental results with the newly generated minimum requirement levels 17000 and 19000 are shown in Fig. 11 (b). We can observe in Fig. 11 that the search of NSGA-II was appropriately directed by the newly generated multiple objectives. We can also observe that good alternative solutions were obtained around the optimal solution of the original scalarizing function in each plot in Fig. 11.

Our approach is applicable not only to the optimization of scalarizing function but also to other optimization problems. For example, let us consider the maximization of $f(\mathbf{x})$. If we have another objective $g(\mathbf{x})$, we can generate two objectives as $f(\mathbf{x}) + w \cdot g(\mathbf{x})$ and $f(\mathbf{x}) - w \cdot g(\mathbf{x})$. In this case, the choice of $g(\mathbf{x})$ is not so important because its effect can be adjusted by the weight w. The direct use of $f(\mathbf{x})$ and $g(\mathbf{x})$ as two objectives is not a good strategy for optimizing $f(\mathbf{x})$ because the optimal solution of $f(\mathbf{x})$ is located at the edge of the Pareto front of the two-objective problem with $f(\mathbf{x})$ and $g(\mathbf{x})$. In Fig. 12, we show experimental results on the optimization of $f_2(\mathbf{x})$ of the 2-500 knapsack problem. We used $f_1(\mathbf{x})$ and $f_2(\mathbf{x})$ as two objectives in Fig. 12 (a), which is not a good strategy. On the other hand, we used $f_2(\mathbf{x}) + 0.3 f_1(\mathbf{x})$ and $f_2(\mathbf{x}) - 0.3 f_1(\mathbf{x})$ in Fig. 12 (b), which is a good strategy as multiobjectivization.

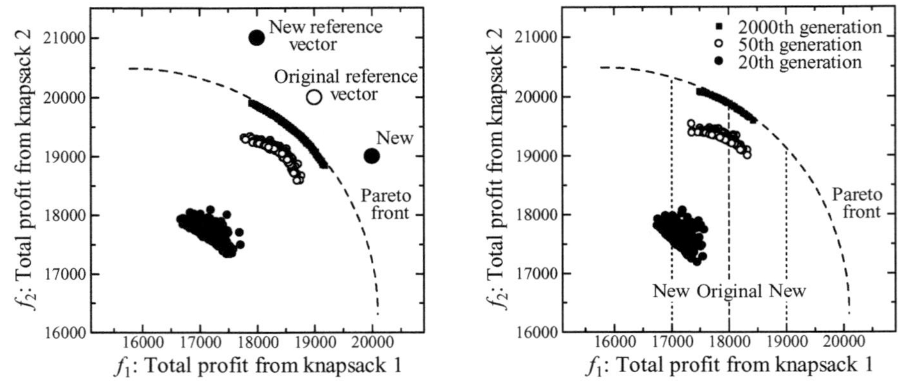

(a) Results based on two reference vectors (b) Results based on two requirement levels

Fig. 11. Experimental results on the 2-500 knapsack problem using other scalarizing functions

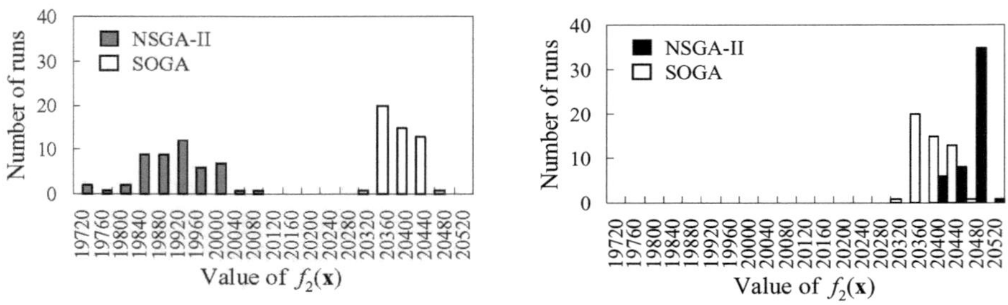

(a) Use of the original 2-500 problem (b) Use of the modified 2-500 problem

Fig. 12. Optimization of $f_2(\mathbf{x})$ of the 2-500 knapsack problem by NSGA-II

6 Conclusions

In this paper, we proposed an idea of using an EMO algorithm to optimize a scalarizing function. Our approach generates multiple objectives, which are similar to but different from the given scalarizing function, so that the location of the optimal solution of the scalarizing function is near the center of the Pareto front of the generated multiobjective problem. The effectiveness of our approach was examined through various computational experiments using NSGA-II. Experimental results showed that the performance of NSGA-II as a single objective optimizer highly depends on the choice of multiple objectives. One interesting observation is that NSGA-II worked very well even when it was applied to a four-objective 0/1 knapsack problem generated by our approach (whereas NSGA-II usually does not work well for many-objective problems). This is because our approach generates similar objectives.

This work was partially supported by Grant-in-Aid for Scientific Research on Priority Areas (18049065) and for Scientific Research (B) (17300075).

References

1. Colombo, G., Mumford, C. L.: Comparing Algorithms, Representations and Operators for the Multi-Objective Knapsack Problem. Proc. of 2005 Congress on Evolutionary Computation (2005) 2241-2247
2. Deb, K.: Multi-Objective Optimization Using Evolutionary Algorithms. John Wiley & Sons, Chichester (2001)
3. Deb, K., Pratap, A., Agarwal, S., Meyarivan, T.: A Fast and Elitist Multiobjective Genetic Algorithm: NSGA-II. IEEE Trans. on Evolutionary Computation 6 (2002) 182-197
4. Deb, K., Sundar, J.: Reference Point Based Multi-Objective Optimization Using Evolutionary Algorithms. Proc. of Genetic and Evolutionary Computation Conference (2006) 635-642
5. Fonseca, C. M., Fleming, P. J.: On the Performance Assessment and Comparison of Stochastic Multiobjective Optimizers. Lecture Notes in Computer Science, Vol. 114: PPSN IV. Springer, Berlin (1996) 584-593
6. Hughes, E. J.: Multiple Single Objective Sampling. Proc. of 2003 Congress on Evolutionary Computation (2003) 2678-2684
7. Hughes, E. J.: Evolutionary Many-Objective Optimization: Many Once or One Many? Proc. of 2005 Congress on Evolutionary Computation (2005) 222-227
8. Ishibuchi, H., Doi, T., Nojima, Y.: Incorporation of Scalarizing Fitness Functions into Evolutionary Multiobjective Optimization Algorithms. Lecture Notes in Computer Science, Vol. 4193: PPSN IX. Springer, Berlin (2006) 493-502
9. Ishibuchi, H., Murata, T.: A Multi-Objective Genetic Local Search Algorithm and Its Application to Flowshop Scheduling. IEEE Trans. on Systems, Man, and Cybernetics - Part C: Applications and Reviews 28 (1998) 392-403
10. Ishibuchi, H., Nojima, Y., Doi, T.: Application of Multiobjective Evolutionary Algorithms to Single-Objective Optimization Problems. Abstract Booklet of 7th International Conference on Multi-Objective Programming and Goal Programming (2006) 4 pages
11. Ishibuchi, H., Yoshida, T., Murata, T.: Balance between Genetic Search and Local Search in Memetic Algorithms for Multiobjective Permutation Flowshop Scheduling. IEEE Trans. on Evolutionary Computation 7 (2003) 204-223
12. Jaszkiewicz, A.: Genetic Local Search for Multi-Objective Combinatorial Optimization. European Journal of Operational Research 137 (2002) 50-71
13. Jaszkiewicz, A.: On the Performance of Multiple-Objective Genetic Local Search on the 0/1 Knapsack Problem - A Comparative Experiment. IEEE Trans. on Evolutionary Computation 6 (2002) 402-412
14. Jaszkiewicz, A.: On the Computational Efficiency of Multiple Objective Metaheuristics: The Knapsack Problem Case Study. European Journal of Operational Research 158 (2004) 418-433
15. Knowles, J. D., Watson, R. A., Corne, D. W.: Reducing Local Optima in Single-Objective Problems by Multi-Objectivization. Lecture Notes in Computer Science, Vol. 1993: EMO 2001. Springer, Berlin (2001) 269-283
16. Mumford, C. L.: A Hierarchical Solve-and-Merge Framework for Multi-Objective Optimization. Proc. of 2005 Congress on Evolutionary Computation (2005) 2241-2247
17. Purshouse, R. C., Fleming, P. J.: Evolutionary Many-Objective Optimization: An Exploratory Analysis. Proc. of 2003 Congress on Evolutionary Computation (2003) 2066-2073
18. Watanabe, S., Sakakibara K.: Multi-Objective Approaches in a Single-Objective Optimization Environment. Proc. of 2005 Congress on Evolutionary Computation (2005) 1714-1721
19. Zitzler, E., Thiele, L.: Multiobjective Evolutionary Algorithms: A Comparative Case Study and the Strength Pareto Approach. IEEE Trans. on Evolutionary Computation 3 (1999) 257-271

Reliability-Based Multi-objective Optimization Using Evolutionary Algorithms

Kalyanmoy Deb[1], Dhanesh Padmanabhan[2], Sulabh Gupta[1], and Abhishek Kumar Mall[1]

[1] Kanpur Genetic Algorithms Laboratory (KanGAL)
Indian Institute of Technology Kanpur
Kanpur, PIN 208 016, India
{deb,sulabhg,akmall}@iitk.ac.in
[2] India Science Laboratory, GM R&D, Bangalore, PIN 560066, India
dhanesh.padmanabhan@gm.com

Abstract. Uncertainties in design variables and problem parameters are inevitable and must be considered in an optimization task including multi-objective optimization, if reliable optimal solutions are to be found. Sampling techniques become computationally expensive if a large reliability is desired. In this paper, first we present a brief review of statistical reliability-based optimization procedures. Thereafter, for the first time, we extend and apply multi-objective evolutionary algorithms for solving two different reliability-based optimization problems for which evolutionary approaches have a clear niche in finding a set of reliable, instead of optimal, solutions. The use of an additional objective of maximizing the reliability index in a multi-objective evolutionary optimization procedure allows a number of trade-off solutions to be found, thereby allowing the designers to find solutions corresponding to different reliability requirements. Next, the concept of single-objective reliability-based optimization is extended to multi-objective optimization of finding a reliable frontier, instead of an optimal frontier. These optimization tasks are illustrated by solving test problems and a well-studied engineering design problem. The results should encourage the use of evolutionary optimization methods to more such reliability-based optimization problems.

1 Introduction

For practical optimization studies, reliability-based techniques are getting increasingly popular, due to their ability to handle uncertainties involved in realizing decision variables and stochasticities involved in various problem parameters. For a canonical deterministic optimization task, the optimum solution usually lies on a constraint surface or at the intersection of more than one constraint surfaces. However, if the design variables or some system parameters cannot be achieved exactly and are uncertain with a known probability distribution of variation, the deterministic optimum (lying on one or more constraint surfaces) will fail to remain feasible in many occasions [5,10]. In such scenarios, a stochastic optimization problem (also known as *chance programming*) is usually formed

S. Obayashi et al. (Eds.): EMO 2007, LNCS 4403, pp. 66–80, 2007.

and solved, in which the constraints are converted into probabilistic constraints meaning that probability of failures (of being a feasible solution) is limited to a pre-specified value (say $(1-R)$) [6], where R is called the reliability of design.

Existing reliability-based optimization techniques vary from each other in the manner they handle the probabilistic constraints. One simple-minded approach would be to use a Monte-Carlo simulation technique to create a number of samples following the uncertainties and stochastitices in the design variables and problem parameters and evaluate them to compute the probability of failure [12,2]. However, such a technique becomes computationally expensive when the desired probability of failure is very small. To alleviate this computational problem, more sophisticated sampling techniques are suggested.

Recently, optimization-based methodologies, instead of sampling methods, are suggested to take care of the probabilistic constraints. In these methods, stochastic variables and parameters are transformed into the standard normal variate space and a separate optimization problem is formulated to compute the largest probability of failure and equate it with the desired value. At least three different concepts – double-loop methods, single-loop methods and decoupled methods – have been followed. In this paper, for the first time, we extend one of these methodologies and apply it with an evolutionary algorithm to solve two different types of optimization problems and demonstrate by solving test problems and an engineering design problem that the evolutionary optimization based reliability consideration is quite appropriate for these problems.

2 Existing Reliability-Based Methodologies

We consider here a reliability-based single-objective optimization problem of the following type:

$$\begin{array}{lll} \underset{(\mathbf{x},\mathbf{d})}{\text{Minimize}} & f(\mathbf{x},\mathbf{d},\mathbf{p}), & \\ \text{Subject to} & g_j(\mathbf{x},\mathbf{d},\mathbf{p}) \geq 0, & j=1,2,\ldots,J, \\ & h_k(\mathbf{d}) \geq 0, & k=1,2,\ldots,K, \\ & \mathbf{x}^{(L)} \leq \mathbf{x} \leq \mathbf{x}^{(U)}, \quad \mathbf{d}^{(L)} \leq \mathbf{d} \leq \mathbf{d}^{(U)}. & \end{array} \tag{1}$$

Here, $\mathbf{x}$ is a set of design variables which are uncertain, $\mathbf{d}$ is a set of deterministic design variables, and $\mathbf{p}$ is a set of uncertain parameters (which are not design variables). Thus, the stochasticity in the optimization problem comes from two sets of variables: $\mathbf{x}$ and $\mathbf{p}$. Here, we only consider inequality constraints. This is because if an equality constraint involves $\mathbf{x}$ or $\mathbf{p}$, there may not exist a solution for any arbitrary desired reliability against failure. All inequality constraints can be classified into two categories: (i) stochastic (or chance) constraints g_j involves at least one random variables ($\mathbf{x}$, $\mathbf{p}$ or both) and (ii) h_k involves no random variables. Figure 1 shows a hypothetical problem with two inequality constraints. Typically, the optimal solution lies on a constraint boundary or at the the intersection of more than one constraints, as shown in the figure. In the event of uncertainties in design variables, as shown in the figure with a probability distribution around the optimal solution, in many instances such a solution will be infeasible.

In order to find a solution which is more reliable (meaning that there is a very small probability of instances producing an infeasible solution), the true optimal solution must be sacrificed and a solution interior to the feasible region may be chosen. For a desired reliability measure R, it is then desired to find that feasible solution which will ensure that the probability of having an infeasible solution instance created through uncertainties from this solution is at most $(1-R)$. To arrive at such a solution, the above optimization problem can be converted to a deterministic optimization problem.

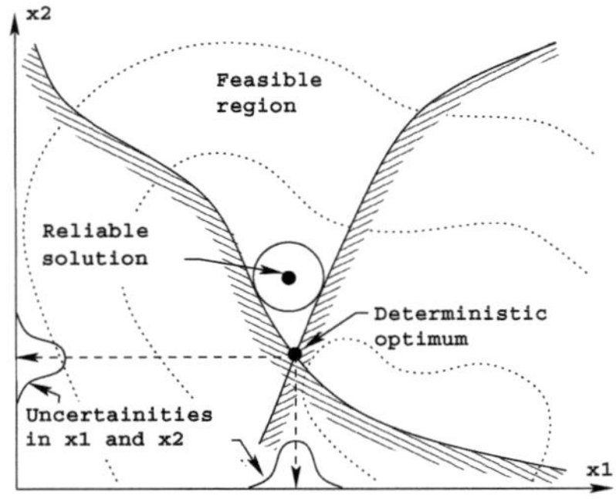

Fig. 1. The concept of reliability-based optimization procedure is illustrated

Since the objective function f and constraints g_j are also random due to the randomness in variables $\mathbf{x}$ and parameter $\mathbf{p}$, usually the following deterministic formulation is made:

$$\begin{array}{lll} \underset{(\mu_{\mathbf{x}},\mathbf{d})}{\text{Minimize}} & \mu_f(\mu_{\mathbf{x}}, \mathbf{d}, \mu_{\mathbf{p}}), & \\ \text{Subject to} & P(g_j(\mathbf{x}, \mathbf{d}, \mathbf{p}) \geq 0) \geq R_j, & j = 1, 2, \ldots, J, \\ & h_k(\mathbf{d}) \geq 0, & k = 1, 2, \ldots, K, \\ & \mathbf{x}^{(L)} \leq \mu_{\mathbf{x}} \leq \mathbf{x}^{(U)}, & \mathbf{d}^{(L)} \leq \mathbf{d} \leq \mathbf{d}^{(U)}. \end{array} \tag{2}$$

where μ_f is the function value computed at the mean of variable vector $\mathbf{x}$ and random parameters $\mathbf{p}$. The quantity R_j is the required reliability (within $[0, 1]$) for satisfying j-th constraint. A computational method is used to estimate the probability term $P()$, which we discuss next. Since the above formulation contains all deterministic expressions, any existing optimization methodology can be used to solve the problem.

The only difficulty the above problem poses is to compute the probability, $P()$. The existing reliability-based design optimization procedures can be classified into four classes [1], mainly based on the way the probability term $P()$ is computed:

1. Simulation methods
2. Double-loop methods
3. Decoupled methods, and
4. Single-loop methods

2.1 Simulation Methods

In this procedure, a set of N different solutions can be created by following the known distribution of variation of $\mathbf{x}$ and $\mathbf{p}$. Thereafter, for each sample each constraint g_j can be evaluated and checked for its violation. If r cases (of N) satisfy all g_j constraints, the probabilistic constraint can be substituted by a deterministic constraint as follows: $\frac{r}{N} \geq R$. Such a method is simple and works well when the desired reliability R is not very close to one [11]. However, a major bottleneck of this approach is that the sample size N needed for finding r must be of the order of at least $O(1/(1-R))$, such that at least one infeasible case is present in the sample. This may be too computationally expensive to be of any practical use. Although better procedures with a biased Monte-Carlo simulation procedure exist [2], the use of reliability techniques [13,14] to evaluate probabilistic constraints is getting increasingly popular in the recent past, which we discuss next.

2.2 Double-Loop Methods

In the double-loop method, a nested optimization is used. To compute the probability of success of each constraint, an optimization procedure (an inner-level optimization) is used. The outer loop optimizes the original objective function and the inner loop finds an equivalent deterministic version of each probabilistic constraint by formulating and solving an optimization problem. There are two approaches used for this purpose: (i) Performance measure approach (PMA) and (ii) Reliability index approach (RIA). Because of the nested optimization procedures, double-loop methods are computationally expensive.

To find whether a given hard constraint (g_j) is satisfied at a design point, we need to compute the following probability of the complementary failure event:

$$P_j = \int_{g_j(\mathbf{x},\mathbf{d},\mathbf{p})<0} f_{\mathbf{X}}(\mathbf{X})d\mathbf{X}, \tag{3}$$

where $f_{\mathbf{X}}$ is the joint probability density function of $\mathbf{X} = (\mathbf{x}, \mathbf{p})$. The reliability can be computed as $R_j = 1 - P_j$. It is usually impossible to find an analytical expression for the above integral. Thus, we first convert the $\mathbf{X}$ coordinate system into an independent standard normal coordinate system $\mathbf{U}$, through the Rosenblatt transformation [15]. The standard normal random variables are characterized by zero mean and unit variance. In this space, we approximate the curve ($g_j(\mathbf{x}, \mathbf{d}, \mathbf{p}) = 0$ or equivalently $G_j(\mathbf{U}) = 0$) by a first-order approximation at a suitable point known as the MPP (most probable point) of failure. In other words, the MPP point corresponds to a reliability index β_j, which makes a first-order approximation of $P_j = \Phi(-\beta_j)$, where $\Phi()$ is the standard normal density function. To compute the MPP (or β_j), we have the following two approaches.

Performance Measure Approach (PMA). To find the MPP in the PMA approach, the following optimization problem is solved:

$$\text{Minimize } G_j(\mathbf{U}), \quad \text{Subject to } \|U\| = \beta_j^r, \tag{4}$$

where β_j^r is the required reliability index, computed from the required reliability R_j as follows: $\beta_j^r = \Phi^{-1}(R_j)$. The above formulation finds a $\mathbf{U}^*$ point which lie on a circle of radius β_j^r and makes $G_j(\mathbf{U})$ minimum. The original probability constraint is replaced by

$$G_j(\mathbf{U}^*) \geq 0. \tag{5}$$

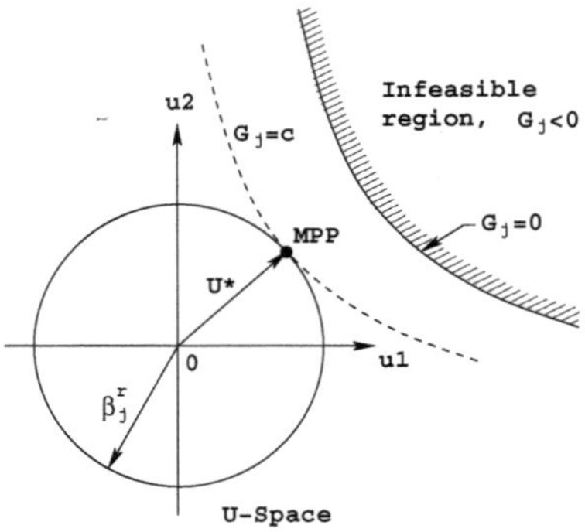

Fig. 2. The PMA approach

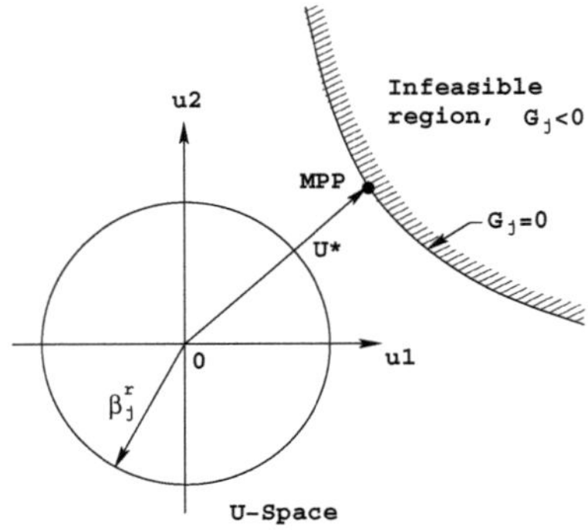

Fig. 3. The RIA approach

Figure 2 illustrates this approach on a hypothetical problem. The figure shows a probabilistic constraint g_j into the **U**-space (for ease of illustration, two variables are considered here). The corresponding constraint $G_j(u_1, u_2)$ and the feasible region are shown. The circle represents **U** solutions which corresponds to a reliability index of β_j^r. Thus, the PMA approach finds a point $\mathbf{U}^*$ on the circle for which the function $G_j(\mathbf{U})$ takes the minimum value. Then, if the corresponding constraint function value is non-negative, the probabilistic constraint $P(g_j(\mathbf{x}, \mathbf{d}, \mathbf{p}) \geq 0) \geq R_j$ is considered to have been satisfied.

Although the procedure involves an equality constraint, a special optimization procedure can be used to consider solutions only on the $\|U\| = \beta_j^r$ surface, thereby making every solution a feasible solution. For multiple such probabilistic constraints, ideally the above problem can be solved for each constraint at a time and the minimum of $G_j(\mathbf{U}^*)$ value for all constraints ($j = 1, 2, \ldots, J$) can be used in equation 5.

Reliability Index Approach (RIA). In this method, the following optimization problem is solved:

$$\text{Minimize } \|\mathbf{U}\|, \quad \text{Subject to } G_j(\mathbf{U}) = 0. \tag{6}$$

Here, the MPP is calculated by finding a point which is on the constraint curve in the **U**-space and is nearest to the origin. The optimum point $\mathbf{U}^*$ is used to replace the original probability constraint as follows:

$$\|\mathbf{U}\| \geq \beta_j^r. \tag{7}$$

Figure 3 illustrates the procedure. During the optimization procedure, the desired reliability index β_j^r is ignored and the minimum **U**-vector on the constraint boundary is found. Thereafter, the minimal $\mathbf{U}^*$ is compared with β_j^r.

This approach also involves an equality constraint. Although this method is computationally inferior compared to the PMA approach, a nice aspect is that the optimization problem does not involve the supplied reliability index value. For multiple such constraints, the above procedure can be applied for each constraint and the minimum $\mathbf{U}^*$ can be considered in equation 7.

2.3 Single-Loop Methods

The single-loop methods [12] combine both optimization tasks together by not exactly finding the optimum of the inner-level optimization task. An approximation procedure is used for the task. As an example, Liang [12] suggested the following replacement of the original probabilistic constraint:

$$g_j(\mathbf{x}, \mathbf{p}, \mathbf{d}) \geq 0, \tag{8}$$

where $\mathbf{x}$ and $\mathbf{p}$ are computed from the derivatives of g_j with respect to $\mathbf{x}$ and $\mathbf{p}$ at the means respectively, as follows:

$$\mathbf{x} = \mu_{\mathbf{x}} - \beta_j^r \sigma \frac{\nabla_{\mathbf{x}} g_j}{\sqrt{\|\nabla_{\mathbf{x}} g_j\|^2 + \|\nabla_{\mathbf{p}} g_j\|^2}}, \quad \mathbf{p} = \mu_{\mathbf{p}} - \beta_j^r \sigma \frac{\nabla_{\mathbf{p}} g_j}{\sqrt{\|\nabla_{\mathbf{x}} g_j\|^2 + \|\nabla_{\mathbf{p}} g_j\|^2}}.$$

Since the above is only an approximation to the double-loop procedure, the single-loop methods often cannot produce accurate results, but are computationally quicker methods than the double-loop methods.

2.4 Decoupled Methods

In the decoupled methods, two optimization (outer and inner-level) approaches are applied one after another in a sequence. Decoupled methods are shown to be the best of the three optimization approaches in a number of recent studies. These methods are started by first finding the best solution in the search space (without considering any uncertainty on design variables $\mathbf{x}$ or parameters $\mathbf{p}$ and using the mean of $\mathbf{x}$ as decision variables). Thereafter, the most-probable point (MPP) for each constraint g_j is found using the PMA or RIA approach. Then in the next iteration, the constraints are shifted according to their MPP points found in the last inner-level optimization. This dual optimization continues in tandem till no further improvement in the current solution is achieved. Figure 4 shows a particular approach (Sequential Optimization and Reliability Assessment (SORA) method) suggested elsewhere [8], in which the PMA approach is used as the second optimization problem.

Next, we discuss and show simulation results of two different problems in which one of the above methodologies coupled with an evolutionary multi-objective optimization algorithm make an efficient and useful search procedure.

3 Optimization for Seeking Multiple Solutions for Different Reliability Values

In most reliability-based optimization studies, the aim is to find the reliable optimum corresponding to a given failure probability (or a given reliability index). However, in the context of design optimization, it would be educative to learn how the reliable solutions would change with different levels of reliability index, as shown in Figure 5. When reliability is not considered, the deterministic optimum

$$\begin{array}{ll} \underset{(\mu_{\mathbf{X}},\mathbf{d})}{\text{Minimize}} & f(\mu_{\mathbf{X}}, \mathbf{d}, \mu_{\mathbf{p}}), \\ \text{Subject to} & g_j(\mu_{\mathbf{X}} - \mathbf{s}_j, \mathbf{d}, \mathbf{p}^k_{MPP,j}) \geq 0, \; j = 1, 2, \ldots, J, \\ & h_k(\mathbf{d}) \geq 0, \quad\quad k = 1, 2, \ldots, K, \\ & \mathbf{x}^{(L)} \leq \mu_{\mathbf{X}} \leq \mathbf{x}^{(U)}, \\ & \mathbf{d}^{(L)} \leq \mathbf{d} \leq \mathbf{d}^{(U)}. \end{array}$$

$(\mu^k_{\mathbf{X}}, \mathbf{d}^k) \downarrow$ $\quad\quad$ $\uparrow k = k + 1$, $\mathbf{s}_j = \mu^k_{\mathbf{X}} - \mathbf{x}^k_{MPP,j}$

For $j = 1, 2, \ldots, J$
PMA Approach to find MPP for j-th constraint $\rightarrow (\mathbf{x}^k_{MPP,j}, \mathbf{p}^k_{MPP,j})$

Fig. 4. A specific decoupled method (SORA) [8]. Initial value of s_j is set equal to zero for all j.

is the desired solution. As discussed earlier, when the optimization is performed for a particular reliability (say $R = 0.9$), an original feasible (inactive) solution becomes the desired reliable solution. As the reliability value is increased, a different solution somewhat more inside into the feasible region is likely to be the desired solution. That is, if we can locate the reliable optimum for small (say 80%) to large value (say 99.99%) of reliability, it would be worthwhile to analyze the solutions and investigate if they all bring out any common design principles (marked by the solid line connecting the reliable solutions). Such multiple optimal solutions can be deciphered by treating the problem as a two-objective optimization problem of optimizing the original objective and in addition maximize the reliability index (R or β) and by finding a number of Pareto-optimal solutions using an evolutionary multi-objective optimization (EMO) strategy to this bi-objective optimization problem. Multiple independent optimization tasks can be eliminated by a single bi-objective optimization task, if a two-objective optimization problem having the original objective function and an additional objective of maximizing reliability R dictated by the solution is formulated and solved for $(\mu_{\mathbf{X}}, \mathbf{d})$:

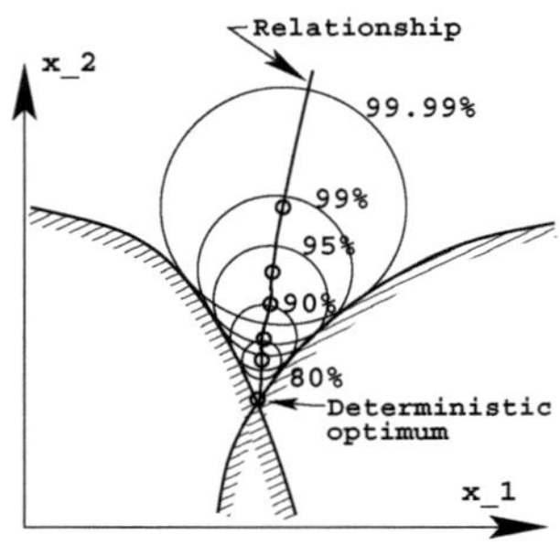

Fig. 5. Different reliability index may provide useful information

$$\begin{array}{l} \text{Minimize} \;\; \mu_f(\mu_{\mathbf{X}}, \mathbf{d}, \mu_{\mathbf{p}}), \;\; \text{Maximize} \; R(\mu_{\mathbf{X}}, \mathbf{d}, \mu_{\mathbf{p}}) = \min_{i=1}^{J} R_j(\mu_{\mathbf{X}}, \mathbf{d}, \mu_{\mathbf{p}}), \\ \text{Subject to} \; h_k(\mathbf{d}) \geq 0, \;\; k = 1, 2, \ldots, K, \;\; \mathbf{x}^{(L)} \leq \mu_{\mathbf{X}} \leq \mathbf{x}^{(U)}, \;\; \mathbf{d}^{(L)} \leq \mathbf{d} \leq \mathbf{d}^{(U)}. \end{array} \tag{9}$$

where $R_j(\mu_{\mathbf{X}}, \mathbf{d}, \mu_{\mathbf{p}}) = P(g_j(\mathbf{x}, \mathbf{d}, \mathbf{p}) \geq 0)$. The evolutionary multi-objective optimization (EMO) procedure is capable of finding multiple Pareto-optimal solutions for such a bi-objective optimization problem, thereby finding multiple reliable solutions corresponding to differing reliability values.

3.1 Reliability-Based Evolutionary Approach

For handling the above problem, the RIA-based approach is used here. This is because, for every solution it is desired to find the reliability value it corresponds. We use a relatively faster yet an approximate approach here for finding the MPP. First, we find the MPP ($\mathbf{U}^*$) on a unit-circle (assuming $\beta_j^r = 1$) based on the above PMA-based fast approach. Thereafter, we perform a uni-directional search along $\mathbf{U}^*$ and locate the point for which $G_j(\mathbf{U}) = 0$. We employ the Newton-Raphson approach for performing the uni-directional search. Here also, we use a tolerance of 0.001 for terminating the uni-directional search. Due to a single-variable search, the computation is usually quick, requiring only a derivative of the constraint function in the **U**-space. However, the MPP point obtained by this procedure may be an approximate solution and more sophisticated methods may be necessary. For handling multiple constraints, the above procedure of finding $\mathbf{U}^*$ and then performing the Newton-Raphson method to locate the point on $G_j(\mathbf{U}) = 0$ is repeated for each constraint one at a time. Thus, the inner-level optimization procedure using all design variables needed in the classical double-loop method is replaced by a fast procedure of finding the $\mathbf{U}^*$ direction and then performing a single-variable line search. Such a technique can be employed with a classical optimization procedure or with any other non-classical methods as well.

3.2 Simulation Results

First, we consider the following two-variable test problem [1]:

$$\left.\begin{array}{ll}\text{Maximize} & x + y,\\ \text{Subject to} & g_1(x,y) \equiv \frac{1}{20}x^2y - 1 \geq 0,\\ & g_2(x,y) \equiv \frac{1}{30}(x+y-5)^2 + \frac{1}{120}(x-y-12)^2 - 1 \geq 0,\\ & g_3(x,y) \equiv \frac{80}{x^2+8y+5} - 1 \geq 0,\\ & 0 \leq x, y \leq 10.\end{array}\right\} \quad (10)$$

Here, $\mathbf{x} = (x, y)$. To find a set of trade-off optimal solutions, we use the elitist non-dominated sorting GA or NSGA-II [4]. Figure 6 shows the non-dominated solutions obtained by NSGA-II with a population size of 100 and run for 100 generations. The SBX recombination operator with a probability of 0.9 and distribution index of 2 and the polynomial mutation with a probability of $1/n$ with a distribution index of 50 are used here [3]. Here, we have restricted the reliability index to vary between 0.5 and 5.0, causing a reliability of 69.14625% to 99.99997%. We have used $\sigma_1 = \sigma_2 = 0.3$ for this problem. It is interesting to note that to achieve a better reliable solution, a compromise of the optimal objective function value must be made. Such a variation of optimal objective value and reliability is important to decision-makers, as this will provide a deeper insight into the trade-off between these two important parameters.

To investigate whether NSGA-II solutions are optimal, we have solved the same problem with single-objective SORA method (decoupled method discussed above) for different β values. These solutions are also shown in Figure 6 with a

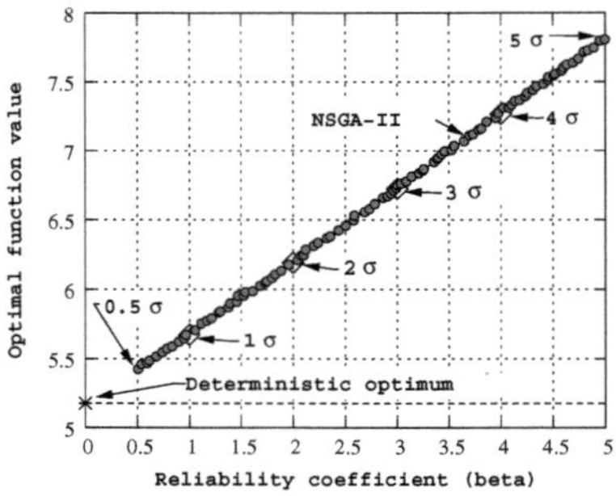

Fig. 6. Trade-off frontier between f^* and reliability index β for the two-variable test problem

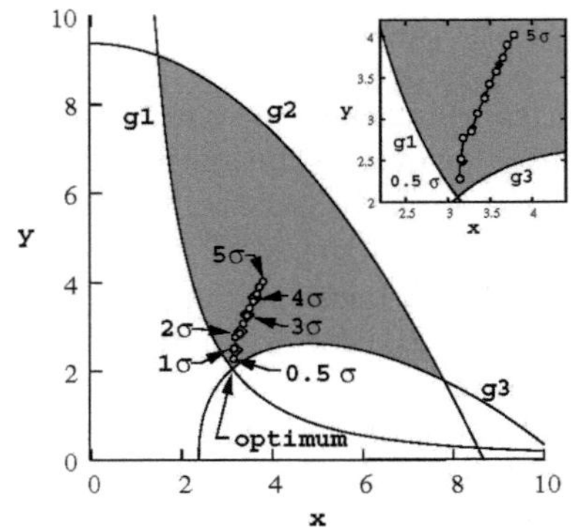

Fig. 7. Reliable solutions move inside the feasible region for the two-variable test problem

'diamond'. Since these solutions lie close to the obtained NSGA-II front, the near-optimality of other NSGA-II solutions can be ensured from the figure. Figure 7 shows how the optimal solutions move away from the deterministic optimum with an increase in the reliability index. The constraints g_1 and g_3 are active at the deterministic optimum. The inset figure shows that even for $\beta^r = 0.5$, the reliable solution is inside the feasible region (marked shaded). Thereafter, as β^r increases up to 5, the reliable solutions move inside the feasible space. The manner in which the solutions move inside also reveals important insights about the dependency of optimal solutions on the reliability index.

Car Side-Impact Problem. Next, we consider the car side-impact problem [9]. A car is subjected to a side impact based on European Enhanced Vehicle-Safety Committee (EEVC) procedures. There are 11 design variables. Their description and the standard deviation of their variations are shown below:

x_1: Thk. of B-Pillar inner (0.03)
x_2: Thk. of B-Pillar reinforcement (0.03)
x_3: Thk. of floor side inner (0.03),
x_4: Thk. of cross members (0.03),
x_5: Thk. of door beam (0.05),
x_6: Thk. of door reinforcement (0.03)
x_7: Thk. of roof rail (0.03),
x_8: Material of B-Pillar inner (0.006),
x_9: Material of floor side inner (0.006),
x_{10}: Barrier height (10),
x_{11}: Barrier hitting position (10).

The problem formulation is as follows:

$$\begin{aligned}
\underset{(x_1,\ldots,x_7)}{\text{Minimize}} \quad & f(\mathbf{x}) = \text{Weight},\\
\text{Subject to} \quad & g_1(\mathbf{x}) \equiv \text{Abdomen load} \leq 1 \text{ kN}, \quad g_2(\mathbf{x}) \equiv V * C_u \leq 0.32 \text{ m/s},\\
& g_3(\mathbf{x}) \equiv V * C_m \leq 0.32 \text{ m/s}, \quad g_4(\mathbf{x}) \equiv V * C_l \leq 0.32 \text{ m/s},\\
& g_5(\mathbf{x}) \equiv D_{ur} \text{ upper rib deflection} \leq 32 \text{ mm},\\
& g_6(\mathbf{x}) \equiv D_{mr} \text{ middle rib deflection} \leq 32 \text{ mm},\\
& g_7(\mathbf{x}) \equiv D_{lr} \text{ lower rib deflection} \leq 32 \text{ mm},\\
& g_8(\mathbf{x}) \equiv F \text{ Pubic force} \leq 4 \text{ kN},
\end{aligned} \tag{11}$$

$g_9(\mathbf{x}) \equiv V_{MBP}$ Velocity of V-Pillar at middle point ≤ 9.9 mm/ms,
$g_{10}(\mathbf{x}) \equiv V_{FD}$ Velocity of front door at V-Pillar ≤ 15.7 mm/ms,
$0.5 \leq x_1 \leq 1.5, \quad 0.45 \leq x_2 \leq 1.35, \quad 0.5 \leq x_3 \leq 1.5,$
$0.5 \leq x_4 \leq 1.5, \quad 0.875 \leq x_5 \leq 2.625, \quad 0.4 \leq x_6 \leq 1.2, \quad 0.4 \leq x_7 \leq 1.2.$

In this problem, we partition the 11-variable vector $\mathbf{x}$ into two sets: uncertain decision variables $\mathbf{x} = (x_1, \ldots, x_7)$ and uncertain parameters $\mathbf{p} = (x_8, \ldots, x_{11})$. Here, all variables/parameters (in mm) are assumed to be stochastic with a standard deviations (in mm) marked above. Problem parameters x_8 to x_{11} are assumed to take a particular distribution with a fixed mean of 0.345, 0.192, 0, and 0 mm, respectively. Thus, the stochastic optimization problem involves seven decision variables, whereas all 11 quantities vary with normal distribution around their mean values and are assumed to be independent. This functional forms of the objective function and constraints are given in the appendix.

We use a population of size 100 and run NSGA-II to optimize two objectives $f(x)$ and R for 100 generations. Figure 8 shows the trade-off between f^* and R. Here, we restrict β to vary between 0.5 and 3 (reliability of 99.865%). The

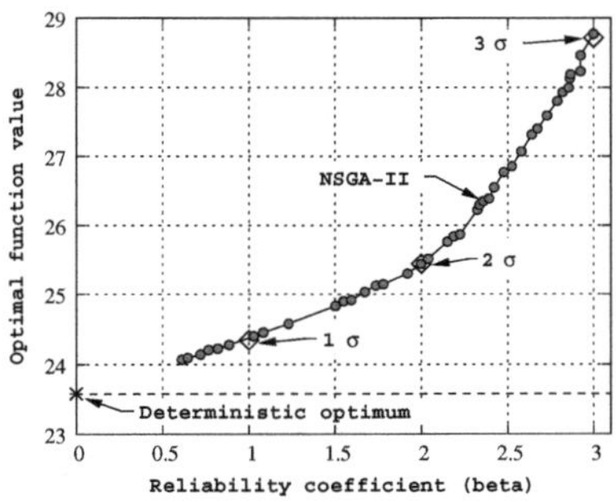

Fig. 8. Trade-off frontier between f^* and reliability index β for the car side impact problem

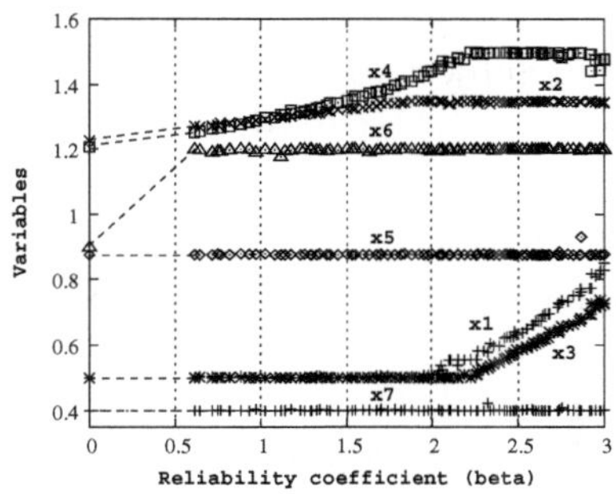

Fig. 9. Reliable solutions move inside the feasible region for the car side impact problem

corresponding SORA solutions, marked with 'diamonds' in Figure 9), confirm the near-optimality of the obtained solutions. Interestingly, in this problem, a larger than linear sacrifice in f^* needs to be made for an increase in β. The deterministic minimum (f^*) – without any variation in variables – is also shown in the figure. Figure 9 also shows how all seven variables vary with the reliability coefficient. Interestingly, x_5, x_6 and x_7 remains fixed for all reliable solutions. Variables x_5 and x_7 are fixed at their lower bounds and x_6 gets fixed at its upper bound. The optimal strategy in all solutions seems to use the smallest dimension for thickness of door beam (x_5) and roof rail (x_7) and the largest possible dimension for the door beltline reinforcement (x_6). For solutions up to around $\beta = 2$ (corresponding to 97.725% reliability), x_1 and x_3 remain fixed to their lower bounds and thereafter they increase with reliability. These variables represent the thickness of B-Pillar inner and floor side inner, respectively. On the other hand, till about this critical reliability requirement, x_2 (thickness of B-Pillar reinforcement) and x_4 (thickness of cross members) take large values and increase with reliability. After a critical reliability index, these values get fixed to their allowed upper limit. Thus, overall it seems that a good recipe to obtain an optimal yet reliable solution is to make the reinforcements stronger, while compromising the weight by using thinner members of other components. The figure also reveals that if no upper bound is used for these variables, the optimal strategy would be to use a monotonically increased dimension of x_2 and x_4 with increased reliability requirement. Such information about the nature of solutions and their interactions with the reliability index are interesting and provide valuable knowledge about the problem to a design engineer.

4 Multi-objective Reliability-Based Optimization

The concept of reliability-based optimization methods can also be applied to solve multi-objective reliability-based optimization problems:

$$\begin{array}{lll} \underset{(\mathbf{x},\mathbf{d})}{\text{Minimize}} & (f_1(\mathbf{x},\mathbf{d},\mathbf{p}),\ldots,f_M(\mathbf{x},\mathbf{d},\mathbf{p})), & \\ \text{Subject to} & g_j(\mathbf{x},\mathbf{d},\mathbf{p}) \geq 0, & j=1,2,\ldots,J, \\ & h_k(\mathbf{d}) \geq 0, & k=1,2,\ldots,K, \\ & \mathbf{x}^{(L)} \leq \mathbf{x} \leq \mathbf{x}^{(U)}, \quad \mathbf{d}^{(L)} \leq \mathbf{d} \leq \mathbf{d}^{(U)}. & \end{array} \qquad (12)$$

In such cases, instead of a single reliable solution, a reliable frontier is the target, as shown in Figure 10. When reliability aspects are considered, the corresponding

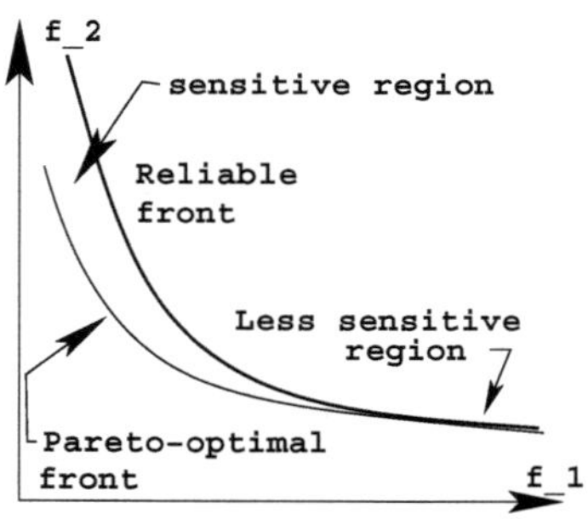

Fig. 10. Reliable front in a multi-objective reliability-based optimization problem

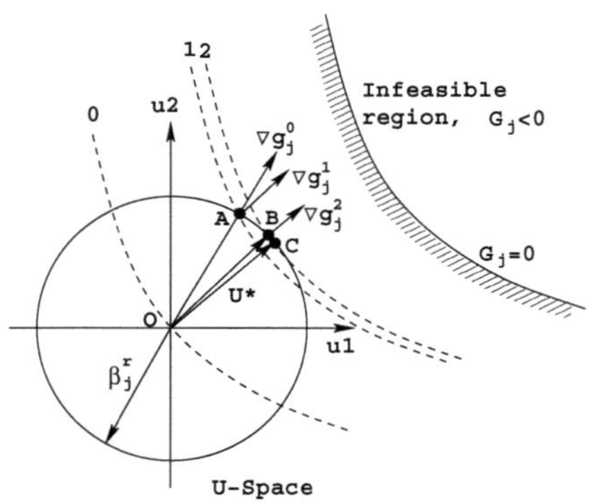

Fig. 11. A fast approach for solving the PMA problem

reliable Pareto-optimal front will be different from the original front and is placed inside the feasible region. As the reliability index is increased (to get more reliable solutions), the front is expected to move further inside the feasible region in the objective space.

An use of EMO procedure can be applied directly on the following deterministic optimization problem:

$$\begin{array}{lll} \underset{(\mu_{\mathbf{X}},\mathbf{d})}{\text{Minimize}} & (\mu_{f_1}(\mu_{\mathbf{X}},\mathbf{d},\mu_{\mathbf{P}}),\ldots,\mu_{f_M}(\mu_{\mathbf{X}},\mathbf{d},\mu_{\mathbf{P}})), & \\ \text{Subject to} & P(g_j(\mathbf{x},\mathbf{d},\mathbf{p})\geq 0)\geq R_j, & j=1,2,\ldots,J, \\ & h_k(\mathbf{d})\geq 0, & k=1,2,\ldots,K, \\ & \mathbf{x}^{(L)}\leq \mu_{\mathbf{X}}\leq \mathbf{x}^{(U)}, & \mathbf{d}^{(L)}\leq \mathbf{d}\leq \mathbf{d}^{(U)}. \end{array} \tag{13}$$

The probability constraint $P()$ can handled as before by using any of the four methods. The advantage of finding the complete reliable frontier is that the relative sensitivity of different regions of the frontier can be established with respect to the uncertainties in design variables and parameters. These information will be useful to the designers and decision-makers in choosing a solution from a relatively insensitive region of the trade-off frontier.

4.1 Reliability-Based Evolutionary Procedure

For the first time, we suggest here a reliability-based optimization procedure using evolutionary optimization algorithms. Here, we suggest a quick procedure of computing the MPP based on the PMA approach. Figure 11 illustrates this procedure. A gradient vector, ∇g_j^0, of each probabilistic constraint g_j is first computed at the origin of the $\mathbf{U}$-space. Its intersection (point A) with a circle of radius β_j^r is computed and a new gradient (∇g_j^1) is recomputed at this point (A). Thereafter, the intersection (point B) of this new gradient direction from the origin with the circle is recomputed and a new gradient vector ((∇g_j^2) is computed at B. This procedure is continued till a convergence of the norm of two consecutive gradient vectors with a predefined tolerance (of 0.001) is met. At this point, we have an approximate solution ($\mathbf{U}^*$) to the PMA approach. Such a procedure is already suggested elsewhere [7]. In our approach, we redo the above procedure for each probabilistic constraint and a deterministic constraint is formulated using equation 5. Thereafter, an EA with a penalty-parameter-less constraint handling approach [3] is used to handle all deterministic constraints. The above procedure can also be used with other EMO approaches and classical multi-objective optimization algorithms.

4.2 Simulation Results

We use the car side-impact problem discussed earlier, but now use an additional objective of minimizing the average rib deflection, calculated by taking the average of three deflections $g_5(\mathbf{x})$, $g_6(\mathbf{x})$ and $g_7(\mathbf{x})$. All 10 constraints are considered. Figure 12 shows the reliable front as a function of β. Once again, with an increase in the reliability index, the optimal frontier gets worse. We observe the following features from the figure:

- The figure indicates the rate at which the front deteriorates. In this problem, the rate of deterioration seems to be faster than linear, as was also discussed

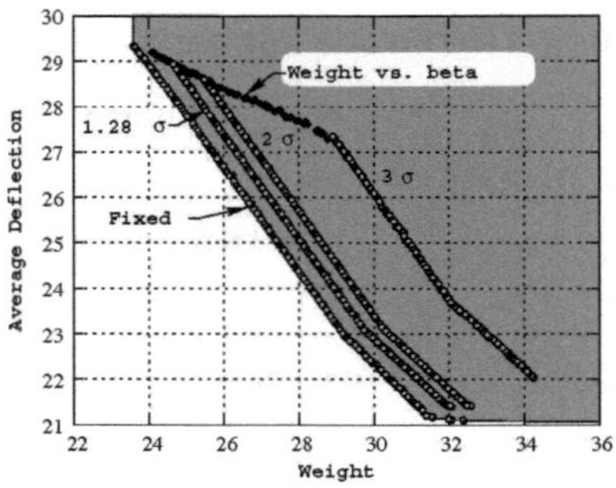

Fig. 12. Trade-off frontiers between f_1 and f_2 for different β for the car side impact problem

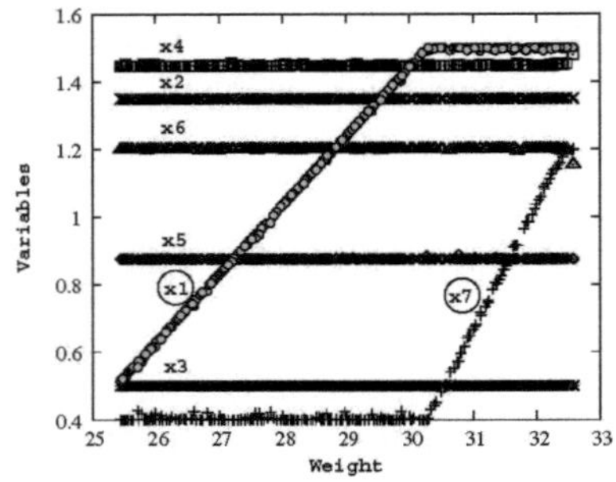

Fig. 13. Reliable solutions corresponding trade-off frontiers for $\beta = 2$ in the car side impact problem

in Section 3.2. Thus, an unnecessary large reliability index corresponds to solutions which are far from being optimum. Designers must carefully set a reliability index to make a good compromise on the optimality of solutions.

- An interesting fact about this problem is that the front moves inside the feasible objective space parallel to each other, indicating that the whole front is uniformly sensitive to a change in the reliability index.
- The minimum-weight solutions are found to be sensitive to the chosen reliability index. The optimal solutions obtained in Figure 9 in Section 3.2 are also plotted in Figure 12 (marked as 'Weight vs. beta'). Interestingly, these solutions mark the boundary to the obtained NSGA-II solutions of this section. This fact provides confidence in the accuracy of the obtained solutions.

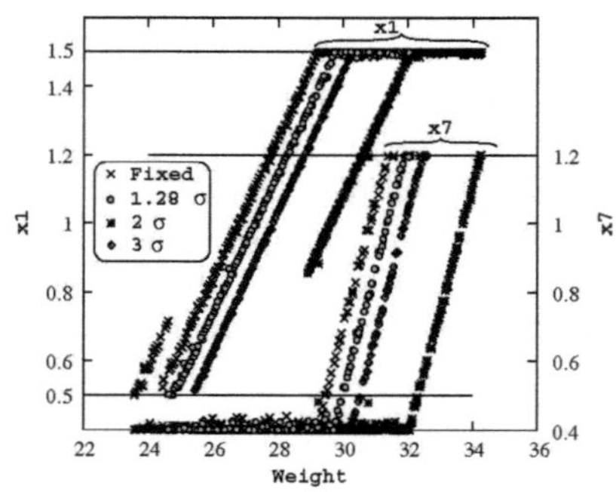

Fig. 14. Variables x_1 and x_7 for different β values

Figure 13 shows the variation of design variables for the solutions of the reliable frontier with $\beta = 2$. It is interesting to note that the optimal way to have the trade-off between the weight and rib-deflection is to have changes in variables x_1 (thickness of B-Pillar inner) and x_7 (thickness of roor rail). All other variables remain fixed to either their lower or their upper bounds. The minimum weight solution corresponds to the lowest allowable values of x_1 and x_7, as were found in section 3.2. For better rib-deflection solutions, the thickness of B-Pillar inner (x_1) must be increased steadily. When its value reaches its upper limit (1.5 mm), the thickness of roof rail (x_7) must be increased by keeping x_1 at its upper limit. The kink in the Pareto-optimal front occurs when x_1 hits its upper bound. Since a different pattern in the variables (now x_1 is kept fixed and x_7 is increased), the Pareto-optimal behavior changes. Figure 14 (above) shows variation of x_1 and x_7 for cases with different β values. In all cases, a similar pattern of changes in these two variables is observed.

Such a study also indicates the differential sensitivities of Pareto-optimal solutions, a matter which is useful for designers and practitioners to make a better multi-criterion decision-making by concentrating on the portion of the Pareto-optimal front which is less sensitive to design variable and parameter uncertainties.

5 Conclusions

In this paper, we have combined the classical reliability optimization techniques with evolutionary multi-objective optimization (EMO) approaches for better handling problems having uncertainties in decision variables and problem parameters. The first approach exploits an EMO approach to find multiple reliable solutions, each corresponding to a different reliability value. The second EMO approach directly finds the reliable frontier, instead of the optimal frontier, in a multi-objective stochastic problem. On a number of problems, the proposed procedures have shown their efficacy in quickly finding the desired reliable solution. In a car side-impact design problem, a number of interesting properties about the optimal and reliable solutions have been revealed. The study would encourage researchers and practitioners in the area of reliability-based design optimization to pay more attention to EA-based search and optimization procedures, a process which may lead to the development of hybrid evolutionary-classical reliability-based approaches in the coming years.

Acknowledgments

The authors acknowledge the funding from India Science Laboratory (ISL), General Motors R&D, Bangalore for executing the study.

References

1. H. Agarwal. *Reliability based design optimization: Formulations and Methodologies.* PhD thesis, University of Notre Dame, 2004.
2. T. R. Cruse. *Reliability-based mechanical design.* New York: Marcel Dekker, 1997.
3. K. Deb. *Multi-objective optimization using evolutionary algorithms.* Chichester, UK: Wiley, 2001.
4. K. Deb, S. Agrawal, A. Pratap, and T. Meyarivan. A fast and elitist multi-objective genetic algorithm: NSGA-II. *IEEE Transactions on Evolutionary Computation*, 6(2):182–197, 2002.
5. K. Deb and H. Gupta. Searching for robust Pareto-optimal solutions in multi-objective optimization. In *Proceedings of the Third Evolutionary Multi-Criteria Optimization (EMO-05) Conference (Also Lecture Notes on Computer Science 3410)*, pages 150–164, 2005.
6. O. Ditlevsen and H. O. Madsen. *Structural Reliability Methods.* New York: Wiley, 1996.

7. X. Du and W. Chen. A most probable point based method for uncertainty analysis. *Journal of Design and Manufacturing Automation*, 4:47–66, 2001.
8. X. Du and W. Chen. Sequential optimization and reliability assessment method for efficient probabilistic design. *ASME Transactions on Journal of Mechanical Design*, 126(2):225–233, 2004.
9. L. Gu, R. J. Yang, C. H. Tho, L. Makowski, O. Faruque, and Y. Li. Optimization and robustness for crashworthiness of side impact. *International Journal of Vehicle Design*, 26(4), 2001.
10. Yaochu Jin and Bernhard Sendhoff. Trade-off between performance and robustness: An evolutionary multiobjective approach. In *Proceedings of the Evolutionary Multi-Criterion Optimization (EMO-2003)*, pages 237–251, 2003.
11. R. T. F. King, H. C. S. Rughooputh, and K. Deb. Evolutionary multi-objective environmental/economic dispatch: Stochastic versus deterministic approaches. In *Proceedings of the Third International Conference on Evolutionary Multi-Criterion Optimization (EMO-2005)*, pages 677–691. Lecture Notes on Computer Science 3410, 2005.
12. J. Liang, Z. Mourelatos, and J. Tu. A single loop method for reliability based design optimization. In *Proceedings of the ASME Design Engineering Technical Conferences*, 2004.
13. S. Mahadevan and A. Chiralaksanakul. Reliability-based design optimization methods. In *Proceedings of the ASME Design Engineering Technical Conferences DETC2004-57456*, 2004.
14. D. Padmanabhan and S. M. Batill. Reliability based optimization using approximations with applications to multi-disciplinary system design. In *Proceedings of the 40th AIAA Sciences & Exhibit,* AIAA-2002-0449, 2002.
15. M. Rosenblatt. Remarks on a multivariate transformation. *Annals of Mathematical Statistics*, 23:470–472, 1952.

A Functions for Car Side Impact Problem

$$
\begin{aligned}
f(\mathbf{x}) &= 1.98 + 4.9x_1 + 6.67x_2 + 6.98x_3 + 4.01x_4 + 1.78x_5 + 0.00001x_6 + 2.73x_7,\\
g_1(\mathbf{x}) &= 1.16 - 0.3717x_2x_4 - 0.00931x_2x_{10} - 0.484x_3x_9 + 0.01343x_6x_{10},\\
g_2(\mathbf{x}) &= 0.261 - 0.0159x_1x_2 - 0.188x_1x_8 - 0.019x_2x_7 + 0.0144x_3x_5 + 0.87570.001x_5x_{10}\\
&\quad +0.08045x_6x_9 + 0.00139x_8x_{11} + 0.00001575x_{10}x_{11},\\
g_3(\mathbf{x}) &= 0.214 + 0.00817x_5 - 0.131x_1x_8 - 0.0704x_1x_9 + 0.03099x_2x_6 - 0.018x_2x_7\\
&\quad +0.0208x_3x_8 + 0.121x_3x_9 - 0.00364x_5x_6 + 0.0007715x_5x_{10} - 0.0005354x_6x_{10}\\
&\quad +0.00121x_8x_{11} + 0.00184x_9x_{10} - 0.018x_2x_2,\\
g_4(\mathbf{x}) &= 0.74 - 0.61x_2 - 0.163x_3x_8 + 0.001232x_3x_{10} - 0.166x_7x_9 + 0.227x_2x_2,\\
g_5(\mathbf{x}) &= 28.98 + 3.818x_3 - 4.2x_1x_2 + 0.0207x_5x_{10} + 6.63x_6x_9 - 7.77x_7x_8 + 0.32x_9x_{10},\\
g_6(\mathbf{x}) &= 33.86 + 2.95x_3 + 0.1792x_{10} - 5.057x_1x_2 - 11x_2x_8 - 0.0215x_5x_{10} - 9.98x_7x_8 + 22x_8x_9,\\
g_7(\mathbf{x}) &= 46.36 - 9.9x_2 - 12.9x_1x_8 + 0.1107x_3x_{10},\\
g_8(\mathbf{x}) &= 4.72 - 0.5x_4 - 0.19x_2x_3 - 0.0122x_4x_{10} + 0.009325x_6x_{10} + 0.000191x_{11}x_{11},\\
g_9(\mathbf{x}) &= 10.58 - 0.674x_1x_2 - 1.95x_2x_8 + 0.02054x_3x_{10} - 0.0198x_4x_{10} + 0.028x_6x_{10},\\
g_{10}(\mathbf{x}) &= 16.45 - 0.489x_3x_7 - 0.843x_5x_6 + 0.0432x_9x_{10} - 0.0556x_9x_{11} - 0.000786x_{11}x_{11}.
\end{aligned}
$$

Multiobjective Evolutionary Algorithms on Complex Networks

Michael Kirley and Robert Stewart

Department of Computer Science and Software Engineering
The University of Melbourne
Parkville, Victoria 3010, Australia
{mkirley,robertls}@csse.unimelb.edu.au

Abstract. Spatially structured populations have been used in evolutionary computation for many years. Somewhat surprisingly, in the multiobjective optimization domain, very few spatial models have been proposed. In this paper, we introduce a new multiobjective evolutionary algorithm on complex networks. Here, the individuals in the evolving population are mapped onto the nodes of alternative complex networks – regular, small-world, scale-free and random. A selection regime based on a non-dominance rating and a crowding mechanism guides the evolutionary trajectory. Our model can be seen as an extension of the standard cellular evolutionary algorithm. However, the dynamical behaviour of the evolving population is constrained by the particular network architecture. An important contribution of this paper is the detailed analysis of the impact that the structural properties of the network – node degree distribution, characteristic path length and clustering coefficient – have on the behaviour of the evolutionary algorithm using benchmark bi-objective problems.

1 Introduction

In many real-world search and optimization tasks, we are often confronted with a problem involving several incommensurable and often conflicting objectives. A family of equivalent non-dominated compromises – the *Pareto-optimal set* – represent solutions for this class of problem [1,2]. These solutions are optimal in the wider sense that no other solution in the search space is superior to them when all objectives are considered. The goal of any multiobjective optimization technique is to generate a diverse set of points distributed along the non-dominated front.

Evolutionary algorithms are now an established technique for solving multiobjective optimization problems [1,2] (such algorithms will be referred to as MOEAs). Well known MOEAs include NSGA-II [3], SPEA2 [4] and PAES [5]. Typically, these models evaluate a population of candidate solutions with respect to each objective. Non-dominated solutions are identified and form the mating pool, which then undergoes evolutionary transformations. As the model is iterated, the non-dominated set converges towards the true Pareto-optimal set.

S. Obayashi et al. (Eds.): EMO 2007, LNCS 4403, pp. 81–95, 2007.

Non-dominated sorting routines, the use of an external archive with appropriate niching and elitism mechanisms are often incorporated into the model to help ensure that the algorithm produces a uniformly distributed non-dominated front at the end of the search.

It has long been recognized that parallelism offers important advantages for evolutionary computation systems [6,7]. However, in the MOEA domain, there has only been a relatively small number of parallel models described as compared with the single objective domain (see [8] for a review). In recent years, there has been an increased interest in the study of complex networks in many areas including communication networks, biological networks and sociology [9,10,11,12]. Recent results reported from evolutionary game theory [13,14] and evolutionary algorithms evolving on both regular and small-world networks for single objective problems [6,15,16], suggest that the topology of the network influences the overall behaviour of the evolving population. In this study, we extend this work into the multiobjective optimization domain.

We present a new MOEA where the individuals of the population are mapped onto the nodes of alternative complex networks – regular lattice, small-world, scale-free and random (see Section 4 for an overview). A key component of the model is that individuals only interact with their local neighbours. Here, the network topology defines the local neighbourhood size. Thus, for different network architectures there will be different local neighbourhood sizes and average path lengths between individuals. The key hypothesis we investigate here, is that the structural characteristics of a given network will influence the quality of solutions generated by a MOEA.

The remainder of the paper is organized as follows: In Section 2, we formally describe multiobjective optimization problems. This is followed by a brief review of parallel evolutionary algorithms in Section 3, with an emphasis on cellular-based models. In Section 4, we describe complex networks, including a description of the particular network architectures used in this study. In Section 5, the new MOEA on complex networks is presented. This is followed by a description of the simulation experiments and results in Section 6. Finally, in Section 7 a discussion of the results is presented in terms of the impact of the underlying network topology and future research directions are identified.

2 Multiobjective Optimization

Multi-objective problems are problems that consist of a set of objective functions to be minimized or maximized subject to specified constraints. A multi-objective optimization problem can be stated generally as follows [2]:

$$\begin{aligned} \text{Minimize} \quad & f(x) = [f_1(x), f_2(x), \ldots, f_k(x)]^T \\ \text{subject to:} \quad & g_i(x) \geq 0 \ , \ i \in [1, \ldots, q] \\ & h_i(x) = 0 \ , \ i \in [1, \ldots, p] \end{aligned} \quad (1)$$

where x is a vector of decision variables, g_i is an inequality constraint, and h_i is an equality constraint. A solution is said to dominate another if, for all objectives,

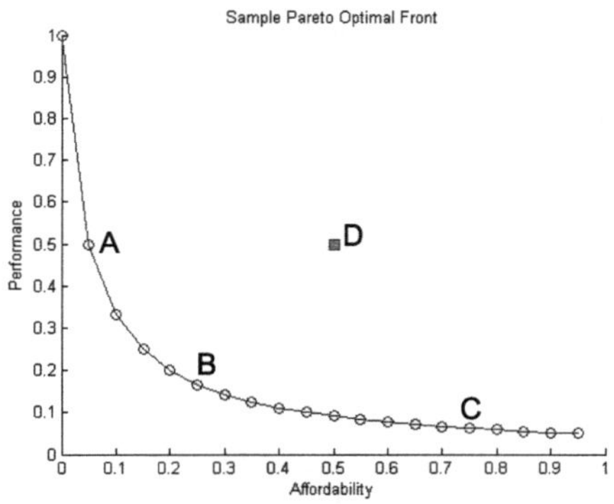

Fig. 1. A sample Pareto-optimal front for a two-objective minimization problem

it is as good as the other solution and better in at least one objective. That is, a solution x^* dominates a solution x (or $x^* \prec x$) iff

$$\forall i \ f_i(x^*) \leq f_i(x) \ \wedge \ \exists j \ f_j(x^*) < f_j(x) \ , i \in [1, \ldots, m] \tag{2}$$

The set of non-dominated solutions with respect to all other feasible solutions form the Pareto-optimal set. Figure 1 illustrates a Pareto-optimal set for a sample two-objective minimization problem. Solutions A, B and C are optimal solutions on the Pareto-optimal set but solution D is a dominated solution. A and B are said to dominate solution D since they are superior in at least one objective and no worse in any of the others.

3 Spatial Evolutionary Models

A range of *structured* or parallel evolutionary algorithms have been proposed where the population is decentralized in some way (see [6,7,8,16] for an overview). The models may be loosely classified into one of four types: single-population master-slaves, multiple populations (island model), cellular (diffusion model) and hierarchical combinations. In this study, the diffusion models are the most relevant, thus we limit our discussion to a brief review of specific applications in both single and multi objective optimization domains.

3.1 Single Objective Models

Recent studies from the single objective domain using cellular-based models are directly related to this study. For example, Sarma and De Jong [17] have described how the shape (and in particular the radius) of the local neighbourhood influences the takeover time of good solutions. Alba and Dorronsoro [6] also provide a comprehensive analysis of the trade-off between exploration and exploitation in dynamic models. In related work, Giacobini and co-workers [16] investigate the impact of selection intensity in cellular models. More recently, they have also compared elite solution takeover times in regular lattice models and small-world networks.

It is clear from the work described above that the topology of the network has an influence on the overall behaviour of the evolving population. In the fine-grained cellular models, the relatively small local neighbourhoods help to maintain population diversity as "good solutions" slowly diffuse across the network.

3.2 Multiobjective Models

Perhaps the first significant cellular-based MOEA was the predator-prey model introduced by Laumanns and co-workers [18], where solutions were mapped to a 2D lattice. In our own previous work, we have developed novel parallel algorithms, which relied on phase shifts in the connectivity of the solution space to provide a balance between exploration (global search) and exploitation (local search) [19]. The metapopulation evolutionary algorithm described in [20] employs these ideas and utilizes a flexible population structure based on a 2D "pseudo landscape". In other work, Mehnen and co-workers [21] introduced a *hypergraph* inspired parallel version of NSGA-II, with the aim of scaling freely between fine-grained and coarse-grained population structures. The flexibility of their model was demonstrated by allowing comparisons between different population structures to be made and an appropriate number of islands chosen for a given problem.

The main motivation of this current study is to combine relevant work from the complex networks domain with evolutionary computation in order to evolve a set of non-dominated solutions for a given multiobjective optimization problem.

4 Complex Networks

4.1 Definitions

A network can be modelled as a graph $G(N, E)$ where N is a finite set of *nodes* (vertices) and E a finite set of *edges* (links) such that each edge is associated with a pair of nodes i and j. G can be represented by an $N \times N$ adjacency (or connection) matrix whose entry a_{ij} is 1 if there is an edge joining node i to node j and is 0 otherwise.

Three measures are typically used to characterize the structural properties of a network: the node degree distribution ($P(k)$), the characteristic path length (L), and the clustering coefficient (C).

The degree k_i of a node i is usually defined to be the total number of edges between node i and all other nodes. The larger the degree, the "more important" the node is in a network. The node degree distribution function $P(k)$ is the probability that a randomly selected node has exactly k edges.

The characteristic path length measures the average separation between any two nodes in the network, and thus represents the "effective size" of the network.

The distance d_{ij} between two nodes, labelled i and j respectively, is defined as the number of edges along the shortest path connecting them, thus:

$$L = \frac{1}{N(N-1)} \sum_{i \neq j} d_{ij} \tag{3}$$

The clustering coefficient is the probability that two nearest neighbours of a node are also nearest neighbours of each other – so called "friends of friends." C_i of node i is then defined as the ratio between the number E_i of edges that actually exist among these k_i nodes and the total possible number $k_i(k_i - 1)/2$, that is:

$$C_i = \frac{2E_i}{k_i(k_i - 1)} \tag{4}$$

The clustering coefficient C of the whole network is the average of C_i over all i. $C = 1$ only in fully connected networks. In all other cases, $C < 1$.

4.2 Network Models

A brief description of the four alternative network models (see Fig. 2) that provide the scaffolding for the MOEA investigated in this study is provide below. A comprehensive discussion of the networks can be found in [10].

Regular networks may be defined as nearest-neighbour coupled networks (regular lattice) in which every node in the network is joined by a few of its neighbours. Regular networks have a high clustering coefficient and characteristic path length. The node degree distribution is constant (*delta* function).

At the opposite end of the spectrum from regular networks are completely *random networks* (see [22] cited in [10]). A random network is created by specifying that each pair of nodes is connected by an edge with uniform probability p. The degree distributions of random networks are approximately Gaussian (Poisson, in the limit of large n) and the shape falls off exponentially away from the peak value $< k >$. Random networks have a relatively low clustering coefficient ($C \approx p$) and have short average path lengths [12].

The transition from a regular lattice to a random graph best describes *small-world networks* [12,23]. Starting from a base regular lattice, a small–world network can be generated by re-wiring each edge with some probability p. For $p = 0$ the result is a completely regular network. For $p = 1$ the result is a completely random network. However, for low but nonzero p, the effect of re-wiring is the substitution of some short-range connections with long-range connections. Consequently, even though there may be relatively few long-range connections, the shortest path length between two individual nodes is likely to be relatively small, hence the "small-world" description. Small-world networks have a high clustering coefficient. The shape of the degree distribution is similar to that of a random graph.

Scale-free networks are characterized by their distinctive connectivity distributions – the probability that a node selected uniformly at random has a certain number of links (degree) follows a power law governed by the relationship

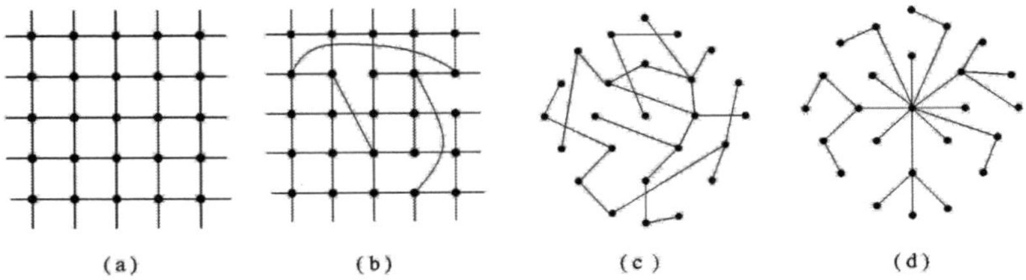

Fig. 2. Complex network models. Here, we illustrate the structure of the alternative networks using 25 nodes. (a) Regular 2D lattice, (b) Small-world, (c) Random, and (d) Scale-free.

$P(k) \sim k^{-\gamma}$. Many real-world networks have been shown to exhibit this "scale-free" behaviour, where γ lies somewhere between 2 and 3 [9]. An important feature of the scale-free model is that most nodes in the network have very few links and yet a small number of nodes are very highly connected. The clustering coefficient of the scale-free model is typically larger than the corresponding value of a random graph and the average path length is smaller than in a corresponding random graph.

5 The Model

The MOEA proposed in this study is an extension of the standard cellular evolutionary algorithm. A key component of the model is the communication topology determined by the network architecture. Here, the individuals are mapped to the nodes of alternative complex networks and interact in their local neighbourhood. Algorithm 1 provides an overview of the key steps.

An important feature of the model is the variation in local neighbourhood size between networks – and within particular networks. The size of the local neighbourhood is determined by the degree k_i of the current node i. Thus, the selection pressure will also vary. The exception to this rule is when a 2D regular lattice with Moore neighbourhood (8 nearest neighbours) is used. In the selection phase, a relative non-dominance ranking mechanism is used to generate a pool of potential mates from the local neighbourhood. A crowding measure is then used to rank individuals in the mating pool. Here, the least crowed individual is viewed as better. This selection regime results in the identification of a "best" mate, j, for the current individual i. After the recombination stage, the resulting offspring are mutated. The parent occupying node i and the resultant offspring are then compared using the dominance ranking mechanism. The non-dominated individual is then copied into the auxilary population. In the event of a tie, one of the children or parent is selected randomly to enter the auxiliary population. After all nodes in the network have been processed, the auixluary population is copied to the main population and the evolutionary cycle continues.

On each iteration of the model, the network is scanned and non-dominated solutions are added to an external archive. A non-dominated sort and crowding

```
initializeModel()
evaluatePopulation()
updateArchive()
while NOT terminationCriteria
    outputParetoFront()
    for individual (node) i in network do in parallel
        j = findBestNeighbour() //using non-domination/crowding
        o* = applyGeneticOperators( i,j )
        evaluate(o*)
        w = compare(i,o*)  //using non-domination
        addToAuxPopulation(w)
    end for
    updateArchive()
    population = AuxPopulation
end while
outputParetoFront()
```

Algorithm 1. Complex network based MOEA

procedure based on the NSGA-II implementation is then used to truncate the archive population size.

The complex network MOEA is fundamentally a parallel algorithm, however, parallel implementation is not a requirement. A sequential implementation can be used in order to take advantage of the spatial properties of the model.

6 Experiments and Results

To examine the impact that the network topology has on the quality of the Pareto-set found, a number of scenarios (network architecture–benchmark problem combinations) were investigated. The details are described below.

6.1 Model Parameters

Six complex networks with different node degree distributions, characteristic path lengths and clustering coefficients were used as the scaffolding for the spatial MOEA. Table 1 lists the networks' structural properties. Figure 3 shows the degree distributions $P(k)$ v k for each of the networks. A range of alternative networks were generated by "tweaking" the parameters used to construct the networks. However, the experimental results obtained were not significantly different between classes of network, thus we report results using the six networks described in the table.

Our network-based MOEA has many features in common with NSGA-II, subsequently we have used real-encoding with SBX crossover and the mutation probability was set to $1/n$, where n is the number of decision variables. The selection regime was descibed in Section 5. Given the emphasis in our model on local interactions, it was necessary to work with population sizes that are larger than typical population sizes used in panmictic models. Here, we have set the number of nodes in the network and archive size to 1024.

Table 1. Complex network models

Network	Description	$< k_i >$	L	C
SF	Scale-free network generated using the preferential attachment model.	3.9	4.4	0.007
M	Regular 2D network with Moore neighbourhood.	8.0	10.7	0.429
SWA	Small-world network, base 2D regular lattice, radius=1, $p = 0.05$.	8.0	5.3	0.355
SWB	Small-world network, base 3D regular lattice, radius=1, $p = 0.05$.	24.0	3.2	0.449
SWC	Small-world network, base 3D regular lattice, $p = 0.05$.	48.0	2.6	0.469
R	Erdös and Rényi random graph with $p = 0.01$.	50.7	2.0	0.049

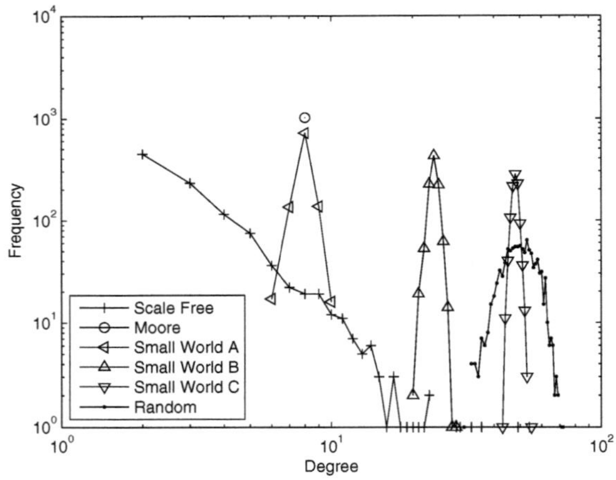

Fig. 3. Complex network degree distributions $P(k)$ v k

Table 2 lists the five benchmark multiobjective problems used in this study. For each network-test problem combination, the model was run for a maximum of 200 time steps. Thirty independent trials were completed for each scenario.

6.2 Results

In order to compare the performance of alternative network architectures, we need to examine both the convergence time and spread of solutions across the Pareto-front. To do this comparison, we have constructed a reference set, R, by merging all of the archival non-dominated solutions found by each of the network-based models and NSGA-II for a given problem across all output Pareto-fronts and trials. We then use two dominance compliance indicators, I_H^-, the

Table 2. Multiobjective test functions from Zitzler et al., [24]

Test function format:	Minimize $\mathcal{T}(\mathbf{x}) = (f_1(x_1), f_2(\mathbf{x}))$ with $f_2(\mathbf{x}) = g(x_2, \ldots, x_m)h(f_1(x_1), g(x_2, \ldots, x_m))$ subject to $\mathbf{x} = (x_1, \ldots, x_m)$	
ZDT1	$f_1(x_1) = x_1$ $g(x_2, \ldots, x_m) = 1 + 9 \cdot \sum_{i=2}^{m} x_i/(m-1)$ $h(f_1, g) = 1 - \sqrt{f_1/g}$	$m = 30$ $x_i \in [0, 1]$
ZDT2	$f_1(x_1) = x_1$ $g(x_2, \ldots, x_m) = 1 + 9 \cdot \sum_{i=2}^{m} x_i/(m-1)$ $h(f_1, g) = 1 - (f_1/g)^2$	$m = 30$ $x_i \in [0, 1]$
ZDT3	$f_1(x_1) = x_1$ $g(x_2, \ldots, x_m) = 1 + 9 \cdot \sum_{i=2}^{m} x_i/(m-1)$ $h(f_1, g) = 1 - \sqrt{f_1/g} - (f_1/g)\sin(10\pi f_1)$	$m = 30$ $x_i \in [0, 1]$
ZDT4	$f_1(x_1) = x_1$ $g(x_2, \ldots, x_m) = 1 + 10(m-1) + \sum_{i=2}^{m}(x_i^2 - 10\cos(4\pi x_i))$ $h(f_1, g) = 1 - \sqrt{f_1/g}$	$m = 30$ $x_1 \in [0, 1]$ $x_2, \ldots, x_m \in [-5, 5]$
ZDT6	$f_1(x_1) = 1 - \exp(-4x_1)\sin^6(6\pi x_1)$ $g(x_2, \ldots, x_m) = 1 + 9 \cdot \left(\left(\sum_{i=2}^{m} x_i\right) / (m-1)\right)^{0.25}$ $h(f_1, g) = 1 - (f_1/g)^2$	$m = 30$ $x_i \in [0, 1]$

hypervolume difference indicator and the I_ϵ unary epsilon indicator. The hypervolume indicator, I_H, measures the portion of the objective space that is weakly dominated by R [24]. Here, we use I_H^-, which measures differences in the value. The I_ϵ indicators gives the minimum factor a solution set must be adjusted to arrive at R [25].

Convergence plots. Figures 4 and 6 show the indicator convergence rates for each of the network models on test problem ZDT6. The data points at time t were found by calculating the I_H^- and I_ϵ values using the archive at time t (across all trials) and the reference set R. The relative ordering of the convergence rate follows the mean degree distributions $< k_i >$ listed in Table 1. That is, networks with higher $< k_i >$ values converge towards R faster than networks with lower $< k_i >$ values. The box plots in Figures 5 and 7 provide snapshots of differences in solution quality in terms of I_H^- and I_ϵ respectively, at time $t = 40$ and $t = 200$ (the end of the run). Similar trends are evident between networks at each time value, however there is significant improvment in the SF network results as the number of generations increases.

The trends in performance evident in the ZDT6 plots were very similar to the other test problems. Space constraints preclude the inclusion of plots for the other test problems. However, performance comparisons at the end of runs ($t = 200$) are listed below.

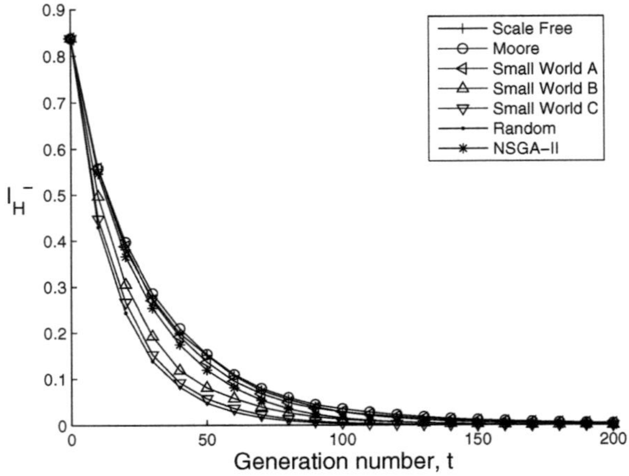

Fig. 4. Convergence of I_H^- for each network on problem ZDT6

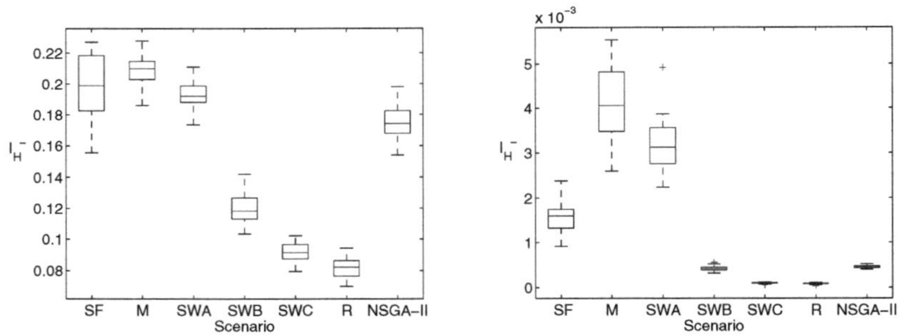

(a) $t = 40$ (b) $t = 200$

Fig. 5. Box plot of I_H^- indicator values for problem ZDT6

Comparing final solution quality. A non-parametric Kruskal-Wallis test was used to test for significance differences between scenarios. Tables 3 and 4 presents a pair-wise statistical comparison between the complex network based MOEAs and NSGA-II for I_H^- and I_ϵ respectively. We test the null hypothesis that the indicator value for the row entry is significantly better than the column entry ($p-$value < 0.05). If the result for a given problem is statistically significant, the problem number is listed in the cell. Box plots of I_H^- and I_ϵ values at the end of runs ($t = 200$) for each scenario are presented in Figure 8. (Note: box plots for ZDT6 were listed previously).

7 Discussion and Conclusion

The main focus of this study was to investigate the relationship between the structural characteristics of complex networks and the evolutionary dynamics of

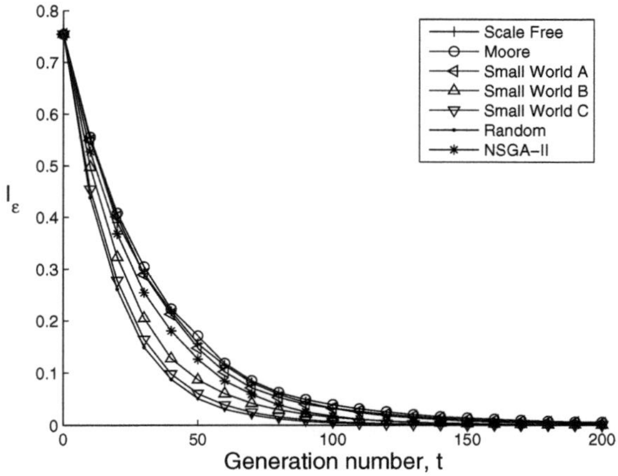

Fig. 6. Convergence of I_ϵ for each network on problem ZDT6

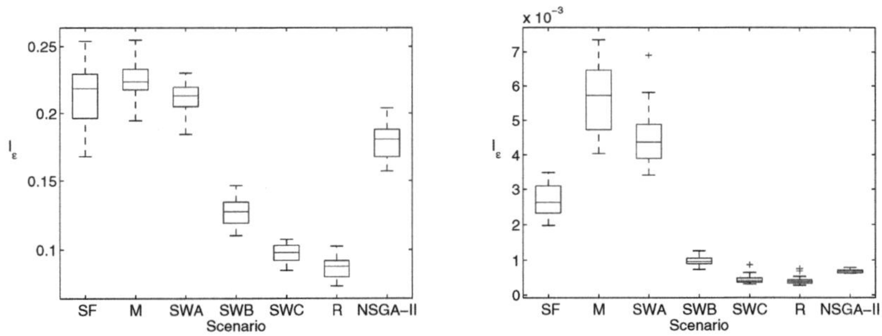

(a) $t = 40$ (b) $t = 200$

Fig. 7. Box plot of I_ϵ indicator values for problem ZDT6

a MOEA. The underlying hypothesis on which this work was based was that as we move away from the regular structure of a 2D lattice, with a constant local neighbourhood size, there would be a corresponding improvement in the performance of the algorithm. However, once the level of randomness exceeded some threshold and/or the network structure included a small number of highly connected nodes the performance may taper off.

In our MOEA, the specific topological features of a network, characterized by its connectivity, influences the evolutionary dynamics. By varying the degree distribution, characteristic path length and clustering coefficient it is possible to control the rate of diffusion across the network. It is to be expected that the population diversity would be greater in spatial models compared with a globally coupled network (base NSGA-II model), which has the smallest average path length and the largest clustering coefficient.

ZDT1:

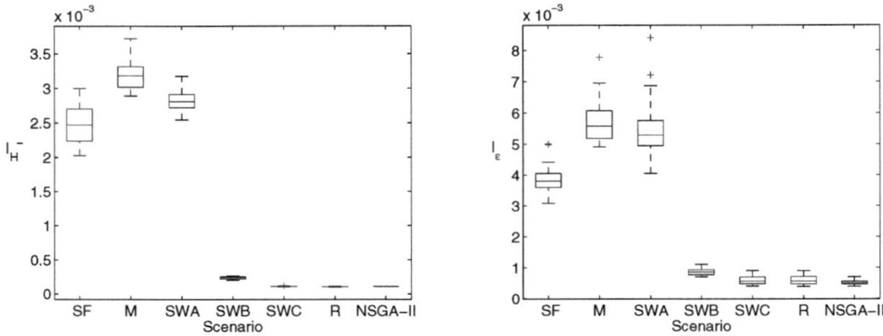

ZDT2:

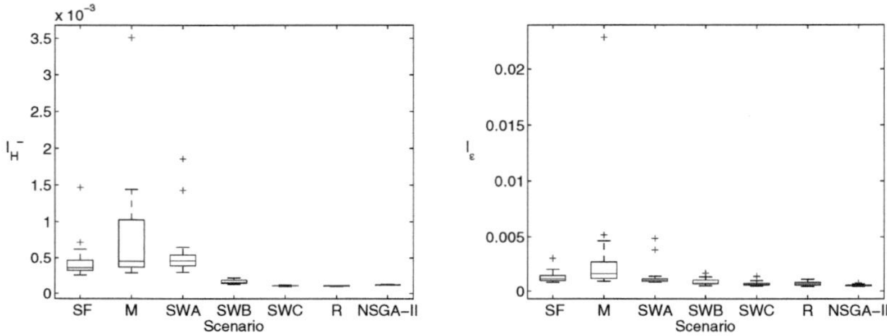

ZDT3:

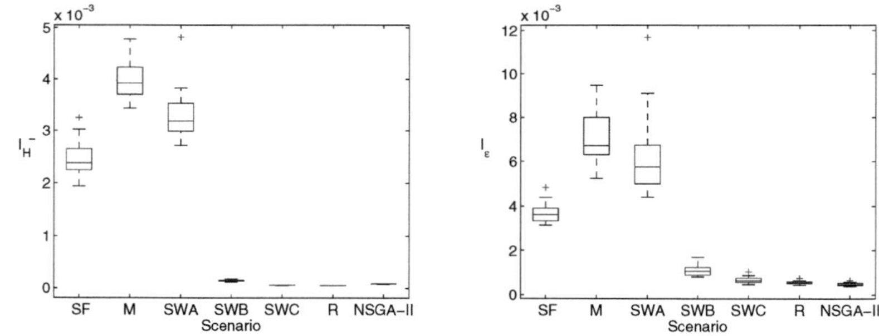

ZDT4:

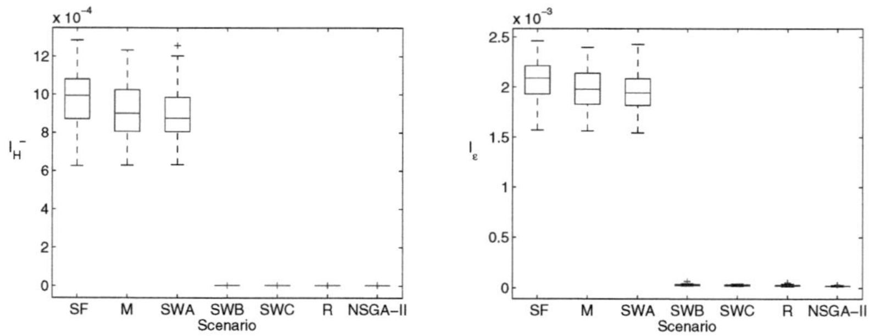

Fig. 8. Box plots for each of the indicators I_H^- and I_ϵ for each of the test problems

Table 3. Pair-wise statistical analysis for I_H^-

	SF	M	**SWA**	**SWB**	**SWC**	**R**	**NSGA-II**
SF	~	1,2,3,6	1,2,3,6				
M		~					
SWA	4	1,3,6	~				
SWB	1,2,3,4,6	1,2,3,4,6	1,2,3,4,6	~			6
SWC	1,2,3,4,6	1,2,3,4,6	1,2,3,4,6	1,2,3,4,6	~		2,3,4,6
R	1,2,3,4,6	1,2,3,4,6	1,2,3,4,6	1,2,3,4,6	1,2,3,4,6	~	1,2,3,4,6
NSGA-II	1,2,3,4,6	1,2,3,4,6	1,2,3,4,6	1,2,3			~

Table 4. Pair-wise statistical analysis for I_ϵ

	SF	M	**SWA**	**SWB**	**SWC**	**R**	**NSGA-II**
SF	~	1,2,3,6	1,3,6				
M		~					
SWA	4	2,3,6	~				
SWB	1,2,3,4,6	1,2,3,4,6	1,2,3,4,6	~			
SWC	1,2,3,4,6	1,2,3,4,6	1,2,3,4,6	1,2,3,6	~		6
R	1,2,3,4,6	1,2,3,4,6	1,2,3,4,6	1,2,3,4,6	3	~	6
NSGA-II	1,2,3,4,6	1,2,3,4,6	1,2,3,4,6	1,2,3,4,6	1,2,3,4	1,2,3,4	~

A systematic analysis of performance of the algorithm in terms of its ability to generate a range of non-dominated solutions distributed along the Pareto-optimal front was conducted. Using alternative complex networks and benchmark test problems, we were able to show that our spatially-structured MOEA was comparable with, and in some cases out-performed, it's panmictic equivalent in terms of I_H^- and I_ϵ. As expected, the results obtained using the regular lattice and small-world models with small re-wiring probabilities were similar. As "shortcuts" were introduced (as a result of link re-wiring), some improvement in performance in I_H^- and I_ϵ can be observed across all problems. However, networks with relatively high $< k_i >$ values tended to produce higher quality solutions.

The convergence rate is obviously an important property of any MOEA. The simulation results indicate that in the early stages of a run ($t < 50$), the quality of the solutions found was directly correlated with the magnitude of $< k_i >$. Large values of $< k_i >$ indicate increased selection intensity. However, as the number of generations increased, the performance differences in terms of I_H^- and I_ϵ was not as significant. This suggests that the rate of diffusion of genetic information – the balance between exploration and exploitation of the search space – is an important quality indicator. In the long run, high $< k_i >$ values may limit the range of solutions found. In addition, for larger values of t, the magnitude of the average path length and clustering level of the network directly effect the ability of the MOEA to find solutions distributed along the Pareto-front. That is, small average path lengths and clustering coefficient values limit the rate of diffusion, thus local (but possibly not global) non-dominated solutions have the opportunity to survive and reproduce. This increased population diversity may result in a wider range of non-dominated solutions being identified by the end of the run.

The indicator values for the scale-free network suggest that this particular architecture does not outperform the other network models. Generally, as the number of generations increased, the performance of the scale-free network also improved. The small number of highly connected nodes may act as "selection amplifiers" for local non-dominated solutions. Similarly, the degree distribution of random networks may also promote the spread of non-dominated solutions. As a consequence of the initial random placement of the individuals across the networks, large spreads in performance metrics are to be expected. In the event that a high quality solution occupies a highly connected node, it is not unreasonable to expect that the individual could take-over and spread its genetic information, which could possibly lead to premature convergence and a degradation in solution quality. Clearly, the outcomes are sensitive to the selection pressure and the choice of niching parameters. It is necessary to conduct further sensitivity analysis of the results before firm conclusions can be drawn.

This study has raised many questions, which will be addressed in future work. For example, how can we exploit the structural properties of the network to guide the search? The highly connected nodes of scale-free networks and the relatively high mean degree of nodes in the random network could act as sinks for migrants (selected from an external archive). Obviously, time-scales associated with fixation would have to be considered. However, the results presented here suggest that by controlling the rate of diffusion we can guide the evolutionary trajectory. It would be interesting to examine how an alternative crowding/niching mechanism could be used to divide the search space (decision/objective) into different geographical regions of the network. Finally, there is a need to investigate the scalability of the algorithm for problems with a larger number of objectives.

Acknowledgments

This work was supported by an Australian Research Council Discovery Grant (DP0664674).

References

1. Coello, C.C., Veldhuizen, D.V., Lamont, G.: EA for Solving Multi-Objective Problems. Kluwer (2002)
2. Deb, K.: Multi-Objective Optimization Using Evolutionary Algorithms. John Wiley and Sons, Ltd., Chichester, England (2001)
3. Deb, K., Pratap, A., Agarwal, S., Meyarivan, T.: A fast and elitist multiobjective genetic algorithm: NSGA-II. IEEE Transactions on Evolutionary Computation **6** (2002) 182–197
4. Zitzler, E., Laumanns, M., Thiele, L.: SPEA2: improving the strength pareto evolutionary algorithm. Technical Report 103, Computer Engineering and Networks Laboratory (TIK), Swiss Federal Institute of Technology (ETH) Zurich, Gloriastrasse 35, CH-8092 Zurich, Switzerland (2001)
5. Knowles, J.D., Corne, D.: Approximating the nondominated front using the pareto archived evolution strategy. Evolutionary Computation **8**(2) (2000) 149–172

6. Alba, E., Dorronsoro, B.: The exploration/exploitation tradeoff in dynamic cellular genetic algorithms. IEEE Transactions on Evolutionary Computation **9** (2005) 126–142
7. Cantu-Paz, E.: Efficient and Accurate Parallel Genetic Algorithms. Kluwer (2000)
8. Veldhuizen, D.A.V., Zydallis, J.B., Lamont, G.B.: Considerations in engineering parallel multiobjective evolutionary algorithms. IEEE Transactions on Evolutionary Computation **7** (2003) 144–173
9. Barabasi, A., Albert, R.: Emergence of scaling in random networks. Science **286** (1999) 509–512
10. Dorogovtsev, S., Mendes, J.: Evolution of Networks: From Biological Nets to the Internet and WWW. Oxford University Press, Oxford (2003)
11. Strogatz, S.: Exploring complex networks. Nature **410** (2001) 268–276
12. Watts, D.: Small Worlds: The Dynamics of Networks between Order and Randomness. Princeton University Press (1999)
13. Kirley, M.: Evolutionary minority games with small-world interactions. Physica A: Statistical Mechanics **365** (2006) 521–528
14. Lieberman, E., Hauert, C., Nowak, M.: Evolutionary dynamics on graphs. Nature **433** (2005) 312–316
15. Giacobini, M., Tomassini, M., Tettamanzi, A.: Takeover time curves in random and small-world structured populations. In: Proceedings of the Genetic and Evolutionary Computation Conference (GECCO'05). (2005) 1133–1340
16. Giacobini, M., Tomassini, M., Tettamanzi, A., Alba, E.: Selection intensity in cellular evolutionary algorithms for regular lattices. IEEE Transactions on Evolutionary Computation **9** (2005) 489–505
17. Sarma, J., De Jong, K.: An analysis of the effects of neighborhood size and shape on local selection algorithms. In Voigt, H., Ebeling, W., Rechenberg, I., eds.: Parallel Problem Solving from Nature – PPSN IV (Berlin, 1996), Berlin, Springer (1996) 236–244
18. Laumanns, M., Rudolph, G., Schwefel, H.: A spatial predator-prey approach to multi-objective optimization: A preliminary study. In: PPSN V. (1998) 241–249
19. Kirley, M.: A cellular genetic algorithm with disturbances: Optimization using dynamic spatial interactions. Journal of Heuristics **8** (2002) 321–342
20. Kirley, M.: M.E.A.: A metapopulation evolutionary algorithm for multi-objective optimisation problems. In: Proceedings of the 2001 Congress on Evolutionary Computation CEC2001. (2001) 949–956
21. Mehnen, J., Michelitsch, T., Schmitt, K., Kohlen, T.: pMOHypEA: Parallel evolutionary multiobjective optimization using hypergraphs. Technical Report Reihe CI-189/04, SFB 531, ISSN 1433-3325, University of Dortmund (2004)
22. Erdös, P., Rényi, A.: On the evolution of random graphs. Publ. Math. Inst. Hung. Acad. Sci., **5** (1959) 17–60
23. Watts, D., Stogatz, S.: Collective dynamics of "small-world" networks. Nature **393** (1998) 440–441
24. Zitzler, E., Thiele, L.: Multiobjective evolutionary algorithms: A comparative case study and the strength pareto approach. IEEE Transactions on Evolutionary Computation **3**(4) (1999) 257–271
25. Zitzler, E., Thiele, L., Laumanns, M., Fonseca, C.M., da Fonseca, G.V.: Performance assessment of multiobjective optimizers: An analysis and review. IEEE Transactions on Evolutionary Computation **7**(2) (2003) 117–132

On Gradient Based Local Search Methods in Unconstrained Evolutionary Multi-objective Optimization

Pradyumn Kumar Shukla

Institute for Numerical Mathematics
Dresden University of Technology
Dresden, D-01062, Germany
pradyumn.shukla@mailbox.tu-dresden.de

Abstract. Evolutionary algorithms have been adequately applied in solving single and multi-objective optimization problems. In the single-objective case various studies have shown the usefulness of combining gradient based classical methods with evolutionary algorithms. However there seems to be limited number of such studies for the multi-objective case. In this paper, we take two classical methods for unconstrained multi-optimization problems and discuss their use as a local search operator in a state-of-the-art multi-objective evolutionary algorithm. These operators require gradient information which is obtained using finite difference method and using a stochastic perturbation technique requiring only two function evaluations. Computational studies on a number of test problems of varying complexity demonstrate the efficiency of resulting hybrid algorithms in solving a large class of complex multi-objective optimization problems. We also discuss a new convergence metric which is useful as a stopping criteria for problems having an unknown Pareto-optimal front.

1 Introduction

Multi-objective optimization is a rapidly growing area of research and application in modern-day optimization. There exist a plethora of non-classical methods which follow some natural or physical principles for solving multi-objective optimization problems, see for example the book by [5]. On the other hand a large amount of studies have been devoted to develop classical methods for solving multi-objective optimization problems ([8]).

Evolutionary algorithms use stochastic transition rules using crossover and mutation search operators to move from one solution to another. In this way global structure of search space is exploited. Classical methods, on the other hand, usually use deterministic (usually gradient based) transition rules to move from one solution to another. Classical methods effectively use local information thus ensuring fast convergence. This however comes up at the cost of requiring gradient or Hessian information which requires a large number of function evaluations. Hence one sees that there is a trade-off between fast convergence

S. Obayashi et al. (Eds.): EMO 2007, LNCS 4403, pp. 96–110, 2007.

and number of function evaluations. Hybrid implementations thus continue to be developed and tested (see for example [1,2,10,11]).

In this contribution we take two classical gradient based Pareto front generating methods and use their search principles as mutation operators in a state-of-the-art multi-objective evolutionary algorithm to create a powerful hybrid multi-objective metaheuristics algorithm. We demonstrate their efficiency in solving real valued differentiable problems of varying complexity.

This paper is structured as follows. The next section present an overview of various classical generating methods and the gradient estimation technique, while the third section presents the simulation results. Conclusions as well as extensions which emanated from this study are presented in the end of this contribution.

2 Classical Generating Methods

For the present study we take two classical algorithms and use their search operators as mutation operator in the elitist non-dominated sorting GA or NSGA-II developed by [6]. The gradients of objective functions are numerically computed by two methods: one-sided finite difference method and a stochastic perturbation method. These gradient estimation methods and the classical search operators used in this study are described in this section.

2.1 Gradient Estimation Methods

In almost all classical algorithms (for both single and multi-objective problems) the gradient of a function (say in general h) are required. The standard approach for estimating the gradient is the Finite Difference (FD) method (one-sided or two-sided). Let $\mathbf{e}_i$ denote a unit vector in the ith direction, then for a variable (say $\mathbf{x}$) of dimension n the one-sided FD method of gradient estimation requires $(n+1)$ function evaluations and is given by

$$g_i(\mathbf{x}) = \frac{f(\mathbf{x} + c\mathbf{e}_i) - f(\mathbf{x})}{c},$$

This is costly in terms of function evaluations (of the order $O(n)$). The Simultaneous Perturbation (SP) method [16] on the other hand requires function evaluation independent of n. The one-sided SP method required only *one* additional function measurement and is thus $O(1)$) as follows

$$g_i(\mathbf{x}) = \frac{f(x + c\Delta) - f(x)}{c\Delta_i},$$

where the ith component of the gradient is denoted by $g_i(\mathbf{x})$, Δ is a n dimensional vector (Δ_i is its ith component) of random perturbations satisfying certain statistical conditions ([16]). A simple (and theoretically valid) choice for each component of Δ is to use a Bernoulli distribution ± 1 with probability of 0.5 for each ± 1 outcome. The step size c at each iteration (denoted by c_k) is given as $c_k = c_0/(k+1)^{\gamma}$. Practically effective (and theoretically valid [16]) values of c_0, γ are 0.001 and 1/6 which are used here.

2.2 Schäffler's Stochastic Method (SSM)

This method [13], is based on the solution of a set of stochastic differential equations. It requires the objective functions to be twice continuously-differentiable. In each iteration, a trace of non-dominated points is constructed by calculating at each point $\mathbf{x}$ a direction $(-q(\mathbf{x}))$ in the decision space which is a direction of descent for *all* objective functions (note that we consider m to be the number of objective functions denoted by f_i for all $i = 1, 2, \cdots, m$). The direction of descent is obtained by solving a quadratic subproblem. Let $\hat{\alpha}$ be the minimizer. Then $q(\mathbf{x}) := \sum_{i=1}^{m} \hat{\alpha}_i \nabla f_i(\mathbf{x})$. A set of non-dominated solution is obtained by perturbing the solution (minimum along the direction of descent) using a Brownian motion concept. The following stochastic differential equation (SDE) is employed for this purpose:

$$d\mathbf{X}_t = -q(\mathbf{X}_t)d(t) + \varepsilon dB_t, \quad \mathbf{X}_0 = \mathbf{x}_0, \tag{1}$$

where $\varepsilon > 0$ and B_t is a n-dimensional Brownian motion. As the first search operator we use Equation 1 to create a child instead of the mutation operator. The gradients are obtained using one-sided FD and one-sided SP method. We name the hybrid algorithm with new mutation operator using FD and SP gradient estimates as S-NSGA-FD and S-NSGA-SP respectively. In all simulations here, to solve the above equation numerically, we employ the Euler's method with a step size σ. The approach needs two parameters to be set properly: (i) the parameter ε which controls the amount of global search and (ii) the step size σ used in the Euler's approach which controls the accuracy of the integration procedure.

2.3 Timmel's Population Based Method (TPM)

As early as in 1980, [17,18] proposed a population-based stochastic approach for finding multiple Pareto-optimal solutions of a differentiable multi-objective optimization problem. In this method, first, a feasible solution set (we call it a population) is randomly created. The non-dominated solutions ($\mathbf{X}_0 = \{\mathbf{x}_1^0, \mathbf{x}_2^0, \ldots, \mathbf{x}_s^0\}$) are identified and they serve as the first approximation to the Pareto-optimal set. Thereafter, from each solution $\mathbf{x}_k^0$, a child solution is created in the following manner:

$$\mathbf{x}_k^1 = -\sum_{i=1}^{M} t_1 u_i \nabla f_i(\mathbf{x}_k^0), \tag{2}$$

where u_i is the realization of a uniformly distributed random number (between 0 and 1) and t_1 is step-length in the first generation. It is a simple exercise to show that the above formulation ensures that not all functions can be worsened simultaneously. The variation of the step-length over iterations must be made carefully to ensure convergence to the efficient frontier. The original study suggested the following strategy for varying the step-length t_j with generation j: $t_j = C/j$ (where C is a positive constant). As the second search operator we use

Equation 2 to create a child instead of the mutation operator. As in S-NSGA-II the gradients are obtained using one-sided FD and SP methods. We name the hybrid algorithm with new mutation operator using FD and SP gradient estimates as T-NSGA-FD and T-NSGA-SP respectively.

3 Simulation Results

In this section, we compare the above two hybrid methods with the elitist non-dominated sorting GA or NSGA-II [6] on a number of unconstrained test problems. The test problems are chosen in such a way so as to systematically investigate various aspects of an algorithm. We consider two-objective ZDT test problems discussed in [5]. In their initial form these test problems are box constrained ones as the Pareto-optimal set lies on box constrains. The test problems are slightly modified so that they become unconstrained multi-objective optimization problems. Also the box constraints are slightly modified so that the functions are twice continuously differentiable in the entire feasible region (required as per SSM). Similar such modifications are also proposed in [14]. Table 1 present these test problems.

For all problems solved, we use a population of size 100. We set the number of function evaluations as 5000 for ZDT1, ZDT2, ZDT3 since in about 5000 function evaluations the population reaches the Pareto-optimal front by the best algorithm. For ZDT4 and ZDT6 we set the number of function evaluation to be 15000 for the same reason. For the NSGA-II, we use a standard real-parameter SBX and polynomial mutation operator with $\eta_c = 10$ and $\eta_m = 10$, respectively [5] (unless otherwise stated). For both T-NSGA-FD and T-NSGA-SP the parameter $C = 10.0$ is used (unless otherwise stated). For S-NSGA-FD and S-NSGA-SP the parameters $\sigma = 1.0$ along with $\epsilon = 0.1$ is used for all the test problems.

Convergence and diversity are two distinct goals in multi-objective optimization. In order to evaluate convergence we use the Inverted Generational Distance (IGD) metric [5]. This measure of convergence indicated how far is the true Pareto-optimal front from the obtained front by each of the algorithms. Diversity of solutions is evaluated using the Spread (denoted by S) metric [5]. Algorithms A is better than Algorithm B in terms of convergence (diversity) if IGD (S) of Algorithm A is less than IGD (S) of Algorithm B. We run each algorithm for 20 times (using same initial population) and the final combined non-dominated solutions are used for calculating the average, best worst and standard deviation of IGD and S metric values.

These unary metrices for convergence and diversity are used together with two binary metrices which can detect whether an approximation set is better than another. We use the multiplicative binary ϵ indicator discussed by Zitzler [20] and the two Set Coverage (SC) ([19]) to assess the performance of the algorithms. Given two outcomes A and B, of different algorithms, the binary ϵ indicator $I_\epsilon(A, B)$ gives the factor by which an approximation set is worse than another with respect to all objectives. The Set Coverage (SC) metric $I_{\mathbb{C}}$ calculates the proportion of solutions produced by Algorithm B, which are weakly dominated by solutions produced by Algorithm A. We use these binary metrices to conclude whether an approximation set produced by an algorithm *strictly dominates, dominates, better, weakly dominates* or *is incomparable* with the approximation set produced by another algorithm (see [20] and Table 2 for definitions). For example, if $I_\epsilon(A, B) \leq 1$ and $I_\epsilon(B, A) > 1$ occurs then we can conclude that Algorithm A is better than Algorithm B. These conditions are quite difficult to satisfy using binary ϵ indicator values. We will use the binary I_ϵ indicator values to conclude *partial* results: we will say that Algorithm A is *relatively better as per I_ϵ metric* than Algorithm B if $I_\epsilon(A, B) < I_\epsilon(B, A)$. For the SC metric we will say that Algorithm A is *relatively better as per SC metric* than Algorithm B if $SC(A, B) > SC(B, A)$. We run each algorithm for 5 times (using same initial population) and the final combined non-dominated solutions are used for calculating the binary performance metrices and for producing the plots shown.

For statistical evaluation we use attaintment surface based statistical metric ([9]). We run each algorithm for 21 times (using same initial population) and the median attaintment surface (11th) plots are shown.

Table 1. Unconstrained test problems used in this study

Name	Objective functions	$g(\cdot)$ function/ Optimal solutions	Variable bounds/ Type
ZDT1	$f_1(\mathbf{x}) = x_1$	$g(x) = 1 + \frac{9}{n-1}\sum_{i=2}^{30} x_i^2$	$[0.01, 1] \times [-1, 1]^{29}$
	$f_2(\mathbf{x}) = g(\mathbf{x})\left[2 - \sqrt{\frac{x_1}{g(\mathbf{x})}}\right]$	$x_1 \in [0.01, 1], x_i = 0 \forall i \neq 1$	convex
ZDT2	$f_1(\mathbf{x}) = x_1$	$g(x) = 1 + \frac{9}{n-1}\sum_{i=2}^{30} x_i^2$	$[0.01, 1] \times [-1, 1]^{29}$
	$f_2(\mathbf{x}) = g(\mathbf{x})\left[2 - \left(\frac{x_1}{g(\mathbf{x})}\right)^2\right]$	$x_1 \in [0.01, 1], x_i = 0 \forall i \neq 1$	non-convex
ZDT3	$f_1(\mathbf{x}) = x_1$	$g(x) = 1 + \frac{9}{n-1}\sum_{i=2}^{30} x_i^2$	$[0.01, 1] \times [-1, 1]^{29}$
	$f_2(\mathbf{x}) = g(x)\left[2 - \sqrt{\frac{x_1}{g(\mathbf{x})}} - \frac{x_1}{g(\mathbf{x})}\sin(10\pi x_1)\right]$	$x_1 \in [0.01, 1], x_i = 0 \forall i \neq 1$	convex, disconnected
ZDT4	$f_1(\mathbf{x}) = x_1$	$g(x) = 1 + 10(n-1) + \sum_{i=2}^{10}(x_i^2 - 10\cos(4\pi x_i))$	$[0.01, 1] \times [-5, 5]^9$
	$f_2(\mathbf{x}) = g(\mathbf{x})\left[2 - \sqrt{\frac{x_1}{g(\mathbf{x})}}\right]$	$x_1 \in [0.01, 1], x_i = 0 \forall i \neq 1$	convex, multi-modal
ZDT6	$f_1(\mathbf{x}) = 1 - \exp(-4x_1)\sin^6(4\pi x_1)$	$g(x) = 1 + 9\left(\frac{\sum_{i=2}^{10} x_i^2}{n-1}\right)^{0.25}$	$[0.01, 1] \times [-1, 1]^9$
	$f_2(\mathbf{x}) = g(\mathbf{x})\left[2 - \left(\frac{f_1(\mathbf{x})}{g(\mathbf{x})}\right)^2\right]$	$x_1 \in [0.01, 1], x_i = 0 \forall i \neq 1$	non-convex, non-uniform density

The unconstrained ZDT1 problem has a convex efficient front for which solutions correspond to $0.01 \leq x_1^* \leq 1$ and $x_i^* = 0$ for $i = 2, 3, \ldots, 30$. In this problem the algorithms face difficulty in tackling a large number of variables. Figure 1 shows the performance of all the algorithms after 5000 function evaluations. From the figure it can be visually concluded that all the algorithms perform well on this problem in maintaining a diverse set of solutions close to the efficient front. Table 3 shows the IGD convergence metric values. In terms of average IGD values T-NSGA-FD performs the best while S-NSGA-FD and S-NSGA-SP along with T-NSGA-FD perform better than NSGA-II. Table 4 shows the S diversity metric values. In terms of average S values T-NSGA-FD

Table 2. Description of relations used to compare algorithms

Relation	Conditions	Description
strictly dominates	$I_\epsilon(A,B) < 1$	every $b \in B$ is strictly dominated by at-least one $a \in A$
dominates	$I_{\mathbb{C}}(A,B) = 1$, $I_{\mathbb{C}}(B,A) = 0$	every $b \in B$ is dominated by at-least one $a \in A$
better	$I_\epsilon(A,B) \le 1$, $I_\epsilon(A,B) > 1$; $I_{\mathbb{C}}(A,B) = 1$, $I_{\mathbb{C}}(B,A) < 1$	every $b \in B$ is weakly dominated by at-least one $a \in A$ and $A \neq B$
weakly dominates	$I_\epsilon(A,B) \le 1$; $I_{\mathbb{C}}(A,B) = 1$	every $b \in B$ is weakly dominated by at-least one $a \in A$
incomparable	$I_\epsilon(A,B) > 1$, $I_\epsilon(B,A) > 1$; $I_{\mathbb{C}}(A,B) \in (0,1)$, $I_{\mathbb{C}}(B,A) \in (0,1)$	neither A weakly dominates B nor B weakly dominates A
relatively better	$I_\epsilon(A,B) < I_\epsilon(B,A)$;$I_{\mathbb{C}}(A,B) > I_{\mathbb{C}}(A,B)$	approximation set A is relatively better than set B
better in terms of convergence	$IGD(A) < IGD(B)$	convergence of approximation set A is better than that of set B
better in terms of diversity	$S(A) < S(B)$	diversity of approximation set A is better than that of set B

still performs the best while S-NSGA-FD, S-NSGA-SP, T-NSGA-SP along with T-NSGA-FD perform better than NSGA-II. Hence as far as convergence and diversity separately are considered almost all the hybrid algorithms perform better than original NSGA-II. The binary performance metrices values of all the algorithms are shown in Table 5. For a particular test problem, an element (i, j) in this table (rows and columns corresponding to different algorithms) represents I_ϵ(algorithm j, algorithm i) while the $I_{\mathbb{C}}$(algorithm j, algorithm i) values are shown in italics. From the table one obtains that the efficient front obtained by T-NSGA-FD *dominates* that of NSGA-II. Moreover the front produced by T-NSGA-FD, S-NSGA-FD and S-NSGA-SP are *relatively better* than that of NSGA-II. Figure 2 shows the median attaintment surface plots of all the algorithms. It can be visually seen that with the possible exception of T-NSGA-SP and S-NSGA-SP all the other algorithms perform better stochastically.

The unconstrained ZDT2 problem has a non-convex efficient front. Figure 3 shows the performance of all the algorithms after 5000 function evaluations. From the figure it can be visually concluded that with the possible exception of S-NSGA-SP all the other algorithms perform better that NSGA-II on this problem in maintaining a diverse set of solutions close to the efficient front. In terms of average IGD values (Table 3) T-NSGA-SP performs the best while all the other algorithms perform better than NSGA-II. Similar results are obtained from Table 4 in terms of S spread metric values. Hence as far as convergence and diversity separately are considered all the hybrid algorithms perform better than original NSGA-II. Next we consider the binary performance metrices on this problem (Table 5). From the table one obtains that the efficient front obtained by T-NSGA-FD and T-NSGA-SP *dominates* that of NSGA-II. Moreover the

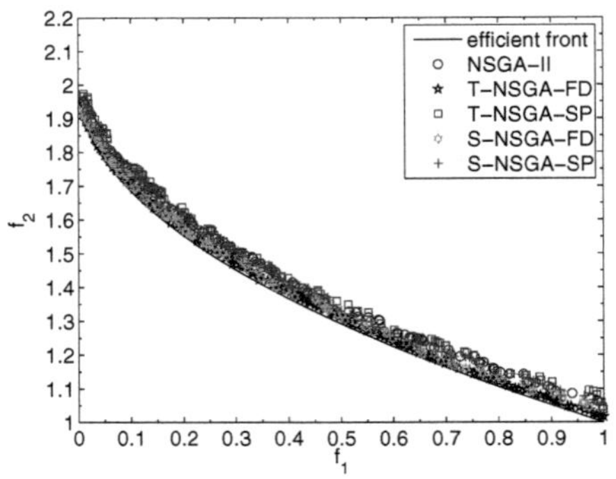

Fig. 1. Performance of the five algorithms on ZDT1

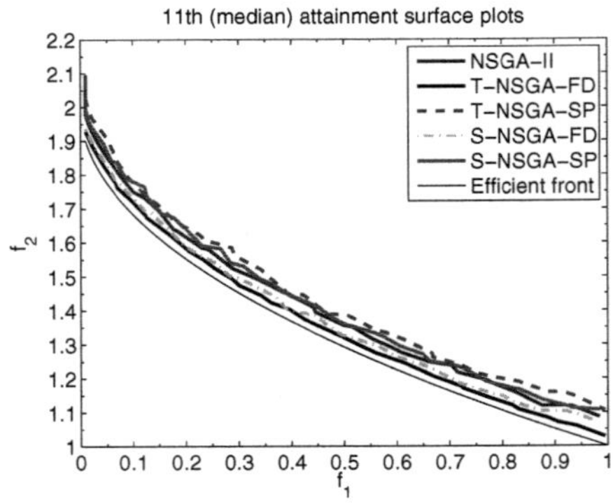

Fig. 2. Median attaintment surface plots of the five algorithms on ZDT1

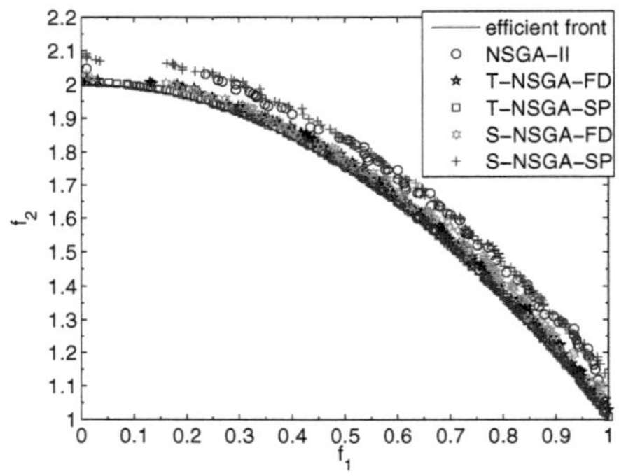

Fig. 3. Performance of the five algorithms on ZDT2

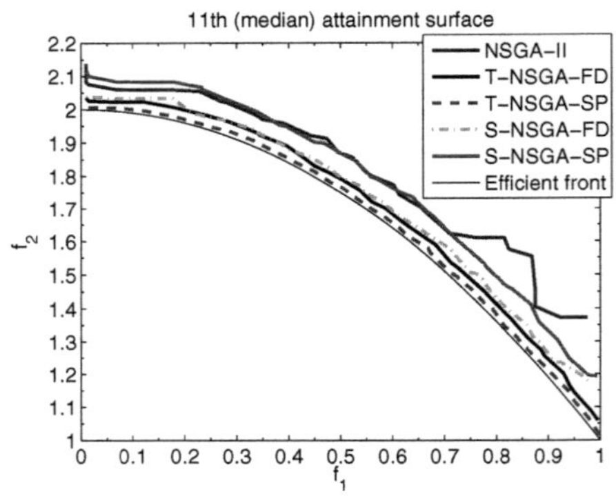

Fig. 4. Median attaintment surface plots of the five algorithms on ZDT2

front produced by S-NSGA-SP is *relatively better* than that of NSGA-II. One may wonder why does some of the algorithms for example T-NSGA-SP does not strictly dominates NSGA-II (i.e. the condition $I_{\epsilon}(A, B) < 1$ is never met). This is due to the presence of some solutions by NSGA-II on the $f_1 = 0.01$ weak efficient front, in this case there cannot exist any feasible points which are strictly better in all objective values. Figure 4 shows the median attaintment surface plots of all the algorithms. It can be visually seen that with the possible exception of T-NSGA-SP all the other algorithms perform better stochastically.

Next we consider unconstrained ZDT3, this problem has a convex discontinuous efficient frontier. Figure 5 shows the performance of all the algorithms after 5000 function evaluations. From the figure it can be visually concluded that all the algorithms perform well on this problem in maintaining a diverse set of solutions close to the efficient front. In terms of average IGD values (Table 3) T-NSGA-SP performs the best while NSGA-II outperforms all the other

Table 3. Inverted generational distance metric values for test problems

ZDT1	NSGA-II	T-NSGA-FD	T-NSGA-SP	S-NSGA-FD	S-NSGA-SP
best	0.0432	0.0162	0.0520	0.0256	0.0349
worst	0.3321	0.0213	0.2625	0.0558	0.0657
average	0.1065	0.0180	0.1071	0.0339	0.0539
std. dev.	0.1264	0.0020	0.8790	0.0124	0.0125
rank.	4	1	5	2	3
ZDT2	NSGA-II	T-NSGA-FD	T-NSGA-SP	S-NSGA-FD	S-NSGA-SP
best	0.0627	0.0241	0.0084	0.0317	0.0718
worst	0.6363	0.0347	0.0103	0.0523	0.0969
average	0.2250	0.0308	0.0093	0.0421	0.0835
std. dev.	0.2420	0.0041	0.0006	0.0085	0.0126
rank.	5	2	1	3	4
ZDT3	NSGA-II	T-NSGA-FD	T-NSGA-SP	S-NSGA-FD	S-NSGA-SP
best	0.0303	0.0178	0.0288	0.0600	0.0277
worst	0.0436	0.0375	0.0501	0.1085	0.0600
average	0.0370	0.0294	0.0409	0.0848	0.0455
std. dev.	0.0055	0.0082	0.0076	0.0212	0.0142
rank.	2	3	1	5	4
ZDT6	NSGA-II	T-NSGA-FD	T-NSGA-SP	S-NSGA-FD	S-NSGA-SP
best	0.1270	0.0098	0.0207	0.2655	0.0193
worst	0.3771	0.0134	0.0382	0.4007	0.0377
average	0.2354	0.0114	0.0259	0.3432	0.0273
std. dev.	0.0951	0.0116	0.0071	0.0485	0.0090
rank.	4	1	2	5	3

Table 4. Spread metric values for test problems

ZDT1	NSGA-II	T-NSGA-FD	T-NSGA-SP	S-NSGA-FD	S-NSGA-SP
best	0.6052	0.0351	0.0520	0.6488	0.4434
worst	0.8349	0.6574	0.7417	0.7259	0.5437
average	0.7022	0.4604	0.6723	0.6997	0.4910
std. dev.	0.0909	0.1182	0.0922	0.0314	0.0482
rank.	5	1	3	4	2
ZDT2	NSGA-II	T-NSGA-FD	T-NSGA-SP	S-NSGA-FD	S-NSGA-SP
best	0.6416	0.6455	0.3924	0.6779	0.6771
worst	1.2529	0.7495	0.4312	0.7465	0.7722
average	0.9059	0.7034	0.4178	0.7232	0.7444
std. dev.	0.2398	0.0401	0.0173	0.0282	0.0404
rank.	5	2	1	3	4
ZDT3	NSGA-II	T-NSGA-FD	T-NSGA-SP	S-NSGA-FD	S-NSGA-SP
best	0.6695	0.6316	0.5738	0.6442	0.5848
worst	0.7271	0.7467	0.6569	0.7665	0.6846
average	0.7006	0.6833	0.6123	0.7072	0.6320
std. dev.	0.0277	0.0568	0.0318	0.0550	0.0360s
rank.	4	3	1	5	2
ZDT6	NSGA-II	T-NSGA-FD	T-NSGA-SP	S-NSGA-FD	S-NSGA-SP
best	0.6267	0.3700	0.3655	0.7304	0.3459
worst	0.8011	0.4844	0.4428	0.8216	0.4334
average	0.7098	0.4304	0.3969	0.7782	0.3931
std. dev.	0.0629	0.0416	0.0284	0.0362	0.0316
rank.	4	3	2	5	1

algorithms. Except S-NSGA-SP all other algorithms perform better than NSGA-II in terms of S spread metric values (Table 4). Next we consider the binary performance metrices on this problem (Table 5). From the table one obtains that the front produced by T-NSGA-FD and S-NSGA-SP is *relatively better* than that of NSGA-II while NSGA-II is *relatively better* than T-NSGA-SP and T-NSGA-FD.

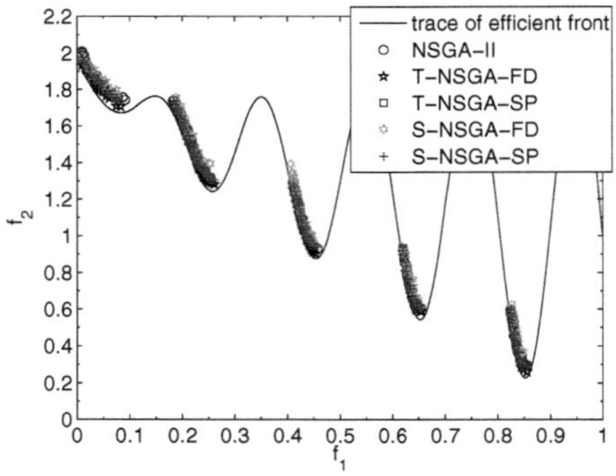

Fig. 5. Performance of the five algorithms on ZDT3

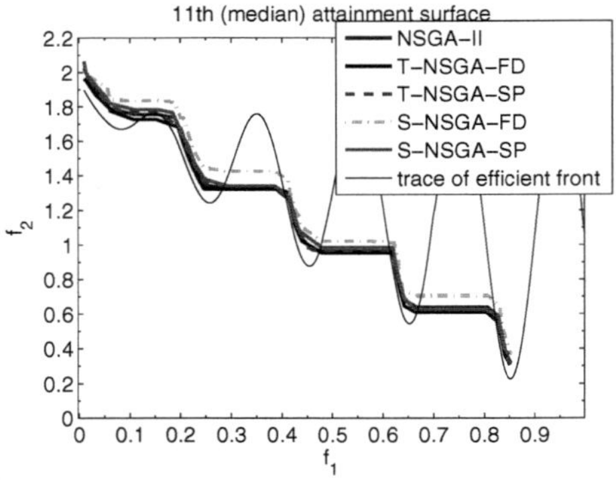

Fig. 6. Median attaintment surface plots of the five algorithms on ZDT3

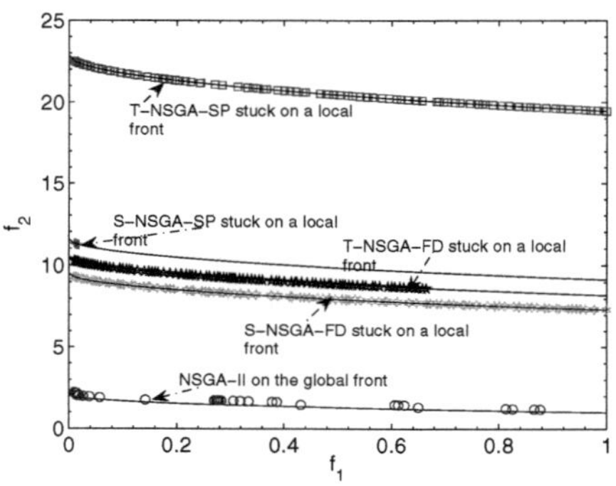

Fig. 7. Performance of the five algorithms on ZDT4

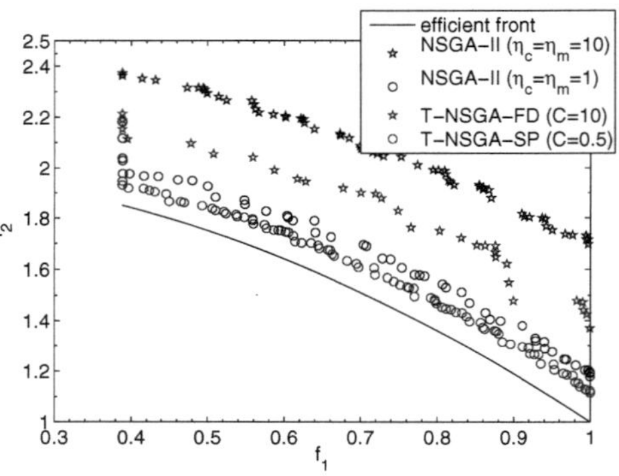

Fig. 8. Influence of parameters of NSGA-II, T-NSGA-FD and T-NSGA-SP on ZDT6

Figure 6 shows the median attaintment surface plots of all the algorithms. It can be visually seen that with the possible exception of S-NSGA-SP all the other algorithms perform almost the same stochastically.

The problem unconstrained ZDT4 has a total of 100 distinct local efficient fronts in the objective space. The global Pareto-optimal solutions correspond to $0.01 \leq x_1^* \leq 1$ and $x_i^* = 0$ for $i = 2, 3, \ldots, 10$. Since ZDT4 is a complex multi-modal problem in this problem all the algorithms are run till 15000 function evaluations. Figure 7 shows the performance of all the algorithms. It can be seen that only NSGA-II is able to overcome many local efficient fronts. All the other algorithms are stuck at some of local efficient fronts. In this problem we can visually conclude that NSGA-II outperforms all other algorithms and thus other performance metric values are not computed.

Next we consider another difficult problem, ZDT6. This problem has a non-convex and non-uniformly spaced Pareto-optimal solutions. The Pareto-optimal

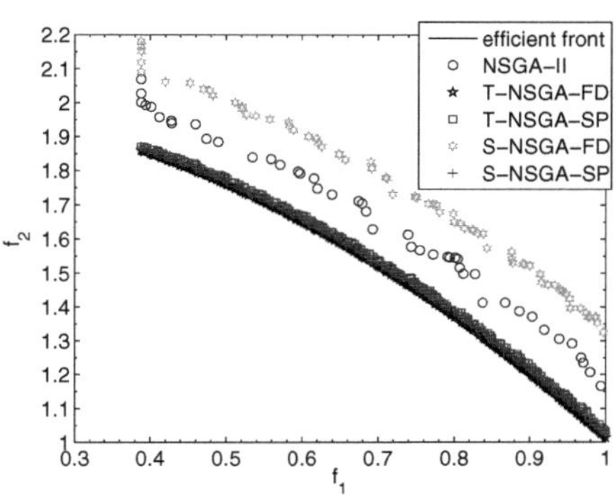

Fig. 9. Performance of the five algorithms on ZDT6

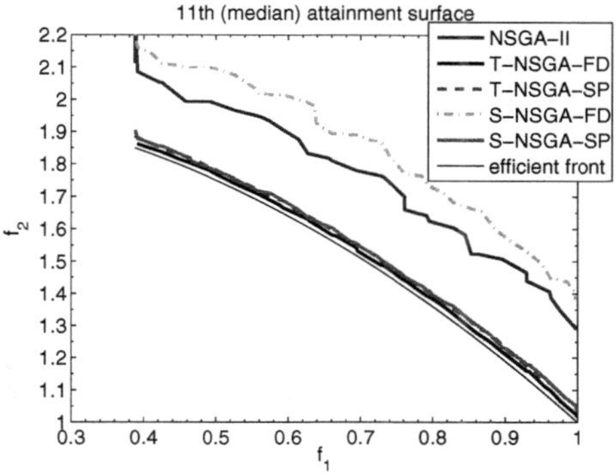

Fig. 10. Median attaintment surface plots of the five algorithms on ZDT6

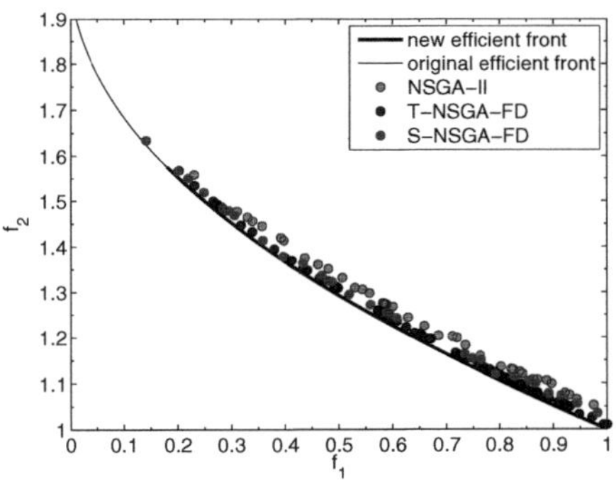

Fig. 11. Performance of three algorithms in the f_1-f_2 objective space

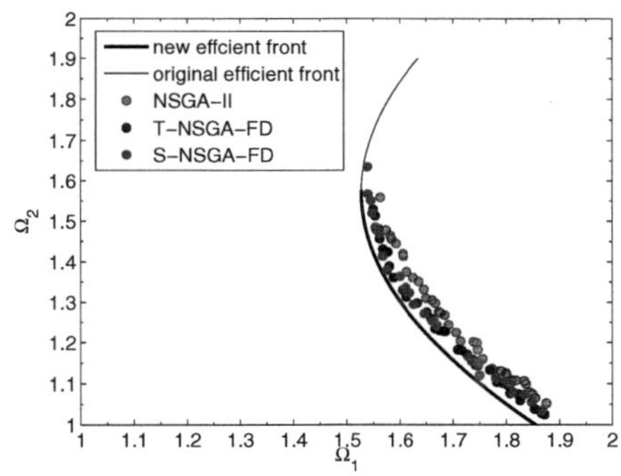

Fig. 12. Performance of three algorithms in the Ω_1-Ω_2 objective space

solutions correspond to $0.01 \leq x_1^* \leq 1$ and $x_i^* = 0$ for $i = 2, 3, \ldots, 10$. In this problem also we set the number of function evaluations to be 15000. A limited parametric study has also been done for this problem and we can see from Figure 8 that in the case of NSGA-II the algorithm performs much better with parameter value as $\eta_c = \eta_m = 1.0$ rather than $\eta_c = \eta_c = 10.0$ and thus these values are used for NSGA-II. Similarly in the case of T-NSGA-FD and T-NSGA-SP it is seen that the parameter $C = 0.5$ performs better than $C = 10.0$ and hence these values are used here. The parameter values does not influence to a large extent the S-NSGA-FD and S-NSGA-SP algorithms and hence we use the original parameter values for these algorithms. Figure 9 shows the performance of all the algorithms after 15000 function evaluations. From the figure it can be visually concluded that with the possible exception of S-NSGA-FD all the other algorithms perform much better that NSGA-II on this problem in maintaining a diverse set of solutions close to the efficient front. In terms of average IGD values (Table 3) T-NSGA-FD performs the best while all the other algorithms

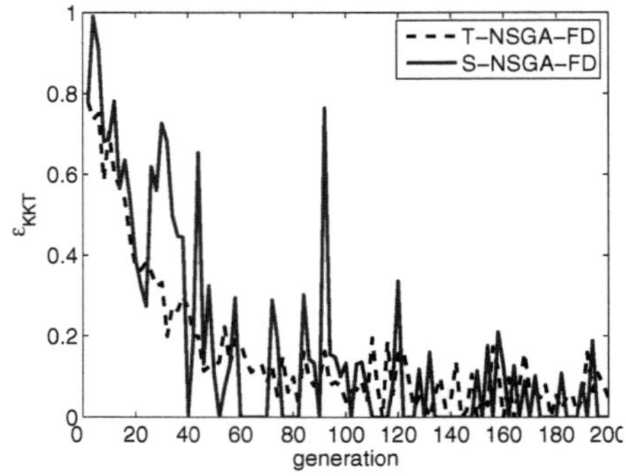

Fig. 13. Plot of ϵ_{KKT} metric with generation for ZDT1

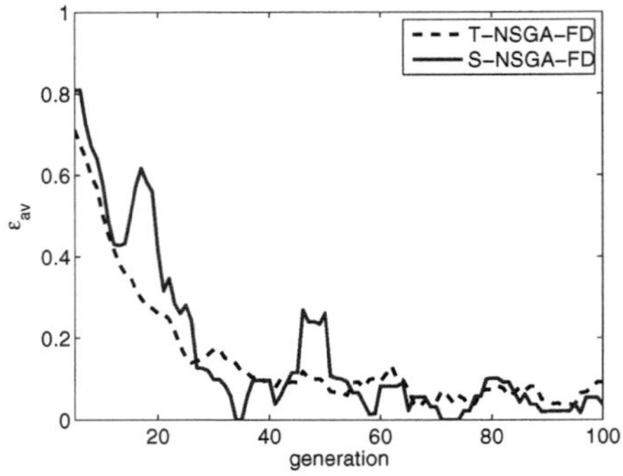

Fig. 14. Plot of ϵ_{av} metric with generation for ZDT1

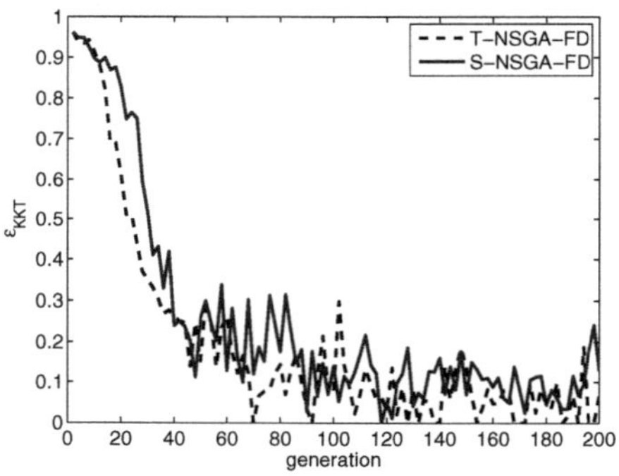

Fig. 15. Plot of ϵ_{KKT} metric with generation for ZDT2

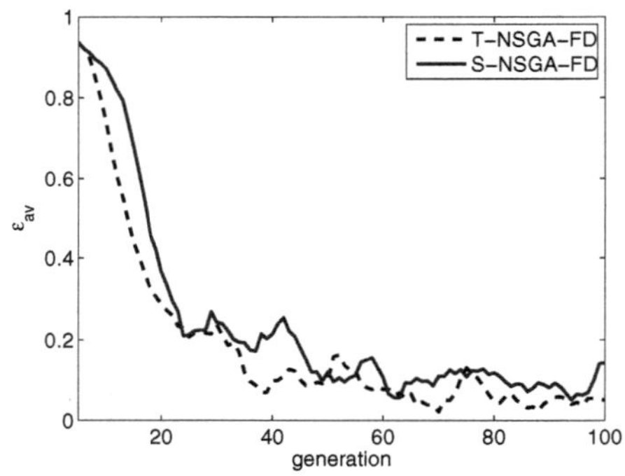

Fig. 16. Plot of ϵ_{av} metric with generation for ZDT2

except S-NSGA-FD perform better than NSGA-II. In terms of average S values (Table 4) S-NSGA-SP performs the best while all the other algorithms except S-NSGA-FD perform better than NSGA-II. Hence as far as convergence and diversity separately are considered all the hybrid algorithms except S-NSGA-FD perform better than original NSGA-II. Next we consider the binary performance metrices on this problem (Table 5). From the table one obtains that the efficient front obtained by T-NSGA-FD, T-NSGA-SP and S-NSGA-SP *dominates* that of NSGA-II while NSGA-II *dominates* S-NSGA-FD in this problem. Figure 10 shows the median attaintment surface plots of all the algorithms. It can be visually seen that with the exception of S-NSGA-FD all the other algorithms perform much better stochastically than NSGA-II.

Next we show how these hybrid methods can be effectively combined in the Guided Domination Approach ([3,4]). In the Guided Dominated Approach, a weighted sum of the objectives is formed. For example in the case of two objectives f_1 and f_2 following two weighted objectives are formed.

Table 5. Epsilon and coverage metric values for test problems

ZDT1	NSGA-II		T-NSGA-FD		T-NSGA-SP		S-NSGA-FD		S-NSGA-SP	
	I_ϵ	$I_\mathbb{C}$	I_ϵ	$I_\mathbb{C}$	I_ϵ	$I_\mathbb{C}$	I_ϵ	$I_\mathbb{C}$	I_ϵ	$I_\mathbb{C}$
NSGA-II	1.00	*1.00*	1.00	*1.00*	1.02	*0.07*	1.01	*0.97*	1.01	*0.46*
T-NSGA-FD	1.04	*0.00*	1.00	*1.00*	1.04	*0.00*	1.02	*0.12*	1.04	*0.00*
T-NSGA-SP	1.02	*0.75*	1.00	*1.00*	1.00	*1.00*	1.00	*0.99*	1.02	*0.84*
S-NSGA-FD	1.03	*0.00*	1.00	*0.66*	1.03	*0.00*	1.00	*1.00*	1.02	*0.00*
S-NSGA-SP	1.02	*0.32*	1.00	*1.00*	1.02	*0.03*	1.00	*0.99*	1.00	*1.00*
ZDT2	NSGA-II		T-NSGA-FD		T-NSGA-SP		S-NSGA-FD		S-NSGA-SP	
	I_ϵ	$I_\mathbb{C}$	I_ϵ	$I_\mathbb{C}$	I_ϵ	$I_\mathbb{C}$	I_ϵ	$I_\mathbb{C}$	I_ϵ	$I_\mathbb{C}$
NSGA-II	1.00	*1.00*	1.00	*1.00*	1.00	*1.00*	1.01	*0.98*	1.01	*0.11*
T-NSGA-FD	1.08	*0.00*	1.00	*1.00*	1.00	*0.97*	1.04	*0.12*	1.10	*0.00*
T-NSGA-SP	1.11	*0.00*	1.02	*0.00*	1.00	*1.00*	1.06	*0.00*	1.12	*0.00*
S-NSGA-FD	1.03	*0.00*	1.01	*0.72*	1.00	*1.00*	1.06	*1.00*	1.12	*0.00*
S-NSGA-SP	1.01	*0.67*	1.00	*1.00*	1.00	*1.00*	1.00	*0.99*	1.00	*1.00*
ZDT3	NSGA-II		T-NSGA-FD		T-NSGA-SP		S-NSGA-FD		S-NSGA-SP	
	I_ϵ	$I_\mathbb{C}$	I_ϵ	$I_\mathbb{C}$	I_ϵ	$I_\mathbb{C}$	I_ϵ	$I_\mathbb{C}$	I_ϵ	$I_\mathbb{C}$
NSGA-II	1.00	*1.00*	1.01	*0.85*	1.02	*0.38*	1.24	*0.04*	1.01	*0.41*
T-NSGA-FD	1.12	*0.05*	1.00	*1.00*	1.10	*0.05*	1.40	*0.00*	1.10	*0.02*
T-NSGA-SP	1.01	*0.46*	1.00	*0.84*	1.00	*1.00*	1.26	*0.02*	1.01	*0.40*
S-NSGA-FD	1.01	*0.94*	1.00	*1.00*	1.00	*0.93*	1.00	*1.00*	1.00	*0.97*
S-NSGA-SP	1.01	*0.34*	1.00	*0.86*	1.01	*0.19*	1.27	*0.00*	1.00	*1.00*
ZDT6	NSGA-II		T-NSGA-FD		T-NSGA-SP		S-NSGA-FD		S-NSGA-SP	
	I_ϵ	$I_\mathbb{C}$	I_ϵ	$I_\mathbb{C}$	I_ϵ	$I_\mathbb{C}$	I_ϵ	$I_\mathbb{C}$	I_ϵ	$I_\mathbb{C}$
NSGA-II	1.00	*1.00*	1.00	*1.00*	1.00	*1.00*	1.13	*0.00*	1.00	*1.00*
T-NSGA-FD	1.15	*0.00*	1.00	*1.00*	1.01	*0.00*	1.30	*0.00*	1.01	*0.00*
T-NSGA-SP	1.13	*0.00*	1.00	*1.00*	1.00	*1.00*	1.28	*0.00*	1.00	*0.51*
S-NSGA-FD	1.00	*1.00*	1.00	*1.00*	1.00	*1.00*	1.00	*1.00*	1.00	*1.00*
S-NSGA-SP	1.13	*0.00*	1.00	*0.99*	1.00	*0.21*	1.28	*0.00*	1.00	*1.00*

$$\Omega_1(f_1, f_2) = f_1 + a_{12} f_2,$$
$$\Omega_2(f_1, f_2) = a_{21} f_1 + f_2.$$

The domination check is now made in the $\Omega = (\Omega_1, \Omega_2)$ space rather than in (f_1, f_2) space. This approach is used to guide an evolutionary algorithm towards some parts of the efficient front for convex problems. We show next that when using the TPM and SSM classical search operators with the Ω function instead of the usual SBX operator in Guided NSGA-II (G-NSGA-II) discussed in ([5,4]) we can improve the algorithms. For this we take the unconstrained ZDT1 test problem and use the Guided NSGA-II to find a region near the minimum of f_2 (with $a_{21} = 0, a_{12} = 1/\arctan(0.75\pi)$ as weights, the same as used in [7]). For T-NSGA-FD and S-NSGA-FD algorithms we use the SSM and TPM search operators on the weighted objectives (i.e. Ω). Figures 11 and 12 show the performance of the algorithms in the f_1-f_2 and Ω_1-Ω_2 space. From the figures it can be visually concluded that T-NSGA-FD and S-NSGA-FD perform better than original Guided NSGA-II finding a portion of the efficient front. The new mutation operators help additionally the Guided NSGA-II by pushing the solutions towards the portion to be discovered by these algorithms.

These use of classical gradient based local search methods have other advantages too. Since now we use gradient information, this can be further used to check the KKT conditions for the test problems without any knowledge of the efficient frontier. We can use the norm of $q(\mathbf{x}) = \sum_{i=1}^{m} \hat{\alpha}_i \nabla f_i(\mathbf{x})$. If the norm equals zero we are sure that the obtained solution lies on the efficient frontier for convex problems and we are more confident in the case of non-convex problems. The norm can be also taken as a measure of deviations from the distance from the unknown efficient front [15]. We propose the following new KKT condition based running convergence metric (which can be obtained from the hybrid algorithms once gradients are known) as follows:

$$\epsilon^i_{KKT} = \sum_{k_1}^{n^i_m} \frac{\|q(\mathbf{x})\|}{n^i_m}$$

where n^i_m if the number of mutations (based on TPM and SSM operators) in generation i. An advantage of this running performance metric is that the efficient front need not be known. For all existing convergence metrices the knowledge of the exact efficient front is required which limits their application to only test problems. Moreover this ϵ_{KKT} can be used as a stopping criteria for differentiable problems having an unknown efficient front. Figure 13 and Figure 15 shows how plot of running performance metric ϵ_{KKT} on ZDT1 and ZDT2 test problem for T-NSGA-FD and S-NSGA-FD algorithms. Since the mutation probability is usually very less only few solutions are used in the computation of the above metric and thus the values fluctuate. For this we suggest taking a running average for last 5 generations and using this values (we call it ϵ_{av}) instead. Figure 14 and Figure 16 shows how plot of running performance metric ϵ_{av} on ZDT1 and ZDT2 test problem for T-NSGA-FD and S-NSGA-SP algorithms. It can be seen that the values quickly go to zero.

4 Conclusions

This study brings into light how the local search operators of two classical generating methods which can be effectively used in a state-of-the-art evolutionary algorithm. These local search methods require gradient information which is numerically evaluated using the Forward Finite-Difference technique and a Stochastic Perturbation method. These local search methods are used instead of the normal mutation operator. The comparison of these methods with NSGA-II on a number of test problems have adequately demonstrated that these methods perform very well when the problem size and search space complexity is large. Among the three hybrid algorithms, the T-NSGA-FD and T-NSGA-SP use a random directional weighted objective gradient which produced a non-dominated or better child. S-NSGA-FD and S-NSGA-SP produce child which is always superior to the parent.

On a number of two-objective test problems, it has been observed that the using the TPM and SSM search operators as local search operators is beneficial for uni-modal problems. In fact they perform much better than NSGA-II on a uni-modal difficult problem with non-uniform density. However for problems with multi-modal fronts these methods in their present form fail. However these classical methods are designed for solving multi-modal problems and thus efficient methods for parameter variation needs to be designed for multi-modal problems. For example better methods of numerically solving the stochastic differential equation, adaptive parameter variation scheme for TPM needs to be designed.

On the other hand, on all other problems considered here, T-NSGA-FD and T-NSGA-SP has performed well. These classical hybrid algorithms are also effectively used in Guided NSGA-II. A convergence metric and stopping criteria is also introduced in the present study. This is effective for differentiable problems having an unknown efficient front. Finally it is to be mentioned that although this study has focussed on only differentiable problems there exists techniques for non-differentiable problems also which needs to be explored ([12]). Some such extensions would be an immediate focus for useful research and application in the area of multi-objective optimization.

References

1. Peter A. N. Bosman and Edwin D. de Jong. Exploiting gradient information in numerical multi–objective evolutionary optimization. In *GECCO '05: Proceedings of the 2005 conference on Genetic and evolutionary computation*, pages 755–762, New York, NY, USA, 2005. ACM Press.
2. Peter A. N. Bosman and Edwin D. de Jong. Combining gradient techniques for numerical multi-objective evolutionary optimization. In *GECCO '06: Proceedings of the 8th annual conference on Genetic and evolutionary computation*, pages 627–634, New York, NY, USA, 2006. ACM Press.
3. J. Branke, T. Kau*β*ler, and H. Schmeck. Guidance in evolutionary multi-objective optimization. *Advances in Engineering Software*, 32:499–507, 2001.
4. Jürgen Branke and Kalyanmoy Deb. Integrating User Preferences into Evolutionary Multi-Objective Optimization. In Yaochu Jin, editor, *Knowledge Incorporation in Evolutionary Computation*, pages 461–477. Springer, Berlin Heidelberg, 2005. ISBN 3-540-22902-7.
5. K. Deb. *Multi-objective optimization using evolutionary algorithms.* Chichester, UK: Wiley, 2001.
6. K. Deb, S. Agrawal, A. Pratap, and T. Meyarivan. A fast and elitist multi-objective genetic algorithm: NSGA-II. *IEEE Transactions on Evolutionary Computation*, 6(2):182–197, 2002.
7. K. Deb, P. Zope, and A. Jain. Distributed computing of pareto-optimal solutions using multi-objective evolutionary algorithms. In *Proceedings of the Second Evolutionary Multi-Criterion Optimization (EMO-03) Conference (LNCS 2632)*, pages 535–549, 2003.
8. M. Ehrgott. *Multicriteria Optimization.* Berlin: Springer, 2000.
9. C. M. Fonesca and P. J. Fleming. On the performance assessment and comparison of stochastic multiobjective optimizers. In *Proceedings of Parallel Problem Solving from Nature IV (PPSN-IV)*, pages 584–593, 1996.
10. Ken Harada, Kokolo Ikeda, and Shigenobu Kobayashi. Hybridization of genetic algorithm and local search in multiobjective function optimization: recommendation of ga then ls. In *GECCO '06: Proceedings of the 8th annual conference on Genetic and evolutionary computation*, pages 667–674, New York, NY, USA, 2006. ACM Press.

11. Ken Harada, Jun Sakuma, and Shigenobu Kobayashi. Local search for multiobjective function optimization: pareto descent method. In *GECCO '06: Proceedings of the 8th annual conference on Genetic and evolutionary computation*, pages 659–666, New York, NY, USA, 2006. ACM Press.
12. K. C. Kiwiel. Descent methods for nonsmooth convex constrained minimization. In *Nondifferentiable optimization: motivations and applications (Sopron, 1984)*, volume 255 of *Lecture Notes in Econom. and Math. Systems*, pages 203–214. Springer, Berlin, 1985.
13. S. Schäffler, R. Schultz, and K. Weinzierl. Stochastic method for the solution of unconstrained vector optimization problems. *Journal of Optimization Theory and Applications*, 114(1):209–222, 2002.
14. P. K. Shukla, K. Deb, and S. Tiwari. Comparing Classical Generating Methods with an Evolutionary Multi-objective Optimization Method. In C. A. Coello Coello, A. Hernández Aquirre, and E. Zitzler, editors, *Evolutionary Multi-Criterion Optimization. Third International Conference, EMO 2005*, pages 311–325, Guanajuato, México, March 2005. Springer. Lecture Notes in Computer Science Vol. 3410.
15. P. K. Shukla, J. Dutta, and K. Deb. Approximate solutions in multiobjective optimization. Technical report, KanGal Report No. 2004009, Indian Institute Of Technology Kanpur, India, 2004.
16. J. C. Spall. Implementation of the Simultaneous Perturbation Algorithm for Stochastic Optimization. *IEEE Transactions on Aerospace and Electronic Systems*, 34(3):817–823, 1998.
17. G. Timmel. Ein stochastisches Suchverrahren zur Bestimmung der optimalen Kompromilsungen bei statischen polzkriteriellen Optimierungsaufgaben. *Wiss. Z. TH Ilmenau*, 26(5):159–174, 1980.
18. G. Timmel. Modifikation eines statistischen Suchverfahrens der Vektoroptimierung. *Wiss. Z. TH Ilmenau*, 28(6):139–148, 1982.
19. E. Zitzler and L. Thiele. Multiobjective optimization using evolutionary algorithms – A comparative case study. In *Parallel Problem Solving from Nature V (PPSN-V)*, pages 292–301, 1998.
20. Eckart Zitzler, Lothar Thiele, Marco Laumanns, Carlos M. Fonseca, and Viviane Grunert da Fonseca. Performance Assessment of Multiobjective Optimizers: An Analysis and Review. *IEEE Transactions on Evolutionary Computation*, 7(2):117–132, April 2003.

Symbolic Archive Representation for a Fast Nondominance Test*

Martin Lukasiewycz, Michael Glaß, Christian Haubelt, and Jürgen Teich

Hardware-Software-Co-Design
Department of Computer Science 12
University of Erlangen-Nuremberg, Germany
{martin.lukasiewycz, glass, haubelt, teich}@cs.fau.de

Abstract. Archives are used in Multi-Objective Evolutionary Algorithms to establish elitism. Depending on the optimization problem, an unconstrained archive may grow to an immense size. With the growing number of nondominated solutions in the archive, testing a new solution for nondominance against this archive becomes the main bottleneck during optimization. As a remedy to this problem, we will propose a new data structure on the basis of Binary Decision Diagrams (BDDs) that permits a nondominance test with a runtime that is independent from the archive size. For this purpose, the region in the objective space weakly dominated by the solutions in the archive is represented by a BDD. We will present the algorithms for constructing the BDD as well as the nondominance test. Moreover, experimental results from using this symbolic data structure will show the efficiency of our approach in test cases where many candidates have to be tested but only few have to be added to the archive.

1 Introduction

Multi-Objective Evolutionary Algorithms [1,2] using elitism to prevent nondominated solutions from being deleted during generations can be proven to converge to the true Pareto front [3]. Moreover, elitism increases the probability of creating better offspring [1]. Hence, keeping nondominated solutions in an archive A is an important issue in multi-objective optimization. In general there are two strategies for handling archives: (1) using so called *constrained archives* requires a method for *limiting* the number of nondominated solutions in the archive. (2) so called *unconstrained archives*, i.e., archives for storing an unlimited number of nondominated solutions, rely on efficient data structures. Constrained archives are afflicted with the problem of *shrinking* the Pareto front or *oscillating* between different approximations of the Pareto front [4]. On the other hand, keeping an archive dominant-free has a large influence on the computational complexity of the optimization and, thus, narrowing the benefits from using huge or even unconstrained archives.

As a remedy to this problem, several data structures have been proposed in the recent years. Most of these data structures are tree-based [4,5,6]. Unfortunately, the worst case behavior of the nondominance test using these data structures is similar to the complexity of using linear lists, i.e., a new solution, called *candidate*, has to be compared with

* Supported in part by the German Science Foundation (DFG), SFB 694.

S. Obayashi et al. (Eds.): EMO 2007, LNCS 4403, pp. 111–125, 2007.

each solution already in the archive A resulting in $|A|$ comparisons. In this paper, we will present a novel data structure on the basis of Binary Decision Diagrams (BDDs) [7]. Instead of explicitly storing all members in the archive, we encode the region in the objective space weakly dominated by the nondominated solutions as a BDD. A new solution can be tested for nondominance by traversing the BDD. This operation returns a value *true* or *false* and is independent from the archive size, thus allowing a faster nondominance test than any up to now reported archive data structure.

The rest of the paper is organized as follows: Section 2 discusses data structures for archive representation and Section 3 will formally state the problem this paper is dedicated to. In Section 4 our novel symbolic data structure based on BDDs together with the most important algorithms will be presented. First experimental results from comparing our symbolic representation with a linear list data structure will be discussed in Section 5 before we conclude the paper in Section 6.

2 Related Work

While implementing an archive A as linear list requires in the worst case $|A|$ tests to check the nondominance of a candidate resulting in a complexity of $\mathcal{O}(|A| \cdot m)$ in a m-dimensional objective space, tree-like data structures showed improved runtimes: In [6], Mostaghim and Teich proposed the use of so called *quad trees* (cf. [8]). A quad tree is a tree-based data structure where each node has at most 2^m successors where m is the number of objectives. A new vector can be inserted in the quad tree if it is not dominated by any node in the tree. Therefore, a nondominance test is done against the root. If it is not dominated by the root it will be tested against all nodes in the k-th subtree of the root. Here, k is the binary encoding of the $\geq$-relation of the vector's components to the root's components. The dominance test is recursive, i.e., the new solution is next tested against the root of the k-th subtree. If the k-th subtree does not exist the new vector will be inserted. A more sophisticated algorithms is needed in order to keep the data structure *dominant-free*, e.g., if the new solution dominates nodes already in the quad tree. Mostaghim and Teich present experimental results from a comparison of quad trees with linear lists. As a result quad trees outperform linear lists in case of large populations and small archives.

In [5], Schütze proposed the use of a data structure based on m-ary trees, called *dominance decision trees*, as well as algorithms to test for dominance and tree update. Each node has at most m successors where again m is the number of objectives. For the k-th successor of a given node the following properties hold: The first $k-1$ objectives fulfill the $\leq$-relation between the k-th successor and the node. The k-th objective of the k-th successor is greater than the k-th objective of the given node. In [5], several experimental results from comparing dominance decision trees with quad trees and linear lists are presented. In many cases, the dominance decision tree outperforms the linear list and quad trees. Considering problem instances with more than three objectives, the quad trees perform better.

Both data structures have some common disadvantages: The worst case computation time is similar to the case using linear lists, i.e., $\mathcal{O}(|A| \cdot m)$. This is due to the fact that the depth of the trees depends on the order in which nondominated candidates are added to the archive.

A combination of two new data structures, called *dominated trees* and *nondominated trees*, avoids the above mentioned problem and was proposed by Fieldsend et al. [4]. Both data structures are based on the notion of so called *composite points* where each composite point represents a set of so called *constituent points* with a maximum cardinality m, where m is the number of objectives. Composite points can be constructed from a set of vectors by successively determining the maximum (minimum) for each dimension (starting with the first objective) and removing the corresponding vector (a constituent point) from the set. Having m objectives, a maximum of m constituent points contribute to a composite point. Dominated trees allow an efficient *nondominance check* whereas nondominated trees permit an efficient computation of dominated points. In the best case, the nondominance test using dominated trees can be done in $\log_2(|A|/m)$ comparisons between the candidate and the composite points with $|A|$ being the cardinality of the archive. However, in general k additional tests with individual solutions in the archive are required, leading to a complexity of $\mathcal{O}(\log_2(|A|/m)+k\,m)$. Hence, the computational complexity is still dependent on the archive size $|A|$. Before we will present our approach for a nondominance test that is independent from the archive size through representing the region in the objective space weakly dominated by the solutions in the archive by binary decision diagrams, we first will start with a formal introduction to the problem.

3 Problem Formulation

Given the following multi-objective optimization problem:[1]

$$\min f : X \subset \mathbb{R}^n \to \mathbb{R}^m \tag{1}$$

The goal in multi-objective optimization is to find all *Pareto-optimal solutions* $X_\text{p} \subseteq X$ [9]. A solution x_1 is said to be Pareto-optimal if it is not *dominated* by any solution $x_2 \in X$.

Definition 1 (Pareto dominance (cf. [10])). *For any two solutions x_1 and x_2,*

$$\begin{aligned}
x_2 \succ x_1 \quad & (x_2 \textit{ dominates } x_1) \textit{ if } \forall i : f_i(x_2) \leq f_i(x_1) \wedge \exists i : f_i(x_2) < f_i(x_1)\\
x_2 \succeq x_1 \quad & (x_2 \textit{ weakly dominates } x_1) \textit{ if } \forall i : f_i(x_2) \leq f_i(x_1)\\
x_2 \sim x_1 \quad & (x_2 \textit{ is indifferent to } x_1) \textit{ if } \forall i : f_i(x_2) = f_i(x_1)\\
x_2 \parallel x_1 \quad & (x_2 \textit{ is incomparable to } x_1) \textit{ if } \exists i,j : f_i(x_2) > f_i(x_1) \wedge f_j(x_2) < f_j(x_1).
\end{aligned}$$

The so called *Pareto-optimal front* is given by $Y_\text{p} = f(X_\text{p}) = \{y \mid y = f(x) \wedge x \in X_\text{p}\}$. Thus the goal in multi-objective optimization can also be stated as: Sort out the nondominated objective vectors $y \in Y_\text{p}$ from a set of all objective vectors $Y = f(X) = \{y \mid y = f(x) \wedge x \in X\}$. X is called the *decision space*. Y is called the *objective space*. In the following, we will limit our discussion to the objective space.

[1] Without loss of generality, we consider minimization problems in this paper.

This so called *nondominance problem* can be divided into two classes (cf. [5]): The *static* nondominance problem is to find the subset of nondominated solutions $Y_{\rm p}$ in a given set Y [11]. The *dynamic* nondominance problem arises during the archive update in Multi-Objective Evolutionary Algorithms: Given a dominant-free archive $A \subseteq Y$ and a sequence of candidate solutions $(y_1, y_2, \ldots, y_l)$. Each candidate solution in this sequence has to be tested for nondominance against the solution in A. If a candidate solution y_i is not weakly dominated by any solution in the archive, y_i has to be added to A. In addition, if y_i dominates solutions from the archive, these solutions have to be removed from the archive keeping it dominant-free.

The most intuitive solution for the dynamic nondominance problem is using a linear list as data structure. In that case, the computational complexity of testing whether a candidate is weakly dominated is linear in the size of the archive, i.e., $\mathcal{O}(|A| \cdot m)$. Removing dominated solutions from A has the same complexity as well.

In the following, we will show how to use Binary Decision Diagrams (BDDs) as data structure to represent the region in the objective space weakly dominated by the solutions in the archive. However, by using BDDs we will not substitute the linear list but rather give a support to it. A Binary Decision Diagram (BDD) is a data structure that can be used to represent Boolean functions [7,12]. By extending the linear list archive with BDDs, the costs for adding a new candidate increase. On the other hand, the test if a candidate is weakly dominated by a solution in the archive and whether it should be added to the archive is independent from the archive size.

4 Using BDDs for a Fast Nondominance Test

In addition to save the nondominated solutions $y \in A$ in the linear, we encode the region that is weakly dominated by these solutions in a single BDD. To test a candidate vector for weak dominance, the binary encoding of its objective values is used to traverse this BDD. This traversal returns *true*, i.e., the BDD is satisfied, if the candidate solution is weakly dominated by at least one solution in the archive A. Otherwise, the traversal returns *false*.

A BDD is a data structure to represent Boolean functions as rooted directed acyclic graphs. Each node in the BDD has either exactly two or none successor. Nodes without successors are called *terminal nodes* and are labeled *true* or *false*. All other nodes are called *decision nodes* and are labeled with a binary variable x of the Boolean function. One edge from the node to one of its successors is labeled 0, the other edge connecting the second successor is labeled 1. This labeling corresponds to the assignment to x. Hence, a path from the root of the BDD to a terminal node is an assignment to the variables of the Boolean function. Moreover, the label of the terminal node determines if the Boolean function is satisfied (*true*) or not satisfied (*false*) under the given variable assignment. On each path from the root to one terminal node each variable appears at most once. The corresponding BDD of the Boolean function $(\overline{x_0} \wedge \overline{x_1} \wedge \overline{x_2}) \vee (x_0 \wedge x_1) \vee (x_1 \wedge x_2)$ is shown in Figure 1.

To clarify the goal of our methodology we assume the example from Figure 2. Given is a three-dimensional problem (x, y, z) in which the values are natural numbers in the range from 0 to 15. Therefore, for each dimension exactly four binary

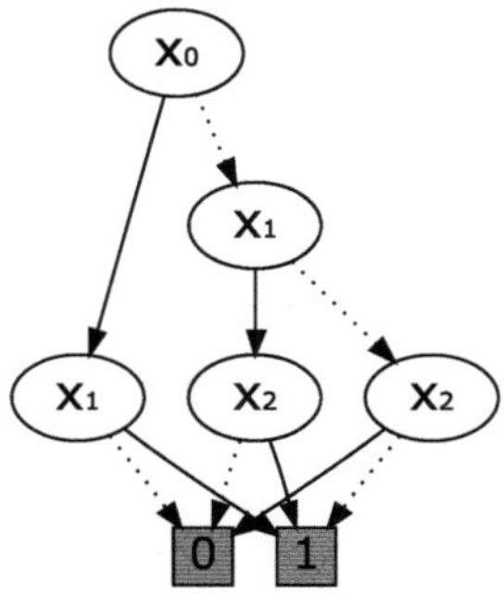

Fig. 1. A Binary Decision Diagram (BDD) of the Boolean function $(\overline{x_0} \wedge \overline{x_1} \wedge \overline{x_2}) \vee (x_0 \wedge x_1) \vee (x_1 \wedge x_2)$. A dotted (solid) edge corresponds the case where the decision variable is 0 (1). The variable order is (x_0, x_1, x_2).

variables are needed to encode a natural number, i.e., for the dimension x we have x_3, x_2, x_1, x_0 where x_3 is the most significant bit and x_0 the least significant bit. For the set $A = \{(15, 12, 4), (6, 12, 8), (2, 2, 14), (2, 5, 13)\}$ of nondominated solutions the BDD is given in Figure 2. To test if a candidate vector is weakly dominated by any vector from A we have to use its binary representation to traverse the BDD. For instance the binary representation of the candidate vector $v = (8, 12, 8)$ is $(x_0 = 0, x_1 = 0, x_2 = 0, x_3 = 1, y_0 = 0, y_1 = 0, y_2 = 1, y_3 = 1, z_0 = 0, z_1 = 0, z_2 = 0, z_3 = 1)$ and by traversing the BDD with that assignment the result is $true$ which means that v is weakly dominated by some vector in A.

By using BDDs, it is mandatory that the objective values are encoded by a binary representation. Although our methodology does not limit the objective values to be natural numbers, we will assume that the objective space is given by $Y \subset {\mathbb{N}_0}^m$. Additionally it is recommended that the upper and lower bounds for the values in each dimension are known, such that the minimal number of variables can be used in the BDD. With the known upper bound hi_i and lower bound lo_i for the values in each dimension $i \in \{1, ..., m\}$ the number of required binary variables is given by:

$$\sum_{i=1}^{m} \lceil \log_2(hi_i - lo_i + 1) \rceil \qquad (2)$$

Before each objective value is converted to its binary representation, it is normalized by subtracting the corresponding lower bound. To decrease the number of required binary variables even more, all objective values occurring in the i-th dimension are divided by their greatest common divisor. An effect on the number of required binary variables will only take place if the greatest common divisor is greater than 1.

Using the binary encoding of the objective values, we start to construct the Binary Decision Diagram which encodes the region in the objective space weakly dominated by the archive A. In the following, we assume that we use k_i variables to encode the i-th objective value. We start by introducing an algorithm that constructs a BDD with k_i binary variables $var_i[k_i - 1], \ldots, var_i[0]$. This BDD returns *true* if the binary encoding

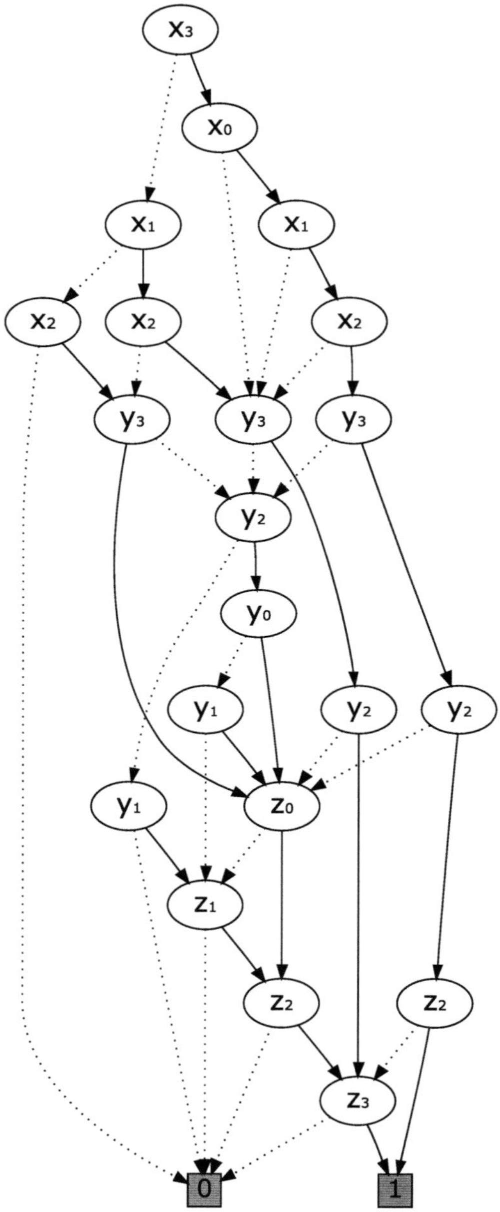

Fig. 2. An example of the BDD that represents the weakly dominated space in a three-dimensional (x, y, z) problem. The variable order is $(x_3, x_0, x_1, x_2, y_3, y_2, y_0, y_1, z_0, z_1, z_2, z_3)$. A dotted (solid) edge corresponds the case where the decision variable is 0 (1).

Algorithm 1. The function bdd_greater&equal constructs a BDD with the variables $var_i[k_i - 1], \ldots, var_i[0]$. The BDD is satisfied, i.e., it returns *true*, if a binary encoded objective value assigned to the variables $var_i[k_i - 1], \ldots, var_i[0]$ is greater than or equal to the given constant value represented by a binary number $c_i = (c_i[k_i - 1], \ldots, c_i[0])$. Otherwise, it will return *false*. If $var_i = (var_i[k_i - 1], \ldots, var_i[0])$ is a list of variables with the length k_i, the first element $var_i[0]$ is the least significant bit and the last element $var_i[k_i - 1]$ is the most significant bit.

The algorithm operates from the least to the most significant bit. The if-condition determines whether the k-th position of c_i is 1 or 0. If the if-condition is *true* and $c_i[k]$ is 1, it is mandatory that the corresponding variable $var_i[k]$ is also 1 in order to fulfill the greater or equal condition. Therefore, the variable is appended by a logical AND ($\wedge$). If $c_i[k]$ is 0 a 1 for the corresponding variable $var_i[k]$ fulfills the greater or equal condition albeit the variable assignment of the less significant bits. Therefore, the variable in the else-branch is appended by a logical OR ($\vee$).

```
bdd_greater&equal(var_i, c_i)
{
    bdd b = true;

    for(k = 0; k < k_i; k++){
        if(c_i[k] == 1) {
            b = b ∧ var_i[k];
        } else {
            b = b ∨ var_i[k];
        }
    }

    return b;
}
```

of the i-th objective value of a candidate solution is assigned to the variables var_i and the value is greater than or equal to a given constant value represented by the binary number $c_i[k_i - 1], \ldots, c_i[0]$. Otherwise, it will return *false*. One can think of c_i being the i-th objective value of a solution stored in the archive. That means the BDD covers the statement

$$2^{k_i-1} var_i[k_i - 1] + \cdots + 2^0 var_i[0] \geq 2^{k_i-1} c_i[k_i - 1] + \cdots + 2^0 c_i[0].$$

The construction of this BDD is shown in Algorithm 1.

By Definition 1, a solution x_1 is weakly dominated by a solution x_2 if $\forall i : f_i(x_2) \leq f_i(x_1)$. For a given solution x_2 it is possible to construct a BDD that returns *true* if a candidate solution x_1 is weakly dominated by x_2. Otherwise, it will return *false*. As we are only interested in improving the set of solutions in the archive A, dominated and moreover weakly dominated candidate vectors can be disregarded. The binary encoding of $f(x_2)$ is given by $c = (c_0, \ldots, c_{m-1})$ with m being the number of

Algorithm 2. The function bdd_weakdominated constructs a BDD that returns *true* if the binary encodings of the objective values $f(x_1)$ of a candidate solution x_1 are assigned to $var = (var_0, \ldots, var_{m-1})$ and x_1 is weakly dominated by x_2 with its objective values $f(x_2) = c = (c_0, \ldots, c_{m-1})$. Otherwise, it will return *false*. m is the number of objectives.

In particular, weak dominance is detected if the candidate vector is greater or equal to c in all m dimensions. Therefore, the greater or equal condition has to be fulfilled in each dimension. This is reached by appending the single dimension conditions with a logical AND ($\wedge$).

```
bdd_weakdominated(var, c)
{
        bdd b = true;

        for(i = 0; i < m; i++){
                b = b ∧ bdd_greater&equal(var_i, c_i);
        }
        return b;
}
```

objectives and $c_i = (c_i[k_i - 1], \ldots, c_i[0])$. The BDD is constructed with the variables $var = (var_0, \ldots, var_{m-1})$ which are the binary encodings of the m objective values where $var_i = (var_i[k_i - 1], \ldots, var_i[0])$. Following Definition 1, if all values $f_i(x_1)$ are greater than or equal to the objective values $f_i(x_2)$ for each dimension x_1 is weakly dominated by x_2. For this reason the BDDs constructed by the bdd_greater&equal function from Algorithm 1 have to be connected by applying the logical AND operation. This is shown in Algorithm 2.

Finally, we can construct a BDD that will validate if a candidate solution x_1 is weakly dominated by any solution in the archive A. This can be easily done by combining the BDDs constructed by Algorithm 2. For each solution in the archive A a BDD is created by Algorithm 2 and connected by a logical OR (see Algorithm 3). If the resulting BDD is interpreted as the entire region in the objective space that is weakly dominated by the solutions in the archive, the OR operation is equivalent to a union of the regions weakly dominated by each solution. Note that the weakly dominated region can only grow monotonously in the dynamic nondominance problem. Thus, we do not need to remove any solutions from the BDD even if they are dominated by new candidate solutions. This is illustrated in Figure 3.

With the ability to iteratively add new solutions to our BDD archive, it is possible to test any candidate solution x_1 if it is weakly dominated by any solution in the archive. If the BDD is satisfied by the assignment the variables of the binary encoding of the objective values $f(x_1)$, x_1 is weakly dominated by at least one solution in the archive A. Testing the satisfiability of a BDD is done by traversing it with the given variable assignment. As in BDDs each variable from the root to terminal node appears at most once, the costs of this operation are linear related to the number of used variables. As the number of used variables only depends on the number of objectives (m) and the

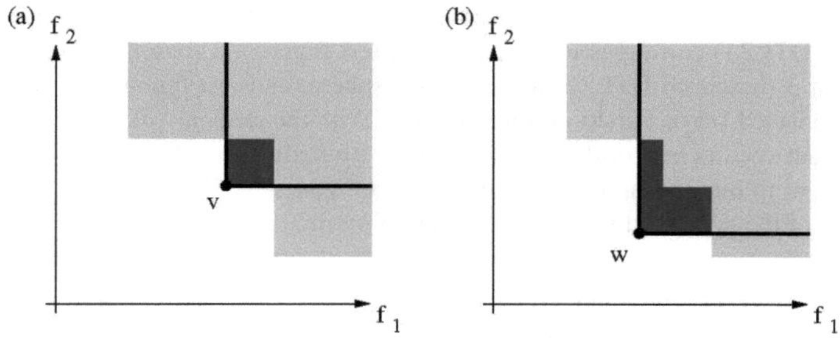

Fig. 3. Objective space with two dimensions. The entire weakly dominated region is built by the union of the regions weakly dominated by each solution vector in the archive. (a) A new candidate solution v incomparable to any other solution in the archive is added. (b) A new candidate solution w which dominates solutions in the archive is added. Note that the dominated solutions need not be removed from the BDD.

ranges of the objective values, this test has a computational complexity $\mathcal{O}(m)$ which is independent from the number of solutions in the archive A. On the other hand, the size of the BDD can grow exponentially in the worst case. But even if it does not, we should expect that constructing the BDD or adding solutions to the BDD will be a time consuming operation. Algorithm 3 shows the two main functions for the BDD archive, i.e., adding and testing a candidate solution with its objective values represented by the vector $v = (v_0, \ldots, v_{m-1})$ with $v_i = (v_i[k_i - 1], \ldots, v_i[0])$.

The BDD archive only encodes the region weakly dominated by the solutions in the archive. There is no easy way to extract the solutions $y \in A$ from the BDD. Thus, it can not completely replace the data structure for storing the solutions. Hence, the BDD gives support to the archive in determining whether a candidate solution should be added or not, and this independently from the archive size. As adding solutions to the BDD causes runtimes much greater than adding solution to a linear list, the rate between added and denied candidates should be small.

In our experiments, we extended a linear list archive by our BDD data structure in such a way, that for each candidate solution a test for weak dominance is carried out with the BDD archive. If the candidate solution is weakly dominated, it is rejected. If the candidate solution is incomparable to or dominates solutions in the archive, it has to be added to the linear list as well as to the BDD archive. Furthermore, the dominated solutions have to be removed from the linear list archive.

5 Experimental Results

In this section, we will present experimental results from using our BDD archive in combination with a linear list archive. One of the key factors for the success of the BDD archive implementation is the performance of the used BDD library. The chosen library for the tests is Buddy 2.4 [13]. In order to create appropriate test functions, we use adapted versions of some DTLZ functions from [14] (compare Table 1).

Table 1. The used test functions are based on the DTLZ test functions from [14]: Test function 1 (based on DTLZ1) converges to a linear front, test function 2 (based on DTLZ2) to a sphere, test function 3 (based on DTLZ2) to an inverse sphere, and test function 4 (based on DTLZ7) has a disconnected set of Pareto-optimal regions. With the random values $z_1, \cdots, z_m$ the functions construct vectors in a m-dimensional space. By scaling the objectives $f_i(z)$ as needed and rounding them to integers appropriate test cases are generated. The function $g(z_m)$ determines the distance of the generated vectors to the Pareto-optimal front. It is scaled with the factor r.

Test Function 1 (TF1)

$$\begin{aligned}
&\min f_1(z) = \tfrac{1}{2}(1 + g(z_m))z_1 z_2 \cdots z_{m-1}\\
&\min f_2(z) = \tfrac{1}{2}(1 + g(z_m))z_1 z_2 \cdots (1 - z_{m-1})\\
&\vdots \quad \vdots\\
&\min f_{m-1}(z) = \tfrac{1}{2}(1 + g(z_m))z_1(1 - z_2)\\
&\min f_m(z) = \tfrac{1}{2}(1 + g(z_m))(1 - z_1)
\end{aligned}$$

Test Function 2 (TF2)

$$\begin{aligned}
&\min f_1(z) = (1 + g(z_m))cos(z_1\pi/2) \cdots cos(x_{m-2}\pi/2)cos(x_{m-1}\pi/2)\\
&\min f_2(z) = (1 + g(z_m))cos(z_1\pi/2) \cdots cos(x_{m-2}\pi/2)sin(x_{m-1}\pi/2)\\
&\min f_3(z) = (1 + g(z_m))cos(z_1\pi/2) \cdots sin(x_{m-2}\pi/2)\\
&\vdots \quad \vdots\\
&\min f_m(z) = (1 + g(z_m))sin(x_1\pi/2)
\end{aligned}$$

Test Function 3 (TF3)

$$\begin{aligned}
&\min f_1(z) = (1 + g(z_m))(1 - cos(z_1\pi/2) \cdots cos(x_{m-2}\pi/2)cos(x_{m-1}\pi/2))\\
&\min f_2(z) = (1 + g(z_m))(1 - cos(z_1\pi/2) \cdots cos(x_{m-2}\pi/2)sin(x_{m-1}\pi/2))\\
&\min f_3(z) = (1 + g(z_m))(1 - cos(z_1\pi/2) \cdots sin(x_{m-2}\pi/2))\\
&\vdots \quad \vdots\\
&\min f_m(z) = (1 + g(z_m))(1 - sin(x_1\pi/2))
\end{aligned}$$

Test Function 4 (TF4)

$$\begin{aligned}
&\min f_1(z) = z_1\\
&\min f_2(z) = z_2\\
&\vdots \quad \vdots\\
&\min f_{m-1}(z) = z_{m-1}\\
&\min f_m(z) = 2m - \textstyle\sum_{i=1}^{m-1}[\frac{f_i(z)}{1+g(z_m)}(1 + sin(3\pi f_i(z)))]/m
\end{aligned}$$

where $z = (z_1, \cdots, z_m)$ and $0 \le z_i \le 1, i = 1, \cdots, m$
and $g(z_m) = z_m \cdot r$

First, we will analyze how the size of the archive affects the size of the BDD focusing on the dynamic variable reordering. The reorder algorithm we used is called Sifting [15] and is activated each time the BDD size doubles. Figure 4 shows that dynamic variable reordering has a huge effect on the BDD archive. By using the reorder algorithm the size of the BDD is halved, but it even has a bigger effect on the runtime. The reordering algorithm itself is time consuming. It can be recognized in Figure 4 on the right as a

Algorithm 3. The main functions that are provided by the BDD archive. var is containing the lists of variables for each dimension. The region that is weakly dominated by the archive A is encoded in the BDD b.

Adding a new nondominated vector equals a union on the weak dominated region. Therefore, new weak dominated regions are added by applying a logical OR where $v = (v_0, \dots, v_{m-1})$ with $v_i = (v_i[k_i - 1], \dots, v_i[0])$ corresponds the added candidate vector.

By traversing the BDD with a binary representation of the candidate vector $v = (v_0, \dots, v_{m-1})$ with $v_i = (v_i[k_i - 1], \dots, v_i[0])$ weak dominance is detected by a resulting *true*. Otherwise, the traversal returns *false*.

```
var;
bdd b=false;

add(v){
      b = b ∨ bdd_weakdominated(var, v);
}

is_weakdominated(v){
      return b.traverse(v);
}
```

vertical characteristic. On the other hand, the runtime of the reordering algorithm is just a fraction of the whole runtime. This is due to the minimization of the BDD size. Thus, it is recommended to use dynamic variable reordering, which is also used in all following test cases.

In the following all four test functions were used and an appropriate average was calculated over 100 test runs. Figure 5 shows that the BDDs in our test cases are never growing exponentially. The difference in BDD size and time consumption is insignificant between the four test functions as they are all in the same order of magnitude. With a growing archive the size of the BDDs shows in fact an increase that is similar to the logarithmic function. Adding solutions to the archive seems to be a constant time operation independent of the archive size if the curves are considered as linear.

In many cases an increasing number of variables in a BDD leads to a growth of the BDD. In our test cases the number of variables increases if the range of one dimension increases or if the number of dimensions is increased. Therefore, we examined the effect of additional bits for the encoding of the objective values. For this purpose, the test functions were analogous scaled. Table 2 shows the increases in size and time consumption. In our testcases the growth of time consumption and BDD size were linear in the number of variables of the BDD, if the number of dimensions is kept constant.

To test which effect will take place with a growing number of dimensions, test cases with a constant number of BDD variables were created. Table 3 shows that an increasing number of dimensions leads to additional runtime and an increased BDD size.

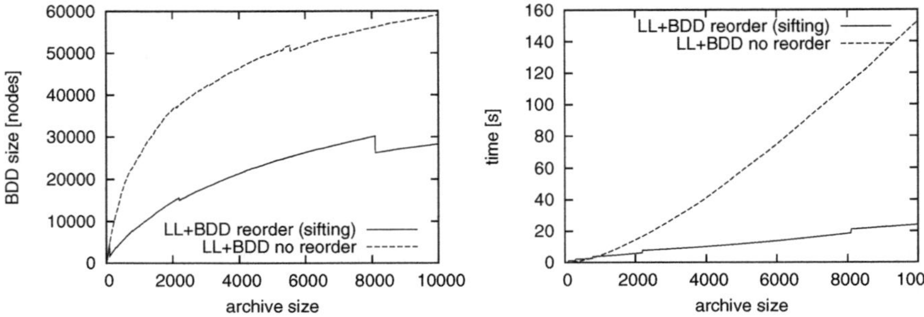

Fig. 4. Example with $m = 3$, $r = 10^{-2}$ and 10 bit encoding per dimension. $10,000$ nondominated solutions from TF1 were added iteratively to the BDD archive. The figures illustrate the size of the BDD and the time consumption with and without dynamic variable reordering.

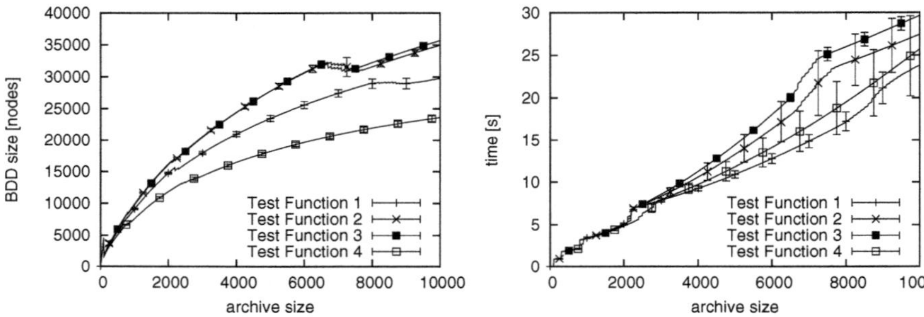

Fig. 5. Example with $m = 3$, $r = 10^{-2}$ and 10 bit encoding per dimension. $10,000$ nondominated solutions from all test functions were added iteratively to the BDD archive. The figures illustrate the size of the BDD and the time required to fill the archive. The vertical bars indicate the standard deviation for 100 runs.

Table 2. Example with $m = 3$, $r = 10^{-2}$, the number of bits per dimension is varied. The number of variables in the BDD is increasing from 27 over 36 to 45. The time consumption and BDD size are listed for adding $2,500$ nondominated solutions from all test functions to the BDD archive. The small numbers indicate the standard deviation for 100 runs.

m	bits per dimension	**TF1** time[s]	**TF2** time[s]	**TF3** time[s]	**TF4** time[s]
3	9	3.87 (0.02)	4.41 (0.04)	4.43 (0.04)	4.15 (0.03)
3	12	12.3 (0.98)	12.9 (0.54)	12.4 (0.23)	13.2 (0.43)
3	15	18.08 (1.69)	20.55 (1.13)	20.47 (0.83)	21.36 (0.70)
m	bits per dimension	**TF1** size[nodes]	**TF2** size[nodes]	**TF3** size[nodes]	**TF4** size[nodes]
3	9	8875 (124)	10829 (88)	10729 (138)	7034 (58)
3	12	32646 (213)	35639 (143)	34765 (184)	30892 (299)
3	15	59167 (363)	62569 (301)	61479 (236)	58427 (224)

Table 3. Example with $r = 10^{-2}$, the number of dimensions m and the the number of bits per dimension was scaled so that the number of variables in the BDD is constantly 36. The time consumption and BDD size are listed for adding 2,500 nondominated solutions from all test functions to the BDD archive. The small numbers indicate the standard deviation for 100 runs.

m	bits per dimension	**TF1** time[s]	**TF2** time[s]	**TF3** time[s]	**TF4** time[s]
2	18	7.9 (0.81)	7.7 (0.77)	7.5 (0.50)	5.7 (1.03)
3	12	12.3 (0.98)	12.9 (0.54)	12.4 (0.23)	13.2 (0.43)
4	9	23.6 (0.92)	25.1 (2.02)	17.5 (0.92)	21.7 (0.93)
m	bits per dimension	**TF1** size[nodes]	**TF2** size[nodes]	**TF3** size[nodes]	**TF4** size[nodes]
2	18	15593 (576)	14861 (174)	14709 (98)	12535 (143)
3	12	32646 (213)	35639 (143)	34765 (184)	30892 (299)
4	9	64929 (761)	91755 (1456)	79888 (1468)	74875 (940)

Therefore, a constant number of variable is not a guarantee for a constant time consumption and BDD size. Hence, a growing number of dimensions leads to a worse than linear growth of time consumption and BDD size.

In the next test, we compared the performance of a simple linear list archive with a linear list archive extended by our BDD archive. As all test functions had similar traits for the BDD archive, all following test cases are based on TF1. The test is separated in adding nondominated solutions to an empty archive and testing random candidate solutions for weak dominance with the filled archive. Although it is known that the added solutions are nondominated, a test for weak dominance is needed before they can be added to the linear list archive. In Figure 6 this is illustrated in comparison to the archive size. As expected, Figure 6 shows that filling the archive is much slower in the case when using the BDD archive extension. However, using the BDD archive extension, also the check for weak dominance turned out to be a constant time operation for our test case, i.e., it is in fact independent of the archive size.

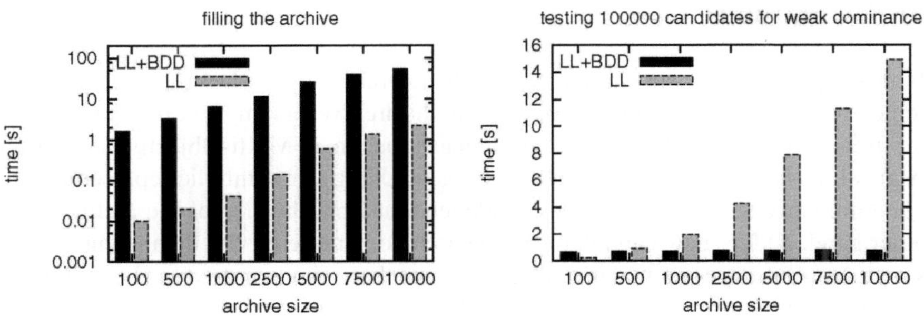

Fig. 6. Example with $m = 3$, $r = 10^{-2}$ and 10 bit encoding per dimension. The used test function is TF1. The archive size is varied. The left figure illustrates the time consumption for filling the archive, the right figure illustrates the time consumption for testing 100,000 candidate solutions for weak dominance. The used archives are a simple linear list archive and a linear list archive extended by our BDD archive.

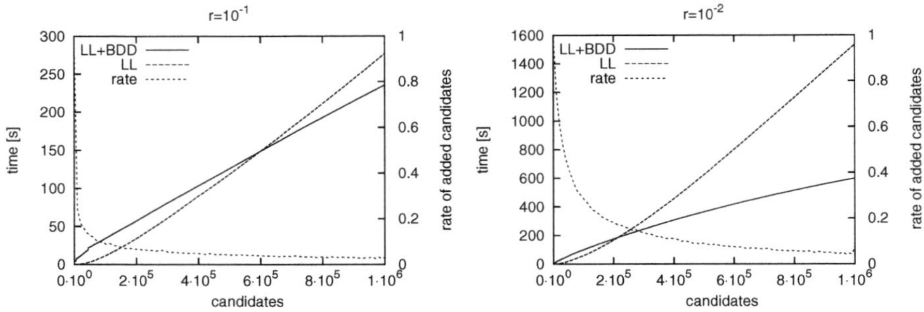

Fig. 7. Example with $m = 3$ and 10 bit encoding per dimension. In the figure on the left the value r is 10^{-1} in the figure on the right $r = 10^{-2}$. $1,000,000$ candidate solutions were generated by TF1 and iteratively tested and, if not dominated, added to the archive. The used archives are a simple linear list archive and linear list archive extended by the BDD archive. Additional to the runtime of the archives the rate of added candidate solutions is stated.

Finally, we created a dynamic nondominance problem with TF1. The used number of dimensions is three while 10 bits for each dimension are used to encode the binary values. With the factor r, the quality of the adopted optimization algorithm is biased. One million candidate solutions are created and iteratively added to both variants of the archive. Figure 7 shows the result. In both cases the BDD extended linear list archive turns out to be the better solution for the chosen test cases on long term run. The reason is that the archive is getting fuller and the weak dominance test is getting more expensive if the simple linear list archive is used. On the other hand, the number of added solutions to the archive decreases. In the case that r is 10^{-1} the archive grows slower compared to the value $r = 10^{-2}$ and the gain of the BDD extended archive is not so clear. But with a growing number of candidate solutions it should get more distinct.

6 Conclusions

In this paper, we have shown that extending an archive by a BDD representation of the region weakly dominated by the solutions in the archive can improve the runtime behavior in the dynamic nondominance problem as it occurs in Multi-Objective Evolutionary Algorithms using archives to establish elitism. Using our symbolic representation, the nondominance test of a candidate is independent from the size of the archive. On the other hand, adding new candidates to the archive is more costly, than using other data structures. Our experimental results have shown that using our proposed nondominance test in case of many candidate tests but only few archive updates clearly outperforms an archive based on a linear list.

In future work, we will combine our symbolic data structure with quad trees [6] and dominance decision trees [5] and integrate it in a Multi-Objective Evolutionary Algorithm.

References

1. Deb, K.: Multi-Objective Optimization using Evolutionary Algorithms. John Wiley & Sons, Ltd., Chichester, New York, Weinheim, Brisbane, Singapore, Toronto (2001)
2. Coello Coello, C.A., Van Veldhuizen, D.A., B.Lamont, G.: Evolutionary Algorithms for Solving Multi-Objective Problems. Kluwer Academic Publishers (2002)
3. Rudolph, G., Agapie, A.: Convergence Properties of Some Multi-Objective Evolutionary Algorithms. In: Proceedings of the 2000 Congress on Evolutionary Computation (CEC00), San Diego, California (2000) 1010–1016
4. Fieldsend, J.E., Everson, R.M., Singh, S.: Using Unconstrained Elite Archives for Multiobjective Optimization. IEEE Transactions on Evolutionary Computation **7**(3) (2003) 305–323
5. Schütze, O.: A New Data Structure for the Nondominance Problem in Multi-Objective Optimization. In: Proceedings of the Second International Conference on Evolutionary Multi-Criterion Optimization, FARO, Portugal (2003) 509–518
6. Mostaghim, S., Teich, J.: Quad-Trees: A Data structure for Storing Pareto-Sets in Multi-Objective Evolutionary Algorithms with Elitism. Advanced Information and Knowledge Processing. In: Evolutionary Multiobjective Optimization – Theoretical Advances and Applications. Springer, London (2005) 81–104
7. Bryant, R.E.: Graph-Based Algorithms for Boolean Function Manipulation. IEEE Transactions on Computers **C-35**(8) (1986) 677–691
8. Habenicht, W.: Quad Trees, a Data Structure for Discrete Vector Optimization Problems. In: Lecture Notes in Economic and Mathematical Systems, Springer-Verlag (1983) 136–145
9. Pareto, V.: Cours d'Économie Politique. Volume 1. F. Rouge & Cie., Lausanne, Switzerland (1896)
10. Laumanns, M.: Analysis and Applications of Evolutionary Multiobjective Optimization Algorithms. PhD thesis, Eidgenössische Technische Hochschule Zürich (2003)
11. Gupta, P., Janardan, R., Smid, M., Dasgupta, B.: The Rectangle Enclosure and Point-Dominance Problems Revisited. International Journal of Computational Geometry and Applications **7**(5) (1997) 437–455
12. Bryant, R.E.: Symbolic Boolean Manipulation with Ordered Binary Decision Diagrams. Technical report, School of Computer Science, Carnegie Mellon University (1992)
13. Buddy: http://sourceforge.net/projects/buddy (2006)
14. Deb, K., Thiele, L., Laumanns, M., Zitzler, E.: Scalable MultiObjective Optimization Test Problems. In: Proceedings of the 2002 Congress on Evolutionary Computation (CEC02), Piscataway, New Jersey (2002) 825–830
15. Rudell, R.: Dynamic Variable Ordering for Ordered Binary Decision Diagrams. In: ICCAD '93: Proceedings of the 1993 IEEE/ACM International Conference on Computer-Aided Design, Los Alamitos, CA, USA, IEEE Computer Society Press (1993) 42–47

Design Issues in a Multiobjective Cellular Genetic Algorithm

Antonio J. Nebro, Juan J. Durillo, Francisco Luna, Bernabé Dorronsoro, and Enrique Alba*

Departamento de Lenguajes y Ciencias de la Computación
E.T.S. Ingeniería Informática
Campus de Teatinos, 29071 Málaga (Spain)
{antonio,durillo,flv,bernable,eat}@lcc.uma.es

Abstract. In this paper we study a number of issues related to the design of a cellular genetic algorithm (cGA) for multiobjective optimization. We take as an starting point an algorithm following the canonical cGA model, i.e., each individual interacts with those ones belonging to its neighborhood, so that a new individual is obtained using the typical selection, crossover, and mutation operators within this neighborhood. An external archive is used to store the non-dominated solutions found during the evolution process. With this basic model in mind, there are many different design issues that can be faced. Among them, we focus here on the synchronous/asynchronous feature of the cGA, the feedback of the search experience contained in the archive into the algorithm, and two different replacement strategies. We evaluate the resulting algorithms using a benchmark of problems and compare the best of them against two state-of-the-art genetic algorithms for multiobjective optimization. The obtained results indicate that the cGA model is a promising approach to solve this kind of problem.

1 Introduction

Most optimization problems in the real world are multiobjective in nature. This feature, along with the facts that function evaluations can require a significant computation time and the search spaces tends to be very large, make metaheuristics popular techniques to solve multiobjective optimization problems (MOPs). Among them, evolutionary algorithms have been investigated by many authors, and some of the most well-known algorithms for solving MOPs belong to this class (e.g. NSGA-II [1], PAES [2], SPEA2 [3]). Nevertheless, in recent years there is a trend to adapt other kinds of metaheuristics (sometimes called "alternative methods", with reference to evolutionary algorithms) to the multiobjective field, such as tabu search [4] or scatter search [5].

Many evolutionary algorithms for solving MOPs are some kind of genetic algorithm (GA). These algorithms work over a set (*population*) of potential solutions

* This work has been partially funded by the Ministry of Science and Technology and FEDER under contract TIN2005-08818-C04-01 (the OPLINK project).

S. Obayashi et al. (Eds.): EMO 2007, LNCS 4403, pp. 126–140, 2007.

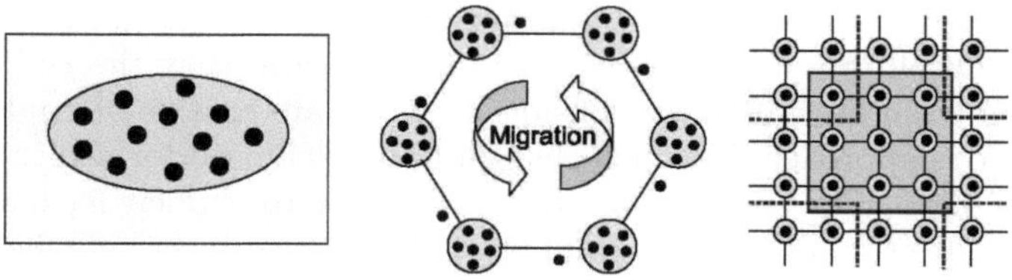

Fig. 1. Panmictic (a), distributed (b), and cellular (c) GAs

(*individuals*) which undergoes stochastic operators in order to search for better solutions. These operators are typically selection, crossover, and mutation. Most GAs use a single population (panmixia) of individuals and apply the operators to them as a whole (see Fig. 1a). Conversely, there exist the so-called structured GAs, in which the population is decentralized somehow. Among the many types of structured GAs, the distributed and cellular models are two popular variants [6,7] (see Fig. 1b and Fig. 1c, respectively). In many cases, these decentralized algorithms provide a better sampling of the search space, resulting in an improved numerical behavior with respect to an equivalent algorithm in panmixia.

In this work, we focus on the cellular model of GAs (cGAs). This kind of GAs uses the concept of (small) *neighborhood* in the sense that an individual may only interact with its nearby neighbors in the breeding loop [8,9,10]. The overlapped small neighborhoods of cGAs help in exploring the search space because the induced slow diffusion of solutions through the population provides a kind of exploration (diversification), while exploitation (intensification) takes place inside each neighborhood by genetic operations. It is worth mentioning that the neighborhood is defined among tentative solutions in the algorithm, with no relation to the geographical neighborhood definition in the problem space.

cGAs have proven to be very effective tools for solving a diverse set of single objective optimization problems from both classical and real world settings [11,12], but little attention has been paid to their use in the multiobjective optimization field [13,14,15,16,17]. In [18] we proposed the MultiObjective Cellular (MOCell) algorithm, which, together with cMOGA [16] is the unique existing adaptation of the canonical cGA model to the multiobjective field. MOCell uses an external archive to store the non-dominated solutions found during the execution of the algorithm, like many other multiobjective evolutionary algorithms do (e.g., PAES, SPEA2).

MOCell is characterized by selecting a fixed number of individuals from the archive to replace the same number of randomly chosen individuals from the population (archive feedback) at the end of each iteration of the algorithm. This is carried out with the hope of taking advantage of the search experience in order to find a Pareto front with good convergence and spread. This new replacement coexists with the typical replacement of a canonical cGA, in which the newly generated individual replaces the current one if the latter is worse than the former. However, there are many other different strategies that could be applied in the context of MOCell. In this paper we explore three possibilities. Firstly, MOCell is a synchronous cGA, in the sense that the cell updates are carried out simultaneously; we consider here the alternative of implementing an asynchronous approach, in which the cell updates are performed in a sequential order [12]. Secondly, we study an alternative way to explicit feedbacking which lies in the selection of an individual from the archive to be matched to the one taken from the neighborhood in the reproductive cycle. Thirdly, in the context of an asynchronous cGA, instead of considering the current individual to be replaced, it is possible to consider all the neighborhood, i.e., the new individual replaces the worst one in the neighborhood.

The contributions of our work can be summarized as follows:

- We explore three different design issues in the MOCell algorithm to study their influences in the accuracy of the algorithm.
- The resulting configurations are validated using two benchmarks: the ZDT family of MOPs [19], which is used in many studies in the field, and a benchmark constructed using the WFG Toolkit [20].
- The experiments are carried out using a rigorous statistical analysis around three performance metrics.
- In order to determine how competitive the resulting algorithm is, we compare the MOCell version yielding the best results against NSGA-II and SPEA2.

The remaining of the paper is organized as follows. In Section 2, we discuss related works concerning cGAs and multiobjective optimization. In Section 3, we describe MOCell and propose six variants of it, resulting from combining three different features. Experimental results are presented in Section 4. Finally, in Section 5 we give some conclusions and lines for future research.

2 Related Work

In this section we discuss related works about cGAs and multiobjective optimization. As mentioned in the introduction, only few works can be found in the literature.

In [13,17], two multiobjective evolution strategies following a predator-prey model are presented. This is a model similar to a cGA, because solutions (preys) are placed on the vertices of an undirected connected graph, thus defining neighborhoods, where they are 'caught' by predators.

Murata and Gen presented in [14] an algorithm in which, for an n-objective MOP, the population is structured in an n-dimensional weight space, and the location of individuals (called cells) depends on their weight vector. Thus, the information given by the weight vector of individuals is used for guiding the search. Notice that in this work the neighborhoods are defined in the objective space, instead of in the location of individuals in the topology of the population, as it is in classical cGAs.

A metapopulation evolutionary algorithm (called MEA) is presented in [15]. This algorithm follows a cellular model with the peculiarity that disasters can occasionally happen in the population, thus killing all the individuals located in the disaster areas (extinction). Additionally, these empty areas produced by disasters can also be occupied by individuals (colonization). Thus, this model allows a flexible population size, combining the ideas of cellular and spatially distributed populations.

Alba et al. proposed in [16] cMOGA, a cellular multiobjective algorithm which, in contrast to the aforementioned works, is based on the canonical cGA model. This algorithm uses an external archive to store the non-dominated solutions found during the search. This feature was included in MOCell [18], which can be considered as an evolution of cMOGA in which a feedback from the archive to the population has been included. MOCell was compared in [18] to the algorithms NSGA-II and SPEA2 using a benchmark of both unconstrained and constrained bi-objective MOPs, obtaining competitive results in convergence and clearly outperforming the other algorithms in terms of diversity.

3 The Algorithm

In this section we first detail a description of a canonical cGA; then, we describe the algorithm MOCell and its different configurations that we intend to study.

3.1 Canonical cGA Model

A canonical cGA follows the pseudo-code included in Algorithm 1. In this basic cGA, the population is usually structured in a regular grid of d dimensions ($d = 1, 2, 3$), and a neighborhood is defined on it. The algorithm iteratively considers as current each individual in the grid (line 3). An individual may only interact with individuals belonging to its neighborhood (line 4), so its parents are chosen among its neighbors (line 5) with a given criterion. Crossover and mutation operators are applied to the individuals in lines 6 and 7, with probabilities P_c and P_m, respectively. Afterwards, the algorithm computes the fitness value of the new offspring individual (or individuals) (line 8), and inserts it (or one of them) into the equivalent place of the current individual in a new auxiliary population (line 9) following a given replacement policy. After each generation (or loop), the auxiliary population is assumed to be the population for the next generation. This loop is repeated until a termination condition is met (line 2). The most usual termination conditions are to reach the optimal value, to perform a maximum number of fitness function evaluations, or a combination of both of them.

According to this canonical cGA, all the cells can be updated in parallel, yielding the so-named *synchronous* cGA. The alternative is the *asynchronous* cGA, in which the cells are updated one at a time in some sequential order. An asynchronous cGA can be easily obtained from Algorithm 1 assuming that the cells are sequentially updated, so the auxiliary population is not needed in the algorithm.

Algorithm 1. Pseudocode for a Canonical cGA

```
proc Steps_Up(cga)      //Algorithm parameters in 'cga'
while not Termination_Condition() do
  for individual ← 1 to cga.popSize do
     n_list←Get_Neighborhood(cga,position(individual));
     parents←Selection(n_list);
     offspring←Recombination(cga.Pc,parents);
     offspring←Mutation(cga.Pm,offspring);
     Evaluate_Fitness(offspring);
     Replace(position(individual),offspring,cga,aux_pop);
  end for
  cga.pop←aux_pop;
end while
end_proc Steps_Up;
```

Algorithm 2. Pseudocode of MOCell

```
1: proc Steps_Up(mocell)      //Algorithm parameters in 'mocell'
2: Pareto_front = Create_Front() //Creates an empty Pareto front
3: while not TerminationCondition() do
4:   for individual ← 1 to mocell.popSize do
5:      n_list←Get_Neighborhood(mocell,position(individual));
6:      parents←Selection(n_list);
7:      offspring←Recombination(mocell.Pc,parents);
8:      offspring←Mutation(mocell.Pm,offspring);
9:      Evaluate_Fitness(offspring);
10:     Replacement(position(individual),offspring,mocell,aux_pop);
11:     Insert_Pareto_Front(offspring);
12:  end for
13:  mocell.pop←aux_pop;
14:  mocell.pop←Feedback(mocell,ParetoFront);
15: end while
16: end_proc Steps_Up;
```

3.2 A Multiobjective cGA: MOCell

In this section we describe MOCell, a multiobjective metaheuristic based on the previously explained cGA model. Its pseudo-code is given in Algorithm 2. We can observe that Algorithms 1 and 2 are very similar. One of the main differences between the two algorithms is the existence of a *Pareto front* in the multiobjective case. The Pareto front is just an additional population (the external archive) composed of the non-dominated solutions found. The archive has a maximum size and, therefore, the insertion of solutions in the Pareto front has to be carefully managed to obtain a diverse set. Hence, a density estimator is needed to remove solutions from the archive when it becomes full.

MOCell starts by creating an empty Pareto front (line 2 in Algorithm 2). Individuals are arranged in a 2-dimensional toroidal grid, and the genetic operators are successively applied to them (lines 7 and 8) until the termination condition is met (line 3). Hence, for each individual, the algorithm selects two parents from its neighborhood, recombines them in order to obtain an offspring, mutates this offspring, and evaluates the resulting individual; then the algorithm decides whether the new offspring replaces the current one (line 10). The next step (line 11) is to insert the offspring into the external archive, if appropriate.

Finally, after each generation, the old population is replaced by the auxiliary one (line 13), and a feedback procedure is invoked to replace a number of randomly chosen individuals by a number of solutions from the archive (line 14).

In this algorithm, the resulting offspring replaces the individual at the current position if the former is better than the later, but, as it is usual in multiobjective optimization, we need to define the concept of "best individual". Our approach is to replace the current individual if it is dominated by the offspring or both the two are non-dominated and the current individual has the worst crowding distance (as defined in NSGA-II) in a population composed of the neighborhood plus the offspring. This criterion is also used to decide whether the offspring solutions are added to the external archive (line 11 in Algorithm 2). For inserting individuals in the Pareto front, the solutions in the archive are ordered according to the crowding distance; then, when inserting a non-dominated solution, if the Pareto front is already full, the solution with the worst (lowest) crowding distance value is removed.

MOCell has been implemented in Java using the jMetal framework [21]. It can be downloaded from: `http://neo.lcc.uma.es/metal/index.html`.

3.3 MOCell Configurations

The MOCell algorithm, as just described, was designed to fit as far as possible into the canonical cGA model. However, we can envision many different variants or configurations, although many of them could lead to a *non-orthodox* cGA. Our interest is not only to design a pure cGA per se, but also an efficient and accurate multiobjective metaheuristic. So, we propose here a number of possible variants with the aim of studying whether they outperform the base cGA model of MOCell.

We focus on three features of MOCell: synchronicity, archive feedback, and replacement. Let us start with the first of these. The algorithm described in Algorithm 2 is a synchronous cGA, but an asynchronous version can be obtained as explained in Section 3.1: the cells can be updated in sequential order, using a unique population. In the context of mono-objective cGAs, asynchronous algorithms can be more efficient (faster) that synchronous ones, while synchronous algorithms can be more effective (in terms of hit rate) [12]; a goal of this paper is to study the influence of synchronicity in MOCell. The asynchronous update policy we consider here is the so called *Line Sweep* [12], the simplest one, which sequentially updates the individuals in the same order as they are placed in the population, line by line.

As is usual in those multiobjective metaheuristics using an external archive, the solutions contained in it are re-used somehow with the idea in mind of progressing towards the Pareto optimal front. In MOCell this is carried out by explicitly selecting a number of individuals from the archive and inserting them into the population, replacing randomly selected cells. An alternative way to use the information in the archive is to use a simple scheme: in the selection method of the algorithm, one parent is taken from the neighborhood, while the other one will be randomly chosen from the archive, thus removing the explicit feedback from the algorithm.

Finally, in the canonical cGA, the replacement policy (line 9 in Algorithm 1) defines whether the offspring individual will be inserted into the population instead of the current one, replacing it. In the context of an asynchronous cGA, a possible variation is to compare the offspring individual not only with the current one but also with its whole neighborhood, thus replacing the worst neighbor.

Taking into consideration these design issues, we propose six new configurations for our algorithm. They are summarized in the following list:

- **sMOCell1**: The original synchronous MOCell algorithm.
- **sMOCell2**: MOCell + archive feedback through parent selection.
- **aMOCell1**: Asynchronous version of sMOCell1.
- **aMOCell2**: aMOCell1 + archive feedback through parent selection.
- **aMOCell3**: aMOCell1 + replacing of the worst neighbor.
- **aMOCell4**: Combination of aMOCell2 and aMOCell3.

4 Computational Results

This section is devoted to the evaluation of MOCell and its variants. For that, we have chosen several test problems taken from the specialized literature, and, in order to assess how competitive MOCell is, we decided to compare it against two algorithms that are representative of the state-of-the-art, namely NSGA-II and SPEA2. Next, we briefly comment on the main features of these two algorithms, including the parameter settings used in the subsequent experiments.

The NSGA-II algorithm was proposed by Deb *et al.* [1]. It is characterized by a Pareto ranking of the individuals and the use of a crowding distance as density estimator. A crossover probability of $p_c = 0.9$ and a mutation probability $p_m = 1/n$ (where n is the number of decision variables) are used. The operators for crossover and mutation are SBX and polynomial mutation [22], with distribution indexes of $\eta_c = 20$ and $\eta_m = 20$, respectively. The population and archive sizes are 100 individuals. The algorithm stops after 25000 function evaluations.

SPEA2 was proposed by Zitler *et al.* in [3]. In this algorithm, each individual has assigned a fitness value that is the sum of its strength raw fitness and a density estimation based on the distance to the k-th nearest neighbor. We have used the following values for the parameters. Both the population and the archive have a size of 100 individuals, and the crossover and mutation operators are the same as used in NSGA-II, using the same values concerning their application probabilities and distribution indexes. As in NSGA-II, the stopping condition is to compute 25000 function evaluations.

Both algorithms have been implemented, as MOCell has, using the jMetal framework. This way, the three techniques share the same internal code, so we can make a fair comparison avoiding problems derived from using different implementations. See [21] for further details.

In Table 1 we show the parameters used by MOCell. A square toroidal grid of 100 individuals has been chosen for structuring the population. The neighborhood used is composed of nine individuals: the considered individual plus those

Table 1. Parameterization used in MOCell

Population Size	100 individuals (10×10)
Stopping Condition	25000 function evaluations
Neighborhood	1-hop neighbors (8 surrounding solutions)
Selection of Parents	binary tournament + binary tournament
Recombination	simulated binary, $p_c = 0.9$
Mutation	polynomial, $p_m = 1.0/n$ (n = number of decision variables)
Replacement	rep_if_better_individual (NSGA-II crowding)
Archive Size	100 individuals
Density estimator	crowding distance (NSGA-II crowding)
Feedback (for sMOCell1 & aMOCell1)	20 individuals

located at its North, East, West, South, NorthWest, SouthWest, NorthEast, and SouthEast (see Fig. 1c). We have also used SBX and polynomial mutation with the same distribution indexes as NSGA-II and SPEA2. Crossover and mutation rates are $p_c = 0.9$ and $p_m = 1/n$, respectively. To set the number of individuals to be inserted from the archive to the population in the feedback procedure in sMOCell1 and aMOCell1, we carried out a number of preliminary experiments; as a result, we choose a value of 20.

We have made 100 independent runs of each experiment, and we have obtained the median, $\tilde{x}$, and interquartile range, IQR, as measures of location (or central tendency) and statistical dispersion, respectively. Since we are dealing with stochastic algorithms and we want to provide the results with confidence, the following statistical analysis has been performed in all this work [23]. Firstly, a Kolmogorov-Smirnov test is performed in order to check whether the values of the results follow a normal (gaussian) distribution or not. If so, an ANOVA test is done, otherwise we perform a Kruskal-Wallis test. We always consider in this work a confidence level of 95% (i.e., significance level of 5% or p-value under 0.05) in the statistical tests, which means that the differences are unlikely to have occurred by chance with a probability of 95%. Successful tests are marked with "+" symbols in the last column in all the tables; conversely, "−" means that no statistical confidence was found (p-value > 0.05). The best result for each problem has a gray colored background.

4.1 Test Problems

We have selected for our tests two benchmarks, the ZDT problems and a set of MOPs defined using the WFG Toolkit. The ZDT benchmark [19] is a family of bi-objective MOPs which have been widely used to assess the performance of metaheuristics for multiobjective optimization; they are formulated in Table 2. The WFG Toolkit allows the user to define benchmarks of MOPs having different properties; in this study, we use the bi-objective version of the nine problems, WFG1 to WFG9, defined in [20]. The properties of these problems are detailed in Table 3.

Table 2. The ZDT test functions

Problem	Objective functions	Variable bounds	n
ZDT1	$f_1(\boldsymbol{x}) = x_1$ $f_2(\boldsymbol{x}) = g(\boldsymbol{x})[1 - \sqrt{x_1/g(\boldsymbol{x})}]$ $g(\boldsymbol{x}) = 1 + 9(\sum_{i=2}^{n} x_i)/(n-1)$	$0 \leq x_i \leq 1$	30
ZDT2	$f_1(\boldsymbol{x}) = x_1$ $f_2(\boldsymbol{x}) = g(\boldsymbol{x})[1 - (x_1/g(\boldsymbol{x}))^2]$ $g(\boldsymbol{x}) = 1 + 9(\sum_{i=2}^{n} x_i)/(n-1)$	$0 \leq x_i \leq 1$	30
ZDT3	$f_1(\boldsymbol{x}) = x_1$ $f_2(\boldsymbol{x}) = g(\boldsymbol{x}) \left[1 - \sqrt{\frac{x_1}{g(\boldsymbol{x})}} - \frac{x_1}{g(\boldsymbol{x})} \sin(10\pi x_1)\right]$ $g(\boldsymbol{x}) = 1 + 9(\sum_{i=2}^{n} x_i)/(n-1)$	$0 \leq x_i \leq 1$	30
ZDT4	$f_1(\boldsymbol{x}) = x_1$ $f_2(\boldsymbol{x}) = g(\boldsymbol{x})[1 - (x_1/g(\boldsymbol{x}))^2]$ $g(\boldsymbol{x}) = 1 + 10(n-1) + \sum_{i=2}^{n}[x_i^2 - 10\cos(4\pi x_i)]$	$0 \leq x_1 \leq 1$ $-5 \leq x_i \leq 5$ $i = 2, ..., n$	10
ZDT6	$f_1(\boldsymbol{x}) = 1 - e^{-4x_1} \sin^6(6\pi x_1)$ $f_2(\boldsymbol{x}) = g(\boldsymbol{x})[1 - (f_1(\boldsymbol{x})/g(\boldsymbol{x}))^2]$ $g(\boldsymbol{x}) = 1 + 9[(\sum_{i=2}^{n} x_i)/(n-1)]^{0.25}$	$0 \leq x_i \leq 1$	10

Table 3. Properties of the MOPs created using the WFG toolkit

Problem	Separability	Modality	Bias	Geometry
WFG1	separable	uni	polynomial, flat	convex, mixed
WFG2	non-separable	f_1 uni, f_2 multi	no bias	convex, disconnected
WFG3	non-separable	uni	no bias	linear, degenerate
WFG4	non-separable	multi	no bias	concave
WFG5	separable	deceptive	no bias	concave
WFG6	non-separable	uni	no bias	concave
WFG7	separable	uni	parameter dependent	concave
WFG8	non-separable	uni	parameter dependent	concave
WFG9	non-separable	multi, deceptive	parameter dependent	concave

4.2 Performance Metrics

For assessing the performance of the algorithms on the test problems, two different issues are normally taken into account: (i) to minimize the distance of the Pareto front generated by the proposed algorithm to the exact Pareto front, and (ii) to maximize the spread of solutions found, so that we can have a distribution of vectors as smooth and uniform as possible. This way, the performance metrics can be classified into three categories depending on whether they evaluate the closeness to the Pareto front, the diversity in the obtained solutions, or both factors [24]. We have adopted one metric of each type: Generational Distance (GD) [25], Spread (Δ) [1], and Hypervolume (HV) [26].

4.3 Comparison of the MOCell Variants

In this section we analyze and compare the results obtained when executing the different versions of MOCell. Let us now start with the GD metric, whose values are included in Table 4. Here, aMOCell4 gets the best (lowest) values in six out of the fourteen problems, and with statistical confidence in five of them (see "+"

Table 4. Different MOCell versions: median and interquartile range of the *GD* metric

MOP	sMOCell1 $\tilde{x}_{IQR}$	sMOCell2 $\tilde{x}_{IQR}$	aMOCell1 $\tilde{x}_{IQR}$	aMOCell2 $\tilde{x}_{IQR}$	aMOCell3 $\tilde{x}_{IQR}$	aMOCell4 $\tilde{x}_{IQR}$	
ZDT1	6.288e-4 $_{1.5e-4}$	2.749e-4 $_{7.5e-5}$	4.207e-4 $_{7.2e-5}$	2.518e-4 $_{4.5e-5}$	2.222e-4 $_{4.1e-5}$	1.753e-4 $_{2.0e-5}$	+
ZDT2	5.651e-4 $_{2.0e-4}$	1.778e-4 $_{1.1e-4}$	2.884e-4 $_{1.9e-4}$	1.111e-4 $_{1.1e-4}$	1.197e-4 $_{5.3e-5}$	5.629e-5 $_{2.5e-5}$	+
ZDT3	3.326e-4 $_{8.5e-5}$	2.493e-4 $_{3.0e-5}$	2.644e-4 $_{4.4e-5}$	2.427e-4 $_{2.8e-5}$	2.077e-4 $_{2.0e-5}$	2.008e-4 $_{1.8e-5}$	+
ZDT4	9.668e-4 $_{6.4e-4}$	3.848e-4 $_{2.9e-4}$	7.847e-4 $_{5.9e-4}$	4.235e-4 $_{3.3e-4}$	6.179e-4 $_{4.0e-4}$	3.293e-4 $_{2.0e-4}$	+
ZDT6	3.963e-3 $_{1.3e-3}$	1.080e-3 $_{2.0e-4}$	2.397e-3 $_{7.8e-4}$	9.334e-4 $_{1.3e-4}$	8.778e-4 $_{1.3e-4}$	6.323e-4 $_{3.4e-5}$	+
WFG1	1.962e-4 $_{8.0e-3}$	1.859e-4 $_{1.9e-5}$	1.906e-4 $_{6.5e-3}$	1.889e-4 $_{2.0e-5}$	1.921e-4 $_{8.1e-3}$	2.052e-4 $_{1.0e-2}$	+
WFG2	4.408e-4 $_{1.4e-4}$	4.339e-4 $_{1.2e-4}$	4.410e-4 $_{1.3e-4}$	4.316e-4 $_{1.3e-4}$	4.337e-4 $_{1.3e-4}$	4.336e-4 $_{7.1e-5}$	+
WFG3	1.372e-4 $_{1.4e-5}$	1.349e-4 $_{1.5e-5}$	1.375e-4 $_{1.8e-5}$	1.340e-4 $_{1.3e-5}$	1.367e-4 $_{1.5e-5}$	1.354e-4 $_{1.4e-5}$	+
WFG4	6.423e-4 $_{2.2e-5}$	6.259e-4 $_{2.6e-5}$	6.396e-4 $_{2.6e-5}$	6.252e-4 $_{2.4e-5}$	6.341e-4 $_{2.6e-5}$	6.253e-4 $_{3.2e-5}$	+
WFG5	2.634e-3 $_{2.6e-5}$	2.633e-3 $_{1.4e-5}$	2.636e-3 $_{3.4e-5}$	2.631e-3 $_{1.4e-5}$	2.633e-3 $_{1.2e-5}$	2.635e-3 $_{1.1e-5}$	+
WFG6	4.984e-4 $_{4.3e-4}$	1.210e-3 $_{2.1e-3}$	5.146e-4 $_{7.1e-4}$	1.268e-3 $_{3.4e-3}$	5.976e-4 $_{7.1e-4}$	1.906e-3 $_{3.4e-3}$	+
WFG7	3.069e-4 $_{2.2e-5}$	3.048e-4 $_{2.3e-5}$	3.025e-4 $_{2.1e-5}$	3.038e-4 $_{2.7e-5}$	3.067e-4 $_{2.4e-5}$	3.011e-4 $_{2.4e-5}$	-
WFG8	1.009e-2 $_{6.6e-3}$	1.460e-2 $_{5.4e-3}$	1.000e-2 $_{6.0e-3}$	1.468e-2 $_{3.3e-3}$	1.434e-2 $_{5.2e-3}$	1.474e-2 $_{4.9e-3}$	+
WFG9	1.072e-3 $_{6.1e-5}$	1.055e-3 $_{5.3e-5}$	1.081e-3 $_{5.5e-5}$	1.067e-3 $_{6.6e-5}$	1.067e-3 $_{5.8e-5}$	1.065e-3 $_{6.0e-5}$	+

Table 5. Different MOCell versions: median and interquartile range of the Δ metric

MOP	sMOCell1 $\tilde{x}_{IQR}$	sMOCell2 $\tilde{x}_{IQR}$	aMOCell1 $\tilde{x}_{IQR}$	aMOCell2 $\tilde{x}_{IQR}$	aMOCell3 $\tilde{x}_{IQR}$	aMOCell4 $\tilde{x}_{IQR}$	
ZDT1	1.541e-1 $_{2.1e-2}$	9.645e-2 $_{1.4e-2}$	1.345e-1 $_{1.9e-2}$	9.161e-2 $_{1.3e-2}$	1.011e-1 $_{1.7e-2}$	7.493e-2 $_{1.3e-2}$	+
ZDT2	1.753e-1 $_{3.8e-2}$	9.907e-2 $_{1.9e-2}$	1.363e-1 $_{3.5e-2}$	9.089e-2 $_{2.4e-2}$	1.003e-1 $_{2.1e-2}$	8.095e-2 $_{1.3e-2}$	+
ZDT3	7.106e-1 $_{7.5e-3}$	7.073e-1 $_{7.4e-3}$	7.091e-1 $_{7.8e-3}$	7.069e-1 $_{7.0e-3}$	7.039e-1 $_{4.0e-3}$	7.054e-1 $_{5.4e-3}$	+
ZDT4	1.964e-1 $_{9.1e-2}$	1.257e-1 $_{3.6e-2}$	1.854e-1 $_{6.3e-2}$	1.324e-1 $_{4.5e-2}$	1.419e-1 $_{3.1e-2}$	1.089e-1 $_{2.5e-2}$	+
ZDT6	3.806e-1 $_{1.1e-1}$	1.513e-1 $_{2.5e-2}$	2.953e-1 $_{7.3e-2}$	1.363e-1 $_{1.9e-2}$	1.536e-1 $_{1.8e-2}$	9.234e-2 $_{1.1e-2}$	+
WFG1	5.469e-1 $_{9.3e-2}$	5.653e-1 $_{7.6e-2}$	5.298e-1 $_{1.0e-1}$	5.571e-1 $_{7.3e-2}$	4.679e-1 $_{1.2e-1}$	5.790e-1 $_{8.6e-2}$	+
WFG2	7.490e-1 $_{1.1e-2}$	7.468e-1 $_{1.0e-2}$	7.474e-1 $_{1.1e-2}$	7.468e-1 $_{9.9e-3}$	7.468e-1 $_{1.0e-2}$	7.471e-1 $_{8.5e-3}$	+
WFG3	3.698e-1 $_{9.8e-3}$	3.657e-1 $_{8.0e-3}$	3.725e-1 $_{8.2e-3}$	3.634e-1 $_{7.3e-3}$	3.684e-1 $_{7.2e-3}$	3.648e-1 $_{8.7e-3}$	+
WFG4	1.349e-1 $_{1.9e-2}$	1.341e-1 $_{1.7e-2}$	1.335e-1 $_{1.7e-2}$	1.336e-1 $_{1.9e-2}$	1.335e-1 $_{1.7e-2}$	1.333e-1 $_{1.7e-2}$	-
WFG5	1.311e-1 $_{2.5e-2}$	1.298e-1 $_{1.7e-2}$	1.377e-1 $_{2.3e-2}$	1.289e-1 $_{2.3e-2}$	1.300e-1 $_{2.3e-2}$	1.293e-1 $_{1.8e-2}$	+
WFG6	1.178e-1 $_{2.1e-2}$	1.339e-1 $_{3.4e-2}$	1.167e-1 $_{2.7e-2}$	1.344e-1 $_{4.6e-2}$	1.190e-1 $_{2.7e-2}$	1.348e-1 $_{4.1e-2}$	+
WFG7	1.059e-1 $_{1.8e-2}$	1.096e-1 $_{1.7e-2}$	1.033e-1 $_{1.6e-2}$	1.069e-1 $_{2.1e-2}$	1.040e-1 $_{1.7e-2}$	1.084e-1 $_{2.1e-2}$	+
WFG8	5.596e-1 $_{6.3e-2}$	5.664e-1 $_{8.4e-2}$	5.710e-1 $_{7.2e-2}$	5.691e-1 $_{5.0e-2}$	5.531e-1 $_{6.4e-2}$	5.703e-1 $_{6.7e-2}$	+
WFG9	1.597e-1 $_{1.8e-2}$	1.449e-1 $_{1.8e-2}$	1.609e-1 $_{2.1e-2}$	1.482e-1 $_{1.8e-2}$	1.606e-1 $_{1.8e-2}$	1.435e-1 $_{1.7e-2}$	+

Table 6. Different MOCell versions: median and interquartile range of the *HV* metric

MOP	sMOCell1 $\tilde{x}_{IQR}$	sMOCell2 $\tilde{x}_{IQR}$	aMOCell1 $\tilde{x}_{IQR}$	aMOCell2 $\tilde{x}_{IQR}$	aMOCell3 $\tilde{x}_{IQR}$	aMOCell4 $\tilde{x}_{IQR}$	
ZDT1	6.543e-1 $_{2.0e-3}$	6.592e-1 $_{7.3e-4}$	6.573e-1 $_{1.1e-3}$	6.595e-1 $_{7.3e-4}$	6.603e-1 $_{5.2e-4}$	6.610e-1 $_{2.7e-4}$	+
ZDT2	3.216e-1 $_{2.8e-3}$	3.265e-1 $_{1.6e-3}$	3.256e-1 $_{2.6e-3}$	3.274e-1 $_{1.7e-3}$	3.276e-1 $_{8.1e-4}$	3.284e-1 $_{5.1e-4}$	+
ZDT3	5.111e-1 $_{2.2e-3}$	5.135e-1 $_{8.2e-4}$	5.132e-1 $_{1.2e-3}$	5.137e-1 $_{8.1e-4}$	5.152e-1 $_{2.8e-4}$	5.152e-1 $_{4.0e-4}$	+
ZDT4	6.487e-1 $_{9.6e-3}$	6.573e-1 $_{4.3e-3}$	6.517e-1 $_{8.4e-3}$	6.568e-1 $_{4.5e-3}$	6.539e-1 $_{5.9e-3}$	6.580e-1 $_{3.2e-3}$	+
ZDT6	3.487e-1 $_{1.7e-2}$	3.885e-1 $_{3.1e-3}$	3.699e-1 $_{1.0e-2}$	3.909e-1 $_{2.0e-3}$	3.920e-1 $_{2.4e-3}$	3.970e-1 $_{8.4e-4}$	+
WFG1	5.491e-1 $_{1.1e-1}$	6.047e-1 $_{5.8e-2}$	5.906e-1 $_{1.2e-1}$	5.983e-1 $_{1.0e-1}$	6.115e-1 $_{1.2e-1}$	5.043e-1 $_{1.7e-1}$	+
WFG2	5.616e-1 $_{2.9e-3}$	5.616e-1 $_{2.7e-3}$	5.616e-1 $_{2.8e-3}$	5.616e-1 $_{2.7e-3}$	5.616e-1 $_{2.7e-3}$	5.616e-1 $_{1.1e-3}$	-
WFG3	4.420e-1 $_{2.0e-4}$	4.420e-1 $_{1.6e-4}$	4.420e-1 $_{3.0e-4}$	4.420e-1 $_{1.6e-4}$	4.420e-1 $_{2.5e-4}$	4.420e-1 $_{1.6e-4}$	+
WFG4	2.187e-1 $_{3.1e-4}$	2.186e-1 $_{3.2e-4}$	2.186e-1 $_{2.8e-4}$	2.186e-1 $_{2.9e-4}$	2.187e-1 $_{2.9e-4}$	2.188e-1 $_{2.6e-4}$	+
WFG5	1.961e-1 $_{7.5e-5}$	1.962e-1 $_{5.4e-5}$	1.961e-1 $_{7.5e-5}$	1.962e-1 $_{6.9e-5}$	1.962e-1 $_{7.4e-5}$	1.962e-1 $_{4.7e-5}$	+
WFG6	2.051e-1 $_{7.0e-3}$	1.949e-1 $_{2.8e-2}$	2.049e-1 $_{1.1e-2}$	1.940e-1 $_{4.3e-2}$	2.036e-1 $_{1.1e-2}$	1.859e-1 $_{4.2e-2}$	+
WFG7	2.104e-1 $_{1.7e-4}$	2.104e-1 $_{1.7e-4}$	2.104e-1 $_{1.6e-4}$	2.104e-1 $_{2.0e-4}$	2.104e-1 $_{2.0e-4}$	2.105e-1 $_{1.6e-4}$	+
WFG8	1.456e-1 $_{2.1e-2}$	1.472e-1 $_{3.0e-3}$	1.459e-1 $_{4.9e-3}$	1.466e-1 $_{2.5e-3}$	1.462e-1 $_{2.8e-3}$	1.479e-1 $_{2.8e-3}$	+
WFG9	2.380e-1 $_{2.2e-3}$	2.389e-1 $_{2.4e-3}$	2.375e-1 $_{3.1e-3}$	2.390e-1 $_{2.0e-3}$	2.380e-1 $_{2.3e-3}$	2.381e-1 $_{3.6e-3}$	+

symbols in the last column). If we compare synchronous *vs.* asynchronous versions, the latter ones are able to compute sets of non-dominated solutions which are closer to the exact Pareto fronts of the MOPs. Indeed, asynchronous versions of MOCell reach the best *GD* value in eleven out of the fourteen problems.

The results of the Δ metric are shown in Table 5. Again, aMOCell4 gets the best values in six (out of fourteen) MOPs. Concerning this metric, asynchronous versions of MOCell also outperform the synchronous ones, reaching in this case the best distribution of non-dominated solutions along the Pareto front in thirteen out of fourteen problems.

Finally, the *HV* metric reinforces the results of the two previous metric (Table 6). Firstly, aMOCell4 achieves the best (highest) values in eight out of the fourteen MOPs of the benchmark and, secondly, asynchronous versions overcome synchronous ones (sMOCell1 in WFG6 is the on exception).

Regarding the feedback policy used, the results show that selecting one parent from the archive yields better results than the original feedback policy of MOCell (in which a number of solutions were copied from the archive into the population) in the cases of both synchronous (sMOCell2 outperforms sMOCell1 in 12, 10, and 9 problems for the *GD*, Δ, and *HV* metrics, respectively) and asynchronous (aMOCell2 is better than aMOCell1 in 11, 10, and 11 problems for the *GD*, Δ, and *HV* metrics, respectively) update policies.

We also want to remark that, even though differences in all the metric values among the optimizers are very small, they are due to the normalization process that the resulting non-dominated sets of solutions undergo before the corresponding metric is computed. In fact, they are rather meaningful and most of the comparisons are supported with statistical confidence ("+" symbols in the last columns of the tables). Consequently, we can state that the combination of the replacement and feedback strategies leads aMOCell4 to be the best algorithm out of the six different proposed configurations over the considered benchmark. Now, in order to determine how competitive this algorithm is, we proceed to compare it against NSGA-II and SPEA2 in the next section.

Table 7. Comparison against NSGA-II and SPEA2. Median and interquartile range of the *GD* metric.

MOP	aMOCell4 $\tilde{x}_{IQR}$	NSGA-II $\tilde{x}_{IQR}$	SPEA2 $\tilde{x}_{IQR}$	
ZDT1	1.753e-4 $_{2.0e-5}$	2.198e-4 $_{4.8e-5}$	2.211e-4 $_{2.8e-5}$	+
ZDT2	5.629e-5 $_{2.5e-5}$	1.674e-4 $_{4.3e-5}$	1.770e-4 $_{4.8e-5}$	+
ZDT3	2.008e-4 $_{1.8e-5}$	2.126e-4 $_{2.1e-5}$	2.320e-4 $_{2.0e-5}$	+
ZDT4	3.293e-4 $_{2.0e-4}$	4.353e-4 $_{3.2e-4}$	5.753e-4 $_{4.4e-4}$	+
ZDT6	6.323e-4 $_{3.4e-5}$	1.010e-3 $_{1.3e-4}$	1.750e-3 $_{2.9e-4}$	+
WFG1	2.052e-4 $_{1.0e-2}$	1.967e-4 $_{8.3e-3}$	6.438e-4 $_{1.0e-2}$	+
WFG2	4.336e-4 $_{7.1e-5}$	5.196e-4 $_{1.7e-4}$	4.474e-4 $_{1.2e-4}$	+
WFG3	1.354e-4 $_{1.4e-5}$	1.553e-4 $_{1.9e-5}$	1.448e-4 $_{1.2e-5}$	+
WFG4	6.253e-4 $_{3.2e-5}$	6.870e-4 $_{1.4e-4}$	6.377e-4 $_{3.0e-5}$	+
WFG5	2.635e-3 $_{1.1e-5}$	2.655e-3 $_{3.2e-5}$	2.718e-3 $_{1.7e-5}$	+
WFG6	1.906e-3 $_{3.4e-3}$	5.539e-4 $_{6.5e-4}$	4.654e-4 $_{6.7e-4}$	+
WFG7	3.011e-4 $_{2.4e-5}$	3.444e-4 $_{4.7e-5}$	3.020e-4 $_{4.5e-5}$	+
WFG8	1.474e-2 $_{4.9e-3}$	1.446e-2 $_{5.3e-3}$	1.569e-2 $_{6.0e-3}$	+
WFG9	1.065e-3 $_{6.0e-5}$	1.223e-3 $_{2.1e-4}$	9.313e-4 $_{9.1e-5}$	+

Table 8. Comparison against NSGA-II and SPEA2. Median and interquartile range of the Δ metric.

MOP	aMOCell4 $\tilde{x}_{IQR}$	NSGA-II $\tilde{x}_{IQR}$	SPEA2 $\tilde{x}_{IQR}$	
ZDT1	7.493e-2 $_{1.3e-2}$	3.753e-1 $_{4.2e-2}$	1.486e-1 $_{1.8e-2}$	+
ZDT2	8.095e-2 $_{1.3e-2}$	3.814e-1 $_{3.9e-2}$	1.558e-1 $_{2.8e-2}$	+
ZDT3	7.054e-1 $_{5.4e-3}$	7.458e-1 $_{2.0e-2}$	7.099e-1 $_{7.7e-3}$	+
ZDT4	1.089e-1 $_{2.5e-2}$	3.849e-1 $_{5.3e-2}$	2.612e-1 $_{1.7e-1}$	+
ZDT6	9.234e-2 $_{1.1e-2}$	3.591e-1 $_{4.6e-2}$	2.268e-1 $_{3.0e-2}$	+
WFG1	5.790e-1 $_{8.6e-2}$	7.170e-1 $_{4.5e-2}$	6.578e-1 $_{7.0e-2}$	+
WFG2	7.471e-1 $_{8.5e-3}$	7.968e-1 $_{1.5e-2}$	7.519e-1 $_{1.1e-2}$	+
WFG3	3.648e-1 $_{8.7e-3}$	6.101e-1 $_{3.8e-2}$	4.379e-1 $_{1.3e-2}$	+
WFG4	1.333e-1 $_{1.7e-2}$	3.835e-1 $_{4.3e-2}$	2.681e-1 $_{3.1e-2}$	+
WFG5	1.293e-1 $_{1.8e-2}$	4.077e-1 $_{4.0e-2}$	2.805e-1 $_{2.7e-2}$	+
WFG6	1.348e-1 $_{4.1e-2}$	3.807e-1 $_{4.2e-2}$	2.506e-1 $_{2.4e-2}$	+
WFG7	1.084e-1 $_{2.1e-2}$	3.836e-1 $_{4.4e-2}$	2.453e-1 $_{2.7e-2}$	+
WFG8	5.703e-1 $_{6.7e-2}$	6.472e-1 $_{5.1e-2}$	6.108e-1 $_{5.7e-2}$	+
WFG9	1.435e-1 $_{1.7e-2}$	3.994e-1 $_{3.9e-2}$	2.945e-1 $_{2.4e-2}$	+

Table 9. Comparison against NSGA-II and SPEA2. Median and interquartile range of the HV metric.

MOP	aMOCell4 $\tilde{x}_{IQR}$	NSGA-II $\tilde{x}_{IQR}$	SPEA2 $\tilde{x}_{IQR}$	
ZDT1	6.610e-1 $_{2.7e-4}$	6.594e-1 $_{4.0e-4}$	6.600e-1 $_{3.5e-4}$	+
ZDT2	3.284e-1 $_{5.1e-4}$	3.261e-1 $_{4.8e-4}$	3.263e-1 $_{7.4e-4}$	+
ZDT3	5.152e-1 $_{4.0e-4}$	5.148e-1 $_{2.7e-4}$	5.141e-1 $_{3.4e-4}$	+
ZDT4	6.580e-1 $_{3.2e-3}$	6.552e-1 $_{4.7e-3}$	6.518e-1 $_{1.0e-2}$	+
ZDT6	3.970e-1 $_{8.4e-4}$	3.887e-1 $_{2.2e-3}$	3.785e-1 $_{4.3e-3}$	+
WFG1	5.043e-1 $_{1.7e-1}$	5.140e-1 $_{1.5e-1}$	4.337e-1 $_{1.4e-1}$	+
WFG2	5.616e-1 $_{1.1e-3}$	5.631e-1 $_{2.9e-3}$	5.615e-1 $_{2.9e-3}$	-
WFG3	4.420e-1 $_{1.6e-4}$	4.411e-1 $_{3.2e-4}$	4.418e-1 $_{2.2e-4}$	+
WFG4	2.188e-1 $_{2.6e-4}$	2.173e-1 $_{5.3e-4}$	2.181e-1 $_{3.4e-4}$	+
WFG5	1.962e-1 $_{4.7e-5}$	1.948e-1 $_{4.8e-4}$	1.956e-1 $_{1.5e-4}$	+
WFG6	1.859e-1 $_{4.2e-2}$	2.033e-1 $_{9.9e-3}$	2.056e-1 $_{1.1e-2}$	+
WFG7	2.105e-1 $_{1.6e-4}$	2.088e-1 $_{4.3e-4}$	2.098e-1 $_{2.7e-4}$	+
WFG8	1.479e-1 $_{2.8e-3}$	1.470e-1 $_{2.3e-3}$	1.469e-1 $_{1.7e-3}$	+
WFG9	2.381e-1 $_{3.6e-3}$	2.372e-1 $_{2.2e-3}$	2.386e-1 $_{2.2e-3}$	+

4.4 Comparison Against NSGA-II and SPEA2

The results of aMOCell4, NSGA-II, and SPEA2 for the GD, Δ, and HV metrics are included in Tables 7, 8, and 9, respectively. If we start by analyzing the closeness to the exact Pareto fronts, Table 7 shows that our cellular approach obtains the best (lowest) values of the GD metric in ten out of the fourteen MOPs. aMOCell4 is especially well suited for solving the ZDT family, where it is the best algorithm on the five problems. In the case of the WFG functions, NSGA-II and SPEA2 obtain the closest approximated fronts in two problems each (out of nine), but aMOCell4 is still able to reach the best results in five of these MOPs. It is therefore clear that, in terms of convergence, the cellular algorithm outperforms both NSGA-II and SPEA2 over the considered benchmark. Note that these claims are supported with statistical confidence ("+" symbols in the last column).

If we turn to compare the distribution of the non-dominated solutions along the Pareto front computed by the optimizers, Table 8 points out that aMOCell4

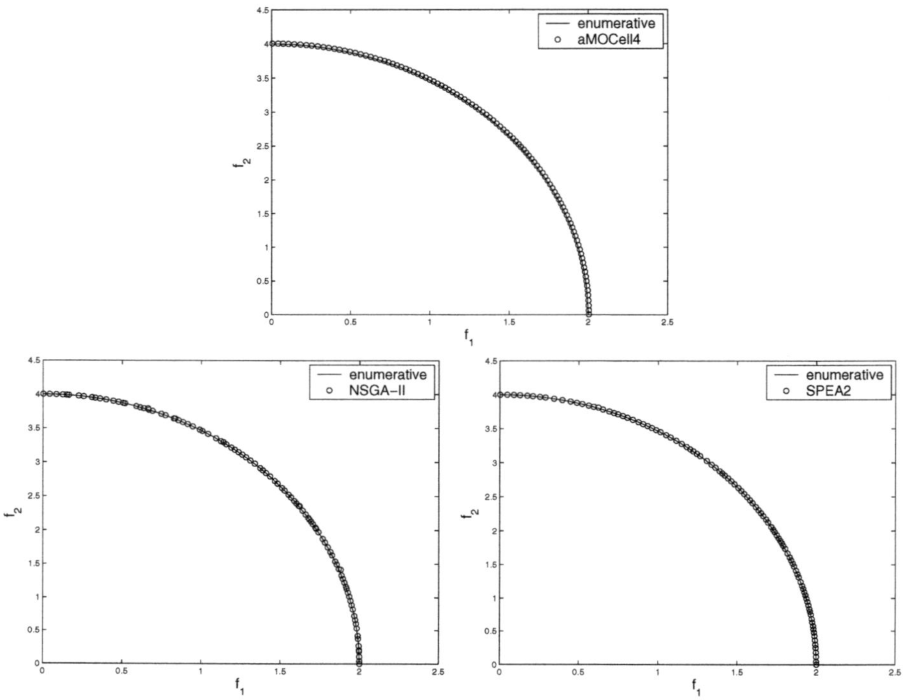

Fig. 2. Approximated fronts of aMOCell4, NSGA-II, and SPEA2 when solving WFG6

clearly outperforms NSGA-II and SPEA2 over all the considered MOPs. The most important differences come out in the ZDT family, where the Δ values of aMOCell4 in ZDT1, ZDT2 and ZDT6 are one order of magnitude lower than those reached by NSGA-II and SPEA2. In order to illustrate this fact, Fig. 2 displays the approximated WFG6 front of each algorithm which reported the best Δ value out of the 100 trials. (We have chosen this MOP because it is one in which aMOCell4 is the best neither in *GD* nor in *HV*.) As it can be seen, aMOCell4 presents an almost perfect distribution of non-dominated solutions along the Pareto front, whereas some gaps exits in the fronts computed by NSGA-II and SPEA2. This is especially important when comparing aMOCell4 against NSGA-II, because they are implemented within the jMetal framework and hence they share the same ranking and crowding procedures. This shows the enhanced search capabilities of our cellular approach.

The results of the *HV* metric are included in Table 9. As in the *GD* metric, aMOCell4 gets the best (now highest) values in ten out of the fourteen considered MOPs. It is clearly the best algorithm for the ZDT family and it reaches the best values in five out of the nine problems of the WFG toolkit (NSGA-II and SPEA2 are the best in two problems each). At this point, we want to remind the reader again that the small differences among the optimizers are because of the normalization process performed before calculating the metrics. However, these small differences lead to discernible differences in the Pareto fronts, as shown in Fig. 2.

5 Conclusions and Future Work

In this work we have proposed and studied several issues in the design of MOCell in order to improve its performance. The basic MOCell algorithm is a multiobjective cGA which uses an external archive to store the non-dominated solutions found during the search. The studied design issues are related to the synchronicity in the update of individuals, the feedback from the archive, and the replacement policy.

Six variants of MOCell have been compared using a standard methodology which is currently used in the evolutionary multiobjective optimization community. We have selected two benchmark of MOPs, the classical ZDT set of problems, usually used in similar studies in the area, and the recent benchmark obtained by using the WFG toolkit. Three metrics were used to assess the performance of the algorithms. The obtained results indicate that an asynchronous version, combined with replacing the worst cell in the neighborhood and using an individual from the archive in the selection operator leads to the best variant of the six analyzed versions of MOCell.

To assess how competitive the most promising variant of MOCell is, we have compared it against two state-of-the-art evolutionary algorithms for solving MOPs, NSGA-II and SPEA2. The results of the comparison reveal that, in the context of the problems, the metrics, and the parameter settings used, aMOCell4 clearly outperforms the other two algorithms.

A deeper study of MOCell, using problems of more than two dimensions, as well as an analysis of other design issues, such as different cell update strategies in the asynchronous versions, are matters for future work.

References

1. Deb, K., Pratap, A., Agarwal, S., Meyarivan, T.: A fast and elitist multiobjective genetic algorithm: NSGA-II. IEEE Trans. on Evol. Computation **6** (2002) 182–197
2. Knowles, J., Corne, D.: The pareto archived evolution strategy: A new baseline algorithm for multiobjective optimization. In: CEC 1999. (1999) 9–105
3. Zitzler, E., Laumanns, M., Thiele, L.: SPEA2: Improving the strength pareto evolutionary algorithm. Technical Report 103, Computer Engineering and Networks Laboratory (TIK), Swiss Federal Institute of Technology (ETH) (2001)
4. Jaeggi, D., Parks, G., Kipouros, T., Clarkson, J.: A multi-objective tabu search algorithm for constrained optimisation problems. In Coello, C., Hernández, A., Zitler, E., eds.: EMO 2005. Volume 3410 of LNCS. (2005) 490–504
5. Nebro, A.J., Luna, F., Alba, E.: New ideas in applying scatter search to multiobjective optimization. In Coello, C., Hernández, A., Zitler, E., eds.: EMO 2005. Volume 3410 of LNCS. (2005) 443–458

6. Alba, E., Tomassini, M.: Parallelism and Evolutionary Algorithms. IEEE Trans. on Evolutionary Computation **6** (2002) 443–462
7. Cantú-Paz, E.: Efficient and Accurate Parallel Genetic Algorithms. Kluwer Academic Publishers (2000)
8. Manderick, B., Spiessens, P.: Fine-grained parallel genetic algorithm. In: Proc. of the Third Int. Conf. on Genetic Algorithms (ICGA). (1989) 428–433
9. Whitley, D.: Cellular genetic algorithms. In Forrest, S., ed.: Proc. of the Fifth International Conference on Genetic Algorithms (ICGA), California, CA, USA, Morgan Kaufmann (1993) 658
10. Tomassini, M.: Spatially Structured Evolutionary Algorithms: Artificial Evolution in Space and Time. Natural Computing Series. Springer-Verlag, Heidelberg (2005)
11. Alba, E., Dorronsoro, B.: The exploration/exploitation tradeoff in dynamic cellular evolutionary algorithms. IEEE Trans. on Evol. Computation **9** (2005) 126–142
12. Alba, E., Dorronsoro, B., Giacobini, M., Tomasini, M.: Decentralized Cellular Evolutionary Algorithms. In: Handbook of Bioinspired Algorithms and Applications, Chapter 7. CRC Press (2006) 103–120
13. Laumanns, M., Rudolph, G., Schwefel, H.P.: A Spatial Predator-Prey Approach to Multi-Objective Optimization: A Preliminary Study. In: PPSN V. (1998) 241–249
14. Murata, T., Gen, M.: Cellular Genetic Algorithm for Multi-Objective Optimization. In: Proc. of the 4th Asian Fuzzy System Symposium. (2002) 538–542
15. Kirley, M.: MEA: A metapopulation evolutionary algorithm for multi-objective optimisation problems. In: CEC 2001, IEEE Press (2001) 949–956
16. Alba, E., Dorronsoro, B., Luna, F., Nebro, A., Bouvry, P., Hogie, L.: A Cellular Multi-Objective Genetic Algorithm for Optimal Broadcasting Strategy in Metropolitan MANETs. Computer Communications (2006) To appear
17. Grimme, C., Schmitt, K.: Inside a predator-prey model for multi-objective optimization: A second study. In Cattolico, M., ed.: GECCO-2006, Seattle, Washington, USA, ACM Press (2006) 707–714
18. Nebro, A.J., Durillo, J.J., Luna, F., Dorronsoro, B., Alba, E.: A cellular genetic algorithm for multiobjective optimization. In Pelta, D.A., Krasnogor, N., eds.: NICSO 2006. (2006) 25–36
19. Zitzler, E., Deb, K., Thiele, L.: Comparison of multiobjective evolutionary algorithms: Empirical results. IEEE Trans. on Evol. Computation **8** (2000) 173–195
20. Huband, S., Barone, L., While, R.L., Hingston, P.: A scalable multi-objective test problem toolkit. In Coello, C., Hernández, A., Zitler, E., eds.: EMO 2005. Volume 3410 of LNCS. (2005) 280–295
21. Durillo, J.J., Nebro, A.J., Luna, F., Dorronsoro, B., Alba, E.: jMetal: A java framework for developing multiobjective optimization metaheuristics. Technical Report ITI-2006.10, Dpto. de Lenguajes y Ciencias de la Computación (2006)
22. Deb, K., Agrawal, R.: Simulated Binary Crossover for Continuous Search Space. Complex Systems **9** (1995) 115–148
23. Demšar, J.: Statistical comparisons of classifiers over multiple data sets. Journal of Machine Learning Research **7** (2006) 1 30
24. Deb, K.: Multi-Objective Optimization Using Evolutionary Algorithms. John Wiley & Sons (2001)
25. Van Veldhuizen, D.A., Lamont, G.B.: Multiobjective Evolutionary Algorithm Research: A History and Analysis. Technical Report TR-98-03, Dept. Elec. Comput. Eng., Air Force Inst. Technol. (1998)
26. Zitzler, E., Thiele, L.: Multiobjective Evolutionary Algorithms: A Comparative Case Study and the Strength Pareto Approach. IEEE Trans. on Evol. Computation **3** (1999) 257–271

FastPGA: A Dynamic Population Sizing Approach for Solving Expensive Multiobjective Optimization Problems

Hamidreza Eskandari[1], Christopher D. Geiger[1], and Gary B. Lamont[2]

[1] Department of Industrial Engineering and Management Systems, University of Central Florida, 4000 Central Florida Blvd., Orlando, FL 32816
{eskandar, cdgeiger}@mail.ucf.edu
[2] Department of Electrical and Computer Engineering, Graduate School of Engineering and Management, Air Force Institute of Technology
Wright Patterson Air Force Base, Dayton, OH 45433
gary.lamont@afit.edu

Abstract. We present a new multiobjective evolutionary algorithm (MOEA), called fast Pareto genetic algorithm (FastPGA). FastPGA uses a new fitness assignment and ranking strategy for the simultaneous optimization of multiple objectives where each solution evaluation is computationally- and/or financially-expensive. This is often the case when there are time or resource constraints involved in finding a solution. A population regulation operator is introduced to dynamically adapt the population size as needed up to a user-specified maximum population size. Computational results for a number of well-known test problems indicate that FastPGA is a promising approach. FastPGA outperforms the improved nondominated sorting genetic algorithm (NSGA-II) within a relatively small number of solution evaluations.

Keywords: multiobjective optimization, evolutionary algorithms, Pareto optimality, fast convergence.

1 Introduction

Most real-world problems often involve multiple conflicting objectives, where improving one objective may degrade the performance of one or more of the other objectives. Traditional approaches for solving MOPs typically try to scalarize the multiple objectives into a single objective using a vector of user-defined weights. This transforms the original multiple objective optimization problem formulation into a single objective optimization problem yielding a single solution. Several disadvantages of using such traditional methods have motivated researchers and practitioners to seek alternative techniques to find a set of Pareto optimal solutions rather than just a single solution [2, 3]. A solution is Pareto optimal if there exists no feasible solution for which an improvement in one objective does not lead to a simultaneous degradation in one (or more) of the other objectives. In other words, the solution is a nondominated solution.

1.1 Evolutionary Algorithms for Multiobjective Optimization

Many Pareto-based heuristic search algorithms have been developed to solve MOPs including simulated annealing, tabu search, scatter search, and evolutionary algorithms

S. Obayashi et al. (Eds.): EMO 2007, LNCS 4403, pp. 141–155, 2007.

(EAs). EAs, the focus of this study, are population-based stochastic search algorithms inspired by Darwinian evolutionary theory (*i.e.*, the survival of the fittest). It has been shown that EAs are able to balance exploration and exploitation of the solution search space [6].

Several variations of multiobjective evolutionary algorithms (MOEAs) have been developed to handle MOPs [2, 3]. Many of the suggested MOEAs have been employed in a variety of real-world applications [1]. Among the existing algorithms, an improved version of the nondominated sorting genetic algorithm (NSGA-II) of Deb *et al.* [4], a newer version of strength Pareto EA (SPEA2) of Zitzler *et al.* [14], an improved version of multiobjective messy GA (MOMGA) of Zydallis *et al.* [16], and Pareto-archived evolution strategy (PAES) of Knowles and Corne [8] are the more popular MOEAs. Given the variations of MOEAs, the idea of using dynamic population sizing has not been thoroughly investigated, and to date only a few studies have explored this idea. For example, Tan *et al.* [9] introduce an incrementing MOEA that uses dynamic population sizing based on the online discovered Pareto front and its desired population distribution density. In another study, Yen and Lu [12] propose a dynamic MOEA, called DMOEA, which incorporates cell-based rank and density estimation strategy to efficiently compute dominance and diversity information when the population size varies dynamically.

After developing many effective MOEAs for solving inexpensive MOPs, there is now a growing need for designing MOEAs capable of dealing with expensive MOPs in that there are computational and financial resource constraints. Few multiobjective optimization algorithms exist for expensive MOPs. Most recently, Knowles [7] introduces a hybrid algorithm with online landscape approximation, called ParEGO, for expensive MOPs where only 100 and 250 solution evaluations are permitted.

1.2 Purpose of Research

In many previous real-world applications of MOEAs, the time to perform a single solution evaluation is typically of the order of minutes or even hours resulting in a limited number of solution evaluations that can be performed. This is especially relevant when implementation of the discovered solutions is time-sensitive. Additionally, many real-world problems involve complicated objective functions making a large number of solutions evaluations computationally-prohibitive [7]. Specifically, our motivation comes from simulation-based optimization research. Computer simulation of real-world systems tend to involve construction of complicated models that capture the complex, nonlinear interrelationships between independent and dependent variables and can report the value of several system performance objectives simultaneously. These models are used to evaluate candidate system design solutions in search of the best solution (or set of solutions) according to several performance objectives. A multiobjective optimization algorithm capable of rapidly finding a diverse set of Pareto optimal solutions would be greatly beneficial in such a situation. The purpose of this research is to propose a multiobjective optimization methodology that finds evenly-distributed Pareto optimal solutions in a computationally-efficient manner.

The remainder of this paper is organized as follows. Section 2 describes the proposed MOEA. This description includes the introduction of a new fitness assignment and ranking strategy and population regulation operator. The proposed MOEA and a benchmark MOEA are used to solve a diverse suite of published test problems. Section 3 and Section 4 present the experimental design and computational results, respectively. Conclusions and future research directions are discussed in Section 5.

2 Proposed Methodology – Fast Pareto Genetic Algorithm (FastPGA)

The proposed framework named Fast Pareto genetic algorithm (FastPGA) is a population-based evolutionary algorithm. A real-coded multiobjective GA is implemented to avoid the difficulties associated with binary representation and bit operations, particularly when dealing with continuous search spaces with large dimension.

The framework of the proposed algorithm incorporates a new fitness assignment and ranking strategy. An elitism operator is implemented to ensure the fast propagation of the Pareto optimal solution set. A population regulation operator is introduced to dynamically adapt the population size as needed up to a user-specified maximum population size, which is the size of the set of nondominated solutions in this study. The population regulation operator improves the proposed algorithm's convergence behavior and reduces the required computational effort. Fig. 1 shows the pseudocode of the logic of FastPGA.

```
Initialize user decision parameters (numvars, numobjs, maxpopsize, maxsoleval,
pc, pm, ...)
    t := 0
    create initial population P_t = {x_1^t, x_2^t, x_3^t, ...}
    evaluate(P_t)
    do while (stopping criterion is not met)
    {
        t := t + 1
        {
            P'_t := select(P_{t-1})  // select pairs of solutions for reproduction
            O_t := crossover( P'_t )
            O_t := mutate(O_t)
            evaluate(O_t)
            CP_t := P_{t-1} ∪ O_t // form composite population
            rank(CP_t)
            regulate(CP_t)
            P_t := generate(CP_t)
        }
    }end do
```

Fig. 1. Pseudocode of fast Pareto genetic algorithm

The major steps of FastPGA are as follows:

1. Initialize all decision parameters to user-specified values;
2. Create an initial population of candidate solution vectors $\mathbf{P}_t$ randomly at the first generation; however, FastPGA can be easily modified to generate the initial

population heuristically, seeded with user-defined solution vectors, or using a combination of these approaches;
3. If it is the first generation, go to Step 5; otherwise, increment the generation number and select pairs of solutions $\mathbf{P}_t'$ as parents from the previous population $\mathbf{P}_{t-1}$ in the reproduction operation using binary tournament selection;
4. Perform the crossover and mutation operations to generate new candidate solutions (offspring) $\mathbf{O}_t$;
5. Evaluate the candidate solution vectors for the m objective functions and record them;
6. Combine generated candidate solutions $\mathbf{O}_t$ with the previous population $\mathbf{P}_{t-1}$ to form a *composite* population $\mathbf{CP}_t$;
7. Rank the composite population of solutions $\mathbf{CP}_t$ based on the new ranking strategy using their fitness values;
8. Regulate the population size according to the number of nondominated solutions and generate a new population $\mathbf{P}_t$ from the composite population $\mathbf{CP}_t$ by discarding the inferior solutions; and
9. Terminate the search if the stopping criterion is met; otherwise, return to Step 3.

A detailed discussion of the primary features of FastPGA is provided in the sections that follow.

2.1 FastPGA Initialization and Solution Evaluation

After initializing the user-specified parameter settings (*e.g.*, number of decision variables, number of objectives, maximum population size, maximum number of solution evaluations, *etc.*), the initial population is created by random sampling of each decision variable within its defined range of variation. The evaluation of new solutions in terms of the objective functions is accomplished by calculating the complicated mathematical, closed-form expressions specified by the published test problems discussed later. At each generation, the obtained solutions with their corresponding objective values are all recorded. If a solution advances to subsequent generations, its corresponding attributes are retrieved and copied to the new generations. In FastPGA, before ranking and fitness assignment is performed, the new solution set $\mathbf{O}_t$ generated by crossover and mutation operations are combined with the previous population $\mathbf{P}_{t-1}$ to form a composite population $\mathbf{CP}_t$, *i.e.*, $\mathbf{CP}_t = \mathbf{P}_{t-1} \cup \mathbf{O}_t$, where $\cup$ denotes the union of the two sets.

2.2 Solution Ranking and Fitness Assignment

The new ranking strategy is based on the classification of candidate solutions of the composite population $\mathbf{CP}_t$ into two different categories (ranks) according to solution dominance. All *nondominated* solutions are identified as the first rank, which implies that there is no solution that is better than these solutions with respect to all objectives simultaneously. All *dominated* solutions are identified as the second rank. These ranks are used to evaluate solution fitness for the purpose of solution reproduction.

The fitness of the nondominated solutions in the first rank is calculated by comparing each nondominated solution with one another and assigning a fitness value. These values are computed using the crowding distance approach suggested by Deb *et al.* [4], which has been shown to help maintain diversity among the nondominated solutions in the Pareto optimal front.

Each dominated solution in the second rank is compared to all other solutions and assigned a fitness value depending on the number of solutions it dominates. The idea here is similar to the strength concept employed in SPEA and SPEA2 [13, 14]; however, in our work, it has been generalized. More precisely, the fitness assignment takes into account both dominating *and* dominated solutions for any dominated solution $\mathbf{x}_i$. Here, each solution $\mathbf{x}_i$ in the composite population $\mathbf{CP}_t$ is assigned a strength value $S(\mathbf{x}_i)$ indicating the number of solutions it dominates, where

$$S(\mathbf{x}_i)=\left|\left\{\mathbf{x}_j \mid \forall \mathbf{x}_j \in \mathbf{CP}_t \wedge \mathbf{x}_i \succ \mathbf{x}_j \wedge j \neq i\right\}\right|. \tag{1}$$

The cardinality of a set is denoted as $|\cdot|$, and the expression $\mathbf{x}_i \succ \mathbf{x}_j$ means that solution $\mathbf{x}_i$ dominates solution $\mathbf{x}_j$. Then, the fitness value of each dominated solution is its net strength, which is calculated as

$$F(\mathbf{x}_i)=\sum_{\mathbf{x}_i \succ \mathbf{x}_j} S(\mathbf{x}_j)-\sum_{\mathbf{x}_k \succ \mathbf{x}_i} S(\mathbf{x}_k),\ \forall \mathbf{x}_j, \mathbf{x}_k \in \mathbf{CP}_t \wedge j \neq i \neq k. \tag{2}$$

In other words, a fitness value assigned to each dominated solution $\mathbf{x}_i$ is equal to the summation of the strength values of all solutions it dominates minus the summation of the strength values of all solutions by which it is dominated. In contrast to SPEA and SPEA2, where the strength values of only solutions that $\mathbf{x}_i$ is dominated by (second term in Eq. 2) is considered, FastPGA takes into account both the dominating and dominated solutions with respect to solution $\mathbf{x}_i$. This strategy provides more information on Pareto dominance and niching relations among solutions in the composite population and reduces the chance that two solutions will have the same fitness value. Thus, no additional diversity preservation mechanism is used among the dominated solutions in the second rank requiring less computation (unlike SPEA2, which requires higher computation for the density estimator). We note that if most solutions do not dominate one another, it is implied that they are in the first rank where the crowding distance operator is invoked to maintain the diversity among them.

After the fitness values of all candidate solutions in $\mathbf{CP}_t$ are calculated, the solutions are compared, where one of three different scenarios occurs. In the first scenario, two solutions with different ranks are selected. In this situation, the solution with the better rank is preferred. In the second scenario, two selected solutions have the same rank but different fitness values. In this case, the solution with the larger fitness value is preferred. Lastly, two solutions may have the same rank and the same fitness value, where one of them is randomly preferred.

2.3 Elitism and Population Regulation

An elitism operator with relatively high intensity is implemented to ensure propagation of the nondominated solutions to subsequent generations. This is accomplished by copying all solutions in the population in the previous generation $\mathbf{P}_{t-1}$ to the composite population $\mathbf{CP}_t$. The combination of $\mathbf{P}_{t-1}$ with generated offspring $\mathbf{O}_t$ provides an opportunity to preserve the superior solutions in the next generation and discard the inferior solutions depending on the number of nondominated solutions obtained in the composite population.

The number of nondominated solutions usually increases over generations resulting in low elitism intensity in early generations if the population size is quite large and kept fixed. Moreover, the fluctuations of the number of nondominated solutions over generations demand an adaptive population sizing strategy to place appropriate emphasis of elitism intensity on nondominated solutions. If elitism intensity is too high, premature convergence may occur and if elitism intensity is too low, convergence may be too slow and computationally-expensive. Therefore, FastPGA employs a regulation operator to dynamically adjust the population size until it reaches a user-specified maximum population size as calculated by

$$\left|\mathbf{P}_t\right| = \min\left\{a_t + \left\lceil b_t \times \left|\left\{\mathbf{x}_i \mid \mathbf{x}_i \in \mathbf{CP}_t \wedge \mathbf{x}_i \text{ is nondominated}\right\}\right|\right\rceil, \mathit{maxpopsize}\right\}, \tag{3}$$

where $|\mathbf{P}_t|$ is the population size at generation t, a_t is a positive integer variable that can change over generations, b_t is a positive real variable that might change over generations, $\lceil x \rceil$ is the smallest integer that is greater than or equal to the real number x, and *maxpopsize* is the user-specified maximum population size. In this study, we set $a_t = 20$, $b_t = 1$ and *maxpopsize* = 100. Thus, we get

$$\left|\mathbf{P}_t\right| = \min\left\{20 + \left|\left\{\mathbf{x}_i \mid \mathbf{x}_i \in \mathbf{CP}_t \wedge \mathbf{x}_i \text{ is nondominated}\right\}\right|, 100\right\}. \tag{4}$$

In other words, the population size at generation t is 20 plus the number of nondominated solutions in the composite population if it is not larger than the user-specified maximum population size. Otherwise, it is kept (truncated) equal to the maximum population size.

FastPGA generates a small number of offspring using crossover and mutation operations. The number of offspring to generate is computed as

$$\left|\mathbf{O}_t\right| = \min\left\{c_t + \left\lceil d_t \times \left|\left\{\mathbf{x}_i \mid \mathbf{x}_i \in \mathbf{CP}_t \wedge \mathbf{x}_i \text{ is nondominated}\right\}\right|\right\rceil, \mathit{maxsoleval}\right\}, \tag{5}$$

where $|\mathbf{O}_t|$ is the number of offspring created at generation t, c_t is a positive integer variable that might change over generations, d_t is a positive real variable that might change over generations, and *maxsoleval* is the user-specified maximum number of solution evaluations at each generation. In this study, we set $c_t = 20$, $d_t = 0$ and *maxsoleval* = 100. Thus, we get $|\mathbf{O}_t| = 20$ meaning that the number of offspring created at each generation is small, but constant during the search process. This feature makes FastPGA capable of saving a significant number of solution evaluations early in the search and utilizes exploitation in a more efficient manner at later generations. This conservative offspring generation scheme makes FastPGA very appropriate for solving computationally-expensive MOPs. Creating a large number of offspring at early generations consumes a considerable number of solution evaluations limiting the total number of generations, which results in no extensive utilization of exploitation. Bear in mind that in expensive MOPs where only a small number of solution evaluations is allowed, more emphasis on exploitation and less emphasis on exploration could be extremely beneficial. During a single run, FastPGA requires, on average, fewer solution evaluations per generation allowing more generations to be used for search exploitation. The values for a_t, b_t, c_t and d_t are obtained by performing several pilot runs. As the intent of this research is to introduce a novel approach that conservatively generates offspring, any serious attempt to determine the best parameter values is left for future study.

2.4 Search Stopping Criterion

Different approaches have been used to stop the search process of EAs including those that consider the landscape of the response surface, the desired solution quality, the specific number of solution evaluations and the required computation time. Designed for dealing with expensive MOPs, FastPGA uses a new stopping criterion that considers the convergence speed towards the true Pareto optimal front. Here, when the number of nondominated solutions reaches the pre-specified maximum population size and, thereafter, no changes are made in the number of nondominated solutions within a certain number of solution evaluations[1], the search stops. For better understanding of the suggested stopping criterion for expensive MOPs, a velocity measure is defined.

Definition 1: The Pareto production ratio (*PPR*) is the ratio of the number of nondominated solutions to the current population size at any given generation t and is calculated as

$$PPR_t = \frac{|\mathbf{NP}_t|}{|\mathbf{P}_t|}, \tag{6}$$

where $\mathbf{P}_t$ is the population at generation t, and $|\mathbf{NP}_t|$ denotes the number of nondominated solutions belonging to population $\mathbf{P}_t$. When PPR_t reaches one (*i.e.*, all solutions in the population are nondominated) and it does not make any changes over a pre-specified number of solution evaluations implying no promising nondominated solutions are found within this period, the search stops. Based upon our experiments on several real-variable test problems, PPR_t consistently reaches one. However, if in a rare case PPR_t does not reach one, the search could easily be stopped after a pre-specified maximum number of solution evaluations.

3 Experimental Study

In this section, we evaluate the performance of FastPGA on a suite of published test problems having two objectives and no coupled constraints. In all test problems, the functions are to be minimized. The results of FastPGA are also benchmarked against one of the state-of-the-art MOEAs – the real-coded NSGA-II of Deb *et al.* [4]. It has been reported that NSGA-II outperforms most of its competitors including SPEA and PAES, and it competes closely with SPEA2 in terms of convergence to the true Pareto optimal front while maintaining solution diversity [*e.g.*, 4, 5, 13]. However, SPEA2 requires higher computational complexity of $O(mN^2\log N)$ [14] compared to that of NSGA-II, $O(mN^2)$, raising the question of whether the computationally-expensive fitness assignment strategy and truncation operator in SPEA2 is of great worth. The m

[1] The expression "solution evaluations" could be replaced by "generations" if the MOEA has identical population size over generations.

and N are the number of objectives and the population size, respectively. Some studies report that there is no statistically significant difference between the performance of SPEA2 and NSGA-II, although SPEA2 requires significantly higher computational time [*e.g.*, 5].

3.1 Test Problems

The suite of test problems consists of four well-known ZDT real-variable problems [15]. Note that the ZDT problems have two objectives and no coupled constraints. The test problems ZDT1 and ZDT3 have 30 decision variables each. The former has a convex Pareto optimal front and the latter has five discontinuous Pareto optimal fronts. The 10-decision variable test problem ZDT4 is a multi-frontal (multi-modal) problem having a large number of local Pareto optimal fronts and a single global Pareto optimal front. The test problem ZDT6 has 10 decision variables and a nonconvex Pareto optimal front. Moreover, the density of solutions across its Pareto optimal front is non-uniform and the density towards the Pareto optimal front gets thin. These test problems possess extremely challenging Pareto optimality characteristics including nonconvex, discontinuous, non-uniformly spaced properties. Many researchers use these problems to evaluate their proposed algorithms [*e.g.*, 4, 5].

3.2 Algorithm Parameter Settings

For both FastPGA and NSGA-II, all of the parameter settings, except the maximum number of solution evaluations, are used according to the suggested values in the original study of Deb *et al.* [4]. They are summarized in Table 1. In order to make better comparisons, the maximum population size for FastPGA is set to the suggested population size used by Deb *et al.* [4]. The number of solution evaluations shown in Table 1 depends on the characteristics and complexity of the underlying problem. In this study, the number of solution evaluations is kept small. This is appropriate

Table 1. Parameter settings for FastPGA and the real-coded NSGA-II

Algorithm Parameter	**FastPGA and Real-Coded NSGA-II**			
Test Problem	ZDT1	ZDT3	ZDT4	ZDT6
No. of Solution Evaluations	6,500	6,000	10,000	10,000
Initial Population Size	100			
Maximum Population Size	100			
Crossover Probability	1			
Mutation Probability	$1/n$ (where n is number of variables)			
Crossover Type	Simulated Binary Crossover (η_c=15)			
Mutation Type	Polynomial Mutation (η_m=20)			
Selection Scheme	Binary Tournament			

since an aim of this research is to evaluate the performance of each algorithm for expensive, real-world MOPs that only allow a small number of solution evaluations.

3.3 Performance Metrics

Generally, when solving MOPs, there are three primary goals: 1) *fast convergence* to the true Pareto frontier solution set in the objective space, 2) *close proximity* to the true Pareto frontier solution set, and 3) *diversity and even dispersion* of the obtained nondominated solutions along the true Pareto optimal front. Many performance metrics have been introduced within the last decade [*e.g.*, 2, 3, 10-12]. Most previous studies emphasize only the closeness and diversity measures. Fast convergence to optimal solutions for computationally-expensive MOPs is very important and fast convergence towards Pareto optimal solutions is a highly desired feature of any promising algorithm. This is especially the case in real-world problems where finding optimal or even near-optimal solutions is often computationally-prohibitive.

In this study, four performance metrics are used to measure the convergence behavior and diversity of FastPGA and NSGA-II. Two metrics are newly introduced in this study. They are the diversity metric and the delineation metric. Two of the four metrics (the delineation and the hypervolume ratio metrics) are employed for the simultaneous evaluation of both the proximity of the obtained solutions to the true Pareto front and the diversity of the obtained solutions. For each test problem, each algorithm is run with 30 different seed values and the mean, standard deviation and 95% confidence interval are computed. The lower and upper bounds of the 95% confidence interval are calculated by $\bar{x} \pm t_{\alpha/2,\eta-1} s/\sqrt{\eta}$, where $\bar{x}$ is the sample mean, s is the sample standard deviation, α is the significance level and is equal to 0.05, and η is the sample size and is equal to 30. Given the fact that in expensive MOPs, the time required for solution evaluations is a significant portion of the actual CPU time of any approach, no attempt is made to measure the computation time needed to run each algorithm. However, the computational complexity of both FastPGA and NSGA-II is $O(mN^2)$ meaning that there should be no appreciable difference between their computation times.

Distance from the Pareto Optimal Front. Distance metric is originally introduced by Van Veldhuizen and Lamont [10], which evaluates the convergence to a known Pareto optimal frontier set. To calculate the distance metric, a set of H evenly-spaced solutions from the true Pareto optimal set in the objective space must be known. The set of H solutions should be large enough such that it well reflects the true Pareto optimal front. In this study, a set of 500 Pareto optimal solutions is used for each of the four test problems. Deb [4] presents a variation of distance metric Υ in which the minimum Euclidean distance from each obtained nondominated solution to the H solutions is calculated and the average of these distances is used as the distance metric. It is important to note that all solutions obtained by an algorithm including those that are dominated are considered for the calculation of this metric. The goal of this metric is to identify how close a set of obtained solutions are to the true Pareto optimal set. The smaller the value of this metric, the closer the solutions are to the true Pareto optimal frontier set.

Diversity of Nondominated Solutions. We define the diversity metric Δ to evaluate both the spread and the uniform spacing of dispersion of the obtained nondominated solutions in the objective space. Here, the goal is to obtain a set of nondominated solutions that are both widely- and uniformly-distributed along the Pareto optimal front at the end of the search. To compute the diversity metric Δ, the Euclidean distance d_i between consecutive nondominated solutions is calculated in the objective space, where $i = 1, \ldots, |\mathbf{NP}_T|-1$ and $|\mathbf{NP}_T|$ is the number of nondominated solutions at the end of the search. Then, the standard deviation of these distances σ_d is calculated representing the degree of non-uniformity of the nondominated solutions. The minimum Euclidean distance of the two extreme Pareto solutions of the true Pareto optimal set from the nondominated solutions, denoted by d_p and d_q, is calculated. Note that the distances d_p and d_q are the distances from the closest nondominated solutions, not necessarily the endpoints of nondominated solutions, to the two extreme Pareto solutions. The diversity of the set of nondominated solutions is

$$\Delta(\mathbf{NP}_T) = d_p + d_q + \sqrt{\frac{1}{|\mathbf{NP}_T|-1} \sum_{i=1}^{|\mathbf{NP}_T|-1} \left(d_i - \bar{d}\right)^2} \; . \tag{7}$$

Delineation of Pareto Optimal Front. The delineation metric Φ is introduced to evaluate the extent of both convergence and diversity to a known Pareto optimal front. The goal of this study is to identify a set of solutions that well represent the Pareto optimal set. The idea behind this metric is how well each solution on the Pareto optimal front is represented by the obtained nondominated solutions. To calculate the delineation metric Φ, a large set of H evenly-spaced solutions from the Pareto optimal set that well reflects the true Pareto optimal front must be known. The same set of H solutions used in calculating the distance metric Υ is used here. The minimum Euclidean distance from each Pareto optimal solution to the obtained solutions l_i is calculated and the average of these distances is used as the delineation metric Φ, *i.e.*,

$$\Phi(\mathbf{P}_T) = \frac{1}{H} \sum_{i=1}^{H} l_i \; . \tag{8}$$

It is important to note that all solutions obtained by an algorithm including those that are dominated are considered for the calculation of this metric.

Hypervolume Ratio. The hypervolume metric HV, originally suggested by Zitzler and Thiele [13], calculates the volume of the objective space dominated by the nondominated solutions having the reference point $\mathbf{R}$. The goal of this measure is to identify the proportion of the volume enclosed by reference point and Pareto optimal front covered by the nondominated solutions obtained at the end of the search. To be consistent with other performance metrics used in this study (*i.e.*, the smaller value of the metric, the better), a modification of hypervolume metric is employed here. We use the hypervolume ratio (HVR). Here, the proportion of the volume enclosed by the reference point $\mathbf{R}$ and true Pareto optimal front that is not covered by the nondominated solutions is of interest, and is given by

$$HVR(\mathbf{NP}_T) = 1 - \frac{HV(\mathbf{NP}_T)}{HV(\mathbf{PF})}. \tag{9}$$

PF is the set of solutions on the true Pareto optimal front. *HVR* returns a value in the range [0, 1].

4 Computational Results

Table 2 and Table 3 show the output statistics including mean, standard deviation and 95% confidence interval (CI) of the four performance metrics obtained from generating 30 random runs for each test problem using FastPGA and NSGA-II. The Υ and Δ metrics are shown in Table 2, and Φ and *HVR* are given in Table 3. Recall that lower values are preferred for all four metrics.

Table 2. Mean, standard deviation and 95% confidence interval of distance and diversity metrics with FastPGA and NSGA-II (over 30 random runs)

Test Problem	Algorithm	Distance Υ			Diversity Δ		
		Avg.	Std. Dev.	95% CI	Avg.	Std. Dev.	95% CI
ZDT1	FastPGA	0.0210	0.0110	[0.0169, 0.0251]	0.0769	0.0296	[0.0659, 0.0879]
	NSGA-II	0.0659	0.0128	[0.0612, 0.0707]	0.1324	0.0220	[0.1242, 0.1406]
ZDT3	FastPGA	0.0200	0.0092	[0.0166, 0.0235]	0.2017	0.1036	[0.1631, 0.2403]
	NSGA-II	0.0297	0.0091	[0.0263, 0.0331]	0.1968	0.0233	[0.1882, 0.2055]
ZDT4	FastPGA	0.0332	0.0262	[0.0234, 0.0430]	0.3812	0.1804	[0.3140, 0.4485]
	NSGA-II	0.7677	0.3414	[0.6404, 0.8950]	1.5111	0.5797	[1.2950, 1.7273]
ZDT6	FastPGA	0.0445	0.0082	[0.0414, 0.0475]	0.1393	0.0256	[0.1297, 0.1488]
	NSGA-II	0.2647	0.0380	[0.2506, 0.2789]	0.7239	0.1063	[0.6843, 0.7636]

Table 3. Mean, standard deviation and 95% confidence interval of delineation and hypervolume ratio metrics with FastPGA and NSGA-II (over 30 random runs)

Test Problem	Algorithm	Delineation Φ			Hypervolume Ratio *HVR*		
		Avg.	Std. Dev.	95% CI	Avg.	Std. Dev.	95% CI
ZDT1	FastPGA	0.0208	0.0097	[0.0172, 0.0244]	0.0443	0.0198	[0.0369, 0.0517]
	NSGA-II	0.0599	0.0111	[0.0557, 0.0640]	0.1259	0.0226	[0.1175, 0.1343]
ZDT3	FastPGA	0.0269	0.0255	[0.0174, 0.0364]	0.0850	0.0345	[0.0722, 0.0979]
	NSGA-II	0.0286	0.0084	[0.0255, 0.0318]	0.1086	0.0252	[0.0992, 0.1180]
ZDT4	FastPGA	0.0701	0.0457	[0.0531, 0.0872]	0.0910	0.0479	[0.0732, 0.1089]
	NSGA-II	0.6557	0.3128	[0.5391, 0.7724]	0.8173	0.2123	[0.7381, 0.8964]
ZDT6	FastPGA	0.0415	0.0079	[0.0385, 0.0444]	0.1083	0.0190	[0.1012, 0.1154]
	NSGA-II	0.2538	0.0396	[0.2391, 0.2686]	0.5731	0.0690	[0.5473, 0.5988]

The results shown in Table 2 indicate that FastPGA significantly outperforms NSGA-II with respect to the convergence to the Pareto front. There is no overlap of the confidence intervals of the distance metric ϒ for FastPGA and NSGA-II in all problems. Compared to FastPGA, NSGA-II exhibits poor convergence in the ZDT4 and ZDT6 test problems. Both MOEAs have acceptable standard deviations for ϒ on most problems. An exception occurs on ZDT4, where NSGA-II has a very high standard deviation for ϒ.

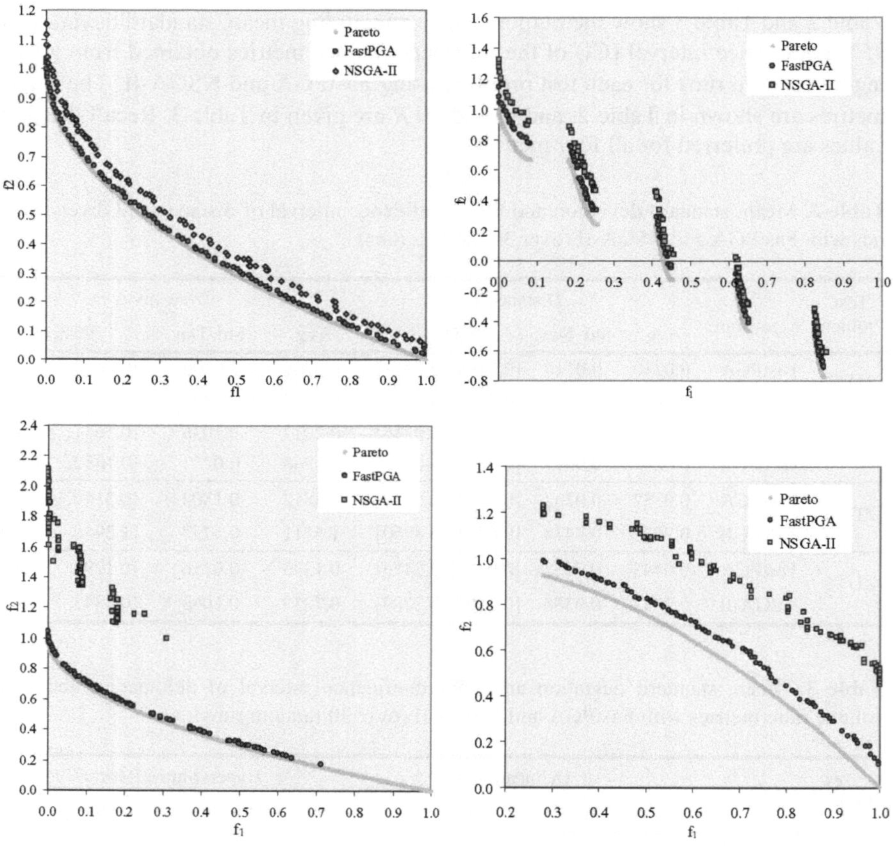

Fig. 2. The populations with FastPGA and NSGA-II on ZDT1 (top left), ZDT3 (top right), ZDT4 (bottom left) and ZDT4 (bottom right)

To illustrate the convergence behavior of FastPGA and NSGA-II, the sample obtained populations at the end of the search together with the Pareto optimal front for ZDT1, ZDT3, ZDT4 and ZDT6 are shown in Fig. 2. This figure shows the superiority of FastPGA over NSGA-II in rapidly converging to the true Pareto optimal solution set while preserving a diverse set of nondominated solutions. Within the given number of solution evaluations, FastPGA obtains the population of nondominated solutions while a significant proportion of solutions in NSGA-II are dominated solutions, indicating that FastPGA converges much faster than NSGA-II. It is interesting to note that all obtained nondominated solutions yielded by NSGA-II at the end of the search are dominated by the nondominated solutions of FastPGA in most problems. The favorable performance of FastPGA is most likely due to high elitism intensity and regulation operator employment. These settings help to improve search space exploitation and to save a considerable number of solution evaluations for further investigation at later generations.

Table 2 shows that FastPGA has significantly better performance than NSGA-II in terms of the diversity metric Δ for most problems. There is no overlap of the confidence intervals of Δ for FastPGA and NSGA-II in ZDT1, ZDT4 and ZDT6 problems. NSGA-II performs only slightly better than FastPGA for ZDT3 with respect to this metric. It is surprising to note that FastPGA has a better Δ than NSGA-II in many replications for ZDT3, but its performance is actually poor in few replications. The reason for this happening is most likely due to the employment of high elitism intensity resulting in biasedness towards some particular regions of the Pareto front in a few of the random runs.

Table 3 indicates that FastPGA has better performance than NSGA-II in terms of the delineation metric Φ for most problems. There is no overlap of the confidence intervals of Φ for FastPGA and NSGA-II in ZDT1, ZDT4 and ZDT6 problems. FastPGA has a slightly better average performance than NSGA-II on ZDT3, but there is a considerable overlap of their confidence intervals. The standard deviations of Φ across all problems for both MOEAs are small, except for FastPGA on ZDT3 (due to the poor diversity in a few replications) and for NSGA-II on ZDT4.

For the hypervolume ratio *HVR*, the reference point $\mathbf{R}$ is set at (1, 1.1) for all test problems, except for KUR where it is set at (-14, 1). Here, $\mathbf{R}$ is selected as a very close point to the Nadir objective vector for each test problem so that a more precise comparative analysis can be performed. The results shown in Table 3 indicate that FastPGA outperforms NSGA-II with respect *HVR*. There is no overlap of the confidence intervals of *HVR* for FastPGA and NSGA-II in all problems. It is interesting to note that, although there is considerable overlap of the confidence intervals of the delineation metric of FastPGA and NSGA-II on ZDT3, FastPGA outperforms NSGA-II with respect to *HVR*. Regarding the obtained results, it is implied that, although the nondominated solutions generated by FastPGA in few replications do not represent the Pareto fronts of ZDT3 well, they dominate a considerable portion of the hypervolumes enclosed by the Pareto fronts and reference point $\mathbf{R}$.

5 Conclusions and Future Work

This research presents a MOEA, called FastPGA, for dealing with MOPs where each solution evaluation is computationally- and/or financially-expensive. This approach incorporates a Pareto-based multiobjective optimization method into a genetic algorithm. Population regulation operator is introduced to enhance the algorithm's performance in finding Pareto optimal solutions while minimizing computational effort. Computational results for a number of well-known test problems with different Pareto optimality characteristics indicate that FastPGA is capable of efficiently and effectively direct the search toward Pareto optimal front. Analysis shows that, within a relatively small number of solution evaluations, FastPGA outperforms NSGA-II in most problems in terms of rapidly converging to the true Pareto optimal solution set while preserving a diverse, evenly-distributed set of nondominated solutions. Adaptive population sizing is most likely one of the main factors resulting in the superiority of FastPGA over NSGA-II in this study.

Future research includes additional testing and benchmarking FastPGA on several other MOPs higher in dimension of objective space. The attempt to find the best FastPGA parameter settings will also be pursued in the next step of this study. Finally, more precise statistical analyses of the results will be performed.

References

1. Coello, C.A.C., and G.B. Lamont. (2004). *Applications of Multi-Objective Evolutionary Algorithms*. Singapore: World Scientific.
2. Coello, C.A.C., D.A. Van Veldhuizen and G.B. Lamont. (2002). *Evolutionary Algorithms for Solving Multi-Objective Problems*. 1st Ed. New York: Kluwer Academic Publishers.
3. Deb, K. (2001). *Multi-Objective Optimization Using Evolutionary Algorithms*. 1st Ed. Chichester, UK: John Wiley & Sons.
4. Deb, K., A. Pratap, S. Agarval and T.A. Meyarivan. (2002). "Fast and Elitist Multiobjective Genetic Algorithm: NSGA-II." *IEEE Transactions on Evolutionary Computing* 6, 182–197.
5. Deb, K., M. Mohan and S. Mishra. (2005). "Evaluating the epsilon-Domination Based Multi-Objective Evolutionary Algorithm for a Quick Computation of Pareto-Optimal Solutions." Evolutionary Computation 13(4), 501–525.
6. Goldberg, D.E. (1989). *Genetic Algorithms in Search, Optimisation and Machine Learning*. Addison Wesley.
7. Knowles, J., (2006). "ParEGO: a hybrid algorithm with on-line landscape approximation for expensive multiobjective optimization problems," *IEEE Transactions on Evolutionary Computation* 10(1), 50-66.
8. Knowles, J. and D. Corne. (1999). "The Pareto Archived Evolution Strategy: A New Baseline Algorithm for Multiobjective Optimization." *Proceedings of the 1999 Congress on Evolutionary Computation* (CEC'1999), Washington, D.C., July 1999, IEEE Service Center, pp. 98–105.
9. Tan, K. C., T.H. Lee, and E.F. Khor. (2001). "Evolutionary algorithm with dynamic population size and local exploration for multi-objective optimization," *IEEE Transactions on Evolutionary Computation* 5(6), 565-588.
10. Van Veldhuizen, D.A., and G.B. Lamont. (2000). "Multiobjective Evolutionary Algorithms: Analyzing the State-of-the-Art," *Evolutionary Computation*, 7(3), 1-26.
11. Van Veldhuizen, D.A., and G.B. Lamont. (2000). "On Measuring Multiobjective Evolutionary Algorithm Performance," *2000 IEEE Congress on Evolutionary Computation*, 1, 204-211.
12. Yen, G.G. and H. Lu (2003). "Dynamic multiobjective evolutionary algorithm: adaptive cell-based rank and density estimation," *IEEE Transactions on Evolutionary Computations* 7(3), 253-274.
13. Zitzler, E. and L. Thiele. (1999). "Multiobjective Evolutionary Algorithms: A Comparative Study and Strength Pareto Approach." *IEEE Transactions on Evolutionary Computation* 3(4), 257–271.

14. Zitzler, E., M. Laumanns and L. Thiele. (2001). "SPEA2: Improving the Strength Pareto Evolutionary Algorithm." Technical Report 103, Computer Engineering and Networks Laboratory (TIK), Swiss Federal Institute of Technology (ETH) Zurich, Gloriastrasse 35, CH-8092 Zurich, Switzerland.
15. Zitzler, E., K. Deb and L. Thiele. (2000). "Comparison of Multiobjective Evolutionary Algorithms: Empirical Results." *Evolutionary Computation* 8(2), 173–195.
16. Zydallis, J.B., D.A. Van Veldhuizen, and G.B. Lamont. (2001). "A Statistical Comparison of Multiobjective Evolutionary Algorithms Including the MOMGA-II," *First International Conference on Evolutionary Multi-Criterion Optimization*, 226-240.

Constraint-Handling Method for Multi-objective Function Optimization: Pareto Descent Repair Operator

Ken Harada, Jun Sakuma, Isao Ono, and Shigenobu Kobayashi

Department of Computational Intelligence and Systems Science,
Tokyo Institute of Technology,
4259 Nagatsuta-cho Midori-ku Yokohama-shi Kanagawa-ken 226-8502, Japan
{ken,jun}@fe.dis.titech.ac.jp,{isao,kobayasi}@dis.titech.ac.jp
http://www.fe.dis.titech.ac.jp/

Abstract. Among the multi-objective optimization methods proposed so far, Genetic Algorithms (GA) have been shown to be more effective in recent decades. Most of such methods were developed to solve primarily unconstrained problems. However, many real-world problems are constrained, which necessitates appropriate handling of constraints. Despite much effort devoted to the studies of constraint-handling methods, it has been reported that each of them has certain limitations. Hence, further studies for designing more effective constraint-handling methods are needed.

For this reason, we investigated the guidelines for a method to effectively handle constraints. Based on these guidelines, we designed a new constraint-handling method, Pareto Descent Repair operator (PDR), in which ideas derived from multi-objective local search and gradient projection method are incorporated. An experiment comparing GA that use PDR and some of the existing constraint-handling methods confirmed the effectiveness of PDR.

1 Introduction

Multi-objective optimization (MOO) has many real-world applications, e.g. portfolio optimization, for which multiple conflicting objective functions are to be simultaneously optimized. MOO whose variables are real-valued is called multi-objective function optimization, which is the subject of this paper. Genetic Algorithms (GA) are known to be relatively efficient and effective MOO methods [1]. GA applies crossover and selection to a set of solutions and converge them to entire Pareto-optimal solutions. Selection for MOO consists of *ranking*, which brings solutions closer to Pareto-optimal solutions, and *sharing*, which enhances the diversity of solutions.

Most MOO methods, including GA, were designed for solving primarily unconstrained problems. However, real-world problems often have constraints, and the handling of them can substantially influence the performance of the optimization methods. When GA is applied to constrained problems, two major difficulties arise.

S. Obayashi et al. (Eds.): EMO 2007, LNCS 4403, pp. 156–170, 2007.

One of them is that some GA require feasible solutions to start with. The most naive way of obtaining feasible solutions is to randomly generate solutions until a prespecified number of them are found. However, this approach fails when the probability of obtaining a feasible solution in such a way is very low. Therefore, feasible solutions must be explicitly searched for, which is one role that constraint-handling methods play.

The other difficulty is that, on problems whose Pareto-optimal solutions lie on feasible region boundaries (boundaries hereafter), GA may not be able to obtain solutions close to the Pareto-optimal solutions. The most commonly used constraint-handling method in GA is death penalty (DP), which simply discards infeasible solutions. The solutions that GA generates can be mostly infeasible on problems whose Pareto-optimal solutions lie on boundaries. Extreme examples of such problems are ZDT1 and ZDT2 [1] whose Pareto-optimal solutions form line segments at which 29 constraint boundaries intersect perpendicularly. When the solutions that GA maintains come near the Pareto-optimal solutions, most of the solutions that GA generates are infeasible and discarded by DP, which implies that GA cannot obtain solutions close to the Pareto-optimal solutions. Therefore, effective constraint-handling methods which facilitate searching for Pareto-optimal solutions on boundaries are necessary.

One class of constraint-handling methods modify solution representation and/or crossover so that infeasible solutions can never be generated [2]. However, these methods are not applicable to general problems. Another class of methods attempt to search for feasible solutions from infeasible solutions by reducing constraint violations. The existing methods of this kind are known to have certain limitations as described in Sect. 2.2.

In order to design an effective constraint-handling method, we first investigate the guidelines for a method to effectively handle constraints. We then explain the concepts and calculations necessary to meet these guidelines and propose them as Pareto Descent Repair operator (PDR).

Section 2 formulates constrained multi-objective function optimization, explains Pareto-optimality, and reviews existing constraint-handling methods. Section 3 presents the guidelines for effective constraint handling and explains the details of PDR. To demonstrate the effectiveness of PDR, Sect. 4 shows the results of experiments comparing PDR and other constraint-handling methods when they are used in GA. Lastly, Sect. 5 summarizes this paper.

2 Constraint Handling in Multi-objective Function Optimization

2.1 Constrained Multi-objective Function Optimization

Formulation. Constrained multi-objective function optimization problem can generally be formulated as

$$\text{Minimize } \boldsymbol{f}(\boldsymbol{x}) \text{ subject to } \boldsymbol{x} \in S \ , \tag{1}$$

where $\boldsymbol{x} \in \mathbb{R}^N$, and $\boldsymbol{f} = (f_1, f_2, \ldots, f_M)^T$ is a vector of M objective functions. *Feasible region* S is the region that satisfies inequality constraints $\boldsymbol{g}(\boldsymbol{x}) \leq \boldsymbol{0}$, where $\boldsymbol{g} = (g_1, g_2, \ldots, g_P)^T$ is a vector of P constraint functions. Solutions that satisfy all constraints are said to be *feasible*, and those that do not, *infeasible*. Objective functions are defined for arbitrary feasible solutions, and constraint functions, for arbitrary solutions. Constraint functions are assumed to be continuously differentiable in this paper, since there are a considerable number of problems for which analytical or approximate gradients of constraint functions are available and continuous.

If $g_j(\boldsymbol{x}) = 0$ holds for solution $\boldsymbol{x}$, $\boldsymbol{x}$ lies on the *boundary* of the corresponding constraint, and the constraint is said to be *active* at $\boldsymbol{x}$. If a direction $\boldsymbol{d} \in \mathbb{R}^N$ satisfies $\boldsymbol{d} \cdot \nabla g_j(\boldsymbol{x}) \leq 0$, $\boldsymbol{d}$ is said to be *feasible* w.r.t. the active constraint. The *constraint violation* of constraint $g_j(\boldsymbol{x}) \leq 0$ at $\boldsymbol{x}$ can be defined as $g_j^+(\boldsymbol{x}) = \max(g_j(\boldsymbol{x}), 0)$. By reducing the positive components of $\boldsymbol{g}^+ = (g_1^+, g_2^+, \ldots, g_P^+)^T$, feasible solutions can be searched for.

Pareto-Optimality and the Objective of MOO methods. If, for $\boldsymbol{x_1}, \boldsymbol{x_2} \in S$,

$$\forall i \in \{1, 2, \ldots, M\},\ f_i(\boldsymbol{x}_1) \leq f_i(\boldsymbol{x}_2) \wedge \exists i \in \{1, 2, \ldots, M\},\ f_i(\boldsymbol{x}_1) < f_i(\boldsymbol{x}_2)$$

holds, $\boldsymbol{x_1}$ is said to be *superior* to $\boldsymbol{x_2}$, which is denoted by $\boldsymbol{x_1} \succ \boldsymbol{x_2}$. If a solution $\boldsymbol{x}$ in a set of solutions is not inferior to any other solution in the set, $\boldsymbol{x}$ is said to be *non-inferior* within the set. If $\boldsymbol{x}' \in S$ such that $\boldsymbol{x}' \succ \boldsymbol{x}$ does not exist, $\boldsymbol{x}$ is said to be *Pareto-optimal*. If a solution $\boldsymbol{x}'$ such that $\boldsymbol{x}' \succ \boldsymbol{x}$ does not exist in the feasible ε-vicinity of $\boldsymbol{x}$, $\boldsymbol{x}$ is said to be *locally Pareto-optimal*. There are often multiple Pareto-optimal and locally Pareto-optimal solutions.

The objective of MOO methods is to find a set of solutions which are close to Pareto-optimal solutions (*proximity*) and which evenly cover entire Pareto-optimal solutions (*diversity*) [1,3].

2.2 Existing Constraint-Handling Methods

This section reviews prominent constraint-handling methods which search feasible solutions by reducing constraint violations, and explains their drawbacks regarding the abilities to find feasible solutions and to search Pareto-optimal solutions on boundaries.

Penalty Methods. A vector of penalty functions $\boldsymbol{P}(\boldsymbol{x}) = (P_1(\boldsymbol{x}), P_2(\boldsymbol{x}), \ldots, P_M(\boldsymbol{x}))^T$, each of whose components represents the degree of overall constraint violation at a solution, is defined, and the unconstrained optimization problem

$$\text{Minimize } \boldsymbol{f}(\boldsymbol{x}) + \boldsymbol{P}(\boldsymbol{x})$$

is solved. It has been pointed out that it is difficult to design appropriate penalty functions [2]. In addition, penalty methods cannot be used when there are infeasible solutions for which objective functions are undefined.

Objectivization of Constraint Violations. Constraint violations are regarded as additional objective functions, and the unconstrained problem

$$\text{Minimize } \tilde{\boldsymbol{f}}(\boldsymbol{x}), \text{ where } \tilde{\boldsymbol{f}} = (f_1, f_2, \ldots, f_M, g_1^+, g_2^+, \ldots, g_P^+)^T \ , \qquad (2)$$

is solved [2]. Many methods that use GA for the optimization of (2) have been proposed [4,2]. We call such methods OCV(e) since constraint violations and objective functions are treated equally. It has also been proposed to redefine superiority relationship in OCV so that the regular superiority relationship is used for feasible solutions, the multi-objective superiority relationship w.r.t constraint violations is used for infeasible solutions, and feasible solutions are always superior to infeasible solutions [5,6]. Such variants of OCV are called OCV(ne) in this paper.

Regardless of whether or not constraint violations and objective functions are treated equally, OCV may not be able to find feasible solutions. Consider the 2-variable-2^2-constraint problem shown in Fig. 1. Assume that the feasible region is sufficiently small, and, when solutions are randomly generated, at least one solution is inside each of the 2^2 dark-shaded areas which violate two constraints. In this situation, at least one non-inferior solution w.r.t. constraint violations exists in each of the dark-shaded areas, and there are at least 2^2 of them in total. On a similar problem with N variables, the number of such solutions is at least 2^N. When N is big, almost all of the randomly-generated solutions are non-inferior, for which ranking does not function. Sharing, in the meantime, attempts to increase the diversity of solutions and, as a result, disperses the solutions. Hence, OCV may not find feasible solutions, which will be demonstrated in Sect. 4.

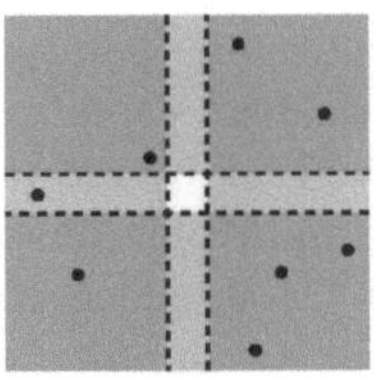

Fig. 1. A problem on which OCV fails when the dimension of the variable space becomes big. Dashed lines denote constraint boundaries. Infeasible regions are shaded, and, the more constraints they violate, the darker they are shaded.

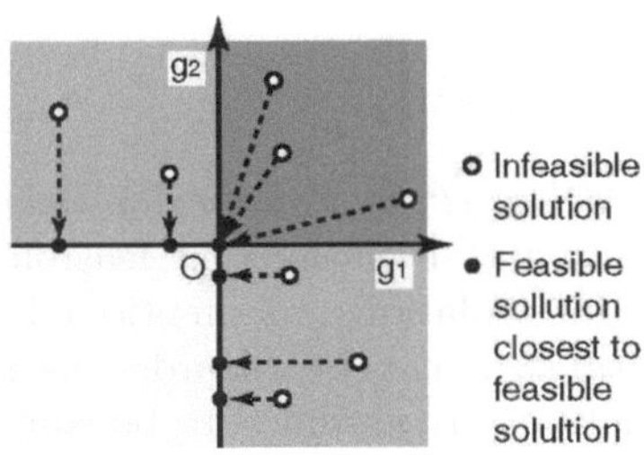

Fig. 2. Feasible solutions closest to infeasible solutions in the constraint function space

Another drawback of OCV(e) is that, on problems whose Pareto-optimal solutions lie on boundaries, infeasible solutions remain in the set of solutions throughout the entire search [2]. Furthermore, in some cases, not even one feasible solution may be found, as demonstrated in Sect. 4. In addition, OCV(e) itself is infeasible when there are some infeasible solutions for which objective functions are undefined. Although OCV(ne) does not have these drawbacks, it can practically reduce to DP: when most of the offspring solutions that GA generates are infeasible, they are simply discarded since their parent solutions are feasible and superior to them. Therefore, OCV(ne) cannot facilitate searching for Pareto-optimal solutions on boundaries, either.

Repair Operators. Repair operators for function optimization search for feasible solutions by reducing constraint violations, *without* considering objective functions. Such repair operators have a great potential since they are applicable to any function optimization problems. However, there have been a very small number of such studies [2]. Although GENOCOP III [7] is proposed as a repair operator, it cannot be used to search for feasible solutions since it assumes that some feasible solutions are available.

3 Pareto Descent Repair Operator

3.1 Guidelines for Effective Constraint Handling

This section gives the guidelines for designing an effective constraint-handling method that circumvents the problems pointed out in Sect. 2.2.

Guideline 1: Take the Repair Operator Approach. It is difficult to define appropriate penalty functions for penalty methods, and OCV may find no feasible solutions, as pointed out in the previous section. In addition, some of these methods themselves are infeasible if objective functions are undefined for some infeasible solutions. These imply that the approach of repair operators is more promising.

Guideline 2: Monotonically Decrease the Number of Violated Constraints and Constraint Violations. A feasible solution can be searched for by reducing constraint violations, as mentioned earlier. Since there are multiple constraint violations, it can be regarded as an MOO problem. Note that, since constraint functions are assumed to be continuous, there is a region, surrounding each feasible region, in which constraint functions can be regarded as unimodal. In fact, when constraint functions are linear or quadratic, constraint functions are unimodal in the entire infeasible region. Note also that infeasible solutions generated during GA's search are often near feasible regions, and constraint functions can be regarded as unimodal around the infeasible solutions. Being able to repair infeasible solutions in such regions is important in terms of both improving the probability of obtaining initial feasible solutions and facilitating GA's search on problems whose Pareto-optimal solutions lie on boundaries. In order to repair infeasible solutions in such regions, it is appropriate to monotonically decrease both the number of violated constraints and constraint violations.

Guideline 3: Search for the Feasible Solution Closest in the Constraint Function Space. Since constraint violations represent the degrees of violation of constraints, it is reasonable to search for the feasible solution at which violated constraint functions are as close to zero as possible, that is, the feasible solution closest to the infeasible solution in the constraint function space. This repairing approach in the case of two constraints is shown schematically in Fig. 2.

3.2 Strategies for Meeting the Guidelines

Guideline 1 implies the use of repair operators for constraint handling. This section explains what are necessary for meeting Guidelines 2 and 3.

Decrease Constraint Violations using Multi-objective Local Search. To monotonically decrease constraint violations, a multi-objective local search can be used with violated constraint functions regarded as objective functions. In this paper, Pareto Descent Method (PDM) [8,9] is used, which, as mentioned in the appendix, calculates appropriate Pareto descent directions and descent directions with relatively small computational cost and efficiently decreases all objective functions simultaneously. PDM consists of search direction calculation and linear search, and a repair operator based on it has a similar structure.

Search for Feasible Solutions on Boundaries. When no violated constraints have been satisfied yet, there is no active unviolated constraints (active constraints hereafter), and the search direction should be a Pareto descent direction of violated constraint functions so that they are decreased efficiently. When there are active constraints, violated constraint functions have to be reduced on the boundaries of the active constraints, since the feasible solutions closest to infeasible solutions in the constraint function space are on the boundaries of initially violated constraints. For this purpose, we can draw on the ideas of gradient projection method [10]. In the constraint-handling context, the search direction must be in the null-space of the gradients of active constraint functions. In order to decrease constraint violations in the null-space, the search direction should be a Pareto descent direction of the violated constraint functions in the null-space if such a direction exists, and a descent direction in the null-space otherwise. Linear search must be conducted while moving solutions in the search direction back onto the boundaries of active constraints.

Even when there are no descent directions in the null-space, the number of violated constraints and constraint violations may be further reduced by regarding some of the active constraints as inactive (*inactivation*). Consider the 2-variable-3-constraint problem shown in Fig. 3. Since there are two active constraints at $\boldsymbol{x}_1$, no descent directions of c_3 exist in the null-space. When c_1 is considered inactive, there are descent directions of c_3 in the null-space that are feasible w.r.t. c_1. Hence, feasible solution $\boldsymbol{x}_2$ can be obtained by inactivating c_1. Note that, when some active constraints are inactivated, not all violated constraint functions can be zero at the resulting feasible solution.

The details of these direction calculations and linear search are described in the following sections.

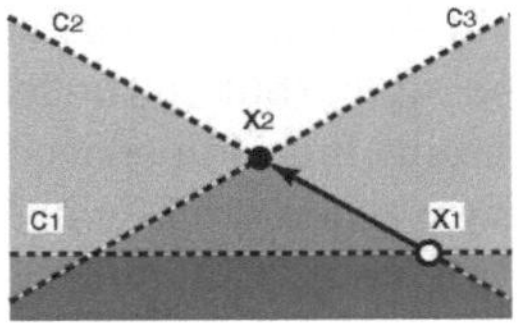

Fig. 3. Inactivation. At infeasible solution $\boldsymbol{x}_1$, there are no descent directions of the violated constraint functions in the null-space of the gradients of the active constraint functions. Inactivation of c_1 allows for obtaining feasible solution $\boldsymbol{x}_2$.

3.3 Search Direction Calculation

When No Active Constraints Exist. Pareto descent directions of violated constraint functions can be obtained using PDM, if they exist. If they do not, PDM detects it [8,9]. In this case, descent directions do not exist either, which implies that constraint violations cannot be locally decreased any further.

When Active Constraints Exist. Denote the active constraint functions by $\hat{g}_j^{\mathrm{u}}$ $(j = 1, 2, \ldots, \hat{P}^{\mathrm{u}})$ and those of violated constraints by g_j^{v} $(j = 1, 2, \ldots, P^{\mathrm{v}})$. In order for a search direction $\boldsymbol{d} \in \mathbb{R}^N$ to be a descent direction of the violated constraint functions in the null-space of the gradients of the active constraint functions, $\boldsymbol{d}$ has to satisfy

$$\hat{G}^{\mathrm{u}T}\boldsymbol{d} = \mathbf{0} \ , \ \text{where } \hat{G}^{\mathrm{u}} = [\nabla \hat{g}_1^{\mathrm{u}}, \ldots, \nabla \hat{g}_{\hat{P}^{\mathrm{u}}}^{\mathrm{u}}] \text{ and} \tag{3}$$

$$G^{\mathrm{v}T}\boldsymbol{d} \leq \mathbf{0} \ , \ \text{where } G^{\mathrm{v}} = [\nabla g_1^{\mathrm{v}}, \ldots, \nabla g_{P^{\mathrm{v}}}^{\mathrm{v}}] \ . \tag{4}$$

The following sections detail the calculations of Pareto descent directions and descent directions in the null-space.

Pareto Descent Directions in the Null-space. The condition for a descent direction $\boldsymbol{d}$ in the null-space to be a Pareto descent direction is that there exists a convex combination weight $\boldsymbol{\alpha} = (\alpha_1, \alpha_2, \ldots, \alpha_{P^{\mathrm{v}}})^T \in \mathbb{R}_+^{P^{\mathrm{v}}}$, where $\mathbb{R}_+$ is the set of non-negative real numbers, such that

$$\boldsymbol{d} = -G^{\mathrm{v}}\boldsymbol{\alpha} \ . \tag{5}$$

Substituting this into (3) gives

$$G\boldsymbol{\alpha} = \mathbf{0} \ , \ \text{where } G = -\hat{G}^{\mathrm{u}T}G^{\mathrm{v}} \ . \tag{6}$$

Denote the rank of G by $r(G)$. When $r(G) = P^{\mathrm{v}}$, the sole solution $\boldsymbol{\alpha} = \mathbf{0}$ of (6) represents $\boldsymbol{d} = \mathbf{0}$, which implies that no Pareto descent directions exist. When $r(G) = 0$, the gradients of violated constraint functions are already in the null-space, and the search direction should simply be a Pareto descent direction of the violated constraint functions. When $0 < r(G) < P^{\mathrm{v}}$, (6) implies that $\boldsymbol{\alpha}$ exists in a subspace of dimension $P^{\mathrm{v}} - r(G)$. Denote the basis vectors of the subspace by $\boldsymbol{u}_1, \boldsymbol{u}_2, \ldots, \boldsymbol{u}_{P^{\mathrm{v}}-r(G)} \in \mathbb{R}^{P^{\mathrm{v}}}$ and the coordinates of $\boldsymbol{\alpha}$ in the subspace by $\boldsymbol{\beta} \in \mathbb{R}^{P^{\mathrm{v}}-r(G)}$. Now $\boldsymbol{\alpha}$ can be expressed as $\boldsymbol{\alpha} = U\boldsymbol{\beta}$, where $U = \left[\boldsymbol{u}_1, \boldsymbol{u}_2, \ldots, \boldsymbol{u}_{P^{\mathrm{v}}-r(G)}\right]$. Substituting this into (5) and then substituting the result into (4) gives

$$-G^{\mathrm{v}T} G^{\mathrm{v}} U \boldsymbol{\beta} \leq \mathbf{0} \ . \tag{7}$$

The constraint $U\boldsymbol{\beta} \geq \mathbf{0}$ that each component of $\boldsymbol{\alpha}$ is non-negative and (7) are homogeneous linear inequalities of $\boldsymbol{\beta}$. This is of the same form as that for calculating Pareto descent directions in PDM. PDM calculates $\boldsymbol{\alpha}$ that maximizes α_i for each $i = 1, 2, \ldots, M$ to obtain Pareto descent directions. Similarly, $\boldsymbol{\beta}$ that maximizes α_i for each $i = 1, 2, \ldots, M$ can be calculated, which give Pareto descent directions in the null-space.[1]

Descent Directions in the Null-space. When $r(\hat{G}^{\mathrm{u}T}) = N$, the sole solution $\boldsymbol{d} = \mathbf{0}$ of (3) implies that no descent directions exist. When $r(\hat{G}^{\mathrm{u}T}) < N$, (3) implies that $\boldsymbol{d}$ exists in a subspace of dimension $N - r(\hat{G}^{\mathrm{u}T})$. Denote the basis vectors of the subspace by $\boldsymbol{e}_1, \boldsymbol{e}_2, \ldots, \boldsymbol{e}_{N-r(\hat{G}^{\mathrm{u}T})} \in \mathbb{R}^N$ and the coordinates of $\boldsymbol{d}$ in the subspace by $\boldsymbol{\gamma} \in \mathbb{R}^{N-r(\hat{G}^{\mathrm{u}T})}$. Now $\boldsymbol{d}$ can be expressed as $\boldsymbol{d} = E\boldsymbol{\gamma}$, where $E = \left[\boldsymbol{e}_1, \boldsymbol{e}_2 \ldots, \boldsymbol{e}_{N-r(\hat{G}^{\mathrm{u}T})}\right]$. Substituting this into (4) gives

$$G^{\mathrm{v}T} E \boldsymbol{\gamma} \leq \mathbf{0} \ . \tag{8}$$

This is a homogeneous linear inequality of $\boldsymbol{\gamma}$ and is the same form as that for calculating descent directions in PDM. Therefore, descent directions in the null-space can be obtained as descent directions are calculated in PDM.

Inactivation. Denote the set of active constraints by $\hat{C}^{\mathrm{u}}$, its subset by $\acute{C}^{\mathrm{u}}$, and the constraint functions of constraints in $\acute{C}^{\mathrm{u}}$ by $\acute{g}^{\mathrm{u}}_j$ ($j = 1, 2, \ldots, \acute{P}^{\mathrm{u}}$). In order for Guideline 2 to be satisfied when $\acute{C}^{\mathrm{u}}$ is inactivated, there must exist descent directions in the null-space of the gradients of the constraint functions of the constraints in $\hat{C}^{\mathrm{u}} \backslash \acute{C}^{\mathrm{u}}$ that are feasible w.r.t. $\acute{C}^{\mathrm{u}}$, i.e.,

$$\acute{G}^{\mathrm{u}T} \boldsymbol{d} \leq \mathbf{0} \ , \text{ where } \acute{G}^{\mathrm{u}} = [\nabla \acute{g}^{\mathrm{u}}_1, \ldots, \nabla \acute{g}^{\mathrm{u}}_{\acute{P}^{\mathrm{u}}}] \ . \tag{9}$$

Equation (9) can be incorporated into the above-mentioned calculations of descent directions and Pareto descent directions, and their existence can be tested by PDM. Hence, the possibility of inactivation of $\acute{C}^{\mathrm{u}}$ can be determined using PDM.[2]

When multiple subsets of $\hat{C}^{\mathrm{u}}$ can be inactivated, the one to be inactivated should be chosen based on the following rules according to Guideline 3:

---

[1] Some of the thus found Pareto descent directions may be redundant. Such redundant directions can be identified and removed as done in the calculation of descent directions in PDM [8,9].

[2] In order to find subsets that can be inactivated, every subset of $\hat{C}^{\mathrm{u}}$ must be examined. When the cardinality of $\hat{C}^{\mathrm{u}}$ is big, however, not all subsets can be examined, and some compromise has to be made.

1. Choose the subset with the smallest cardinality, and
2. If there are more than one such subsets, choose the one for which there exist Pareto descent directions of the violated constraints in the null-space of the gradients of the active constraint functions that are feasible w.r.t. the inactivated constraints.

3.4 Linear Search over Active Constraint Boundaries

Moving Solutions back onto Active Constraint Boundaries. In order to move a solution $\boldsymbol{y}$ in a search direction back onto active constraint boundaries, we can search for a solution that satisfies active constraints by small margins using $\boldsymbol{y}$ as the initial solution.[3] Consider using golden section method for the linear search in the optimization. The length of closed linear search interval and the number of iterations determine the maximum error ε that can transpire in the linear search. Minimizing $\sum_{j=1}^{\hat{P}^{\mathrm{u}}} \left(d_{\hat{g}_j^{\mathrm{u}}}(\boldsymbol{x}) - (-\varepsilon) \right)^2$ gives a solution that satisfies active constraints by the distance of at most 2ε, where $\hat{g}_j^{\mathrm{u}}(\boldsymbol{x})$ $(j = 1, 2, \ldots, \hat{P}^{\mathrm{u}})$ are active constraint functions, and $d_{\hat{g}_j^{\mathrm{u}}}(\boldsymbol{x})$ is the signed distance of $\boldsymbol{x}$ from the j-th boundary. Since $d_{\hat{g}_j^{\mathrm{u}}}(\boldsymbol{x})$ cannot usually be calculated precisely in practice, it has to be approximated. Applying Taylor expansion to $\hat{g}_j^{\mathrm{u}}(\boldsymbol{x})$ and ignoring the terms of order greater than two, $d_{\hat{g}_j^{\mathrm{u}}}(\boldsymbol{x})$ can be approximated [11] by

$$\tilde{d}_{\hat{g}_j^{\mathrm{u}}}(\boldsymbol{x}) = \frac{\nabla \hat{g}_j^{\mathrm{u}}(\boldsymbol{x}) \cdot \boldsymbol{x} + \hat{g}_j^{\mathrm{u}}(\boldsymbol{x})}{||\nabla \hat{g}_j^{\mathrm{u}}(\boldsymbol{x})||} \ . \tag{10}$$

Linear Search. Since the number of violated constraints and constraint violations must be monotonically decreased according to Guideline 2, the step-size must be chosen so that the solution in the search direction is 1) just before any of the unviolated constraints is violated or 2) just before any of the violated constraint functions increases. Additionally, the step-size must be chosen so that the solution in the search direction is 3) just after any of the violated constraints is satisfied, since the next iteration searches over the boundaries of active constraints including the one just satisfied.

3.5 Proposal of Pareto Descent Repair Operator

We propose the repair operator consisting of the above-mentioned search direction calculations and linear search as Pareto Descent Repair operator (PDR). PDR efficiently decreases constraint violations by calculating an appropriate search direction for each case it may encounter: active constraints may or may not exist, and Pareto descent directions and descent directions may or may not exist. The most computationally intense part of PDR is that of solving linear

[3] We can alternatively search for a solution which minimizes the distance to each boundary. The optimum, however, may violate the active constraints by small margins since the linear search used in that optimization always transpires a small error.

programming problems for direction calculations. Since computationally efficient linear programming solvers such as simplex method [12] can be used, the computational complexity of PDR is accordingly small.

3.6 Use of PDR in GA

When a repair operator is used in GA, infeasible solutions can be replaced by their corresponding feasible solutions (Lamarckian). They can also be stored and used in crossover, and their corresponding feasible solutions are used for evaluations of objective functions (non-Lamarckian) [2]. Consider solving a problem with a single feasible region using a crossover operator such as UNDX [13] which generates offspring solutions around the center of mass of parent solutions. Note that, when GA generates infeasible offspring solutions, Pareto-optimal solutions are likely to lie on boundaries. When Lamarckian PDR (PDR(l) hereafter) is used, parent solutions are either inside the feasible region or on boundaries, and their offspring solutions therefore are inside the feasible region. When non-Lamarckian PDR (PDR(nl) hereafter) is used, parent solutions are both inside and outside the feasible region, and their offspring solutions are more likely to be generated near the boundaries. This difference becomes prominent on problems such as ZDT2 on which a number of boundaries intersect at the Pareto-optimal solutions. Therefore, Pareto-optimal solutions on boundaries are expected to be obtained with higher precision when PDR(nl) is used than when PDR(l) is used. Even if there are multiple feasible regions, a similar argument applies when mating restriction is imposed so that solutions close to each other, which often belong to the same feasible region, are chosen for mating.

4 Experiments

In order to verify the effectiveness of PDR, GA that use PDR(l), PDR(nl), OCV(e), and OCV(ne) are compared on some well-known multi-objective benchmark problems. The results of death penalty (DP) will also be shown just for a reference, since it is the standard constraint-handling method for GA.

4.1 Experiment Setup

Performance Metrics. In order to evaluate the proximity and diversity of solutions, we use generational distance (GD) and D1R, which are used in many existing studies. GD is defined as the mean of the distances from each solution to its nearest Pareto-optimal solution in the normalized objective space [1] and measures proximity. D1R is defined as the mean of the distances from each Pareto-optimal solution to its nearest solution in the normalized objective space [14] and measures both proximity and diversity.

Pareto-optimal solutions are necessary to evaluate GD and D1R. We assume that the solutions obtained by running GA with a large population size and many generations are Pareto-optimal, as existing studies do. Note that, when OCV is used, the set of solutions may contain infeasible solutions. Since the number of non-inferior solutions, which are necessarily feasible, is sometimes used as a performance metric [1], obtaining more feasible solutions is better than obtaining less. Since infeasible solutions only deteriorate both proximity and diversity, GD and D1R are calculated using all the solutions in the solution set.

Benchmark Problems. Since the methods being compared are applicable to problems with arbitrary numbers of objective functions and feasible regions, the benchmark problems in Table 1 are used, each of which has two objective functions and a single feasible region. These problems with relatively simple constraints were chosen so that the behaviors of the constraint-handling methods can be examined in detail.

Table 1. The properties of the benchmark problems used in the experiment [1]

Name	N	M	**# const.**	**Linearity of const.**	**Local Pareto-opt.**	**Pareto-optimal solutions in the variable space**
BNH	2	2	6	Linear and non-linear	No	A kinked line partly on a boundary
TNK	2	2	2	Non-linear	Yes	Multiple curves on a boundary
ZDT2	30	2	60	Linear	No	A line at which 29 boundaries intersect

GA. Population size is 100, which is commonly used for MOO. Initial solutions are generated uniformly at random in $[-100, 100]^N$. For DP, however, initial solutions are generated uniformly at random in feasible regions. 50 parent pairs are formed at each generation. Since it has been reported that proximity is improved as the number of offspring solutions for each parent pair is increased [15,16], 20 offspring solutions are generated for each pair. Since it has also been reported in [15,16] that, although the best-performing crossover is problem dependent, UNDX [13] performs relatively well on many problems, UNDX is used in the experiment. SPEA2 [17] is known to exhibit good performance as a survival selection [18]. However, since the original SPEA2 requires substantial computation and memory space, modified SPEA2 [15,16] is used, which approximates crowdedness around a solution with the Euclidean distance from the solution to the other solution nearest to it in the normalized objective space.

PDR. Gradients are approximated by forward difference with the difference of 10^{-4}. To move a solution in a search direction back onto active constraint boundaries, steepest descent method is used. For linear search, golden section method is used, with the closed linear search interval length of 10^{-2}, the maximum number of extension of the interval of 20, and the basic number of iterations of 20. When active constraints are $\hat{C}^{\mathrm{u}} = \{\hat{c}^{\mathrm{u}}_1, \ldots, \hat{c}^{\mathrm{u}}_{\hat{P}^{\mathrm{u}}}\}$, $\hat{C}^{\mathrm{u}}$ and $\{\hat{c}^{\mathrm{u}}_i\}$ for each $i = 1, 2, \ldots, \hat{P}^{\mathrm{u}}$ are considered for inactivation. In order to accommodate a solution violating 30 linear constraints, search direction calculation and linear search are applied at most 30 times. Unrepairable infeasible solutions are discarded. Infeasible initial solutions are repaired in the Lamarckian way for PDR(nl) as well.

OCV. Since diversity w.r.t. constraint violations is unnecessary, sharing is applied in the original objective space.

4.2 Results

Figure 4 shows the transitions of GD and D1R, averaged over 10 trials, against the number of objective function evaluations when GA with constraint-handing methods are applied to the benchmark problems. For the methods that diverged GD, the lines of both GD and D1R are omitted.

BNH and TNK. Regarding GD, PDR(nl) performed the best, followed by PDR(l). PDR(nl) performed better because part or all of the Pareto-optimal solutions of these problems lie on boundaries, and GA's search for the Pareto-optimal solutions was better facilitated by PDR(nl), as explained in Sect. 3.6. No performance difference regarding D1R can be observed between PDR(l) and PDR(nl).

DP and OCV(ne) performed worse than PDR w.r.t. GD since it is difficult for DP to search for solutions on boundaries, and OCV(ne) behaves practically the same as DP, as explained in Sect. 2.2. Regarding D1R, PDR performed no worse than DP and OCV(ne).

On BNH, OCV(e) performed the worst. This is because OCV(e) maintains infeasible solutions throughout the entire search. OCV(e) performed poorly in D1R because its GD is not good. OCV(e) diverged GD on TNK. The entire third quadrant of TNK is Pareto-optimal w.r.t. constraint violations and objective functions, and solutions in the second and fourth quadrants can also be non-inferior. Therefore, ranking did not function on TNK, and crossover and sharing dispersed solutions.

ZDT2. Again, PDR(nl) performed the best regarding both GD and D1R since PDR(nl) better facilitates the search of the Pareto-optimal solutions on boundaries than DP and PDR(l) do, as explained in Sect. 1 and Sect. 3.6, respectively. OCV diverged GD as predicted in Sect. 2.2. Since this was observed despite the strong interpolating property of UNDX, similar results are expected to be observed when other less interpolative crossovers are used.

On the Whole. Experimental results confirmed that GA performs the best when PDR is used, which requires additional computational complexity comparable to that of linear programming solvers. They have also shown that OCV(ne) exhibits performance similar to that of DP on low dimensional problems, and OCV(e) and OCV(ne) can disperse solutions on problems with many constraints. In addition, it has been confirmed that, on problems whose Pareto-optimal solutions lie on boundaries, GA's search is better facilitated and solutions are obtained with higher precision when PDR is applied in the non-Lamarckian way.

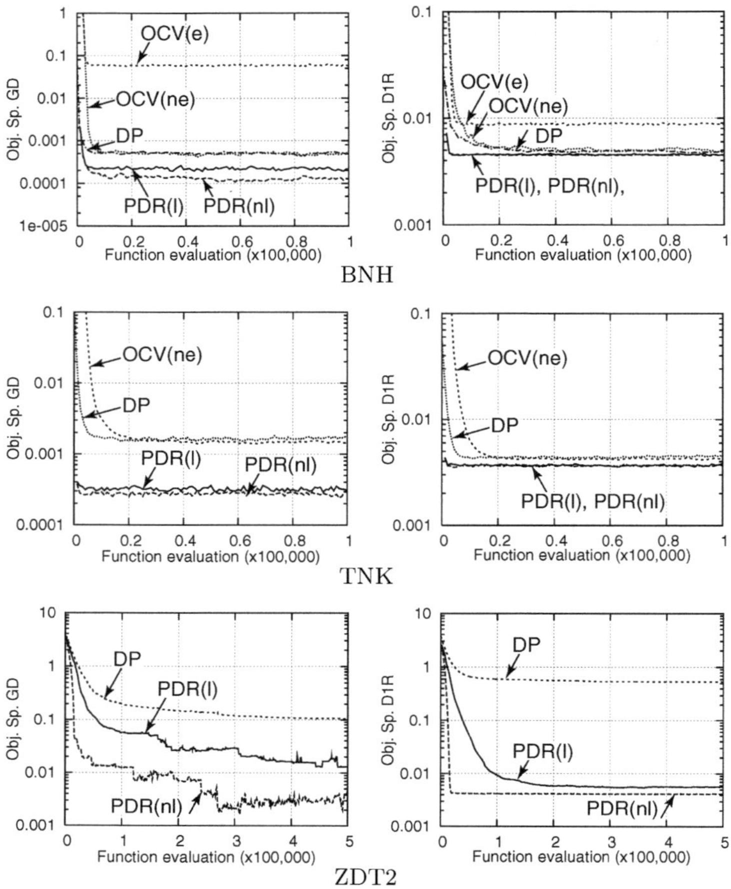

Fig. 4. Transitions of GD and D1R, averaged over 10 trials, when GA combined with constraint-handling methods are applied to the benchmark problems

5 Conclusions

This paper first presented the guidelines for designing effective constraint-handling methods. It then proposed Pareto Descent Repair operator (PDR) that meets these guidelines. PDR's effectiveness was verified through experiments comparing it with other constraint-handling methods. It was also confirmed that Pareto-optimal solutions on the boundaries are obtained with higher precision when PDR is applied in the non-Lamarckian way.

Although this paper proposed PDR as a repair operator for MOO, it can also be applied to single-objective optimization problems. Hence, it remains to investigate the effectiveness of PDR on single-objective optimization problems.

References

1. Deb, K.: Multi-Objective Optimization Using Evolutionary Algorithms. John Wiley & Sons, Chichester (2001)
2. Coello, C.A.C.: Theoretical and numerical constraint handling techniques used with evolutionary algorithms: A survey of the state of the art. Computer Methods in Applied Mechanics and Engineering **191**(11-12) (2002) 1245–1287
3. Knowles, J.D., Corne, D.W.: Memetic algorithms for multiobjective optimization: issues, methods and prospects. In Krasnogor, N., Smith, J.E., Hart, W.E., eds.: Recent Advances in Memetic Algorithms. Springer (2004) 313–352
4. Coello, C.A.C.: Treating constraints as objectives for single-objective evolutionary optimization. Engineering Optimization **32**(3) (2000) 275–308
5. Deb, K.: An efficient constraint handling method for genetic algorithms. Computer Methods in Applied Mechanics and Engineering **186** (2000) 311–338
6. Oyama, A., Shimoyama, K., Fujii, K.: New constraint-handling method for multi-objective multi-constraint evolutionary optimization and its application to space plane design. In Schilling, R., Haase, W., Periaux, J., Baier, H., Bugeda, G., eds.: Evolutionary and Deterministic Methods for Design, Optimization and Control with Applications to Industrial and Societal Problems (EUROGEN 2005). (2005) 416–428
7. Michalewicz, Z., Nazhiyath, G.: Genocop III: A co-evolutionary algorithm for numerical optimization problems with nonlinear constraints. In: Proceedings of the 2nd IEEE International Conference on Evolutionary Computation. Volume 2. (1995) 647–651
8. Harada, K., Sakuma, J., Ikeda, K., Ono, I., Kobayashi, S.: Local search for multiobjective function optimization: Pareto descent method (in Japanese). Transactions of the Japanese Society for Artificial Intelligence **21**(4) (2006) 340–350
9. Harada, K., Sakuma, J., Ikeda, K., Ono, I., Kobayashi, S.: Local search for multiobjective function optimization: Pareto descent method. In: Proceedings of the Genetic and Evolutionary Computation Conference (GECCO-2006), New York, NY, ACM Press (2006) 659–666
10. Luenberger, D.G.: Linear and Nonlinear Programming. Second edn. Addison-Wesley, Reading, MA (1984)
11. Gellert, W., Gottwald, S., Hellwich, M., Kästner, H., Künstner, H., eds.: VNR Concise Encyclopedia of Mathematics. Van Nostrand Reinhold, New York (1989)
12. Cormen, T.H., Leiserson, C.E., Rivest, R.L., Stein, C.: Introduction to Algorithms. Second edn. MIT Press (2001)
13. Ono, I., Kobayashi, S.: A real-coded genetic algorithm for function optimization using unimodal normal distribution crossover. In: 7th International Conference on Genetic Algorithms (ICGA7). (1997) 246–253
14. Knowles, J.D., Corne, D.W.: On metrics for comparing non-dominated sets. In: Proceedings of the 2002 Congress on Evolutionary Computation Conference (CEC02), IEEE Press (2002) 711–716
15. Harada, K., Sakuma, J., Kobayashi, S., Ikeda, K., Ono, I.: Hybridization of genetic algorithm and local search in multiobjective function optimization: Recommendation of GA then LS. In: Proceedings of the Genetic and Evolutionary Computation Conference (GECCO-2006), New York, NY, ACM Press (2006) 667–674
16. Harada, K., Ikeda, K., Sakuma, J., Ono, I., Kobayashi, S.: Hybridization of genetic algorithm with local search in multiobjective function optimization: Recommendation of GA then LS (in Japanese). Transactions of the Japanese Society for Artificial Intelligence **21**(6) (2006) 482–492

17. Zitzler, E., Laumanns, M., Thiele, L.: SPEA2: Improving the strength pareto evolutionary algorithm for multiobjective optimization. In Giannakoglou, K., et al., eds.: EUROGEN 2001, Evolutionary Methods for Design, Optimization and Control with Applications to Industrial Problems. (2001) 12–21
18. Deb, K., Thiele, L., Laumanns, M., Zitzler, E.: Scalable multi-objective optimization test problems. In: Proceedings of the Congress on Evolutionary Computation (CEC-2002). (2002) 825–830
19. Fliege, J., Svaiter, B.F.: Steepest descent methods for multicriteria optimization. Mathematical Methods of Operations Research **51**(3) (2000) 479–494

Appendix: Pareto Descent Method

Denote the normalized gradients of objective functions at solution $\boldsymbol{x}$ by $\bar{\nabla} f_i(\boldsymbol{x})$ $(i = 1, 2, \ldots, M)$. If a direction $\boldsymbol{d} \in \mathbb{R}^N$ satisfies

$$\boldsymbol{d} \cdot (-\bar{\nabla} f_i(\boldsymbol{x})) \geq 0 \ (i = 1, 2, \ldots, M) \ , \tag{11}$$

all objective functions can be simultaneously decreased by moving $\boldsymbol{x}$ in direction $\boldsymbol{d}$. Such directions are called *descent directions* for MOO. There are often multiple descent directions. The descent directions to which no other descent directions are superior in improving all objective functions are called *Pareto descent directions* [8,9]. There are often multiple Pareto descent directions. A descent direction is a Pareto descent direction if it can be expressed as a convex combination of the steepest descent directions of objective functions. Descent directions and Pareto descent directions of a 2-variable-2-objective problem are shown in Fig. 5.

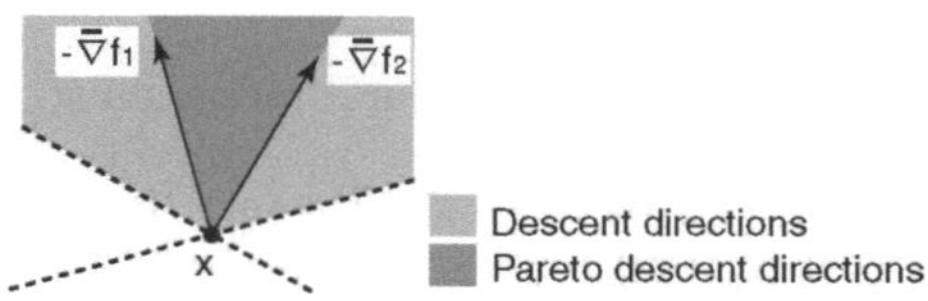

Fig. 5. Descent directions and Pareto descent directions of a 2-variable-2-objective problem

Since objective functions can be efficiently decreased by searching in Pareto-descent directions, several methods that calculates such directions were proposed in recent years, which include Multi-objective Steepest Descent Method (MSDM) [19] and Pareto Descent Method (PDM) [8,9]. PDM calculates feasible Pareto descent directions or descent directions, as appropriate, by solving linear programming problems, which has less computational complexity than MSDM does. Therefore, PDM can both effectively and efficiently decrease all objective functions simultaneously.

Steady-State Selection and Efficient Covariance Matrix Update in the Multi-objective CMA-ES

Christian Igel[1], Thorsten Suttorp[1], and Nikolaus Hansen[2]

[1] Institut für Neuroinformatik, Ruhr-Universität Bochum, 44780 Bochum, Germany
{christian.igel, thorsten.suttorp}@neuroinformatik.rub.de
[2] Institute of Computational Science, ETH Zurich, 8092 Zurich, Switzerland
nikolaus.hansen@inf.ethz.ch

Abstract. The multi-objective covariance matrix adaptation evolution strategy (MO-CMA-ES) combines a mutation operator that adapts its search distribution to the underlying optimization problem with multi-criteria selection. Here, a generational and two steady-state selection schemes for the MO-CMA-ES are compared. Further, a recently proposed method for computationally efficient adaptation of the search distribution is evaluated in the context of the MO-CMA-ES.

1 Introduction

Evolution strategies (ES) for real-valued optimization rely on Gaussian random variations. Appropriately adapting the covariance matrices of these mutations during optimization allows for learning a variable metric for the search distribution. It is well known that such an automatic adaptation of the mutation distribution drastically improves the search performance on non-separable and/or badly scaled single-objective functions [1,2,3,4].

In [5], we incorporated the step size and covariance matrix adaptation from the covariance matrix adaptation ES (CMA-ES, [3]) into a multi-objective framework. The resulting MO-CMA-ES used generational selection based on [6] combined with the sorting criterion proposed in [7,8]. We chose generational selection in order to make our performance comparisons with alternative methods easier to interpret. However, in [7,8] steady-state selection is used with good results and the question arises whether the MO-CMA-ES would profit from this selection scheme. In [9], we presented a new, computationally efficient update scheme for covariance matrices. The complexity reduction from $\mathcal{O}(n^3)$ to $\mathcal{O}(n^2)$ per update of the mutation distribution, where n is the dimensionality of the search space, comes at the cost of slower adaptation rates. However, as in the MO-CMA-ES many mutation distributions need to be traced, this approach seems to be particularly promising for the MO-CMA-ES.

In this work, we first investigate the computationally efficient update proposed in [9] within the framework of the MO-CMA-ES. Second, we compare variants of the MO-CMA-ES with different steady-state selection schemes and generational selection, respectively.

S. Obayashi et al. (Eds.): EMO 2007, LNCS 4403, pp. 171–185, 2007.

2 Covariance Matrix Adaptation

Let us consider an additive mutation $\boldsymbol{v}_i^{(g)} \in \mathbb{R}^n$ of individual i in generation g. The mutation $\boldsymbol{v}_i^{(g)}$ is a realization of an n-dimensional random vector distributed according to a zero-mean Gaussian distribution with covariance matrix $\boldsymbol{C}_i^{(g)}$, that is, $\boldsymbol{v}_i^{(g)} \sim \mathcal{N}\left(\mathbf{0}, \boldsymbol{C}_i^{(g)}\right)$. To sample this mutation distribution, n independent standard normally distributed random numbers are drawn to generate a realization of an n-dimensional normally distributed random vector $\boldsymbol{z}_i^{(g)} \sim \mathcal{N}(\mathbf{0}, \boldsymbol{I})$ with unit covariance matrix and zero mean. Then this random vector is rotated and scaled by a linear transformation $\boldsymbol{A}_i^{(g)} \in \mathbb{R}^{n \times n}$ such that

$$\boldsymbol{A}_i^{(g)} \boldsymbol{z}_i^{(g)} \sim \mathcal{N}(\mathbf{0}, \boldsymbol{C}_i^{(g)}) \text{ for } \boldsymbol{z}_i^{(g)} \sim \mathcal{N}(\mathbf{0}, \boldsymbol{I}) \ .$$

Thus, for sampling the mutation distribution the covariance matrix $\boldsymbol{C}_i^{(g)}$ has to be decomposed into Cholesky factors $\boldsymbol{C}_i^{(g)} = \boldsymbol{A}_i^{(g)} \boldsymbol{A}_i^{(g)T}$. One of the decisive features of ES is that the covariance matrices are subject to adaptation. The general policy is to alter the covariance matrices such that steps promising larger fitness gain are sampled more often. Here we consider matrix updates of the form $\boldsymbol{C}^{(g+1)} = \alpha \boldsymbol{C}^{(g)} + \beta \boldsymbol{V}^{(g)}$, where $\boldsymbol{V}^{(g)} \in \mathbb{R}^{n \times n}$ is positive definite and $\alpha, \beta \in \mathbb{R}^+$ are weighting factors (e.g., see [3,10]). Let $\boldsymbol{v}^{(g)} \in \mathbb{R}$ be a step in the search space promising large fitness gain. To increase the probability that $\boldsymbol{v}^{(g)}$ is sampled in the next iteration, the rank-one update

$$\boldsymbol{C}^{(g+1)} = \alpha \boldsymbol{C}^{(g)} + \beta \boldsymbol{v}^{(g)} \boldsymbol{v}^{(g)T} \tag{1}$$

can be used. This update rule shifts the mutation distribution towards the line distribution $\mathcal{N}(\mathbf{0}, \boldsymbol{v}^{(g)} \boldsymbol{v}^{(g)T})$, which is the distribution with the highest probability to generate $\boldsymbol{v}^{(g)}$ among all normal distributions with zero mean [3].

In general, each factorizing of a covariance matrix requires $O(n^3)$ operations. Thus, in an ES with additive covariance matrix update the Cholesky factorization of the covariance matrix is the computationally dominating factor apart from the fitness function evaluations. In [9] we therefore proposed not to factorize the covariance matrix, but to use an incremental rank-one update rule for the Cholesky factorization. This reduces the computational complexity to $O(n^2)$. The idea is not to compute the covariance matrix explicitly, but to operate on Cholesky factors only. Setting $\boldsymbol{v}^{(g)} = \boldsymbol{A}^{(g)} \boldsymbol{z}^{(g)}$ with $\boldsymbol{z}^{(g)} \sim \mathcal{N}(\mathbf{0}, \boldsymbol{I})$ we can rewrite the rank-one update of the covariance matrix equation (1) as

$$\boldsymbol{C}^{(g+1)} = \alpha \boldsymbol{C}^{(g)} + \beta \boldsymbol{A}^{(g)} \boldsymbol{z}^{(g)} \left[\boldsymbol{A}^{(g)} \boldsymbol{z}^{(g)}\right]^T \ . \tag{2}$$

Using the following theorem, we turn this update for $\boldsymbol{C}^{(g)}$ into an update for $\boldsymbol{A}^{(g)}$.

Theorem 1 ([9]). *Let $\boldsymbol{C}_t \in \mathbb{R}^{n \times n}$ be a symmetric nonnegative definite matrix with Cholesky factorization $\boldsymbol{C}_t = \boldsymbol{A}_t \boldsymbol{A}_t{}^T$. Assuming that $\boldsymbol{C}_t$ is updated using*

$$\boldsymbol{C}_{t+1} = \alpha \boldsymbol{C}_t + \beta \boldsymbol{v}_t \boldsymbol{v}_t^T ,$$

with $\boldsymbol{v}_t = \boldsymbol{A}_t \boldsymbol{z}_t$*, where* $\boldsymbol{z}_t$ *is a column vector and* $\alpha, \beta \in \mathbb{R}^+$*. Then, the Cholesky factorization* $\boldsymbol{C}_{t+1} = \boldsymbol{A}_{t+1} \boldsymbol{A}_{t+1}{}^T$ *is given by*

$$\boldsymbol{A}_{t+1} = \sqrt{\alpha}\boldsymbol{A}_t + \frac{\sqrt{\alpha}}{\|\boldsymbol{z}_t\|^2}\left(\sqrt{1 + \frac{\beta}{\alpha}\|\boldsymbol{z}_t\|^2} - 1\right)[\boldsymbol{A}_t \boldsymbol{z}_t]\, \boldsymbol{z}_t^T \ .$$

The new update rule guarantees a positive-definite covariance matrix. The numerical stability of the new update is likely to be better than an update requiring decompositions (e.g., see the discussion in [11, chapter 6]).

3 Generational and Steady-State Multi-objective Selection

Our multi-objective selection schemes is based on the non-dominated sorting approach used in NSGA-II [12,6] and the selection scheme used in SMS-EMOA [7,8].

First of all, the elements in a population A of candidate solutions are ranked according to their level of non-dominance. Let the non-dominated solutions in A be denoted by $\text{ndom}(A) = \{a \in A \,|\, \nexists a' \in A : a' \prec a\}$, where $a' \prec a$ means that a' dominates a. The Pareto front of A is then given by $\{(f_1(a), \ldots, f_M(a)) \,|\, a \in \text{ndom}(A)\}$, where the f_i are the M real-valued objective functions. The elements in $\text{ndom}(A)$ get rank 1. The other ranks are defined recursively by considering the set without the solutions with lower ranks (cf. [6,9]). Formally, let $\text{dom}_0(A) = A$, $\text{dom}_l(A) = \text{dom}_{l-1}(A) \setminus \text{ndom}_l(A)$, and $\text{ndom}_l(A) = \text{ndom}(\text{dom}_{l-1}(A))$ for $l \in \{1, \ldots\}$. For $a \in A$ we define the level of non-dominance $\text{r}(a, A)$ to be i iff $a \in \text{ndom}_i(A)$.

A second sorting criterion is needed to rank the solutions having the same level of non-dominance. This criterion is very important, as usually (in particular in real-valued optimization of continuous objective functions) after some generations there are more non-dominated solutions in the population than solutions to be selected. We consider the contributing hypervolume as second sorting criterion, which gave better results than the crowding-distance [6] in the experiments in [9]. The hypervolume measure or $\mathcal{S}$-metric was introduced by [13] in the domain of evolutionary MOO. It can be defined as the Lebesgue measure Λ (i.e., the volume) of the union of hypercuboids in the objective space:

$$\mathcal{S}_{a_{\text{ref}}}(A') = \Lambda\left(\bigcup_{a \in \text{ndom}(A')} \{(f_1(a'), \ldots, f_M(a')) \,|\, a \prec a' \prec a_{\text{ref}}\}\right) ,$$

where a_{ref} is an appropriately chosen reference point. The contributing hypervolume of a point $a \in \text{ndom}(A')$ is given by

$$\Delta_{\mathcal{S}}(a, A') := \mathcal{S}_{a_{\text{ref}}}(A') - \mathcal{S}_{a_{\text{ref}}}(A' \setminus \{a\}) \ .$$

The rank $s(a, A')$ of an individual a can be defined recursively based on its contribution to the hypervolume, where ties are broken at random. The individual contributing least to the hypervolume of A' gets the worst rank. The individual contributing least to the hypervolume of A' without the individual with the worst rank is assigned the second worst rank and so on. We call $a \in A'$ a boundary element if $\Delta_{\mathcal{S}}(a, A')$ depends on the choice of the reference point a_{ref}. We choose a_{ref} such that all elements in A' dominate a_{ref} and that for any boundary element $a \in A'$ and any non boundary element $a' \in A'$ we have $\Delta_{\mathcal{S}}(a, A') > \Delta_{\mathcal{S}}(a', A')$. That is, the individuals at the "boundaries" of the Pareto front of A' are preferably selected. Let a lower rank be worse. Formally (assuming that argmin breaks ties randomly), for $a \in \text{ndom}(A')$ we have $s(a, A') = 1$ if $a = \text{argmin}_{a' \in A'}\{\Delta_{\mathcal{S}}(a', A')\}$ and $s(a, A') = k$ if $a = \text{argmin}_{a' \in A'}\{\Delta_{\mathcal{S}}(a', A' \setminus \{a'' \mid s(a'', A') < k\})\}$. Based on this ranking and the level of non-dominance we define the relation

$$a \prec_A a' \Leftrightarrow \text{r}(a, A) < \text{r}(a', A) \text{ or} \\ \left[(\text{r}(a, A) = \text{r}(a', A)) \wedge (s(a, \text{ndom}_{\text{r}(a,A)}(A)) > s(a', \text{ndom}_{\text{r}(a',A)}(A)))\right] ,$$

for $a, a' \in A$. That is, a is better than a' when compared using $\prec_A$ if either a has a better level of non-dominance or a and a' are on the same level but a contributes more to the hypervolume when considering the points at that level of non-dominance.

In the following, we consider three reproduction and selection schemes based on this ranking. First, in *generational* selection $(\mu+\mu)$ as described in [9] each of the μ parents generates one offspring per generation. The resulting 2μ individuals are sorted as described above and the μ best form the next parent population.

Then we consider two *steady-state* [14,15] selection schemes, in which only a single parent creates one offspring per generation. If this offspring a is better than the worst individual in the parent population A w.r.t. $\prec_{A \cup \{a\}}$, a replaces the worst individual. Otherwise the offspring is discarded. The two steady state variants differ in the way the parent of the offspring is selected. In the first variant $(\mu+1)$, the parent is chosen uniformly at random from A. In the second version $(\mu_{\prec}+1)$, the parent is chosen uniformly at random from $\text{ndom}(A)$. The idea behind the second approach, which can be regarded as more greedy, is that it is more promising to allow reproduction of non-dominated individuals than of dominated. Because the individuals $A \setminus \text{ndom}(A)$ do not influence the evolutionary process anymore, they can be discarded. That is, the second variant is an algorithm with varying population (or archive) size. Only non-dominated individuals remain in the population while the size of the population is still upper bounded by μ.

4 MO-CMA-ES

In the MO-CMA-ES with standard covariance matrix update the kth individual in generation g is a 5-tuple denoted by $a_k^{(g)} = [\boldsymbol{x}_k^{(g)}, \overline{p}_{\text{succ},k}^{(g)}, \sigma_k^{(g)}, \boldsymbol{p}_{c,k}^{(g)}, \boldsymbol{C}_k^{(g)}]$. Here, $\boldsymbol{x}_k^{(g)} \in \mathbb{R}^n$ is the point in the search space, $\overline{p}_{\text{succ},k}^{(g)} \in \mathbb{R}_0^+$ the average success rate,

$\sigma_k^{(g)} \in \mathbb{R}^+$ the global step size, $\boldsymbol{p}_{c,k}^{(g)} \in \mathbb{R}^n$ the evolution path, and $\boldsymbol{C}_k^{(g)} \in \mathbb{R}^{n \times n}$ the covariance matrix.

The standard version of the generational MO-CMA-ES reads as follows (ignoring lines 5b and 10b for a moment):

Algorithm 1. generational MO-CMA

1 $g = 0$, initialize $a_k^{(g)}$ for $k = 1, \ldots, \mu$
2 **repeat**
3 **for** $k = 1, \ldots, \mu$ **do**
4 $a'^{(g+1)}_k \leftarrow a_k^{(g)}$
5a $\boldsymbol{x}'^{(g+1)}_k \sim \mathcal{N}\left(\boldsymbol{x}_k^{(g)}, {\sigma_k^{(g)}}^2 \boldsymbol{C}_k^{(g)}\right)$
5b $\boldsymbol{x}'^{(g+1)}_k \sim \mathcal{N}\left(\boldsymbol{x}_k^{(g)}, {\sigma_k^{(g)}}^2 \boldsymbol{A}_k^{(g)} {\boldsymbol{A}_k^{(g)}}^T\right)$
6 $Q^{(g)} = \left\{a'^{(g+1)}_k, a_k^{(g)} \mid 1 \leq k \leq \mu\right\}$
7 **for** $k = 1, \ldots, \mu$ **do**
8 updateStepSize $\left(a_k^{(g)}, \mathbb{1}[a'^{(g+1)}_k \prec_{Q^{(g)}} a_k^{(g)}]\right)$
9 updateStepSize $\left(a'^{(g+1)}_k, \mathbb{1}[a'^{(g+1)}_k \prec_{Q^{(g)}} a_k^{(g)}]\right)$
10a updateCovariance $\left(a'^{(g+1)}_k, \frac{\boldsymbol{x}'^{(g+1)}_k - \boldsymbol{x}_k^{(g)}}{\sigma_k^{(g)}}\right)$
10b updateCholesky $\left(a'^{(g+1)}_k, \frac{\boldsymbol{x}'^{(g+1)}_k - \boldsymbol{x}_k^{(g)}}{\sigma_k^{(g)}}\right)$
11 **for** $i = 1, \ldots, \mu$ **do** $a_i^{(g+1)} \leftarrow Q_{\prec:i}^{(g)}$
12 $g \leftarrow g + 1$
until stopping criterion is met

Each of the μ parents generates one offspring (lines 3–5) . Parents and offspring form the set $Q^{(g)}$ (line 6). The step sizes of a parent and its offspring are updated depending on whether the mutations were successful (lines 7–9), that is, whether the offspring is better than the parent according to the relation $\prec_{Q^{(g)}}$ (the indicator function $\mathbb{1}[\cdot]$ is 1 if its argument is true and 0 otherwise).

The covariance matrix of the offspring (line 10a) is adjusted taking into account the mutation that has led to its genotype. Both step size and covariance matrix update are the same as in the single-objective (1+1)-CMA-ES, see [5,9] for details. The best μ individuals in $Q^{(g)}$ sorted by $\prec_{Q^{(g)}}$ form the next parent generation (line 11, where $Q_{\prec:i}^{(g)}$ is the ith best offspring in $Q^{(g)}$ w.r.t. $\prec_{Q^{(g)}}$).

The update rule for the global step size is rooted in the 1/5-success-rule proposed in [1] and is an extension from the rule proposed in [4]:

Procedure `updateStepSize`$(a = [\boldsymbol{x}, \overline{p}_{\text{succ}}, \sigma, \boldsymbol{p}_c, \boldsymbol{C}], p_{\text{succ}})$

$\overline{p}_{\text{succ}} \leftarrow (1 - c_p)\,\overline{p}_{\text{succ}} + c_p p_{\text{succ}}$

$\sigma \leftarrow \sigma \cdot \exp\left(\frac{1}{d}\frac{\overline{p}_{\text{succ}} - p_{\text{succ}}^{\text{target}}}{1 - p_{\text{succ}}^{\text{target}}}\right)$

This rule implements the well-known heuristic that the step size should be increased if the success rate of mutation is high, and the step size should be decreased if the success rate is low. The damping parameter d controls the rate of the step size adaptation.

Then the covariance matrices are adapted (see main routine line 10a):

Procedure `updateCovariance`$(a = [\boldsymbol{x}, \overline{p}_{\text{succ}}, \sigma, \boldsymbol{p}_c, \boldsymbol{C}], \boldsymbol{x}_{\text{step}} \in \mathbb{R}^n)$

1 **if** $\overline{p}_{\text{succ}} < p_{\text{thresh}}$ **then**
2 $\quad \boldsymbol{p}_c \leftarrow (1 - c_c)\boldsymbol{p}_c + \sqrt{c_c(2 - c_c)}\,\boldsymbol{x}_{\text{step}}$
3 $\quad \boldsymbol{C} \leftarrow (1 - c_{\text{cov}})\boldsymbol{C} + c_{\text{cov}} \cdot \boldsymbol{p}_c {\boldsymbol{p}_c}^T$
4 **else**
5 $\quad \boldsymbol{p}_c \leftarrow (1 - c_c)\boldsymbol{p}_c$
6 $\quad \boldsymbol{C} \leftarrow (1 - c_{\text{cov}})\boldsymbol{C} + c_{\text{cov}} \cdot \left(\boldsymbol{p}_c {\boldsymbol{p}_c}^T + c_c(2 - c_c)\boldsymbol{C}\right)$

The update of the evolution path $\boldsymbol{p}_c$ depends on the value of $\overline{p}_{\text{succ}}$. If the smoothed success rate $\overline{p}_{\text{succ}}$ is high, that is, above $p_{\text{thresh}} < 0.5$, the update of the evolution path $\boldsymbol{p}_c$ is stalled. This prevents a too fast increase of axes of $\boldsymbol{C}$ when the step size is far too small, for example, in a linear surrounding. If the smoothed success rate $\overline{p}_{\text{succ}}$ is low, the update of $\boldsymbol{p}_c$ is accomplished with exponential smoothing. The constants c_c and c_{cov} $(0 \leq c_{\text{cov}} < c_c \leq 1)$ are learning rates for the evolution path and the covariance matrix, respectively. The factor $\sqrt{c_c(2 - c_c)}$ normalizes the variance of $\boldsymbol{p}_c$ viewed as a random variable (see [3]). The evolution path $\boldsymbol{p}_c$ is then used to update the covariance matrix. The new covariance matrix is a weighted mean of the old matrix and the outer product of $\boldsymbol{p}_c$. In the second case (line 5), the second summand in the update of $\boldsymbol{p}_c$ is missing and the length of $\boldsymbol{p}_c$ shrinks. Although of minor relevance, the term $c_c(2 - c_c)\boldsymbol{C}$ (line 6) compensates for this shrinking in $\boldsymbol{C}$.

The (external) strategy parameters are the population size, target success probability $p_{\text{succ}}^{\text{target}}$, step size damping d, success rate averaging parameter c_p, cumulation time horizon parameter c_c, and covariance matrix learning rate c_{cov}. Default values, as given in [9] and used in this paper, are: $d = 1 + n/2$, $p_{\text{succ}}^{\text{target}} = (5 + \sqrt{1/2})^{-1}$, $c_p = p_{\text{succ}}^{\text{target}}/(2 + p_{\text{succ}}^{\text{target}})$, $c_c = 2/(n + 2)$, $c_{\text{cov}} = 2/(n^2 + 6)$, and $p_{\text{thresh}} = 0.44$.

The elements of the initial individual, $a_{\text{parent}}^{(0)}$ are set to $\overline{p}_{\text{succ}} = p_{\text{succ}}^{\text{target}}$, $\boldsymbol{p}_c = \boldsymbol{0}$, and $\boldsymbol{C} = \mathbf{I}$. The initial candidate solution $\boldsymbol{x} \in \mathbb{R}^n$ and the initial $\sigma \in \mathbb{R}^+$ must be chosen problem dependent. The optimum should presumably be within the cube $\boldsymbol{x} \pm 2\,\sigma\,(1,\ldots,1)^T$.

4.1 Cholesky Update

In the standard generational MO-CMA-ES, denoted by $(\mu{+}\mu)$ in the following, there are up to μ covariance updates (in an efficient implementation only covariance matrices of those offspring that will be in the next parent population are updated). Therefore, computation time could be significantly reduced using the concepts described in [9] and Section 2—if objective function evaluation is fast and the dimensionality n of the search space is large.

In the generational MO-CMA with "Cholesky update", denoted by $(\mu{+}\mu)_{\text{chol}}$ in the following, the kth individual in generation g consists of a 4-tuple $a_k^{(g)} = [\boldsymbol{x}_k^{(g)}, \bar{p}_{\text{succ},k}^{(g)}, \sigma_k^{(g)}, \boldsymbol{A}_k^{(g)}]$, where the Cholesky factor $\boldsymbol{A}_k^{(g)} \in \mathbb{R}^{n \times n}$ is stored instead of the covariance matrix. The update of the Cholesky factor is given by applying Theorem 1:

Procedure `updateCholesky`$(a = [\boldsymbol{x}, \bar{p}_{\text{succ}}, \sigma, \boldsymbol{p}_c, \boldsymbol{A}], \boldsymbol{x}_{\text{step}} \in \mathbb{R}^n)$

---

if $\bar{p}_{\text{succ}} < p_{\text{thresh}}$ **then**

$$\boldsymbol{A} \leftarrow \sqrt{1 - c_{\text{cov}}}\,\boldsymbol{A} + \frac{\sqrt{1 - c_{\text{cov}}}}{\|\boldsymbol{x}_{\text{step}}\|^2}\left(\sqrt{1 + \frac{c_{\text{cov}}\|\boldsymbol{x}_{\text{step}}\|^2}{1 - c_{\text{cov}}}} - 1\right)\boldsymbol{A}\boldsymbol{x}_{\text{step}}\boldsymbol{x}_{\text{step}}^T$$

Because this update rule does not work with an evolution path (see [5]), the covariance adaptation usually slows down in terms of the number of generations needed to learn the metric of the underlying problem [9]. However, how strong this effect is depends on the optimization problem.

The algorithmic description of the $(\mu{+}\mu)$ is obtained from Algorithm 1 using lines 5b and 10b instead of 5a and 10a. The replacement allows for a simple implementation of the $(\mu{+}\mu)$, because it avoids the otherwise necessary matrix decomposition of the standard generational MO-CMA-ES.

4.2 Steady-State Selection

In steady-state selection schemes only one offspring is generated per generation. Here, we consider two different variants of steady-state selection. The first one, denoted by $(\mu_{\prec}{+}1)$ in the remainder of this article, selects the parent among the non-dominated solutions in the population. As the dominated solutions in the parent population do not influence the evolutionary dynamics, they can be removed from the population. Thus, this variant can be viewed as an evolutionary algorithm with adaptive population size, where the number of individuals equals the number of non-dominated solutions upper bounded by μ.

The second steady-state algorithm, denoted by $(\mu{+}1)$ in the following, considers all μ members of the population as potential parents and hence is less greedy than the first variant. This corresponds to the selection scheme used in [7,8].

Both variants are described in Algorithm 2 and are quite similar to the generational MO-CMA. The main difference is the selection of the parent for reproduction in line 4a for the $(\mu_{\prec}+1)$ and in line 4b for the $(\mu+1)$, respectively.

Algorithm 2. steady-state MO-CMA

1 $g = 0$, initialize $a_k^{(g)}$ for $k = 1, \ldots, \mu$

2 **repeat**

3 $\quad Q^{(g)} = \left\{ a_k^{(g)} \mid 1 \leq k \leq \mu \right\}$

4a $\quad i \leftarrow \mathcal{U}(1, |\operatorname{ndom}(Q^{(g)})|)$

4b $\quad i \leftarrow \mathcal{U}(1, |Q^{(g)}|)$

5 $\quad a^{(g+1)} \leftarrow Q^{(g)}_{\prec:i}$

6 $\quad a'^{(g+1)} \leftarrow a^{(g+1)}$

7 $\quad \boldsymbol{x}'^{(g+1)} \sim \mathcal{N}\left(\boldsymbol{x}^{(g+1)}, {\sigma^{(g)}}^2 \boldsymbol{C}^{(g)}\right)$

8 $\quad Q^{(g)} \leftarrow Q^{(g)} \cup \{a'^{(g+1)}\}$

9 $\quad$ `updateStepSize` $\left(a^{(g)}, \mathbb{1}[a'^{(g)} \prec_{Q^{(g)}} a^{(g)}]\right)$

10 $\quad$ `updateStepSize` $\left(a'^{(g+1)}, \mathbb{1}[a'^{(g)} \prec_{Q^{(g)}} a^{(g)}]\right)$

11 $\quad$ `updateCovariance` $\left(a'^{(g+1)}, \frac{\boldsymbol{x}'^{(g+1)} - \boldsymbol{x}^{(g)}}{\sigma^{(g)}}\right)$

12 $\quad$ **for** $i = 1, \ldots, \mu$ **do** $a_i^{(g+1)} \leftarrow Q^{(g)}_{\prec:i}$

13 $\quad g \leftarrow g + 1$

until stopping criterion is met

5 Experiments

In the following, we empirically evaluate the different variants of the MO-CMA-ES presented in the previous section. In [9] we compared the generational MO-CMA-ES with other multi-objective evolutionary algorithms, namely NSGA-II and the multi-criteria differential evolution algorithm NSDE [16]. Because of the good performance of the MO-CMA-ES in [9], we do not consider any other reference algorithm in the present study.

5.1 Evaluating the Performance of MOO Algorithms

Many ways of measuring the performance of MOO algorithms have been proposed. Here we follow recommendations in [17] and use unary quality indicators, for a detailed description of the methods we refer to [18,17,5].

An unary quality indicator assigns a real valued quality to a set of solutions. Here, the hypervolume indicator [13] and the ϵ-indicator [18] are measured. We use the performance assessment tools contributed to the PISA [19] software package with standard parameters. The hypervolume indicator w.r.t. reference set A_{ref} (see below) is defined as $\mathcal{I}_{\mathcal{S}, A_{\text{ref}}}(A) = \mathcal{S}_{a_{\text{ref}}}(A_{\text{ref}}) - \mathcal{S}_{a_{\text{ref}}}(A)$ where a_{ref} denotes a (hypothetical) reference point having in each objective an objective function value worse than all considered individuals. The additive unary ϵ-indicator $\mathcal{I}_{\epsilon,A_{\text{ref}}}$ w.r.t. reference set A_{ref} is defined as the smallest offset by which the fitness values of the elements in A have to be shifted such that the resulting set dominates A_{ref}. Both a small $\mathcal{I}_{\mathcal{S},A_{\text{ref}}}$ and a small $\mathcal{I}_{\epsilon,A_{\text{ref}}}$ are preferable.

Before the performance indicators are computed, the data are normalized. We want to compare k algorithms on a particular optimization problem after g_1 and g_2 fitness evaluations (here, 25000 and 50000) and we assume that we have conducted t trials. We consider the non-dominated individuals of the union of all $2 \cdot k \cdot t$ populations after g_1 and g_2 evaluations. These individuals make up the reference set A_{ref}. Their objective vectors are normalized such that for every objective the smallest and largest objective function value are mapped to 1 and 2, respectively, by an affine transformation. The mapping to $[1, 2]^M$ is fixed and applied to all objective vectors under consideration. The reference point a_{ref} is chosen to have an objective value of 2.1 in each objective. Note that the set A_{ref} is comprised of rather well performing individuals, whereas the point a_{ref} has bad objective function values.

5.2 Benchmark Functions

We consider three groups of test functions. The first group comprises six common benchmark problems taken from the literature, namely the function FON proposed in [20] and the test functions ZDT1, ZDT2, ZDT3, ZDT4, and ZDT6 proposed in [21]. All functions have box constraints also given in the table. As most components of the optimal solution lie on the boundary of these box constraints, we question the general relevance of these test functions. In accordance with [22,23], we believe that "rotated" functions, which are less aligned with the coordinate system of the search space, are more appropriate. This led to the definition of the two other groups of benchmark functions, see [5] for details.

Table 1. Unconstrained benchmark problems to be minimized, with $a = 1000$, $b = 100$, $\boldsymbol{y} = \boldsymbol{O}_1\boldsymbol{x}$, and $\boldsymbol{z} = \boldsymbol{O}_2\boldsymbol{x}$, where $\boldsymbol{O}_1$ and $\boldsymbol{O}_2$ are orthogonal matrices

Problem	n	Initial region	Objective functions
ELLI_1	10	$[-10, 10]$	$f_1(\boldsymbol{y}) = \frac{1}{a^2 n} \sum_{i=1}^{n} a^{2\frac{i-1}{n-1}} y_i^2$
			$f_2(\boldsymbol{y}) = \frac{1}{a^2 n} \sum_{i=1}^{n} a^{2\frac{i-1}{n-1}} (y_i - 2)^2$
ELLI_2	10	$[-10, 10]$	$f_1(\boldsymbol{y}) = \frac{1}{a^2 n} \sum_{i=1}^{n} a^{2\frac{i-1}{n-1}} y_i^2$
			$f_2(\boldsymbol{z}) = \frac{1}{a^2 n} \sum_{i=1}^{n} a^{2\frac{i-1}{n-1}} (z_i - 2)^2$
CIGTAB_1	10	$[-10, 10]$	$f_1(\boldsymbol{y}) = \frac{1}{a^2 n} \left[y_1^2 + \sum_{i=2}^{n-1} a y_i^2 + a^2 y_n^2\right]$
			$f_2(\boldsymbol{y}) = \frac{1}{a^2 n} \left[(y_1 - 2)^2 + \sum_{i=2}^{n-1} a\,(y_i - 2)^2 + a^2 (y_n - 2)^2\right]$
CIGTAB_2	10	$[-10, 10]$	$f_1(\boldsymbol{y}) = \frac{1}{a^2 n} \left[y_1^2 + \sum_{i=2}^{n-1} a y_i^2 + a^2 y_n^2\right]$
			$f_2(\boldsymbol{z}) = \frac{1}{a^2 n} \left[(z_1 - 2)^2 + \sum_{i=2}^{n-1} a\,(z_i - 2)^2 + a^2 (z_n - 2)^2\right]$

The second group of benchmarks are functions where for each objective the objective function is quadratic (a quadratic approximation close to a local optimum is reasonable for any smooth enough fitness function), see Table 1. They are of the general form $f_m(\boldsymbol{x}) = \boldsymbol{x}^T\boldsymbol{Q}\boldsymbol{x} = \boldsymbol{x}^T\boldsymbol{O}_m^T\boldsymbol{A}\boldsymbol{O}_m\boldsymbol{x}$, where $\boldsymbol{x} \in \mathbb{R}^n, \boldsymbol{Q}, \boldsymbol{O}_m, \boldsymbol{A} \in \mathbb{R}^{n\times n}$ with $\boldsymbol{O}_m$ orthogonal and $\boldsymbol{A}$ diagonal and positive definite. There are two types of functions, ELLI and CIGTAB, which differ in the eigenvalue spectrum of $\boldsymbol{Q}$. In each optimization run the coordinate system of the objective functions is changed by a random choice of $\boldsymbol{O}_m$ (see [9] for details). In the case of the test functions ELLI_1 and CIGTAB_1 the same rotation is used for both objective functions (i.e., $\boldsymbol{O}_1 = \boldsymbol{O}_2$). In the more general case of ELLI_2 and CIGTAB_2 two independent rotation matrices $\boldsymbol{O}_1$ and $\boldsymbol{O}_2$ are generated, which are applied to the first and second objective function, respectively.

Table 2. New benchmark problems to be minimized, $\boldsymbol{y} = \boldsymbol{O}\boldsymbol{x}$, where $\boldsymbol{O} \in \mathbb{R}^{n\times n}$ is an orthogonal matrix, and $y_{\max} = 1/\max_j(|o_{1j}|)$. In the case of ZDT4', $o_{1j} = o_{j1} = 0$ for $1 < j \leq n$ and $o_{11} = 1$. The auxiliary functions are defined as $h : \mathbb{R} \to [0,1], x \mapsto \left(1 + \exp\left(\frac{-x}{\sqrt{n}}\right)\right)^{-1}$, $h_f : \mathbb{R} \to \mathbb{R}, x \mapsto \begin{cases} x & \text{if } |y_1| \leq y_{\max} \\ 1 + |y_1| & \text{otherwise} \end{cases}$, and $h_g : \mathbb{R} \to \mathbb{R}_0^+, x \mapsto \frac{x^2}{|x|+0.1}$.

Problem	n	Variable bounds	Objective function
ZDT4'	10	$x_1 \in [0,1]$	$f_1(\boldsymbol{x}) = x_1$
		$x_i \in [-5,5]$	$f_2(\boldsymbol{x}) = g(\boldsymbol{y})\left[1 - \sqrt{x_1/g(\boldsymbol{y})}\right]$
		$i = 2,\ldots n$	$g(\boldsymbol{y}) = 1 + 10(n-1) + \sum_{i=2}^{n}\left[y_i^2 - 10\cos(4\pi y_i)\right]$
IHR1	10	$[-1,1]$	$f_1(\boldsymbol{x}) = \|y_1\|$
			$f_2(\boldsymbol{x}) = g(\boldsymbol{y})\, h_f\!\left(1 - \sqrt{h(y_1)/g(\boldsymbol{y})}\right)$
			$g(\boldsymbol{y}) = 1 + 9\left(\sum_{i=2}^{n} h_g(y_i)\right)/(n-1)$
IHR2	10	$[-1,1]$	$f_1(\boldsymbol{x}) = \|y_1\|$
			$f_2(\boldsymbol{x}) = g(\boldsymbol{y})\, h_f(1 - (y_1/g(\boldsymbol{y}))^2)$
			$g(\boldsymbol{y}) = 1 + 9\left(\sum_{i=2}^{n} h_g(y_i)\right)/(n-1)$
IHR3	10	$[-1,1]$	$f_1(\boldsymbol{x}) = \|y_1\|$
			$f_2(\boldsymbol{x}) = g(\boldsymbol{y})\, h_f\!\left(1 - \sqrt{h(y_1)/g(\boldsymbol{y})} - \frac{h(y_1)}{g(\boldsymbol{y})}\sin(10\pi y_1)\right)$
			$g(\boldsymbol{y}) = 1 + 9\left(\sum_{i=2}^{n} h_g(y_i)\right)/(n-1)$
IHR4	10	$[-5,5]$	$f_1(\boldsymbol{x}) = \|y_1\|$
			$f_2(\boldsymbol{x}) = g(\boldsymbol{y})\, h_f\!\left(1 - \sqrt{h(y_1)/g(\boldsymbol{y})}\right)$
			$g(\boldsymbol{y}) = 1 + 10(n-1) + \sum_{i=2}^{n}\left[y_i^2 - 10\cos(4\pi y_i)\right]$
IHR6	10	$[-1,1]$	$f_1(\boldsymbol{x}) = 1 - \exp(-4\,\|y_1\|)\sin^6(6\pi y_1)$
			$f_2(\boldsymbol{x}) = g(\boldsymbol{y})\, h_f(1 - (f_1(\boldsymbol{x})/g(\boldsymbol{y}))^2)$
			$g(\boldsymbol{y}) = 1 + 9\left[\left(\sum_{i=2}^{n} h_g(y_i)\right)/(n-1)\right]^{0.25}$

The third group of problems shown in Table 2 are new benchmarks that generalize the ZDT problems to allow a rotation of the search space as in the second group. In the first function ZDT4' the rotation is applied to all but the first coordinate. That is, we consider $\boldsymbol{y} = \boldsymbol{O}\boldsymbol{x}$, where $\boldsymbol{O} \in \mathbb{R}^{n \times n}$ is an orthogonal matrix with $o_{1j} = o_{j1} = 0$ for $1 < j \leq n$ and $o_{11} = 1$. In the other functions the rotation matrices are not restricted. Compared to the ZDT functions, the search space is expanded and the Pareto front is not completely located on the boundaries anymore. The lower end $y_1 = 0$ of the Pareto front is induced by the absolute value in the definition of f_1. The ends $y_1 = \pm y_{\max}$ of the Pareto front are determined by h_f. The value $y_{\max}$ can be chosen between 1 and $1/\max_j(|o_{1j}|)$, and in the latter case the Pareto optimal solution $y_1 = y_{\max}$ lies on the search space boundary. The function $h : \mathbb{R} \to [0, 1]$ is monotonic and emulates the original variable boundary $x_1 \in [0, 1]$. Similar, the function $h_g : \mathbb{R} \to \mathbb{R}_0^+$ emulates the original lower variable boundary of $x_i \geq 0$ for $i = 2, \ldots, n$.

Table 3. Median results over 100 trials on standard benchmark functions after 25000 and 50000 evaluations, respectively. Superscripts indicate significant differences; I: $(\mu+\mu)$, …, IV: $(\mu+1)$, two-sided Wilcoxon rank sum test, normal font $p < 0.001$, slanted $p < 0.01$.

hypervolume indicator						
algorithm	FON	ZDT1	ZDT2	ZDT3	ZDT4	ZDT6
			25000 evaluations			
$(\mu+\mu)$	0.00480^{II}	0.00377	0.00483	0.00139	0.17444^{III}	0.00052
$(\mu+\mu)_{\mathrm{chol}}$	0.00482	0.00377	0.00484	0.00140	0.16218^{III}	0.00052
$(\mu_{\prec}+1)$	$0.00448^{\mathrm{I,II},\mathit{IV}}$	$0.00357^{\mathrm{I,II,IV}}$	$0.00451^{\mathrm{I,II,IV}}$	$0.00129^{\mathrm{I,II,IV}}$	0.39748	$0.00050^{\mathrm{I,II}}$
$(\mu+1)$	$0.00448^{\mathrm{I,II}}$	$0.00363^{\mathrm{I,II}}$	$0.00466^{\mathrm{I,II}}$	0.00137	0.17113^{III}	$0.00050^{\mathrm{I,II}}$
			50000 evaluations			
$(\mu+\mu)$	0.00473^{II}	0.00365	0.00472	0.00132	0.12979^{III}	0.00052
$(\mu+\mu)_{\mathrm{chol}}$	0.00476	0.00365	0.00473	0.00134	0.13068^{III}	0.00052
$(\mu_{\prec}+1)$	$0.00448^{\mathrm{I,II}}$	$0.00349^{\mathrm{I,II}}$	$0.00445^{\mathrm{I,II,IV}}$	$0.00125^{\mathrm{I,II},\mathit{IV}}$	0.35486	$0.00050^{\mathrm{I,II}}$
$(\mu+1)$	$0.00448^{\mathrm{I,II}}$	$0.00350^{\mathrm{I,II}}$	$0.00452^{\mathrm{I,II}}$	$0.00129^{\mathit{I},\mathrm{II}}$	0.15571^{III}	$0.00050^{\mathrm{I,II}}$
			ϵ-indicator			
algorithm	FON	ZDT1	ZDT2	ZDT3	ZDT4	ZDT6
			25000 evaluations			
$(\mu+\mu)$	0.00698	0.00624	0.00707	0.00345	0.16344^{III}	0.00147
$(\mu+\mu)_{\mathrm{chol}}$	0.00695	0.00615	0.00703	0.00342	0.15132^{III}	0.00149
$(\mu_{\prec}+1)$	$0.00497^{\mathrm{I,II}}$	$0.00491^{\mathrm{I,II}}$	$0.00541^{\mathrm{I,II,IV}}$	$0.00285^{\mathrm{I,II,IV}}$	0.35884	$0.00106^{\mathrm{I,II}}$
$(\mu+1)$	$0.00501^{\mathrm{I,II}}$	$0.00501^{\mathrm{I,II}}$	$0.00569^{\mathrm{I,II}}$	$0.00306^{\mathrm{I,II}}$	0.15800^{III}	$0.00106^{\mathrm{I,II}}$
			50000 evaluations			
$(\mu+\mu)$	0.00694	0.00607	0.00699	0.00354	0.13596^{III}	0.00149
$(\mu+\mu)_{\mathrm{chol}}$	0.00690	0.00613	0.00697	0.00355	0.12269^{III}	0.00150
$(\mu_{\prec}+1)$	$0.00487^{\mathrm{I,II}}$	$0.00460^{\mathrm{I,II}}$	$0.00514^{\mathrm{I,II},\mathit{IV}}$	$0.00269^{\mathrm{I,II}}$	0.33336	$0.00101^{\mathrm{I,II}}$
$(\mu+1)$	$0.00487^{\mathrm{I,II}}$	$0.00460^{\mathrm{I,II}}$	$0.00526^{\mathrm{I,II}}$	$0.00273^{\mathrm{I,II}}$	0.14155^{III}	$0.00101^{\mathrm{I,II}}$

5.3 Experiments

We used the same parameters in the MO-CMA-ES as in [9]. For the functions FON, ZDT1, ZDT2, ZDT3, ZDT4, and ZDT6 we set $\sigma^{(0)}$ equal to 60 % of the feasible region $x_2^u - x_2^l$ (we rescaled the first component of ZDT4 to $[-5, 5]$). In the unconstrained problems, Table 1, we set $\sigma^{(0)}$ equal to 60 % of the initialization range of one component. In all algorithms the population size μ was set to 100. For each test problem, 100 trials were conducted per algorithm.

6 Results and Discussion

The results are summarized in Tables 3 to 5. In general, the MO-CMA with Cholesky update performed worse than the MO-CMA using an evolution path, although the differences are often not significant. After 50000 evaluations the methods differ significantly in at least one indicator on FON, ELLI_2, CIGTAB_1, CIGTAB_2, IHR1, and IHR6. On the multi-modal problems, where the results are dominated by the global search performance, the $(\mu+\mu)_{\mathrm{chol}}$ results look slightly better than those with $(\mu+\mu)$, while the differences are not apparent in our

Table 4. Median results over 100 trials on rotated quadratic benchmark functions after 25000 evaluations and 50000 evaluations, respectively

		hypervolume indicator		
algorithm	ELLI_1	ELLI_2	CIGTAB_1	CIGTAB_2
		25000 evaluations		
$(\mu+\mu)$	0.003931	0.000037	0.003466	0.000037
$(\mu+\mu)_{\mathrm{chol}}$	0.004050	0.000037	0.003486	0.000042
$(\mu_{\prec}+1)$	$\underline{0.003771}^{II}$	$\underline{0.000018}^{\mathrm{I,II,IV}}$	$\underline{0.003092}^{\mathrm{I,II,IV}}$	$\underline{0.000015}^{\mathrm{I,II,IV}}$
$(\mu+1)$	0.003963	0.000039	$0.003132^{\mathrm{I,II}}$	$0.000031^{I,\mathrm{II}}$
		50000 evaluations		
$(\mu+\mu)$	$\underline{0.003468}^{IV}$	0.000012^{II}	0.003382^{II}	0.000004
$(\mu+\mu)_{\mathrm{chol}}$	0.003611	0.000019	0.003400	0.000004
$(\mu_{\prec}+1)$	0.003575	$\underline{0.000005}^{\mathrm{I,II,IV}}$	$\underline{0.003068}^{\mathrm{I,II}}$	$\underline{0.000002}^{\mathrm{I,II,IV}}$
$(\mu+1)$	0.003592	0.000012^{II}	$0.003077^{\mathrm{I,II}}$	0.000004
		ϵ-indicator		
algorithm	ELLI_1	ELLI_2	CIGTAB_1	CIGTAB_2
		25000 evaluations		
$(\mu+\mu)$	0.006015	0.000120	0.005835	0.000196
$(\mu+\mu)_{\mathrm{chol}}$	0.005981	0.000134	0.005897	0.000214
$(\mu_{\prec}+1)$	$\underline{0.004717}^{\mathrm{I,II,IV}}$	$\underline{0.000060}^{\mathrm{I,II,IV}}$	$\underline{0.004294}^{\mathrm{I,II,IV}}$	$\underline{0.000145}^{\mathrm{I,II,IV}}$
$(\mu+1)$	$0.005191^{\mathrm{I,II}}$	0.000122	$0.004463^{\mathrm{I,II}}$	$0.000185^{\mathrm{I,II}}$
		50000 evaluations		
$(\mu+\mu)$	0.005742	0.000056^{II}	0.005779	0.000149^{II}
$(\mu+\mu)_{\mathrm{chol}}$	0.005823	0.000073	0.005759	0.000152
$(\mu_{\prec}+1)$	$\underline{0.004261}^{\mathrm{I,II}}$	$\underline{0.000045}^{\mathrm{I,II,IV}}$	$\underline{0.004030}^{\mathrm{I,II}}$	$\underline{0.000142}^{\mathrm{I,II,IV}}$
$(\mu+1)$	$0.004313^{\mathrm{I,II}}$	0.000047^{II}	$0.004030^{\mathrm{I,II}}$	0.000148^{II}

Table 5. Median results over 100 trials on new rotated benchmark functions after 25000 evaluations and 50000 evaluations, respectively

			hypervolume indicator			
algorithm	ZDT4'	IHR1	IHR2	IHR3	IHR4	IHR6
			25000 evaluations			
$(\mu+\mu)$	$\underline{0.18487}^{\mathrm{III}}$	0.00750	0.04023	0.02678	$\underline{0.00484}^{\mathrm{III}}$	0.17635
$(\mu+\mu)_{\mathrm{chol}}$	0.19488^{III}	0.00759	0.03960	0.02686	0.00496^{III}	0.18078
$(\mu_{\prec}+1)$	0.48206	$\underline{0.00119}^{\mathrm{I,II,IV}}$	$\underline{0.03877}^{\mathrm{I,II,IV}}$	$\underline{0.02634}^{IV}$	0.01753	$\underline{0.03198}^{\mathrm{I,II,IV}}$
$(\mu+1)$	0.21022^{III}	0.00813	$0.03927^{\mathrm{I,II}}$	0.02654	0.00521^{III}	$0.13856^{\mathrm{I,II}}$
			50000 evaluations			
$(\mu+\mu)$	$\underline{0.14438}^{\mathrm{III},IV}$	0.00161^{II}	0.03799	0.02633	$0.00415^{\mathrm{III},IV}$	0.03522^{II}
$(\mu+\mu)_{\mathrm{chol}}$	0.16716^{III}	0.00658	0.03789^{I}	0.02633	$\underline{0.00402}^{\mathrm{III},IV}$	0.03928
$(\mu_{\prec}+1)$	0.42775	$\underline{0.00082}^{\mathrm{I,II,IV}}$	$\underline{0.03785}^{\mathrm{I,II}}$	$\underline{0.02633}^{IV}$	0.01746	$\underline{0.02749}^{\mathrm{I,II}}$
$(\mu+1)$	0.18228^{III}	$0.00115^{\mathrm{I,II}}$	$0.03787^{I,II}$	0.02633	0.00501^{III}	$0.03034^{\mathrm{I,II}}$
			ϵ-indicator			
algorithm	ZDT4'	IHR1	IHR2	IHR3	IHR4	IHR6
			25000 evaluations			
$(\mu+\mu)$	$\underline{0.18533}^{\mathrm{III}}$	0.01440	0.14304	0.04360	$\underline{0.00526}^{\mathrm{III}}$	0.18192
$(\mu+\mu)_{\mathrm{chol}}$	0.19352^{III}	0.01446	0.14269	0.04367	0.00533^{III}	0.18782
$(\mu_{\prec}+1)$	0.45294	$\underline{0.00444}^{\mathrm{I,II,IV}}$	$\underline{0.14268}^{\mathrm{I,II,IV}}$	$\underline{0.04321}^{IV}$	0.01666	$\underline{0.06542}^{\mathrm{I,II,IV}}$
$(\mu+1)$	0.20867^{III}	0.01515	$0.14284^{\mathrm{I,II}}$	0.04340	0.00537^{III}	$0.15154^{\mathrm{I,II}}$
			50000 evaluations			
$(\mu+\mu)$	$\underline{0.14594}^{\mathrm{III},IV}$	0.00553^{II}	0.14066	0.04320	$\underline{0.00393}^{\mathrm{III},IV}$	0.05598^{II}
$(\mu+\mu)_{\mathrm{chol}}$	0.16683^{III}	0.01345	0.14046	0.04320	0.00408^{III}	0.06493
$(\mu_{\prec}+1)$	0.37796	$\underline{0.00362}^{\mathrm{I,II,IV}}$	$\underline{0.14043}^{I,II}$	$\underline{0.04320}$	0.01665	0.06490
$(\mu+1)$	0.17945^{III}	$0.00446^{\mathrm{I,II}}$	0.14046	0.04320	0.00474^{III}	$\underline{0.05340}^{\mathrm{II}}$

statistics. After 25000 evaluations, where the covariance matrix adaptation did not pay off yet, the differences are not significant, except for the FON function, where covariance matrix adaptation is faster due to the low dimensionality. These results are in accordance with those in [9].

The newly developed covariance matrix update rule reduces the computational complexity of the rank-one covariance matrix adaptation from $O(n^3)$ to $O(n^2)$. This is a significant improvement on high dimensional, but fast computable fitness functions. However, in practice it is not necessary to perform the covariance matrix decomposition, as required in the original covariance matrix adaptation, each generation, but only every τ generations. Then the computational complexity becomes $O(n^3/\tau + n^2)$. For $\tau = o(n)$ the Cholesky approach is still faster for large n, while $\tau = \omega(n)$ is not advisable. Apart from that, the new update rule is much simpler to implement (e.g., allowing for easy implementations in hardware and in low level programming languages) and is completely specified without any hidden, numerically involved procedures such as a singular value decomposition.

On the unimodal problems, the steady-state algorithms perform better than the generational MO-CMA-ES. Here the greedy steady-state ($\mu_{\prec}$+1)-MO-CMA-ES performs best. But on the multi-modal problems, the generational algorithms are superior. However, the (μ+1)-MO-CMA-ES is not significantly worse, whereas the performance of the greedy ($\mu_{\prec}$+1)-MO-CMA-ES is so bad that it should not be considered as an alternative to the generational MO-CMA despite its good performance on the other test problems.

Thus, we recommend the (μ+1)-MO-CMA-ES. The selection strategy of this variant is equal to the strategy in the SMS-EMOA proposed in [7,8], only the variation operators and the strategy adaptation differ between SMS-EMOA and (μ+1)-MO-CMA.

References

1. Rechenberg, I.: Evolutionsstrategie: Optimierung Technischer Systeme nach Prinzipien der Biologischen Evolution. Werkstatt Bionik und Evolutionstechnik. Frommann-Holzboog, Stuttgart (1973)
2. Schwefel, H.P.: Evolution and Optimum Seeking. Sixth-Generation Computer Technology Series. John Wiley & Sons (1995)
3. Hansen, N., Ostermeier, A.: Completely derandomized self-adaptation in evolution strategies. Evolutionary Computation **9** (2001) 159–195
4. Kern, S., Müller, S., Hansen, N., Büche, D., Ocenasek, J., Koumoutsakos, P.: Learning probability distributions in continuous evolutionary algorithms – A comparative review. Natural Computing **3** (2004) 77–112
5. Igel, C., Hansen, N., Roth, S.: Covariance matrix adaptation for multi-objective optimization. Evolutionary Computation (2006) Accepted.
6. Deb, K., Agrawal, S., Pratap, A., Meyarivan, T.: A fast and elitist multiobjective genetic algorithm: NSGA-II. IEEE Transactions on Evolutionary Computation **6** (2002) 182–197
7. Emmerich, M., Beume, N., Naujoks, B.: An EMO algorithm using the hypervolume measure as selection criterion. In Coello Coello, C.A., Zitzler, E., Hernandez Aguirre, A., eds.: Third International Conference on Evolutionary Multi-Criterion Optimization (EMO 2005). Volume 3410 of LNCS., Springer-Verlag (2005) 62–76
8. Beume, N., Naujoks, B., Emmerich, M.: SMS-EMOA: Multiobjective selection based on dominated hypervolume. European Journal of Operational Research (2007) In press.
9. Igel, C., Suttorp, T., Hansen, N.: A computational efficient covariance matrix update and a (1+1)-CMA for evolution strategies. In: Proceedings of the Genetic and Evolutionary Computation Conference (GECCO 2006), ACM Press (2006) 453–460
10. Hansen, N., Müller, S.D., Koumoutsakos, P.: Reducing the time complexity of the derandomized evolution strategy with covariance matrix adaptation (CMA-ES). Evolutionary Computation **11** (2003) 1–18
11. Grewal, M.S., Andrews, A.P.: Kalman Filtering: Theory and Practice. Prentice-Hall (1993)
12. Deb, K.: Multi-Objective Optimization Using Evolutionary Algorithms. Wiley (2001)

13. Zitzler, E., Thiele, L.: Multiobjective optimization using evolutionary algorithms – a comparative case study. In Eiben, A.E., Bäck, T., Schoenauer, M., Schwefel, H.P., eds.: Fifth International Conference on Parallel Problem Solving from Nature (PPSN-V), Springer-Verlag (1998) 292–301
14. Whitley, L.D.: The GENITOR algorithm and selection pressure: Why rank–based allocation of reproductive trials is best. In Schaffer, J.D., ed.: Proceedings of the Third International Conference on Genetic Algorithms (ICGA'89), Morgan Kaufmann Publishers (1989) 116–121
15. Syswerda, G.: Uniform crossover in genetic algorithms. In Schaffer, J.D., ed.: Proceedings of the Third International Conference on Genetic Algorithms (ICGA'89), Morgan Kaufmann Publishers (1989) 2–9
16. Iorio, A., Li, X.: Solving rotated multi-objective optimization problems using differential evolution. In Webb, G.I., Yu, X., eds.: Proceeding of the 17th Joint Australian Conference on Artificial Intelligence. Volume 3339 of LNCS., Spriner-Verlag (2005) 861–872
17. Knowles, J., Thiele, L., Zitzler, E.: A tutorial on the performance assessment of stochastic multiobjective optimizers. TIK-Report 214, Computer Engineering and Networks Laboratory (TIK), Swiss Federal Institute of Technology (ETH) Zurich (2005)
18. Zitzler, E., Thiele, L., Laumanns, M., Fonseca, C.M., Grunert da Fonseca, V.: Performance assesment of multiobjective optimizers: An analysis and review. IEEE Transactions on Evolutionary Computation **7** (2003) 117–132
19. Bleuler, S., Laumanns, M., Thiele, L., Zitzler, E.: PISA – A platform and programming language independent interface for search algorithms. In Fonseca, C.M., Fleming, P.J., Zitzler, E., Deb, K., Thiele, L., eds.: Evolutionary Multi-Criterion Optimization (EMO 2003). Volume 2632 of LNCS., Springer-Verlag (2003) 494 – 508
20. Fonseca, C.M., Fleming, P.J.: Multiobjective optimization and multiple constraint handling with evolutionary algorithms—Part II: Application example. IEEE Transactions on Systems, Man, and Cybernetics, Part A: Systems and Humans **28** (1998) 38–47
21. Zitzler, E., Deb, K., Thiele, L.: Comparison of multiobjective evolutionary algorithms: Empirical results. Evolutionary Computation **8** (2000) 173–195
22. Deb, K., Sinha, A., Kukkonen, S.: Multi-objective test problems, linkages, and evolutionary methodologies. In: Proceedings of the Genetic and Evolutionary Computation Conference (GECCO 2006), ACM Press (2006) 1141–1148
23. Iorio, A.W., Li, X.: Rotated test problems for assessing the performance of multiobjective optimization algorithms. In: Proceedings of the Genetic and Evolutionary Computation Conference (GECCO 2006), ACM Press (2006) 683–690

A Multi-tiered Memetic Multiobjective Evolutionary Algorithm for the Design of Quantum Cascade Lasers*

Mark P. Kleeman[1], Gary B. Lamont[1], Adam Cooney[2], and Thomas R. Nelson[3]

[1] Air Force Institute of Technology, Dept of Electrical and Computer Engineering, Graduate School of Engineering & Management, Wright-Patterson AFB (Dayton) OH, 45433, USA
[2] Air Force Research Laboratory, Materials and Manufacturing Directorate Wright-Patterson AFB (Dayton) OH, 45433, USA
[3] Air Force Research Laboratory, Sensors Directorate Wright-Patterson AFB (Dayton) OH, 45433, USA

Abstract. Recent advances in quantum cascade lasers (QCLs) have enabled their use as (tunable) emission sources for chemical and biological spectroscopy, as well as allowed their demonstration in applications in medical diagnostics and potential homeland security systems. Finding the optimal design solution can be challenging, especially for lasers that operate in the terahertz region. The production process is prohibitive, so an optimization algorithm is needed to find high quality QCL designs. Past research attempts using multiobjective evolutionary algorithms (MOEAs) have found good solutions, but lacked a local search element that could enable them to find more effective solutions. This research looks at two memetic MOEAs that use a neighborhood search. Our baseline memetic MOEA used a simple neighborhood search, which is similar to other MOEA neighborhood searches found in the literature. Alternatively, our innovative multi-tiered memetic MOEA uses problem domain knowledge to change the temporal focus of the neighborhood search based on the generation. It is empirically shown that the multi-tiered memetic MOEA is able to find solutions that dominate the baseline memetic algorithm. Additional experiments suggest that using local search on only non-dominated individuals improves the effectiveness and efficiency of the algorithm versus applying the local search to dominated individuals as well. This research validates the importance of using multiobjective problem (MOP) domain knowledge in order to obtain the best results for a real world solution. It also introduces a new multi-tiered local search procedure that is able to focus the local search on specific critical elements of the problem at different stages in the optimization process.

* The views expressed in this article are those of the authors and do not reflect the official policy of the United States Air Force, Department of Defense, or the United States Government.

S. Obayashi et al. (Eds.): EMO 2007, LNCS 4403, pp. 186–200, 2007.

1 Introduction

In 2000, Zhores Alferov and Herbert Kroemer received a share of the Nobel Prize in Physics for their work in developing a semiconductor laser using interband transitions in a double heterostructure. These types of lasers are now quite common, and can be found in everyday devices such as laser printers, compact disk players, and laser pointers. Unfortunately, these devices can only operate in a limited range of wavelengths, a fundamental limitation imposed by the bandgap of the constituent materials. For wavelengths greater than 2 microns, suitable semiconductor materials have yet to be developed that can enable interband lasing at room temperature [1].

Quantum cascade lasers (QCL) are semiconductor lasers that are not based on the heterostructure design, but on quantum mechanics. In these devices, lasing is based upon intersubband transitions with properties that are tailored through the careful epitaxial growth different semiconductor layers. Therefore, a QCL does not have the same limitations as the traditional double heterostructure laser. As such, QCLs are used in applications where the standard double heterostructure cannot be utilized [2].

This research focuses on developing good QCL designs in the terahertz frequency range. A terahertz QCL can have potential applications in spectroscopy, astronomy, medicine, free-space communication, near-space radar, and possibly chemical/biological detection [3]. Of particular interest is its potential use as a sensor for security purposes, particularly in the realm of homeland security.

Our previous QCL research [4,3] using multiobjective evolutionary algorithms (MOEAs) found valid designs, which had to be manually adjusted in order to find more stable designs. This is a tedious process that can be remedied by adding a local search technique. We initially implemented a neighborhood search into our MOEA, but this failed to provide us with the type of results we were looking for. We then implemented an innovative multi-tiered neighborhood search that utilizes problem domain knowledge. This new local search technique focuses its search on a specific region based on the stage of the algorithm. The results detailed below empirically show that our multi-tiered memetic MOEA is more effective than a memetic MOEA designed without domain knowledge.

This paper also compares the results of implementing the multi-tiered memetic MOEA on only the non-dominated solutions, the top 10 ranked solutions, and on all current population members. The results show that the multi-tiered memetic MOEA works best when the local search is applied to only the current non-dominated solutions.

Section 2 presents the basics of QCL design. Section 3 discusses generic memetic MOEAs and describes the ones created for the QCL problem. Section 4 describes the algorithm that was extended with local search. Section 5 presents an analysis of the results.

2 Quantum Cascade Laser Overview

The first QCL was experimentally realized in 1994 by researchers from Bell Laboratories [5]. A QCL uses only one type of charge carrier, typically negatively charged electrons and not more massive positively charged holes, so it is classified as a unipolar laser. The laser name comes from its operation. It operates using quantum mechanics and a cascading electronic waterfall, hence the name. The semiconductor crystals are grown in such a way that identical energy steps are created for the electrons to cascade down. At each energy step, the electrons emit photons. A normal diode laser can only emit one photon in per cycle where a QCL can emit many more. In fact, a QCL operating at the same wavelength can outperform a diode laser by a factor greater than 1000 in terms of power because of both the cascading effect and its ability to carry large currents [6]. Further, the QCL can can be designed to emit wavelengths over a broad spectrum of frequencies using the same combination of materials in the active region.

Since QCLs have demonstrated near Watt-level output powers at room temperature in the mid-infrared portion of the spectrum, they are ideal candidates to be used as sensors. Many pollutants, explosives, industrial chemicals, and medical substances can only be detected with high accuracy with mid-infrared lasers [7]. Given the wide range of capabilities listed, QCLs can be applied in the environmental, military, security, and medical fields.

Lasers work by controlling the photon emissions of atoms as electrons move from higher energy states to lower ones. The wavelength of a laser is determined by the electrons change in energy state. In the case of a QCL, such energy states are formed through careful design of one or more quantum wells (QWs), which may form a coupled state system. In such a case, the ensemble properties of the coupled QWs can differ from the individual structures. In general, QWs are formed in semiconductors by placing a thin layer of narrow bandgap semiconductor between two potential barriers with a higher energy bandgap. In one stage of a QCL laser an electron is captured and makes a transition from a higher energy state to a lower energy state through the emission of a photon. The electron is subsequently collected and injected into the next stage, performing the same processes, giving rise to similar photon emissions. This is a continuous process and is illustrated in Figure 1. This cascading, which causes emission of photons and in turn lasing, is the attribute of the quantum *cascade* laser that gives it its name, QCL [8]. In addition to these attributes, QCLs are unique because their performance is not directly related to the properties of the specific semiconductor used, but rather is governed by the thickness of the fabricated layer. In essence, this means a QCL is tunable to a desired terahertz frequency.

We are currently investigating QCL designs that have five QWs per period. Banerjee et al [10] have investigated QCL designs that utilize three wells, and Friedrich [11] investigates five well designs, but neither discuss how they developed their design parameters. This research, along with that from [4,3], are the first to describe how QCL designs can be discovered using an MOEA.

The QCL problem domain description utilizes two fitness functions that attempt to model two of the most important properties of a QCL. The first fitness

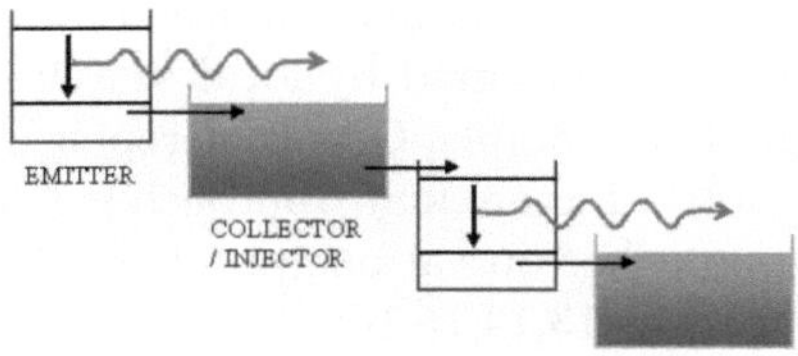

Fig. 1. Cascading Scheme of a QCL [9]

function determines how well the energy levels are lining up. The goal is to have good injection of electrons at the top of each quantum well, but at the same time, have good drainage at the bottom of the well. If a laser has good injection, but poor drainage, then the electrons at the top of the well won't be able to jump to the next energy state since it drains slower than the injection process. The second fitness function determines the overlap ratios. This describes how electrons jump from one state to another. In essence, the fitness function is a measure of how close states are and the ability of the electron to transfer between the states.

3 Memetic MOEAs

GAs typically have difficulty fine-tuning chromosomes that are close to the optimal solution [12]. Memetic MOEAs (also called hybrid MOEAs, genetic local search, and cultural algorithms) are designed as an attempt to find better solutions in these instances. The algorithm is typically a combination of an MOEA and a local search algorithm. By balancing the genetic global search and local search, researchers can improve their results for some problem instances. Permutation problems are one example of where memetic MOEAs have performed well. The landscape of the QCL design problem also suggests that infusing local search into the algorithm would generate better solutions. This section discusses how researchers have integrated local search into their MOEAs and how we integrated local search into our MOEA. Additionally, Table 1 lists a few of the varying approaches that researchers have applied a local search technique to an MOEA.

3.1 Lamarckian vs. Baldwinian

The local search technique used by researchers typically hinges on one of two approaches: Lamarckian search and Baldwinian search. The Lamarckian search technique uses a local search operator on the chromosomes generated by a GA in an effort to find a better solution in the genotype neighborhood. If a better solution is found in a nearby chromosome, that chromosome takes the place of the original. The Baldwinian search technique performs a local search around each chromosome and the best fitness from the local search results is recorded with the chromosome, but the chromosome remains the same. So the fitness value represents search neighborhoods that have the most promise instead of actual good solutions.

Researchers typically apply the Lamarckian search approach for memetic algorithms: If a better solution is found by local search, that chromosome takes the place of the original. For example, GAs that repair chromosomes in an effort to satisfy constraints apply a Lamarckian approach [13]. Several researchers have examined these two approaches on various problems, but there is no consensus on which method is superior [13,14,15,16]

For this research effort, the focus is on finding good solutions with a an approximation model of a QCL so a more accurate model can be used to validate that the solution is good. We use an approximation model because it produces solutions much faster than the accurate model, which must make many compound loops in the code until it reaches a steady state. Using the approximation model allows us to search many more areas in the landscape. Since the goal is finding the best solutions to analyze further, a Lamarckian approach is the most appropriate method to use. With the Lamarckian method, the best solution from the local search is saved, and this is the solution that we analyze in depth with the more accurate model.

3.2 Application of Local Search Approaches

An evolutionary algorithm can generally implement local search in three different ways: after each generation, on the final generation, and after a predefined number of generations. Goel and Deb [17] compared how a purely posterior approach fared against an online (local search after every generation) approach. The methods were run for a fixed number of evaluations using NSGA-II as the guiding MOEA and a bit-wise hill-climbing strategy as the local search technique. The algorithm is used to find the optimal engineering shape design for a cantilever plate and a simply-supported plate. The research showed that the posterior approach found better solutions than the online approach.

The posterior approach is able to generate better diversity and convergence. The results, in [17], show that applying local search after each generation applies too much emphasis on local search and not enough on the evolutionary search.

Based on the results found in [17], this research applies local search after predefined generations throughout the search process. While their research only addresses applying local search techniques at the two extremes (all generations or the last generation only), we take the middle road and apply local search after predetermined intervals. By applying local search at set intervals, good early solutions can be improved multiple times by the algorithm through local search techniques as opposed to only once at the end. Since a stochastic local search method is used in this research, the multiple local searches have a better chance of finding good solutions in a very rugged search landscape.

3.3 Type of Local Search

A researcher can apply several different types of local search techniques to an algorithm. Forms of simulated annealing (SA) and neighborhood search (NS) have been successful local search enhancements to MOEAs[18,19,20,21,22]. Neighborhood search has been successfully applied by numerous researchers in the MOEA community [18,19,20,21]. They found that adding a neighborhood search in the decision space to their MOEA improved their results. Table 1 lists a few of the memetic MOEAs that have been implemented.

Simulated annealing (SA) allows for a larger local search area early in the run and slowly decreases the area as the algorithm continues. Leiva et al [22] compared a simulated annealing approach to a neighborhood search and found that neighborhood search performed better than SA for their problem instance. They based these results on how many nondominated points one algorithm found compared to the other. But their research in no way suggests that neighborhood search should be used in all instances.

The local search (LS) algorithm used in this research is a "multi-tiered" neighborhood search algorithm. The neighborhood search selects an individual from the global search algorithm and uses that individual as a baseline for the neighborhood. The LS algorithm only varies a specified allele in the chromosome. After the results are gathered for the associated neighborhood, another allele is varied. This continues until all specified alleles have been varied. We call this a multi-tiered neighborhood search because the alleles that are varied is dependent upon how many generations the MOEA has run. This type of neighborhood search was selected based our knowledge of the problem domain and landscape. We have found the QCL landscape to be very rugged and noted that certain design parameters affect the solution more than others, specifically the electric field.

Therefore, the experiments in this research only vary the electrical field allele in the earlier runs of the algorithm. So if our algorithm runs for 200 generations, our neighborhood search would only vary the electrical field allele at the 50th and 100th generations. At the 150th and 200th generation, the neighborhood search would vary the barrier wall and width first, and then vary the electrical field. This multi-tiered approach is used because the electrical field has many possible solutions and it can play a major role in finding good solutions. So we focus our earlier searches solely on the electrical field in order to narrow the search region. In the later generations, we fine tune our search by attempting to match up the best physical characteristics with the electrical field. Then we slightly tweak the electrical field in order to find the best solutions in that region. The multi-tiered neighborhood searches are limited to how far they can search from the original allele value.

3.4 How Local Search Is Applied

Researchers have attempted to apply local search to the population of individuals in a variety of ways. Many have applied it to all the individuals in the population [18,23,19,17], while others have applied it to only the non-dominated individuals [22,24]. This research looks at how the these two methods compare. We also compare the results of taking the top 10 ranked individuals (all nondominated individuals and all individuals dominated by 9 points or less). By doing this one is gaining additional insight into their problem domain. For example, if there are individuals who are slightly dominated, they may contain good genetic material in their chromosome that the local search may be able to use and create a new individual that could not be created using a nondominated point.

3.5 Method of Selecting Individuals from Local Search

Researchers predominately use two different methods for selecting individuals from the local search to continue on to the next generation. Many use a weighted vector method [23,25,19,24,17,21], where a weighting is applied to each objective function and the local search attempts to find the best solution that fits that weighting. After a solution is found the weighting is typically changed in a random fashion and the new results are plotted. The goal is to have enough variation in the weighting process to obtain solutions across the Pareto front.

The dominance method selects an new individual only if it is nondominated. There are many ways a researcher can handle the occurrence of multiple nondominated points. They can either add all of them into the population or use some method to determine which individual should be picked. We chose to use the dominance method because we didn't want the possibility of missing good solutions because we picked the wrong weighting for an individual. Plus, we knew that a local search in the solution space can generate diverse points in the objective space. Therefore keep all nondominated solutions that were found. This problem domain typically generates a limited number of nondominated points, so we knew that there wouldn't be too many solutions if we kept all of the solutions.

4 Algorithm Selection

For the QCL problem, the general multi-objective parallel (GENMOP) algorithm was selected because it incorporates some of the major operators of the NSGA-II [27], SPEA2 [28], and GENOCOP [29] algorithms. The algorithm is extended to include both local search procedures - the neighborhood search and the multi-tiered neighborhood search. GENMOP has been applied successfully to a broad range of problems ranging from in-situ bioremediation of contaminated groundwater [30] to solving the aircraft engine maintenance scheduling problem [31]. The initial MOEA was applied to the QCL problem twice before [4,3], but with mixed results. The solutions received in the earlier research obtained good solutions, but they required tedious tweaking in order to find better, more stable solutions. This was due to our algorithm providing a good global search, but inadequate local search. To alleviate this problem, a local search technique was added to the algorithm as indicated.

This section first discusses the particulars of the GENMOP algorithm. Then the two local search algorithms are discussed in more detail.

4.1 GENMOP Description

GENMOP is a Pareto-based algorithm that utilizes real values for crossover and mutation operators. The algorithm also employs fitness sharing through a niche radius and a ranking structure that is similar to the one employed in NSGA-II.

Table 1. Memetic MOEAs

Algorithm	Local Search Used	Where applied	How applied	Method
MSPC-LS1 [22]	Simulated Annealing (SA)	After MOEA	Non-dominated individuals	Dominance
MSPC-LS2 [22]	SA	Each Generation	All individuals	Dominance
MSPC-LS3 [22]	Neighborhood Search (NS)	After MOEA	Non-dominated individuals	Dominance
M-PAES [18]	NS	Continually	All individuals	Dominance (archive)
Thomson EA [20]	NS	Mutation operator	Random individuals	Single Objective
Polar Dominance [26]	Polar dominance	Each generation	All Subpopulations	Dominance
S-MOGLS [23]	NS	Each Generation	All individuals	Weighted vector
C-MOGLS [25]	NS	Each Generation	All individuals	Weighted vector
PGS-WLS [19]	NS	Each Generation	All individuals	Weighted vector
WGS-PLS [19]	NS	Each Generation	All individuals	Dominance
PGS-PLS [19]	NS	Each Generation	All individuals	Dominance
M-NSGA-II [24]	NS	After MOEA	Non-dominated individuals	Weighted vector
M-NSGA-II [17]	NS	Each Generation	All individuals	Weighted vector
EDWA [21]	NS	Each Generation	All individuals	Weighted vector
GENMOP-MTLS	Tiered NS	After x generations	Top 10 ranked individuals	Dominance

For the QCL problem, the individual chromosomes are encoded with values denoting the physical size of the multiple barriers and wells for the cascading region of the semiconductor, as well as the electrical field that is applied to the laser. Auxiliary genes are also associated with the individual chromosomes to define fitness values and Pareto ranking. All GA operations, to include mutation and crossover, are performed solely on the chromosome without interaction from the auxiliary genes.

There are six parameters that the user has the ability to specify with GENMOP (parameter values used in this paper are shown): mutation probability, $p_m = .25$, initial population size, $Pop_0 = 25$, number of generations, $N = 200$, mating pool size, $MP = 10$, the niche radius, $\sigma_{share} = .6$, and save generations.

If no input file is specified to begin GENMOP execution a population of size Pop_0 is randomly initialized. Instead of utilizing a repair function after new individuals are created, all parameters have minimum and maximum values that constrain the chromosome construction. These initial chromosomes are stored in the cumulative population, Pop_{cum}. Each individual within this population is evaluated for its fitness and then these fitnesses are given a Pareto ranking. This rank corresponds to the number of chromosomes that dominate the particular individual. A non-dominated chromosome would hold the Pareto rank of zero. This is the Pareto ranking scheme as developed by Fonseca and Fleming [32].

Selection for the mating pool occurs after the Pareto ranking has terminated. Individuals are selected first based on their Pareto rank. When more individuals are similarly ranked than the spaces left in the mating pool, defined by *MP*, then the equivalence class sharing technique [33] is used to measure crowding within the objective space. Chromosomes from less crowded areas of the objective space are chosen for the mating pool to help preserve diversity within the population.

The GENMOP software flows from the required input to population initialization, through a preliminary evaluation, ranking and normalization of this population. If the maximum number of user specified generations has not been reached, then GENMOP fills the mating pool with individuals from the cumulative population maintaining the highest rank. Crossover and mutation are performed on these individuals, followed by an evaluation. Once all the individuals are returned to the cumulative population, ranking takes place. The children are saved in an output file. If the maximum number of generations are reached, the program terminates and writes all the individuals in the cumulative population to an additional output file. If the maximum number of generations has not been reached, the GA loops back, refills the mating pool, performs crossover and mutation, evaluates the new individuals, places them back in the cumulative population and ranks the whole population. This loop continues until the maximum number of generations is reached and the program is terminated.

Crossover. The entire mating pool is now subjected to crossover and mutation operators developed in GENOCOP [34]. Crossover occurs in one of four ways $\forall\ x_i \in$ the mating pool, where $i = \{1,2,3,\ldots,|MP|\}$. For the first three types of crossover described below a second individual, x_r, is chosen at random from *MP* to be crossed with x_i.

1) *Whole Arithmetical Crossover*: All genes of x_i and x_r are linearly combined to form chromosomes x_1 and x_2. GENMOP retains x_1 and discards x_2.

2) *Simple Crossover*: One gene is chosen in both x_i and x_r and swapped to form chromosomes x_1 and x_2. GENMOP retains x_1 and discards x_2.

3) *Heuristic Crossover*: Individuals x_i and x_r are combined to form one individual $\ni x_1 = R \cdot (x_r - x_i) + x_r$, where R is a uniform random number between zero and one and the rank of $x_r \leq x_i$.

4) *Pool Crossover*: Randomly chooses genes from individuals in the mating pool and combines them to create x_1.

The type of crossover to be performed on the two individuals is chosen based upon an adaptive probability distribution. Each of the four crossover types described above begins with the same probability of being chosen. As the algorithm progresses through generations these probabilities are adapted through the fitness of the individuals they create. If the newly created individual dominates x_i, then the fitness of the newer individual was increased over the previous through use of this particular crossover operator. Consequently, because of the success of the new individual the crossover operator's selection probability increases [34].

Mutation. The new individuals created through a crossover operation are now subject to mutation with a probability defined by the user, p_m. If a number, n is randomly selected from a uniform distribution, so that $0 < n < 1$ and $n < p_m$, then one of three mutation operators described below is chosen to perform on the individual.

1) *Uniform Mutation*: Chooses a gene existing in the chromosome to reset to a random value within its specified ranges.

2) *Boundary Mutation*: Chooses a gene existing in the chromosome to reset to either its maximum or minimum value.

3) *Non-uniform Mutation*: Chooses a gene to modify by some random value decreases probabilistically, until it equals zero, as the generation number approaches the maximum generations.

The mutation operator is selected using the same adaptive probability distribution described previously for crossover operations. Between these two operators a new population is developed, Pop_{new} which is equal to $|MP|$. Each individual in Pop_{new} is evaluated for fitness and then placed in Pop_{cum}. This is an archive that contains the chromosomes created in previous generations.

4.2 Local Search Description

In the research, a baseline local search procedure was added to the algorithm. This initial approach, which is not much different from the neighborhood searches listed in Table 1, was intended to be used as a baseline method for comparison with the innovative multi-tiered MOEA. In this baseline local search procedure, the algorithm explores for allele values in the neighborhood of the initial value. Specifically, the procedure limits its search to the area that is within $\alpha = 0.1$ of the total values that the allele can take on. The algorithm stochastically selects an equal number of neighbors that are above the current allele value and below. The local search is applied at set generations. It is applied after every generation, every 20 generations, every 50 generations, and at the end of the algorithm. This setting is changed to see the effect. The GENMOP-LS algorithm is typically run on only the nondominated points, but it is run once to include some of the best dominated points in order to determine if that method performs better.

4.3 Multi-tiered Local Search Description

The multi-tiered local search addition to GENMOP, nicknamed GENMOP-MLTS, focuses the first search on the electrical field. The neighborhood search in this tier, uses a neighborhood size that is 4% the size of the actual electrical field search space. This neighborhood size was chosen because it balances efficiency (limit the number of fitness evaluations) with effectiveness. This first tier of local search is run after generations 50 and 100. We chose to run the local search at these intervals because it provided the algorithm with a good balance of local search and global search. The goal in this first tier of local search is to focus the attention, and fitness evaluations, on the search region that has the most influence on creating good solutions. Through previous experimentation and problem domain knowledge, we determined that the electrical field had the biggest impact on creating good solutions.

The second tier of GENMOP-MLTS applies the neighborhood search first to the width of the quantum wells and then to the electrical field. The neighborhood for the wells is 20% of the size of the wells. This neighborhood is larger because the actual search area for the quantum wells is much smaller than the electrical field. The electrical field is then varied with a much smaller neighborhood (1%) than used in the first tier. This second tier is run after generations 150 and 200. The second tier of neighborhood searches are more for fine-tuning the solutions that are generated in the earlier portion of the algorithm. So the first tier is used for larger adjustments in the solution and the second tier is used to fine tune our results. This is similar to the principles of simulated annealing, but this approach also directs the search to different regions of the decision space based on what stage the algorithm has reached.

5 Results and Analysis

Each implementation of memetic GENMOP is run 100 times in order to be able to effectively compare the results with previous results found in [4,3]. The Pareto front generated by the best results are compared to the other runs. Each implementation is run for 200 generations and starts with 25 individuals. We again chose these numbers in order to better compare these research results with previous results. To compare GENMOP-LS to the multi-tiered GENMOP-MTLS, each MOEA is run with local search applied every 50 generations (local search applied a total of 4 times during the run). In all instances, the GENMOP-MTLS was able to find high quality solutions (which are designs that can be effectively fabricated into a stable QCL). Figure 2 shows graph comparing GENMOP-LS with GENMOP-MTLS.

The values in the graph are overlap ratios and energy level differences. These are described in further detail in Section 2. The overlap ratios are unitless while the energy level difference is measured in Angstroms. A zero level is considered the best and a one is the worst. We are interested in finding solutions that are at roughly 0.25 or less. Any values that reach 0.1 are highly desirable. As Figure 2 shows, numerous values from both algorithms are considered acceptable solutions. But it is easy to note that all 11 points generated by GENMOP-MTLS were nondominated while only 1 out of 34 point were nondominated using the baseline GENMOP-LS. This figure empirically confirms that the GENMOP-MTLS is more effective at finding higher quality solutions. One disadvantage is

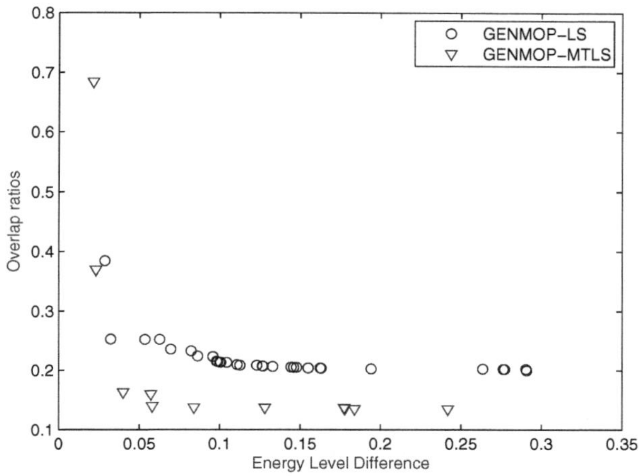

Fig. 2. Comparison of GENMOP-LS with a local search applied every 50 generations and the improved, multi-tiered, GENMOP-MTLS

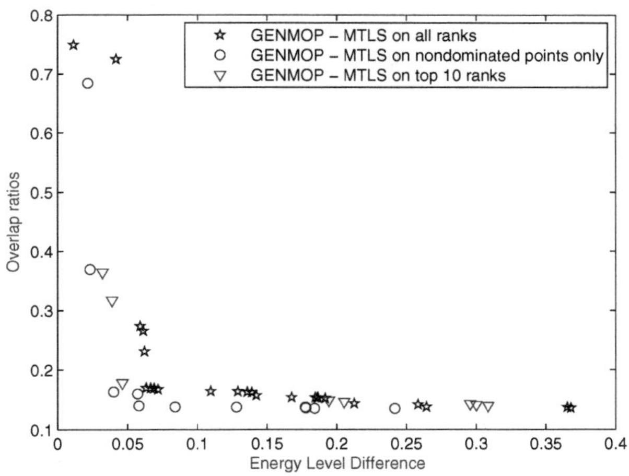

Fig. 3. Comparison of results based on the number of individuals having the multi-tiered memetic GENMOP applied to them

that fewer points are found along the GENMOP-MTLS Pareto front. This is probably due to the landscape, where multiple solutions may lead to the same nondominated point.

Next, we compared the runs that the local search was applied to all individuals, nondominated individuals, and only individuals that were dominated by 10 points or less. These experiments were run in the same fashion as the previous experiments (200 generations, 25 individuals in the initial population, local search applied every 50 generations). Figure 3 shows the results. It appears that for this problem domain, applying GENMOP-MTLS to only the nondominated solutions works the best. The results are all very similar, but applying GENMOP-MTLS to the nondominated points only appears to generate better compromise solutions, while applying it to all ranks appears to do a better job at finding solutions on the edges of the Pareto front. While many of these values fall out of our 0.25 criteria for acceptable solutions, it is an interesting phenomenon that should be investigated in the future.

6 Conclusion

Memetic MOEAs are a key tool for researchers to use to apply more local search to MOP domains. In this paper, we empirically validate that our innovative, multi-tiered memetic MOEA, GENMOP-MTLS, is capable of generating better solutions than a standard memetic MOEA - much like those listed in Table 1. By applying problem domain knowledge, and changing the local search focus in stages, we are able to more effectively generate solutions. This multi-tiered approach can be useful in other problem domains, where several alleles play a major role in determining the fitness of an individual. Various other alleles can be used to "fine tune" the results. Understanding the problem domain search landscape can play a major role in how a memetic algorithm should be implemented.

The authors wish to acknowledge former Air Force Institute of Technology student Traci Keller and Researcher Andrés F. Rodríguez , whose initial work on this problem enabled us to have better insight on the problem domain and enabled us to develop better fitness functions and improve our algorithm.

References

1. Capasso, F., Gmachl, C., Sivco, D.L., Cho, A.Y.: Quantum cascade lasers. Physics Today (2002)
2. Tihov, M.I.: Chemical sensors based on distributed feedback quantum cascade laser for environmental monitoring. Master's thesis, Ecole Polytechnique (2003)
3. Rodríguez, A.F., Keller, T.A., Lamont, G.B., Nelson, T.R.: Using a Multiobjective Evolutionary Algorithm to Develop a Quantum Cascade Laser Operating in the Terahertz Frequency Range. In: 2005 IEEE Congress on Evolutionary Computation (CEC 2005). Volume 1., Edinburgh, Scotland (2005) 9–16
4. Keller, T.A., Lamont, G.B.: Optimization of a Quantum Cascade Laser Operating in the Terahertz Frequency Range Using a Multiobjective Evolutionary Algorithm. In: 17th International Conference on Multiple Criteria Decision Making (MCDM 2004). Volume 1. (2004)
5. Capasso, F., Faist, J., L., S.D., Sirtori, C., L., H.A., Chu, S.N.G., Cho, A.Y.: Quantum cascade laser: A unipolar intersubband semiconductor laser. In: 14th IEEE International Semiconductor Laser Conference, 1994. (1994) 71–72
6. Capasso, F., Sivco, D.L., Cho, A.Y., Gmachl, C.: Quantum cascade lasers. Physics World **12** (1999) 27
7. Ouellette, J.: Quantum cascade lasers turn commercial. The Industrial Physicist **7** (2001) 8–13

8. Menon, V.: Design, fabrication and characterization of quantum cascade terahertz emitters. (2001) PhD Dissertation.
9. Ram-Mohan, L.R.: Quantum Cascade Terahertz Lasers (2003) Private Communications, Air Force Research Labs, Wright Patterson AFB, OH.
10. Banerjee, S., Spencer, P.S., Shore, K.A.: Design of a tunable quantum cascade laser with enhanced optical non-linearities. In: IEE Proceddings on Optoelectronics. Volume 153. (2006) 40–42
11. Friedrich, A., Boehm, G., Amann, M.C.: Low-threshold quantum-cascade lasers without injector regions, emitting at /spl lambda/ /spl sim/ 6.7 /spl mu/m. In: Proceddings of the 2006 International Conference on Indium Phosphide and Related Materials. (2006) 19–22
12. Goldberg, D.E.: Genetic Algorithms in Search, Optimization and Machine Learning. Addison-Wesley Publishing Company, Reading, Massachusetts (1989)
13. Julstrom, B.A.: Comparing Darwinian, Baldwinian, and Lamarckian search in a genetic algorithm for the 4-cycle problem. In Brave, S., Wu, A.S., eds.: Late Breaking Papers at the 1999 Genetic and Evolutionary Computation Conference, Orlando, Florida, USA (1999) 134–138
14. Ishibuchi, H., Kaige, S., Narukawa, K.: Comparison between lamarckian and baldwinian repair on multiobjective 0/1 knapsack problems. In: Proceedings of the Third International Conference onEvolutionary Multi-Criterion Optimization, (EMO 2005) Guanajuato, Mexico, March 9-11. Volume 3410 of Lecture Notes in Computer Science., Springer (2005) 370–385
15. Ku, K.W.C., Mak, M.W.: Exploring the effects of lamarckian and baldwinian learning in evolving recurrent neural networks. In Fogel, D.B., ed.: Proceedings of the Fourth IEEE Conference on Evolutionary Computation (ICEC'97), Piscataway, New Jersey, IEEE Press (1995) 617–621
16. Gruau, F., Whitley, L.D.: Adding learning to the cellular development of neural networks: Evolution and the baldwin effect. Evolutionary Computation **1** (1993) 213–233
17. Goel, T., Deb, K.: Hybrid Methods for Multi-Objective Evolutionary Algorithms. In: Proceedings of the 4th Asia-Pacific Conference on Simulated Evolution and Learning (SEAL'02). Volume 1., Singapore (2002) 188–192
18. Knowles, J., Corne, D.: M-PAES: A Memetic Algorithm for Multiobjective Optimization. In: 2000 Congress on Evolutionary Computation. Volume 1., Piscataway, New Jersey, IEEE Service Center (2000) 325–332
19. Ishibuchi, H., Narukawa, K.: Performance evaluation of simple multiobjective genetic local search algorithms on multiobjective 0/1 knapsack problems. In: 2004 Congress on Evolutionary Computation (CEC'2004). Volume 1., Portland, Oregon, USA, IEEE Service Center (2004) 441–448
20. Thomson, R., Arslan, T.: The evolutionary design and synthesis of non-linear digital vlsi systems. In: EH '03: Proceedings of the 2003 NASA/DoD Conference on Evolvable Hardware, Washington, DC, USA (2003) 125–134
21. Jin, Y., Okabe, T., Sendhoff, B.: Adapting Weighted Aggregation for Multiobjective Evolution Strategies. In Zitzler, E., Deb, K., Thiele, L., Coello, C.A.C., Corne, D., eds.: First International Conference on Evolutionary Multi-Criterion Optimization. Springer-Verlag. Lecture Notes in Computer Science No. 1993 (2001) 96–110
22. Leiva, H.A., Esquivel, S.C., Gallard, R.H.: Multiplicity and local search in evolutionary algorithms to build the pareto front. In: SCCC '00: Proceedings of the XX International Conference of the Chilean Computer Science Society (SCCC'00), Washington, DC, USA, IEEE Computer Society (2000) 7

23. Ishibuchi, H., Murata, T.: Multi-Objective Genetic Local Search Algorithm. In Fukuda, T., Furuhashi, T., eds.: Proceedings of the 1996 International Conference on Evolutionary Computation, Nagoya, Japan, IEEE (1996) 119–124
24. Deb, K., Goel, T.: A Hybrid Multi-Objective Evolutionary Approach to Engineering Shape Design. In Zitzler, E., Deb, K., Thiele, L., Coello, C.A.C., Corne, D., eds.: First International Conference on Evolutionary Multi-Criterion Optimization. Springer-Verlag. Lecture Notes in Computer Science No. 1993 (2001) 385–399
25. Murata, T., Nozawa, H., Tsujimura, Y., Gen, M., Ishibuchi, H.: Effect of Local Search on the Performance of Cellular Multi-Objective Genetic Algorithms for Designing Fuzzy Rule-based Classification Systems. In: Congress on Evolutionary Computation (CEC'2002). Volume 1. (2002) 663–668
26. Sato, H., Aguirre, H.E., Tanaka, K.: Local Dominance Using Polar Coordinates to Enhance Multiobjective Evolutionary Algorithms. In: 2004 Congress on Evolutionary Computation (CEC'2004). Volume 1., Portland, Oregon (2004) 188–195
27. Deb, K., Agrawal, S., Pratab, A., Meyarivan, T.: A Fast Elitist Non-Dominated Sorting Genetic Algorithm for Multi-Objective Optimization: NSGA-II. KanGAL report 200001, Indian Institute of Technology, Kanpur, India (2000)
28. Zitzler, E., Laumanns, M., Thiele, L.: SPEA2: Improving the Strength Pareto Evolutionary Algorithm. In: EUROGEN 2001. Evolutionary Methods for Design, Optimization and Control with Applications to Industrial Problems. (2002) 95–100
29. Michalewicz, Z.: Evolutionary computation techniques for nonlinear programming problems. International Transactions in Operational Research **1** (1994) 175
30. Knarr, M.R., Goltz, M.N., Lamont, G.B., Huang, J.: *In Situ* Bioremediation of Perchlorate-Contaminated Groundwater using a Multi-Objective Parallel Evolutionary Algorithm. In: Congress on Evolutionary Computation (CEC'2003). Volume 1., Piscataway, New Jersey, IEEE Service Center (2003) 1604–1611
31. Kleeman, M.P., Lamont, G.B.: Solving the aircraft engine maintenance scheduling problem using a multi-objective evolutionary algorithm. In: EMO. (2005) 782–796
32. Fleming, P.J., Fonseca, C.M.: Genetic algorihtms for multi-objective optimization: Formulation, discussion and generalization. In: 5th International Congress on Genetic Algorithms (ICGA). (1993) 416–423
33. Horn, J., Nafpliotis, N., Goldberg, D.E.: A niched pareto genetic algoirthm for multiobjective optimization. (1994) 82–87
34. Michalewicz, Z., Janikow, C.Z.: Genocop: a genetic algorithm for numerical optimization problems with linear constraints. Commun. ACM **39** (1996) 223–240

Local Search in Two-Fold EMO Algorithm to Enhance Solution Similarity for Multi-objective Vehicle Routing Problems

Tadahiko Murata[1,2] and Ryota Itai[1]

[1] Department of Informatics, Kansai University
[2] Policy Grid Computing Laboratory, Kansai University
2-1-1 Ryozenji, Takatsuki 569-1095, Osaka, Japan
murata@res.kutc.kansai-u.ac.jp
http://www.res.kutc.kansai-u.ac.jp/~murata/

Abstract. In this paper, we propose a memetic EMO algorithm that enhances the similarity of two sets of non-dominated solutions. We employ our algorithm in vehicle routing problems (VRPs) where the demand of customers varies. We consider two periods of different demand in a problem that are Normal Demand Period (NDP) and High Demand Period (HDP). In each period, we can find a set of non-dominated solutions with respect to several objectives such as minimizing total cost for delivery, minimizing maximum cost, minimizing the number of vehicles, minimizing total delay to the date of delivery and so on. Although a set of non-dominated solutions can be searched independently in each period, drivers of vehicles prefer to have similar routes in NDP and HDP in order to reduce their fatigue to drive on a different route. In this paper, we propose a local search that enhance the similarity of routes in NDP and HDP. Simulation results show that the proposed memetic EMO algorithm can find a similar set of non-dominated solutions in HDP to the one in NDP.

Keywords: memetic algorithm, local search, solution similarity, vehicle routing problem.

1 Introduction

Although we have various approaches in EMO (Evolutionary Multi-criterion Optimization) community [1-3], there are few research works that investigate the similarity among obtained sets of non-dominated solutions. Deb considered topologies of several non-dominated solutions in Chapter 9 of his book [4]. He examined the topologies or structures of three-bar and ten-bar truss. He showed that neighboring non-dominated solutions on the obtained front are under the same topology, and NSGA-II can find the gap between the different topologies. While he considered the similarity of solutions in a single set of non-dominated solutions from a topological point of view, there is no research work relating to EMO that considers the similarity of solutions in different sets of non-dominated solutions. In this paper, we propose a memetic EMO algorithm that enhances the similarity of solutions in different sets of non-dominated solutions.

S. Obayashi et al. (Eds.): EMO 2007, LNCS 4403, pp. 201–215, 2007.

We employ Vehicle Routing Problems (VRPs) to consider the similarity in different sets of solutions. The VRP is a complex combinatorial optimization problem that can be seen as a merge of two well-known problems: Traveling Salesman Problems (TSP) and Bin Packing Problems (BRPs). This problem can be described as follows: Given a fleet of vehicles, a common depot, and several customers scattered geographically. Find the sets of routes for the fleet of vehicles. As for objective functions considered in VRPs, many research works [5-9] on the VRP try to minimize the total route cost that is calculated using the distance or the duration between customers. Among them the research works in [7-9] are related to multi-objective optimization. Tan *et al.* [7] and Saadah *et al.* [8] employed the travel distance and the number of vehicles to be minimized. Chitty and Hernandez [9] tried to minimize the total mean transit time and the total variance in transit time.

In this paper, we employ three objectives. One is to minimize maximum routing time and another is to minimize the number of vehicles in VRPs. It should be noted that we don't employ the total routing time of all the vehicles, but use the maximum routing time among the vehicles. We employed it in order to minimize the active duration of the central depot. Even if the total routing time is minimized, the central depot should be opened until the last vehicle comes back to the depot. In order to minimize the active duration of the central depot, the maximum routing time should be minimized.

As for the third objective, we consider the similarity of solutions. In this paper, we suppose two periods with different demands. One period has a normal demand of customers. The other has a higher demand. We refer the former period and the latter period as Normal Demand Period (NDP) and High Demand Period (HDP), respectively. We define the demand in the HDP as an extended demand of the NDP in this paper. For example, we assume that the demand in the HDP is a demand occurring in a high season such as Christmas season. In that season, the depot may have an extra demand as well as the demand in the normal season. In order to avoid big changes of each route from the depot, a solution (i.e., a set of route) in HDP should be similar to a solution in NDP. This situation requires us to consider the similarity of solutions on different non-dominated solutions in multi-objective VRPs.

In order to find a set of non-dominated solutions in the HDP that is similar to a set of non-dominated solutions in the NDP, we apply a two-fold EMO algorithm [10] to the problem. In a two-fold EMO algorithm, first we find a set of non-dominated solutions for the NDP by an EMO algorithm. In order to enhance the similarity between sets of non-dominated solutions in NDP and HDP, we showed the effectiveness of utilization of a solution set in NDP for population initialization in HDP [10]. In this paper, we propose a local search method in a memetic EMO algorithm to enhance the solution similarity in HDP to a solution set in NDP.

We organize this paper as follows: Section 2 gives the problem model for multi-objective VRPs. We define a measure of the similarity between solutions in Section 3. The outline of our two-fold EMO algorithm is described in Section 4. In section 5, we propose a local search algorithm that enhances the solution similarity introduced in the second phase of the two-fold memetic EMO algorithm. Section 6 describes simulation results that show the effectiveness of the proposed local search algorithm. Conclusions are drawn in Section 7.

2 Multi-objective Vehicle Routing Problems

The domain of VRPs has large variety of problems such as capacitated VRP, multiple depot VRP, periodic VRP, split delivery VRP, stochastic VRP, VRP with backhauls, VRP with pick-up and delivering, VRP with satellite facilities, VRP with time windows and so on. These problems have the basic architecture of the VRP except their own constraints. Those constraints are arisen in practical cases. Please see for the detail of the VRP problem in [11].

A solution of the VRPs is represented by a permutation of N customers, and we split it into M parts as shown in Figure 1. It shows eight customers that are served by three vehicles. The first vehicle denoted v_1 in the figure visits three customers in the order of Customers 1, 2, and 3. Each solution is divided by a closed triangle. Therefore the driving duration for v_1 is calculated by $c_{D,1} + c_{1,2} + c_{2,3} + c_{3,D}$. Figure 2 shows an example of three routes depicted on the map of eight customers and the depot. It should be noted that, we consider only problems with symmetric cost where $c_{1,2} = c_{2,1}$ in this paper.

The objective employed in many VRPs is to minimize a total cost is described as follows:

$$\text{Min. } \sum_{k=1}^{M} c_k \text{ ,} \tag{1}$$

where M is the number of vehicles that start from the depot and is routed by a sequence of customers, then return to the depot. The cost of k-th vehicle is denoted by c_k and described as follows:

$$c_k = c_{D,1} + \sum_{i=1}^{n_k - 1} c_{i,i+1} + c_{n_k,D} \text{ ,} \tag{2}$$

where $c_{i,j}$ means the cost between Customers i and j. Let us denote D as the index for the depot in this paper. Equation (2) indicates the sum of the cost between the

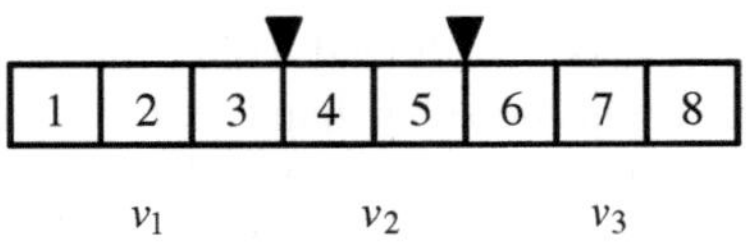

Fig. 1. An example of eight customers visited by three vehicles. Each triangle shows the split between the routes for vehicles.

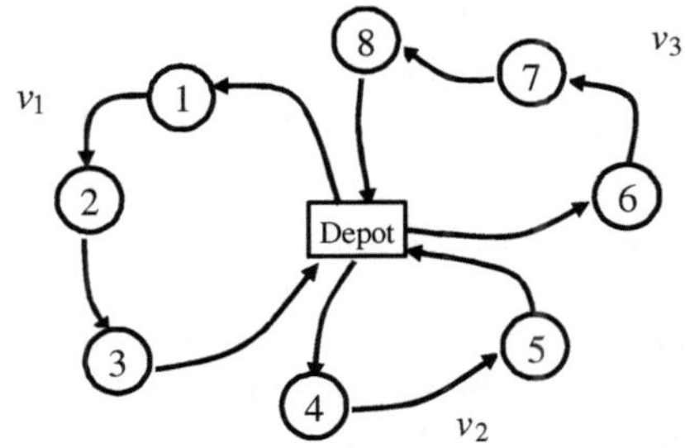

Fig. 2. An example of eight customers visited by three vehicles

depot and the first customer assigned to the k-th vehicle (i.e., $c_{D,1}$), the total cost from the 1st customer to the n_k -th customer (i.e., $\sum_{i=1}^{n_k-1} c_{i,i+1}$), and the cost between the final customer n_k and the depot. Each vehicle is assigned to visit n_k customers, thus we have $N = \sum_{k=1}^{M} n_k$ customers in total. The aim of this VRP is to find a set of sequences of customers that minimizes the total cost. Each customer should be visited exactly once by one vehicle.

While the total cost of all the vehicles is ordinarily employed in the VRP, we employ the maximum cost to be minimized in this paper. When the cost $c_{i,j}$ is related to the driving duration between Customers i and j in Equation (2), the total cost c_k for the k-th vehicle means the driving duration from the starting time from the depot to the returning time to the depot. In order to minimize the activity duration of the depot, the maximum duration of the vehicles should be minimized since the depot should wait until all the vehicles return to the depot. We also consider the minimization of the number of vehicles in our multi-objective VRP. The objectives in this paper can be described as follows:

$$\text{Min. } \max_k c_k , \tag{3}$$

$$\text{Min. } M . \tag{4}$$

When we have a solution with $M = 1$, our problem becomes the traveling salesman problem, the TSP. In that case, the other objective, to minimize the maximum driving duration in Equation (3), becomes just to minimize the total driving duration by one vehicle. On the other hand, the maximum driving duration becomes minimum when the number of vehicles equals to the number of customers (i.e., $M = N$). In that case, each vehicle visits only one customer. The driving duration for each vehicle in (2) can be described as follows:

$$c_k = c_{D,[1]_k} + c_{[1]_k, D} , \tag{5}$$

where $[1]_k$ denotes the index of the customer visited by the k-th vehicle. The maximum driving duration in (5) over M vehicles becomes the optimal value of that objective in the case of $M = N$. Therefore we face the trade off between these two objectives: the minimization of the maximum driving duration and the minimization of the number of vehicles.

We consider two periods with different demands: NDP and HDP. In NDP, a normal demand of customers should be satisfied. On the other hand, extra demands should also be satisfied in HDP. In this paper, we increase the number of customers in HDP. That is, $N_{NDP} < N_{HDP}$, where N_{NDP} and N_{HDP} are the number of customers in NDP and HDP, respectively. We can obtain a set of non-dominated solutions for each problem. We refer a set of non-dominated solutions for NDP as Ψ_{NDP} , and that for HDP as Ψ_{HDP} . These two sets of non-dominated solutions can be obtained by applying one of EMO algorithms such as NSGA-II [12]. But if we apply the algorithm to each of NDP and HDP independently, we can not expect to obtain a set of solutions with similar routes in HDP to that obtained for NDP.

3 Similarity Between Sets of Non-dominated Solutions

We define a similarity measure between a non-dominated solution Ψ_{HDP} obtained for HDP and Ψ_{NDP} for NDP. Since the aim of measuring the similarity is to find a solution in HDP that is similar to one in NDP, we measure the similarity of a solution in HDP to the one in NDP. We measure it by a ratio of the number of the same edges to the number of all edges in a solution of NDP.

We define the similarity of solution x in HDP is as follows:

$$similarity(x) = \max_{y \in \Psi_{NDP}} (similarity(x, y)) = \max_{y \in \Psi_{NDP}} \left(\frac{sames(x, y)}{edges(y)}\right), \ x \in \Psi_{HDP}, \quad (6)$$

where $similarity(x, y)$ is the similarity of the solution x to the solution y, that is calculated by $sames(x, y)$ (i.e., the number of the same edges) and $edges(y)$ (i.e., the number of edges in a solution y). Figure 3 shows an example to calculate the similarity of solution x (5, 1, 2, 3, 6, 4 with four vehicles) to three non-dominated solutions (2, 1, 3 with one vehicle, 1, 2, 3 with two, and 1, 3, 2 with three) obtained for NDP. The similarity of solutions x becomes the maximum similarity 0.8 through the calculation.

4 Two-Fold EMO Algorithm for Multi-objective VRPs

In this section, we show a Two-Fold EMO algorithm for our multi-objective VRPs [10]. Then we show how we apply a two-fold EMO algorithm to obtain a similar set of solutions in NDP and HDP.

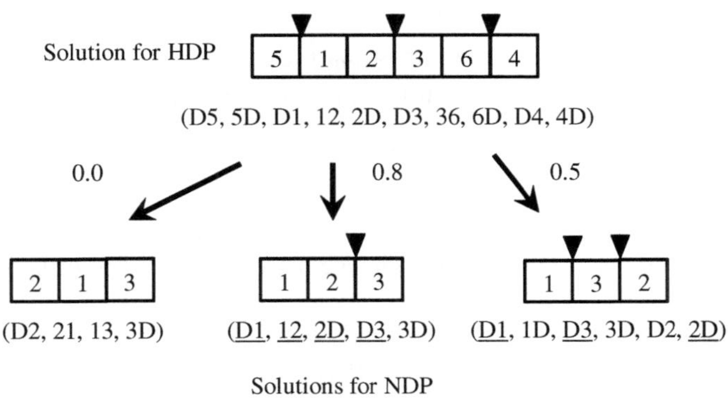

Fig. 3. The similarity of a solution for HDP that is calculated as the maximum similarity among three similarities

4.1 Genetic Operators

[Crossover]

We employ the edge exchange crossover (EXX) [13] as a crossover operator. This crossover produces offspring only by exchanging edges in parents chromosome, where an edge means a segment between two customers. Therefore offspring chromosomes preserve segments between customers well. The following is the algorithm of this crossover:

Step 1) Select an edge randomly from one parent (Parent 1), and let i_1 be the position of the edge. Let i_2 be the position of the edge of the other parent (Parent 2) whose origin customer is the same as that of the i_1-th edge in Parent l.

Step 2) Let j_2 be the position of the edge of Parent 2 whose origin customer is the same as the destination customer of the i_1-th edge in Parent l, and j_1 be the position of the edge of Parent 1 whose origin city is the same as the destination customer of the i_2-th edge in Parent 2.

Step 3) Exchange the i_1-th edge of Parent 1 and the i_2-th edge of Parent 2. If the destination customers of them are the same, terminate the algorithm.

Step 4) Invert the order of the edges and their origin and destination customers of Parent 1 between the positions i_1 and j_1, and those of Parent 2 between the positions i_2 and j_2.

Step 5) Let $i_1 = j_1$ and $i_2 = j_2$ and go to Step 2.

Figure 4 shows the above procedure between the following parents with one vehicle:

Parent 1: (1 2 3 4 5 6 7 8), and Parent 2: (2 5 4 1 6 7 3 8).

Their edges can be represented as follows:

Parent 1: (12 23 34 45 56 67 78 81), and Parent 2: (25 54 51 16 67 73 38 82).

As an example where the edge 23 of Parent l is taken as the starting edge in Step 1 of the above procedure. We have the following offspring after the crossover operation

Offspring 1: (1 2 5 4 3 8 7 6), and Offspring 2: (7 3 2 8 1 4 5 6).

In this paper, we consider any chromosome with multiple vehicles as that with one vehicle. Thus the following cases have the same result in the order of the customers while their positions of Depot do not change between parent and offspring.

Case A: Parent 1: (1 2 | 3 4 5 6 | 7 8), and Parent 2: (2 5 4 1 | 6 7 3 8).
Case B: Parent 1: (1 2 3 | 4 5 6 7 8), and Parent 2: (2 | 5 4 | 1 6 7 | 3 8).

Case A: Offspring 1: (1 2 | 5 4 3 8 | 7 6), and Offspring 2: (7 3 2 8 | 1 4 5 6).
Case B: Offspring 1: (1 2 5 | 4 3 8 7 6), and Offspring 2: (7 | 3 2 | 8 1 4 | 5 6).

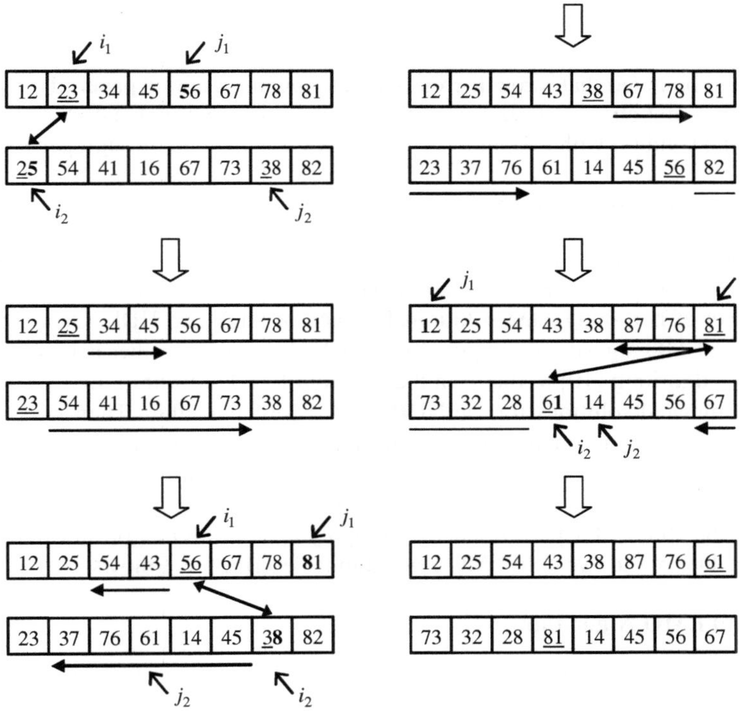

Fig. 4. Examples of Edge Exchange Crossover [13]

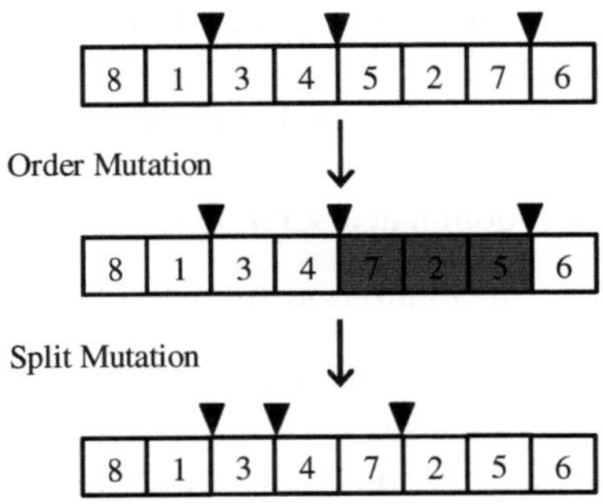

Fig. 5. Examples of two mutation operators. In the order mutation, a selected route is inversed its order of customers. In the split mutation, locations of splits are changed randomly.

[Mutation]

As for the mutation, we employ two kinds of operators in order to modify the order of customers and the locations of splits in a selected route. Figure 5 shows examples of these mutations. It should be noted that the order mutation itself does not affect the two objectives (i.e., the maximum driving duration and the number of vehicles). But it can be useful to increase the variety of solutions when it is used with the crossover and the split mutation.

It should be noted that through crossover and mutation in this paper, the number of vehicles does not change. Therefore if there is no individual with a certain number of vehicles, no solution with that number of vehicles is generated through genetic search.

4.2 Two-Fold EMO Algorithm

In our multi-objective VRP, we have two periods, NDP and HDP. Since HDP has extra demands of customers with the demands of NDP, we have two approaches to search a set of non-dominated solutions for each of NDP and HDP. One approach is to apply an EMO algorithm individually to each of them. The other is to apply a two-fold EMO algorithm [10] to them. In the two-fold EMO algorithm, first we find a set of non-dominated solutions for the NDP by an EMO algorithm. Then we generate a set of initial solutions for the HDP from the non-dominated solutions for the NDP. We apply an EMO algorithm to the HDP with initial solutions that are similar to those of the NDP problem. In our former study [10], we showed that the two-fold EMO algorithm has the better performance than applying an EMO algorithm individually. The procedure of the two-fold EMO algorithm is described as follows:

[Two-Fold EMO Algorithm]

Step 1: Initialize a set of solutions randomly for the NDP. The number of vehicles and the order of customers in each solution are defined randomly.
Step 2: Apply an EMO algorithm to find a set of non-dominated solutions until the specified stopping condition is satisfied.
Step 3: Obtain a set of non-dominated solutions for the NDP.
Step 4: Initialize a set of solutions for the HDP using a set of non-dominated solutions of the NDP.
Step 5: Apply an EMO algorithm to find a set of non-dominated solutions until the specified stopping condition is satisfied.
Step 6: Obtain a set of non-dominated solutions for the HDP.

In Step 4, we initialize a set of solutions as follows:

Step 4.1: Obtain a set of non-dominated solutions of the NDP.
Step 4.2: Specify a solution of the set.
Step 4.3: Insert new customers randomly into the solution.
Step 4.4: Repeat Steps 4.2 and 4.3 until all solutions in the set of non-dominated solutions of the NDP are modified.

It should be noted that the number of vehicles of each solution is not changed by this initialization. The number of vehicles of each solution is changed by the crossover operation. Using this initialization method, we found that the similarity between non-dominated solutions for the NDP and those for the HDP can be increased [10].

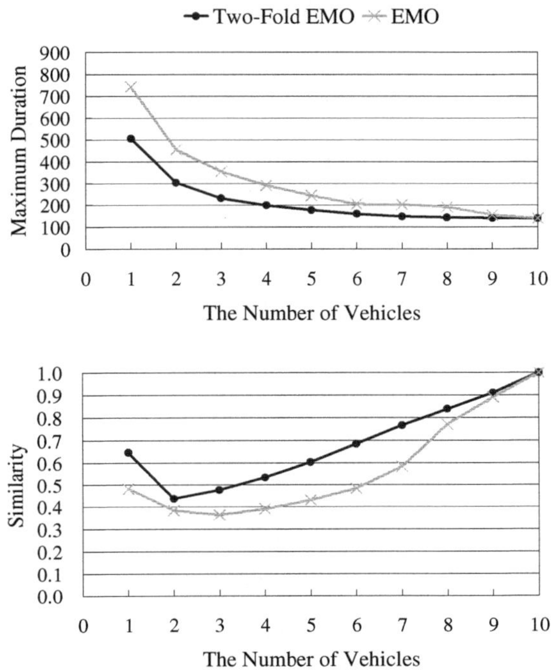

Fig. 6. The obtained non-dominated solutions in the HDP with ten customers. The similarity is to be maximized, the maximum duration to be minimized, and the number of vehicles to be minimized.

We applied the two-fold EMO algorithm to a VRP that has five customers in NDP and ten customers in HDP. As for an EMO algorithm, we employed NSGA-II [12]. Figure 6 shows the results of the two-fold EMO, and the EMO applied the HDP with a population initialized randomly. In this problem, we consider only two objectives: the maximum duration and the number of vehicles. We obtained the average maximum duration of a set of non-dominated solutions over 100 trials. We calculate the average similarity after obtaining a set of non-dominated solutions for HDP. In the first figure of Figure 6, we can find that the two-fold EMO can find better non-dominated solutions with respect to the minimization of the maximum duration and the number of vehicles. From the second figure, we can find that the similarity of non-dominated solutions obtained by the two-fold EMO algorithm is better than that obtained by the EMO algorithm. In this experiment, we can see that improving solutions with respect to the maximum duration does not lead to deterioration of the similarity of non-dominated solutions to those in NDP. Therefore we can say that the two-fold EMO could find the better solutions compared to the EMO for HDP without initial solutions from NDP.

5 Two-Fold Memetic EMO Algorithm

In this paper, we propose a local search that enhances the similarity of non-dominated solutions for HDP. In order to increase the similarity of a solution for HDP, we incorporate segments between customers from a solution of NDP to a solution of HDP. Therefore we introduce this procedure in an EMO search for HDP not for NDP. The algorithm of the proposed local search can be described as follows:

[Local Search Algorithm]

Step 1: Select a solution x from the current Ψ_{HDP}.

Step 2: Select a non-dominated solution y from Ψ_{NDP} that is used for the calculation of the maximum similarity of a solution x.

Step 3: Select an edge between two customers in y. Therefore the edge should be selected within a vehicle.

Step 4: Find a start customer of the selected edge in x.

Step 5: Incorporate the edge to x at the position of the first customer in x. Since the following customer to the first customer in x is replaced by the second customer in the edge, a repairing process should be followed. Find the second customer in x, and replace that with the following customer.

Step 6: Return to Step 3 until all edges in y is incorporated in x.

Figure 7 shows an example of this local search. We apply this local search to each solution of the current set of non-dominated solutions. Since this local search process is introduced to an EMO search in HDP, the two-fold memetic EMO algorithm can be depicted as Figure 8.

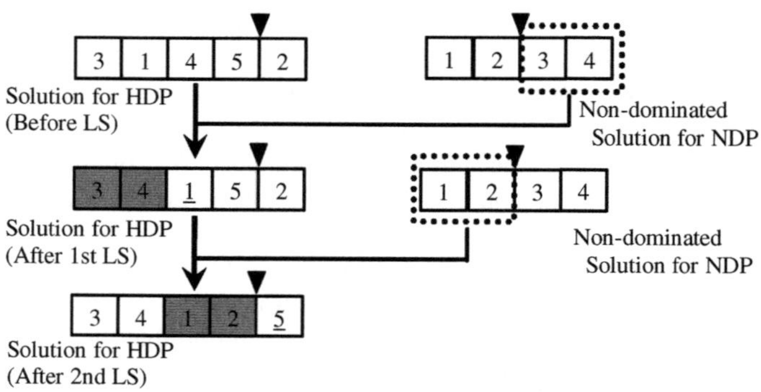

Fig. 7. Local search applied to a solution for HDP. An edge (3,4) in a non-dominated solution for NDP is incorporated to a solution for HDP. Since (3,1) in the solution for HDP is replaced with (3,4) the customers 1 and 4 in the solutions for HDP should be exchanged in a repairing process. Since the solution for NDP has two edges in its string, the local search process terminates at the second time.

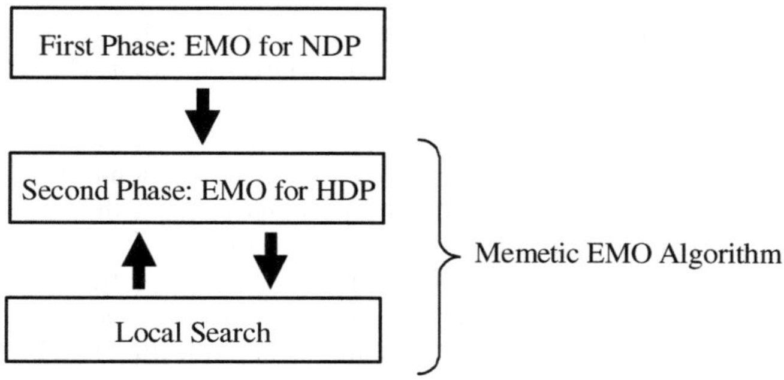

Fig. 8. Two-Fold Memetic EMO Algorithm. A local search is introduced in the second phase of EMO search for HDP.

Table 1. The parameter specifications in EMO algorithms

# of population	30
Crossover rate	1.0
Order mutation rate	0.04
Split mutation rate	0.02
Terminal Generation	2000

6 Simulation Results by Two-Fold Memetic EMO Algorithm

We show the simulation result on a multi-objective VRP with NDP and HDP. In that problem, there are five customers in NDP, and ten customers in HDP. Table 1 shows the parameter specifications in our two-fold memetic EMO algorithm. We apply our two-fold memetic EMO algorithm to the problem with 100 different initial solution sets. That is, we obtain average results over 100 trials in a problem. In this section, first we examine the effect of introducing the similarity as third objective. Then we show the effectiveness of the proposed local search to enhance the similarity.

6.1 Effect of Similarity

We apply two EMO algorithms to a problem in HDP. One is the two-fold EMO algorithm with three objectives (2F-EMO-3). The other is the two-fold EMO algorithm with two objectives (2F-EMO-2). We don't employ the proposed local search in this section. Therefore both algorithms have an initial population generated by a non-dominated solutions for NDP. The same result is obtained by Two-Fold EMO in Figure 6 by 2F-EMO-2. We calculate the similarity of non-dominated solutions obtained by 2F-EMO-2 after the search. Figure 9 shows the simulation results obtained by these algorithms. Since the 2F-EMO-3 finds non-dominated solutions on the surface of three objectives, we project them onto the two objective space in Figure 9. Therefore they are projected between two lines. We depicted two lines of extreme cases. That is, the lowest and the highest similarity on the space with the maximum duration and the number of vehicles. On the other hand, the shortest and the longest maximum duration on the space with the similarity and the number of vehicles. From Figure 9, we can see that slightly better solutions are obtained by the 2F-EMO-2 with respect to the minimization of the maximum duration. On the other hand, the 2F-EMO-2 finds worse solutions with respect to the maximization of the similarity. As for the 2F-EMO-3, it produces slightly better non-dominated solutions in the similarity when their maximum duration becomes near to those of the 2F-EMO-2. On the other hand, when the 2F-EMO-3 sacrifices the minimization of the maximum duration, the similarity of non-dominated solutions becomes much better than the 2F-EMO-2. Through this figure, we can find that the similarity of non-dominated solutions has the trade-off relationship with the maximum duration. Therefore the introduction of the similarity as third objective is needed for those who wants to have similar routes in HDP to NDP.

6.2 Effect of Local Search to Enhance the Similarity in HDP

In this section, we examine the effectiveness of the proposed local search to enhance the similarity of non-dominated solutions for HDP. We compare the 2F-EMO-3 and the two-fold memetic EMO algorithm (2F-mEMO). From Figure 10, We can see that the 2F-EMO-3 could find better solutions with respect to the maximum duration when

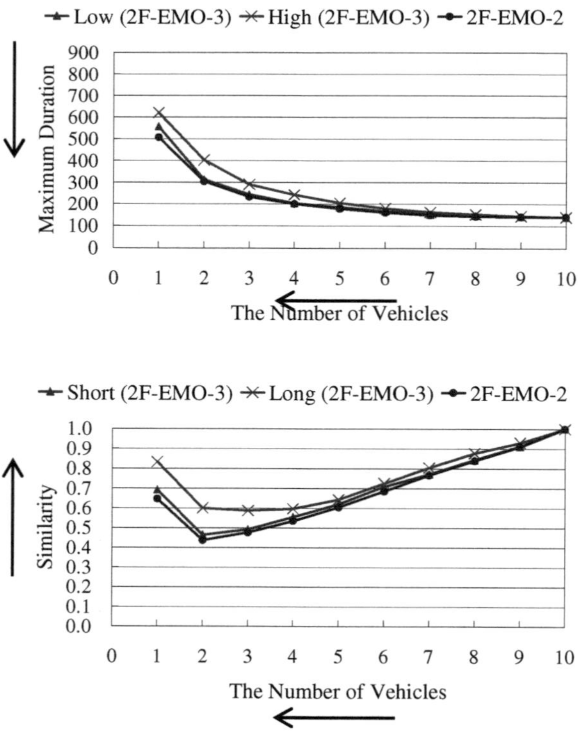

Fig. 9. The effect of the similarity

it sacrifices the similarity. On the other hand, almost similar maximum durations are obtained by both the algorithms when they seek to maximize the similarity. Although both the algorithms have similar maximum durations in the case of high similarity, the degree of the similarity of these algorithms is quite different in the latter figure of Figure 10. Using the proposed local search, the 2F-mEMO could enhance the similarity especially in non-dominated solutions with two through six vehicles. As for the solutions with more than seven vehicles the similarity is not improved well. This is because each vehicle should not visit so many customers when the number of vehicles is similar to the number of customers. Similar routes are required when each vehicle has several customers to visit. From Figure 10, we can see that the proposed local search is very much effective in enhancing the similarity with a slight deterioration in the maximum duration.

7 Conclusion

In this paper, we proposed a local search that can be used in a two-fold memetic EMO algorithm for multiple-objective VRPs with different demands. The simulation results show that the proposed method have the fine effectiveness to enhance the similarity of

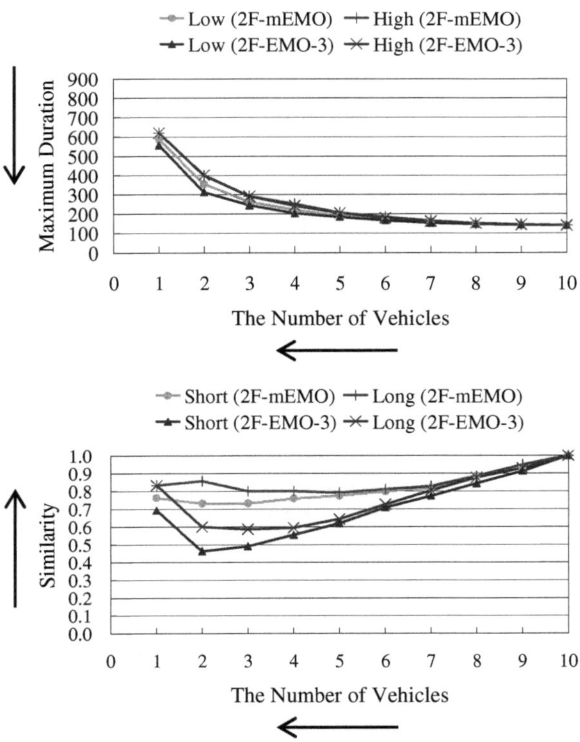

Fig. 10. The effect of the local search in HDP

obtained routes for vehicles. Although the local search slightly deteriorates the maximum duration, we can see the effectiveness of the similarity of the routes because the possibility of getting lost the way of drivers may be decreased. If drivers get lost their ways during their delivery, the cost of his routes may increase. The enhancing the similarity of set of non-dominated solutions seems important when we apply EMO algorithms to practical problems.

Since the algorithm of the proposed local search to enhance the similarity depends on the problem specifications, we should make further research on the similarity of a set of non-dominated solutions with different problems. We may define similarity on the genotype, and it on the phenotype. Since the similarity on the phenotype may depend on problems, we should research further on the similarity on the genotype of various problems.

Acknowledgement

This work was supported in part by the Japan MEXT (Ministry of Education, Culture, Sports, Science and Technology) under Collaboration with Local Communities Project for Private Universities starting 2005.

References

1. Zitzler, E., Deb, K., Thiele, L., Coello Coello, C. A., Corne, D. (eds.): *Proc. of First International Conference on Evolutionary Multi-Criterion Optimization* (EMO 2001).
2. Fonseca, C. M., Fleming, P. J., Zitzler, E., Deb, K., Thiele, L. (eds.): *Proc. of Second International Conference on Evolutionary Multi-Criterion Optimization* (EMO 2003).
3. Coello Coello, C. A., Hernandez Aguirre A., Zitzler E. (eds.): *Proc. of Third International Conference on Evolutionary Multi-Criterion Optimization* (EMO 2005).
4. Deb, K.: Applications of multi-objective evolutionary algorithms. In: *Multi-objective Optimization Using Evolutionary Algorithms.* John Wiley & Sons, England (2001) 447-479.
5. Tavares, J., Pereira, F. B., Machado, P., Costa, E.: Crossover and diversity: A study about GVR, In *Proc. of AdoRo (Workshop held at GECCO 2003)*, 7 pages.
6. Berger, J., and Barkaoui, M.: A hybrid genetic algorithm for the capacitated vehicle routing problem," *Proc. of GECCO 2003* (2003) 646-656.
7. Tan, K. C., Lee, T. H., Chew, Y. H., Lee, L. H.: A hybrid multiobjective evolutionary algorithms for solving vehicle routing problem with time windows, In *Proc. of IEEE International Conf. on SMC 2003* (Washington D.C., U.S.A., Oct. 5-8, 2003) 361-366.
8. Saadah, S., Ross, P., Paechter, B.: Improving vehicle routing using a customer waiting time colony, In *Proc. of 4th European Conf. on Evolutionary Computation in Combinatorial Optimization* (2004) 188-198.
9. Chitty, D. M., Hernandez, M. L.: A hybrid ant colony optimisation technique for dynamic vehicle routing, In *Proc. of GECCO 2004* (2004) 48-59.
10. Murata, T., Itai, R.: Multi-objective vehicle routing problems using two-fold EMO algorithms to enhance solution similarity on non-dominated solutions, In *Proc. of Third International Conference on Evolutionary Multi-Criterion Optimization* (EMO 2005), 885-896.

11. Lenstra, J. K., Rinnooy Kan, A.H.G.: Complexity of vehicle routing and scheduling problems. In *Networks*, **11** (1981) 221-227.
12. Deb, K., Pratap, A., Agarwal, S., Meyarivan, T.: A fast and elitist multiobjective genetic algorithm: NSGA-II, In *IEEE Trans. on Evolutionary Computation*, **6** (2) (2002) 182-197.
13. Maekawa, K., Mori, N., Tamaki, H., Kita, H., Nishikawa, H.: A Genetic Solution for the Traveling Salesman Problem by Means of a Thermodynamical Selection Rule. In *Proc. 1996 IEEE Int. Conf. on Evolutionary Computation* (1996) 529-534.

Mechanism of Multi-Objective Genetic Algorithm for Maintaining the Solution Diversity Using Neural Network

Kenji Kobayashi[1], Tomoyuki Hiroyasu[2], and Mitsunori Miki[2]

[1] Graduate Student, Department of Knowledge Engineering and Computer Sciences, Doshisha University ,1-3 Tatara Miyakodani,Kyo-tanabe, Kyoto, 610-0321, Japan
kkobayashi@mikilab.doshisha.ac.jp
[2] Department of Knowledge Engineering and Computer Sciences, Doshisha University ,1-3 Tatara Miyakodani,Kyo-tanabe, Kyoto, 610-0321, Japan
tomo@is.doshisha.ac.jp, mmiki@mail.doshisha.ac.jp

Abstract. When multi-objective genetic algorithms are applied to real-world problems for deriving Pareto-optimal solutions, the high calculation cost becomes a problem. One solution to this problem is to use a small population size. However, this often results in loss of diversity of the solutions, and therefore solutions with sufficient precision cannot be derived. To overcome this difficulty, the solutions should be replaced when they have converged on a certain point. To perform this replacement, inverse analysis is required to derive the design variables from objects as the solutions are located in the objective space. For this purpose, an Artificial Neural Network (ANN) is applied. Using ANN, the solutions concentrating on certain points are replaced and the diversity of the solutions is maintained. In this paper, a new mechanism using ANN to maintain the diversity of the solutions is proposed. The proposed mechanism was introduced into NSGA-II and applied to test functions. In some functions, the proposed mechanism was useful compared to the conventional method. In other numerical experiments, the results of the proposed algorithm with large populations are discussed and the effectiveness of the proposed mechanism is also described.

1 Introduction

Since Schaffer developed the genetic algorithm for multi-objective optimization problems [14], many evolutionary multi-objective algorithms that can derive good solutions have been introduced in this field [7,19]. Recently, these algorithms have been applied to real-world problems and effective results have been obtained. [3,16] One of the most important points to obtain solutions in real-world problems is to derive a superior solution within a reasonable time. Usually, it takes a large amount of time to evaluate one parameter set in a real-world problem. Therefore, even with a strong algorithm, satisfactory results cannot be

S. Obayashi et al. (Eds.): EMO 2007, LNCS 4403, pp. 216–226, 2007.

derived if the calculation time is of insufficient length. In this case, an algorithm that can derive reasonable solutions with a small number of evaluation calls should be used. There are two approaches to develop such algorithms.

The first approach is to use the response surface methodology [17], which is a technique for approximating objective functions. This method reduces the calculation costs by generating approximations of objective function and treating these approximations as objective functions for each evaluation. There are several response surface methodologies, such as the quadratic polynomial model [1], neural network model [2,6,11], and Kriging model [12]. Among these, the quadratic polynomial model is commonly used, because it is the simplest and has low calculation costs for approximation. Although the costs associated with the other models for approximation are greater than those for the quadratic model, the neural network model and Kriging model allow approximation of more complicated objective functions [18].

On the other hand, the method discussed here involves a search with a small number of individuals. For MOGA search, it is critical to search the Pareto-optimal solutions with keeping the diversity of individuals, because it is more likely that solutions with high accuracy and diversity will be obtained. This approach can reduce the calculation cost, but the solutions often converge on a certain point in the search process and the diversity of the solutions may be lost. In this paper, we propose a mechanism that eases the reduction of diversity of solutions during the search process by using an Artificial Neural Network (ANN). This mechanism is expected to reduce the calculation cost, and provide a good set of Pareto-optimum solutions with a high degree of diversity and accuracy, even when the search is performed with a small number of individuals.

In this paper, we discuss in detail a mechanism to ease the reduction of diversity using ANN. The proposed mechanism is introduced into NSGA-II [7], a typical MOGA, and its effectiveness and influence on the search are investigated for mathematical test functions.

2 Problem of Multi-Objective Genetic Algorithms with a Small Population

The advantage of multi-objective optimization using a MOGA is that it can derive several Pareto-optimum solutions in one calculation trial. However, a number of function evaluations are required before the Pareto-optimum solution can be obtained. The calculation cost can be reduced when using a small number of individuals or a small number of generations. However, with this approach, the diversity of individuals is often lost. This may have a negative influence on the progress of the search, as in MOGA it is important to maintain the diversity during the search. In some cases, it may become difficult to obtain non-dominated solutions with high accuracy and diversity. To overcome this difficulty, whenever solutions are converged they are relocated evenly on an interpolated line.

However, it is very difficult to determine the parameters of these target solutions, as the target solutions exist in the objective field but not in the design field.

Therefore, the problem of how to determine the design variable values of the solutions is very important. For example, the objective function values of the relocated solutions are known, and the design variable values are not. Thus, the design variable values must be determined through inverse analysis. Here, ANN is used for this inverse analysis. ANN is a powerful approach to modeling stochastic and noisy patterns of data to produce predicted values of unknown systems [13]. In recent years, there have been many studies on multi-objective optimization using ANN. They are mainly classified into two types: methods that reduce the calculation cost for each evaluation by obtaining an approximation function of the objective function [15], and those to obtain the approximation function that is the inverse of the objective function and apply it for local searches [5,13]. Multilayer perceptrons, i.e., feedforward neural networks that use back propagation [2,6,10] for the learning algorithm, are often used for ANN.

3 Maintaining the Solution Diversity Mechanism Using Neural Network

In this paper, a method of maintaining the solution diversity using a neural network is proposed. The concept of the proposed mechanism is shown in Fig. 1.

In this mechanism, whenever the solutions are converged, they are relocated on the Pareto front line and MOGA search is performed using these relocated solutions. Iterating this process is expected to maintain the diversity and obtain good Pareto-optimal solutions with a smaller number of individuals.

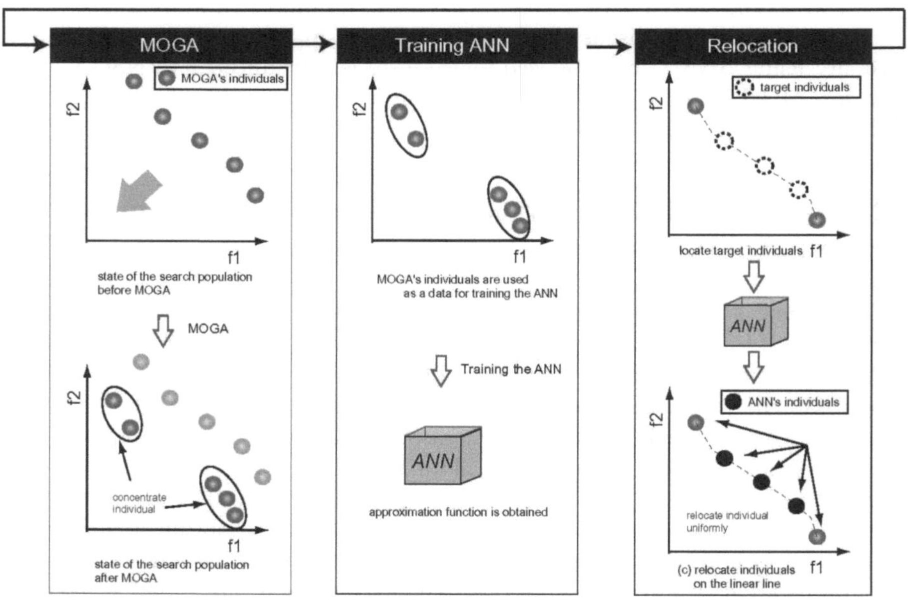

Fig. 1. Concept of the proposed method

The proposed method using both MOGA and ANN aims to reduce the calculation cost and to obtain solutions with high accuracy by maintaining diversity during the search process, even in a search with a small number of individuals. The algorithm of the proposed method is as follows:

N : Number of executions of ANN.
t_{max} : Max number of generations .
t : Number of generations .
k : Number of non-dominated solutions.

Step 1: NSGA-II search is performed up to $i \times t_{max}/N$ generations.(i=1)
Step 2-1: A set of non-dominated solutions is obtained, and a linear line passing through the set of non-dominated solutions is obtained through interpolation.
Step 2-2: A set of non-dominated solutions is used as a data set for training the ANN, and an approximation function is created.
(Input: objective function values; Output: design variable values)
Step 2-3: In a set of n non-dominated solutions , all individuals are removed except those on both ends , then $n-2$ target individuals are created so that the distances regarding f1 between adjacent individuals are equal.
Step 2-4: The approximation function created by ANN is used to obtain the design variable values corresponding to the objective function values of the target individuals.
Step 2-5: The design variable values obtained by ANN are evaluated using the real objective function, not the approximation function.
Step 2-6: Individuals and archives obtained from ANN are combined and the archive update mechanism of NSGA-II is executed.
Step 2-7: End if all the end conditions are satisfied. If not, return to Step 1($i = i + 1$).

The processes using interpolation mentioned in Steps 2-3 and 2-4 are discussed. An approximation function that is the inverse of the objective function is created by the ANN, which is trained based on individual data derived by the MOGA. Using the approximation function, all non-dominated solutions will be relocated except those on both ends, as it is better to obtain as many individuals as possible at equal distances to the Pareto front.

Next, we describe how to determine the objective function values of the target individuals. There are two steps to obtain these values. In the first step, a linear interpolation line is obtained (Step 2-1). A linear interpolation method is adopted, because it showed more positive results in preliminary experiments than two-dimensional interpolation. In the second step, target individuals are relocated so that they satisfy the following two conditions: 1) individuals lie on the interpolation line; 2) individuals are at equal distances with regard to f1. As it is difficult to set the target individuals for interpolation in many objectives (more than three), this paper focuses on two-objective optimization problems. Many objectives problems (more than three) will be examined in future studies. The scheme of the process is shown in Fig. 2. The left diagram in Fig. 2 shows the Pareto-optimal solution obtained from the MOGA, and the relocation concept is illustrated on the right in Fig. 2.

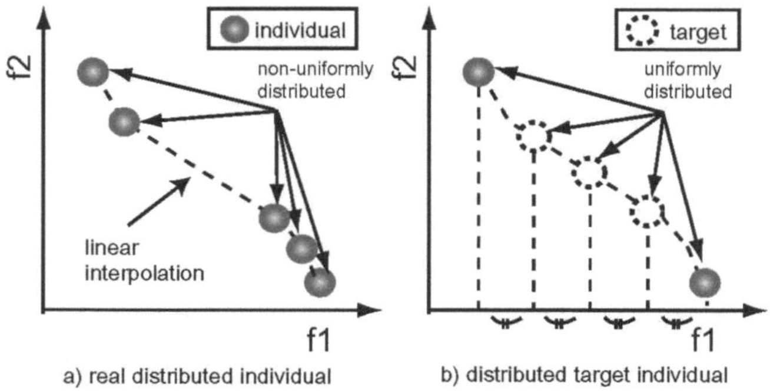

Fig. 2. Linear Interpolation

4 Effectiveness of Diversity Maintenance Mechanism Using ANN

4.1 Examination Environment

In this section, we report testing of the proposed mechanism to examine its effectiveness. Here, NSGA-II is applied for the basic MOGA and the ANN mechanism is attached to NSGA-II.

The proposed hybridized method with MOGA and ANN is designed for real-world problems that require large computational cost for each evaluation. As an initial study, we chose a problem where the landscape of the function is relatively smooth. ZDT6 [4] unimodal test problems with a non-convex Pareto front are selected as a test function. The equation of test functions is shown in Table 1.

In this experiment, the ratio of non-dominated individuals (RNI) [9] is used to evaluate a set of non-dominated solutions obtained using various methods. RNI measures the accuracy in objective function space. This method compares two sets of non-dominated solutions, and counts the number of solutions that are inferior to those obtained by the other method. This method evaluates items with regard to accuracy. The method used by Tan and colleagues [8] is expanded to compare two sets of non-dominated solutions to create this method. The comparison procedure of this method is as follows. The union of the solution sets X and Y obtained by the two methods is set as S^U. Next, solutions not

Table 1. Test Problem(ZDT6)

Problem	Functions
ZDT6	min $f_1 = 1 - \exp(-4x_1)sin^6(6\pi x_1)$
	min $f_2 = g \times h$
	$g = 1 + 9[(\sum_{i=2}^{n} x_i)/9]^{0.25}$
	$h = 1 - (f_1/g)^2$
	$x_i \in [0,1]$, n=2

dominated by any solution are selected from S^U, and the selected set of solutions is set as S^P. Then, the ratio of S^P of each method is derived as RNI(X,Y). The closer this ratio is to the maximum, 100%, the better it is compared to the other method, indicating that a solution that is closer to the true solution is being obtained.

4.2 Assessment of Approximation Ability of ANN

An archive obtained from a MOGA is used as a training data set for ANN. There are two methods of training: one is to use all of the archive, and the other is to use only the non-dominated solutions from the archive as training data. The two methods are compared. The test data for ANN are shown in Fig. 3. In this data, the non-dominated and dominated solutions are mixed. The results from the method using the whole archive are shown in Fig. 4(a), while the results using only the non-dominated solutions are shown in Fig. 4(b).

As shown in Fig. 4(b), the developed ANN has sufficient efficiency to derive the solutions, as the derived solutions are very close to the Pareto front. At the same time, as shown in Fig. 4, when all the solutions in the archive are used, those solutions except the non-dominated solutions become noise, and thus solutions

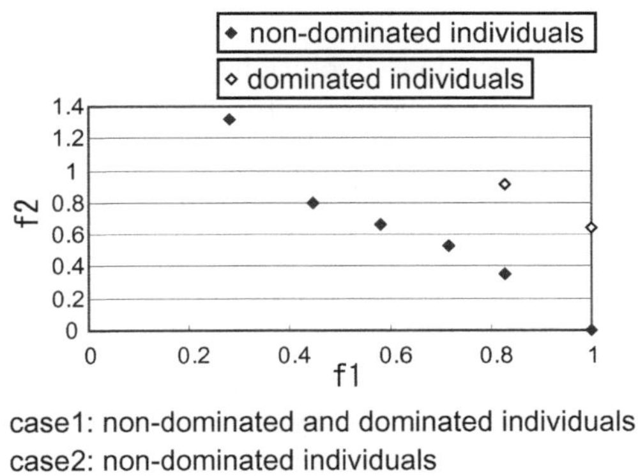

Fig. 3. Test data for the ANN training

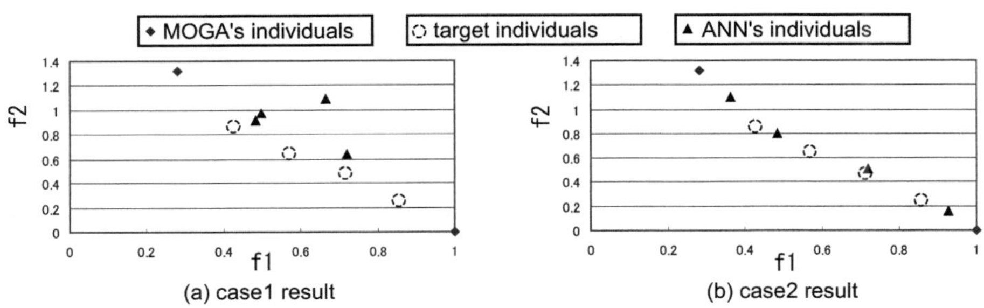

Fig. 4. Result of the ANN training

are derived in the area away from the target Pareto front. On the other hand, when only non-dominated solutions are used in the training data set, target solutions are derived appropriately on the Pareto front. Thus, we use only the non-dominated solutions in the archive as a training data set for ANN.

4.3 Examination of Diversity Improvement Using ANN

We next examine whether the reduced diversity of solutions can be improved by hybridized NSGA-II. The archive size is set to 10, which showed good results in a preliminary experiment. The parameters used are shown in Table 2.

Table 2. Parameter settings in examination of diversity improvement using ANN

Population size	6
Number of generations	60
Archive size	10
Number of dimensions	2
Crossover rate	1.0
Method of crossover	Two-point crossover
Number of times ANN applied	2
Number of trials	30

Fig. 5(a) shows the non-dominated solutions obtained by NSGA-II, and Fig. 5(b) shows the target individuals relocated according to these non-dominated solutions. Individuals obtained using the proposed mechanism are shown in Fig. 5(c).

Comparison of Fig. 5(a) and (c) indicates that a more uniform solution distribution was achieved after application of the diversity maintenance mechanism using ANN.

In the next experiment, conventional NSGA-II and hybridized NSGA-II are compared to examine their effectiveness. The parameters used are shown in Table 3. ANN is applied evenly during the search (e.g., 20th and 40th generations).

RNI of conventional NSGA-II and hybridized NSGA-II are shown in Fig. 6. In addition, plots of each method are shown.

Table 3. Parameter settings in examination of diversity improvement using ANN

Search technique	Hybrid	NSGA-II
Number of generations	60	61
Number of evaluations	368	366
Number of times ANN applied	2	None

The results shown in Fig. 6 indicate that RNI of hybridized NSGA-II is higher than that of conventional NSGA-II. In addition, hybridized NSGA-II can derive solutions close to the region of the Pareto-optimum solutions in more trials than the conventional NSGA-II. The above observations indicate that the issue of reduced diversity by the conventional NSGA-II with a small number of individuals

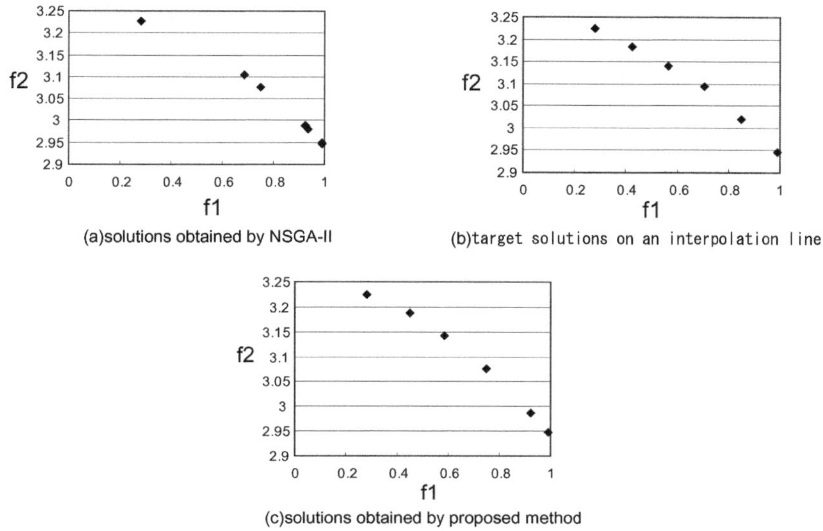

Fig. 5. Results of solutions relocation using ANN

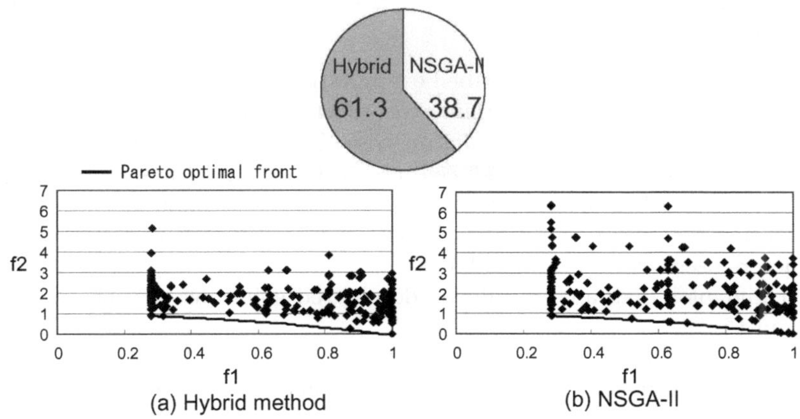

Fig. 6. Comparison of RNI and search result

can be resolved using the proposed method and it is possible to execute a search while maintaining its diversity.

4.4 Examination of Number of Times ANN Is Applied

The diversity mechanism using ANN is introduced into NSGA-II, and the influence of the number of times ANN is applied is examined. The parameters used are shown in Table 4. The results of this experiment are shown in Fig. 7. RNI is obtained by comparing the various number of times ANN is applied with the same number of evaluations.

Table 4. Parameter settings in examination of number of times ANN applied

Number of generations	30
Number of times ANN applied	0,3,6,10,15,30

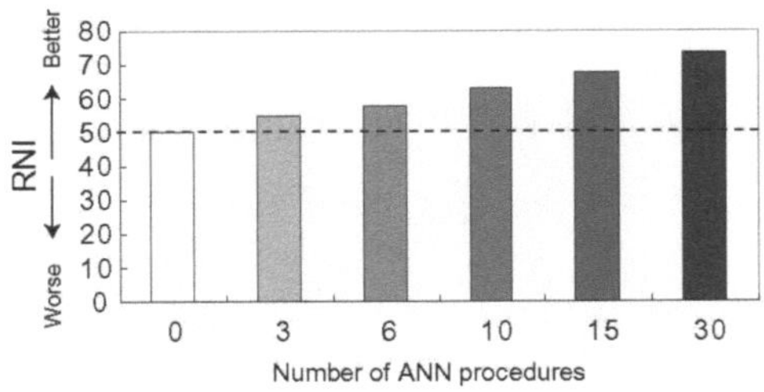

Fig. 7. RNI comparison

Fig. 7 shows that RNI is better when ANN is applied often during the search. This indicates that the diversity maintenance mechanism using ANN has a positive effect on the search.

4.5 Comparison of a Search with Small and Large Numbers of Individuals

In the previous sections, we described examination of searches with a small number of individuals. Here, we examine the effectiveness of the proposed method related to the reduction of calculation cost. In this experiment, the diversity maintenance mechanism is introduced into NSGA-II and the search is executed with a small number of individuals. The results are compared with those obtained using a conventional NSGA-II with a large number of individuals. The parameters used are shown in Table 5.

RNI and plot diagrams of all non-dominated solutions of each method are shown in Fig. 8.

Fig. 8 shows that the performance of the hybridized NSGA-II with a smaller number of individuals is equivalent to conventional NSGA-II with a larger population size. The hybridized NSGA-II used 300 evaluations, while the conventional NSGA-II required 540. In this numerical example, a simple test function was used and the calculation cost was very small. Our target is real-world problems,

Table 5. Parameter settings in examination of comparison of a search with small and large number of individuals

Search technique	Hybrid	NSGA-II
Population size	6	18
Number of generations		30
The number of evaluations	300	540
Archive size	10	18
Number of times ANN applied	30	None

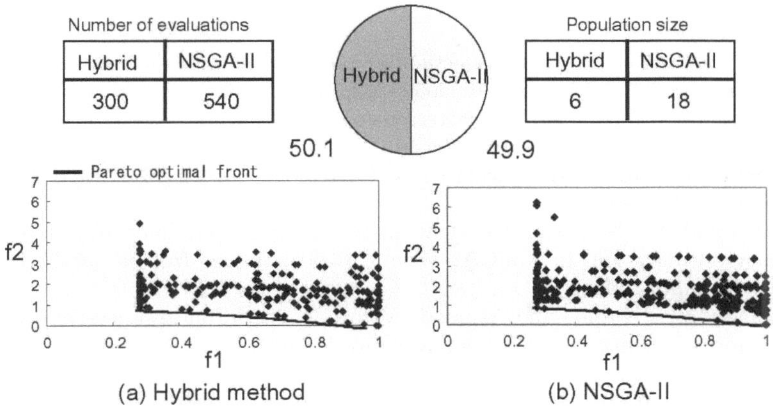

Fig. 8. Comparison of RNI and search result

and these often take a huge amount of time. For example, if evaluation takes around 1 h at each fitness, our method can save about 240 h. These observations indicated that when a conventional MOGA search with a small number of individuals is performed, the search ability can be improved by using our diversity maintenance mechanism. This is comparable to a conventional MOGA search with a larger number of individuals when the number of evaluations is small. Therefore, this mechanism allows reduction of the calculation cost. With this search, the calculation cost required by the ANN must be considered. When large computational cost for evaluation in GA is needed, the relative calculation cost of ANN is negligible; examples of such problems include airplane design and automobile collision analysis.

5 Conclusions

In this paper, a mechanism is proposed for a MOGA search that maintains its diversity during the search process, even when the search is performed with a small number of individuals. It can restore the reduced diversity of solutions by using ANN, whenever the diversity is lost. The mechanism of maintaining diversity by ANN creates an approximation function, which is used to obtain the design variable values corresponding to the objective function values of target individuals. The target objective values are determined by relocating the individuals so that they are equally distanced. The proposed method is expected to derive individuals with a high degree of accuracy and diversity. Here, the proposed mechanism was introduced into NSGA-II, and its effectiveness was examined for mathematical test functions. The results of numerical experiments indicated that the search performance of the proposed hybrid method is comparable to that of the conventional method. Using the hybrid method, it is possible to derive solutions with high accuracy and diversity with a small number of evaluations, even when performing the search with a small number of individuals.

References

1. Julio l, hierarchical variable selection in polynomial regression models. http:// www.jstor.org/view/00031305/di020596/02p0164o/0.
2. C.Bishop. Neural networks for pattern recognition. *Oxford University Press*, 1997.
3. Kazuhisa Chiba and Shigeru Obayashi. High-fidelity multidisciplinary design optimization of aerostructural wing shape for regional jet. In *23rd Applied Aerodynamics Conference*, 2005.
4. K. Deb and T. Meyarivan. Constrained test problems for multi-objective evolutionary optimization. In KanGAL report 200005, *Indian Institute of Technology, Kanpur, India,* 2000.
5. Antonio Gaspar-Cunha and Armondo Vieira. A hybrid multi-objective evolutionary algorithm using an inverse neural network. 2004.
6. HENSELER J. Back propagation.*Lect Notes Comput Sci,* 931:37 – 66, 1999.
7. Amrit Pratab Kalyanmoy Deb, Samir Agrawal and T. Meyarivan. A fast elitist non-dominated sorting genetic algorithm for multi-objective optimization: Nsga-ii. In *KanGAL report 200001, Indian Institute of Technology, Kanpur, India,* 2000.
8. K.C.Tan, T.H.Lee, and E.F.Khor. Increamenting multi-objective evolutionary algorithms: Performance studies and comparisons. In *First International Conference on Evolutionary Multi-Criterion Optimization,* pages 111 – 125, 2001.
9. J. D. Knowles and D. W. Corne. Approximating the nondominated front using the pareto archived evolution strategy. volume 8, pages 149 – 172, 2000.
10. L.Bull. On modelbased evolutionary computation. 1999.
11. HAYMAN S. The mcculloch-pitts model.*IEEE Int Conf Neural Netw Proc,* 6:4438 – 4439, 1999.
12. J. et al. Sacks.*Statistical Science,* 4:pp.409 – 435, 1989.
13. Ian Griffin Salem F. Adra, Ahmed I. Hamody and Peter J. Fleming. A hybrid multi-objective evolutionary algorithm using an inverse neural network for aircraft control system designa. *2005 IEEE Congress on Evolutionary Computation (CEC'2005),* 1:1 – 8, 2005.
14. J. D. Shaffer. Multiple objective optimization with vector evaluated genetic algorithms. pages pp.93 – 100, 1985.
15. Hirotaka Nakayama Takaharu Sirai, Masao Arakawa. Application of rbf network in approximate optimization in satisfcing method. In *Design and Systems,* pages Vol14th page 85 – 86, 2004.
16. J.Kamiura Shinya Watanabe H.Hiroyasu T.Hiroyasu, M.Miki. Multi-objective optimization of diesel engine emissions and fuel economy using genetic algorithms and phenomenological model. In*SAE 2002 Powertrain and Fluid Systems Conference,* 2002.
17. Akira Todoroki. Response surface method, http:// orida.mes.titech.ac.jp/responsesurface.pdf.
18. Kikuo Fujita Yusuke Kounoe. Third-order polynominal response surface with optimal selection of interaction terms for global approximation. In*Design and Systems,* volume 15, pages 31 – 34.
19. E. Zitzler, M. Laumanns, and L. Thiele. Spea2: Improving the performance of the strength pareto evolutionary algorithm. In*Technical Report 103, Computer Engineering and Communication Networks Lab (TIK), Swiss Federal Institute of Technology (ETH) Zurich,* 2001.

Pareto Evolution and Co-evolution in Cognitive Game AI Synthesis

Yi Jack Yau, Jason Teo, and Patricia Anthony

School of Engineering and Information Technology
Universiti Malaysia Sabah, Locked Bag No. 2073,
88999 Kota Kinabalu, Sabah, Malaysia
yijackyau@gmail.com, {jtwteo,panthony}@ums.edu.my

Abstract. The Pareto-based Differential Evolution (PDE) algorithm is one of the current state-of-the-art Multi-objective Evolutionary Algorithms (MOEAs). This paper describes a series of experiments using PDE for evolving artificial neural networks (ANNs) that act as game-playing agents. Three systems are compared: (i) a canonical PDE system, (ii) a co-evolving PDE system (PCDE) with 3 different setups, and (iii) a co-evolving PDE system that uses an archive (PCDE-A) with 3 different setups. The aim of this study is to provide insights on the effects of including co-evolutionary techniques on a well-known MOEA by investigating and comparing these 3 different approaches in evolving intelligent agents as both first and second players in a deterministic zero-sum board game. The results indicate that the canonical PDE system outperformed both co-evolutionary PDE systems as it was able to evolve ANN agents with higher quality game-playing performance as both first and second game players. Hence, this study shows that a canonical MOEA without co-evolution is desirable for the synthesis of cognitive game AI agents.

Keywords: Game AI, Co-Evolution, Evolutionary Artificial Neural Networks, Pareto Differential Evolution, Evolutionary Multi-Objective Optimization.

1 Introduction

Artificial intelligence for games (game AI) represents one of the most useful and practical platforms for studying evolutionary computation systems. Game has well defined rules that make them easier to simulate on a computer and its applications to many real world problems in economics, politics, biology, and countless other areas. Zero-sum board games provide a simple yet interesting testing-bed to study both the machine learning and the optimization aspects of soft computing systems. Firstly, these games have perfectly defined sets of rules that limit the possible behaviors of the players, thereby simplifying the problem at hand. Furthermore, games have a clear objective for the players to reach. In addition, games have enough information to allow a wide range of possible behaviors to emerge as represented by the various strategies of the game players.

S. Obayashi et al. (Eds.): EMO 2007, LNCS 4403, pp. 227–241, 2007.

In spite of the additional complexities of co-evolutionary models, they hold some significant advantages that have been exploited within the context of EAs to support the generation of solutions to a series of complex problems. Co-evolutionary techniques have been successfully applied to a number of games, for instance, Awari [1], Pong [2], Nim [3], and Go [4,5]. Evolutionary Programming (EP) has also been used to create ANNs that are capable of playing Tic-Tac-Toe (TTT) [6,7]. Although many single-objective evolutionary techniques have been successfully applied to many different kinds of games, a large number of research issues and questions still remain for multi-objective evolutionary techniques when applied to games.

In previous work, an enhanced version of a hybrid co-evolutionary implementation using the Pareto Differential Evolution (PDE) algorithm known as the Pareto Co-evolutionary Differential Evolution (PCDE) algorithm [8]. This algorithm was reported to be able to automatically synthesize neural network game-playing agents both as the first and second players with reasonable playing strength through the introduction of Pareto multi-objective evolution [8]. In this study, the main objective is to look into the effects of the introduction of the co-evolution technique and whether it is actually beneficial or otherwise to the Pareto evolutionary optimization process. A comprehensive empirical comparison of performance between the previous system of PCDE, a new archived-based version of PCDE called PCDE-A and the canonical PDE without co-evolution is carried out. All of the above implementations do not require an explicit evaluation function for the purpose of automatically generating the game AI for TTT since the scoring from playing against a rule-based player is used as the objective evaluation method during evolution. Finally, the performance of the respective approaches will be measured according to the playing strength of the evolved ANN game-playing agents pitted against three different levels of players (expert, medium and random players).

1.1 Tic-Tac-Toe

TTT is a standard two-player zero-sum game, in which two players alternately put crosses and circles in one of the compartments of a 3 by 3 board. The objective of the game is to get a row of 3 crosses or 3 circles before the opponent does. Player one is the player that moves first, making a cross, followed by player two, making a circle. If at the end of the game both players cannot meet the objective, it means that a draw is awarded to both players. There are 4 player types in TTT. The novice player makes random moves, the intermediate player will block their opponent from winning, the experienced player knows that playing in certain first squares will lose the game, and the expert player will never lose. When both players are at the expert level, the purpose of a TTT first player is to force a win or a draw; however a second player should force a draw by blocking the winning moves of first player. This is because if the first player starts the game with an optimal first move, it will never lose if no mistake is made for following moves, so the second player can only force a tie. The only chance for a second player to force a win is when the first player did not make a best first move, or making a mistake during subsequent moves.

2 Methods

2.1 Pareto Differential Evolution (*PDE*)

The multi-objective optimization method used here was an adaptation of a well known multi-objective evolutionary algorithm (MOEA) which is the PDE algorithm proposed in [9]. Evolved artificial neural networks (ANNs) act as the cognitive game AI agents in the game. The system was initialized with a population of 100 ANNs, each one having its weight connections and bias term value set at random in a uniform distribution ranging over $[-0.5, 0.5]$. Parent created an offspring through mutation of each weight and bias term value by adding a Gaussian random variable with zero mean and a standard deviation of 1 ($GaussianF(0,1)$). Based on the mutation rate, the number of nodes in hidden layer was allowed to vary, subject to the constraints on the maximum and minimum number of nodes. All new added node weights and the bias terms are set to 0.0. A crossover function that implements the DE concept was used in the creation of new offspring. The scaling factor (F) is a random real value between $[0.0, 0.5]$. All layers (l) of ANN (the input layer, hidden layer and output layer) and all the synapses (connection between input layer and hidden layer, and also between hidden layer and output layer, bl) are involved in both operations of reproduction (mutation and crossover). Criteria for marking non-dominated solutions will be done directly based on the scores gained from these two payoff functions (see sub-section 2.1). The algorithm works as described in the following section.

Pseudocode of PDE

1. Randomly initialize population of 100 ANNs, each with its weight (W) connections and bias term (B) value set at random in a uniform distribution ranging over $[-0.5, 0.5]$.
2. Repeat:
   a. Evaluate individuals in the population and mark non-dominated ANNs.
   b. If the number of non-dominated ANNs is less then 50, repeat the following until the number of non-dominated ANNs is greater than or equal to 50:
      i. Find the next layer of non-dominated solutions among those marked ANNs.
      ii. Re-mark the ANN as non-dominated.
   c. Delete all dominated ANNs from the population.
   d. Repeat:
      i. Randomly pick an ANN as the main parent p1, and two ANNs, p2, p3, as supporting parents.
      ii. Crossover:
         If within the probability of CrRate, Then do

$B_l^{child} \leftarrow B_l^{p1} + F(B_l^{p2} - B_l^{p3})$
$W_{bl}^{child} \leftarrow W_{bl}^{p1} + F(W_{bl}^{p2} - W_{bl}^{p3})$
Else
$B_l^{child} \leftarrow B_l^{p1}$
$W_{bl}^{child} \leftarrow W_{bl}^{p1}$

iii. Mutation:
If within the probability of MtRate, Then do
$B_l^{child} \leftarrow B_l^{p1} + GaussianF(0, 1)$
$W_{bl}^{child} \leftarrow W_{bl}^{p1} + GaussianF(0, 1)$
Else
$B_l^{child} \leftarrow B_l^{p1}$
$W_{bl}^{child} \leftarrow W_{bl}^{p1}$
And With probability of MtRate, do
$NumberNode \leftarrow Vary(NumberNode)$

iv. Evaluate the child, if the child dominates p1, place it into the population.

e. Until the population size is maximum (100).

3. If termination conditions are satisfied then end, else return to (2).

Evaluation of individuals (in PDE). Each ANN will compete with the same rule-based procedure as the first player "X" in 2 sets of 8 games and the second player "O" in 2 sets of 9 games. The first move of the rule-based player will not be repeated in each set of games, based on all possible moves being stored in an array at the beginning of the particular set of games. Two different payoff functions were used to reward each agent performance as the first and second player. For grading the performance of the ANN as the first player, the payoff function $\{+1, -10, 0\}$ are the rewards for winning, losing, and drawing, respectively (as in [6]). However, for grading the performance of the ANN as the second player, the payoff function $\{+2, -5, 3\}$ are the rewards for winning, losing, and drawing, respectively (which was obtained from preliminary testing). Marking non-dominated solutions will be done directly based on the scores gained from these two payoff functions.

2.2 Pareto Co-evolutionary Differential Evolution (*PCDE*)

The PCDE algorithm introduced co-evolutionary techniques into PDE (see 2.1), where the force of evolution is from the competition among the (evolving) ANNs. The main difference between PDE and PCDE was on the evaluation of each individual. For implementation of PCDE, after completing the evaluation using the rule-based agent and grading with payoff function similar with PDE, each ANN will then be compared with a constant number of randomly picked ANNs from the population of the current generation. If the score of the ANN was greater than or equal to its opponent (the randomly picked ANN), it will receive a win. Furthermore, the ranking of first Pareto layer (by marking non-dominated solutions) will be based on the number of wins as the main evaluation criterion.

2.3 Pareto Co-evolutionary Differential Evolution with an Archive (*PCDE-A*)

Similar to PCDE, after grading each ANN using the payoff functions, a second competition is held. However, PCDE-A has an extra archive, which is used to store Pareto solutions at every 50th generation. Consequently, each ANN is compared to a minimum number of randomly picked ANNs (without repetition) from the archive. Only if the number of ANNs in the archive is less than the minimum required number of random opponents, then the opponent list will be filled with randomly picked ANNs from the population. Similarly, an ANN will receive a win, if its score was greater than or equal to its competitor. The number of wins will then be used as the main evaluation criterion for marking non-dominated solutions to rank the first Pareto layer. This evaluation value will be less dependent on luck for marking of dominated solutions because of the bounded set of evaluators.

2.4 Adaptive Evolution

In adaptive evolution, direction and/or magnitude of the strategy parameters' modification is decided using some form of feedback from the EA. However, in self-adaptive evolution, self-adaptation of parameters is the implementation of the **evolution of evolution** idea. A hybrid adaptive/self-adaptive evolution combines two adaptation methods mentioned above. Parameters consisting of the mutation rate and crossover rate are encoded into the chromosomes of individuals. Instead of undergoing genetic operations of mutation and recombination, these parameters will be varied by some deduction within a range. The highest ANN's score value of the current generation is used as the feedback from the EA. If the feedback comprises of a non-negative value, a deduction within a range will be applied to the strategy parameters. For all the experiments in this study, the probability of the initialization value and deduction range (for mutation rate and crossover rate) are 1 and between 0.0001 and 0.0005, respectively.

2.5 Cognitive Game AI Representation

The cognitive game AI is represented by a standard multi-layered feed-forward ANN. A board pattern is received as the input of ANN, and the output of ANN is a position of the board as the corresponding move. Each node of the hidden layer and output layer performs a sum of the weighted input strengths, subtracts off an adaptable bias term and passes the result through a sigmoid filter as shown in (1),

$$\frac{1}{1+e^{-x}} \tag{1}$$

where x is the sum of the weighted input strengths.

The ANN's input layer consisted of nine input nodes (with an additional bias unit), a hidden layer of varying size (between 1 and 10 hidden nodes with an additional bias unit) and the output layer consisted of nine nodes, where each of the input and output nodes corresponded to a square in the TTT grid. The

3 by 3 matrix board state is represented as a 2-dimensional 3 by 3 array of nine values. A blank open space was denoted by the value 0.0, an "X" was denoted by the value 1.0, and an "O" was denoted by the value (minus) -1.0. The two-dimensional array represents the current board pattern and is presented to the ANN to determine the move of the opposing player and correspondingly, the relative strengths of the nine output nodes were examined to determine the equivalent counter-move by the game AI system. An empty square's position with the maximum output strength was chosen as the output. This is to ensure only legal moves are made. Placed squares were ignored and selection pressure was not applied to force the output to zero [6].

3 Experimental Setup

This series of experiments was designed to examine and observe the effects of synthesizing TTT agents with and without the introduction of co-evolutionary techniques into the PDE algorithm. Settings of implementations of PCDE and PCDE-A involved in all experiments were directly adopted from the PDE setup (see section 2.1). Each experiment was repeated for 50 trial runs, each run was run for 800 generations (as in [6]). Table 1 shows the details of the experimental setup, the number and target location of randomly picked opponents.

Table 1. Experimental setup details. The main differences between each system were the number of opponents (that will compete with each candidate ANN in the population) and where the opponents were picked from.

System	No. of opponents	Origin of Opponent
PDE	0	None
PCDE1	30	Population
PCDE2	50	Population
PCDE3	70	Population
PCDE-A1	30	Archive & Population
PCDE-A2	50	Archive & Population
PCDE-A3	70	Archive & Population

After completing all of the experiments, all agents representing Pareto solutions at the 800th generation from each experiment are selected to be the representatives of each system. Non-dominated ANNs from all the Pareto multi-objective optimization systems mention above are able to synthesize both first and second players in a single run. Consequently, each selected ANN has to compete with three different levels of evaluation players, that are a near-perfect expert player (*VS Expert*), an average player (*VS Medium*) and a random player (*VS Random*). A competition consists of a set of 20 games for each level of player. A selected ANN will thus play a total of 60 games firstly as a first player and secondly as a second player.

4 Experimental Results and Discussion

4.1 Overall Performance of All Experiments

Figure 1 shows the overall performance of all experiments and is summarized by focusing on the number of lost games only. It clearly shows all experiments successfully produced agent(s) that never lost any games to all three level of players as the first player. Nevertheless, only PDE, PCDE2, PCDE3 and PCDE-A1 successfully produced intelligent agent(s) that never lost any game to the expert level of player as the second player. Figure 2 shows the global Pareto solution(s) layer for all the experiments in this study. Figures 3, 4, 5, 6 and Table 2 show details of the overall performances of the selected agents from each system involved in the competition against all three level of players as the first and second players.

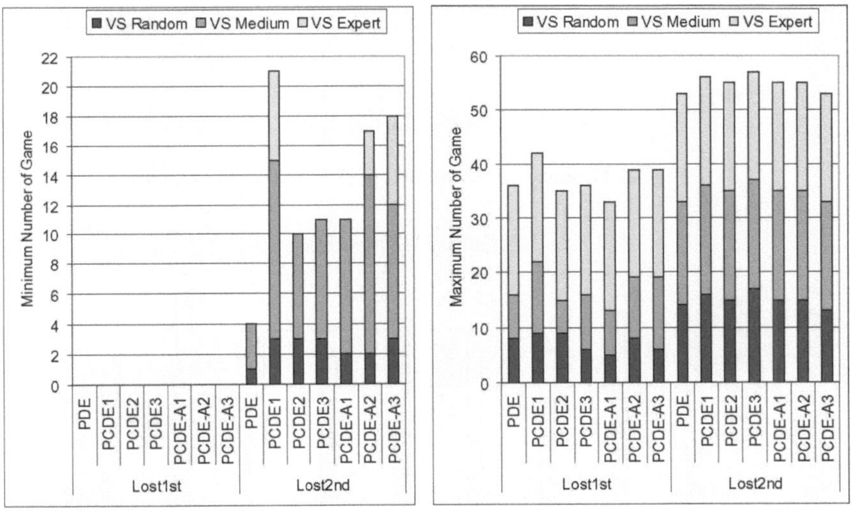

Fig. 1. This figure shows minimum and maximum number of games lost as the first player (*Lost1st*) and the second player (*Lost2nd*), respectively

4.2 The Introduction of Co-evolution

Figures 3 and 4 show the details of the overall performances of implementations from the PCDE system competing as the first and second players against three different levels (expert, medium and random) of evaluation players. Overall, PCDE2 was successful in outperforming PCDE1 and PCDE3 in terms of producing good performing agents both as first and second players.

Based on the results of the competition against the expert-level player, PCDE2 and PCDE3 were successful in producing agent(s) that never lost to the expert-level player as the first and second players. PCDE1 was successful in producing

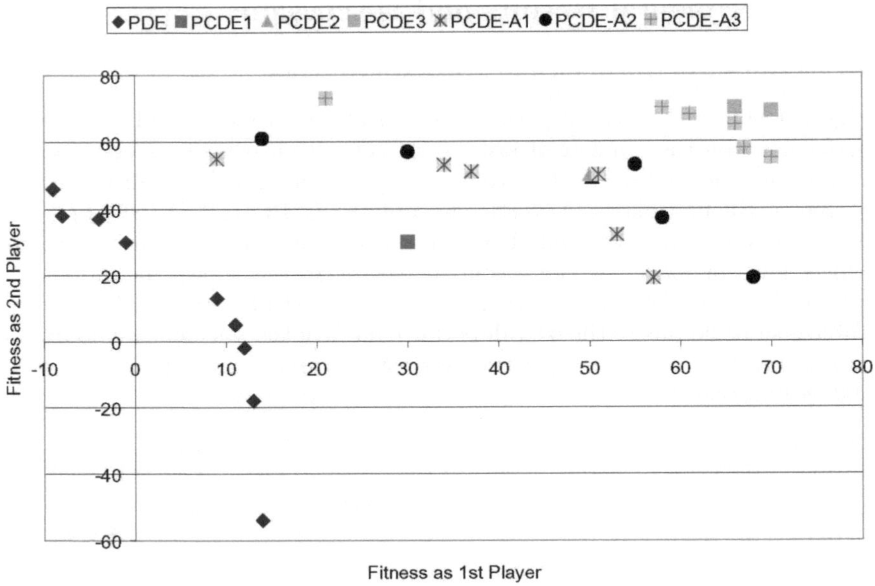

Fig. 2. Global Pareto Solution(s) of each experiment

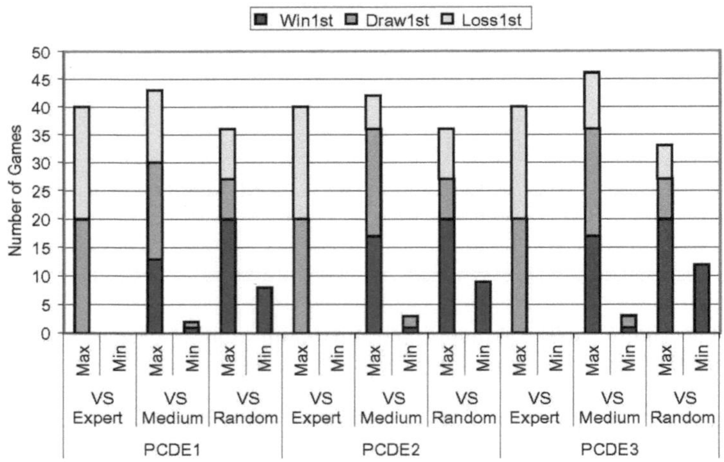

Fig. 3. Minimum (*Min*) and maximum (*Max*) number of games won (*Win1st, Win2nd*), drawn (*Draw1st, Draw2nd*) and lost (*Loss1st, Loss2nd*) as the first player, respectively from ANNs of PCDE

agent(s) that never lost any game to the expert-level player as the first player but its best second player lost 6 games to the expert-level player.

Based on the results obtained after competing with the medium-level player, PCDE2 performed slightly better than PCDE1 and PCDE3. Each selected agent

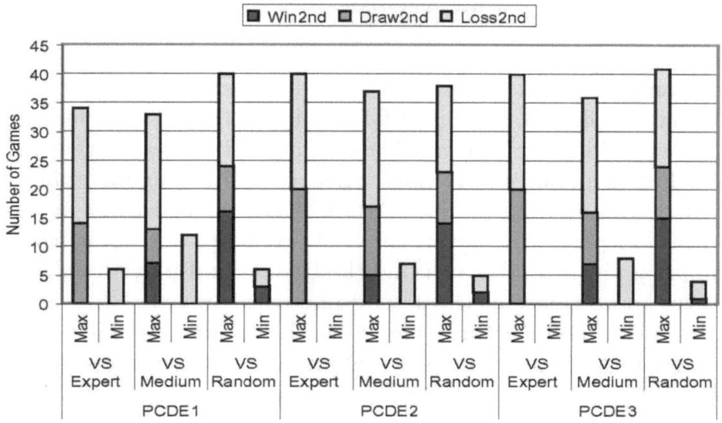

Fig. 4. Minimum (*Min*) and maximum (*Max*) number of games won (*Win1st, Win2nd*), drawn (*Draw1st, Draw2nd*) and lost (*Loss1st, Loss2nd*) as the second player from ANNs of PCDE

from PCDE2 had the lowest maximum number of lost games as the first player, and the lowest minimum number of lost games as the second player. Similarly, PCDE2 and PCDE3 were performing better then PCDE1, where both had the highest maximum number of games won as the first player. Nevertheless, PCDE3 was performing slightly better then PCDE1 and PCDE2 when competing against the random player. The minimum number of wins as the first and second player of the selected agent(s) from PCDE3 was slightly higher then other agents. In addition, the maximum number of losses as the first player of selected networks from PCDE3 was also slightly lower than other representatives as well.

The only difference between each implementation of this algorithm (PCDE) is the number of randomly picked opponent. The increase (from 30 to 50 comparisons) successfully improved the performance of PCDE but the performance decreased for PCDE3, which is closer to a round-robin (70 comparisons). The marking of dominated agents in PCDE is based on the number of wins obtained from the evaluation against a constant number of randomly picked opponents from the population of the current generation. This evaluation is only presenting "performance of the current generation" with a high probability of luck involved. All implementations of PCDE did not have a good spread of global non-dominated solution. Furthermore, PCDE3 and the other two implementations of PCDE converged into two and a single non-dominated point(s) respectively (see Figure 2). Hence, it proves that the PCDE systems were not able to fully exploit the range of good solutions between the first and second players offered by the Pareto evolutionary optimization process. Furthermore, although it may appear that the other systems' solutions dominate the canonical PDE solutions, post-evolution empirical evaluation proves otherwise (see Section 4.5).

4.3 Co-evolution with an Archive

Figure 5 shows details of the overall performances of implementations from PCDE-A competing as the first and second players against three different levels of evaluation players. Overall, PCDE-A1 was the best implementation from PCDE-A, outperforming PCDE-A2 and PCDE-A3. Selected agents from PCDE-A1 were able to perform very well when competing against the expert-level player. Furthermore, the best agent from PCDE-A1 never lost any game as the first and second player. However, the best second-player agent from PCDE-A2 lost 3 games as the second player and the best second-player agent from PCDE-A3 lost even more games as the second player (6 games).

The discussion will continue based on the results obtained from the competition against the medium-level evaluator. Representatives from PCDE-A1 successfully outperformed representatives from the other setups. They have the lowest minimum number of losses as the second player (9 games) and the lowest maximum number of losses as first player (8 games). Similarly, agents from PCDE-A3 have the lowest minimum number of losses as the second player. However, PCDE-A2 did not perform well, since its agents have the highest minimum number of losses as the second player (12 games). Moving the focus of this discussion to the competition against the random player, PCDE-A1 again had the best performance compared with the other two implementations of PCDE-A. Representatives from PCDE-A1 have the lowest number of minimum losses as the second player (as well as PCDE-A2, 2 games) and the lowest maximum number of losses as the first player (5 games). The maximum number of losses as the second player of PCDE-A3's agent(s) was slightly lower (13 games), whereas the maximum number of losses as the second player of PCDE-A1's agents and PCDE-A2's agent(s) were both 15 games.

The PCDE-A has an embedded archive, that stores Pareto solution(s) of every 50th generation. At generation of 800th, the number of agents stored in the archive can approach 50. The addition of an archive is to have a better quality of evaluation in terms of play strength representation and fairness by having the comparison against a similar set of opponents. Overall, PCDE-A1 was the best implementation of PCDE-A, outperforming the other two PCDE-A systems. PCDE-A2 and PCDE-A3 were initialized with a larger number of randomly picked opponents (50 and 70 opponents, respectively). Hence, PCDE-A2 and PCDE-A3 may still be randomly picking opponents from the population of current generation for comparison until the 800th generation. However, since PCDE-A1 was initialized with a smaller number of comparisons (30 opponents), the archive will contain more non-dominated opponents compared to randomly picked opponents much earlier in the evolutionary run compared to the other two systems with larger archives. Thus, the most important observation in the success of the evolution of PCDE-A1 is that the number of randomly picked opponent was lower. The number of randomly picked opponent of PCDE-A1 was almost zero after the 500th generation.

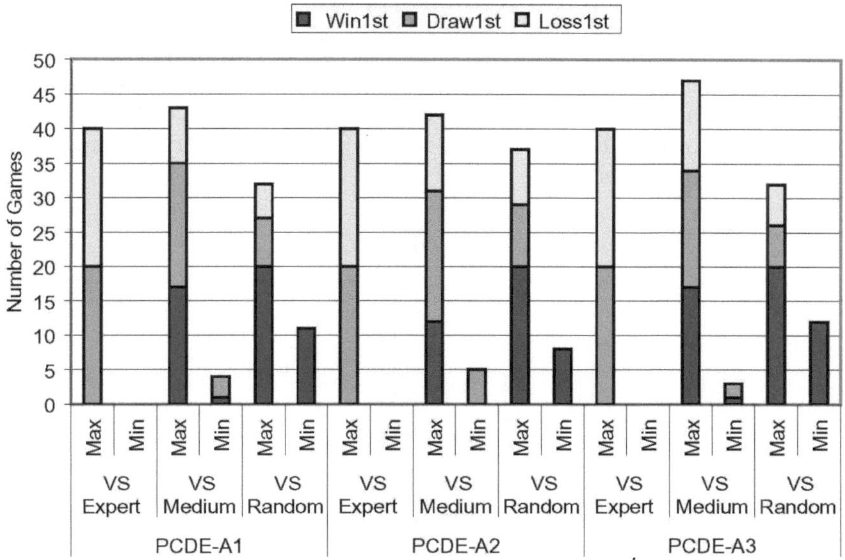

(a) Play as the 1st player.

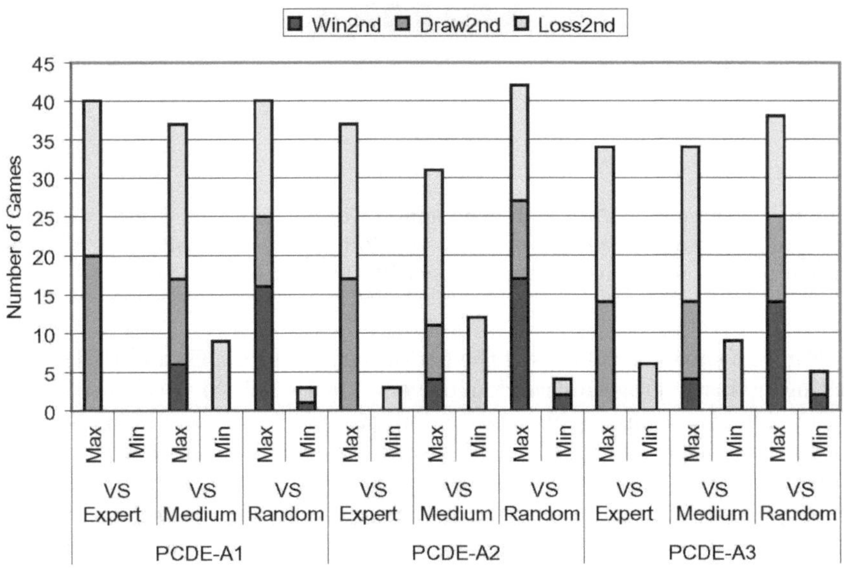

(b) Play as the 2nd player.

Fig. 5. Selected ANNs from implementations of PCDE-A competing against a near-perfect expert player (*VS Expert*), an average player (*VS Medium*) and a random player (*VS Random*). This figure shows the minimum (*Min*) and maximum (*Max*) number of games won (*Win1st, Win2nd*), drawn*Draw1st, Draw2nd*) and lost (*Loss1st, Loss2nd*) as the first and second players, respectively.

4.4 Performance With/Without the Additional Archive

PCDE2 and PCDE-A1 were the best systems among implementations of its own algorithm. To look into the effects of the additional archive, the discussion continues with the comparison between PCDE-A1 and PCDE2 only, because the performance of PCDE2 and PCDE-A1 were the same in the competition with the expert-level player.

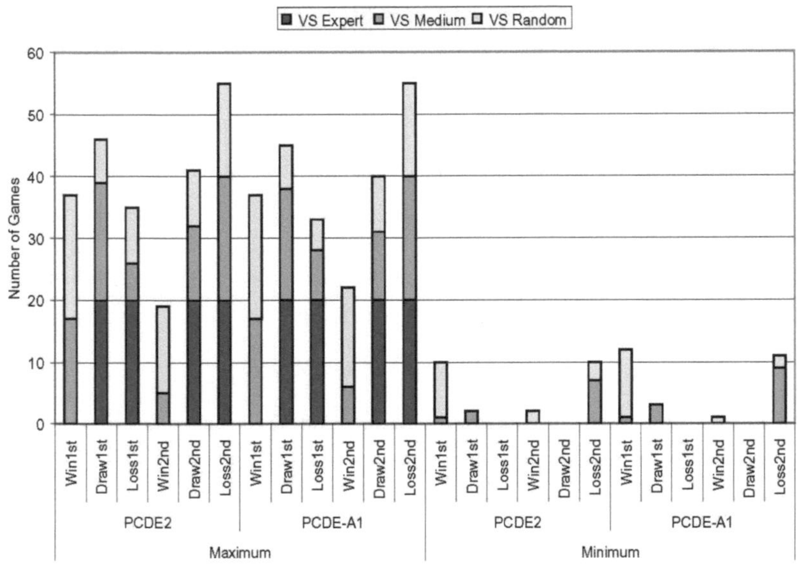

Fig. 6. Minimum (*Min*) and maximum (*Max*) number of games won (*Win1st, Win2nd*), drawn (*Draw1st, Draw2nd*) and lost (*Loss1st, Loss2nd*) as the first and second players, respectively from ANNs of PCDE2 and PCDE-A1

Figure 6 shows the details of the performances of PCDE2 and PCDE-A1 competing against average and random players. The PCDE-A1 performed better than PCDE2 in competing against the random player. The number of maximum losses as the first player of PCDE-A1 (5 games) was smaller than PCDE2 (9 games). Furthermore for the competition against the random player, PCDE-A1 also has a slightly higher minimum number of wins and maximum number of wins as the first and second players, respectively. However for the competition against the medium-level player, PCDE2 performed slightly better than PCDE-A1.

Continues with the focus on the suitability of both algorithms in terms of Pareto multi-objective optimization. Figure 6 clearly shows the suitability of PCDE-A1 was significantly higher than the PCDE2. Implementations of PCDE-A (especially PCDE-A1) have a significantly better spread of global non-dominated solution compared to implementations of PCDE. Marking of dominated solutions in PCDE-A was based on the number of wins obtained from evaluation against agent(s) from the archive (as well as randomly picked agent(s) from population of the current generation, only if the size of the archive is less then the required minimum number of comparisons). This evaluation method is thus presenting more than the "performance of the current generation", more global (over the whole process of co-evolution) and less elements of luck, since the evaluation is using the same set of Pareto solutions. PCDE-A1 was randomly picking agent(s) from the population before the 500th generation approximately. However, for the rest of the evolution, PCDE-A1 was evaluated using agents from the archive only, since the size of the archive had already exceeded the minimum number of comparisons.

4.5 Performance Without Co-evolution

Table 2 shows the overall performances of PDE, PCDE2 and PCDE-A1 competing against three different levels of evaluation players. PCDE2 and PCDE-A1 were the best systems among implementations of its own algorithm. Overall, PDE was successful in outperforming PCDE2 and PCDE-A1 in terms of producing good performing agents as both the first player and second player. Compared to agents from PCDE2 and PCDE-A1, agent(s) from PDE were performing better as a second player, since they have the lowest minimum and maximum number of losses when competing against the medium and random players as the second player. However, compared to representatives of PDE, agents of PCDE2

Table 2. Minimum (*Min*) and maximum (*Max*) number of games won (*Win1st, Win2nd*), drawn (*Draw1st, Draw2nd*) and lost (*Loss1st, Loss2nd*) as the first and second players, respectively from ANNs of PDE, PCDE2 and PCDE-A1

			As 1st Player			**As 2nd Player**		
			Win	Draw	Lost	Win	Draw	Lost
Maximum	PDE	VS Expert	0	20	20	0	20	20
		VS Medium	17	18	8	6	17	19
		VS Random	20	7	8	14	9	14
	PCDE2	VS Expert	0	20	20	0	20	20
		VS Medium	17	19	6	5	12	20
		VS Random	20	7	9	14	9	15
	PCDE-A1	VS Expert	0	20	20	0	20	20
		VS Medium	17	18	8	6	11	20
		VS Random	20	7	5	16	9	15
Minimum	PDE	VS Expert	0	0	0	0	0	0
		VS Medium	2	0	0	0	0	3
		VS Random	8	0	0	2	0	1
	PCDE2	VS Expert	0	0	0	0	0	0
		VS Medium	1	2	0	0	0	7
		VS Random	9	0	0	2	0	3
	PCDE-A1	VS Expert	0	0	0	0	0	0
		VS Medium	1	3	0	0	0	9
		VS Random	11	0	0	1	0	2

and PCDE-A1 performed slightly better as the first player when competing against the medium-level player and random player, respectively.

Furthermore, PDE uses the fitness values from the payoff functions directly to mark dominated solutions, which is more suitable in Pareto multi-objective optimization as justified by the results of the comprehensive testing above. The PDE has the best spread of global non-dominated solutions on the Pareto frontier compared with frontiers from other systems.

5 Conclusion

This paper reports the first comprehensive study in evolving cognitive systems for game AI of TTT using Pareto evolution as well as co-evolutionary techniques. Overall, the canonical PDE system was able to automatically synthesize ANN as game-playing TTT agents successfully both as the first and second players without the introduction of co-evolution. Furthermore, PDE was successful in outperforming all the other systems in terms of producing high play-strength game agents as the second player. Moreover, using an archive in PCDE-A did not produce significantly better results as first expected. Using an evaluation value which represents a bigger picture as PDE (not the current generation only) and is less dependent on luck for marking of dominated solutions should introduce significant effects on the performance of the Pareto multi-objective optimization process.

The poor performance of the PCDE systems, even those utilizing an archive, in producing a good spread of solutions along the Pareto front is further proof that co-evolutionary methods are not particularly beneficial for synthesizing intelligent agents for game AI in Pareto evolution. In terms of playing strength, PCDE2 was successful in outperforming PCDE1 and PCDE3, similarly PCDE-A1 was the best system compared with PCDE-A2 and PCDE-A3. Only PDE, PCDE2 and PCDE-A1 were successful in outperforming all other systems in terms of producing high play-strength game agents that never lost any game to the expert-level player both as the first and second players. Based on the performances of PCDE and PCDE-A systems, the co-evolution process was very sensitive to the number of randomly picked opponents. Initializing the co-evolution process with a suitably small number to limit the number of randomly picked opponents early enough in the evolutionary process can cause significant effects on the performance of co-evolutionary process.

Lastly, it is also shown clearly here that all the implementations of the co-evolutionary algorithms faced one problem, that is the "forgetting" problem. Although the majority of agents can win or draw against an expert-level player, they "forgot" how to win or draw against medium-level and random players. This is one of the known problems in co-evolutionary techniques. This problem is particularly obvious when the focus is on the synthesis of intelligent agents that act as the second player. As such, this should be investigated further in future work and may be utilized at the same time as a suitable test bed for evaluating new methods proposed for solving the "forgetting" problem in co-evolutionary algorithms.

Acknowledgments. The authors wish to thank MIMOS Berhad for providing time on their computational cluster to conduct the evolutionary runs. In particular, we wish to thank J.Y. Luke, H. Kauthary and W.K Ng for their time and assistance. We are also grateful to the anonymous reviewers for their helpful comments.

References

1. J. Davis and G. Kendall. An investigation, using coevolution, to evolve an awari player, 2002. In *Proc. of the IEEE CEC '02*, pp. 1408-1413,2002
2. G. A. Monroy, K. O. Stanley, and R. Miikkulainen. Coevolution of neural networks using a layered pareto archive. In *Proc. of GECCO '06*, pp. 329–336, New York, NY, USA, 2006. ACM Press.
3. L. Panait and S. Luke. A comparison of two competitive fitness functions. In W. B. L. et al, editor, *Proc. of GECCO '02* , pp. 503–511. 2002.
4. H. A. Mayer and P. Maier. Coevolution of neural Go players in a cultural environment. In *Proc. of the IEEE CEC '05*, Vol. 2, pp. 1017–1024, 2005.
5. A. Lubberts and R. Miikkulainen. Co-evolving a go-playing neural network. In R. K. Belew and H. Juillè, editors, *Coevolution: Turning Adaptive Algorithms upon Themselves*, pp. 14–19, San Francisco, California, USA, 2001.
6. K. Chellapilla and D. B. Fogel. Evolution, neural networks, games, and intelligence. In *Proceedings of IEEE*, 87(9):1471–1496, 1999.
7. Y. J. Yau and J. Teo. An empirical comparison of non-adaptive, adaptive and self-adaptive co-evolution for evolving artificial neural network game players. In *Proc. IEEE CIS '06*, pp. 405–410, 2006.
8. Y. J. Yau and J. Teo. Co-evolving Neural Agents for Game AI: From Single-objective to Pareto Multi-objective Evolution In *The 22nd ACM Symposium on Applied Computing (SAC'07)*. 2007, Seoul, Korea. (Submitted).
9. H. A. Abbass and R. Sarker. The Pareto Differential Evolution Algorithm. In *International Journal on Artificial Intelligence Tools*, 11(4):531–552, 2002.

The Development of a Multi-threaded Multi-objective Tabu Search Algorithm

Peter Dawson[1], Geoff Parks[1], Daniel Jaeggi[1], Arturo Molina-Cristobal[2], and P. John Clarkson[1]

[1] Engineering Design Centre, Department of Engineering, University of Cambridge, Trumpington Street, Cambridge CB2 1PZ, UK
[2] Electrical Machines and Drives Group, Department of Electronic and Electrical Engineering, University of Sheffield, Mappin Street, Sheffield S1 3JD, UK

Abstract. The reliance of Tabu Search (TS) algorithms on a local search leads to a logical development of algorithms that use more than one search concurrently. In this paper we present a multi-threaded TS algorithm employing a number of threads that share information. We assess the performance of this algorithm compared to previous multi-objective TS algorithms, via the results obtained from applying the algorithms to a range of standard test functions. We also consider whether an optimal number of threads can be found, and what impact changing the number of threads used has on performance. We discover that, contrary to the popular belief that multi-threading is usually beneficial, performance only improves in a few special cases.

1 Introduction

Whilst there has been much research into other types of algorithms for multi-objective optimization [1], Tabu Search remains an option under-represented in the volume of research in this area. Following the recent development of a multi-objective TS algorithm, presented in [2], there is an opportunity to investigate variants of this algorithm. We are specifically interested in a multi-threading TS algorithm. The integral part played by a local search in the TS algorithm points towards potential benefit from using many local searches at once. These parallel searches are known as threads, and together they form a multi-threaded algorithm. Single-objective multi-threading strategies have been heavily researched. An overview is given in [3] (see particularly Type 3 parallelisation), and examples of specific implementations of multi-threaded single-objective TS are presented in [4,5]. Research in multi-objective meta-heuristics is currently following the pattern of studies previously undertaken for single-objective problems, and the fact that multi-threaded single-objective TS implementations exist provides us with more motivation to develop multi-objective variants of such algorithms.

2 Background

2.1 Multi-objective Optimization

The problem we seek to solve is that of multi-objective optimization. We wish to find a vector $\mathbf{x}$ that minimizes all the components of the vector $F(\mathbf{x}) =$

S. Obayashi et al. (Eds.): EMO 2007, LNCS 4403, pp. 242–256, 2007.

$(f_1(\mathbf{x}), f_2(\mathbf{x}), f_3(\mathbf{x}), ...)$. The components of $\mathbf{x}$ are known as design variables and each $f_i(\mathbf{x})$ is an objective function. We usually seek values for design variables only in a predetermined range, and there may also be functional constraints on the variables.

Each set of vectors, those of design variables and those of objective functions, forms a vector space. We call these variable/decision/design space and objective space respectively. When we talk of searching a space we mean variable space, whilst when we consider the outcomes of our optimization algorithm we almost always refer to objective space.

Our aim is to minimize all our objective functions. However, we almost always work with situations where different points in variable space minimize different objective functions. We therefore must use the idea of Pareto dominance. Consider two points in variable space, $\mathbf{x}$ and $\mathbf{y}$, with objective functions $F(\mathbf{x})$ and $F(\mathbf{y})$. We say $\mathbf{x}$ *dominates* $\mathbf{y}$ if no component of $F(\mathbf{x})$ is greater than the corresponding component in $F(\mathbf{y})$ and at least one component is smaller. If neither $\mathbf{x}$ dominates $\mathbf{y}$ nor $\mathbf{y}$ dominates $\mathbf{x}$ the two points are said to be Pareto equivalent.

Now our aim is clearer – we wish to find a set of Pareto equivalent points not dominated by any point outside the set. This represents the trade-off surface between objective functions, and it is impossible to say that any one point is better than another within this set without weighting the objectives.

2.2 Tabu Search

The basic idea of Tabu Search is that of a local search looking at one point at a time before moving to the next, and maintaining three memories along the way. These are termed respectively the Short. Medium and Long Term Memories (STM, MTM, LTM). TS was developed for single-objective optimization problems and this is reflected in the original use of these memories. The STM is used to store the points that have been recently visited and then stops the search from revisiting these points, hence the method's name. The MTM stores optimal and near-optimal points found during search. The LTM stores information about all points visited. An important feature of TS is that an 'uphill' move will be made in preference to converging at a local minimum, as "a bad strategic choice can yield more information than a good random choice" [6].

The implementation of a multi-objective TS algorithm in this paper follows closely that presented by Jaeggi et al. in [2]. The functions of the three memories change, and a fourth – an Intensification Memory (IM) – is added. These changes are detailed in Sect. 3.1. The functions of the memories must change slightly to be compatible with a multi-threading algorithm which shares information, but all changes endeavour to remain consistent with the basic tenets of TS.

3 Multi-threaded Multi-objective TS Implementation

The algorithm presented in [2] has a single thread starting at a given point. The search then proceeds in an iterative manner; at each iteration a new point is selected via a variety of methods. The standard method is the Hooke and Jeeves

(H&J) move [7]. After each of these, a pattern move is attempted. The other types of move occur less often. If a certain numbers of iterations pass without an addition to the MTM, intensification, diversification or step-size reduction (SSR) moves occurs. Each of these moves will be explained in more detail in the following sections.

A multi-threaded multi-objective TS algorithm can intuitively be considered as a number of these local searches working in parallel. However, this on its own gives very little advantage over the single-threaded search. Therefore we must change the structure of the algorithm such that information collected by one search thread is available to all of the threads thereafter. We do this by introducing a master program, which keeps some of the memories globally. Each thread retains its own STM and an individual list of points it has ever visited. The MTM, IM and LTM are kept by the master program, and in this way a thread may intensify, diversify or reduce step-size with knowledge of the progress of all threads. It also prevents interference between the threads – we now can be sure a thread will not diversify, for example, to an area already heavily searched by another thread. The new implementation still has, within each local search thread, the possibility of H&J, pattern, intensification, diversification and SSR moves. The difference lies in the memory usage, and the consequent sharing of information between threads.

3.1 The Memories

During the search, the algorithm maintains four memories: the STM, MTM, LTM and IM. Our use of these memories differs notably from that in single-objective TS and will now be described.

The STM retains its functionality as a list (of user-specified size) of points recently visited by the search thread, updated on a first-in, first-out basis. The thread is not allowed to revisit these points.

The MTM is an unbounded set of globally non-dominated solutions found by all threads. As new points are evaluated they become candidates for addition to this set. Points dominated by newly added points are removed. The MTM is used by the search during a SSR move, and its final contents represent the main output from the optimization.

The LTM records the regions of the search space which have been explored, and is used on diversification, directing the thread to regions which are under-explored. Information is recorded by dividing the allowed range for each design variable into a certain number of regions and counting the number of solutions evaluated by all threads in those regions.

Like the MTM, the IM is an unbounded set of Pareto equivalent points. All points that are added the MTM are also added to the IM (and IM points dominated by them removed). However, if a point in the IM is then chosen to be the next point in a thread as a result of any type of move (H&J, intensification, SSR etc.), then it is removed from the IM. Thus the IM maintains a list of non-dominated points found by the threads but not subsequently explored, which are then available for further exploration when intensification occurs.

There is also an important difference made to the way in which points are rendered tabu. On the consideration of any point, we deem it tabu if it is in the STM of any thread. In this way two threads are prevented from searching in the same area simultaneously.

3.2 The Hooke and Jeeves Move and Pattern Move

Each local search at each stage has a current position in variable space, and moves are used to change this position, and to consider new points for addition to the MTM and IM. In the algorithm the standard move used is a H&J move [7]. This is performed by creating a set of neighbouring points, each determined through either adding or subtracting a given step-size from the value of one variable. This new set is then checked for points that violate constraints or are tabu, and these are removed. Having evaluated the objectives for these remaining points, the Pareto-optimal set (POS) among them is identified, and a random point from this set is chosen as the next point for the local search thread, while the other set members are considered as candidates for addition to the IM. Note that this means that even if the POS of the neighbours does not dominate the current point, we still move, in an 'uphill' direction.

This basic move is, in our implementation, augmented by a roulette wheel sampling strategy. At initialization, after an intensification, diversification or SSR, and also every so many iterations (as a user-set parameter), the distance of each neighbour from the current POS in the MTM is calculated, as in the Path Relinking strategy outlined in [2]. Then each neighbour is assigned a probability inversely proportional to this distance, and a random neighbour is chosen with these probabilities. This process is repeated until we have chosen the desired number of neighbours to sample.

This means that at each move, a random sample of a given size is taken from the set of neighbouring points, and objective function evaluations are only performed on these points. If the POS of sampled neighbours does not dominate the current point then another sample is taken, and so on. If all neighbours are considered and none are found to dominate the current point, then an 'uphill' move is made, as before, by considering the POS of the entire set of neighbours. This sampling strategy uses at worst the number of objective function evaluations employed in the standard H&J move, and at best many fewer evaluations. This serves to improve the performance of the algorithm in terms of speed by sacrificing the best direction for simply a better one, in an intelligent manner.

After each H&J move, a pattern move is performed. This checks only the point found by incrementing the variable changed in the preceding move by the same amount again. This point's objective functions are evaluated. If it dominates the current point, it becomes the new current search point; if it does not, the previous point is retained. This serves to accelerate motion once a 'downhill' direction has been found.

3.3 Intensification, Diversification and Step-Size Reduction

If specified numbers of iterations pass without a successful addition to the MTM by an individual thread (each thread has its own independent counter for this purpose), we conclude that the thread's current strategy needs to be changed. This is the rationale behind the intensification, diversification and SSR moves.

When intensification occurs, a random point is chosen from the IM as the next point of search. This point is removed from the IM so that it is not chosen again or by a different thread. Tabu points are not allowed to be chosen.

Diversification is slightly more complicated. The search space is divided into distinct regions, with the number of solutions from all threads evaluated in each region being tallied in the LTM. At diversification a number of candidate next points are generated by choosing regions, with the probability of selection being proportionately higher for less explored regions, and then choosing a point randomly within the chosen region. Tabu and infeasible points are removed and the remaining points evaluated. The POS of these solutions is identified and one of these randomly chosen as the next position for the thread.

The SSR move simply multiplies the step-sizes used for finding the neighbours in a H&J move by some given number between 0 and 1. The thread then proceeds anew from a point randomly selected from the MTM.

3.4 Parallelisation Strategy

As real-world problems are likely to have computationally expensive objective functions and many design variables, optimizations are likely to be run on clustered processors to ensure a viable run-time. We have therefore designed our algorithm for such use, using an MPI/LAM environment.

There are now two possibilities for parallelisation: functional decomposition or separation of threads (Type 1 or 3 parallelisation [3]). The second approach could follow the structure of the algorithm very closely: the threads could work not only independently but on separate processors, sending back information to the master program. However, despite offering a large potential increase in speed, this requires a complicated system of information exchange. In this study we consider the algorithm parallelised via functional decomposition, and suggest parallelisation of threads as holding much potential for future study.

Functional decomposition effectively means that at each move, there is a specific point where all the possible positions to which each thread may move are known by the master program. At this time, all of the function evaluations are requested at once, and farmed out to all the slave processors (and to the processor on which the master program runs). This allows an approximately linear increase in speed with the addition of more slave processors, up to $n_{\text{threads}} \times n_{\text{sample}}$ processors – the maximum number of requested points at each move.

3.5 Constraint Handling

Throughout the investigations in this study, we use a method of binary constraint handling – where points are either allowed or not. This is very suitable to the kinds of constraints considered in our test problems, which are simple and solely serve to bound variable space. However, TS lends itself well to highly constrained problems due to its use of a local search, which reduces the likelihood of violating constraints once feasible space has been located, and our algorithm is easily adapted to the use of penalty functions or modified dominance relations necessary for handling complex constraints.

4 Test Procedures

To assess the performance of the multi-threading version of our multi-objective TS algorithm, we undertook a rigorous series of tests on benchmark problems.

4.1 Test Functions

The problems chosen are Zitzler, Deb and Thiele's suite of six problems [8], excluding number 5. The design rationale behind these problems is given in [9]. They are chosen here as they are easy to implement and relatively fast to compute, whilst still providing a challenge for the optimizers. Their relevance to real-world problems is debatable, but they serve as a useful starting point for performance assessment.

4.2 Performance Assessment Using Unary Indicators

The key idea in assessing the performance of an optimizer is assigning a measure of quality to an approximation set (the Pareto front found in a single run). We therefore wish to reduce a run of our algorithm to a single number representing how 'good' it was, which we can then compare to other algorithms or variants run on the same problem. We use two such indicators in this study, the *unary epsilon indicator*, and the *hypervolume indicator*. These both require a reference set with respect to which we measure the quality of the approximation set.

In this study we reject the idea of using an ideal set as a reference set, as it is rare that this can be generated analytically in any real-world situation – if the answer is already known there is little point solving the problem! Instead we use a 50% attainment surface of 1000 runs of a random search. The random search simply picks a single point in the search space and evaluates the objective functions at it, so after 1000 runs we have 1000 points each forming an approximation set on their own. An $n\%$ attainment surface is formed of those points that are dominated by $n\%$ of the approximation sets under consideration. It is possible to represent an attainment surface using a finite set of points at its corners. As each approximation set is finite – this is always the case with any set generated by an optimization algorithm taking a finite amount of time, but is particularly easy to see in this case as each set has size 1 – and there are a finite number of them, any attainment surface will take the form of a stepped line in objective space with a finite number of vertices. These are then the points taken as the reference set for use in calculating unary indicators for all algorithms under test. The 50% attainment surface we are using converges quickly with respect to the

Table 1. Bounds chosen for the calculation of indicator values

Problem	Upper Bound	Lower Bound
ZDT1	(1,7)	(0,0)
ZDT2	(1,7)	(0,0)
ZDT3	(1,7)	(-1,0)
ZDT4	(1,700)	(0,0)
ZDT6	(1,10)	(0,0)

number of runs considered [11], so 1000 runs should be enough to render our results comparable to others taken using this reference set.

In order to eliminate weighting of the objective functions in the calculation of indicator values, we must normalize our approximation set. We now think of our set as points in objective space. We must perform operations which leave all coordinates of all points in our reference set and approximation sets lying between 1 and 2. In this way each coordinate contributes a comparable amount to the distances or volumes in objective space used in calculating indicators. In order to normalize the coordinates in this way we must choose bounds in objective space with respect to which we transform the space. These are the points that will become (1,1,...) and (2,2,...). In this study we explicitly choose these points globally, instead of normalizing with respect to the bounds of the approximation sets given by runs of a specific algorithm, in order to ensure that comparisons between indicators assigned to runs of different algorithms are the same. See Table 1 for the values chosen for each test problem.

Unary Epsilon Indicator. The additive unary epsilon indicator, proposed by Zitzler et al. [10], makes direct use of Pareto dominance, and hence is highly intuitive. For two approximation sets A and B, the epsilon indicator can be thought of as a measure of the minimum distance we must shift set B by so that A only just dominates it. If set A dominates set B then the indicator value will be negative, and if B dominates A then the value will be positive. In the case of Pareto equivalence it could be either.

We transform this binary indicator into a unary indicator by fixing one of the sets compared, set B. We then call this set a reference set and label it R. We use the 50% attainment surface described above as our reference set. As for any reference set, a lower indicator value indicates a (possibly) better approximation set. However, we must be careful, as we expect that our approximation sets will be a definite improvement upon our reference set, and therefore our indicator values will be negative. We consider lower, and so larger, values to be better.

Hypervolume Indicator. For a given point in an approximation set and a given reference point, considering both in objective space, we can define a unique hyper-rectangle with these two points at opposite vertices (defining its orientation as parallel to our axes). If we do this for all points in the approximation set, and take the hypervolume of the union of the hyper-rectangles given, this gives us an indicator of the quality of the approximation set [10]. In this study we choose the point (2,2,...) as our reference point. This indicator can also be used in conjunction with a reference set – the hypervolume for each set is found and then the difference is taken. We make use of this with our 50% attainment surface of the random search as the reference set.

The Kruskal-Wallis Test. In order to compare indicator values obtained from the different algorithms rigorously, we use a statistical test called the Kruskal-Wallis test. This is an extension of the Mann-Whitney test which allows more than two sets of data to be compared at once. This test is chosen as we wish to make no assumptions about the distribution of the approximation sets and so must use a nonparametric test. There are two main types of nonparametric tests – rank tests and permutation tests [11]. Here we choose a rank test as it is computationally cheaper. The output of the test is in the form of p-values – these represent the certainty with which we can say that one algorithm's indicator values are better than another's. We take our null hypothesis to be that all of the algorithms tested perform equally well. We only accept that one is better than another if the p-values are in the ranges [0,0.05] and [0.95,1], thus carrying out a two-tailed test at a 10% significance level. Full details of this test (and the Mann-Whitney test) can be found in [12].

4.3 Details of the Procedure

Each algorithm was run 45 times on each test problem for each stopping criterion. The random number generator for each run was initialised with a different seed.

The algorithms tested were the original single-threaded multi-objective TS algorithm (termed MOTS), based on that in [2], and the multi-threaded variant presented here with between 1 and 6 threads (MTMOTS1-6).

MOTS and MTMOTS1 have been extensively tested using seeding of the random number generators to ensure that they perform identically. They have been both included in the results tables to illustrate the level of uncertainty to which we are exposed in comparing the results of different algorithms, due to the random selection occurring throughout the implementation.

The stopping criterion used was a limit on the number of points in variable space for which the (two in the case of each of our test problems) objective functions are evaluated. Runs were performed with this limit set at 1000, 5000 and 10000 evaluations. (The other MOTS and MTMOTS parameter settings are detailed in Table 2. Advice about appropriate choices for these parameter settings can be found in [13] and [2].)

This means that, for the multi-threaded algorithms, each thread may only perform $1000/n_{\text{threads}}$ evaluations. We are therefore setting a high target for a multi-threaded algorithm to outperform the original MOTS algorithm – it must find a better POS in the same overall number of evaluations, meaning that each thread has many fewer. We rely on the benefits of intercommunication between threads and the advantage of considering more of the search space at once to gain performance improvements.

Table 2. MOTS and MTMOTS parameter settings

Parameter	Description	Value
n_{stm}	Size of the STM	10
n_{regions}	Search space split into $n_{\mathrm{var}} \times n_{\mathrm{regions}}$ regions in the LTM	4
n_{sample}	Number of points randomly sampled at each move	6
n_{tasks}	Number of diversify points considered	6
$i_{\mathrm{intensify}}$	Number of iterations after a successful addition to the MTM at which an intensification move is performed	25
$i_{\mathrm{diversify}}$	Number of iterations after a successful addition to the MTM at which a diversification move is performed	45
i_{reduce}	Number of iterations after a successful addition to the MTM at which a SSR move is performed	90
i_{select}	Re-evaluate the distance of neighbours to the Pareto front every i_{select} iterations	20
h	Step-size (as a percentage of the range of each variable)	8%
δh	Step-size reduction factor	0.5

Table 3. Indicator value summary for tests at 1000 evaluations

Problem			ZDT1	ZDT2	ZDT3	ZDT4	ZDT6
MOTS	hyp	mean	-0.47189861	-0.48296787	-0.24807474	-0.47282969	-0.58526265
		std dev	0.03379669	0.00163838	0.00059581	0.01395013	0.04665498
	eps	mean	-0.57978382	-0.48296787	-0.24807474	-0.47282969	-0.58526265
		std dev	0.04564560	0.00163838	0.00059581	0.01395013	0.04665498
MTMOTS1	hyp	mean	-0.46170811	-0.63888054	-0.28958053	-0.51768275	-0.31132050
		std dev	0.03994537	0.04726613	0.02531064	0.02899083	0.04785066
	eps	mean	-0.56829806	-0.44377244	-0.23126602	-0.13501031	-0.38764545
		std dev	0.05329170	0.03683993	0.01066563	0.02314708	0.06473030
MTMOTS2	hyp	mean	-0.42155799	-0.66066378	-0.29790510	-0.53065500	-0.28508967
		std dev	0.02523835	0.03463798	0.01657247	0.02953229	0.05478347
	eps	mean	-0.52871405	-0.46555037	-0.23336463	-0.48376550	-0.34037449
		std dev	0.04117396	0.01258062	0.01120566	0.01470215	0.07162364
MTMOTS3	hyp	mean	-0.39276954	-0.61003153	-0.25719158	-0.52037129	-0.27124040
		std dev	0.02672701	0.04793333	0.03163013	0.03058630	0.04381376
	eps	mean	-0.48826261	-0.43763577	-0.20646111	-0.48149464	-0.32478213
		std dev	0.04041251	0.05131813	0.04161739	0.01979420	0.04722428
MTMOTS4	hyp	mean	-0.37342153	-0.51284643	-0.23662673	-0.53146441	-0.26822716
		std dev	0.02737189	0.05098687	0.02015291	0.02437211	0.04058641
	eps	mean	-0.46254002	-0.38467187	-0.21076098	-0.14486690	-0.30225486
		std dev	0.03886867	0.05561409	0.02543820	0.02349137	0.06412030
MTMOTS5	hyp	mean	-0.35064797	-0.45925508	-0.20942130	-0.53204800	-0.26747420
		std dev	0.02056303	0.04259167	0.01490424	0.02360550	0.03895329
	eps	mean	-0.42753125	-0.34071154	-0.18955525	-0.14440422	-0.31403610
		std dev	0.02972445	0.04594956	0.02425446	0.02128838	0.05028005
MTMOTS6	hyp	mean	-0.35326050	-0.43212912	-0.19853758	-0.53037486	-0.28331800
		std dev	0.01618564	0.04293036	0.01228371	0.02805780	0.03053868
	eps	mean	-0.42756207	-0.31251449	-0.17569703	-0.14612598	-0.33820096
		std dev	0.02755102	0.04515031	0.02442164	0.02239561	0.04578612

5 Results and Discussion

The values of both the epsilon and hypervolume indicators for all runs are summarised in Tables 3, 4 and 5 through their means and standard deviations over

Table 4. Indicator value summary for tests at 5000 evaluations

Problem			ZDT1	ZDT2	ZDT3	ZDT4	ZDT6
MOTS	hyp	mean	-0.75771113	-0.77604662	-0.34128927	-0.47282969	-0.47189861
		std dev	0.01080879	0.02368686	0.01946946	0.01395013	0.03379669
	eps	mean	-0.48614000	-0.47885986	-0.24595268	-0.47282969	-0.57978382
		std dev	0.02130317	0.00532947	0.00202096	0.01395013	0.04564560
MTMOTS1	hyp	mean	-0.74546818	-0.77465641	-0.34000247	-0.57349963	-0.46170811
		std dev	0.03029480	0.02610753	0.01995945	0.01783768	0.03994537
	eps	mean	-0.49272115	-0.47796028	-0.24539430	-0.17291130	-0.56829806
		std dev	0.00535743	0.00677950	0.00272025	0.01751066	0.05329170
MTMOTS2	hyp	mean	-0.74021378	-0.74698593	-0.33980378	-0.57489295	-0.42155799
		std dev	0.01804256	0.02304431	0.01810125	0.01645429	0.02523835
	eps	mean	-0.48075048	-0.47302125	-0.24577017	-0.17716111	-0.52871405
		std dev	0.01352368	0.01042583	0.00216076	0.01626620	0.04117396
MTMOTS3	hyp	mean	-0.73204179	-0.74413296	-0.33247867	-0.57260769	-0.39276954
		std dev	0.02057182	0.02588503	0.01353899	0.02004265	0.02672701
	eps	mean	-0.47460730	-0.47336821	-0.24367137	-0.49707953	-0.48826261
		std dev	0.01864253	0.00992169	0.00452546	0.00406652	0.04041251
MTMOTS4	hyp	mean	-0.72444517	-0.73243387	-0.32650285	-0.56678358	-0.37342153
		std dev	0.01835684	0.03057232	0.01676906	0.01419364	0.02737189
	eps	mean	-0.46866599	-0.47382256	-0.24438498	-0.49591891	-0.46254002
		std dev	0.01626027	0.01159641	0.00448835	0.00635217	0.03886867
MTMOTS5	hyp	mean	-0.71613364	-0.72588713	-0.32269983	-0.56905458	-0.35064797
		std dev	0.02223144	0.03955106	0.01418449	0.01640497	0.02056303
	eps	mean	-0.46199537	-0.47443139	-0.24352935	-0.17478283	-0.42753125
		std dev	0.02137072	0.01098106	0.00502267	0.01606452	0.02972445
MTMOTS6	hyp	mean	-0.71892105	-0.73547479	-0.32559845	-0.57060564	-0.35326050
		std dev	0.01653063	0.03553801	0.01313763	0.01452726	0.01618564
	eps	mean	-0.46492066	-0.47583449	-0.24502034	-0.17650711	-0.42756207
		std dev	0.01614469	0.00977956	0.00387710	0.01440561	0.02755102

each set of 45 runs. The results of the Kruskal-Wallis tests for each set of indicators are presented in Tables 6, 7 and 8. For compactness, in the latter set of tables the algorithms MTMOTS1-6 are notated MT1-6. The epsilon and hypervolume indicator values are all negative, indicating that in general the approximation set is better than the reference set, as we would expect. A larger, more negative value in both cases indicates that the approximation set is further from the reference set and therefore closer to the true Pareto front.

As we are using the same reference set and bounding points for every run on a given problem, we may freely compare indicator values for different numbers of evaluations or algorithms on that problem. The testing has been designed to be as transparent as possible, for ease of comparison of our results with others'.

For the most part, either MOTS or MTMOTS1 seem to be the best algorithms. However, there are significant exceptions to this.

Although we know the MOTS and MTMOTS1 algorithms to be functionally identical, the Kruskal-Wallis test nevertheless chooses a winner between the two each time. We notice that, in almost all of such cases, the p-value that MOTS is better than MTMOTS1 is not significant for a 10% two-tailed test between them. This reassures us that the parameters we have set for our tests are sensible.

We find two cases where MOTS does win significantly over MTMOTS1, and one where MTMOTS1 wins significantly over MOTS. The two former cases are both with 1000 evaluations, and the latter is at 5000 evaluations. As the two algorithms under comparison are identical, this points to the fact that the random seed used in initialization is more significant to the end result than which

Table 5. Indicator value summary for tests at 10000 evaluations

Problem			ZDT1	ZDT2	ZDT3	ZDT4	ZDT6
MOTS	hyp	mean	-0.75804522	-0.79038349	-0.34670252	-0.49205272	-0.49205272
		std dev	0.02861861	0.02150583	0.02161375	0.04517364	0.04517364
	eps	mean	-0.48611454	-0.48296787	-0.24807474	-0.58526265	-0.58526265
		std dev	0.02668620	0.00163838	0.00059581	0.04665498	0.04665498
MTMOTS1	hyp	mean	-0.76420495	-0.78934388	-0.34562586	-0.49752378	-0.49752378
		std dev	0.02078507	0.02216047	0.01986500	0.03068751	0.03068751
	eps	mean	-0.49191725	-0.48309933	-0.24794868	-0.58502840	-0.58502840
		std dev	0.01920746	0.00164277	0.00086166	0.04035133	0.04035133
MTMOTS2	hyp	mean	-0.76031847	-0.78424399	-0.35241522	-0.49742877	-0.49742877
		std dev	0.01792981	0.02376584	0.01324688	0.01285745	0.01285745
	eps	mean	-0.49053026	-0.48050908	-0.24649687	-0.59839771	-0.59839771
		std dev	0.01688889	0.00380593	0.00211902	0.00357564	0.00357564
MTMOTS3	hyp	mean	-0.75603023	-0.76955898	-0.34772538	-0.46111812	-0.46111812
		std dev	0.00971037	0.02164517	0.01103327	0.01554581	0.01554581
	eps	mean	-0.49163574	-0.47937156	-0.24628508	-0.58113050	-0.58113050
		std dev	0.00609305	0.00403372	0.00202930	0.02302866	0.02302866
MTMOTS4	hyp	mean	-0.75185899	-0.76907325	-0.34513998	-0.43699462	-0.43699462
		std dev	0.00790369	0.02727055	0.01070044	0.02216875	0.02216875
	eps	mean	-0.49026121	-0.47715134	-0.24571855	-0.55088889	-0.55088889
		std dev	0.00485044	0.00731959	0.00288883	0.03212537	0.03212537
MTMOTS5	hyp	mean	-0.74732437	-0.76383152	-0.34381879	-0.41948442	-0.41948442
		std dev	0.01005336	0.02892708	0.00789131	0.01741343	0.01741343
	eps	mean	-0.48585004	-0.47740937	-0.24469911	-0.52484087	-0.52484087
		std dev	0.01040263	0.00731432	0.00300585	0.02962981	0.02962981
MTMOTS6	hyp	mean	-0.74413543	-0.76351363	-0.34045766	-0.40218758	-0.40218758
		std dev	0.01238293	0.02310450	0.01094006	0.01701943	0.01701943
	eps	mean	-0.48465399	-0.47673877	-0.24553331	-0.49944897	-0.49944897
		std dev	0.01090706	0.00678134	0.00251925	0.02569089	0.02569089

Table 6. Kruskal-Wallis test values for tests at 1000 evaluations

Test Indicator	ZDT1		ZDT2		ZDT3		ZDT4		ZDT6	
	hyp	eps	hyp	eps	hyp	eps	hyp	eps	hyp	eps
p(MOTS>MT1)	0.105556	0.346889	0.620349	0.816576	0.816786	0.992924	H0	0.993142	0.274042	0.381567
p(MOTS>MT2)	0.080925	0.130265	0.005423	0.000766	0.107537	0.935704		0.000000	0.982873	0.999889
p(MOTS>MT3)	1.000000	1.000000	0.999349	0.824997	1.000000	1.000000		0.000000	0.999806	1.000000
p(MOTS>MT4)	1.000000	1.000000	1.000000	1.000000	1.000000	1.000000		0.672640	0.999944	1.000000
p(MOTS>MT5)	1.000000	1.000000	1.000000	1.000000	1.000000	1.000000		0.711272	0.999969	1.000000
p(MOTS>MT6)	1.000000	1.000000	1.000000	1.000000	1.000000	1.000000		0.534900	0.992000	0.999852
p(MT1>MT2)	0.440777	0.232023	0.002195	0.000026	0.016295	0.173034		0.000000	0.996629	0.999966
p(MT1>MT3)	1.000000	1.000000	0.998227	0.512864	1.000000	0.999997		0.000000	0.999982	1.000000
p(MT1>MT4)	1.000000	1.000000	1.000000	1.000000	1.000000	1.000000		0.021546	0.999996	1.000000
p(MT1>MT5)	1.000000	1.000000	1.000000	1.000000	1.000000	1.000000		0.027817	0.999998	1.000000
p(MT1>MT6)	1.000000	1.000000	1.000000	1.000000	1.000000	1.000000		0.008697	0.998646	0.999954
p(MT2>MT3)	1.000000	1.000000	1.000000	0.999977	1.000000	1.000000		0.497769	0.927409	0.937168
p(MT2>MT4)	1.000000	1.000000	1.000000	1.000000	1.000000	1.000000		1.000000	0.962719	0.997312
p(MT2>MT5)	1.000000	1.000000	1.000000	1.000000	1.000000	1.000000		1.000000	0.972951	0.990474
p(MT2>MT6)	1.000000	1.000000	1.000000	1.000000	1.000000	1.000000		1.000000	0.616193	0.470359
p(MT3>MT4)	1.000000	0.998119	1.000000	1.000000	0.999982	0.670996		1.000000	0.628719	0.897106
p(MT3>MT5)	1.000000	1.000000	1.000000	1.000000	1.000000	1.000000		1.000000	0.681716	0.793954
p(MT3>MT6)	1.000000	1.000000	1.000000	1.000000	1.000000	1.000000		1.000000	0.122520	0.054209
p(MT4>MT5)	0.999979	0.998450	0.999991	0.999977	1.000000	0.999999		0.543778	0.557281	0.327748
p(MT4>MT6)	1.000000	1.000000	1.000000	1.000000	1.000000	1.000000		0.359552	0.068170	0.002141
p(MT5>MT6)	0.996554	0.978707	0.990596	0.973026	0.997049	0.974458		0.319332	0.051255	0.007806
Winning Algorithm	MT2	MT2	MT2	MT2	MT2	MOTS		MT3	MT1	MT1

algorithm is used! This applies more so for fewer evaluations. It is significant that there are no cases where either MOTS or MTMOTS1 significantly win over each other at 10000 evaluations. We therefore must be careful when drawing conclusions about results for runs with fewer evaluations.

The most striking results are for ZDT4. For the epsilon indicators, we see that for 10000 evaluations MTMOTS6 is a clear winner, for 5000 evaluations MTMOTS3 is the winner closely followed by MTMOTS4, and for 1000 evaluations MTMOTS3 is also the winner. The p-values for these indicator values show that

Table 7. Kruskal-Wallis test values for tests at 5000 evaluations

Test Indicator	ZDT1		ZDT2		ZDT3		ZDT4		ZDT6	
	hyp	eps	hyp	eps	hyp	eps	hyp	eps	hyp	eps
p(MOTS>MT1)	0.991387	0.031571	0.628211	0.669757	0.659320	H0	0.733933	0.770235	0.945943	0.941797
p(MOTS>MT2)	0.999999	0.999897	1.000000	0.998999	0.715541		0.533915	0.268905	1.000000	1.000000
p(MOTS>MT3)	1.000000	1.000000	1.000000	0.998923	0.999269		0.761078	0.000000	1.000000	1.000000
p(MOTS>MT4)	1.000000	1.000000	1.000000	0.981008	0.999998		0.995667	0.000000	1.000000	1.000000
p(MOTS>MT5)	1.000000	1.000000	1.000000	0.952175	1.000000		0.961834	0.560861	1.000000	1.000000
p(MOTS>MT6)	1.000000	1.000000	1.000000	0.876979	1.000000		0.932913	0.415977	1.000000	1.000000
p(MT1>MT2)	0.988556	1.000000	0.999999	0.996086	0.563216		0.294699	0.087829	0.999995	0.999886
p(MT1>MT3)	0.999992	1.000000	1.000000	0.995822	0.997283		0.533915	0.000000	1.000000	1.000000
p(MT1>MT4)	1.000000	1.000000	1.000000	0.949436	0.999991		0.977687	0.000000	1.000000	1.000000
p(MT1>MT5)	1.000000	1.000000	1.000000	0.890514	1.000000		0.875077	0.278727	1.000000	1.000000
p(MT1>MT6)	1.000000	1.000000	1.000000	0.764752	0.999999		0.809236	0.170687	1.000000	1.000000
p(MT2>MT3)	0.981167	0.969488	0.647191	0.491073	0.995652		0.733933	0.000000	0.999997	0.999985
p(MT2>MT4)	0.999982	0.999973	0.988984	0.151308	0.999981		0.994476	0.000000	1.000000	1.000000
p(MT2>MT5)	1.000000	1.000000	0.997682	0.074720	1.000000		0.954283	0.779084	1.000000	1.000000
p(MT2>MT6)	1.000000	1.000000	0.946050	0.025783	0.999998		0.921197	0.656915	1.000000	1.000000
p(MT3>MT4)	0.981921	0.986244	0.972323	0.156602	0.938325		0.972813	0.528685	0.999806	0.998888
p(MT3>MT5)	0.999870	0.999923	0.993052	0.077913	0.997748		0.856757	1.000000	1.000000	1.000000
p(MT3>MT6)	0.999739	0.999762	0.890984	0.027142	0.980736		0.785270	1.000000	1.000000	1.000000
p(MT4>MT5)	0.943992	0.946603	0.708884	0.340109	0.905503		0.194266	1.000000	0.999999	0.999993
p(MT4>MT6)	0.919277	0.905525	0.245649	0.178685	0.702795		0.127573	1.000000	0.999992	0.999997
p(MT5>MT6)	0.424625	0.381849	0.108000	0.305484	0.216979		0.391154	0.357436	0.346757	0.562926
Winning Algorithm	MOTS	MT1	MOTS	MOTS	MOTS		MOTS	MT3	MOTS	MOTS

Table 8. Kruskal-Wallis test values for tests at 10000 evaluations

Test Indicator	ZDT1		ZDT2		ZDT3		ZDT4		ZDT6	
	hyp	eps	hyp	eps	hyp	eps	hyp	eps	hyp	eps
p(MOTS>MT1)	0.121011	0.152397	0.509236	0.261613	0.884989	0.629323	H0	0.518490	0.249132	0.714321
p(MOTS>MT2)	0.973477	0.982249	0.967079	0.999965	0.303948	0.999999		0.233605	0.545664	0.153195
p(MOTS>MT3)	1.000000	0.999868	1.000000	1.000000	0.993803	1.000000		0.177433	1.000000	0.999999
p(MOTS>MT4)	1.000000	1.000000	0.999998	1.000000	0.999947	1.000000		0.120832	1.000000	1.000000
p(MOTS>MT5)	1.000000	1.000000	1.000000	1.000000	1.000000	1.000000		0.639767	1.000000	1.000000
p(MOTS>MT6)	1.000000	1.000000	1.000000	1.000000	1.000000	1.000000		0.000000	1.000000	1.000000
p(MT1>MT2)	0.998991	0.999073	0.965356	0.999998	0.043557	0.999997		0.219665	0.785758	0.056297
p(MT1>MT3)	1.000000	0.999998	1.000000	1.000000	0.904892	1.000000		0.165661	1.000000	0.999986
p(MT1>MT4)	1.000000	1.000000	0.999998	1.000000	0.996581	1.000000		0.111794	1.000000	1.000000
p(MT1>MT5)	1.000000	1.000000	1.000000	1.000000	0.999946	1.000000		0.622284	1.000000	1.000000
p(MT1>MT6)	1.000000	1.000000	1.000000	1.000000	0.999999	1.000000		0.000000	1.000000	1.000000
p(MT2>MT3)	0.999497	0.942294	0.999377	0.906584	0.998670	0.785605		0.421331	1.000000	1.000000
p(MT2>MT4)	1.000000	0.999508	0.997536	0.996922	0.999994	0.923291		0.328256	1.000000	1.000000
p(MT2>MT5)	1.000000	0.999999	0.999859	0.990395	1.000000	0.999852		0.860856	1.000000	1.000000
p(MT2>MT6)	1.000000	1.000000	0.999959	0.997988	1.000000	0.979562		0.000000	1.000000	1.000000
p(MT3>MT4)	0.988419	0.959294	0.335161	0.923915	0.920252	0.738280		0.402726	0.999995	0.999999
p(MT3>MT5)	0.999832	0.999573	0.660612	0.848289	0.995273	0.997780		0.900078	1.000000	1.000000
p(MT3>MT6)	0.999996	0.999762	0.767400	0.941867	0.999741	0.895889		0.000000	1.000000	1.000000
p(MT4>MT5)	0.909995	0.946850	0.799401	0.343022	0.884741	0.986682		0.936634	0.999348	0.998805
p(MT4>MT6)	0.988004	0.962282	0.875984	0.555472	0.981639	0.732908		0.000000	1.000000	1.000000
p(MT5>MT6)	0.822077	0.565092	0.624153	0.706661	0.814676	0.054740		0.000000	0.999348	0.996217
Winning Algorithm	MT1	MT1	MOTS	MT1	MT2	MOTS		MT6	MT1	MT2

we may regard these wins as highly significant. These are the only cases where the wins over both MOTS and MTMOTS1 are "significantly" significant.

Interestingly, although the conclusions drawn for ZDT4 from the epsilon indicators are clear, the hypervolume indicators disagree – for 1000 and 10000 evaluations, in fact, the null hypothesis is accepted. This points to the fact that the Pareto fronts found by the multi-threading algorithms are better in some ways but not in others. Inspecting the plots of the Pareto fronts found in Figs. 1 and 2, we see that, in general, those for the winning algorithms find the *whole* of a global or local Pareto front. The shape of these fronts in objective space is such that partial convergence to a front can still give a high hypervolume value, and thus algorithms are hard to distinguish on this basis. We also see that, in general, local fronts closer to the global front are found by the winning algorithms. We conclude that for ZDT4 multi-threading is beneficial. Our results indicate that the optimal number of threads increases with the number of evaluations allowed, and this would provide a good avenue for further study. We must note,

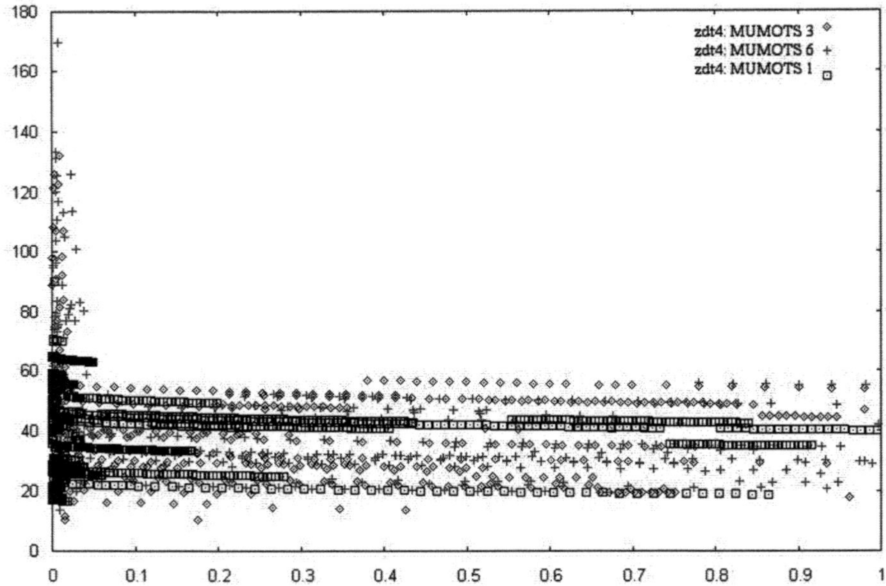

Fig. 1. All 45 Pareto fronts found by various MTMOTS algorithms for the test problem ZDT4, 5000 evaluations

however, that the true Pareto front is still not consistently found, even with the improved results of the multi-threaded algorithm.

The reason why the new algorithm makes such a difference on ZDT4 may be attributed to the fact that it is a deceptive problem with many local fronts. It makes sense that a multi-threaded algorithm would be advantageous here – as soon as one thread finds that the front in which it was stuck is only local, all threads are alerted, and then, if another thread becomes stuck, intensification will occur, moving it onto the new front. On a non-deceptive problem the sharing of information may not help the individual threads very much; they have to find their way to the Pareto front independently.

We might also expect multi-threading to improve results for problems with discontinuous Pareto fronts, on the basis that different threads could more efficiently find and explore different parts of the front. Of the problems tested in this study ZDT3 has this characteristic, but the results are not very compelling. At 10000 evaluations and 1000 evaluations MTMOTS2 wins, but not significantly, for the hypervolume indicators only, and elsewhere MOTS or MTMOTS1 win. It is hard to draw any conclusions from the results, except that the higher numbers of threads (above 4 threads) definitely make things worse.

The other notable feature in the results is that for 1000 evaluations MTMOTS2 wins more often than not. These tests are significant over MTMOTS1,

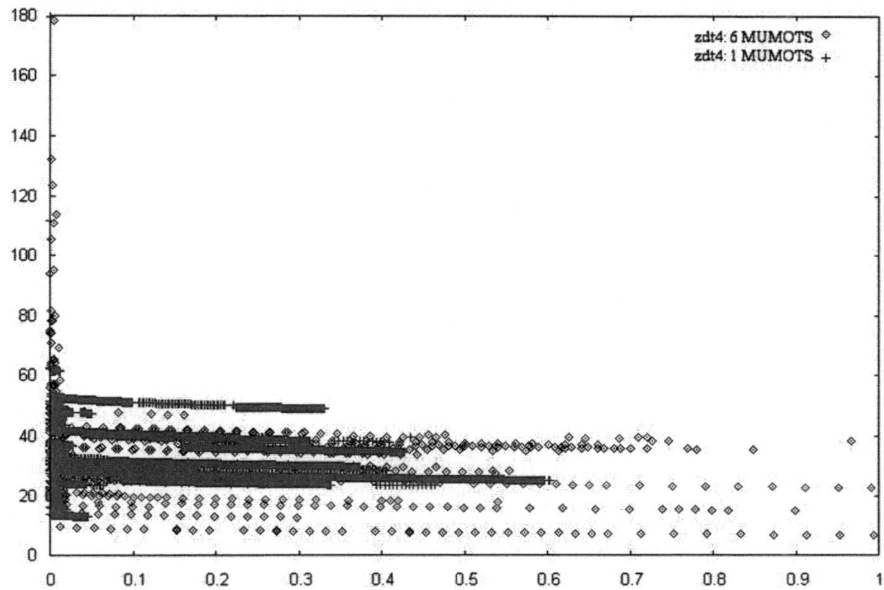

Fig. 2. All 45 Pareto fronts found by various MTMOTS algorithms for the test problem ZDT4, 10000 evaluations

but not MOTS. We must therefore treat them with caution as MOTS and MTMOTS1 are the same but for their random number seeds each run. However, we may tentatively suggest that adding an extra thread may improve initial progress and may therefore be particularly beneficial if, for some reason, operation of the algorithm is limited to a small number of evaluations. Again, this is an area which could be further investigated.

6 Conclusions

The performance of seven algorithms, MOTS and MTMOTS1-6, on five benchmark test problems, ZDT1-4 and ZDT6, was investigated. The results show a clear improvement with multi-threading on ZDT4, indicating that this may be beneficial in general for problems with deceptive local Pareto fronts. Our results indicate that there may be a relationship between the optimal number of threads to use and the number of evaluations allowed. It also seems that, if only a small numbers of evaluations are permitted, adding one thread may provide small advantages. However, overall we must conclude that on most problems the implementation of a multi-threaded algorithm, as specified in this paper, does not result in an improvement in optimiser performance.

References

1. Jones, D.F., Mirrazavi, S.K., Tamiz, M.: Multi-objective meta-heuristics: An overview of the current state-of-the-art. European Journal of Operational Research **137**(1) (2002) 1–9
2. Jaeggi, D.M., Parks, G.T., Kipouros, T., Clarkson, P.J.: The development of a multi-objective Tabu Search algorithm for continuous optimisation problems. European Journal of Operational Research, in press (2006)
3. Crainic, T.G., Toulouse, M.: Parallel strategies for meta-heuristics. Parallel Computing **30**(1) (2002) 57–79
4. Taillard, E.D.: Parallel Taboo Search techniques for the job shop scheduling problem. ORSA Journal of Computing **6**(2) (1994) 108–117
5. Connor, A.M.: A multi-thread Tabu Search algorithm. Design Optimization **1**(3) (1999) 293–304
6. Glover, F., Laguna, M.: Tabu Search. Boston, MA: Kluwer Academic Publishers (1997)
7. Hooke, R., Jeeves, T.A.: "Direct search" solution of numerical and statistical problems. Journal of the ACM **8**(2) (1961) 212–229
8. Zitzler, E., Deb, K., Thiele, L.: Comparison of Multiobjective Evolutionary Algorithms. TIK-Report **70** (1999)
9. Deb, K.: Multi-objective Optimization using Evolutionary Algorithms (Section 8.3). Chichester, UK: Wiley & Sons Ltd. (2001)
10. Zitzler, E., Thiele, L., Laumanns, M., Fonseca, C.M., Grunert da Fonseca, V.: Performance assessment of multiobjective optimizers: An analysis and review. IEEE Transactions on Evolutionary Computation **7**(2) (2003) 117–132
11. Knowles, J.D., Thiele, L., Zitzler, E.: A Tutorial on the Performance Assessment of Stochastic Multiobjective Optimizers. TIK-Report **214** (2006)
12. Conover, W.J.: Practical Nonparametric Statistics, 3rd Edition. Chichester, UK: Wiley & Sons Ltd. (1999)
13. Jaeggi, D.M., Parks, G.T., Kipouros, T., Clarkson, P.J.: A multi-objective Tabu Search algorithm for constrained optimisation problems. In Coello Coello, C.A., Hernandez Aguirre, A., Zitzler, E., eds.: Proc. 3rd Int. Conf. Evolutionary Multi-Criterion Optimization – EMO 2005. Lecture Notes in Computer Science, Vol. 3410, Springer-Verlag, Berlin, Heidelberg, New York (2005) 490–504

Differential Evolution Versus Genetic Algorithms in Multiobjective Optimization

Tea Tušar and Bogdan Filipič

Department of Intelligent Systems, Jožef Stefan Institute
Jamova 39, SI-1000 Ljubljana, Slovenia
tea.tusar@ijs.si, bogdan.filipic@ijs.si

Abstract. This paper presents a comprehensive comparison between the performance of state-of-the-art genetic algorithms NSGA-II, SPEA2 and IBEA and their differential evolution based variants $\text{DEMO}^{\text{NS-II}}$, DEMO^{SP2} and DEMO^{IB}. Experimental results on 16 numerical multiobjective test problems show that on the majority of problems, the algorithms based on differential evolution perform significantly better than the corresponding genetic algorithms with regard to applied quality indicators. This suggests that in numerical multiobjective optimization, differential evolution explores the decision space more efficiently than genetic algorithms.

1 Introduction

Differential Evolution (DE) [1] is a simple yet powerful algorithm that outperforms Genetic Algorithms (GAs) on many numerical singleobjective optimization problems [2]. In this paper we show that DE can achieve better results than GAs also on numerical multiobjective optimization problems (MOPs). To this end, we compare three state-of-the-art Multiobjective Evolutionary Algorithms (MOEAs), namely NSGA-II [3], SPEA2 [4] and IBEA [5], to their counterparts – algorithms that use the same environmental selection, but DE instead of GAs for exploring the decision space. While DE-based algorithms for multiobjective optimization have already been proposed in the past (see Related Work in Section 3), comparisons between these approaches and GA-based algorithms lack: (a) a wide choice of difficult test problems with more than two objectives, (b) performance assessment with Pareto compliant indicators, and (c) inferences about algorithm performance based on statistical tests. The comparison in this paper includes all these usually omitted features.

The paper is further organized as follows. Section 2 introduces the basic GA as the underlying algorithm for NSGA-II, SPEA2 and IBEA, while the proposed algorithm DEMO is explained in detail in Section 3. Section 4 outlines the experiments, whose results are presented and discussed in Section 5. Section 6 concludes the paper with a summary of the results.

2 Multiobjective Optimization with the Basic GA

Most of the efforts spent on adapting GAs to multiobjective optimization have been focusing on finding new approaches for environmental selection. These

S. Obayashi et al. (Eds.): EMO 2007, LNCS 4403, pp. 257–271, 2007.

approaches try to produce good approximations of the Pareto optimal front by incorporating different preferences. For example, the environmental selection in NSGA-II [3] first ranks the individuals using nondominated sorting. To distinguish between individuals with the same rank, the crowding distance metric is used, which prefers individuals from less crowded regions of the objective space. SPEA2 [4] works similarly, calculating the raw fitness of the individuals according to Pareto dominance relations between them and using a density measure to break the ties. The individuals that reside close together in the objective space are discouraged from entering the archive of best solutions. IBEA [5], on the other hand, uses a different approach. The fitness of individuals is determined only according to the value of a predefined indicator. This indicator has to be dominance preserving and no other explicit diversity preserving mechanism (such as crowding in NSGA-II or density in SPEA2) is applied.

While directing all attention to environmental selection, the popular algorithms NSGA-II, SPEA2 and IBEA use practically the same algorithm for exploring the decision space. It is therefore possible to describe all three algorithms using a unifying framework, which will be called *Basic Genetic Algorithm* in the remainder of this paper. This algorithm is presented in Fig. 1. After initialization of the populations $\mathcal{P}$ and $\mathcal{Q}$, which is slightly different in NSGA-II, SPEA2 and IBEA[1], the evolutionary steps of selection, crossover and mutation are repeated until a stopping criterion is met. In environmental selection, one of the previously described approaches is used to calculate the fitness of the individuals. This fitness is used again when comparing individuals in tournament selection. Figure 2 shows the variation operators on individuals encoded as real vectors. In case of combinatorial MOPs, different operators need to be used.

Basic Genetic Algorithm for Multiobjective Optimization

1. Initialize populations $\mathcal{P}_0$ and $\mathcal{Q}_0$.
2. Set $t = 0$.
3. Repeat:
   - 3.1. Set $t = t + 1$.
   - 3.2. Calculate the objectives for new individuals from $\mathcal{P}_{t-1}$ and $\mathcal{Q}_{t-1}$.
   - 3.3. Get $\mathcal{P}_t$ from $\mathcal{P}_{t-1}$ and $\mathcal{Q}_{t-1}$ with environmental selection.
   - 3.4. If stopping criterion met, return nondominated individuals from $\mathcal{P}_t$.
   - 3.5. Fill the mating pool $\mathcal{M}_t$ using tournament selection on $\mathcal{P}_t$.
   - 3.6. Apply variation to individuals from $\mathcal{M}_t$ to get $\mathcal{Q}_t$ (see Fig. 2).

Fig. 1. Outline of the basic genetic algorithm

[1] While NSGA-II initializes the population $\mathcal{P}_0$ with randomly created individuals and sets $\mathcal{Q}_0$ to be empty, in SPEA2, $\mathcal{P}_0$ represents the archive of best solutions and is therefore initially empty, while $\mathcal{Q}_0$ is filled with randomly created individuals. IBEA originally uses a single population of variable size instead of two separate populations. Without altering its performance, we can assume that IBEA uses two populations, which are initialized in the same way as in NSGA-II.

Variation
Input: Mating pool $\mathcal{M}_t$
1. Create empty population $\mathcal{Q}_t$.
2. For each pair of individuals I_i, I_{i+1} ($i = 1, 3, \dots$) from $\mathcal{M}_t$ do:
   2.1. Modify the individuals I_i, I_{i+1} with uniform crossover.
   2.2. Modify the individuals I_i, I_{i+1} with simulated binary crossover.
   2.3. Modify the individual I_i with polynomial mutation.
   2.4. Modify the individual I_{i+1} with polynomial mutation.
   2.5. Add individuals I_i and I_{i+1} to $\mathcal{Q}_t$.

Output: Population $\mathcal{Q}_t$

Fig. 2. Variation of real-coded individuals

3 Multiobjective Optimization with DE

DE [1] is a simple evolutionary algorithm that encodes solutions as vectors and uses operations such as vector addition, scalar multiplication and exchange of components (crossover) to construct new solutions from the existing ones. When a new solution, also called *candidate*, is constructed, it is compared to its parent. If the candidate is better than its parent, it replaces the parent in the population. Otherwise, the candidate is discarded. As a steady-state algorithm, DE implicitly incorporates elitism, i.e. no solution can be deleted from the population unless a better solution is found. While being a very successful optimization method, DE's greatest limitation originates in its encoding. As no vector representation of solution exists for combinatorial problems, DE can only be applied in numerical optimization.

3.1 Related Work

DE has been adapted to solve MOPs in several ways. In the early approaches (PDE [6] and GDE [7]), only the concept of Pareto dominance was used to compare the individuals. The candidate replaced its parent only if it (weakly) dominated it. Otherwise, it was discarded. This is a rather strict demand, especially when the number of objectives is high. Many subsequent approaches (PDEA [8], MODE [9], NSDE [10], GDE2 [11], DEMO [12], GDE3 [13] and NSDE-DCS [14]) used nondominated sorting and/or the crowding distance metric to calculate the fitness of individuals. Only recently, new algorithms that do not follow the environmental selection of NSGA-II were proposed (ε-MyDE [15] and DEMORS [16]). To our best knowledge, no algorithms that combine DE with the environmental selection of SPEA2 or IBEA have been presented so far.

3.2 DEMO$^{\text{NS-II}}$, DEMO$^{\text{SP2}}$ and DEMO^{IB}

The idea presented here is to use DE for exploring the decision space and environmental selection from either NSGA-II, SPEA2 or IBEA to select the best individuals for the next population. This idea is implemented in the algorithm called DEMO (Differential Evolution for Multiobjective Optimization)[2].

The outline of DEMO is presented in Figs. 3 and 4. In the main loop, the candidate replaces the parent if it dominates it. If the parent dominates the candidate, the candidate is discarded. Otherwise (when the candidate and parent are nondominated with regard to each other), the candidate is added to the population. This step is repeated until *popSize* number of candidates are created. After that, we get a population of size between *popSize* and $2 \times popSize$. If the population has enlarged, it is truncated to *popSize* using environmental selection.

Differential Evolution for Multiobjective Optimization

1. Evaluate the initial population $\mathcal{P}$ of random individuals.
2. While stopping criterion not met, do:
   2.1. For each individual P_i ($i = 1, \dots, popSize$) from $\mathcal{P}$ repeat:
      (a) Create candidate C from parent P_i (see Fig. 4).
      (b) Calculate the objectives of the candidate.
      (c) If the candidate dominates the parent, the candidate replaces the parent. If the parent dominates the candidate, the candidate is discarded. Otherwise, the candidate is added in the population.
   2.2. If the population has more than *popSize* individuals, apply environmental selection to get the best *popSize* individuals.
   2.3. Randomly enumerate the individuals in $\mathcal{P}$.
3. Return nondominated individuals from $\mathcal{P}$.

Fig. 3. Outline of DEMO

Candidate creation

Input: Parent P_i

1. Randomly select three individuals $P_{i_1}, P_{i_2}, P_{i_3}$ from $\mathcal{P}$, where i, i_1, i_2 and i_3 are pairwise different.
2. Calculate candidate C as $C = P_{i_1} + F(P_{i_2} - P_{i_3})$, where F is a scaling factor.
3. Modify the candidate with binary crossover with the parent P_i.
4. Repair the candidate if it falls out of bounds of the decision space.

Output: Candidate C

Fig. 4. Candidate creation using scheme *DE/rand/1/bin*

Note that the newly created candidates that enter the population (either by replacement or by addition) instantly take part in the creation of subsequent candidates. This helps achieving fast convergence to the Pareto optimal front. Moreover, it resembles very closely the steady-state mechanism of DE. This is why we prefer the described approach to a somewhat more straightforward way to use DE

[2] DEMO is a generalization of the DEMO/parent variant presented in [12], which used the *DE/rand/1/bin* scheme [2] and environmental selection from NSGA-II.

in the basic GA, which consists of replacing the variation phase (see Fig. 2) with candidate creation (as in Fig. 4) for each individual from the mating pool.

In candidate creation, the use of vector addition can result in candidates that fall out of bounds of the decision space. In such cases, many repair methods are possible. We address this problem by replacing the candidate value violating the boundary constraints with the closest boundary value. In this way, the candidate becomes feasible with as few alterations to it as possible and there is no need for making a new candidate. It is important to note, however, that this repair method may yield more boundary individuals and is biased for problems where the Pareto optimal set lies on one of the bounds of the decision space.

DEMO, as described in Fig. 3, can incorporate arbitrary environmental selection. In the remainder of the paper we will use $\text{DEMO}^{\text{NS-II}}$, DEMO^{SP2} and DEMO^{IB} to denote the variants of DEMO that use environmental selection from NSGA-II, SPEA2 and IBEA, respectively.

4 Experimental Setup

To compare the presented algorithms, extensive experiments on 16 test problems were performed. The focus of the experiments was on comparing $\text{DEMO}^{\text{NS-II}}$ to NSGA-II, DEMO^{SP2} to SPEA2, and DEMO^{IB} to IBEA, and not on comparing algorithms with different environmental selection among themselves. Such a comparison can be found, for example, in [5].

4.1 Test Problems

Two test problem suits were used in the experiments. The first consists of the first seven DTLZ test problems from [17] and the second of the nine WFG test problems presented in [18]. Both suits comprise difficult problems that present many challenges for multiobjective optimizers, such as the existence of many local Pareto optimal fronts, uneven distribution of points on the Pareto optimal front, nonseparable variables etc.

Let n and m denote the dimensionality of the decision space and variable space, respectively. Each of the 16 problems was used three times – each time with a different number of objectives (m = 2, 3 and 4). The other parameters were set as follows:

- The parameters of DTLZ problems were set as recommended in [17], i.e. $n = m + k - 1$, where $k = 5$ for DTLZ1, $k = 10$ for problems DTLZ2 to DTLZ6 and $k = 20$ for DTLZ7.
- Parameters of the WFG test suite are: the number of position related parameters k, number of distance related parameters l and number of objectives m. The number of decision variables is calculated as $n = k + l$. Because of some additional requirements (l must be an even number for WFG2 and WFG3, and k must be divisible by $m - 1$), we used the following setting: $k = 6$ and $l = 4$ (consecutively $n = 10$), which satisfies all the requirements for $m = 2$, 3 and 4.

All test problems suppose minimization of all objectives.

4.2 Parameters of the Algorithms

The experiments with NSGA-II, SPEA2 and IBEA were performed using the PISA environment [19]. The parameter settings for the basic GA, used by all three algorithms, are the same as the ones used in the comparison between NSGA-II and SPEA2 on the DTLZ1 problem in [17]:

- population size = 100,
- number of generations = 300,
- tournament size = 2,
- size of the mating pool = 100,
- individual crossover probability = 1,
- variable probability for SBX crossover = 1,
- distribution index for crossover $\eta_c = 15$,
- variable uniform crossover probability = 0.5,
- individual mutation probability = 1,
- variable polynomial mutation probability = $1/n$,
- distribution index for mutation $\eta_m = 20$.

The parameters of all three variants of DEMO were set as in [12] (except for the number of generations, which equals the number of generations used by the basic GA):

- population size = 100,
- number of generations = 300,
- DE selection scheme = *DE/rand/1/bin*,
- scaling factor $F = 0.5$,
- probability used in binary crossover = 0.3.

DEMO^{IB} and IBEA used additional parameters: indicator = I_{HD}[3], scaling factor $\kappa = 0.05$ and reference point for the hypervolume calculation $\rho = (2, \ldots, 2) \in \mathbb{R}^m$. Each algorithm was run on each problem 30 times.

4.3 Performance Assessment

The performance assessment was carried out using PISA and the guidelines from [20] and [21]. Consider for example the comparison between $\text{DEMO}^{\text{NS-II}}$ and NSGA-II on one problem. Firstly, the bounds of approximation sets of both algorithms were calculated so that the approximation sets could be normalized to the interval $[1, 2]$. After that, a dominance rank was calculated for each of the 60 approximation sets by simply counting the number of approximation sets that are better than the observed one. The Mann-Whitney rank sum test was used to discover if there are significant differences between the dominance ranks of the two algorithms.

[3] The same set of experiments was performed also with the $I_{\varepsilon+}$ indicator. Because of space limitations, we report only the results obtained using I_{HD} as they are less favorable for DEMO^{IB}. The interested reader can access all results from `http://dis.ijs.si/tea/EMO2007/demo.htm`.

Additional assessment was carried out using unary quality indicators. From the approximation sets of both algorithms, the set containing only nondominated solutions was computed and used as the reference set for the unary indicators $I^1_{\varepsilon+}$ and I^1_{R2}. Other parameters of the I^1_{R2} indicator were: $\rho = 0.01$, $(0.9, \ldots, 0.9)$ and $(2.1, \ldots, 2.1) \in \mathbb{R}^m$ served as the ideal and Nadir points, and 501, 496 and 455 uniformly spread parameter vectors were used for the problems with two, three and four objectives, respectively. The hypervolume indicator I_H used the point $(2.1, \ldots, 2.1) \in \mathbb{R}^m$ as the reference point. All three indicators were calculated for each approximation set of both algorithms. The significance of these outcomes was tested independently with the Fisher's independent permutation test. Because we used dominance ranking and three indicators on the same data, the significance level α for all significance tests was reduced from 0.05 to 0.0125 using the Bonferroni correction.

The same procedure was repeated in comparing DEMO^{SP2} to SPEA2 and DEMO^{IB} to IBEA. The outcomes of these comparisons are presented in the next section.

5 Results and Discussion

Looking at the outcomes of dominance ranking (Tables 1, 3 and 5) we can observe that on many problems, approximation sets of DEMO achieve significantly better domination ranks than the approximation sets of the basic GA. Only rarely (see Subsection 5.3), the basic GA outperforms DEMO. On the majority of problems, however, there are no significant differences between the two algorithms with regard to dominance ranking.

As expected, when dominance ranking shows a significant difference between two algorithms, so do the three applied indicators (an exception is again explained in Subsection 5.3). On the majority of problems, DEMO achieves significantly better results with regard to the chosen indicator (see Tables 2, 4 and 6). Note that on a few problems (see for example DTLZ5 for $m = 4$ in Table 2), DEMO is significantly better than the basic GA with regard to one indicator (I^1_{r2}) and significantly worse with regard to another indicator (I_H). This suggests that the outcomes of DEMO and the basic GA are incomparable on such problems.

Besides these results, we also investigated the plots of approximation sets (for $m = 2$ and 3) and plots of attainment surfaces (for $m = 2$) [22] to gain further insight into the comparison between DEMO and the basic GA. Despite statistical tests show that there is almost always a significant difference in indicator values of the two algorithms, in general no noticeable distinction was *visible* between the approximation sets (and attainment surfaces) of DEMO and the basic GA on problems DTLZ2, DTLZ4, DTLZ5, DTLZ7, WFG3, WFG4, WFG5, WFG8 and WFG9. On problems DTLZ1, DTLZ3 and DTLZ6, where it is very difficult to converge to the Pareto optimal front, and on the non-separable WFG6 problem, DEMO generally attained the Pareto optimal front more efficiently than the basic GA. On problems DTLZ3, WFG1, WFG2 and WFG7, DEMO achieved better spread of solutions along the Pareto optimal front than the basic GA.

Table 1. Outcomes of the Mann-Whitney rank sum test ($\alpha = 0.0125$) on dominance ranking for DEMO$^{\text{NS-II}}$ and NSGA-II. The '▲ p-value' ('▽ p-value') denotes the problems, on which DEMO$^{\text{NS-II}}$ is significantly better (worse) than NSGA-II, while '-' indicates there are no significant differences between the two algorithms.

	$m = 2$	$m = 3$	$m = 4$
DTLZ1	-	▲ $3.9{\times}10^{-13}$	▲ $7.9{\times}10^{-15}$
DTLZ2	-	-	-
DTLZ3	▲ $2.0{\times}10^{-11}$	▲ $3.9{\times}10^{-13}$	▲ $3.5{\times}10^{-12}$
DTLZ4	▲ 0.0052	-	-
DTLZ5	-	-	-
DTLZ6	▲ $4.1{\times}10^{-14}$	▲ $7.9{\times}10^{-15}$	▲ $3.9{\times}10^{-13}$
DTLZ7	-	-	▲ $1.5{\times}10^{-4}$
WFG1	-	▲ $1.6{\times}10^{-7}$	-
WFG2	-	-	-
WFG3	-	-	-
WFG4	-	-	-
WFG5	-	-	-
WFG6	▲ $1.5{\times}10^{-4}$	-	-
WFG7	-	-	-
WFG8	-	-	-
WFG9	-	-	-

Table 2. Outcomes of the Fisher-independent test ($\alpha = 0.0125$) on indicator values for DEMO$^{\text{NS-II}}$ and NSGA-II. A '▲' ('▽') under the indicator I means that DEMO$^{\text{NS-II}}$ is significantly better (worse) than NSGA-II regarding indicator I, while '-' indicates there are no significant differences between the two algorithms regarding indicator I

	$m = 2$			$m = 3$			$m = 4$		
	$I^1_{\varepsilon+}$	I_{H}	I^1_{R2}	$I^1_{\varepsilon+}$	I_{H}	I^1_{R2}	$I^1_{\varepsilon+}$	I_{H}	I^1_{R2}
DTLZ1	▲	▲	▲	▲	▲	▲	▲	▲	▲
DTLZ2	▲	▲	▲	▲	▲	▲	▲	▲	▲
DTLZ3	▲	▲	▲	▲	▲	▲	▲	▲	▲
DTLZ4	▲	▲	▲	▲	▲	▲	▲	▲	▲
DTLZ5	▲	▲	▲	▲	▲	▲	-	▽	▲
DTLZ6	▲	▲	▲	▲	▲	▲	▲	▲	▲
DTLZ7	▲	▲	▲	▲	▲	▲	▲	▲	▲
WFG1	▲	▲	▲	▲	▲	▲	▲	▲	▲
WFG2	▲	▲	▲	▲	▲	▲	▲	▲	▲
WFG3	-	▲	▲	▲	▲	▲	-	▽	▽
WFG4	-	▲	-	-	▲	▲	▲	▲	▲
WFG5	▽	-	▽	-	-	-	-	-	▲
WFG6	▲	▲	▲	▲	▲	▲	-	▲	▲
WFG7	▲	▲	▲	▲	▲	▲	-	▲	-
WFG8	▲	-	▲	-	-	-	-	-	▽
WFG9	-	▲	-	-	▲	▲	-	-	-

Table 3. Outcomes of the Mann-Whitney rank sum test ($\alpha = 0.0125$) on dominance ranking for DEMO^{SP2} and SPEA2. The '▲ p-value' ('▽ p-value') denotes the problems, on which DEMO^{SP2} is significantly better (worse) than SPEA2, while '-' indicates there are no significant differences between the two algorithms.

	$m=2$	$m=3$	$m=4$
DTLZ1	-	▲ 9.7×10^{-13}	▲ 3.2×10^{-13}
DTLZ2	-	-	-
DTLZ3	▲ 2.2×10^{-11}	▲ 2.0×10^{-14}	▲ 3.2×10^{-13}
DTLZ4	-	-	-
DTLZ5	-	-	-
DTLZ6	▲ 2.0×10^{-14}	▲ 7.9×10^{-15}	▲ 1.7×10^{-12}
DTLZ7	-	-	-
WFG1	-	▲ 2.7×10^{-9}	-
WFG2	-	-	-
WFG3	-	-	-
WFG4	-	-	-
WFG5	-	-	-
WFG6	▲ 2.6×10^{-6}	-	-
WFG7	-	-	-
WFG8	-	-	-
WFG9	-	-	-

Table 4. Outcomes of the Fisher-independent test ($\alpha = 0.0125$) on indicator values for DEMO^{SP2} and SPEA2. A '▲' ('▽') under the indicator I means that DEMO^{SP2} is significantly better (worse) than SPEA2 regarding indicator I, while '-' indicates there are no significant differences between the two algorithms regarding indicator I.

	$m=2$			$m=3$			$m=4$		
	$I^1_{\varepsilon+}$	I_H	I^1_{R2}	$I^1_{\varepsilon+}$	I_H	I^1_{R2}	$I^1_{\varepsilon+}$	I_H	I^1_{R2}
DTLZ1	▲	▲	▲	▲	▲	▲	▲	▲	▲
DTLZ2	▲	▲	▲	▲	▲	▲	▲	▲	▲
DTLZ3	▲	▲	▲	▲	▲	▲	▲	▲	▲
DTLZ4	▲	▲	▲	▲	▲	▲	▲	▲	▲
DTLZ5	▲	▲	▲	▲	▲	▲	▽	-	▲
DTLZ6	▲	▲	▲	▲	▲	▲	▲	▲	▲
DTLZ7	▲	▲	▲	▲	▲	▲	▲	▲	▲
WFG1	▲	▲	▲	▽	-	-	▲	▲	▲
WFG2	▲	▲	▲	▲	▲	▲	▲	▲	▲
WFG3	▲	▲	▲	-	-	-	▽	-	-
WFG4	-	▲	-	-	▲	▲	-	▲	▲
WFG5	▽	▽	▽	▲	▲	▲	▲	▲	▲
WFG6	▲	▲	▲	▲	▲	▲	-	▲	▲
WFG7	▲	▲	▲	▲	▲	▲	▲	▲	▲
WFG8	▲	-	▲	▲	▲	▲	-	▲	▲
WFG9	-	-	-	▲	▲	▲	-	▲	▲

Table 5. Outcomes of the Mann-Whitney rank sum test ($\alpha = 0.0125$) on dominance ranking for DEMO^{IB} and IBEA. The '▲ p-value' ('▽ p-value') denotes the problems, on which DEMO^{IB} is significantly better (worse) than IBEA, while '-' indicates there are no significant differences between the two algorithms.

	$m = 2$	$m = 3$	$m = 4$
DTLZ1	-	-	-
DTLZ2	-	-	-
DTLZ3	-	▲ 3.0×10^{-12}	▲ 9.7×10^{-12}
DTLZ4	▲ 0.0013	▲ 0.0104	-
DTLZ5	-	-	-
DTLZ6	▲ 1.2×10^{-13}	▲ 2.4×10^{-11}	-
DTLZ7	▽ 0.0023	▽ 1.3×10^{-7}	▽ 1.8×10^{-7}
WFG1	-	▽ 0.0058	-
WFG2	-	-	-
WFG3	-	-	-
WFG4	-	-	-
WFG5	-	-	-
WFG6	▲ 6.1×10^{-6}	-	-
WFG7	-	-	-
WFG8	-	-	-
WFG9	-	-	-

Table 6. Outcomes of the Fisher-independent test ($\alpha = 0.0125$) on indicator values for DEMO^{IB} and IBEA. A '▲' ('▽') under the indicator I means that DEMO^{IB} is significantly better (worse) than IBEA regarding indicator I, while '-' indicates there are no significant differences between the two algorithms regarding indicator I.

	$m = 2$			$m = 3$			$m = 4$		
	$I^1_{\varepsilon+}$	I_{H}	I^1_{R2}	$I^1_{\varepsilon+}$	I_{H}	I^1_{R2}	$I^1_{\varepsilon+}$	I_{H}	I^1_{R2}
DTLZ1	▲	▲	▲	▲	▲	▲	▲	▲	-
DTLZ2	-	-	▲	-	▲	▲	-	-	▲
DTLZ3	-	▽	▽	▲	▲	▲	▲	▲	▲
DTLZ4	▲	▲	▲	▲	▲	▲	▲	▲	▲
DTLZ5	-	-	▲	▽	▽	▲	▲	▲	▲
DTLZ6	▲	▲	▲	▲	▲	▲	▲	▲	▲
DTLZ7	▽	▽	▽	▽	▽	▽	▽	▽	-
WFG1	▲	▲	▲	▲	▲	▲	▲	▲	▲
WFG2	▲	▲	▲	▲	▲	▲	-	▲	▲
WFG3	-	▲	▲	-	▲	▲	▽	▽	▽
WFG4	-	▽	▽	-	▲	▲	-	▲	▲
WFG5	▽	▽	▽	-	-	▲	-	▽	▲
WFG6	▲	▲	▲	▲	▲	▲	▲	▲	▲
WFG7	▲	▲	▲	-	▲	▲	-	▲	▲
WFG8	-	-	-	-	-	▽	-	-	▽
WFG9	▲	▲	-	-	▲	▲	-	-	▲

In the following subsections, we review the performance of DEMO and basic GA on selected problems in more detail.

5.1 DEMO$^{\text{NS-II}}$ vs. NSGA-II

The comparison between DEMO and the basic GA is very favorable to DEMO, when nondominated sorting is used for environmental selection. Let us explore in more detail the outcomes of both algorithms on the DTLZ6 problem. The difficulty of this problem reflects in poor convergence of certain algorithms to the Pareto optimal front. Figure 5 shows that DEMO$^{\text{NS-II}}$ reaches the Pareto optimal front for $m = 2$ and $m = 3$, while NSGA-II does not. The most probable cause for such behavior is the repair method used by DEMO, since in this problem, the Pareto optimal set lies at the bounds of the decision space and boundary points are likely to be found after applying DEMO's repair method.

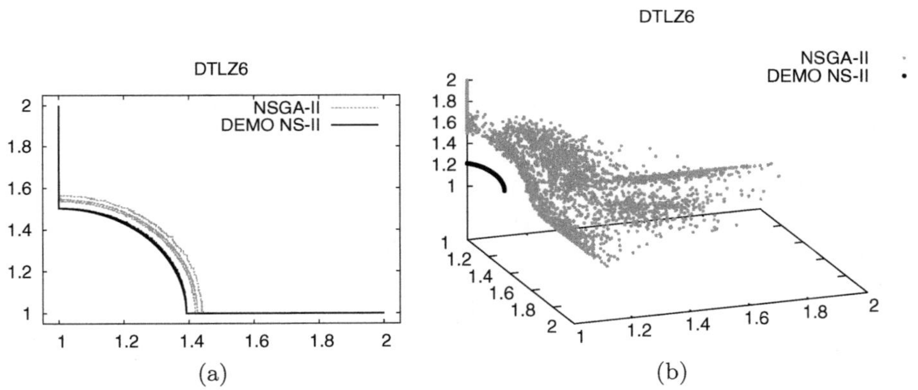

Fig. 5. Plots of normalized attainment surfaces and approximation sets of DEMO$^{\text{NS-II}}$ and NSGA-II on the DTLZ6 problem: (a) the best, worst and 50%-attainment surfaces for each algorithm on the problem with two objectives; (b) 30 approximation sets for each algorithm on the problem with three objectives

It is interesting to note, however, that on the only other problem (DTLZ7), where the Pareto optimal set lies on the bounds of the decision space, no big differences between approximation sets could be noticed. This is probably because on this problem, none of the algorithms has difficulties in reaching the Pareto optimal front.

5.2 DEMO$^{\text{SP2}}$ vs. SPEA2

Using the strength Pareto approach for environmental selection yields very similar results in the comparison between DEMO and the basic GA as the use of nondominated sorting. The findings from the previous subsection (on problems DTLZ6 and DTLZ7) hold also for DEMO$^{\text{SP2}}$ and SPEA2. Similarly, some of

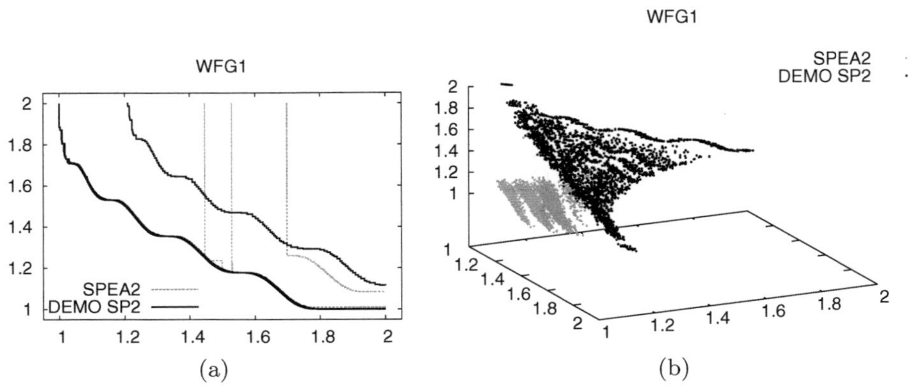

Fig. 6. Plots of normalized attainment surfaces and approximation sets of DEMO^{SP2} and SPEA2 on the WFG1 problem: (a) the best, worst and 50%-attainment surfaces for each algorithm on the problem with two objectives; (b) 30 approximation sets for each algorithm on the problem with three objectives

the characteristics of the comparison between DEMO^{SP2} and SPEA2 on the WFG1 problem, which will be discussed shortly, are true also when comparing DEMO$^{NS\text{-}II}$ and NSGA-II.

Consider now the WFG1 problem for $m = 2$. From the plot of attainment surfaces (see Fig. 6) we can see that DEMO^{SP2} reaches a wider portion of the Pareto optimal front than SPEA2, while having comparable convergence properties in the best and average case (50%-attainment surface) and a little worse in the worst case. When this problem is tackled in three objectives, DEMO^{SP2} loses some of its convergence power while keeping the good coverage. SPEA2, on the other hand, still covers only a small part of the whole front, while achieving much better convergence than DEMO^{SP2}. Although this is not visible from the plots, we wish to point out that neither of the algorithms reached the Pareto optimal front for this problem.

There is an additional interesting aspect of the results on this problem, which is related to the performance assessment using dominance ranking and quality indicators. Note that Tables 3 and 4 show that DEMO^{SP2} is significantly better than SPEA2 with regard to dominance ranking, and significantly worse than SPEA2 with regard to the $I^1_{\varepsilon+}$ indicator. This happens because the approximation sets of DEMO^{SP2} are never dominated, while the approximation sets of SPEA2 sometimes dominate each other. As a result, dominance ranking prefers DEMO^{SP2} to SPEA2 although approximation sets of SPEA2 are closer to the Pareto optimal front than approximation sets of DEMO^{SP2}.

5.3 DEMO^{IB} vs. IBEA

From Tables 5 and 6 it is obvious that using indicator based environmental selection brought DEMO less improvement over the basic GA than using the other two approaches. For the first time, DEMO was outperformed with regard

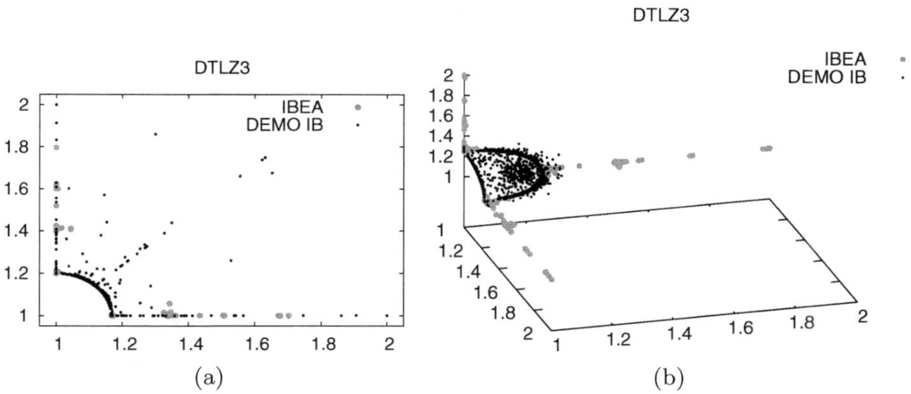

Fig. 7. Plots of normalized approximation sets of DEMO^{IB} and IBEA on the DTLZ3 problem: (a) 30 approximation sets for each algorithm on the problem with two objectives; (b) 30 approximation sets for each algorithm on the problem with three objectives

to dominance ranking. The DTLZ7 problem with 2^{m-1} disconnected Pareto optimal regions proved to be very hard for DEMO^{IB}. While the convergence to the Pareto optimal front was not difficult, maintaining diverse solutions was hard for DEMO^{IB}. Out of 30 runs for each objective space dimensionality, DEMO converged to a single point 29 times for $m = 2$, 26 times for $m = 3$ and 25 times for $m = 4$. Note, however, that in combination with all other approaches to environmental selection (including using $I^1_{\varepsilon+}$ instead of I_{HD} in indicator based selection), DEMO could always maintain diverse solutions.

Let us analyze in more detail also the DTLZ3 problem, where the main difficulty rises from its $3^{10} - 1$ local Pareto optimal fronts. As shown in the plots in Fig. 7, IBEA has more difficulties in reaching the Pareto optimal front than DEMO^{IB}. In the case of two objectives, DEMO^{IB} performs worse than IBEA in the worst case while achieving a much better spread in the best case. On the three-objective problem, DEMO^{IB} achieves good results in all 30 runs, while IBEA still gets stuck in local optima and has a poor spread of solutions.

6 Conclusion

This paper compared the well-known multiobjective evolutionary algorithms NSGA-II, SPEA2 and IBEA to their DE-based variants DEMO$^{\text{NS-II}}$, DEMO$^{\text{SP2}}$ and DEMO^{IB} on 16 state-of-the-art benchmark problems (each with 2, 3 and 4 objectives). The results have shown that on 20% of the problems, DEMO achieved significantly better dominance ranks than the basic GA, while significantly worse dominance ranks were obtained on only 3% of the problems. Furthermore, DEMO outperformed the basic GA with regard to the used quality indicator on the majority (almost 83%) of the problems.

On the basis of these results we conclude that DE explores the decision space more efficiently than GAs also when multiple objectives need to be optimized. It is important to note, however, that DE and, consequently, DEMO are limited to vector representation of solutions and can therefore only be used in numerical optimization.

Acknowledgment

The work presented in the paper was supported by the Slovenian Research Agency under Research Program P2-0209 *Artificial Intelligence and Intelligent Systems.* The authors wish to thank Eckart Zitzler for useful comments on algorithm performance assessment.

References

1. Price, K.V., Storn, R.: Differential evolution – A simple evolution strategy for fast optimization. Dr. Dobb's Journal **22**(4) (1997) 18–24
2. Price, K., Storn, R.M., Lampinen, J.A.: Differential Evolution: A Practical Approach to Global Optimization. Springer-Verlag New York, Inc. (2005)
3. Deb, K., Pratap, A., Agrawal, S., Meyarivan, T.: A fast and elitist multiobjective genetic algorithm: NSGA–II. IEEE Transactions on Evolutionary Computation **6**(2) (2002) 182–197
4. Zitzler, E., Laumanns, M., Thiele, L.: SPEA2: Improving the strength Pareto evolutionary algorithm. In: Proceedings of Evolutionary Methods for Design, Optimization and Control with Applications to Industrial Problems – EUROGEN 2001. (2001) 95–100
5. Zitzler, E., Künzli, S.: Indicator-based selection in multiobjective search. In: Proceedings of the Eighth International Conference on Parallel Problem Solving from Nature – PPSN VIII. (2004) 832–842
6. Abbass, H.A., Sarker, R., Newton, C.: PDE: A Pareto-frontier differential evolution approach for multi-objective optimization problems. In: Proceedings of the 2001 Congress on Evolutionary Computation – CEC 2001. Volume 2. (2001) 971–978
7. Lampinen, J.: DE's selection rule for multiobjective optimization. Technical report, Lappeenranta University of Technology (2001)
8. Madavan, N.K.: Multiobjective optimization using a Pareto differential evolution approach. In: Proceedings of the 2002 Congress on Evolutionary Computation – CEC 2002. Volume 2. (2002) 1145–1150
9. Xue, F., Sanderson, A.C., Graves, R.J.: Pareto-based multi-objective differential evolution. In: Proceedings of the 2003 Congress on Evolutionary Computation – CEC 2003. Volume 2. (2003) 862–869
10. Iorio, A.W., Li, X.: Solving rotated multi-objective optimization problems using differential evolution. In: Proceedings of the 17th Australian Joint Conference on Artificial Intelligence – AI 2004. (2004) 861–872
11. Kukkonen, S., Lampinen, J.: An extension of generalized differential evolution for multi-objective optimization with constraints. In: Proceedings of Parallel Problem Solving from Nature – PPSN VIII. (2004) 752–761

12. Robič, T., Filipič, B.: DEMO: Differential evolution for multiobjective optimization. In: Proceedings of the Third International Conference on Evolutionary Multi-Criterion Optimization – EMO 2005. (2005) 520–533
13. Kukkonen, S., Lampinen, J.: GDE3: The third evolution step of generalized differential evolution. In: Proceedings of the 2005 Congress on Evolutionary Computation – CEC 2005. Volume 1. (2005) 443–450
14. Iorio, A.W., Li, X.: Incorporating directional information within a differential evolution algorithm for multi-objective optimization. In: Proceedings of the 2006 Genetic and Evolutionary Computation Conference – GECCO 2006. Volume 1. (2006) 675–682
15. Santana-Quintero, L.V., Coello Coello, C.A.: An algorithm based on differential evolution for multiobjective problems. In: Smart Engineering System Design: Neural Networks, Evolutionary Programming and Artificial Life. (2005) 211–220
16. Hernández-Díaz, A.G., Santana-Quintero, L.V., Coello Coello, C., Caballero, R., Molina, J.: A new proposal for multi-objective optimization using differential evolution and rough sets theory. In: Proceedings of the 2006 Genetic and Evolutionary Computation Conference – GECCO 2006. Volume 1. (2006) 675–682
17. Deb, K., Thiele, L., Laumanns, M., Zitzler, E.: Scalable test problems for evolutionary multi-objective optimization. In Abraham, A., Jain, R., Goldberg, R., eds.: Evolutionary Multiobjective Optimization: Theoretical Advances and Applications. Springer (2005) 105–145
18. Huband, S., Barone, L., White, L., Hingston, P.: A scalable multi-objective test problem toolkit. In: Proceedings of the Third International Conference on Evolutionary Multi-Criterion Optimization – EMO 2005. (2005) 280–295
19. Bleuler, S., Laumanns, M., Thiele, L., Zitzler, E.: PISA – A platform and programming language independent interface for search algorithms. In: Proceedings of the Second International Conference on Evolutionary Multi-Criterion Optimization – EMO 2003. (2003) 494–508
20. Knowles, J.D., Thiele, L., Zitzler, E.: A tutorial on the performance assessment of stochastic multiobjective optimizers. TIK-Report No. 214, Computer Engineering and Networks Laboratory, ETH Zürich, Switzerland (2006)
21. Zitzler, E., Thiele, L., Laumanns, M., Fonseca, C.M., da Fonseca, V.D.: Performance assessment of multiobjective optimizers: An analysis and review. IEEE Transactions on Evolutionary Computation **7**(2) (2003) 117–132
22. Fonseca, C.M., Fleming, P.J.: On the performance assessment and comparison of multiobjective optimizers. In: Proceedings of Parallel Problem Solving from Nature – PPSN IV. (1996) 584–593

EMOPSO: A Multi-Objective Particle Swarm Optimizer with Emphasis on Efficiency

Gregorio Toscano-Pulido[1], Carlos A. Coello Coello[2], and Luis Vicente Santana-Quintero[2]

[1] Universidad Autónoma de Nuevo León, AP 34 - F
Cd. Universitaria, San Nicolás de los Garza, NL 66450, Mexico*
[2] CINVESTAV-IPN (Evolutionary Computation Group)
Depto. de Computación, Av. IPN No 2508
Col. San Pedro Zacatenco, México, D.F., 07360, Mexico
gtoscano@gmail.com
ccoello@cs.cinvestav.mx, lvspenny@hotmail.com

Abstract. This paper presents the Efficient Multi-Objective Particle Swarm Optimizer (EMOPSO), which is an improved version of a multi-objective evolutionary algorithm (MOEA) previously proposed by the authors. Throughout the paper, we provide several details of the design process that led us to EMOPSO. The main issues discussed are: the mechanism to maintain a set of well-distributed nondominated solutions, the turbulence operator that avoids premature convergence, the constraint-handling scheme, and the study of parameters that led us to propose a self-adaptation mechanism. The final algorithm is able to produce reasonably good approximations of the Pareto front of problems with up to 30 decision variables, while performing only 2,000 fitness function evaluations. As far as we know, this is the lowest number of evaluations reported so far for any multi-objective particle swarm optimizer. Our results are compared with respect to the NSGA-II in 12 test functions taken from the specialized literature.

1 Introduction

Particle swarm optimization (PSO) has been found to be a very effective engine for multi-objective optimization, and several multi-objective particle swarm optimizers (MOPSOs) have been proposed in the last few years [1]. Nevertheless, very few researchers have studied the basic mechanisms of a MOPSO, aiming to design a more efficient search engine, which achieves competitive performance at a low number of objective function evaluations (see for example [2,3]). This paper reports a detailed study of a MOPSO previously proposed by the authors in [4]. Such study led us to propose new mechanisms that produced a new MOPSO that only performs 2,000 fitness function evaluations, while solving test problems

* This author is currently affiliated to CINVESTAV-Tamaulipas Laboratorio de Tecnologías de la Información, Carretera a Monterrey Km. 6, Cd. Victoria, Tamps 87261, MEXICO.

S. Obayashi et al. (Eds.): EMO 2007, LNCS 4403, pp. 272–285, 2007.

of up to 30 decision variables. To the best of the authors' knowledge, this is the lowest number of fitness function evaluations ever reported for any MOPSO in the specialized literature. Our results are compared with respect to the NSGA-II [5], which is a multi-objective evolutionary algorithm (MOEA) representative of the state-of-the-art in the area.

2 Towards an Efficient MOPSO

In [4], we proposed the use of clustering techniques to improve the performance of a MOPSO. In order to improve the performance of our original algorithm, we performed several modifications to it. As a first step, we incorporated a mechanism to distribute the nondominated solutions obtained by the algorithm. Next, we used a turbulence operator, in order to avoid premature convergence. After that, and in order to maximize the subswarms performance, we performed an experiment to fix the number of subswarms to be adopted.Then, we incorporated a mechanism to handle constraints. Finally, we performed an empirical study of the influence of the C_1, C_2 and W parameters, and we proposed a simple methodology to self-adapt these parameters. Each of these componentes will be briefly described in the following subsections.

2.1 Handling Well-Distributed Solutions

The MOPSO proposed in [4] does not impose a bound on the total number of nondominated solutions that it can store. This makes it difficult for the decision maker to choose one of them and also complicates the definition of a baseline to perform fair comparisons with respect to other MOEAs. Researchers have proposed several mechanisms to reduce the nondominated solutions generated by a MOEA (most of them applicable to external archives): clusters [6], adaptive grids [7], crowding [5] and relaxed forms of Pareto dominance [8]. In our case, we implemented two mechanisms: (1) an adaptive grid and (2) a relaxed form of Pareto dominance (ε-dominance). These two mechanisms are described next:

- **Adaptive Grid:** Proposed by Knowles & Corne [7], the adaptive grid is really a space formed by hypercubes. Such hypercubes have as many components as objective functions has the problem to be solved. Each hypercube can be interpreted as a geographical region that contains an n number of individuals. The adaptive grid allows us to store nondominated solutions and to redistribute them when its maximum capacity is reached.
- **ϵ-dominance:** This is a relaxed form of dominance proposed by Laumanns et al. [8]. The so-called ε-Pareto set is an archiving strategy that maintains a subset of generated solutions. It guarantees convergence and diversity according to well-defined criteria, namely the value of the ε parameter, which defines the resolution of the grid to be adopted for the secondary population. The general idea of this mechanism is to divide objective function space into boxes of size ε. Each box can be interpreted as a geographical region that

contains a single solution. The approach accepts a new solution into the ε-Pareto set if 1) it is the only solution in the box which it belongs to, 2) it dominates other(s) solution(s) or 3) it competes against other nondominated solution inside the box, but it is closer to the origin vertex of the box. This algorithm is very attractive both from a theoretical and from a practical point of view. However, in order to achieve the best performance, it is necessary to provide the size of the box (the ε parameter) which is problem-dependent, and it's normally not known before executing a MOEA.

Additionally, we also propose a mechanism to distribute nondominated solutions, which is called **Hyper-plane distribution**. The core idea of this proposal is to perform a good distribution of the hyper-plane space defined by the minima (assuming minimization) from the objectives, and use such distribution to select a representative subset from the whole set of nondominated solutions. The algorithm works as follows: First, it requires as input, a set of nondominated solutions and the quantity n of desirable final solutions. Then, the algorithm selects those solutions which have the minima value on each objective. A hyperplane among all minima solutions is thus computed. Next, the algorithm divides such space into $n-1$ regions. Therefore, on the vertex of each region, a perpendicular line to the hyper-plane is computed. Finally, the algorithm only accepts those solutions which are closest to each line. This algorithm has a complexity $O(N^2)$, because each individual is compared against everybody else with respect to distance.

In Figure 1, we can see an example that aims to clarify the algorithm's description. In this example, two objective functions are used. Five nondominated

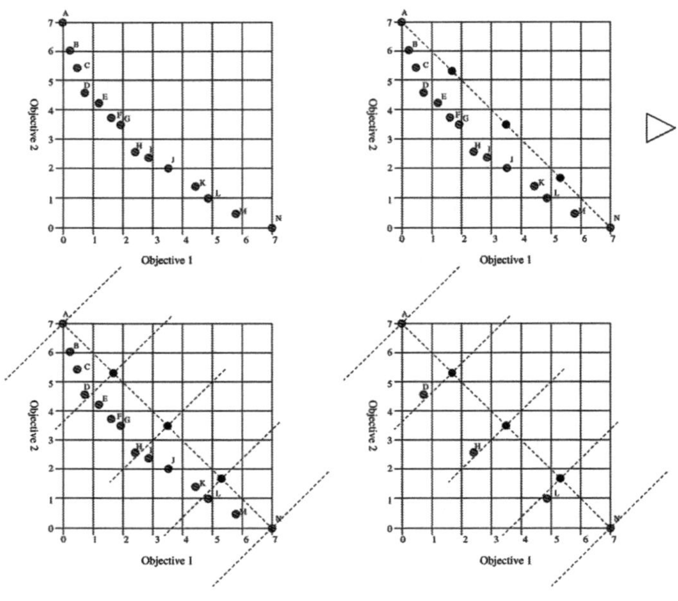

Fig. 1. Graphical representation of the hyper-plane distribution

solutions need to be selected. So, the hyper-plane (a line in this case) formed by the minima of objective 1 and objective 2 is divided into 4 line segments. Then, each vertex is projected towards the Pareto front. Finally, the solutions closest to those projected points are selected.

A First Comparative Study. The three approaches previously described to select the best distributed nondominated individuals were implemented and called at each generation. Therefore, on each generation we will have the set of nondominated solutions with a bounded and well-distributed subset. The three algorithms were requested to select 50 nondominated solutions each. The scheme based on the ε-Pareto set needs an extra parameter (ε). This parameter was computed in each test function as follows: the algorithm was executed using a total number of iterations of 200. Then, the ε values were manually fine-tuned to find an average of 40 nondominated solutions in each of the 30 executions. Since the aim of this experiment was to observe if there was an approach which performed best, we adopted the Inverted Generational Distance (IGD) metric [9], as a way to estimate how far is the true Pareto front from the solutions obtained. This measure is defined as:

$$GD = \frac{\sqrt{\sum_{i=1}^{n} d_i^2}}{n} \tag{1}$$

where n is the number of nondominated vectors found by the algorithm being analyzed and d_i is the Euclidean distance (measured in objective space) between each of these and the nearest member of the true Pareto front. It should be clear that a value of $GD = 0$ indicates that all the elements generated are in the true Pareto front of the problem. Therefore, any other value will indicate how "far" we are from the global Pareto front of our problem. This metric, also, penalizes when the solutions obtained do not cover completely the true Pareto front.

For this first study, we adopted 8 test functions taken from the specialized literature: **ZDT1**, **ZDT2**, **ZDT3**, **ZDT4**, **ZDT6** proposed by *Zitzler et al.* in [10], **Kursawe**'s problem proposed in [11], and **Deb** and **Deb2**, proposed in [12]. In Table 1, we can see that the adaptive grid was the worst choice, and that the Hyper-plane Distribution outperformed the ε-Dominance approach mainly because of its property of selecting a fixed number of solutions (while we can only estimate the total number of solutions when using ε-dominance). Therefore, we will use the Hyper-plane distribution in our MOPSO. Table 2 shows the average of nondominated solutions found in the last generation by each approach. We can see in this table, that ε-dominance was the algorithm that had more problems to reach the target (40 nondominated solutions).

2.2 Avoiding Premature Convergence

Frans van den Bergh [13] discovered a potentially dangerous property in PSO: if the position of the particles coincides with gbest, then it will move away from the gbest if the inertia or the current velocity is non-zero. This may lead to premature convergence (i.e. all the particles will converge to the gbest particle, which is

Table 1. Comparison of results of three approaches to maintain a good distribution of nondominated solutions (adaptive grid, ε-dominance and hyper-plane distribution) with respect to the Inverted Generational Distance metric

	Approach		
Function	**Adaptive-Grid**	**ϵ-Dominance**	**Hyper-plane Distribution**
ZDT1	0.002855236	**0.00149249**	0.002029507
ZDT2	0.031790172	0.033058091	**0.023006512**
ZDT3	0.007237485	0.005816599	**0.005786199**
ZDT4	2.763645000	2.903810666	**2.715304000**
ZDT6	0.000224374	0.000186435	**0.000197410**
Kursawe	0.009648260	0.008716252	**0.007975956**
Deb	0.001454523	0.001606147	**0.000475256**
Deb2	**0.00942435**	0.009346694	0.009427080

Table 2. Comparison of the number of reported solutions at the end of the executions by each of the three approaches

	Approach		
Function	**Adaptive-Grid**	**ε-Dominance**	**Hyper-plane Distribution**
ZDT1	39	37	40
ZDT2	24	19	27
ZDT3	38	29	37
ZDT4	2	2	2
ZDT6	40	42	39
Kursawe	40	33	39
Deb	40	36	40
Deb2	39	34	37

usually a local minimum). To determine how often this behavior occurred in our algorithm, we counted how many times a particle's position was the same than the gbest position in all the executions for all the test functions adopted. In Table 3, we can see the results of this experiment. Frans van den Bergh proposed a new parameter to address this issue, but his proposal is hard to adopt in a multiobjective approach based on Pareto ranking, since we would have to deal with several "best" solutions. Therefore, in [14], we proposed the use of a turbulence operator. This turbulence operator consists of an alteration to the flight velocity of a particle.[1] This modification is performed in all the dimensions (i.e., in all the decision variables), such that the particle can move to a completely isolated region. The turbulence operator acts based on a probability factor that considers the current generation and the total number of iterations to be performed. The idea is to have a much higher probability to perturb the flight of the particles at the beginning of the search, and decrease it as we progress in the search. The turbulence can be seen as a mutation operator and it is based on the following expression:

[1] This mechanism is inspired on the one proposed in [15].

$$temp = current\_generation / total\_generations \quad (2)$$
$$prob_{turbulence} = temp^{1.7} - 2.0 \times temp + 1.0 \quad (3)$$

where *temp* is used as a temporary variable, *current_generation* is the current generation number, *total_generations* is the total number of generations and $prob_{turbulence}$ refers to the probability of affecting the flight of a particle using the turbulence operator. The values used for this expression were empirically derived after a set of experiments. The details of the experiments that led us to this setup may be found in [2].

2.3 Maximizing the Spread

The appropriate selection of leaders is essential for the good performance of a MOPSO. If the particle chooses an inappropriate leader (i.e., a leader who is too far away in the search space) then most of the flight will be fruitless because the particle will not be visiting promissory regions of the search space. In [4], we proposed to use not one but several swarms to avoid this type of problem (in order to make a difference between the use of the word *swarm* from traditional PSO's approaches, and the use of several swarms, from our approach, we introduce the word *subswarm*, which means a set of particles that has its own PSO's behavior. The subswarms share information among them by interchanging their leaders with certain probability). However, we did not provide any statistical analysis related to the number of subswarms needed. Thus, we proceeded to perform such an analysis. Table 4 summarizes the results obtained, showing the mean of 30 independently executions of the algorithm using 2,000 fitness function evaluations. Each execution was tuned to perform $2,000$ fitness function evaluations. We found that by using 8 subswarms, the algorithm exhibited its best performance in 5 out of 8 test functions. We think, that the use of this value can be beneficial most of the time. Therefore, we adopted it as the default value for the number of subswarms. It is important to note that this experiment was performed using 40 particles, which means that each subswarm will have 5 particles. Because of the results, we suggest to use 8 subswarms as a fixed parameter in our algorithm.

Table 3. Statistical results obtained from the counting of the cases in which the particle was equal to its gbest

Statistics	Particle = GBest
Mean	104.81192
Best	2
Worst	729
St.dev.	110.59983
Median	74

Table 4. Results of the impact of the number of clusters on each test function, with respect to Inverted Generational Distance metric

	Subswarms (clusters)				
Function	**1**	**2**	**4**	**8**	**20**
ZDT1	0.004820567	0.005645242	0.004388993	**0.002853130**	0.003191553
ZDT2	0.034759800	0.013606517	0.017762557	**0.012582413**	0.013187440
ZDT3	0.016337620	0.017927403	0.017630486	**0.008787392**	0.011176582
ZDT4	4.536191000	3.445736000	**3.27696600**	3.7399510000	3.762421666
ZDT6	0.003337423	0.000509384	0.000472352	**0.000334933**	0.000431701
Kursawe	0.069521400	**0.04149130**	0.043504603	0.0498330600	0.042813556
Deb	0.034178800	0.009745156	0.007927125	**0.007114026**	0.008663098
Deb2	0.009801385	0.009445994	**0.00889158**	0.008950303	0.009510023

2.4 A Constraint-Handling Mechanism

Since the approach proposed in [4] does not use any special mechanism to deal with constrained search spaces, we incorporated the mechanism proposed in [14]. This approach does not require any user-defined parameters and it performs less objective function evaluations than any of the other approaches with respect to which it was compared, while obtaining similar results (see [14] for further details).

2.5 Analyzing the Impact of the PSO's Parameters

The PSO algorithm has three parameters that play a key role in the algorithm's behavior:

1. **W**: velocity inertia
2. **C1**: cognitive component
3. **C2**: social component

The fact that a MOEA converges to a set of solutions rather than to a single value, makes it difficult to perform an statistical analysis such as the analysis of variance, which can determine how sensitive is an algorithm to its parameters. Nevertheless, we performed a very thorough analysis of parameters (similar to an analysis of variance), with the aim of finding the best possible parameter configuration for our approach (considering the set of test functions adopted). It is worth indicating that the parameters settings that have been previously proposed (see for example [16]) for the original (unconstrained single-objective) PSO, do not provide a good performance in the context of multiobjective optimization and therefore the motivation to perform the thorough analysis reported in [2]. In order to analyze the impact of the parameters on our proposed approach, we considered several configurations, and performed an exhaustive analysis. The configurations adopted are:

$$\begin{aligned} W &= \{0.0, 0.1, 0.2, 0.3, 0.4, 0.5, 0.6, 0.7, 0.8, 0.9, 1.0\} \\ C1 &= \{1.0, 1.2, 1.4, 1.6, 1.8, 2.0, 2.2, 2.4, 2.6, 2.8, 3.0\} \\ C2 &= \{1.0, 1.2, 1.4, 1.6, 1.8, 2.0, 2.2, 2.4, 2.6, 2.8, 3.0\} \end{aligned} \tag{4}$$

For all our experiments, we adopted 40 particles. However, we analyzed three different performance scenarios:

1. **Experiment 1:** Use of a randomly generated initial population and a low number of fitness function evaluations (we used $Gmax = 25$, which gives us a total of 1000 fitness function evaluations).
2. **Experiment 2:** Use of a good approximation of the Pareto front in the initial population. In all the experiments performed in this case, the same approximation was fed to the algorithm in its initial generation. In this case, we only performed 600 fitness function evaluation ($Gmax = 15$), since we were interested in analyzing the capability (or possible difficulties) of our algorithm to reach the true Pareto front of a problem once a good (and sufficiently close) approximation had been produced.
3. **Experiment 3:** Use of a large number of fitness function evaluations ($Gmax = 250$, which gives a total of $10,000$ fitness function evaluations) in order to assess the performance of our approach in the long term.

We adopted eleven test functions for **Experiment 1** and **Experiment 2**. Due to the high CPU time required by each run, we only adopted six test functions for **Experiment 3**. Since we needed to assess performance in each case, we chose the inverted generational distance metric, since it can measure both closeness to the true Pareto front and spread of solutions. An obvious problem with so many experiments was how to present the results in a compact form. For that sake, we adopted a set of squares (called "mosaics"), such that each of them has a gray scale that corresponds to the mean value of the inverted generational distance over 30 independent runs, produced from one combination of {W,C1,C2} (all the possible combinations were adopted, considering the sets of possible values previously defined for these three parameters). The mean results were normalized between zero and 255 (where zero is the best possible value and 255 is the worst). The results of each of the three above experiments are briefly discussed next, but the mosaics themselves could not be included due to space restrictions; however, they can be found in [2].

Conclusions from the Second Series of Experiments. After combining the results from the three experiments, we concluded the following:

- The best range for C1 and C2 is from 1.2 to 2.0. From within this range, we can say that C1=C2=1.4 provides the best overall performance.
- The best values for W are those less or equal to 0.5.
- Note, however, that the above values do not produce the best possible performance in all cases (as expected, because of the No Free Lunch Theorem [17]). This is precisely what motivated us to propose a mechanism to self-adapt the parameters of our approach so that it automatically adjusts the parameters according to the characteristics of the search space being explored.

2.6 Self-adaptive Mechanism

The results obtained from the experiments reported in the previous subsection led us to propose a self-adaptive mechanism which is described in this

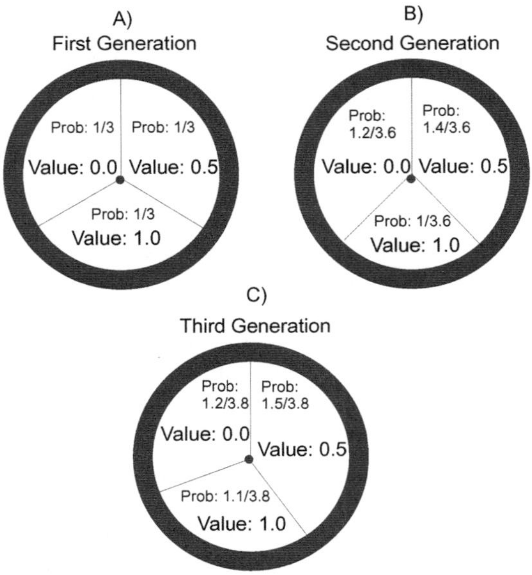

Fig. 2. Roulette wheel selection example at the a) first, b) second and c) third generation

subsection. We proposed the use of a traditional proportional selection mechanism to select the most appropriate combination of parameters values to be adopted. We adopted roulette-wheel selection for that sake [18]. Rather than computing fitness for each combination of values, we count the number of nondominated solutions generated by each combination of values. Let's assume that the parameter W can take 3 possible values: 0, 0.5, 1.0. So, at the beginning of the search, each value has a 33% probability of occurring (see Figure 2 (a)). At generation zero, each possible value has a "fitness" of one. Now, let's assume that after one generation, $W = 0$ generates 2 new nondominated solution, $W = 0.5$ generate 4 new nondominated solution and $W = 1$ did not contribute with any new nondominated solution. Thus, we reward the "fitness" of W=0.5 by increasing its value in 0.4 (we increase fitness in 0.1 for each new nondominated solution produced). So, $W = 0.5$ now has a fitness of 1.4. Analogously, $W = 0$ has a fitness of 1.2 and $W = 1$ remains with a fitness of 1.0. The new share in the roulette wheel for each value is shown in Figure 2 (b). Since the total fitness is now 3.6 $(1.4+1.2+1.0)$, the share of each value is: $W = 0.5$ $(1.4/3.6)$, $W = 0$ $(1.2/3.6)$ and $W = 1$ $(1.0/3.6)$. After generation two, $W = 0.5$ generated one new nondominated solution, the same happened with $W = 1.0$ and $W = 0$ did not produce any new nondominated solutions. So, $W = 0.5$ now has a fitness of 1.5, $W = 1.0$ has a fitness of 1.1 and $W = 0$ remains with a fitness of 1.2. Thus, the total fitness is now 3.8, and we have the following: $W = 0.5$ has a share of $1.5/3.8$, $W = 1$ has a share of $1.1/3.8$ and $W = 0$ has a share of $1.2/3.8$ (see Figure 2 (c)).

To validate our proposal, we compare it against two other parameter selection mechanisms: a) deterministic selection and b) random selection.

Deterministic selection : $C1$, $C2$ and W are deterministically set to 1.4, 1.4 and 0.2, respectively (these values are the center of the region which performed best from the experiments reported in the previous subsection).

Random selection : $C1$, $C2$ and W pick their values randomly from an interval from 1.2 to 2 for $C1$ and $C2$ and from the range from 0.0 to 0.4 for W (this is the range of values which performed best in the experiments reported in the previous subsection).

The detailed results from this experiment can be seen in [2]. Although the results of this study seem inconclusive, we argue that the use of a self-adaptive mechanism for adjusting the parameters gives a better performance in a wider range of functions, and avoids that the user has to setup the parameters of the algorithm by hand. In this experiment, most of the time, the approach which performed best for 25 generations, was also the best performer when adopting 50 and 250 generations. So, we argue, that the proposal to self-adapt the parameters improved the overall performance of the algorithm. After introducing this last component (i.e., the self-adaptive mechanism), the final version of our proposed algorithm, which we call Efficient Multi-Objective Particle Swarm Optimizer (EMOPSO) is shown in flowchart 3. This approach does not require any manual fine-tuning of its PSO parameters.

3 Test Problems

Deb *et al.* [19] proposed a set of constrained multiobjective problems in which the difficulty can be controlled by varying a set of parameters. In this study, we use $g(\boldsymbol{x}) = 1 + x_1$ for the test problems CTP1 to CTP7, $a_1 = 0.858$, $b_1 = 0.541$, $a_2 = 0.728$, and $b_2 = 0.295$ for CTP1 and the parameters shown in Table 5 for CTP2 to CTP7.

Table 5. Parameters chosen for CTP2 to CTP7

Problem	θ	a	b	c	d	e	constraints	boundaries
CTP2	-0.2π	0.2	10.0	1.0	6.0	1.0	2	$0 \leq x_1, x_2 \leq 1$
CTP3	-0.2π	0.1	10.0	1.0	0.5	1.0	1	$0 \leq x_1, x_2 \leq 1$
CTP4	-0.2π	0.75	10.0	1.0	0.5	1.0	1	$0 \leq x_1, x_2 \leq 1$
CTP5	-0.2π	0.1	10.0	2.0	0.5	1.0	1	$0 \leq x_1, x_2 \leq 1$
CTP6	-0.1π	40.0	0.5	1.0	2.0	-2.0	1	$0 \leq x_1 \leq 1, 0 \leq x_2 \leq 10$
CTP7	-0.05π	40.0	0.5	1.0	2.0	-2.0	1	$0 \leq x_1 \leq 1, 0 \leq x_2 \leq 10$

4 Comparison of Results

In order to allow a quantitative assessment of results, we adopted the following performance measures: Inverted Generational Distance [9], Two Set

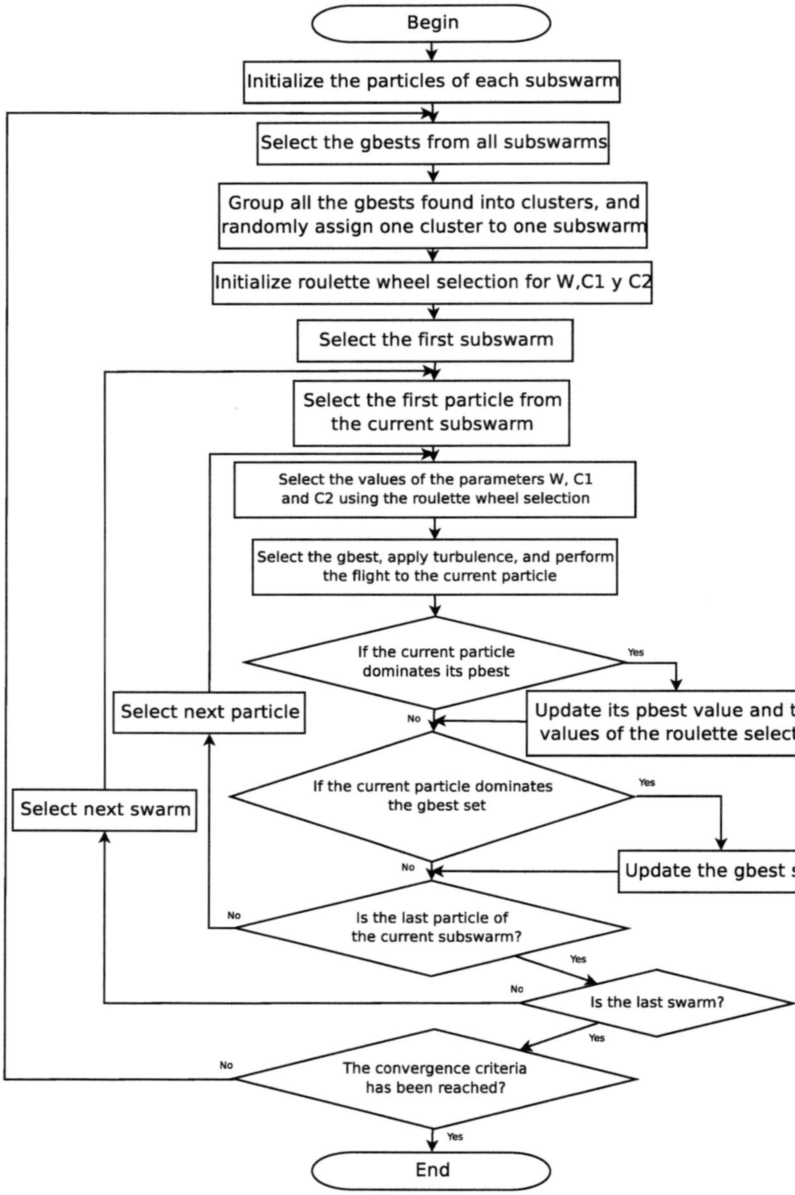

Fig. 3. Flowchart of the proposed algorithm

Coverage [10] and the Spread metric [20]. We compare results with respect to the NSGA-II using 12 standard test functions: 7 from the Constrained Test Problems (CTP1 to CTP7) proposed by Deb *et al.* [19], which were previously discussed; and 5 from the ZDTs test problems proposed by *Zitzler et al.* in

Table 6. Comparison of results between our approach (EMOPSO) and the NSGA-II

	IGD				Set Coverage				Spread			
Function	EMOPSO		NSGA-II		EMOPSO		NSGA-II		EMOPSO		NSGA-II	
	Mean	σ	Mean	σ	Mean	σ	Mean	σ	Mean	σ	Mean	σ
CTP1	**0.0004**	0.0001	0.0013	0.0014	**0.0868**	0.0393	0.2133	0.0657	**0.1957**	0.0104	0.2859	0.0386
CTP2	**0.0014**	0.0003	0.0094	0.0056	**0.3208**	0.0698	0.3266	0.0811	**0.3932**	0.0682	0.4210	0.0320
CTP3	**0.0084**	0.0016	0.0100	0.0035	**0.5275**	0.1788	0.6242	0.1223	**0.5959**	0.1253	0.6191	0.080
CTP4	**0.0327**	0.0163	0.0417	0.0121	**0.1465**	0.2039	0.7294	0.1814	**0.7585**	0.2583	0.9909	0.1401
CTP5	**0.0060**	0.0012	0.0102	0.0064	**0.3933**	0.1404	0.4808	0.1135	0.5126	0.1054	**0.4925**	0.0770
CTP6	**0.002**	0.0006	0.0096	0.0108	**0.1113**	0.0643	0.2041	0.0612	0.8679	0.0575	**0.3067**	0.0707
CTP7	**0.0002**	0.0006	0.0017	0.0027	**0.0042**	0.0132	0.0741	0.0480	**0.7112**	0.1874	0.9998	0.0633
ZDT1	**0.0022**	0.0001	0.0740	0.0177	**0**	0	1	0	**0.0870**	0.0164	0.5627	0.0756
ZDT2	**0.0007**	0.0003	0.1938	0.0632	**0**	0	1	0	**0.0755**	0.0980	0.7585	0.0757
ZDT3	**0.0020**	0.0005	0.061	0.011	**0**	0	1	0	**0.5011**	0.0290	0.7033	0.0609
ZDT4	8.1128	5.1184	**5.5563**	1.8295	**0.0531**	0.2018	0.4438	0.3703	0.9978	0.0081	**0.9856**	0.0128
ZDT6	**0.0980**	0.1191	0.6098	0.1447	**0**	0	1	0	**0.4380**	0.2798	0.8799	0.0775

[10]. The detailed description of these 12 test functions was omitted due to space restrictions. The CTPs problems all have 2 decision variables each, and the ZDTs functions are unconstrained and have between 10 (ZDT4 and ZDT6) and 30 (ZDT1, ZDT2 and ZDT3) decision variables. ZDT5 is not included in our study because it is a binary function, and we only adopted test functions in which the decision variables are real numbers. In the following examples, the NSGA-II was run using a population size of 40, a crossover rate of 0.9, tournament selection, a mutation rate of $1/N$, where N = number of variables (real numbers representation was adopted), a distribution index of 15 for SBX, and a distribution index of 20 for its parameter-based mutation operator. Our EMOPSO used 40 particles and a total of 8 swarms. The total number of fitness function evaluations was set to $2,000$ for the two algorithms compared (50 generations).

Table 6 shows that the results obtained by our EMOPSO were superior to those generated by the NSGA-II. Our EMOPSO outperformed the NSGA-II in all the test problems with respect to the set coverage metric, and in all but one (ZDT4) with respect to the inverted generational distance metric. Regarding spread, our approach outperformed the NSGA-II in most problems (except for CTP5, CTP6 and ZDT4). Obviously, if allowed to perform a higher number of fitness function evaluations, the NSGA-II would be able to converge to the true Pareto front of most of these test functions, but our main purpose was to show that our EMOPSO is a viable choice when dealing with objective functions whose computational cost is very high (e.g., in aerospace engineering). Due to space restrictions, the Pareto fronts obtained in each case are not included in the paper.

5 Conclusions and Future Work

Our main conclusions are the following:

- We found that the use of subswarms promotes local search as an emergent behavior in our EMOPSO. Consequently, the performance of our approach

was improved by the use of subswarms, particularly in the presence of disconnected Pareto fronts.
- We have proposed a mechanism called "hyper-plane distribution", to distribute nondominated solutions.
- The use of a perturbation mechanism in our multiobjective particle swarm optimizer was found to be critical to control its high selection pressure, as to avoid premature convergence.
- In general, we found that it is quite difficult to find fixed values for the three most significant parameters of our approach (W, C1 and C2). It is worth indicating that the comprehensive study of parameters that we performed is, as far as we know, the first of its type (in the context of multiobjective particle swarm optimization). Based on the results of this study, we designed a self-adaptive mechanism for these parameters, and we found this to be a good alternative to facilitate the use of our approach.

Some possible paths to extend this work are the following:

- Experiment with other PSO's models and with different interconnection topologies.
- Study alternative methods for the survivor selection mechanism.
- Study alternative (perhaps more elaborate) constraint-handling mechanisms.
- Study alternative mechanisms to accelerate convergence while keeping the same quality of results achieved by our EMOPSO.
- Study alternative mechanisms to distribute nondominated solutions.
- To assess the performance of our EMOPSO in a real-world problem in which the cost of evaluating the objective functions is very high (computationally speaking).

Acknowledgments

The second author acknowledges support from CONACyT through project number 45683-Y. The third author acknowledges support from CONACyT through a scholarship to pursue graduate studies at the Computer Science Department of CINVESTAV-IPN.

References

1. Reyes-Sierra, M., Coello Coello, C.A.: Multi-Objective Particle Swarm Optimizers: A Survey of the State-of-the-Art. International Journal of Computational Intelligence Research **2**(3) (2006) 287–308
2. Toscano Pulido, G.: On the Use of Self-Adaptation and Elitism for Multiobjective Particle Swarm Optimization. PhD thesis, Computer Science Section, Department of Electrical Engineering, CINVESTAV-IPN, Mexico (2005)
3. Branke, J., Mostaghim, S.: About Selecting the Personal Best in Multi-Objective Particle Swarm Optimization. In Runarsson, T.P., Beyer, H.G., Burke, E., Merelo-Guervós, J.J., Whitley, L.D., Yao, X., eds.: Parallel Problem Solving from Nature - PPSN IX, 9th International Conference. Springer. Lecture Notes in Computer Science Vol. 4193, Reykjavik, Iceland (2006) 523–532

4. Toscano Pulido, G., Coello Coello, C.A.: Using Clustering Techniques to Improve the Performance of a Particle Swarm Optimizer. In et al., K.D., ed.: Genetic and Evolutionary Computation–GECCO 2004. Proceedings of the Genetic and Evolutionary Computation Conference. Part I, Seattle, Washington, USA, Springer-Verlag, Lecture Notes in Computer Science Vol. 3102 (2004) 225–237
5. Deb, K., Pratap, A., Agarwal, S., Meyarivan, T.: A Fast and Elitist Multiobjective Genetic Algorithm: NSGA–II. IEEE Transactions on Evolutionary Computation **6**(2) (2002) 182–197
6. Zitzler, E., Thiele, L.: Multiobjective Evolutionary Algorithms: A Comparative Case Study and the Strength Pareto Approach. IEEE Transactions on Evolutionary Computation **3**(4) (1999) 257–271
7. Knowles, J.D., Corne, D.W.: Approximating the Nondominated Front Using the Pareto Archived Evolution Strategy. Evolutionary Computation **8**(2) (2000) 149–172
8. Laumanns, M., Thiele, L., Deb, K., Zitzler, E.: Combining Convergence and Diversity in Evolutionary Multi-objective Optimization. Evolutionary Computation **10**(3) (2002) 263–282
9. Veldhuizen, D.A.V.: Multiobjective Evolutionary Algorithms: Classifications, Analyses, and New Innovations. PhD thesis, Department of Electrical and Computer Engineering. Graduate School of Engineering. Air Force Institute of Technology, Wright-Patterson AFB, Ohio (1999)
10. Zitzler, E., Deb, K., Thiele, L.: Comparison of Multiobjective Evolutionary Algorithms: Empirical Results. Evolutionary Computation **8**(2) (2000) 173–195
11. Kursawe, F.: A Variant of Evolution Strategies for Vector Optimization. In Schwefel, H.P., Männer, R., eds.: Parallel Problem Solving from Nature. 1st Workshop, PPSN I. Volume 496 of Lecture Notes in Computer Science Vol. 496., Berlin, Germany, Springer-Verlag (1991) 193–197
12. Deb, K.: Multi-Objective Genetic Algorithms: Problem Difficulties and Construction of Test Problems. Evolutionary Computation **7**(3) (1999) 205–230
13. van den Bergh, F.: An Analysis of Particle Swarm Optimization. PhD thesis, Faculty of Natural and Agricultural Science, University of Petoria, Pretoria, South Africa (2002)
14. Toscano-Pulido, G., Coello Coello, C.A.: A Constraint-Handling Mechanism for Particle Swarm Optimization. In: Proceedings of the Congress on Evolutionary Computation 2004 (CEC'2004). Volume 2., Piscataway, New Jersey, Portland, Oregon, USA, IEEE Service Center (2004) 1396–1403
15. Fieldsend, J.E., Singh, S.: A Multi-Objective Algorithm based upon Particle Swarm Optimisation, an Efficient Data Structure and Turbulence. In: Proceedings of the 2002 U.K. Workshop on Computational Intelligence, Birmingham, UK (2002) 37–44
16. Kennedy, J., Eberhart, R.C.: Swarm Intelligence. Morgan Kaufmann Publishers, San Francisco, California (2001)
17. Wolpert, D.H., Macready, W.G.: No Free Lunch Theorems for Optimization. IEEE Transactions on Evolutionary Computation **1**(1) (1997) 67–82
18. Goldberg, D.E.: Genetic Algorithms in Search, Optimization and Machine Learning. Addison-Wesley Publishing Company, Reading, Massachusetts (1989)
19. Deb, K., Pratap, A., Meyarivan, T.: Constrained Test Problems for Multi-objective Evolutionary Optimization. In Zitzler, E., Deb, K., Thiele, L., Coello, C.A.C., Corne, D., eds.: First International Conference on Evolutionary Multi-Criterion Optimization. Springer-Verlag. Lecture Notes in Computer Science No. 1993 (2001) 284–298
20. Deb, K.: Multi-Objective Optimization using Evolutionary Algorithms. John Wiley & Sons, Chichester, UK (2001) ISBN 0-471-87339-X.

A Novel Differential Evolution Algorithm Based on ϵ-Domination and Orthogonal Design Method for Multiobjective Optimization

Zhihua Cai[1,2], Wenyin Gong[1], and Yongqin Huang[1]

[1] School of Computer Science
China University of Geosciences, Wuhan 430074, P.R. China
zhcai@cug.edu.cn, cug11100304@yahoo.com.cn
[2] Dept. of computer science
The University of Western Ontario,London,Ontario,N6A 5B7,Canada

Abstract. To find solutions as close to the Pareto front as possible, and to make them as diverse as possible in the obtained non-dominated front is a challenging task for any multiobjective optimization algorithm.ϵ-dominance is a concept which can make genetic algorithm obtain a good distribution of Pareto-optimal solutions and has low computational time complexity,and the orthogonal design method can generate an initial population of points that are scattered uniformly over the feasible solution space.In this paper, combining ϵ-dominance and orthogonal design method, we propose a novel Differential Evolution (DE) algorithm for multiobjective optimization .Experiments on a number of two- and three-objective test problems of diverse complexities show that our approach is able to obtain a good distribution with a small computational time in all cases. Compared with several other state-of-the-art evolutionary algorithms, it achieves not only comparable results in terms of convergence and diversity metrics, but also a considerable reduction of the computational effort.

1 Introduction

Evolutionary Algorithms (EAs) (including genetic algorithms, evolution strategies, evolutionary programming, and genetic programming) are heuristics that have been successfully applied in a wide set of areas. In real-world optimization applications, it is often hard to formulate the optimization goal as a scalar function. Typically, there are several criteria or objectives, and not unusually, these objectives stay in conflict with each other. Simply combining the different associated objective functions in a linear way is usually unsatisfactory. Instead, one is interested in a so-called Pareto optimal set of solutions, i.e., any solution that cannot be improved with respect to one objective without worsening the situation with respect to the other objectives. Consequently, there are two goals in multiobjective optimization: (i) to find solutions as close to the Pareto front as possible, and (ii) to find solutions as diverse as possible in the obtained non-dominated front. Satisfying the two goals is a challenging task for any multiobjective optimization algorithm. Special strategies are therefore needed to deal with such multiobjective optimization problems. Since EAs work on populations of candidate solutions, they represent a promising basic framework for multiobjective optimization.

S. Obayashi et al. (Eds.): EMO 2007, LNCS 4403, pp. 286–301, 2007.

In the last few years, many variants and extensions of classical EAs have been developed for Multiobjective Optimization Problems (MOPs). Such as Nondominated Sorting GA (NSGA-II) [1], Strength Pareto EA (SPEA2) [2], Vector Evaluated GA (VEGA) [3], Hajela and Lins GA (HLGA) [4], Pareto-based Ranking Procedure(FFGA) [5], Niched Pareto GA (NPGA) [6], Pareto Archived Evolution Strategy (PAES) [7], and so on. Among these, the NSGA-II by Deb *et al.* [1] and SPEA2 by Zitzler *et al.* [2] are the most popular approaches.

Differential evolution (DE) [8] is a novel evolutionary algorithm for faster optimization, which mutation operator is based on the distribution of solutions in the population. And DE has won the third place at the first International Contest on Evolutionary Computation on a real-valued function test-suite. Unlike Genetic Algorithm (GA) that uses binary coding to represent problem parameters, DE is a simple yet powerful population based, direct search algorithm with the generation-and-test feature for globally optimizing functions using real valued parameters. Among the DE's advantages are its simple structure, ease of use, speed and robustness. Price & Storn [8] gave the working principle of DE with single strategy. Later on, they suggested ten different strategies of DE [9]. It has been successfully used in solving single-objective optimization problems [10]. Hence, several researchers have tried to extend it to handle MOPs. Such as Pareto DE (PDE) [11,12], Pareto DE Approach (PDEA) [13], Multiobjective DE (MODE) [14], and DE for Multiobjective Optimization (DEMO) [15].

Combing orthogonal array (OA) and factor analysis (such as the statistical optimal method),Orthogonal design method [16] is developed to sample a small and representative set for all possible combinations to obtain good combinations. Recently, some researchers applied the orthogonal design method incorporated with EAs to solve optimization problems. Leung and Wang [17] incorporated orthogonal design in genetic algorithm for numerical optimization problems and found such method was more robust and statistically sound than the classical GAs. OMOEA [18] and OMOEA-II [19] presented by Sangyou Zeng *et al.* adopted the orthogonal design method to solve the MOPs. Numerical results demonstrated the efficiency of the two tools.

ϵ-MOEA [20] is a steady-state Multiobjective EA (MOEA) based on the ϵ-dominance concept introduced in [21]. Also, it incorporated efficient parent and archive update strategies to obtain a good distribution of Pareto-optimal solutions within less computational time. The ϵ-dominance does not allow two solutions with a difference ϵ_i in the i-th objective to be nondominated to each other, thereby allowing a good diversity to be maintained in the population. Besides, the method is quite pragmatic, because it allows the user to choose a suitable ϵ_i depending on the desired resolution in the i-th objective [20].

Inspired by the ideas from OGA/Q [17] and ϵ-MOEA [20], in this paper, we propose an extension of DE algorithm based on the ϵ-dominance concept and orthogonal design method. Our proposed DE algorithm is named ϵ-ODEMO. Our algorithm has three novelties. Firstly, the proposed approach adopts orthogonal design method with quantization technique to generate an initial population of points. And then, it uses the DE/rand/1/exp strategy to produce new candidate solutions. Thirdly, ϵ-dominance concept and efficient parent and archive update strategies introduced in [20] are used to update the archive and population. To evaluate the efficiency of the proposed ϵ-ODEMO, we test it on a number of two- and three-objective problems.

The rest of this paper is organized as follows. In Section 2, we briefly introduce the background of ϵ-MOEA. In Section 3, we describe the function optimization problems in conventional DE. A detailed description of the proposed ϵ-ODEMO algorithm is provided in Section 4. In Section 5, we test our algorithm through a number of two- and three-objective problems. This is followed by results and discussions of the optimization experiments for ϵ-ODEMO in Section 6. The last section, Section 7, is devoted to conclusions and future studies.

2 Background of ϵ-MOEA

ϵ-MOEA [20] is a new approach for MOPs, which is a steady-state MOEA based on the ϵ-dominance concept [21]. In ϵ-MOEA, the search space is divided into a number of grids (or hyper-boxes) and diversity is maintained by ensuring that a grid or hyper-box can be occupied by only one solution. There are two co-evolving populations: (i) an EA population $P(t)$ and (ii) an archive population $E(t)$ (where t is the iteration counter). The archive population stores the nondominated solutions and is updated iteratively with the ϵ-dominance concept. And the EA population is updated iteratively with the usual domination. ϵ-MOEA is described as follows:

Algorithm 1. ϵ-MOEA algorithm

Generate an initial population $P(0)$ uniform randomly
Create the archive $E(0)$ with the ϵ-nondominated solutions of $P(0)$
while the halting criterion is not satisfied **do**
  Choose one solution each from $P(t)$ and $E(t)$ for mating
  Use crossover and mutation to produce λ offspring solutions
  Compare each of these offspring solutions with the archive and the EA population to update them respectively
end while

As mentioned above, The archive population is updated by the offspring solutions iteratively with the ϵ-dominance concept. In ϵ-MOEA, every solution in the archive is assigned an identification array ($\boldsymbol{B}$) which can be calculated by:

$$B_j(\boldsymbol{f}) = \begin{cases} \left\lfloor (f_j - f_j^{min})/\epsilon_j \right\rfloor, \text{ for minimizing} f_j \\ \left\lceil (f_j - f_j^{min})/\epsilon_j \right\rceil, \text{ for maxmizing} f_j \end{cases} \tag{1}$$

where f_j^{min} is the minimum possible value of the j-th objective (if the decision-makers don't know the minimum possible value exactly, use $f_j^{min} = 0$) and ϵ_j is the allowable tolerance in the j-th objective beyond which two values are significant to the user. This ϵ_j value is similar to the ϵ used in the ϵ-dominance definition [21]. The identification arrays make the whole search space into grids having ϵ_j size in the j-th objective [20]. The ϵ-domination is used first when the archive is updated with the offspring solutions. More details about the ϵ-MOEA can be found in [20].

3 Function Optimization by Conventional DE

A general MOP includes a set of n parameters (decision variables), a set of k objective functions, and a set of m constraints. Objective functions and constraints are functions of the decision variables. The optimization goal is to

$$\begin{aligned} minimize: \boldsymbol{y} &= f(\boldsymbol{x}) = (f_1(\boldsymbol{x}), f_2(\boldsymbol{x}), \cdots, f_k(\boldsymbol{x})) \\ subject\ to: \boldsymbol{e}(\boldsymbol{x}) &= (e_1(\boldsymbol{x}), e_2(\boldsymbol{x}), \cdots, e_m(\boldsymbol{x})) \geq 0 \\ where: \boldsymbol{x} &= (x_1, x_2, \cdots, x_n) \in \boldsymbol{X} \\ \boldsymbol{y} &= (y_1, y_2, \cdots, y_k) \in \boldsymbol{Y} \end{aligned} \tag{2}$$

and $\boldsymbol{x}$ is the decision vector, $\boldsymbol{y}$ is the objective vector, $\boldsymbol{X}$ denotes as the decision space, and $\boldsymbol{Y}$ represents the objective space. Generally, for each variable x_i it satisfies a constrained boundary

$$l_i \leq x_i \leq u_i, i = 1, 2, \cdots, n \tag{3}$$

The constraints $\boldsymbol{e}(\boldsymbol{x}) \geq 0$ determine the set of feasible solutions.

The algorithm of DE with the DE/rand/1/exp strategy [8] is as follows:

1. Generate the initial population with *NP* individuals, and set current iteration $k = 1$. Each individual is taken as a real valued vector $X_i, \forall i \in \{1, 2, \cdots, NP\}$, where X_is are objective variables.
2. Evaluate the fitness score for each individual $X_i, \forall i \in \{1, 2, \cdots, NP\}$, of the population based on the objective function, $f(X_i)$.
3. Stop if the halting criterion such as $k = MAX\_GEN$ is satisfied; otherwise, continue.
4. For each individual i, $\forall i \in \{1, 2, \cdots, NP\}$, select r_1, r_2, r_3 uniform randomly from $O \in \{1, 2, \cdots, NP\}$ with $r_1 \neq r_2 \neq r_3 \neq i$.
5. Generate the offspring using DE crossover-mutation operator as following:
   Mutation:
   $$X_i' = X_{r_1} + F \times (X_{r_2} - X_{r_3}) \tag{4}$$
   where $F > 0$ is a scaling factor, and x_{r_1} is known as the base vector. The trial point Y_i is found from its parents X_i and X_i' using the following crossover rule:
   Crossover:
   $$Y_i^j = \begin{cases} X_i'^j & \text{if } R^j \leq CR \text{ or } j = t \\ X_i^j & \text{if } R^j > CR \text{ and } j \neq t \end{cases} \tag{5}$$
   where t is a randomly chosen integer in the set $Q \in \{1, 2, \cdots, n\}$; the superscript j represents the j-th component of corresponding vectors; $R^j \in (0, 1)$, drawn uniformly for each j. And $CR > 0$ is the user defined probability of the crossover operator.
6. Select each trial vector Y_i for the $k+1$ iteration using the acceptance criterion:replace $X_i \in S$ with Y_i if $f(Y_i) \prec f(X_i)$, otherwise retain X_i. Set $k = k + 1$ and go to Step 3.

4 Our Approach: ϵ-ODEMO

Here, we propose an extension DE algorithm called ϵ-ODEMO for MOPs.

4.1 Orthogonal Initial Population

Before solving an optimization problem, we usually have no information about the location of the global minimum. It is desirable that an algorithm starts to explore those points that are scattered evenly in the feasible solution space. In our presented manner, the algorithm can evenly scan the feasible solution space once to locate good points for further exploration in subsequent iterations. As the algorithm iterates and improves the population of points, some points may move closer to the global minimum.We apply the quantization technique and the orthogonal design to generate this initial population.

4.1.1 Design of the Orthogonal Array

Although the proposed algorithm may require different orthogonal arrays (OAs) for different optimization problems. We will only need a special class of OAs. To design an OA, in this research, we use $L_R(Q^C)$ to denote the OA with different level Q, where Q is odd and use $R = Q^J$ to indicate the number of the rows of OA, where J is a positive integer fulfilling

$$C = \frac{Q^J - 1}{Q - 1} \tag{6}$$

C denotes the number of the columns in the above equation. The orthogonal array needs to find a proper J to satisfy

$$\begin{aligned} &minimize : R = Q^J \\ &subject\ to : C \geq n \end{aligned} \tag{7}$$

where n is the number of the variables. In this study, we adopt the algorithm described in [17] to construct an orthogonal array. In particular, we use $L(R, C)$ to indicate the orthogonal array; and $a(i, j)$ to denote the level of the jth factor in the ith combination in $L(R, C)$. If $C > n$, we delete the last $C - n$ columns to get an OA with n factors.

4.1.2 Quantization

For one decision variable with the boundary $[l, u]$, we quantize the domain into Q levels $\alpha_1, \alpha_2, \cdots, \alpha_Q$, where the design parameter Q is odd and α_i is given by

$$\alpha_i = \begin{cases} l & i = 1 \\ l + (i-1)(\frac{u-l}{Q-1}) & 2 \leq i \leq Q-1 \\ u & Q \end{cases} \tag{8}$$

In other words, the domain $[l, u]$ is quantized $Q - 1$ fractions, and any two successive levels are same as each other.

4.1.3 Generation of Initial Population

After constructing a proper OA and quantizing the domain of each decision variable, we can generate the initial population which can scatter uniformly over the feasible solution space. The algorithm for generating the initial population is omitted here, please refer [17] for details. Regularly, the number of the rows of the OA is larger than the population size NP, so we first create the archive with ϵ-nondominated solutions, And then we generate the initial EA population from the archive and the orthogonal solutions. If $ar\_size > NP$, we select NP solutions from the archive randomly; or all of the ar_size solutions in the archive are inserted the EA population, and the remainder $NP - ar\_size$ solutions are selected from the orthogonal solutions randomly.

4.2 Producing New Solutions with DE/rand/1/exp Strategy

In this study, we use DE/rand/1/exp strategy described in Section 3 to produce the offspring solutions. Firstly, when the size of the archive $ar\_size \geq 5$, we select the mating parents from the archive to generate new solutions, which are used to update the archive with ϵ-domination and EA population with usual domination respectively. Secondly, we get the mating parents from the EA population to generate new solutions and update the archive and EA population.

4.3 Procedure of ϵ-ODEMO

The procedure of ϵ-ODEMO is similar to the ϵ-MOEA with the exception that in ϵ-ODEMO we generate the initial population with orthogonal design method and use DE/rand/1/exp to produce new solutions. The algorithm is followed by

Algorithm 2. The procedure of the proposed ϵ-ODEMO

```
Generate a proper OA and generate the orthogonal solutions OS
Create the archive E(0) with the ϵ-nondominated solutions of OS
Create the orthogonal initial population P(0) from E(0) and OS
while The maximum number of the fitness function evaluations (NFE) does not reach do
  if ar_size ≥ 5 then
    for i = 1 to ar_size do
      Produce the new solution with DE/rand/1/exp with archive members
      Update the archive using ϵ-dominance concept
      Update the EA population with usual domination
    end for
  end if
  for i = 1 to NP do
    Produce the new solution with DE/rand/1/exp with EA population members
    Update the archive using ϵ-dominance concept
    Update the EA population with usual domination
  end for
end while
```

5 Simulation Results

In order to test the performance of ϵ-ODEMO a number of two- and three-objective problems were used, where two-objective test problems (ZDT1, ZDT2, ZDT3, ZDT4 and ZDT6) are introduced in [22], and also have been used in [1,13,14,15,18]. And three-objective test problems (DTLZ1 and DTLZ6) are introduced in [23]. The brief information of the test problems is described in Table 1, where k is the number of the objective functions and n is the dimension of the decision vector. We also test these problems with three other approaches: (i) ϵ-DEMO, which is similar to ϵ-ODEMO except

Table 1. Brief information of the test problems in this study

Problem	k	n	Property
ZDT1	2	30	high dimensionality, convex Pareto front
ZDT2	2	30	high dimensionality, non-convex Pareto front
ZDT3	2	30	high dimensionality, discontinuous Pareto front
ZDT4	2	10	many (21^9) local Pareto fronts
ZDT6	2	10	low density of solutions near Pareto front
DTLZ1	3	7	many ($11^5 - 1$) local Pareto fronts, linear hyper-plane Pareto front
DTLZ6	3	22	high dimensionality, 2^{19} disconnected Pareto-optimal regions

using random initial population, (ii) ϵ-MOEA proposed in [20], and (iii) ϵ-OMOEA, which uses orthogonal initial population mentioned above instead of the random initial population in ϵ-MOEA.

5.1 Performance Measures

There are three metrics used in this study. The smaller the value of these metrics, the better the performance of the algorithm.

- **Convergence metric** γ [1]: It measures the distance between the obtained nondominated front Q and the set P^*. of Pareto-optimal solutions.
- **Diversity metric** Δ [1]: It measures the extent of spread achieved among the non-dominated solutions.
- **Generational distance** GD [15]: It is similar to the convergence metric. It measures the distance between the obtained nondominated front Q and the set P^*. of Pareto-optimal solutions.

For all the three metrics, we need to know the true Pareto front for a problem. Since we are dealing with artificial test problems, the true Pareto front is not difficult to be obtained. In our experiments we use uniformly spaced Pareto-optimal solutions as the approximation of the true Pareto front (For ZDT test problems, they were made available online at http://www.scis.ecu.edu.au/research/wfg/datafiles.html. And for DTLZ test problems, they were made available online at http://www.lania.mx/~ccoello/EMOO/testfuncs/.).

5.2 Experimental Setup

For all experiments, we used the following parameters:

- Population size: *NP* = 100;
- Number of fitness function evaluations: *NFE* = 20,000, which is less than the compared approaches (NSGA-II, SPEA, PAES, PDEA, MODE, and DEMO/parent), where the *NFE* of them is 25,000;
- Probability of crossover: *CR* = 0.5;
- Scaling factor: *F* = 0.5;
- Positive integer in orthogonal design: *J* = 2;
- Number of the quantization levels: if $n > 29$, $Q = 29$; else $Q = 21$;
- The ϵ values for different problems are described in Table 2 in order to get roughly 100 solutions in the archive after 20,000 fitness function evaluations.

Table 2. The ϵ values for different test problems

	ZDT1	ZDT2	ZDT3	ZDT4	ZDT6	DTLZ1	DTLZ6
ϵ	[0.0075,0.0075]	[0.0075,0.0075]	[0.0025,0.0035]	[0.0075,0.0075]	[0.0075,0.0075]	[0.02,0.02,0.05]	[0.05,0.05,0.05]

5.3 Experimental Results

The performance of the four methods are compared on all the test problems over ten independent trials each respectively. The average execution time (in seconds) in each run using a PC with an Intel Celeron IV 2.53 GHz processor and 512MB memory running Microsoft Windows XP operating system is shown in the column labeled Time(s).

5.3.1 Two-Objective Test Problems

Tables 3 and Table 4 present the mean and variance of the values of the convergence and diversity metric, averaged on ten runs. Results of other algorithms are taken from the literature (see [1] for the results and parameter settings of both versions of NSGA-II, SPEA and PAES, [13] for PDEA, [14] for MODE, and [15] for DEMO/parent). Although in [15] they proposed three variants of DE algorithm for MOPs, DEMO/parent obtained almost the same performance compared with the other two variants. So we only select the results of DEMO/parent to compare with our approaches.

We also present the additional comparison of generational distance results in Tables 5 for PDEA [14] and DEMO/parent [15]. Once more, we present the mean and variance of the values of generational distance, averaged over 30 runs.

Fig. 1 shows the nondominated fronts obtained by a single run of ϵ-ODEMO. Table 6 summarizes the values of the convergence and diversity metrics for the nondominated fronts from Fig. 1.

5.3.2 Three-Objective Test Problems

With respect to three-objective problems DTLZ1 and DTLZ2, we only make comparisons on ϵ-ODEMO, ϵ-DEMO, ϵ-OMOEA, ϵ-MOEA. Table 7 shows the convergence metric values of the four approaches, averaged on 10 runs.

Fig. 2 and Fig. 3 shows the nondominated fronts obtained by a single run of ϵ-ODEMO and ϵ-OMOEA. Table 8 summarizes the values of the convergence metric for the nondominated fronts from Fig. 2 and Fig. 3.

6 Results Analysis

6.1 Two-Objective Test Problems

For the five two-objective problems, [15] obtained superior results on these problems and gave a good discussion of the comparison between DEMO with NSGA-II, SPEA, PAES, PDEA and MODE. Hence, here we only discuss the comparison between our approaches and DEMO. From Table 3 - 6 and Fig. 1, we can see that

- ϵ-ODEMO and ϵ-OMOEA can get very good results on all of the five test problems in both goals of multiobjective optimization (convergence to the true Pareto

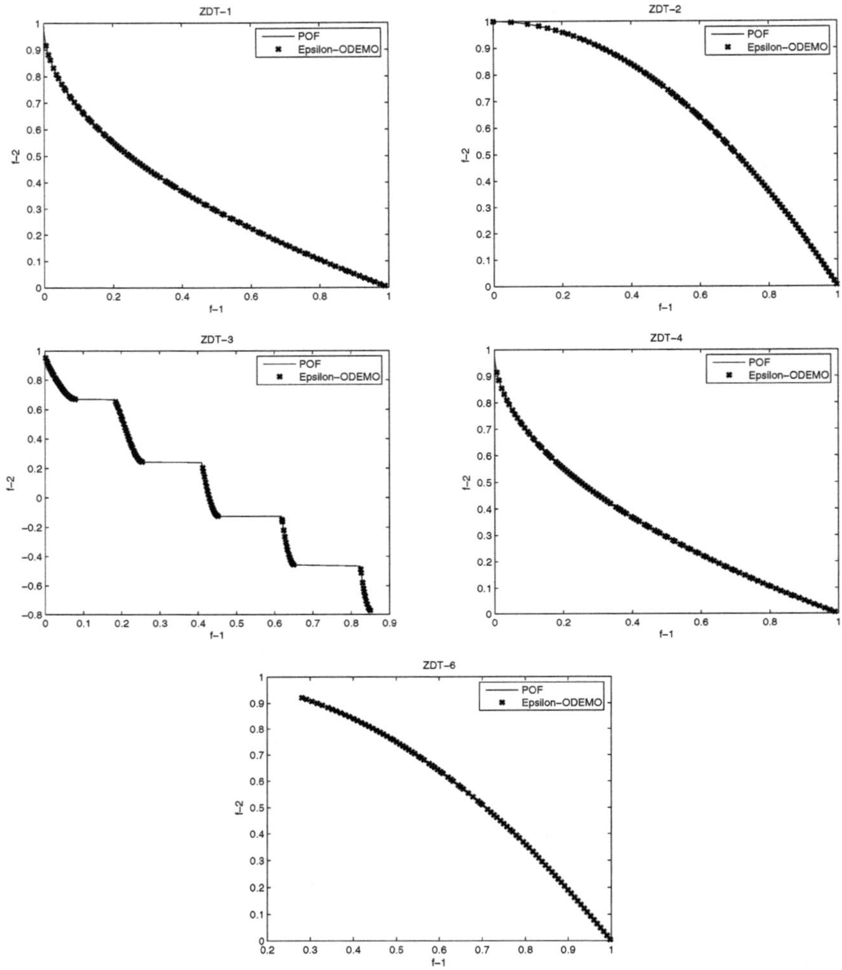

Fig. 1. Non-dominated solutions of the final archive obtained by ϵ-ODEMO on five ZDT test problems (see Table 6 for more details on these fronts). Where POF means Pareto-optimal front. The presented fronts are the outcome of a single run of ϵ-ODEMO.

front and uniform spread of solutions along the front). However, the performance of ϵ-DEMO and ϵ-MOEA are not good. This indicates that the orthogonal initial population can evenly scan the feasible solution space once to locate good points for further exploration in subsequent iterations.

- For ZDT1, ZDT2 and ZDT3, they have high-dimensionality, but many MOEAs have achieved very good results on these problems in both goals of multiobjective optimization. The results for ZDT1, ZDT2 and ZDT3 shown in Table 3 demonstrate that ϵ-ODEMO and ϵ-OMOEA can obtain better convergence metrics than DEMO/parent. But the diversity metrics for ZDT1 and ZDT3 are slightly worse than DEMO due to the absence of extreme solutions in the Pareto-optimal front (POF). Since the ϵ-dominance concept is used in ϵ-ODEMO and ϵ-OMOEA, the extreme solutions usually get dominated by solutions within ϵ and which are better in other objectives [20]. ϵ-OMOEA outperforms ϵ-ODEMO on ZDT1 and ZDT2. But ϵ-ODEMO gets better results on ZDT3.

- ZDT4 has 21^9 local Pareto fronts, which is difficult for many optimizers. Our proposed ϵ-ODEMO can find the true POF in all of the ten runs and get better results than DEMO/parent. Another three approaches (ϵ-OMOEA, ϵ-MOEA and ϵ-DEMO) are trapped into a local Pareto front sometimes.
- For the results from ZDT6 in Table 4, ϵ-ODEMO, ϵ-OMOEA and DEMO/parent obtain similar results on convergence metric. However, ϵ-ODEMO and ϵ-OMOEA get better results of diversity metric than DEMO/parent. But ϵ-DEMO and ϵ-MOEA get worse results in both goals of multiobjective optimization.
- With respect to the generational distance metric, Table 5 shows that among PDEA, DEMO/parent and ϵ-ODEMO, ϵ-ODEMO can find the best results on all of the test problems.
- From Fig. 1, we can see that all of the solutions obtained by ϵ-ODEMO are scattered on the true POF on all of five test problems.
- Regarding computational cost, we don't run NSGA-II, SPEA, PAES, PDEA, MODE and DEMO in my PC, but from the analysis in [20], all of them are more time consuming than the approaches only used the ϵ-dominance concept. Although ϵ-DEMO requires the least computational time on all of five test problems, it obtains worse results on the convergence and diversity results. However, ϵ-ODEMO can get very good results on all of five test problems in both goals of multiobjective optimization with less time.

6.2 Three-Objective Test Problems

With respect to three-objective problems DTLZ1 and DTLZ2, we summarize the results in Table 7. And the nondominated fronts obtained by a single run of ϵ-ODEMO and ϵ-OMOEA are illustrated in Fig. 2 and Fig. 3. The results show that

- ϵ-OMOEA obtains the best results of convergence metric on two test problems, followed by ϵ-ODEMO. ϵ-DEMO and ϵ-MOEA find worse results, especially for DLTZ1.
- On DLTZ1, ϵ-OMOEA can get a better result than ϵ-ODEMO with more computational time. They can all find the true POF. But ϵ-DEMO and ϵ-MOEA find many local Pareto solutions in the final archive ($ar\_size \gg 100$), hence, they get the worst results of convergence.
- With respect to DTLZ6, there is an interesting result from Table 8 and Fig. 3 that although ϵ-OMOEA finds a better result of convergence than ϵ-ODEMO, ϵ-OMOEA can obtain only one Pareto-optimal region, however, ϵ-ODEMO can find the true POF in all disconnected Pareto-optimal front regions. Therefore, ϵ-ODEMO was able to get a better distribution of the Pareto solutions than ϵ-OMOEA.

Table 3. Statistics of the average results on test problems ZDT1, ZDT2, ZDT3 and ZDT4 over 30 independent runs. The results obtained by the proposed ϵ-ODEMO are shown in **boldface**. NA = Not Available.

Algorithm	ZDT1		
	Convergence γ	Diversity Δ	Time (s)
NSGA-II (real-code) [1]	0.033482±0.004750	0.390307±0.001876	NA
NSGA-II (binary-code) [1]	0.000894±0.000000	0.463292±0.041622	NA
SPEA [1]	0.001799±0.000001	0.784525±0.004440	NA
PAES [1]	0.082085±0.008679	1.229794±0.004839	NA
PDEA [13]	NA	0.298567±0.000742	NA
MODE [14]	0.005800±0.000000	NA	NA
DEMO/parent [15]	0.001083±0.000113	0.325237±0.030249	NA
ϵ-ODEMO	**0.000761±0.000058**	**0.360154±0.011059**	**2.553**
ϵ-DEMO	0.040202±0.018254	0.387340±0.040272	0.925
ϵ-OMOEA	0.000721±0.000023	0.358369±0.012453	6.201
ϵ-MOEA	0.020125±0.012800	0.364210±0.020230	6.188

Algorithm	ZDT2		
	Convergence γ	Diversity Δ	Time (s)
NSGA-II (real-code) [1]	0.072391±0.031689	0.430776±0.004721	NA
NSGA-II (binary-code) [1]	0.000824±0.000000	0.435112±0.024607	NA
SPEA [1]	0.001339±0.000000	0.755148±0.004521	NA
PAES [1]	0.126276±0.036877	1.165942±0.007682	NA
PDEA [13]	NA	0.317958±0.001389	NA
MODE [14]	0.005500±0.000000	NA	NA
DEMO/parent [15]	0.000755±0.000045	0.329151±0.032408	NA
ϵ-ODEMO	**0.000764±0.000035**	**0.276872±0.007013**	**2.502**
ϵ-DEMO	0.190147±0.081243	0.615820±0.051986	0.621
ϵ-OMOEA	0.000760±0.000015	0.283013±0.045573	6.567
ϵ-MOEA	0.034273±0.020354	0.402345±0.077234	6.342

Algorithm	ZDT3		
	Convergence γ	Diversity Δ	Time (s)
NSGA-II (real-code) [1]	0.114500±0.007940	0.738540±0.019706	NA
NSGA-II (binary-code) [1]	0.043411±0.000042	0.575606±0.005078	NA
SPEA [1]	0.047517±0.000047	0.672938±0.003587	NA
PAES [1]	0.023872±0.000010	0.789920±0.001653	NA
PDEA [13]	NA	0.623812±0.000225	NA
MODE [14]	0.021560±0.000000	NA	NA
DEMO/parent [15]	0.001178±0.000059	0.309436±0.018603	NA
ϵ-ODEMO	**0.000915±0.000050**	**0.534329±0.018301**	**2.453**
ϵ-DEMO	0.008754±0.003127	0.632701±0.025327	0.924
ϵ-OMOEA	0.006453±0.007956	0.687538±0.032879	6.583
ϵ-MOEA	0.005689±0.003357	0.673541±0.012586	6.037

Table 3. *(continued)*

Algorithm	ZDT4		
	Convergence γ	Diversity Δ	Time (s)
NSGA-II (real-code) [1]	0.513053±0.118460	0.702612±0.064648	NA
NSGA-II (binary-code) [1]	3.227636±7.307630	0.479475±0.009841	NA
SPEA [1]	7.340299±6.572516	0.798463±0.014616	NA
PAES [1]	0.854816±0.527238	0.870458±0.101399	NA
PDEA [13]	NA	0.840852±0.035741	NA
MODE [14]	0.638950±0.500200	NA	NA
DEMO/parent [15]	0.001037±0.000134	0.359905±0.037672	NA
ϵ-ODEMO	**0.000712±0.000056**	**0.354847±0.003956**	**2.325**
ϵ-DEMO	0.856829±0.702439	0.679368±0.120357	0.903
ϵ-OMOEA	0.010389±0.009354	0.180321±0.531570	6.237
ϵ-MOEA	8.137894±5.27689	0.927910±0.025221	5.832

Table 4. Statistics of the results on test problem ZDT6 over 30 independent runs. The results obtained by the proposed ϵ-ODEMO are shown in **boldface**. NA = Not Available.

Algorithm	ZDT6		
	Convergence γ	Diversity Δ	Time (s)
NSGA-II (real-code) [1]	0.296564±0.013135	0.668025±0.009923	NA
NSGA-II (binary-code) [1]	7.806798±0.001667	0.644477±0.035042	NA
SPEA [1]	0.221138±0.000449	0.849389±0.002713	NA
PAES [1]	0.085469±0.006664	1.153052±0.003916	NA
PDEA [13]	NA	0.473074±0.021721	NA
MODE [14]	0.026230±0.000861	NA	NA
DEMO/parent [15]	0.000629±0.000044	0.442308±0.039255	NA
ϵ-ODEMO	**0.000581±0.000030**	**0.204142±0.005012**	**1.559**
ϵ-DEMO	0.875253±0.0573254	1.530470±0.046340	0.387
ϵ-OMOEA	0.000618±0.000102	0.180214±0.000864	6.189
ϵ-MOEA	0.675321±0.537001	0.795721±0.068538	6.057

Table 5. Generational distance achieved by PDEA, DEMO and ϵ-ODEMO on the test problems ZDT1, ZDT2, ZDT3, ZDT4 and ZDT6. The results obtained by the proposed ϵ-ODEMO are shown in **boldface**.

Algorithm	Generational distance				
	ZDT1	ZDT2	ZDT3	ZDT4	ZDT6
PDEA [13]	0.00062±0.00000	0.00065±0.00000	0.00056±0.00000	0.61826±0.82688	0.02389±0.00329
DEMO/parent [15]	0.00023±0.00005	0.00009±0.00001	0.00016±0.00001	0.00020±0.00005	0.00007±0.00001
ϵ-ODEMO	**0.00010±0.00001**	**0.00009±0.00000**	**0.00013±0.00001**	**0.00009±0.00001**	**0.00007±0.00000**

Table 6. Metric values for the nondominated fronts shown in Fig. 1 by ϵ-ODEMO

Problem	Convergence γ	Diversity Δ
ZDT1	0.000799	0.372004
ZDT2	0.000753	0.268825
ZDT3	0.000974	0.383188
ZDT4	0.000750	0.353701
ZDT6	0.000613	0.193624

Table 7. Statistics of the results on test problems DTLZ1 and DTLZ6 over 30 independent runs. The results obtained by the proposed ϵ-ODEMO are shown in **boldface**.

Algorithm	DTLZ1		DTLZ6	
	Convergence γ	Time(s)	Convergence γ	Time(s)
ϵ-ODEMO	**0.004389±0.000204**	**1.872**	**0.020387±0.000789**	**3.427**
ϵ-DEMO	1.732468±0.182287	17.320	0.039879±0.025680	1.199
ϵ-OMOEA	0.003790±0.000158	6.251	0.013768±0.002786	7.350
ϵ-MOEA	7.789134±1.867357	67.035	0.063201±0.031879	6.044

Table 8. Convergence metric values for the nondominated fronts shown in Fig. 2 and Fig. 3

Algorithm	DTLZ1	DTLZ6
ϵ-ODEMO	0.004370	0.020155
ϵ-OMOEA	0.003997	0.013433

From the comparison above, we can conclude that our proposed approach, ϵ-ODEMO, produced competitive results based on quality with respect to many other techniques representative of the state-of-the-art in multiobjective optimization. ϵ-ODEMO can deal with two- and three-objective problems of diverse complexities; problems with low (ZDT4, ZDT6 and DTLZ1) and high (ZDT1, ZDT2, ZDT3 and DTLZ6) dimensionality, with different types of Pareto fronts (convex, non-convex, discontinuous, thin density and non-uniform spread) and with many local Pareto fronts (ZDT4 and DTLZ1). Furthermore, the approach is very fast in terms of the computational time on each of the test problems.

7 Conclusion

In this paper, we proposed a novel DE algorithm based on ϵ-dominance concept and orthogonal design method for MOPs. ϵ-ODEMO implies the orthogonal design method with quantization technique to generate the initial population of points that are scattered uniformly over the feasible solution space, so that the algorithm can evenly scan the feasible solution space once to locate good points for further exploration in subsequent iterations. And it uses DE/rand/1/exp strategy to produce offspring solutions. Meanwhile, in order to find good distribution Pareto solutions with less computational time, ϵ-dominance concept and efficient parent and archive update strategies are adopted to

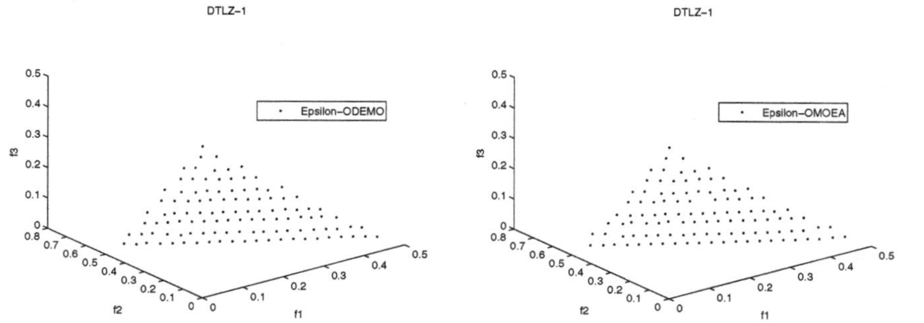

Fig. 2. Non-dominated solutions of the final archive obtained by ϵ-ODEMO and ϵ-OMOEA on DTLZ1 (see Table 8 for more details on these fronts). The presented fronts are the outcome of a single run.

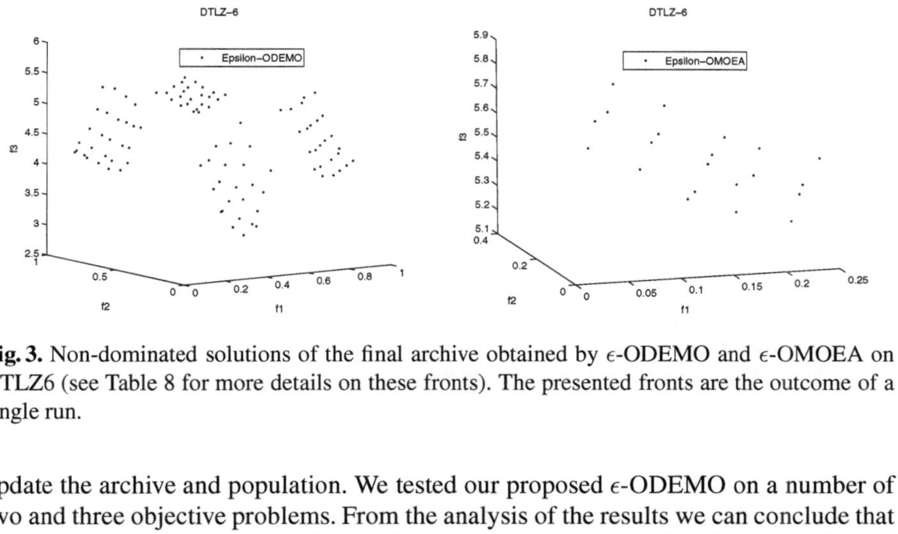

Fig. 3. Non-dominated solutions of the final archive obtained by ϵ-ODEMO and ϵ-OMOEA on DTLZ6 (see Table 8 for more details on these fronts). The presented fronts are the outcome of a single run.

update the archive and population. We tested our proposed ϵ-ODEMO on a number of two and three objective problems. From the analysis of the results we can conclude that ϵ-ODEMO can obtain a good distribution Pareto solutions on all of the test problems. Moreover, it requires small computational time. Although ϵ-OMOEA got slightly better results on some test problems (ZDT1, ZDT2, ZDT6 and DTLZ1) than ϵ-ODEMO, it needed more computational time and it obtained worse results on ZDT3, ZDT4 and DTLZ6. Hence, we recommend our proposed ϵ-ODEMO be used in future experimentation. Our future work consists on using the proposed ϵ-ODEMO to solve the constrained MOPs and dynamic MOPs.

Acknowledgments

The authors would like to thank the anonymous reviewers for their constructive advice and Kanpur Genetic Algorithms Laboratory (KanGAL) for making the source code of ϵ-MOEA available online at http://www.iitk.ac.in/kangal/codes.shtml. This work is supported by the Humanities Base Project of Hubei Province of China(Grant No. 2004 B0011) and the Natural Science Foundation of Hubei Province of China(Grant No. 2003ABA043).

References

1. Deb, K., Pratap, A., Agarwal, S., Meyarivan, T.: A fast and elitist multiobjective genetic algorithm: NSGACII. IEEE Transactions on Evolutionary Computation 6 (2002) 182 - 197
2. Zitzler, E., Laumanns, M., Thiele, L.: SPEA2: Improving the strength pareto evolutionary algorithm. Technical Report 103, Computer Engineering and Networks Laboratory, (2001)
3. Schaffer, J. D.: Multiple Objective Optimization with Vector Evaluated Genetic Algorithms. Ph. D. thesis, Vanderbilt University. Unpublished. (1984)
4. Hajela, P. and Lin, C. Y.: Genetic search strategies in multicriterion optimal design. Structural Optimization 4 (1992) 99 - 107
5. Fonseca, C. M. and Fleming, P. J.: Genetic algorithms for multiobjective optimization: Formulation, discussion and generalization. In S. Forrest (Ed.), Proceedings of the Fifth International Conference on Genetic Algorithms, (1993) 416 - 423
6. Horn, J. and Nafpliotis,N.: Multiobjective optimization using the niched pareto genetic algorithm. IlliGAL Report 93005, Illinois Genetic Algorithms Laboratory, University of Illinois, Urbana, Champaign. (1993)
7. Knowles,J.D., Corne, D.W.: Approximating the Nondominated Front Using the Pareto Archived Evolution Strategy. Evolutionary Computation 8(2) (2000) 149 - 172
8. Storn, R. and Price, K.: Differential evolution–A simple and efficient heuristic for global optimization over continuous spaces. Journal of Global Optimization. 11 (1997) 341 - 359
9. Storn, R. and Price, K.: Home Page of Differential Evolution. Available online at http://www.ICSI.Berkeley.edu/~storn/code.html. (2003)
10. Lampinen, J.: A bibliography of differential evolution algorithm. Available online at http://www2.lut.fi/~jlampine/debiblio.htm
11. Abbass, H.A., Sarker, R., Newton, C.: PDE: A pareto-frontier differential evolution approach for multi-objective optimization problems. In: Proceedings of the Congress on Evolutionary Computation 2001 (CEC2001). Volume 2, Piscataway, New Jersey, IEEE Service Center (2001) 971 - 978
12. Abbass, H.A.: The self-adaptive pareto differential evolution algorithm. In: Congress on Evolutionary Computation (CEC2002). Volume 1, Piscataway, New Jersey, IEEE Service Center (2002) 831 - 836
13. Madavan, N.K.: Multiobjective optimization using a pareto differential evolution approach. In: Congress on Evolutionary Computation (CEC2002). Volume 2, Piscataway, New Jersey, IEEE Service Center (2002) 1145 - 1150
14. Xue, F., Sanderson, A.C., Graves, R.J.: Pareto-based multi-objective differential evolution. Proceedings of the 2003 Congress on Evolutionary Computation (CEC2003). Volume 2, Canberra, Australia, IEEE Press (2003) 862 - 869
15. Robič, T. and Filipič, B.: DEMO: Differential Evolution for Multiobjective Optimization. proceedings of the Third International Conference on Evolutionary Multi-Criterion Optimization (EMO-05), (2005) 520 - 533
16. Fang, K. T. and Ma, C. X.: Orthogonal and Uniform Design (in Chinese). Science Press, (2001)
17. Leung, Y. W. and Wang, Y.: An Orthogonal Genetic Algorithm with Quantization for Global Numerical Optimization. IEEE Transactions on Evolutionary Computation, 5(1) (2001) 41 - 53

18. S. Zeng, L. Kang, L. Ding: An Orthogonal Multiobjective Evolutionary Algorithm for Multi-objective Optimization Problems with Constraints. Evolutionary Computation, 12(1) (2004) 77 - 98
19. S. Zeng, S. Yao, L., and Y. Liu: An Efficient Multi-objective Evolutionary Algorithm: OMOEA-II. In C. A. Coello Coello *et al.* (Eds.), Third International Conference on Evolutionary Multi-Criterion Optimization (EMO-05), (2005) 108 - 119
20. Deb, K., Mohan, M. and Mishra, S.: Towards a quick computation of well-spread Pareto-optimal solutions. Second Evolutionary Multi-Criterion Optimization (EMO-03),Springr LNCS 2632(2003) 222 - 236
21. Laumanns, M., Thiele, L., Deb, K., and Zitzler, E.: Combining convergence and diversity in evolutionary multi-objective optimization. Evolutionary Computation, 10(3) (2002) 263 - 282
22. Zitzler, E., Deb, K., Thiele, L.: Comparison of multiobjective evolutionary algorithms: Empirical results. Evolutionary Computation 8 (2000) 173 - 195
23. Deb, K., Thiele, L., Laumanns, M., Zitzler, E.: Scalable Multi-Objective Optimization Test Problems. In: Congress on Evolutionary Computation (CEC2002), (2002) 825 - 830

Molecular Dynamics Optimizer

Swee Chiang Chiam[1], Kay Chen Tan, and Abdullah Al Mamun

[1] Department of Electrical and Computer Engineering
National University of Singapore
4 Engineering Drive 3
Singapore 117576
g0500055@nus.edu.sg

Abstract. Molecular system possesses two main characteristics that seem to be applicable for the contrary goals of proximity and diversity in multiobjective optimization, namely the converging pressure in potential fields as dictated by the Maxwell-Boltzmann distribution and the inherent drift to a homogenous and uniform equilibrium with maximum entropy, even without any prior knowledge on the geometry and state of the enclosure. Inspired by this association, this paper explores the notion of exploiting molecular motion to solve multiobjective problems. By adapting the algorithmic structure of molecular dynamics, which essentially represents a technique for the computer simulation of molecular motion, a molecular system that is relevant for multiobjective optimization is proposed, known as molecular dynamics optimizer (MDO). The performance of MDO was subsequently compared with other conventional multiobjective optimizers, specifically EA and PSO, and the experimental results demonstrated that MDO is indeed a viable and practical approach for multiobjective optimization.

Keywords: Multiobjective optimization, molecular dynamics.

1 Introduction

The development of computational techniques for multiobjective optimization (MOO) has significantly grown in the last few years due to their success in satisfying the optimization goals of attaining near-optimal, diverse and uniformly distributed solution sets for various multiobjective problems (MOP). The basic requirement for multiobjective optimizers is to balance between the proximity and diversity goals of MOO, providing sufficient convergence pressure without compromising diversity. Hence, regardless of the configurations of the initial solution, the ultimate solutions generated by the multiobjective optimizer should ideally be well distributed within the optimal region for the MOP.

Many of these optimizers were actually inspired from nature, for instance evolutionary algorithm (EA) from evolutionary biology, ant colony optimization (ACO) from the behavior of real ant colony and particle swarm optimization (PSO) from the swarm behavior of birds. Actually, one could find a resemblance between the evolving solutions generated by evolutionary optimizers and the dynamics of ideal gas molecules in an enclosure. In the presence of unequal potential fields, there will

S. Obayashi et al. (Eds.): EMO 2007, LNCS 4403, pp. 302–316, 2007.

be a higher probability for molecules to reside in the lower potential regions. Furthermore, these molecules will always tend to a homogenous and uniform equilibrium with maximum entropy [1], even without any prior knowledge on the geometry and state of the enclosure. The convergence and diversity nature of molecular motions appears to be inherently well suited for MOO. Thus, it might be possible to develop a multiobjective optimizer based on this natural phenomenon.

Actually, the notion of exploiting the dynamics of molecules for MOO is not new. Kita *et al* [2] and Cui *et al* [3] formulated strategies based on the thermodynamic notion of energy and entropy that will prevent premature convergence by maintaining an appropriate level of diversity. Similarly, Sobieski *et al.* [4] designed variation operators that project solutions based on a bell-curve distribution. This idea was realized in [5], where an expansion operator designed based on the thermodynamic behavior of ideal gas expanding in an enclosure was used to complement the crossover and mutation operation. Of more recent issue, a selection strategy and stopping conditions were designed by using the natural principles of a dynamical system approaching its equilibrium state [6]. Closely related to molecular motion is a well established theoretical framework, statistical mechanics, which studies the overall behavior of many molecules in terms of their actual motion and interactions. While statistical mechanics cannot chart out the life history of one molecule in a system, it is able to describe the macro behavior of the molecular system as a whole. Because MOO usually involve manipulating populations of solutions, the value of this framework is obvious. In fact, statistical mechanics had been used on several occasions to analyze the dynamics of EA [7].

Nevertheless, analytical solutions in statistical mechanics are restricted only for simple and ideal cases. For intractable systems, molecular simulation is used instead for more accurate results and has played an important role as a bridge connecting models to theory. Molecular simulation is a general term for the use of computer models to describe physical systems at a molecular level of detail, particularly the individual position and orientation of every molecule and from that, both thermodynamic and kinetic properties of the system could be derived. Molecular simulation is commonly done by employing either Monte Carlo or molecular dynamics [8], with the latter being the research subject in this paper. Molecular dynamics numerically solves Newton's equations of motion on molecular system and update their state properties correspondingly.

Since molecular dynamics is essentially a computer simulation of molecular motion and considering the inherent diverse behavior of molecular system and their converging drift pressure in the presence of unequal potential fields, molecular dynamics appears to be a possible computational platform to apply molecular motion for MOO purposes. As such, this idea will be explored in this paper. By adapting the algorithmic structure of molecular dynamics, a molecular system that is relevant for multiobjective optimization is proposed, known as molecular dynamics optimizer (MDO). It should be highlighted that this paper aims to show that the basic principles derived from the dynamics of molecules can be applied for MOO purposes, instead of modeling exactly any real-life molecular phenomenon. Thus, the model proposed will be as generic as possible and the proof of principle study will focus mainly on the investigation of its feasibility and characteristics.

The remainder of this paper is organized as such. First, a formal introduction of molecular dynamics will be given, which will be followed by a discussion on how it can be relevant to MOO. The algorithmic framework of MDO and its implementation issues will then be highlighted in the next section. Subsequently, the results of the experimental study will be presented to show the viability of applying molecular dynamics for MOO purposes before the conclusions are drawn in section 4.

2 The Etiology of Molecular Dynamics Optimizer

This section presents the conceptual development of MDO, which is crucial for a better understanding of its algorithmic structure in the next section. First, a brief introduction to molecular dynamics will be given, followed by a discussion on how it could be relevant to the proximity and diversity goals in MOO.

2.1 Molecular Dynamics

Molecular dynamics refers to a computer simulation technique where the time evolution of a set of interacting molecules is simulated by integrating their equations of motion. For that purpose, it specifies a collection of *N* molecules, $\Phi = (m_1, \ldots, m_N)$ of equal mass, *M* in a simulation cell and their locations are represented by a position vector, $\vec{x}$. The trajectory of the various molecules follows the law of classical mechanics, most notably Newton's second law of motion:

$$\vec{f} = m\ddot{\vec{x}} \tag{1}$$

where $\vec{f}$ denotes the net force acting on each molecule due to the interactions with other molecules and $\ddot{\vec{x}} = \frac{d^2\vec{x}}{dt^2}$ is its corresponding acceleration. The equations of motion are used to calculate step by step the next sequence of coordinates. In more pictorial terms, the various molecules will "fly" around in the confined enclosure, bump into each other and alter their course of motion accordingly, in a way very similar to how molecules in the real world behave.

$$\vec{f} = -\frac{\partial}{\partial \vec{x}} U \tag{2}$$

$$U = \sum_i u_1(m_i) + \sum_i \sum_{j>i} u_2(m_i, m_j) + \sum_i \sum_{j>i} \sum_{k>j>i} u_3(m_i, m_j, m_k) + \ldots \tag{3}$$

The overall moment of the molecules is governed by the potential function, *U* and the formal relationship between $\vec{f}$ and *U* is given in (2). The calculation of $\vec{f}$ and *U* is non-trivial as *U* can be split into 1-body, 2-body, 3-body… terms as spelt out in (3). The first term denotes the effect of externally applied potential fields and the other terms represent the interaction between the molecules. Among them, u_2 represents the potential between pairs of molecules, while u_3, u_4... is the potential between triplets, quadruplets... respectively.

To put everything into context, a general algorithm for molecular dynamics is shown in fig. 1. It begins with the creation of an initial configuration. Subsequently, if the time has not reached its maximum limit, the configuration of the systems will be updated by solving the equation of motions. Of course, a substantial degree of implementation complexity has been concealed for brevity. For example, there are many possible methods for solving the equation of motions, like the Gear predictor-corrector or Verlet algorithms. Interested readers are referred to [8], [9] for more details.

```
Create an Initial Configuration
Loop
  Calculate Potential Forces
  Solve the Equation of Motions
  Update Molecular Trajectory
End loop
```

Fig. 1. General outline of a molecular dynamics algorithm

2.2 Applying Molecular Dynamics for Multi Objective Optimization

The work in this paper is motivated by how the contrary goals of proximity and diversity in multiobjective optimization can be met by the converging drift pressure of molecular systems in potential fields and their inherent drift to a homogenous and uniform equilibrium with maximum entropy. For the former, it is actually governed by the Maxwell-Boltzmann distribution. To illustrate this clearer, let's consider an assembly of *N* molecules in an isolated system of volume *V*. Their energy is limited at certain discrete values and summing their energy will yield the total energy of the system, *E*. The state or condition of each molecule at any time is specified by its position $\vec{x}$ and momentum $\vec{p}$ at that instant. Without any loss in generality, a six-dimensional space denoted as phase-space is defined, and a six variables coordinates system represents the state of one molecule at any given time, where the first three variables represent the spatial coordinates, $\vec{x}$ and the subsequent three being the momentum, $\vec{p}$.

The *N* molecules will be distributed almost continuously throughout the phase space, and at equilibrium, the number of molecule in a volume element $d\Omega$, $d\Omega = dx_1 dx_2 dx_3 dp_1 dp_2 dp_3$ of the phase space, at time *t* is given by (4),

$$n = \int N \rho A e^{-\beta E} d\Omega \tag{4}$$

In fact, *n* depends on the energy of the system. For brevity, the derivation has been left out. Interested readers are referred to standard statistical mechanics textbooks for details.

Considering a more complicated situation where they may be some potential energy present, let's suppose that the phase space consists of two regions of equal volume, R_1 and R_2, where the former has zero potential and the latter has a constant potential energy, $V_o > 0$. The number of molecule in region R_1 and R_2 is given in (5) and (6) respectively, where E_k refers to the kinetic energy of the molecules.

$$n_1 = \int N \rho A e^{-\beta E_k} d\Omega \tag{5}$$

$$n_2 = \int N \rho A e^{-\beta (E_k + V_o)} d\Omega \tag{6}$$

Since V_o is a constant and R_1 and R_2 has the same volume, the ratio of n_2/n_1 can be easily evaluated in (7). This ratio is just the Boltzmann constant with the energy difference V_o appearing in the exponent.

$$\frac{n_2}{n_1} = e^{-\beta V_o} \tag{7}$$

The direct implication is that when the system is in dynamic equilibrium, there will be a higher proportion of molecules residing in the lower potential region of the field with the ratio being related to the difference in energy level. Hence, if the optimal region in the search space of a MOP could be associated with low potential region in the molecular simulation cell, most molecules should converge to the optimum. Fig. 2 suggests how this could be done.

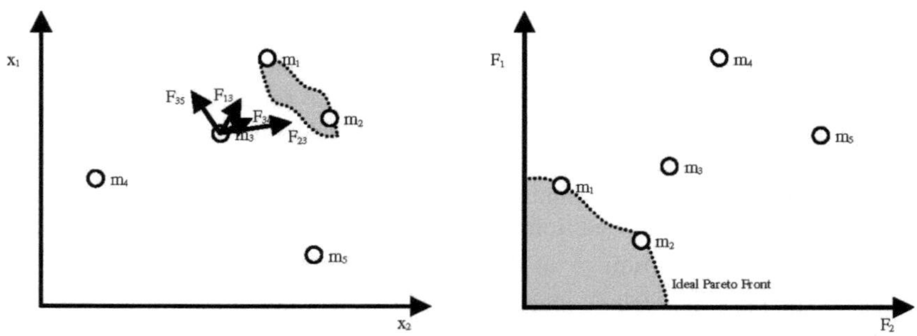

(a) (b)

Fig. 2. (a) Search space and (b) objective space of a 5-molecules system

Fig. 2a shows five solutions situated in the search space of a hypothetical MOP. Fig. 2b shows their positions in the objective space after they have been evaluated. Obviously, they have different degree of optimality, depending on their proximity to the Pareto optimal front. Clearly, molecules m_1 and m_2 are the most optimal solutions while m_4 and m_5 are the least optimal.

Ideally, molecules should be accelerated towards the lower potential region delineated by the gray region. However in practice, the complete potential landscape can never be charted. Thus, what can be done instead is to accelerate each molecule towards those that dominate it and away from those that it dominates. Figure 2a illustrates this for the simple example considered where m_3 is accelerated towards molecules m_1 and m_2 and away from m_4 and m_5. Thus, the resultant force vector for m_3 points towards the optimal region. As governed by (7), this simple system will tend to dynamic equilibrium with a higher proportion of the molecules settling in the region of optimality, hence satisfying the proximity goals of MOO.

As for diversity, it can be achieved by introducing intermolecular potential into the MDO, since they are responsible for the uniform distribution of molecules in actual molecular systems. To illustrate, let's consider the Lennard-Jones potential, which is the most widely used pair-wise potential for molecular dynamics due to its realistic description for the intermolecular interaction and computational simplicity. Its formulation is given in (8), where $\left|\vec{x}_{ij}\right|$ is the inter-molecular separation, ε is the depth of the potential well at the minimum in $u_2(m_i, m_j)$ and σ is the collision diameter, which is the separation of the molecules at $u_2(m_i, m_j) = 0$.

$$u_2(m_i, m_j) = 4\varepsilon\left[\left(\frac{\sigma}{\left|\vec{x}_{ij}\right|}\right)^{12} - \left(\frac{\sigma}{\left|\vec{x}_{ij}\right|}\right)^{6}\right] \tag{8}$$

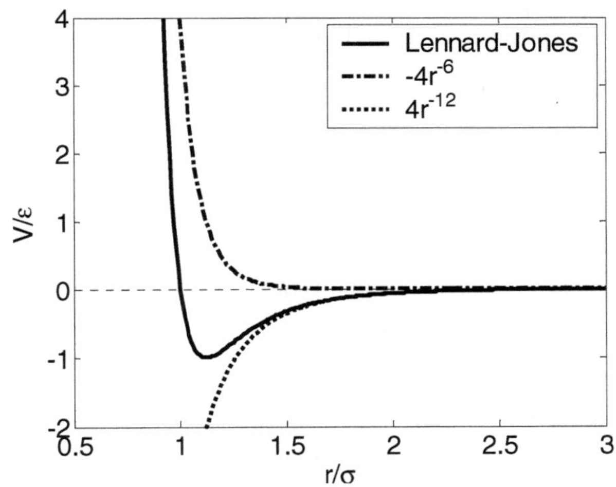

Fig. 3. Lennard-Jones potential

The graphical plot of the Lennard-Jones potential and the two distinct forces that constituted it are illustrated in fig 3. The positive potential is of repulsive nature and the other is of attractive nature. The former prevents the molecules from congregating to a single point, while the latter maintains their cohesion. Both of these forces vary inversely with the distance between the molecules. Also, the repulsive potential has a higher value initially as compared to the attractive potential but it converges to zero faster. As a result, molecular interaction is strongly repulsive in close proximity, becomes mildly attractive at intermediate range and vanishes at long distance. The effect is that molecules in the system will be uniformly spaced from each other at a distance roughly equivalent to the collision diameter, σ. Hence extending this concept to MOO, the molecules in the optimal region will be spaced evenly for a well-defined Pareto front.

3 Molecular Dynamics Optimizer

Having discussed the affinity between molecular dynamics and MOO, the development and implementation of MDO is straightforward. The main aspects of molecular motion taken into account to develop the algorithm are the converging drift pressure in the presence of unequal potential fields and the inherent diverse behavior of molecular system.

```
Generate an Initial Configuration Randomly
Loop t = 1 … T
  Evaluate each Molecule's Optimality Level
  Determine the Potential
   Calculate the Potential Field
   Calculate the Intermolecular Potential
  Solve the Equation of Motions
  Update Molecular Trajectory
   Account for Boundary Constraints
End loop
```

Fig. 4. Pseudo code of MDO

Fig. 4 describes the pseudo code of MDO and the algorithmic details are briefly described as follows:

1. *Generate:* The population initialization process create a molecular system $\Phi = (m_1, \ldots, m_N)$ consisting of N modules at t = 0. A random position vector $\vec{x} = (x_1, \ldots, x_l)$ and velocity vector $\vec{v} = (v_1, \ldots, v_l)$ will be assigned to each of them. The coordinates of each molecule represents a solution of the MOP, $F(\vec{x})$ at hand and its values will be confined within the dimension of the simulation cell, corresponding to the search space of $F(\vec{x})$.
2. *Evaluate:* All the molecules are evaluated based on $F(\vec{x})$. From their objective vectors, their optimality levels are determined, which will affect the type of potential forces acting on it.
3. *Determine:* The potential of each molecules are determined via the potential function outlined in (3). The two main potentials considered will be potential field and the intermolecular potential which respectively relates to the convergence and divergence property of molecular motions that are essential for MOO purposes.

4. *Solve:* With the potential determined, the resultant force acting on each molecules can be computed. Subsequently, by solving the equations of motion, the position and velocity of every molecule at the next time step can be calculated.
5. *Update:* The velocity and position of the molecules are updated based on the values attained earlier. Here, the boundary constraints are checked to ensure that each molecule is still confined within the simulation cell.

After these steps, the termination criterion will be checked and if it is not met, the whole algorithm will repeat again. In this paper, the termination criterion is a predefined maximum number of time steps, T.

3.1 Implementation of MDO

With the generic algorithmic framework of MDO in place and the discussion of the various implementation issues, we are in a better position to evaluate the capability of MDO to satisfy the goals of proximity and diversity in MOO. For this purpose, a simple instance of MDO will be implemented and the algorithm details are described as follows.

Since the equilibrium behavior of the system is independent on the initial conditions, all reasonable initial configurations are, in principle, acceptable. The coordinates for each molecule are assigned values that are drawn from a uniform distribution which ranges within the limits of the search space. Likewise, their velocity is determined in similar fashion.

The concept of optimality level in MDO is analogous to fitness level in the context of evolutionary optimizers and the optimality difference between molecules is used to determine the nature of their intermolecular potential. Since the solutions of MOP are all partially ordered, using the technique of Pareto ranking or aggregation function can generate a total ordering within the solutions, and hence determining their optimality level. However, since we are only concerned with the optimality difference, this implementation will simply use Pareto dominance to compare solutions in pairs and determine the nature of their intermolecular potential.

The potential function outlined in (3) is too complex even for the actual molecular simulation of realistic systems. A simpler alternative will be to ignore higher order interactions and just focus on the two-body interaction or pair-wise potential which contributes bulk of the system energy. Also, there are many types of intermolecular potential function used in molecular simulation. Beside the Lennard Jones potential mentioned earlier, there is also the Weeks-Chandler-Anderson potential [8], [9], a truncated version of the Lennard Jones potential so as to keep computation at a reasonable level, and the Barker-Fisher-Watts potential [8], [9], where the repulsion between molecules have an exponential dependence on distance. While implementing those potential functions might be a more direct approach, this will only introduce additional parameters and complexity to the model, which this implementation tries to avoid.

Hence, linear potential functions will be considered instead and to simplify the model complexity further, the force acting on each molecule due to the intermolecular potential is derived directly. Each molecule will be compared to each and every other molecule. Based on their difference in optimality level, different type of force will be assigned to each molecular pairing. Altogether, there are three cases.

1. If m_i is compared to a superior molecule, m_j, m_i will experience a attractive force given by (10) that will accelerate it towards m_j. This force varies directly with the difference in ranking and inversely with the intermolecular separation. The attractive potential force constant, C_a vary its influence on the dynamics of the molecules. A random number was considered to introduce stochasticity into this model.

$$\vec{f}_{ij} = C_a \cdot rand() \cdot {(r_{\max} - r_j)}/{r_{\max}} \cdot \left(|\vec{x}_{max}| - |\vec{x}_{ij}|\right) \cdot \hat{\vec{x}}_{ij} \tag{9}$$

2. Conversely, if m_i is compared to a inferior molecule, a repulsive force will be experienced by m_i to accelerate itself from m_j, given by (11). This formulation is similar to (10), except for the negative sign which reflects the different nature. The repulsive potential force constant, C_r has similar function as C_a.

$$\vec{f}_{ij} = C_r \cdot -rand() \cdot {(r_{\max} - r_j)}/{r_{\max}} \cdot \left(|\vec{x}_{max}| - |\vec{x}_{ij}|\right) \cdot \hat{\vec{x}}_{ij} \tag{10}$$

3. In the event where both molecules are of the same optimality, which implies that the intermolecular forces should be applied to improve their diversity. To minimize the model complexity, a simplistic linear force that varies inversely with the intermolecular separation will be considered instead of the Lennard-Jones potential. The force experienced by m_i is given in (12) is given as below, which is actually a simplified version of the Lennard-Jones potential. The constant R_0 is analogous to the depth of the potential well in (8). It allows adjustment of the difference in magnitude between this short ranged force and the long-ranged forces in (10)-(11) respectively. Apart from that, its sign will determine the nature of the force i.e. attractive or repulsive. x_{niche} is the radius within which this force will act. It is similar to σ in (8) which define the intermolecular separation where no force is experienced.

$$\vec{f}_{ij} = \begin{cases} 0 & \text{if } |\vec{x}_{ij}| \geq x_{niche} \\ R_0 \cdot rand() \cdot \left(x_{niche} - |\vec{x}_{ij}|\right) \cdot \hat{\vec{x}}_{ij} & \text{if } |\vec{x}_{ij}| < x_{niche} \end{cases} \tag{11}$$

The resultant force acting on m_i will be given by

$$\vec{f}_i = \sum_j^N \vec{f}_{ij} \tag{12}$$

After the force for each molecule has been determined, the velocity and position of the molecules can be updated. They are related to the previously calculated force vector as shown,

$$\vec{v}_{i,t} = I \cdot \vec{v}_{i,t-1} + \vec{f}_{i,t} \tag{13}$$

$$\vec{x}_{i,t} = \vec{x}_{i,t-1} + \vec{v}_{i,t} \tag{14}$$

where I is the inertia constant. I will determine the responsiveness of the system to new changes and is also being widely adopted in PSO also [11].

If the updated position of the molecules is outside the boundary of the search space, the corresponding value will be simply truncated to the boundary value accordingly. Although this method is biased towards the boundary and rather unrealistic, it is simple in implementation and adequate for this preliminary application of MDO for MOO. Furthermore, the optimal region of the problems considered in the experimental study later is far away from the boundary. Hence, this type of strategy will not be biased towards MDO in any means. The more common technique used in actual molecular simulation is the periodic boundary conditions [9].

The time counter will then be incremented by one and if the maximum time had not been reached, the algorithm will re-evaluate the solutions and repeat itself again.

3.2 Comparison of MDO with Conventional Evolutionary Optimizers

MDO represents a computational implementation of molecular system, simulating the dynamics of molecules trapped in a fixed enclosure with uneven potential field. The molecules will fly around the simulation cell, which represents the search space of the investigated MOP. Each molecule will be attracted by those that Pareto-dominate it and repelled from those they Pareto-dominate. Regardless of the initial configurations, the ensemble of molecules will settle into dynamic equilibrium, where a larger proportion of molecules having a higher degree of optimality. The intermolecular forces help to space the molecules evenly within the optimal region.

Noticeably, MDO and PSO have some similarities, where both used the concept of flying molecules (particles in the case for PSO) within a search space. However, their difference is that the particles' trajectories in PSO are updated based on principles of social psychology, emulating the socio-cognitive behavior of human and animals, whereas in MDO, the overall movement of the molecules is governed by the potential function dictating intermolecular interaction in nature. MDO performs its search through the molecular interaction and dynamics, balancing the exploration of the search space with the exploitation of the best solutions.

Furthermore, it should be mentioned that no form of archiving is being considered in MDO, unlike for conventional multiobjective optimizers, where archiving is deemed necessary to maintain a stable population in the vicinity of the optimal region, as there is a non-zero probability that the optimum solutions will be dropped from the evolving population during the selection process. The convergence witnessed in simulation studies is actually due to the algorithmic trick of keeping a record of Pareto optimal solutions throughout the evolutionary progress [12]. The importance of archiving for evolutionary optimizers, specifically EA and PSO, in MOO will be explored further during the experimental study. Of course, there is no doubt that including archiving will most probably improve the algorithmic performance of MDO. But considering the main objective of the paper, which is to explore the possibility of exploiting molecular dynamics for MOO, the model developed should be kept as generic as possible without any additional features. Furthermore, the implementation of archiving will garner issues like the level of elitism, where solutions from the solutions are thrown back into the evolving population, and the type of solutions to be archived or eliminated.

Lastly, due to the complexity of the model that the conventional evolutionary optimizers are emulating, it is difficult to establish a complete theoretical framework that could explain their behavior and characteristics. Therefore, current theoretical works are mostly based on simplified versions of the original model instead. As such, evolutionary optimizers are often regarded as a black box; where its capability is well understood but the underlying dynamics still remain a mystery. As a result, design of evolutionary optimizers for practical MOO purposes is normally based on intuition and guesswork. While theoretical works in molecular motions are restricted also to simpler systems, they are more developed and comprehensive as compared to those of the evolutionary optimizers, which provide another motivating factor to apply molecular motion for MOO purposes. Furthermore, the convergence properties of molecular motion are actually being proven in one of the related theoretical works.

4 Experimental Study

MDO is implemented in this section to investigate whether it can satisfy the goals of proximity and diversity in MOO. Since a mathematical analysis of the model had not been established, which could guide the parameter setting in various scenarios, parameters for MDO will be arbitrarily chosen. Its performance will be compared against other evolutionary optimizers so as to quantify the significance of the experimental results. But before the experimental studies, a formal introduction to the metrics used to quantify the algorithmic performance of MDO will be presented.

4.1 Performance Metrics

As mentioned earlier, there are several goals in MO optimization [13], [14] including proximity and diversity where the former describes the accuracy of the solution set and the latter measures how well the solution set is defined. In this paper, the generational distance, *GD*, metric is used to measure proximity. It quantifies how "close" the set of PF_{known} is from PF_{true} [15]. A low *GD* signifies that PF_{known} is very close to the PF_{true}.

The measure of diversity depends on factors such as spread and spacing of the solution set. The former can be measured by the maximum spread, *MS*, metric [14], which indicates how well the PF_{true} is covered by PF_{known} through the hyper-boxes formed by the extreme function values observed in both fronts. The greater the value of *MS* is, the more the area of PF_{true} is covered by PF_{known}. The spacing metric, *S*, is used to measure how "evenly" solutions in PF_{known} are distributed [15]. A low value of *S* indicates that the members in PF_{known} are evenly distributed.

Collectively, these three metrics assess how well the optimization goals are achieved by MDO. Apart from these metrics, a simple metric that measures the number of non-dominated solutions found, *N*, is included in the analysis also. In general, a higher *N* corresponds to a better defined PF_{known}, providing the decision-maker more choices and thus increases the likelihood of a better final decision.

4.2 Experimental Results

It should be highlighted that the objective in this paper is not to develop a full fledged algorithm that is on-par with state-of-the-art multiobjective optimizers in real life MOPs, rather it aims to explore the concept of applying molecular dynamics for the purpose of MOO. Hence, baseline version of EA and PSO was adopted instead. Also, the benchmark problem used is a simple MOP that could highlight the relevant characteristics under investigation, and its formulation is given in (15). Typical benchmark problem like the Zitzler series [14] will be reserved for future studies.

$$
\begin{aligned}
&\min\ f_1(x_1,x_2) = x_1^2 + x_2^2 \\
&\min\ f_2(x_1,x_2) = (x_1 - 5)^2 + (x_2 - 5)^2 \\
&-5 \le x_1, x_2 \le 10
\end{aligned}
\tag{15}
$$

A generic Pareto-based EA was considered and its parameter configurations are listed in table 1. The selection criterion is based on Pareto ranking and in the event of a tie, the niche count will be employed. The mechanism of niche sharing is used in the tournament selection, as well as diversity maintenance in the archive. Two version of EA was considered where one adopted elitism, and one without. For the former, elitism was implemented by selecting individuals to the mating pool through a binary tournament selection of the combined archive and evolving population. As for the latter, no archive was implemented. As such, there is a non-zero probability that the good solutions will be eliminated from the evolving population.

Table 1. Parameter Settings for EA

Chromosome	Binary coding , 30 bits per decision variables
Population	Population size of 100. Archive size of 100
Selection	Binary Tournament Selection
Crossover Probability	0.8
Mutation Probability	1 / Chromosome_length
Ranking scheme	Pareto ranking
Diversity Operator	Niche count with radius 0.01 in the normalized objective space
Generation	100

Table 2. Parameter Settings for PSO

Population	Population size of 100. Archive size of 100
Selection	Binary Tournament Selection
Inertia weight	0.4
Social learning factor	1.0
Cognitive learning factor	1.0
Ranking scheme	Pareto ranking
Diversity Operator	Niche count with radius 0.01 in the normalized objective space
Generation	100

As for PSO, a baseline version which adopt similar selection scheme and diversity operator was used. Similarly, two version of PSO was considered where one adopted elitism, and one without. The global best will be randomly chosen from the archive and from the evolving population (for the former version) or just from the evolving population for the latter. The personal best for each particle is updated whenever its pervious location is Pareto-dominated by its current location. If they are of the same level of optimality, the personal best will be randomly chosen between them.

The parameters settings for MDO were arbitrarily chosen and are summarized below in table 3. The different algorithms and their corresponding index and notation are shown in Table 4.

Table 3. Parameter Settings for MDO

Number of molecules, N	100
Total Time, T	100
Strength of attractive potential field, C_a	1
Strength of repulsive potential field , C_r	1
Strength of intermolecular potential, R_0	100
Inertia constant, I	0.9
Niche radius, xniche	0.02

Table 4. Notations for the various algorithms

Index	Algorithm	Notation
1	EA without elitism	EA-I
2	EA with elitism	EA-II
3	PSO without elitism	PSO-I
4	PSO with	PSO-II
5	Molecular Dynamics Optimizer	MDO

30 independent simulation runs were performed altogether. Fig. 5 illustrates the experimental results for the different algorithms when applied to BINH. Despite their differences in *GD*, statistical tests revealed that these differences are not statistically significant except for the few cases where EA-I failed to converge to PF_{true}. This is expected due to the simplicity of the problem. As for *MS*, MDO is able to sustain a wide spread of solution even without any form of archive. In contrast, PSO-I just managed to attain a single point on the PF_{true}, while EA attained a better spread only after elitism was employed in EA-II. After archive was implemented, higher value of *N* was obtained as in the case for EA. Though PSO-I attained high values for *N* on average, all the solutions attained are very near as reflected by the low *MS*.

The PF_{known} of the various algorithms from a randomly chosen experimental run are shown in fig. 6. These plots correspond to the box plots, where PF_{known} attained by PSO-I indeed comprised of only one single point. Generally, the implementation of elitism improves the performance of EA and PSO with PSO-II having the performance amongst the various test algorithms. Nevertheless, even without any form of archiving or elitism, the performance of MDO is on par with PSO-II and was able to attain a wider spread among the solutions.

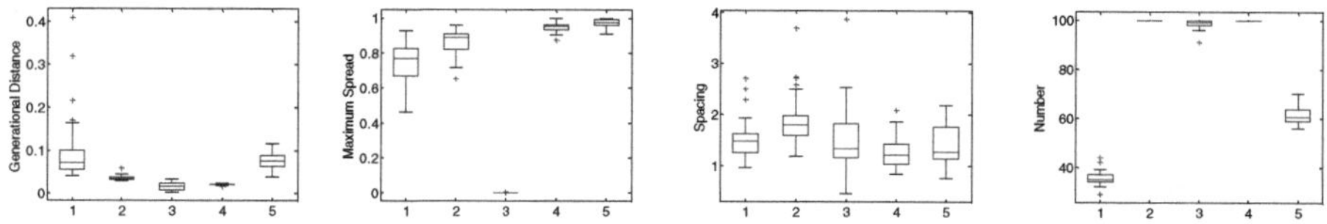

(a) (b) (c) (d)

Fig. 5. Performance metrics of (a) *GD*, (b) *MS*, (c) *S* and (d) *N* when applied to BINH

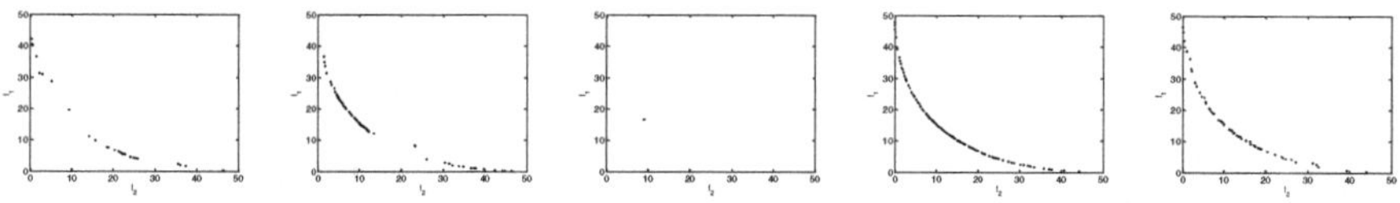

(a) EA-I (b) EA-II (c) PSO-I (d) PSO-II (e) MDO

Fig. 6. PF_{known} attained by the different algorithms

The experimental results conformed to the theoretical hypothesis that conventional evolutionary optimizer required the incorporation of archive before reasonable performance in MOO could be derived. This is in stark contrast to MDO where, the inherent convergence property of MDO allows the solution to converge to PF_{true} even without elitism. Also, the diversity nature of the intermolecular potential allows the solutions to be evenly spread along $PF_{true.}$

Nevertheless, this simple experimental study is insufficient, if the performance of MDO is to be properly evaluated and compared with conventional optimizers of modern day standards. A comprehensive proper evaluation study against existing evolutionary optimizers is currently in progress and the results will be reported in due course. But, the positive performance of MDO does suggest that molecular dynamics is indeed a viable tool for multiobjective optimization.

5 Conclusion

In this paper, a multiobjective optimizer based on the simulation of molecular motion has been presented. This algorithmic framework draws inspiration from the flow of molecules to lower potential region in the presence of uneven potential fields and its inherent drift to a homogenous and uniform equilibrium, which make it suitable for the contrary goals of proximity and diversity in MOO. A simple instance of MDO was implemented and the experimental results demonstrated the viability and practicality of the approach.

Nevertheless, there is still plenty avenues in which future research can proceed, most notably is to consider a fully-operational version of MDO, which is on par with state-of-the-art multiobjective optimizers in real life problems. As the implementation considered in the experimental study is actually a very simplified form of the molecular dynamics algorithm, ongoing work includes the implementation of a more realistic version that considers the existing techniques discussed earlier and conducting a comprehensive performance analysis based on statistical tests and a proper suite of benchmark problem and real life problems.

References

1. Fay, J. A.: Molecular Thermodynamics. Adison Wesley (1965)
2. Kita, H., Yabumoto, Y., Mori, N., Nishikawa, Y.: Multi-objective optimization by means of the thermodynamical genetic algorithm optimization. Lecture Notes in Computer Science (1996) 504–512
3. Cui, X., Li, M., Fang, T.: Study of population diversity of multiobjective evolutionary algorithm based on immune and entropy principles. Congress on Evolutionary Computation, No. 2 (2001) 1316–1321
4. Sobieski, J.S., Laba, K., Kincaid, R.: Bell-curved based evolutionary optimization algorithms. 7th IAA/USAF/NASA/ISSMO Symp. on Multidisciplinary Analysis and Optimization (1998) 2083–2096
5. Farhang-Mehr, Azarm, S.: Entropy-based multi-objective genetic algorithm for design optimization. Structural and Multidisciplinary Optimization, Vol. 24, No. 5 (2002) 351–361
6. Li, X., Zou, X.F., Kang, L.S., Michalewicz, Z.: A new dynamical evolutionary algorithm based on statistical mechanics. Journal of Computer Science and Technology, vol. 18, no. 3 (2003) 361–368
7. Prugel-Bennett, A., Shapiro, J.: An analysis of genetic algorithms using statistical mechanics. Physical Review Letter, Vol. 72 (1994) 305–1309
8. Allen, M.P.: Introduction to Molecular Dynamics Simulation, Lecture Notes in Computational Soft Matter: From Synthetic Polymers to Proteins, Vol. 23 (2004) 1–28
9. Frenkel, D., Smit, B.: Understanding Molecular Simulation: From Algorithms to Applications. Academic Press (1996)
10. Ercolessi, F.: A molecular dynamics primer. http://www.fisica.uniud.it/ercolessi/md/md.pdf (1997)
11. Hu, X.H., Shi, Y.H., Eberhart, R.: Recent advances in particle swarm. Congress on Evolutionary Computation, Vol. 1 (2004) 90–97
12. Mario V.A., Coello, C.A.C., Onésimo, H.L.:Asymptotic Convergence of some Metaheuristics used for Multiobjetive Optimization. Foundations of Genetic Algorithms, Vol. 3469 (2005) 95–111
13. Deb, K.: Multi-objective Optimization Using Evolutionary Algorithms. John Wiley & Sons, New York (2001)
14. Zitzler, E., Deb, K., Thiele, L.: Comparison of multiobjective evolutionary algorithms: empirical results. Evolutionary Computation, Vol. 8, No. 2 (2000) 173–195
15. Veldhuizen, D.A.V., Lamont, G.B.: Multiobjective evolutionary algorithms: Analyzing the state-of-the-arts. Evolutionary Computation, Vol. 8, No. 2 (2000) 125–147

Sequential Approximation Method in Multi-objective Optimization Using Aspiration Level Approach

Yeboon Yun[1], Hirotaka Nakayama[2], and Min Yoon[3]

[1] Kagawa University, Kagawa 761-0396, Japan
yun@eng.kagawa-u.ac.jp
[2] Konan University, Kobe 658-8501, Japan
nakayama@konan-u.ac.jp
[3] Konkuk University, Seoul 143-701, Republic of Korea
myoon515@konkuk.ac.kr

Abstract. One of main issues in multi-objective optimization is to support for choosing a final solution from Pareto frontier which is the set of solution to problem. For generating a part of Pareto optimal solution closest to an aspiration level of decision maker, not the whole set of Pareto optimal solutions, we propose a method which is composed of two steps; i) approximate the form of each objective function by using support vector regression on the basis of some sample data, and ii) generate Pareto frontier to the approximated objective functions based on given the aspiration level. In addition, we suggest to select additional data for approximating sequentially the forms of objective functions by relearning step by step. Finally, the effectiveness of the proposed method will be shown through some numerical examples.

1 Introduction

Many decision making problems are formulated as multi-objective optimization problem so as to satisfy the diverse demands of decision maker. Usually, there does not necessarily exist an optimal solution which minimizes all objective functions simultaneously, because of the trade-off among the objective functions. And then, Pareto optimal solution is introduced, and the set of them in the objective function space is called Pareto frontier. Generally, there exist a number of Pareto optimal solutions, which are considered as the candidates of a final decision making solution. Therefore, it is one of main issues in multi-objective optimization how to obtain Pareto optimal solutions, and how to choose one solution from many Pareto optimal solutions. To the end, the aspiration level methods have been developed. These methods search a decision making solution by processing the following two stages repeatedly: 1) solving auxiliary optimization problem to obtain the closest Pareto optimal solution to a given aspiration level of decision maker, and 2) revising her/his aspiration level by making the trade-off analysis. For the cases with many objective functions, it is difficult to visualize Pareto frontier, and also to depict the trade-off among many objective functions. In this case, the conventional interactive optimization methods

S. Obayashi et al. (Eds.): EMO 2007, LNCS 4403, pp. 317–329, 2007.

are useful, although these approaches give one Pareto optimal solution with a single-optimization run. On the other hand, it may be the best way to depict Pareto frontier in the cases with two or three objective functions, since visualizing Pareto frontier helps to grasp trade-off among the objective functions. For that purpose, genetic algorithm (GA) has been applied for solving a multi-objective optimization problem, and multi-objective GA (MOGA) has been shown to be effective for generating Pareto optimal solutions. However, MOGA has some problems as follows; i) it is difficult to treat many objective functions, ii) so many function evaluations are needed in generating the whole Pareto frontier. In particularly, the number of function evaluations is very important when applying MOGA as well as conventional multi-objective optimization methods to the real problems such engineering design problem which have black-box objective functions whose forms are not explicitly known in terms of design variables. Under this circumstance, the value associated with each design variable is given by sampled real/computational experiments such as structural analysis, fluid-mechanical analysis, thermodynamic analysis, and so on. These analyses take long execution time and high cost. Therefore, it is essential to reduce the number of function evaluations as few as possible which is needed in finding an optimal solution.

In multi-objective optimization considering the number of function evaluations, it would rather be practicable to generate a necessary part, not the whole of Pareto frontier. In this research, we propose a new method which is composed of two stages; i) the first stage is to approximate the form of each objective function by using support vector regression on the basis of some sample data, ii) in the second stage using MOGA, we generate Pareto frontier to the approximated objective functions based on a given aspiration level of decision maker. Furthermore, we discuss the way how to select additional data for revising the forms of objective functions by relearning step by step. Finally, we illustrate the effectiveness of proposed method through a numerical example.

2 Support Vector Regression

In this section, to begin with, we introduce support vector regression (SVR), which is a kind of Support vector machine (SVM) for function approximation.

SVM has been recognized as a powerful machine learning technique. SVM was originally developed for pattern classification and later extended to regression, [2], [3], [10], [12]. Therefore, we review briefly SVM for classification.

Let $\mathcal{F}$ be a space of conditional attributes. For binary classification problems, the value of $+1$ or -1 is assigned to each pattern $\boldsymbol{x}_i \in \mathcal{F}$ according to its class $\mathcal{A}$ or $\mathcal{B}$. The aim of machine learning is to predict which class newly observed patterns belong to on the basis of the given training data set $(\boldsymbol{x}_i, y_i)$ $(i = 1, \ldots, \ell)$, where $y_i = +1$ or -1. This is performed by finding a discriminant function $h(\boldsymbol{x})$ such that $h(\boldsymbol{x}) \geqq 0$ for $\boldsymbol{x} \in \mathcal{A}$ and $h(\boldsymbol{x}) < 0$ for $\boldsymbol{x} \in \mathcal{B}$. Linear discriminant functions, in particular, can be expressed by a linear form

$$h(\boldsymbol{x}) = \boldsymbol{w}^T\boldsymbol{x} + b$$

with the property

$$\begin{aligned} \boldsymbol{w}^T\boldsymbol{x} + b &\geqq 0 \quad \text{for} \quad \boldsymbol{x} \in \mathcal{A} \\ \boldsymbol{w}^T\boldsymbol{x} + b &< 0 \quad \text{for} \quad \boldsymbol{x} \in \mathcal{B} \end{aligned}$$

In cases where training data set $\mathcal{F}$ is not linearly separable, we map the original data set $\mathcal{F}$ to a feature space Z by some nonlinear map ϕ. Increasing the dimension of the feature space, it is expected that the mapped data set becomes linearly separable. We try to find linear classifiers with maximal margin in the feature space.

Letting $\boldsymbol{z}_i = \phi(\boldsymbol{x}_i)$, the separating hyperplane with maximal margin can be given by solving the following problem with the normalization $\boldsymbol{w}^T\boldsymbol{z} + b = \pm 1$ at points with the minimum distance:

$$\begin{aligned} &\underset{\boldsymbol{w},b}{\text{minimize}} && \frac{1}{2}||\boldsymbol{w}||_2^2 && (\text{hard-SVC})_P \\ &\text{subject to} && y_i\left(\boldsymbol{w}^T\boldsymbol{z}_i + b\right) \geqq 1,\ i = 1,\ldots,\ell. \end{aligned}$$

Using the kernel function $K(\boldsymbol{x},\boldsymbol{x}') = \phi(\boldsymbol{x})^T\phi(\boldsymbol{x}')$, the dual problem for (hard-SVC)$_P$ can be obtained as follows:

$$\begin{aligned} &\underset{\alpha_i}{\text{maximize}} && \sum_{i=1}^{\ell}\alpha_i - \frac{1}{2}\sum_{i,j=1}^{\ell}\alpha_i\alpha_j y_i y_j K(\boldsymbol{x}_i,\boldsymbol{x}_j) && (\text{hard-SVC})_D \\ &\text{subject to} && \sum_{i=1}^{\ell}\alpha_i y_i = 0, \\ &&& \alpha_i \geqq 0,\ i = 1,\ldots,\ell. \end{aligned}$$

Although several kinds of kernel functions have been suggested, typical kernel functions are Gaussian kernel function and $p-$order polynomial kernel function, and so on [10]:

$$\begin{aligned} K(\boldsymbol{x},\boldsymbol{x}') &= \exp\left(-\frac{\|\boldsymbol{x}-\boldsymbol{x}'\|_2^2}{2\sigma^2}\right), \\ K(\boldsymbol{x},\boldsymbol{x}') &= (\langle\boldsymbol{x}\cdot\boldsymbol{x}'\rangle + 1)^p \end{aligned}$$

In our simulation of this paper, we use Gaussian kernel.

The above hard-SVM is the most basic model, and various SVMs have been suggested. (See the references [2,3,9] about the details.)

Later, SVM has been extended to regression by introducing the $\varepsilon-$insensitive loss function by Vapnik [12]:

$$L^{\varepsilon}(\boldsymbol{z},y,h) = |y - h(\boldsymbol{z})|_{\varepsilon} = \max(0, |y - h(\boldsymbol{z})| - \varepsilon).$$

For a given insensitivity parameter ε,

$$\begin{aligned} \underset{\boldsymbol{w},b,\xi_i,\acute{\xi}_i}{\text{minimize}} \quad & \frac{1}{2}\|\boldsymbol{w}\|_2^2 + C\Big(\frac{1}{\ell}\sum_{i=1}^{\ell}(\xi_i + \acute{\xi}_i)\Big) && \text{(soft}-\text{SVR)} \\ \text{subject to} \quad & \big(\boldsymbol{w}^T\boldsymbol{z}_i + b\big) - y_i \leqq \varepsilon + \xi_i, \ \ i = 1,\dots,\ell, \\ & y_i - \big(\boldsymbol{w}^T\boldsymbol{z}_i + b\big) \leqq \varepsilon + \acute{\xi}_i, \ \ i = 1,\dots,\ell, \\ & \varepsilon, \ \ \xi_i, \ \ \acute{\xi}_i \geqq 0, \end{aligned}$$

where C is the trade-off parameter between the norm of $\boldsymbol{w}$ and ξ_i ($\acute{\xi}_i$). In order to decide an insensitivity parameter ε automatically, $\nu-$SVR was proposed by Schölkopf and Smola [9].

$$\begin{aligned} \underset{\boldsymbol{w},b,\varepsilon,\xi_i,\acute{\xi}_i}{\text{minimize}} \quad & \frac{1}{2}\|\boldsymbol{w}\|_2^2 + C\Big(\nu\varepsilon + \frac{1}{\ell}\sum_{i=1}^{\ell}(\xi_i + \acute{\xi}_i)\Big) && (\nu-\text{SVR}) \\ \text{subject to} \quad & \big(\boldsymbol{w}^T\boldsymbol{z}_i + b\big) - y_i \leqq \varepsilon + \xi_i, \ \ i = 1,\dots,\ell, \\ & y_i - \big(\boldsymbol{w}^T\boldsymbol{z}_i + b\big) \leqq \varepsilon + \acute{\xi}_i, \ \ i = 1,\dots,\ell, \\ & \varepsilon, \ \ \xi_i, \ \ \acute{\xi}_i \geqq 0, \end{aligned}$$

where C and ν are trade-off parameters between the norm of $\boldsymbol{w}$ and ε and ξ_i ($\acute{\xi}_i$).

Using multi-objective programming and goal programming techniques [6], [15], the authors have developed several varieties of SVM and extended the family of SVM for classification to regression [7] including the primal problem

$$\begin{aligned} \underset{\boldsymbol{w},b,\varepsilon,\xi,\acute{\xi}}{\text{minimize}} \quad & \frac{1}{2}\|\boldsymbol{w}\|_2^2 + \nu\varepsilon + \mu(\xi + \acute{\xi}) && (\mu - \nu-\text{SVR}) \\ \text{subject to} \quad & \big(\boldsymbol{w}^T\boldsymbol{z}_i + b\big) - y_i \leqq \varepsilon + \xi, \ \ i = 1,\dots,\ell, \\ & y_i - \big(\boldsymbol{w}^T\boldsymbol{z}_i + b\big) \leqq \varepsilon + \acute{\xi}, \ \ i = 1,\dots,\ell, \\ & \varepsilon, \ \ \xi, \ \ \acute{\xi} \geqq 0, \end{aligned}$$

where ν and μ are the trade-off parameters between the norm of $\boldsymbol{w}$ and ε and ξ ($\acute{\xi}$).

Here, we show the effectiveness of $\mu - \nu-$SVR comparing the performance through the following simple problem:

$$y = (x^2 - 1)(x^2 - 9), \ \ -3 \leqq x \leqq 3.$$

The parameters in each SVR are given by

i) soft$-$SVR : $C = 100, \ \varepsilon = 0.1$
ii) $\nu-$SVR : $C = 100, \ \nu = 0.5$
iii) $\mu - \nu-$SVR : $\mu = 50, \ \nu = 50$

and the training data is 30 points randomly with noise.

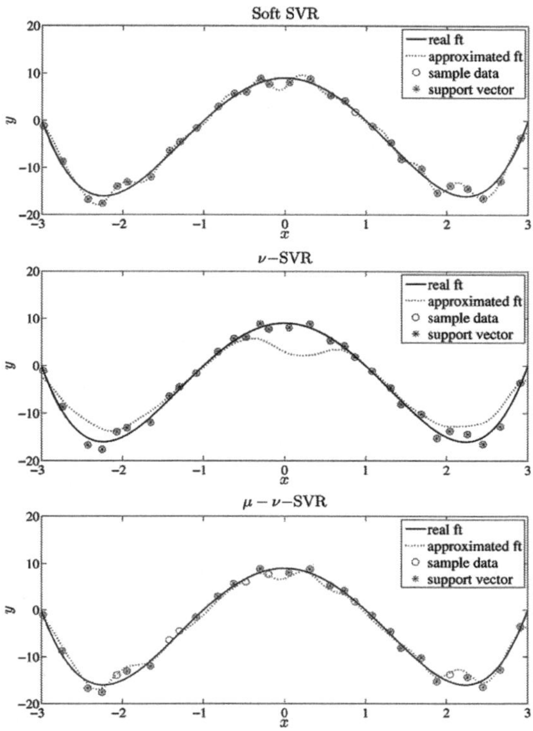

Fig. 1. Comparison of Several SVR

As shown in **Fig. 1**, it can be observed that $\mu - \nu-$SVR provides the least number of support vectors while keeping a reasonable error rate, compared with soft−SVR and $\nu-$SVR, that is, $\mu - \nu-$SVR is promising for sparse approximation. One of the most prominent features in $\mu - \nu-$SVR is that it provides relatively less support vectors, which means the computation is less expensive. The fact that $\mu - \nu-$SVR yields good function approximation with reasonable accuracy and with less support vectors, is important in practice in engineering design. For some kinds of engineering design problems, approximation functions should be realized on the basis of as few data points as possible. Such practical engineering design problems are considering as the application, and therefore, we adopt $\mu - \nu-$SVR as a useful tool of function approximation in this research.

3 Multi-objective Optimization

Next, we explain the concept of solution in multi-objective optimization. Consider a multi-objective optimization problem with m-objective functions and n-dimensional decision variable as follows:

$$\begin{array}{lll} \underset{\boldsymbol{x}}{\text{minimize}} & \boldsymbol{f}(\boldsymbol{x}) = (f_1(\boldsymbol{x}), f_2(\boldsymbol{x}), \ldots, f_m(\boldsymbol{x}))^T & \text{(MP)} \\ \text{subject to} & \boldsymbol{x} \in X = \{\, \boldsymbol{x} \in \mathbb{R}^n \mid g_j(\boldsymbol{x}) \leqq 0,\ j = 1, \ldots, l \,\}, & \end{array}$$

where X is the set of all feasible solution set in design variable space.

The objective function space is partially ordered, thus for convenience, we use the following notations for two vectors $\boldsymbol{y}^1$ $(= (y_1^1, \ldots, y_m^1)^T)$ and $\boldsymbol{y}^2$ $(= (y_1^2, \ldots, y_m^2)^T)$ in $\mathbb{R}^m$.

$$\begin{array}{lll} \boldsymbol{y}^1 > \boldsymbol{y}^2 & \Longleftrightarrow & y_i^1 > y_i^2,\ i = 1, \ldots, m, \\ \boldsymbol{y}^1 \geqq \boldsymbol{y}^2 & \Longleftrightarrow & y_i^1 \geqq y_i^2,\ i = 1, \ldots, m, \\ \boldsymbol{y}^1 \geq \boldsymbol{y}^2 & \Longleftrightarrow & y_i^1 \geqq y_i^2,\ i = 1, \ldots, m \text{ but } \boldsymbol{y}^1 \neq \boldsymbol{y}^2. \end{array}$$

Generally, unlike traditional optimization problem with a single objective function, there seldom exists an optimal solution that minimizes all objective functions $f_i(\boldsymbol{x})$, $i = 1, \ldots, m$, simultaneously in the problem (MP). Using on the above partial order relation, hence Pareto optimal solution is defined as follows [8]:

Definition 1 (Pareto optimal solution). A point $\hat{\boldsymbol{x}} \in X$ is said to be *Pareto optimal solution* if there exists no $\boldsymbol{x} \in X$ such that

$$\boldsymbol{f}(\boldsymbol{x}) \leq \boldsymbol{f}(\hat{\boldsymbol{x}}).$$

Also, Pareto optimal solution $\hat{\boldsymbol{x}} \in X$ in the objective space is called as *Pareto optimal value*, and the set of them is *Pareto frontier*. (See **Fig. 2**.)

Usually, Pareto optimal solutions are considered as the candidates of a final solution to the problem (MP). Therefore, main issues in multi-objective optimization are how to obtain Pareto optimal solutions and how to choose one solution from the set of Pareto optimal solutions. To the end, many interactive optimization methods , for example, aspiration level approach [8] have been developed. In the aspiration level approach, for a given aspiration level $\overline{\boldsymbol{f}} := (\overline{f}_1, \ldots, \overline{f}_m)$,

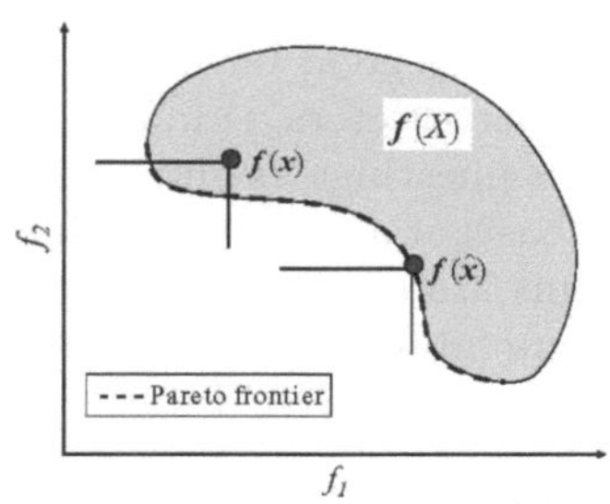

Fig. 2. Pareto frontier

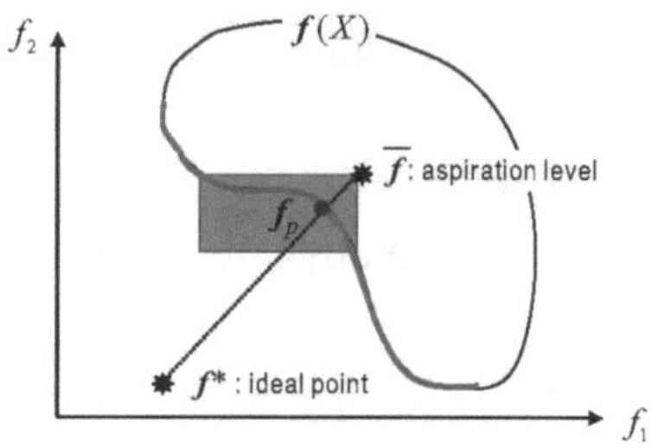

Fig. 3. Aspiration level method

we consider the following problem with the augmented Tchebyshev scalarization function:

$$\begin{array}{lll} \underset{\boldsymbol{x},\delta}{\text{minimize}} & \delta + \lambda \sum_{i=1}^{m} \omega_i \left(f_i(\boldsymbol{x}) - \overline{f}_i \right) & \text{(AP)} \\ \text{subject to} & \omega_i \left(f_i(\boldsymbol{x}) - \overline{f}_i \right) \leqq \delta, \ \ i = 1, \ldots, m, & \\ & \boldsymbol{x} \in X, & \end{array}$$

where $\omega_i = \dfrac{1}{\overline{f}_i - f_i^*}$, $i = 1, \ldots, m$, $\boldsymbol{f}^* := (f_1^*, \ldots, f_m^*)$ is an ideal point, and λ $(= 10^{-7})$ is sufficiently small number.

By solving the problem (AP), as shown in **Fig. 3**, we can obtain the closest Pareto optimal value $\boldsymbol{f}_p$ where is an intercept of Pareto frontier and the line passing the aspiration level point $\overline{\boldsymbol{f}}$ and the ideal point $\boldsymbol{f}^*$.

4 Generation of Pareto Frontier by the Proposed Method

As was mentioned in Section 1, GA has been applied for solving a multi-objective optimization problem. Especially, GA can generate a set of Pareto optimal solutions, since GA is a kind of multi-point searching methods. As a results, GA has been proved to be a very effective method in multi-objective optimization, and several multi-objective GA (MOGA) have been researched [1], [4]. Main efforts in MOGA are made for the diversity and the accuracy of solutions. Consequently, when applying MOGA to actual problems, so many function evaluations (= the number of population $\times$ generation) are needed. However, in the problems such as engineering design problems, because the value of objective function cannot be obtained so easily, it is very important to reduce the number of function evaluations as few as possible in finding an explicit/implicit optimal solution to the problem. In order to reduce the number of function evaluations, we have proposed several algorithms, [13], [14]. Especially, the paper [14] suggested the method combining MOGA and the aspiration level approach to choose one solution from the set of Pareto optimal solutions.

In this research, we suggest a hybrid algorithm that SVR approximates the objective functions and MOGA searches Pareto optimal solutions to the approximate objective functions. We summarize the proposed method as follows:

Step 1. (Real Calculation)
Calculate actually the values of objective functions $\boldsymbol{f}(\boldsymbol{x}_1), \boldsymbol{f}(\boldsymbol{x}_2), \ldots, \boldsymbol{f}(\boldsymbol{x}_\ell)$ for sampled data $\boldsymbol{x}_1, \ldots, \boldsymbol{x}_\ell$.

Step 2. (Approximation)
Approximate each objective function $\hat{f}_1(\boldsymbol{x}), \ldots, \hat{f}_m(\boldsymbol{x})$ by the learning of $\mu - \nu-$SVR on the basis of real sample data set $\{(\boldsymbol{x}_i, \boldsymbol{f}(\boldsymbol{x}_i)), \ i = 1, \ldots, \ell\}$.

Step 3. (Finding of Solution Closest to Aspiration Level and Generation of Pareto Frontier)
Find the closest Pareto optimal solution to the given aspiration level by minimizing the problem (AP), which is solved by genetic algorithm. In parallel, generate the whole set of Pareto optimal solutions by using MOGA.

Step 4. (Choice of Additional Data)
Choose the additional ℓ_0-data for relearning. Go to Step 1. (Set $\ell \leftarrow \ell + \ell_0$.) Here, we choose the additional data by the method as described in the below.

Stage 1. First, add the closest Pareto optimal solution to the given aspiration level which was found in Step 3. This is for well approximating the neighborhood of Pareto optimal solution to the aspiration level) ($\leftarrow$ local information)

Stage 2. Evaluate the ranks $R(\boldsymbol{x}_1), R(\boldsymbol{x}_2), \ldots, R(\boldsymbol{x}_\ell)$ for sampled data $\boldsymbol{x}_1, \boldsymbol{x}_2, \ldots, \boldsymbol{x}_\ell$ by the ranking method [5]: if an individual $\boldsymbol{x}_i$ is dominated by another $n-$individuals, then the rank of $\boldsymbol{x}_i$ is given by $R(\boldsymbol{x}_i) = n + 1$.

Stage 3. Approximate a ranking function $\hat{R}(\boldsymbol{x}_i)$ on the basis of data set $\{(\boldsymbol{x}_i, R(\boldsymbol{x}_i)), \ i = 1, \ldots, \ell\}$ by $\mu - \nu-$SVR.

Stage 4. Calculate the values of ranking function for the whole Pareto optimal solutions obtained in Step 3.

Stage 5. Among them, add the points with high value of ranking function. This is to grasp the configuration of Pareto frontier. ($\leftarrow$ global information)

5 Numerical Examples

For illustrating the proposed method, we show the results of a simple example with one design variable and two objective functions. **Fig. 4** (a) and **Fig. 5** (a) shows the approximated objective functions on the basis of some real sample data. **Fig. 4** (b) and **Fig. 5** (b) shows the population generated at the final generation. Finally, **Fig. 4** (c) and **Fig. 5** (c) represents Pareto optimal values accumulated in the whole generations.

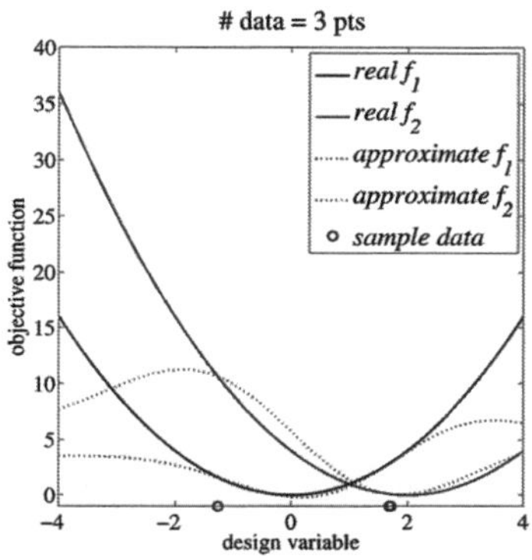

(a) approximation

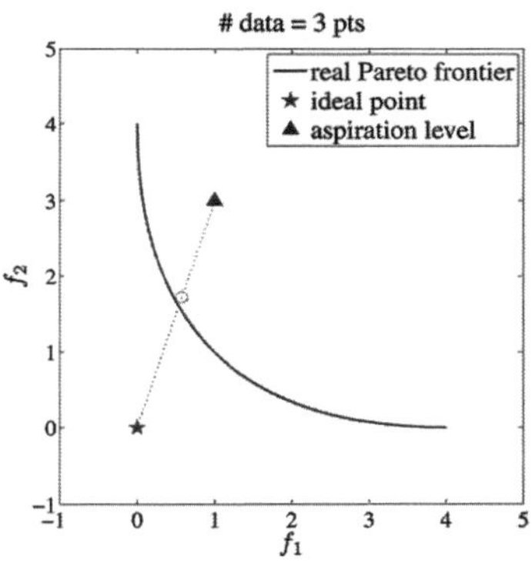

(b) population at the final generation

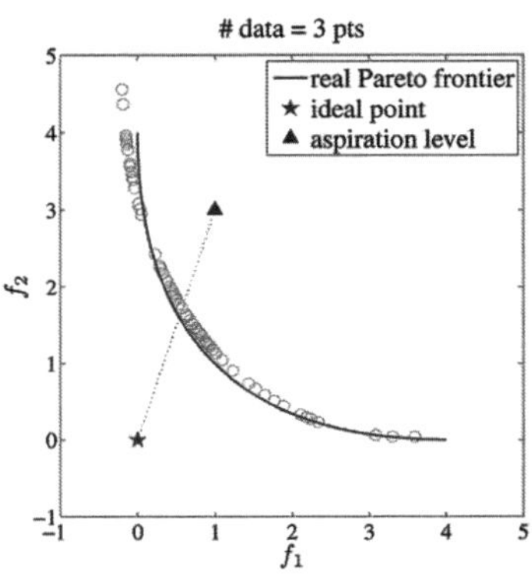

(c) Pareto optimal values

Fig. 4. # sample data = 3 points

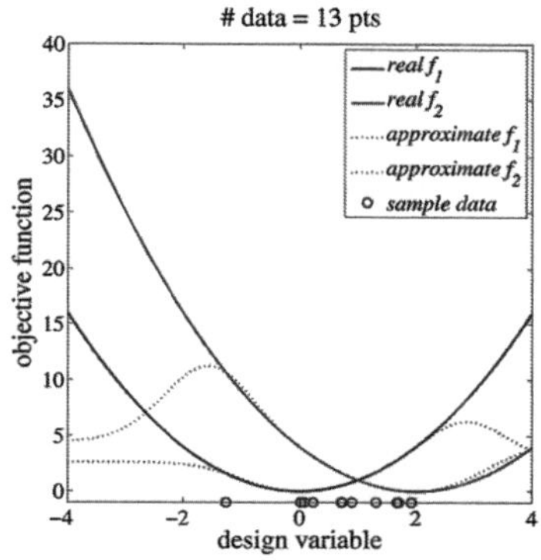

(a) approximation

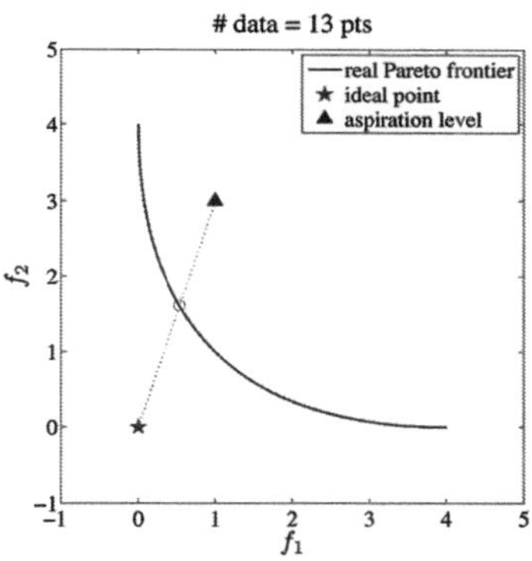

(b) population at the final generation

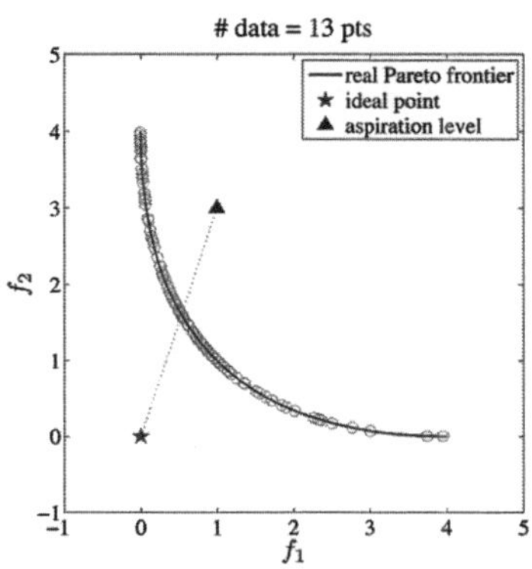

(c) Pareto optimal values

Fig. 5. # sample data = 13 points

Next, we consider the following problem:

$$\begin{aligned} \text{minimize} \quad & f_1(\boldsymbol{x}) = x_1 + x_2 \\ & f_2(\boldsymbol{x}) = 20\cos(15x_1) + (x_1 - 4)^4 + 100\sin(x_1 x_2) \\ \text{subject to} \quad & 0 \leqq x_1,\ x_2 \leqq 3. \end{aligned} \tag{Ex-1}$$

The true function of each objective function f_1 and f_2 in the problem (Ex-1) are shown in **Fig. 6**.

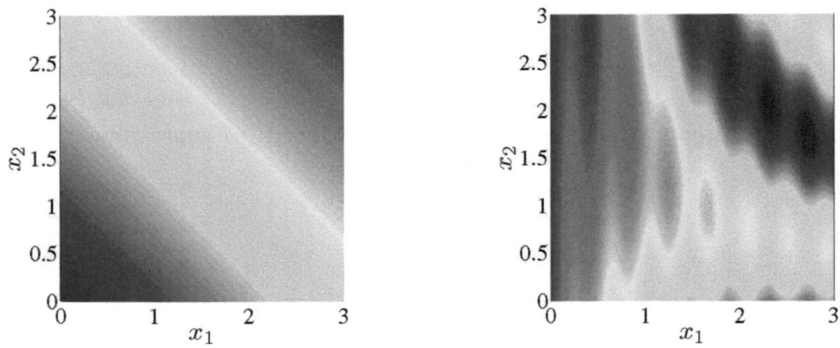

(a) f_1 (b) f_2

Fig. 6. The True contours to the problem (Ex-1)

In our simulation, the ideal point and the aspiration level is given by

$$\begin{aligned} (f_1^*,\ f_2^*) &= (0,\ -120), \\ (\overline{f}_1,\ \overline{f}_2) &= (3,\ 200), \end{aligned}$$

and the closest Pareto optimal solution to the above condition is as follows:

$$\begin{aligned} \text{exact optimal solution } (\hat{x}_1,\ \hat{x}_2) &= (1.413,\ 0) \\ \text{exact optimal value } (\hat{f}_1,\ \hat{f}_2) &= (1.413,\ 30.742) \end{aligned}$$

Starting with initial data 10 points randomly, we obtained the following approximate solution by proposed method:

$$\begin{aligned} \text{approximate solution } (x_1,\ x_2) &= (1.405,\ 0) \\ \text{approximate value } (f_1,\ f_2) &= (1.405,\ 29.853) \end{aligned}$$

and also, we show the results in **Fig. 8** – **Fig. 11**. As can be seen from **Fig. 6** (b) of the true function, using just 40 sample points, the proposed method can generate the approximate solution with almost same exact one, although it may be so difficult to approximate the second objective function f_2.

Additionally, the proposed method provides many Pareto optimal solutions on the neighborhood of the exact solution as well as a rough configuration of Pareto frontier as shown in **Fig. 7**.

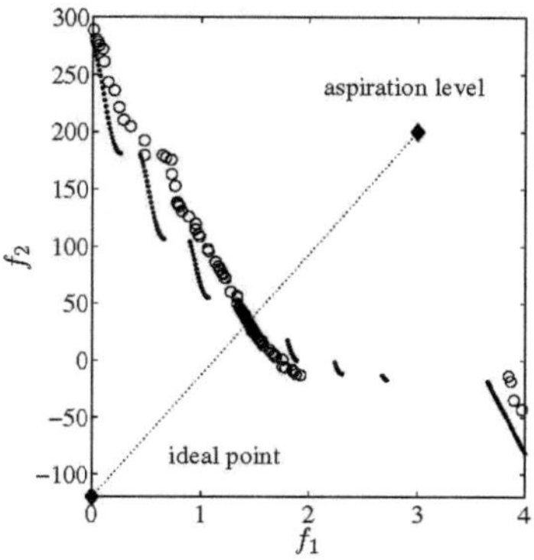

Fig. 7. The Whole Pareto frontier with 40 sample points

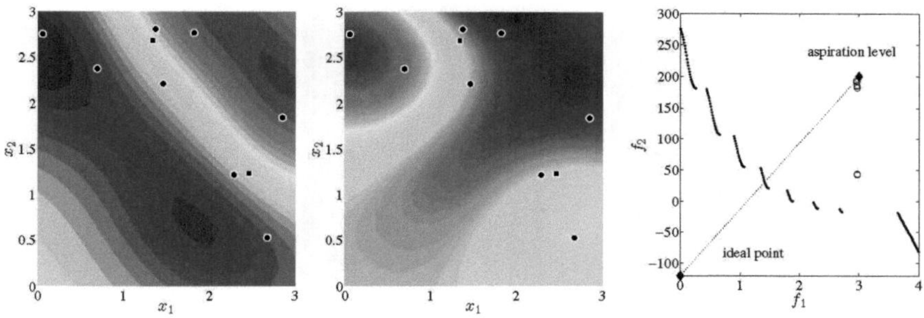

(a) contour of f_1 (b) contour of f_2 (c) population at the final generation

Fig. 8. Initial training : 10 points (Ex-1)

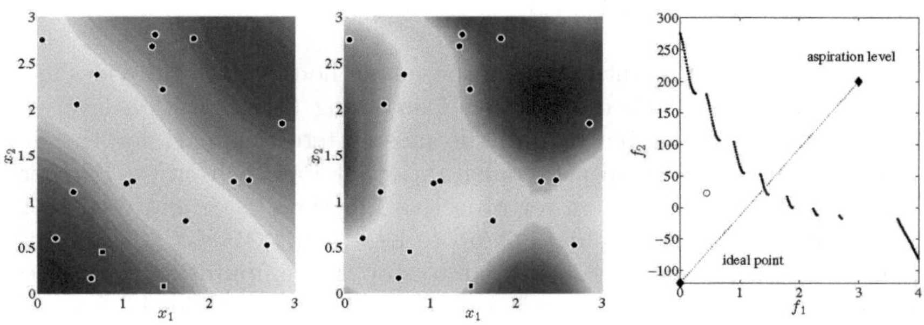

(a) contour of f_1 (b) contour of f_2 (c) population at the final generation

Fig. 9. # sample data : 20 points (Ex-1)

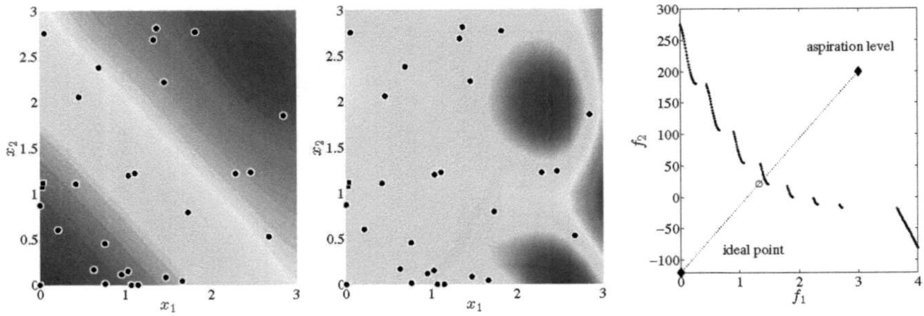

(a) contour of f_1 (b) contour of f_2 (c) population at the final generation

Fig. 10. # sample data : 30 points (Ex-1)

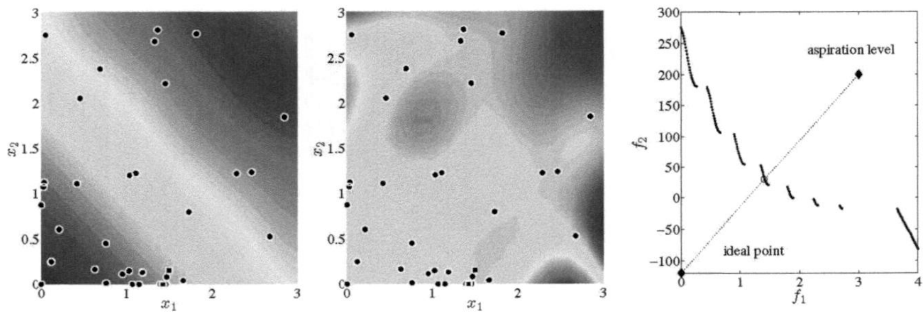

(a) contour of f_1 (b) contour of f_2 (c) population at the final generation

Fig. 11. # sample data : 40 points (Ex-1)

6 Concluding Remarks

The most prominent feature in the proposed method is that it provides local and global information of Pareto frontier. Combining the aspiration level method and MOGA, it is possible to find the most interesting part for the decision maker as well as to grasp the configuration of Pareto frontier. Furthermore, employing the approximation for objective functions, it is expected to reduce the number of function evaluations up to less than 1/100 to 1/5 of using only MOGA. Usually, we do not know when to stop the computation in advance, and the computation is terminated relatively early by the limitation of time and cost in practical problem. In practical engineering design problems, it is desirable that the number of function evaluations is less than 1000 (100 if possible). Therefore, we can conclude that the proposed method is capable of wide application.

References

1. C.A. Coello Coello, D.A. Van Veldhuizen and G.B. Lamont, "Evolutionary Algorithms for Solving Multi-Objective Problems", *Kluwer Academic Publishers*, 2002
2. N. Cristianini and J. Shawe-Taylor, *An Introduction to Support Vector Machines and Other Kernel-based Learning Methods*, Cambridge University Press, 2000
3. C. Cortes and V.N. Vapnik, Support Vector Networks, *Machine Learning*, Vol. 20, pp. 273–297, 1995
4. K. Deb, "Multi-Objective Optimization using Evolutionary Algorithms", *John & Wiley Sons, Ltd.*, 2001
5. C.M. Fonseca and P.J. Fleming, "Genetic Algorithms for Multi-objective Optimization: Formulation, Discussion and Generalization", *Proceedings of the Fifth International Conference on Genetic Algorithms*, pp. 416–426, 1993
6. H. Nakayama and Y.B. Yun, Generating Support Vector Machines using Multiobjective Optimization and Goal Programming, *Multi-Objective Machine Learning*, Springer Series on Studies in Computational Intelligence, pp. 173–198, 2006
7. H. Nakayama and Y.B. Yun, Support Vector Regression based on Goal Programming and Multiobjective Programming, *Proceedings of World Congress on Computational Intelligence*, 2006
8. Y. Sawaragi, H. Nakayama and T. Tanino, "Theory of Multiobjective Optimization", *Academic Press Inc.*, 1985
9. B. Schölkopf and A.J. Smola, New Support Vector Algorithms, *NeuroCOLT2 Technical report Series*, NC2-TR-1998-031, 1998
10. B. Schölkopf and A.J. Smola, "Learning with Kernels", *MIT Press*, 2002
11. V.N. Vapnik, "The Nature of Statistical Learning Theory", *Springer Verlag*, 1995
12. V.N. Vapnik, *Statistical Learning Theory*, John Wiley & Sons, New York, 1998
13. Y.B. Yun, H. Nakayama, M. Arakawa, "Multiple criteria decision making with generalized DEA and an aspiration level method", *European Journal of Operational Research*, Vol. 158, No. 1, pp. 697–706, 2004
14. Y.B. Yun, H. Nakayama, T. Tanino, M. Arakawa, "Generation of Efficient Frontiers in Multi-Objective Optimization Problems by Generalized Data Envelopment Analysis", *European Journal of Operational Research*, Vol. 129, No. 3, pp. 586–595, 2001
15. M. Yoon, Y.B. Yun and H. Nakayama, "Total Margin Algorithms in Support Vector Machines", *IEICE Transactions on Information and Systems*, Vol. E87, No. 5, pp. 1223-1230, 2004

Multi-objective Optimisation of a Hybrid Electric Vehicle: Drive Train and Driving Strategy

Robert Cook[1], Arturo Molina-Cristobal[2], Geoff Parks[1], Cuitlahuac Osornio Correa[3], and P. John Clarkson[1]

[1] Engineering Design Centre, Department of Engineering, University of Cambridge, Trumpington Street, Cambridge CB2 1PZ, UK
[2] Electrical Machines and Drives Group, Department of Electronic and Electrical Engineering, University of Sheffield, Mappin Street, Sheffield S1 3JD, UK
[3] Department of Engineering, Iberoamericana University, Prolongacion Paseo de la Reforma 880, Lomas de Santa Fe, C.P. 01210, Mexico City, Mexico

Abstract. The design of a Hybrid Electric Vehicle (HEV) system is an energy management strategy problem between two sources of power. Traditionally, the drive train has been designed first, and then a driving strategy chosen and sometimes optimised. This paper considers the simultaneous optimisation of both drive train and driving strategy variables of the HEV system through use of a multi-objective evolutionary optimiser. The drive train is well understood. However, the optimal driving strategy to determine efficient and opportune use of each prime mover is subject to the driving cycle (the type of dynamic environment, e.g. urban, highway), and has been shown to depend on the correct selection of the drive train parameters (gear ratios) as well as driving strategy heuristic parameters. In this paper, it is proposed that the overall optimal design problem has to consider multiple objectives, such as fuel consumption, reduction in electrical energy stored, and the 'driveability' of the vehicle. Numerical results shows improvement when considering multiple objectives and simultaneous optimisation of both drive train and driving strategy.

1 Introduction

A current environmental issue is the reduction of the total energy consumption of a passenger car. Despite their higher manufacturing cost, HEVs have been shown to be an effective way to substantially reduce fuel consumption [1]. Combining an electric motor and internal combustion engine to propel a vehicle results in an energy management problem. The fundamental issue of seeking for an effective and optimal strategy to split the power between thermal and electrical paths is addressed in this paper.

Guzzella and Sciarretta [2] have classified the optimisation of a HEV system in three layers, as follows: 1) Structural optimisation, where the objective is to find the best possible structure (arrangement of power train and prime movers);

S. Obayashi et al. (Eds.): EMO 2007, LNCS 4403, pp. 330–345, 2007.

2) Parametric optimisation, where the objective is to find the best possible parameters for a fixed power train structure; and 3) Control system optimisation, where the objective is to find the best possible supervisory control algorithm and best parameters thereof. Guzzella and Sciarretta identify that these stages are not independent. However, due to the limitations of conventional optimisation techniques (nonlinear programming, dynamic programming) they have yet to be considered simultaneously. In this paper, the parametric optimisation, via an Evolutionary Algorithm (EA), for a fixed power train structure and for a supervisory control algorithm is simultaneously considered.

This problem could be seen as an optimisation problem in Dynamic Environments (DEs). Using Branke's criteria [3], it clearly has the characteristics of a DE: the change in optimum value (optimal distribution of the power between the thermal and electrical paths depends on time and energy usage of the vehicle), frequency of the change and severity of the change depend on the driving cycle. There is some degree of predictability of change: there are three main types of driving cycle. In the United States, the Urban Dynamometer Driving Schedule (UDDS – also known as the US federal test procedure, FTP-72) represents a city driving cycle. The federal highway driving cycle (FUDS) represents extra-urban and high speed driving, and in Europe the urban motor vehicle expert group (MVEG-95) represents a combined cycle. These driving cycles are standard profiles of speed and serve as test cycles for performance comparison among different vehicles on the same basis.

It has been discussed in [3] that the solution of an optimisation problem in a DE, solved via an EA, doesn't need to cope with the dynamics (the optimum doesn't change over time) if a controller strategy is involved, and then the problem becomes static for the optimiser. Thus, the success of the design relies on the effective selection and parameter optimisation of the driving strategy. Guzzella and Sciarretta [2] have classified the driving controller strategies as follows: 1) Heuristic control strategies where rules are set up to meet torque demand and vehicle speed [4,5,6]; 2) Optimal control strategies where minimisation of the fuel energy use over the entire cycle is sought subject to a constraint over the final state of charge (SOC) of the battery. This strategy needs detailed knowledge of the future driving condition, so its use is impractical. However, it serves as a basis of comparison for evaluating the quality of other control strategies [7,8,9]; and 3) Suboptimal control strategies or real-time control strategies consider that some *a priori* knowledge of future driving conditions is available during the actual operation and that the self-sustainability of the electrical path has to be guaranteed. The idea then is to perform online optimisation to find the optimal distribution of power. This strategy assumes that some instantaneous state variables are available to evaluate a cost function, which is in terms of the fuel consumption and the SOC variation [9,10].

In Sect. 2, the HEV model used for the optimal transmission design and optimal tuning of the driving control strategy is described. This is a sketched outline of the model used, and the interested reader is referred to [6].

In Sect. 3, a multi-objective Genetic Algorithm (MOGA) to automate the process of finding the optimal parameters for the transmission design is introduced. In this case we use the MOGA proposed in [11,12] to carry out the optimisation. This model was used to evaluate the variation of vehicle performance with various parameters of the transmission. Results from a multi-objective optimisation of fuel consumption and electrical energy use as well as cycle adherence factor are presented.

Although recent optimisation techniques applied to the energy management problem for HEVs have shown promising results, these techniques haven't been able to include the optimisation of discrete motor-engine options. In Sect. 4, the overall design optimisation is considered, it includes: HEV transmission design, tuning of the driving control strategy parameters (heuristic control strategy) and twenty combinations of motor-engine (four different engines and five motors were available) as decision variables.

2 The Model of the Hybrid Vehicle

A HEV model from [6] is used in this study. In [6], a framework for designing a flexible transmission and the complete power train configuration was developed. No optimisation technique is applied. However, the results show an improvement over the original conventional vehicle. The design methodology is based on heuristic selection (engineering assumptions) using intensive evaluations of the most representative design variables. It also was found that the control strategy is intimately bound to the transmission design and the driving cycle.

2.1 Vehicle Dynamics

Ignoring energy dissipated in the power train and used to accelerate the components inside the vehicle, the elementary equation that describes the longitudinal dynamics of a road vehicle takes the following form:

$$m_v \frac{\mathrm{d}}{\mathrm{d}t} v(t) = F_t(t) - \underbrace{\frac{1}{2}\rho_a v(t)^2 C_D A_f}_{\text{air resistance}} \quad - \quad \underbrace{m_v g C_{rr}}_{\text{rolling resistance}} \tag{1}$$

The required tractive force F_t is composed of a sum of resistive and acceleration requirements. $C_D (= 0.29)$ is the coefficient of aerodynamic drag, and $A_f (= 3.2\,\mathrm{m}^2)$ the effective frontal area of the vehicle. No gradients were used in the simulations for this paper. In this analysis, the coefficient of rolling resistance, C_{rr}, is assumed a constant, valued 0.013, and the vehicle mass m_v for all simulations was 2800 kg.

Using the quasistatic paradigm, velocity and acceleration are defined at each time interval. In this way the demanded power is also defined via $P_t(t) = F_t(t)v(t)$. Torques and angular velocities are then defined at the exit of the transmission via the wheel size.

In this study, the simulations are performed using the backward-forward modeling approach used in [4,7]. Given the torque and speed required at the wheels, and going 'backwards' (in the thermal path and in the electric path) component by component this calculation is carried through the drive train, until the fuel use or electrical energy use that would be necessary to meet the trace is computed. The weakness of using only the backward-facing approach is the assumption that the vehicle is able to meet the trace at all times, making it inappropriate for calculating best-effort performance. Since in this optimisation many contrasting designs are considered, some of which may not be capable of matching the trace, the forward calculation must also be carried out. This consists of applying limits to torque, speed and power in the transmission and prime mover calculations, and then following the causality path back to the wheels where the *true* acceleration is calculated.

In Sect. 3, an objective is proposed that measures the difference between demanded and actual velocity. This is introduced to the optimisation to address the problem of designs which do not meet the desired trace. Rather than impose a hard constraint this objective will encourage the algorithm to find solutions which are both efficient, *and* are able to produce torque and power in a variety of situations when needed. In order to introduce the vehicle model to the optimiser, some minor changes to the model in [6] were necessary.

2.2 Hybrid Power Train and Driving Control Strategy

The architecture of the parallel hybrid power train is sketched in Fig. 1a. This architecture allows effective flexibility in operation of the vehicle from an internal combustion engine (ICE) only mode, through a parallel hybrid configuration, to a purely electric motor (EM) powered vehicle. This flexibility is due to the implementation of a planetary gear set (PGS) in the transmission, which provides two mechanical degrees of freedom. In Fig. 1b, the continuously variable transmission (CVT) regulates the speed of the ICE, and the torque of EM is added to that of the ICE to impel the vehicle in critical conditions. In hybrid parallel mode, the ICE is connected directly to the sun gear of a PGS as well as to the ring of PGS through the CVT and a simple gear train (ST). Finally, the torque at the wheels is transmitted through the final drive (FD) gear box, and by the conservation of power we have that: $P_{total}(t) = P_{EM}(t) + P_{ICE}(t)$.

In Fig. 1b, the geometric variables of the transmission are the fixed gear ratio of the PGS g_{rs}, the simple gear train g_c, and the range limits on the CVT g_{cvt}. The final drive ratio is not used as a variable, and takes a value of $g_{fd} = 3.523$ throughout. The degree of hybridisation (DH) in a standard driving cycle is given by $DH(t) = P_{EM}(t)/(P_{EM}(t) + P_{ICE}(t))$, and is determined (in this case indirectly) by the driving control strategy. The value $DH(t)$=1 therefore means that all power needed at the wheels is provided by the electrical path or that all negative power available at the wheels from regenerative braking is driven entirely to the electrical path. When $DH(t)$=0, it means that all the power needed at the wheels is provided by the fuel path.

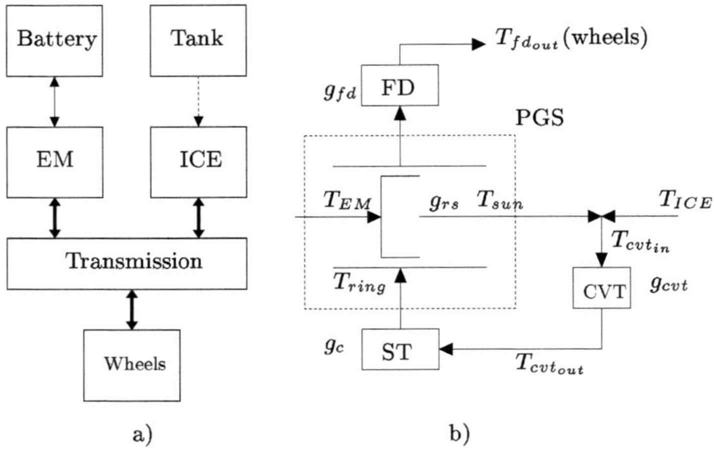

Fig. 1. a) Schematic of the HEV power flows, bold lines: mechanical link, solid lines: electrical link, dashed line: fuel link; b) Schematic of torque flux in the transmission

Driving Control Strategy. The strategy used here is that suggested in [6] and is a heuristic strategy with the following attributes:

- The ICE remains turned off below a minimum connection speed $v_{cutoff_{ICE}}$. Below this speed the EM provides all of the motive power.
- Once $v_{cutoff_{ICE}}$ is reached the ICE then attempts to supply all power needed to drive the vehicle, within some limits defined by the engine's efficiency map. Figure 2 shows the control of the operating point for the ICE. When running, the ICE is ideally kept on the optimal line (dashed) and between the curves of maximum (bold-dotted) and minimum (bold-dashdot) nominal constant power. The ICE may leave the optimal line if forced to do so by limits on the CVT. However, operation outside the region between the power lines is allowed for battery charging only. The curves of maximum and minimum nominal power are calculated automatically for each engine by finding the intersection of the optimal line (dotted) with the point where efficiency falls to 94% *of best*, marked "*".
- Any additional power needed to be supplied/drained is produced/absorbed by the EM by applying a suitable voltage across its windings.
- The state of discharge of the battery is monitored, and falling below a determined lower limit of charge, bat_{lo}, will cause different behaviour from the ICE. At this stage the ICE will run at maximum nominal power until the battery reaches a suitable charge, bat_{hi}, regardless of the power demand from the driving cycle. (The exception is if the vehicle stops, since the ICE cannot then continue to operate.) If the battery level reaches a lower *critical* limit of 50%, the ICE is allowed to operate up to maximum rated power to recover battery charge.

– On deceleration, the EM recovers as much energy as possible (within torque limits) as a generator. Any additional deceleration is provided by friction (dissipative) braking.

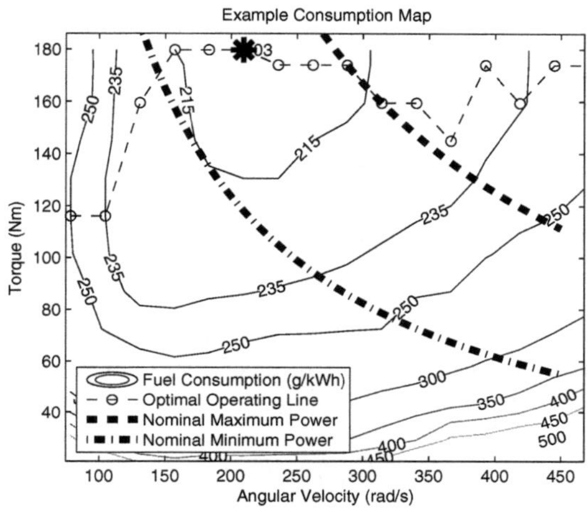

Fig. 2. Selection of the engine operating point in the ICE

3 Multi-objective Optimisation of the Parallel Hybrid Power Train

By considering how the design of transmission parameters (g_c, g_{rs} and g_{cvt}) affect the performance of the vehicle over a pre-defined driving cycle, useful information is obtained about choosing these parameter values.

Three objectives are introduced for simultaneous minimisation: to consider the total energy use from each path, and the adherence of the vehicle to the chosen cycle. Other research studies on multi-objective optimisation of HEV fuel economy have considered reduction of CO and NOx emissions as another objective [13,14]. However, as discussed in [7], novel control systems and catalytic converter technology have reduced pollutant emissions in diesel and gasoline engines to almost negligible levels. Therefore, pollutant emissions are not considered in this study. The multi-objective problem MOHEV-I solved is:

$$\text{minimise} \begin{cases} O_1(g_c, g_{rs}, g_{cvt_{offset}}) \\ O_2(g_c, g_{rs}, g_{cvt_{offset}}) \\ O_3(g_c, g_{rs}, g_{cvt_{offset}}) \end{cases} \tag{2}$$

Objective 1 is simply the total ICE fuel consumption f_c over the driving cycle. This is interpolated from a consumption map at each time step (see Fig. 2) so that:

$$O_1 = \frac{1}{\rho_f} \sum_{t=0}^{t=t_{final}} \frac{f_{c_{inst}}(T(t), \omega(t))}{d_{inst}(t)} \tag{3}$$

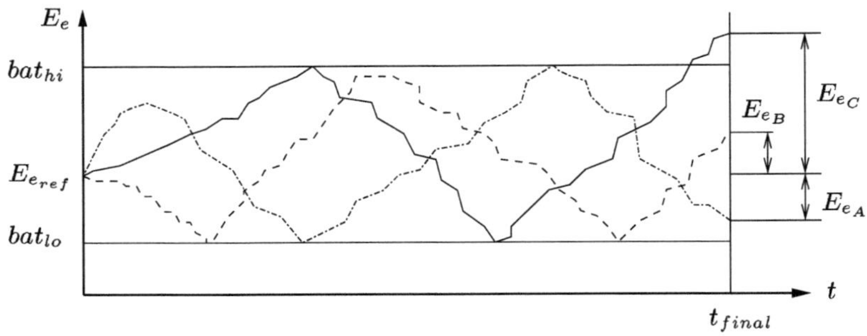

Fig. 3. Battery charge diagram

where $d_{\text{inst}}(t)$ is the distance traveled in each time step and ρ_f the fuel density.

Objective 2 is the overall *reduction* in electrical energy stored in the battery over the complete cycle. Thus:

$$O_2 = E_{e_{ref}} - E_e(t_{final}) \tag{4}$$

Here it is important to note that the vehicle model is that of a *charge sustaining* vehicle. Whilst in the long term this means battery energy is maintained, over the time scale of the UDDS cycle (1369 s) it is a variable outcome. Different vehicle designs with varying final battery energy levels cannot be fairly compared without introducing it as an objective. Whilst in [2] an 'equivalence factor' is introduced for the combination of O_1 and O_2, here the two energy sources are kept separate to provide more information and retain the true complexity of the problem. Figure 3 clarifies this calculation. $E_{e_{ref}}$ is the initial energy state of the battery and is constant in each simulation. E_{e_A}, E_{e_B} and E_{e_C} represent example designs. Here example A would have a larger O_2 than B and C.

As mentioned, with some designs the trace cannot be met when the acceleration of the trace exceeds the capabilities of the power train. Therefore, objective 3 is effectively a measure of how well the vehicle will respond to acceleration demand in different situations. The adherence to the cycle is calculated as:

$$O_3 = \sum_{t=0}^{t=t_{final}} \frac{|v_{cycle}(t) - v_{actual}(t)|}{t_{final}} \tag{5}$$

What this allows us to effectively present is a trade-off study between vehicle efficiency and vehicle 'driveability'. This brings more information than simply defining hard constraints, such as the time taken for a best effort 0-60 km/hr acceleration or similar [13,14].

In order to investigate the effects of different motors and engines in the parametric optimisation of gear ratios, a second multi-objective problem with an additional variable, *MotEng*, was considered. *MotEng* is the selection number for the combination of prime movers to be used in the vehicle. In [13], a 'scale factor' took a similar role. However, the resulting engines are not off-the-shelf

Table 1. MOHEV-I: decision variables for the model (*left*) and the input parameters for the optimiser (*right*)

Variable	Allowed Range
g_{rs}	2.0–4.0
g_c	0.5–1.0
$g_{cvt_{offset}}$	(0.6649–3.611) $\pm$ 0.9

Parameter	Value
Individuals per generation	50
Maximum number of generations	50
Precision	32bits/variable
Selective pressure	2
Probability of mutation	$\approx$ 0.0073
Crossover rate	0.7

designs, and are unrealistic. In [14], independent optimisations with different manufactured motors and engines were considered, but the different combinations were not allowed to compete in a single optimisation.

Here four different engines and five motors were available from ADVISOR [4]. The aim is to identify the global optimum, not just for a specific engine or motor. The motors are: MCI_95, PRIUS_43, INSIGHT, MCI_102, FC_C167; and engines are: PM_49, PM_58, PRIUS_JPN, AC_59. The multi-objective problem MOHEV-II solved is:

$$\text{minimise} \begin{cases} O_1(g_c, g_{rs}, g_{cvt_{offset}}, MotEng) \\ O_2(g_c, g_{rs}, g_{cvt_{offset}}, MotEng) \\ O_3(g_c, g_{rs}, g_{cvt_{offset}}, MotEng) \end{cases} \tag{6}$$

The objective functions are evaluated over a full quasistatic simulation of operation using the UDDS driving cycle, using a resolution step size of 0.2 s. The computational system was a 9 node, Pentium 4 2.8GHz Linux cluster, and a complete 50 generation optimisation took around 10 hours.

The simple heuristic driving strategy described in Sect. 2 determines when each prime mover operates and the flow of power and torque in the drive train. In this case, values selected in [6] are used for the control strategy parameters: $v_{cutoff_{ICE}} = 6.0\,\text{m/s}$, $bat_{lo} = 60\%$, $bat_{hi} = 80\%$.

Table 1 lists the input variables for the model that were used and the ranges allowed. These are chosen to correspond to the model [6] and enable considerable change in the geometry of the transmission path. The ranges chosen are those suggested by Osornio's PhD thesis and reflect realistic scales for manufacture. Note that the CVT will change its ratio as part of the control strategy as the driving cycle progresses. Thus, it is only in fact the range limits which may be set at design time. Here we use a variable $g_{cvt_{offset}}$ which offsets the range used in [6] (0.6649–3.611) by up to 0.9, allowing a thorough search of suitable values. Obviously the simulation did not allow negative values of the ratio to occur. For this first optimisation, the prime movers used were the FC_CI67, a 67 kW diesel engine, and the PM_58, a 58 kW electric motor. In Table 1 the settings used for the MOGA are also detailed.

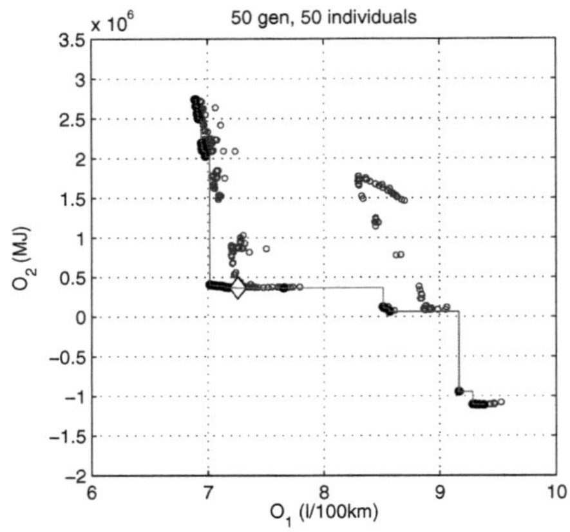

Fig. 4. MOHEV-I: 2D projection of the objective space

3.1 Optimisation Results: Drive Train

The objective and decision spaces of the final generation for the problem are shown in Figs. 4 and 5. The Pareto-front in Fig. 4 is marked with "•", and a line shows the attainment surface. The datum based on evaluation of the model using data from [6] is also shown, marked with a "⋄".

In this case it seems that the range of variation has not been sufficient to produce *any* individuals unable to cope with the cycle. Whilst some random errors produce a very small difference between demanded and actual velocities *at every time step*, they affect every design in the same way, and cause O_3 to be constant across the solutions. It is non-competitive in this case. Hence only the useful O_1 vs O_2 trade-off is shown here.

Whilst not immediately apparent at this magnification level, the algorithm has been able to find some solutions which marginally dominate the datum. It is clear though that the main effect of the evolution is a widening of the Pareto-front, giving a decision-maker a larger range of suitable solutions. In fact the solution picked manually by Osornio is close to Pareto-optimal given the ranges allowed here for the chosen variables.

Figure 5 shows the design variables plotted in three dimensions. Again, the •'s are the variables which generate solutions found to be Pareto-optimal in 3D objective space. These nondominated solutions appear to lie in three clusters, one around $g_c = 0.5$, another at $g_c = 1.0$ and a smaller one around the datum point. The suggestion is that without using the extremes of the variable range, only a small region around the datum point generates competitive solutions. A physical explanation is suggested by analysing the link between objective and decision spaces. The cluster around $g_c = 0.5$ is found to correspond to solutions lying at both extremes of the Pareto-front. This may be because such a low value of g_c causes the CVT range to be more limiting on the engine operating point, and produce more extreme behaviour from the electric motor. The points

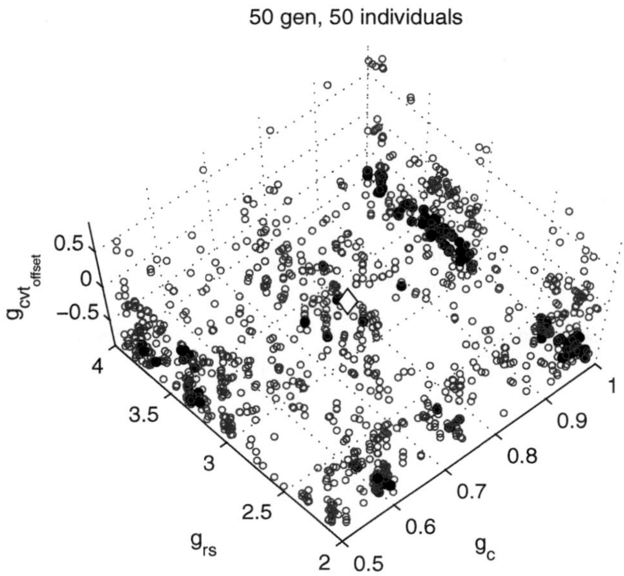

Fig. 5. MOHEV-I: 3D visualisation of the decision space

which lie around $g_c = 0.7$ to 0.9 cluster near to the datum in objective space, presenting very similar solutions.

Figure 6 shows the full 3D objective space for the MOHEV-II optimisation. In this case there has been more evolutionary progress with a widening of the Pareto-front reflecting the wider range of model inputs available. A trade-off has also begun to appear in O_1 vs O_3. This reflects vehicles which may save fuel but are less able to respond to power demands. It is important to note though that the improvement over the datum design is still very limited.

4 Multi-objective Optimisation of the Drive Train and Driving Control Strategy

4.1 The Introduction of Control Strategy Variables

This section presents the *simultaneous* optimisation of drive train and driving control strategy. As mentioned in the introduction and discussed in [2], this optimisation hasn't previously been considered due to the limitations of traditional optimisation techniques. Although optimisation via metaheuristic techniques, such as MOGA, has allowed the incorporation of complex objectives and parameters into the HEV optimisation problem [13,14], to the best of the authors' knowledge this is the first time that *simultaneous* optimisation of drive train

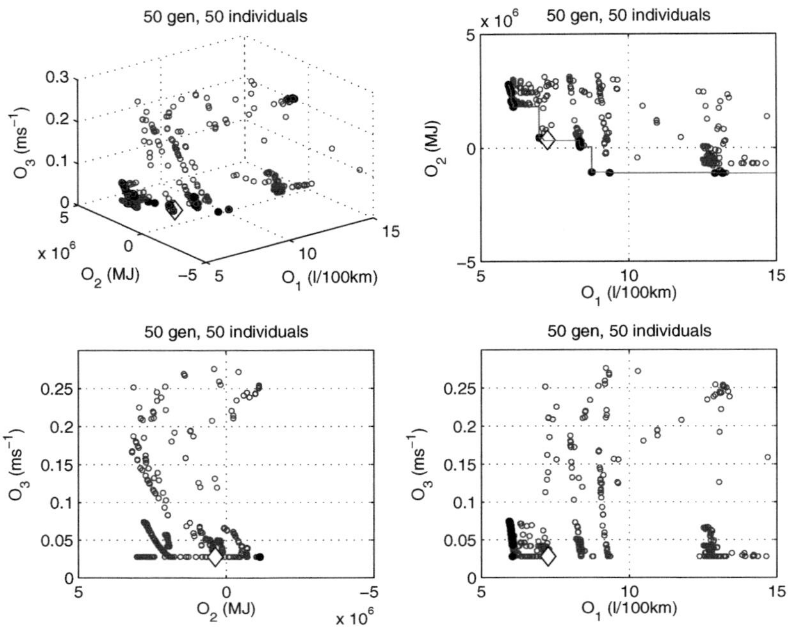

Fig. 6. MOHEV-II: multi-objective representation of the feasible objective space. O_1 *is the fuel consumed;* O_2 *the reduction in battery energy;* O_3 *is a factor of the total difference between demanded and actual velocity traces.*

and driving control strategy has been considered. The multi-objective problem MOHEV-III solved is:

$$\text{minimise} \begin{cases} O_1(g_c, g_{rs}, g_{cvt_{offset}}, v_{cutoff_{ICE}}, bat_{hi}, bat_{lo}, MotEng) \\ O_2(g_c, g_{rs}, g_{cvt_{offset}}, v_{cutoff_{ICE}}, bat_{hi}, bat_{lo}, MotEng) \\ O_3(g_c, g_{rs}, g_{cvt_{offset}}, v_{cutoff_{ICE}}, bat_{hi}, bat_{lo}, MotEng) \end{cases} \quad (7)$$

Notice that the optimisation is over the transmission parameters, driving strategy parameters and the selection of motor-engine pairs. Table 2 lists the range of input variables now used. The variables bat_{hi}, bat_{lo} and $v_{cutoff_{ICE}}$ are described in Sect. 2.2.

4.2 Optimisation Results: Drive Train and Driving Strategy

Figure 7 shows the 3D objective space for the problem, and includes projections of each objective pair. Plotted points are all those visited by the algorithm throughout the 50 generation run.

Note in the 'top view' the linear nature of the Pareto-optimal set. The suggestion here is that a point has been reached where the total *actual* energy used for the cycle is close to constant across all solutions (the line may be defined by

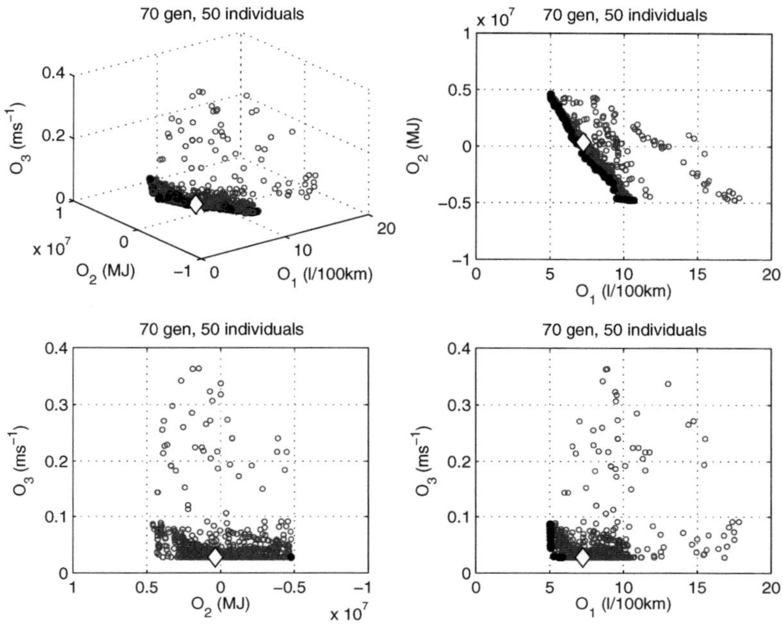

Fig. 7. MOHEV-III: multi-objective representation of the feasible objective space. O_1 *is the fuel consumed;* O_2 *the reduction in battery energy;* O_3 *is a factor of the total difference between demanded and actual velocity traces.*

Table 2. MOHEV-III: input variables for the model (*left*) and the optimiser (*right*)

Variable	Allowed Range
Transmission	
g_{rs}	2.0–4.0
g_c	0.5–1.0
$g_{cvt_{offset}}$	(0.6649–3.611) ± 0.9
ICE/EM selection no.	0–19
Control Strategy	
$v_{cutoff_{ICE}}$	5.0–10.0 m/s
bat_{lo}	50%–70%
bat_{hi}	70%–100%

Parameter	Value
Individuals per generation	50
Maximum number of generations	50
Precision	32bits/variable
Selective pressure	2
Probability of mutation	≈ 0.0031
Crossover rate	0.7

$E_e = E_{const} - kE_f$). These solutions are then a true trade-off with each being equally favoured by the first two objectives.

In the 'left' and 'right' views it is clear that the algorithm has been successful in finding points which significantly dominate the datum, for only two objectives in each case. In this case O_3 and O_1 compete, but O_3 and O_2 are still not in competition. A possible reason for this is that reducing fuel consumption will tend to reduce the overall power capability of the vehicle, whilst a lower rate

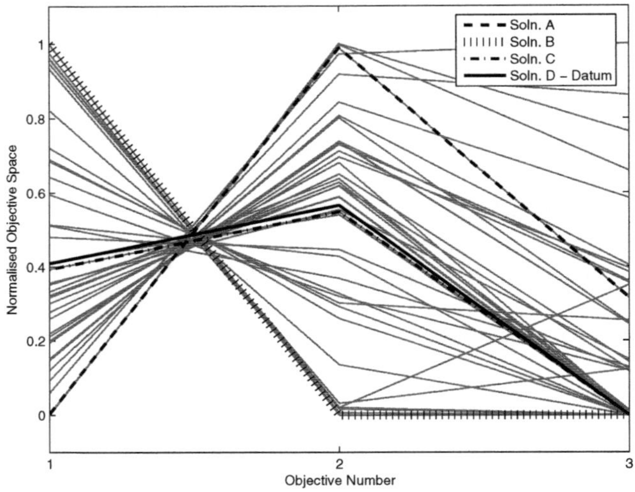

Fig. 8. Parallel coordinates, selected solutions

of battery usage is possible simply by using more fuel and cycle adherence will then not suffer.

The result of a multi-objective optimisation is a Pareto front – a set of competing solutions. However, a real-life design problem will ultimately require a single solution. In order to analyze solutions along the Pareto front, three solutions are chosen using a parallel coordinates visualisation. In Fig. 8 each line joining the three objectives represents a solution. The x-axis represents the objectives and the y-axis shows normalised performance in the interval $[0, 1]$. Crossing lines represent conflicting objectives, with less competitive tradeoffs having more parallel lines.

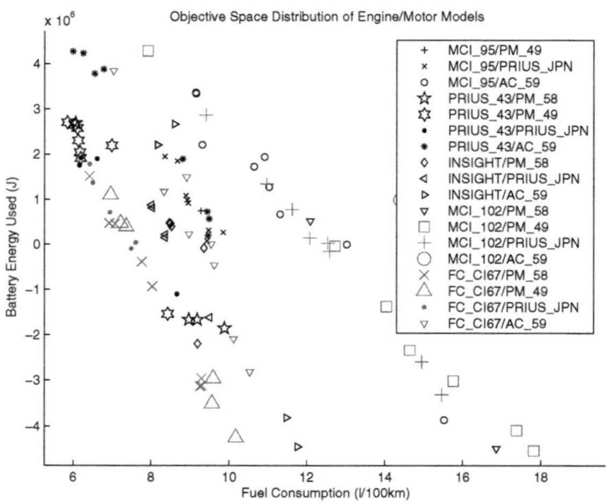

Fig. 9. Analysis of the best engine/motor combinations identified by the algorithm

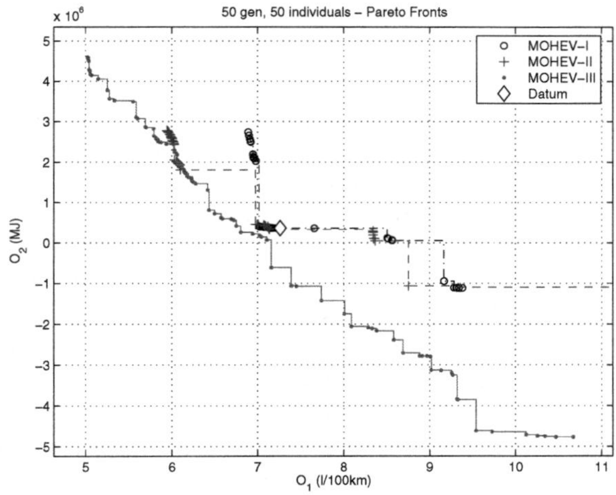

Fig. 10. Comparison of the three optimisation runs

In Fig. 8, solution A has the lowest fuel consumption, the biggest reduction in stored electrical energy, and it doesn't adhere to the cycle. Solution B is the opposite, having the worst fuel consumption, but the best reduction in electrical energy, while adhering to the cycle completely. A good compromise will be solution C. Here a design is found which has sufficient power and torque to follow all parts of the trace, and is able to outperform the datum (D) in both of the energy path objectives. The corresponding decision variable and objective values are given in Table 3.

Figure 9 gives an idea of the spread of results in the 2D objective space with respect to the engine-motor-pairs. The data are selected from different generations of the algorithm, and clearly show some pairs (e.g. □ - MCI_102/PM_49) which did not survive to form part of the final Pareto-front.

In Fig. 10 a comparison of the attainment surfaces for each of the optimisations is overlayed. It is clear that the introduction of the *simultaneous* optimisation has brought about a tangible broadening and improvement of the Pareto-surface from a decision-maker's viewpoint.

Table 3. Decision variable and objective values for selected designs

Solution	Variable							Objectives		
	g_c ratio	g_{rs} ratio	$V_{cutoff_{ICE}}$ m/s	bat_{lo} %	bat_{hi} %	$g_{cvt_{offset}}$ $\Delta ratio$	$MotEng$ $number$	O_1 $l/100Km$	O_2 MJ	O_3 avem/s
A	0.605	3.466	9.299	54.9	96	-0.734	PRIUS_43/PM_49	5.058	4.187	0.0458
B	0.680	3.900	6.092	54.9	96.8	0.445	FC_CI67/PM_58	10.465	-4.739	0.0277
C	0.930	3.933	5.339	68.5	95.6	-0.562	FC_CI67/PM_58	7.176	0.206	0.0277
D-datum	0.736	3.000	6.000	60	80	0	FC_CI67/PM_58	7.260	0.365	0.0278

5 Conclusions

This paper has shown that the optimisation of HEV drive train and driving control strategy is not independent. A comparison of three different multi-objective optimisations has demonstrated that *simultaneous* optimisation leads to a wider exploration of solutions and/or a more diversified and optimal Pareto front. Although it was found that D, the *datum* design, is close to the Pareto-front, significant engineering experience in HEV systems and a cumbersome design procedure was needed to find it. The introduction of multi-objective optimisation means the trade-offs can be investigated in a single run, giving the design engineer an immediate choice from among the family of Pareto-optimal solutions.

Future work should include consideration of other combinations of motor-engine, since in this case the exploration of motor-engines other than diesel engines was limited. There is also a need for more research into how MOEAs handle dynamic environments, in this case driving cycles. This will facilitate the exploration of alternative driving control strategies.

References

1. Voelcker, J.: Top 10 tech cars [fuel efficient cars]. IEEE Spectrum **43** (2006) 34–35
2. Guzzella, L., Sciarretta, A.: Vehicle Propulsion Systems: Introduction to Modeling and Optimization. Berlin Heidelberg, New York: Springer (2005)
3. Branke, J.: Evolutionary Optimization in Dynamic Environments. Norwell, MA: Kluwer Academic Publishers (2001)
4. Wipke, K.B., Cuddy, M.R., Burch, S.D.: ADVISOR 2.1: a user-friendly advanced powertrain simulation using a combined backward/forward approach. IEEE Transactions on Vehicular Technology **48**(6) (1999) 1751–1761
5. Johnson, V., Wipke, K.B., Rausen, D.: HEV control strategy for real-time optimization of fuel economy and emissions. SAE paper no. 2000-01-1543 (2000)
6. Osornio Correa, C.: Caracterización de una transmisión flexible y dimensionamiento del tren transmisión de potencia de un vehículo eléctrico híbrido para máxima eficiencia. PhD thesis, Faculty of Engineering, Universidad Nacional Autonoma de Mexico (2006)
7. Guzzella, L., Onder, C.: Past, present and future of automotive control. In: Control of Uncertain Systems: Modeling, Approximation and Design, LNCIS **329** (2006) 163–182
8. Uthaichana, K., Bengea, S., DeCarlo, R.: Suboptimal supervisory level power flow control of a hybrid electric vehicle. In: Proceedings of the IFAC World Congress on Automatic Control, Prague, Czech Republic (2005)
9. Koot, M., Kessels, J.T.B.A., de Jager, B., Heemels, W.P.M.H., van den Bosch, P.P.J., Steinbuch, M.: Energy management strategies for vehicular electric power systems. IEEE Transactions on Vehicular Technology **54**(3) (2005) 771–782
10. Sciarretta, A., Back, M., Guzzella, L.: Optimal control of parallel hybrid electric vehicles. IEEE Transactions on Control Systems Technology **12**(3) (2004) 352–363
11. Fonseca, C.M., Fleming, P.J.: Genetic algorithms for multiobjective optimisation: Formulation, discussion and generalization. In: Proceedings of the Fifth International Conference on Genetic Algorithms, San Mateo, USA (1993) 416–423

12. Fonseca, C.M., Fleming, P.J.: Multiobjective optimization and multiple constraint handling with evolutionary algorithms I: A unified formulation. IEEE Transactions on Systems, Man and Cybernetics Part A: Systems and Humans **28**(1) (1998)26–37
13. Hu, X., Wang, Z., Liao, L.: Multi-objective optimization of HEV fuel economy and emissions using evolutionary computation. In: Proceedings of the Society of Automotive Engineering World Congress, Electronics Simulation and Optimization (SP-1856), Detroit, USA (2004)
14. Molyneaux, A., Leyland, G., Favrat, D.: Multi-objective optimisation of vehicle drivetrains. In: Proceedings of the 3rd Swiss Transport Research Conference, Ascona, Switzerland (2003)

Multiobjective Evolutionary Neural Networks for Time Series Forecasting

Swee Chiang Chiam[1], Kay Chen Tan, and Abdullah Al Mamun

[1] Department of Electrical and Computer Engineering
National University of Singapore
4 Engineering Drive 3
Singapore 117576
g0500055@nus.edu.sg

Abstract. This paper will investigate the application of multiobjective evolutionary neural networks in time series forecasting. The proposed algorithmic model considers training and validation accuracy as the objectives to be optimized simultaneously, so as to balance the accuracy and generalization of the evolved neural networks. To improve the overall generalization ability for the set of solutions attained by the multiobjective evolutionary optimizer, a simple algorithm to filter possible outliers, which tend to deteriorate the overall performance, is proposed also. Performance comparison with other existing evolutionary neural networks in several time series problems demonstrates the practicality and viability of the proposed time series forecasting model.

Keywords: Time Series Forecasting, Multiobjective Evolutionary Neural Network.

1 Introduction

Time series forecasting (TSF), the forecast of a chronologically ordered variable, is an important tool in the modeling of complex systems, where the primary aim is to predict the system's behavior without the need to understand its underlying mechanism. The importance of TSF has motivated development in the field of operational research, statistics and computer science for more advanced methodologies and techniques to handle more realistic time series with nonlinear and noisy components.

Neural networks (NN), connectionist models that mimic the central nervous systems, are excellent candidates for TSF due to their capabilities like nonlinear generalization, input-output mapping and noise tolerance. There is currently a wide variety of NN available of different architecture, learning paradigm and etc [1], and studies have shown that NN generally perform better than classical econometric models in TSF. Nevertheless, despite the differences in approach between NN and econometric models i.e. nonlinear generalization versus explicit regression modeling, the ultimate objective in TSF remains in getting the most accurate forecast of the time series i.e. minimizing the error between the forecasted and actual values [2].

S. Obayashi et al. (Eds.): EMO 2007, LNCS 4403, pp. 346–360, 2007.

Designing NN of suitable architecture and connection weights for a given time series is itself a difficult combinatorial optimization problem. This has thus motivated the incorporation of evolutionary optimizers due to their efficacy in dealing with large and complex search spaces. However, the usual approach of evolving NN via training data often results in the over-fitting phenomena, where there is over concentration on the peculiarities of the training set at the expense of losing the regularities needed for good generalization [3]. Consequently, this has inspired the application of evolutionary multiobjective optimization that can simultaneously balance the accuracy and generalization of the evolved NN. The optimization of NN via multiobjective evolutionary optimizers is also known as multiobjective evolutionary neural network (MOENN). Specifically, accuracy and generalization refer to the algorithmic performance of the NN with respect to the training and test data respectively.

As such, this paper investigates the application of MOENN in TSF. While there had been several related works, a more direct approach to balance accuracy and generalization is adopted here, by considering the training and validation accuracy as the objective functions. For this purpose, a TSF model comprising of multiobjective evolutionary algorithm hybridized with particle swarm optimization (PSO) and backpropagation (BP) is proposed. The remainder of the paper is organized as such. Preliminary concepts of TSF and MOENN will be introduced over the next two sections. Following that, experiments to examine the validity of the proposed multiobjective approach will be conducted, before the TSF model is formally presented. Lastly, the performance of the proposed model will be evaluated and compared with other existing algorithms.

2 Time Series Forecasting

Time series is a sequence of observation values of a physical or financial variable ordered at equally spaced time intervals, Δt and is represented as a set of discrete values $x_1, x_2, x_3, x_4, ...etc$. Basically, a TSF model assumes that past patterns will occur in the future and thus, predictions can be done by identifying and generalizing patterns among the past data. As such, the usual training methodology is to split the given time series into two parts, namely the training data, where the learning is performed and the test set, where the performance of the resulting TSF model is measured. The underlying principle is that the TSF model will attune to the training data via the training algorithm and its performance on the test data will approximate its actual performance in real life implementation. The performance in the two separate data is known as training and test accuracy respectively.

Due to the impossibility of deriving a perfect model, there will always be some deviation between the actual and forecasted values and this error, e_t is defined as such,

$$e_t = x_t - \hat{x}_t \tag{1}$$

where $\hat{x}_t$ is the forecasted value for time, t.

There are many error measures available to quantify the time series prediction performance [4], but the more commonly adopted forecasting accuracy measure are namely the Sum Squared Error (SSE) and the Root Mean Squared Error (RMSE)

$$SSE = \sum_{t=1}^{L} e_t^{\,2} \tag{2}$$

$$RMSE = \frac{\sqrt{SSE}}{L} = \frac{\sqrt{\sum_{t=1}^{L} e_t^{\,2}}}{L} \tag{3}$$

where L represents the number of forecasts.

Theoretically, time series consists of four main (or a combination) characteristics, namely stationarity, linearity, trend and seasonality. Traditional TSF methodologies will differentiate and segregate these various components and identify the extent in which they are present in the given time series. Subsequently, after extracting these partial data, regression model like autoregressive moving average (ARMA), autoregressive integrated moving average (ARIMA) will be applied to obtain the TSF model. However, real-life time series are corrupted by noise interference that results in statistical fluctuations around the original values. Although exponential smoothing (ES), a conventional yet popular regression model, has been quite successful for such cases, its capability is only restricted to seasonal time series [5]. In lieu of these limitations, MOENN represents a promising alternative for TSF, as its generalization ability need not take into account the different form of nonlinearity present in the time series.

3 Multiobjective Evolutionary Neural Networks

NN refers to interconnected groups of artificial neurons that use a mathematical model for information processing based on a connectionist approach to computation. Practically, NN are non-linear data modeling tools, capable of finding complex relationships between inputs and outputs and/or patterns within data. Amongst the different NN available [1], the most common model and training methodology is the feedforward NN and the backpropagation (BP) training algorithm. Fig. 1 illustrates a feedforward NN with 1 hidden layer. There are a total of n0 inputs, n1 hidden nodes and 1 output layer. This is also referred to as an n0-n1-1 feedforward NN and it will be used as the basic NN architecture in this paper.

The fundamental design issue in NN is to determine the appropriate architecture and weights for the given time series. In the absence of proper guidelines and framework, the architecture is often constructed through the process of trial and error or via some heuristics [6], [7]. However, the former is time consuming and tedious while the latter fail when the problem specifications changes. As for the training of weights, conventional methodologies like BP are inherently gradient-based approach, which are generally susceptible to local optimum, especially in noisy, multimodal and large environment. Miller *et al.* [8] highlighted that evolutionary optimizers will perform better in such landscape, as they are less prone to be trapped in local optima due to their non-dependence on gradient information. This has thus motivated the development of evolutionary NN, the optimization of NN via evolutionary optimizers. For brevity, the interested reader is referred to [9] for a detailed discussions on this topic.

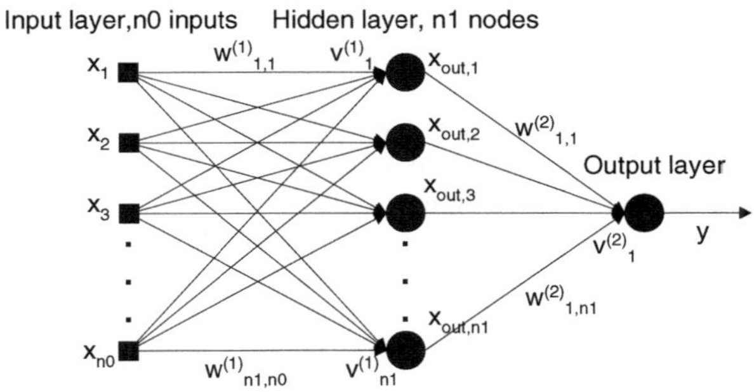

Fig. 1. Feed forward NN

Conventional evolutionary NN model adopted a single objective approach, where the optimization of NN is solely based on training accuracy. Unfortunately, this approach is prone to overfitting, where the evolved NN failed to perform equally well in the test data. As such, multiobjective approaches, which consider additional objectives to control the extent of generalization, have been proposed, so as to balance the accuracy and generalization of the NN evolved. The various approaches can be broadly categorized into three different types:

I. Tradeoff between more than one type of error measures
II. Tradeoff between network complexity and error measure
III. Tradeoff between network diversity and error measure

Table 1 provides some examples of the different approaches.

Table 1. Different objective functions and their classification

Type	Objective Functions
I	Average Euclidean Error and Maximum Euclidean Error [2]
I	Root Mean Squared Error, Correlation Coefficient, Maximum Absolute Percentage Error and Mean Absolute Percentage Error [10]
I	Error for Training Data with and without noise [11]
I	Error between two subsets of Training Data [11]
II	Training Error and Number of Hidden Units [12], [13]
II	Mean Squared Error and Norm of Weight Vector [14]
II	Error rate and Number of Features [15]
II	Mean Squared Error and Weight Decay [16]
III	Mean Squared Error and Correlation Penalty Function [17]

Even though the various approaches are inherently different, their underlying motivation is to avoid overfitting in MOENN and maintain a certain degree of generalization. For instance, type I makes use of additional error measures that can quantify the extent of generalization for NN. Likewise, based on the observation that the architecture complexity of a NN is related to its generalization ability [12], where specifically, more complex NN are more prone to over fitting, type II maintains generalization by minimizing the complexity of the evolved NN. Lastly, unlike in single objective optimization, MOENN will obtain a set of solutions at the end of the algorithmic run. For better algorithmic performance, it is important to preserve the diversity in the ensemble of NN [18].

The fact that MOENN will obtain a set of solutions naturally led to the question of which NN should be selected eventually for the forecasting of the time series. Current approach simply considers the simple average of all their forecasted values or selects the NN with the best training accuracy. Instead of directly examining the selection issue, this paper will investigate how the filtering of outliers can improve the overall algorithmic performance of MOENN.

Recent studies advocated the use of validation data in the optimization of MOENN. Basically, the training data will be further subdivided into two separate set, namely the training set, which is used in the actual training, and the validation set, which serves as a pseudo test set to evaluate the quality of NN during training [19]. Such an evaluation is also termed as cross validation and is frequently used in single objective approaches to avoid the over-fitting phenomenon. Due to the fact that only a minority of the Pareto optimal solutions with respect to the training set is Pareto optimal with respect to the test data [2], the use of the validation data has been used instead to select the relevant NN from the eventual pool of solutions [15] or as an archiving criteria [2], [20]. However, in the literature reviewed, none has attempted to consider training and validation accuracy simultaneously as the objective functions for MOENN, which seems to a more direct approach to balance accuracy and generalization for the evolved NN.

4 Preliminary Investigation

The feasibility of using training and validation accuracy as objective functions in MOENN will be examined in this section. The time series considered in this experiment is the SUNSPOT data, which consists of the annual Wolf's sunspot numbers collected from 1700 until 1989 [21]. The 1st 260 data will be used as training data. A training ratio of 90% was used so as to synchronize with [22], [23], as their result will be used subsequently in the comparative study. The training data is further divided into a training set, consisting of the 1st 208 (80%) data and the remaining will be the validation set.

For this purpose, a simple evolutionary NN (SENN) is considered. The architecture used is a feed forward NN with 1 hidden layer as shown earlier in Fig. 1. The number of inputs ($n0$) and hidden units ($n1$) are limited to 20 and 10 respectively. Binary representation is adopted to specify the connections within this architecture, where an $n0 \times n1$ and $n1$ binary matrix, C is used to describe the first layer and second layer respectively. $Cij = 1$ indicates a connection between node i and j while $Cij = 0$ represents no connection. This coding scheme is suitable for precise and local fine-tuning search of a compact NN architecture, as a single connection can be added or removed from the NN easily. Also, the inputs to the NN will be evolved and adapted as well. This, hence, avoids the problem of choosing the appropriate size and type of sliding window, which is itself a fundamental design issue in TSF. Essentially, a larger sliding window will increase the system complexity and affect the learning capabilities of the model, while a smaller window may lead to insufficient information for the TSF model [22]. The binary representation will be evolved using uniform crossover and bit-wise mutation. As for the weights of the NN, they are represented by real number matrixes, with values of range [-1, 1]. The weight matrixes will be evolved solely by Gaussian mutation.

To investigate the effects of adopting training and validation accuracy as the objective functions, three single objective approaches are considered. Their description and notation are summarized in table 2. SENN_TRAIN and SENN_VALID will minimize the training and validation accuracy respectively, while the SENN_SUM will minimize their sum. Lastly, SENN_MULTI represents the multiobjective approach of considering training and validation accuracy simultaneously.

Table 2. Different objective functions for SENN and their notation

Objective function	Notation
Training Accuracy	SENN_TRAIN
Validation Accuracy	SENN_VALID
Sum of Training and Validation Accuracy	SENN_SUM
Tradeoff between Training and Validation Accuracy	SENN_MULTI

Table 3. Algorithmic configuration for SENN

Selection	Binary Tournament Selection
Crossover	Uniform Crossover with Probability 0.9
Mutation	Bitwise Mutation with Probability $1/\, n0 \times n1$
	Gaussian Mutation of Mean 0 & Standard Deviation 0.1
Population	100

The evolutionary platform adopted is a generic elitist Pareto-based evolutionary algorithm. This algorithm maintains a fixed-size population for evolution and an archive to store the non-dominated solutions discovered during the evolution. Elitism is implemented by selecting individuals to a mating pool through a binary tournament selection of the combined archive and evolving population. For the single objective approaches, the selection criterion is very straightforward, as there is only one objective function. However, for SENN_MULTI, it will be based on Pareto dominance with respect to the training and validation accuracy. In the event of a tie, the niche count will be employed. The mechanism of niche sharing is used in the tournament selection as well as diversity maintenance in the archive. The algorithmic settings for SENN are summarized in Table 3.

For all the algorithms, their accuracy are measured by RMSE (3) based on the different data. Fig. 2 shows the various RMSE trace for the different algorithms in a single run. For the single objective approaches, if the RMSE is used as the objective function, its corresponding trace will decrease monotonically. Unfortunately, this does not necessarily correspond to a decrease in other RMSE values, in particularly the RMSE for the test data. In fact, improvements often occurred at the expense of

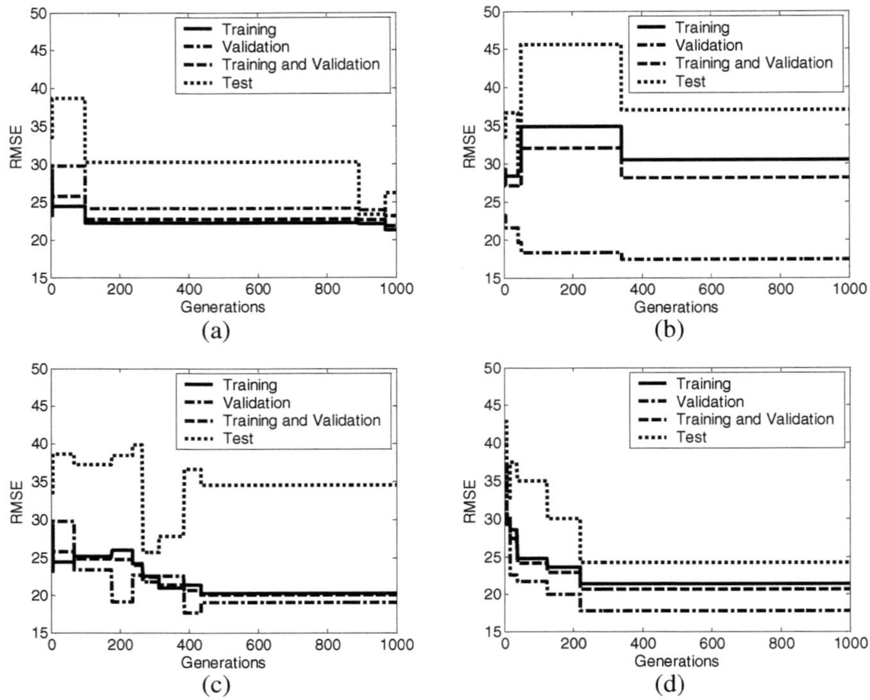

Fig. 2. RMSE trace for (a) SENN_TRAIN (b) SENN_VALID, (c) SENN_SUM and (d) SENN_MULTI

other RMSE. For instance in Fig. 2(a), there is a sudden dip in training RMSE at the late generations around 1000, which coincides with a corresponding increase in test RMSE, signifying the overfitting phenomena. Conversely when the multiobjective approach is adopted, optimizing the training and validation accuracy simultaneously results in a progressive decrease in all RMSE during the evolutionary progress.

Due to the stochastic nature of evolutionary optimization, performance comparison should be based on multiple experimental runs. As such, 20 runs were conducted and the average result is summarized in Table 4. Contrary to the single objective approaches, SENN_MULTI will obtain multiple solutions instead of one. Hence, simple averaging is performed and the RMSE reported is the average of all the solutions in the archive. The lowest value attained for each RMSE is bold.

Similar to the observations earlier, the RMSE value will be the lowest if it is used as the objective function. However, this does not necessarily mean that higher test accuracy could be obtained. The lowest test RMSE is attained by SENN_MULTI, where the training and validation accuracy is balanced throughout the evolutionary progress. However, the difference is not at all significant. This observation is not surprising, as the mean result is based on the simple averaging of all the solutions found in the archive. Very often, the solution set will contain outliers that will significantly deteriorate the average test accuracy.

Table 4. RMSE attained by the various algorithms

	Training RMSE	Validation RMSE	Sum RMSE	Test RMSE
SENN_TRAIN	**20.50**	22.72	21.09	26.96
SENN_VALID	28.93	**16.71**	26.83	37.33
SENN_SUM	20.60	20.98	**20.77**	26.93
SENN_MULTI	21.00	16.97	20.29	**26.33**

To investigate this further, Fig. 3 depicts a Pareto front obtained by SENN_MULTI in one of the experimental run. The Pareto front clearly illustrates the trade off between training and validation accuracy. Furthermore, each solution is labeled with its corresponding test RMSE and the NN with the best test accuracy is highlighted. Clearly, the best NN is situated at the region where training and validation accuracy are balanced. Also, the outliers are observed to have lower test accuracy and their inclusion will generally deteriorate the average algorithmic performance.

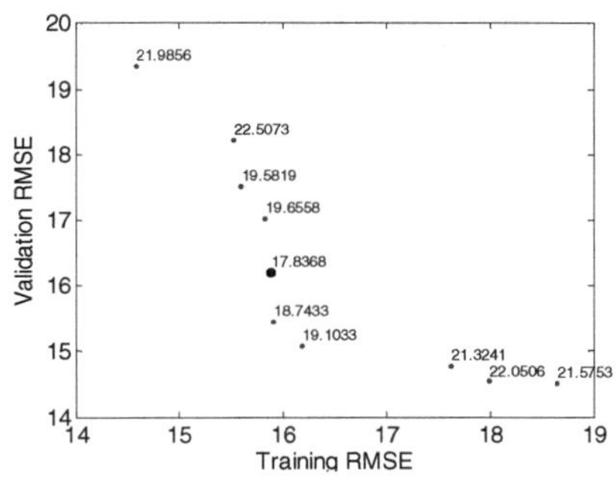

Fig. 3. Pareto front obtained by SENN_MULTI

However, identification of the outliers is not as straightforward as choosing the solutions at the extreme ends of the Pareto front found. Fig. 4 illustrates two other Pareto fronts obtained by SENN_MULTI. In Fig. 4(a), even though filtering of outliers will eliminate solutions with low test accuracy, it will eliminate better solutions as well. As for Fig. 4(b), the best solution is the outlier itself. Ultimately, the position of the outliers will depend on the location of the evolved Pareto front on the optimal Pareto front. Unfortunately, this information is unattainable in real world optimization, as the optimal Pareto front is never known. Furthermore, the identification of the outliers should be conducted without any prior knowledge on the test accuracy of any of the NN found.

Nevertheless, it might be interesting to analyze the distribution of better solution in the objective space of training and validation RMSE. Fig. 5(a) plots all the points found by SENN_MULTI during the 20 runs, while Fig. 5(b) and Fig. 5(c) highlights the solutions of test RMSE lesser then 26.3 and 22 respectively. From the plots, the better solutions seem to be clustered in a particular region of the Pareto front where the training and validation RMSE are approximately equal.

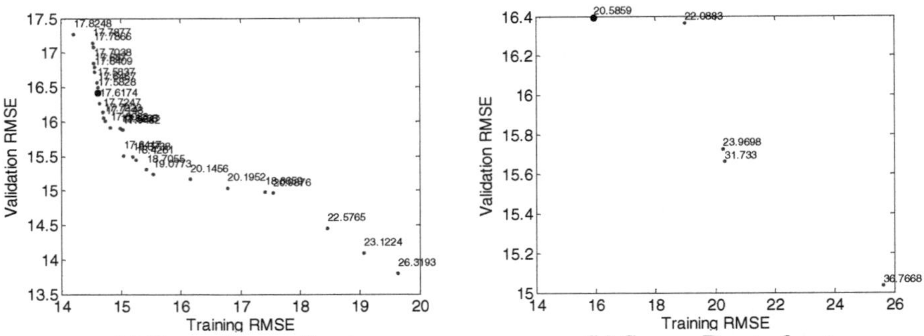

(a) Dense Pareto front (b) Sparse Pareto front

Fig. 4. Other instances of Pareto front obtained by SENN_MULTI

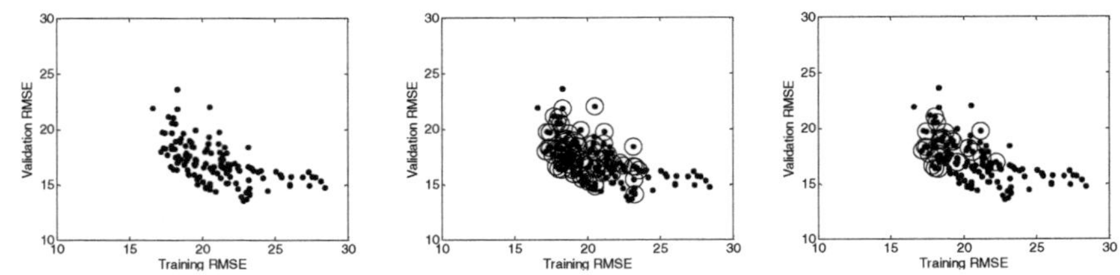

(a) (b) (c)

Fig. 5. Objective space with (a) all solutions, (b) solutions with test RMSE of less than 26.3 highlighted and (c) solutions with test RMSE of less than 22 highlighted

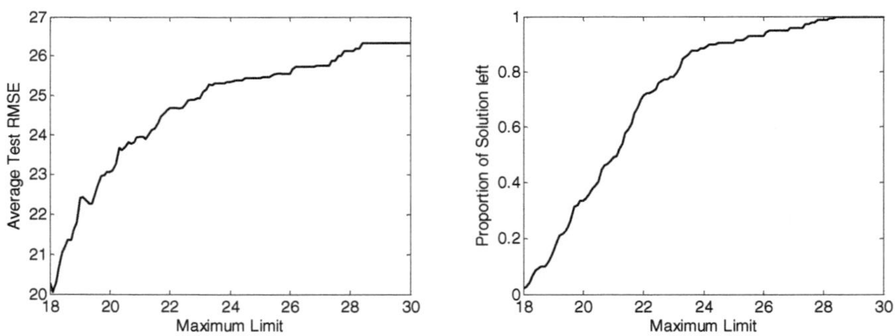

(a) (b)

Fig. 6. Effects of Maximum limit on (a) average test RMSE and (b) proportion of solutions left

Based on this observation, a simple rule to filter the outliers is proposed. Basically, a value which corresponds to the maximum limit allowable for both training and validation RMSE will be defined and solutions that fail to meet this simple criterion will be removed. The effects of varying this limit value on the resultant average test RMSE and the proportion of solution are illustrated in Fig. 6. A smaller limit value will correspond to a tighter criterion, which will result in a larger proportion of solutions being filtered. Consequently, the average test RMSE attained will be smaller. This seems to be a viable approach for the filtering of outliers, where the limit value could be defined such that a certain proportion of the solutions are removed. This will naturally lead to an improvement in the average test accuracy via simple averaging of the remaining solutions. From the gradient variation in Fig. 6(a), the average test RMSE is more sensitive to changes in the limits values for smaller limit values. It might be instructive to investigate whether this simple filtering rule can be generalized to other time series.

5 Algorithm Description

Results from the preliminary investigation suggest that the multiobjective approach of training and validation accuracy is indeed viable. For further validation purposes, a MOENN model for TSF is proposed and its performance will be evaluated with other existing TSF model on several benchmark time series in the next section. The general algorithmic flow for the proposed MOEN is illustrated in Fig. 7.

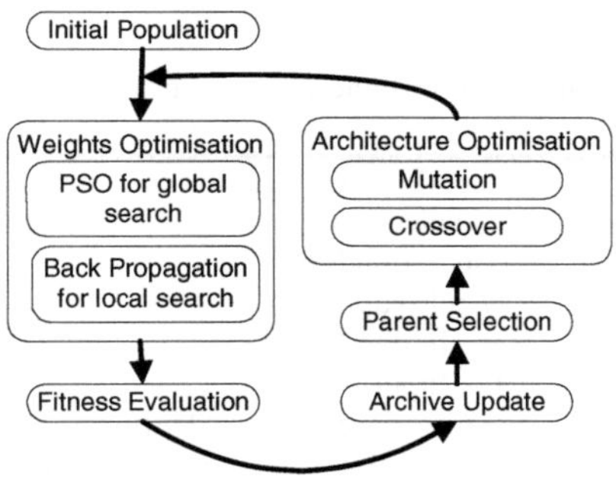

Fig. 7. Algorithmic flow of MOENN

The algorithm will independently optimize the architecture and the connection weight and their representation is identical to SENN described earlier. Also, the architecture will be optimized in similar fashion as SENN. The main difference is that the various weights will be optimized by a hybrid PSO-BP training approach instead. PSO can handle global search better in a vast, complex, multimodal and non-differentiable surface and search for the globally optimal solution, while BP can exploit local gradient information and find a local optimum in the vicinity of the initial solution. Ideally, the synergy of these two search operators will result in a more effective and efficient search of optimal solutions.

As illustrated in Fig. 7, the algorithm will begin with the random initialization of a population of individuals of specified architecture. For each individual, a sub-population of connection weights will be generated at random and optimized by the hybrid PSO-BP operator. The best weight vector for each individual in terms of training accuracy will continue with the evolutionary progress. Subsequently, these individuals will be assessed based on validation accuracy and the solutions of superior Pareto optimality in terms of training and validation accuracy will be stored in the archive. From the combined pool of archive and evolving population, parent solutions will be selected to breed new offspring of different architecture via the variation operation. The weights of these newly created offspring will then be optimized by the hybrid PSO-BP operator. The algorithm will repeat until the stopping criterion is met.

6 Experimental Results

The performance of the proposed MOENN model will be compared with ENN [22] and the Meta-GA [23], as they had been applied to different types of time series in a comprehensive study on the application of evolutionary NN on TSF [22], [23]. Furthermore, the study also reports the performance of other TSF models, like ARIMA and ES. This is contrary to most MOENN related works, which focus mainly on classification problems. The algorithmic parameter configuration for MOENN is illustrated in Table 5. It should be highlighted that the total number of function evaluation for MOENN is only 4 million, as compared to 50 million for ENN and Meta-GA, so as to reflect the efficacy and efficiency of the proposed MOENN.

Table 5. Algorithmic configuration for MOENN

General	Terminal Generation	100
	Selection	Binary tournament selection
	Population Size	10
	Archive Size	10
	Max Hidden Nodes	4
	Max Input Nodes	13
Architecture Optimization	Crossover	Uniform crossover with probability 0.9
	Mutation	Bitwise mutation with probability $1/n0\times n1$
Weight Optimization (BP)	Learning Epoch	10
	Learning Rate	0.5
Weight Optimization (PSO)	Population size	20
	Learning Time	20
	Inertia	0.7
	Individual weights	1.49
	Sociality weight	1.49

MOENN is applied to the SUNSPOT and MAXTEMP time series, where the latter is a meteorological time series of the monthly maximum temperature (in Celsius degrees) measured in Melbourne, Australia, from January 1971 to December 1990 [24]. Similar training to test ratio was adopted to ensure a fair comparison [22], [23].

Figure 11 shows the deviation between the predicted and actual time series attained by MOENN in a randomly chosen run for both the time series. The forecasted time series are similar to the actual time series, except for the discrepancies occurring at the peak values.

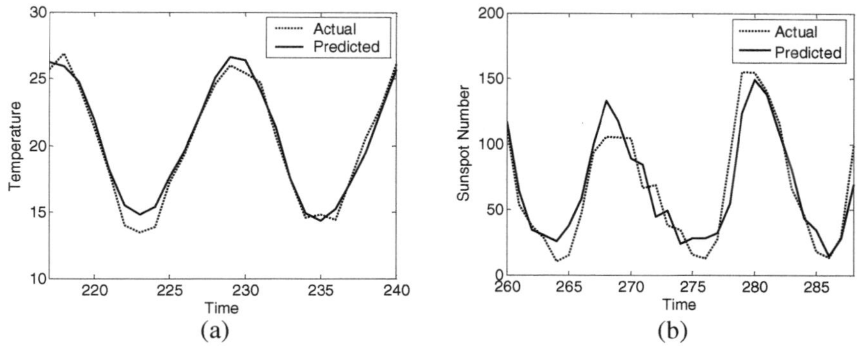

Fig. 8. Predicted time series for (a) MAXTEMP and (b) SUNSPOT

The average test RSME obtained by the various algorithms are summarized in table 6. ES [22] obtained the lowest RMSE for MAXTEMP. This is not surprising, as ES was developed specifically for this particular type of time series, i.e. seasonal. Apart from ES, MOENN obtained a lower RMSE as compared to the rest of the algorithms at a lower computational cost. Of course, a statistical test like ANOVA will be essential to test the significance of the mean differences. This unfortunately could not be performed here, as the authors only revealed the mean RMSE in their comparative studies.

Table 6. Average test RMSE for the various algortihms

Algorithm description	Mean RMSE	
	MAXTEMP	SUNSPOT
ES [22]	0.72	28.4
ARIMA [22], [23]	1.07	21.4
ENN [22]	0.93	17.4
Meta-GA [23]	0.87	17.6
MOENN	0.82	18.56

For the SUNSPOT time series, MOENN has a higher average test RMSE as compared to ENN and Meta-GA. However, observing the evolutionary trace of the test RMSE in Fig. 8, the value is actually still decreasing. The exact statistical information is shown in table 7. Clearly, as the generation increases, the mean will decrease as well. This is accompanied by a decrease in the standard deviation also, reflecting convergence behavior. At generation 250, MOENN is comparable to the two other evolutionary based methods, despite having 5 times lesser function evaluations. Lastly, the improvement of test RMSE from 26.33 for SENN_MULTI to 18.56 for MOENN demonstrated the efficacy of the hybrid PSO-BP operator for weight optimization.

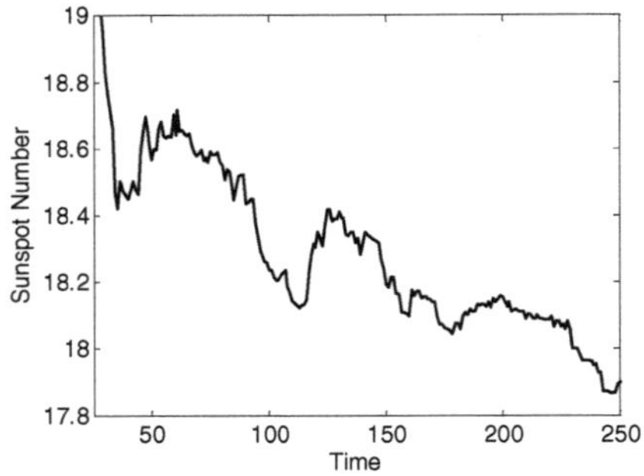

Fig. 9. Closer examination of the mean RMSE for the SUNSPOT series

Table 7. Statistics of RMSE attained by MOENN at different generations

Generations	50	100	150	200	250
Mean	18.56	18.23	18.19	18.15	17.90
Minimum	16.80	17.18	16.62	16.88	16.46
Maximum	20.07	19.65	20.18	20.38	19.04
Standard Deviation	0.82	0.70	0.87	0.75	0.62

Improving algorithmic performance via higher computational cost is impractical at times. Alternatively, the algorithmic performance could be improved by the removal of outliers as discussed earlier in the preliminary investigations. Fig. 10 plots all the solutions found from the experimental runs for MOENN and highlights those having test accuracy below the mean value of 18.56. Again, similar observations can be made where better solutions are situated in the region where validation and training accuracy are equal.

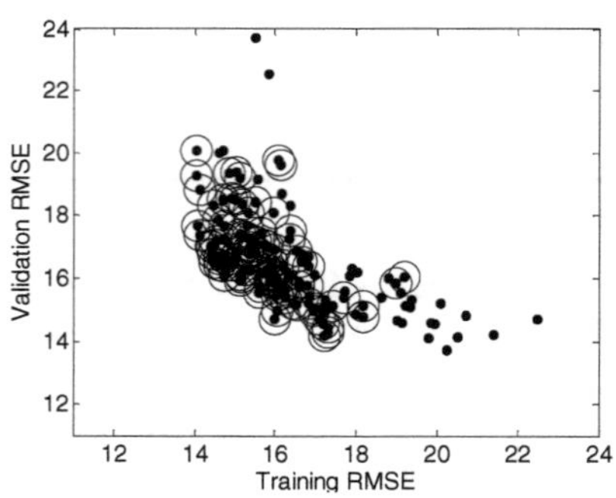

Fig. 10. Objective space with solutions with test RMSE of less than 25 highlighted

Again, the filtering rule is applied to the experimental results for SUNSPOT. In Fig. 11, if the limit value is set to 17.1, which will filter away 50% of the solutions, this will correspond to a test RMSE of 17.19, lower then the value obtained by increasing computational time. Of course, the average test RMSE can be further decreased by using a smaller limit, but the robustness of the solution will be affected.

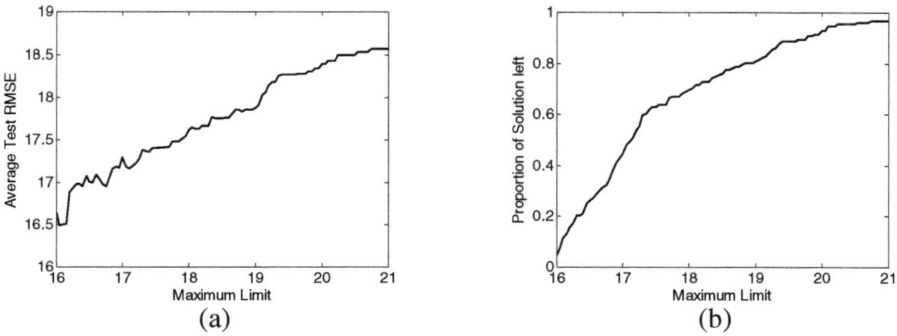

Fig. 11. Effects of Maximum limit on (a) average test RMSE and (b) proportion of solutions left

7 Conclusion

The preliminary investigation has demonstrated the viability of using a multiobjective approach comprising of training and validation accuracy to balance the accuracy and generalization of the evolved NN. Also, a simple algorithm to filter possible outliers from the Pareto optimal solutions found is proposed. Using a MOENN model hybridized with PSO and BP, this approach is compared with other existing TSF techniques and its performance is comparable, if not better than the other existing evolutionary NN models.

However, there are still plenty of avenues for future work. One important area is to investigate further the feasibility of the rule that filter potential outliers, as this simple technique can enhance algorithmic performance significantly, without incurring any additional computational burden. Also, MOENN has a lot of parameters due to the incorporation of several different operators. Parameter sensitivity analysis should be conducted to examine the robustness of MOENN with respect to its parameters. Alternatively, adaptation techniques could be considered to lessen the burden of parameter tuning for MOENN. Lastly, a more comprehensive test on a wider range of TSF problems is needed to further affirm the practicality of the proposed TSF model.

References

1. Haykin, S.: Neural Networks: A Comprehensive Foundation. Prentice Hall (1999)
2. Fieldsend, J., Singh, S.: Pareto Evolutionary Neural Networks. IEEE Transactions on Neural Networks, Vol. 16, No. 2, (2005) 338-354
3. Geman, S., Bienenstock, E., Doursat, R.: Neural Networks and the Bias/Variance Dilemma. Neural Computation, Vol. 4 (1992) 1-58
4. Armstrong, J., Collopy, F.: Error measures for generalizing about forecasting methods-Empirical Comparisons. International Journal of Forecasting, Vol. 8, No. 1 (1992) 69-80

5. Makridakis, S.G., Wheelwright, S.C.: Forecasting methods and applications. New York, Wiley (1998)
6. Ash, T.: Dynamic node creation in backpropagation networks. Connection Science, Vol. 1, No. 4 (1989) 365–375
7. Lehtokangas, M.: Modeling with constructive backpropagation. Neural Networks, Vol. 12 (1999) 707–716
8. Miller, G., Todd, P., Hegde, S.: Designing neural networks using genetic algorithms. Third International Conference on Genetic Algorithms (2000) 379-384
9. Yao, X.: Evolving artificial neural networks. Proceedings of the IEEE, Vol. 87, No. 9 (1999) 1423-1447
10. Abraham, A., Grosan, C., Han, S.Y., Gelbukh, A.: Evolutionary Multiobjective Optimization Approach for Evolving Assemble of Intelligent Paradigms for Stock Market Modeling. Advances in Artificial Intelligence, Lecture Notes in Computer Science, Springer, Vol. 3789 (2005) 673-681
11. Abbass, H.A.: Pareto Neuro-Evolution: Constructing Ensemble of Neural Networks Using Multi-objective Optimization. IEEE Congress on Evolutionary Computation, Vol. 3 (2003) 2074-2080
12. Abbass, H.A.: A Memetic Pareto Evolutionary Approach to Artificial Neural Networks. Australian Joint Conference on Artificial Intelligence (2001) 1-12
13. Abbass, H.A.: An evolutionary artificial neural networks approach for breast cancer diagnosis. Artificial Intelligence in Medicine, Vol. 25, No. 3 (2002) 265-281
14. Teixeira, R.D.A., Braga, A.D.P., Takahashi, R.H.C., Saldanha, R.R.: A Multi-Objective Optimization Approach for Training Artificial Neural Networks. Brazilian Symposium on Neural Networks (2000) 168-172
15. Oliveira, L.S., Morita, M., Sabourin, R.: Multi-Objective Genetic Algorithms to Create Ensemble of Classifiers. 3rd International Conference on Evolutionary Multi-Criterion Optimization (2005) 592-606
16. Jin, Y., Okabe, T., Sendhoff, B.: Neural network regularization and ensembling using multi-objective evolutionary algorithms. Congress on Evolutionary Computation (2004) 1-8
17. Chandra, A., Yao, X.: DIVACE: Diverse and Accurate Ensemble Learning Algorithm. Conference on Intelligent Data Engineering and Automated Learning, Lecture Notes in Computer Science, Springer, Vol. 3177 (2004) 619-625
18. Brown, G., Wyatt, J., Harris, R., Yao, X.: Diversity creation methods: a survey and categorization. Journal of Information Fusion, Vol. 6, No. 1 (2005) 5-20
19. Prechelt, L.: Proben1-A Set of Neural Network Benchmarking Problems and Benchmarking Rules. Technical Report Nr. 21/ 94, Faculty of Computer Science, University of Karlsruhe, Germany (1994)
20. Radtke P.V.W., Wong T., Sabourin R.: An evaluation of Over-Fit Control Strategies for Multi-Objective Evolutionary Optimization. IEEE Congress on Evolutionary Computation (2006) 6359-6366.
21. Tong, H.: Threshold models in non-linear time series analysis. Springer lecture notes in Statistics, Vol. 21 (1983)
22. Cortez, P., Rocha, M., Neves, J.: Evolving Time Series Forecasting Neural Network Models. International Symposium on Adaptive Systems: Evolutionary Computation and Probabilistic Graphical Models (2001)
23. Cortez, P., Rocha, M., Neves, J.: A Meta-Genetic Algorithm for Time Series Forecasting. Workshop on Artificial Intelligence Techniques for Financial Time Series Analysis, 10th Portuguese Conference on Artificial Intelligence (2001) 21-31
24. Hyndman, R.J.: Time Series Data Library, http://www.robhyndman.info/TSDL/. Accessed on 14102006.

Heatmap Visualization of Population Based Multi Objective Algorithms

Andy Pryke[1], Sanaz Mostaghim[2], and Alireza Nazemi[3]

[1] Cercia, School of Computer Science / [3]Department of Civil Engineering
University of Birmingham
Birmingham, UK
{A.N.Pryke,A.Nazemi}@cs.bham.ac.uk
[2] AIFB Institute, University of Karlsruhe,
Karlsruhe, Germany
mostaghim@aifb.uni-karlsruhe.de

Abstract. Understanding the results of a multi objective optimization process can be hard. Various visualization methods have been proposed previously, but the only consistently popular one is the 2D or 3D objective scatterplot, which cannot be extended to handle more than 3 objectives. Additionally, the visualization of high dimensional parameter spaces has traditionally been neglected. We propose a new method, based on heatmaps, for the simultaneous visualization of objective and parameter spaces. We demonstrate its application on a simple 3D test function and also apply heatmaps to the analysis of real-world optimization problems. Finally we use the technique to compare the performance of two different multi-objective algorithms.

Keywords: Visualization; Multi-objective optimization; Multi-objective algorithms; Evolutionary algorithms; Real-world applications.

1 Introduction

Visualization of the optimal solutions plays a very important role in multi-objective optimization (MO). In MO with conflicting objectives there is no single optimum, and search methods return a set of solutions from which one must be selected. In order to select the solution, a decision maker usually needs to visualize the discovered solutions in the objective space. This can be done using scatterplots of the objective space if there are only 2 or 3 objectives.

Visualization is also used to show the quality of the solutions. A good set of optimal solutions should contain well-distributed converged solutions along the Pareto front. Almost every new algorithm in MO is tested on several 2- and 3-objective problems, and beside numerical measurements, the obtained solutions are illustrated in objective space plots. This illustration is very valuable for many applications in science and industry as domain experts get information about the whole set of optimal solutions. Also, understanding the algorithm behavior is easier with this view.

S. Obayashi et al. (Eds.): EMO 2007, LNCS 4403, pp. 361–375, 2007.

Visualizing the objective space directly is possible for 2 and 3 objective spaces whereas for higher number of objectives, the solutions are only be evaluated by metrics. Metrics and numerical measures hide too much information and only consider the objective space. On the other hand, in almost all of the proposed methods and applications in MO, there is a high (>3) number of parameters which are not being visualized. However, showing the parameter values and visualizing them has a great impact on the decision making process.

Here, we investigate several visualization methods in order to visualize parameters and objective values of a set of optimal solutions. We note previous application of objective plots, Self Organizing Maps (SOM) [2], and Distance and Distribution Charts[1].

We also introduce a new application of pre-existing visualization methods to population based algorithms: Heatmap visualization, which has previously been used mainly for the visualization of biological data. Heatmaps typically make full use of color, but this isn't available in this printed paper. A pre-print with full color figures is available as a technical report[16].

When evaluating visualization methods, it is wise to consider what features of the system we may wish to reveal. The possibilities for population based MOEAs include: the diversity and convergence of the solutions in both objective and parameter space; the relationship between parameter and objective values; A comparison between different runs or algorithms; Identification of clusters of solutions in either the parameter or objective spaces; Dynamically illustrating the progress of an algorithm towards its optimization goals.

This introductory section gives some background on Multi-objective Optimization and previous visualization methods which have been applied to it.

To familiarize the reader with our methods, we make use of a 3-objective test problem defined in Section 2. We then apply heatmaps to this problem in sections 3. Sections 4 applies heatmaps on a real world application in mineralogy. A second application in hydrological modeling shows the use of our methods for the comparison of the behavior of algorithms (Section 5).

1.1 Multi-objective Optimization Problems (MOPs)

A Multi-objective Optimization Problem (MOP) contains several objective functions, which are to be optimized at the same time:

$$\begin{aligned}&\text{minimize}\, \vec{f}(\vec{x}) = (f_1(\vec{x}), \cdots, f_m(\vec{x}))\\&\text{subject to}\ \ \vec{e}(\vec{x}) < 0\\&\vec{x} \in S\end{aligned}$$

involving $m \geq 2$ (normally conflicting) objective functions $f_i : \Re^n \rightarrow \Re^m$ that we want to minimize simultaneously. The parameters $\vec{x} = (x_1, \cdots, x_n)$ belong to the feasible region *S*. The feasible region is formed by constraint functions $\vec{e}(\vec{x})$. We call the image of the feasible region feasible objective region. Its elements are called objective vectors and they consist of objective values.

Many MOP have conflicting objectives, i.e. it is not possible to find a single solution that would be *optimal* for all the objectives simultaneously. In this case, we aim to find some optimal solutions where none of their objective values can be improved without deterioration of at least one of the other objective values.

These solutions are called *Pareto-optimal* solutions. A solution $\vec{x}_1$ is called *Pareto-optimal,* if there is no other solution that dominates[1] it. The Pareto-optimal set in the objective space is called *Pareto-optimal front*. The *Pareto-rank* of a solution is a count of the number of other solutions which dominate it.

We note that many of the visualizations presented below can be applied either to a complete solution set, or to a subset of solutions selected using Pareto-rank.

1.2 Visualization of MOPs

Objective space plots: The most commonly used visualization of the obtained solutions is to plot the objective values in the objective space plots. Figure 1 shows an example of the solutions of a 2-objective problem. The dotted line and small circles show the Pareto-front and the obtained solutions respectively. A good algorithm must obtain solutions with both good diversity and good convergence.

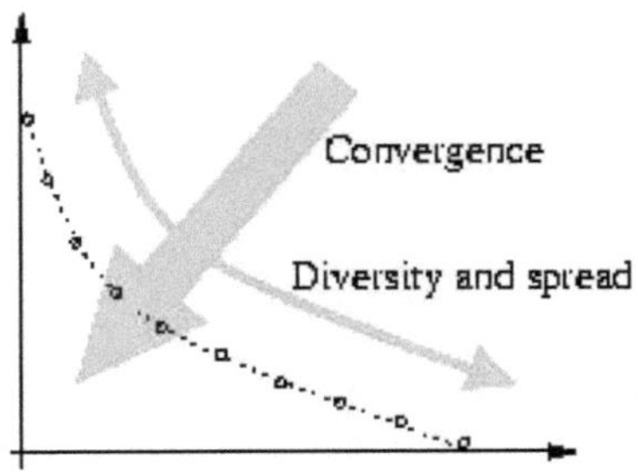

Fig. 1. An example of a 2-objective plot

This method is very useful, but cannot deal with more than 3 objectives. For a large set of optimal solutions, these plots are not accurate enough and numerical measures are required.

Distance and distribution (DD) charts: Ang et al. [1] propose two separate charts which plot a set of non-dominated solutions using their distance to an approximate Pareto front and the distance between each other. This method requires the approximate Pareto front to be found, which is not always straightforward or even possible. These plots are based solely on the objective values and parameters values are not considered.

[1] A parameter $\vec{x}_1$ is said to *dominate* $\vec{x}_2$ if $\vec{x}_1$ is not worse than $\vec{x}_2$ in all objectives and it is strictly better in at least one objective. Among a set of solutions, the non-dominated set of solutions contains those solutions that are not dominated by any member of the set.

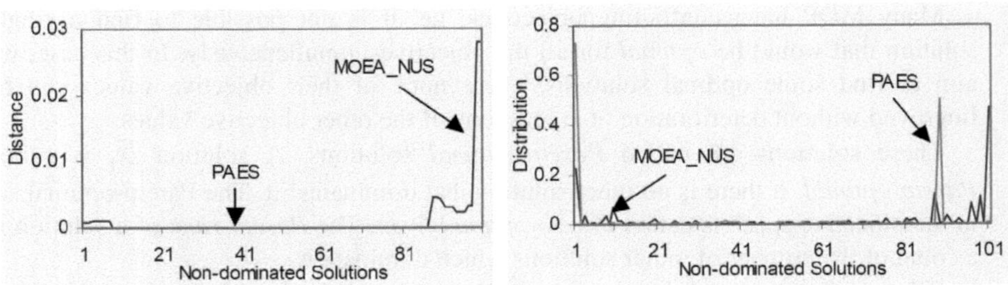

Fig. 2. Distance and Distribution Charts for solution sets generated by two algorithms

Self-Organizing Map (SOM) method: Obayashi and Sasaki [2] use SOM to reduce the dimensionality of parameters and objectives for visualization. SOM is an unsupervised neural network method which generates a mapping of thehigh dimension data into cells in fewer, usually 2, dimensions. This has been used to visualize a set of relatively large set of non-dominated solutions. Figure 3 shows an example. To facilitate the analysis of SOM and the data, similar cells on the map are clustered into groups. In this approach, the parameter space itself is not visualized, but information about parameters is provided by examples of designs overlaid at appropriate points on the 2D map.

Fig. 3. Self Organizing Map for aircraft part designs evolved using multiple objectives

2 An Example Multi-objective Optimization Problem

In order to introduce the visualization techniques, they are tested on a 3-objective version of the test function known as DTLZ2, the m objective sphere problem. This function is defined in table 1. [3]. In order to observe the quality of solutions, we produce solutions with different diversity and convergence using a multi-objective optimization method from [4].

We produce multiple solution sets, known as *good*, *bad-conv*, *bad-div-1* to *bad-div-5*. *bad-conv* refers to a set of solutions with bad convergence and *bad-div-1* to *bad-div-5* refer to sets with bad diversity and spread of solutions along the Pareto front. Figure 4 Shows the objective plot of the results of DTLZ2, in 3 dimensions.

We have chosen this 3-objective function to allow us to illustrate the other visualization methods using consistent data, before applying the techniques to examples of real optimization problems.

Table 1. Test functions

Test	function	constraints
DLTZ2	$f_1(\vec{x}) = (1+g)\cos(\vec{x}_1 \frac{\pi}{2})\cos(\vec{x}_2 \frac{\pi}{2})$	$x_i \in [0,1]$
	$f_2(\vec{x}) = (1+g)\cos(\vec{x}_1 \frac{\pi}{2})\sin(\vec{x}_2 \frac{\pi}{2})$	$n = 10$
	$f_3(\vec{x}) = (1+g)\sin(\vec{x}_1 \frac{\pi}{2})$	
	$g = \sum_{i=m}^{n} (\vec{x}_i - 0.5)^2$	

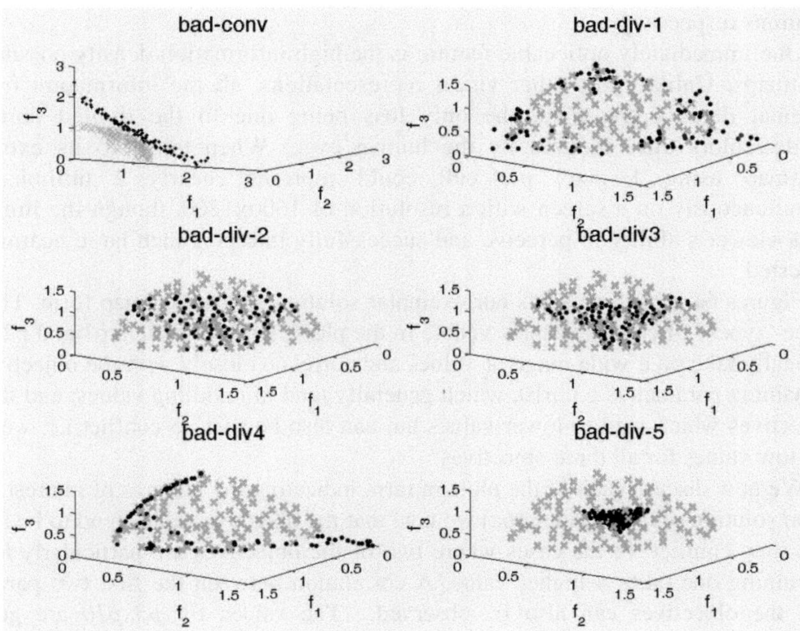

Fig. 4. Results of the DLTZ2 test problem with different diversities. *Crosses* indicatesolutions from the good solution set while *circles* are solutions from the six "bad" sets.

3 Heatmaps

Heatmap visualization is a technique which is most often applied to data gathered from microarrays. Microarray analysis [5] is a biological technique used to investigate the activation levels of large numbers of genes within cell samples. A typical dataset in this application consists of perhaps a dozen samples and hundreds if not thousands of genes. However, there is nothing intrinsically biological about heatmaps, and the technique can be applied to other data with a few modifications.

The results of a MOEA are a population of solutions, with each solution consisting of values for a set of parameters and an associated scores on multiple objectives. In computing terms this is a two-dimensional array, the dimensions being solution ID and parameter/objective. In the heatmap in figure 6 each row is a solution and each column is a parameter or objective. The color/shade of the cells represents the value of a parameter or objective for a particular solution. In figure 6 columns show parameters in numerical order, followed by objectives also in numerical order. The solutions have been ordered by using a hierarchical clustering method which keeps solutions with similar objective values together. The computational complexity of calculating a heatmap depends on the clustering method, this is typically O(Nr2+Nc2) with Nc and Nr being the number of rows and columns respectively.

One immediately noticeable feature is the high information density possible with heatmaps. Unlike most other visual representations, all the information from the original data is presented, the only loss being due to the limited number of shades/colors differentiable by the human eye. When taken to its extreme, a heatmap using 1 pixel per cell could represent nearly 2 million values simultaneously on a screen with a resolution of 1600x1200, though the limitations on a viewer's ability to perceive and successfully interpret such large heatmaps are untested.

Figures 6a and 6b presents our exemplar solution sets in heatmap form. There are three "types" of column clearly visible in the plots. The parameters p1 and p2, which typically take on a wide range of values and correlate closely with the objectives; the remaining parameters p3..p10, which generally tend to middling values; and the three objectives which tend to lower values but can also be seen to conflict i.e. we do not get low values for all three objectives.

We now discuss each of the plots in turn, indicating the features of interest. In the *good* solution set plot (Figure 6a) we note that the objective values tend to be low, but that they conflict. In the cases where two of the objectives are particularly low, the remaining one takes a higher value. A correlation between the first two parameters and the objectives can also be observed. The values for *p3..p10* are generally middling and quite similar, as we'd expect from examining the equations defining the objectives, which indicate the optimum value of *p3..p10* is always 0.5, independent of other parameters.

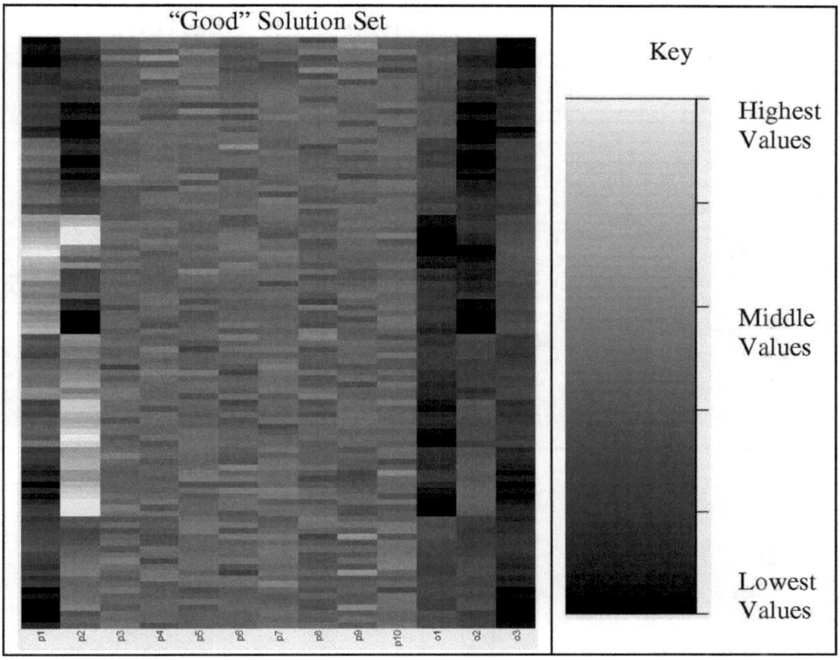

Fig. 5a. Heatmap of Good Solution Set. The 1st ten columns are parameter values, the final three are objective values. Data is normalized in such a way that shades are comparable with Fig 5b.

Comparing Figure 5a to Figure 5b, the *bad convergence* solution set shows much less consistency in *p3..p10*, than the *good* set. However, it is interesting to note that the correlation between *p1,p2* and the objectives is visible, indicating that the optimization process has begun to work on those parameters. As we'd expect in a badly converged solution set, there are some quite high values for the objectives, particularly in cases where the other objectives are low.

Bad Div 1 solutions can be seen to optimize two objectives at once, but the third objective always takes a high value. In comparison with the *good* set, there are more extreme values for *p1* and *p2*, and *p3..p10* show less convergence.

The *bad div 2* and *bad div 3* heatmaps shows the only one of the objectives is optimized at a time. Where one objective has low values, the other two have high values. This behavior can also be seen in the 3d scatterplot in Figure 4

The bias towards *o1* and *o3* can be see for *Bad Div 4*. Another interesting feature is that the values of *p4..p8* have a high variance in solutions where *p2* is very high (bottom of diagram). In these case, *o2* also takes on a high value.

In *bad div 5* we observe mainly middling values with little diversity for both parameters and objectives. Again, this can be seen to tally with Figure 4.

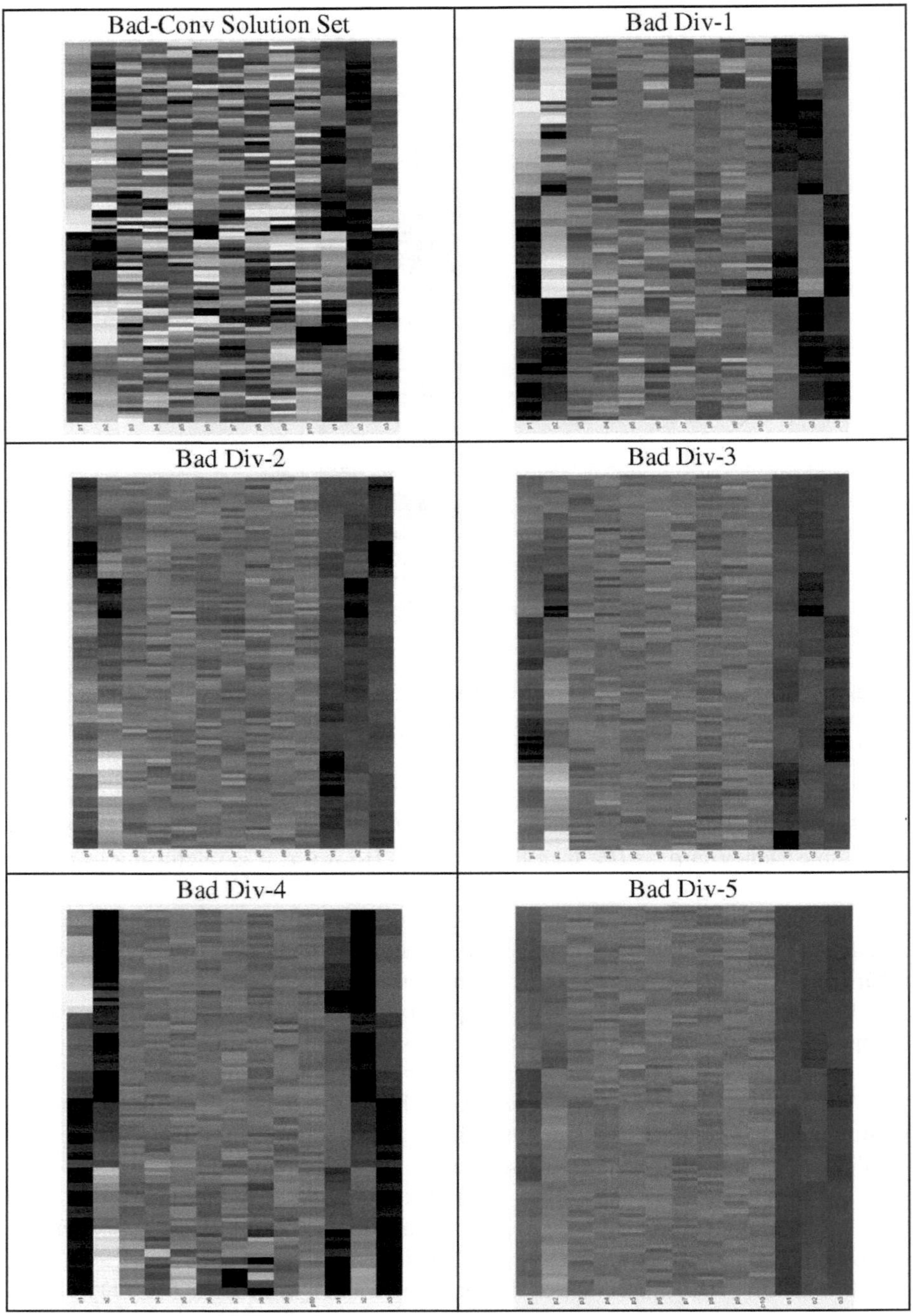

Fig. 5b. Heatmaps of "bad" solution sets. The key can be seen in Figure 5a

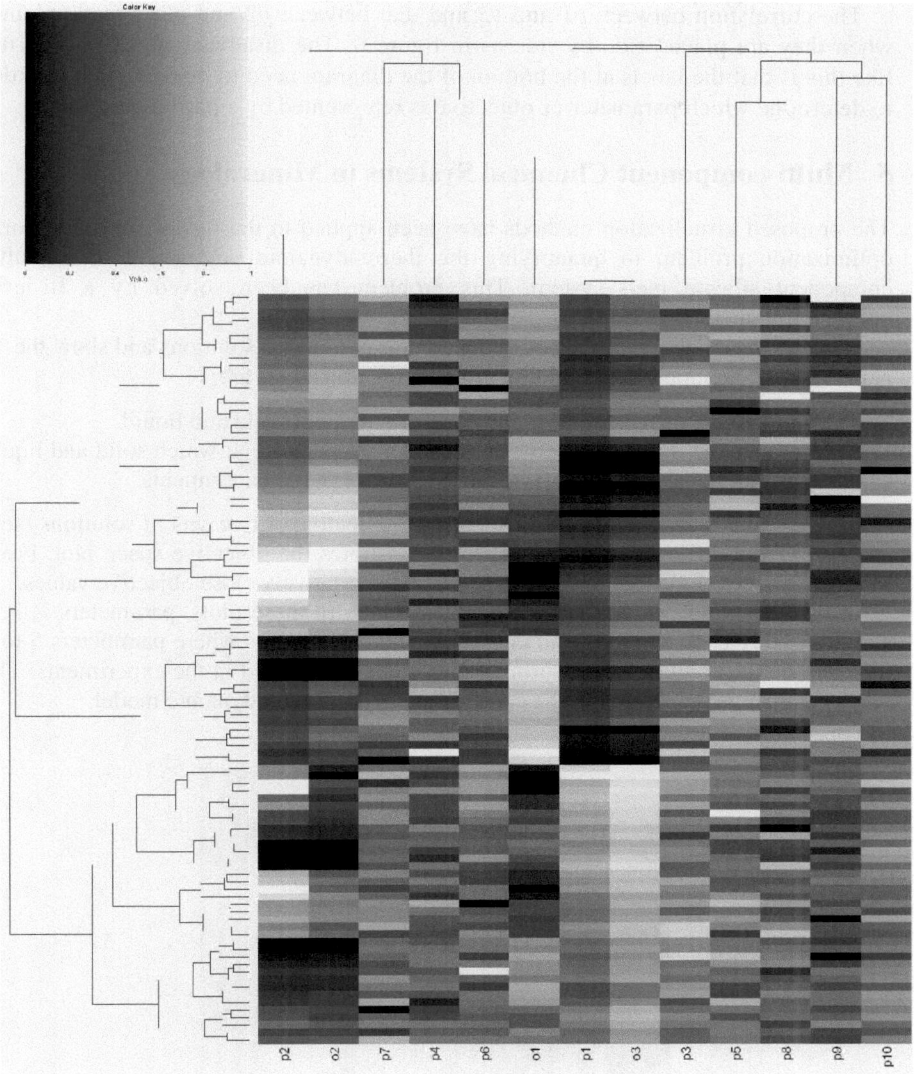

Fig. 6. Heatmap of "good" solution set. Trees indicating variable and solution clustering canbeseen at the top and left of the diagram respectively. The key shown indicates the shades associated with values in the cells. In order to make full use of the shade information the data has been normalized across the *good* set. Shades cannot therefore be compared directly with those in Figure 5a and 5b.

The correlation between *p1* and *o3* and that between *p2* and *o2* is made plainer when they are placed side by side as in figure 6. The disadvantage of re-ordering like this is that the labels at the bottom of the diagram need to be consulted in order to determine which parameter or objective is represented by a particular column.

5 Multi-component Chemical Systems in Mineralogy

The proposed visualization methods have been applied to the results of a real-world optimization problem in quantifying the thermodynamic parameters of a multi-component silicate melt system. This problem has been solved by a Bi-level optimization method in [6].

Here we visualize the last set of obtained non-dominated solutions and show the 13 parameters of the upper-level. In this problem the objectives are:

1) Minimize the difference between the free energy of solid and liquid.

2) Minimize the difference between obtained temperature at which solid and liquid can coexist and the absolute temperature T recorded in the experiments.

Figure 7 and 8 show the plain-heatmaps-by-objective of two sets of solutions (set1 and set2) from different runs and Figure 9 illustrates the objective space plot. From Figure 9, we conclude that both of these sets have relatively close objective values.

In the following, we analyze the heatmap plots. In these plots, parameters 1, 3, 9 and 2, 4, 10 refer to enthalpy and entropy of the components where parameters 5 to 8 indicate the uncertainty in measuring the free energy of solid in the experiments. The other parameters are related to the coefficients in the thermodynamic model.

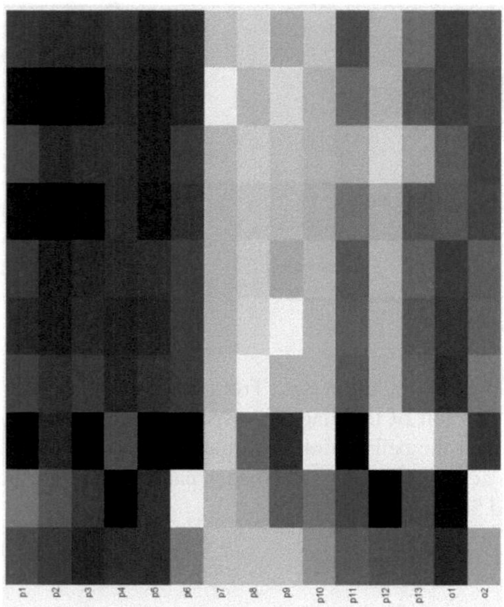

Fig. 7. Volcano Model. Solution Set 1. See Fig. 5a for key

Fig. 8. Volcano model. Solution Set 2. See Fig. 5a for key

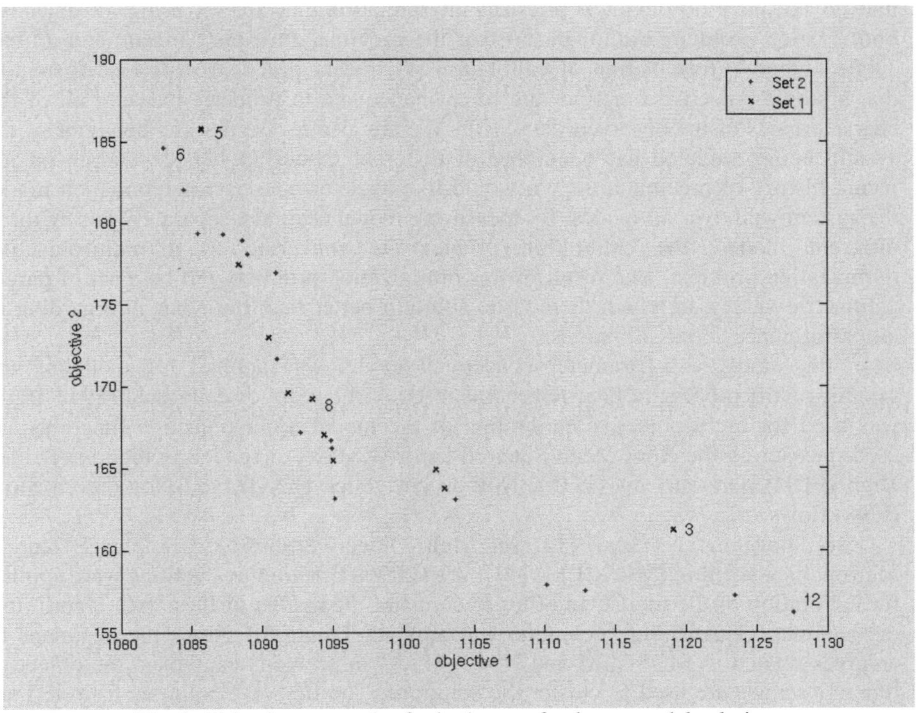

Fig. 9. Objective space plot for both sets of volcano model solutions

Figure 8 shows the heatmap of set2 and indicates if we select a certain constant value for uncertainties, almost all of the solutions on the front have the same enthalpy and entropy values. This can be observed by comparing the rows for the extreme solutions 12 and 6. This indicates that that the solutions, despite having good objective values, are located in a local optimum, where the uncertainty parameters are all equal and constant.

But for lower values of uncertainties like shown in Figure 7 (heatmap of set1), there is a distinct difference between the parameters of the extreme solutions on the front (solutions 3 and 5). This is valuable knowledge if we want to select a solution from the middle of the front, we have to select the parameters in the middle of their ranges (solution 8).

This analysis indicates, although the objective values of the two sets are very close to each other, parameters are located in different areas of the search space. This is very important for designing a proper thermodynamic model for mineralogists and for selecting an appropriate solution from the non-dominated set.

6 Application in Multi-objective Calibration of Hydrologic Models

The conversion of rain and snow to runoff has long been studied by engineers to design hydraulic systems and by scientist to develop an understanding of the process involved [7]. Most of the practical rainfall-runoff models contains some parameters that do not have any explicit physical interpretation and are set using an objective optimization problem. Although most of the previous attempts concentrated around single objective formulation of calibration [9], recent practical experiences suggest that a single objective functions are often inadequate to properly measure all of the characteristics of the observed data [10]. We are also aware that the behavior of the system being modeled has a number of different modes[11,12], depending on the recent history of precipitation. We would like to determine parameters which model the system well over all modes. By measuring model accuracy separately during these different modes, the calibration problem is converted to a multi-objective optimization problem. The result of this optimization problem will be a set of pareto parametric values, in which there is no solution better than the other in regarding to all performance across all modes.

In this study, a 5-parameter conceptual model was applied for modeling the rainfall-runoff process in Leaf River Basin, USA. 11 years of daily data (1948-1958) was used for multi-objective calibration of the model. Four objective functions are used, measuring the Root Mean Squared Error (RMSE) of model predictions Driven High (FDH), Driven Low (FDL), Non-driven Quick (FNQ), and Non-driven Slow (FNS) flows.

Two algorithms, NSGA-II [14] and Multi-Objective Shuffled Complex Evolution Metropolis algorithm (MOSCEM) [13], with 20000 function evaluations were applied for calibration of the model. In order to compare the results of these two algorithms, the heatmap visualization technique was applied. Figure 10 shows the heatmaps of archives related to MOSCEM and NSGA-II. As in previous examples, the objective function values are used to cluster the solutions. The first five columns from left are represent parametric values and the next four are represent objective functions.

Comparing the number of solutions in the archives (941 solutions for NSGA-II and 88 for MOSCEM), initially it was supposed that NSGA-II would have a finer texture. However as it appears, the textures of both heatmaps are almost the same, showing that NSGA-II converged to limited number of distinct solutions for each model parameter. Additionally, we note that the two algorithms have converged to different regions of the parameter space. For the first parameter, MOSCEM generally converges to values which are smaller and more diverse than NSGA-II. For parameter 2 MOSCEM has again used a wider parametric region than NSGA-II. For parameter 3 on the other hand, NSGA-II has found some low values unused by MOSCEM, though MOSCEM covers other values which NSGA-II did not find. For parameter 4, it is quite obvious that NSGA-II converges almost completely to two extreme values, whereas MOSCEM produces a broad spectrum of results. For parameter 5, the differences are not so great, but NSGA-II has located some higher values. Looking to the columns related to objective functions, they reveal that for the first objective function both algorithms can find many good results, but NSGA-II allowed some bad results which trade off against objective 4. For second objective function, convergence is similar, however NSGA-II is again more tolerant of bad results. For third objective function the results of both algorithms are very similar. For the final objective function, it is quite obvious that NSGA-II can produce much better results than MOSCEM. However, the best results that NSGA-II could find for objective 4 are some of the worst for the other objectives. Looking at the four objectives together it is clear that the calibration process of the applied hydrologic model is inherently a multi-objective task, and that the applied model can not represent the whole hydrologic behavior of the catchment with a single parametric set or even a parametric region.

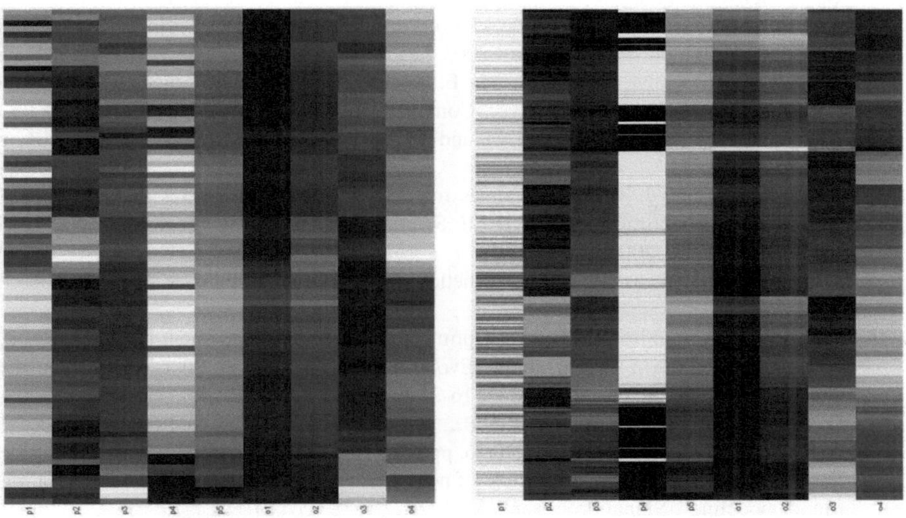

(a) MOSCEM (b) NSGA-II

Fig. 10. The heatmaps of solution archives for hydrological model parameterization

7 Conclusion and Future Work

The heatmap is a novel and interesting visualization method which can provide detailed insight into the multiple solutions generated by population based multi objective algorithms. The high information density of heatmaps allows whole populations of solutions to be visualized. One application for heatmaps is to gain greater insight into the behavior of particular MO algorithms, and to compare the performance of algorithms. This is of particular interest to the MO algorithm community. Another is the exploratory analysis of the parameter space and its relationship to the objective space. Domain experts and those applying MO algorithms to real-world problems will find this aspect particularly useful.

Future work includes: the developments of color-scales and grayscales appropriate for colorblind users and grayscale printing; Ordering rows consistently between datasets to ease comparison; experimentation to understand the relationship between heatmaps representations and Pareto ranking; and development of an easy to use toolkit for MO researchers.

References

1. Kiam Heong Ang, Gregory Chong and Yun Li. Visualization Technique for Analyzing Non-Dominated Set Comparison, in Lipo Wang, Kay Chen Tan, Takeshi Furuhashi, Jong-Hwan Kim and Xin Yao (editors), Proceedings of the 4th Asia-Pacific Conference on Simulated Evolution and Learning (SEAL'02), pp. 36--40, Vol. 1, Nanyang Technical University, Orchid Country Club, Singapore, November 2002.
2. Shigeru Obayashi and Daisuke Sasaki, Visualization and Data Mining of Pareto Solutions Using Self-Organizing Map, Second International Conference on Evolutionary Multi-Criterion Optimization (EMO 2003), Faro, Portugal, LNCS 2632, Springer-Verlag Berlin Heidelberg 2003, pp. 796-809, April 2003.
3. K. Deb, L. Thiele, M. Laumanns, and E. Zitzler, "Scalable multi-objective optimization test problems," in Proc. Congr. Evol. Comput., D. B. Fogel, M. A. El-Sharkawi, X. Yao, G. Greenwood, H. Iba, P. Marrow, and M. Shackleton, Eds., May 2002, vol. 1, pp. 825–830
4. S. Mostaghim and J. Teich., Strategies for finding good local guides in multi-objective particle swarm optimization. In IEEE Swarm Intelligence Symposium, pages 26–33, Indianapolis, USA, 2003.
5. The Chipping Forecast II, Nature Genetics Special Issue, December 2002, Volume 32, No 4s
6. Halter, W.; Mostaghim, S., "Bilevel Optimization of Multi-Component Chemical Systems Using Particle Swarm Optimization," Evolutionary Computation, 2006. CEC 2006. IEEE Congress on , vol., no.pp. 1240- 1247, 16-21 July 2006
7. O' Loughlin, G., Huber, W., Chocat, B., "Rainfall-runoff process and modeling", Journal of Hydraulic Research, VOL. 34, NO. 6, pp. 733-751, 1996.
8. Furundzic, D.,: "Application example of neural networks for time series analysis: rainfall-runoff modeling", Signal Processing, VOL. 64, pp. 383-396, 1998.
9. Gan, T., Y., Biftu, G., F., "Automatic calibration of conceptual rainfall-runoff models: optimization algorithms, catchment conditions, and model structure", Water resources research, VOL. 32, NO. 12, pp. 3513-3524, 1993.

10. Vrugt, J. A., Gupta H. V., Bouten, W., and Sorooshian, S., "A shuffled complex evolution metropolis algorithm for optimization and uncertainty assessment of hydrologic model parameters", Water Resources Research, VOL. 39 (8), 2003.
11. Boyle, D. P., Gupta, H. V., and Sorooshian S., "Toward improved calibration of hydrological models: Combination the strengths of manual and automatic methods, Water Resources Research, VOL. 36(12), 3663-3674, 2000.
12. Wagener, T., Wheater, H. S., "On the evaluation of conceptual rainfall-runoff models using multiple-objectives and dynamic identifiability analysis", In Littlewood, I. (ed.), Continuous river flow simulation: methods, applications and uncertainty, British Hydrological Society, Occasional paper, No. 13, Wallingford, UK, pp.45-51, 2002.
13. Vrugt, J. A., Gupta H. V., Bastidas, L. A., Bouten, W., and Sorooshian, S., "Effective and efficient algorithm for multi-objective optimization of hydrologic models", Water Resources Research, VOL. 39 (8), 2003.
14. Deb, K., Pratap, A., Agrawal, S. and Meyarivan, T. (2000). A fast and elitist multiobjective genetic algorithm: NSGA-II. Technical Report No. 2000001. Kanpur: Indian Institute of Technology Kanpur, India. http://citeseer.ist.psu.edu/article/deb00fast.html
15. Pryke A., Mostaghim S., Nazemi A. (2006),"Heatmap Visualization of Population Based Multi Objective Algorithms", Technical Report Number CSR-06-14, University of Birmingham, School of Computer Science, ftp://ftp.cs.bham.ac.uk/pub/tech-reports/2006/CSR-06-14.pdf

Multiplex PCR Assay Design by Hybrid Multiobjective Evolutionary Algorithm

In-Hee Lee, Soo-Yong Shin, and Byoung-Tak Zhang

Biointelligence Laboratory
School of Computer Science and Engineering
Seoul National University, Seoul 151-742, Korea
{ihlee,syshin,btzhang}@bi.snu.ac.kr

Abstract. Multiplex Polymerase Chain Reaction (PCR) assay is to amplify multiple target DNAs simultaneously using different primer pairs for each target DNA. Recently, it is widely used for various biology applications such as genotyping. For sucessful experiments, both the primer pairs for each target DNA and grouping of targets to be actually amplified in one tube should be optimized. This involves multiple conflicting objectives such as minimizing the interaction of primers in a group and minimizing the number of groups required for the assay. Therefore, a multiobjective evolutionary approach may be an appropriate approach. In this paper, a hybrid multiobjective evolutionary algorithm which combines ϵ-multiobjective evolutionary algorithm with local search is proposed for multiplex PCR assay design. The proposed approach was compared with another multiobjective method, called MuPlex, and showed comparative performance by covering all of the given target sequences.

1 Introduction

The Polymerase Chain Reaction (PCR) is a very powerful biological technique which is widely used to amplify DNA and plays a key role in biotechnology and biology research. In standard protocol, PCR can amplify only one target DNA at a time (Fig. 1(a)). But the biological or clinical assay usually involves multiple target DNAs, it is much more desirable to amplify these DNAs simultaneously. The multiplex PCR is an extension of PCR in which multiple target DNAs are amplified at the same time (Fig. 1(b)). It has a wide variety of applications in biology and is recently spotlighted as a core tool for high throughput single nucleotide polymorphism (SNP) genotyping [1,2,3].

For successful experimental results, a careful design of multiplex PCR assay is important. A multiplex PCR assay design is a complex problem composed of two optimization processes: optimizing primers for each target while minimizing the number of partition. First, the primers for each target DNA should be optimized so that the interactions between primers and non-target DNAs are minimized. Since the multiple targets are amplified in one tube at the same time, it is important that the primers for one target do not interact another targets or primers. If such an unwanted interaction happens, some of the target DNAs

S. Obayashi et al. (Eds.): EMO 2007, LNCS 4403, pp. 376–385, 2007.

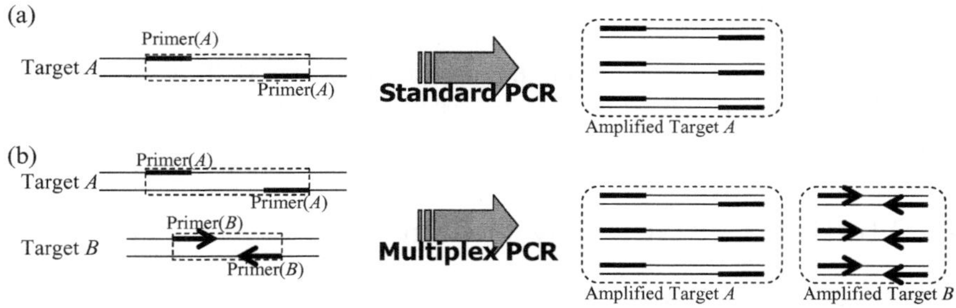

Fig. 1. (a) The concept of standard polymerase chain reaction (PCR). The region between primers in A (the dashed box) is amplified by PCR. (b) The concept of multiplex PCR. Multiple targets A and B are amplified simultaneously by primers A and B, respectively, in one experiment.

might not be amplified. Last, the grouping of target DNAs which will be amplified together should be decided. It would be ideal when all of the target DNAs can be amplified together. Unfortunately, it is very likely that the primers for some targets can not be chosen to prevent unwanted interactions. In such cases, the targets should be put to different groups for separate multiplex PCR runs. However, the number of such separate groups should be minimized.

There have been many studies to tackle this problem [4,5,6,7,8,9,10,11,12] . Most of these works assumed only one group and the targets that do not fit to be amplified together were discarded [5,6,7,8,11,12]. In [4,9,10], on the other hand, the partitioning of targets into multiple groups was handled. First, a set of primer candidates for each target is selected according to predefined conditions. Then, the targets are partitioned into appropriate groups in deterministic way while selecting optimal primers from the candidate sets. From the methodological point of view, most of the previous researches used a deterministic search and only a few evolutionary approaches are published [7,8,11].

Rachlin et al. formulated the design of multiplex PCR assay as finding cliques in graph to optimize both objectives [13]. According to their formulation, the nodes in graph G represent the target DNAs and edges connect two targets(nodes) which can be put into the same group. Each node has multiple states (candidate primers) and the state of two nodes determines whether they can be connected or not. They empirically showed that there is a tradeoff relationship between the specificity of each primer pair to their target and the overall degree of multiplexing. Moreover, it is well known that finding a clique in a given graph is a hard computational problem [14]. Considering these properties of multiplex PCR assay design, a multiobjective evolutionary approach with local search is suggested here.

The suggested multiobjective evolutionary algorithm is based on ϵ-MOEA which was originally suggested in [15]. The algorithm is modified to perform local search after the generation of every new offspring and a genotypical niching is adopted to keep the population diversity.

The rest of the paper is organized as follows. Section 2 describes the multiplex PCR assay design and the strategy taken here in more detail. The hybrid multiobjective evolutionary algorithm used in multiplex PCR assay design is explained in Section 3. The experimental results are shown and compared to other methods in Section 4 and the conclusions are drawn in Section 5.

2 Multiplex PCR Assay Design

As already mentioned in Section 1, two problems should be solved for a successful multiplex PCR assay. One is to select primers for each target. Those primers must have uniform experimental conditions and minimize the interaction with other targets and primers. The other is to divide the set of targets into multiple subsets. For example, some target DNAs share similar subsequences. In that case, it is very likely that one target's primer can interact with other target and vice versa. Therefore, these targets should be separated into different subsets. But the number of different subsets needs to be minimized to simplify the experimental process. However, we can not put all compatible targets together because there is a limitation on the number of targets that can be amplified in one tube for technical reasons. Hence we should minimize the number of different subsets while keeping the maximum size of each subset.

The primers having uniform experimental conditions can be chosen independently for each target. But the minimization of the interaction between other targets and primers depends on the partitioning of targets. The partitioning of targets is also dependent on the selection of primers. Therefore, the selection of primer candidates can be separated from the optimization process. From now on, we assume that a set of primer candidates for each target is given *a priori* and concentrate on primer assignment on each target from the candidates and partition of targets.

In [13], it is more formally described using the concept of multi-node graph. In a multi-node graph, each node has its own set of states it can take. In multiplex PCR context, a node corresponds to a target and a node's state means the primers for the target. If a node u is in state i, the i-th primer candidate is assigned for target u. Each edge in the graph is associated with variable weight depending on the states on the two nodes it connects (see Fig. 2). We denote the

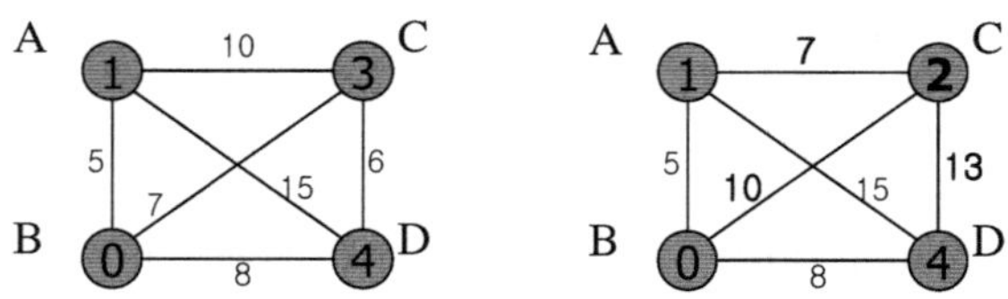

Fig. 2. An example of multi-node graph (left). After changing the state of node C, the weights on edges connected to C are changed as the second graph (right).

weight values for edge between nodes u and v by the matrix W_{uv}. For multiplex PCR assay design, the weight matrix denotes the compatibility between two targets, which means whether the targets can be put together in one tube. The elements of $W_{uv}[i][j]$ represents how compatible the targets u and v are when they are in states i and j respectively. For the two targets to be compatible, they should satisfy two conditions: minimizing the undesirable hybridization and the sequence similarity among targets and their primers. Minimizing the undesirable hybridization alone is not enough because similar sequences can also reduce the hybridization chance. For example, the sequences 'AAAA' and 'CCCC' do not hybridize each other, but 'AAAA' and 'AAAA' do not, either. Hence, we decompose the compatibility between targets (W_{uv}) by two values: H-measure (H_{uv}) and Similarity (S_{uv}) defined in [16]. H_{uv} is a matrix whose element $H_{uv}[i][j]$ denotes how much undesirable hybridization can occur among targets u and v and their primers i and j. Similarly, S_{uv} is a matrix whose element $S_{uv}[i][j]$ denotes how similar the targets u and v and their primers i and j are.

Given a set of n targets T and the set of primer candidates C_i for each target i, a multi-node graph G and its associated matrices H and S can be constructed. Then, formally, the multiplex PCR assay design is to find,

1. The partition $S_1, \ldots, S_M$ such that $\bigcup S_i = T$ and $S_i \cap S_j = \varnothing$ for each i, j and
2. The state assignment A for each node in G from $C_1 \times C_2 \times \cdots \times C_n$,

while satisfying the following:

1. Minimize $\sum_i \sum_{u,v \in S_i} H_{uv}[A(u)][A(v)]$ and
2. Minimize $\sum_i \sum_{u,v \in S_i} S_{uv}[A(u)][A(v)]$ and
3. Minimize the number of partitions M,

where $A(u)$ denotes the state of node u under assignment A.

According to the above definition, two different variable spaces should be searched: state assignment and partition. We explored both space by combining a multiobjective evolutionary algorithm and local search. Here, the main evolutionary algorithm searches the space of the state assignment. During local search, the space of partition is explored. Detailed description of the evolutionary search procedure will be given in the next section.

Our approach to multiplex PCR assay design is summarized in Fig. 3. First, candidate primers are generated by Primer3 [17]. We used an external program at this step from two reasons: one is that the candidate primers can be selected independently from the main evolutionary optimization and the other is that there exist many open softwares for primer selection since it is a fundamental tool in biology. Among various open softwares, we chose the most popular program, Primer3. Next, the partition of targets and primer assignment for each target are optimized by hybrid multiobjective evolutionary algorithm. At the end, a variety of multiplex PCR assays will be presented to user.

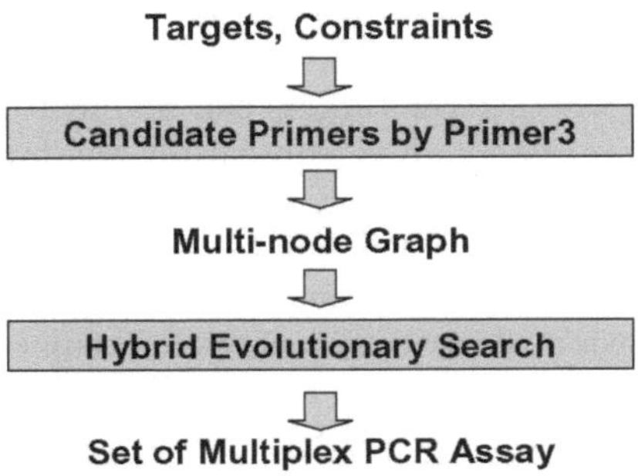

Fig. 3. The flow chart for Multiplex PCR assay design

3 Hybrid Multiobjective Evolutionary Algorithm for Multiplex PCR Assay Design

3.1 The Preprocessing

The first step is to select primer candidates for each target which having similar chemical properties. We used Primer3 program to choose candidate primers [17]. Considering the size of the search space and running time, five candidate primers are selected for each target. And for each pair of target u and v, the elements of $H_{uv}[i][j]$ and $S_{uv}[i][j]$ are calculated using the H-measure and Similarity function described in [16].

3.2 The Hybrid Multiobjective Evolutionary Algorithm

Our next step is the evolutionary search for optimal partition of groups and primer assignments. We used a variation of ϵ-MOEA [15] combined with local search.

Each individual is a concatenation of partition part and state assignment part. The state assignment is a vector from $C_1 \times C_2 \times \cdots \times C_n$ which determines the configuration of multi-node graph. The i-th value denotes which primer candidate from C_i is assigned for target i. The partition part means the set of cliques from the multi-node graph configured by the state assignment part. It is a vector from $[1, \ldots, M]^n$ where i-th value determines which partition the node i belongs to.

For each generation of ϵ-MOEA,

1. One parent P_1 is chosen at random from the archive and the other parent P_2 is chosen from the population by tournament selection.
2. Generate two offsprings O_1 and O_2 from P_1 and P_2 by genetic operators. The uniform crossover and 1-bit mutation operators are used here.
3. Apply the local search to O_1 and O_2 for a predefined number of times, L.

   (a) O'_1 and O'_2 is produced through local search operators. One of the two local search operators which will be explained in Section 3.2 is applied in random.
   (b) Replace O_1 with O'_1 if O'_1 dominates O_1. If O_1 and O'_1 non-dominates each other, O'_1 replaces O_1 with probability of 0.5. Otherwise, O'_1 is discarded.
   (c) Similar procedure with O_2 and O'_2.
4. Update the archive.
   (a) The offspring is accepted to the archive if it is in the same front as the archive members or it dominates one or more of them. The dominated archive members are removed.
   (b) If the archive reaches its maximum size, the nearest archive member in objective space is replaced.
5. Update the population.
   (a) If the offspring dominates one of the population, the dominated member is replaced with the offspring.
   (b) If the offspring and the population do not dominate each other, the nearest population member in variable space is replaced with probability of 0.5.
6. Repeat to Step 1 until the termination condition is satisfied.

In original ϵ-MOEA, Step 4 followed after Step 2 and there was no restriction on maximum size of the archive. Theoretically, the number of archive members that ϵ-dominate the population is not infinite [15]. However, the size of the archive often grows very large. So we put a limit on the maximum size of the archive.

Another difference between the suggested approach and the original ϵ-MOEA is the niching method. In original ϵ-MOEA, there was no explicit niching method for the archive except the ϵ-domination concept and the new offsprings replaced random individual from the population. However, the ϵ-domination concept is not enough to keep the size of the archive in reasonable size and the random replacement in the population can lead to the quick loss of genetic diversity. As can be seen from Step 4 and 5, we tried to handle the problem by using different distance measures. In archive, the distance in objective space is used to provide diverse solutions. In contrast, the distance in variable space is used in population update. This is to keep the genetic diversity of population and to prevent premature convergence.

3.3 Local Search

When generating new offsprings in Step 2 of the main loop, the crossover and mutation operator targets the entire chromosome. Both the state assignment vector and the partition vector are treated as one string of size $2n$ and undergo uniform crossover or 1-bit mutation.

On the other hand, the local search operator targets the partition vector only. Since every candidate primers generated by Primer3 guarantee minimum level

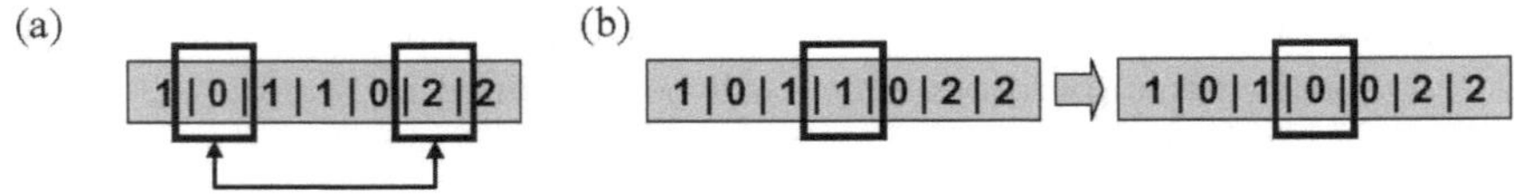

Fig. 4. The two local search operators. (a) The swapping operator exchanges two targets between two different partitions. (b) The migration operator moves one target from a partition to another.

of quality, the change of primers assigned to a target does not result in drastic change in objective values. Hence, we concentrate on finding optimal partition during local search. As can be seen in Fig. 4, two local search operators are adopted. One is the swapping operator that exchanges two targets from two different partitions. The other is the migration operator which moves one target from a partition to another.

Local search proceeds in Lamarckian way. At each cycle of local search, the individual produced by local search which dominates the previous replaces the original individual. If the individuals before and after the local search non-dominates each other, one is chosen at random.

4 Experimental Results

We tested our approach on 52 target sequences from Arabidopsis multigene family and compared the result with an open program for multiplex PCR assay design, MuPlex [10]. Among the MULTIPCR [4], MultiPLX [9] and MuPlex [10] that can handle multiple partition, MuPlex was chosen because MULTIPCR was not open to public and MultiPLX could not find any acceptable result for the given problem.

MuPlex uses an agent-based multi-objective optimization. The agents encapsulating specific algorithm either create new solutions from scratch, improve or modify existing solutions, or remove unpromising solutions from further consideration [10]. By the interactions between agents, MuPlex implements similar approach as evolutionary algorithm. The solutions in MuPlex are evaluated in similar terms as our approach.

We set the size of population and archive as 100 and 200, respectively. The maximum generation was set to 100,000 and local search was performed 100 times for each offspring. The probability for crossover and mutation was 0.9 and 0.01, respectively. The maximum number of partition and maximum size of a partition is set to 10. These experimental parameters were chosen empirically.

The results are evaluated from three perspectives. One is the sum of total cross-hybridization within a partition. This is to estimate total experimental errors. The DNA-DNA hybridization simulator NACST/Sim is used to calculate this value [16]. It checks all possibilities of cross-hybridization between two given sequences. Others are the number of groups and the average number of targets per group. These are to estimate the efficiency of multiplex PCR assay.

Table 1. The comparison of designed Multiplex PCR Assay. Solution 1, Solution 2 and Solution 3 are generated by the hybrid multiobjective evolutionary algorithm suggested in this paper. The solution named MuPlex is generated using MuPlex program.

	MuPlex	Solution 1	Solution 2	Solution 3
The total cross-hybridization	13719	10915	13269	10683
The number of groups	5	9	8	10
The average number of targets	9.4	5.7	6.5	5.2

Table 2. The design examples from MuPlex and the proposed approach. The Solution 2 from Table 1 is shown here. The columns group ID and group size denote each partition of target DNAs and the number of targets in each partition, respectively. # of cross-hyb. value means the total undesirable hybridization value calculated by NACST/Sim.

MuPlex			Hybrid ϵ-MOEA		
Group ID	Group Size	# of cross-hyb.	Group ID	Group Size	# of cross-hyb.
1	10	4926	1	6	1746
2	10	2397	2	7	1385
3	10	2789	3	6	1655
4	10	2032	4	6	2295
5	7	1575	5	6	1234
			6	7	1493
			7	6	1912
			8	8	1549
Total	47	13719	Total	52	13269

The best run of our algorithm output only three solutions in the final archive. These solutions are compared with the only solution from MuPlex in Table 1. From the three solutions produced by our approach, the tradeoff between primer optimization and partition efficiency is clear. As the number of group increases, the average number of targets in each group decreases. And if the number of targets in a group is small, there is little chance to the cross-hybridization. In that sense, all of the four solutions in Table 1 form a tradeoff front.

The design examples from MuPlex and the proposed approach are compared in detail in Table 2. The columns group ID and group size denote each partition of target DNAs and the number of targets in each partition, respectively. NACST/Sim value means the total undesirable hybridization value calculated by NACST/Sim. In MuPlex, some target can be discarded if it is hard to find a partition for that target. Therefore, as can be seen in Table 2, only 47 of 52 targets were partitioned. In contrast, every target belongs to a partition in our approach. But this is dependent on the purpose of the user. In some case as high throughput screening, users need a design which is efficient but do not cover every target. But in cases of clinical assay, the coverage becomes critical. Also, the constraint of perfect coverage upon the proposed approach can be relaxed.

5 Conclusions

The problem of multiplex PCR assay design is addressed and formulated as a multiobjective optimization problem. A hybrid multiobjective evolutionary search is applied to the problem and compared with other similar program. The suggested approach combines a variant of existing algorithm and two simple local search operators and shows a reasonable performance. This is a preliminary result and further work is required.

Acknowledgements

This research was supported in part by the Ministry of Education & Human Resources Development under the BK21-IT Program, the Ministry of Commerce, Industry and Energy through MEC project, and the NRL Program from Korean Ministry of Science and Technology. The ICT at Seoul National University provides research facilities for this study. Soo-Yong Shin was supported by the Korea Research Foundation Grant funded by the Korean Government (MOEHRD) (KRF-2006-214-D00140).

References

1. A. L. L. Cortez, A. C. Carvalho, A. A. Ikuno, K. P. Bürger, and A. M. C. Vidal-Martins. Identification of *Salmonella* spp. isolates from chicken abattoirs by multiplex-PCR. *Research in Veterinary Science*, 81(3):340–344, 2006.
2. Christina L. Aquilante, Taimour Y. Langaee, Peter L. Anderson, Issam Zineh, and Courtney V. Fletcher. Multiplex PCR-pyrosequencing assay for genotyping CYP3A5 polymorphisms. *Clinica Chimica Acta*, 372(1–2):195–198, 2006.
3. Anne J. Jääskeläinen, Heli Piiparinen, Marija Lappalainen, Marjaleena Koskiniemi, and Vaheri Antti. Multiplex-PCR and oligonucleotide microarray for detection of eight different herpesviruses from clinical specimens. *Journal of Clinical Virology*, 37(2):83–90, 2006.
4. Pierre Nicodème and Jean-Marc Steyaert. Selecting optimal oligonucleotide primers for multiplex PCR. In *Proceedings of the 5th International Conference on Intelligent Systems for Molecular Biology*, pages 210–213, 1997.
5. Eric C. Rouchka, Abdelnaby Khalyfa, and Nigel G. F. Cooper. MPrime: efficient large scale multiple primer and oligonucleotide design for customized gene microarrays. *BMC Bioinformatics*, 6(175), 2005.
6. Richard Schoske, Pete M. Vallone, Christian M. Ruitberg, and John M. Butler. Multiplex PCR design strategy used for the simultaneous amplification of 10 Y chromosome short tandem repeat (STR) loci. *Analytical and Bioanalytical Chemistry*, 375:333–343, 2003.
7. Hong-Long Liang, Chungnan Lee, and Jain-Shing Wu. Multiplex PCR primer design for gene family using genetic algorithm. In *GECCO '05*, pages 67–74, Washington, DC, USA, 2005. ACM.
8. Feng-Mao Lin, Hsien-Da Huang, His-Yuan Huang, and Jorng-Tzong Horng. Primer design for multiplex PCR using a genetic algorithm. In *GECCO '05*, pages 475–476, Washington, DC, USA, 2005. ACM.

9. Lauris Kaplinski, Peidar Andreson, Tarmo Puurand, and Maido Remm. MultiPLX: automatic grouping and evaluation of PCR primers. *Bioinformatics*, 21(8):1701–1702, 2005.
10. John Rachlin, Chunming Ding, Charles Cantor, and Simon Kasif. MuPlex: multi-objective multiplex PCR assay design. *Nucleic Acids Research*, 33:W544–W547, 2005.
11. Chungnan Lee, Jain-Shing Wu, Yow-Ling Shiue, and Hong-Long Liang. MultiPrimer: Software for multiple primer design. *Applied Bioinformatics*, 5(2):99–109, 2006.
12. Tomoyuki Yamada, Haruhiko Soma, and Shinichi Morishita. PrimerStation: a highly specific multiplex genomic PCR primer design server for the human genome. *Nucleic Acids Research*, 34:W665–W669, 2006.
13. John Rachlin, Chunming Ding, Charles Cantor, and Simon Kasif. Computational tradeoffs in multiplex PCR assay design for SNP genotyping. *BMC Genomics*, 6(102), 2005.
14. M. R. Garey and D. S. Johnson. *Computers and Intractability: A Guide to the Theory of NP-completeness*. W. H. Freeman and company, 1979.
15. Marco Laumanns, Lothar Thiele, Kalyanmoy Deb, and Eckart Zitzler. Combining convergence and diversity in evolutionary multi-objective optimization. *Evolutionary Computation*, 10(3):263–282, 2002.
16. Soo-Yong Shin, In-Hee Lee, and Byoung-Tak Zhang. Multi-objective evolutionary optimization of DNA sequences for reliable DNA computing. *IEEE Transactions on Evolutionary Computation*, 9(2):143–158, 2005.
17. Steve Rozen and Helen J. Skaletsky. Primer3 on the www for general users and for biologist programmers. In S. Krawetz and S. Misener, editors, *Bioinformatics Methods and Protocols: Methods in Molecular Biology*, pages 365–386. Humana Press, Totowa, NJ, 2000. Source code available at http://fokker.wi.mit.edu/primer3/.

ParadisEO-MOEO: A Framework for Evolutionary Multi-objective Optimization

Arnaud Liefooghe, Matthieu Basseur, Laetitia Jourdan, and El-Ghazali Talbi

INRIA Futurs, Laboratoire d'Informatique Fondamentale de Lille (LIFL), CNRS
Bât. M3, Cité Scientifique, 59655 Villeneuve d'Ascq cedex, France
{liefooga,basseur,jourdan,talbi}@lifl.fr

Abstract. This paper presents ParadisEO-MOEO, a white-box object-oriented generic framework dedicated to the flexible design of evolutionary multi-objective algorithms. This paradigm-free software embeds some features and techniques for Pareto-based resolution and aims to provide a set of classes allowing to ease and speed up the development of computationally efficient programs. It is based on a clear conceptual distinction between the solution methods and the multi-objective problems they are intended to solve. This separation confers a maximum design and code reuse. ParadisEO-MOEO provides a broad range of archive-related features (such as elitism or performance metrics) and the most common Pareto-based fitness assignment strategies (MOGA, NSGA, SPEA, IBEA and more). Furthermore, parallel and distributed models as well as hybridization mechanisms can be applied to an algorithm designed within ParadisEO-MOEO using the whole version of ParadisEO. In addition, GUIMOO, a platform-independant free software dedicated to results analysis for multi-objective problems, is briefly introduced.

Keywords: object-oriented frameworks, design and code reuse, multi-objective optimization, evolutionary algorithms.

1 Introduction

Nowadays, the usefulness of Multi-Objective Optimization (MOO) is globally established in the whole operational research community. Furthermore, evolutionary algorithms (EAs) are commonly used to solve multi-criterion problems since they naturally found a well-diversified set of good-quality solutions. EAs [12] are stochastic optimization processes based on an iterative improvement of a population of solutions (called individuals). As discussed later in the paper, several frameworks such as MOEA [20], MOMHLib++, Open BEAGLE [9], PISA [2], TEA [7] (to quote only them) already attempt to simplify and accelerate the development process of evolutionary MOO applications. We here propose a new library, called ParadisEO-MOEO (MOEO for short), that aims to produce efficient programs while having a minimal programming effort and a maximum code reuse. MOEO (Multi-Objective Evolving Objects) is an extension of the Evolving Objects framework [15]. It includes a broad range of reusable features and

S. Obayashi et al. (Eds.): EMO 2007, LNCS 4403, pp. 386–400, 2007.

techniques related to Pareto-based MOO such as performance metrics, elitism, fitness sharing and the most common Pareto-based fitness assignment schemes: MOGA, NSGA, NSGA-II, SPEA, SPEA2, IBEA, ... The fine-grained components of MOEO confer a high genericity, flexibility, adaptability and extensibility. Thus, a genuine conceptual effort has been done in order to allow the user to write only the minimum problem-specific code and to incrementally adapt an algorithm rather than entirely re-implementing it. Moreover, MOEO is itself extented to compose the full ParadisEO framework which is devoted to hybridization and parallel/distributed computing. Besides, MOEO has already been used to solve various academic problems likewise real-world applications.

The remainder of the paper is organized as follows. Section 2 gives the necessary background about MOO. Sections 3 and 4 describe the aims, the implementation and the provided features of the MOEO framework. In section 5, we present ParadisEO as well as the most common parallel/distributed models and hybridization mechanisms for multi-objective problems. In section 6, we survey some existing MOEO-designed applications and we introduce a Graphical User Interface for Multi-Objective Optimization (GUIMOO). Finally, the last section concludes the paper and highlights several perspectives about this work.

2 Multi-objective Optimization

Widely investigated since the end of the 1980's, multi-objective optimization concerns many areas of the industry (telecommunication, transport, aeronautics, etc). In this section, we briefly present some required notions about Pareto-based multi-objective optimization such as the formulation of a multi-objective optimization problem (MOOP) and some concepts relating to Pareto optimality (the reader is referred to [4,5] for more details).

Multi-objective optimization problem. A MOOP is defined by a decision space D, an objective space Z, and $n \geq 2$ objective functions $f_1, f_2, \ldots, f_n$. Each objective function can be either minimized or maximized. A solution $x = (x_1, x_2, \ldots, x_k)$ is represented by a vector of k decision variables. To each solution $x \in D$ is assigned exactly one objective vector $z \in Z$ on the basis of a vector function $F : X \rightarrow Z$ with $z = F(x) = (f_1(x), f_2(x), \ldots, f_n(x))$.

Pareto optimality. A multi-objective algorithm aims to approximate the set of Pareto optimal solutions according to F. A solution $x_a \in D$ is Pareto optimal if there exists no solution $x_b \in D$ that dominates x_a. For a minimization problem, the Pareto dominance relation is defined as follows:

Definition 1. *A solution $x_a \in D$ dominates a solution $x_b \in D$ if and only if $\forall i \in [1..n]$, $f_i(x_a) \leq f_i(x_b)$ and $\exists i \in [1..n]$ such as $f_i(x_a) < f_i(x_b)$.*

The overall goal is then to find a well-converged and well-diversified set of Pareto optimal solutions.

These basic notions already emphasize the most important points to consider for the design of a library devoted to evolutionary multi-objective optimization.

3 ParadisEO-MOEO Motivations

The 'EO' part of MOEO stands for *Evolving Objects*. EO is a C++ LGPL open-source object-oriented framework for evolutionary computation[1] that has been developed through an European joint work [15]. This library aims to provide a set of evolving objects dedicated to the flexible design of EAs. Furthermore, EO integrates many services including visualization facilities, on-line definition of parameters, application checkpointing, etc. MOEO is an extended version of the EO framework that includes some features related to Pareto-based multi-objective optimization. In this section, we present the goals of MOEO and we review some existing multi-objective optimization frameworks.

3.1 Goals

A framework is usually intended to be generic and could then be useful only if some important criteria are satisfied. Thence, the main goals of the MOEO framework are:

- *Services.* The framework must cover a wide range of features relating to Pareto-based multi-objective optimization.
- *Design and generic components.* MOEO must provide a whole architecture design of the solution method. This objective requires a clear and maximal conceptual distinction between the method and the problem representations. Therefore, the designer might only write the minimal problem-specific code, and the development process should be done in an incremental way.
- *Maximum code reuse.* The framework must allow the programmer to rewrite as little code as possible. Everthing that is already coded might be reusable. Then, it must be a commonplace to extend a problem from the mono-objective (and the EO framework) to the multi-objective case (and the MOEO framework), and from the classical to the parallel or the hybridizing case (and the whole version of the paradisEO framework, see section 5) without re-implementing the whole algorithm. For instance, it should not be necessary to re-code variation operators or solutions initialization.
- *Extensibility, flexibility and adaptability.* Some new features must easily be added or modified without implicating other components. Furthermore, existing components must be adaptable, as, in practice, existing problems evolve and new ones arise. Thence, MOEO must be a *white-box* framework (and not a black-box one); users must have access to source-code and must use inheritance or specialization to derive new components from base or abstract classes.

3.2 Existing Multi-objective Optimization Frameworks

Many frameworks dedicated to combinatorial optimization have been proposed. However, very few reached the whole goals stated above. A non-exhaustive comparative study between some existing multi-objective optimization frameworks

[1] EO is available at `http://eodev.sourceforge.net`

is given in table 1. These frameworks are distinguished according to the following criteria: the available metaheuristic(s), the framework type (black-box or white-box), the licence (open-source or not), the available metrics, the availability of hybridization and parallel features and the programming language.

Table 1. Existing frameworks for multi-objective optimization (hybrid. stands for hybridization features, // for parallel features, lang. for programming language, ref. for reference, EA for Evolutionary Algorithm, LS for Local Search, SA for Simulated Annealing, TS for Tabu Search, ACO for Ant Colony Optimization and PSO for Particle Swarm Optimization).

name	available metaheuristic(s)	black box	open source	metrics	hybrid.	//	lang.	ref.
MOEA for Matlab	EA	X	X / -	-	-	X	Matlab	[20]
MOMHLib++	EA, LS, SA	-	X	R, coverage, O_w, O_s, O_c	X	-	C++	-
Open BEAGLE	EA	-	X	-	-	-	C++	[9]
PISA	EA	X	X	I_ϵ, $I_{\epsilon+}$, R_2, R_3, S	-	-	-	[2]
TEA	EA	-	X	-	-	X	C++	[7]
ParadisEO-MOEO	EA (+ LS, SA, TS)	-	X	$I_{\epsilon+}$, I_{HD}, entropy, contribution	X	X	C++	-

As we can see, the whole presented frameworks are open-source (only MOEA, although open-source, is based on Matlab which is not). Moreover, a large part of these frameworks are white-box frameworks, that is to say that the source-code can easily be extended or adapted in order to offer the most possible flexibility. Even so, only a thin part includes all the major metaheuristics and some metrics for performance evaluation or comparison. Also, parallel models and hybridization mechanisms are both provided at once only within ParadisEO-MOEO. Furthermore, ParadisEO is portable on distributed-memory machines and shared-memory multi-processors, it offers a high flexibility and, to our knowledge, is the only one that is portable on grid computing.

4 ParadisEO-MOEO Implementation and Deployment

Using EO and MOEO, it is possible to build a complete multi-objective evolutionary computation application. Two major contributions of the MOEO framework refer to i) archive-related features and ii) multi-objective fitness assignment techniques. On each level of its architecture, a set of classes, devoted to the first or the second point, is provided. First, the general implementation of a multi-objective EA is shown in order to see how simple is to code a whole algorithm and to add or to change features. The implementation is conceptually divided into components so that different operators can be experimented without engendering significative modifications. A wide range of components are already provided, but new ones can easily be developed by the user with minimum code writing as MOEO is a white-box framework that tends to be flexible.

4.1 A General Eolutionary Algorithm Implementation

Here is a general implementation of a multi-objective EA:

```
unsigned N;                          /* population size              */
eoPop<EOT> population;               /* population initialization    */
moeoeoArchive<EOT> archive;          /* archive declaration          */
eoEvalFunc<EOT> eval;                /* raw fitnesses evaluation     */
eoInit<EOT> init;                    /* solutions initialization     */
eoTransform<EOT> transform;          /* variation operators          */
eoContinue<EOT> stop;                /* stopping criteria            */
eoCheckPoint<EOT> checkpoint;        /* application checkpointing    */
eoPerf2Worth<EOT,double> p2w;        /* multi-objective ranking      */
eoSelectOne<EOT> selectOne;          /* selector (built using p2w)   */
/* N-element selector */
eoSelect<EOT> select = eoSelectNumber<EOT>(selectOne, N);
eoReplacement<EOT> replace;          /* replacement                  */
/* algorithm definition */
eoEasyEA<EOT> algo(stop, eval, select, transform, replace);
algo(population);                     /* run the algorithm           */
```

All evolution-related objects are templatized[2] regarding to the type of individuals (`EOT`). And, the `eoEasyEA` class is used to define the algorithm.

4.2 Archive-Related Features

An essential point of Pareto-based optimization is the concept of archive. An archive is a secondary population that stores non-dominated solutions. Its main goal is to prevent that these solutions are not lost during the (stochastic) optimization process. As a consequence, it must be updated at each generation with newly found non-dominated individuals:

```
moeoArchiveUpdater<EOT> updater (archive, pop);
checkpoint.add (updater);
```

Moreover, it is possible to save the fitnesses of the archive's members at each generation into a file `fileName` in order to study the evolution of the non-dominated set:

```
moeoArchiveFitnessSavingUpdater<EOT> fitness (archive, fileName);
checkpoint.add (fitness);
```

[2] A template is a generic description of a class or a function created as an instance of the template at compile time.

Performance metrics. Commonly, analyzing Pareto set approximations is done using performance metrics. The entropy [1] and the contribution [16] are both already provided within MOEO, but other ones can easily be implemented and a few will be soon. For instance, it is possible to save the progression of the entropy measured on the archive at every generation into a file `fileName`:

```
moeoEntropyMetric<EOT> entropy;
moeoBinaryMetricSavingUpdater<EOT>
                        metricUpdater (entropy, archive, fileName);
checkpoint.add (metricUpdater);
```

Elitist selection. Another major use of an archive is elitism [4,5]. It consists in choosing individuals in the external population as well as in the current EA population during the selection phase of the algorithm, so that non-dominated solutions also contribute to the definition of variation operators.

```
eoSelectOne<EOT> popSelectOne;
eoSelectOne<EOT> archSelectOne;
selectOne = moeoSelectOneFromPopAndArch<EOT>
                  (popSelectOne, archSelectOne, archive, ratio);
```

At last, MOEO aims to constantly evolve and then, if need be, to provide further archive-related features in order to reflect the advances of the literature.

4.3 Implemented Multi-objective Fitness Assignment Strategies

In EO/MOEO, the fitness of a solution is represented by a vector of real numbers for which must be specified, for each criterion, if it is to be minimized or maximized. For multi-objective problems, fitness functions must convert raw fitnesses into fitness for selection. Various Pareto-based fitness assignment schemes are already implemented in EO and MOEO (see fig. 1), but this list is not exhaustive as the framework perpetually evolves and provides all that is necessary to easily implement new ones without a significant development effort.

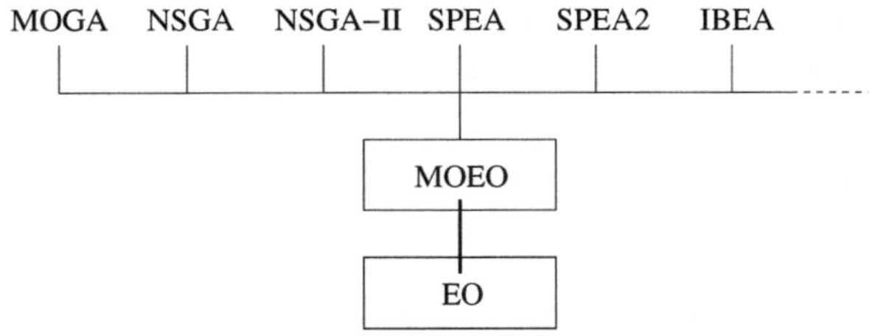

Fig. 1. Pareto-based multi-objective fitness assignment strategies proposed in EO and MOEO: MOGA, NSGA, NSGA-II, SPEA, SPEA2, IBEA, ...

Pareto-based fitness assignment was first proposed by Goldberg [10] to solve the problems of Schaffer's approach [17]. He suggested to use the Pareto dominance relation for ranking and selection. We here present all the fitness assignment strategies provided within MOEO as well as the code that is necessary to add or to modify in order to use them.

Pareto ranking (MOGA). In [8], Fonseca and Fleming proposed a variation of Goldberg's fitness assignment where a solution's rank corresponds to the number of solutions in the current population by which it is dominated (see fig. 2). Then, non-dominated individuals are all assigned the same rank, while dominated ones are penalized according to the population density in the corresponding region of the trade-off surface. The algorithm proceeds, first, by sorting the population according to the ranks previously determined. Then, fitness is assigned to solutions by interpolating from the best to the worst individuals in the population. Finally, fitnesses are averaged between solutions with the same rank.

```
eoDominanceMap<EOT> dominanceMap;
p2w = new eoParetoRanking<EOT> (dominanceMap);
```

Pareto sharing. As Goldberg and Deb noticed in [11], a fitness assignment like the previous one tends to produce premature convergence, what does not guarantee a uniformly sampled final Pareto approximation set. To avoid that, Fonseca and Fleming [8] modified the strategy above by implementing fitness sharing in the objective space to distribute the population over the Pareto-optimal region.

```
double nicheSize;
p2w = new moeoParetoSharing<EOT> (nicheSize);
```

NSGA. Srinivas and Deb [18] introduced another variation of Goldberg's fitness assigment in a similar way than [8], but based on Goldberg's version of Pareto-ranking. This algorithm, called *Non-dominated Sorting Genetic Alorithm*, classifies the solutions into several classes (or fronts). A solution that belongs to a class does not dominate another one from the same class. Then, individuals from the first front all belong to the best non-dominated set of the population; individuals from the second front all belong to the second best non-dominated set; and so on (see fig. 3). Logically, the best fitness value is assigned to solutions of the first class, because they are closest to the true Pareto-optimal front of the problem. This tends to search for solutions located in non-dominated regions. Additionally, a fitness sharing procedure helps to distribute the population over these regions.

```
double nicheSize;
p2w = new eoNDSorting_I<EOT> (nicheSize);
```

NSGA-II. In [6], Deb et al. introduced a modified version of NSGA. This new algorithm, called NSGA-II, is computationnaly more efficient, uses elitism, and keeps diversity without specifying any parameters by using a crowded tournament selection operator.

```
p2w = new eoNDSorting_II<EOT> ();
```

SPEA. Zitzler and Thiele [23] proposed an elitist algorithm called the *Strength Pareto Evolutionary Algorithm.* It maintains an external population (an archive) that stores a fixed number of non-dominated solutions found during the optimization process. For each member of the archive, a *strength value*, proportional to the number of solutions this member dominates, is computed. Then, the fitness of a solution is obtained according to the strength values of the archive's individuals that dominate it (see fig. 4). Moreover, a clustering method is used to keep diversity.

```
unsigned archiveSize;
select = moeoSPSelect_I<EOT> (N, archiveSize);
```

SPEA2. An improved version of SPEA, namely SPEA2, has been introduced by Zitzler et al. [22]. The three main differences of SPEA2 in comparison to its predecessor are that it incorporates: (i) a fine-grained fitness assignment strategy that takes into account the number of individuals that a solution dominates and is dominated by; (ii) a density estimation technique that leads the search process more precisely; (iii) an enhanced archive truncation method that ensures the preservation of boundary solutions.

```
unsigned archiveSize;
select = moeoSPSelect_II<EOT> (N, archiveSize);
```

IBEA. Introduced by Zitzler and Künzli [21], the *Indicator-Based Evolutionary Algorithm* (IBEA) has the characteristic to compute fitness values by comparing individuals on the basis of an arbitrary binary quality indicator I (also called binary performance metric). Thereby, no particular diversification mechanisms, such as fitness sharing, is necessary. The indicator, determined according to the decision maker preferences, denotes the overall goal of the optimization process. Thus, the fitness of a solution measures its usefulness according to the optimization goal. In MOEO, two binary quality indicators are proposed: the additive ϵ-indicator [24] and the I_{HD}-indicator [24] that is based on the hypervolume concept [23] (see fig. 5). However, everything is implemented to easily develop other indicators to be used with IBEA (see [24] for an overview about quality indicators).

```
moeoSolutionVsSolutionBM<EOT> I;
double kappa;        /* scaling factor */
p2w = new moeoIBSorting<EOT> (I, kappa);
```

New fitness assignment strategies. MOEO aims to be extensible, flexible and easily adaptable. All its components are generic in order to provide a modular architecture design that allows the user to quickly and conveniently develop a new fitness assignment scheme with a minimum code writing. The aim is to follow the new strategies coming from the litterature and, if need be, to provide any additional components required for their implementation.

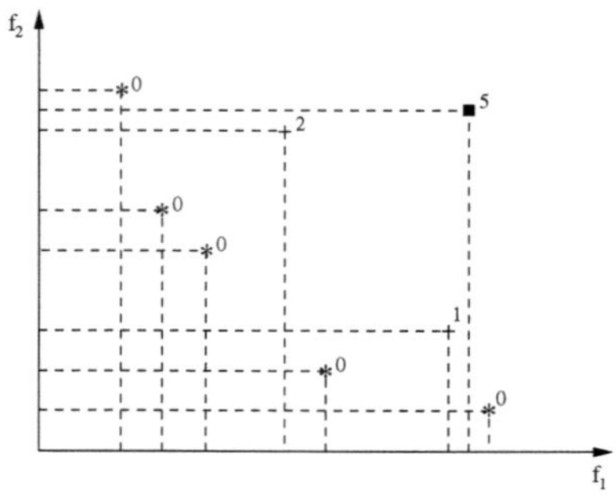

Fig. 2. The Pareto ranking fitness assignment scheme on a two-objective minimization problem

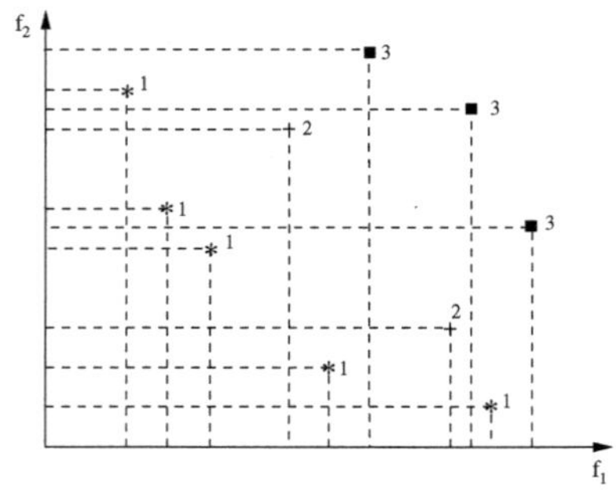

Fig. 3. The NSGA fitness assignment scheme on a two-objective minimization problem

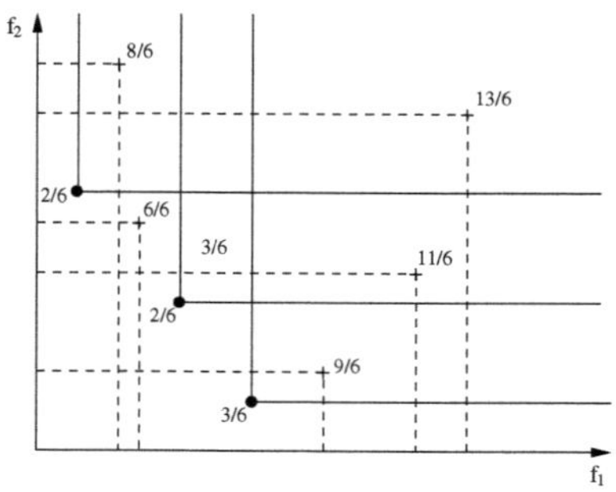

Fig. 4. The SPEA fitness assignment scheme on a two-objective minimization problem (the external population members are shown by circles and the EA population members are shown by crosses)

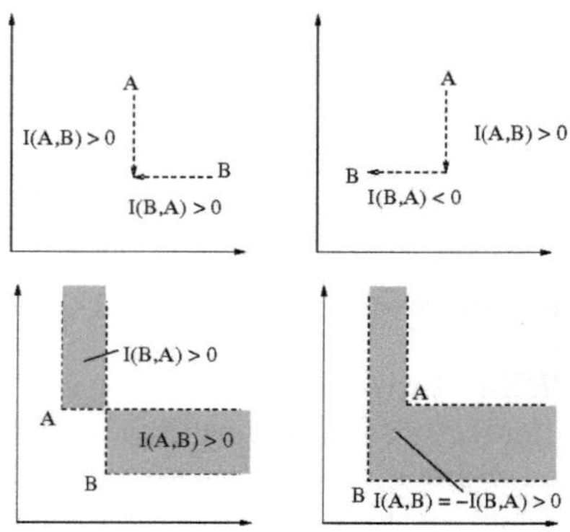

Fig. 5. Illustration of binary quality indicators for a two-objective minimization problem (top: $I_{\epsilon+}$-indicator, bottom: I_{HD}-indicator)

5 Parallelism and Hybridization Design for Multi-objective Problems Using the ParadisEO Framework

In practice, multi-objective optimization problems are varied, they perpetually evolve (with regards to the needs, the constraints, the objectives, etc), they handle a high number of decision variables and they have to deal with instances of increasing size. Despite that, the overall goal is still to find near Pareto-optimal solutions in a tractable time. Then, classical approaches are not sufficient, and hybridization features as well as large scale parallelism must be considered to tackle this kind of problem. As shown in figure 6, in addition to parallel and distributed environments, the ParadisEO framework embeds a broad range of features, including evolutionary algorithms (as it is based on the EO framework), various local searches (as it is based on the MO framework) and multi-objective mechanisms (as it is based on the MOEO framework). Furthermore, its generic aspect allows to easily add some of its features to a MOEO-designed problem.

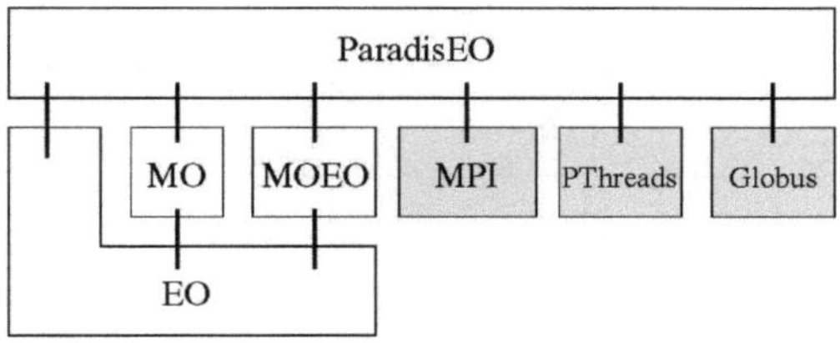

Fig. 6. ParadisEO architecture: Evolving Objects (EO) for the design of population-based metaheuristics, Moving Objects (MO) for the design of solution-based metaheuristics and their hybridization with EAs, Multi-Objective EO (MOEO) for multi-objective optimization, and ParadisEO for parallel and distributed models. These models are portable on many execution platforms thanks to the MPI, PThreads and Globus standard libraries.

5.1 Parallel Distributed Evolutionary Algorithms

Basically, three major parallel models can be distinguished [3]: the cooperative island model, the parallel population evaluation model, and the distributed single-solution evaluation model. These models are all illustrated in figure 7.

The cooperative island model. A number of EAs are simultaneously deployed to cooperate with the aim of improving the solutions's robustness. Each of them performs a search on a sub-population. Then, exchanges of genetic materials are performed in an asynchronous way to diversify the search into the target sub-populations. This allows to delay the global convergence, especially when the EAs are heterogeneous with respect to the variation operators. Individuals migrations are conducted by various parameters and are performed in a regular or irregular way.

The parallel population evaluation model. The evaluation step of an EA is generally the most time-consuming. Therefore, in order to speed up the search, this centralized parallel model distributes the evaluation of the evolving population. As they require a global management of the population, the selection, the transformation and the replacement operations are applied by a master process. At each generation, it distributes the set of newly generated solutions between different workers that evaluate and return back these solutions as well as their fitness values. A particularly efficient execution is often obtained when the ratio between communication and computation is high.

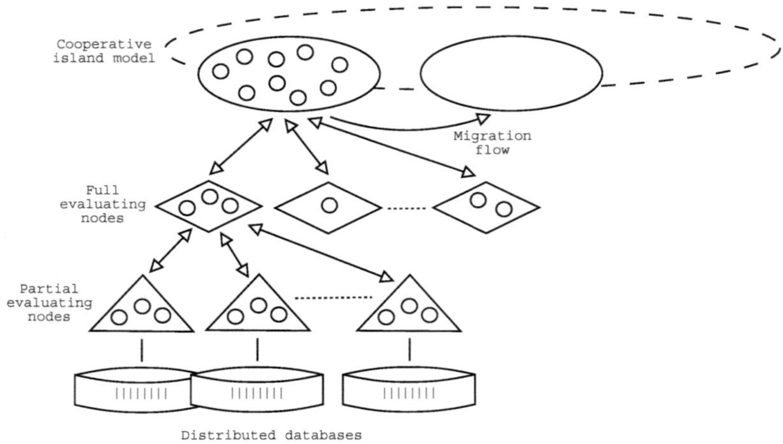

Fig. 7. Three major parallel models for multi-objective EAs

The distributed single-solution evaluation model. The fitness of each solution is evaluated in a parallel centralized way. Such a model is especially interesting when the evaluation of a solution can be itself parallelized as it requires, for instance, an access to large databases distributed among various processing nodes.

Those three parallel and distributed models are all provided within ParadisEO [3]. They are implemented using MPI, PThreads and Globus standard libraries, and are thus portable on different execution platforms such as parallel computing, cluster computing, internet computing and grid computing. Moreover, their deployement is transparent as the user does not need to manage the communications and threads-based concurrency.

5.2 Hybridization

Hybridization have acquired a considerable interest in the field of optimization these last years [19]. A wide variety of hybrid approaches exists in the literature. And, for many academic and real-world applications, best found solutions are obtained by hybrid algorithms. In the multi-objective context, EAs are generally hybridized with local search methods in order to apply the local search algorithm on a selected individual, or to find non-dominated solutions in the neighborhood of an interesting region of the objective space. In [19], two levels (low and high) and two modes (relay and cooperative) of hybridization are distinguished.

The low-level hybrid algorithms address the functional composition of a single optimization method. A given function of a metaheuristic is replaced by another metaheuristic. For high-level hybridization, the different metaheuristics are self-contained. There is no direct relationship between the internal workings of a metaheuristic. Besides, for relay hybridization, a set of metaheuristics is applied the one after another in a pipeline way, each one using the output of the previous one as its input. Contrarily, cooperative hybridization represents a teamwork optimization model in which parallel cooperating agents perform a search in a solution space and exchange solutions with the others.

The ParadisEO framework provides all these most common hybridization mechanisms [3] that can thus directly be applied to a MOEO-designed application in a fast and simple way. They can naturally be exploited to make cooperating metaheuristics belonging either to the same or to different families.

6 Applications

ParadisEO and MOEO have been applied to many areas where multi-objective optimization is required. Before presenting three examples of applications drawn from varied fields that have been implemented within MOEO, we will introduce some prerequisites concerning performance evaluation and results analysis.

6.1 Preliminaries: GUIMOO

In multi-objective optimization, a fundamental part is the performance comparison of various algorithms. Therefore, the question arises on the way of evaluating the quality of Pareto set approximations. To achieve that, let us introduce GUIMOO[3] (a Graphical User Interface for Multi-Objective Optimization). This platform-independant free software is dedicated to the analysis of results for multi-objective optimization and is able to handle different input and output formats. Its main features are:

- The on-line and off-line visualization (in 2 or 3 dimensions) of Pareto set approximations. A Pareto set approximation can be characterized by its (dis)continuity, (dis)convexity, multi-modality, ... Such an information can be useful to help an expert to build more efficient metaheuristics.
- A number of metrics for quantitative and qualitative performance evaluation or comparison [24] (contribution, entropy, generational distance, spacing, coverage of two sets, coverage difference, S-metric, D-metric and R-metrics).

Furthermore, GUIMOO aims to be generic. Its architecture allows to easily customize it in order to provide more functionalities to tackle specific applications (telecom, genomics, engineering design, etc).

6.2 Examples

MOEO has been experimented on different academic and industrial problems. In this section, we present three applications that show the wide range of potential of this framework as it has been applied to scheduling problems, continuous optimization and data-mining applications.

---

[3] GUIMOO is available at `http://guimoo.gforge.inria.fr`

A bi-objective flow-shop scheduling problem. The flow-shop is one of the most widely investigated scheduling problem of the literature. But, the majority of studies considers it on a single-criterion form. However, other objectives than minimizing the makespan can be taken into account, like, *e.g.*, minimizing the total tardiness.

Electromagnetic properties of conducting polymer composites in the microwave band. Due to the proliferation of electromagnetic interferences, designing protecting material for high frequencies equipments has become an important problem. In [14], a new multi-objective model is proposed to design the different layers of a conducting polymer. To solve this model, a multi-objective continuous genetic algorithm is used. This algorithm offers several solutions with different physical properties and different costs.

Knowledge discovery in biological data from microarray experiments. The problem of analyzing microarray data is actually a major issue in genomics. Often used techniques are clustering and classification. In [13], the authors propose to analyze those data through association rules. The problem is modeled as a multi-objective rule mining problem and a genetic algorithm is used to explore the large search space associated. Thence, MOGA permitted to present previously undiscovered knowledge.

7 Conclusion and Perspectives

In this paper, we introduced ParadisEO-MOEO, a framework dedicated to the reusable design of evolutionary multi-objective optimization applications[4]. It provides the most common Pareto-based multi-objective fitness assignment shemes (MOGA, NSGA, NSGA-II, SPEA, SPEA2, IBEA, ...) as well as fitness sharing and a wide range of archive-related features such as non-dominated solutions storage, elitism and performance metrics computation. Moreover, the whole version of ParadisEO, a complete framework for the design of parallel/distributed and hybrid metaheuristics, and GUIMOO, a software for the analysis and the comparison of Pareto set approximations, have both been presented. These frameworks have all been applied to many type of applications, from academic to real-world problems.

ParadisEO-MOEO is an open-source white-box object-oriented framework that aims to simplify and speed up the incremental implementation of a whole efficient multi-objective optimization program. In order to confer a maximum design and code reuse, it is based on a clear conceptual distinction between the metaheuristics and the problem representations. This separation is expressed at the implementation level, and the hierarchical classes that are provided allow the designer to extend the framework by inheritance or specialization. Furthermore, the fine-grained components of ParadisEO-MOEO confer a high flexibility compared to other frameworks. Modifying existing components or adding

[4] ParadisEO-MOEO is available at `http://paradiseo.gforge.inria.fr`

new ones can easily be done without impacting the whole application. Besides, ParadisEO-MOEO is a part of the ParadisEO framework that covers the most common parallel/distributed models and hybridization mechanisms. The user can thus directly include some ParadisEO features into an application designed using ParadisEO-MOEO in a fast and simple way.

In the future, ParadisEO-MOEO needs to constantly evolve in order to reflect the advances of the literature. New Pareto-based fitness assignment strategies as well as new performance metrics for Pareto set approximations should also be proposed before long. Moreover, a major extension of ParadisEO-MOEO would be to allow the design of exact methods as well as their hybridization with already provided metaheuristics. Besides, it would be interesting to introduce new specific concepts emerging from multi-criterion optimization such as the consideration of uncertainty through stochastic or fuzzy multi-objective problems.

References

1. Basseur M., Seynhaeve F., Talbi E.-G.: Design of multi-objective evolutionary algorithms: Application to the flow-shop scheduling problem. In Congress on Evolutionary Computation (CEC'02), Honolulu, Hawaii, USA (2002) 1151–1156
2. Bleuler S., Laumanns M., Thiele L., Zitzler E.: PISA — A Platform and Programming Language Independent Interface for Search Algorithms. In Fonseca C. M. et al. (eds.): Evolutionary Multi-Criterion Optimization, Second International Conference (EMO 2003). Lecture Notes in Computer Science **2632**, Springer, Faro, Portugal (2003) 494–508
3. Cahon S., Melab N., Talbi E.-G.: ParadisEO: A Framework for the Reusable Design of Parallel and Distributed Metaheuristics. Journal of Heuristics **10(3)** (2004) 357–380
4. Coello Coello C. A., Van Veldhuizen D. A., Lamont G. B.: Evolutionary Algorithms for Solving Multi-Objective Optimization Problems. Kluwer Academic Publishers, New York, USA (2002)
5. Deb K.: Multi-Objective Optimization using Evolutionary Algorithms. Wiley, Chichester, UK (2001)
6. Deb K., Agrawal S., Pratap A., Meyarivan T. A Fast Elitist Non-Dominated Sorting Genetic Algorithm for Multi-Objective Optimization: NSGA-II. In Proc. of the 6th International Conference on Parallel Problem Solving from Nature (PPSN VI), Paris, France (2000) 849–858
7. Emmerich M., Hosenberg R.: TEA - A Toolbox for the Design of Parallel Evolutionary Algorithms in C++. Technical report CI-106/01, SFB 531, University of Dortmund, Germany (2001)
8. Fonseca C. M., Fleming P. J.: Genetic Algorithms for Multiobjective Optimization: Formulation, Discussion and Generalization. In Forrest S. (ed.): Proc. of the 5th International Conference on Genetic Algorithms, Morgan Kaufmann publishers, San Mateo, California (1993) 416–423
9. Gagné C., Parizeau M.: Genericity in Evolutionary Computation Software Tools: Principles and Case Study. International Journal on Artificial Intelligence Tools **15(2)** (2006) 173–194
10. Goldberg D. E.: Genetic Algorithms in Search, Optimization and Machine Learning. Addison-Wesley, Boston, MA, USA (1989)

11. Goldberg D. E., Deb K.: A Comparative Analysis of Selection Schemes Used in Genetic Algorithms. In Rawlins G. (ed.): Foundations of Genetic Algorithms, Morgan Kaufmann, San Mateo, CA, USA (1991) 69–93
12. Holland J. H.: Adaptation in Natural and Artificial Systems. University of Michigan Press, Ann Abor, MI, USA (1975)
13. Jourdan L., Khabzaoui M., Dhaenens C., Talbi E.-G.: A Hybrid Evolutionary Algorithm for Knowledge Discovery in Microarray Experiments. In Olariu S. et al. (eds.): Handbook of Bioinspired Algorithms and Applications, CRC Press, USA, chapter 28 (2005) 489–505
14. Jourdan L., Legrand T., Talbi E.-G., Wojkiewicz, J.-L.: Mono and Multi-objective continuous optimization for conducting polymer composites. ECCO/CO 2006, Porto, Portugal (2006)
15. Keijzer M., Merelo J.-J., Romero G., Schoenauer M.: Evolving Objects: a general purpose evolutionary computation library. In Proc. of the 5th International Conference on Artificial Evolution (EA'01), Le Creusot, France (2001) 231–244
16. Meunier H., Talbi E.-G., Reininger P.: A multiobjective genetic algorithm for radio network optimization. In Proc. of the 2000 Congress on Evolutionary Computation (CEC'00). IEEE Press (2000) 317-324
17. Schaffer J. D.: Multiple Objective Optimization with Vector Evaluated Genetic Algorithms. In Grefensette J. J. (eds.): Proc. of the 1st International Conference on Genetic Algorithms, Pittsburgh, PA, USA (1985) 93–100
18. Srinivas N., Deb K.: Multiobjective Optimization Using Nondominated Sorting in Genetic Algorithms. Evolutionary Computation **2(3)** (1994) 221–248
19. Talbi E.-G.: A Taxonomy of Hybrid Metaheuristics. Journal of Heuristics **8(5)** (2002) 541–564
20. Tan K. C., Lee T. H., Khoo D., Khor E. F., Kannan R. S.: MOEA Toolbox for Computer Aided Multi-Objective Optimization. In Proc. of the 2000 Congress on Evolutionary Computation (CEC'00). IEEE Press (2000) 38–45
21. Zitzler E., Künzli S.: Indicator-Based Selection in Multiobjective Search. In Parallel Problem Solving from Nature (PPSN VIII). Lecture Notes in Computer Science **3242**, Springer, Birmingham, UK (2004) 832–842
22. Zitzler E., Laumanns M., Thiele L.: SPEA2: Improving the Strength Pareto Evolutionary Algorithm. TIK Report Nr. 103, Computer Engineering and Networks Lab (TIK), Swiss Federal Institute of Technology (ETH) Zurich, (2001)
23. Zitzler E., Thiele L.: Multiobjective Evolutionary Algorithms: A Comparative Case Study and the Strength Pareto Approach. IEEE Transactions on Evolutionary Computation **3(4)** (1999) 257–271
24. Zitzler E., Thiele L., Laumanns M., Fonseca C. M., Grunert da Fonseca V.: Performance Assessment of Multiobjective Optimizers: An Analysis and Review. IEEE Transactions on Evolutionary Computation **7(2)** (2003) 117–132

Multi-objective Evolutionary Algorithms for Resource Allocation Problems

Dilip Datta[1], Kalyanmoy Deb[1], and Carlos M. Fonseca[2]

[1] Indian Institute of Technology Kanpur, Kanpur - 208 016, India
{ddatta,deb}@iitk.ac.in
[2] Universidade do Algarve, Campus de Gambelas, 8000-117 Faro, Portugal
cmfonsec@ualg.pt

Abstract. The inadequacy of classical methods to handle *resource allocation problems* (RAPs) draw the attention of evolutionary algorithms (EAs) to these problems. The potentialities of EAs are exploited in the present work for handling two such RAPs of quite different natures, namely (1) university class timetabling problem and (2) land-use management problem. In many cases, these problems are over-simplified by ignoring many important aspects, such as different types of constraints and multiple objective functions. In the present work, two EA-based multi-objective optimizers are developed for handling these two problems by considering various aspects that are common to most of their variants. Finally, the similarities between the problems, and also between their solution techniques, are analyzed through the application of the developed optimizers on two real problems.

1 Introduction

A *resource allocation problem* (RAP) involves the allotment of limited amount of resources to certain number of competitive events for achieving the most effective allotment of resources. It is a combinatorial optimization problem, and encountered in a variety of areas in operations research and management science, such as load distribution, production planning, computer scheduling, portfolio selection, apportionment, and so on. An RAP usually contains huge number of integer variables and constraints, a discrete search space, and multiple objectives, which make classical methods, such as linear and integer programming approaches, inadequate to handle RAPs. These inadequacy of classical methods draw the attention of non-classical techniques towards RAPs, among which evolutionary algorithms (EAs) are the widely preferred non-classical techniques. The potentialities of EAs are exploited in the present work for handling two such RAPs of quite different natures, namely (1) university class timetabling problem and (2) land-use management problem. The class timetabling problem involves the scheduling of classes[1], students, teachers and rooms at a fixed number of time-slots. Traditionally, the problem is solved manually by *trial and hit* method, where a valid solution is not guaranteed. Even if a valid solution

[1] A class is a meeting of a group of students and a teacher in a room for a lecture.

S. Obayashi et al. (Eds.): EMO 2007, LNCS 4403, pp. 401–416, 2007.

is found, it is likely to miss far better solutions. These uncertainties motivate for the scientific study of the problem, and to develop an automated solution technique for it. Despite multiple criteria to be met simultaneously, the problem is generally tackled as a single-objective optimization problem. Moreover, most of the earlier works are concentrated on school timetabling, and only a few on university class timetabling. On the other hand, in many cases, the problem is over-simplified by skipping many complex class-structures, such as multi-slot, split, combined and group classes. Land-use management is another scheduling problem, where different competitive land uses, such as agriculture, forest or industries, are to be allocated to different units of a landscape to meet the desired objectives of land managers [29]. The problem has emerged today as a problem of great concern. Due to increasing human activities on land to meet various demands, land and its resources have been under tremendous pressure, which are causing significant transformations of land for a variety of land uses. Most of the land use changes occur without any logical planning to their long-term environmental impacts. Global warming, soil degradation, deforestation, loss of biodiversity, are all consequences of mismanagement of land and its resources. Thus, various land-use management practices are to be understood for developing an integrated land-use policy framework for improving soil quality, ensuing biomass production and food security, maintaining environmental stability, and extending socio-economic benefits [14]. These incommensurable objectives can be achieved only through optimization tools. Owing to the difficulty of deploying field experiments for direct assessment, it is important to enhance the knowledge by developing mechanistic models through extensive study. However, the problem is very new to the computational community, and only a little work has been done so far in this area. NSGA-II-UCTO and NSGA-II-LUM [8,9], two versions of EA-based multi-objective optimizer NSGA-II [11], are developed in the present work for optimizing university class timetabling problem and land-use management problem, respectively. NSGA-II-UCTO is applied for scheduling the classes of Indian Institute of Technology Kanpur, where much better solutions are obtained than a manually prepared solution which is in use. On the other hand, NSGA-II-LUM is applied to a Mediterranean landscape from Southern Portugal. However, due to non-availability of any existing solution for this landscape, the performance of NSGA-II-LUM could not be compared.

2 Related Works

The class timetabling problem drew the attention of the researchers starting with the study of Gotlieb [15], who formulated the problem by considering that each lecture contained one group of students, one teacher, and any number of time-slots which could be chosen freely. Since then the problem is being studied using different methods under different conditions. Initially it was mostly applied to schools. Since the problem in schools is relatively simple because of their simple class structures, classical methods, such as linear or integer programming approaches [19,30], could be used easily. However, the gradual consideration of different cases of universities, which contain different types of complex classes, is increasing the complexity of the problem. As a result, classical methods become inadequate to handle the problem, particularly the huge number of integer variables, a discrete search space, and multiple objective functions, involved with it. These inadequacy of classical methods draw the attention of non-classical techniques to this problem. Worth mentioning non-classical techniques, that are being used to the problem, are genetic algorithms (GAs) [1,5], neural network [22], and tabu search algorithm [6]. However, compared to other non-classical methods, the widely used are the GAs/EAs. This might be due to the reason that once the objectives and constraints are defined, EAs appear to offer good solutions by evolving without a problem solving strategy [2]. A few worth mentioning EAs, used to solve this problem, are found in [1,4,8]. In case of land-use management problem, Bhadwal and Singh [3] made a comparative estimate of land-use and carbon sequestration potential under land-use and relative biomass changes. Other works on the carbon sequestration potential can be found in [17,21]. Liu and Bliss [20] simulated the influences of rainfall-induced soil erosion and deposition on carbon dynamics in soil profiles. Due to the same reasons as in the case of class timetabling problem, classical methods are not fully capable to handle this problem also. Hence, non-classical techniques, particularly GAs/EAs, are motivated to handle the problem [13]. Matthews et al. [23,24] explored the potential of applying GAs to spatially integrated land-use management problem. A bi-objective GA is developed in one work [25] to define the trade-off structure of the objective functions. Stewart et al. [29] used another bi-objective GA to a spatially integrated problem. Seixas et al. [27] also proposed a bi-objective EA to study future land-use configuration.

3 Class Timetabling and Land-Use Management Problems as Multi-objective Optimization Problems

In the present work, both the class timetabling and land-use management problems are modelled as multi-objective optimization problems, subject to a number of constraints which are common to most of their variants.

3.1 Objective Functions in University Class Timetabling Problem

The class timetabling problem involves several criteria that must be satisfied simultaneously, such as compliance with regulations, proper utilization of resources, and satisfaction of people's preferences [28], which are generally known as *soft constraints*. Many researchers transform all such constraints into a single one by using a pre-defined weightage to each constraint, and then treat it as the only objective function of the problem [1,4,26]. However, this approach is likely to miss good solutions as the weightages of different constraints are usually not known beforehand. Hence, the problem essentially becomes a multi-objective optimization problem, where each constraint can be treated as an objective function with an automatically adjusted weightage. However, though the imposition of excess objectives would produce a greater preferred timetable, but it will increase the computational complexity of a problem. Hence, the number of objectives should be as less as possible. Only the following two conflicting objectives are considered in the present work:

1. Minimize the average number of weekly free time-slots between two classes of a student (f_1), and
2. Maximize the weekly average span of time-slots of classes of a teacher (f_2).

The objective function f_1 implies a compact timetable, whereas f_2 conflicts with it, and seeks a well-spread timetable.

3.2 Objective Functions in Land-Use Management Problem

In this case also, the problem was handled earlier as a single-objective optimization problem with the only aim of either increasing the productivity or fulfilling the immediate need. This has caused a numerous damage to the environment in long term, such as global warming, soil degradation, loss of biodiversity, and so on. Hence, to safeguard the environment from destroying further, it has become urgent need to address different issues, such as improving soil quality, ensuing biomass production and food security, maintaining environmental stability, and extending socio-economic benefits [14]. These different issues often conflict with each other, and require the problem to be treated as a multi-objective optimization problem. Realizing the urgent needs of the current society, the following three objective functions are considered in the present work:

1. Maximize net present economic return (f_1),
2. Maximize net amount of carbon sequestration (f_2), and
3. Minimize net amount of soil erosion (f_3).

The objective functions f_2 and f_3 are burning issues to today's researchers as the remedies to global warming and soil degradation.

3.3 Constraints in University Class Timetabling Problem

The number and type of constraints in this problem vary from university to university. The following six types of constraints, which are known as *hard constraints* and must be satisfied by a solution, are considered in the present work.

1. A student should have only one class at a time.
2. A teacher should have only one class at a time.
3. A room should be booked only for one class at a time (a set of combined classes[2] may be treated as a single class).
4. A course[3] should have only one class on a day.
5. A class should be scheduled only in a specific room, if required, otherwise in any room which has sufficient sitting capacity for the students of the class.
6. A class should be scheduled only at a specific time-slot, if required.

[2] Six different types of classes, namely single-slot, multi-slot, split, combined, open and group classes, are considered in the present work [8].

[3] A course is a subject to be studied, e.g. *Theory of Optimization*, or *Fluid Mechanics*.

3.4 Constraints in Land-Use Management Problem

Constraints in this problem arise from various factors. For instance, the productivity of a land-use depends on spatial and temporal conditions of a landscape, availability of water, its influence on runoff, and so on [16]. Landscape ecology and biodiversity are also major issues to be taken into account during any management planning [13]. Based on such requirements, the constraints in this problem can be classified as below [27]:

- Physical constraints on geomorphological structure:
  1. A unit should be assigned a land-use, only if the soil in it is permitted to hold that land-use,
  2. The slope of a unit should be within the permitted range of slope for the land-use applied in that unit,
  3. The aridity index of a unit should be within the permitted range of aridity index for the land-use applied in that unit, and
  4. Topographic soil wetness index (TSWI) of a unit should be within the permitted range of TSWI for the land-use applied in that unit.

- Ecological constraints on spatial coherence:
  1. Area in a patch[4] of a land-use should be within the permitted range of area in a patch for that land-use, and
  2. Total area under a land-use in a landscape should be within the permitted range of total area for that land-use.

4 NSGA-II-UCTO and NSGA-II-LUM

Two chromosome representations and a number of EA operators (crossover and mutation operators) are proposed in the present work for class timetabling and land-use management problems [8,9]. Few guidance are also proposed to speed up EA search by satisfying some constraints. Then those are incorporated in NSGA-II [11], an EA-based multi-objective optimizer, and named them as NSGA-II-UCTO (NSGA-II as university class timetable optimizer) and NSGA-II-LUM (NSGA-II in land-use management). The proposed chromosome representations, EA operators and guidance, along with the salient features of NSGA-II-UCTO and NSGA-II-LUM, are stated briefly in the following subsections.

4.1 Chromosome Representations

The chromosome representation, proposed in the present work for handling class timetabling problem, is a two-dimensional matrix. Each column of the matrix represents a time-slot, and a row represents a room, i.e. a chromosome is a vector of time-slots (genes), and a time-slot is a vector of rooms. Hence, as shown in Fig. 1, the value of each cell of the matrix represents the class scheduled in the corresponding room and time-slot. In this representation, it is not required to

[4] A patch in a landscape is a set of contiguous units under the same land-use.

go through the entire chromosome to check if, at any particular time, more than one class were scheduled in a room, or to a teacher or student. The column of the matrix, representing that time-slot, is sufficient for checking any such possibility. Moreover, complex classes, such as multi-slot, split, combined and group classes, can also be easily scheduled and checked. For example, Fig. 1 shows the scheduling of 2-slot class $C35$ in room R_3 at time-slots T_2 and T_3, and combine classes $C17$ and $C36$ in room R_r at time-slot T_3. On the other hand, the

R/T	$\mathbf{T}_1$	$\mathbf{T}_2$	$\mathbf{T}_3$	..	$\mathbf{T}_j$	..	$\mathbf{T}_t$
$\mathbf{R}_1$	C20	C11	C39	...	C05	...	C16
$\mathbf{R}_2$	C33	C21	C15	...	C40	...	C12
$\mathbf{R}_3$	C01	C35		...	C07	...	C08
..	...	...	...	...	...	...	C27
$\mathbf{R}_k$	C13	C02	C14	...	C22	...	C38
..	...	...	...	...	...	...	C18
$\mathbf{R}_r$	C06	C04	C17 C36	...	C28	...	C31

Fig. 1. Representation of class timetable

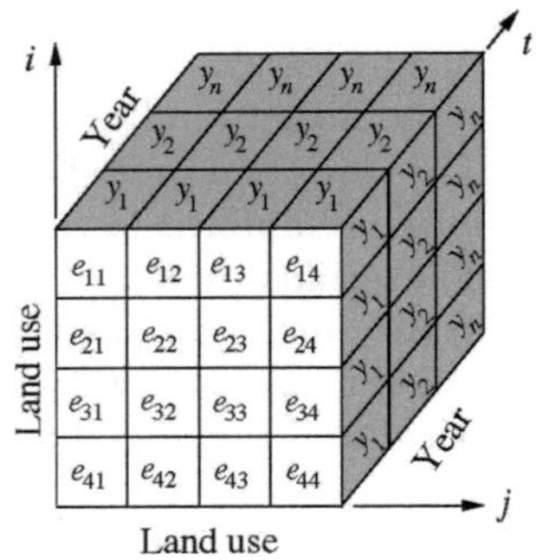

Fig. 2. Representation of a landscape

chromosome representation, proposed here for land-use management problem, is a two-dimensional grid of genes. In this representation, the position of each gene represents a unit of a landscape, and its value determines the land-use for that unit. In addition to the two-dimensional grid, a third dimension is also used in this representation to represent the dynamics of a landscape over a planning horizon, i.e. a chromosome is a two-dimensional grid of genes, where each gene is again a vector of years of the planning horizon. This representation is shown diagrammatically in Fig. 2, where the position of a gene in ij-plane represents a unit and its value gives the land-use for that unit, and the t-th axis represents the years of the planning horizon.

4.2 Crossover Operators

Unlike traditional crossover operators, where few random genes are exchanged between two parent solutions to generate two offspring, a special crossover operator (XVRA) is developed in the present work for university class timetabling problem. On the other hand, a problem independent crossover operator (XTD) is adopted from [7] for land-use management problem.

- **Crossover for Valid Resource Allocation (XVRA):** In XVRA, developed for class timetabling problem, a random feasible portion of an offspring is first generated by two parent solutions. For instance, in generating the first offspring, some genes are randomly selected from two parents, and the information in them are transferred to the corresponding genes of the offspring.

However, for maintaining the feasibility of the offspring, only those information from the second parent should be transferred for which no constraint is violated. Finally, the remaining portion of the offspring is completed by a heuristic approach (HA) (addressed in Sect. 4.4). The HA is used for generating only feasible offspring. Similarly, the second offspring is generated from the second parent, assisted by the first parent and the HA.

- **Two-Dimensional Crossover (XTD):** XTD is originally developed by Datta and Deb [7] for topology optimization of structures. It is a problem independent operator, and based just on two-dimensional structure of a problem. The working procedure of XTD is also very simple. A chromosome is first divided into four blocks by a randomly selected pair of row and column, and then a random block is exchanged with a similar one from another chromosome. In land-use management problem, it works in exchanging one or more patches/parts of patches. However, XTD can not preserve the feasibility of solutions of this problem.

4.3 Mutation Operators

Like XVRA, a special mutation operator (MRRA) is developed in the present work for class timetabling problem. Similarly, two spacial mutation operators (MBC and MSIS) are developed here for land-use management problem. MRRA and MBC generate offspring in the respective problems, and MSIS steers infeasible offspring of land-use management problem towards feasible region.

- **Mutation for Reshuffling Resource Allocation (MRRA):** In MRRA, allotment of resources to events are reshuffled. In case of class timetabling problem, it swaps classes at two slots. Two random time-slots are first taken, and then the classes in each room at those time-slots are swapped, provided no constraint is violated. If any multi-slot class appears in any room, the range of the time-slots are expanded accordingly. Similarly, the number of rooms to be handled at a time is expanded if any group class appears in any room. If the classes of any room cannot be swapped due to the violation of any constraint, an attempt may be made for that room by taking another pair of random time-slots.
- **Mutation on Boundary Cells (MBC):** The main function of an optimizer, in land-use management problem, is to alter the size of a patch of a land-use to meet the objectives of a problem. When the size of a patch is reduced, the size(s) of one or more of its adjacent patches is(are) increased. The size of a patch can be reduced by replacing the land uses of its boundary cells with those of its adjacent cells. MBC is developed by exploiting this problem information. All the boundary cells of a solution are first sorted out. Then a boundary cell is chosen randomly with some probability, and its land-use is replaced as above, provided the new land-use is applicable in that boundary cell.
- **Mutation for Steering Infeasible Solution (MSIS):** MSIS is developed here for steering infeasible solutions towards feasible region. It is not explicitly a repairing mechanism, but mutation is performed only in those patches

which violate minimum patch-size constraints. If the area of the patch, in which a chosen boundary cell belongs, is less than its minimum requirement, the adjacent cells of the boundary cell, having different land uses than in the patch, are merged in the patch, provided the new cells satisfy the physical constraints for the land-use of the patch.

4.4 Guidance to Speed Up the Search for Optimum Solutions

In general, EAs are capable enough to handle infeasible solutions. However, sometimes they suffer from huge computational time, or may even fail, in handling infeasible solutions of many complex problems [8]. Though MSIS is developed to take care of infeasible solutions in land-use management problem, its progress is very slow. Hence, to speed up EA search, one heuristic approach for generating feasible solutions in class timetabling problem, and two other guidance for generating near feasible solutions in land-use management problem are proposed here.

1. **Heuristic Approach (HA) for University Class Timetabling Problem:** It is observed in [8] that, in order to get a feasible solution, random scheduling of classes may not work. Rather classes should be scheduled in some *order*, based on the complexities of classes. Hence, a sequential *Heuristic Approach* (HA) is developed here, where classes are first sorted in descending order of their complexities, and then rooms and time-slots are assigned to them. The detail of the approach is as below:
   (a) All classes are first sorted in the following order:
      i. Ascending order of number of specific time-slots. If not mentioned, this number for a class is the number of total available time-slots.
      ii. If numbers of specific time-slots are equal, descending order of number of time-slots per class.
      iii. If numbers of specific time-slots as well as numbers of time-slots/class are equal, ascending order of number of specific rooms. If not mentioned, this number for a class is the number of total available rooms.
      iv. If numbers of specific time-slots, numbers of time-slots/class as well as numbers of specific rooms are equal, preference to group/split classes (group classes are not supposed to be split).
   (b) Once the sorting of classes is over, they are taken in order to assign rooms and time-slots to them, respecting hard constraints. Since sitting capacity of a room also plays an important role in this problem, a class may be first scheduled in a random room, and then an exhaustive search may be performed among the remaining rooms for finding a suitable smaller one. This will avoid the possibility of occupation of an over-size room by a smaller class, which can be used for a bigger class.
2. **Guidance-1 for Land-Use Management Problem:** During initialization, an attempt may be made for satisfying, if possible, the patch-size constraints by scheduling a land-use in sufficient number of contiguous units.

3. **Guidance-2 for Land-Use Management Problem:** During optimization process, a patch, having less area than the specified one, may be deleted. This can be made by merging the cells of the patch in its adjacent patches, where the physical constraints for the merging cells are satisfied.

4.5 Salient Features of NSGA-II-UCTO and NSGA-II-LUM

1. One of the chromosome representations of Sect. 4.1 (as applicable to NSGA-II-UCTO/NSGA-II-LUM) is used to form an EA population of N solutions.
2. The HA or Guidance-1 (as applicable to NSGA-II-UCTO/NSGA-II-LUM), addressed in Sect. 4.4, is used to initialize the solutions of the population.
3. *Crowded tournament selection operator* is used to form a *mating pool* of N solutions. It is done by randomly selecting two solutions from the population, and sending a copy of the best one to the mating pool. The process is continued until the mating pool is filled up with N solutions. The mating pool is later used by EA operators for generating offspring [11].
4. XVRA/XTD (as applicable to NSGA-II-UCTO/NSGA-II-LUM), addressed in Sect. 4.2, is used for generating a new population of N offspring.
5. MRRA/MBC (as applicable to NSGA-II-UCTO/NSGA-II-LUM), addressed in Sect. 4.3, is used for mutating the offspring of the new population.
6. MSIS, addressed in Sect. 4.3, is used in NSGA-II-LUM for steering an infeasible offspring, if any, towards the feasible region.
7. Next, the Guidance-2, addressed in Sect. 4.4, is used in NSGA-II-LUM for satisfying patch-size constraints, as much as possible.
8. Both the populations, obtained so far, are combined to form a combined population of 2N solutions.
9. Based on ranks and crowding distances [11], the best N solutions from the combined population are picked up to form a single population.
10. Steps (3)-(9) are repeated for required number of generations.
11. Results of the final population are accepted as the optimum results.

To enhance the probabilities of true convergence of NSGA-II-LUM, a local search strategy [12] is also proposed to use to the final Pareto front of NSGA-II-LUM. The search is applied to each boundary unit of a solution, and the land-use in it is replaced by the one in one of its adjacent units. However, the change is accepted only if some improvement is found in the solution.

5 Two Case Studies (IITK2 and LBAP)

Two real problems are considered in the present work, one is the class timetabling problem of Indian Institute of Technology Kanpur (IIT-Kanpur) and the other is the land-use management for a Mediterranean landscape, located in Baixo Alentejo, Southern Portugal. The class timetable of IIT-Kanpur is composed of two phases. The first phase contains the common compulsory classes of its all under-graduate programmes (B.Tech and integrated M.Sc), and the timetable of this phase is prepared by a central team. Then the available time-slots and rooms are allotted to individual departments to prepare the second phase of the timetable, where departments schedule their all compulsory and open classes. The scheduling of even-semester classes is considered in the present work, and it is named here as IITK2 in short (detail of IITK2 is available in [10]). In addition to the classes of the first phase, slots for departmental compulsory and open classes of B.Tech programme of one department are also included in IITK2. This inclusion is only for illustrative purpose to show that both the phases of a timetable can be scheduled by a single team only. Class-structures of IITK2 are very complex, which include different types of classes, such as single-slot, multi-slot, split, and group classes. Laboratory classes are spanned over 3 consecutive time-slots, and many of them are split up to 5 parts. Most of the tutorial classes are grouped, and few groups contain up to 20 classes. Total number of only common compulsory classes is 242, where there are 11 simple single-slot, 12 three-slot split, and 219 single-slot group classes. These classes, spanning over 266 time-slots, are to be taught to around 2000 students by 103 teachers, and to be scheduled in 40 rooms (including laboratories) in 5 days/week, where each day has 8 time-slots. The objective functions and constraints in the problem are the same with those of Sect. 3.1 and 3.3, respectively. Hence, as per the formulation in [8], the total number of constraints, involved only with common compulsory classes, is (S+M+R)TD+CD+3E = 47071, where S, M, R, T, D, C and E represent, respectively, the total numbers of students, teachers, rooms, time-slots/day, days/week, courses, and classes. The land-use management problem of the landscape of Baixo Alentejo is named here as LBAP in short. The latitude and longitude at the centroid of the landscape are $38^0 0' 50.3''$ N and $7^0 51' 56.94''$ W, respectively. According to the available data of LBAP (detail of LBAP is available in [9,27]), the landscape is divided into 100×100 units, each unit of which is subject to a number of physical constraints as mentioned in Sect. 3.4. Five different land uses, namely annual agriculture, permanent agriculture, mixed agriculture, forest, and shrubs, are allowed in the landscape, which are subject to two ecological constraints as mentioned in Sect. 3.4. Hence, as per the formulation in [9], there are total 100×100 = 10000 physical constraints, and $2(\sum_{e=1}^{5} N_e + 5)$ ecological constraints, where N_e is the number of patches under land-use e. The number of ecological constraints is not fixed, but varies with the number of patches under a land-use in a particular instance.

5.1 NSGA-II-UCTO to IITK2

It is shown elsewhere [8] that, even if NSGA-II-UCTO is able to handle infeasible solutions, its progress is very slow. Hence, no attempt is made here to test whether NSGA-II-UCTO can be applied with less guidance by allowing constraint violation. Instead of that, it is applied directly maintaining the feasibility of solutions. Presently IITK2 is solved manually with the only aim of producing a feasible timetable. Applying NSGA-II-UCTO, not only feasible, but much better solutions than a manually prepared one could be obtained. A comparison plot of the results for common compulsory classes of IITK2 is shown

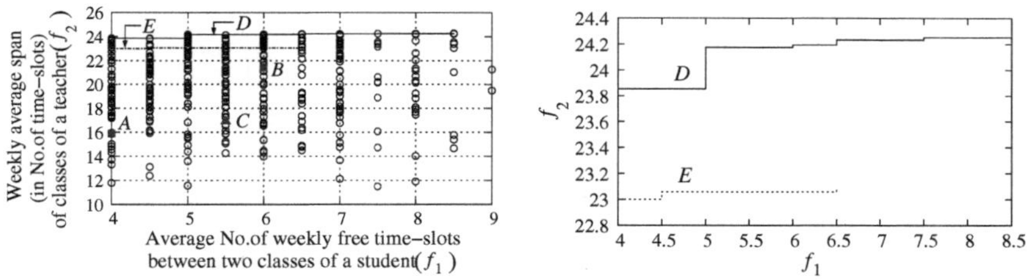

(a) Comparison of different solutions (b) Pareto fronts

Fig. 3. Manual and different simulated solutions of IITK2

in Fig. 3(a). Points A and B represent, respectively, single-objective optimization of f_1 and f_2 under crossover probability $(p_c) = 0.90$ and evolving mutation probability (p_m). Curve D is the Pareto front of multi-objective optimization under $p_c = 0.90$ and evolving p_m, while curve E is that under evolving p_m only (no crossover). Point C is the manually prepared result which is in use. It is observed that the solutions, obtained from both the single-objective and multi-objective optimization, are better than a manually prepared solution. In case of multi-objective optimization, Pareto fronts of which are shown separately in Fig. 3(b), variation in f_2 is very small. This might be due to the fact that most of the classes of IITK2 are grouped, for which it is very tough to shift classes to different slots. IITK2 is tested separately with combined XVRA and MRRA, and MRRA alone. Plots of one run with combined XVRA and MRRA are shown in Fig. 4. It is observed in Fig. 4(b) that on an average of 5 minutes 35 seconds per

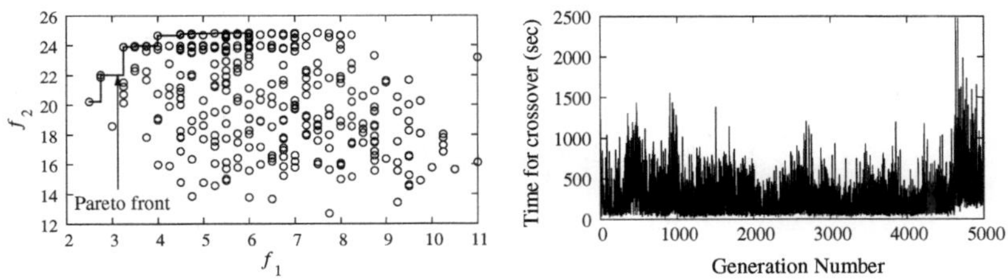

(a) Obtained all solutions (b) Time taken by XVRA

Fig. 4. Multi-objective solutions of IITK2 ($p_c = 0.90$ and $p_m = 0.01$)

generation is required by XVRA. Total execution time of 465 hours 14 minutes 39 seconds, in Linux environment in a Pentium IV machine with 1.0 GB RAM and 2.933 GHz processor, is required for 5000 generations. When solved IITK2 using MRRA alone, total execution time for 5000 generations comes down to 11 hours 31 minutes 42 seconds only. However, both visual observation and *statistical comparison test* [18], conducted on a number of runs, show that the overall performance of NSGA-II-UCTO is better with combined XVRA and MRRA than with MRRA alone. Compared Pareto fronts of both the cases for three different runs are shown in Fig. 5. While studying different cases of IITK2, it

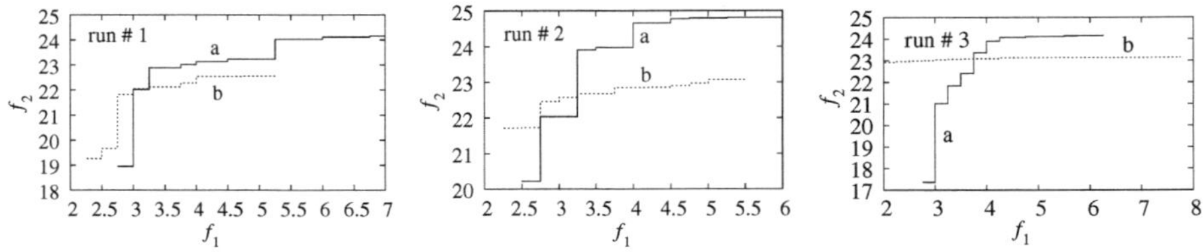

Fig. 5. Comparison of Pareto fronts of IITK2. Curve (a): combined XVRA and MRRA with $p_c = 0.90$ and $p_m = 0.01$, and curve (b): only MRRA with $p_m = 0.90$.

is also noticed that NSGA-II-UCTO has the tendency for scheduling particular classes in particular slots in different solutions. The percentage of classes in the same slots increases with the increasing complexities of classes.

5.2 NSGA-II-LUM to LBAP

LBAP is considered under XTD, MBC, MSIS, and the two guidance of Sect. 4.4. EA parameters, p_c and p_m, are chosen as 0.9 and 0.05, respectively. Then NSGA-II-LUM is executed for 5000 generations, with 50 solutions in its population, which take 82 hours 3 minutes 52 seconds. Local search strategy takes another 56 minutes 59 seconds for 50 solutions of the final Pareto front of NSGA-II-LUM. It is observed in Fig. 6(a) that significant improvements in both values and spread

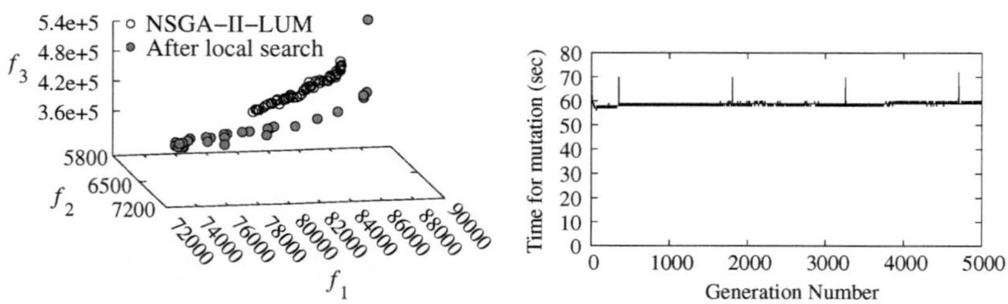

(a) Pareto fronts (b) Mutation time (sec)

Fig. 6. Solution of LBAP under $p_c = 0.90$ and $p_m = 0.010$

of the Pareto front are obtained from the local search. Fig. 6(b) shows that most of the computational time is required by MBC, which takes around 59 seconds per generation. While studying the distribution of land uses in different solutions of LBAP (two solutions are shown in Fig. 7), it is observed that NSGA-II-LUM has the tendency for allocating particular land uses in particular locations of the landscape. Permanent agriculture is always allocated over the entire landscape, except the South-West corner where shrubs are allocated. Annual agriculture is also mostly preferred in the South-West corner. However, there is no preference of specific locations for mixed agriculture and forest, and these are allocated over the entire landscape.

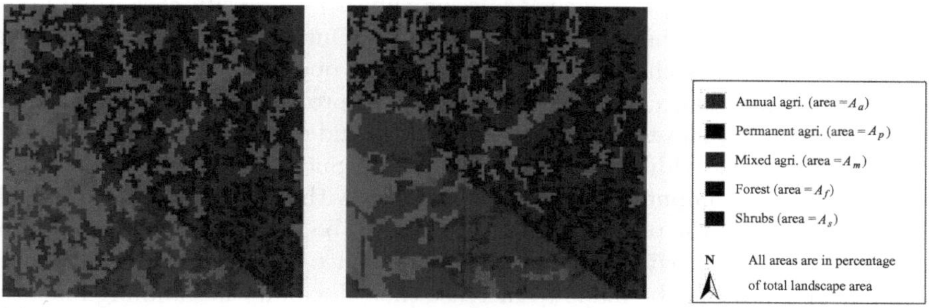

(a) f_1=84495, f_2=6698, f_3=511877, A_a=17.58, A_p=29.32, A_m=13.73, A_f=29.33, A_s=10.04

(b) f_1=84495, f_2=6698, f_3=511877, A_a=17.58, A_p=29.32, A_m=13.73, A_f=29.33, A_s=10.04

(c) Color-code for different land uses of LBAP

Fig. 7. Distribution of land uses in LBAP

6 Similarity Among RAPs

During the above study on class timetabling and land-use management problems, a number of similarities between the problems, and also between their solution techniques, are encountered, which are summarized below:

- **Similarities between the problems:**
  1. Both are highly constrained, and spatial and temporal-based multi-objective combinatorial optimization problems.
  2. In class timetabling problem, each event (class) is required to be scheduled exactly once, while an event (land-use) in land-use management problem may be scheduled multiple times within a certain range. These requirements slightly differ the problems from one another.
- **Similarities between their solution techniques:**
  1. Classical methods are not fully capable to handle the problems, and hence, non-classical techniques are motivated for these problems.
  2. Both NSGA-II-UCTO and NSGA-II-LUM need some guidance to speed up the search for optimum solutions, or even for feasible solutions.
  3. Individual problem information can be exploited for designing similar chromosome representations and EA operators for both the problems.

7 Conclusions

It is learnt that classical methods, such as linear and integer programming approaches, are not fully capable to handle resource allocation problems (RAPs), particularly the huge number of integer variables and constraints, a discrete search space, and multiple objectives, involved with an RAP. Hence, non-classical heuristic techniques, such as evolutionary algorithms (EAs), are motivated for these problems. In the present study, the potentialities of multi-objective EAs are exploited for handling two such RAPs, namely (1) university class timetabling problem and (2) land-use management problem. During the study, it is observed that similar chromosome representations and EA operators, based on individual problem information, can be used in both the problems. Future reseach may attempt to design more general representations and operators, by reducing their dependency on individual problem information. Such operators would possibly reduce the huge computational time as involved with the present operators. The EAs, developed in the present study, are either to be fed by feasible/near feasible solutions or to be provided some run-time guidance to speed up the EA search. An attempt may also be made to increase their flexibility in handling infeasible solutions. On the other hand, the present study is based on NSGA-II only. Other evolutionary multi-objective optimizers (EMOs) may be attempted to compare the performance of different EMOs on RAPs.

Acknowledgement. This work was partially supported by an Indo-Portuguese scientific and technical cooperation grant from DST, India, and GRICES, Portugal.

References

1. Abramson, D., & Abela, J., A parallel genetic algorithm for solving the school timetabling problem. In Proceedings of 15 Australian Comp. Sc. Conference, Hobart, 1992, 1-11.
2. Al-Attar, A., White Paper: A hybrid GA-heuristic search strategy. AI Expert, USA, 1994.
3. Bhadwal, S., & Singh, R., Carbon sequestration estimates for forestry options under different land-use scenarios in India, Current Sc., 2002, 83(11), 1380-1386.
4. Blum, C., Correia, S., Dorigo, M., Paechter, B., Rossi-Doria, O., & Snoek, M., A GA evolving instructions for a timetable builder. In Proceedings of the Practice and Theory of Automated Timetabling (PATAT'02), 2002, 120-123.
5. Colorni, A., Dorigo, M., & Maniezzo, V., Genetic algorithms and highly constrained problems: The time-table case. In Proceedings of the first International Workshop on Parallel Problem Solving from Nature (PPSN-1, 1990), Lecture Notes in Computer Science (1991), Springer-Verlag, 1990, 496, 55-59.
6. Costa, D., A tabu search algorithm for computing an operational timetable. European Journal of Op. Res., 1994, 76(1), 98-110.
7. Datta, D., & Deb, K., Design of optimum cross-sections for load-carrying members using multi-objective evolutionary algorithms. Int. J. of Systemics, Cybernetics and Informatics (IJSCI, Jan. 2006), 2006, 57-63.
8. Datta, D., Deb, K., & Fonseca, C.M., Multi-objective evolutionary algorithm for university class timetabling problem, In Evolutionary Scheduling, Springer-Verlag (in Press), 2006.
9. Datta, D., Deb, K., Fonseca, C.M., Lobo, F., & Condado, P. Multi-objective evolutionary algorithm for land-use management problem, KanGAL Technical Report No. 2006005, IIT-Kanpur, 2006.
10. Datta, D., Deb, K., & Fonseca, C.M., Solving class timetabling problem of IIT Kanpur using multi-objective evolutionary algorithm, KanGAL Technical Report No. 2006006, IIT-Kanpur, 2006.

11. Deb, K., Multi-Objective Optimization using Evolutionary Algorithms. John Wiley & Sons Ltd, Chichester, England, 2001.
12. Deb, K., & Goel, T., A hybrid multi-objective evolutionary approach to engineering shape design, In Proceedings of the First International Conference on Evolutionary Multi-Criterion Optimization (EMO-2001), Zürich, Switzerland, 2001, 385-399.
13. Ducheyne, E., Multiple objective forest management using GIS and genetic optimisation techniques, PhD Thesis, Faculty of Agricultural and Applied Biol. Sc., University of Ghent, Belgium, 2003.
14. Gautam, N.C., & Raghavswamy, V., Preface, In Land use/land cover and management practices in India, BS Publications, Hyderabad (India), 2004.
15. Gotlieb, C.C., The construction of class-teacher timetables. In Proceedings of IFIP Congress, North-Holland Pub. Co., Amsterdam, 1962, 73-77.
16. Huston, M., The need for science and technology in land management, Online Book - The International Development Research Centre, http://www.idrc.ca/en/ev-29587-201-1-DO_TOPIC.html, January-2006.
17. Kerr, S., Liu, S., Pfaff, A.S.P., & Hughes, R.F., Carbon dynamics and land-use choices: building a regional-scale multidisciplinary model, Journal of Environmental Management, 2003, 69, 25-37.
18. Knowles, J.D., & Corne, D.W., Approximating the nondominated front using the Pareto archived evolution strategy. Evo. Comput., 2000, 8(2), 149-172.
19. Lawrie, N.L., An integer programming model of a school timetabling problem. The Computer Journal, 1969, 12, 307-316.
20. Liu, S., & Bliss, N., Modeling carbon dynamics in vegetation and soil under the impact of soil erosion and deposition, Global Biochem. Cycles, 2003, 17(2), 1074.
21. Liu, S., Liu, J., & Loveland, T.R., Spatial-temporal carbon sequestration under land use and land cover change, In Proceedings of 12th International Conference on Geoinformatics - Geospatial Information Research: Bridging the Pacific and Atlantic University of Gävle, Sweden, 2004, 525-532.
22. Looi, C., Neural network methods in combinatorial optimization. Computers and Op. Res., 1992, 19(3/4), 191-208.
23. Matthews, K.B., Craw, S., MacKenzie, I., Elder, S., & Sibbald, A.R., Applying genetic algorithms to land use planning, In Proc. of the 18th Workshop of the UK Planning and Scheduling Special Interest Group, University of Salford, UK, 1999, 109-115.
24. Matthews, K.B., Sibbald, A.R., & Craw, S., Implementation of a spatial decision support system for rural land use planning: integrating GIS and environmental models with search and optimisation algorithms, Comp. & Electr. in Agri., 1999, 23, 9-26.
25. Matthews, K.B., Buchan, K., Sibbald, A.R., & Craw, S., Using soft-systems methods to evaluate the outputs from multi-objective land use planning tools, In Integrated Assessment and Decision Support: Proceedings of the 1st biennial meeting of the International Environmental Modelling and Software Society, University of Lugano, Switzerland, 2002, 3, 247-252.
26. Piola, R., Evolutionary solutions to a highly constrained combinatorial problem. In Proceedings of IEEE Conference on Evol. Comput. (First World Congress on Computational Intelligence), Orlando, Florida, 1994, 1, 446-450.
27. Seixas, J., Nunes, J.P., Lourenço, P., Lobo, F., & Condado, P., GeneticLand: Modeling land use change using evolutionary algorithms, In Proc. of the 45th Congress of the European Regional Sc. Asso., Land Use and Water Management in a Sustainable Network Society, Vrije Universiteit Amsterdam, 2005, 23-27.

28. Silva, J.D.L., Burke, E.K., & Petrovic, S., An introduction to multiobjective metaheuristics for scheduling and timetabling. In Proceedings of Metaheuristic for Multiobjective Optimisation, Lecture Notes in Economics and Math. Systems, Springer, 2004, 535, 91-129.
29. Stewart, T.J., Janssen, R., & Herwijnen, M.V., A genetic algorithm approach to multiobjective land use planning, Comp. & Op. Res., 2004, 31, 2293-2313.
30. Tripathy, A., School timetabling - A case in large binary integer linear programming. Management Science, 1984, 30(12), 1473-1489.

Multi-objective Pole Placement with Evolutionary Algorithms

Gustavo Sánchez, Minaya Villasana, and Miguel Strefezza

Universidad Simón Bolívar. Venezuela
gsanchez@usb.ve

Abstract. Multi-Objective Evolutionary Algorithms (MOEA) have been succesfully applied to solve control problems. However, many improvements are still to be accomplished. In this paper a new approach is proposed: the Multi-Objective Pole Placement with Evolutionary Algorithms (MOPPEA). The design method is based upon using complex-valued chromosomes that contain information about closed-loop poles, which are then placed through an output feedback controller. Specific cross-over and mutation operators were implemented in simple but efficient ways. The performance is tested on a mixed multi-objective $\mathcal{H}_2/\mathcal{H}_\infty$ control problem.

Keywords: Multi-objective control; Pole placement; Evolutionary Algorithms.

1 Introduction

Most control design problems can be solved using numerical optimization. Thus, Multi-Objective Evolutionary Algorithms (MOEA) have been successfully applied for this purpose, provided the problem is non-convex in the optimization parameters and cannot be efficiently solved by conventional local optimization algorithms [1]. To illustrate this point, let's review five previous related publications. Note that an excellent survey can be found in [2].

In 1995, Fonseca and Fleming [3] developed an approach to multiple objective and constraint handling with genetic algorithms, with application to control system design. A Multiple Objective Genetic Algorithm (MOGA) was proposed, which is still frequently used in many applications.

In 1995, Whidborne *et al* [4] compared the performance of three search methods. Two were based on hill-climbing techniques: Nelder-Mead Dynamic Min-Max (NMDM) and Moving Boundaries Process (MBP). The third was precisely MOGA. The three were found to be useful for interactive multi-objective controller design. Besides, the author introduced MODCONS: a MATLAB toolbox for Multi-Objective Design of Control Systems.

In 2000, Herreros [5] proposed an algorithm for Multi-objective Robust Control Design (MRCD). It was tested against a Linear Matrix Inequalities (LMI) approach for mixed multi-objective $\mathcal{H}_2/\mathcal{H}_\infty$ control problems. An adaptive search space was proposed, motivated by two reasons: the selection of the initial

S. Obayashi et al. (Eds.): EMO 2007, LNCS 4403, pp. 417–427, 2007.

population and the delimitation of the search space. In fact, these are still open problems in the field.

In 2005, Liu and Ishihara [6] discussed the use of multi-objective genetic algorithms and the method of inequalities. The performance of the proposed design method was tested on a special set of benchmark control problems.

In 2006, Molina-Cristobal *et al* [7] compared MOGA against a LMI approach to find the trade-off of a multi-objective $\mathcal{H}_2/\mathcal{H}_\infty$ control problem. The author asserted that MOGA could find an improved Pareto-optimal front compared to the LMI approach.

Despite all this important work, many improvements remain still to be accomplished. This includes elements like parameters tuning, space search adaptation, performance assessment and controller coding. Particularly, regarding the latter, a new approach for solving the design problem is proposed in this work: the Multi-Objective Pole Placement with Evolutionary Algorithms (MOPPEA) technique.

The main idea is using complex-valued chromosomes, containing information about closed-loop poles. This representation allows poles be placed through a classical observer-based feedback controller, based on the information contained within each chromosome. Note that, unlike this representation, usually controllers are coded in terms of real parameters [8]. Specific cross-over and mutation operators were implemented in simple but efficient ways.

The exposition is organized as follows. In section **2**, we formulate the controller design problem. The proposed solution method is described in section **3**. In section **4**, it is applied to solve a mixed $\mathcal{H}_2/\mathcal{H}_\infty$ control problem. Finally, conclusions are drawn in section **5**.

2 Problem Formulation

2.1 Preliminaries

Let $L_2^{p\times m}$ be the set of $p \times m$ matrix functions $f : \mathcal{R}_0^+ \rightarrow \mathcal{R}^{p\times m}$ such that:

$$\int_0^{+\infty} trace[f^T(t)f(t)]dt < \infty \tag{1}$$

Let $\mathcal{R}(s)^{p\times m}$ be the set of $p \times m$ rational complex matrix functions G : $\overline{\mathfrak{C}}^+ \rightarrow \mathfrak{C}^{p\times m}$ such that:

$$G_{ij}(s) = \frac{b_{nu}s^{nu} + b_{nu-1}s^{nu-1} + .. + b_1 s + b_0}{s^{nd} + a_{nd-1}s^{nu-1} + .. + a_1 s + a_0} \quad i = 1, 2, ..., p \quad j = 1, 2, ..., m \tag{2}$$

and $nd \geq nu$, $\forall G_{ij}(s)$; $a_k, b_h \in \mathcal{R}$, $k = 1, 2, ..., nd$ and $h = 1, 2, ..., nu$.

2.2 Control Topology

We focus on the problem of designing a linear controller $Kc \in \mathcal{R}(s)^{n_u \times n_y}$ for the continuous-time model shown in figure 1. Matrices $A \in \mathcal{R}^{n \times n}$, $B \in \mathcal{R}^{n \times n_u}$ and $C \in \mathcal{R}^{n_y \times n}$ denote the given plant state matrices.

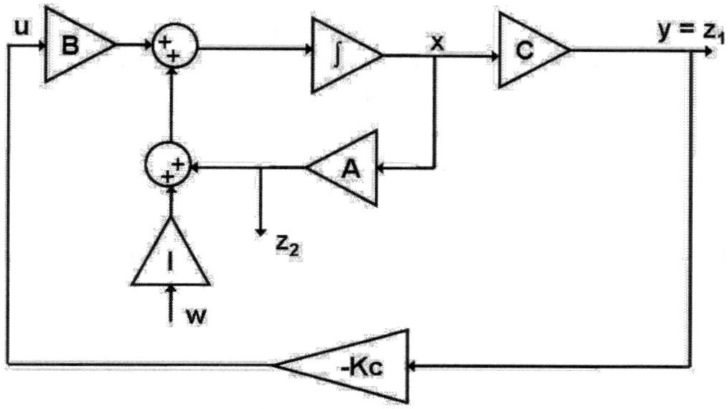

Fig. 1. Continuous-time closed-loop design model

As usual, $w \in L_2^{n_w \times 1}$denote the exogenous input, $z_1 \in L_2^{n_{z_1} \times 1}$ and $z_2 \in L_2^{n_{z_2} \times 1}$ represent the outputs to be regulated, while $u \in L_2^{n_u \times 1}$ and $y \in L_2^{n_y \times 1}$ represent the control input and the measured output respectively. It is assumed that $G(s) = C(sI - A)^{-1}B$ is strictly proper, stabilizable from u and detectable from y. The open-loop state-space equations are:

$$\begin{cases} \dot{x} = Ax + Iw + Bu \\ z_1 = y = Cx \\ z_2 = Ax \end{cases} \tag{3}$$

The state-space equations of the controller are:

$$\begin{cases} \dot{x}_c = A_c x_c - Ly \\ u = Kx_c \end{cases} \tag{4}$$

with $K \in \mathcal{R}^{n_u \times n}, L \in \mathcal{R}^{n \times n_y}$ and

$$Kc(s) = -K(sI - A_c)^{-1}L \tag{5}$$

Let

$$T_1(K_c) = T_{z_1 w}(K_c) \tag{6}$$
$$T_2(K_c) = T_{z_2 w}(K_c) \tag{7}$$

be the closed-loop transfer function from w to z_1 and z_2 respectively.

2.3 The Control Problem

The mixed $\mathcal{H}_2/\mathcal{H}_\infty$ Objective Control Problem (MOCP) can now be stated as:

$$\min_{Kc \in \mathcal{K}^n \subset \mathcal{R}(s)^{n_u \times n_y}} \begin{pmatrix} \|T_1(K_c)\|_2 \\ \|T_2(K_c)\|_\infty \end{pmatrix} \tag{8}$$

where $\mathcal{K}^n \subset \mathcal{R}(s)^{n_u \times n_y}$ is the set of all stabilizing controllers of degree n. Note that in this problem the term "*min*" means finding a solution which give the values of the objective functions acceptable to the designer [9]. Note also that control law specifications (i.e. controller structure) are not considered in this formulation.

3 MOPPEA: A Linear Controller Design Method

A flowchart of the proposed design method is shown in figure 2.

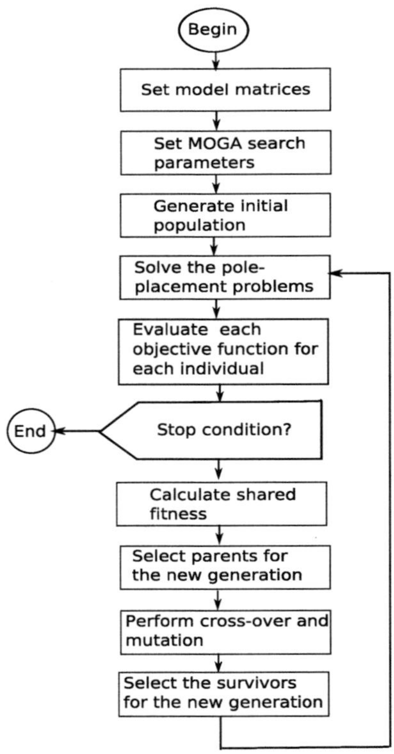

Fig. 2. MOPPEA flowchart

3.1 The Pole Placement Method

An output feedback controller can be designed by combining a full information controller (i.e. a controller which has immediate access to the information about all states) with a state observer (a subsystem which attempts to reconstruct the current states using the information about past measurements and control inputs). The resulting output feedback sub-system is called "observer-based controller" and has the following state-equations:

$$\begin{cases} \dot{\widehat{x}} = (A + BK + LC)\,\widehat{x} - Ly \\ u = K\widehat{x} \end{cases} \tag{9}$$

where $\widehat{x}$ is the estimated state. Thus, in our framework the system closed-loop state equations, using the estimation error $\widehat{e} = x - \widehat{x}$ as state variable are:

$$\begin{cases} \begin{pmatrix} \dot{x} \\ \dot{\widehat{e}} \end{pmatrix} = \begin{pmatrix} A + BK & -BK \\ 0 & A + LC \end{pmatrix} \begin{pmatrix} x \\ \widehat{e} \end{pmatrix} + \begin{pmatrix} I \\ I \end{pmatrix} w \\ z_1 = y = (C\ 0) \begin{pmatrix} x \\ \widehat{e} \end{pmatrix} \\ z_2 = (A\ 0) \begin{pmatrix} x \\ \widehat{e} \end{pmatrix} \end{cases} \tag{10}$$

Let $pk \in \mathfrak{C}^{n_k}$ and $pl \in \mathfrak{C}^{n_l}$ be the eigenvalues of $A + BK$ and $A + LC$ respectively. To assure closed-loop system stability, the gain matrix K and L must be calculated in such way that pk and pl belong to $\mathfrak{C}^-$ (open left-half complex plan). Several algorithms to compute K and L from pk, pl, B and C have been proposed [10]. In this work, the MATLAB *place* function has been used.

3.2 Problem Reformulation

The key concept of the proposed design method is using an evolutionary process in order to evolve pk and pl, moving across $\mathfrak{C}^-$ in order to find the best feasible closed-loop poles locations. Thus, the MOCP problem (see equation 8) can be stated again as:

$$\min_{pk,pl \in \mathfrak{C}^-} \begin{pmatrix} \|T_1(pk, pl)\|_2 \\ \|T_2(pk, pl)\|_\infty \end{pmatrix} \tag{11}$$

Note that, in this case, the stability restriction $Kc \in \mathcal{K}^n$ has disappeared. This is the main advantage of the proposed method when compared to previous works (see [5] and [7]).

3.3 Representation of Individuals

Different representations have been proposed for controllers (see [5], [7] and [8]). In this work, chromosomes are complex-valued vectors containing the concatenation of pk and pl (see figure 3). A recursive algorithm has been implemented for randomly generating the initial population (see appendix A). Parameters *supr* and *supi* determine the size of the initial search space.

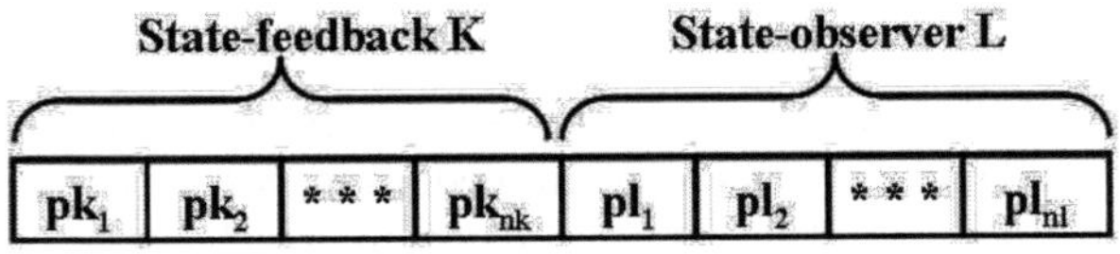

Fig. 3. Chromosome structure

3.4 Variation Operators

Two cross-over operators were implemented. The first (see figure 4) performs a "block" exchange between *pk* and *pl*, belonging to different individuals. The second (see figure 5) performs an uniform random cross-over [11].

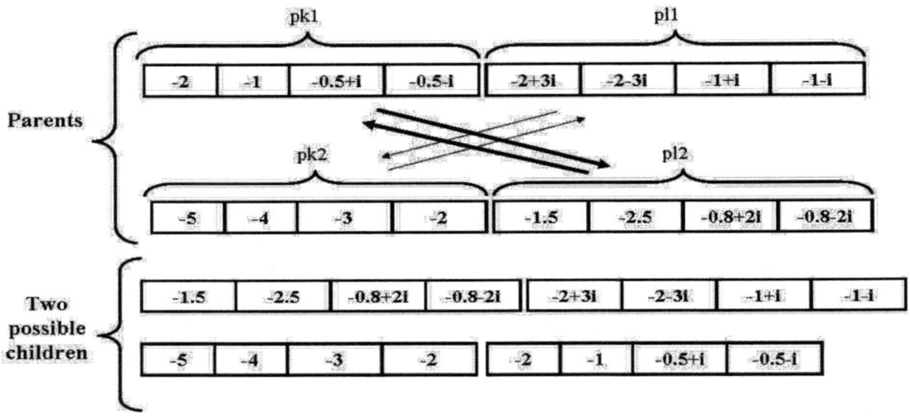

Fig. 4. First cross-over operator

Moreover, two mutation operators were implemented. The first (see figure 6) works like cross-over operator #1, but this time between *pk* and *pl* belonging to the same individual. The second (see figure 7) slightly moves the poles in random directions.

3.5 Multi-Objective Genetic Algorithm (MOGA)

MOGA has been widely applied to solve a number of practical applications [3]. This algorithm assigns the smallest rank value for all non-dominated individuals. The dominated ones are ranked according to the number of individuals that dominated them. The fitness value of each individual is computed by implementing a mapping inversely related to its rank. This fitness will be degraded based upon a sharing function, according to the distribution density in the feature space. The parameter α regulates the shape of the sharing function.

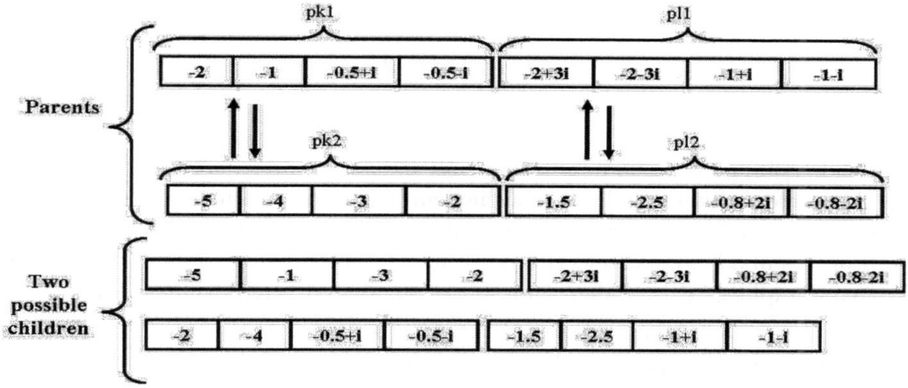

Fig. 5. Second cross-over operator

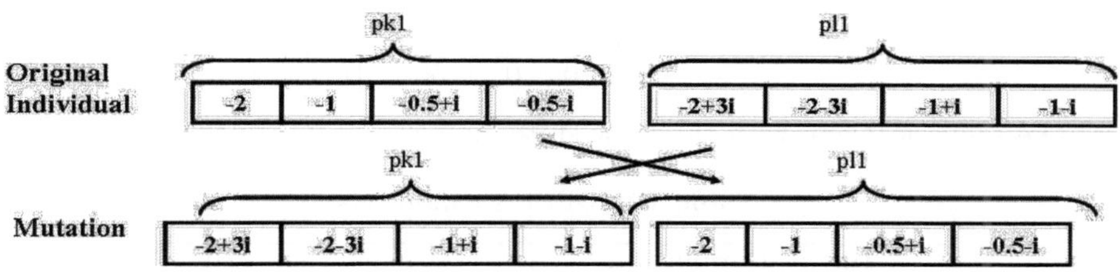

Fig. 6. First mutation operator

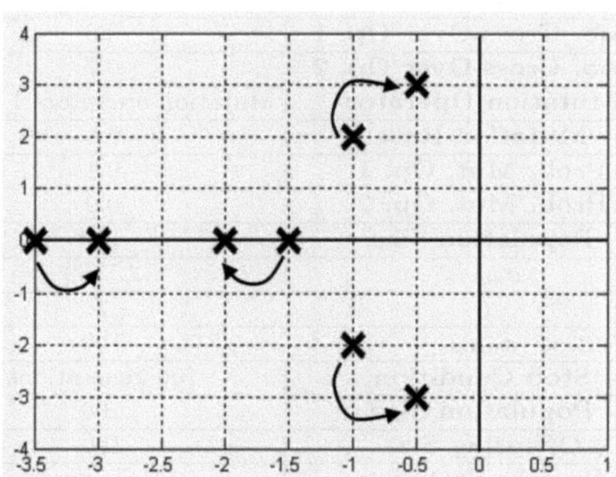

Fig. 7. Second mutation operator

The sharing distance σ_{share} determines the extent of the sharing region for each individual. In this work (two objectives case), the following equation was used:

$$\sigma_{share} = \frac{d}{2 \times N} \qquad (12)$$

where d is the diameter of the trade-off curve (estimated from previous works [5] and [7]) and N is the population size.

A simple mating restriction mechanism was also implemented: only pairs of individuals that lie within a distance of σ_{mate} were allowed for mating [1].

Parents and survivor selection were implemented using the "Stochastic Universal Sampling" (SUS) algorithm, based on shared fitness values.

4 Design Example: A Mixed $\mathcal{H}_2/\mathcal{H}_\infty$ Control Problem

The proposed design method was applied to solve the mixed $\mathcal{H}_2/\mathcal{H}_\infty$ control problem (see equation 11) with the following state matrices:

$$\left(\begin{array}{c|c} A & B \\ \hline C & D \end{array}\right) = \left(\begin{array}{ccc|c} -21 & -120 & -100 & 1 \\ 1 & 0 & 0 & 0 \\ 0 & 1 & 0 & 0 \\ \hline 0 & 0 & 150 & 0 \end{array}\right) \qquad (13)$$

The same problem, with the same state matrices, was tackled in [5] and [7].

Table 1. MOPPEA parameters

Initial Population	Randomly generated
Representation	Complex vectors
supr	10
supi	10
Cross-Over Recombination	Cross-over operators 1 and 2
Cross-Over Rate	0.8
Prob. Cross-Over Op. 1	0.3
Prob. Cross-Over Op. 2	0.7
Mutation Operator	Mutation operators 1 and 2
Mutation Rate	0.1
Prob. Mut. Op. 1	0.1
Prob. Mut. Op. 2	0.9
Population Size	100
σ_{share}	2.7733
α	2
σ_{mate}	100
Stop Condition	100 generations
Population Size	100
Offspring Size	100
Parents Selection	SUS $(s = 2)$
Survivor Selection	SUS $(s = 2)$

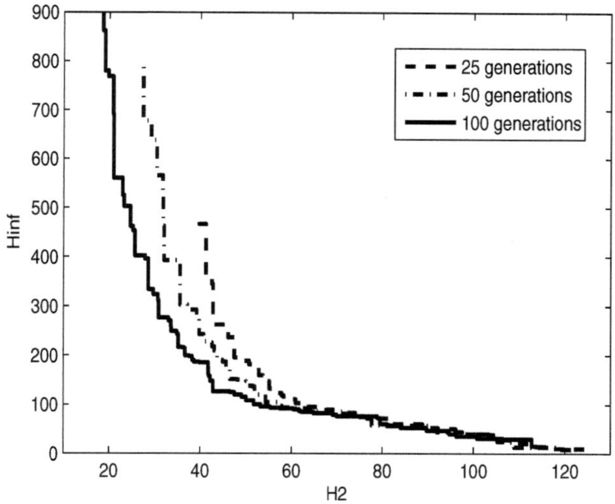

Fig. 8. Evolution of a typical attainment surface after 25, 50 and 100 generations

4.1 Experimental Results

All algorithms were coded in MATLAB, taking advantage of its control toolbox. Table 1 shows the parameters used during tests. Figure 8 shows the evolution of a typical attainment surface after 25, 50 and 100 generations. An attainment surface is the family of tightest goal vectors known to be attainable during the optimization process. Note that these results are comparable to those presented in [7].

5 Conclusions and Future Work

The proposed approach was able to find attainment surfaces as good as previous works. Its main advantage is that the stability restriction disappears from the optimization problem. However, the controller order is high and its structure cannot be optimized.

The goal of this paper was just to show experimentally that the proposed method is efficient enough. The focus was not really put on improving the optimization tool. Moreover, no other performance measure has been tested: it is well-known that assessing the performance of a multi-objective optimization algorithm is also a multi-objective problem [12].

Despite the simplicity of the proposed method, problems have been reported when using the pole placement technique. In high-order systems, certain pole locations result in very large gains [13]. This fact suggests caution during the optimization evolutionary process: a penalty mechanism can be used in order to avoid such locations.

Finally, it is clear that more tests are needed. The authors are currently developing a MATLAB toolbox, which allows using "state-of-the art" MOEA, in order to solve more complex control problems.

References

1. K.C Tan, E.F Khor, and T.H Lee. *Multiobjective Evolutionary Algorithms and Applications.* Springer-Verlag, 2005.
2. Peter J. Fleming and R.C. Purshouse. Evolutionary algorithms in control systems engineering: a survey. *Control Engineering Practice (10)*, pages 1223–1241, 2002.
3. Carlos Fonseca. *Multiobjective Genetic Algorithms with Application to Control Engineering Problems.* PhD thesis, Departement of Automatic Control and Systems Engineering. University of Sheffield, 1995.
4. J.F Whidborne, D.W Gu, and I. Postlethwaite. Algorithms for the method of inequalities: A comparative study. *American Control Conference, Washington*, 1995.
5. Alberto Herreros. Diseo de controladores robustos multiobjetivo por medio de algoritmos geneticos. *Tesis Doctoral, Departamento de Ingenieria de Sistemas y Automatica. Universidad de Valladolid, Espana.*, 2000.
6. V Zakian. *Control systems design: a new framework.* Springer-Verlag London Limited, 2005.
7. A Molina-Cristobal, I.A Griffin, P.J Fleming, and D.H Owens. Linear matrix inequalities and evolutionary computation. *International Journal of Systems Science. Vol. 37, No. 8*, pages 513–522, 2006.
8. Shinn-Jang Ho, Shinn-Ying Ho, Ming-Hao Hung, Li-Sun Shu, and Hui-Ling Huang. Designing Structure-Specified Mixed H2/Hinf Optimal Controllers Using an Intelligent Genetic Algorithm. *IEEE Transactions on Control System Technology, Vol. 13, No. 6*, pages 1119–1124, 2005.
9. A. Osyczka. *Multicriteria Optimization for Engineering Design.* Academic Press, Cambridge, MA, 1985.
10. J. Kautsky and N.K. Nichols. Robust Pole Assignment in Linear State Feedback. *Int. J. Control, 41*, 1985.
11. A.E Eiben and J.E Smith. *Introduction to Evolutionary Computing.* Springer-Verlag, 2003.
12. E. Zitzler, K. Deb, and L. Thiele. Comparison of Multiobjective Evolutionary Algorithms on Test Functions of Different Difficulty. In Annie S. Wu, editor, *Proceedings of the 1999 Genetic and Evolutionary Computation Conference. Workshop Program*, pages 121–122, Orlando, Florida, 1999.
13. A.J Laub and M. Wette. Algorithms and Software for Pole Assignment and Observers. *UCRL-15646 Rev. 1, EE Dept., Univ. of Calif., Santa Barbara, CA*, 1984.

Appendix A. A MATLAB Recursive Function for Randomly Generating the Initial Population

```
function [ pol,nr,ni ] = genpol( n, supr, supi )
   %supr and supi are the limits of the complex region
    if n == 1,
       pol = -supr*[rand]; nr = 1; ni = 0;
       return
   end
    if n == 2,
        if rand <= 0.5,
           pol = -supr*[rand;rand];
           nr = 2;ni = 0;
        else
           im = 2*rand-1;re = -rand;
           pol = [complex(supr*re,supi*im);complex(supr*re,-supi*im)];
            nr = 0;ni = 1; end
   return
   end
   if n >= 3,
   est = rand;
       if est <= 0.5,
            polaux = -supr*[rand;rand];
            nr = 2;
            ni = 0;
        else
           im = 2*rand-1;
            re = -rand;
    polaux = [complex(supr*re,supi*im);complex(supr*re,-supi*im)];
           nr = 0;
           ni = 1;
       end
    [polaux1,nraux,niaux] = genpol(n-2,supr,supi);
    if est <= 0.5,
        pol = [ polaux ; polaux1];
    else
        pol = [ polaux1 ; polaux];
    end

    nr = nr + nraux;
    ni = ni + niaux;
    return
    end
```

A Multi-objective Evolutionary Approach for Phylogenetic Inference

Waldo Cancino and Alexandre C.B. Delbem

Institute of Mathematics and Computer Science
University of Sao Paulo
Sao Carlos, SP. Brazil 13560–970
{wcancino,acbd}@icmc.usp.br

Abstract. The phylogeny reconstruction problem consists of determining the most accurate tree that represents evolutionary relationships among species. Different criteria have been employed to evaluate possible solutions in order to guide a search algorithm towards the best tree. However, these criteria may lead to distinct phylogenies, which are often conflicting among them. In this context, a multi-objective approach can be useful since it could produce a spectrum of equally optimal trees (Pareto front) according to all criteria. We propose a multi-objective evolutionary algorithm, named PhyloMOEA, which employs the maximum parsimony and likelihood criteria to evaluate solutions. PhyloMOEA was tested using four datasets of nucleotide sequences. This algorithm found, for all datasets, a Pareto front representing a trade-off between the criteria. Moreover, SH-test showed that most of solutions have scores similar to those obtained by phylogenetic programs using one criterion.

Keywords: Phylogenetic Inference, Multi-Objective Optimization, Genetic Algorithms.

1 Introduction

In a recent paper, Handl et al [1] discussed applications of multi-objective optimization in several bioinformatics and computational biology problems. Phylogenetic inference, which searches for the best explanation for evolutionary events from input data, is one of the central problems in this area. It is often modeled as a single objective optimization problem using one criterion for evaluating possible solutions. Moreover, several researches [2,3,4] have shown that the employment of different reconstruction methods can lead to unequal trees for the same input data. Thus, a multi-objective approach, which can search for phylogenies using more than one criterion, can be a relevant contribution since it can produce solutions which are consistent with all employed criteria.

Rokas et al [5] pointed out that there are several sources of incongruence in phylogenetic analysis: optimality criterion employed, data used and evolutionary assumptions about data. Moreover, Poladian and Jermiin [6] suggested that multi-objective optimization can be applied to phylogenetic inference from several conflicting input data. The authors showed that this approach can reveal sources of such conflicts and provide useful information for a robust inference.

S. Obayashi et al. (Eds.): EMO 2007, LNCS 4403, pp. 428–442, 2007.

We propose a multi-objective approach for phylogenetic inference using maximum parsimony [7] and likelihood [8] criteria. The algorithm developed to solve such problem, named PhyloMOEA, is a multi-objective evolutionary algorithm based on NSGA-II model proposed by Deb et al [9]. The output of PhyloMOEA is a solution set representing a trade-off between the criteria considered.

This paper is organized as follows. Section 2 provides relevant background information about phylogenetic inference. Section 3 presents the main concepts of Genetic Algorithms and their application to phylogeny. Section 4 discusses multi-objective optimization problems and shows how Genetic Algorithms can contribute to solve this kind of problems. Section 5 presents the PhyloMOEA algorithm. Section 6 describes the experiments involving four nucleotide datasets and discusses the main results. Finally, Section 7 presents conclusions and proposes future works.

2 Phylogenetic Inference Problem

Phylogenetic analysis investigates evolutionary relationships among species. Sequence data from actual species (nucleotide or aminoacid sequences) are frequently employed for this purpose, although other types of data can be used [10]. Evolutionary relationships can be illustrated as a leaf-labelled tree, named phylogenetic tree. In such tree, external nodes refer to actual species in data, internal nodes refer to hypothetical ancestors and branches represent relations among species. Since sequence data used in phylogenetic analysis are obtained from contemporary species, a phylogenetic tree is a hypothesis (of many possible trees) about the evolutionary events in the history of species.

A phylogenetic tree can be rooted or unrooted. In a rooted tree, there is a special node named root that defines the direction of the evolution, allowing the determination of ancestral relationships among nodes. An unrooted tree shows only the relative positions of nodes without an evolutionary direction. Additionally, tree branches may have an associated length showing genetic distances between connected nodes. Figures 1 and 2 show a rooted and unrooted tree, respectively.

The main goal of the phylogenetic inference is the determination of a tree that best explains the evolutionary events of species under analysis. Swofford et al [11] classified phylogenetic inference methods into two categories: algorithmic and optimality criterion methods. The former follows a sequence of well-defined steps to generate a tree. Clustering methods, like Neighboor Joining [12] are in this category. These algorithms go directly to the final answer without examining manyS alternatives in the search space, quickly producing a phylogenetic tree. Optimally criterion methods include two components: an optimality criterion and a search mechanism. The optimality criterion defines an objective function that scores every possible solution. Using this criterion, the search mechanism should determine the best scored solution in the search space. However, finding the optimal solution requires exhaustive or exact strategies, which are only applicable to small datasets. Since the tree search space increases exponentially

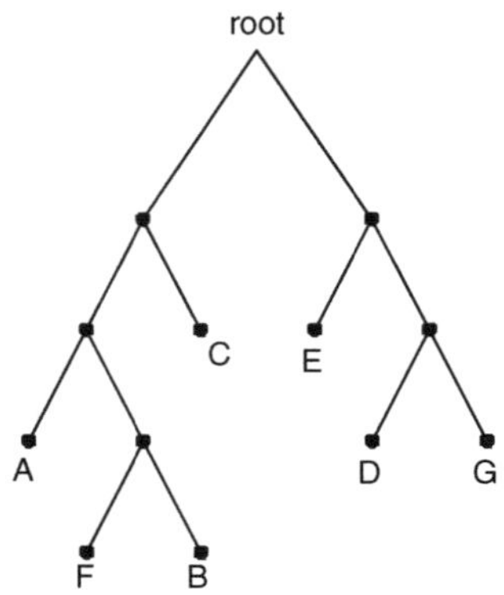

Fig. 1. A rooted tree

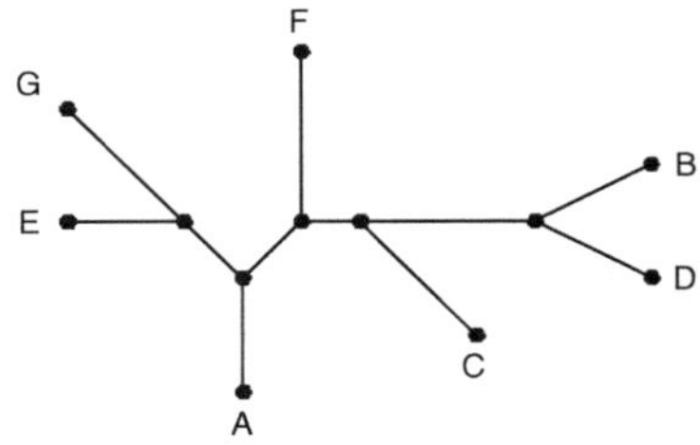

Fig. 2. An unrooted tree

with the number of species from data, only heuristic search is feasible for moderate and large datasets. Examples of optimality criterion methods are maximun parsimony [7], maximun likelihood [8] and least squares [13]. The following sections present a brief review of maximun parsimony and maximun likelihood since they are employed in PhyloMOEA.

2.1 Maximum Parsimony

The parsimony principle states that the simplest hypothesis which explains an observed phenomenon must always be preferred. In phylogenetic inference, parsimony methods search for a tree that minimizes the number of character state changes in its topology [10]. Such tree is named maximum parsimony tree, which refers to the simplest hypothesis.

Let D be a dataset containing n species. Each species has N sites, where d_{ij} is the character state of species i at site j. Given a tree T with a node set $V(T)$ and a branch set $E(T)$, the parsimony score of T is defined as:

$$PS(T) = \sum_{j=1}^{N} ps_j \ , \tag{1}$$

where ps_j, defined by counting character changes along branches in T, is the parsimony score for site j, which can be formulated as:

$$ps_j(T) = \sum_{(v,u)\in E(T)} C(v_j, u_j) \ , \tag{2}$$

where v_j and u_j are the character states of nodes v and u at site j for each branch (u, v) in T, C is the cost matrix such that $C(v_j, u_j)$ is the cost of changing from v_j to u_j. The leaves of T are labelled by character states of species from D, i.e. a leaf representing $k-th$ species has character state d_{kj} for position j.

The problem of finding the most parsimonious tree can be divided into two sub-problems:

- Small parsimony problem, which determines the character states of internal nodes minimizing PS for a given tree T;
- Large parsimony problem, which finds the most parsimonious tree from the tree search space.

The small parsimony problem can be easily solved using the Sankoff algorithm [14] for any cost matrix C. When C satisfies $C(x, y) = 1$ if $x \neq y$ and $C(x, y) = 0$, the Fitch algorithm [7] can be employed. The large parsimony problem was proved to be NP-hard [10].

2.2 Maximum Likelihood

Likelihood is a widely-used statistical measurement that indicates the conditional probability of data given a hypothesis [10]. The likelihood of a phylogenetic tree, denoted by $L = P(D|T)$, is the conditional probability of the sequence data D given a tree T and stochastic evolution model. Two assumptions are necessary to compute likelihoods:

1. Evolution at different sites is independent;
2. Evolution from different tree lineages is independent, i.e. each subtree evolves separately.

Let D_i be the sequence data set D at site i. L is calculated from the product of partial likelihoods from all sites:

$$L = \prod_{i}^{N} L_i \ , \tag{3}$$

where $L_i = P(D_i/T)$ is the likelihood at site i. An efficient method to calculate L was proposed by Felsenstein [8] using a dynamic programming approach, where L is obtained by traversing from leaves to root. Usually, it is necessary to work with logarithmic values of L, then Equation 3 results in:

$$\ln L = \sum_{i=1}^{n} L_i \ . \tag{4}$$

In order to maximize L, it is necessary to optimize the branch lengths of T and the parameters of the employed evolutionary model. This can be achieved using classical optimization methods [10].

Several heuristic techniques were proposed to find the best parsimony tree or the best likelihood tree [15,16,17]. One of such approaches is Genetic Algorithms (GAs), which are important for the purposes of this paper. Next section presents a review of GAs and their application to phylogenetic inference.

3 Genetic Algorithms in Phylogenetic Inference

Genetic Algorithms are search and machine learning techniques inspired by natural selection principles. They have been applied to a wide range of problems of science and engineering [18]. A GA uses a set of individuals, named population, which respresent feasible solutions for a given optimization problem. Each individual has an associated fitness value, which is based on the problem objective function. GAs search for an optimal or near-optimal solution in the fitness landscape. Individuals use an internal encoding that must be able to store all relevant problem variables and codify all feasible solutions.

First, a GA creates a initial population and calculates the fitness of its individuals. Then, a new population is generated using three genetic operators: selection, crossover and mutation [19]. The selection operator uses individuals'fitness to choose candidates to generate the next population. Features of the selected solutions are combined by the crossover operator and new offspring solutions are generated. Then, small modifications are performed by the mutation operator at a very low rate. While crossover is useful to explore the search space, mutation allows scaping from local optima. The average fitness of the new population is expected to be better than average fitness of the previous population. This process is repeated until a stop criterion has been reached. The solutions found by the GA are in the final population.

Several papers have described the application of GAs to the phylogeny problem focused on one criterion. In general, they use maximun likelihood [20,21,22], parsimony [23] or distance-based [24] criterion. Experimental results have shown that GAs have better performance and accuracy when compared to heuristics implemented in widely-used phylogenetic software like PHYLIP [25] and PAUP* [26]. Moreover, GAs can also work with several criteria in order to solve multi-objective optimization problems (MOOP). The following section briefly describes MOOPs and the application of GA to these problems.

4 Multi-objective Optimization

A MOOP deals with two or more objective functions that must be simultaneously optimized. In this problem, the *Pareto dominance* concept is commonly used to compare two solutions. A solution $\mathbf{x}$ dominates a solution $\mathbf{y}$ if $\mathbf{x}$ is not worse than $\mathbf{y}$ in all objectives and if it is better for at least one. Solving an MOOP implies calculating the Pareto optimal set whose elements, named Pareto optimal solutions, represent a trade-off among objective functions. These solutions are not dominated by any other in the search space. The curve formed by plotting these solutions in the objective function space is named Pareto front. If there is no additional information regarding the relevance of the objectives, all Pareto optimal solutions have the same importance. Deb [18] pointed out two fundamental goals in MOOP:

1. To find a set of solutions as close as possible to the Pareto optimal front;
2. To find a set of solutions as diverse as possible.

Several classical optimization techniques have been proposed to deal with MOOPs [18]. However, these methods have limitations. A possible approach is the conversion of an MOOP into a single optimization problem by a weighted sum of objectives. This strategy only finds a single point in the Pareto front for each weight combination. Thus, several runs using different weight values are required to obtain a reasonable number of Pareto optimal solutions. Nevertheless, this method does not guarantee solution diversity in the frontier. Other methods need *a priori* knowledge of the problem, as for example, target values; which are not always available.

On the other hand, evolutionary algorithms for multi-objective optimization (MOEA) have been successfully applied to both theoretical and practical MOOPs [18]. In general, MOEAs are capable of finding a distributed Pareto optimal set in a single run.

The folowing sections describe PhyloMOEA, the proposed MOEA to solve the phylogenetic inference problem using maximum parsimony and likelihood criteria.

5 A Multi-objective Approach to Phylogenetic Inference

One difficulty in the phylogenetic problem is the occurrence of conflicting alternative solutions. There are various researches in the literature that compare several aspects of main reconstruction methods and their variants [2, 27, 3, 4]. These analyses have shown that reconstruction methods often lead to different topologies for the same input data. It occurs if the method's requirements are violated or if there is a bias in the experiment.

This paper formulates the phylogenetic inference problem as an MOOP with two optimality criteria: maximum parsimony and likelihood. In order to solve such problem, we have propose an MOEA algorithm named PhyloMOEA, which is based on the NSGA-II [9] algorithm. The main goal of PhyloMOEA is the determination of a set of non-dominated solutions (trees), which represents a trade-off between parsimony and likelihood scores. The following subsections describe the proposed algorithm in more details.

5.1 Internal Encoding

There are several possible data structures that can be used to represent a phylogenetic tree, although phylogenetic programs commonly employ graph data structures [10]. PhyloMOEA employs graph structures provided by the Graph Template Library (GTL) [28] to represent unrooted trees. GTL allows for an easy implemention of genetic operators and facilitates the storage of additional information, like branch lengths.

5.2 Initial Solutions

PhyloMOEA uses two populations as NSGA-II: one to retain non-dominated solutions and the other to store new solutions generated by genetic operators [9].

These populations are, respectively, denoted by P_i and Q_i, where i refers to generation i. In the first generation, initial solutions are created for P_1. Then, solutions in Q_1 are obtained by applying selection, crossover and mutation operators to P_1.

PhyloMOEA can generate initial random trees in P_1; however, these trees are often far from maximum parsimony and likelihood trees. In order to overcome this drawback, additional trees, provided by maximum likelihood, parsimony or bootstrap analysis, can be included in the initial population. This strategy is often used in GA-based phylogenetic programs [20, 21].

5.3 Fitness Evaluation

PhyloMOEA evaluates trees in P_i and Q_i using the parsimony and likelihood criteria. The parsimony and likelihood scores are calculated using Fitch [7] and Felsenstein [8] algorithms, respectively. The fitness of a tree requires two values: a rank and a crowding distance [18].

The rank is calculated using a non-dominated sorting algorithm [18] applied to both populations for all generations. This algorithm divides $R = P_i \cup Q_i$ into several frontiers, denoted by $\mathcal{F}_1, \mathcal{F}_2, \ldots, \mathcal{F}_j$. The first frontier ($\mathcal{F}_1$) is formed by non-dominated solutions from R. Thus, the second frontier $\mathcal{F}_2$ is by non-dominated solutions from $R - \mathcal{F}_1$. This process is repeated to $R - \mathcal{F}_1 - \mathcal{F}_2$, and so on, until R is empty. The rank value of an individual is the index of the frontier it belongs to.

The crowding distance of a solution reflects the density of solutions around its neighborhood. This value is calculated from a perimeter defined by the nearest neighbors in each objective and used to maintain the population diversity.

PhyloMOEA uses a tournament selection which picks two individuals at random and choose the best one, which has the lowest rank. If both solutions have the same rank, the solution with the longest crowding distance is preferred.

5.4 Crossover Operator

The crossover operator implemented in PhyloMOEA is the same of [22]. It combines a subtree from two parent trees and creates two new offspring trees. Given trees T_1 and T_2, this operator performs the following steps:

1. Prune a subtree s from T_1;
2. Remove all leaves in s from T_2;
3. Offspring T_1' results from regrafting s in T_2.

Similarly, a second offspring T_2' is created. The operator prunes a subtree from T_2 and regrafts it in T_1. Figure 3 illustrates this operator.

5.5 Mutation Operator

There are three well-known topological modifications used in phylogenetic inference [11]: Nearest Neighboor Interchange (NNI), Sub-tree Pruning and Regrafting (SPR) and Tree Bisection and Reconnection (TBR). NNI was employed in

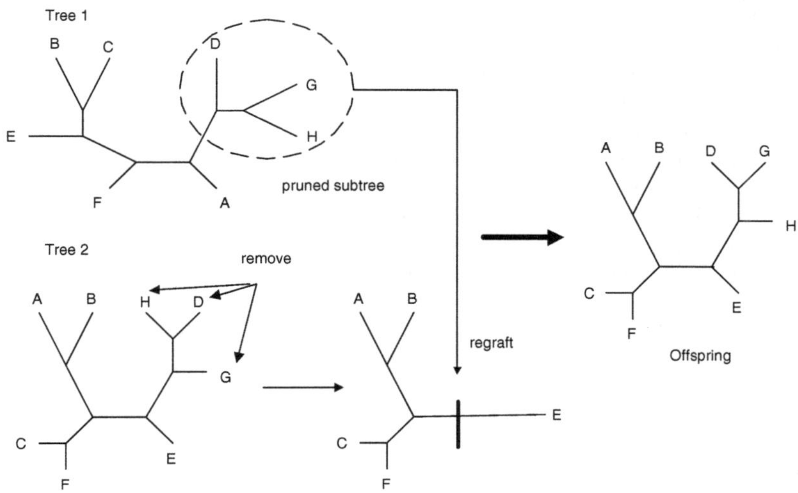

Fig. 3. Example of the crossover operator

PhyloMOEA, since it performs minimal tree modifications. This operator carries out the following steps:

1. Choose an interior branch whose connected nodes i, j define two pairs of neighbors: w, x adjacent to i ($w, x \neq j$) and y, z adjacent to j ($y, z \neq i$).
2. Execute a swap of nodes between each pair of neighboors.

Figure 4 illustrates the NNI operator. The mutation operator also modifies branch lengths in order to improve the tree likelihood value. A branch length is multiplied by a factor obtained from a gamma distribution [22]. In each mutation, some branch lengths chosen at random are modified.

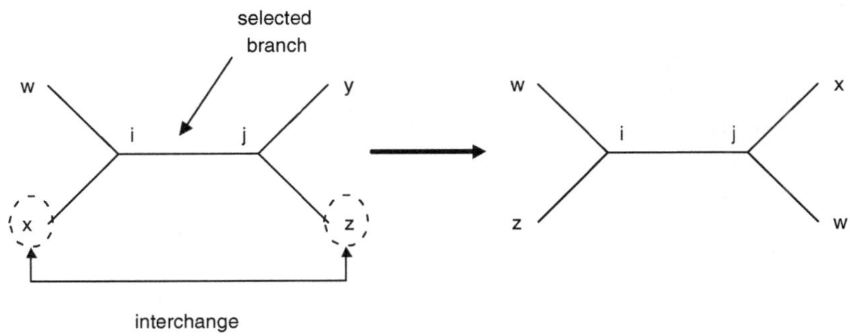

Fig. 4. Example of NNI mutation operator

Algorithm: PhyloMOEA
begin
  Create an initial population P_1 containing N solutions
  Perform non-dominated Sorting in $R = P_1$
  Calculate crowding distance values of P_1
  Apply selection, crossover and mutation operators in P_1 and generate a new population Q_1
  foreach *generation* $t = 2, \ldots, n$ **do**
    Perform non-dominated sorting in $R = P_t \cup Q_t$
    Calculate crowding distance values of R
    Calculate Pareto frontiers $\mathcal{F}_1, \mathcal{F}_2, \ldots \mathcal{F}_j$ from R
    Store the N best solutions from $\mathcal{F}_k$ in $P_{t+1}, |\mathcal{F}_k| \leq N, \quad k = 1 \ldots l$
    Create a new population Q_{t+1} by applying selection, crossover and mutation operators in P_{t+1}
  end
  Perform branch length optimization of solutions in P_n.
end

Fig. 5. PhyloMOEA algorithm

After a PhyloMOEA execution, branch lengths of trees in the final population are optimized using a non-decreasing Newton-Raphson method described by Yang [29]. Figure 5 shows the PhyloMOEA algorithm.

6 Experiments

PhyloMOEA was tested using four nucleotide data sets. The *rbcL_55* dataset has sequences of the rbcL chloroplast gene from 55 species of green plants (1314 sites) [22]. The *mtDNA_186* dataset contains 186 human mitochondrial DNA sequences (16608 sites) taken from The Human Mitochondrial Genome Database (mtDB) [30]. The *RDPII_218* dataset comprises 218 prokaryotic RNA sequences (4182 sites) taken from the Ribosomal Database Project II [31]. Finally, the *ZILLA_500* dataset includes 500 rbcL sequences (1428 sites) from plant plastids [16].

Maximum parsimony and likelihood analyses were performed in each dataset using programs NONA [15] and RAxML-V [17], respectively. These programs include sophisticated heuristics that produces satisfactory results quickly. Table 1 shows the parsimony and likelihood results obtained from these programs.

Trees in the initial population were generated from bootstrap data applied to each dataset. The boostrap analysis was performed by PHYML [16], which employs the BIONJ algorithm [32] to infer trees. The parsimony and likelihood scores of these trees approximate the scores shown in Table 1. However, for *RDPII_218* and *ZILLA_500* datasets, bootstrap tree scores are not close enough to the scores obtained by NONA and RAxML-V. Consequently, it slows the PhyloMOEA's convergence. In order to overcome this drawback, solutions from Table 1 are included in the initial population.

Table 1. Parsimony and Likelihood results found by NONA and RAxML-V

Solution	NONA		RAxML-V	
	Parsimony	Likelihood	Parsimony	Likelihood
*rbcL*_55	4874	-24627.8480	4894	-24583.3313
*mtDNA*_186	2438	-41049.7677	2450	-40894.5497
*RDPII*_218	41534	-170831.1213	42631	-156595.8725
*ZILLA*_500	16219	-87361.4841	16276	-86993.8264

Table 2 shows the parameters of PhyloMOEA used for the experiments. It can be noted that *ZILLA*_500 dataset requires the largest number of generations and population size. Furthermore, HKY85 [33] nucleotide model was employed in likelihood calculations since it is often used in the literature [16, 21, 22, 17].

Table 2. Parameters used by PhyloMOEA in the experiments

Parameter	Value
Generations	500 (*rbcL*_55, *mtDNA*_186, *RDPII*_218), 2000 (*ZILLA*_500)
Population size	50 (*rbcL*_55, *mtDNA*_186, *RDPII*_218), 100 (*ZILLA*_500)
Crossover rate	0.8
Mutation rate	0.05
Mutation operator	NNI
Evolution model	HKY85

At the end of a PhyloMOEA execution, duplicate trees are removed from the final population. Finally, the Pareto optimal solutions are calculated, although this may eliminate adequate topologies from the perspective of parsimony criterion. If two solutions have an equal parsimony score, only the solution with the best likelihood remains in the Pareto-set. Thus, all non-duplicated topologies, named Final Solutions, are maintained.

Table 3 presents a summary of the experiment results. The second and third columns indicate the number of trees found by PhyloMOEA in the Pareto optimal and Final Solutions, respectively. The parsimony and likelihood scores for the best trees found, which represent extreme points in the Pareto front, are also shown. It can be noted that there are 5 Pareto optimal solutions for *rbcL*_55 dataset. This reduced number of solutions is due to the proximity of the extreme points. Moreover,

Table 3. Parsimony and Likelihood scores of the extreme points found by PhyloMOEA

	Pareto	Final	Best Parsimony Tree Scores		Best Likelihood Tree Scores	
Dataset	Trees	Trees	Parsimony	Likelihood	Parsimony	Likelihood
*rbcL*_55	5	46	4874	-24626.4337	4884	-24583.3297
*mtDNA*_186	12	44	2438	-41004.3018	2450	-40894.3433
218_*RDPII*	21	82	41534	-158724.2803	42631	-156595.8224
500_*ZILLA*	16	107	16219	-87275.2812	16276	-86993.8250

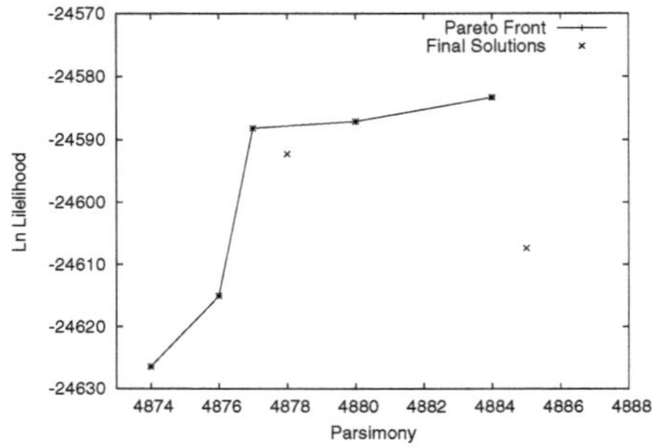

Fig. 6. Final Solutions and Pareto front for *rbcL_55* dataset

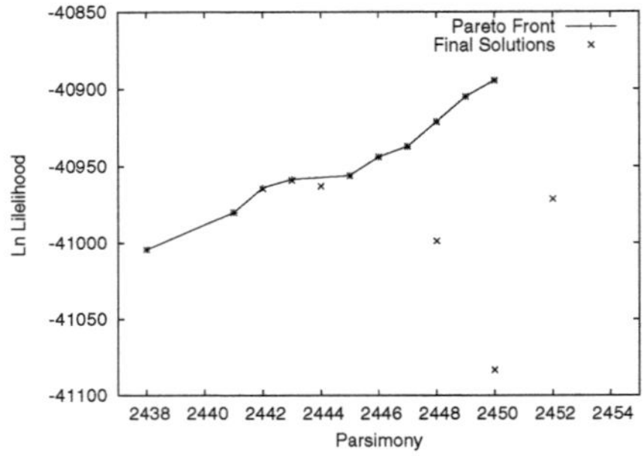

Fig. 7. Final Solutions and Pareto front for *mtDNA_186* dataset

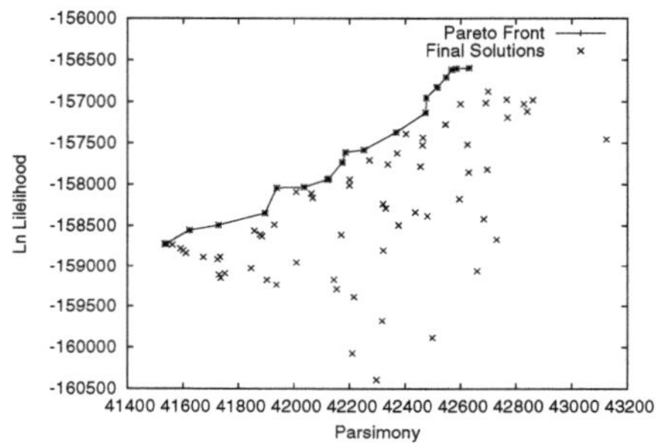

Fig. 8. Final Solutions and Pareto front for *RDPII_218* dataset

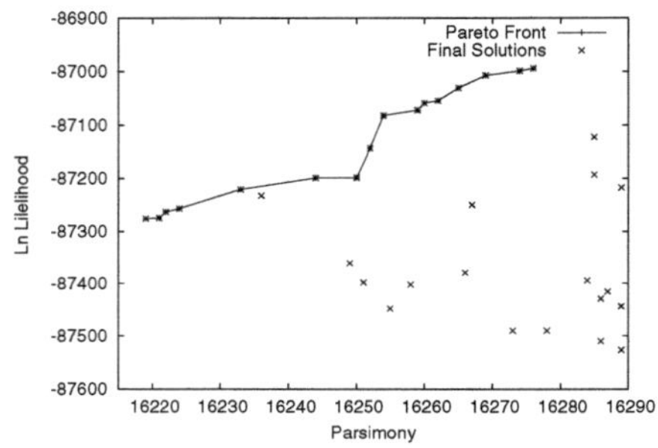

Fig. 9. Final Solutions and Pareto front for *ZILLA_500* dataset

parsimony scores are integer values restricting the number of possible solutions in a small interval. For the other datasets, intervals are greater and, consequently, the number of Pareto optimal trees increases. However, for all datasets, PhyloMOEA found a relatively large number of Final Solutions. On the other hand, Table 3 also shows improvements in likelihood scores for the best found parsimony tree when compared with values from Table 1. Nevertheless, likelihood scores for the best likelihood trees shown in Table 1 are only slightly improved.

Figures 6, 7, 8 and 9 show the Pareto fronts for *rbcL_55*, *mtDNA_186*, *RDPII_218* and *ZILLA_500* respectively. Due parsimony scores are integer values, the resulting Pareto Front is a discontinuous set of points connected by lines. These figures also show Final Solutions near to Pareto front. For *RDPII_218* and *ZILLA_500* datasets, there is larger number of Final Solutions displayed since there is a large interval between extreme points, mainly for the *RDPII_218* dataset.

Pareto optimal and Final Solutions were evaluated using Shimodaira-Hasegawa test (SH test) [34]. For each tree, SH test calculates a $P-$value, which indicates if a tree is significantly worse than the best tree according to a criterion. When a tree has a $P-$value lower than a given bound (usually 0.05), it can

Table 4. Summary of SH-test results for Pareto Solutions

Dataset	Trees	SH-Test Parsimony		SH-Test Likelihood	
		$P \geq 0.05$	$P < 0.05$	$P \geq 0.05$	$P < 0.05$
rbcL_55	5	5	0	5	0
mtDNA_186	12	10	2	9	3
RDPII_218	21	4	17	6	15
ZILLA_500	16	12	4	9	7

Table 5. Summary of SH-test results for Final Solutions

Dataset	Trees	SH-Test Parsimony		SH-Test Likelihood	
		$P \geq 0.05$	$P < 0.05$	$P \geq 0.05$	$P < 0.05$
rbcL_55	46	15	31	13	33
mtDNA_186	44	36	8	19	25
RDPII_218	82	14	68	10	72
ZILLA_500	107	29	78	21	86

be rejected. For the experiments, SH-tests for parsimony and likelihood criteria were performed using programs PHYLIP [25] and PAML [35], respectively.

Tables 4 and 5 show the results of SH-test applied to Pareto optimal and Final solutions, respectively. The second column indicates the number of trees tested and the others show the number of non-rejected ($P \geq 0.05$) and rejected trees ($P < 0.05$), according to parsimony and likelihood criteria. It can be seen from Table 4 that none of the optimal solutions were rejected for *rbcL_55*, while 2 Pareto optimal solutions were rejected for *mtDNA_186* dataset. In both datasets, the extreme points are close, thus, intermediate solutions cannot be rejected. On the other hand, the extreme points in *RDPII_218* and *ZILLA_500* are distant from each other, therefore several intermediate solutions are rejected.

Table 5 shows that most of the Final Solutions were rejected in SH-test for both criteria, meaning that the scores of these solutions are far from the extreme point scores. The only exception is *mtDNA_186* dataset, where 8 out of 44 and 25 out of 44 trees are rejected for the parsimony and likelihood criteria, respectively. Consequently, there are more solutions whose scores are in the neighborhood of the maximum parsimony score.

We pointed out that SH-test must be applied for one criterion separately, even though, our results provide an insight of the score distribution of the Pareto optimal trees. Moreover, SH-test shows that some of the Pareto optimal solutions are not significantly worse than the best found trees obtained from a separate analysis. However, SH-test is not useful to describe the nature of Pareto-front since it tends to eliminate intermediate solutions when Pareto front extreme points are distant from each other. Thus, SH-test was not use for this pourposes.

7 Conclusions and Future Works

In this paper, we proposed a multi-objective evolutionary algorithm to solve the phylogenetic inference problem using both parsimony and likelihood criteria. This proposal was motivated by the literature in the area [2,27,3,4], which points out that several phylogenetic inference methods leads to different solutions. This fact was verified for all datasets analysed in the experiments.

The proposed algorithm, named PhyloMOEA, was designed on the basis of the NSGA-II model. The crossover and mutation take into account heuristics specifically designed for phylogetic inference (Section 5). In the experiments, PhyloMOEA was able to find a set of trees that represents a trade-off between the parsimony and likelihood criteria. The Pareto-front obtained is formed by discontinuos points along the maximum parsimony and maximum likelihood scores obtained. Several trees found by PhyloMOEA were not rejected in SH-test applied, which indicates alternative solutions for both criteria have been found.

Despite the relevant results found by PhyloMOEA, there are some aspects that should be addressed in order to improve the algorithm and corresponding results:

- PhyloMOEA requires several hours to find acceptable Pareto-solutions if initial trees are poorly estimated. Performance can be improved using advanced genetic operators that take into account local search strategies [16, 17];
- Likelihood calculations performed by PhyloMOEA do not consider the rate heterogeneity among sites. In real datasets, sites frequently evolve at different rates. Thus, if the rate heterogeneity is considered, the accuracy of likelihood analysis is often improved [36].
- This research has not investigated metrics for convergence and diversity of the obtained Pareto front. Measurements for convergence are difficult to obtain since the Pareto front is unknown in this case. However, several diversity metrics found in the literature [18] can be employed.

In summary, preliminary results have shown that PhyloMOEA can make relevant contributions to phylogenetic inference. Moreover, there are several aspects that can be investigated to improve the current approach.

Acknowledgments

The authors would like to thank the Sao Paulo State Research Foundation (FAPESP) for the financial support provided (Grant Nº 01/13846-0).

References

1. Handl, J., Kell, D., Knowles, J.: Multiobjective Optimization in Computational Biology and Bioinformatics. IEEE Transactions on Computational Biology and Bioinformatics (2006) To appear.
2. Huelsenbeck, J.: Performance of Phylogenetic Methods in Simulation. Systematic Biology **44** (1995) 17–48

3. Kuhner, M., Felsenstein, J.: A Simulation Comparison of Phylogeny Algorithms under Equal and Unequal Evolutionary Rate. Molecular Biology and Evolution **11** (1994) 459–468
4. Tateno, Y., Takezaki, N., Nei, M.: Relative Efficiences of the Maximum-Likelihood, Neighbor-Joining, and Maximum Parsimony Methods when Substitution Rate Varies with Site. Molecular Biology and Evolution **11** (1994) 261–267
5. Rokas, A., Wiliams, B., King, N., Carroll, S.: Genome-Scale Approaches to Resolving Incongrounce in Molecular Phylogenies. Nature **425**(23) (2003) 798–804
6. Poladian, L., Jermiin, L.: Multi-Objective Evolutionary Algorithms and Phylogenetic Inference with Multiple Data Sets. Soft Computing **10**(4) (2006) 359–368
7. Fitch, W.: Toward Defining the Course of Evolution: Minimum Change for a Specific Tree Topology. Systematic Zoology **20**(4) (1972) 406–416
8. Felsenstein, J.: Evolutionary Trees from DNA Sequences: A Maximum Likelihood Approach. Journal of Molecular Evolution **17** (1981) 368–376
9. Deb, K., Agrawal, S., Pratab, A., Meyarivan, T.: A Fast Elitist Non-Dominated Sorting Genetic Algorithm for Multi-Objective Optimization: NSGA-II. KanGAL report 200001, Indian Institute of Technology, Kanpur, India (2000)
10. Felsenstein, J.: Inferring Phylogenies. Sinauer, Sunderland, Massachusetts (2004)
11. Swofford, D., Olsen, G., Waddell, P., Hillis, D.: Phylogeny Reconstruction. In: Molecular Systematics. 3 edn. Sinauer (1996) 407–514
12. Saitou, N., Nei, M.: The Neighbor-Joining Method: A New Method for Reconstructing Phylogenetic Trees. Molecular Biology and Evolution **4**(4) (1987) 406–425
13. Cavalli-Sforza, L., Edwards, A.: Phylogenetic Analysis: Models and Estimation Procedures. Evolution **21**(3) (1967) 550–570
14. Sankoff, D.: Simultaneous Solution of the RNA Folding, Alignment and Proto-Sequence Problems. SIAM Journal on Applied Mathematics **45**(5) (1985) 810–825
15. Goloboff, P.: NONA (no name) ver. 2 (1999) Published by the author.
16. Guindon, S., Gascuel, O.: A Simple, Fast, and Accurate Algorithm to Estimate Large Phylogenies by Maximum Likelihood. Systematic Biology **5**(52) (2003) 696–704
17. Stamatakis, A., Meier, H.: New Fast and Accurate Heuristics for Inference of Large Phylogenetic Trees. In: 18th IEEE/ACM International Parallel and Distributed Processing Symposium (IPDPS2004). (2004)
18. Deb, K.: Multi-Objective Optimization using Evolutionary Algorithms. John Wiley & Sons, New York (2001)
19. Goldberg, D.: Genetic Algorithms in Search, Optimization, and Machine Learning. Addison-Wesley Publishing Company, Inc., Reading, MA (1989)
20. Katoh, K., Kuma, K., Miyata, T.: Genetic Algorithm-Based Maximum-Likelihood Analysis for Molecular Phylogeny. Journal of Molecular Evolution **53** (2001) 477–484
21. Lemmon, A.R., Milinkovitch, M.C.: The Metapopulation Genetic Algorithm: An Efficient Solution for the Problem of Large Phylogeny Estimation. In: Proceedings of the National Academy of Sciences. Volume 99. (2002) 10516–10521
22. Lewis, P.O.: A Genetic Algorithm for Maximum-Likelihood Phylogeny Inference Using Nucleotide Sequence Data. Molecular Biology and Evolution **15**(3) (1998) 277–283
23. Moilanen, A.: Searching for Most Parsimonious Trees with Simulated Evolutionary Optimization. Cladistics **15** (1999) 39–50

24. Cotta, C., Moscato, P.: Inferring Phylogenetic Trees Using Evolutionary Algorithms. In Merelo, J., ed.: Parallel Problem Solving From Nature VII, Springer-Verlag (2002) 720–729
25. Felsenstein, J.: PHYLIP (Phylogeny Inference Package) (2000)
26. Swofford, D.: PAUP* Phylogenetic Analisys Using Parsimony (2000) CSIT Florida State University.
27. Jin, L., Nei, M.: Limitations of the Evolutionary Parsimony Method of Phylogenetic Analysis. Molecular Biology and Evolution **7** (1990) 82–102
28. Forster, M., Pick, A., Raitner, M., Bachmaier, C.: GTL - Graph Template Library Documentation. University of Pasdau. (2004)
29. Yang, Z.: Maximum-Likelihood Estimation of Phylogeny from DNA Sequences when Substitution Rates Differ over Sites. Molecular Biology and Evolution **10**(6) (1993) 1396–1401
30. Ingman, M., Gyllensten, U.: mtDB: Human Mitochondrial Genome Database, a Resource for Population Genetics and Medical Sciences. Nucleic Acids Research **34** (2006) D749–D751
31. Cole, J., Chai, B., Farris, R., Wang, Kulam, S., McGarrell, D., Garrity, G., Tiedje, J.: The Ribosomal Database Project (RDP-II): Sequences and Tools for High-throughput rRNA Analysis. Nucleic Acids Research **33** (2005) D294–D296
32. Gascuel, O.: BIONJ: An Improved Version of the NJ Algorithm Based on a Sample Model of Sequence Data. Molecular Biology and Evolution **14**(7) (1997) 685–695
33. Hasegawa, M., Kishino, H., Yano, T.: Dating of the Human–Ape Splitting by a Molecular Clock of Mitochondrial DNA. Journal of Molecular Evolution **22** (1985) 160–174
34. Shimodaira, H., Hasegawa, M.: Likelihood-Based Tests of Topologies in Phylogenetics. Molecular Biology and Evolution **16**(8) (1999) 111–1116
35. Yang, Z.: PAML: A Program Package for Phylogenetic Analysis by Maximum Likelihood. Computer Applications in Biosciences **13**(5) (1997) 555–6
36. Yang, Z.: Maximum Likelihood Estimation on Large Phylogenies and Analysis of Adaptative Evolution in Human Influenza Virus A. Journal of Molecular Evolution **51**(51) (2000) 423–432

On Convergence of Multi-objective Pareto Front: Perturbation Method

Raziyeh Farmani, Dragan A. Savic, and Godfrey A. Walters

Centre for Water Systems, School of Engineering, Computer Science and Mathematics, University of Exeter, Exeter, EX4 4QF, UK
{r.farmani, d.savic, g.a.walters}@exeter.ac.uk
http://www.ex.ac.uk/cws

Abstract. A perturbation method is proposed to detect convergence of the Pareto front for multi-objective algorithms and to investigate its effect on the rate of convergence of the optimization. Conventionally, evolutionary algorithms are allowed to run for a fixed number of trial solutions which can result in a premature convergence or in an unnecessary number of calls to a computationally intensive real world problem. Combination of evolutionary multi-objective algorithms with perturbation method will improve the rate of convergence of the optimization. This is a very important characteristic in reducing number of generations and therefore reducing the computational time which is important in real world problems where cost and time constraint prohibit repeated runs of the algorithm and the simulation. The performance of the method will be examined by its application to two water distribution networks from literature. The results will be compared with previously published results from literature and those generated by evolutionary multi-objective algorithm. It will be shown that the method is able to find the Pareto optimal front with less computational effort.

Keywords: Multiple objective, Pareto front, convergence, water distribution.

1 Introduction

Water Distribution network design involves conflicting objectives that each needs to be optimized. Optimal performance according to one objective often implies low performance in one or more of the other objectives. Evolutionary algorithms (EAs) have demonstrated unique ways of handling multi-objective optimization problems. Farmani et al. [1] investigated the application of different evolutionary multi-objective optimization methods in the search for the non-dominated (Pareto) set of solutions to the water distribution network problem. Two non-elitist methods, Multi-Objective Genetic Algorithms (MOGA) [2] and Niched Pareto Genetic Algorithms (NPGAs) [3] and three elitist methods, Non-Dominated Sorting Genetic Algorithm (NSGAII) [4], Pareto-Archived Evolution Strategy (PAES) [5] and Strength Pareto Evolutionary Algorithm (SPEA2) [6] were investigated through application to two test cases for the comparative study.

S. Obayashi et al. (Eds.): EMO 2007, LNCS 4403, pp. 443–456, 2007.

SPEA2 performed better than the other methods. NSGAII method performed slightly worse than SPEA2 but outperformed other three methods. In this work the NSGAII constrained optimization method will be considered as an evolutionary multi-objective optimization technique for the developed hybrid method. Farmani et al. [1] also showed that although most of the methods have managed to converge to the region of optimal solution but Pareto front was sub optimal. Evolutionary algorithms have good initial convergence characteristics, but slow down considerably once the region of the optimal solution has been identified. Combining local search with evolutionary algorithm, hybrid methods, have been very successful in the context of single objective optimization [7]. Talbi et al. [8] proposed a hybrid two-phase approach for multicriteria optimization problems as follows: run an MOEA for a fixed number of generations; then for each Pareto optimal solution, compute the neighborhood and store any non-dominated solutions found; update the list of Pareto front solutions and again recompute all the neighborhoods; iterate the procedure until no improvement occurs. Goel and Deb [4] presented two local search strategies to enhance the probability of NSGAII's true convergence. In the first method, the posteriori approach, the obtained non-dominated solutions of a multi-objective evolutionary algorithm run were modified using a local search method (The local search strategy was suggested from each obtained solution of NSGAII to find a better solution. A weighted objective function was used to convert multiple objectives into a single objective for local search). In the second method, the online approach, a local search method was applied to each solution obtained by genetic operation in a MOEA run. Goel and Deb [4] concluded that the posteriori approach is better than the online approach, the main reason being that in the online approach more emphasis is allocated to the local search method.

To achieve optimal non-dominated front in less computational time, the posteriori approach was applied to two benchmark water systems. Two issues might be raised by doing so, first the order of variables might affect the final result and secondly the weighted method might result in solutions that are dominated by original Pareto front and also it might cluster solutions, which is contrary to second goal of multi-objective optimization (diversity among the solutions). Preliminary research, by Farmani et al. [9], into effect of order of variables in local search method made it clear that final solution is not effected considerably by the order of design variables. The main reason for this is that the local search is done on individuals on Pareto front at final generation of NSGAII, and usually there are only few small changes possible. However, study into the choice of weighting method for decision making in accepting or rejecting of changes by local search showed that solutions are usually clustered in a small region and sometimes they are dominated by the original Pareto set.

In this paper, perturbation of variables of the individuals on the Pareto front is presented to assess the sensitivity of the solutions. The perturbation method is implemented in two ways, random and deterministic, and judgement on efficiency of newly generated solution is done based on Pareto non-domination rather than weighting method. Two different runs are carried out. The first run involves online implementation of the perturbation method, where perturbation is applied for each solution at each generation on Pareto front and in the second run perturbation is only applied to solutions on the Pareto front at the final generation. The performance of the perturbation method in detecting convergence of the Pareto front is illustrated by applying the method at different stages of the run until application of the perturbation method does not result in any further improvement in the Pareto front. The evolutionary algorithm will then be allowed to run further to demonstrate the lack of improvement in the Pareto front.

In what follows, description of the algorithm is presented and the efficiency of the method is illustrated by application to two benchmark water systems. The results are compared to those of Goel and Deb [4] to show the effect of random and deterministic perturbation and also Pareto non-domination over weight non-domination. Also a comparison is made to demonstrate the effect on the computational time.

2 Perturbation Method

In this study, the efficiency of NSGAII was improved by developing a hybrid optimization method which is based on perturbation of individuals on the Pareto front. Ultimately, this results in better or global Pareto set and reduces the computational time which is important in real world problems where cost and time constraint prohibit repeated runs of the algorithm and hydraulic evaluation. Perturbation of individuals on the Pareto front is done based on random or a deterministic method. In the random method design variables are chosen randomly for perturbation and the number of perturbations is equal to the total design variables for each solution. The deterministic approach is applied in two ways; in the first approach all the design variables are subject to perturbation (order-based perturbation) while in the second approach perturbation is implemented on certain design variables that are identified (based on knowledge related to the problem) to have direct effect on improving the objective function values (knowledge-based perturbation).

The knowledge-based perturbation involves introduction of a set of rules generated based on characteristics of each system and the nature of design constraints. In this method, first a non-dominated solution is chosen for possible perturbation and then the pipe or node that has the maximum infeasibility (violating the minimum or maximum velocity constraint in the pipe or the minimum individual surplus head constraint in the node) is identified. Depending on the nature of the violation, the design variable related to the identified node or pipe is perturbed to reduce or eliminate the level of infeasibility. The order-based perturbation algorithm involves implementation of the following procedure: first a search direction is determined by taking a small step in the positive direction (i.e., decision variable value will be increased to the next high value) or if required, in the negative direction; if the objective function value is improved (the generated solution dominate the existing solution), further steps are implemented in the same direction until no improvement in the optimal objective values are achieved.

The sensitivity analysis of Pareto front for water distribution is investigated here through the analysis of the expansion of the New York water supply system (Fig. 1) and the new water distribution trunk network in Hanoi (Fig. 2).

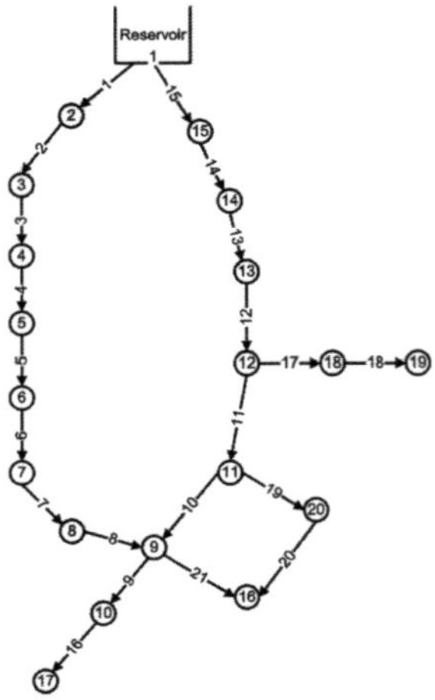

Fig. 1. Existing New York City Water Supply Tunnels

3 Test Cases

3.1 New York Tunnels (NYT) Problem

The New York pipe network has been studied using evolutionary techniques by a large number of researchers in the past ([10]; [11]; [12]; [13]; [14]; [9]). The objective of the NYT problem was to determine the most economically effective design for addition to the existing system of tunnels that constituted the primary water distribution system of the city of New York. Because of age and increased demands, the existing gravity flow tunnels were found to be inadequate to meet the pressure requirements (at nodes 16,17,18,19 and 20) for the projected consumption level [11]. The construction of additional gravity flow tunnels parallel to the existing was considered. The node and link data are from Murphy et al. [10].

Tunnel (pipe) diameters are considered as design variables. There are 15 available discrete diameters [36, 48, 60, 72, 84, 96, 108, 120, 132, 144, 156, 168, 180, 192, 204 inches] and one extra possible decision which is the 'do nothing' option. All twenty one existing tunnels are considered for duplication. Supplying demand at an adequate pressure to consumers is the main constraint (system performance indicator) in the design of water distribution systems. The performance of each candidate design solution is evaluated through simulation of the network flows. EPANET2 computer program [15] is the network solver used in this work.

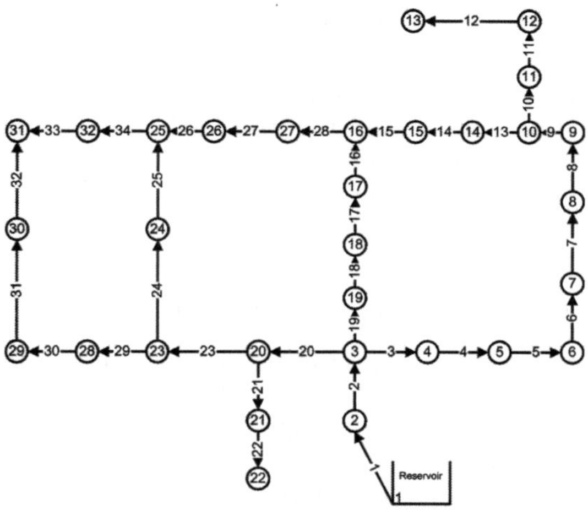

Fig. 2. The Hanoi Network

The cost function is non-linear, $C = 1.1D_{ij}^{1.24}L_{ij}$, in which cost C is in dollars, diameter D_{ij} is in inches, and length L_{ij} is in feet. The optimization problem was set based on capital expenditure as an objective function and the minimum pressure as constraints.

3.2 The Hanoi Network

The configuration of the water distribution trunk network in Hanoi, Vietnam, is shown in figure 2. This pipe network has 32 nodes and 34 pipes, organized in 3 loops. No pumping facilities are considered since only a single fixed head source at elevation of 100 m is available. The minimum head requirement at all nodes is fixed at 30 m. The set of commercially available diameters (in inches) is [12, 16, 20, 24, 30, 40]. The node data and link data are from Fujiwara and Khang [16]. The cost function is non-linear, $C = 1.1D_{ij}^{1.5}L_{ij}$, in which cost C is in dollars, diameter D_{ij} is in inches, and length L_{ij} is in meters.

4 Solutions and Analysis

The performance of the NSGAII has been investigated here for the solution of New York Tunnel and Hanoi optimization problems. In the optimization problems, the pay-off is investigated between the total cost and the maximum head deficiency.

Different runs were carried out for each network; in the first run the trade off between the capital cost and the maximum head deficiency was analyzed using NSGAII, and in the second run the analysis was based on pay off curves generated by the perturbation method or the posteriori method. In order to determine the pay off characteristic between the capital cost and the head deficiency all the methods were run with a population of 200 sample solutions, and were allowed to run for (1000, 100, 50) generations for NYT network and (1000, 500, 100) generations for Hanoi network. Early results showed that out of the three perturbation methods (random, order-based and knowledge-based), the random perturbation method had the worst efficiency. In this method, targeting the right design variables sometimes resulted in a large number of function evaluations for small improvements. Therefore, only deterministic methods were used in perturbation of solutions on the Pareto fronts.

4.1 Performance in Finding Pareto Front

Figure 3 illustrates the Pareto front characteristic between the capital cost and the maximum head deficiency for the NYT network obtained through application of standard NSGAII run for 1000 generations and the Pareto set after perturbation and posteriori methods for Pareto set of 100 generations. It can be seen from the figure that the solutions generated by the perturbation and posteriori methods for 100 generations are in a close match to those obtained by NSGAII run for 1000 generations. The above results are presented to illustrate the performance of the algorithms over the whole range of both objectives. However, in practice, only those solutions which are marginally infeasible would be of interest. Figure 4 shows the result for the New York Tunnels network for an individual head deficit in the range of 0-20 m, showing the trade-off between capital cost and head deficiency in the region of practical design interest.

Figure 5 illustrates the same plot for Hanoi water system for 1000 generations of NSGAII and Pareto sets obtained using the perturbation and posteriori methods. The efficiency of the algorithm is apparent from the number of solutions obtained along the pay-off curve. Figure 6 shows the results for the Hanoi network for an individual head deficit in the range of 0-18 m, showing the trade-off between capital cost and head deficiency in the region of practical design interest. This figure illustrates one of the main drawbacks of the posteriori method, some individuals in the Pareto set generated by the posteriori method have been dominated by the main pay off curve. Therefore, using the weighting method in judgement on acceptance or rejection of the newly generated individual in the Pareto front might not always result in a non-dominated set.

4.2 Comparison of Results

Figures 7 and 8 illustrate the Pareto front for New York Tunnel problem generated from 1000, 100 and 50 generations by NSGAII and those by the perturbation method and the posteriori method respectively. Detailed inspection of Pareto front generated by different algorithms shows that perturbation of the Pareto front of 100 generations resulted in a pay off curve that dominates the Pareto front generated by NSGAII after 1000 generations. This is a very important characteristic in reducing the number of generations and therefore reducing the computational time which is important in real world problems where cost and time constraints prohibit repeated runs of the algorithm and hydraulic evaluation. The number of hydraulic evaluations in the knowledge based and

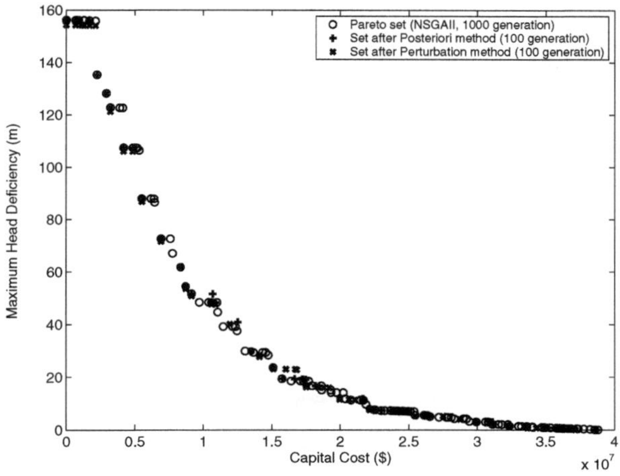

Fig. 3. Cost vs. Deficiency pay-off for NSGAII, Posteriori method and Perturbation method (New York Tunnel)

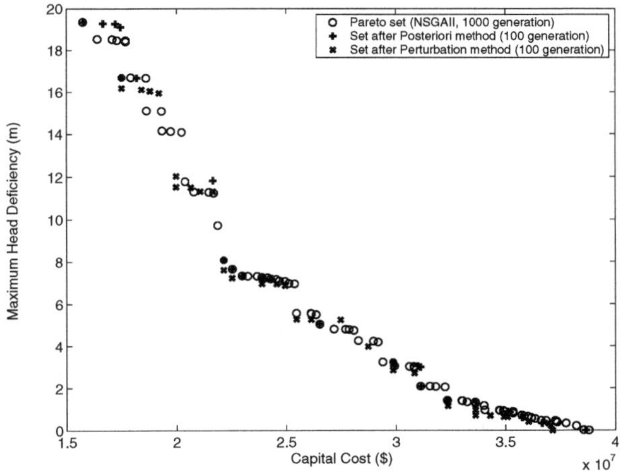

Fig. 4. Cost vs. Deficiency pay-off for NSGAII, Posteriori method and Perturbation method (New York Tunnel)

order based perturbation are 830 and 1760 respectively that is 4 and 8 times the number of hydraulic evaluations in one generation of NSGAII. This leads to a reduction of about 90 percent in the number of evaluations.

Figures 9 and 10 illustrate the Pareto front for Hanoi generated from 1000, 500 and 100 generations by NSGAII and those by the perturbation and posteriori methods respectively. Due to complexity of the problem as shown in figure 9 the pay off curve resulted from 100 generations is far from the one generated

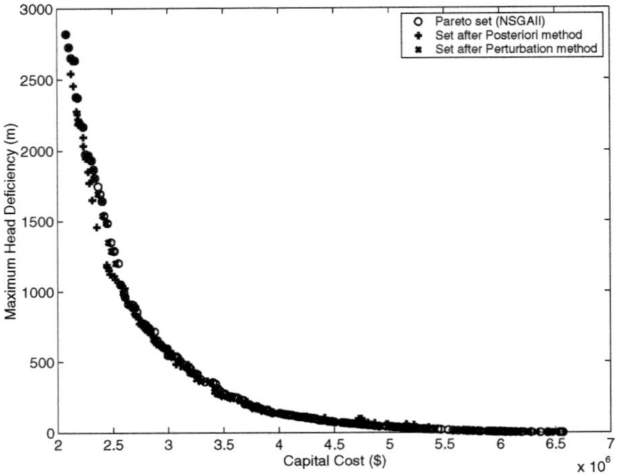

Fig. 5. Cost vs. Deficiency pay-off for NSGAII, Posteriori method and Perturbation method (Hanoi Network)

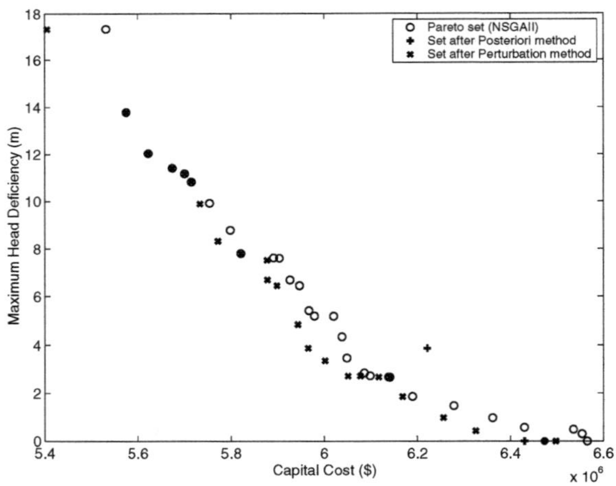

Fig. 6. Cost vs. Deficiency pay-off for NSGAII, Posteriori method and Perturbation method (Hanoi Network)

by 1000 generations, and the perturbation and the posteriori methods did not manage to close the gap, however the perturbation method still performed better than the posteriori method. Perturbation of the individuals on the pay off curve generated by 500 generations resulted in a Pareto front that often dominates the one generated by 1000 generations except in a small part of the curve. Improvements of the Pareto front generated by 1000 generations after implementation of perturbation method indicate that the set was not the optimal set and further runs would be required to find a near optimal Pareto front. This is

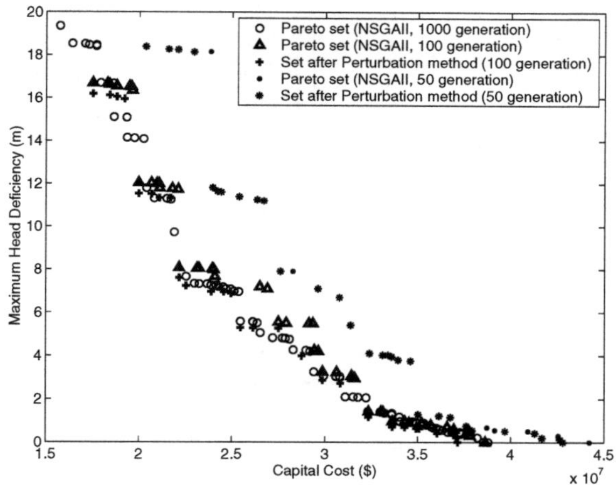

Fig. 7. Cost vs. Deficiency pay-off for Perturbation method (New York Tunnel)

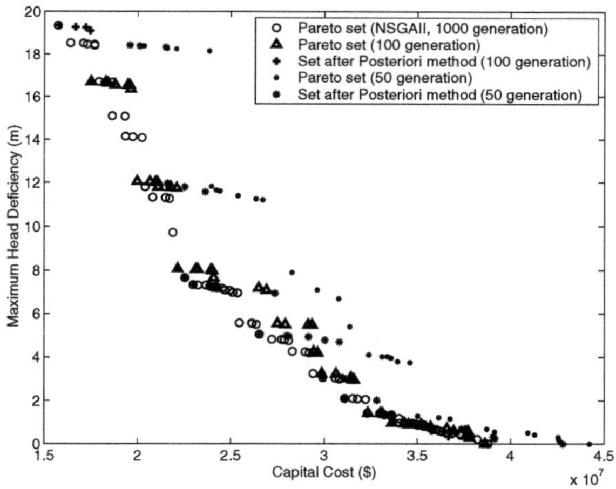

Fig. 8. Cost vs. Deficiency pay-off for Posteriori method (New York Tunnel)

another important characteristic of the method in identifying the termination criteria for evolutionary algorithms. No improvement due to the perturbation method in pay off curve means that individuals are not sensitive to changes in variable values. In another word, they have reached their limit in the feasible region. Figure 10 illustrates the poor performance of the posteriori method in the region of practical design interest. This figure illustrates another drawback of the posteriori method that there are only few individuals in Pareto set generated by the posteriori method that clustered in a very small part of the search space. The perturbation method consistently generated solutions that either dominate individuals on the main pay off curve or overlap them.

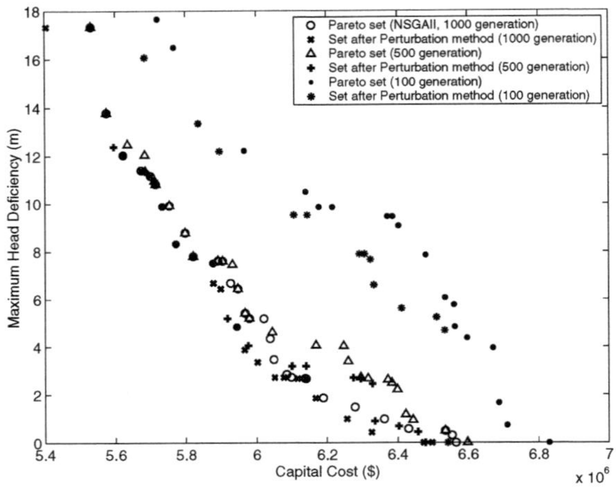

Fig. 9. Cost vs. Deficiency pay-off for Perturbation method (Hanoi Network)

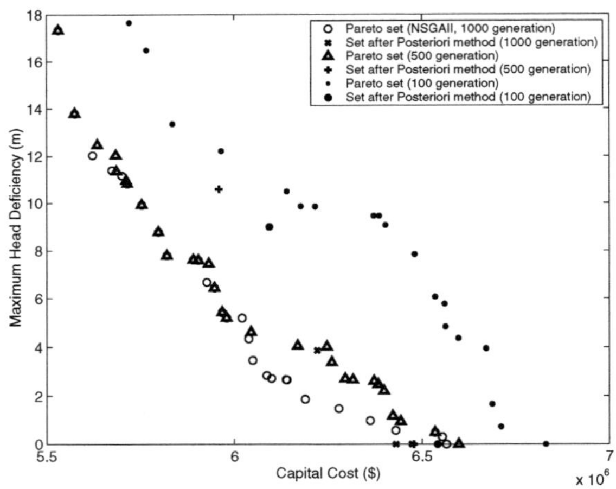

Fig. 10. Cost vs. Deficiency pay-off for Posteriori method (Hanoi Network)

Figure 11 shows a comparison of the non-dominated fronts obtained by the three algorithms for Hanoi network in three ways, using graphics, using a coverage metric (I_c) and using an epsilon metric (I_{eps}). Quantitative comparison of the performance of the different evolutionary multi-objective algorithms is an important issue. Zitzler et al. [17] proposed the binary $\epsilon-$indicator (I_ϵ), which gives a factor by which an approximation set is worse than another with respect to all objectives. I_ϵ can be calculated as:

$$I_\epsilon(A,B) = \max_{z^2\in B} \min_{z^1\in A} \max_{1\le i\le n} \frac{z_i^1}{z_i^2} \quad (1)$$

where A and B are two approximation sets; z^1 and z^2 are objective vectors; and n is the number of objectives.

Zitzler and Thiele [18] suggested the coverage indicator I_c, which gives the fraction of solutions in one set that are weakly dominated by at least one solution in another set. I_c can be calculated as:

$$I_c(A, B) = \frac{\left|\left\{z^2 \in B\,;\, \exists z^1 \in A : z^1 \leq z^2\right\}\right|}{|B|} \tag{2}$$

They considered the relationships "strictly dominates, $\gg$; "dominates, $>$; "better, $\rhd$; "weakly dominates, $\geq$; and "incomparable, $\|$ on objective vectors and approximation sets. The summary of equivalence of these dominance relationships in approximation sets is given in Table 1.

Table 1. Dominance relationships and their definition in approximation sets

Relation	Approximation set
$A \gg B$	every $z^2 \in B$ is strictly dominated by at least one $z^1 \in A$
$A > B$	every $z^2 \in B$ is dominated by at least one $z^1 \in A$
$A \rhd B$	every $z^2 \in B$ is weakly dominated by at least one $z^1 \in A$ and $A \neq B$
$A \geq B$	every $z^2 \in B$ is weakly dominated by at least one $z^1 \in A$
$A \parallel B$	neither A weakly dominates B nor B weakly dominates A

Binary indicators can be interpreted as different dominance relationships according to their values. Table 2 gives a summary of these interpretations by Zitzler et al. [17] for I_ϵ and I_c.

Table 2. Overview of binary indicators and dominance relationships

Binary indicator	compatible and complete with respect to relation					
	$\gg$	$>$	$\rhd$	$\geq$	$=$	$\parallel$
I_ϵ	$I_\epsilon(A, B) < 1$		$I_\epsilon(A, B) \leq 1$ $I_\epsilon(B, A) > 1$	$I_\epsilon(A, B) \leq 1$	$I_\epsilon(A, B) = 1$ $I_\epsilon(B, A) = 1$	$I_\epsilon(A, B) > 1$ $I_\epsilon(B, A) > 1$
I_c		$I_c(A, B) = 1$ $I_c(B, A) = 0$	$I_c(A, B) = 1$ $I_c(B, A) < 1$	$I_c(A, B) = 1$	$I_c(A, B) = 1$ $I_c(B, A) = 1$	$0 < I_c(A, B) < 1$ $0 < I_c(B, A) < 1$

The binary coverage indicator values were evaluated for different combinations of the NSGAII, posteriori and perturbation algorithms for the Hanoi water distribution network for the Pareto front generated from 100, 500 and 1000 generations. Most of the values for different combinations were greater than zero and less than one. According to table 2 this can be interpreted as "incomparable, $\|$" relationship. However, the coverage indicator values for the combination of the perturbation method with the other two methods were higher than other combinations and usually equal to 1.0. This means the Pareto front generated by the perturbation method has better coverage of the Pareto fronts generated by the other methods. The posteriori method has the lowest value of the coverage indicator therefore has the worst coverage of the other Pareto fronts.

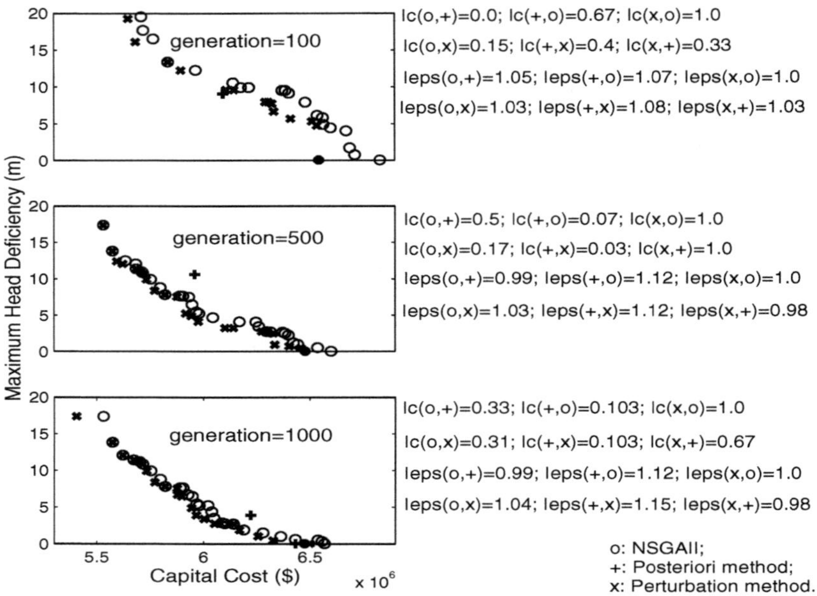

Fig. 11. Cost vs. Deficiency pay-off for NSGAII, Posteriori method and Perturbation method (Hanoi Network)

The binary ϵ-indicator was also evaluated for different combinations of the NSGAII, posteriori and perturbation algorithms for the Hanoi water distribution network. All the values for different combinations from 100 generations were greater than one, except for four combinations. According to table 2 this can be interpreted as "incomparable, $\|$" relationship. $I_\epsilon(perturbation, NSGAII)$ was equal to 1.0 for one of the combinations for the Hanoi network and $I_\epsilon(NSGAII, perturbation)$ for that combination was equal to 1.03. According to table 2 ($I_\epsilon(perturbation,\ NSGAII) \leqq 1$ and $I_\epsilon(NSGAII, perturbation) > 1$) indicating that the perturbation performed better "$\triangleright$" than NSGAII. It can be concluded that, based on the binary ϵ values for combinations of the perturbation and NSGAII and posteriori methods for the Pareto fronts generated by 500 and 1000 generations, the perturbation performed better than NSGAII and posteriori methods. It also can be concluded that, NSGAII performed better than the posteriori method. This also confirms the conclusions drawn from the graphical presentation.

5 Conclusion

A perturbation method was introduced that is a combination of multi-objective evolutionary algorithms (NSGAII) and local search methods to speed up the convergence rate of the algorithm in finding optimal Pareto set. The perturbation method was implemented in three different ways, random, order-based and knowledge-based deterministic, and judgment on efficiency of the newly generated solutions was made based on Pareto non-domination. The method was applied to two benchmark water distribution networks (New York Tunnel and Hanoi water distribution system). The results were compared to those obtained by posteriori method of Goel and Deb [4] to show the effect of random and deterministic hybrid methods and also Pareto non-domination over weighting method. Also a comparison was made to demonstrate the effect on computational time.

The search results indicate that combination of the NSGAII approach with local search has the potential to find Pareto optimal solutions. Preliminary research into effect of order of variables in local search method indicated that the final solution is not affected a lot by order of design variables. The main reason for this is that the local search is done on individuals on Pareto front at final generation of NSGAII, and usually there are only few small changes possible. Out of three methods used for perturbation of individuals on the Pareto front, the random method performed the worst, in terms of the number of function evaluations and in finding better Pareto front. The knowledge-based perturbation method outperformed the order-based method by having slightly lower number of function evaluations. However, it is expected that the knowledge-based method would perform much better in problems with larger number of design variables by targeting the most effective design variables and therefore, reducing the number of function evaluations. Perturbation of pay off curve resulted in a Pareto front that not only dominates the original pay off curve but also dominates the one generated by 10 times and twice the number of the generations for the original pay off curve (for New York Tunnel and Hanoi network respectfully). This is a very important characteristic in reducing the number of generations and therefore reducing the computational time which is important in real world problems where cost and time constraint prohibit repeated runs of the algorithm and hydraulic evaluation. It can be concluded that the hybrid NSGAII multi-objective evolutionary algorithm shows potential for the optimization of water systems.

References

1. Farmani, R., Savic, D.A., Walters, G.A.: Evolutionary multi-objective optimization in water distribution network design. Journal of Engineering Optimisation, Vol. 37.**2** (2005) 167–185
2. Fonseca, C.M., Fleming, P.J.: Genetic Algorithms for multi-objective Optimization: Formulation, Discussion and Generalization. Proceedings of the International Conference on Genetic Algorithms, Vol. 5. (1993) 416–423
3. Horn, J., Nafploitis, N., Goldberg, D.E.: A niched Pareto genetic algorithm for multi-objective optimisation. Proceeding of the first IEEE Conference on Evolutionary Computation, New York. IEEE Press. Vol. 1. (1994) 82–87
4. Goel, T., Deb, K.: Hybrid methods for multi-objective evolutionary algortihms. KanGAL Report (2001004) (2001)
5. Knowles, J.D., Corne, D.W.: Approximating the non-dominated front using the Pareto archived evolution strategy. Evolutionary Computation Journal, Vol. 8.**2** (2000) 149–172

6. Zitzler, E., Laumanns, M., Thiele, L.: SPEA2: Improving the strength pareto evolutionary algorithm for multiobjective optimization, Evolutionary methods for design, Optimisation and control, Giannakoglou, K., Tsahalis, D., Periaux, J., Papailiou, K., and Fogarty, T. (eds.), Barcelona, Spain, (2002)
7. Van zyl, J., Savic, D.A., Walters, G.A.: Operational Optimization of Water Distribution Systems Using a Hybrid Genetic Algorithm Method, Journal of Water Resources Planning and Management, ASCE, Vol. 130.**2** (2005) 160–170
8. Talbi El-Ghazali, Rahoual, M., Mabed, M.H., Dhaenens, C.: A hybrid evolutionary approach for multicriteria optimization problems: Application to the flow shop. First international conference on Evolutionary Multi-Criterion Optimization, Zitzler, E., Deb, K., Thiele, L., Coello Coello, C.A. and Corne D. (Eds.), Springer-Verlag. Lecture Notes in Computer Science No. 1993, (2001) 416–428
9. Farmani R., Prasad, T.Devi, Walters, G.A., Savic, D.A.: Mutli-criteria decision making in water systems. Proceedings of the 17th European Junior Scientist workshop(EJSW), Rehabilitation management of urban infrastructure networks, Dresden, September 2003, Baur R., Kropp I. and Herz, R. (eds.), (2003) 103–112
10. Murphy, L.J., Simpson, A.R., Dandy, G.C.: Pipe network optimisation using an improved genetic algorithm. Research Report No. R109, Dept of Civil and Environment Engineering, University of Adelaide, Australia. (1993)
11. Savic, D.A., Walters, G.A.: Genetic Algorithms for least cost design of water distribution networks. Journal of water resources planning and management, ASCE, Vol. 123.**2** (1997) 67–77
12. Lippai, I., Heaney, J.P., Laguna, M.: Robust water system design with commercial intelligent search optimisers. Journal of Computing in Civil Engineering, Vol. 13.**3** (1999) 135–143
13. Vairavamoorthy, K., Ali, M.: Optimal design of water distribution systems using genetic algorithms. Computer Aided Civil and Infrastructure Engineering, Vol. 15. (2000) 374–382
14. Wu, Z.Y., Boulos, P.F., Orr, C.h., Ro, J.J.: Using genetic algorithm to rehabilitate distribution systems. Journal AWWA, Vol. 93.**11** (2001) 74–85.
15. Rossman, L.A.: EPANET, users manual. U.S. Envir. Protection Agency, Cincinnati, Ohio, (2000)
16. Fujiwara, O., Khang, D.B.: A two Phase decomposition method for optimal design of looped water distribution networks. Water resources research, Vol. 26.**4** (1990) 539–549
17. Zitzler, E., Thiele, L., Laumanns, M., Fonseca, C.M., Fonseca, V.G.: Performance assessment of multiobjective optimizers: An analysis and review. IEEE Transactions on Evolutionary Computation, Vol. 7.**2** (2003) 117–132
18. Zitzler, E., Thiele, L.: Multiobjective evolutionary algorithms: A comparative case study and the strength Pareto approach. IEEE Transactions on Evolutionary Computation, Vol. 3. (1999) 257–271

Combinatorial Optimization of Stochastic Multi-objective Problems: An Application to the Flow-Shop Scheduling Problem

Arnaud Liefooghe, Matthieu Basseur, Laetitia Jourdan, and El-Ghazali Talbi

INRIA Futurs, Laboratoire d'Informatique Fondamentale de Lille (LIFL), CNRS
Bât. M3, Cité Scientifique, 59655 Villeneuve d'Ascq cedex, France
{liefooga,basseur,jourdan,talbi}@lifl.fr

Abstract. The importance of multi-objective optimization is globably established nowadays. Furthermore, a great part of real-world problems are subject to uncertainties due to, *e.g.*, noisy or approximated fitness function(s), varying parameters or dynamic environments. Moreover, although evolutionary algorithms are commonly used to solve multi-objective problems on the one hand and to solve stochastic problems on the other hand, very few approaches combine simultaneously these two aspects. Thus, flow-shop scheduling problems are generally studied in a single-objective deterministic way whereas they are, by nature, multi-objective and are subjected to a wide range of uncertainties. However, these two features have never been investigated at the same time.

In this paper, we present and adopt a proactive stochastic approach where processing times are represented by random variables. Then, we propose several multi-objective methods that are able to handle any type of probability distribution. Finally, we experiment these methods on a stochastic bi-objective flow-shop problem.

Keywords: multi-objective combinatorial optimization, stochasticity, evolutionnary algorithms, flow-shop, stochastic processing times.

1 Introduction

A large part of concrete optimization problems are subject to uncertainties that have to be taken into account. Therefore, many works relate to optimization in stochastic environments (see [10] for an overview), but very few deal with the multi-objective case where Pareto dominance is used to compare solutions. Thus, Hughes [8] and Teich [18] independently suggested to extend the concept of Pareto dominance in a stochastic way by replacing the rank of a solution by its probability of being dominated; but both studies make an assumption on probability distributions. In [1], another ranking method, based on an average value per objective and on the variance of a set of evaluations, is presented. Likewise, Deb and Gupta [5] proposed to apply standard deterministic multi-objective optimizers using an average value, determined over a sample of objective vectors, for each dimension of the objective space. Finally, Basseur and Zitzler [2]

S. Obayashi et al. (Eds.): EMO 2007, LNCS 4403, pp. 457–471, 2007.

recently extended the concept of multi-objective optimization using quality indicators [21] to take stochasticity into account. However, even if existing methods are generally adaptable to the combinatorial case, most of them were only tested on continuous mathematical test functions. Thence, it is not obvious that the performances of these algorithms are similar for combinatorial and continuous problems. Furthermore, a large part of these algorithms exploits problem knowledge that may not be available in real-world applications.

The deterministic indicator-based approach [21] consists in assigning each Pareto set approximation a real value reflecting its quality, using a function I [20]. The goal is then to identify a Pareto set approximation that optimizes I. As a result, I induces a total order into the set of aproximation sets in the objective space, and gives rise to a total order into the corresponding objective vectors. The interest of this perception is that no additional diversity preservation mechanisms are required, the concept of Pareto dominance not being directly used for fitness assignment. To extend this approach to the stochastic case, we must consider that every solution can be associated to an arbitrary probability distribution over the objective space.

In this paper, we propose various models to represent stochasticity for a bi-objective flow-shop scheduling problem. Then, we introduce different ways to handle uncertainty, insisting on the various technical aspects. And, we apply the resulting methods to the concrete case of a flow-shop scheduling problem with stochastic processing times, that have, to our knowledge, never been investigated in a multi-objective form. Each approach has advantages and drawbacks and is adapted from indicator-based optimization.

The paper is organized as follows. In section 2, we formulate a bi-objective flow-shop scheduling problem with stochastic processing times (SFSP). In section 3, we present three different approaches dedicated to stochastic multi-objective optimization and apply them on a SFSP. Section 4 presents experimental results. And finally, the last section draws conclusion and suggests further topics in this research area.

2 A Bi-objective Flow-Shop Scheduling Problem with Stochastic Processing Times

The flow-shop is one of the numerous scheduling problems. It has been widely studied in the literature (see, for example, [6] for a survey). However, the majority of works dedicated to this problem considers it on a deterministic single-criterion form and mainly aims at minimizing the makespan, which is the completion time of the last job. Following the formulation of the deterministic model of a bi-objective flow-shop scheduling problem, this section presents various sources of uncertainty that have to be taken into account and introduces different probability distributions to model stochastic processing times. Note that, although this part focuses on the flow-shop, it can easily be generalizable to other types of problem.

2.1 Deterministic Model

Solving the flow-shop problem consists in scheduling N jobs $J_1, J_2, \ldots, J_N$ on M machines $M_1, M_2, \ldots, M_M$. Machines are critical resources, *i.e.* two jobs cannot be assigned to one machine simultaneously. A job J_i is composed of M consecutive tasks $t_{i1}, t_{i2}, \ldots, t_{iM}$, where t_{ij} is the jth task of the job J_i, requiring the machine M_j. A processing time p_{ij} is associated to each task t_{ij} and a job J_i must be achieved before its due date d_i. For the permutation flow-shop, the operating sequences of the jobs are identical and unidirectional on every machines.

In this study, we focus on minimizing both the makespan (C_{max}) and the total tardiness ($\overline{T}$), which are two of the most studied objectives of the literature . For each task t_{ij} to be scheduled at the time s_{ij}, we can compute the two considered objectives as follows:

$$C_{max} = \max_{i \in [1..N]} [s_{iM} + p_{iM}] \; . \tag{1}$$

$$\overline{T} = \sum_{i=1}^{N} \max(0, s_{iM} + p_{iM} - d_i) \; . \tag{2}$$

In the Graham et al. notation [7], this problem is noted $F/permu, d_j/(C_{max}, \overline{T})$. Besides, the interested reader is referred to [14,16] for a review on multi-objective scheduling.

2.2 Sources of Uncertainty

In real-world scheduling situations, uncertainty can occur from many sources such as release or due date variations, machine breakdowns, unexpected arrival or cancellation of orders, variable processing times, ... According to the literature, in the particular case of our permutation flow-shop scheduling problem, the uncertainty mainly stem from due dates and processing times.

Firstly, in the deterministic model, the due date of a job J_i is given by a fixed number d_i. However, it looks difficult to determine it without ambiguity. So, it seems more natural to determine it using an interval $[d_i^1, d_i^2]$ during which the human satisfaction for the completion of the job J_i decreases between d_i^1 and d_i^2. Moreover, a due date d_i may change dynamically since a less important job today may be of high importance tomorrow, and vice versa. Secondly, a processing time may vary from an execution to another and some unexpected events may occur during the process. So, the processing time p_{ij} of a task t_{ij} rarely corresponds to a constant value. To conclude, it is obvious that no parameter can be regarded as an exact and precise data and that non-classical approaches are required to solve concrete scheduling problems. Thus, we decide to adopt a proactive stochastic approach where processing time values are regarded as uncertain and are represented by random variables.

2.3 Stochastic Models

Widely studied in its single-criterion form, the stochastic flow-shop scheduling problem has, to our knowledge, never been investigated in a multi-objective way. Furthermore, as soon as historic data about processing times are available, it seems quite easy to determine which probability distribution is associated to those parameters. Following an analysis, we propose four different general distributions a processing time may follow. Of course, a rigorous statistical analysis, based on real data, is imperative to determine the concrete and exact distribution associated to a certain processing time p_{ij} of a real-world problem.

Uniform distribution. A processing time p_{ij} can uniformly be included between two values a and b. Then, p_{ij} follows a uniform distribution over the interval $[a,b]$ and its probability density function is:

$$f(x) = \begin{cases} \frac{1}{b-a} & \text{if } x \in [a,b] \\ 0 & \text{otherwise} \end{cases} . \tag{3}$$

This kind of distribution is used to provide a simplified model of real industrial cases. For example, it has already been used by Kouvelis et al. [12].

Exponential distribution. A processing time p_{ij} may follow an exponential distribution $\mathcal{E}(\lambda, a)$. Thus, its probability density function is:

$$f(x) = \begin{cases} \lambda\, e^{-\lambda(x-a)} & \text{if } x \geq a \\ 0 & \text{otherwise} \end{cases} . \tag{4}$$

Exponential distributions are commonly used to model random events that may occur with uncertainty. This is typically the case when a machine is subject to unpredictable breakdowns. For example, processing times have been modeled by an exponential distribution in [3,13].

Normal distribution. A processing time p_{ij} may follow a normal distribution $\mathcal{N}(\mu, \sigma)$ where μ stands for the mean and σ stands for the standard deviation, in which case its probability density function is:

$$f(x) = \frac{1}{\sigma\sqrt{2\pi}}\, exp\Big(-\frac{1}{2}\Big(\frac{x-\mu}{\sigma}\Big)^2\Big) . \tag{5}$$

This kind of distribution is especially usual when human factors are observed. A process may also depend on unknown or uncontrollable factors and some parameters can be described in a vague or ambiguous way by the analyst. Therefore, processing times vary according to a normal distribution.

Log-normal distribution. A random variable X follows a log-normal distribution with parameters μ and σ if $\log X$ follows a normal distribution $\mathcal{N}(\mu, \sigma)$. Its probability density function is then:

$$f(x) = \begin{cases} \frac{1}{\sigma\sqrt{2\pi}}\, \frac{1}{x}\, exp(-\frac{1}{2}(\frac{\log x-\mu}{\sigma})^2) & \text{if } x > 0 \\ 0 & \text{otherwise} \end{cases} . \tag{6}$$

The log-normal distribution is often used to model the influence of uncontrolled environmental variables. In our case, a processing time p_{ij} following a log-normal distribution takes into account simultaneously the whole observed uncertainties. For example, this modeling has already been used in [4].

3 Indicator-Based Evolutionary Methods

This section contains a brief presentation of the indicator-based approach introduced in [21] (the interested reader will refer to this article for more details). Then, we present its extension to the stochastic case and propose three multi-objective methods that result from this extension.

3.1 Indicator-Based Multi-objective Optimization

Let us consider a generic multi-objective optimization problem defined by a decision space X, an objective space Z, and n objective functions $f_1, f_2, \ldots, f_n$. Without loss of generality, we here assume that $Z \subseteq \mathbb{R}^n$ and that all n objective functions are to be minimized. In the deterministic case, to each solution $x \in X$ is assigned exactly one objective vector $z \in Z$ on the basis of a vector function $F : X \rightarrow Z$ with $z = F(x) = f_1(x), f_2(x), \ldots, f_n(x)$. The mapping F defines the 'true' evaluation of a solution $x \in X$, and the goal of a deterministic multi-objective algorithm is to approximate the set of Pareto optimal solutions according to F[1]. However, generating the entire set of Pareto optimal solutions is usually infeasible, due to, *e.g.*, the complexity of the underlying problem or the large number of optima. Therefore, in many applications, the overall goal is to identify a good approximation of the Pareto optimal set. The entirety of all Pareto set approximations is represented by Ω.

Different interpretations of what a good Pareto set approximation is are possible, and the definition of approximation quality strongly depends on the decision maker preferences and the optimization scenario. As proposed in [21], we here assume that the optimization goal is given in terms of a binary quality indicator $I : \Omega \times \Omega \rightarrow \mathbb{R}$. Then, $I(A, B)$ quantifies the difference in quality between two sets A and $B \in \Omega$, according to the decision maker preferences. So, if R denotes the set of Pareto optimal solutions, the overall optimization goal can be formulated as $\operatorname{argmin}_{A \in \Omega} I(F(A), F(R))$. Binary quality indicators represent a natural extension of the Pareto dominance relation and can thus directly be used for fitness assignment. Therefore, the fitness of a solution x contained in a set of solutions can be determined using the indicator values obtained by x compared to the whole set. It measures the usefulness of x according to the optimization goal.

3.2 Handling Stochasticity

In the stochastic case, the objective values are different each time a solution is evaluated. So, the vector function F does not represent a deterministic mapping from X to Z, because an infinite set of different objective vectors is now assigned to a solution $x \in X$. Note that we consider that the 'true' objective vector of a solution is absolutely unknown before the end of the optimization process.

---

[1] A solution $x_1 \in X$ is Pareto optimal if and only if there exists no $x_2 \in X$ such that (i) $F(x_2)$ is component-wise smaller than or equal to $F(x_1)$ and (ii) $F(x_2) \neq F(x_1)$.

3.3 Proposed Methods

To tackle the optimization of stochastic multi-objective problems, we here propose three different adaptations inspired by a multi-objective evolutionary algorithm designed for deterministic problems and recently introduced by Zitzler and Künzli [21], namely $IBEA$ (Indicator-Based Evolutionary Algorithm). For each algorithm, we will use the additive ϵ-indicator [20,21] as the binary performance measure needed in the selection process of $IBEA$. This indicator seems to be efficient [21] and obtained significantly better results on our problem in its deterministic form than the I_{HD}-indicator (that is based on the hypervolume concept introduced in [19]). The additive ϵ-indicator ($I_{\epsilon+}$) gives the minimum ϵ-value by which B can be moved in the objective space such that A is at least as good as B. For a minimization problem, it is defined as follows [20,21]:

$$I_{\epsilon+}(A, B) = \inf_{\epsilon \in \mathbb{R}} \{\forall x_2 \in B, \exists x_1 \in A : f_i(x_1) - \epsilon \leq f_i(x_2), i \in \{1, ..., n\}\} . \quad (7)$$

Single evaluation-based estimate. The first method, $IBEA_1$, consists in preserving the approach used in the deterministic case. A solution is evaluated only once and its fitness is estimated using this single evaluation (see fig. 1). Actually, most of the methods proceed like that since they are based on constant parameters and do not take uncertainties into account. The advantage of this method is its low computation cost, but the estimation error may be large since the evaluation used is not necessarily representative of the potential evaluation space.

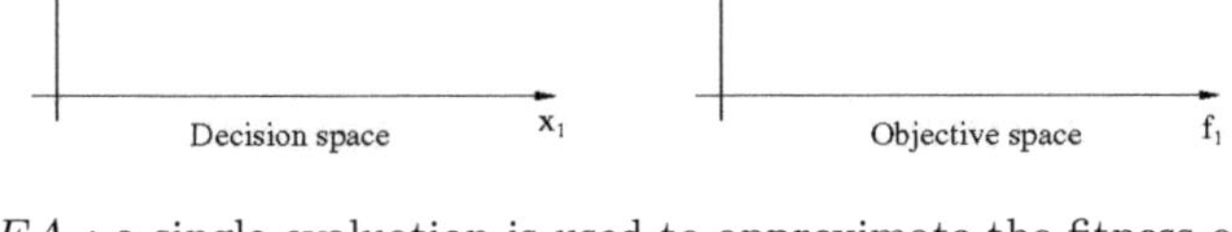

Fig. 1. $IBEA_1$: a single evaluation is used to approximate the fitness of a solution

Average estimate. The second method, called $IBEA_{avg}$, follows the idea commonly used in the single-criterion form and suggested in several multi-objective studies (such as, *e.g.*, [1,5]). It consists in doing several evaluations of the same solution, and then in calculating the average value of these evaluations on each objective function. Next, the deterministic approach is applied using those average values (see fig. 2). This method also has the advantage of having a low computation cost if the evaluation of a solution is not too expensive (what is the case for our problem). However, losses of information may occur during the average estimate, like, e.g., the potential evaluations distribution in the research space.

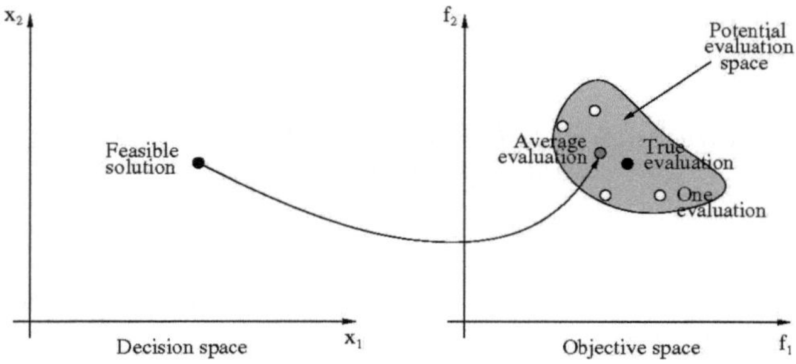

Fig. 2. $IBEA_{avg}$: the average evaluation values are used to approximate the fitness of a solution

Probabilistic estimate. The last method consists in estimating the fitness of a solution in a probabilistic way. Here, contrary to some other approaches [10], we do not assume that there is a 'true' objective vector per solution which is blurred by noise, but we consider that a probability distribution is associated to each solution on the objective space (see fig. 3). In order to allow the comparison of potential values of solutions, this extension of $IBEA$, called $IBEA_{stoch}$, consists in modifying the performance assessment procedure as proposed by Basseur and Zitzler in [2].

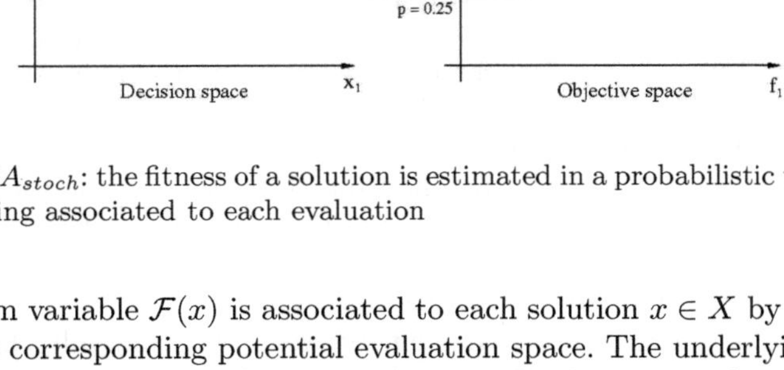

Fig. 3. $IBEA_{stoch}$: the fitness of a solution is estimated in a probabilistic way, a quality indicator being associated to each evaluation

A random variable $\mathcal{F}(x)$ is associated to each solution $x \in X$ by the range of which is its corresponding potential evaluation space. The underlying probability distribution is usually unknown and may differ for other solutions. Thus, in practice, for a binary quality indicator I, the fitness of a solution x is computed using an estimation of expected I-values on a finite set of evaluations. Hence, for a population $P = \{x_1, x_2, \dots, x_m\}$ and a finite set of evaluation $S(x)$, the fitness of an individual x is defined as the estimated loss of quality if x is removed from the population, *i.e.*:

$$\begin{aligned} Fitness(x) &= \hat{E}(I(\mathcal{F}(P \setminus \{x\}), \mathcal{F}(P)) \\ &= \hat{E}(I(\mathcal{F}(P \setminus \{x\}), \mathcal{F}(\{x\})) \\ &= \frac{1}{|S(x)|} \sum_{z \in S(x)} \hat{E}(I(\mathcal{F}(P \setminus \{x\}, \{z\}))) \; . \end{aligned} \tag{8}$$

To compute the estimated expected $I_{\epsilon+}$-value between a multiset $A \in \Omega$ and a reference objective vector $z^\star$, we consider all pairs (x_j, z_k) where $x_j \in A$ and $z_k \in S(x_j)$ and sort them in the increasing order according to the indicator values $I_{\epsilon+}(\{z_k\}, \{z^\star\})$. Suppose the resulting order is $(x_{j_1}, z_{k_1}), (x_{j_2}, z_{k_2}), \ldots, (x_{j_l}, z_{k_l})$, the estimated expected $I_{\epsilon+}$-value is then:

$$\begin{aligned} \hat{E}(I_{\epsilon+}(\mathcal{F}(A), \{z^*\})) = \; & I_{\epsilon+}(\{z_{k_1}\}, \{z^*\}) \times \hat{P}(\mathcal{F}(\{x_{j_1}\}) = \{z_{k_1}\}) \quad + \\ & I_{\epsilon+}(\{z_{k_2}\}, \{z^*\}) \times \\ & \quad \hat{P}(\mathcal{F}(\{x_{j_2}\}) = \{z_{k_2}\}) \,|\, \mathcal{F}(\{x_{j_1}\}) \neq \{z_{k_1})\}) \quad + \\ & \ldots \\ & I_{\epsilon+}(\{z_{k_l}\}, \{z^*\}) \times \\ & \quad \hat{P}(\mathcal{F}(\{x_{j_l}\}) = \{z_{k_l}\} \,|\, \forall_{1 \le i < l} \mathcal{F}(\{x_{j_i}\}) \neq \{z_{k_i}\}) \; . \end{aligned} \tag{9}$$

The running time complexity of an estimated expected $I_{\epsilon+}$-value computation is of order $\mathcal{O}(n(Ns)^2 \log(Ns))$, where N stands for the population size, n for the number of objectives and s for the number of evaluations per solution. Nevertheless, note that all l sums do not necessarily need to be computed and, thereby, the real computation time can be reduced.

3.4 Implementation

To implement our algorithms, we use the EO framework [11] linked to its extension dedicated to multi-objective optimization ParadisEO-MOEO[2]. First, it was necessary to extend this framework by defining the pareto fitness notion for stochastic problems. Then, we implemented those methods in the same way as existing deterministic methods. All these new concepts are now available within the ParadisEO-MOEO framework.

The three approaches differ the one from the others by the way their fitness function is defined. The common points of the different algorithms are:

- *Initialization*: randomly generated individuals.
- *Selection*: deterministic tournament between two randomly chosen individuals.
- *Crossover*: two-point crossover [9].
- *Mutation*: shift mutation [9].
- *Replacement*: elitist.

[2] ParadisEO-MOEO is available at `http://paradiseo.gforge.inria.fr`.

4 Simulation Results

4.1 Benchmarks

To test our algorihms, we propose differents benchmark suites[3] built from Taillard's instances [17]. These instances contain processing times for problems whose size varies from 20 to 500 jobs and from 5 to 20 machines.

Deterministic bi-objective benchmarks. First, we need to extend the Taillard's instances for the bi-objective deterministic case by adding a due date for every job. These dates were fixed using a random value chosen between $\overline{p} \times M$ and $\overline{p} \times (N + M - 1)$, where N stands for the number of jobs, M for the number of machines and $\overline{p}$ for the mean of previously generated processing times. Thus, a due date d_i lies between the average completion date of the first scheduled job and the average completion date of the last scheduled job. Moreover, in addition to Taillard's instances, we propose some instances with intermediate sizes. Each benchmark's name is composed on the same way: the first number represents the number of jobs to schedule, the second one the number of machines and the last one the index of the instance among the instances of same size.

Stochastic bi-objective benchmarks. To generate stochasticity on a deterministic instance, the four probability distributions a processing time may follow can be applied over initial data using a configuration file. We choose to allow to configure this uncertainty over the machines only, by specifying, for each machine, a probability distribution and its parameters or some proportions depending on its central tendency. Thus, as in real-world problems, each time stochasticity is generated on an initial deterministic instance using the same configuration file, processing times contained in the obtained stochastic instance will be different.

4.2 Optimization Runs

For the optimization runs, we generate stochasticity over the processing times of some deterministic benchmarks. Thus, for a given instance, we carry out 10 different evaluations into which processing times follow a uniform, an exponential, a normal, a log-normal or various distributions in the following way:

- uniform distribution: $p_{ij} \sim \mathcal{U}(a = 0.85 \times p_{ij}, b = 1.15 \times p_{ij})$;
- normal distribution: $p_{ij} \sim \mathcal{N}(\mu = p_{ij}, \sigma = 0.15 \times p_{ij})$;
- exponential distribution: $p_{ij} \sim \mathcal{E}(a = p_{ij}, \lambda = \frac{1}{0.15 \times p_{ij}})$;
- log-normal distribution: $p_{ij} \sim log\text{-}\mathcal{N}(\mu = \log p_{ij}, \sigma = 0.15 \times \log p_{ij})$;
- various distributions: the distribution of the processing times differs on every machine.

The population size is set to 50. For each kind of distribution, we perform 10 runs per instance and per algorithm using 10 evaluations per solution (except

[3] All benchmarks are available at `http://www.lifl.fr/~liefooga/benchmarks/`.

for $IBEA_1$ where only the first evaluation is used). The different methods are tested using the same initial populations and the same number of generations. The crossover and mutation probabilities are set to 0.05 and 1.00 respectively. The scaling factor $\mathcal{K}$, required in $IBEA_1$ [21], is set to 0.05.

4.3 Performance Assessment

To our knowledge, no protocol fully adapted to evaluate the effectiveness of multi-objective optimization methods for stochastic problems exists by now. Consequently, we choose to revalue each final set of solutions on the reference benchmark (the one from which stochasticity was generated) and to regard this evaluation as the 'true' evaluation. Then, we only keep the non-dominated solutions (according to this 'true' evaluation) obtained by each algorithm. Therefore, we are able to apply traditional metrics, used in the deterministic case, to assess the quality of the obtained Pareto set approximations.

Here, we use two measures to compare the obtained Pareto fronts: the contribution metric [15] and the S metric [19]. The contribution metric evaluates the proportions of Pareto optimal solutions given by each front, whereas the S metric measures the size of the objective space dominated by a non-dominated set. For the contribution metric, the performance comparison is carried out using a reference set R determined by merging all the solutions found during the whole optimization runs of every algorithm into a single set and keeping only the non-dominated solutions. For the S metric, the required reference point Z is composed of the worst objective values observed on the whole optimization runs for the benchmark under consideration, multiplied by 1.1.

4.4 Computational Results and Discussion

To significantly compare all the algorithms, we choose to perform a Wilcoxon rank test for every instances and every kind of probability distribution. On each of the following results tables, the 'T' columns give the result of the test for a p-value lower than 5%, *i.e.*:

- according to the metric under consideration, the results of the algorithm located at the specific row are significantly better than those of the algorithm located at the specific column (+);
- according to the metric under consideration, the results of the algorithm located at the specific row are significantly worse than those of the algorithm located at the specific column (−);
- according to the metric under consideration, there is no significant difference between the results of the two algorithms (≡).

Results for uniformly, exponentially, normally, log-normally and variously distributed processing times are respectively given in tables 1, 2, 3, 4 and 5.

In a general way, according to the performance metrics used, $IBEA_{avg}$ globally outperforms $IBEA_1$ and $IBEA_{stoch}$ for all probability distributions (except for the first benchmark whose processing times are log-normally distributed, where

$IBEA_1$ performs significantly better according to the S metric, see table 4). However, as $IBEA_{avg}$ uses average values, we could have expected such results for probability distributions whose central tendency is the mean (*i.e.* uniform and normal distributions), but these results are more surprising for the other distributions. For uniformly and normally distributed processing times, $IBEA_1$ is significantly more efficient than $IBEA_{stoch}$ according to the S metric (see tables 1 and 3). But, according to the contribution metric, there is no significant difference between these two algorithms (except for the uniform distribution where $IBEA_{stoch}$ performs significantly better on the last benchmark, see table 1). For the exponential distribution, there is globally no significant difference between $IBEA_1$ and $IBEA_{stoch}$ (see table 2). Even so, according to the contribution metric, $IBEA_{stoch}$ is more effective on the last benchmark. And, according to the S metric, $IBEA_1$ is more effective on the second one. Finally, according to the contribution metric, for log-normally and variously distributed processing times, there is no significant difference between all the algorithms. On the contrary, according to the S metric, $IBEA_{stoch}$ generally outperforms $IBEA_1$ for the log-normal distribution (except for the first benchmark) whereas it outperforms $IBEA_1$ only on the last benchmark for variously distributed processing times (see tables 4 and 5).

Table 1. Comparison of the quality assessment values obtained by $IBEA_1$, $IBEA_{avg}$ and $IBEA_{stoch}$ for uniformly distributed processing times using the Wilcoxon rank test

		contribution metric				S metric			
		$IBEA_1$		$IBEA_{avg}$		$IBEA_1$		$IBEA_{avg}$	
		p-value	T	p-value	T	p-value	T	p-value	T
20_5_01	$IBEA_{avg}$	$> 5\%$	$\equiv$			0.002	$+$		
	$IBEA_{stoch}$	$> 5\%$	$\equiv$	0.009	$-$	0.002	$-$	0.001	$-$
20_5_02	$IBEA_{avg}$	0.013	$+$			0.002	$+$		
	$IBEA_{stoch}$	$> 5\%$	$\equiv$	0.006	$-$	0.005	$-$	0.001	$-$
20_10_01	$IBEA_{avg}$	0.002	$+$			0.001	$+$		
	$IBEA_{stoch}$	0.005	$+$	0.002	$-$	0.001	$-$	0.001	$-$

Table 2. Comparison of the quality assessment values obtained by $IBEA_1$, $IBEA_{avg}$ and $IBEA_{stoch}$ for exponentially distributed processing times using the Wilcoxon rank test

		contribution metric				S metric			
		$IBEA_1$		$IBEA_{avg}$		$IBEA_1$		$IBEA_{avg}$	
		p-value	T	p-value	T	p-value	T	p-value	T
20_5_01	$IBEA_{avg}$	$> 5\%$	$\equiv$			0.001	$+$		
	$IBEA_{stoch}$	$> 5\%$	$\equiv$	$> 5\%$	$\equiv$	$> 5\%$	$\equiv$	0.033	$-$
20_5_02	$IBEA_{avg}$	$> 5\%$	$\equiv$			$> 5\%$	$\equiv$		
	$IBEA_{stoch}$	$> 5\%$	$\equiv$	$> 5\%$	$\equiv$	0.002	$-$	0.001	$-$
20_10_01	$IBEA_{avg}$	0.045	$+$			0.001	$+$		
	$IBEA_{stoch}$	0.036	$+$	$> 5\%$	$\equiv$	$> 5\%$	$\equiv$	0.005	$-$

Table 3. Comparison of the quality assessment values obtained by $IBEA_1$, $IBEA_{avg}$ and $IBEA_{stoch}$ for normally distributed processing times using the Wilcoxon rank test

		contribution metric				S metric			
		$IBEA_1$		$IBEA_{avg}$		$IBEA_1$		$IBEA_{avg}$	
		p-value	T	p-value	T	p-value	T	p-value	T
20_5_01	$IBEA_{avg}$	> 5 %	≡			0.002	+		
	$IBEA_{stoch}$	> 5 %	≡	> 5 %	≡	0.021	–	0.001	–
20_5_02	$IBEA_{avg}$	> 5 %	≡			0.001	+		
	$IBEA_{stoch}$	> 5 %	≡	> 5 %	≡	0.002	–	0.001	–
20_10_01	$IBEA_{avg}$	0.004	+			0.002	+		
	$IBEA_{stoch}$	> 5 %	≡	0.004	+	0.001	–	0.001	–

Table 4. Comparison of the quality assessment values obtained by $IBEA_1$, $IBEA_{avg}$ and $IBEA_{stoch}$ for log-normally distributed processing times using the Wilcoxon rank test

		contribution metric				S metric			
		$IBEA_1$		$IBEA_{avg}$		$IBEA_1$		$IBEA_{avg}$	
		p-value	T	p-value	T	p-value	T	p-value	T
20_5_01	$IBEA_{avg}$	> 5 %	≡			0.032	–		
	$IBEA_{stoch}$	> 5 %	≡	> 5 %	≡	0.001	–	0.001	–
20_5_02	$IBEA_{avg}$	> 5 %	≡			0.001	+		
	$IBEA_{stoch}$	> 5 %	≡	> 5 %	≡	0.001	+	0.001	–
20_10_01	$IBEA_{avg}$	> 5 %	≡			0.001	+		
	$IBEA_{stoch}$	> 5 %	≡	> 5 %	≡	0.001	+	0.020	–

Table 5. Comparison of the quality assessment values obtained by $IBEA_1$, $IBEA_{avg}$ and $IBEA_{stoch}$ for variously distributed processing times using the Wilcoxon rank test

		contribution metric				S metric			
		$IBEA_1$		$IBEA_{avg}$		$IBEA_1$		$IBEA_{avg}$	
		p-value	T	p-value	T	p-value	T	p-value	T
20_5_01	$IBEA_{avg}$	> 5 %	≡			0.001	+		
	$IBEA_{stoch}$	> 5 %	≡	> 5 %	≡	0.019	–	0.001	–
20_5_02	$IBEA_{avg}$	> 5 %	≡			0.014	+		
	$IBEA_{stoch}$	> 5 %	≡	> 5 %	≡	> 5 %	≡	0.010	–
20_10_01	$IBEA_{avg}$	> 5 %	≡			0.001	+		
	$IBEA_{stoch}$	> 5 %	≡	> 5 %	≡	0.003	+	> 5 %	≡

The poor overall effectiveness of $IBEA_{stoch}$ can be partly explained by a low diversity among the Pareto set approximation obtained during the evaluation on the reference benchmark. Then, using an indicator that would increase the diversity in the decision space could give better results. Besides, the whole of the experimental results can be discussed as the evaluation protocol is not fully adapted to stochastic multi-objective problems.

5 Conclusion and Perspectives

In this paper, we investigated a bi-objective flow-shop scheduling problem with stochastic processing times as well as general combinatorial optimization algorithms applied to its resolution. First, we saw that, in real-world situations, none of the parameters related to this kind of problem is deprived of uncertainty. Thus, non-deterministic models are required to take this uncertainty into account. To this end, a proactive approach, where processing times are represented by random variables, have been taken and several general stochastic models have been proposed. Next, we introduced three different indicator-based methods for stochastic multi-objective problems that are able to handle any type of uncertainty. The first method, called $IBEA_1$, consists in preserving the deterministic approach by computing the fitness of a solution on a single evaluation. The second method, namely $IBEA_{avg}$, is based on average objective values. At last, the $IBEA_{stoch}$ method consists in estimating the quality of a solution in a probabilistic way. The latter, already investigated in [2] on continuous problems, has never been applied neither to the combinatorial case nor to the stochastic models proposed here. All these algorithms and the fitness concept for multi-objective stochastic problems are now available within the ParadisEO-MOEO framework; new methods can thus easily be implemented in order to compare their effectiveness with those presented in this paper. To test these algorithms on our stochastic flow-shop problem, we initially had to build benchmark suites, first for the deterministic bi-objective case, then for the stochastic case. According to the experimental protocol we formulated, we concluded that $IBEA_{avg}$ was overall more efficient than $IBEA_1$ and $IBEA_{stoch}$. Even so, from a purely theoretical point of view, $IBEA_{stoch}$ seems to be more representative of the quality associated to a solution than the two other methods. In spite of that, the results it obtained are a little disappointing in comparison with the continuous case [2]; even if, in that last case, its effectiveness is especially significant for more than two objectives. This can be explained by the fact that the final solutions found by this algorithm are relatively close the ones from the others in the decision space, which generally implies, for our problem, that they are also close in the objective space. As a result, this method cannot contest the others in term of diversity. However, all these results should be moderated. No experimental protocol fully adapted to the combinatorial optimization of stochastic multi-objective problems exists up to now. The one we proposed, although simple and fast, is not really natural and is imperfectly adapted to this kind of problem. Moreover, considering the evaluation on the deterministic benchmark as the 'true' evaluation advantages a lot $IBEA_{avg}$, especially for probability distributions whose central tendency is the mean.

Different perspectives emerge from this work. First of all, other sources of uncertainty that processing times could be taken into account for the flow-shop problem. Furthermore, a reactive approach could be linked to our proactive approach in order to largely improve the effectiveness of all the algorithms. Additionally, as all the results obtained during our experiments reveal a weakness of convergence, hybridizing our evolutionary algorithms with local searches could be beneficial, especially to accelerate the exploration near potentially interesting solutions. Moreover, even if most of quality indicators to be used with $IBEA$ take diversity into account, they only deal with diversity on the objective space, and not on the decision space. Then, to fill the low-level of diversity of $IBEA_{stoch}$, it could be interesting to create an indicator that would allow the decision maker to obtain diversified solutions in the decision space and that would not be completely focused on the Pareto dominance relation. Also, perhaps $IBEA_{stoch}$ is simply to fine-grained compared to $IBEA_{avg}$, and so presents a landscape with a complex structure whereas $IBEA_{avg}$ provides a reasonable guidance and uses stochasticity to pass over such a structure. Studying the lanscape more precisely could then be helpful. Besides, the population size as well as the number of evaluations per solution are two parameters that influence a lot the effectiveness of all the algorithms. Studying more finely the way of determining them and analyzing how to make them evolve in a more efficient way during the optimization process could be profitable. Lastly, working out an experimental protocol adapted to the combinatorial optimization of stochastic multi-objective problems proves to be essential to evaluate the results in a more rigorous way. As well, this study could be extended in order to consider problems with more than two ojectives.

References

1. Babbar M., Lakshmikantha A., Goldberg D. E.: A Modified NSGA-II to Solve Noisy Multiobjective Problems. In Foster J. et al. (eds.): Genetic and Evolutionary Computation Conference (GECCO'2003). Lecture Notes in Computer Science **2723**, Springer, Chicago, Illinois, USA (2003) 21–27
2. Basseur M., Zitzler E.: Handling Uncertainty in Indicator-Based Multiobjective Optimization. International Journal of Computational Intelligence Research **2(3)** (2006) 255–272
3. Cunningham A. A., Dutta S. K.: Scheduling jobs with exponentially distributed processing times on two machines of a flow shop. Naval Research Logistics Quarterly **16** (1973) 69–81
4. Dauzère-Pérès S., Castagliola P., Lahlou C.: Niveau de service en ordonnancement stochastique. In Billaut J.-C. et al. (eds.): Flexibilité et robustesse en ordonnancement. Hermès, Paris (2004) 97–113
5. Deb K., Gupta H.: Searching for Robust Pareto-Optimal Solutions in Multi-Objective Optimization. In Coello Coello C. A. et al. (eds.): Evolutionary Multi-Criterion Optimization, Third International Conference (EMO 2005). Lecture Notes in Computer Science **3410**, Springer, Guanajuato, Mexico (2005) 150–164
6. Dudek R. A., Panwalkar S. S., Smith M. L.: The Lessons of Flowshop Scheduling Research. Operations Research **40(1)**, INFORMS, Maryland, USA (1992) 7–13
7. Graham R. L., Lawler E. L., Lenstra J. K., Rinnooy Kan A. H. G.: Optimization and Approximation in Deterministic Sequencing and Scheduling: A Survey. Annals of Discrete Mathematics **5** (1979) 287–326
8. Hughes E. J.: Evolutionary Multi-Objective Ranking with Uncertainty and Noise. In Zitzler E. et al. (eds.): Evolutionary Multi-Criterion Optimization, First International Conference (EMO 2001). Lecture Notes in Computer Science **1993**, Springer, London, UK (2001) 329–343

9. Ishibuchi H., Murata T.: A Multi-Objective Genetic Local Search Algorithm and Its Application to Flowshop Scheduling. IEEE Transactions on Systems, Man and Cybernetics **28** (1998) 392–403
10. Jin, Y., Branke, J.: Evolutionary Optimization in Uncertain Environments - A Survey. IEEE Transactions on Evolutionary Computation **9** (2005) 303–317
11. Keijzer M., Merelo J. J., Romero G., Schoenauer M.: Evolving Objects: A General Purpose Evolutionary Computation Library. Proc. of the 5th International Conference on Artificial Evolution (EA'01), Le Creusot, France (2001) 231–244
12. Kouvelis P., Daniels R. L., Vairaktarakis G.: Robust scheduling of a two-machine flow shop with uncertain processing times. IIE Transactions **32(5)** (2000) 421–432
13. Ku P. S., Niu S. C.: On Johnson's Two-Machine Flow Shop with Random Processing Times. Operations Research **34** (1986) 130–136
14. Landa Silva J. D., Burke E. K., Petrovic S.: An Introduction to Multiobjective Metaheuristics for Scheduling and Timetabling. In Gandibleux X. et al. (eds.): Metaheuristics for Multiobjective Optimisation. Lecture Notes in Economics and Mathematical Systems **535**, Springer, Berlin (2004) 91–129
15. Meunier H., Talbi E.-G., Reininger P.: A multiobjective genetic algorithm for radio network optimization. In Proc. of the 2000 Congress on Evolutionary Computation (CEC'00). IEEE Press (2000) 317–324
16. T'kindt V., Billaut J.-C.: Multicriteria Scheduling - Theory, Models and Algorithms. Springer, Berlin, 2002.
17. Taillard E. D.: Benchmarks for Basic Scheduling Problems. European Journal of Operational Research **64** (1993) 278–285
18. Teich J.: Pareto-Front Exploration with Uncertain Objectives. In Zitzler E. et al. (eds.): Evolutionary Multi-Criterion Optimization, First International Conference (EMO 2001). Lecture Notes in Computer Science **1993**, Springer, London, UK (2001) 314–328
19. Zitzler E., Thiele L.: Multiobjective Evolutionary Algorithms: A Comparative Case Study and the Strength Pareto Approach. IEEE Transactions on Evolutionary Computation **3(4)** (1999) 257–271
20. Zitzler E., Thiele L., Laumanns M., Fonseca C. M., Grunert da Fonseca V.: Performance Assessment of Multiobjective Optimizers: An Analysis and Review. IEEE Transactions on Evolutionary Computation **7(2)** (2003) 117–132
21. Zitzler E., Künzli S.: Indicator-Based Selection in Multiobjective Search. In Parallel Problem Solving from Nature (PPSN VIII). Lecture Notes in Computer Science **3242**, Springer, Birmingham, UK (2004) 832–842

Evolutionary Algorithm Based Corrective Process Control System in Glass Melting Process

Hosang Jung[1] and F. Frank Chen[2]

[1] Samsung Economic Research Institute, Hangangno 2-Ga, Yongsan-Gu, Seoul 140-702, South Korea
hsjung@seri.org
[2] University of Texas at San Antonio, Department of Mechanical Engineering, San Antonio, TX 78249, USA
FF.chen@utsa.edu

Abstract. This paper presents the corrective process control system for achieving a target quality level in glass melting processes. Since automated data collection devices would monitor and log process attributes that are assumed to correlate to a quality level in the glass melting process, appropriate process control logics utilizing the collected data are definitely needed. In this paper, an evolutionary algorithm based search logic is newly proposed. The objective of the proposed logic is to find the best process condition composed of the process attributes which can generate the target quality level. The proposed logic tries to find the best process condition that needs to satisfy the following two criteria: 1) a process condition should require minimal changes from the current setting of the process attributes; and 2) a process condition can generate the exact or closest value against the target quality level. A case study and a developed process control system are presented.

Keywords: Evolutionary algorithm, corrective process control, glass melting process.

1 Introduction

One of the major activities for improving quality in manufacturing processes is to control the processes correctly. The product defects typically come from even small changes during manufacturing processes that cause big vibrations in the product that in turn result in out of the acceptable limits. This paper is concerned with finding the desirable process conditions that result in achieving a target quality in glass melting processes. Glass melting processes are composed of three parts: a furnace, a refinery, and a forehearth as shown in Fig. 1. Raw materials such as mostly sand and many kinds of chemicals are melted into glass at high temperature in the furnace. The refinery makes the temperature of the molten glass uniform at every spot. Then the temperature of the melted glass is adjusted to a suitable temperature through the forehearth for the glass forming process to follow. Due to flow dynamics, the flow of melted glass is not uniform, especially at the bottom of the furnace. As a matter of fact, each spot of the melted glass has a different temperature profile. In addition to temperature, other attributes also affect the thermal characteristics of melted glass.

S. Obayashi et al. (Eds.): EMO 2007, LNCS 4403, pp 472–485, 2007.

Raw material composition, thermal conductivity, viscosity, fuel/air feed rate, glass color, and the amount of the glass in a furnace are some of the process attributes. Because there are so many – either known or unknown – attributes and the process is so dynamic, it is very difficult to control the behavior of molten glass.

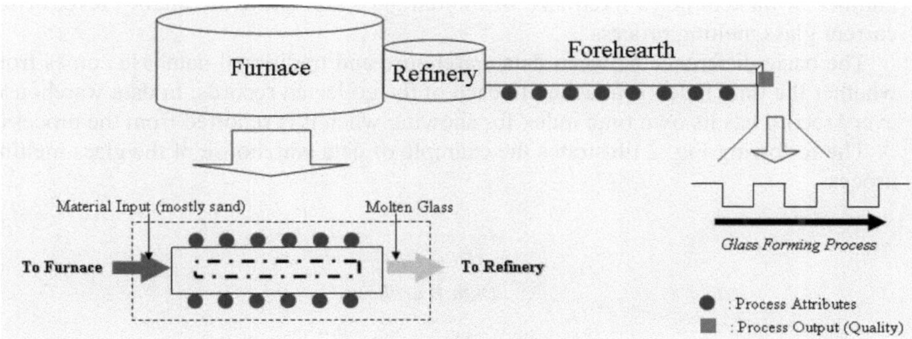

Fig. 1. Glass melting process

In real glass melting processes, there exists the automated data collection devices to monitor and log the process attributes on the assumption that they might be related to the quality level of the process.

In this paper, we propose the corrective process control system for achieving a target quality level in glass melting processes. The proposed corrective process control system has the evolutionary algorithm based search logic (EASL) as a main engine in order to utilize the past data collected from the automated data collection device. The main objective of the EASL is to find the best process condition composed of the process attributes which can generate the target quality level. The proposed EASL tries to find the best process condition that needs to satisfy the following two criteria: 1) a process condition should require minimal changes from the current setting of the process attributes; and 2) a process condition can generate the exact or closest value against the target quality level.

This paper is organized as follows. Section 2 describes data collection procedure, and in Section 3, the EASL is explained in detail. Section 4 presents the implementation of the proposed corrective process control system with a real glass melting process case. Finally, conclusions and future research directions are discussed in Section 5.

2 Data Collection

In glass melting processes, one of the most difficult jobs is to control the process attributes. The automated data collection devices constantly monitor operating parameters (i.e., process attributes) and the quality level throughout the individual facilities. When the quality level becomes out of tolerance, these systems will either respond with an alarm to notify the operation personnel that an unexpected quality level has been found, or some are designed to automatically correct the situation by modifying the process attributes properly.

The collected data on the process attributes and the quality level are stored in data warehouse. As mentioned in the previous section, various factors including temperature and fuel/air feed rate are regarded as the process attributes, while the number of blisters per a fixed size of molten glass can show the quality level of the current glass melting process.

The basic difference between data warehouse and traditional database comes from whether the time index is attached to each of the collected records. In data warehouse, every record has its own time index for showing when it is reported from the process.

The following Fig. 2 illustrates the example of data warehouse of the glass melting process.

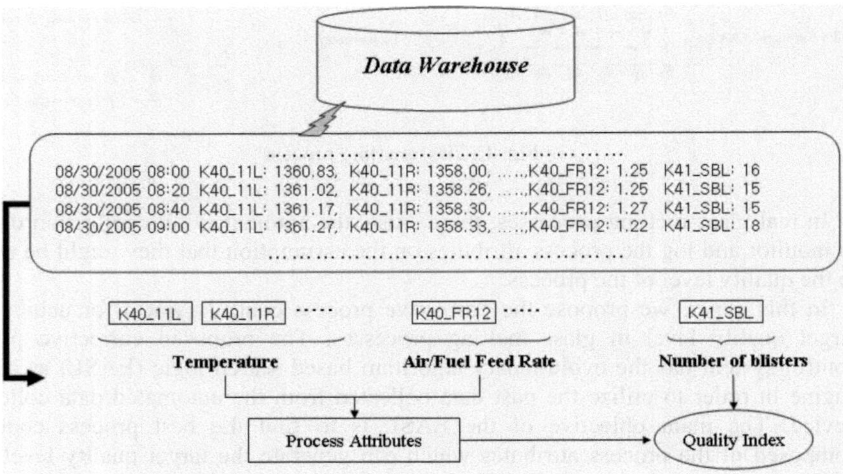

Fig. 2. Data warehouse of the glass melting process

How frequent each record is collected from the process depends on the pre-determined policy of the company. In our real world case, the record containing the status of the process attributes and the target quality is automatically saved every 20 minutes.

Using the collected data in data warehouse, the EASL of the corrective process control system can be performed. In the following section, the corrective process control system is presented focusing on the EASL.

3 Corrective Process Control System

For the corrective process control, many theoretical and application research works are found in diagnostic knowledge representation, diagnostics reasoning, expert systems, neural networks, etc [1]-[4]. Also, although case-based reasoning can be another alternative to find the causes of the defects and furthermore determine the good process conditions [5], little is known in the application of manufacturing processes.

In general, how to set each of the process attributes could be directly related to the process quality level. Suppose that we want to achieve any target quality level that is represented as the average number of blisters for final glass products in glass melting process. Then the control problem is how to set the process attributes appropriately for achieving that quality level.

As stated in Section 1, we propose the EASL for the corrective process control in the glass melting process. Before proceeding to the details of the proposed EASL, a basic pick-up logic (BPL) for the corrective process control is presented. This basic pick-up logic can give us a basic view of the corrective process control.

The following table 1 shows the properties of both the BPL and the EASL.

Table 1. The properties of two corrective process control logics

	BPL	EASL
Logic Type	Pick-up from the past records stored in the data warehouse	Search from both the past records and the randomly generated records
Computational Time	Very short	Relatively long
Expected Performance	Not good	Relatively good

The BPL can be running for a stable process where relationship between the process attributes and the process quality level seems certain and fixed. Under this stable process environment, one of the past records in data warehouse can be chosen for the corrective process control. However, most processes are not repeatable and just picking-up the past record cannot be a proper way to achieve the target quality level.

To figure out the problem of the BPL, the EASL is proposed. The search engine is designed to increase the possibility of finding a better process condition composed of the process attributes which can lead us to achieve the target quality level. While the BPL only searches for the past records, the EASL can explore more various process conditions by considering both the past records and the randomly generated records together. The details of both the BPL and the EASL will be investigated in the following sub sections.

3.1 Basic Pick-Up Logic (BPL)

The assumption for using this logic is that the target process is very stable and repeatable enough. In other words, in this environment, if the operator modifies the values of the process attributes according to one of the past records, the quality level of the process should be close to that of the chosen past record.

The proposed basic pick-up logic has three subsequent steps as follows:

Step (1) Set the target quality level of the process, q (e.g., set the allowable average number of blisters in the glass melting process).

Step (2) From the data warehouse, find a set of past records generating the target quality level.
Step (3) Pick-up the most desirable record which requires the minimal changes or set-up from the current setting of the process attributes using the following equation.

$$e = \min_{j \in J}\{e_j\} = \min_{j \in J}\left\{\sum_{i \in I} w_i \left(X_i^0 - X_i^j\right)^2\right\} \tag{1}$$

where e is the minimal difference between the current process condition and the past process conditions; X_i^0 is the current value of the process attribute i; X_i^j is the value of the process attribute i of the past record j; I is the set of all the process attributes; w_i is the relative weight of the process attribute i; and J is the set including all the past records which can generate the target quality level, q.

The following Fig. 3 briefly shows how the proposed BPL works.

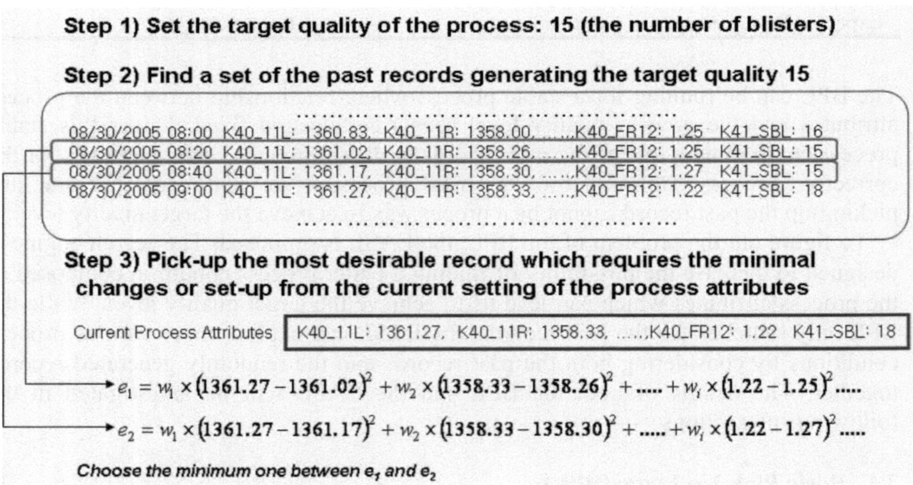

Fig. 3. Example of the BPL

In Step 1, the decision-maker or the operator set the target quality of the process to 15 (i.e. the average number of blisters) since the current average number of blisters is 18. In the following Step 2, two different past records are chosen from the data warehouse. Using Equation (1), one out of the two past records is chosen for the current process control in Step 3.

3.2 Evolutionary Algorithm Based Search Logic (EASL)

Unless the target process is stable and repeatable, we cannot simply pick up the desirable record from the data warehouse for setting the current process attributes to achieve the target quality level. Also, sometimes we cannot find any past records generating the target quality level from the data warehouse.

Unlike the BPL, the proposed EASL can explore various process conditions composed of the process attributes by considering both the past records and the randomly generated records together. This means the EASL attempts to increase the possibility of finding a better solution (i.e., the process condition composed of the process attributes which can generate the exact or closest value against the target quality level).

One of the representative evolutionary algorithms is a genetic algorithm (GA). The GA is a search and optimization algorithm based on the principles of natural evolution [6]. Due to its ease of applicability, numerous applications of the GA are found in the area of business, scientific, and engineering optimization problems. Especially for process control problems, we can find researches in various application areas such as general plant controlling [7], control of a phosphate processing plant [8], control optimization of chemical engineering process [9], and control of a fibre-yarn production process [10]. However, all these GA schemes are case specific so that unfortunately they cannot fit to our glass melting process at all without considerable modifications. In addition, to the best of our knowledge, the GA based process control issues in our particular industry, glass melting process, have not been found at all.

The basic procedure of the proposed EASL is as follows.

(Step 1) Set the target quality level of the process, q (e.g., set the allowable average number of blisters in the glass melting process).
(Step 2) Build a linear causal model using regression method and past data on the process conditions of glass melting process. To simplify the model which represents a causal relationship between process conditions and the target quality level, a linear causal model is adopted here. But, it can be replaced with various models for representing a more complex relationship later.
(Step 3) Using the EA based on the modified GA, search the best process condition that can satisfy the following two criteria: 1) the process condition should require minimal changes from the current attribute status; and 2) the process condition could generate the exact or closest value against the target quality level.

Step (1) of the EASL:

This step is the same as that of the BPL.

Step (2) of the EASL:

In Step 2, the linear causal model is created by a linear regression, and it is used for a basic structure for the EASL.

The proposed linear causal model is as follows:

$$y(t) = b_0 + b_1 x_1(t - l_1) + b_2 x_2(t - l_2) + \ldots + b_n x_n(t - l_n) + e(t) \tag{2}$$

where $y(t)$: the estimated quality value at time t

$x_i(t-l_i)$: the value of the process attribute x_i at time $t-l_i$

l_i : the lag time between the estimated y and the process attribute x_i

$e(t)$: the error term at time t

In the above model, lag time, l, is defined as the time difference between process quality and process attributes at specified observation time. This means that the values of the process attributes observed at time t account for the qualities measured at time $t+l$. To find an accurate lag time is very difficult, especially in dynamic business processes or continuous-flow-type manufacturing environments, because it requires the comprehensive tracking of the process attributes and also the lag time itself has somewhat probabilistic nature. For simplicity, in model (2) we assume that there exists only a single lag between y and x_i. In other words, no multiple lags are distributed over time. Even though it is reported that polynomial distributed lags have been utilized in measuring the lag effect of advertising at retail sales [11], we do not consider the cumulative effects of process attributes at process quality level over time. In this paper, we propose a simple method to determine lag times in glass melting process, which adopts a graphical trend analysis [12].

In order to identify relationship between two different variables (i.e., the process quality and the process attribute), we analyze the trend of two variables simultaneously. Our trend analysis is to measure the similarity of trend between a quality level and a process attribute under a specified lag time l. The similarity, S_l, is computed as follows:

$$S_l = \sum_{t=1}^{T} \left| d_{tl} - \overline{d_l} \right| \Big/ T \tag{3}$$

where $d_{tl} = x_t - y_{t+l}$

$$\overline{d_l} = \sum_{t=1}^{T} \left(x_t - y_{t+l} \right) \Big/ T$$

x_t : the observed value of a process attribute at time t

y_{t+l} : the observed value of a quality level at time $t+l$

S_l is interpreted as an average deviation of differences between a quality level and a process attribute. When S_l approaches zero, $\boldsymbol{x}$ and $\boldsymbol{y}$ have exactly the same trend with a lag time l. Then by plotting S_l along l, we can estimate how the similarity changes according to lag time. If the S_l graph has a convex shape, we might conclude that the lag time at the minimal point should be selected because it gives higher similarity than any other lag time. Here, the convex graph means that the $\boldsymbol{x}$ and $\boldsymbol{y}$ behave proportionally in the same direction and have very similar trend. If the S_l graph has a concave shape, we might select the lag time at the maximal point with the same reason as in the convex case. In this case, the $\boldsymbol{x}$ and $\boldsymbol{y}$ behave in the opposite direction and have reverse trends. What if the graph has no particular shape or multiple convex or concave shapes? Then we can think that the process attribute is not directly related to the quality level and instead it is affected by other process attributes. This means that it is not possible to determine the lag time for this process attribute. This method seems to be somewhat intuitive but very simple and efficient. From the extensive experiments, Jeong et al. [12] reported that this graphical trend analysis is useful for determining lag time in continuous manufacturing environments although they just adopted this approach for building a forecasting system. The lag times need to be updated regularly or every time the process environment changes.

Step (3) of the EASL

In Step 3, once we prepared the linear causal model using regression, the best process condition composed of the process attributes should be searched using the EASL.

First, we can make an initial set of alternative process conditions (i.e., population) composed of U process conditions (i.e., chromosomes). In other words, the size of population is U, and it should be an even number in this research. Any process condition $x(u)$ is composed of n process attributes $x_i(u)$, i=1,…,n. Fig. 4 shows how the initial population of the EASL is created.

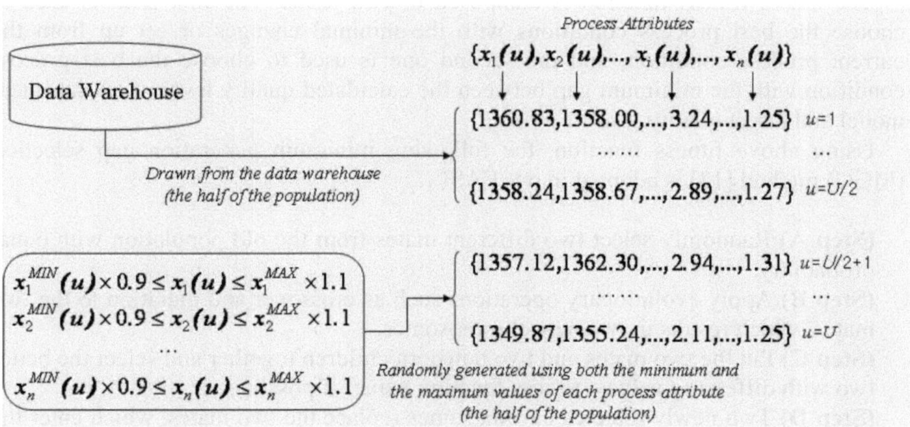

Fig. 4. Population creation in the EASL

As shown in Fig. 4, the half of the initial population is composed of the past records drawn from the data warehouse, and the other half of the initial population is composed of the randomly generated records in order for the EASL to search more various process conditions. To randomly generate the records, we use the minimum and maximum values (i.e., $x_i^{MIN}(u)$ and $x_i^{MAX}(u)$) of the process attribute i which can be found from the data warehouse. Also, the searching range can be extended by multiplying the minimum and maximum values by 0.9 or 1.1, an adjustable extension rate. Once the searching range is fixed, each value of the process attributes $x_i(u)$ can be randomly chosen from the range.

Next, the fitness function of the EASL is needed for evaluating each of the process conditions in the population. Selection using the fitness function is very important to develop and maintain better solutions in the population. At each iteration (i.e., generation) of the EASL, we have to select chromosomes that survive in a way that achieves the objectives for the concerned problem. Our suggested fitness function is given as follows:

$$f = \alpha \sum_{i=0}^{n} \left(x_i^0 - x_i(u)\right)^2 + \beta \left| q_t - y_{x(u)} \right| \tag{4}$$

where x_i^0 : the current process condition (value) of process attribute i

q_t : target quality

$y_{x(u)}$: the estimated quality level using the process condition $x(u)$

it can be obtained by inserting $x(u)$ into Equation (2)

α, β : weight factors

The above fitness function is composed of two terms. The first term is used to choose the best process conditions with the minimal changes or set up from the current process condition, and the second one is used to choose the best process condition with the minimum gap between the calculated quality level of causal linear model and target quality level.

Using above fitness function, the following minimum generation gap selection (MGG) method [13] is adopted in our EASL.

(Step A) Randomly select two different mates from the old population with equal probability.
(Step B) Apply evolutionary operations such as crossover and mutation to the two mates, which results in two new chromosomes.
(Step C) Put the two mates and two newborn children together and select the better two with different f values (fitness function value) if possible.
(Step D) Two newly selected chromosomes replace the two mates, which enter the next generation.
(Step E) Repeat steps A-D until next population is fully built.

One of the strengths of the MGG selection method is that it tries to preserve the good parents in the next generation while other selection methods construct a set of candidates consisting of only child chromosomes so that even parents with good f value cannot enter the next generation.

Finally, we need to define the evolutionary operations of the EASL which diversify the solution alternatives. In this paper, crossover and mutation are applied on individual chromosomes in the population. Crossover is used to avoid destroying parent's characteristics and helps a child chromosome inherit a good sequence from the two parents. While there are so many crossover techniques in various fields, the two-point crossover techniques were introduced in our GA because of generality and the ease of adaptability. Also mutation makes the solution space be searched in various directions using random insertion or swapping techniques. In this research, if any process attribute is chosen for mutation, we insert the minimum or maximum value of the chosen process attribute into the chromosome. Both the minimum and maximum values of the process attributes can be found from the data warehouse as shown in Fig. 4.

Crossover occurs on the two randomly selected chromosomes with probability P_c, and mutation occurs on each of the two mates with probability P_m, which is artificially adjustable.

Fig. 5 represents the overall procedure of the EASL. In Figure 5, the number of generations is set to 50.

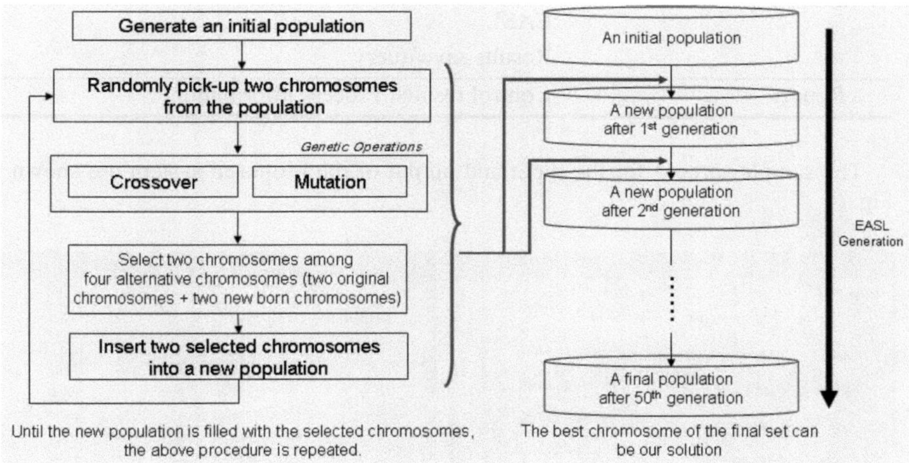

Fig. 5. Overall procedure of the proposed EASL

4 Implementation and Case Study

4.1 System Implementation

The proposed corrective process control system for glass melting process was implemented on a PC Pentium IV-M 2.20 GHz platform using Delphi and C++ Language. Table 2 shows all the functions included in the system.

The lag time determination function is to generate the proper set of lag time between all the process attributes and the quality level by means of the graphical trend analysis. After finding the desirable process condition by the BPL or the EASL, we can report final results via table and graph. In a report table, for a user-friendly view, the founded process conditions can be sorted by either the degree of e_j in the BPL or the degree of fitness function f in the EASL.

Table 2. The functions of the developed system

Main Function	Sub Function
Data Management	Process quality or attributes input/delete/change
	Data validation and filtering
	Lag time determination
	Data query/view
	File save/retrieve
Control Logic	Regression Analysis
	BPL
	EASL
	Results save/query
Report	Control results/ Process information

The sample screens for the input and output of the proposed system are shown in Fig. 6.

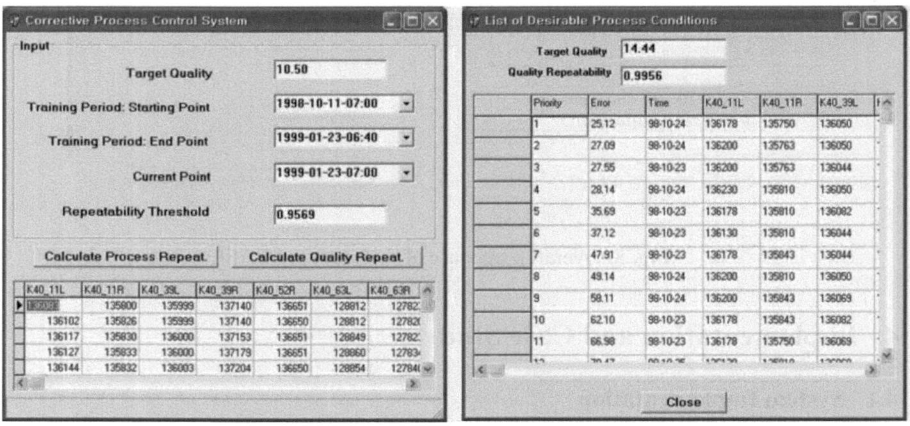

Fig. 6. The sample screen shots for the input and output of the system

4.2 Case Study

This case study was based on a research project conducted at a real glass panels manufacturing company. We gathered and used a set of real process data and quality (i.e., the average number of blisters) data that were collected every 20 minutes for about 135 days in a glass manufacturing line where the glass panels for CRT TV are manufactured. The total number of records is 9793, and there exist 125 process attributes and 1 quality measure. As for the EASL, since the population size is set to 100 in this case study, 50 past records generating the exact or closest target quality level are chosen first, and 50 randomly generated records are additionally generated for the initial population.

Using the pre-experiments, we have fixed the following experimental setup: population size = 100; crossover rate = 0.9; mutation rate = 0.19; number of generations (i.e., stopping criterion of the EASL) = 100; and weight factors of the fitness function $\alpha, \beta = 0.5$, respectively.

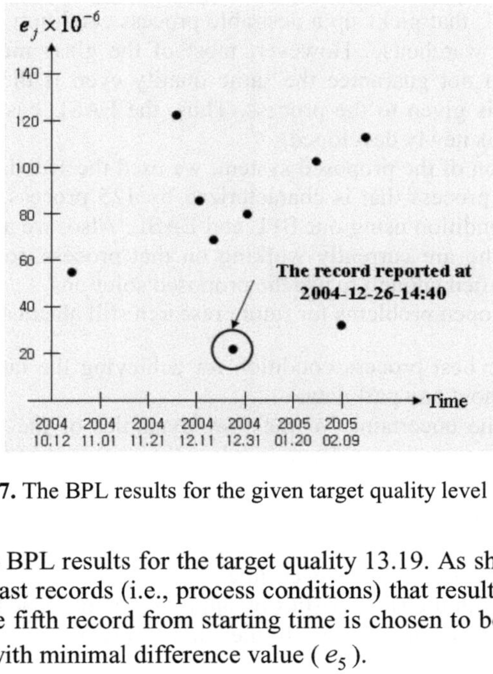

Fig. 7. The BPL results for the given target quality level 13.19

Fig. 7 shows the BPL results for the target quality 13.19. As shown in Fig. 7, there are nine different past records (i.e., process conditions) that result in the same quality level 13.19, and the fifth record from starting time is chosen to be the most desirable process condition with minimal difference value (e_5).

Table 3. The EASL results for four different target quality levels

Experiment No.	Target Quality Level	Estimated Quality Level from the EASL	Computational Time
1	7.92	8.14	28.43
2	18.42	18.02	28.94
3	19.02	19.98	28.09
4	9.65	9.23	29.63

The four different experimental results of the EASL are summarized in Table 3. Table 3 contains the computational time, the target quality level, and the final estimated quality level after the 100 generation running of the EASL. As shown in Table 3, it just takes under 30 seconds for 100 generations to search the best process condition.

5 Conclusions

In this paper we proposed a corrective process control system for the glass melting process, which is based on two different logics to find the best process condition for achieving the given target quality. For a stable process where the same quality is almost guaranteed if we insert the same setting of the process attributes to the process, we can use the BPL that picks up a desirable process condition from the past records stored in the data warehouse. However, most of the glass melting processes have uncertainty and do not guarantee the same quality even if the same setting of the process attributes is given to the process. Thus, the EASL based on a linear causal model and the EA is newly developed.

For the validation of the proposed system, we used the 135 days real data from the real glass melting process that is characterized by 125 process attributes. We found the best process condition using our BPL and EASL. Also, we asked for the group of process analyst, who are currently working on that process, to review those results and they were satisfied enough to use the proposed solutions.

Finally, several open problems for future research still ahead are as follows:

- How to find the best process condition for achieving the target quality in a new process line without any past data.
- How to apply the uncertainty of the flow dynamics of the melted glass into the current proposed system in order to increase the accuracy of the corrective process control.
- How to determine the exact lag time if the multiple lags between the multiple process attributes and the quality level, and how to determine the cumulative effects of the process attributes over time.
- To find the algorithm characteristics by applying the proposed algorithm to several different situations (it may require further case studies).

Acknowledgments. This research is partially supported from the Center for High Performance Manufacturing (CHPM) at Virginia Tech in 2005-2006.

References

1. Fink, P., Lusth, J. C.: Expert Systems and Diagnostic Expertise in the Mechanical and Electrical Domains. IEEE Transactions on Systems, Man, and Cybernetics 17(3) (1987) 340-349
2. Li, C. C.: Path Analysis- A Primer. Boxwood Press (1975)
3. Peng, Y., Reggia, J. A.: A Connectionist Model for Diagnostic Problem Solving. IEEE Transactions on Systems, Man, and Cybernetics 19(2) (1989) 285-298
4. Shortliffe, E. H.: Computer-Based Medical Consultations: MYCIN. American Elsevier, New York (1976)

5. Kolodner, J. L.: Case-Based Reasoning. Morgan Kaufmann (1993)
6. Holland, J.: Adaptation in Natural and Artificial Systems. University of Michigan Press, Ann Arbor (1975)
7. Acosta, G., Todorovich, E.: Genetic Algorithm and Fuzzy Control: a Practical Synergism for Industrial Applications. Computers in Industry 52 (2003) 183-195
8. Karr, C. L.: Control of a Phosphate Processing Plant via a Synergistic Architecture for Adaptive, Intelligent Control. Engineering Applications of Artificial Intelligence 16 (2003) 21-30
9. Pham, Q. T.: Dynamic Optimization of Chemical Engineering Processes by an Evolutionary Method. Computers and Chemical Engineering 22 (1998) 1089-1097
10. Sette, S., Boullart, L., Langenhove, L. V.: Using Genetic Algorithms to Design a Control Strategy of an Industrial Process. Control Engineering Practice 6 (1998) 523-527
11. Chase Jr., C. W.: The Rrole of the Demand Planner in Supply Chain Management. The Journal of Business Forecasting Fall (1998) 23-25
12. Jeong, B., Jung, H., Park, N.: A Computerized Causal Forecasting System using Genetic Algorithms in Supply Chain Management. The Journal of Systems and Software 60 (2002) 223-237
13. Shi, G.: A Genetic Algorithm Applied to A Classic Job Shop Scheduling Problem. International Journal of Systems Science 28 (1997) 25-32

Bi-objective Combined Facility Location and Network Design

Eduardo G. Carrano[1,3], Ricardo H.C. Takahashi[2],
Carlos M. Fonseca[3], and Oriane M. Neto[1]

[1] Universidade Federal de Minas Gerais - Department of Electrical Engineering
[2] Universidade Federal de Minas Gerais - Department of Mathematics
Av. Antonio Carlos, 6627, Belo Horizonte, MG, 31270-010, Brazil
[3] Universidade do Algarve - Centre for Intelligent Systems
Campus de Gambelas, 8005-139 FARO, Portugal
carrano@cpdee.ufmg.br, taka@mat.ufmg.br, cmfonsec@ualg.pt,
oriane@cpdee.ufmg.br

Abstract. This paper presents a multicriterion algorithm for dealing with joint facility location and network design problems, formulated as bi-objective problems. The algorithm is composed of two modules: a multiobjective quasi-Newton algorithm, that is used to find the location of the facilities; and a multiobjective genetic algorithm, which is responsible for finding the efficient topologies. These modules are executed in an iterative way, to make the estimation of whole Pareto set possible. The algorithm has been applied to the expansion of a real energy distribution system. The minimization of financial cost and the maximization of reliability have been considered as the design objectives in this case.

1 Introduction

The problem of combined facility location and network topology design constitutes a difficult task that arises in several contexts, frequently when the facility is the source of a good that must be distributed via some network. Reference [1] discusses in detail this general class of optimization problems, which involves the simultaneous choice of continuous and discrete variables. This problem becomes even harder in the context of multi-objective formulations, since its intrinsic computational burden will affect the generation of a large number of solutions that are contained in the efficient solution set (the Pareto-set solutions). It should be noted that, very often, the real-world problems can be suitably cast in terms of a bi-objective formulation, for instance with the system cost competing with the system reliability, or the service quality. Up to the authors' knowledge, this problem has not been addressed in true multiobjective fashion in the literature, yet.

This paper discusses the structure of such an optimization problem, in the setting of nonlinear multiobjective optimization. For brief reference, the *Multiobjective Joint Facility Location and Network Design* problem is referred to here as the MJFLND problem. An algorithm is proposed here which combines

S. Obayashi et al. (Eds.): EMO 2007, LNCS 4403, pp. 486–500, 2007.

a line search routine for the facility location sub-problem with a genetic algorithm (GA) for the network design sub-problem. Both routines are adapted to generate Pareto-set solutions in their own variables. A heuristic iterative procedure performs the switching between these routines, so that the Pareto-optimal solutions may be approached considering all variables. A convergence criterion based on the stabilization of the various Pareto-set "islands" that characterize this problem is proposed.

The proposed approach is discussed here through the analysis of a case study which is taken from a real joint electric distribution network design and one substation (SS) location problem. Two objectives have been considered: the system reliability and the sum of installation and operation financial costs. The multiobjective genetic algorithm employed here has been adapted from that presented in [2], in which several specific genetic operators are proposed for network design problems. The approach of combining a genetic algorithm with a line search optimization routine for the joint problem of substation location and network design has been used in [3], in an easier mono-objective setting. Here the problem is extended to a multiobjective setting.

The paper is structured as follows. The problem structure and the conceptual algorithm are discussed in sections 2 and 3. The statement of the real problem and the algorithm proposed to solve it are presented in sections 4 and 5. Finally, the numerical results gained for the real case are discussed, and some concluding remarks are drawn.

2 The Structure of Pareto-Sets in MJFLND

2.1 Multiobjective Optimization

Conventional mono-objective optimization is stated as the problem of finding the point, in a space of optimization variables, in which a certain function (the objective function) reaches its minimum (or maximum) value. The multiobjective optimization problem, instead of looking for a single point, searches for a set of points, the *Pareto-optimal set*, which is the set of optimal solutions of a problem with more than one objective functions [4]. The Pareto-optimal set, $\mathcal{X}^*$, is defined as follows.

Consider the minimization of a vector function $f(\cdot) : \mathcal{F} \mapsto \mathbb{R}^m$ (the vector of m objective functions of the problem) in which the set $\mathcal{F}$ represents the problem feasible set. In general, there may not be a single point $x \in \mathcal{F}$ in which $f(\cdot)$ reaches the minimum value for all its components. Then:

$$\begin{aligned} \mathcal{X}^* = \{x^* \in \mathcal{F} \mid \; \nexists\, z \in \mathcal{F} \text{ such that} \\ f(z) \leq f(x^*) \text{ and } f(z) \neq f(x^*)\} \end{aligned} \quad (1)$$

in which the relational operators $\leq$ and $\neq$ are defined for vectors $u, v \in \mathbb{R}^m$, as:

$$\begin{aligned} u \leq v \Leftrightarrow u_i \leq v_i \; \forall\, i = 1, \ldots, m \\ u \neq v \Leftrightarrow u_i \neq v_i \text{ for some } i = 1, \ldots, m \end{aligned} \quad (2)$$

The points $x \in \mathcal{F}$ that do not belong to the set $\mathcal{X}^*$ are said to be *dominated*, since there are some other points, $z \in \mathcal{F}$, such that $f(z) \leq f(x)$ and $f(z) \neq f(x)$, which means that $f(z)$ is better than $f(x)$ in at least one coordinate, without being worse in any other coordinate. In this case, z *dominates* x. The solutions x^* that belong to the set $\mathcal{X}^*$ are called the *efficient solutions*, since they are not dominated by any other point. The space of objective function vectors is also defined, being denoted by $\mathcal{Y}$, and the Pareto-front in this space is denoted by $\mathcal{Y}^*$. Multiobjective optimization looks for the efficient solution set (both $\mathcal{X}^*$ and $\mathcal{Y}^*$) of a vector optimization problem.

2.2 MJFLND Problem

The MJFLND problem can be stated, in a general context, as follows:

$$\mathcal{N}^* = \arg\min_{\mathcal{N}} \begin{cases} f_1 \\ f_2 \\ \vdots \\ f_m \end{cases}$$

$$\text{subject to:} \quad \mathcal{N} \in \mathcal{F}_N$$

where:
$\mathcal{N}^*$ is the set of non-dominated solutions (networks);
f_i is the $i - th$ objective function;
$\mathcal{F}_N$ is the feasible set of solutions.

It is important to note that each one of possible solutions $\mathcal{N}$ is composed of two sets of variables: the facility position (which is a set of continuous variables) and the network topology (which is discrete). Therefore, the multiobjective MJFLND can be decomposed into two sub-problems:

- Finding the set of non-dominated topologies (discrete problem);
- Finding the set of non-dominated facility positions for each topology (continuous problem).

However, these problems cannot be treated separately since they are strongly coupled: changes in position of the nodes usually imply in changes of the topology, and vice-versa.

On the other hand, the difference in the nature of the sub-problems justifies the use of different optimization methods to solve them. A short description of the methods employed and the coupling strategy used is given in the following paragraphs.

The Discrete Variable Sub-problem: In the space of objectives, the Pareto-front $\mathcal{Y}^*$ of a problem with discrete variables consists of a set of isolated points. Figure 1(a) illustrates the Pareto-front for a discrete-variable multiobjective problem with two objective functions. The solutions I to IV compose the Pareto set for this example. One can note that all the other solutions are dominated by at least one of the efficient solutions.

The algorithm for dealing with the discrete-variable optimization sub-problem is chosen to be a *genetic algorithm*.

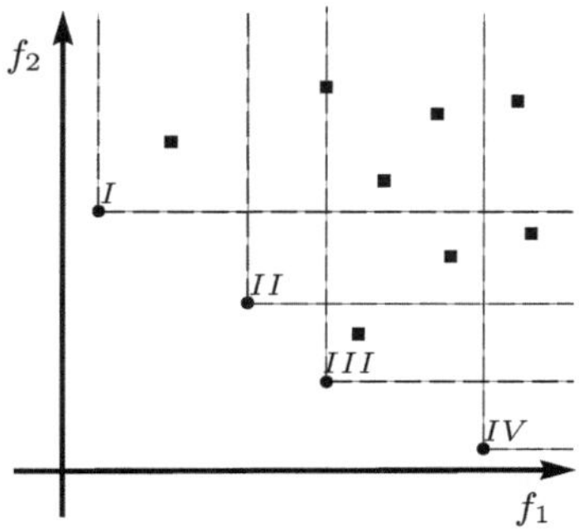

(a) Discrete problem

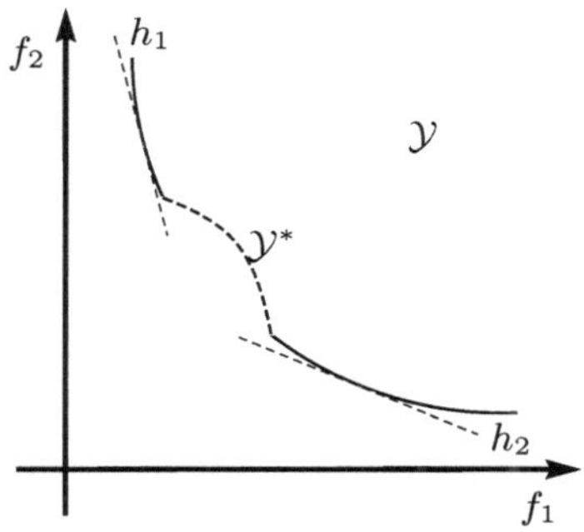

(b) Continuous problem

Fig. 1. Example of Pareto-fronts

The Continuous Variable Sub-problem: Continuous variable multiobjective optimization problems often lead to Pareto-fronts $\mathcal{Y}^*$ that are continuous or, at least, piecewise continuous (but not necessarily connected) in the space of objectives. An instance of such a continuous set, for two objective functions, can be seen in figure 1(b).

Deterministic algorithms for nonlinear optimization, such as BFGS (Broyden-Fletcher-Goldfarb-Shanno) or the Ellipsoid Algorithm [5] can be employed, in the case of multiobjective optimization with continuous variables, after a *scalarization* procedure [4,6]. A very intuitive scalarization technique is P_λ, which uses weighted sums of the objectives. This procedure searches for each point in the Pareto-set through the variation of the weighting vector λ. It should be noted that the algorithm must be executed at least N times for mapping N Pareto-optimal solutions[1]. The P_λ problem formulation is shown in (3).

$$x = \arg\min_x \sum_{i=1}^{m} \lambda_i f_i(x)$$
$$\text{subject to: } \begin{cases} x \in \mathcal{F}_x \\ \lambda_i \geq 0 \\ \sum_{i=1}^{m} \lambda_i = 1 \end{cases} \tag{3}$$

A favorable property of the P_λ scalarization is that it does not affect the nature of the problem. However, this kind of approach is not suitable for achieving the

[1] The algorithm must be executed *at least* N times because different λ vectors can lead to the same solution.

whole Pareto-set in some problems, since it can map only the solutions which belong to the boundary of the convex hull of the efficient solution set (Figure 1(b))[2]. Therefore, using P_λ, the whole Pareto-set can be achieved only in convex problems.

The whole Pareto-set can be mapped in non-convex problems using an enhanced scalarization method, such as the *ϵ-constrained approach* [4] (denoted by P_ϵ). In P_ϵ, one of the problem objectives is considered as the objective function of the scalar problem, and the other objectives are treated as constraints. The solutions of Pareto-set are achieved through variations in the constraint thresholds (ϵ_i). The mathematical formulation of this scalarization technique is shown in (4).

$$x = \arg\min_x f_i(x)$$
$$\text{subject to:} \quad \begin{cases} x \in \mathcal{F}_x \\ f_j(x) \leq \epsilon_j \ , \ j = 1, \ldots, m \ (j \neq i) \end{cases} \tag{4}$$

One limitation of the P_ϵ approach is that this technique changes the structure of problem, adding new constraints to it. It can make the problem considerably harder, mainly when the functions are non-linear.

A viable way of mapping the whole Pareto-set in non-convex problems, with a reasonable computational cost, is the association of P_λ with the P_ϵ. First, P_λ is employed to map an initial estimate of Pareto-set. In a second step, this estimate is analyzed: if it contains significant non-contiguous parts, then, P_ϵ is used in order to fill those "holes", to find the Pareto solutions that are not reachable via the P_λ procedure. This approach is employed here, with the BFGS algorithm being used as the optimization machinery.

2.3 Expected Characteristics of a MJFLND Pareto-Front

Considering the characteristics of the continuous and discrete multiobjective problems mentioned previously, the structure of Pareto-front $\mathcal{Y}^*$ of a MJFLND problem is likely to be composed of a set of disjoint continuous sets in the space of objectives. The "discrete part" of the problem gives rise to isolated solution subsets, while the continuous part of problem leads to continuous surfaces associated with the isolated solutions. Figure 2 shows an example of what the Pareto-front may look like for a two objective MJFLND problem.

Each continuous part of the Pareto-front is generated with a specific network topology (discrete variables) and with the facility position (continuous variables) being continuously varied. Note that the displacement of the facility in a specific topology eventually creates solutions which are not efficient, being dominated by other solutions associated with other topologies. In this case, the Pareto-front "jumps" to the continuous surface associated with another topology. This phenomenon can be seen in Figure 2, at the points where the topologies change.

[2] The parts that do not have a supporting hyperplane, and therefore they do not lie in the boundary of the efficient set convex hull.

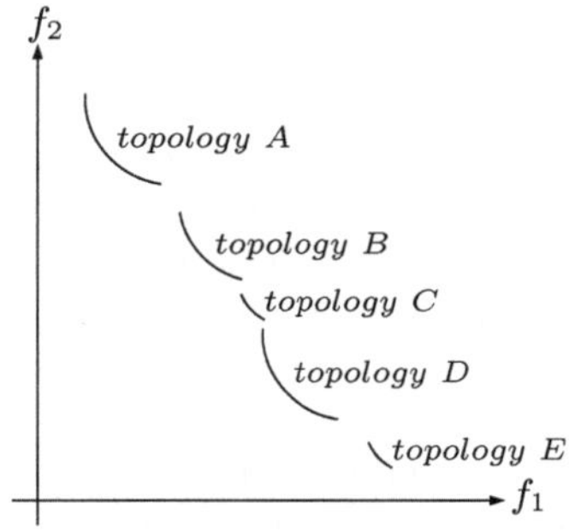

Fig. 2. Example of a Pareto-front for a two objective MJFLND problem

3 Conceptual Algorithm for Finding Pareto-Sets in MJFLND

The algorithm for finding Pareto-sets in MJFLND must execute the following steps:

1. Choose an initial position for the facility; set the counter $i \leftarrow 1$.
2. Find a discrete Pareto-set of the efficient topologies, keeping the current position of facility fixed. This results in the set of topologies $\mathcal{T}_i$, with the j-th topology in the set denoted by $\mathcal{T}_i(j)$.
3. For each such efficient topology in $\mathcal{T}_i$ find local sets of efficient positions for the facility (with each topology being kept fixed), with the k-th point of the j-th local set denoted by $\mathcal{L}_i(j,k)$. Consider that the points in such local sets are ordered in objective f_1, such that $\mathcal{L}_i(j,1)$ is associated to the smallest value of f_1 among the points in this set, and $\mathcal{L}_i(j,\alpha_j)$ is associated to the greatest value of f_1 in this set.
4. Considering the extreme points of each such local set, $\mathcal{L}_i(j,1)$ and $\mathcal{L}_i(j,\alpha_j)$, perform a local search of efficient topologies around such points, keeping each facility location fixed.
5. Perform a dominance analysis considering the Pareto achieved up to the moment and the new solutions, to update the current Pareto-set;
6. If the Pareto-set has not changed after steps 4 and 5, then stop the algorithm. Else, make $i \leftarrow i+1$, define the new set of topologies as $\mathcal{T}_i$ and return to step 3.

Step 2 estimates an initial approximation of the Pareto-set, composed of disjoint points. This is the only point in which a global search is executed in the algorithm[3]. The other searches for enhanced topologies are executed with local algorithms, that are computationally cheaper.

[3] The global algorithm has a higher computational cost, since its search space is greater than the search space of a local algorithm.

Step 3 is responsible for finding the continuous parts of the Pareto-front for each topology.

In step 4, only the extreme points for each topology are analyzed. Figure 3(a) illustrates this process for a two objective example. This is an acceptable approximation, since those are the solutions with greater chance of finding new topologies for their positions. When a new topology is found, it can dominate some solutions of other topologies, changing the extreme points of both. This is illustrated in Figure 3(b).

If the extreme points[4] of a certain topology do not change in two consecutive iterations, this topology is considered stabilized (local stabilization). This topology is not analyzed in the next iterations, unless it becomes affected by the dominance analysis, due to changes that occur in other nearby topologies. The convergence criterion is reached when all topologies have stabilized.

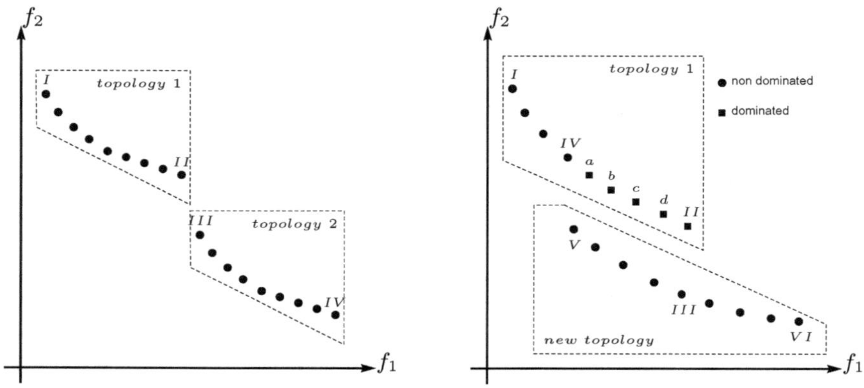

(a) In this example, only the points I and II are analyzed for topology 1, and the points III and IV are analyzed for topology 2.

(b) Suppose that the local algorithm has found the topology III from the topology II, then the facility positions V to VI have been found for topology III. These solutions dominate the solutions a, b, c, d, and II. In this new situation, the extreme points of topology 1 are I and IV and the extreme points of topology 2 are V and VI.

Fig. 3. Solutions analyzed by the algorithm

One should note that this algorithm can be applied to any type of MJFLND problem, provided that the corresponding specific genetic algorithm suitable to the specific problem is available. The generalization for the case of any number of objectives can be performed too, with few adaptations.

[4] The objective f_1 is used as reference to find these extreme points.

In the next sections, the application of this algorithm is illustrated on a practical problem: the expansion of an 8-node distribution network with the location of a new generation facility.

4 Problem Description

The conceptual algorithm which has been been proposed for a general context, can be particularized for dealing with the expansion of power distribution systems. The two sub-problems of the MJFLND can be easily identified in this case:

- Find the optimal substation (SS) position (facility location);
- Find the optimal power distribution network layout (topology design).

This sort of expansion is very common in places which have been growing both in economic activity and population. The existing facilities become unable to supply the new load levels, and new feeder facilities and more robust cables to supply the demand are required. A real case is used here to illustrate the proposed approach [3]. This real case has arisen within Electric Energy Utility of Minas Gerais (CEMIG), in the interior part of Brazil. A detailed description of this case is given in [3]

4.1 Problem Statement

In the optimization approach proposed here, four aspects have been considered as desirable in an optimal distribution system:

- Minimization of energy losses;
- Minimization of investment in new facilities and distribution lines;
- Minimization of the average number of faults;
- Minimization of average interruption time in faults.

It can be seen, from references [7,2,3], that those desirable aspects are related to two "quality indices" of the distribution systems: cost and reliability. From the same references, it can also be seen that those aspects can be aggregated in two objective functions. Equation (5) represents the cost of the network, including the installation (fixed costs) and the maintenance and loss costs (variable costs). Equation (6) represents the cost of system failures (which is accepted as a good index for measuring the reliability of the system).

$$f_c(tpl, SS_c) = \sum_{(i,j)\epsilon S^{FC}} \sum_{a\epsilon S^{PB}} \left\{ \left(FC_{(i,j)}\right)_a + \left(MC_{(i,j)}\right)_a + \left(LC_{(i,j)}\right)_a \right\} \tag{5}$$

$$f_r(tpl, SS_c) = \sum_{(i,j)\epsilon S^{FC}} \sum_{a\epsilon S^{PB}} \left\{ C^{FF} \cdot \left(FF_{(i,j)}\right)_a + C^{FH} \cdot \left(FD_{(i,j)}\right)_a \right\} \tag{6}$$

in which:
tpl is the topology;
SS_c are the SS coordinates;
S^{FC} is the set of feasible connections;
S^{PB} is the set of possible branch types;
$\left(FC_{(i,j)}\right)_a = l_{(i,j)} \cdot instcst_a$ is the fix cost of connection (i,j);
$\left(MC_{(i,j)}\right)_a = l_{(i,j)} \cdot mnntcst_a$ is the maintenance cost of connection (i,j);
$\left(LC_{(i,j)}\right)_a = 8760 \cdot lf \cdot entax \cdot PL_{(i,j)}$ is the loss cost of connection (i,j);
$l_{(i,j)}$ is the length of a branch from i to j (km);
$instcst_a$ is the installation cost of a branch of type a $(R\$/km)$;
$mnntcst_a$ is the maintenance cost of a branch of type a $(R\$/km/\text{year})$;
lf is the loss factor;
$entax$ is the energy tax $(R\$/kW \cdot hour)$;
$PL_{(i,j)}$ are the power losses in the line (i,j) (kW);
C^{FF} is the energy cost per fault;
$\left(FF_{(i,j)}\right)_a = l_{(i,j)} \cdot \lambda_a$ is the fail frequency;
C^{FH} is the energy cost per hour of fault;
$\left(FD_{(i,j)}\right)_a = r_a \cdot P_{(i,j)}$ is the fail duration;
λ_a is the failure rate of a branch of type a $(faults/km)$.
r_a is the average duration of fault of a branch of type a (h);
$P_{(i,j)}$ is the active power in the line (i,j) (kW);
($R\$$ is the Brazilian currency unit).

Besides, some constraints must be considered:

g_1: $x_{MIN} \leq x_{SS} \leq x_{MAX}$;
g_2: $y_{MIN} \leq y_{SS} \leq y_{MAX}$;
g_3: $(x,y) \in \{x_{iIMP}, y_{iIMP}\}$;
g_4: $(x,y) \in \mathbb{R}$;
g_5: $P_l \leq P_{MXl}$;
g_6: $0.92 \leq V_{nd} \leq 1.08$;
g_7: Connectivity and radiality (tree structure) of the network;
g_8: Quality and reliability indexes.

in which:
x_{MIN} and x_{MAX} are the minimal and maximal admissible values for x coordinates;
y_{MIN} and y_{MAX} are the minimal and maximal admissible values for y coordinates;
(x_{SS}, y_{SS}) is the SS coordinate;
$\{x_{iIMP}, y_{iIMP}\}$ is the set of impossible connection (geographical accidents);
P_l is the active power in the line;
P_{MXl} is the maximal admissible power in the line;
V_{nd} is the voltage in load busses;
n_B is the number of branches;
n_N is the number of nodes.

A brief historical review of the use of multiobjective approaches in the design of power distribution systems is presented next.

4.2 Multiobjective Design of Power Distribution Systems

The usual approaches employed in the design of distributions networks just consider the cost minimization as a relevant objective, ignoring the reliability of the system [8,9,10]. Some recent studies have considered the reliability as a relevant aspect to be optimized in distribution systems [11,12]. However, in those approaches the cost and reliability functions have been reduced to a single cost index using a weighted sum of objectives. This objective aggregation is not the most adequate for this kind of problem since, as discussed in section 2.1, it cannot find the whole Pareto set in non-convex problems. Up to now, references [7] and [2] seem to be the only ones that seek the Pareto-optimal solutions using a true multiobjective environment for dealing with the problem.

The approach presented in this paper is a true multiobjective one, including the multiobjective location of new feeder nodes, which extends the other approaches found in the literature.

Here, as it can be seen from reference [2], a small set of initially inactive branches can be placed with the *active network*. These inactive branches define alternative paths that are to be used when a failure breaks down an active branch. This can avoid, at least partially, the energy supply interruption that would occur in such cases. These "extra" connections are called *reserve branches.*

5 The Multiobjective GA-BFGS Algorithm for Power Distribution Systems

The algorithm employed here is a multiobjective extension of the GA-BFGS algorithm proposed in reference [3]. The algorithm is composed of four modules:

- Mono-objective GA-BFGS algorithm;
- Global NSGA-II;
- Multiobjective Quasi-Newton BFGS algorithm;
- Local NSGA-II.

Each one of these modules is briefly described in the next section. They are executed jointly, following the flowchart shown in Figure 4. The main characteristics of the conceptual algorithm (section 3) such as convergence analysis, local and global stabilizations, points that must be analyzed, etc, are valid for this specific case.

5.1 Modules Description

GA-BFGS: The GA-BFGS algorithm proposed in reference [3] has been employed here to find an initial position for the SS. It is composed of a Quasi-Newton BFGS algorithm and a Genetic Algorithm. Jointly, the GA-BFGS can find the optimal topology and the best SS position, but only considering the minimization of costs as a relevant aspect.

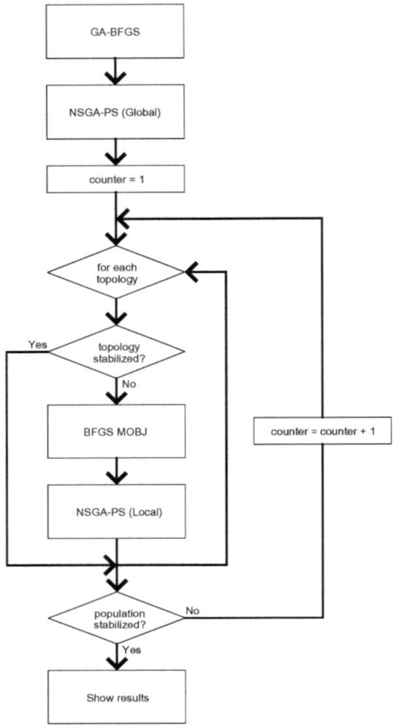

Fig. 4. Multiobjective GA-BFGS algorithm

NSGA-PS: The NSGA-II [13] algorithm has been employed here to find an initial sampling of the Pareto-set and uses problem-specific operators, such those presented in reference [2]: such a combination is called here the NSGA-PS algorithm. The NSGA-PS algorithm appears in two versions:

a Global algorithm: in this version the NSGA-PS searches for the Pareto-set of topologies over the whole search space. Consequently, it requires a larger population and a greater number of generations.

b Local algorithm: in this version the NSGA-PS searches for efficient topologies near a considered topology. The initial population is obtained through perturbations in the analyzed topology. Since the crossover operators have strong heritability and the mutation operators have high locality, it is reasonable to accept that this algorithm performs a local search around the current topology. The size of population and the number of generations are reduced in this version of algorithm.

Multiobjective BFGS: A Quasi-Newton BFGS algorithm [5] has been employed here to solve the SS location sub-problem. The constraints g_1 and g_2 are handled directly by the algorithm which considers the minimum and maximum coordinates as the outer limits of the golden section procedure [5]. The geographical constraints (g_3) are handled by a penalty method. Since the deterministic algorithms cannot deal directly with multiobjective problems, the additional strategy presented in section 2.1 has been used here to map the continuous parts of the Pareto-set. The technique used to assign λ in order to sample the Pareto-set is described following.

λ Assignment. The use of uniformly spaced λ's can lead to Pareto-set samples which are not representative. The samples may present high density of solutions in some small parts and the absence of solutions in other significant parts of that set.

To avoid this sort of trouble, a more efficient way to define the set of λ's is proposed here. The following steps composed the proposed strategy:

1. Set $\lambda(1) = 0$, $\lambda(2) = 1$ and $i = 1$;
2. Solve the SS location sub-problem for $\lambda(1)$ and $\lambda(2)$ respectively;
3. While $i \leq$ the number of desired points,
   (a) Calculate the distance, in the solution space, between each mapped solution and the next one: $d(i) = norm\left(\begin{bmatrix} f_r(i) \\ f_c(i) \end{bmatrix} - \begin{bmatrix} f_r(i+1) \\ f_c(i+1) \end{bmatrix}\right)$;
   (b) Find the solution with the biggest distance: $i^* = \arg\max_i \; d(i)$;
   (c) The λ in next run (λ^n) will be the arithmetic mean of the λ's of solutions with the biggest distance: $\lambda^n = \dfrac{\lambda(i^*) + \lambda(i^*+1)}{2}$;
   (d) Update the λ list: $\begin{cases} \lambda'(j) = \lambda(j) & \forall\, j < i \\ \lambda'(j) = \lambda^n & \forall\, j = i \\ \lambda'(j) = \lambda(j-1) & \forall\, j > i \end{cases}$
   (e) Make $\lambda = \lambda'$, $i = i + 1$ and go to step 3;

Figures 5(a) and 5(b) show the resulting Pareto-fronts associated with the uniform λ assignment and the proposed λ-assignment procedure, respectively. Both experiments have been performed for SS positions, obtained for a randomly generated topology, with 50 multiobjective BFGS runs. It's noticeable that the Pareto-front achieved with the proposed λ-assignment is more representative than the one achieved with the uniform λ set.

6 Numerical Results

As it has been mentioned in section 4, the proposed algorithm has been used to optimize a real system in the interior Brazil. A five years time horizon has been considered and the maximum number of reserve branches has been limited to five. Four types of cables can be used in the design: two uncovered cables and two covered cables (the covered cables are more reliable). The table 1 shows the cable specifications.

Figure 6(a) shows the Pareto-front which has been obtained in a sample run of the proposed algorithm. This run took approximately 36 hours on a Athlon XP 1700+ with 512MB of RAM using Matlab 7[5] and has produced 4,145 solutions, corresponding to 57 different topologies. Figure 6(c) shows the minimum cost

[5] Matlab is a trademark of MathWorks.

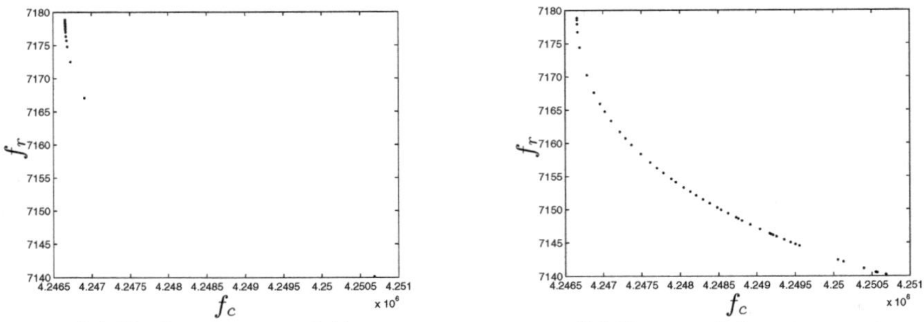

(a) Uniformly spaced λ's (b) Proposed λ assignment

Fig. 5. Differences between λ assignment techniques (sample run)

Table 1. Conductor specifications

C1	C2	C3	C4	C5	C6	C7	C8
1	208	0.7394	0.2682	35000.00	975.00	0.20000	0.33
2	415	0.2469	0.2417	50000.00	975.00	0.20000	0.33
3	180	0.8220	0.3037	65000.00	100.00	0.00625	0.01
4	380	0.2646	0.2567	80000.00	100.00	0.00625	0.01

in which:
C1 - Conductor index;
C2 - Nominal current (A);
C3 - Resistance (Ω);
C4 - Reactance (Ω);
C5 - Installation cost ($R\$/km$);
C6 - Maintenance cost ($R\$/km/year$);
C7 - Fail rate ($faults/year$);
C8 - Average duration of fault ($hours$).

solution (which is the same achieved in [3] for the mono-objective statement of this problem).

It can be seen that some parts of the Pareto-set are almost continuous, as is the case of region a (Figure 6(b) shows a zoom of this region). This region shows various Pareto-optimal SS positions which have been obtained by the Multiobjective BFGS algorithm for the minimal cost topology.

However, this "continuous part" does not occur for some topologies, as is the case of region b. This is expected in some cases, since, for some specific topologies, the network costs and the fault costs are strictly dependent on cable length. Therefore, in these cases both functions would have the same optimal coordinate, resulting in a single non-dominated point. In other topologies, the losses can have a relevant effect in the network cost, shifting the SS from the "minimal cable length" position and resulting in a multi-point Pareto-set.

It is important to note that small changes in the network (which usually imply in a small increment in network financial cost) can result in great improvements

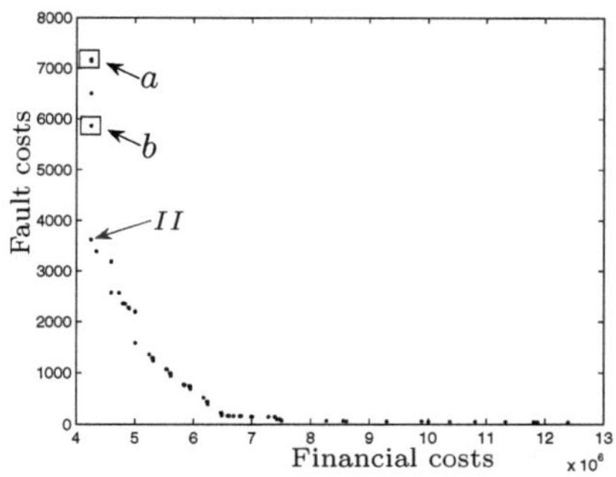

(a) Pareto-front of real case

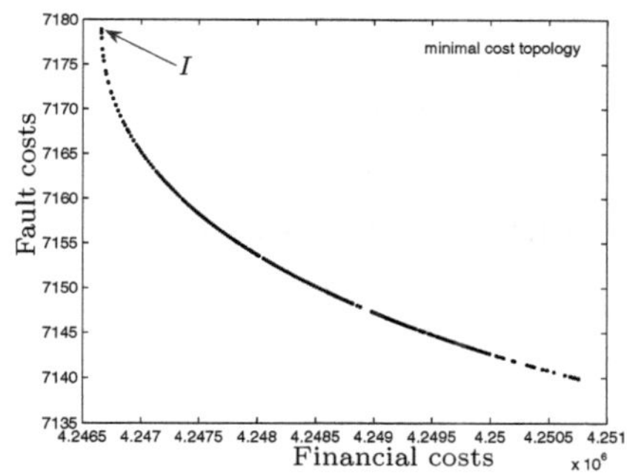

(b) Zoom of region a

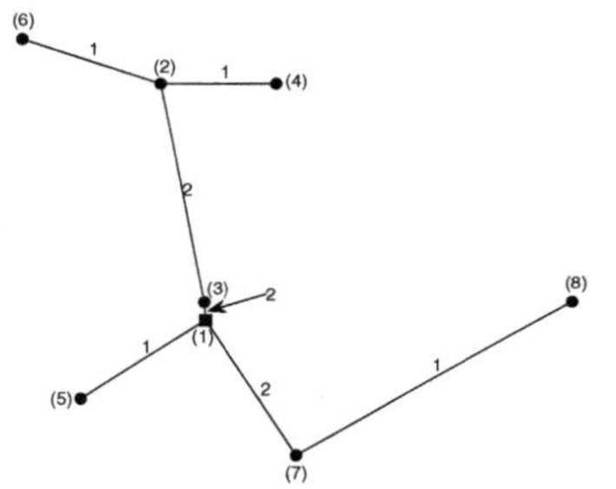

(c) Minimum cost solution (*sol. I*)

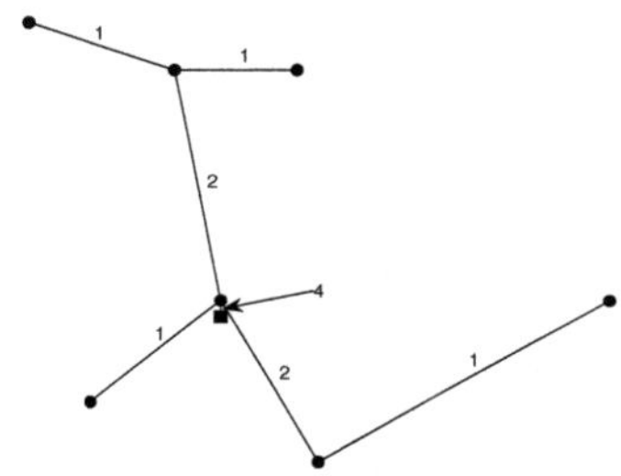

(d) Alternative solution (*sol. II*)

Fig. 6. Two solutions of Pareto-set

in the system reliability. For example, the solution II (Figure 6(d)) has a financial cost just 0.1% greater than the minimal cost solution (solution I). On the other hand, its fault cost is 50% lower than the fault cost of solution I.

7 Conclusions

This paper has presented a multicriterion optimization algorithm for dealing with the multiobjective joint facility location and network design problems. The algorithm is composed of a multiobjective BFGS algorithm and a multiobjective GA, which are responsible for the facility location and topology design, respectively. An iterative structure have been proposed to make the full estimation of the Pareto-front possible.

This algorithm has been applied to the expansion of a power distribution system, considering the minimization of financial costs and the minimization of fault costs as the relevant aspects. The algorithm has produced a diverse set of non-dominated solutions for the problem, which involved optimizing the topology of the network and the new facility position. The results obtained show that great improvements in the reliability are possible at the expense of small increases in financial cost.

Acknowledgments

The authors would like to thank the Brazilian agencies, CAPES and CNPQ for the financial support.

References

1. C. S. ReVelle and H. A. Eiselt, "Location analysis: A synthesis and survey," *European Journal of Operational Research*, vol. 165, no. 1, pp. 1–19, 2005.
2. E. G. Carrano, L. A. E. Soares, R. H. C. Takahashi, R. R. Saldanha, and O. M. Neto, "Electric distribution multiobjective network design usign a problem-specific genetic algorithm," *IEEE Transactions on Power Delivery*, vol. 21, no. 2, pp. 995–1005, 2006.
3. E. G. Carrano, R. H. C. Takahashi, E. P. Cardoso, O. M. Neto, and R. R. Saldanha, "Optimal substation location and energy distribution network design using an hybrid GA-BFGS algorithm," *IEE Proc. on Generation, Transmission and Distribution*, vol. 152, no. 6, pp. 919–926, 2005.
4. V. Chankong and Y. Y. Haimes, *Multiobjective decision making: Theory and methodology*. North-Holland (Elsevier), 1983.
5. M. S. Bazaraa, H. D. Sherali, and C. M. Shetty, *Nonlinear programming: Theory and algorithms*, 2nd ed. John Wiley & Sons, 1993.
6. M. Ehrgott, *Multicriteria optimization*. Springer, 2000.
7. J. L. Bernal-Agustín, "Aplicación de algoritmos genéticos al diseño óptimo de sistemas de distribución de energa eléctrica," Ph.D. dissertation, Zaragoza University, Zaragoza, Spain, 1998.
8. I. Ramírez-Rosado and J. Bernal-Agustín, "Genetic algorithms applied to the design of large power distribution systems," *IEEE Transactions on Power Systems*, vol. 13, no. 3, pp. 696–702, 1998.
9. P. Carvalho, L. Ferreira, F. Lobo, and L. Barruncho, "Optimal distribution network expansion planning under uncertainly by evolutionary decision convergence," *Electrical Power and Energy Systems*, vol. 20, pp. 125–129, 1998.
10. G. Duan and Y. Yu, "Problem-specific genetic algorithm for power transmission system planning," *Electric Power Systems Research*, vol. 61, pp. 41–50, 2002.
11. V. Miranda, J. V. Ranito, and L. M. Proença, "Genetic algorithms in optimal multistage distribution network planning," *IEEE Transactions on Power Systems*, vol. 9, no. 4, pp. 1927–1933, 1994.
12. Y. Tang, "Power distribution system planning with reliability modelling and optimization," *IEEE Transactions on Power Systems*, vol. 11, no. 1, pp. 181–189, 1996.
13. K. Deb, A. Pratap, S. Agrawal, and T. Meyarivan, "A fast and elitist multiobjective genetic algorithm: NSGA-II," *IEEE Transactions on Evolutionary Computation*, vol. 6, no. 2, pp. 182–197, 2002.

Local Search Guided by Path Relinking and Heuristic Bounds

Joseph M. Pasia[1,3], Xavier Gandibleux[2], Karl F. Doerner[3], and Richard F. Hartl[3]

[1] Department of Mathematics, University of the Philippines-Diliman, Quezon City, Philippines
jmpasia@up.edu.ph

[2] Laboratoire d' Informatique de Nantes Atlantique, Université de Nantes Nantes, France
Xavier.Gandibleux@univ-nantes.fr

[3] Department of Management Science, University of Vienna, Vienna, Austria
{richard.hartl, karl.doerner}@univie.ac.at

Abstract. In this paper we present three path relinking approaches for solving a bi-objective permutation flowshop problem. The path relinking phase is initialized by optimizing the two objectives using Ant Colony System. The initiating and guiding solutions of path relinking are randomly selected and some of the solutions along the path are intensified using local search. The three approaches differ in their strategy of defining the heuristic bounds for the local search, i.e., each approach allows its solutions to undergo local search under different conditions. These conditions are based on local nadir points. Several test instances are used to investigate the performances of the different approaches. Computational results show that the decision which allows solutions to undergo local search has an influence in the performance of path relinking. We also demonstrate that path relinking generates competitive results compared to the best known solutions of the test instances.

1 Introduction

Multiobjective optimization (MO) is a field that has been extensively applied to various disciplines. It has many applications in the areas of science and engineering, medicine, finance, operations research and many others. This is one reason why, over the last decades, many researchers have devoted their resources to developing and improving the theories and methodologies of MO.

MO involves solving problems having more than one objective. For example, in the permutation flowshop scheduling problem, where n jobs have to be sequentially processed on m machines, possible objectives are (i) to minimize makespan and (ii) to minimize total tardiness. Given the release date r_i of job i, due date d_i, and processing time p_{ij} on machine j, makespan is defined as $f_1 = \max_i \{C_{i,m}\}$ and total tardiness $f_2 = \sum_i^n \max(C_{i,m} - d_i, 0)$ where $C_{i,m}$ is

S. Obayashi et al. (Eds.): EMO 2007, LNCS 4403, pp. 501–515, 2007.

the completion time of job i at the last machine m. In this paper, we try to find job sequences such that the above objectives are accomplished. We call this problem as bi-objective permutation flowshop scheduling problem (BPFSP).

In general, there is no single solution that simultaneously accomplishes the objectives of a bi-objective optimization problem. Hence, the Pareto optimal solutions or sometimes called the set of efficient solutions are considered. We say that a solution x is an efficient solution if there exists no other feasible solution y such that $f_k(y) \leq f_k(x)$, for $k = 1, 2$ and $f_k(y) < f_k(x)$ for some k. Otherwise, we say that x is dominated by y and we denote this by $y \prec x$.

BPFSP is an $\mathcal{NP}$-hard problem since makespan minimization has been proven $\mathcal{NP}$-hard for more than two machines [1]. Furthermore, the minimization of total tardiness for one machine has been proven $\mathcal{NP}$-hard as well [2]. Therefore, the use of metaheuristics is appropriate.

Path relinking is a population-based heuristic first developed as an intensification strategy for elite solutions obtained by tabu search or scatter search [3]. A path relinking operation starts by selecting an initiating solution I_A and a guiding solution I_B from a set $\mathcal{G}$ of initial solutions. It then generates a path $P : I_A - I_1 - I_2 - \ldots - I_B$ through a given neighborhood space $\mathcal{N}_{PR}$ where the distance $d(I_i, I_B)$ between I_i and I_B is decreasing monotonically in i, i.e., $d(I_i, I_B) < d(I_j, I_B)$ $\forall i > j$. In most cases, there exists a huge number of paths P to be evaluated. Hence, a path selection mechanism is used to choose the preferred path. An intensification phase or local search may be used to improve the quality of the solutions along the path. Finally, PR requires a strategy for updating the set $\mathcal{G}$ where the next I_A and I_B will be selected from.

In general, PR is composed of the following:

* Initial population $\mathcal{G}$
* Neighborhood structure $\mathcal{N}_{PR}$
* Distance measure d
* Selection criteria for initiating and guiding solutions
* Path selection criteria
* Local search for the solutions generated
* Update strategy for set $\mathcal{G}$

In this study, we solve the BPFSP using a PR approach. Our PR uses the ant colony system (ACS) to generate the starting solutions. We will also investigate how the different strategies for computing the heuristic bounds for local search affect the performance of PR. We consider three definitions for our heuristic bounds i.e., for deciding whether local search is applied or not. The three heuristic bounds are defined by local nadir points. A local nadir point corresponds to the worst objectives of two given efficient solutions. Finally we will also demonstrate that our PR approach is competitive with respect to the other existing metaheuristics for BPFSP.

This paper is organized as follows. Section 2 describes the implementation of path relinking in BPFSP. Section 3 presents the numerical results of the study and Sect. 4 provides a short conclusion of the study.

2 Path Relinking for BPFSP

Our path relinking approach follows the two basic steps given by Algorithm 1. In the following sections, we describe the details how we implemented these basic steps.

Algorithm 1. Basic algorithmic framework of PR

```
STEP 1: Create initial solutions for the path relinking /*Algorithm 2*/
STEP 2: While (condition is satisfied) do     /*iteration loop*/
            Path relinking with local search /*Algorithms 3 & 4*/
```

2.1 Initial Solutions for the Path Relinking

Being a population-based heuristic, the starting solutions of PR is important when solving a biobjective optimization problem. It was demonstrated in [4,5,6] that spending time in creating a good initial population improves the convergence in optimization. This concept was applied in the PR approach developed in [7,8], where the starting solutions of PR were obtained from the efficient solutions generated by genetic algorithm.

In this study, we also use a set of good starting solutions for our PR. Our strategy for creating the initial solutions is a two-step process. The first step is to generate two pools of solutions where one pool contains solutions that are good with respect to makespan and the other contains solutions that are good with respect to total tardiness. These pools are generated by using a straightforward implementation of the ant colony system (ACS) [9].

As one of the ant colony optimization (ACO) algorithms, ACS is a constructive metaheuristic that selects the next unscheduled job to be appended in the partial schedule based on the cost of adding the job (heuristic information) and the desirability of the job (pheromone). Unlike the other ACO algorithms, ACS uses a transition rule that balances exploration and exploitation of solutions.

We run the ACS twice where the first run minimizes the makespan and the second run minimizes the total tardiness (see Fig. 1). We consider only the ten best solutions of every run as components of the pools.

The second step is to apply a local search (LS) to all the solutions in the pools. The move of the local search is based on random insertion, i.e., it randomly selects a job and inserts it in another position. If the solution created after the move is not dominated by the set $\mathcal{L}$ of all nondominated solutions found by this local search, this solution becomes the incumbent solution. The process of move is repeated until a maximum number (M) of iterations has been reached.

The set $\mathcal{L}$ updates the external set $\mathcal{G}$ which contains all efficient solutions found so far. The solutions in $\mathcal{L}$ that are not dominated by the solutions in $\mathcal{G}$ are stored in $\mathcal{G}$ and the solutions in $\mathcal{G}$ that are dominated by $\mathcal{L}$ are removed. Algorithm 2 provides the pseudocode for this method.

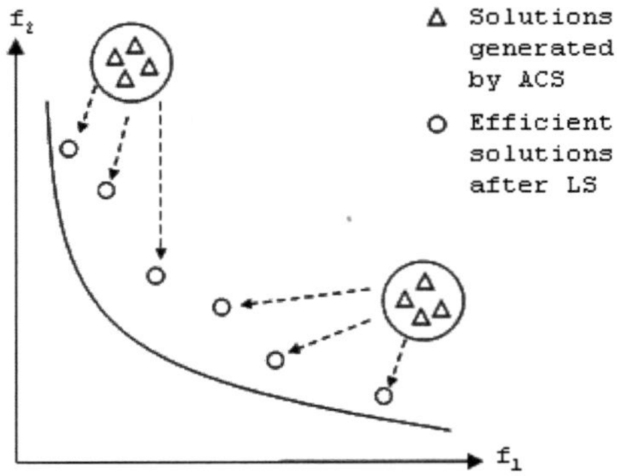

Fig. 1. Initial solutions for PR

Algorithm 2. Create starting solutions of PR

Starting solutions $\mathcal{G}$ /*`generated by ant colony system`*/
Forall $z_i \in \mathcal{G}$
  Pareto set $\mathcal{L} \leftarrow z_i$
  While (condition is satisfied) /*`local search loop`*/
    $y \leftarrow$ Move (z_i)
    If $y \notin \mathcal{L}$ and y is not dominated by $\mathcal{L}$ then
    Set $\mathcal{L} =$ nondominated solutions of $(\mathcal{L} \cup \{y\})$ and $z_i = y$
  Update $\mathcal{G}$ by $\mathcal{L}$

2.2 Path Relinking

Algorithm 3. Path relinking

STEP 1: Select initiating and guiding solutions
STEP 2: Apply path relinking operator
STEP 3: Apply local search /*`Algorithm 4`*/

Selection Criteria. The path relinking is applied on set $\mathcal{G}$ - the initiating solution x and guiding solution y are randomly selected from the set $\mathcal{G}$. These solutions may only be selected once in Algorithm 3, i.e. if solution x is selected as initiating/guiding solution in the current iteration, it will not be selected again as initiating/guiding solution in the current iteration.

Neigborhood and Distance Measure. The insertion operator serves as the neighborhood structure of PR and the Hamming distance (HD) is used to guide the construction of the paths that connect solution x to solution y. Hence the distance between solution x and solution y is the number of positions in which the two solutions have different assigned jobs.

Path Selection Criteria. There are several possible paths that connect one solution to another. Instead of exploring all these paths, only the paths that generate solutions which are nondominated by the entire neighborhood are exploited. However, there are still plenty of these paths (see Fig. 2). Thus a random aggregation of the objectives is applied to select a single path. This technique was also implemented in [8]. However, in that study the solutions along the path are guided by the concept of greatest common substring (GCS) and not by HD. The metric HD is slightly favored over GCS in [10].

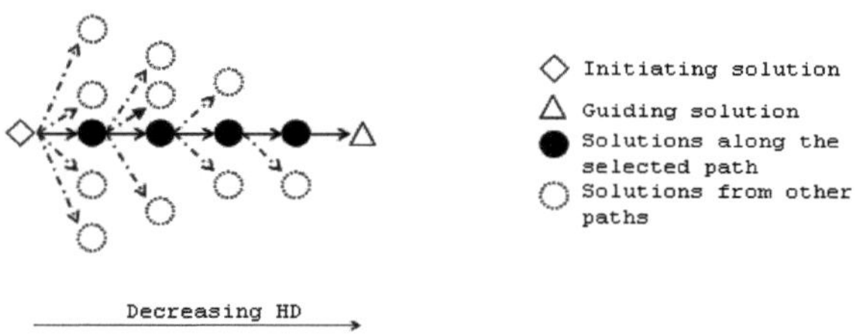

Fig. 2. The initiating solution (◇) is linked to the guiding solution (△) through the path traversed by solutions in solid circles. Each of the solution in the path is selected from among the other efficient solutions (dotted circles) via the random aggregation of the objectives. Note that the solutions along the path have decreasing values of HD.

Update Strategy. Each iteration of PR is completed when no pair of initiating and guiding solutions is possible. The set $\mathcal{G}$ is then updated by the set containing all nondominated solutions generated in the iteration. The next set of pairs of initiating and guiding solutions is then realized on this set. In [8], the set $\mathcal{G}$ is constantly updated, i.e., a new Pareto set is computed after the linking of two solutions.

2.3 Local Search Within PR

We apply local search on some of the solutions along the path. We propose three strategies for defining the heuristic bounds for the local search. The first strategy is to apply the local search if solution z is not dominated by set $\mathcal{G}$. This solution lies in Region A of Fig. 3(a). The second strategy is to apply the local search if solution z lies in the Region B (see Fig. 3(b)) which is the area not dominated by some local nadir points. In this study, we assume the number of local nadir points to be four. In the last strategy, we apply the local search if the solutions lie in Region C which is the area not dominated by three local nadir points described in Fig. 4. This strategy is designed to be biased towards the extreme points. It is worth mentioning that the last two criteria cover an area that is dominated by $\mathcal{G}$. At this point, it is important to note that in the succeeding discussions, the different approaches may sometime be referred to the kind of region they use, i.e., the term "Region A" may refer to the region or the PR that uses Region A.

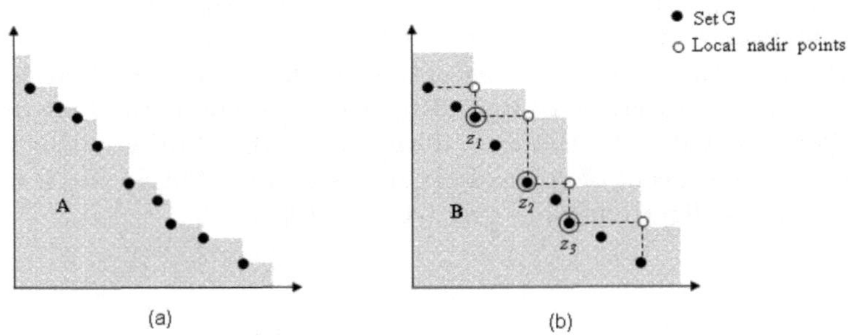

Fig. 3. (a) Region A is the area not dominated by $\mathcal{G}$. (b) Region B is the area not dominated by the local nadir points. The local nadir points are defined by the extreme points and three other points that are evenly distributed in the Pareto front of $\mathcal{G}$.

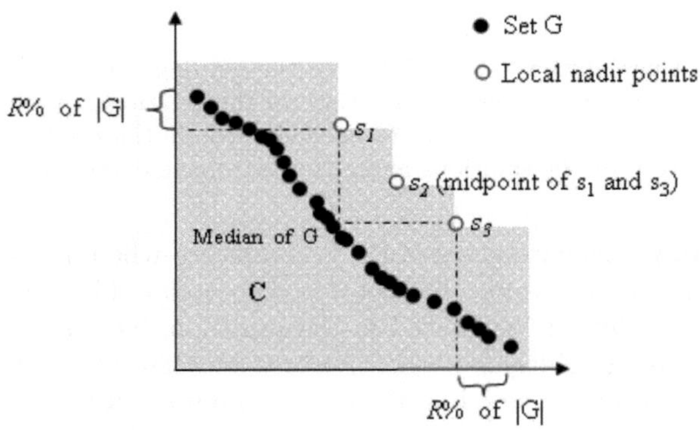

Fig. 4. Region C is the area not dominated by three local nadir points

The local search applied inside the path relinking is illustrated in Algorithm 3. This local search also uses insertion as neighborhood structure. Starting from the selected solution z, the whole neighborhood $\mathcal{N}$ of z is explored. All neighbor solutions are compared and the dominated ones are removed. Each of the remaining efficient solution will undergo the same process as z, i.e., its entire neighborhood is searched and all the dominated solutions are removed. We repeat the entire process of searching the whole neighborhood and removing the dominated solutions until all solutions in the neighborhood are dominated. In this study, the neighborhood $\mathcal{N}$ is also defined by the insertion operator.

Algorithm 4. Local search within PR

Pareto set $\mathcal{S} \leftarrow z$
While $\mathcal{S}$ is non-empty
  Pareto set $C = \emptyset$
  Forall $z \in \mathcal{S}$ do
    Pareto set $\mathcal{L} \leftarrow z$
    Forall $w \in \mathcal{N}(z)$ do
      If $w \notin \mathcal{L}$ and w is not dominated by $\mathcal{L}$ then
      Set $\mathcal{L} =$ nondominated solutions of $(\mathcal{L} \cup \{w\})$ and
      Set $C =$ nondominated solutions of $(C \cup \{w\})$
  $\mathcal{S} = C$
  Update Pareto set $\mathcal{P}$ by C
Return $\mathcal{P}$

3 Numerical Results

The proposed algorithms are tested using nine instances with 20, 50, and 100 jobs having 5, 10, and 20 machines. These instances are taken from some Taillard benchmarks [11] extended into bi-objective case in [12]. We performed all our methods on a personal computer with 3.2 Ghz processor; the algorithms were coded in C++ and compiled using GCC 4.1.0 compiler.

3.1 Evaluation Metrics

The use of unary quality indicators has become one of the standard approaches in assessing the performance of different algorithms for bi-objective problems. It complements the traditional approach of using graphical visualization which may provide information on how the algorithm works [13]. This study considered three unary quantitative measures namely, the *hypervolume indicator*, *unary epsilon indicator*, and *R3 indicator*.

Hypervolume Indicator I_H. This indicator measures the hypervolume of the objective space that is weakly dominated by an approximation set [14]. This is calculated using a boundary point that is dominated by all approximation sets. It has a desirable property that whenever an approximation set A is better than approximation set B, then the hypervolume of A is greater than B.

Unary Epsilon Indicator I_ε. The indicator $I_\varepsilon(A, X)$ gives the minimum factor ε such that if every point in reference set X is multiplied by ε, then the resulting approximation set is weakly dominated by A. For minimization problem, this indicator is formally defined by:

$$I_\varepsilon(A) = I_\varepsilon(A, X) = \inf_{\varepsilon \in \mathrm{R}} \{\forall z^2 \in X \exists z^1 \in A : z^1 \preceq_\varepsilon z^2\} \quad (1)$$

where the ε-dominance relation is defined as $z^1 \preceq_\varepsilon z^2 \Leftrightarrow \forall i \in 1, 2, \ldots, n : z_i^1 \leq \varepsilon \cdot z_i^2$. Note that a small ε value is preferable.

***R*3 Indicator I_{R3}.** The I_{R3} indicator used in this study is one of R indicators proposed in [15]. Given a set of weight vectors Λ, this indicator is defined as:

$$I_{R3}(A) = I_{R3}(A, X) = \frac{\sum_{\lambda \in \Lambda} \left[u^*(\lambda, X) - u^*(\lambda, A)\right] / u^*(\lambda, A)}{|\Lambda|} \tag{2}$$

where u^* is the maximum value attained by a utility function u_λ with weight λ. In this study, the utility function is given by:

$$u_\lambda(z) = -\left(\max_{j \in 1..n} \lambda_j |z_j^* - z_j| + \rho \cdot \sum_{j \in 1..n} |z_j^* - z_j| \right) \tag{3}$$

where z^* is the ideal point and ρ is a sufficiently small positive real number. The values of I_{R3} range from -1 to 1 where values close to -1 are superior.

3.2 Analysis

Ten runs with different random seeds were performed for each of the three PR algorithms and each test instance. The PR algorithms use the same computational time when applied to the same test instance. The computational time ranges from 32 to 36000 seconds. For Region C, the value of R is assumed to be 10. The value of M in the local search ranges from 1200 to 50000. Before applying the different unary indicators, all approximation sets are normalized between 1 and 2. The boundary point used in the hypervolume indicator is (2.1, 2.1), and the ρ and $|\Lambda|$ in I_{R3} are 0.01 and 500 respectively [13]. The point z^* is (1,1) and the reference set for each test instance consists of the points that are not dominated by any of the approximation sets generated by all algorithms under consideration.

Population Plots. Figure 5(a) plots the Pareto approximation sets generated by the 10 runs of the three PR approaches for the T_50_20_01 test instance and Fig. 5(b) graphs their 50% approximation sets[1]. The *k% approximation set* contains all the goals that have been attained independently in *k*% of the runs [13]. Based on these figures, one may say that Region B and Region C outperform Region A for equal computation times. This may be due to the fact that there are very few solutions that fall in Region A, i.e., local search is seldom applied. Table 1 provides the average number of path relinking iterations for all test instances. Note that Region A has high average values, suggesting that local search is rarely used. On the other hand, the average values for Region B and Region C are close to each other.

[1] These sets are computed using the tool in PISA described in [13].

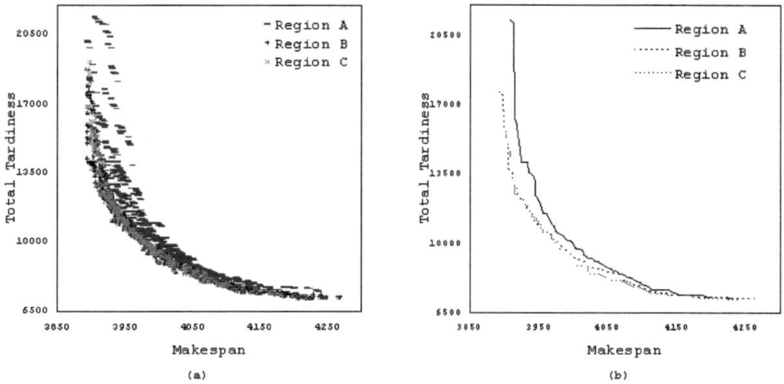

Fig. 5. (a) Population plots of the ten runs and (b) 50% attainment surface generated by three approaches of path relinking for the $T\_50\_20\_01$

Table 1. Average number of PR iterations

Test instance	Region A	Region B	Region C
T_20_5_01	12100.7	1506.6	1499.0
T_20_5_02	11280.2	475.7	475.1
T_20_10_01	1935.6	11.2	9.4
T_20_10_02	4159.3	50.5	34.7
T_20_20_01	2561.8	12.0	8.2
T_50_5_01	10935.8	11.2	9.4
T_50_10_01	24245.0	41.0	31.3
T_50_20_01	19140.1	11.0	8.9
T_100_5_01	1102.2	1.3	1.3

Unary Quality Indicators. Figure 6 provides the boxplots of the different unary quality indicators for all test instances. From these figures, the following can be observed: First, the observation that Region B and Region C outperform Region A in T_50_20_01 is confirmed by the unary indicators. The median of unary indicators of the former algorithms are better than the median of the latter algorithm. This finding is also valid to all other instances. This suggests that accepting more solutions to undergo local search in PR is important. This can be achieved by defining a suitable acceptance region like Region B or C.

Second, in test instances T_20_5_01 and T_100_5_01, the performances of Region B and C are almost the same. This may be explained by the fact that we allow them to have a single nadir point to define their region if the set $|\mathcal{G}|$ does not have enough solutions. For the two instances, the resulting approximation sets do not exceed 12.

Third, Region C performed best with respect to all unary quality indicators in T_20_5_02, T_20_20_01 and T_50_5_01 while Region B has a slight advantage over Region C in T_20_10_01, T_50_10_01 and T_50_20_01. This last observation suggests that there is no solid evidence that defining our region of interest using

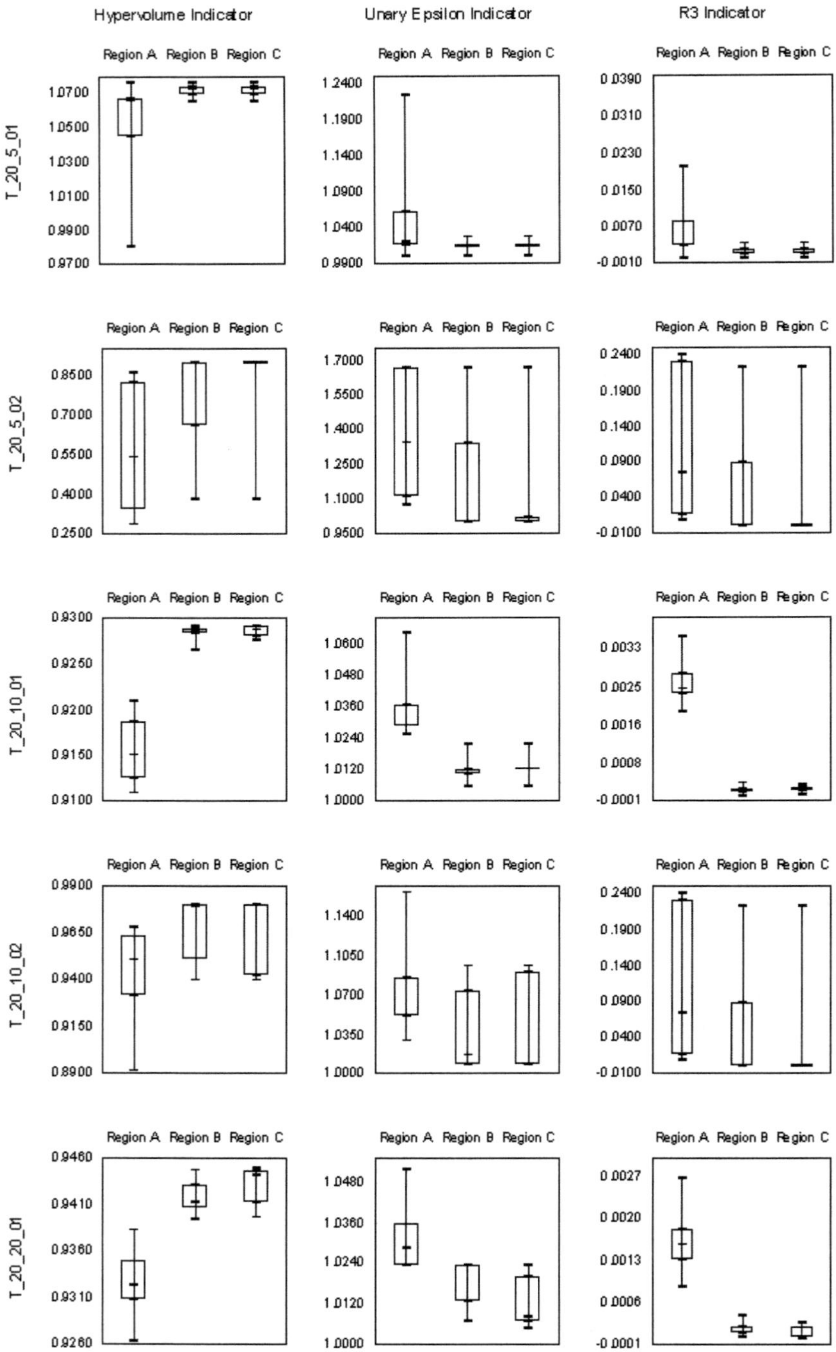

Fig. 6. Boxplots of the different unary indicators for all test instances

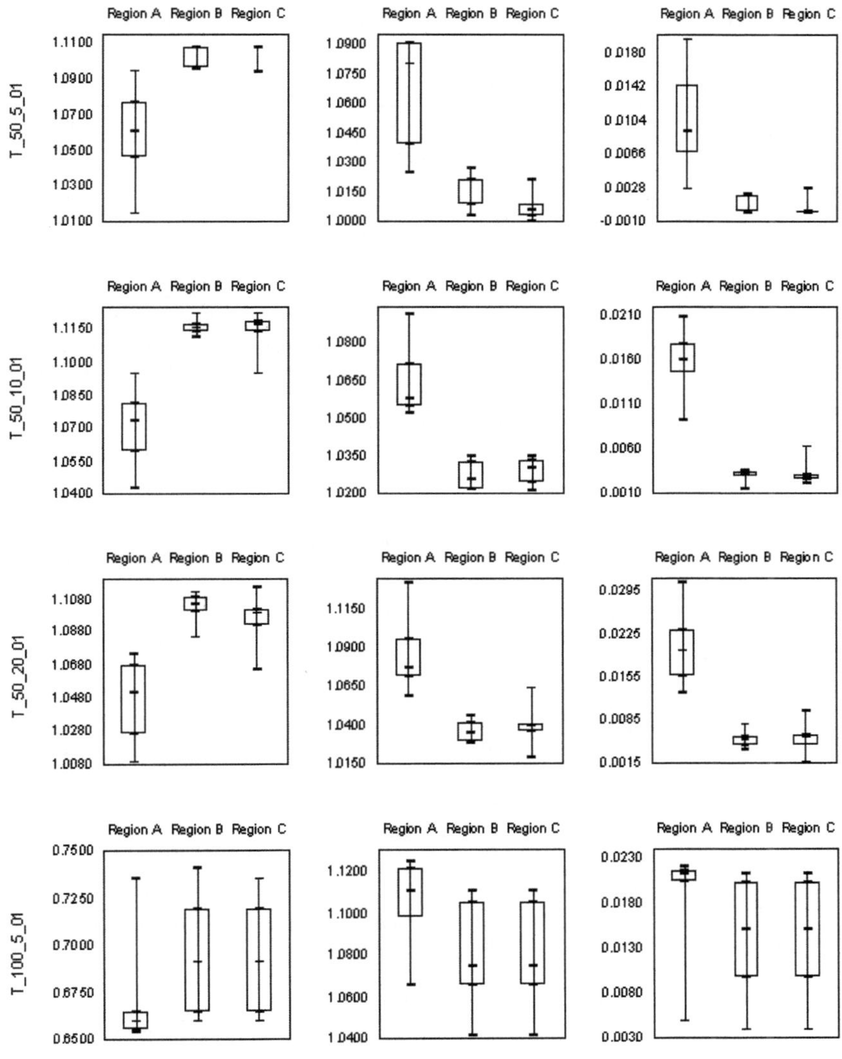

Fig. 6. *(continued)*

three nadir points that are biased towards the extreme solutions is advantageous than defining the regions by simply using four nadir points.

To determine the effectiveness of path relinking as an approach in solving the BPFSP, we compare Region B and Region C to the best known solutions of each of the test instances. These solutions are derived from [8,16] and we refer to them as the benchmarks. These algorithms are based on adaptive genetic/memetic algorithm and hybrid of adaptive genetic algorithm and path relinking. The comparison is done by comparing the unary quality indicators of the benchmarks

Table 2. Unary quality indicators of path relinking and benchmarks

T_20_5_01	Region B			Region C		
k%	Hypervolume	Unary Epsilon	R3	Hypervolume	Unary Epsilon	R3
10	**1.0763**	**1.0000**	**0.0000**	**1.0763**	**1.0000**	**0.0000**
50	1.0744	1.0127	0.0006	1.0744	1.0127	0.0006
100	1.0623	1.0265	0.0040	1.0623	1.0265	0.0040
Benchmark	**1.0763**	**1.0000**	**0.0000**			

T_20_5_02	Region B			Region C		
k%	Hypervolume	Unary Epsilon	R3	Hypervolume	Unary Epsilon	R3
10	**0.8967**	**1.0000**	**0.0000**	**0.8967**	**1.0000**	**0.0000**
50	**0.8967**	**1.0000**	**0.0000**	**0.8967**	**1.0000**	**0.0000**
100	0.3653	1.6667	0.2263	0.3653	1.6667	0.2263
Benchmark	**0.8967**	**1.0000**	**0.0000**			

T_20_10_01	Region B			Region C		
k%	Hypervolume	Unary Epsilon	R3	Hypervolume	Unary Epsilon	R3
10	**0.9296**	**1.0000**	**0.0000**	**0.9296**	**1.0000**	**0.0000**
50	0.9284	1.0120	0.0001	0.9284	1.0120	0.0001
100	0.9255	1.0216	0.0005	0.9264	1.0216	0.0003
Benchmark	**0.9296**	**1.0000**	**0.0000**			

T_20_10_02	Region B			Region C		
k%	Hypervolume	Unary Epsilon	R3	Hypervolume	Unary Epsilon	R3
10	0.9806	1.0074	**0.0000**	0.9806	1.0074	**0.0000**
50	0.9805	1.0074	**0.0000**	0.9805	1.0074	**0.0000**
100	0.9373	1.0956	0.0071	0.9395	1.0956	0.0063
Benchmark	**0.9814**	**1.0000**	**0.0000**			

T_20_20_01	Region B			Region C		
k%	Hypervolume	Unary Epsilon	R3	Hypervolume	Unary Epsilon	R3
10	**0.9452**	**1.0000**	**0.0000**	**0.9452**	**1.0000**	**0.0000**
50	0.9442	1.0067	**0.0000**	0.9449	1.0047	**0.0000**
100	0.9394	1.0233	0.0003	0.9442	1.0067	**0.0000**
Benchmark	**0.9452**	**1.0000**	**0.0000**			

T_50_5_01	Region B			Region C		
k%	Hypervolume	Unary Epsilon	R3	Hypervolume	Unary Epsilon	R3
10	**1.1085**	**1.0000**	**0.0000**	**1.1085**	**1.0000**	**0.0000**
50	1.0968	1.0217	0.0019	1.1080	1.0059	0.0002
100	1.0953	1.0272	0.0022	1.0937	1.0217	0.0029
Benchmark	**1.1085**	**1.0000**	**0.0000**			

T_50_10_01	Region B			Region C		
k%	Hypervolume	Unary Epsilon	R3	Hypervolume	Unary Epsilon	R3
10	1.1273	1.0207	0.0006	1.1282	1.0173	0.0004
50	1.1164	1.0217	0.0030	1.1186	1.0251	0.0029
100	1.1058	1.0348	0.0052	1.0895	1.0348	0.0077
Benchmark	**1.1290**	**1.0087**	**0.0002**			

Table 2. (*continued*)

T_50_20_01	Region B			Region C		
k%	Hypervolume	Unary Epsilon	R3	Hypervolume	Unary Epsilon	R3
10	**1.1214**	1.0203	0.0010	1.1195	1.0177	0.0012
50	1.1080	1.0292	0.0049	1.1015	1.0343	0.0043
100	1.0757	1.0460	0.0097	1.0569	1.0640	0.0121
Benchmark	1.1120	**1.0160**	**0.0005**			

T_100_5_01	Region B			Region C		
k%	Hypervolume	Unary Epsilon	R3	Hypervolume	Unary Epsilon	R3
10	**0.7556**	**1.0003**	**0.0000**	0.7508	1.0133	0.0012
50	0.6924	1.0694	0.0151	0.6924	1.0694	0.0151
100	0.6596	1.1111	0.0213	0.6596	1.1111	0.0213
Benchmark	0.6600	1.1111	0.0213			

to the unary quality indicators of the 10%-, 50%-, and 100%-approximation sets. Note that the 10%-approximation set of the PR algorithm corresponds to the nondominated solutions found in any of the 10 runs. This is also how the benchmarks were derived. Table 2 summarizes these results . From these tables, we can say that Region B and Region C are good alternatives for solving the BPFSP.

The 10%-approximation sets of the two approaches of PR, Region B and Region C, have the same quality indicator values as the benchmarks in 5 of 9 test instances. The benchmarks performed better than Region B with respect to all quality indicators in only one instance (T_50_10_01) and two instances (T_50_10_01 and T_50_20_01) against Region C. However, since unary epsilon and R3 indicators of the benchmarks in T_50_10_01 and T_50_20_01 are not equal to 1 and 0 respectively, one may deduce that the path relinking approaches have found new efficient solutions. Finally, the two approaches of path relinking are better than the benchmark for T_100_5_01 with respect to all three quality indicators. In fact, even the 100%-approximation sets of the two approaches are almost as good as the benchmark.

4 Conclusion

In this study, we proposed a path relinking approach to solve a biobjective flowshop scheduling problem. Our PR approach uses two pools of good starting solutions. One pool contains some best solutions for one objective and the other pool contains some best solutions for the other objective. The solutions in the pools are obtained by applying ACS.

In our PR approach, we allowed some solutions generated by PR to undergo local search. The decision whether a solution will undergo local search is governed by some heuristic bounds. In this study, we also examined three strategies in defining these bounds. These strategies are based on local nadir points. The idea is to allow the solutions to undergo local search if they are not dominated by these points.

Computational results showed that the different heuristic bounds performed differently. Heuristic bounds that allow more solutions to undergo local search have higher chances of attaining better efficient frontiers. We have also demonstrated that for all the test instances, our PR approach was able to reproduce almost all of the best nondominated solutions known so far. In addition, the methods have also generated new efficient solutions for some of the test instances.

References

1. Lenstra, J., Kan, A., Brucker, P.: Complexity of machine scheduling problems. Annals of Discrete Mathematics **1** (1977) 343–362
2. Du, J., Leung, J.: Minimizing total tardiness on one machine is np-hard. Mathematics of operations research **15** (1990) 483–495
3. Glover, F.: Tabu search and adaptive memory programming advances, applications and challenges. In Barr, R., Helgason, R., Kennington, J., eds.: Interfaces in Computer Science and Operations Research. Kluwer Academic Publishers (1996) 1–75
4. Gandibleux, X., Morita, H., Katoh, N.: The supported solutions used as a genetic information in a population heuristic. In: Evolutionary Multi-criterion Optimization, EMO'2001. Volume 1993 of Lecture Notes in Computer Science., Zurich, Switzerland (2001) 429–442
5. Morita, H., Gandibleux, X., Katoh, N.: Experimental feedback on biobjective permutation scheduling problems solved with a population heuristic. Foundations of Computing and Decision Sciences **26**(1) (2001) 23–50
6. Haubelt, C., Gamenik, J., Teich, J.: Initial population construction for convergence improvement of moeas. In Coello, C., Aguirre, A., Zitzler, E., eds.: Evolutionary Multi-criterion Optimization, EMO'2005. Volume 3410 of Lecture Notes in Computer Science. (2005) 191–205
7. Gandibleux, X., Morita, H., Katoh, H.: Evolutionary operators based on elite solutions for bi-objective combinatorial optimization. In Coello, C.C., Lamont, G., eds.: Applications of Multi-Objective Evolutionary Algorithms. Volume 1 of Advances in Natural Computation. World Scientific (2004) 555–579
8. Basseur, M., Seynhaeve, F., Talbi, E.: Path relinking in pareto multi-objective genetic algorithms. In Coello, C., Aguirre, A., Zitzler, E., eds.: Evolutionary Multi-criterion Optimization, EMO'2005. Volume 3410 of Lecture Notes in Computer Science. (2005) 120–134
9. Dorigo, M., Di Caro, G.: The ant colony optimization meta-heuristic. In Corne, D., Dorigo, M., Glover, F., eds.: New Ideas in Optimization. McGraw-Hill, New York (1999) 11–32
10. Pasia, J., Hartl, R., Doerner, K.: Solving bi-objective flowshop problem using pareto-ant colony optimization. In Dorigo, M., et al., eds.: Proceedings of the 5th International Workshop, ANTS 2006. Volume 4150 of Lecture Notes in Computer Science., Brussels, Belgium (2006) 294–305
11. Taillard, E.: Benchmarks for basic scheduling problems. European Journal of Operational Research **64** (1993) 278–285

12. Talbi, E., Rahoual, M., Mabed, M., Dhaenens, C.: A hybrid evolutionary approach for multicriteria optimization problems: Application to the flowshop. In Zitzler, E., et al., eds.: Evolutionary Multi-criterion Optimization. Volume 1993 of Lecture Notes in Computer Science. (2001) 416–428
13. Knowles, J., Thiele, L., Zitzler, E.: A tutorial on the performance assessment of stochastic multiobjective optimizers. Technical Report TIK-Report No. 214, Computer Engineering and Networks Laboratory, ETH Zurich, Gloriastrasse 35, ETH-Zentum, 8092 Zurich, Switzerland (2006)
14. Zitzler, E., Thiele, L.: Multiobjective evolutionary algorithms: a comparative case study and the strength pareto approach. IEEE Trans. Evolutionary Computation **3**(4) (1999) 257–271
15. Hansen, M., Jaszkiewicz, A.: Evaluating the quality of approximations to the non-dominated set. Technical Report Technical Report IMM-REP-1998-7, Technical University of Denmark (1998)
16. Basseur, M., Seynhaeve, F., Talbi, E.: Adaptive mechanisms for multi-objective evolutionary algorithms. In: IMACS multiconference, Computational Engineering in Systems Applications (CESA'03), Piscataway, IEEE Service Center (2003) S3-R-00-222.

Rule Induction for Classification Using Multi-objective Genetic Programming

A.P. Reynolds and B. de la Iglesia

School of Computing Sciences,
University of East Anglia,
Norwich

Abstract. Multi-objective metaheuristics have previously been applied to partial classification, where the objective is to produce simple, easy to understand rules that describe subsets of a class of interest. While this provides a useful aid in descriptive data mining, it is difficult to see how the rules produced can be combined usefully to make a predictive classifier. This paper describes how, by using a more complex representation of the rules, it is possible to produce effective classifiers for two class problems. Furthermore, through the use of multi-objective genetic programming, the user can be provided with a selection of classifiers providing different trade-offs between the misclassification costs and the overall model complexity.

1 Introduction

Earlier work by the authors [1,2,3,4] described the application of multi-objective metaheuristics to the problem of partial classification [5]. This problem is the search for simple rules, that represent 'strong' or 'interesting' descriptions of a specified class, or subsets of the specified class, even when that class has few representative cases in the data. These rules are of the form

- `if age` $\geq$ `28 and firstDegree = mathematics and attendance` $\geq$ `90% then result = distinction`

where the antecedent is a conjunction of simple attribute tests and the consequent, describing the class of interest, is the same for all rules generated.

Such simple rules may have high confidence, in that the rule produces few false positives. They may have high coverage, in that they describe a high proportion of the class of interest. Multi-objective metaheuristics can be used to produce different trade-offs between confidence and coverage. However, this simple rule representation is insufficiently descriptive to produce an individual rule with both high confidence and coverage.

In other work, Ghosh and Nath [6] used a multi-objective genetic algorithm for association rule mining, optimizing the accuracy, comprehensibility and interestingness of the rules produced. Association rules are similar to that shown above, but with tests usually limited to equalities and with an unconstrained consequent that may be any conjunction of such tests.

S. Obayashi et al. (Eds.): EMO 2007, LNCS 4403, pp. 516–530, 2007.

Both partial classification and association rule mining fall primarily into the category of *descriptive* data mining. A natural question is, can this work with simple descriptive rules be extended or modified to create *understandable* and highly *predictive* models that can classify previously unseen records? There are two approaches to take to this task: select a subset of the simple rules created to act as a classifier or increase the expressiveness of the rule representation.

Ishibuchi et al. [7,8] take the first of these approaches. In their work, a multi-objective algorithm is used to select a small subset of association rules produced by another algorithm, minimizing rule set complexity and error rate. However, this approach has a number of disadvantages:

- A good rule set may contain individuals that are far from the Pareto-front, according to the objectives of rule confidence and support [8]. Hence a very large set of simple rules must be created. For example, from a small training set of 342 records, 17070 classification rules were extracted, from which the multi-objective metaheuristic selected no more than 25 [8].
- Ideally, a record should assigned to a class if *any* of the rules in the rule set make this prediction. Then each rule provides a useful description of a subset of the data. However, in practice rules may make conflicting predictions. To handle conflicts, Ishibuchi et al. essentially create a decision list rather than a simple rule set [9], with rules lower on the list being used only when none of the higher rules apply. Converting such a list to a simple rule set reveals added complexity hidden in the decision list representation.
- A set of rules may not be the simplest way in which to represent a model of the data. This is illustrated by the example given in section 2.

In this paper we take the alternative approach of using a more expressive rule representation, specifically by using expression trees. While there is much literature on the use of genetic programming to optimize trees for the purposes of classification, this mostly concentrates on the optimization of decision trees, e.g. [10,11,12,13]. In particular, Mugambi and Hunter [14] apply multi-objective genetic programming to decision tree induction, optimizing both tree accuracy and tree simplicity. However, decision trees are different to the expression trees developed in this paper, with internal nodes that define partitions of the data and leaf nodes that indicate class membership. While rules may easily be extracted from such decision trees, we concentrate in this paper on the direct production and optimization of rules.

The work of Setzkorn and Paton [15,16] is perhaps more relevant to this paper, applying multi-objective genetic programming directly to fuzzy rule induction. However, internal nodes are restricted to two fuzzy forms of the boolean 'and' operation and the algorithm optimizes sets of these rules.

Section 2 provides details of the expression tree representation used. This representation is manipulated by a multi-objective metaheuristic to produce models with different trade-offs between model complexity and model accuracy (section 3). Section 4 describes the experiments performed and presents the results of using this approach. Finally, section 5 presents some conclusions and section 6 describes areas of further research.

2 Rule Representation and Manipulation

2.1 Attribute Tests

The algorithm described in this paper manipulates rules of the form

$$antecedent \rightarrow consequent,$$

where both antecedent and consequent are constructed from *attribute tests.* Three different types of attribute test (AT) are used:

Value: e.g. `colour = red`,
Inequality: e.g. `colour` $\neq$ `green`,
Binary partition: e.g. `age` $\geq$ `42` or `height` $\leq$ `156`.

Value and inequality tests are used exclusively on categorical fields, while binary partition tests can only be used with a numeric field. A more detailed description of ATs and some alternative AT types may be found in previous work [4].

2.2 Attribute Test Representation

Values occurring in each field are stored in reference arrays. The index values, rather than the values from the database, are used in the representation of the ATs as shown in figure 1. Each AT type is represented and mutated as follows:

Value/Inequality: Represented by the categorical field number and the category index. A mutation changes the category index to random value.

Binary partition: Represented by the numeric field number, the index of the bound value and a flag indicating the type of bound. A mutation changes the index of the bound by up to 20% of the number of values that occur in the database, while ensuring that the AT does not become trivial or impossible to satisfy. The type of the bound is not changed.

The algorithm manipulates rule antecedents constructed from combinations of different ATs. The consequent is fixed, representing the *class of interest.* Any unseen record that matches the rule antecedent is predicted to belong to the class of interest. However, in contrast to previous work, any record that does not match the rule antecedent is predicted to belong to some other class.

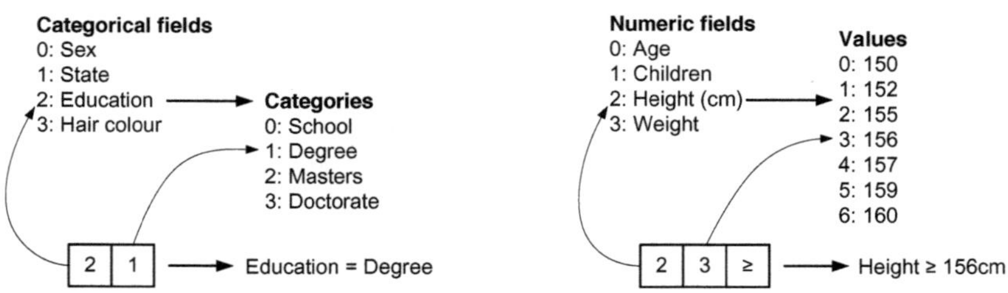

Fig. 1. Representation of a categorical value AT and a binary partition AT

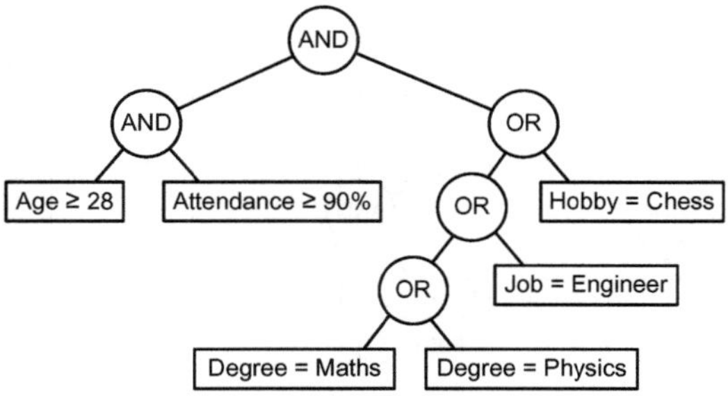

Fig. 2. A binary boolean expression tree

2.3 Rule Trees

ATs are combined in expression trees that represent the rule antecedent, as shown in figure 2. Leaf nodes contain ATs, while internal nodes contain a boolean operator. These operators have been restricted to be either 'or' or 'and' in the experiments reported here, though it is easy to include additional boolean operators if desired. Notice that the tree contains only 6 ATs, one per leaf node, rather than the 12 required to represent this antecedent as a set of simple rules. In order to simplify the genetic operators used, binary trees are used. This increases the number of internal nodes, but leaves the number of leaf nodes unaltered. Note that such trees may easily be converted into rule sets if this is how the client wishes to view the models produced.

2.4 Genetic Operators

Initialization: The population is initialized with randomly generated balanced trees of depth two, where the root node is considered to be at depth zero.

Mutation: During mutation, there is a 50% probability that a random AT is mutated, a 25% probability that an AT and its parent node is removed and a 25% probability that a random AT with a new internal node is added.

Crossover: Subtree crossover [17] proceeds by selecting a node at random in each tree and swapping the subtrees headed by these nodes. As is commonplace in genetic programming, a choice between crossover and mutation is made when creating new solutions, rather than both being applied probabilistically.

2.5 Bloat and Rule Simplification

It is well established that solutions generated during genetic programming tend to suffer from *bloat*, i.e. they grow excessively, often without any great improvements in fitness [17,18]. Such solutions usually contain redundant sections, referred to as *introns*. It has been suggested that the occurrence of such introns

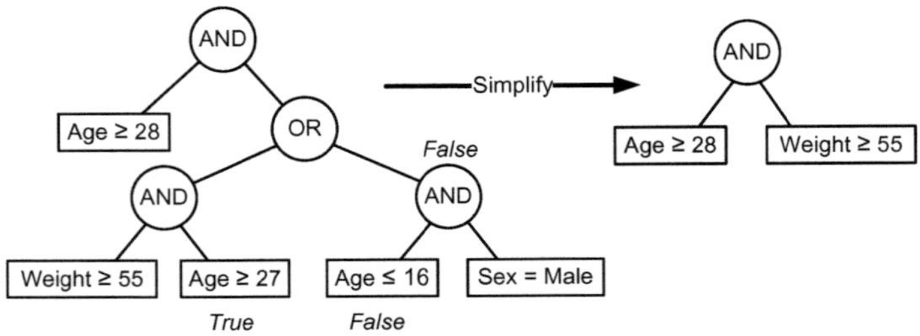

Fig. 3. Simplifying the right hand subtree by assuming a value of 'true' for the left hand subtree

should not be hindered [19], as they protect solutions from the more destructive effects of crossover. However, bloat leads to an increase in evaluation times and may also interfere with finding better solutions, since time is spent manipulating the introns rather than useful code[20]. Langdon and Poli [18] suggest that "since no clear benefits offset these detrimental effects, practical solutions to the code bloat phenomenon are necessary to make GP and related search techniques feasible for real-world applications".

In this paper, bloat is counteracted in three ways. Firstly, although the simplicity of a rule is already considered as an objective of the problem, counteracting bloat provides an additional reason for using this objective. Using rule simplicity in this way has been found to be effective in reducing code bloat in the literature [20]. Secondly, rule simplification is performed, removing redundant sections from rules. Finally, since these measures alone are insufficient to eliminate bloat, a simple limit on rule size is imposed. If, after simplification, a rule exceeds this AT limit, ATs and their parent nodes are removed until the constraint is satisfied. In this paper, this limit has been set to 20 ATs, in order to demonstrate the effect of rule size on misclassification costs. In practice, this limit is likely to be set to a smaller value, since 20 AT rules are too large for easy human comprehension and smaller rules can be evaluated more quickly.

Figure 3 illustrates the rule simplification performed. Here, the right hand subtree need only be evaluated if the left hand subtree evaluates to 'true'. Therefore, assuming that the left hand subtree is 'true', we determine which nodes in the right hand subtree must be 'false' or must be 'true', simplifying as shown. Similarly, if the root node of a tree contains the boolean operation 'or', the right hand subtree need only be evaluated if the left evaluates to false. Note that such simplifications can still be made if the left hand subtree has more than one node. All simplifications of this form are made at every internal node in the rule tree.

Note that this does not ensure that the rule is as simple as possible. For example, we do not currently use the distributivity law to simplify rules. Similarly, given three colours, red, green and blue, the rule antecedent `colour` $\neq$ `red and colour` $\neq$ `blue` could be simplified to `colour = green`.

3 Rule Evaluation

If the understandability of the rule is not a concern, then the overall aim is to produce a rule that, when applied to *previously unseen* data, minimizes the expected costs of misclassification. In practice, we minimize total misclassification costs on the training data and rule complexity.

There are two reasons for minimizing rule complexity:

- The more complex a rule is permitted to be, the more likely it is that *overfitting* [9] will occur, making accuracy on training data an unreliable measure of accuracy on unseen data.
- Simple rules are easier to understand. If part of the aim of classification is the extraction of knowledge, for example when attempting to discover patterns in scientific data, it has been argued [21] that the classifier *must* be comprehensible to a human expert. Also, a client is more likely to use a classifier if he understands it.

3.1 Misclassification Costs

Rule antecedents generated by our algorithm describe the class of interest, with any record not matching the antecedent assumed to be not in the class. A 'false positive' occurs when the rule predicts that a record belongs to the class of interest when it does not, and a 'false negative' occurs when the rule predicts that a records does not belong to the class of interest when it does.

Using equal false positive and false negative costs results in the minimization of the simple error rate. This commonly used measure of performance is not always appropriate, for example when a false positive results in additional labour while a false negative results in injury or death. Also, if the class of interest is small, minimizing the simple error rate may merely result in the rule that predicts that all records are not in the class. If the class is truly of special interest, the false negative cost must be increased to discover of patterns of interest.

In the experiments reported, we have also used the balanced error rate:

$$BER = \frac{1}{2}\left(\frac{\text{No. of false positives}}{\text{Total number of positives}} + \frac{\text{No. of false negatives}}{\text{Total number of negatives}}\right)$$

Other experiments were performed with a false positive cost of 1 and a false negative cost of 10.

3.2 Measuring Rule Complexity

In most of our experiments, the complexity of a rule is given by the number of ATs in the rule tree. This simplifies the comparison of results obtained with different parameter settings and ensures that the complement of the rule, describing records that *do not* belong to the selected class, has the same complexity. (Note that the same could not be said if the categorical inequality AT type was not

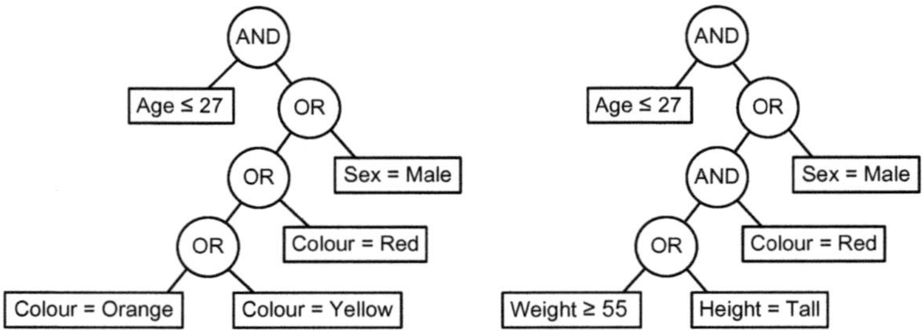

Fig. 4. Rules of the same size need not be equally understandable to the human reader

used.) While counting the ATs has the advantage of simplicity, it may not accurately portray the ease with which a rule can be understood. In figure 4 the first tree is easier to comprehend than the second, due to the repeated use of the both the same operator and the same attribute in the right hand subtree.

In practice, rule complexity also depends upon the client and upon his or her preferences regarding rule presentation. For example, the client may prefer to see the second rule in figure 4 presented as the following rule set.

- `if age` $\leq$ `27 and sex = male then...`
- `if age` $\leq$ `27 and colour = red and weight` $\geq$ `55 then...`
- `if age` $\leq$ `27 and colour = red and height = tall then...`

In this case, the rule complexity may be given as eight ATs rather than the five in the original rule tree. Fortunately, the algorithm used can easily be adapted to use this and other measures of rule complexity and an example is given at the end of section 4.

4 Experimentation and Results

4.1 Data

Rules were extracted from five datasets from the UCI machine learning repository [22]: the Adult, Forest Cover Type, Contraception, Breast Cancer (Wisconsin) and the Pima Indians Diabetes datasets. Any records containing missing data were removed prior to applying the algorithm. Table 1 describes the datasets in more detail.

The Adult dataset and on 10,000 records selected at random from the Cover Type dataset were used to tune algorithm parameters (section 4.2), before running the algorithm on all five of the datasets.

4.2 Algorithm and Parameter Tuning

The multi-objective metaheuristic selected for the optimization of rule trees was NSGA II [23,24], using an external store to hold the best solutions found. This

Table 1. Datasets used and classes of interest

Name	Records	Fields: Numeric	Fields: Categorical	Classes	Class of Interest	Class Prevalence
Adult	45222	6	8	2	Salary > $50,000	24.8%
Cover type	581012	10	2	7	Spruce-fir	36.5%
Contraception	1473	5	4	3	No contraceptive use	42.7%
Breast cancer	683	10	0	2	Malignant	35.0%
Pima Indians	532	7	0	2	Test positive	33.3%

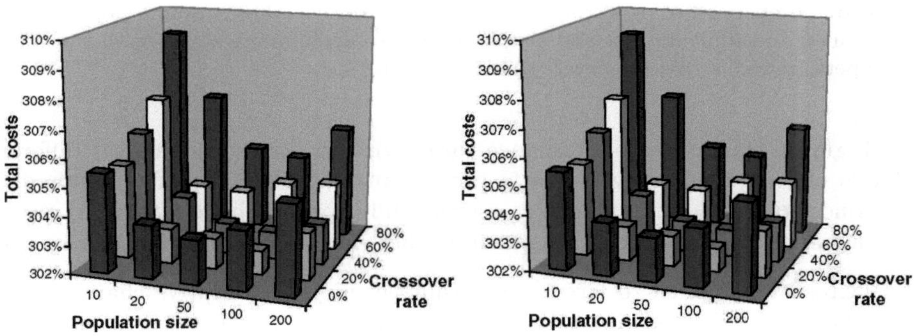

Fig. 5. Comparison of different crossover rates and population sizes for the Adult and Cover Type datasets, using equal false positive and false negative costs

algorithm has been shown to be an effective multi-objective optimizer, both in general and when optimizing rules [1,2,3,4]. Parameter tuning was performed on the Adult dataset and on 10,000 records selected at random from the Cover Type dataset, minimizing the simple error rate and the number of ATs.

The number of different parameters and potential variations of the algorithm made the cost of exhaustive experimentation with parameter settings prohibitive. Instead, effort was focused on finding the best values for crossover rate and population size only. Experiments were performed with six population sizes, 10, 20, 50, 100, 200 and 500, and six crossover rates, 0%, 20%, 40%, 60%, 80% and 100%. Each experiment consisted of 30 runs of the algorithm, with 200,000 rule evaluations per run. Results were compared by summing the error rates of the best rule at each level of rule complexity, up to 20 ATs. This is equivalent to comparing on the dominated area in the objective space [25]. Mean results are shown in figure 5. Results with a population size of 500 or a crossover rate of 100% are omitted since these were considerably worse than the results displayed. Similar graphs were obtained when minimizing the balanced error rate and when the false negative cost was increased to ten times the false positive cost. In each case, best performance was obtained using a crossover rate of 20% or 40% and a population size of 100 for the Adult dataset and 50 for the Cover Type dataset.

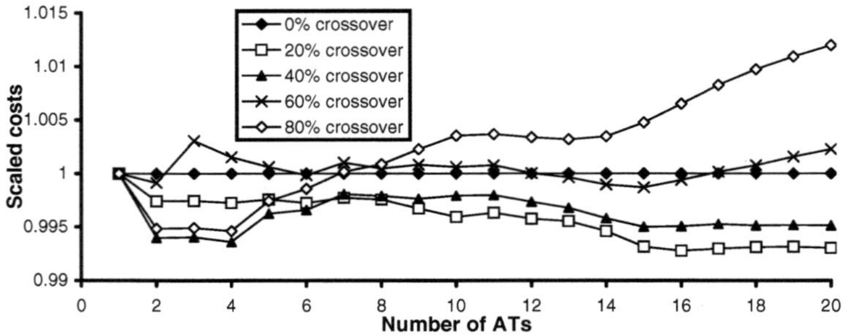

Fig. 6. Comparison of performance at different crossover rates, using a population of 100 rules, the adult dataset and simple error rate. Results are scaled with respect to the performance at 0% crossover, 100% mutation.

Figure 6 shows how performance varies with the crossover rate, at different levels of rule complexity. While a certain amount of crossover is required to produce good results, too much crossover (and hence too little mutation) results in degraded performance for large rules. There are two possible reasons for this:

- Crossover applied to large rules results in major, disruptive changes when subtle modifications may be more appropriate. Adapting the crossover operator to be biased towards smaller changes is a matter for further research.
- A loss of diversity in the population, early in the search process, requires the use of mutation to reintroduce useful ATs and subtrees. This would explain the very poor performance obtained when using crossover only.

4.3 Training, Validation, Selection and Testing

Evaluating the performance of a new classification algorithm often consists of two stages:

Training: The classifier is first trained using a set of training data to create a model to be used for prediction.

Testing: The overall aim is to achieve high performance on *unseen* data, which is not necessarily implied by high performance on the training data. Therefore a set of test data is used to evaluate the model.

Suppose our algorithm is applied to data provided by a hypothetical client. First the algorithm produces a range of rules from the training data with differing trade-offs between misclassification cost and rule complexity. In order to give the client some idea as to how well the rules produced generalize, the rules are re-evaluated on new validation data. At this point, the client selects a rule. To compare with other algorithms that produce just one model, we assume that the client selects the rule with minimum misclassification costs on the validation data, though in practice the client may elect to choose a simpler rule. To ensure a fair comparison, this rule must be re-evaluated again on further test data.

Table 2. Misclassification costs on test data and run times

Dataset	Cost	Select best Mean	Select best StdDev	Select best ATs	Select 5AT Mean	Select 5AT StdDev	Time (s)
Adult	Simple	14.42%	0.10%	19.6	15.64%	0.12%	1032
	Balanced	17.82%	0.14%	18.0	19.42%	0.07%	957
	1-10	12.45%	0.13%	15.8	12.90%	0%	898
Cover type	Simple	21.00%	0.19%	19.5	23.35%	0.08%	8217
	Balanced	21.71%	0.31%	19.5	23.57%	0.16%	9168
	1-10	10.58%	0.17%	19.2	11.20%	0.09%	7988
Contraceptive	Simple	29.45%	3.48%	8.8	29.46%	3.53%	55
	Balanced	31.97%	3.78%	7.9	32.22%	4.18%	55
Breast cancer	Simple	4.14%	2.25%	4.2	4.10%	2.28%	37
	Balanced	4.97%	2.53%	4.8	5.09%	2.61%	30
Pima Indians	Simple	24.38%	4.47%	4.9	24.48%	4.65%	47
	Balanced	26.35%	6.25%	8.0	26.84%	6.33%	41

Both the Adult dataset and the Cover Type dataset were partitioned once only, since the Adult dataset is already split into training and test data (30162 and 15060 records) and the size of the Cover Type dataset limits the amount of experimentation that can be performed. The Adult training set provided was split again at random, with approximately 80% forming the new training set and 20% providing a validation set. The Cover Type data was split 50–25–25 at random, with 50% forming the training set. Each experiment consisted of 30 runs of the algorithm, using a crossover rate of 30% and a population size of 100 for the Adult dataset and 50 for the Cover Type data.

The smaller datasets were split into ten approximately equal parts to be used as the test sets. For each test set, the remaining 90% of the records were split at random into two roughly equal parts to be used as training and validation sets for two experiments, resulting in 20 experiments for each dataset. This use of ten fold cross validation allows for fair comparison with both the results of Ishibuchi et al. [7,8] and with the results of 33 classifiers provided by Lim et al. [26]. The algorithm was run 30 times for each experiment, resulting in 600 runs. The parameter settings found to perform well for the Adult data (a crossover rate of 30% and a population of 100) were used in these experiments.

4.4 Results

Table 2 shows the mean quality and size of the rule selected by the client if he selects the best rule according to error rate on the validation data, breaking ties on the training error rate. Mean error rates for the best 5 AT rule and the approximate run time for one run on a 2GHz processor are also given.

The results obtained when the client chooses not to sacrifice rule accuracy can be roughly compared with those provided in the UCI machine learning repository [22] for the Adult dataset and those provided by Lim et al. [26] for the Breast

Cancer and Pima Indians datasets. In the first case, the UCI repository provides results for 16 algorithms, with error rates of between 14.05% and 21.42% and our algorithm ranks 3rd out of 17 algorithms. The 33 algorithms evaluated by Lim et al. have error rates varying between 2.78% and 8.48% for the Breast cancer dataset and between 22.1% and 31.0% for the Pima Indians dataset. Our algorithm ranks 17th and 22nd out of 34 algorithms respectively.

The trade-off between misclassification costs, on both the training and validation data, and rule simplicity for the Adult dataset is shown in figure 7. Here, the error bars give the standard deviation at each level of rule complexity.

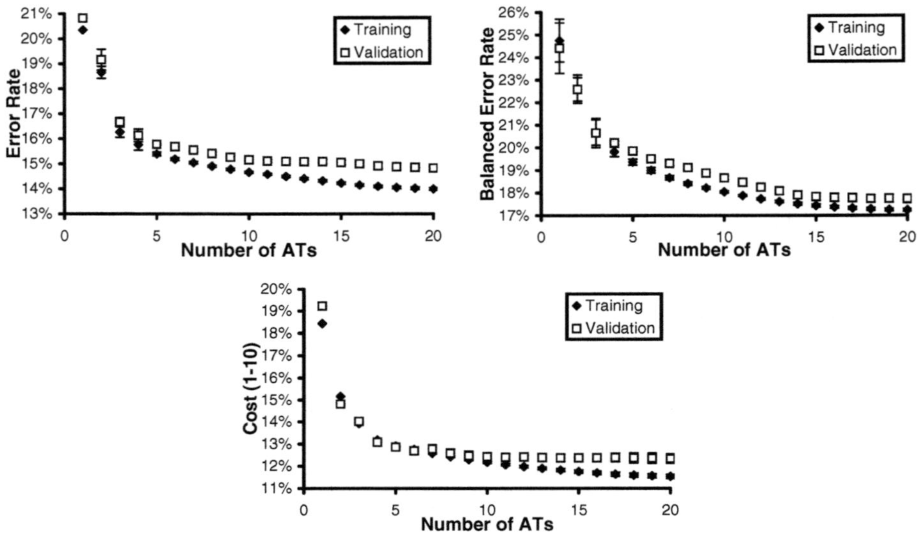

Fig. 7. Rule quality for the adult dataset. Misclassification costs are given by the simple error rate, the balanced error rate and by setting the false negative cost to be ten times the false positive cost respectively.

Typical 5 AT rules, with associated confusion matrices, are shown in figures 8 to 10. The first rule is fairly accurate when predicting that a record belongs to the class of interest, but it identifies little more than half of this class, since more emphasis is placed on accuracy on the larger set of uninteresting records. Subsequent rules have increases in the number of false positives and decreases in the number of false negatives, as would be expected. The rules become less restrictive as illustrated by the greater use of the boolean 'or' operation in figure 10, describing more of the class of interest but with decreased confidence.

Figure 11 shows the results obtained when optimizing the simple error rate for the Breast cancer and Pima Indians diabetes datasets, allowing comparison with the results obtained by Ishibuchi et al. [7,8]. The results obtained by the

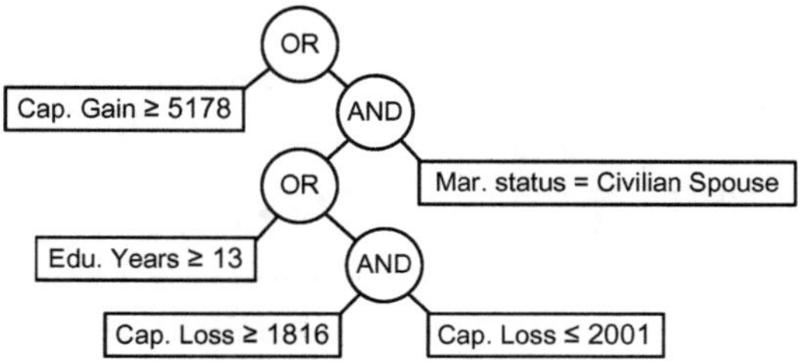

Confusion matrix

		Predicted -	Predicted +
Actual	-	10732	628
Actual	+	1710	1990

Fig. 8. A typical five AT rule antecedent, predicting a high salary, produced when minimizing the simple error rate, with its confusion matrix on test data

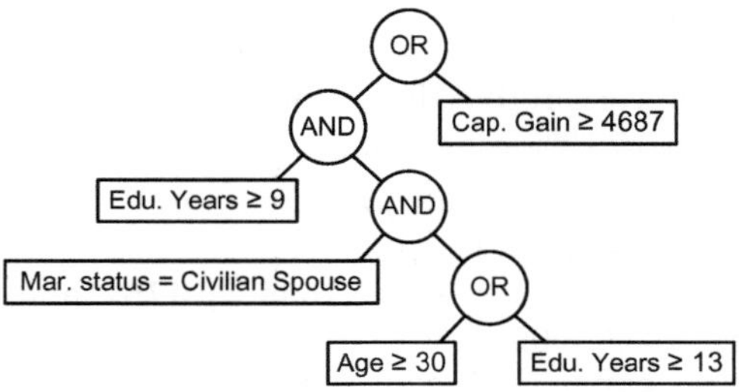

Confusion matrix

		Predicted -	Predicted +
Actual	-	8631	2729
Actual	+	549	3151

Fig. 9. A five AT rule antecedent produced when minimizing the balanced error rate, with its confusion matrix on test data

two approaches on the Breast cancer data are broadly similar. Results for the Pima Indians diabetes datasets are similar in testing — once ten or more rules are permitted in the rule set, Ishibuchi et al. achieve error rates of approximately 24.8% — but on the training data Ishibuchi et al. only reach an error rate of approximately 16% even when 20 rules of up to 3 tests each are permitted. Note that both approaches result in rules that generalize poorly, as indicated by the large disparity between training and testing error rates, though even the best classifier evaluated by Lim et al. [26] has an error rate of 22.1% on test data.

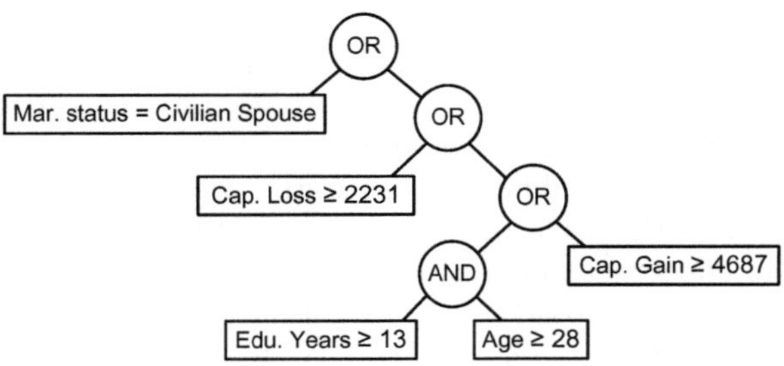

Confusion matrix

		Predicted -	Predicted +
Actual	-	6524	4836
Actual	+	140	3560

Fig. 10. A five AT rule antecedent produced when the cost of a false negative is ten times that for a false positive

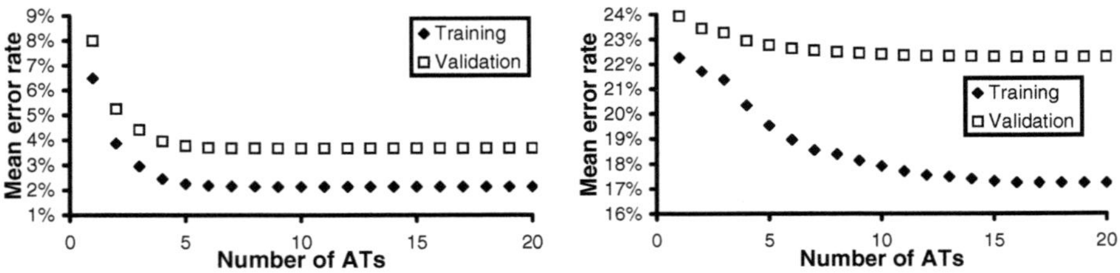

Fig. 11. Error rates for the Breast cancer and Pima Indians datasets

Finally, the algorithm was modified to suit a user with a preference for viewing rule sets, by changing the rule complexity objective to the number of ATs after conversion to such a rule set. Applying this modified algorithm to the Adult dataset and selecting an 8 AT rule set produced the following rules, with an error rate of 15.01% in training and 15.19% in testing:

- If cap. gain $\geq$ 5178 then salary $\geq$ $50,000
- If cap. loss $\geq$ 2392 then salary $\geq$ $50,000
- If mar. status = civilian spouse and cap. loss $\geq$ 1762 and cap. loss $\leq$ 1980 then salary $\geq$ $50,000
- If mar. status = civilian spouse and edu. years $\geq$ 13 and hours per week $\geq$ 31 then salary $\geq$ $50,000
- Otherwise salary < $50,000

5 Conclusions

The results presented illustrate that the approach can produce reasonable results on two class problems, though there is room for improvement. However, the algorithm provides additional benefits. The most obvious is that the client can be provided with a range of models with different trade-offs between rule complexity and misclassification costs. This allows the client to select a rule that is accurate enough while also being comprehensible.

The overall approach is also flexible. Rules may be presented to the client in a number of ways and the measure of rule complexity can easily be adapted to match the method of rule presentation and the client's concept of rule comprehensibility. Different measures of misclassification cost can easily be used. There is no restriction on the data types of the fields of the dataset and no need to discretize numeric fields as required by other algorithms.

While the efficiency of the algorithm could be improved, the approach appears to be able to handle larger datasets than that of Ishibuchi et al. While no timings are given that would allow an effective comparison, Ishibuchi et al. report that their approach has a large computational load [7] and the largest dataset to which their code has been applied — the Pima Indians diabetes data — contains only 768 records. This is understandable, as even when applied to small datasets, 17070 simple rules are generated from which the genetic algorithm must select a set. Many of these simple rules are unlikely to be useful in a full classifier.

6 Further Research

Three or more classes: While the algorithm has been shown to be effective for two class problems an obvious improvement would be to extend the approach to handle three or more classes.

Three objectives: In practice, the client may only be able to approximate the costs of false negatives and false positives. In this case, the problem may be modeled as having three objectives to be minimized: the rule complexity, the number of false positives and the number of false negatives.

Diversity management: Preliminary investigations revealed a major loss of diversity in the population of rules early in the search process, even when no simplification routines were applied. If techniques can be found that provide better management of population diversity, the algorithm may be able to produce better results in a fraction of the time.

Efficiency improvements: There is scope for a number of efficiency improvements. For example, early in the search it makes little sense to evaluate the rules on the entire dataset, when evaluation on just a sample will provide enough information to guide the search at this stage.

Rule types: Other AT types (see previous work [2,4]) and operators may be used to further enhance the expressiveness of the rules produced.

References

1. B. de la Iglesia, M. S. Philpott, A. J. Bagnall, and V. J. Rayward-Smith. Data Mining Rules Using Multi-Objective Evolutionary Algorithms. In *Proceedings of 2003 IEEE Congress on Evolutionary Computation*, pages 1552–1559, 2003.
2. Beatriz de la Iglesia, Alan Reynolds, and Vic J Rayward-Smith. Developments on a Multi-Objective Metaheuristic (MOMH) Algorithm for Finding Interesting Sets of Classification Rules. In *Evolutionary Multi-Criterion Optimization: Third International Conference, EMO 2005*, volume 3410 of *Lecture Notes in Computer Science*, pages 826–840, March 2005.
3. B. de la Iglesia, G. Richards, M. S. Philpott, and V. J. Rayward-Smith. The application and effectiveness of a multi-objective metaheuristic algorithm for partial classification. *European Journal of Operational Research*, 169(3):898–917, 2006.
4. Alan Reynolds and Beatriz de la Iglesia. Rule induction using multi-objective metaheuristics: Encouraging rule diversity. In *Proceedings of the 2006 IEEE World Congress on Computational Intelligence*, pages 6375–6382. IEEE, 2006.
5. K. Ali, S. Manganaris, and R. Srikant. Partial Classification using Association Rules. In *Proceedings of the Third International Conference on Knowledge Discovery and Data Mining*, pages 115–118. The AAAI Press, 1997.
6. Ashish Ghosh and Bhabesh Nath. Multi-objective rule mining using genetic algorithms. *Information Sciences*, 163:123–133, 2004.
7. Hisao Ishibuchi and Yusuke Nojima. Accuracy-Complexity Tradeoff Analysis by Multiobjective Rule Selection. In *Proceedings of the ICDM 2005 Workshop on Computational Intelligence in Data Mining*, pages 39–48, 2005.
8. Hisao Ishibuchi, Isao Kuwajima, and Yusuke Nojima. Multiobjective Association Rule Mining. In *Proceedings of the PPSN Workshop on Multiobjective Problem Solving from Nature*, 2006.

9. Ian H. Witten and Eibe Frank. *Data Mining: Practical Machine Learning Tools and Techniques*. Elsevier, second edition, 2005.
10. T. Tanigawa and Q. Zhao. A Study on Efficient Generation of Decision Trees Using Genetic Programming. In *Proceedings of the Genetic and Evolutionary Computation Conference (GECCO 2000)*, pages 1047–1052. Morgan Kaufmann, 2000.
11. R. K. DeLisle and S. L. Dixon. Induction of Decision Trees via Evolutionary Programming. *Journal of Chemical Information and Modelling*, 44(3):862–870, 2004.
12. J. Eggermont, J. N. Kok, and W. A. Kosters. Detecting and Pruning Introns for Faster Decision Tree Evolution. In *Parallel Problem Solving from Nature VIII*, volume 3242, pages 1071–1080. Springer-Verlag, 2004.
13. M. C. J. Bot. Improving Induction of Linear Classification Trees with Genetic Programming. In *Proceedings of the Genetic and Evolutionary Computation Conference (GECCO 2000)*, pages 403–410. Morgan Kaufmann, 2000.
14. E. M. Mugambi and A. Hunter. Multi-Objective Genetic Programming Optimization of Decision Trees for Classifying Medical Data. In *Proceedings of the 7th International Conference on Knowledge-Based Intelligent Information and Engineering Systems (KES 2003)*, volume 2773 of *Lecture Notes in Computer Science*, pages 293–299, 2003.
15. C. Setzkorn and R. C. Paton. MERBIS - A Multi-Objective Evolutionary Rule Base Induction System. Technical Report ULCS-03-016, Department of Computer Science, University of Liverpool, 2003.
16. C. Setzkorn and R. C. Paton. MERBIS - A Self-Adaptive Multi-Objective Evolutionary Rule Base Induction System. Technical Report ULCS-03-021, Department of Computer Science, University of Liverpool, 2003.
17. John Koza. *Genetic programming: on the programming of computers by means of natural selection.* MIT Press, 1992.
18. William B. Langdon and Riccardo Poli. *Foundations of Genetic Programming.* Springer, 1998.
19. P. J. Angeline. Genetic Programming and Emergent Intelligence. In K. E. Kinnear, editor, *Advances in Genetic Programming*, pages 75–97. MIT Press, 1994.
20. A. Ekárt and S. Z. Németh. Selection Based on the Pareto Nondomination Criterion for Controlling Code Growth in Genetic Programming. *Genetic Programming and Evolvable Machines*, 2:61–73, 2001.
21. J. R. Quinlan. Simplifying decision trees. *International Journal of Man-Machine Studies*, 27(3):221–234, 1987.
22. C. J Merz and P. M. Murphy. UCI repository of machine learning databases. Univ. California, Irvine, 1998.
23. Kalyanmoy Deb. *Multi-Objective Optimization using Evolutionary Algorithms.* John Wiley & Sons Ltd, 2001.
24. Kalyanmoy Deb, Samir Agrawal, Amrit Pratab, and T. Meyarivan. A Fast Elitist Non-Dominated Sorting Genetic Algorithm for Multi-Objective Optimization: NSGA-II. In *Proceedings of PPSN VI*, volume 1917 of *Lecture Notes in Computer Science*, pages 849–858. Springer, 2000.
25. Eckart Zitzler and Lothar Thiele. Multiobjective Optimization Using Evolutionary Algorithms: A Comparative Case Study and the Strength Pareto Approach. *IEEE Transactions on Evolutionary Computation*, 3(4):257–271, 1999.
26. T.-S. Lim, W.-Y. Loh, and Y.-S. Shih. A Comparison of Prediction Accuracy, Complexity, and Training Time of Thirty-three Old and New Classification Algorithms. *Machine Learning*, 40:203–229, 2000.

Combining Linear Programming and Multiobjective Evolutionary Computation for Solving a Type of Stochastic Knapsack Problem

Fermín Mallor-Gímenez, Rosa Blanco, and Cristina Azcárate

Department of Statistics and Operations Research
Edificio Los Magnolios
Public University of Navarre
31006 Pamplona – Spain
{mallor,rosa.blanco,cazcarate}@unavarra.es

Abstract. In this paper, the design of systems using mechanical or electrical energy-transformation devices is treated as a knapsack problem. Due to the well-known NP-hard complexity of the knapsack problem, a combination of integer linear programming and evolutionary multicriteria optimization is presented to solve this real problem with promising experimental results.

1 Introduction

Many real-world problems involve the simultaneous optimization of multiple objectives. These objectives often conflict, that is, an optimal performance in one objective implies a low performance in some of the remaining objectives. Therefore, a compromise solution must therefore be reached.

Several mathematical programming methods have been developed to deal with multi-objective problems through different approaches [1]. From a mathematical point of view, a multi-objective problem is solved by finding a set of non-dominated solutions. For large and complex problems, however, there are practical difficulties involved in using mathematical programming methods to obtain this set of Pareto-optimal solutions. In recent years, techniques based on meta-heuristics have been proposed to increase efficiency when solving real multi-objective problems [2,3].

Evolutionary computational approaches are used successfully to solve hard problems. Inspired in natural evolution, evolutionary approaches are considered powerful stochastic search methods [4]. Evolutionary strategies evolve a population of potential solutions in a single run by means of chromosomal encoding representation of solutions and crossing, mutation and selection operators. Approaches of this type are therefore of interest in attemps to solve multi-objective problems. For an extended review [5,6,7,8,9].

The knapsack problem is a well-known NP-hard combinatorial optimization problem [10] that can be formulated as follows: given a number of item types

S. Obayashi et al. (Eds.): EMO 2007, LNCS 4403, pp. 531–545, 2007.

with corresponding unit profit p_i and unit weights w_i, and a knapsack with fixed capacity c, determine the number of x_i of each item type that maximizes profit without exceeding the capacity limit. Mathematically:

$$\max \sum_i p_i x_i$$
$$\text{subject to}$$
$$w_i x_i \leq c$$
$$x_i \geq 0$$

Due to its application to many real problems, several meta-heuristic methods are proposed to solve combinatorial problems [4,11,12]. Evolutionary techniques are also applied to multi-objective knapsack problems [13,14].

In this paper, a combination of integer programming and multi-objective optimization is presented to handle a complex real problem in the energy sector. The real problem is related to the design of systems using mechanical or electrical devices. More specifically, the number and type of electrical or mechanical devices required to make up the system must be determined.

Due to the complexity of the real problem, the proposed methodology is based on the *divide and conquer* philosophy. In this way, integer linear programming and evolutionary multi-criteria optimization are combined to solve the real problem.

The paper is organized as follows. The stochastic knapsack real problem and its mathematical formulation is described in Section 2. Section 3 presents the proposed methodology to solve the problem. Section 4 analyses the experimental results. Finally, a brief set of conclusions and suggestions for future work are presented in Section 5.

2 Problem Statement and Mathematical Modelling

2.1 Problem Statement

This work addresses the design of mechanical or electrical equipment to achieve a given target. In this case, the target is related to a transformation process. The system or equipment receives an input and delivers an output. The design task is to determine the number and type of electrical and/or mechanical devices required to build the system. There are different types of devices performing the same task. Each device has a set work capacity, efficiency curve, and cost. The capacity is the highest input the device is able to handle. The efficiency curve describes the output-to-input ratio. Each configuration of the system (i.e. a set of devices) is therefore assessed by its total cost and total efficiency.

The transformation process is performed through time. In this problem, it is assumed that the *total input* I is a random quantity whose probability distribution is known or can be estimated.

Specifically, in order to state the problem, there are K different types of devices, $D_1, D_2, \ldots, D_K$. The type D_i is characterized by the following parameters:

- Capacity: L_i is defined as the minimum partial input required to start up the device D_i. U_i is defined as the maximum partial input for device D_i, that is, D_i transforms a partial input I_i ranging from L_i to U_i.
- Efficiency function $Ef_i(partial\_input)$: a function relating the partial input I to device D_i to its output $O = Ef_i(I)$.
- Cost C_i: the cost of integrating a unit of type D_i into the system.

Therefore, the problem is to fix the number $n_1, \ldots, n_k$ of devices of types $D_1, \ldots, D_k$, respectively, that provide the system with maximum efficiency at the minimum cost. Maximum efficiency and minimum cost turn the real problem into a multi-objective problem. The total cost TC of a system is obtained as:

$$TC = \sum_{i=1}^{K} C_i n_i.$$

Due to the random nature of the input, efficiency is not easy to set. Given a configuration of the system, for each quantity of input I a different level of efficiency is reached. Then, efficiency could be set in terms of several statistical measures. Expected efficiency is the objective function considered in this study. Therefore, efficiency is calculated as follows:

$$\overline{Ef}(system) = \frac{\sum_{j=1}^{P} Ef(system, total\_input_j) freq(total\_input_j)}{\sum_{j=1}^{P} total\_input_j freq(total\_input_j)} \quad (1)$$

where $Ef(system, total\_input_j)$ denotes system efficiency (output) when input is fixed to $total\_input_j$, $freq(total\_input_j)$ denotes frequency of input $total\_input_j$, and $\overline{Ef}(system)$ denotes expected efficiency of the system. This formulation is presented for a discrete distribution of input. When a continuous distribution of input is considered, then, an analogous formulation can be proposed, in which the sums of the previous expression have to be replaced by integrals and the frequency by a density function.

The calculation of $Ef(system, total\_input_j)$ is not trivial. It is required to determine how the $total\_input_j$ is divided among the devices in the system– see Figure 1 –. Furthermore, the assignment must be done in an optimal way.

2.2 Modelling Using Stochastic Multi-objective Nonlinear Integer Programming

The proposed real problem can be mathematically represented in several ways. It can be formulated as a stochastic nonlinear multi-objective integer program, in which case the following notation is used:

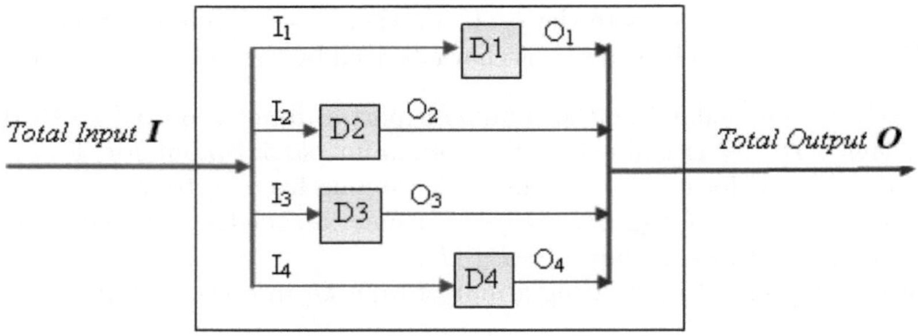

Fig. 1. Equipment consisting of 4 devices: D_1, D_2, D_3 and D_4. $I = I_1 + I_2 + I_3 + I_4$. $0 = 0_1 + 0_2 + 0_3 + 0_4$.

- Indices:
  - k, $k = 1, \dots, r$, is related to devices.
  - j, $j = 1, \dots, P$, is related to system random input with frequency $freq(total\_input_j)$.
- Decision variables:
  - $partial\_input_{kj}$ represents the partial input of device D_k when the system receives $total\_input_j$.
  - w_{kj} and z_k are binary variables defined as:

$$w_{kj} = \begin{cases} 1 \text{ if } partial\_input_{kj} > 0 \\ 0 \text{ if } partial\_input_{kj} = 0 \end{cases} \qquad z_k = \begin{cases} 1 \text{ if } \sum_j w_{kj} > 0 \\ 0 \text{ if } \sum_j w_{kj} = 0 \end{cases}$$

  Note that w_{kj} equals 1 if and only if device D_k is used when input equals $total\_input_j$. Then, z_k equals 1 if and only if device D_k is used for at least one $total\_input_j$.
- Parameters:
  - $total\_input_j$ denotes the system random input, $j = 1, \dots, P$.
  - U_k is the maximum partial input of device D_k, and L_k is the minimum partial input required for the start-up of D_k, $k = 1, \dots, r$.
  - $Ef_k(device\_input)$ is the efficiency curve for device D_k, that is, the output of device D_k when the input is $device\_input$, $k = 1, \dots, r$. Usually, it is a nonlinear function of the $device\_input$.
  - $freq(total\_input_j)$ is the frequency of the system random input when it takes the value $total\_input_j$, $j = 1, \dots, P$.
  - c_k denotes the cost of including device D_k in the system, $k = 1, \dots, r$.

Once all the required parameters and decision variables are defined, the multi-objective nonlinear integer programming can be formulated as in Figure 2.

Constraints. The sets of constraints 2 and 3 verify that $partial\_input_{kj}$ equals zero or a value between L_k and U_k. Constraints 4 ensure that the amount of

$$\min \sum_{k=1}^{r} c_k z_k$$

$$\max \sum_{j=1}^{P} freq(total\_input_j) \sum_{k=1}^{r} Ef_k(partial\_input_{kj})$$

subject to

$$partial\_input_{kj} \geq L_k w_{kj} \tag{2}$$

$$partial\_input_{kj} \leq U_k w_{kj} \tag{3}$$

$$\sum_{k=1}^{r} partial\_input_{kj} \leq total\_input_j \tag{4}$$

$$w_{kj} \leq z_k \tag{5}$$

$$partial\_input_{kj} \geq 0$$

$$w_{kj}, z_k \text{ binary variables}$$

$$\forall k = 1, \ldots, r; j = 1, \ldots, P$$

Fig. 2. Mathematical formulation of the multi-objective nonlinear integer programming

input through all devices, $D_1, \ldots, D_r$ is no higher than the total system input. Finally, constaints 5 check the relation between variables w_{kj} and z_k. Note that the number n_i of devices D_i is fixed by: $n_i = \sum_{k \text{ of type } D_i} z_k$.

Knapsack problem. The mathematical problem presented in this paper can be considered as a generalization of the knapsack problem. Observe that if $L_k = U_k$, the attained constraints are the classical capacity constraints of the knapsack problem. The general situation where $L_k < U_k$ is considered, thus, a knapsack problem where weights w_i are unknown is reached. Moreover, the weights turn into the decision variables of the problem.

Because of the maximization objective, the multi-objective model becomes a stochastic optimization problem. In this real problem scenario, expected efficiency is used as the efficiency measure, assuming that *total_input* is a discrete random variable with P different values. The nonlinearity is also due to the maximization objective, because the efficiency functions of devices are usually considered to be non linear functions. To sum up, the presented real problem is a knapsack multi-objective stochastic nonlinear problem.

Implementation is complicated due to the size of the multi-objective nonlinear problem, which involves rP continuous variables, $r(P+1)$ binary variables and $P(3r+1)$ constraints. Then, the implementation of the nonlinear multi-objective problem has thousands of continuous variables, thousands of binary variables and thousands of constraints.

$$\min \sum_{k=1}^{r} c_k z_k$$

$$\max \sum_{j=1}^{P} freq(total\_input_j) \sum_{k=1}^{r} (y_{1kj}\lambda_{1kj} + y_{2kj}\lambda_{2kj} + y_{3kj}\lambda_{3kj} + y_{4kj}\lambda_{4kj})$$

subject to

$$\begin{aligned}
partial_input_{kj} &\geq L_k w_{kj} \\
partial_input_{kj} &\leq U_k w_{kj} \\
\sum_{k=1}^{r} partial_input_{kj} &\leq total_input_j \\
w_{kj} &\leq z_k \\
partial_input_{kj} &= x_{1kj}\lambda_{1kj} + x_{2kj}\lambda_{2kj} + x_{3kj}\lambda_{3kj} + x_{4kj}\lambda_{4kj} \quad (6)\\
\lambda_{1kj} &\leq bin_{1kj} \\
\lambda_{2kj} &\leq bin_{1kj} + bin_{2kj} \\
\lambda_{3kj} &\leq bin_{2kj} + bin_{3kj} \\
\lambda_{4kj} &\leq bin_{3kj} \\
\lambda_{1kj} + \lambda_{2kj} + \lambda_{3kj} + \lambda_{4kj} - ww_{kj} &= 0 \\
bin_{1kj} + bin_{2kj} + bin_{3kj} - ww_{kj} &= 0 \\
partial_input_{kj} &\geq 0 \\
w_{kj}, ww_{kj}, bin_{1kj}, bin_{2kj}, bin_{3kj} &\text{ binary variables} \\
\forall k = 1, \ldots, r; j &= 1, \ldots, P
\end{aligned}$$

Fig. 3. Mathematical formulation of the multi-objective linear integer programming

2.3 Modelling Using Stochastic Multi-objective Linear Integer Programming

In order to avoid the difficulties caused by the size of the nonlinear problem, a linear piecewise approach is used to analyze the efficiency curve of each device. Figure 3 represents the stochastic linear approach. Parameters (x_{1k}, y_{1k}), (x_{2k}, y_{2k}), (x_{3k}, y_{3k}), and (x_{4k}, y_{4k}) are points defining three linear piecewise segments of the efficiency curve for device D_k, and $\lambda_{ik}, bin_{ik}, ww_k$ are decision variables introduced for modelling purposes.

The linear approach increases the number of continuous and binary variables and, thus, the number of constraints. For example, replacing a nonlinear efficiency function with a three linear piecewise function multiplies the number of continuous variables by 4. Then, $2r(P + 2)$ binary variables and $6rP$ new constraints have to be added. Therefore, a multi-objective linear integer problem with tens of thousands of constraints, tens of thousands of continuous variables and tens of thousands of binary variables has to be carried out. Note that the integer linear model of Figure 3 can be simplified by introducing the equality constraints 6 in suitable places. This simplified model is used to determine the model dimension.

Nevertheless, the size of the problem is still too large to find the exact solution. This means that exact methods are not applicable. Hence, in order to attain good solutions with a feasible computational cost, heuristic algorithms are considered.

2.4 Modelling Using Linear Programming and Evolutionary Computation

Due to all the difficulties described in the previous sections, an iterative two-stage method is proposed. In *stage A*, the expected efficiency of a given configuration for the system is evaluated. *Stage B* evolves a population of configurations according to two objective functions.

Stage A resolves a set of P integer linear programming models of smaller size than discussed above. *Stage B* develops a multi-criteria evolutionary algorithm with Pareto elitist selection to approach the cost-efficient Pareto optimal surface for devices.

3 Hybrid Algorithm Based on Integer Linear Programming and Multi-objective Evolutionary Computation

The proposed hybrid algorithm to solve the real problem presented above is described in depth.

3.1 Stage A: Efficiency Evaluation. An Integer Linear Programming Approach

Let $N = (n_1, \ldots, n_k)$ be a feasible system configuration where n_i is the number of devices of type i and $s = \sum_i n_i$. For each system configuration, the expected efficiency function is evaluated by equation 1. It requires resolution of P integer linear problems, with several values of $total\_input_j, j = 1, \ldots, P$. Figure 4 shows the mathematical formulation of these problems.

The P linear problems have fewer variables and constraints than formulated in the previous section. Specifically, each problem requires $4s$ continuous variables, $5s$ binary variables, and $8s + 1$ constraints. Furthermore, to solve the P linear problems, a unique configuration of the system is considered, thus, $s \ll r$.

3.2 Stage B. Population Evolution. A Multi-criteria Evolutionary Algorithm

A multi-criteria evolutionary Pareto elitist algorithm is proposed to attain a good approximation to the cost-effiency Pareto optimal surface of the real problem. Figure 5 shows the main steps in the proposed method.

$$\max Ef(system, total\_input_j) = \sum_{device\ k} (y_{1k}\lambda_{1k} + y_{2k}\lambda_{2k} + y_{3k}\lambda_{3k} + y_{4k}\lambda_{4k})$$

subject to

$$\begin{aligned}
partial_input_k &= x_{1k}\lambda_{1k} + x_{2k}\lambda_{2k} + x_{3k}\lambda_{3k} + x_{4k}\lambda_{4k} \\
\lambda_{1k} &\le bin_{1k} \\
\lambda_{2k} &\le bin_{1k} + bin_{2k} \\
\lambda_{3k} &\le bin_{2k} + bin_{3k} \\
\lambda_{4k} &\le bin_{3k} \\
\lambda_{1k} + \lambda_{2k} + \lambda_{3k} + \lambda_{4k} - ww_k &= 0 \\
bin_{1k} + bin_{2k} + bin_{3k} - ww_k &= 0 \\
partial_input_k &\ge L_k w_k \\
partial_input_k &\le U_k w_k \\
\sum_{device\ k} partial_input_k &\le total_input_j \\
w_k, ww_k, bin_{ik} &\text{ binary variables} \\
\lambda_{ik}, partial_input_k &\ge 0 \\
\forall k = 1, \ldots, s; i &= 1, \ldots, 4
\end{aligned}$$

Fig. 4. Formulation of integer linear problems to measure expected efficiency

---

Step 1: Generate an initial population pop_0.
  Create an empty archive $archive_0$.
  Set $t = 0$.
Step 2: Evaluate individuals of pop_t (cost and expected efficiency):
  resolve P integer linear programming problems for each individual.
Step 3. Copy all non-dominated solutions in $pop_t + archive_t$ to $archive_{t+1}$.
Step 4. If stopping criterion is met: plot nondominated solutions in $archive_{t+1}$.
Step 5. Select μ individuals from $archive_{t+1}$.
Step 6. Generate λ individuals from μ selected individuals.
  $pop_{t+1} = \{\lambda_{offspring}\}$.
  $t = t + 1$.
  Go to Step 2.

Fig. 5. Main steps of the proposed multi-criteria evolutionary algorithm

Now, each step is explained more in detail.

Step 1: Generation of the initial population. In order to generate the initial population, k different devices are considered. A solution or system configuration is represented by $N = (n_1, \ldots, n_k)$, where n_i is the number of devices D_i. It is assumed that $cost(D_i) \le cost(D_{i+1})$ and $U_i \le U_{i+1}$.

First, the following configurations are evaluated:

$$\begin{aligned}
N_1^* &= (n_1^*, 0, \ldots, 0) \\
N_2^* &= (0, n_2^*, \ldots, 0) \\
&\ldots \\
N_i^* &= (0, \ldots, n_i^*, \ldots, 0) \\
&\ldots \\
N_k^* &= (0, 0, \ldots, n_k^*)
\end{aligned}$$

where n_i is fixed as the integer part of $\frac{total\_input}{U_i}$ and $total\_input = \max_j\{total\_input_j\}$. Note that $N_1^*, \ldots, N_k^*$ are extreme points of a simplex. This simplex approaches the feasible set of configurations for the multi-objective problem. Through the search process, the proposed method can find solutions that, while not contained in the simplex, are not far from the frontier.

Any system solution or configuration is a linear combination of the initial solutions, such that: $N^* = \sum_{i=1}^{k} p_i N_i^*$, with $\sum_{i=1}^{k} p_i \approx 1$, where p_i represents the percentage input assigned to device type i. By setting $\sum_{i=1}^{k} p_i = 1$ the simplex is exactly reached. Therefore, by allowing $\sum_{i=1}^{k} p_i \approx 1$, the simplex neighbourhood is visited.

The initial population, pop_0, contains the k vertices of the simplex and a set of solutions attained as a linear combination of pairs of these vertices. In order to achieve diversification, vertex N_i^* is combined with vertex N_{i+l}^*, where l depends on the total number m of required pairs in the following way:

$$m = (k-l) + (k-l-1) + \ldots + 1 = \sum_{n=1}^{k-l}, \text{ then, } m = \frac{k-l+1}{2}(k-l).$$

From this equation, $l \approx= \frac{(2k+1)-\sqrt{8m+1}}{2}$, thus, l can be considered the round off of this expression.

For every pair of selected solutions, N_r^* and N_s^*, a new solution is generated as the convex combination. Then, the new solution is: $N_t^* = pN_r^* + (1-p)N_s^*$, $p \in (0,1)$. In this way, for every pair, h new solutions are generated by selecting $p = \frac{1}{h+1}, \frac{2}{h+1}, \ldots, \frac{h}{h+1}$. Finally, the number of solutions in the initial population pop_0 is $k + hm$.

Step 2: Evaluation of generated configurations. The configuration cost is directly evaluated depending on the cost of each selected device. The expected efficiency measure requires the evaluation of equation 1. This implies carrying out of *stage A*, that is, the resolution of P integer programming problems –see Figure 4.

Step 3: The evolutionary strategy and the Pareto elitist selection approach. The Pareto elitist selection approach chooses nondominated solutions from pop_t and $archive_t$, and adds them to $archive_{k+1}$. The archive is used to store the current approximation to Pareto optimal surface. It must be remarked that the number of individuals in the archive changes over time.

for $(p = 1; p = \mu; p++)$
  for $(m = 1; m = \beta; m++)$
    1. with $prob\_mut_1$ do $n_i = n_i - 1 \; \forall i \in CD$
       $S = \{i \in CD | n_i \text{ value has been decreased}\}$
    2. $quantity\_decreased = \sum_{i \in S} U_i$
    3. if $(quantity\_decreased == 0)$ repeat 1 and 2
       otherwise $A = \{j | j \in NCD + CD - S\}$
    4. while $(quantity\_decreased > 0)$ do
       $\forall j \in A, prob\_select_j = \frac{L_j U_j}{\sum_{i \in A} L_i U_i}$
       with $prob\_select_j$, select $s \in A$
       if $quantity\_decreased \geq U_s$
         $n_s = n_s + 1$
         $quantity\_decreased = quantity\_decreased - U_s$
       otherwise with probability $\frac{quantity\_decreased}{U_s}$
         $n_s = n_s + 1$
         $quantity\_decreased = 0$
return offspring

Fig. 6. Pseudo-code of the proposed mutation method

Step 4: Stopping condition. Several stopping criteria can be applied. It is usual to implement a given number of generations and stagnation of the estimated Pareto surface are implemented.

Step 5: Selection method. In order to evolve the current population, μ individuals are selected from the non-dominated configurations stored in $archive_{t+1}$. A cluster method is carried out to preserve diversity in the selection. If $card(archive_{k+1}) < \mu$ then all non-dominated solutions are selected.

Step 6: Generation of next population. Each selected solution is a parent of β new individuals. Then, $\lambda = \mu\beta_{offspring}$.

Let $(n_1, \ldots, n_k)$ be a selected parent to create β offspring. The parent components are classified into two sets, depending on the value of $n_i, i = 1, \ldots, k$:

$$CD = \{\text{component } i | n_i > 0\}$$

$$NCD = \{\text{component } i | n_i = 0\}.$$

The evolution process is a mutation operator that creates new individuals. First, a subset of device types is selected with a mutation probability. This subset of device types decreases its number in the new solution. Then, the required decrease in input is evaluated. Finally, this input is randomly reassigned to other types of devices, which increase in number in the new configuration. The pseudo-code of this mutation operator is given in Figure 6.

Table 1. Frequencies of the synthetic experimental framework

input	20	40	60	80	100
frequency	0.30	0.25	0.20	0.15	0.10

Table 2. Characteristics of synthetic devices

device	D_1	D_2	D_3	D_4	D_5	D_6	D_7	D_8
U_k	50	35	20	15	10	5	2	1
L_k	10	7	4	3	2	1	0.4	0.2
unitary cost	15	14	10	9	7	4	1.8	1

4 Industrial Application

4.1 Origin of the Problem

As stated ealier, the problem presented herein is a real world situation. It is part of a broader study of a company in the energy sector that needed an analysis of the feasibility and dimension of an energy-storage and transformation system.

Due to the strict confidentiality clause required by the company, neither details of the workings of the energy system nor the characteristics and specifications of the devices can be revealed. Each device performs an energy transformation task, in such a way that the input of one type of energy produces the output of another type of energy that is esier to store. The transformation process is not completely efficient, that is, output is lower that input. The relation between these two values is given by an efficiency function, usually, a nonlinear function and of a typical shape is shown in Figure 7. This efficiency function gives non-linearity to the initial optimization problem.

The *randomness of the input* is one of the most significant characteristics of the real problem. Obviously, when the energy input comes from nature, i.e, sun, wind, tide, earth's heat, etc, it is impossible to avoid this random behaviour. Although their stochastic behaviour can be analyzed, these natural resources can not be controlled.

In order to select the best configuration of the system, two criteria are imposed by the company: system global efficiency, and, the cost of creating and running the system.

4.2 Experimental Framework of the Multi-criteria Problem

In order to illustrate the performance of the proposed method, a synthetic problem statement is developed. This synthetic experimental framework uses invented values but has a similar structure to the real problem.

Table 1 shows the energy input frequencies considered, which ranges from 20 to 100. There are 8 energy-transforming devices with the characteristics shown in Table 2.

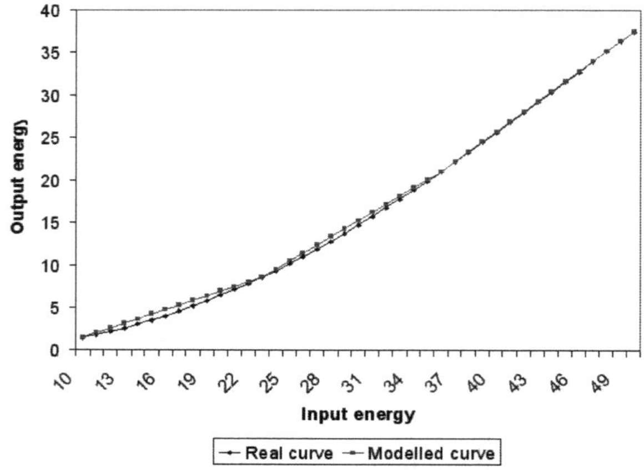

Fig. 7. Real and modelled efficiency curves for the D_1 synthetic device

For simplicity, the same shaped efficiency function is assumed for all the devices (it is in factthe one that occurs the real problem). The efficiency curve is calculated by:

$$output = Ef(input) = \frac{2}{1+3^{1-input/U}} 5^5 10^{-5} \left(-input + 50\frac{input^2}{U} - 25\frac{input^3}{U^2} \right)$$

where U is the maximum input per device, that is, its total capacity U_k. Figure 7 shows the real efficiency curve for D_1. Figure 7 also shows the modelled three linear piecewise function curve, where it must be noted that the loss on the loss-of-efficiency function is barely perceptible.

Therefore, in order to attain maximum expected efficiency and minimum cost it is necessary to determine how many devices of each type should be installed.

4.3 Experimental Results

The main objective of the multi-criteria optimization is to find a set of solutions that provide appreciable efficiency at a reasonable cost. The main drawback of using this kind of models to solve integer problems is the high computational time required. The decision maker agrees to design a method to provide solutions in less than 12 hours. This time constraint guides the parametrization of the evolutionary strategy proposed. More specifically, initial population size is fixed at $m = 25$, mutation probability to $prob\_mut_1 = 0.2$, $\mu = 5$ parents, and $\beta = 25$ offspring.

Then, once these parameters are fixed, the number of evaluated solution in the evolutionary search is around 500. This implies solving around 500 integer linear problems, for which the LINDO API 4.1 is used. In order to control the computational time, the resolution time for each integer linear problem is fixed

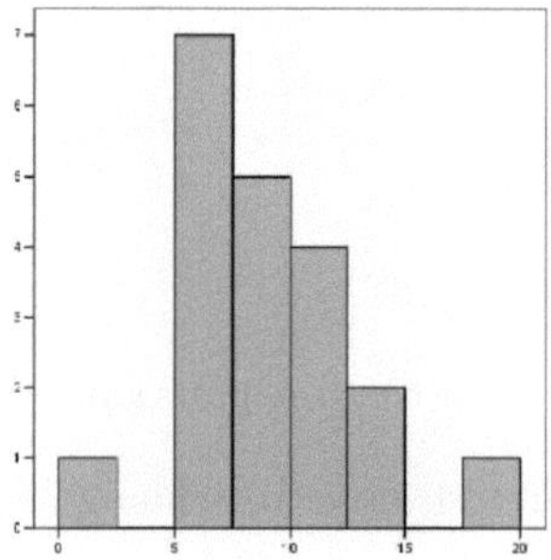

Fig. 8. Histogram of number of populations reached in 20 runs

at 90 seconds. It must be remarked that only in a very few cases this time limit was reached (when considering the size of problems found in practice), and, furthermore, the study of these cases showed that solutions provided in 90 seconds were not far from the optimal ones.

This parametrization is used in an attempt to satisfy standard solution quality measures, such as proximity, diversity, and pertinence. In such a way that:

- proximity: the estimated Pareto frontier is close to the real one
- diversity: a good distribution of solutions in both objectives (efficiency and cost)
- pertinence: the solution set should contain configurations in the area of interest of the decision maker

Since the solution set presented to the decision maker covers a wide range of values for both objectives, the grid method used to obtain the initial population, and the cluster method used to evolve the Pareto frontier assure diversity and pertinence.

In order to check the performance of the proposed method, 20 independent runs of the multi-objective evolutionary method are carried out with the described synthetic devices. For these runs, the stagnation of the Pareto frontier is fixed as the stopping criterion. As a result, 8.65 ± 3.73 (mean $\pm$ st. deviation) populations are required to stagnate the Pareto frontier after 327.62 ± 63.49 seconds. Figure 8 shows the histogram of the number of population reached when the evolutionary algorithm stops.

The results of these 20 independent runs provide us with 20 independent Pareto frontiers. Therefore, a disimilarity measure between two Pareto frontiers, F_r and F_s, is defined as:

$$dif(F_r, F_s) = \frac{\sharp \text{ solutions in } F_r \text{ and not in } F_s + \sharp \text{ solutions in } F_s \text{ and not in } F_r}{\sharp \text{ solutions in } F_r + \sharp \text{ solutions in } F_s}$$

This measure varies between 0 and 1 –the closer to 0, the more similar. The election of this disimilarity measure is based on the fact that not all the Pareto frontiers have the same number of solutions.

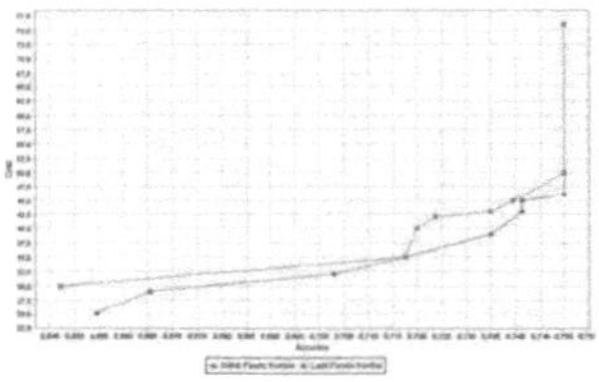
(a) 5 populations

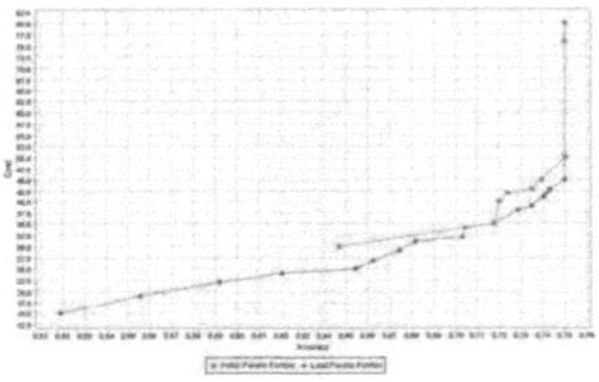
(b) 10 populations

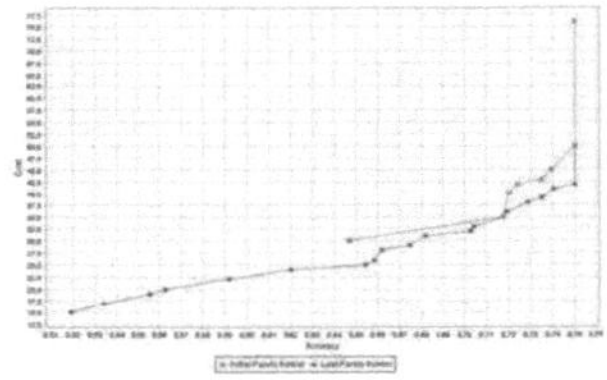
(c) 15 populations

Fig. 9. Spread of Pareto frontier depending on number of population

As a result, 190 disimilarity values are attained where the Pareto frontiers differ at 0.45 ± 0.16 (mean $\pm$ st. deviation). This implies that, by mean, around the 50% of solutions in one Pareto frontier appears in another independent Pareto frontier.

In order to study the spread of the Pareto frontier, 3 indepent runs of the multi-objective evolutionary algorithm is carried out. Now, the stopping criterion is the number of evolved populations. Figure 9 is a graphic representation of the estimated Pareto frontier for these 3 independent runs. The initial and final Pareto frontier are depicted when the algorithm stops evaluating 5, 10 and 15 populations. It can be shown how the Pareto frontier spreads with the number of populations.

It can be observed in Figure 9 that the proposed algorithm evolves the efficient Pareto frontier, including better solutions at each step.

5 Conclusions and Future Work

In this paper, a combination of integer linear programming and evolutionary multi-criteria optimization is proposed to solve a stochastic knapsack problem. The presented methodology has proven to be useful for identifying a set of good feasible configurations. When the decision maker validates the results his intuition about the structure of the system composition is confirmed. Another useful feature of the proposed methodology is that it can usually be implemented in real applications for operational management purposes.

In this work, a new type of stochastic knapsack problem is illustrated by means of an energy transformation process (because of its origins). Nevertheless, similar problems can also be found in other areas.

The design of the algorithm provides a satisfactory answer to a practical problem, while its practical implementation using a restricted parametrization meets key constraints, such as computational time. More research should be done to assess the relationship between the set of parameters and the estimated Pareto frontier, when a wider range of knapsack problems of this type is considered.

A detailed and extensive comparison of the proposed multi-objective evolutionary method with classical first-order evolutionary algorithms (such as genetic algorithms or estimation of distributions algorithms) should be done to check its viability and reliability.

References

1. Steuer, R.: Multiple Criteria Optimization: Theory, Computation and Application. John Wiley and sons (1986)
2. Gandibleux, X., Sevaux, M., Sorensen, K., T'kindt, V., eds.: Metaheuristics for multiobjective optimization. Volume 535 of Lectures Notes in Economics and Mathematical Systems. Springer (2004)
3. Jones, D., Mirrazavi, S., Tamiz, M.: Multi-objective meta-heuristics: An overview of the current state-of-the-art. European Journal of Operational Research **137**(1) (2002) 1–9
4. Michalewicz, Z.: Genetic Algorithms + Data Structures = Evolution Programs. Springer (1999) 3rd, revised and extended edition.
5. Coello, C.A.C.: A comprehensive survey of evolutionary-based multiobjective optimization. Knowledge and Information Systems **1**(3) (1999) 269–308
6. Coello, C.A.C.: Evolutionary multi-objective optimization: A critical review. In Sarker, R., Mohammadian, M., Yao, X., eds.: Evolutionary Optimization. Kluwer Academic Publishers (2002) 117–146
7. Fonseca, C.M., Fleming, P.J.: An overview of evolutionary algorithms in multiobjective optimization. Evolutionary Computation **3**(1) (1995) 1–16
8. Horn, J.: Multicriterion decision making. In Back, T., Fogel, D., Michalewicz, Z., eds.: Handbook of Evolutionary Computation. IOP Publishing and Oxford University Press (1997)
9. Zitzler, E., Deb, K., Thiele, L.: Comparison of multiobjective evolutionary algorithms: Empirical results. Evolutionary Computation **8**(2) (2000) 173–195
10. Pisinger, D., Toth, P.: Knapsack problems. In Du, D., Pardalos, P.M., eds.: Handbook of Combinatorial Optimization. Volume 1. Kluwer Academic Publishers (1998) 299–428
11. Dorigo, M., Stutzle, T.: Ant Colony Optimization. The MIT Press (2004)
12. Glover, F., Laguna, M.: Tabu Search. Kluwer Academic Publishers (1997)
13. Jaskiewicz, A.: On the performance of multiple-objective genetic local search on the 0/1 knapsack problem: a comparative experiment. IEEE Transactions on Evolutionary Computation **6**(4) (2002) 402–412
14. Jaskiewicz, A.: On the computational efficiency of multiple objective metaheuristics. the knapsack problem case study. European Journal of Operational Research **158**(2) (2004) 418–433

Hybridizing Cellular Automata Principles and NSGAII for Multi-objective Design of Urban Water Networks

Yufeng Guo, Edward C. Keedwell, Godfrey A. Walters, and Soon-Thiam Khu

School of Engineering, Computer Science and Mathematics, University of Exeter, Exeter, EX4 4QF, United Kingdom
{Yufeng.Guo,E.C.Keedwell,G.A.Walters,S.T.Khu}@exeter.ac.uk

Abstract. Genetic algorithms are one of the state-of-the-art metaheuristic techniques for optimal design of capital-intensive infrastructural water networks. They are capable of finding near optimal cost solutions to these problems given certain cost and hydraulic parameters. Recently, multi-objective genetic algorithms have become prevalent due to the conflicting nature of these hydraulic and cost objectives. The Pareto-front of solutions obtained enables water engineers to have more flexibility by providing a set of design alternatives. However, multi-objective genetic algorithms tend to require a large number of objective function evaluations to achieve an acceptable Pareto-front. This paper describes a novel hybrid cellular automaton and genetic algorithm approach, called CAMOGA for multi-objective design of urban water networks. The method is applied to four large real-world networks. The results show that CAMOGA can outperform the standard multi-objective genetic algorithm in terms of optimization efficiency and quality of the obtained Pareto fronts.

Keywords: Multi-Objective Optimization, Pipe Networks, Cellular Automata, Genetic Algorithms.

1 Introduction

The optimal design of urban water networks, both water distribution and sewer networks, is of central importance to water industries due to the vast capital investments associated with these underground assets. The problem is normally interpreted as a pipe-sizing problem. A large variety of optimization techniques have been developed for this NP-hard task, including linear and dynamic programming, and recently meta-heuristic algorithms, such as genetic algorithms (GAs), simulated annealing and tabu search.

Amongst these meta-heuristic algorithms, genetic algorithms are the most prevailing algorithm explored [1], [2], [3], [4] tracing back to the mid-nineties. More recently, inline with advances in optimization techniques and also due to the multi-criterion nature of the problem, research in this area has moved from searching for a single "optimal" solution to using multi-objective techniques [5], [6], [7] for more alternatives. Throughout the optimization, a GA is combined with a hydraulic simulator, which evaluates hydraulic performance of each solution attained at every generation of the GA. Although having given promising results, this approach can be

S. Obayashi et al. (Eds.): EMO 2007, LNCS 4403, pp. 546–559, 2007.

time-consuming, especially when designing large networks. One reason is that the hydraulic simulation time is often significantly increased for large networks. Furthermore, as a drawback common to all genetic algorithms, GAs require a large number of objective function evaluations to attain sound solutions, which increases with the complexity of the problem. Since water networks typically have a large number of pipes, the prohibitively high computation cost has become the bottleneck in optimal design practices.

Some attempts have recently been made to overcome this difficulty by accelerating the optimization by way of employing a quicker hydraulic simulator or applying a more efficient optimizer [5], [8], [9], [10]. In Jourdan's work [5], a hybrid optimization method is developed, which integrates machine learning to boost the convergence of a multi-objective genetic algorithm by intermittently intensifying the search on promising areas of the search space. The approach eventually achieves a speed up of the optimization process ranging from 20% to 36%. Keedwell and Khu [10] and Guo et al. [9] developed an efficient cellular automata (CA) based optimizer for designing, respectively, water distribution networks and sewer networks. These approaches can obtain good solutions in a very small number of function evaluations, typically one hundred or less. However as CA are primarily driven by the principle of localism, the cellular automata based approaches cannot guarantee to find global optimal Pareto-fronts.

In this paper, as an extension to our previous work, the authors describe a hybrid optimization method for multi-objective design of urban water networks, called CAMOGA which combines the CA based approaches and a multi-objective GA, specifically a constrained non-dominated sorting genetic algorithm (NSGAII) [11], [12]. The rationale behind this is to exploit the strengths of both approaches, namely the efficiency through localism in the CA and the ability to search for a global optimum from the GA. Two stages can be identified during the optimization. In the first stage, the CA based approach is applied to efficiently obtain a set of preliminary solutions, which are then used as high-quality seeds for the NSGAII implemented in the second stage. In this way, it is expected that a large amount of the initial computation cost incurred at the early stage of GA execution can be saved, and the optimization progress towards global optima can still be assured.

The remainder of the paper is organized as follows. In Section 2, a brief introduction to water network optimization is given. In Section 3, Cellular Automata and CA based optimization methods are described; then the methodology of CAMOGA is explained in detail; In Section 5, case studies are carried out on four real-world problems (two sewer networks and two water distribution networks), and the optimization results are compared with those of NSGAII. A general conclusion is drawn at the end.

2 Multi-objective Water Systems Design

2.1 Water Systems Design

Water systems design has primarily focused on finding a cost-effective solution which minimizes the system cost whilst achieving the requisite system serviceability, such as, sufficient water pressure for high rise buildings and fire fighting in a water distribution system or no flooding under a certain return period storm in a drainage system. The pipe layout, the node connectivity, and the serviceability of the network system are typically assumed to be known. Therefore the problem is simply interpreted as a pipe-sizing problem, namely the decision variables are the diameters of pipes within the network.

2.2 Multi-objective Optimization

As is commonly done, we consider two optimization objectives: the cost and the hydraulic performance of the network, which uses the total head deficit in distribution systems and the total flood volume in sewer systems as a performance indicator. The network cost is calculated based on the cost per unit length associated with the diameter and the length of the pipe.

Since both head deficits in distribution systems and flooding in sewer systems are not generally desirable, a specific value is introduced as a constraint on the hydraulic objective of the Pareto-front. No constraint is applied to the cost objective. Constrained dominance [11] is used to determine the dominance of one solution over another during the optimization.

Finally, the problem is generally expressed by the following formulas:

$$\min_{x\in S} F(x) = (H(x), C(x))^T \quad (1)$$

$$S = \{x : h(x) = 0, g(x) \le 0, a \le x \le b\} . \quad (2)$$

Where $F(x)$ is the objective vector consisting of two objective functions: hydraulic performance $H(x)$ and network cost $C(x)$. S denotes the feasible set constrained by equality and inequality constraints and explicit variable bounds.

There are a large variety of available MOGA algorithms. However, no firm evidence shows that one algorithm can generally outperform the others [13]. As one of the state-of-the-art MOGA algorithms applied for engineering problems, NSGA-II is chosen as the multi-objective genetic algorithm to use in this study. The main advantages of using NSGA-II over other MOGAs are: 1) a strong elitist approach leads to fast convergence; 2) a crowded comparison operator is used to keep diversity and a uniformly spread Pareto-optimal front; 3) an efficient ranking scheme reduces the overall complexity of the algorithm 4) It can easily handle optimization constraints.

3 Cellular Automata Based Optimization

In the last two decades, a heuristic kind of algorithm, called cellular automata (CA) has attracted much attention and has been widely applied to problems in almost all research fields.

3.1 Cellular Automata

The concept of CA was first conceived by von Neumann in the late 1940s [14], [15]. A cellular automaton is a spatially and temporally discrete dynamic system, which consists of a regular lattice of cells. Each cell can have cell states with a finite number of possible values. The states of all cells are updated synchronously at every modeling step according to a set of predefined transition rules. For each cell, the updating only involves the previous state of the cell and of its surrounding neighbors, which are selected according to a neighborhood scheme. The literature suggests that the transition rules and the neighborhood scheme are critical to the performance of a CA model [14], especially the transition rules, a slight alteration in which may result in significantly different behaviors. These two elements are usually problem specific.

CA have historically mostly been utilized as a simulation environment for spatially distributed problems. Recently, the use of CA has been extended to optimization. Although research on CA based optimization is still at its early stage, a number of successful applications have been proposed, such as: for estimating shortest path [16] or trip distribution problems [17], for structural design [18], for computer networks [19], for water distribution system design [10] and for sewer system design [9].

3.2 Cellular Automata Based Water Network Design

In a recent publication [8], [10], a cellular automaton inspired approach (Cellular Automaton for Network Design Algorithm, or CANDA) was developed for the design of water distribution systems. Similarly, we also proposed a CA based approach called CASiNO (Cellular Automata for Sewers in Network Optimization) for the design of sewer systems. For both CANDA and CASiNO, the mechanism was developed on the basis of the definition of cellular automata and hydraulic theory. Therefore they have many features in common but also with some exceptions mainly due to the inherent differences between the two types of systems: (1) distribution systems only deliver pressured flows, and sewer systems mainly deliver gravity flows varying from free-surface to pressured flows; (2) Distribution networks can have loop network connections, but sewer networks can only have a dendritic structure. In general, the mechanism of the CA based approaches for water network optimization is introduced below:

a. Lattice Structure: Each node (network junction) is regarded as one cell, the basic unit of a cellular automaton. The spatial structure of the lattice fully follows the layout of the water network. Because of variable lengths of pipes, instead of having a regular lattice structure as in traditional CA, an irregular two-dimension cellular automata network is constructed.

b. Cell states: Considering the optimization objectives, two variables are set to represent the states of each cell. One is the diameter of the pipe, which reflects the cost information associated with each cell. The other reflects the hydraulic situation at the node, which is pressure head at the node for water distribution systems or flooding volume at the node for sewer systems.

c. Neighborhood scheme: Subject to network layout and hydraulic processes, there is little scope for adopting alternative neighborhood schemes. Hence it involves only the direct surrounding cells.

d. Transition rules: As a critical factor for the quality of CA modeling, the transition rules are defined in a heuristic way which reflects the decisions a water expert might make based on his experience, knowledge and intuition when facing similar hydraulic situations in reality. In line with the optimization objectives, the underling principle of designing those rules is to reduce pipe diameters, namely reduce the network cost, but not violate the constraints on hydraulic performance.

e. Termination criteria: Two criteria are set for terminating the optimization execution: (a) a predefined maximum number of evolving steps of the Cellular Automata model is reached; (b) a homogeneous or periodic behavior is detected. Periodic repetition will start when the solution at the current step is identical to the solution at any previous step.

Despite some exceptions to the definition of a standard cellular automaton, both CANDA and CASiNO preserve all key characteristics of CA: simplicity, homogeneity, parallelism, localism and discreteness both in temporal and in spatial terms.

Unlike traditional optimization algorithms driven by global performance, the CA based approaches do not have an objective function to direct the optimization. Therefore they are not run with any concept of domination, trade-off or even objective function values. Experts' knowledge and judgments have been pre-embedded into the transition rules to direct the optimization towards design objectives, and these rules is executed at a local level, which evolve the cellular automata in a deterministic and self-reproducing pattern. It is this fact which makes it so distinct from the standard optimization approach.

Primarily benefiting from the features of *homogeneity* and *parallelism*, these approaches are proved to be efficient in the optimization of water networks. During the optimization, the heuristic transition rules are homogeneously applied to all cells. At each evolving step, all actions to update the diameters of cells are executed in parallel. It is *parallelism* that also enables the computation cost of the optimization not necessarily to increase with increasing network size. Normally, the CA based approach can attain good solutions requiring only a small number of system evaluations, typically less than 100 evaluations. Furthermore, CANDA and CASiNO are computationally economic in their execution because their transition rules are all designed into a simple "if-then" rule format and no complicated manipulation is associated with the calculations. However, driven by *localism*, the CA based approaches make every update decision based on an exploration limited within the scope of the corresponding local neighborhood and consequently cannot guarantee the solutions to be global optima. This issue is shown to be the major weakness of the CA based approaches for optimization problems.

4 Methodology

4.1 Principles

As discussed in the previous sections, in the optimization of water networks, a GA or a CA based approach each has its own notable advantage but also its own weakness. GAs are directed and sophisticated in finding global optimal solutions, but may entail an unaffordably high computation cost. In contrast, the CA based approach exhibits remarkable optimization efficiency, but inevitably lacks the ability to target the global optima.

Some recent research [20], [21] revealed that seeding GAs with good initial estimates may generate good or even better solutions with faster convergence. However, any algorithm used in the seeding process must be computationally efficient enough not to incur more network simulations than the genetic algorithm itself, or the computational benefits of seeding will be nullified. In light of these findings, Keedwell and Khu [8], [22] described a successful attempt to seed a GA by using CANDA for water distribution network design problems. As an extension to this work, the hybrid approach is here applied to all urban water systems including sewer networks. The approach combines the best features of CA and GA in a multi-objective optimization, hence the name CAMOGA.CA and GA are executed in two consecutive stages during the optimization. The basic CA approach is applied in the first stage to obtain a set of preliminary solutions, which are utilized to seed NSGAII in the second stage. In this way, the CA based approach greatly reduces the high computation cost which is required for NSGAII to reach a similar optimization level for preliminary solutions, and the following GA execution ensures that global-optimal solutions are eventually achieved.

4.2 Execution Process

In the section thus far, general issues relating to optimal design practice have been discussed. The major steps to execute CAMOGA are listed below, and a flowchart of the modeling procedures is illustrated in Fig. 1.

Stage I:

1. Set up the hydraulic simulation model and initialize the system with a given pipe-sizing scenario. A sensible initial scenario is one with minimum pipe diameters for all pipes, as the optimization starting from this scenario can construct a curve analogous to the Pareto front. Theoretically any initial scenario can be used.
2. Simulate the hydraulic model and evaluate the hydraulic performance and capital cost of the system.
3. Check Termination Criteria 1 to decide whether or not to stop the current run of the CA based optimization starting from a certain initial condition. If neither of the criteria is satisfied, the execution goes to Step 4, otherwise Step 5.
4. Implement CANDA/CASiNO to modify the design variables (pipe diameters). Steps 2-4 are repeated until any of Termination Criteria 1 has been satisfied.
5. Record the current solution along with any performance criteria, such as cost and hydraulic performance.
6. Check Termination Criteria 2 to ensure that each pre-defined initial scenario has been employed in an execution of the CA based approach. This is considered as a trigger to switch the optimization from Stage I to Stage II. Otherwise Steps 1-5 are repeated.
7. Create a set of preliminary solutions by selecting non-dominated solutions from all recorded solutions.

Although the process of the CA based optimization is deterministic, it is not possible to predict in advance the number of solutions yielded as it progresses from the initial scenario to meet any of Termination Criteria 1. This is due to the fact that unpredictable and very complex behaviours can be generated by Cellular Automata.

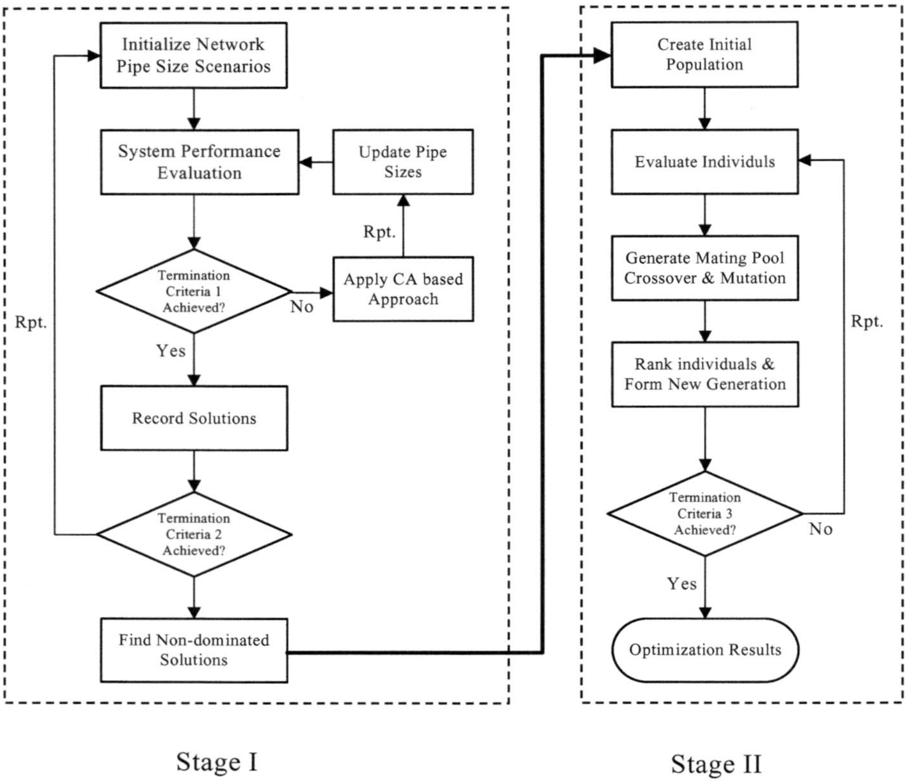

Fig. 1. Execution flow chart of CAMOGA

Stage II:

The preliminary solutions yielded at Stage I are used to seed the initial population of NSGA-II in the second stage. Since only non-dominated solutions are passed from Stage I to Stage II, the number of preliminary solutions is normally smaller than the population size of the GA. The rest of the initial population are therefore random solutions as is normal for a GA. It is in this way that CAMOGA use the same number of network simulations for an optimization as that of the standard NSGA-II. In Stage II, the NSGA-II is executed until Termination Criteria 3 (the required number of generations) is reached.

5 Experiments

5.1 Networks

To illustrate the use of the proposed approach in practice, it is here applied to four large and real world water networks, including two sewer networks and two water

Table 1. Features of the water networks

Network	Pipe No.	Diameter No.	Network Type	Problem Complexity
Case I	112	12	Sewer	12^{112}
Case II	265	10	Sewer	10^{265}
Case III	632	20	Water distribution	20^{632}
Case IV	1277	20	Water distribution	20^{1277}

(a) Sewer Network (Case II) (b) Water Distribution Network (Case IV)

Fig. 2. Layout of the designed network

distribution networks. Detailed information about these designed networks is given in Table 1; and the outlines of two sample networks are illustrated in Fig. 2.

The task for the optimal design is to discover an optimal trade-off surface between cost and constrained hydraulic performance, such as limited head deficit in water distribution network or limited flooding in sewer networks. In order to have an impartial and valid comparison between the hybrid approach and the traditional NSGAII, experiments are implemented in exactly the same fashion for all networks, and as follows:

a. The GA random seed is set to be one of 5 alternatives.

b. Both the traditional NSGAII and CAMOGA use exactly the same parameter configurations (see Table 2), such as, population size, crossover rate and mutation probability, which are determined via a small number of trial and error pre-experiments.

Table 2. Parameter configurations for the NSGAII

Network	Population size	Generation number	Crossover rate	Mutation rate
Case I	150	800	0.90	0.02
Case II	150	1000	0.90	0.02
Case III	100	1000	0.90	0.90
Case IV	100	1000	0.90	0.90

c. The same constraint values on the hydraulic objective of the design problem are applied. The impact of these constraint values on the optimization performance is outside the scope of this study and hence has not been investigated here.
d. Both approaches are run for a same given number of generations.

Since a significantly increased simulation time is required for the big networks in Case III and IV, a relatively smaller number of function evaluations were implemented considering the size of the problems.

In Case III and IV, the mutation rate for the GA is high in comparison with others due to the integer representation used. Variables represented in the more traditional binary form are mutated by the crossover operator which can cut the binary string at any point, thereby creating two new decoded variables. This is not possible with the integer representation, and therefore this is addressed by using a high mutation rate.

5.2 Performance Evaluation Measures

Zitzler, *et al.* [23] indicated that the quality of multi-objective optimization is substantially complex because the optimization goal itself consists of multiple objectives: (a) a minimized distance of the resulting non-dominated set to the Pareto-optimal front (b) a good distribution of solutions on the non-dominated front (c) a wide extent of values covered by the non-dominated solutions for each objective. Therefore it is conceptually difficult to compare the optimality of the results of different methods. As a visual inspection remains most effective and straightforward, a graphical comparison by plotting the final results of both algorithms is firstly made to have a qualitative assessment. Following recent recommendations in the literature, the S-Metric developed by Zitzler and Thiele [24] is employed as a numerical performance indicator. This measurement calculates the size of dominated area in the objective space and in general, combines all three criteria (distance, distribution and extent) into one [23]. The S-Metric requires that a point "W" be defined in objective space which represents the maximum values for each objective. This may not be possible for some optimizations, but for water network optimizations, we define the point W on the cost scale to be where all pipes in the network have maximum diameter and on the hydraulic scale where the hydraulic constraint applies. The S-Metric then computes the union of the rectangles formed between points on the Pareto-curve and the point W. When normalized to the maximum objective values, the S-Metric figure will be in the range 0.0-1.0 where 1.0 is essentially a theoretical rather than practical maximum. This gives a very good objective measure of the optimality of the current solution across the entire optimization, not merely at the end.

5.3 Results and Discussions

During the experiments, each algorithm was run separately with 5 different random seeds for each network. Fig. 3 shows the development trace of the S-Metric values for each of the optimization methods. The numerical comparisons between the S-Metric values achieved by CAMOGA and NSGAII are presented in Table 3.

As can be seen in Table 3 and Fig. 3, in general, more optimal results can be obtained by CAMOGA after the same number of model evaluations. For Case I, the smaller sewer network, the superiority of CAMOGA over the standard NSGAII seems marginal. A similar situation is also experienced for the smaller water distribution network of Case III, in which on an average basis, the two approaches give the same performance. This phenomenon may imply that with the relatively small nature of the constrained problem, CAMOGA and NSGAII converge to the global optima over the given number of model evaluations. However, in all situations, during the optimization up to a considerable number of network simulations, the benefits of using the hybrid approach can be clearly seen. For designing larger networks, NSGAII requires a greater computational effort, probably a significantly larger number of generations with a larger population size, in order to equal the optimality obtained by the CA based approach in Stage I, and to eventually reach global optimal solutions. However, CAMOGA removes onerous computations required at the early stage by NSGAII, and can fully focus on polishing the high quality preliminary solutions. For this reason, the optimization can be more efficient in converging to the global optima. This fact also implies that the CAMOGA methodology actually suits large network problems more than comparatively small ones.

The results show that through a very small number of network simulations, the CA based approach offers a set of solutions with a high level of S-Metrics to seed the following NSGA execution, which is presented in Table 4. For all cases, the CA based approach provides more than 92% of the best performance found using less than 0.1% of the computation cost. This point is most significant for Case I, the smallest network in this study. Its preliminary solutions obtained at Stage I, are only 0.55% worse than the solutions on the final Pareto Front, which is probably the available headroom for the improvement at Stage II. It is also noticed that the increase in complexity of the problems has not adversely influenced the run of the CA based approaches, as they discovered preliminary solutions within a similar level of required computations for all cases. However, the increase in problem complexity does

Table 3. Comparisons of the performance between CAMOGA and NSGAII

Network	Network Simulations	NSGAII Average Normalized S-Metrics	CAMOGA Average Normalized S-Metrics
Case I	120000	0.908	0.911
Case II	150000	0.917	0.946
Case III	100000	0.913	0.913
Case IV	100000	0.886	0.896

Case I

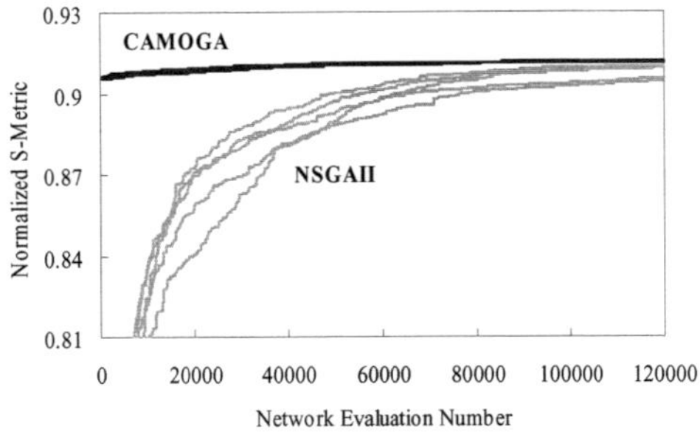

Case II

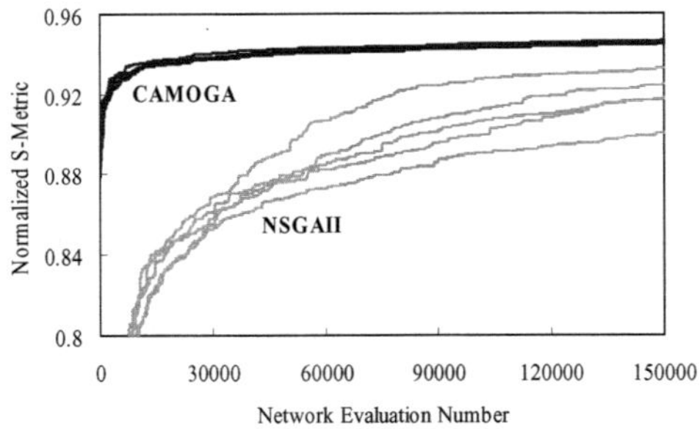

Case III

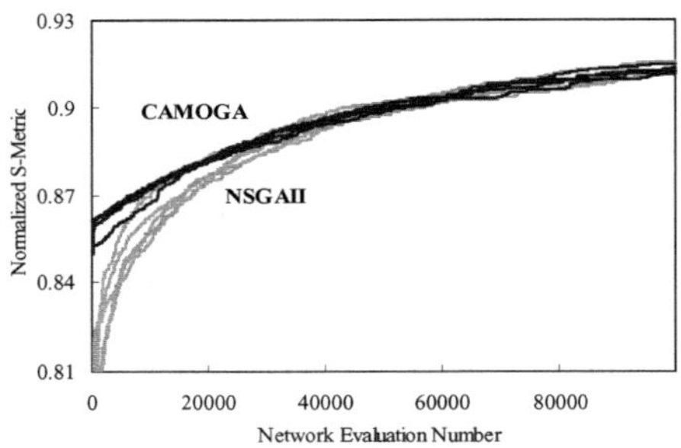

Case IV

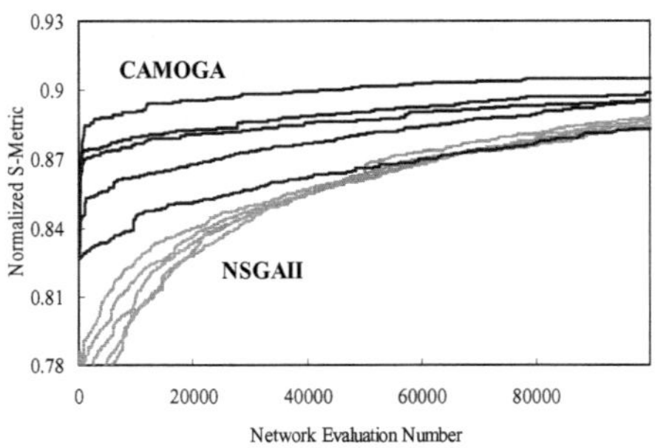

Fig. 3. S-Metric Comparisons of CAMOGA and NSGA-II over 5 Random Seeds

Table 4. Average achievements of the CA based approach at Stage I of CAMOGA

Network	Normalized S-Metrics	Network Evaluation Number
Case I	0.906	39
Case II	0.875	77
Case III	0.850	36
Case IV	0.844	64

amplify the diversity of the S-Metric tracks of different CAMOGA runs, which can be clearly observed from Fig. 3. This is due to a combined effect of the smaller set of preliminary solutions with relatively worse quality, larger headroom available for performance improvement when designing large networks, and the random nature of the evolutionary algorithm.

Multi-objective optimization allows an unprecedented view of the potential cost-benefit trade-offs, and hence offers more freedom to water network designers to have a tailored solution. It also provides the unambiguous evidence that a small shift in the defined hydraulic performance may gain large savings in terms of cost.

Based on the fact that CAMOGA can give an enhanced performance on the water network design problem with no extra computational cost, it will be applicable to networks of high complexity similar to those found in industry. The approach can conceivably tackle difficult multi-objective real world applications within a realistic timeframe, a possibility that is not available with current multi-objective GAs and other formal optimization techniques. This is almost certainly the most attractive aspect of the proposed approach.

6 Conclusions

In this paper, we show that, as an extension to the work previously performed, the CA based optimization approach can be used to seed a modern multi-objective algorithm (NSGA-II) and achieve large-scale computational savings for a given level of performance, or higher quality results for a given computational time.

The approach is clearly proved to be promising for designing large networks as it can obtain reliable results with affordable computation. Its advantage in optimization efficiency and quality over the traditional GA approach seems to increase with the problem size which fits well the actual needs of water industries to design very large scale water networks.

Acknowledgments. This research was partially funded by a UK Engineering and Physical Science Research Council grant GR/R73393 and their support is greatly appreciated. The authors wish to thank Dr. Slobodan Djordjevic and Ewan Group plc. for the provision of the network data in this study, and also Dr. Prasad Tumula for his assistance in network modeling.

References

1. Cembrowicz, R.G.: Evolution Strategies and Genetic Algorithms in Water Supply and Waste Water Systems Design. In: Blain, W.R. et al. (eds.): Water Resources and Distribution, Southampton, United Kingdom (1994) 27-39
2. Dandy, G. C., Simpson, A. R., Murphy, L. J.: An Improved Genetic Algorithm for Pipe Network Optimization. Water Resour. Res., Vol. 32, No. 2, (1996) 449-458
3. Diogo, A.F., Walters, G.A., de Sousa, E.R., Graveto, V.M.: Three-Dimensional Optimization of Urban Drainage Systems. Computer-Aided Civil and Infrastructure Engineering, Vol. 15, (2000) 409-426
4. Savic, D. A., Walters, G.A.: Genetic Algorithms for Least-cost Design of Water Distribution Networks. J. Water Resour. Plann. Mgmt., ASCE, Vol. 123, NO. 2, (1997) 67-77
5. Jourdan, L., Corne, D., Savic, D., Walters, G.: Preliminary Investigation of the `Learnable Evolution Model' for Faster/Better Multiobjective Water Systems Design. In: Coello, C.A.C., et al. (eds.): Evolutionary Multi-Criterion Optimization. Third International Conference, EMO 2005, Springer. Lecture Notes in Computer Science, Vol. 3410, Guanajuato, México, (2005) 841- 855
6. Keedwell, E., Khu, S.T.: A Novel Cellular Automata Approach to Optimal Water Distribution Network Design. J. Computing in Civil Engineering, ASCE, Vol. 20, No. 1, (2006) 49-56
7. Khu, S.T., Keedwell, E.C.: Introducing Flexibility in Upgrading of Water Distribution Network: The New York City Tunnel Network Example. Engineering Optimization, Vol. 37, No. 3, (2005) 291-305
8. Savic, D.A.: Single-objective vs. Multiobjective Optimization for Integrated Decision Support. In: Rizzoli, A.E., et al. (eds.): Integrated Assessment and Decision Support, Proceedings of the First Biennial Meeting of the International Environmental Modelling and Software Society, Lugano, Switzerland, Vol. 1, (2002) 7-12
9. Keedwell, E., Khu, S.T.: A Novel Evolutionary Metaheuristic for the Multi-objective Optimisation of Real-world Water Distribution Networks. Engineering Optimization, Vol. 38, No. 3, (2006) 319-333
10. Guo, Y., Walters, G.A., Khu, S.T., Keedwell, E.C.: A Novel Cellular Automata Based Approach to Storm Sewer Design. Engineering Optimization (in press) (2007)
11. Deb, K., Agrawal, S., Pratap, A., Meyarivan, T.: A Fast Elitist Non-Dominated Sorting Genetic Algorithm for Multi-Objective Optimization: NSGA-II. In: Schoenauer, M. et al. (eds.): Parallel Problem Solving from Nature – PPSN VI, Berlin, Germany (2000) 849–858
12. Deb, K.: Constrained Test Problems for Multi-Objective Evolutionary Optimization. In: The 1st International Conference on Evolutionary Multi-Criterion Optimization, (2001) 284-298
13. Farmani, R., Savic, D.A., Walters, G.A.: Evolutionary Multi-Objective Optimization in Water Distribution Network Design. Engineering Optimization, Vol. 37, No. 2, (2005) 167-185
14. Chopard, B., Droz, M.: Cellular Automata Modelling of Physical Systems. Cambrige University Press, United Kingdom (1998)
15. Ilachinski, A.:Cellular Automata: A Discrete Universe. World Scientific, Singapore (2001)
16. Adamatzky, A.I.: Computation of Shortest Path in Cellular Automata. Mathematical and Computer Modelling, Vol. 23, No. 4, (1996) 105-113

17. El Dessouki, W.M., Fathi, Y., Rouphail, N.: Meta-optimization Using Cellular Automata with Application to The Combined Trip Distribution and Assignment System Optimal problem. Computer-Aided Civil and Infrastructure Engineering, Vol. 16, No. 6, (2001) 384-396
18. Kita, E., Toyoda, T.: Structural Design Using Cellular Automata. Structural and Multidisciplinary Optimization, Vol. 19, (2000) 64-73
19. Shuai, D., Zhao, H. A New Generalized Cellular Automata Approach to Optimization of Fast Packet Switching. Computer Networks, Vol. 45, (2004) 399-419
20. Harik, G.R., Goldberg, D. E.: Linkage Learning Through Probabilistic Expression. Computer Methods in Applied Mechanics and Engineering, Vol. 186, No. 2, (2000) 295-310
21. Neppalli, V.R., Chen, C.L., Gupta, J.N.D.: Genetic Algorithms for the Two Stage Bicriteria Flowshop Problem. European Journal of Operational Research, Vol. 95, (1996) 356-373
22. Keedwell, E., Khu, S.T.: A Hybrid Genetic Algorithm for the Design of Water Distribution Networks, Engineering Applications of Artificial Intelligence, Vol. 18, No. 4, (2005) 461-472
23. Zitzler, E., Deb, K., Thiele, L. Comparison of Multiobjective Evolutionary Algorithms: Empirical Results. Evolutionary Computation, Vol. 8, No. 2, (2000) 173-195
24. Zitzler E., Thiele L.: Multiobjective Optimization Using Evolutionary Algorithms - A Comparative Case Study. Study. In: Parallel Problem Solving from Nature (PPSN-V), Amsterdam, The Netherlands, (1998) 292-301

Using Multiobjective Evolutionary Algorithms to Assess Biological Simulation Models

Rié Komuro[1], Joel H. Reynolds[2], and E. David Ford[3]

[1] Bioengineering Institute, University of Auckland
Level 6, 70 Symonds Street, Auckland, New Zealand
r.komuro@auckland.ac.nz
[2] Division of Natural Resources, U.S. Fish & Wildlife Service
1011 E. Tudor Road, MS 221, Anchorage, Alaska 99503, USA
joel_reynolds@fws.gov
[3] College of Forest Resources, University of Washington
Box 352100, Seattle, Washington 98195-2100, USA
edford@u.washington.edu

Abstract. We introduce an important general Multiobjective Evolutionary Algorithm (MOEA) application – assessment of mechanistic simulation models in biology. These models are often developed to investigate the processes underlying biological phenomena. The proposed model structure must be assessed to reveal if it adequately describes the phenomenon. Objective functions are defined to measure how well the simulations reproduce specific phenomenon features. They may be continuous or binary-valued, e.g. constraints, depending on the quality and quantity of phenomenon data. Assessment requires estimating and exploring the model's Pareto frontier. To illustrate the problem, we assess a model of shoot growth in pine trees using an elitist MOEA based on Nondominated Sorting in Genetic Algorithms. The algorithm uses the partition induced on the parameter space by the binary-valued objectives. Repeating the assessment with tighter constraints revealed model structure improvements required for a more accurate simulation of the biological phenomenon.

Keywords: Multiobjective optimization, Pareto frontier, binary discrepancy measures, process model, mechanistic model, model assessment, structural inference, elitism.

1 Introduction

Simulation models are often used to investigate the processes or mechanisms underlying biological or ecological phenomena. For example, the competition process among tree crowns in a dense forest stand was explored by building a spatially-explicit model of crown growth based on simple rules of resource acquisition and utilization at the foliage and branch level [1]. Developing such models involves a considerable degree of uncertainty in the selection of both the model's components and their representation detail. Merely fitting the model to data

S. Obayashi et al. (Eds.): EMO 2007, LNCS 4403, pp. 560–574, 2007.

does not guarantee that the proposed model structure adequately describes the phenomenon [2,3]. Rather, the model must be assessed to reveal which aspects of the phenomenon it can produce and which it cannot [2].

Model assessment involves solving a multiobjective optimization problem (MOOP). The model developers select the phenomenon features that the model must reproduce to be considered an adequate system description, then they define an objective function for each feature [2,4]. For example, the crown competition model was assessed for its ability to simultaneously reproduce specific features of tree crown shape and stand level growth rates. The Pareto frontier is then estimated to reveal the model performance tradeoffs among objectives.

This problem can differ from the usual engineering design MOOP in many ways:

- The decision maker is the model developer and the goal is to reveal the inadequacies in the proposed model structure's reproduction of selected features.
- Conceptually, the model is a black-box transforming a parameterization (possible solution) X to a point in the objective space. Model complexity prevents expressing the objective functions directly in terms of X. Rather, the objective functions are expressed in terms of features of the model predictions, $M(X)$. Evaluating a solution X requires running the simulation model, which may be quite computationally intensive.
- Depending on the quality and quantity of data available regarding the phenomenon, an objective function may be a continuous measure of discrepancy between a feature of the model predictions and a target value or simply a binary-valued function evaluating whether the predicted feature falls within an acceptable range [2]. An assessment of a model of stem cell development in cats used six continuous objective functions [5] while the crown competition model was assessed with respect to ten binary objectives due to data limitations [2]. Depending on the objectives, the assessment might be viewed as either a MOOP or a constraint-satisfaction problem (CSP) [6]. However, since the objectives can only be directly expressed in terms of $M(X)$ rather than X, standard CSP methods [6] are inapplicable. We use constraint in the remainder to refer to our nonstandard setting more properly defined as a binary-valued objective function.
- Even if continuous objective functions are used, the model developer will likely summarize the final Pareto frontier by defining an acceptable threshold value for each objective, thus partitioning the Pareto optimal set into sections of the parameter space producing identical constraint satisfaction [2,5].
- The existence of model deficiencies is revealed by (i) constraints that are never satisfied (features not adequately reproducible by the model) or (ii) combinations of constraints that can never be simultaneously satisfied. Maximizing the number of satisfied constraints does not reveal the information required for model assessment.
- Computational demands generally limit model assessments to 4 - 15 objectives and models for which 100,000 complete simulations can be conducted on the order of days to weeks [2,4,5,7].

The models of interest generally are dynamic nonlinear multivariate functions of varying complexity. While the crown competition model used simple equations to predict resource acquisition and branch growth at the foliage unit level (cube 10 cm on a side) at each time step, the resulting stand dynamics were a complex synthesis of the spatially and temporally varying conditions [1,2].

Following a heuristic detailed in [2], deficiency sources are located through detailed investigation of how model performance varies across the Pareto optimal set. For example, the assessment of the canopy competition model revealed (i) a set of solutions that satisfied all constraints except two that focused on the growth rate of dominant trees in the stand, and (ii) six other sets of solutions that satisfied one or the other of the growth rate constraints along with differing subsets of the other constraints. The model simulations revealed that the growth rate constraints were only satisfied due to a mathematical artifact in the objective function formulation - the simulations did not reproduce the intended feature [2]. Rather, the simulations revealed an unnatural clustering of growth rates due to the model assumption that certain foliage characteristics varied at the scale of the tree rather than that of the foliage unit [2]. Revising the model to eliminate this structural deficiency greatly improved its ability to reproduce the phenomenon (see [2] for further details).

Model assessment's focus on tradeoffs and unachieved objectives make multiobjective Evolutionary Algorithms (MOEAs) the most promising approach to these MOOPs. Motivated by the success of MOEAs in engineering and other fields [8,9,10,11,12,13,14], one was developed in 1997 for the assessment of ecological models [2] based on the NSGA algorithm [11]. Recognizing the importance of *elitism* for maintaining fit parameterizations across generations [15,16,17,18,19], which we believe increases with increasing numbers of objectives, the original algorithm was recently revised to include elitism [7].

We illustrate the model assessment process and its features using a model of hourly increments of extension of the leading shoot of a conifer tree [4]. All objectives in this application are binary valued, e.g., constraints, so we introduce an elitist algorithm based on the partition the constraints induce over the parameter space [7]. Fairly wide initial threshold values for the objectives returned a Pareto frontier with one set of solutions satisfying all objectives (constraints). The assessment was repeated with smaller thresholds until the Pareto frontier no longer consisted of a single group, ultimately revealing a hysteresis effect in the phenomena that the proposed model structure could not reproduce [4].

1.1 The Pareto Frontier in Model Assessment

Model assessment is built around investigating the Pareto frontier of the multiobjective minimization problem defined by the trio [2]:

$M(\,)$	the simulation model, producing multivariate output,
$\mathbf{X} \subseteq \mathbb{R}^m$	the parameter space, and
$\mathbf{F}() = (F_1(M()), \ldots, F_n(M())) \in \mathbb{R}^n$	the vector objective function, $\mathbb{R}^m \to \mathbb{R}^n$, measuring n distinct features of model performance.

Each model parameterization X is a potential solution of the simultaneous minimization problem defined by $\mathbf{F}(\ )$. The technical task is to reveal if any parameterization can simultaneously minimize every component of $\mathbf{F}(\ )$; generally none can. In such a case, model assessment investigates the model predictions from the solutions in the Pareto optimal set for insight into the sources of the proposed model structure's deficiencies [2].

Definition 1. *parameterization* X dominates X' $(X \succ X' \text{ or } X' \prec X) \Longleftrightarrow$

$$\begin{aligned} &\forall i, \quad 1 \leq i \leq n, \quad F_i(M(X)) \leq F_i(M(X')) \quad \textit{and} \\ &\exists i, \quad 1 \leq i \leq n, \quad \textit{such that } F_i(M(X)) < F_i(M(X')). \end{aligned}$$

X *is* non-dominant *to* X' $(X \parallel X') \Longleftrightarrow$

$$\exists i, j, \ i \neq j \ \textit{ such that } F_i(M(X)) < F_i(M(X')) \ \textit{ and } \ F_j(M(X')) < F_j(M(X)).$$

The *Pareto optimal set*, $\mathcal{P}_{\mathbf{F}}(\mathbf{X}) \subseteq \mathbf{X}$, is the set of all non-dominated solutions with respect to the vector of objective functions $\mathbf{F}$, i.e. the set of solutions which are mutually non-dominated and are not dominated by any other X in the search range. The *Pareto frontier*, $\mathcal{F}_{\mathbf{F}}$, is the associated set of their objective vectors,

$$\mathcal{F}_{\mathbf{F}}(\mathbf{X}) = \{ \ (F_1(M(X)), \ldots, F_n(M(X))) \mid X \in \mathcal{P}_{\mathbf{F}}(\mathbf{X}) \ \} \in \mathbb{R}^n.$$

1.2 Model Assessment Objective Functions

The objective functions may be continuous measures of discrepancy between a model prediction and a target value. Often a lack of quantitative data on the phenomena limits the discrepancy measure to a simple binary function [20,2] - does the prediction fall within an acceptable range of values? E.g., the objective becomes a constraint. We will use the term constraint below, but in the context of model assessment we view this as simply one end of the spectrum of objective function definitions, a spectrum driven by data availability and the nature of the feature and model, e.g., deterministic or stochastic. Fuzzy objective functions would be possible. In general, a model assessment could use both continuous objectives and constraints, but for now we assume all objectives have similar quality - all are continuous or all are constraints. We are specifically interested in MOEA algorithms that can handle either type of objective function.

When the objective functions are all constraints, the objective vector produced by a given model parameterization is termed an *assessment vector* [2]. In this case the Pareto optimal set is partitioned into *Pareto groups*, sets of parameterizations that generate the same assessment vector in the Pareto frontier. Consider assessing six model parameterizations with respect to four constraints, $\mathbf{F} = (F_1, F_2, F_3, F_4)$, producing the results in Table 1. Model predictions using parameterization X_1 satisfy all objectives except F_4, i.e. the predictions fall within the target ranges for (F_1, F_2, F_3). Those from X_2 fail to satisfy F_3 and F_4 so $X_1 \succ X_2$; similarly $X_1 \succ X_4$. For these six parameterizations, $\mathcal{P}_{\mathbf{F}}(\mathbf{X}) = \{X_1, X_3, X_5, X_6\}$ and consists of three Pareto groups, $\{X_1\}, \{X_3\}$,

Table 1. Assessment vectors for six different model parameterizations, where performance was assessed with regard to the four binary valued objectives: $F_1, \ldots, F_4$. The values are discrepancy measures or distances, so an objective with value 0 was satisfied.

Parameterizations	Criteria $F_1(M(X))$	$F_2(M(X))$	$F_3(M(X))$	$F_4(M(X))$
X_1	0	0	0	1
X_2	0	0	1	1
X_3	1	1	0	0
X_4	1	1	0	1
X_5	0	1	1	0
X_6	0	1	1	0

and $\{X_5, X_6\}$. It can be represented in terms of these groups: $\mathcal{P}_{\mathbf{F}}(\mathbf{X}) = \mathcal{G}_{\mathbf{F}} = \{\ \{X_1\}, \{X_3\}, \{X_5, X_6\}\ \}$. Likewise, the Pareto frontier $\mathcal{F}_{\mathbf{F}}$ can be represented in terms of its unique assessment vectors as $\mathcal{F}_{\mathbf{F}} = \mathcal{A}_{\mathbf{F}} = \{\ (0,0,0,1), (1,1,0,0), (0,1,1,0)\ \}$.

Selection and elitism components of MOEAs can be defined either directly in terms of the parameterizations or in terms of the Pareto groups. Operating on the groups allows for random subsampling of parameterizations associated with each group, reducing the storage and computing costs associated with elites.

If the parameterizations in Table 1 were the full parameter space, from $\mathcal{A}_{\mathbf{F}} = \{\ (0,0,0,1), (1,1,0,0), (0,1,1,0)\ \}$ we'd conclude that the model structure M was deficient as it could not simultaneously satisfy all four objectives. The model assessment, following the process in [2], would explore the simulations of the parameterizations in $\mathcal{P}_{\mathbf{F}}(\mathbf{X})$ for insight into (i) the details of *how* $M(X_1)$ failed to satisfy criterion F_4 - overestimate? underestimate? etc.; (ii) the details of how $M(X_3)$, $M(X_5)$, and $M(X_6)$ satisfied F_4 but never F_2, etc.. The process eventually leads to insight into the *why* underlying the deficiency, i.e., its source [2].

We introduce the general optimization problem arising in model assessment, introduce an MOEA developed for this problem, and use the MOEA to assess a process model of shoot extension in conifer trees. A more detailed presentation of the full assessment process is given in [2,4].

2 MOEA Algorithm for Model Assessment

General model assessment software, called Pareto_Evolve [21] (`http://faculty.washington.edu/edford/software.html`), was developed in 1997 based on a modification of the Non-Dominated Sorting Genetic Algorithm [11] (`www.iitk.ac.in/kangal/deb.shtml`). At each of up to $g_{\max}$ generations, the non-dominated rank of each of N individuals (real-valued parameterizations) was calculated with respect to the vector of objective functions $\mathbf{F}$. An individual's chromosome representation was a vector of double precision variables, each with a constrained feasible space (Figure 2b). Each individual's fitness was initially assigned based on their non-dominated rank then modified by their niche count in the objective space. Rank 1 individuals were assigned initial fitness of 100; rank i+1 individuals were assigned initial fitness of $w \times \min\{\text{fitness of rank } i \text{ individuals}\}$, where w was a user defined value in $(0, 1]$ [21]. The fitness values were used in roulette selection to select with replacement N parents from the current population [22]. Multi-point crossover [23] and non-uniform mutation [24] genetic operators were used to stochastically generate N offspring. The probability of a parent being assigned the crossover operator linearly decreased from 0.67 to 0.0 as $g \to g_{\max}$; $Prob(\text{mutation}) = 1 - Prob(\text{crossover})$. The process repeated while $g \leq g_{\max}$ and none of the population members simultaneously satisfied all of the criteria.

An archive was maintained of all non-dominated individuals and their assessment vectors discovered during the search. Specifically, at generation g the archive contained the Pareto optimal set of all previous generations' Pareto optimal sets, as well as its corresponding Pareto frontier:

$$\text{Archive}_g^{\mathcal{P}} = \mathcal{P}_{\mathbf{F}} \left(\bigcup_{k=0}^{g} \mathcal{P}_{\mathbf{F}}(\text{Pop}_k) \right), \tag{1}$$

$$\text{Archive}_g^{\mathcal{F}} = \mathcal{F}_{\mathbf{F}} \left(\text{Archive}_g^{\mathcal{P}} \right), \tag{2}$$

where Pop_k denotes the search population at generation k.

2.1 Elitism

The original algorithm was revised to allow elites to participate in breeding (Figure 1). Rather than select parents from the archive, a secondary population of elites was created from which parents could be selected. The use of binary valued objectives allowed elites to be selected from the current generation's non-dominated individuals using a two-stage process: Pareto groups in the current generation's Pareto optimal set were selected following the rules below, then a subsample of individuals were randomly selected from each of these elite Pareto groups. The two-stage rules for elite selection maintained diversity at the Pareto group level, controlled growth of the total number of elites, and limited the variation among Pareto groups in terms of the number of individuals representing each group. An individual remained in the elite population until dominated by a new elite.

A Pareto group was selected as an elite if it satisfied one of the following conditions (phrased in terms of binary valued objectives or constraints):

1. Its assessment vector satisfied the most constraints of any Pareto group in $\mathcal{P}_{\mathbf{F}}(\text{Pop}_g)$.

Randomly generate Pop_0 from feasible space
$g := 0$
repeat
  $\text{Pop}_g = \{\mathcal{P}^1, \mathcal{P}^2, \ldots\} :=$ pareto_ranking(Pop_g)
  for all i **do**
    fitness_assignment($\mathcal{P}^i$)
  end for
  $\text{E}_g :=$ select_elites($\mathcal{P}^1$)
  $\text{P}_g :=$ select_parents(E_g)
  if $|\text{P}_g| < p$ **then**
    $\text{P}_g := \text{P}_g \cup$ select_parents(Pop_g)
  end if
  $\text{Pop}_{g+1} :=$ breed(P_g)
  $g := g + 1$
until all criteria are achieved or $g = g_{\max}$

Fig. 1. Pseudo-code for elitist Pareto_Evolve. $\mathcal{P}^i$ is the set of individuals with non-dominated rank i, and E_g and P_g are the elite and parent pools at generation g, respectively.

2. It dominated at least one individual in $\mathcal{P}_{\mathbf{F}}(\text{Pop}_{g-1})$.
3. Its assessment vector satisfied a constraint unsatisfied by any other Pareto group in the current elite population.

Rather than limit the size of the elite population, the maximum number of individuals selected from each elite Pareto group was limited. If a Pareto group selected as an elite had fewer than $l = N/10$ individuals then they all became elites; otherwise, l individuals were randomly selected as elites.

At each generation, a total of N parents were selected from the elite population, $\mathbf{E}_g$, and the current search population, Pop_g. Let $\nu = |\mathbf{E}_g|$, the number of elites at generation g. A random draw of $h \sim$ binomial (ν, prob $= 0.5 \times N/\max(\nu, N)$) determined the number of parents randomly selected from the elites; if $h \geq N$, h was set at N. All elites were equally likely to be selected as parents. Note that this selected parents from elites without replacement. If $h < N$, then $N - h$ parents were selected from Pop_g using roulette selection as a function of their fitness, calculated as in the original algorithm.

Parents selected from Pop_g were assigned genetic operators as in the original algorithm. Parents selected from $\mathbf{E}_g$ were all assigned crossover; the companion parent did not have to be an elite. This provided a more efficient exploration of the frontier when using constraints as the nonuniform mutation operator of the original algorithm, in the later generations, tended to produce offspring with assessment vectors identical to those of their elite parents [7].

3 Assessing a Model of Shoot Growth

The elitist algorithm was used to estimate the Pareto frontier of a model of shoot growth in pine trees.

3.1 Ecological Phenomenon and Observations

The growing tip of a conifer branch extends and contracts from hour to hour as influenced both by current and recent environmental conditions, such as temperature and solar radiation. However, the relationship between growth and these environmental conditions is not fully known. Extension and contraction of a Sitka spruce (*Picea sitchensis*) shoot were automatically measured every hour using an electro-mechanical sensor (Figure 3) [25]. A model (below) was proposed for the functional dependence of the observed expansion and contraction on solar radiation, temperature, and transpiration calculated using the Penman-Monteith equation [26]. We summarize the assessment of how well the model reproduced the phenomenon's features over six days [4].

3.2 Process Model

An earlier investigation of daily measurements, rather than hourly, identified a time series model using both daily temperature, at one and two day lags, and daily solar radiation, at two and three day lags, as regressors [27]. The hourly observations revealed a more detailed pattern of extension, particularly contraction of the shoot during daylight hours on sunny days followed by expansion at night (Figure 3). This raised important questions on whether contraction was more rapid than re-expansion [4]. To investigate these questions the daily effects model was revised to an hourly scale and expanded to include a component for contraction and expansion based on development and release of a water deficit [28,29]. Letting S_t denote the shoot growth at time t (other terms are defined in Table 2), the initial model was:

$$
\begin{aligned}
S_t = {} & x_1 \cdot \left(\sum_{k=1}^{24} T_{t-k} \right) \Big/ 24 \; - \; x_2 \cdot \left(\sum_{k=25}^{48} T_{t-k} \right) \Big/ 24 \\
& + \; x_3 \cdot \sum_{k=25}^{48} R_{t-k} \; + \; x_4 \cdot \sum_{k=49}^{72} R_{t-k} \\
& - \begin{cases} x_5 \cdot \Delta_t D \cdot \sum_{k=1}^{24} S_k^* & (*) \\[2ex] x_7 \cdot \Delta_t D \cdot \sum_{k=1}^{24} S_k^* & (**) \end{cases}
\end{aligned}
$$

with

$$\Delta_t D \equiv D_t - D_{t-1} = W_t - U_t = W_t - x_6 \cdot D_{t-1}.$$

The first and second unknown coefficients, x_1, x_2, control the influence on hourly extension of the average temperatures for the 24 hour periods starting 24 and 48 hours prior, respectively. The third and fourth coefficients control the influence

Table 2. Variables in the conifer shoot growth model

t	hour(s) from start of simulation
S_t	hourly simulated growth rate (mm/hour) at time t ($S^*_{t'}$ is rate at time t' on previous day)
T_t	temperature (^oC) at time t;
R_t	hourly solar radiation (MJ/m^2/hour) at time t;
D_t	hourly water deficit (mm/hour) at time t,
W_t	hourly water transpiration (mm/hour) at time t;
U_t	hourly water uptake (mm/hour) from the soil at time t;

of the total solar radiation for the 24 hour periods starting 48 and 72 hours prior, respectively. Details are in Ford et al. [27].

The term marked (*) describes the change in contraction or expansion due to change in plant water status. Estimated hourly transpiration, W_t, accumulates a water deficit D_t. The water deficit decreases, in turn, by uptake of water from the tree bole and soil at a rate of $x_6 \cdot D_{t-1}$. The change in deficit is then applied to the estimated amount of soft tissue available for contraction on the expanding shoot, represented by the sum of growth over the previous 24 hours. The initial model assumed the same effect of changing plant water status on expansion and contraction, x_5, regardless of whether water deficit was increasing (plant tending toward contraction) or decreasing (plant tending toward expansion) [7]. The assessment, summarized below, revealed a need for hysteresis. The model was revised to let the rates vary: term (*) was used with decreasing deficit, term (**) was used with increasing deficit.

3.3 Assessment Objective Functions

Standard methods of calibrating and assessing time series models focus strictly on univariate summaries of overall fit. Process models, though, should be able to simultaneously simulate different features of the series, thus the need for an assessment using multiple objectives. The goal is to develop a model of contraction and expansion that adequately explains the data [30].

The initial model (without the (**) term) was assessed for how well it reproduced the observed mean shoot extension for four periods each day: early morning (hours 2 – 6), late morning (8 – 12), afternoon (12 – 16), and late evening (20 – 24). These represent, respectively, the pre-dawn expansion period when water deficit is lowest; the late morning period of maximum contraction; the afternoon when recovery from contraction starts; and late evening when maximum expansion occurs (Figure 3). Comparisons were quantified as |Predicted mean growth – Observed mean growth| during each four hour period, thus assessing both timing and magnitude. The objective functions were binary-valued. The same threshold value was used for each objective as there was no basis for treating the periods differently; in general this is not required.

The model was assessed separately for the days 179–181 and 182–184 as the latter period exhibited lower minima and greater amplitude of variation. There was an increase in soil moisture tension and a decrease in fine root length over this period [31], suggesting additional processes may have come into play. The water deficit equilibrated during day 178, so those observations were not used in the assessment.

The assessment process employed here differed somewhat from the more general process laid out in [2] since the objectives were conceptually and quantitatively similarly, allowing use of a common threshold value. A large initial threshold value was chosen so that the Pareto frontier would contain a single Pareto group satisfying all the objectives: 1.5 mmh^{-1}. The assessment was repeated with smaller threshold values, each time resulting in an approximate Pareto frontier using the new threshold, until the Pareto frontier no longer contained a single group satisfying all the objectives. That boundary was located by reducing thresholds by 0.1 mmh^{-1}. The whole process was repeated for days 182–184.

Each search used a population size of 100 and maximum generation number of 1000. The maximum generation number was deemed sufficient for convergence based on experiments using the initial model and the original non-elitist algorithm [7]. The search terminated early if all twelve objectives were achieved by a single Pareto optimal solution. Five realizations of each search were conducted. The threshold value was defined to be unachievable when none of the realizations discovered a single Pareto optimal solution satisfying all objectives.

4 Results

4.1 Model Assessment for Days 179–181

The smallest threshold value at which all objectives were satisfied was 0.41 mmh^{-1} (Figure 2a). With a value of 0.40 mmh^{-1}, each realization of the Pareto frontier contained five or six groups; four were common to every realizations in terms of their assessment vectors, the others differed only slightly across realizations. The four common groups each failed to satisfy only a single objective: one overestimated pre-midnight expansion on day 180, the period with the relatively extreme hourly growth rate (Figure 3), the others underestimated early a.m. expansion on 181, underestimated late a.m. contraction on 181, or underestimated early p.m. expansion on 181. That is, the model either failed to predict the extreme expansion late in day 180 but adequately predicted everything else, or it adequately predicted the extreme expansion on 180 at the cost of failing to predict observed growth in one of the three subsequent time periods [4]. All realizations had one or two other groups that adequately predicted the extreme expansion period by failing to predict growth in an immediately prior period on day 180: overestimating late a.m. contraction and/or early p.m. expansion.

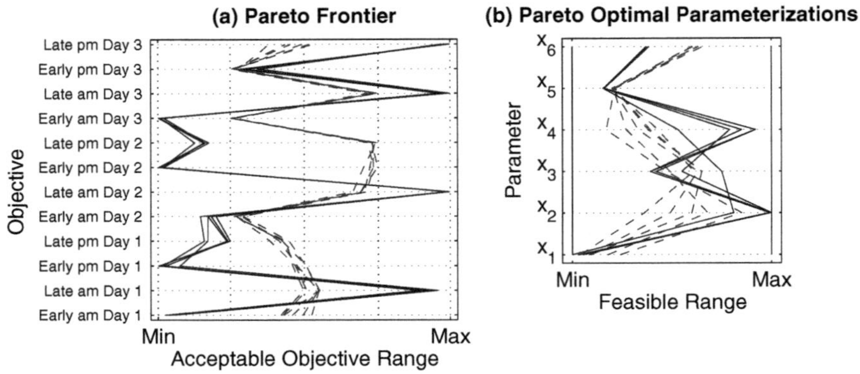

Fig. 2. (a) Parallel plot of the initial model's 12-dimensional Pareto frontier for days 179–181 (dashed curves) and 182–184 (solid curves) using the smallest threshold values (dashed vertical lines) at which all objectives were satisfied: $[-0.41, 0.41]$ for 178–181 (second vertical lines from each side), $[-0.81, 0.81]$ for 182–184 (outermost vertical lines). The center vertical marks zero, perfect prediction. For each period, the independent search realizations discovered approximately identical Pareto frontiers (curves of same line style). (b) Parallel plot of the relative parameter values of the initial model's Pareto optimal set for days 179–181 (dashed) and 182–184 (solid). For each period, the five independent searches discovered approximately identical associated Pareto optimal sets (curves of each line style). Parameters x_1 through x_4 control the overall trend in growth while x_5 and x_6 influence the diurnal cycles of expansion and contraction. All parameter ranges have a minimum of 0; maximum feasible value for parameters x_1 through x_6 were, respectively, 0.4, 0.1, 0.1, 0.1, 1, and 1.

4.2 Model Assessment for Days 182–184

The smallest threshold value at which all objectives were satisfied was 0.81 mmh^{-1}, almost twice the value required for days 179–181. With a value of 0.80 mmh^{-1}, each realization of the Pareto frontier contained five or six groups; three were common to every realization in terms of their assessment vectors, the others were approximately identical across realizations. Every realization revealed one group that only failed to satisfy one objective: it underestimated the late morning contraction on day 183, the period with the most extreme hourly contraction (Figure 3). The other groups all adequately predicted this period of extreme contraction but at the cost of adequately predicting any early morning expansions and most of the afternoon or late evening expansions.

All realizations also had one or two groups that underestimated the afternoon and/or overestimated evening expansion rate on the last day, which was less rapid than on previous days (Figure 3). Further investigation suggested a change in the eco-physiological processes began at this point in the study [4].

4.3 Model Revision

The parameter values associated with approximate Pareto frontiers were consistent across the five realizations within each period but differed between the periods, mainly in x_4 and x_6 (Figure 2b). Differences in parameters $x_1, \ldots, x_4$ cause change in the overall trend of growth but do not contribute directly to the diurnal cycles of contraction and expansion. Thus the differences of parameter values of x_6 (water uptake per unit deficit) between periods demonstrated how the process differences between the periods were accommodated by the model fitting rather than the model structure. Detailed investigations of each period's approximate Pareto frontier when using thresholds that were too small revealed that the inability to adequately predict extreme expansion or contraction rates without causing poor prediction of the preceding or following objectives was due to (i) a need for different growth rates depending on whether water deficit was increasing or decreasing, (ii) a need to change the period of calculation for "recent growth available for contraction" (ΣS_k^*) from the previous day to the period 23 h of previous day to 6h on day of interest, and (iii) a need to smooth the rate of water uptake over three hours [4]. The inability to capture the pre-midnight expansion on day 184 led to further investigation of the longer data series, which revealed a change in extension pattern from that point onward; further assessments did not use this last objective [4].

The model was revised to address each of the identified needs and the assessment process repeated using eleven objectives. For days 179–181 the revised model produced a Pareto frontier with a single group satisfying all objectives using threshold values as low as 0.40 mmh^{-1}. For days 182–184 a single group was found using threshold values as lows as 0.42 mmh^{-1}. The revised model reproduces the observed process better than the original model, especially for the contraction periods of the last three days (Figure 3). However, it does not do as well on the contraction period of day 181, suggesting existence of further deficiencies (see Komuro et al. [4] for further revision). The threshold values suggest the revised model structure accommodated both periods equally well, though this may be due to either the revision or the removal of the last objective.

5 Discussion

Model assessment is an essential phase in the development of any mechanistic simulation model [2,30]. The key task of assessment, estimating the Pareto frontier defined by the proposed model structure and the selected objective functions, is technically challenging but eminently well suited to the flexibility of MOEA methods. This has been demonstrated by successful application of the algorithm presented here, and its precursor, in assessing a variety of models [2,4,5].

Model assessment, in general, is not identical to constraint satisfaction. While the example here was presented in terms of constraint satisfaction, a model assessment can use continuous objective functions rather than binary-valued ones. This simply requires sufficient phenomenon observations to justify defining a specific target value for each feature / objective. For example, the shoot growth

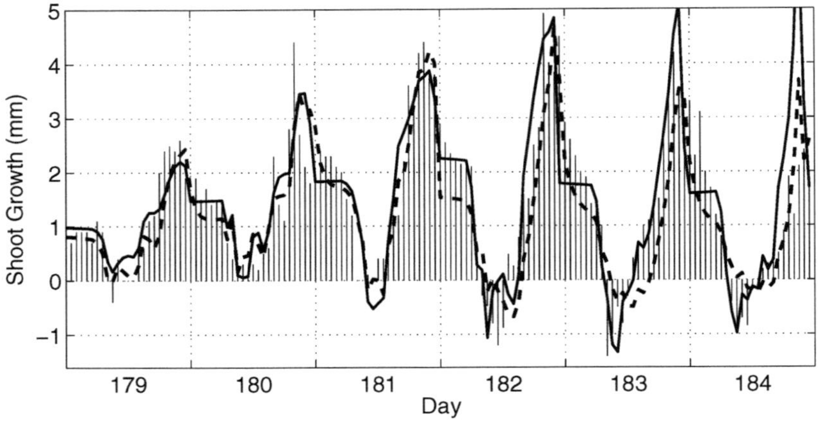

Fig. 3. Examples of simulated results using the original (dashed curve) and revised (solid curve) models with the assessment vector achieving all the twelve criteria. The bar graph shows the measured growth.

model assessment could have estimated the smallest threshold allowing complete constraint satisfaction by defining each objective as a continuous measure of discrepancy from the observed target value, running a single optimization search to produce the Pareto frontier, then a posteriori determining the smallest threshold supporting complete satisfaction; see Reynolds and Golinelli [5] for an example. In most assessments the objectives measure conceptually distinct features that are incommensurable, disallowing the approach employed here.

The example illustrated how improvements from model assessment stem from investigating the objectives, or combinations of objectives, that the model does a poor job achieving. Simply maximizing the number of constraints satisfied, or otherwise aggregating the MOOP to a univariate problem, eliminates the exact information model assessment requires.

The elitism mechanism presented here requires further investigation. Defining elites in terms of Pareto groups rather than individual parameterizations, with individuals randomly selected from each elite group, makes it easier to maintain diversity of representation of the current generation's Pareto optimal set among the next generation's breeding population and reduces some of the memory burden of a growing elite population. The effectiveness of the different rules must be investigated, as well as its performance when continuous objective functions are used. This would likely greatly increase the number of Pareto groups, greatly reduce their cardinalities, and likely reduce the effectiveness of the elitism component as currently defined.

We hope this example serves to interest the MOEA community into this very interesting, and quite general, problem in scientific inference.

Acknowledgments. This work was supported in part by a grant from the U.S. National Science Foundation DBI-0129042 and in part by the United States

Environmental Protection Agency through agreement GR82517301-0 to the University of Washington.

References

1. Sorrensen-Cothern, K.A., Ford, E.D., Sprugel, D.: A process based model of competition for light incorporating plasticity through modular foliage and crown development. Ecological Monographs **63** (1993) 277–304
2. Reynolds, J.H., Ford, E.D.: Multi-criteria assessment of ecological process models. Ecology **80**(2) (1999) 538–553
3. Wood, S.N., Thomas, M.B.: Super-sensitivity to structure in biological models. Proceedings of the Royal Society of London B **266** (1999) 565–570
4. Komuro, R., Ford, E.D., Reynolds, J.H.: The use of multi-criteria assessment in developing a process model. Ecological Modelling **197**(3-4) (2006) 320–330
5. Reynolds, J.H., Golinelli, D.: Multi-criteria inference for process models: structural and parametric inference for a stochastic model of feline hematopoeisis. In: 2004 JSM Proceedings, American Statistical Association (2004) 2978–2986
6. Miguel, I., Shen, Q.: Hard, flexible and dynamic constraint satisfaction. The Knowledge Engineering Review **14**(3) (1999) 199–220
7. Komuro, R.: Multi-Objective Evolutionary Algorithms for Ecological Process Models. PhD thesis, University of Washington, Seattle, Washington (2005)
8. Kursawe, F.: A variant of evolution strategies for vector optimization. In Schwefel, H.P., Männer, R., eds.: Parallel Problem Solving from Nature, Springer-Verlag, Berlin, Germany (1991) 193–197
9. Fonseca, C.M., Fleming, P.J.: Genetic algorithms for multiobjective optimization: Formulation, discussion and generalization. In Forrest, S., ed.: Proceedings of the Fifth International Conference on Genetic Algorithms, Morgan Kaufmann Publishers Inc., San Mateo, California (1993) 416–423
10. Horn, J., Nafpliotis, N.: Multiobjective optimization using the niched Pareto genetic algorithm. IlliGAL Report 93005, Illinois Genetic Algorithm Laboratory (IlliGAL), University of Illinois at Urbana-Champaign, Urbana, Illinois (1993)
11. Srinivas, N., Deb, K.: Multiobjective optimization using nondominated sorting in genetic algorithms. Evolutionary Computation **2**(3) (1994) 221–248
12. Zitzler, E., Thiele, L.: Multiobjective evolutionary algorithms: A comparative case study and the strength Pareto approach. IEEE Transactions on Evolutionary Computation **3**(4) (1999) 257–271
13. Deb, K.: Multi-Objective Optimization Using Evolutionary Algorithms. John Wiley & Sons, Inc., New York, New York (2001)
14. Coello, C.A.C.: Guest editorial: Special issue on evolutionary multiobjective optimization. IEEE Transactions on Evolutionary Computation **7**(2) (2003) 97–99
15. Parks, G.T., Miller, I.: Selective breeding in a multiobjective genetic algorithm. In Eiben, A.E., Bäck, T., Schoenauer, M., Schwefel, H.P., eds.: Parallel Problem Solving from Nature - PPSN V, Springer-Verlag, Berlin, Germany (1998) 250–259
16. Knowles, J., Corne, D.: The Pareto archived evolution strategy: A new baseline algorithm for pareto multiobjective optimisation. In Angeline, P.J., Michalewicz, Z., Schoenauer, M., Yao, X., Zalzala, A., eds.: Proceedings of the Congress on Evolutionary Computation. Volume 1., IEEE Press, Piscataway, New Jersey (1999) 98–105

17. Deb, K., Agrawal, S., Pratap, A., Meyarivan, T.: A first elitist non-dominated sorting genetic algorithm for multi-objective optimization: NSGA-II. In Schoenauer, M., Deb, K., Rudolph, G., Yao, X., Lutton, E., Merelo, J.J., Schwefel, H.P., eds.: Parallel Problem Solving from Nature - PPSN VI, Springer-Verlag, Berlin, Germany (2000) 849–858
18. Zitzler, E., Laumanns, M., Thiele, L.: SPEA2: Improving the strength pareto evolutionary algorithm. TIK Report 103, Computer Engineering and Communication Networks Lab, Swiss Federal Institute of Technology, Zurich, Switzerland (2001)
19. Costa, L., Oliveira, P.: An adaptive sharing elitist evolution strategy for multiobjective optimization. Evolutionary Computation **11**(4) (2003) 417–438
20. Hornberger, G.M., Spear, R.C.: An approach to the preliminary analysis of environmental systems. Journal of Environmental Management **12** (1981) 7–18
21. Ford, E.D., Turley, M., Reynolds, J.H.: User's Manual: The Pareto Optimal Model Assessment Cycle Using Evolutionary Computation. National Research Center for Statistics and the Environment, Seattle, Washington. (2001)
22. Fogel, D.B.: Evolutionary Computation: Toward a New Philosophy of Machine Intelligence. 2nd edn. IEEE Press, Piscataway, New Jersey (2000)
23. Syswerda, G.: Uniform crossover in genetic algorithms. In Schaffer, J.D., ed.: Proceedings of the Third International Conference on Genetic Algorithms, Morgan Kaufmann Publishers Inc., San Mateo, California (1989) 2–9
24. Michalewicz, Z.: Genetic Algorithms + Data Structures = Evolutionary Programs. 3rd edn. Springer-Verlag, London, UK (1996)
25. Milne, R., Smith, S.K., Ford, E.D.: An automatic system for measuring shoot length in Sitka spruce and other plant species. Applied Ecology **14** (1977) 523–529
26. Monteith, J.L.: Evaporation and environment. In Fogg, G.E., ed.: The State and Movement of Water in Living Organisms (Symposia of the Society for Experimental Biology, Number XIX), Cambridge University Press, UK (1965) 205–234
27. Ford, E.D., Deans, J.D., Milne, R.: Shoot extension in *Picea sitchensis* I. seasonal variation within a forest canopy. Annals of Botany **60** (1987) 531–542
28. Milne, R., Ford, E.D., Deans, J.D.: Time lags in the water relations of Sitka spruce. Forest Ecology and Management **5** (1983) 1–25
29. Milne, R., Deans, J.D., Ford, E.D., Jarvis, P.G., Leverenz, J., Whitehead, D.: A comparison of two methods of estimating transpiration rates from a Sitka spruce plantation. Boundary-Layer Meteorology **32** (1985) 155–175
30. Wiegand, T., Jeltsch, F., Hanski, I., Grimm, V.: Using pattern-oriented modeling for revealing hidden information: a key for reconciling ecological theory and conservation practice. Oikos **100** (2003) 209–222
31. Deans, J.D.: Fluctuations of the soil environment and fine root growth in a young Sitka spruce plantation. Plant Soil **52**(2) (1979) 195–208

Improving Computational Mechanics Optimum Design Using Helper Objectives: An Application in Frame Bar Structures

David Greiner, José M. Emperador, Gabriel Winter, and Blas Galván

Institute of Intelligent Systems and Numerical Applications in Engineering (IUSIANI), 35017, University of Las Palmas de Gran Canaria, Spain
{dgreiner, jemperador}@iusiani.ulpgc.es, {gabw, bgalvan}@step.es

Abstract. Considering evolutionary multiobjective algorithms for improving single objective optimization problems is focused in this work on introducing the concept of helper objectives in a computational mechanics problem: the constrained mass minimization in real discrete frame bar structures optimum design. The number of different cross-section types of the structure is proposed as a helper objective. It provides a discrete functional landscape where the non-dominated frontier is constituted of a low number of discrete isolated points. Therefore, the population diversity treatment becomes a key point in the multiobjective approach performance. Two different-sized test cases, four mutation rates and two codifications (binary and gray) are considered in the performance analysis of four algorithms: single-objective elitist evolutionary algorithm, NSGAII, SPEA2 and DENSEA. Results show how an appropriate multiobjective approach that makes use of the proposed helper objective outperforms the single objective optimization in terms of average final solutions and enhanced robustness related to mutation rate variations.

Keywords: Helper objectives, Multiobjectivization, Structural optimization, Evolutionary multiobjective optimization, population diversity.

1 Introduction

The recently developed evolutionary multiobjective algorithms [7][10] have shown a capacity to solve optimization problems in countless fields in science and engineering, and frequently without any increase in cost compared to single objective optimization [9]. Moreover, new possibilities recently opened up by evolutionary multiobjective optimization tools have been proposed to improve the search in single objective problems [26][33]. They include concepts such as 'multiobjectivization' or 'helper objectives'.

In Knowles et al. [32], two ways are proposed to diminish the number of local optima in a search (multiobjectivization): 1) By adding new objectives which allow the problem to be solved as a multiobjective one; 2) By decomposing the problem into simpler sub-problems, whose solutions are optimizing objectives in the multiobjective problem, with the purpose of increasing the number of paths to the

S. Obayashi et al. (Eds.): EMO 2007, LNCS 4403, pp. 575–589, 2007.

global optimum, which are not opened in the single-criteria optimization. Therefore, the non-dominated solutions are coincident with the optima of the original problem. Two examples are solved in the referenced article: the hierarchical-if-and-only-if-function and the travelling salesman problem. They are solved using the first and second aforementioned strategies, respectively, showing improvements with respect to the single objective strategy.

In Abbass and Deb [1], the introduction of a new criterion helping the population diversity is proposed. The following additional criteria are suggested and analyzed: maximization of the objective function inverse, the maximization or minimization of a random value assigned to each individual in its creation and the maximization of the age of the chromosome. With no mutation, the multiobjective approach maintains better the genetic diversity and surpasses the single criteria results. Also, in E. de Jong et al. [31] and in S. Bleuer et al. [4], an additional criterion is added to genetic programming optimization for increasing the population diversity, in order to solve simultaneously the minimization of the tree size and the resolution of the n-parity problem. Other authors have also focused on using diversity-based new objectives in the context of dynamic environments [5][6].

In M. Jensen [29][30], it is emphasized how the inclusion of new additional objectives or criteria called 'helper objectives' and the solving of the problem as a multiobjective one can lead to the decrease or even the disappearance of certain difficulties inherent to the single objective optimization, such as: 1º) avoidance of local minima; 2º) maintenance of diversity at suitable levels; 3º) identification of good building blocks. Jensen applies this new concept to the job shop scheduling problem, where minimizations of individual jobs are generated dynamically with the algorithm evolution, as helper objectives. Using this strategy, the obtained solution is improved in a set of 18 test cases compared with the single criteria optimization.

This new perspective that evolutionary multiobjective optimization algorithms can provide is applied to different types of problems, such as mathematical and classical test functions, combinatorial problems (job shop, travelling salesman) or computational biology and bioinformatics [27]. Developed helper objectives are focused on the diversity item, as well as specific ad-hoc objectives, inherent to each particular problem.

In this work, we introduce the application of the 'helper objectives' concept to a real design optimization problem belonging to the field of computational mechanics. A new type of helper objective is proposed with advantages for the problem resolution, concretely applied to the bar frame optimum design problem of constrained mass minimization. The single criteria optimization is compared with three multiobjective evolutionary algorithms in two different-sized well referenced test cases, considering four different mutation rates and two distinct chromosome codifications.

The organization of this paper is as follows: First, the frame structural optimum design problem is described. Section 3 explains the new proposal of helper objectives. Section 4 sets out the evolutionary multiobjective algorithms considered in the analysis. Following this, both test cases are shown in section 5, continuing with the results and discussion in section 6. Finally, the paper ends with the conclusions section.

2 Frame Structural Optimum Design

The optimization of bar structures has been performed using evolutionary algorithms since the origins of their application to optimum engineering design [15]. The use of real cross section types as variables was introduced in [37], and one of the pioneering works of frame bar structures is explained in [24].

The handled frame structural design problem has one single objective: the minimization of the constrained mass. It is considered in order to minimize the raw material cost of the final structural design. The constraints take into account those conditions that allow the designed frame to carry out its task without collapsing or deforming excessively. They are as follows, taking into account the Spanish design code recommendations:

a) Stresses of the bars: where the limit stress depends on the frame material and the comparing stress takes into account the axial and shearing stresses by means of the shear effort, and also the bending effort (a common value for steel is of 260 MPa), for each bar:

$$\sigma_{co} - \sigma_{\lim} \leq 0 \tag{1}$$

b) Compressive slenderness limit: where the $\lambda_{\lim}$ value is 200 (in order to include the buckling effect, the evaluation of the β factor is based on Julian and Lawrence criteria). For each bar:

$$\lambda - \lambda_{\lim} \leq 0 \tag{2}$$

c) Displacements of joints (in each of the three possible degrees of freedom) or middle points of bars. In the test cases, the maximum vertical displacement of each beam is limited (in the multiobjective test case the maximum vertical displacement of the beams is L / 500):

$$u_{co} - u_{\lim} \leq 0 \tag{3}$$

The objective function is the constrained mass. It is shown in equation 4:

$$ObjectiveFunction_1 = \left[\sum_{i=1}^{Nbars} A_i \cdot \rho_i \cdot l_i \right] [1 + k \cdot \sum_{j=1}^{Nviols} (viol_j - 1)] \tag{4}$$

where:

A_i = bar i cross-section area; ρ_i = bar i density; l_i = length of bar i; k = constant that regulates the equivalence between mass and constraints; $viol_j$ = for each of the violated constraints (stress, displacement or slenderness), is the quotient between the value that violates the constraint limit (violated constraint value) and its reference limit. The constraints reference limits are chosen according to the Spanish design codes. So, constraints if violated are integrated into the mass of the whole structure as a penalty depending on the amount of the violation (for each constraint violation the total mass is incremented):

$$viol_j = \frac{Violated\ Constraint\ Value}{Constraint\ Limit} \tag{5}$$

Handling with frames implies that many bar geometric magnitudes have to be considered: area, the modulus of section and the relation of beam height are required to evaluate the normal stresses; web area is required to evaluate shearing stresses; moment of inertia is required to evaluate the medium span displacement; the radius of gyration is required to consider the buckling effect. Real design is performed using real cross-section types -developing a discrete optimization problem with direct real application-, whose magnitudes are stored in a vector associated to each cross-section type, constituting a database. Therefore, the codification of each solution implies for each bar a discrete value that is assigned to the cross-section type order in the database. It contains the corresponding geometric magnitudes of the cross-section needed to perform the structural calculation, implying the resolution of a finite element modelling -with Hermite approximation functions-, and its associated linear equation system.

3 Helper Objective: A New Proposal

The evolutionary multiobjective optimization of bar structures was first handled in [25]. Other related pioneering works in this field are [12][14][38]. Nevertheless, the focus was mainly centered on the simultaneous minimization of constrained mass and the displacement in certain points of the structure as objectives.

The purpose of this work is the presentation and analysis of a new type of 'helper objective' and its application in the field of computational mechanics optimum design, concretely, in the single-criteria constrained mass minimization problem of real cross-section type frame bar structures design. Helper objectives should accomplish the requirement of being in conflict with the principal objective; otherwise, the performed search would be equivalent to a single criteria optimization. The proposed helper objective should act by increasing the population diversity and guaranteeing the inclusion of the single criteria optimum in the Pareto frontier. With this in mind, we propose as helper objective the number of different Cross-Section Types (CST) of the bar structure. With this objective, each structure is classified in terms of the quantity of different CST included in it. It could also be considered as a measure of the contained structural diversity in terms of the considered cross-section type database. Therefore, the minimum number of CST (absence of diversity in the variable space) is one, and the maximum number of CST is the minimum value between the structural bars number and the database item number. Although here the problem belongs to the field of computational mechanics, the suggested helper objective is also valid for other kinds of optimization problems where the variables are coded through a sorted database whose items selection defines the chromosome individual.

The multiobjective problem of simultaneous minimization of constrained mass and the number of different CST is in itself of interest. It was solved using a weighted approach in [13] and has been successfully solved recently using evolutionary multiobjective algorithms, e.g. in [17]. This second objective poses a condition of constructive order with special relevancy in structures with a high number of bars [25]. It helps towards a better quality control during the execution of the building site and is a factor that has also been recently related to the life cost cycle optimization of steel structures [34].

What we introduce in this work is the comparison of the results of single criteria optimization of constrained mass versus the multiobjective problem considering the different number of CST as helper objective. Since our aim is the minimization of constrained mass, in this case will it be advantageous to use the multiobjective approach instead? Both objectives accomplish the requirement to be in conflict together: as the global optimum (minimum constrained mass structure) has a concrete number of different CST, a decrease in this value is due to an increase in the structural mass. Moreover, it is also guaranteed that the single-criteria global optimum is included in the Pareto frontier. The fact that additional solutions are maintained in the population during the multiobjective algorithm evolution provides richer population diversity than the single criteria approach. Further analysis of the proposed multiobjective approach effects follows in the next sections.

4 Evolutionary Multiobjective Algorithms

Three different evolutionary multiobjective algorithms are included in the analysis. The first two are among the most efficient algorithms used in many fields of sciences and engineering optimum design problems: SPEA2 [41] and NSGAII [11]. They belong to the called 'second generation' [8] of evolutionary multiobjective algorithms, and excel due to their characteristics of elitist algorithms, governed by the Pareto non-dominance criterion, and with operators that homogeneously distribute the solutions along the non-dominated surface or front.

The multiobjective approach handled in this work has been previously solved using evolutionary algorithms. The associated functional space is discrete with a low number of points which are determined by the number of different CST, and therefore the importance of an adequate diversity is enhanced. So, the inclusion of duplicate elimination has been shown to be a key factor in the multiobjective algorithm performance, as can be seen for example in [17][18][22][23]. It seems to be necessary to emphasize the creation and maintenance of population diversity to avoid the scarce number of non-dominated solution saturating the population, obstructing and lengthening the evolution and leading to a premature convergence. This duplicate deletion operator has also been applied to other kinds of problems and has been seen to have certain advantages. The following examples in particular are noteworthy: the SEAMO algorithm [35][39] –which includes duplicate elimination-, applied to the multiobjective knapsack problem; the MNK-landscape problem in [2], using duplicate elimination in the NSGAII; the knapsack problem in [36], where the overlapped solutions are deleted; and also the Combative Accretion Model suggested in [3], which includes the duplicate elimination operator and was compared with NSGAII in a set of multiobjective test functions with promising results.

There are problems (such as the multiobjective approach we are dealing with here), where the number of Pareto optimum solutions is smaller than the population size. This kind of problem usually has one or more discrete fitness functions with reduced values, limiting the capacity to cover the functional space homogeneously by the solutions. A discrete space with a limited and low number of solutions has influence not only on the final non-dominated front obtained, but also during the whole algorithm progress. In this case, the performance of the operators that distribute solutions along the front (crowding distance, clustering, etc.) is drastically reduced, because an accumulation of solutions in these discrete points happens without discrimination possibility. So, in the multiobjective structural problem dealt with here, the second objective function, the number of cross-section types, is a discrete one, having as limits 1 and the minimum value between the number of bars of the truss and the number of different cross-section types considered in the reference database.

With the above in mind, the DENSEA Algorithm (*D*uplicate *E*limination *N*on-dominated *S*orting *E*volutionary *A*lgorithm) was presented and applied successfully to bar frame optimum design in [16] and to bar truss optimum design in [19]. It is the third evolutionary multiobjective algorithm considered in the analysis of helper objective performance. DENSEA emphasizes the creation and maintenance of population diversity and is conceived as an improvement medium for the previously explained difficulties of having a non-dominated solution reduced set in functional space with respect to the population size. DENSEA is based on the non-domination sorting criterion selection and has incorporated some elitism, but it is characterized by offering population diversity maintenance based on various characteristics: 1) Deletion of duplicate solutions; 2) Replacement of these duplicate solutions; 3) Replacement selection of population of next generation. After the initial population creation, individuals' fitnesses are evaluated. The obtained population is ordered by the non-domination criterion, with the distribution operator along the front (secondary criterion when solutions belong to the same front) explained later. After this sorting, each individual has a linear selection probability by Roulette Wheel Selection, which determines the individuals that are selected for crossover and mutation. These constitute the offspring population. This population is also ordered by the non-domination criterion after evaluation. DENSEA has specifically a deletion operator for duplicate solutions: the algorithm deletes the accumulated duplicate solutions due to the reduced non-dominated solutions quantity in the functional space. The replacement of each deleted solution is performed by inserting the individual that has the same ordering in the second half of the population until the completion of 50% of the population size (N/2). In that way, the inclusion of diverse solutions replacing duplicates is fostered, helping to maintain the population diversity. This filtering process is implemented both in the parent and offspring population. Merging both filtered populations (each of size N/2) generates the parent population of the next generation, size N. So, the individuals belonging to the parent population of the next generation are selected by using an elitism that also guarantees the renovation of 50% of the individuals each generation, promoting population diversity and without losing genetic information of the best solutions (because the non-dominated solutions number is small due to the problem characteristics). The operator that distributes the solutions, after the non-dominated ordering calculation, along the non-dominated surface has been implemented with consideration for the specific characteristics of the problem handled and with a search for simplicity. However, any other second generation evolutionary multiobjective algorithms scheme could have been adopted. Considering the discrete property of the helper objective (number of different cross-section types), the considered operator acts as follows: A crowding distance is calculated taking into account only the distances of the discrete fitness function, simplifying its calculation and computational cost (it only requires the subtraction of natural numbers). This operator is based on the discrete nature of the functional search space, and considers as crowding distance among solutions the difference of the previous and next values of the fitness function number of different cross section types of each individual.

The results and their analysis, having taken into account the three previously mentioned evolutionary multiobjective algorithms (NSGAII, SPEA2 and DENSEA), are shown in section 6.

5 Test Cases

Two different sized frame bar structures are considered as test cases. The smaller one is a four bar sized frame which is called test case X. The second one is a fifty-five sized frame named test case Y. In both cases, the buckling effect and the own gravitational load of the structure bars have been considered. They are based on a reference problem in [28], where a continuous optimization without buckling effect is handled. The density and elasticity modulus are the typical values of steel: 7850 kg/m^3 and $2.1 \cdot 10^5$ MPa, respectively.

Test Case X is graphically represented in figure 1. The spot lengths are in meters, and there is a maximum displacement constraint in the second bar middle point, of length/300. All the bars belong to the IPE cross-section type series (consecutively from IPE-80 to IPE-500), with admissible stresses of 240 MPa.

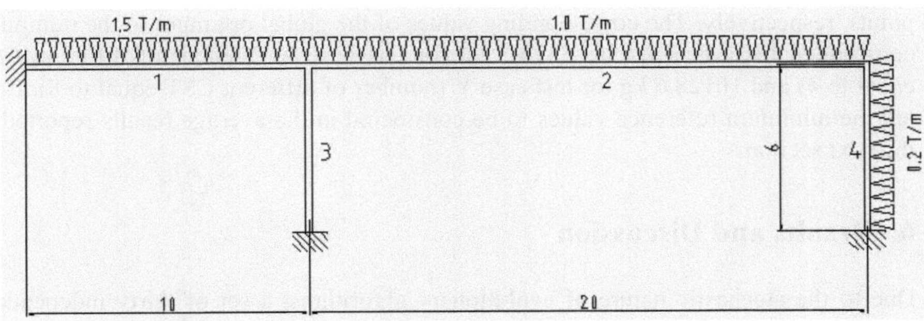

Fig. 1. Frame Test Case X

Test case Y is represented in figure 2. The figure includes the numbering of the bars and the precise loads in Tons. In addition, in every beam there is a uniform load of 39945 N/m. The lengths of the beams are 5.6 m and the heights of the columns are 2.80 m. The columns belong to the HEB cross-section type series (consecutively from HEB-100 to HEB-450), and the beams belong to the IPE cross-section type series (consecutively from IPE-80 to IPE-500); with admissible stresses of 200 MPa and 220 MPa, respectively. The maximum vertical displacement in the middle point of each beam is established at $l/300 = 1.86 \cdot 10^{-2}$ m.

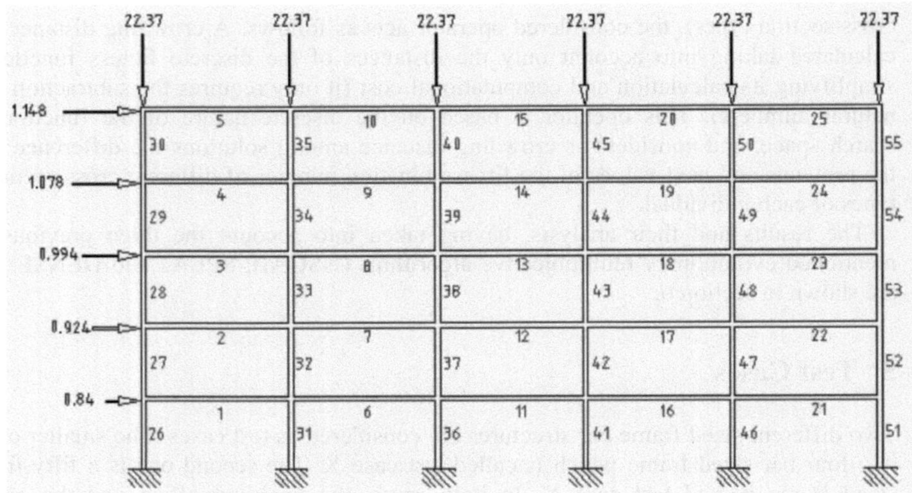

Fig. 2. Frame Test Case Y

The optimal reported solutions of each test case can be consulted in the following references: [16][23] for test case X and [16][21] for test case Y. In particular, the number of solutions that the Pareto frontiers could integrate in our test cases is between 1 and 4 (number of bars of the structure) in the first test case X and between 1 and 32 (number of different cross-section types included in the reference database) in the second test case Y. Their Pareto frontiers comprise 4 and 8 isolated single points, respectively. The corresponding values of the global optimum of the minimum constrained mass designs are 3324.3 kg for test case X (number of different CST equal to 4) and 10128.6 kg for test case Y (number of different CST equal to 8). Both are the minimum reference values to be considered in the average results reported in the next section.

6 Results and Discussion

Due to the stochastic nature of evolutionary algorithms, a set of thirty independent runs has been performed to compare four different algorithms: NSGAII, SPEA2, DENSEA and the single-objective approach. The single objective evolutionary algorithm is an elitist approach, where the best two individuals are mandatorily maintained and whose selection is an order-based approach with roulette wheel assignation of probabilities.

In every case, a population size of 50 individuals and uniform crossover with 1.0 crossover rate are fixed. Four different mutation rates have been analyzed in every test case: 0.8%, 1.5%, 3% and 6% for test case X; 0.4%, 0.8%, 1.5% and 3% for test case Y. Both have been selected in relation to the chromosome length of each structure and following previous results [20]. Two different codifications have also been analyzed, both based on 0s and 1s due to the discrete nature of the problem variables: 1) the standard binary code and 2) the standard binary reflected gray code as described in [40]. The gray codification has shown better behaviour in previous works, both in the single-criteria [23] and multiobjective approaches [21].

The main purpose of the analysis is the comparison of algorithm performances, and especially between the single and multiobjective (that includes the proposed helper objective: number of different CST) approaches, focusing on the minimization of the constrained structural mass; therefore, results are presented in terms of constrained mass in kg.

The average results over 30 runs are presented graphically in figures 3 to 6. The number of fitness function evaluations is represented along the x-axis, and along the y-axis the averaged constrained mass during the whole convergence of the algorithm. Algorithms Single approach, NSGAII, SPEA2 and DENSEA are represented in the upper left, upper right, lower left and lower right parts of each figure, respectively. Each part contains the evolution of the four mutation rates with, from lower to higher in each test case, the continuous line (lower), circle, lined asterisk and square (higher). The numerical averaged values at the stop criterion ($2 \cdot 10^4$ and $2 \cdot 10^5$ fitness function evaluations for test case X and Y, respectively) are shown in tables 1 (case X) and 2 (case Y), where the best constrained mass value among algorithms corresponding to each codification and mutation rate is bold typed.

The importance of an adequate diversity treatment is more noticeable in the first test case X, where both the Pareto frontier and the search space (associated with the chromosome length) are of smaller size than in test case Y. Therefore, we can observe how, in figures 3 and 4 and in table 1, DENSEA achieves remarkable results versus the other algorithms, with certain difficulties being seen in the other multiobjective approaches. However, since the search space and Pareto frontier are bigger in test case Y, the diversity treatment is not so critical, and the three multiobjective approaches show a more homogeneous behaviour, clearly differentiated from the single objective approach.

Analyzing the numerical results of tables 1 and 2, DENSEA is the best multiobjective approach in all the cases in X and in 5 of 8 in Y, being second in another 2. It is also the best overall approach in all the cases in X and in 3 of 8 in Y, being second in another 4. The multiobjective approach is always better than the single-objective approach in test case X, where the single objective algorithm is the worst in 7 of 8 cases. The multiobjective approach is better than the single-objective in test case Y in 6 of 8 cases, where the single objective algorithm is the worst in those 6 cases.

The best averaged values are obtained in both test cases by the combined use of gray coding and DENSEA: In test case X, the achieved value equals the global optimum reported in section 5 (3324.3 kg) and is obtained in the four tested mutation rates. In test case Y, the achieved value is obtained with a mutation rate of 0.8% (10214 kg.), which is 0.8% higher than the global optimum value.

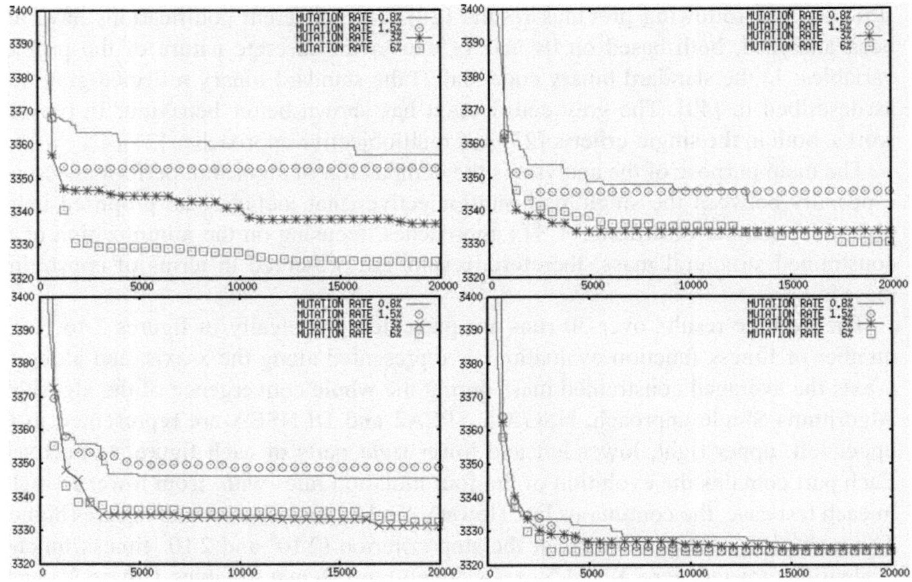

Fig. 3. Case X, Binary Coding. Single objective (upper left), NSGAII (upper right), SPEA2 (lower left) and DENSEA (lower right).

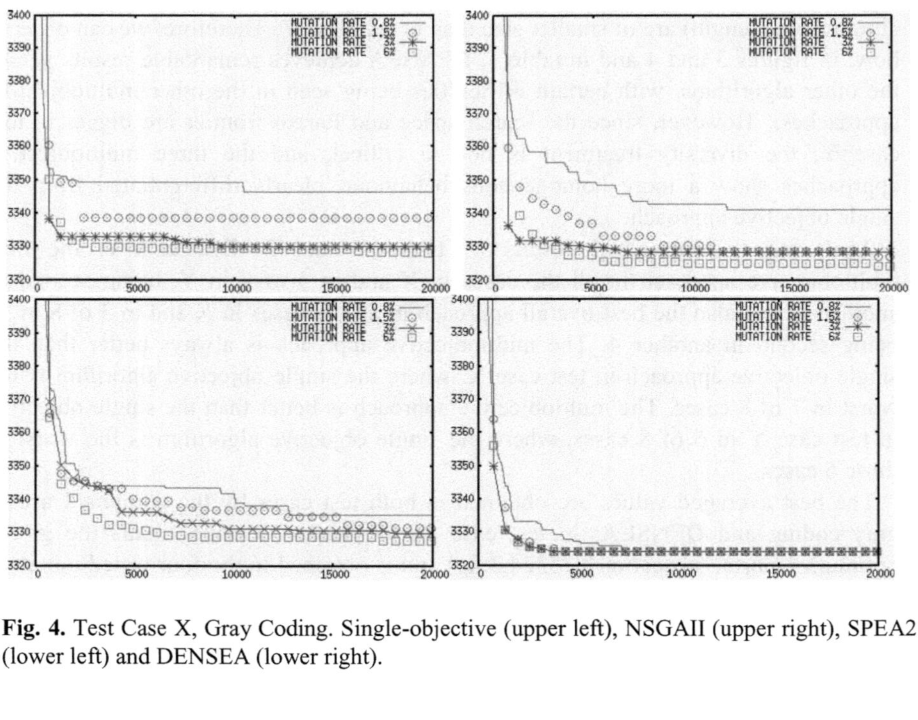

Fig. 4. Test Case X, Gray Coding. Single-objective (upper left), NSGAII (upper right), SPEA2 (lower left) and DENSEA (lower right).

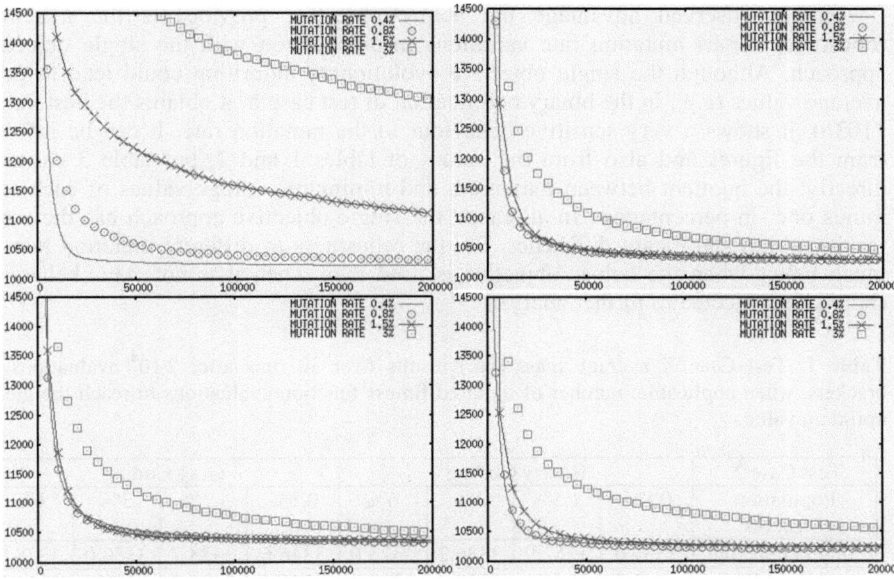

Fig. 5. Test Case Y, Binary Coding. Single-objective (upper left), NSGAII (upper right), SPEA2 (lower left) and DENSEA (lower right).

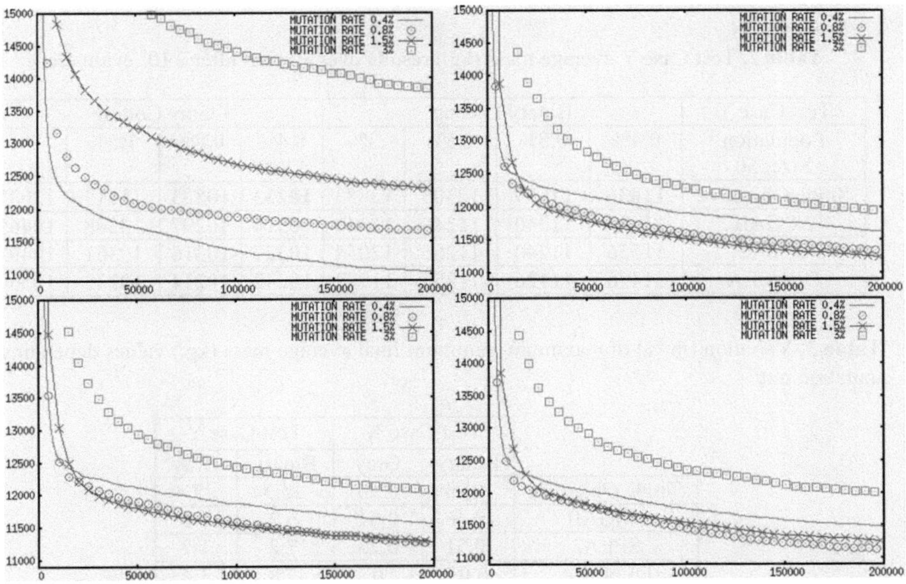

Fig. 6. Test Case Y, Gray Coding. Single-objective (upper left), NSGAII (upper right), SPEA2 (upper left) and DENSEA (upper right).

Another observed advantage the helper objective provides is the increased robustness versus mutation rate variations in comparison with the single objective approach. Although the single objective evolutionary algorithm could lead to good average values (e.g., in the binary codification of test case Y it obtains the best value: 11036), it shows a very sensitive behaviour to the mutation rate. It can be inferred from the figures and also from the values of tables 1 and 2, but table 3 shows it directly: the quotient between maximum and minimum average values of each case minus one -in percentages-. In all cases, the single objective approach has the worst results with a significant difference. So, the robustness to different mutation rates is much better when the helper objective is used than when it is not. This behaviour should be subjected to further analysis.

Table 1. Test Case X average mass (kg.) results over 30 runs after $2 \cdot 10^4$ evaluations. In brackets, when applicable: number of required fitness function evaluations to reach the global optimum value.

Test Case X	Binary Coding				Gray Coding			
Population Size: 50	0.8%	1.5%	3%	6%	0.8%	1.5%	3%	6%
SingleObjective	3357.0	3352.9	3336.2	3325.0	3346.6	3338.7	3330.0	3328.0
NSGAII	3342.4	3345.6	3333.7	3330.4	3341.2	3326.9	3328.4	**3324.3** (19000)
SPEA2	3344.8	3348.7	3331.5	3332.5	3335.5	3331.4	3329.8	3327.6
DENSEA	**3325.9**	**3324.3** (11250)	**3324.9**	**3324.3** (2800)	**3324.3** (7150)	**3324.3** (3850)	**3324.3** (3300)	**3324.3** (3550)

Table 2. Test Case Y average mass (kg.) results over 30 runs after $2 \cdot 10^5$ evaluations

Test Case Y	Binary Coding				Gray Coding			
Population Size: 50	0.4%	0.8%	1.5%	3%	0.4%	0.8%	1.5%	3%
SingleObjective	**11036**	11659	12303	13833	**10233**	10333	11113	13037
NSGAII	11623	11340	**11245**	**11944**	10314	10297	10308	**10465**
SPEA2	11556	11240	11266	12054	10327	10316	10361	10497
DENSEA	11476	**11126**	11255	11992	10242	**10214**	**10242**	10560

Table 3. Variation (in %) of maximum-minimum final average mass (kg.) values depending on mutation rate

	Test Case X		Test Case Y	
	Binary	Gray	Binary	Gray
Single Objective	*0.96*	*0.56*	*25.3*	*27.4*
NSGAII	0.46	0.51	6.2	1.6
SPEA2	0.51	0.23	7.2	1.7
DENSEA	0.05	0	7.8	3.4

The Gray codification also clearly outperforms the standard binary. Only in one case out of 64 is the behaviour of the binary codification algorithm better (the single-objective algorithm with mutation rate 6% in test case X). In test case Y, a more complex problem due to its greater search space, the advantage of its use is increased, with the mass averages differences between both codifications evident.

7 Conclusions

Focusing on the helper objective concept, a new type of helper objective has been introduced in this work applied to a computational mechanics field problem: the number of cross-section types of the bar structure.

The effect of this multiobjective approach has been analyzed in the structural constrained mass optimum design problem, comparing a single objective optimization and three multiobjective algorithms: NSGAII, SPEA2 and DENSEA in two test cases. The simultaneous combination of a suitable evolutionary multiobjective algorithm (considering an appropriate population diversity treatment) and the helper objective as second fitness function has shown advantages in the resolution of the problem: the best overall average results are achieved using this multiobjective approach which also provides a more robust behaviour compared to the mutation rate variation. Therefore, this article describes another application where evolutionary multiobjective algorithms can be seen as a tool that helps to improve single objective optimization.

This kind of helper objective (*number of different cross-section types of the structure*) could also be generalized to other kinds of problems where the chromosome variables are composed of a set of database orderings, whose members contain the information required to compute the fitness function (*number of different items of the database contained in the candidate chromosome or solution*). An analysis of its performance in other fields could provide more light about the practical use and generalization of this multiobjective approach.

References

1. Abbass H., Deb, K.: Searching under multi-evolutionary pressures. Second Evolutionary Multi-Criterion Optimization EMO 2003, Lecture Notes in Computer Science, Vol. 2632 (2003) 391-405
2. Aguirre, H., Tanaka, K.: Selection, Drift, Recombination, and Mutation in Multiobjective Evolutionary Algorithms on Scalable MNK-Landscapes. Third International Conference on Evolutionary Multi-Criterion Optimization EMO 2005, Lecture Notes in Computer Science, Vol. 3410 (2005) 355-369
3. Berry, A., Vamplew, P.: "The combative accretion model – multiobjective optimisation without explicit Pareto ranking", Third International Conference on Evolutionary Multi-Criterion Optimization 2005, Lecture Notes in Computer Science, Vol. 3410 (2005) 77-91
4. Bleuer, S., Brack, M., Thiele, L., Zitzler, E.: Multiobjective Genetic Programming: Reducing Bloat using SPEA2. Proceedings of Congress on Evolutionary Computation (2001) 536-543
5. Bui, L., Branke, J., Abbass, H.: Multiobjective optimization for dynamic environments. 2005 IEEE Congress on Evolutionary Computation, Vol. 3, (2005) 2349-2356
6. Bui, L., Branke, J., Abbass, H.: Diversity as a Selection Pressure in Dynamic Environments. Proceedings of the Genetic and Evolutionary Computation Conference GECCO, Vol. 2, ACM Press (2005) 1557-1558
7. Coello, C., Van Veldhuizen, D., Lamont, G.: Evolutionary Algorithms for solving multi-objective problems. Kluwer Academic Publishers - GENA Series (2002)

8. Coello, C.: Evolutionary Multiobjective Optimization: A Historical View of the Field. IEEE Computational Intelligence Magazine, Vol. 1, No. 1, February (2006) 28-36
9. Corne, D., Deb, K., Fleming, P., Knowles, J.: The Good of the Many Outweighs the Good of the One: Evolutionary Multi-Objective Optimization. IEEE coNNections, 1-1 (2003) 9-13
10. Deb, K.: Multiobjective Optimization using Evolutionary Algorithms. John Wiley & Sons -Series in Systems and Optimization- (2001)
11. Deb, K., Pratap, A., Agrawal, S., Meyarivan, T.: A fast and elitist multiobjective genetic algorithm NSGAII. IEEE Transactions on Evolutionary Computation 6(2), (2002) 182-197
12. Dhingra, A.K., Lee, B.H.: A Genetic Algorithm Approach to Single and Multiobjective Structural Optimization with Discrete-Continuous Variables. International Journal for Numerical Methods in Engineering, 37 (1994) 4059-4080
13. Galante, M.: Genetic Algorithms as an approach to optimize real-world trusses. International Journal Numerical Methods Engineering, 39 (1996) 361-382
14. Gero, J.S., Louis, S.J.: Improving Pareto Optimal Designs using Genetic Algorithms. Microcomputers in Civil Engineering, 10-4 (1995) 241-249
15. Goldberg, D.E., Samtani, M.P.: Engineering Optimization via genetic algorithm. Proceedings Ninth Conference on Electronic Computation, ASCE, New York, NY (1986) 471-482
16. Greiner, D.: Multiobjective optimization of metallic frames using evolutionary algorithms. phD Thesis, Departments of Computer Science, Applied Mathematics and Civil Engineering, University of Las Palmas de Gran Canaria, Spain (2005), (in Spanish)
17. Greiner, D., Emperador, J.M., Winter, G.: Single and Multiobjective Frame Optimization by Evolutionary Algorithms and the Auto-adaptive Rebirth Operator. Computer Methods in Applied Mechanics and Engineering, Elsevier, 193 (2004) 3711-3743
18. Greiner, D., Emperador, J.M., Winter, G.: Multiobjective Optimisation of Bar Structures by Pareto-GA. European Congress on Computational Methods in Applied Sciences and Engineering, CIMNE (2000)
19. Greiner, D., Emperador, J.M., Winter, G.: Enhancing the multiobjective optimum design of structural trusses with evolutionary algorithms using DENSEA. 44th AIAA (American Institute of Aeronautics and Astronautics) Aerospace Sciences Meeting and Exhibit, paper AIAA-2006-1474, (2006)
20. Greiner, D., Winter, G., Emperador, J.M.: A comparative study about the mutation rate in multiobjective frame structural optimization using evolutionary algorithms", in: Schilling, R., et al. (eds): Proceedings of the 6th Conference on Evolutionary and Deterministic Methods for Design, Optimization and Control with Applications to Industrial and Societal Problems, Munich, September (2005)
21. Greiner, D., Winter, G., Emperador, J.M., Galván, B.: Gray coding in evolutionary multicriteria optimization: Application in frame structural optimum design. Third International Conference on Evolutionary Multi-Criterion Optimization EMO 2005. Lecture Notes in Computer Science, Vol. 3410 (2005) 576-591
22. Greiner, D., Winter, G., Emperador, J.M.: Searching for an efficient method in multiobjective frame optimisation using evolutionary algorithms. Computational Solid and Fluid Mechanics, Massachusetts Institute of Technology Conference on Computational Fluid and Solid Mechanics 2003, Elsevier Science (2003) 2285-2290
23. Greiner, D., Winter, G., Emperador, J.M.: Optimising Frame Structures by different strategies of GA, Finite Elements in Analysis and Design, Elsevier, 37-5 (2001) 381-402
24. Grierson, D.E., Pak, W.H.: Optimal sizing, geometrical and topological design using a genetic algorithm. Structural Optimization, 6-3 (1993) 151-159

25. Hajela, P., Lin, C.Y.: Genetic search strategies in multicriterion optimal design. Structural Optimization, 4 (1992) 99-107
26. Handl, J., Knowles, J.: Bias, Proxies and Solution Selection in Multiobjective Optimization. PPSN Workshop on Multiobjective Problem Solving from Nature, September, Reykjavik, Iceland (2006)
27. Handl, J., Knowles, J., Kell, D.: Multiobjective optimization in computational biology and bioinformatics. IEEE Transactions on computational biology and bioinformatics, in press
28. Hernández Ibáñez, S.: Structural Optimum Design Methods. Colección Seinor. Colegio de Ingenieros de Caminos, Canales y Puertos, Madrid, (1990), (in Spanish)
29. Jensen, M. T.: Guiding Single-Objective Optimization using Multi-Objective Methods. Proceedings of EvoCOP 2003. Lecture Notes in Computer Science, Vol. 2611 (2003) 268-279
30. Jensen, M. T.: Helper-Objectives: Using Multi-Objective Evolutionary Algorithms for Single-Objective Optimisation. Journal of Mathematical Modelling and Algorithms, Vol. 3, Nº 4 (2004) 323-347
31. de Jong, E., Watson, R., Pollack, J.: Reducing bloat and promoting diversity using multi-objective methods. Proceedings of the Genetic and Evolutionary Computation Conference GECCO (2001) 11-18
32. Knowles, J., Watson, R., Corne, D.: Reducing Local Optima in Single-Objective Problems by Multiobjectivization. First International Conference on Evolutionary Multi-Criterion Optimization EMO 2001. Lecture Notes in Computer Science, Vol. 1993 (2001) 269-283
33. Landa Silva, J. D., Burke, E. K.: Using Diversity to guide the search in Multi-Objective Optimization. In: Coello Coello C.A., Lamont G.B. (eds.): Applications of Multi-Objective Evolutionary Algorithms, Advances in Natural Computation, Vol. 1, World Scientific (2004) 727-751
34. Liu, M.: Seismic design of steel moment-resisting frame structures using multi objective optimization. Earthquake Spectra, Vol. 21-2 (2005) 389-414
35. Mumford, C.: Simple population replacement strategies for a steady-state multiobjective evolutionary algorithm. Genetic and Evolutionary Computation Conference GECCO (2004) 1389-1400
36. Nojima, Y., Narukawa, K., Kaige, S., Ishibushi, H.: Effects of Removing Overlapping Solutions on the Performance of the NSGAII Algorithm. Third International Conference on Evolutionary Multi-Criterion Optimization 2005. Lecture Notes in Computer Science, Vol. 3410 (2005) 341-354
37. Rajeev, S., Krishnamoorthy, C.S.: Discrete Optimization of Structures using Genetic Algorithms. Journal of Structural Engineering, Vol. 118-5 (1992) 1233-1250
38. Rao, S.S.: Genetic Algorithmic Approach for Multiobjective Optimization of Structures. ASME Annual Winter Meeting, Structures and Controls Optimization, vol. AD-38 (1993) 29-38
39. Valenzuela, C.: A simple evolutionary algorithm for multi-objective optimization (SEAMO). Congress on Evolutionary Computation-CEC (2002) 717-722
40. Whitley, D., Rana, S., Heckendorn, R.: Representation Issues in Neighborhood Search and Evolutionary Algorithms. In: Quagliarella, D., Périaux, J., Poloni, C., Winter G. (eds.): Genetic Algorithms and Evolution Strategies in Engineering and Computer Science. John Wiley & Sons (1997) 39-57
41. Zitzler, E., Laumanns, M., Thiele, L.: SPEA2: Improving the Strength Pareto Evolutionary Algorithm for Multiobjective Optimization. Evolutionary Methods for Design, Optimization and Control with Applications to Industrial Problems, CIMNE (2002) 95-100

A Multi-objective Approach to the Design of Conducting Polymer Composites for Electromagnetic Shielding

Oliver Schütze[1], Laetitia Jourdan[1], Thomas Legrand[1], El-Ghazali Talbi[1], and Jean Luc Wojkiewicz[2]

[1] INRIA Futurs, LIFL, CNRS Bât M3, Cité Scientifique, 59655 Villeneuve d'Ascq Cedex, France
{schuetze,jourdan,legrand,talbi}@lifl.fr
[2] Ecole des Mines de Douai, 941, rue Charles Bourseul, BP 10838 - 59508 Douai Cedex, France
wojkiewicz@ensm-douai.fr

Abstract. This work deals with the design of new shielding materials for the protection of electrical devices. Since there are many different requirements for modern materials, we have chosen a multi-objective approach to this problem. As material under consideration we chose conducting polymer composites due to their excellent electromagnetic properties in the microwave band and their high potential for the optimization process. In this paper, we start this process with the formulation of a novel model, deal further with the approximation of these solution sets, and finally consider the decision support related to this problem.

1 Introduction

Electromagnetic interferences have become an important problem due to the proliferation of commercial, military, and scientific electrical devices and equipments in high frequencies. Electronic devices must be shielded to be protected against the incoming and potentially disturbing radiation.

Conducting polymer composites (CPCs) like Polyaniline Polyurethane (PAni/PU) are very promising for applications in electromagnetic interference shielding ([9]). These materials are e.g. characterized by relatively high conductivities and permittivities. Since these properties can easily be tuned via chemical processes in the making of these composites, CPCs are well-suited for the demanding optimization in this field. Further, these materials are lighter, more flexible and offer better environmental stability compared to the classical shielding materials subjected to corrosion which make them an interesting potential alternative.

In this work, we are particularly interested in the design of new high-protecting and light-weight materials which are realisable for reasonable prices. In search of

S. Obayashi et al. (Eds.): EMO 2007, LNCS 4403, pp. 590–603, 2007.

these materials, we propose in the following a new multi-objective optimization model, address the numerical treatment of these problems, and present possible techniques which are designed to support the decision maker (DM) to find the preferred solution according to the specific problem.

The remainder of this work is organized as follows: in Section 2 we state the background required for the understanding of the particular design problem which is proposed in Section 3. Section 4 deals with the approximation of the Pareto sets of the resulting MOPs, and in Section 5 we show how these sets can be visualized according to the preference of the DM. Finally we make a conclusion in Section 6.

2 Background

In this section we briefly summarize the background required for this work: we introduce the electromagnetic properties which are interesting in our context, present a theoretical model for these properties which serves as the basis for further considerations, and finally address the concept of multi-objective optimization.

2.1 Electromagnetic Properties

Since our aim is to design new protecting materials we are particularly interested in what happens when an electromagnetic wave (EM) arrives at the surface of a material. In that case, three physical phenomena can occur: *absorbtion*, *reflection* and *transmission* of the incidental wave (see Figure 1).

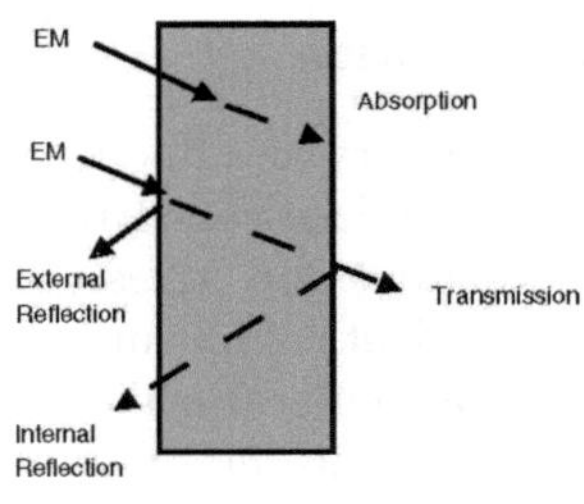

Fig. 1. The three kinds of physical wave interaction

For our purpose it is sufficient just to consider the reflection and the transmission. In [17] a theoretical model for these two wave interactions was proposed which will be used in this work and which will be described in the following. For this, we consider a compound consisting of N layers and assume each layer to be homogeneous and isotropic. The design parameters of the i-th layer, $i = 1, \ldots, N$, are the conductivity σ_i, the permittivity ϵ_i, and the thickness d_i of the material of each layer.

The *characteristic matrix* $M_i \in \mathbb{C}^{2\times 2}$ of the i-th layer is given by:

$$M_i = \begin{bmatrix} cos(A_i) & -jZ_i sin(A_i) \\ -\frac{j}{Z_i} sin(A_i) & cos(A_i) \end{bmatrix},$$

where

$$A_i = \omega d_i \sqrt{\mu_0 \epsilon_0 \left[\epsilon_i - j\frac{\sigma_i}{\omega \epsilon_0}\right]}, \qquad Z_i = \sqrt{\frac{\mu_0}{\epsilon_0 \left[\epsilon_i - j\frac{\sigma_i}{\omega\epsilon_0}\right]}},$$

with $\omega = 2\pi f$, where f is the frequency of the electromagnetic wave, and j denotes the imaginary unit. Z_i is the impedance of the i-th layer. Due to their contact to air media, the impedances of the outer layers are set to $Z_0 = Z_{N+1} = 377(\Omega)$.

The characteristic matrix of the entire compound is given by the product of the characteristic matrices for each layer, i.e.

$$M = M_1 \cdot M_2 \cdot \ldots \cdot M_N = \begin{bmatrix} M_{11} & M_{12} \\ M_{21} & M_{22} \end{bmatrix}$$

Now we are in the position to state the coefficients for the reflection R and the transmission T:

$$R = \frac{(M_{11}Z_0 - M_{12}) - Z_0(M_{22} - M_{21}Z_0)}{(M_{11}Z_0 - M_{12}) + Z_0(M_{22} - M_{21}Z_0)}, \tag{1}$$

and

$$T = \frac{2[M_{22}(M_{11}Z_0 - M_{12}) + M_{12}(M_{22} - M_{21}Z_0)]}{(M_{11}Z_0 - M_{12}) + Z_0(M_{22} - M_{21}Z_0)}. \tag{2}$$

2.2 Multi-objective Optimization

In a variety of applications in industry and finance a problem arises that several objective functions have to be optimized concurrently. One important feature of these problems are that the different objectives typically contradict each other and therefore certainly not have identical optima. Thus, the question arises how to approximate one or several particular 'optimal compromises' (e.g., by *interactive methods* [16]) or how to compute *all* optimal compromises of this *multi-objective optimization problem* (MOP). For this, for instance a huge variety of evolutionary strategies have been proposed during the last years (see e.g. [5] or [3] for an overview on existing methods).

Mathematically speaking, an MOP can be stated in its general form as follows:

$$\min_{x\in\mathcal{S}}\{F(x)\}, \quad \mathcal{S} = \{x \in \mathbb{R}^n : h(x) = 0,\ g(x) \le 0\},$$

where F is defined as the vector of the objectives, i.e.

$$F : \mathbb{R}^n \to \mathbb{R}^k, \quad F(x) = (f_1(x), \ldots f_k(x)),$$

with $f_1, \dots, f_k : \mathbb{R}^n \to \mathbb{R}$, $h : \mathbb{R}^n \to \mathbb{R}^m$, $m \le n$, and $g : \mathbb{R}^n \to \mathbb{R}^q$. A vector $v \in \mathbb{R}^k$ is said to be *dominated* by a vector $w \in \mathbb{R}^k$ if $w_i \le v_i$ for all $i \in \{1, \dots, k\}$ and $v \ne w$ (i.e., there exists a $j \in \{1, \dots, k\}$ such that $w_j < v_j$). A vektor v is called *nondominated* with respect to a set P, if none of the vectors $p \in P$ dominate v.

A point $x \in \mathcal{S}$ is called optimal or *Pareto optimal*[1], if $F(x)$ is not dominated by any vector $F(y), y \in \mathcal{S}$. The solution set – the so-called *Pareto set* – consists typically not of finitely many points as for scalar optimization problems, but forms a $(k-1)$-dimensional object.

3 The Design Problem

In this section we propose a novel multi-objective model for the design of conducting polymer composites for shielding electrical devices. We aim in particular at high-shielding *and* light-weight materials since there seems to be a growing interest in alternatives to classical materials like metals, which are too heavy e.g. for aeronautic applications.

The electromagnetic shielding of conducting polymer composites and the related optimization problem have been considered in some works so far (e.g., [17], [4], [11]). Albeit this is of course the most important feature of this material, the mono-objective approach reveals some limitations since it does not consider other physical properties of the compound which are getting more and more important for commercial products (as weight and cost).

Now we propose the objectives – formulated as minimization problems – which have to be considered in search for modern conducting polymer composites for the shielding of electronic devices.

The first objective is the *electromagnetic shielding*, i.e. the 'classical' objective, which can be expressed as follows ([17]):

$$f_s(x) = 20 \log(|T|), \tag{3}$$

where T is the transmission coefficient defined in (2).
Alternatively, it can be desirable to aim in particular for a high reflection coefficient (see (1)), which leads to the objective

$$f_r = -|R|. \tag{4}$$

Since reflection and transmission of an electromagnetic wave are closely related, only one of these two objectives – depending on the preference of the DM – is required for the formulation of the MOP.

Next, we propose to take the *mass percentage* of a particular material (*Polyaniline*) inside the polymer compound in each layer into account since this value is highly responsible for the (relatively high) cost of these composites. Thus, to be

[1] Named after the economist Vilfredo Pareto, 1848–1923.

efficient and realisable, the materials must have a small mass percentage, which leads to the following minimization problem (e.g., [9]):

$$f_p = \log\left(\sum_i^N p_i\right), \quad p_i = \left(\frac{\sigma_i}{\sigma_0}\right)^{\frac{1}{t}} + pc,\ i = 1, \ldots, N, \tag{5}$$

where σ_0 is a reference conductivity, pc the percolation threshold, and t a critical exponent. p_i is the mass percentage of the i-th layer.

Finally, we propose to take the thickness of the compound into account since this has a direct influence on the weight and the cost of the resulting material. Thus, the 4th objective reads as follows:

$$f_t = \sum_i^N d_i \tag{6}$$

There exist of course other possible goals as well as other models which could be interesting for particular applications and which cannot be stated all here. However, the objectives presented above seem to be the most generic ones.

4 Approximation of the Pareto Fronts

In this section we shortly introduce the two methods which were used and adapted to compute the Pareto fronts of the MOPs which grew out of the design problem under consideration. Since so far mainly compounds with few layers are being studied, we are faced with low or moderate dimensional models which do not represent a challenge to state of the art (EMO) algorithms. Consequently, the approximation of the solution sets is not the main contribution in this work, but, however, this is and will be one important task in multi-objective optimization, and has to be accomplished thoroughly.

In [14] a MOEA is proposed which is designed for the present context. The genetic algorithm used for the optimization of electromagnetic shielding properties allows to obtain diversified and pertinent results. For all the steps of the algorithm, a satisfying diversity of the population was maintained which allowed to present a large number of different solutions to the DM. In the first step of this algorithm, the components of each individual belonging to the initial population are generated at random from subdivided intervals in order to gain homogeneity. The selection step combines two populations, the current one and another one which is stored in a Pareto archive in an elitist manner. In this algorithm, the crossover and mutation operators which are presented in ([2]) are used as they have proven their efficiency on continuous optimization problems. Excellent results have been obtained on different benchmarks with genetic algorithm using these operators their flexible configuration represents a advantage to get varied components. Furthermore, these operators have not increased the time of computation. Also, the obtained individuals had diversified components and the results in the Pareto fronts offered a large palette of solutions to the decision

Algorithm 1. MOEA for shielding design problems

choose initial population P_0
set $A_0 :=$ nondominated points of P_0 ▷ archive
$i := 0$
repeat
  compute P_{i+1} from P_i and A_i by the following steps
    (a) perform NSGA selection from P_i and A_i as proposed in [20]
    (b) perform Crossover and Mutation as proposed in [2]
    (c) perform the generational replacement
  $A_{i+1} :=$ nondominated points of $A_i \cup P_{i+1}$.
  $i := i + 1$
until (stopping criteria fulfilled)

makers. The algorithm has been developed using the platforms EO ([13]) and its extension ParadisEO ([1]).

In order to compare the results obtained by the method described above, we have alternatively used and adapted *subdivision techniques* ([18], [7]) for this problem. These techniques have been primarily designed for unrestricted MOPs and work particularly well for moderate dimensions, i.e. when few layers are considered in the compound.

The algorithms of this type start with a compact subset $Q \subset \mathcal{D}$ of the domain, represented by a collection of n-dimensional boxes (where n is the dimension of $\mathcal{D}$). Each box gets subdivided into smaller sub-boxes and after certain conditions it is decided if a box is promising – i.e., if it could contain a part of the Pareto set – or not. The 'unpromising' boxes are deleted from the collection while the process – subdivision and selection – is continued successively on the remaining boxes until the desired granularity of the boxes is reached. In our design problem the minimal radii of the boxes are given in a natural way by the manufacturing accuracy of the material (which in turn results in a certain accuracy for the parameters ϵ and σ).

The approach is of global nature, i.e. in principle capable of detecting the entire Pareto set. However, it is restricted to moderate dimensions and not rigorous. That is, boxes which are deleted once from the collection but contain a part of the Pareto set will not be reconsidered in further iteration steps. In [19] a variant is described which hybridizes with a MOEA to reduce this problem, which allows to attack higher dimensional and more complicated models, and which was used to compute the Pareto sets of the present design problems. It is planned to integrate the MOEA described above into the subdivision techniques in order to unite the strengths of both algorithms.

Example: A 3-layered Material

As a general test example which will serve for the remainder of this work we have chosen a 3-layered material which is a compound of *Polyaniline Polyurethane* (PAni/PU) and *Kapton*[2] (see Figure 2). We have decided to include Kapton

[2] Kapton is a registered trademark of DuPont.

into the compound since it is a polymer which offers high chemical resistance and good interaction with the Polyaniline - Polyurethane solution in the chemical production process ([10]). Using this compound and the model described above this leads to a design problem with four free parameters (see Table 1). Due to the low dimensionality of the parameter the corresponding model is easy to handle with our (and other) algorithms, and we are able to include all four objectives proposed above into the design problem. Doing so, this leads to the MOP

$$\min_{x} F_c : Q \subset \mathbb{R}^4 \to \mathbb{R}^4, \tag{7}$$

where Q is the hyper-rectangle which is given by the box constraints shown in Table 1.

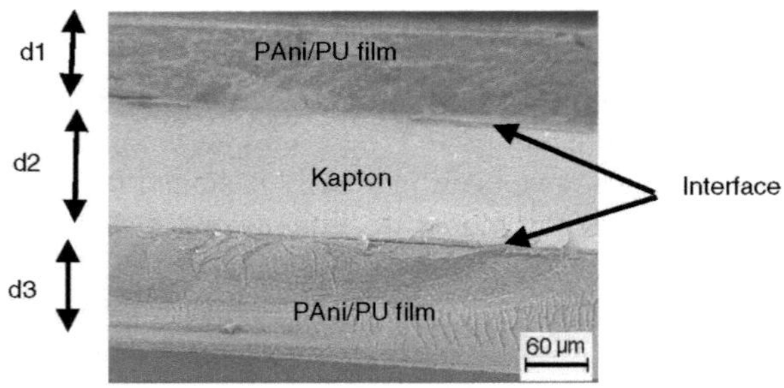

Fig. 2. Electron microscope image of the three layered material under consideration

Figure 3 shows two projections of a front as well as a short discussion. We have chosen $f = 50$ MHz for the frequency of the incoming wave[3]. The results are certainly highly satisfying – from the point of view of the developer of the optimization algorithm. However, it is ad hoc more than doubtful if and how this huge amount of data can help the DM to find the 'right' material according to the given problem. Therefore, the next section deals with the problem specific visualization of these solution sets.

Since this multi-objective approach to the shielding problem is novel and since every application has its special environmental peculiarities, the result of this optimization can hardly be compared to existing materials documented in literature (but is a task for future work). However, the results seem to be promising regarding (a) the large portion of the front where international standards for the shielding efficiency ([12], [8]) are satisfied, and (b) the significant diversity with respect to f_p and f_t, which influence cost and weight of the material. An example for the latter can be seen in Table 2. A motivation for the choice of the points – we have chosen $x_i \in B_i, i = 1, 2$ – is given in the next section. The

[3] Since the frequency of the incidental wave may vary, it is desired in this situation to have a high shielding efficiency in the entire frequency range between $5.0 \cdot 10^6$ and $1.0 \cdot 1.0^9$ Hz. See discussion below.

Table 1. Parameters MOP 7. Since the permittivities of the outer layers are fixed and we consider Kapton as the 2nd layer (see [11] for a motivation of this choice), merely four free design parameters have to be considered.

No. of layer	Material	ϵ	$\sigma(S/m)$	$d(\mu m)$
1	PAni/PU	0.0	from 30 to 10^4	from 0 to 300
2	Kapton	3.1	0	125
3	PAni/PU	0.0	from 30 to 10^4	from 0 to 300

material related to point x_1 reaches the maximal value for the thickness and is further relatively expensive due to the high mass percentage ($p_1 = 31.27$ %m and $p_3 = 31.30$ %m for the 1st and 3rd layer respectively) but offers in turn a shielding efficiency of 61.20 dB. The material which is given by x_2 is much thinner and less expensive due to a significant lower mass percentage ($p_1 = 22.39$ %m and $p_3 = 2.92$ %m). Practically, this indicates that in fact the 3rd layer is not required in case the function values of x_2 have been selected which would lead to further reduction of the cost due to a simplification of the making process.

Table 2. Two selected solutions of MOP (7) with different properties

Point	f_s (dB)	f_r (dB)	f_p (%m)	f_t (μm)	d_1 (μm)	σ_1 (S/m)	d_3 (μm)	σ_3 (S/m)
x_1	-61.21	-0.9991	1.80	598.4	298.6	9945.5	299.8	9965.7
x_2	-40.86	-0.9909	1.40	127.9	126.5	4587.8	1.3	37.3

5 Selecting the Preferred Material

In this section we present two possible ways to present the Pareto fronts of a given design problem in a way that allows the DM to obtain a suitable, problem specific, and maybe subjective overview of the available possibilities, and thus, to help to find the preferred solution. In the following we report on the applicability of an existing visualization tool which offers an unbiased overview on the entire front and propose alternatively a new way of the visualization of 2-dimensional solution sets which can involve preferences of the DM.

The PARETO FRONT VIEWER[4] is based on the interactive decision maps technique ([15]) and was particularly developed for the exploration of Pareto fronts with more than two objectives. In this approach, a non-negative cone is added to every point of the approximation of the Pareto front. The combination of these cones approximates the Edgeworth-Pareto Hull (EPH). By displaying various *decision maps* – i.e. collections of two-dimensional slices of the EPH – which are depicted on the value of a third objective, the decision maps help to understand the influence of this objective. The influence of further objectives

[4] For further description and a demo version of the software tool see `http://www.ccas.ru/mmes/mmeda/mcdm.htm`.

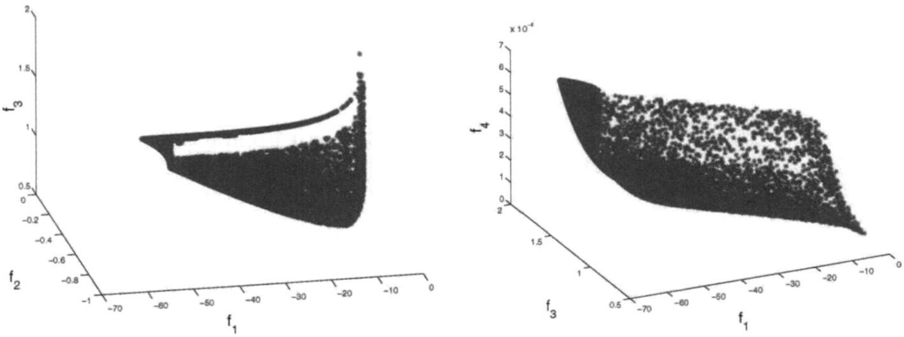

(a) $f_s - f_r - f_p$ (b) $f_s - f_p - f_t$

Fig. 3. Two projections of a Pareto front of MOP (7). In order to obtain this set we have used the algorithm *EA-Subdivision* proposed in [19] which has used 724,600 fitness evaluations. Since the objective values are given in short analytical form the computation took less than one minute on a standard computer. Using the MOEA described in Section 4 very similar results were obtained.

can be experienced by using sliders, which move the efficiency frontiers described above according to the values of these objectives. Though 2-dimensional maps are used for the visualization of the fronts, this approach allows the DM to handle multiple objective values. See [15] for various applications where up to nine objectives are involved.

Figure 4 shows a snapshot of a decision map which displays a Pareto front of MOP (7). The values of f_m and f_p are plotted in the x-axis and y-axis respectively. The map shows an amount of 15 such efficiency frontiers using different values of the shielding efficiency. Finally, these maps can be moved according to the value of the reflection coefficient. Using this tool, a good understanding of the criterion tradeoff the can be obtained.

One important requirement for a proper visualization of a Pareto front is certainly to tame the complexity of the huge amount of data, in particular when more than two objectives are under consideration. For this, the authors of this work share the opinion that it makes sense for this application – and certainly for others as well – to introduce an additional, problem specific indicator which can (hopefully) help the DM to identify parts of the Pareto front which are potentially promising for the current situation.

Let us consider one example. The multi-objective model presented above fixes the frequency of the incoming wave. However, since it is desired to have a high shielding efficiency in an entire frequency range $[f_{min}, f_{max}]$, one could e.g. consider a 'shielding indicator' as follows:

$$I_s(x) := \int_{f_{min}}^{f_{max}} f_s(x, f)df, \tag{8}$$

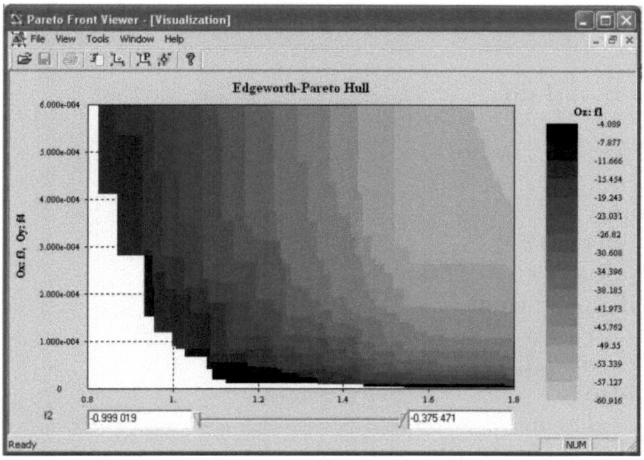

Fig. 4. Snapshot of the PARETO FRONT VIEWER which shows a visualization of a Pareto front which is similar to the one displayed in Figure 3

where $x = (\epsilon, \sigma, d) \in \mathbb{R}^{3N}$ and $f_s(\cdot, \cdot)$ denotes the shielding efficiency analogue to (3). If the value of $I_s(x)$ is high, it does obviously not follow that $f_s(x, f)$ is high for all relevant frequencies, but the underlying heuristic is that in this case the values of f_s must be high in at least one sub-region.

Further examples for possible indicators in the current design problem are e.g. the value of the reflection, the cost or the weight of the material, the robustness of the material against possible errors in the production process, etc. See Figures 5 and 6 for examples.

For the visualization of the 2-dimensional fronts in combination with a property indicator (or without), we propose to use *boxes* since they bring the required 3D effect to the appearance of the fronts[5]. Further, they can be used to reduce the complexity of the data since several 'neighboring' points are collected in one box, depending on the location and the size of the boxes which can be both adjusted according to the problem.

A k-dimensional box B can be represented by a center $c \in \mathbb{R}^k$ and a radius $r \in \mathbb{R}^k$:

$$B = B(c, r) = \{x \in \mathbb{R}^k \; : \; |x_i - c_i| \leq r_i \; \forall i = 1, \ldots, k\}$$

In order to obtain a clear view on these new designed fronts it is certainly advantageous to build a box collection where the interiors of its boxes are mutually non-intersecting. Algorithm 2 represents one possible way to construct such a collection $\mathcal{B}$ given a set of points P, a domain $Q = [a_1, b_1] \times \ldots \times [a_k, b_k]$, and a number *sd* of subdivision steps. This algorithm does not treat adequately the fact that the same boxes may be constructed several times. For this, we refer to [6], where the same data structure is used for a different purpose.

[5] For the visualization we have used MATLAB, see `http://www.mathworks.com`.

Algorithm 2. $\mathcal{B} := Build\_BC(P = \{p_1, \ldots, p_l\}, Q = [a_1, b_1] \times \ldots \times [a_k, b_k], sd)$

```
B := ∅
for all i = 1,...,l do
    y := p_i
    for all j = 1,...,k do
        l := a_j
        r := b_j
        c := (l + r)/2
        for all s = 1,...,sd do
            if y_j ≤ c then
                r := c
                c := (l + c)/2
            else
                l := c
                c := (c + r)/2
            end if
            c_j := c
            r_j := (b_j − a_j)/2
        end for
    end for
    B := B ∪ B(c, r)
end for
```

Figure 5 shows a box collection where the Pareto front was used which is displayed in Figure 3. For the shading of the boxes the shielding indicator (8) was used.

Using this example we want to demonstrate on two (hypothetical) settings how this visualization form can be of advantage for the DM. First, we assume that we are aiming at a high-shielding material where the cost is of minor interest (e.g., in a military application). The image of the Pareto front displays one connected component which is shaded in white (corresponding to the highest value of the indicator). Thus, a point in this section can be chosen – maybe in Box B_1 –, or the search can be continued in this region, e.g. in an interactive manner. Second we assume we want to design a material for a 'standard' device. Thus, it is sufficient to fulfil the required norm for the shielding efficiency while it is desired to minimize cost and weight of the material. Using the shielding indicator (e.g., by looking at the boxes which correspond to a value of $I_s(x) \geq SE_{norm}(f_{max} - f_{min})$) this could lead to the conclusion that points inside B_2 have to be examined for possible realization.

The results demonstrate that the two visualization techniques are well-suited to screen and filter the available possibilities offered by the multi-objective approach in their own way. Both approaches achieved to reduce the complexity of the incoming data in the required amount for this particular application, which motivates that this can also be possible for other design problems.

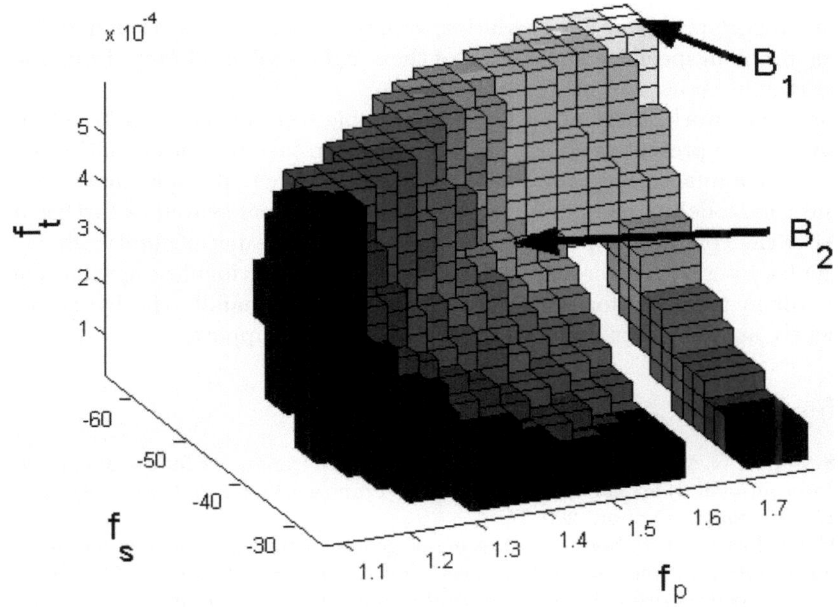

Fig. 5. Classification of a Pareto front consisting of 15987 nondominated solutions (see Figure 3) into 469 different boxes which are grayed into 15 different tones from black (corresponding to the smallest value of the indicator) to white (highest value)

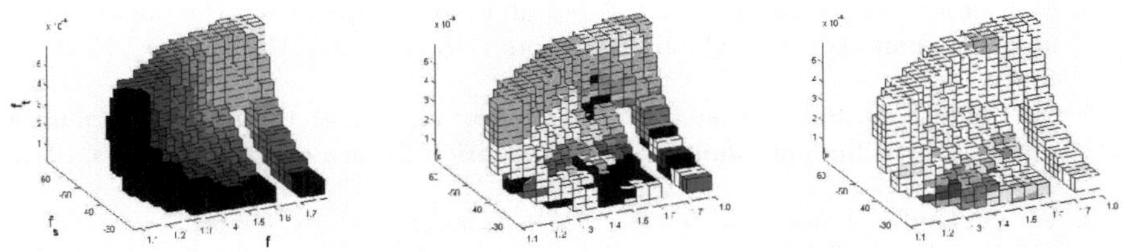

(a) Indicator 'Shielding' (b) Indicator 'Robustness' (c) Indicator 'Reflection'

Fig. 6. Further examples of possible shading of the Pareto front leading to different results: in (b) the value of $\frac{\partial SE}{\partial d_1} + \frac{\partial SE}{\partial d_3}$ is taken as an indicator for the robustness of the production with respect to a manufacturing error in the thickness of the PAni/PU compounds, while for (c) the value of the reflection coefficient f_r is used

6 Conclusions and Future Work

We have presented a multi-objective approach to the design of conducting polymer composites for the shielding of modern devices which demand for high-shielding and light-weight materials for reasonable prices. For this, we have proposed a novel model, have shown the applicability of EMO algorithms (which were tuned for this purpose) to these MOPs, and have addressed the related decision support problem. For the latter we have proposed a particular technique for the problem specific visualization of these 2-dimensional Pareto fronts, which can certainly be used in other applications.

For future work, there are a lot of interesting topics which can be addressed to advance the present work. For instance, one can take the uncertainties coming from the manufacturing process into account. Further, it could make sense to extend the model by including e.g. the number of layers as well as further properties of the conducting polymer composites (e.g., the permeability which is of particular interest for aeronautic applications) as additional design parameter. This will in turn call for new techniques which can handle the hybrid model efficiently as well as for more sophisticated decision support.

References

1. S. Cahon, N. Melab, and E-G. Talbi. ParadisEO: a framework for metaheuristics. In *International Workshop on Optimization Frameworks for Industrial Applications (ICOPI 2005)*, October 2005.
2. R. Chelouah and P. Siarry. A continuous genetic algorithm designed for the global optimization of multimodal functions. *Journal of Heuristics*, 6(2):191–213, 2000.
3. C. A. Coello Coello, D. A. Van Veldhuizen, and G. B. Lamont. *Evolutionary Algorithms for Solving Multi-Objective Problems.* Kluwer Academic Publishers, 2002.
4. N. F. Colaneri and L. W. Shacklette. EMI shielding measurements of conductive polymer blends. *IEEE transactions on instrumentation and measurement*, 4(2):291–297, 1992.
5. K. Deb. *Multi-Objective Optimization Using Evolutionary Algorithms.* Wiley, 2001.
6. M. Dellnitz and A. Hohmann. A subdivision algorithm for the computation of unstable manifolds and global attractors. *Numerische Mathematik*, 75:293–317, 1997.
7. M. Dellnitz, O. Schütze, and T. Hestermeyer. Covering Pareto sets by multilevel subdivision techniques. *Journal of Optimization Theory and Applications*, 124:113–155, 2005.
8. W. G. Duff. *A handbook series of electromagnetic interference and compatibility, Fundamentals of electromagnetic compatibility. Volume 1 Interference Control Technologies.* Gainesville, (VA), 1998.
9. N. N. Hoang. *Réalisation et caractérisation de structure composite polyaniline-polyuréthane dans le domaine de micro-ondes. Modélisation et optimisation de blindage électromagnétique multicouche en utilisant un algorithme génétique.* PhD thesis, Ecole Doctorale Des Sciences Physiques et de l'Ingenieur. Université de Bordeaux 1, France, 2005.
10. N. N. Hoang, J.-L. Wojkiewicz, and R. S. Biscaro. Genetic algorithm applied to the optimisation of the electromagnetic properties of conducting polymer composites in the microwave band. In *Proceedings of ICMS'05*, 2005.
11. N.N. Hoang, J-L. Wojkiewicz, and J-L. Miane. Optimization of electromagnetic shielding effectiveness of composites. In *8th International Workshop on Optimization and Inverse Problem in Electromagnetism IOPE 2004*, pages 61–62, 2004.
12. B. Kaiser. *Principles of Electromagnetic Compatibility.* Artech House, Norwood, MA, 1987.

13. M. Keijzer, J. J. Merelo, G. Romero, and M. Schoenauer. Evolving Objects: a general purpose evolutionary computation library. In *5. conférence internationale sur l'Évolution Artificielle (EA'01)*, October 2001.
14. T. Legrand. Algorithmes génétiques et composites conducteurs. Master's thesis, Polytech Lille, France, 2006. <http://www2.lifl.fr/OPAC/index.php?page=Publications/publiRec>.
15. A. Lotov, V. A. Bushenkov, and G. Kamenev. *Interactive Decision Maps. Approximation and Visualization of Pareto Frontier.* Kluwer Academic Publishers, 2004.
16. K. Miettinen. *Nonlinear Multiobjective Optimization.* Kluwer Academic Publishers, 1999.
17. K. Naishadham. Shielding effectiveness of conductive polymers. *IEEE Transaction on Electromagnetic Compatibility*, 34(1):47–50, February 1992.
18. O. Schütze. *Set Oriented Methods for Global Optimization.* PhD thesis, University of Paderborn, 2004. <http://ubdata.uni-paderborn.de/ediss/17/2004/schuetze/>.
19. O. Schütze, S. Mostaghim, M. Dellnitz, and J. Teich. Covering Pareto sets by multilevel evolutionary subdivision techniques. In C. M. Fonseca, P. J. Fleming, E. Zitzler, K. Deb, and L. Thiele, editors, *Evolutionary Multi-Criterion Optimization*, Lecture Notes in Computer Science, 2003.
20. N. Srinivas and Kalyanmoy Deb. Multiobjective optimization using nondominated sorting in genetic algorithms. *Evolutionary Computation*, 2(3):221–248, 1994.

Evolutionary Multiobjective Optimization of Steel Structural Systems in Tall Buildings

Rafal Kicinger[1], Shigeru Obayashi[2], and Tomasz Arciszewski[1]

[1] George Mason University, The Volgenau School of Information Technology & Engineering, 4400 University Drive MS 6C1, Fairfax, VA 22030, USA
{rkicinge, tarcisze}@gmu.edu
[2] Tohoku University, Institute of Fluid Science, 2-1-1 Katahira Aoba-ku Sendai, 980-8577, Japan
obayashi@ifs.tohoku.ac.jp

Abstract. This paper presents results of extensive computational experiments in which evolutionary multiobjective algorithms were used to find Pareto-optimal solutions to a complex structural design problem. In particular, Strength-Pareto Evolutionary Algorithm 2 (SPEA2) was combined with a mathematical programming method to find optimal designs of steel structural systems in tall buildings with respect to two objectives (both minimized): the total weight and the maximum horizontal displacement of a tall building. SPEA2 was employed to determine Pareto-optimal topologies of structural members (topology optimization) whose cross-sections were subsequently optimized by the mathematical programming method (sizing optimization). The paper also presents the shape of the Pareto front in this two-dimensional objective space and discusses its dependence on the building's aspect ratio. The results reported provide both qualitative and quantitative knowledge regarding the relationship between the two objectives. They also show the trade-offs involved in the process of conceptual and detailed design of complex structural systems in tall buildings.

Keywords: evolutionary multiobjective optimization, structural design, Pareto front, tall buildings.

1 Introduction

Finding solutions for many structural engineering problems involves multiple and often conflicting objectives. Traditionally, however, due to lack of efficient multiobjective optimization methods, a single 'most important' criterion was selected and treated as the objective with respect to which structural designs were optimized. The remaining objectives were usually converted into constraints which were subsequently used to determine the feasibility of generated structural designs [1]. In the vast majority of structural design applications, the total weight of a structural system was employed as the objective of choice mainly because it can be regarded as a good estimate of structural system's cost [2]. In several other studies, including authors' previous research [3], multiple design objectives were combined into an

S. Obayashi et al. (Eds.): EMO 2007, LNCS 4403, pp. 604–618, 2007.

aggregate fitness function using a linear combination of weights [4]. This approach, however, has its obvious limitations. They include the necessity of conducting a large number of design optimization runs for each combination of weights in order to determine the shape of the Pareto front as well as inability to produce proper Pareto optimal solutions when the design spaces are non-convex [5].

Thus, in this paper, we extend the previous aggregate function approach to multiobjective optimization of steel structural systems in tall buildings by using a 'truly' multiobjective optimization algorithm, namely the Strength-Pareto Evolutionary Algorithm 2 (SPEA2) [6]. In our study SPEA2 was used to optimize topologies of steel structures in tall buildings with respect to two objectives (both to be minimized): the total weight and the maximum horizontal displacement of a tall building. SPEA2 was integrated with a mathematical programming method which was utilized to determine optimal cross-sections of structural members. Thus, multiobjective topology and sizing optimization of steel structural systems in tall buildings was achieved.

A large number of computational structural design experiments was conducted in order to determine the shape of the Pareto front in this two-dimensional objective space and its dependence on the building's aspect ratio. In order to achieve this goal, the reported experiments were performed for two classes of steel structural systems in tall buildings: one with 30 stories and 5 bays with a relatively small aspect ratio and the other with 36 stories and 3 bays with a relatively high aspect ratio. The qualitative and quantitative relationships between the two objectives have been investigated and identified for both classes of structural design problems. Also, as in the authors' previous study [3], the qualitative changes among topologies of structural systems located in various regions of Pareto front were analyzed.

The paper is organized as follows. First, a brief review of the history of evolutionary optimization in structural engineering is presented together with a short introduction to the structural design problem investigated in this paper. Next, the problem of topological optimum design of steel structural systems in tall buildings is formalized and its representation introduced. Further, the structure and parameters of conducted multiobjective optimization experiments are described and followed by discussion of obtained results. Finally, initial conclusions are provided.

2 Background

2.1 Evolutionary Computation in Structural Design

The concept of evolutionary-based optimization in structural design is not new. It has a relatively long history dating back to the 1980s and to the initial applications of evolutionary algorithms to sizing and shape optimization of relatively simple structural systems (e.g., trusses [7, 8] and frames [9]). The progress in the fields of evolutionary computation and computing resulted in applications of evolutionary methods to more complex and computationally intensive structural design problems, including the topology optimization of discrete-member trusses [10], topology optimization of truss structures in pylons [11], and topology, shape, and sizing optimization of truss structures [12]. Evolutionary-based topological optimum design of steel structural systems in tall buildings was initially studied in [13, 14] and later extended in [1].

Initial applications of evolutionary algorithms in structural design considered only a single-objective fitness function (usually the total weight). Later, however, several multiobjective evolutionary design problems were studied. Early applications of multiobjective optimization methods include the conceptual design of airframes [15]. The weighted min-max algorithm was also used to optimize a 10-bar plane truss [16], and to optimize I-beams [17] and truss designs [18]. Multiobjective Genetic Algorithm (MOGA) has been used in many engineering design applications including gas turbine controller [19] and supersonic wings [20, 21]. A variation of MOGA (called MGA) was applied to conceptual design of office buildings [22]. NSGA-II has been recently applied to a topological optimum design problem [23]. In this approach, both the weight and the maximal displacement of a cantilever plate were minimized. A hybrid approach, NSGA-II and a hill climber, was employed to solve several engineering shape optimization problems [24]. Initial exploration of multiobjective optimization of steel structural systems in tall buildings using an aggregate function approach was reported in [3].

A comprehensive survey of evolutionary computation in structural design, including a discussion on multiobjective structural optimization methods, can be found in [2].

2.2 Steel Structural Systems in Tall Buildings

Topological optimum design of steel structural systems in tall buildings is considered amongst the most complex problems in structural engineering. Its complexity can be compared to such complex structural design problems as the design of large span bridges or of space structures. Steel structural systems are designed to provide structural support for tall buildings. They have to satisfy numerous requirements regarding the building's stability, transfer of loads (i.e., gravity, wind and earthquake loads), displacements, vibrations, etc. For this reason, the design of structural systems in tall buildings requires the analysis of their behavior under various combinations of loading and the determination of an optimal configuration of structural members. It is difficult, complex, and still not fully understood domain of structural engineering, particularly as the generation of novel structural concepts is concerned.

Usually, steel structural systems in tall buildings are designed as a system of vertical members called "columns", horizontal members called "beams", and various diagonal members called "wind bracings". Finding optimal configurations of these structural members is the subject of the topological optimum design problem (also known as a "conceptual design problem") while finding optimal dimensions of cross-sections of members for a given topology is the subject of sizing optimization (also known as a "detailed design problem").

3 Multiobjective Optimization of Tall Buildings

3.1 Topological Optimum Design of Steel Structures in Tall Buildings

The paper investigates multiobjective topological optimum design of steel structural systems in tall buildings. Hence, its goal is to determine optimal topologies (i.e., configurations) of structural members (column, beams, and wind bracings) in a steel structural system. The fitness of the produced design topologies (design concepts) is determined by two evaluation criteria (the total weight and the maximum horizontal displacement of a structural system), which are both minimized.

In order to determine the total weight of a structural system and its maximum horizontal displacement, not only the topology of a given design must be established but also the dimensions of structural members' cross-sections need to be determined. Hence, the design optimization process considered in this paper consists of two stages. In the first stage, a multiobjective evolutionary algorithm produces the topology of a structural system (a design concept), which is understood here as a complete description of a configuration of the following members of a structural system: wind bracings, beams, and column supports. The configurations of individual columns are assumed fixed (the locations and types of columns do not change) and hence are not evolved. In the second stage of the design process, sizing optimization of all structural members, including wind bracings, beams, and columns, is conducted for the topology determined in the first stage.

The sizing optimization is conducted by SODA. It is a commercial computer program for the analysis of internal forces, dimensioning and numerical optimization of steel structural systems. In the project, a modified SODA program developed by the Waterloo Systems in Waterloo, Ontario, Canada, was used. The optimization method used in SODA is described in [25]. In the structural analysis conducted by SODA, dead, live, and wind loads as well as their combinations were considered. The structural elements were designed using several groups of sections for beams, columns, and wind bracings. In the performed experiments the first order analysis was used (P-Delta effects were not considered).

As stated earlier, the designs were optimized with respect to the total weight of a steel structural system and its maximum horizontal displacement. The former provides a good estimate of the cost of a steel structural system while the latter gives a good approximation of the structure's stiffness. These two objectives are usually conflicting as the reduction of the weight of a steel structure may increase its maximum horizontal displacement (and thus reduce its stiffness) and vice versa. The goal of the reported design experiments was to both qualitatively and quantitatively analyze the trade-offs between the two objectives and in doing so determine, or provide a good approximation to, the shape of the Pareto front for this complex structural design problem.

3.2 Representations of Steel Structural Systems in Tall Buildings

In the design experiments reported in this paper, a simplified two-dimensional model of a three-dimensional structural system of a tall building was considered. The representation space for this 2D structural design problem was developed using a direct (1-to-1) mapping between design attributes determining the types of structural members and their corresponding symbolic values encoded in the genome.

Design attributes represented the following types of structural members: bracings, beams, and supports. Fig. 1 shows the values of design attributes representing wind bracing elements in a steel structural system and their phenotypic and genotypic representation. Each such attribute can have up to seven possible values encoding various types of wind bracings: no bracing, diagonal bracing \, diagonal bracing /, K bracing, V bracing, simple X bracing, and X bracing. Each wind bracing attribute was encoded in the genome as an integer value from 0 to 6.

In Fig. 2 values of design attributes describing beams and supports are presented. Each design attribute representing a beam in a steel structural system had two possible values (binary attributes) encoding two types of beams: a pinned beam or a fixed beam. Similarly, each design attribute representing a support in a steel structural system was binary and encoded two types of supports: a pinned support, or a fixed support.

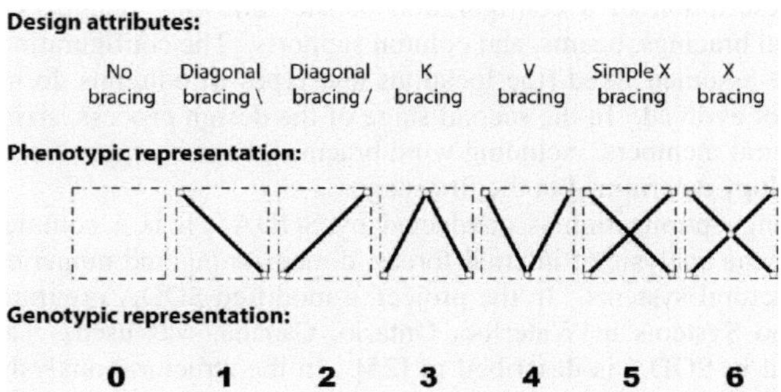

Fig. 1. Design attributes representing types of wind bracing members and their phenotypic and genotypic representations

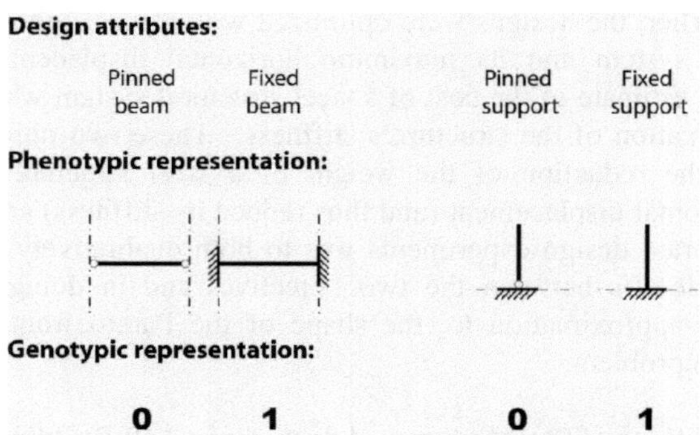

Fig. 2. Design attributes representing types of wind bracing members and their phenotypic and genotypic representations

Thus, the actual design representation manipulated by the multiobjective evolutionary algorithm was encoded in the form of a non-homogeneous genome composed of integer-valued genes. In the design experiments reported in this paper, fixed-length genomes were used as representations of steel structural systems. However, the length of a genome depended on the class of the structural design problem. The genomes were composed of 306 genes for tall buildings with 30 stories and 5 bays (Problem I) and of 220 genes for tall buildings with 36 stories and 3 bays (Problem II).

In the process of evaluation of each genome, symbolic attributes encoded in the genome were mapped to the corresponding types of structural members. At this point, a complete description of the *topology* of a steel structural system was specified. This specification, together with applicable loads and load combinations, was converted into a SODA input file. SODA was subsequently run to perform the second level of design optimization, i.e., sizing optimization, during which optimal cross-sections of structural members were determined. The *detailed design* created in this way was regarded feasible when it satisfied the requirements of the relevant steel design code (here AISC-LRFD-93). The results produced during the SODA run were saved in an output file from which the total weight of the structural system and its maximum horizontal displacement were extracted and assigned as genome's fitness values.

4 Experimental Design

Extensive computational experiments were designed to answers the following research questions:

1. What is the shape of the Pareto front for the topological optimum design problem investigated in this paper?
2. What are the qualitative and quantitative changes between the Pareto fronts when the aspect ratio of the structural system changes?
3. What are the qualitative and quantitative differences among designs located in various regions of the Pareto front?

In order to answer research question No. 1, SPEA2 algorithm was used to optimize steel structural systems in tall buildings with respect to two-objectives: the total weight and the maximum horizontal displacement of a structural system. Experimental results were subsequently analyzed and non-dominated solutions identified. Also, extensive sensitivity analyses of several key evolutionary computation parameters were conducted at this stage (i.e., mutation rate, population size, and SPEA2 archive size) to determine the ones that produces the best results, i.e., the ones that generated the largest numbers of non-dominated solutions. The termination criterion for evolutionary runs was based on the number of generations. It was varied with the population size in order to keep a comparable budget of about 50,000 fitness evaluations per run. All evolutionary parameters and their values used in the reported experiments are shown in Table 1.

Table 1. Evolutionary computation parameters and their values used in the experiments

EC Parameter	Value
EA	SPEA2
Population sizes	100, 200, or 500
SPEA archive sizes	10% of the population size
Mutation (type, rate)	(random reset, 0.05, 0.1, 0.3, or 0.5)
Crossover (type, rate)	(uniform, 0.2)
Genome length	306 genes for Problem I, and 220 genes for Problem II
Objectives	1. The total weight of the structural system 2. The maximum horizontal displacement of the structural system ('sway')
Initialization method	Random
Termination criterion	500 generations (for pop. size 100) 250 generations (for pop. size 200) 100 generations (for pop. size 500)

In order to answer research question No. 2, two classes of structural design problems were investigated. The first class of problems (i.e., Problem I) involved structural systems with 30 stories and 5 bays. These tall buildings have a relatively low value of the aspect ratio. The second class of problems (i.e., Problem II) involved 36 story tall buildings with 3 bays. For these structures the aspect ratio is much higher. For both classes of problems, the height of each story was equal to 14 ft (4.27 m) while the bay widths were equal to 20 ft (6.01 m). As discussed in the previous section, 7 types of wind bracings (see Fig. 1), 2 types of beams, and 2 types of supports were considered (see Fig. 2). All parameters of the design problems and their values are presented in Table 2.

Table 3 shows the magnitudes of dead, live, and wind loads used in the structural analysis conducted by SODA. Five load combinations were considered, following the design specifications for steel, concrete, and composite structures in tall buildings provided in [26]. They included the following combinations of loads:

- Dead + Live
- 0.75(Dead + Live + Wind)
- 0.75(Dead + Live – Wind)
- 0.75(Dead + Wind)
- 0.75(Dead – Wind)

Table 2. Design problem parameters and their values

Problem Parameter	Value
Number of stories	30 (for Problem I), and 36 (for Problem II)
Number of bays	5 (for Problem I), and 3 (for Problem II)
Bay width	20 feet (6.01 m)
Story height	14 feet (4.27 m)
Distance between transverse systems	20 feet (6.01 m)
Types of bracing elements	No, Diagonal \, Diagonal /, K, V, Simple X, and X
Types of beam elements	Pinned-Pinned, and Fixed-Fixed
Types of column elements	Fixed-Fixed (only)
Types of supports	Pinned, and Fixed

Table 3. Magnitudes of dead, live, and wind loads

Load Parameter	Value
Dead load magnitude	50 psf (2.39 kN/m^2)
Live load magnitude:	
- building	100 psf (4.78 kN/m^2)
- roof	30 psf (1.43 kN/m^2)
Wind load:	
- Wind speed	100 mph (160.9 km/h)
- Wind importance factor	1.0
- Wind exposure category	C

The negative sign placed in front of the wind loads indicates that the wind forces considered in a given load combination act in the opposite direction, i.e. wind pressure is replaced by wind suction and vice versa, when compared to the case when the plus sign is used.

Finally, the research questions No. 3 was answered by analyzing the values of both objectives for designs located in various parts of the Pareto front. Qualitative changes among Pareto optimal designs located in various parts of the front were inspected visually and analyzed statistically. Basic statistical analysis was conducted to detect variation in frequencies of occurrence of specific types of structural members at specific locations in the structural system.

The experimental results are presented in the following section.

5 Experimental Results

5.1 Sensitivity Analysis

The sensitivity analysis was conducted to determine optimal evolutionary computation parameters. It revealed that the multiobjective optimization progress (measured in terms of the number of non-dominated solutions found) depends mostly on values of two parameters: the rate of mutation and the size of the population. Fig. 3 shows that low mutation rates (i.e., 0.05 and 0.1) produced the largest number of non-dominated solutions for both problems. The number of all non-dominated solutions shown in Fig. 3 includes all non-dominated solutions generated during the entire run as opposed to only those solutions contained in the final population or in the archive.

Similar pattern is presented in Fig. 4. This time, however, the relationship between the number of non-dominated solutions and the size of the population is shown for Problem II only. Fig. 4 clearly shows that the SPEA2 with the population size of 100 produced the largest number of non-dominated solutions regardless of the mutation rate used. Thus, the sensitivity analysis of evolutionary computation parameters identified the most crucial ones and their corresponding optimal values. These values (i.e., mutation rate of 0.05 and population size of 100) were subsequently used in further computational explorations reported in this paper.

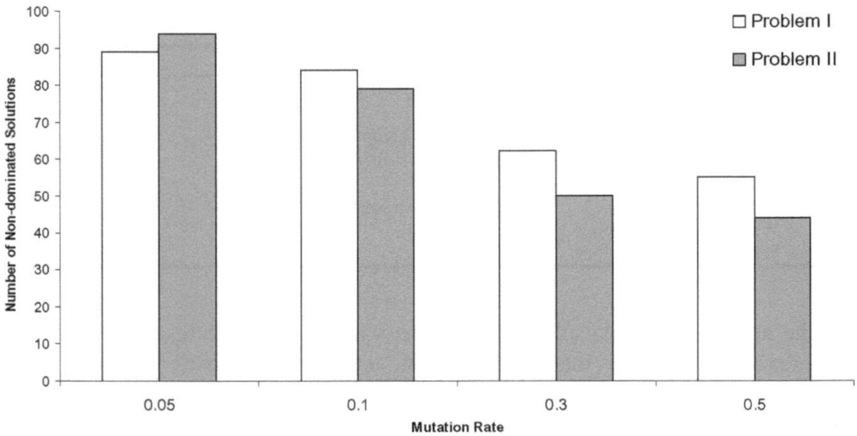

Fig. 3. Impact of the mutation rate on the number of non-dominated solutions for both classes of structural design problems

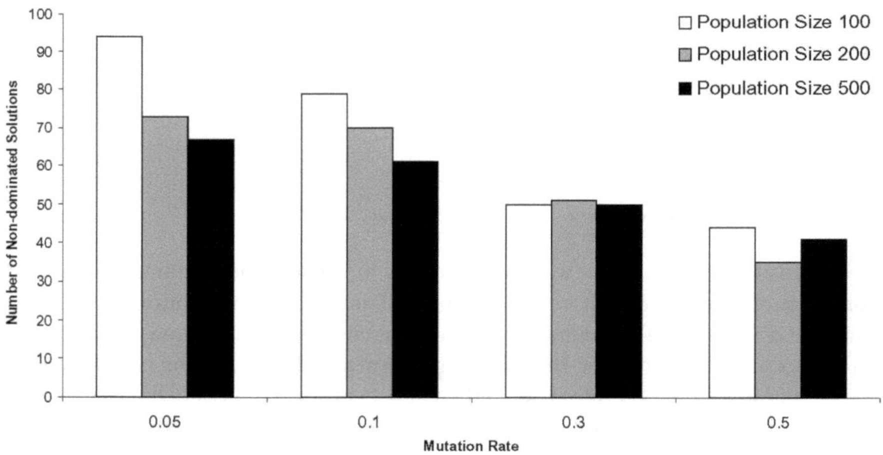

Fig. 4. Impact of the population size and mutation rate on the number of non-dominated solutions

5.2 Shape of the Pareto Front

Fig. 5 shows typical results obtained during multiobjective topological optimum design experiments for Problem II. It presents about 50,000 solutions generated by SPEA2 with population size equal to 100 and mutation rate equal to 0.05. The solutions form three distinct regions:

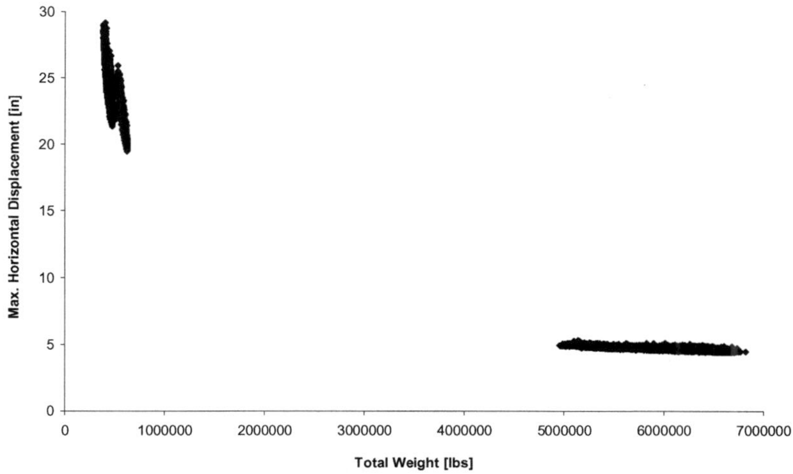

Fig. 5. Typical results produced during a single run using SPEA2 for Problem II

- One region located to the right corresponds to small values of the maximum horizontal displacements (between 4.3 and 5.3 in) and large values of the total weight (between 4,900,000 and 6,700,000 lbs).
- Two adjacent regions to the left correspond to relatively large values of the maximum horizontal displacements (between 19 and 30 in) and small values of the total weight (between 370,000 and 620,000 lbs).

In order to determine the shapes of Pareto fronts for both investigated problems, the results of the corresponding computational experiments were analyzed and non-dominated solutions identified. Fig. 6 shows the approximate Pareto front for Problem I formed by its 133 non-dominated solutions. The leftmost part of the front has been additionally magnified in a zoom window to more clearly present its structure.

The structure of the identified Pareto front is essentially similar to the results presented in Fig. 5. It forms two separate regions which show significant trade-offs between the two objectives. A relatively small reduction in the total weight of the structural system causes large increase in the value of the maximum horizontal displacement and vice versa. Fig. 6 also provides upper and lower bounds on the ranges of variability of the total weight and maximum horizontal displacement for Problem I. They are analyzed in more detail in the next section.

Similar results were obtained for Problem II. Fig. 7 shows the structure of the Pareto front formed by 160 non-dominated solutions found in the conducted experiments. As before, the front is divided into two separated regions located in the opposite corners of the two-dimensional objective spaces. Again, the zoom window was used to magnify the leftmost part of the Pareto front and show more clearly its structure.

The discussion of the similarities and differences between the shapes of Pareto fronts for Problems I and II is presented in the following section.

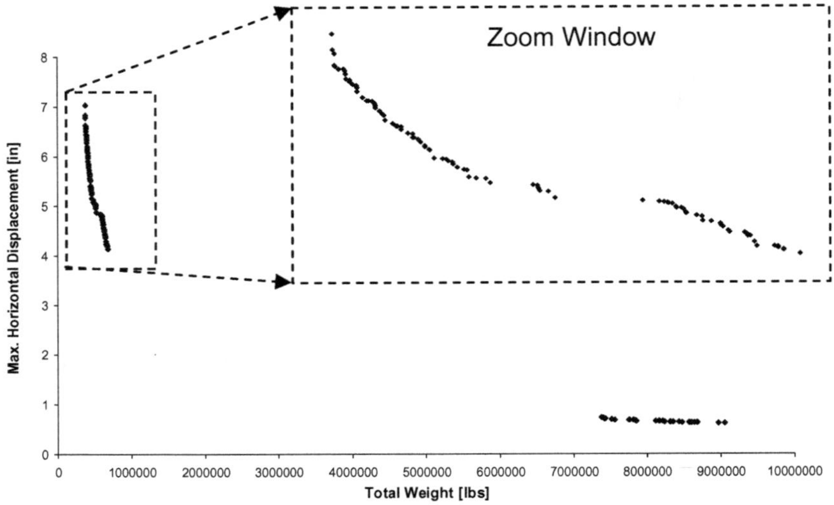

Fig. 6. Approximate shape of the Pareto front for Problem I (leftmost part of the front has been magnified in the zoom window for greater clarity)

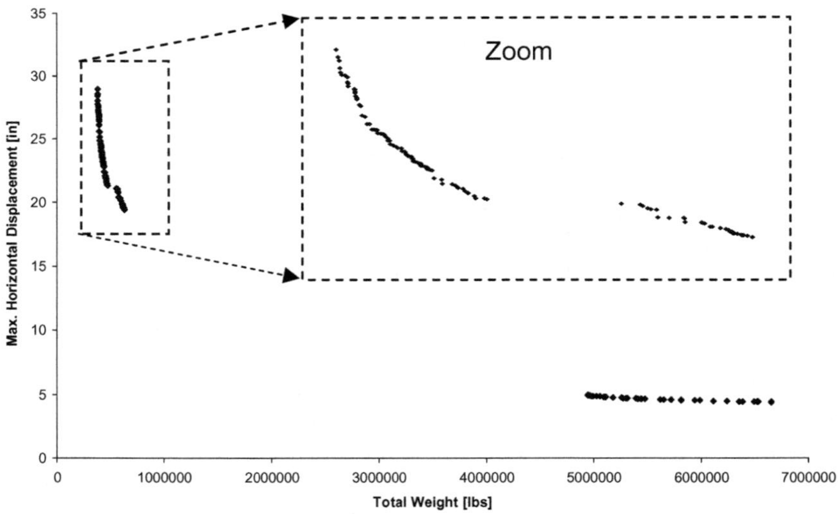

Fig. 7. Approximate shape of the Pareto front for Problem II (leftmost part of the front has been magnified in the zoom window for greater clarity)

5.3 Impact of the Aspect Ratio on the Pareto Front

In Fig. 8, the shapes of obtained Pareto fronts for Problems I and II are compared. This illustrates the changes in trade-offs between the two objectives when the aspect ratio of the structural system in a tall building is varied. Fig. 8 shows that the aspect

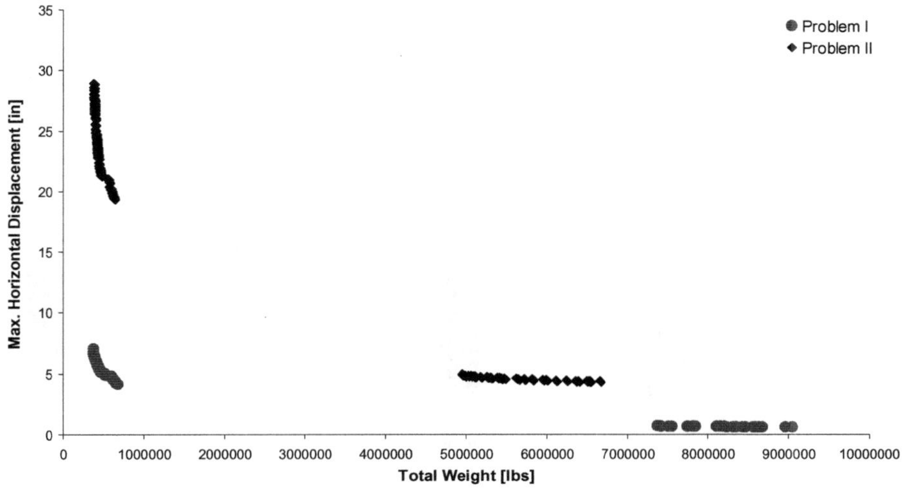

Fig. 8. Comparison of the shapes of Pareto fronts for Problems I and II

ratio has a significant impact on the location of the Pareto front in two-dimensional objective spaces. When the ratio is high (Problem I), the leftmost part of the front achieves significantly higher values of maximum horizontal displacements (between 19 and 29 in) than the values obtained for the low value of the aspect ratio. In the latter case, maximum horizontal displacements corresponding to non-dominated solutions in the left part of the Pareto front varied between 4 and 7 in.

One can also observe significant shift to the right of the right part of the Pareto front for Problem I, when compared to the corresponding part of the Pareto front for Problem II. In the former case, non-dominated solutions had the total weight between 7,300,000 and 8,900,000 lbs with the maximum horizontal displacements from 0.6 to 0.75 in. In the latter case, the total weight ranged 4,900,000 to 6,700,000 lbs and the corresponding displacements varied from 4.4 to 5 in.

5.4 Optimal Structural Topologies Along the Pareto Front

Variations in optimal topologies along the Pareto front were analyzed visually and statistically. The statistical analysis focused on the leftmost part of the Pareto fronts corresponding to structural designs of feasible total weight (acceptable cost from the practical point of view). It was discovered that major differences occur in the frequencies of occurrence of K or V and simple X or X bracings. Fig. 9 shows the frequencies (in percent) of occurrence of these four types of wind bracings for two designs for each problem (a total of four designs) located in the left part of the Pareto front:

- Top - corresponding to the minimum total weight and maximum horizontal displacement
- Bottom – corresponding to the maximum total weight and minimum horizontal displacement (within the leftmost part of the Pareto front)

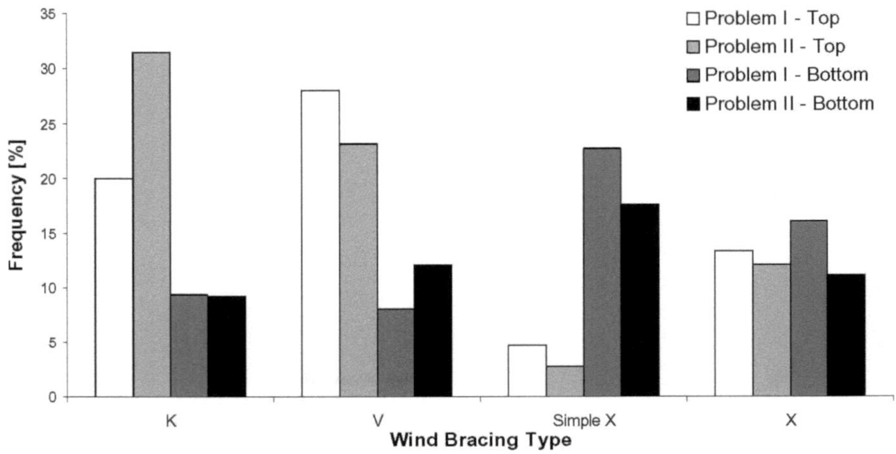

Fig. 9. Frequencies of occurrence of various types of wind bracings along the Pareto front for Problems I and II

Fig. 9 clearly shows that the designs characterized by the minimum total weight and relatively large horizontal displacements use mostly K or V bracings instead of simple X or X bracings. The opposite is true for the designs with minimum horizontal displacements and relatively large total weights. As Fig. 9 shows, these findings are universal for Problems I and II and hence seem to be independent of the aspect ratio of the structural system.

6 Conclusions

In this paper, results of extensive multiobjective topological optimum design experiments were presented. SPEA2 algorithm was integrated with a mathematical programming method to optimize topologies and cross-sections of structural members in steel structures of tall buildings with respect to two objectives (both minimized): the total weight and the maximum horizontal displacement of a tall building. The experiments identified the structure of the Pareto front in the two-dimensional objective space as well its relationship to the aspect ratio of the tall building. It was further discovered that there are significant differences in the frequencies of occurrence between two groups of bracings among non-dominated designs located in various parts of the Pareto front. The first group, containing K- and V-type bracings, was mostly found in the regions of the front corresponding to low values of the total weight and the large values of horizontal displacements. On the other hand, the group, containing simple X- or X-type bracings, occurred mostly in the regions with large values of total weight and relatively small values of horizontal displacements.

Acknowledgements

The first author would like to thank the 21 Century Center of Excellence of Flow Dynamics, Institute of Fluid Science, Tohoku University, Sendai, Japan, for providing financial support in the form of an International Internship Award. It allowed him to spend six weeks at Tohoku University in the early 2005 and conduct the initial part of the reported research. He also acknowledges current support from the National Science Foundation under CMMI grant #0632507.

References

1. Kicinger, R., T. Arciszewski, and K.A. De Jong, *Evolutionary design of steel structures in tall buildings.* Journal of Computing in Civil Engineering, 2005. **19**(3): p. 223-238.
2. Kicinger, R., T. Arciszewski, and K.A. De Jong, *Evolutionary computation and structural design: A survey of the state of the art.* Computers & Structures, 2005. **83**(23-24): p. 1943-1978.
3. Kicinger, R. and T. Arciszewski. *Multiobjective evolutionary design of steel structures in tall buildings.* in *Proceedings of the AIAA 1st Intelligent Systems Technical Conference, Chicago, IL, September 20-23, 2004.* 2004. Reston, VA: American Institute of Aeronautics and Astronautics Press.
4. Liu, X., D.W. Begg, and R.J. Fishwick, *Genetic approach to optimal topology/controller design of adaptive structures.* International Journal for Numerical Methods in Engineering, 1998. **41**: p. 815-830.
5. Coello Coello, C.A., *A comprehensive survey of evolutionary-based multi-objective optimization techniques.* Knowledge and Information Systems, 1999. **1**(3): p. 269-308.
6. Zitzler, E., M. Laumanns, and L. Thiele, *SPEA2: improving the strength Pareto evolutionary algorithm.* 2001, Computer Engineering and Networks Laboratory (TIK), Swiss Federal Institute of Technology (ETH): Zurich, Switzerland.
7. Goldberg, D.E. and M. Samtani. *Engineering optimization via genetic algorithm.* in *Ninth Conference on Electronic Computation.* 1986. University of Alabama, Birmingham: American Society of Civil Engineers.
8. Hajela, P., *Genetic search - an approach to the nonconvex optimization problem.* AIAA Journal, 1990. **26**: p. 1205-1212.
9. Grierson, D.E. and W. Pak, *Discrete optimal design using a genetic algorithm*, in *Topology Design of Structures*, M.P. Bendsoe and C.A. Mota Soares, Editors. 1993, Kluwer Academic Publishers: Dordrecht, The Netherlands. p. 89-102.
10. Shankar, N. and P. Hajela, *Heuristics driven strategies for near-optimal structural topology development*, in *Artificial Intelligence and Structural Engineering*, B.H.V. Topping, Editor. 1991, Civil-Comp Press: Oxford, UK. p. 219-226.
11. Bohnenberger, O., J. Hesser, and R. Männer. *Automatic design of truss structures using evolutionary algorithms.* in *Second IEEE International Conference on Evolutionary Computation (ICEC'95).* 1995. Perth, Australia: IEEE Press.
12. Rajan, S.D., *Sizing, shape, and topology design optimization of trusses using genetic algorithm.* Journal of Structural Engineering, 1995. **121**: p. 1480-1487.
13. Arciszewski, T., K.A. De Jong, and H. Vyas. *Inventive design in structural engineering: evolutionary computation approach.* in *Fifth International Conference on the Applications of AI to Civil and Structural Engineering.* 1999. Oxford, England.

14. Murawski, K., T. Arciszewski, and K.A. De Jong, *Evolutionary computation in structural design.* Engineering with Computers, 2001. **16**(3-4): p. 275-286.
15. Cvetkovic, D., I.C. Parmee, and E. Webb, *Multi-objective optimisation and preliminary airframe design*, in *The Integration of Evolutionary and Adaptive Computing Technologies with Product/System Design and Realisation*, I.C. Parmee, Editor. 1998, Springer-Verlag: Plymouth, UK. p. 255-267.
16. Hajela, P. and C.-Y. Lin, *Genetic search strategies in multicriterion optimal design.* Structural Optimization, 1992(4): p. 99-107.
17. Coello Coello, C.A. and A.D. Christiansen, *Two new GA-based methods for multiobjective optimization.* Civil Engineering Systems, 1998. **15**(3): p. 207-243.
18. Coello Coello, C.A. and A.D. Christiansen, *Multiobjective optimization of trusses using genetic algorithms.* Computers & Structures, 2000. **75**(6): p. 647-660.
19. Chipperfield, A.J. and P.J. Fleming. *Gas turbine engine controller design using multiobjective genetic algorithms.* in *First IEE/IEEE International Conference on Genetic Algorithms in Engineering Systems: Innovations and Applications (GALESIA'95).* 1995. Halifax Hall, University of Sheffield, UK: IEEE Press.
20. Obayashi, S., *Pareto genetic algorithm for aerodynamic design using the Navier-Stokes equations*, in *Genetic algorithms and evolution strategies in engineering and computer science: recent advances and industrial applications*, D. Quagliarella, et al., Editors. 1998, John Wiley & Sons: Chichester, England. p. 245-266.
21. Obayashi, S., *Pareto solutions of multipoint design of supersonic wings using evolutionary algorithms*, in *Adaptive Computing in Design and Manufacture V*, I.C. Parmee, Editor. 2002, Springer-Verlag: London. p. 3-16.
22. Grierson, D.E. and S. Khajehpour, *Method for conceptual design applied to office buildings.* Journal of Computing in Civil Engineering, 2002. **16**(2): p. 83-103.
23. Hamda, H., O. Roudenko, and M. Schoenauer. *Multi-objective evolutionary topological optimum design.* in *Fifth International Conference on Adaptive Computing Design and Manufacture (ACDM 2002).* 2002. University of Exeter, Devon, UK: Springer-Verlag.
24. Deb, K. and T. Goel. *A hybrid multi-objective evolutionary approach to engineering shape design.* in *First International Conference on Evolutionary Multi-Criterion Optimization (EMO'2001).* 2001. Zurich, Switzerland: Springer-Verlag.
25. Grierson, D.E. *Computer-automated optimal design for structural steel frameworks.* in *NATO ASI Conference on Optimization and Decision Support Systems in Civil Engineering.* 1989. Edinburgh, UK: Kluwer Academic Publishers.
26. Taranath, B.S., *Steel, concrete and composite design of tall buildings.* 2nd ed. 1998, New York: McGraw-Hill.

Multi Criteria Decision Aiding Techniques to Select Designs After Robust Design Optimization

Mattia Ciprian, Valentino Pediroda, and Carlo Poloni

Department of Mechanical Engineering, University of Trieste, Trieste, 34100, Italy
{pediroda, poloni, mciprian}@units.it

Abstract. Robust Design Optimization is the most appropriate approach to face problems characterized by uncertainties on operating conditions, which are peculiarity of aeronautical research activities. The Robust Design methodology illustrated in this paper is based on multi-objective approach. When a Pareto approach is used, a Multi Criteria Decision Method is needed for selecting the final optimal solution. This method is tested on an aeronautic case: the design of a transonic airfoil with uncertainties on free Mach number and angle of attack. The final solution is compared with a well known airfoil: the new design performs as the original one , especially concerning lift and drag stability.

Keywords: Robust Design, uncertainties, Multi-Objective, Multi Criteria Decision Making, airfoil.

1 Introduction

Multidisciplinary Design Optimization (MDO) is getting more and more important, especially in the aerospace community. The AIAA Association (American Institute of Aeronautics and Astronautics) has organized several sessions dedicated to the MDO (last session, AIAA 2004) and recently the First Session of MDO for specialists (AIAA, 2005). Consequently the development of numerical methodologies to solve this kind of problems is increasing in importance, in order to help industry during the phases of complex design. It seems useful to remark that designs, in particular in aeronautic fields, are extremely complex, because of the physic model and the huge number of input and output parameters.

One important aspect in industrial design is the management of uncertainties, to find solutions which are not sensitive to stochastic fluctuations of parameters. The name of this design model is Robust Design.

The need of Robust Design method appears in many contests, especially in Multi Disciplinary Design. In fact it is possible to find uncertainties in many different cases. During the preliminary design process the exact value of some input parameters could be unknown or the input parameters could change in the next design phases. Consequently the aim is to look for a solution as less dependent as possible on unknown parameters. Other concerns are to find out solutions which are insensitive to the tolerance manufacturing parameters, to fluctuations in operative conditions or numerical fluctuations in the high fidelity simulation models.

S. Obayashi et al. (Eds.): EMO 2007, LNCS 4403, pp. 619–632, 2007.

The present paper shows a new optimization method that look for solutions which are insensitive to fluctuations, any source they are caused by. The method refers to the statistical definition of stability and is based on a multi objective approach, in particular on Game Theory. It is able to find good solutions for stability and performances.

Finally an important aspect of multi disciplinary optimization is presented: the selection of the most interesting design among those obtained by a Multi Objective Optimization. It is known that a co-operative Game Theory gives a set of solutions (Pareto frontier). In order to choose the final design a Multi Criteria Decision Making (MCDM) algorithm has been adopted.

MCDM is both an approach and a set of techniques, with the goal of providing an overall ordering of designs, starting from the most preferred to the least preferred one. Obviously no one alternative design will be the best in achieving all objectives; in addition, some conflict or trade-off is usually evident among the objectives: costs and benefits typically conflict. MCDM is a way of looking at (and solving) complex problems that are characterized by both costs and benefits. In this work we used an out-ranking method that allows ranking the alternatives from the best to the worst one.

2 The Idea of Robust Design in Aeronautics

The study of uncertainties in engineering begins with Taguchi (Taguchi, 1978), who codified the methodology for the quality engineering. Taguchi divides the design in three different phases: the firsts one, called system design, determinates the most feasible region for the following numerical optimizations, the second phase, called robust design, determinates the optimal parameter for maximizing the final quality of considered system, and in the third final phase, called tolerance design, a parameter tuning is performed to reach the best possible final solution.

The necessity to study uncertainty is well known in aeronautics; in fact it is possible to cite the definition of uncertainty given on AIAA Guideline (AIAA, 1998):

Definition 1.2. *Uncertainty: A potential deficiency in any phase or activity of the modeling process that is due to lack of knowledge.*

Notice that the uncertainty is defined how a lack of knowledge, which obviously leads to the need of a different approach for studying the model.

From a numerical point of view the study of a model affected by uncertainties could be defined as:

$f: A \times B \rightarrow \Re$ where

- $a \in A$ represents the design parameters chosen by the designers
- $b \in B$ represents the input parameters permeated by uncertainties consequently not controllable by designers.

In (Trosset, 2005) the common uncertainties (parameter b) for external aerodynamic are well described, in particular in the case of two dimensional airfoil design:

1. Uncertainties on geometry parameters due to manufacturing tolerance ε which modifies the geometry parameters in a-ε This situation is deeply explained in (Welch, 1989), where an airfoil with minimum drag over geometrical uncertainties is designed.

2. Uncertainties on operative conditions (design point): usually the fluctuations of free stream Mach number [M_{min}, M_{max}] are considered. For important references see (Drela, 1989, Wu Li 2003).

In (Hicks 1977) it is well demonstrated why the study of fluctuations is of primary importance in external aeronautics. It is shown how minimizing the drag coefficient of an airfoil with fixed operative conditions, in particular the free Mach number, the final solution has good performance at the design point but poor off-design characteristics (Fig. 1); this concept is known as over-optimization.

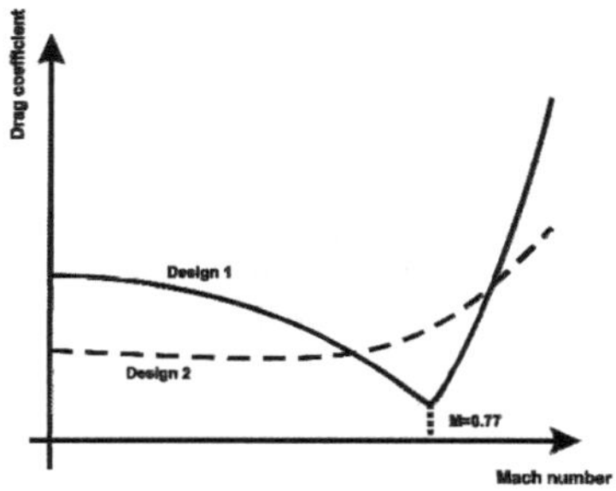

Fig. 1. Drag profile for stable (Design 2) e not stable (Design 1) solution respect to Mach number

This behavior becomes more evident for supercritical airfoils where the relationship between drag and free stream velocity is highly nonlinear because of the fluctuations of shock wave position on the airfoil surface.

Consequently the possibility to determine solutions with good performances over a range of operative conditions appears attractive, also to avoid sudden changes in the behavior of the system; it is necessary to remember that a stable behavior minimizes the operative risk of the system.

Many numerical methods have been developed to optimize a system under uncertainties, in particular in the case of fluctuations of operative conditions.

In (Wu Li 2003) a methodology based on multi-point optimization has been proposed, using a sum-weighted formulation; in the examined example, the minimization of the drag coefficient of a lift-constrained airfoil is performed with uncertainties on cruise Mach number. The proposed formulation is:

$$\min_{\alpha, d \in D} \sum_{i=1}^{n} w_i c_d(d, \alpha_i, M_i) \tag{1}$$

subjected to

$$c_l(d, \alpha_i, M_i) \geq c_l^* \quad \text{for } 1 \leq i \leq n \tag{2}$$

where w_i are arbitrary weights, d is a set of geometric design variables that define the airfoil, C_d and C_l are the drag and lift coefficient defined as function of free stream Mach number M_i and angle of attack α_i which can fluctuate around the design point values, and C_l^* is the required lift. The problem of the Eq.1 is that the final result depends on the choice of the weights w_i, too arbitrary to define.

In (Huyse 2001) a new concept is introduced for the Robust Design and an approach different from the multi-point optimization is used. The innovative idea is the formulation of a risk ρ to minimize:

$$\min \rho = \int_{M_\infty} C_d(d, M_\infty) f(M_\infty) dM_\infty \tag{3}$$

The term $f(M_\infty)$, which appears in Eq. 3, is the probability density of the cruise Mach number. The risk (from Bayesian theory) represents the mean value of drag coefficient inside the fluctuations of operative conditions. The author proposes for the solution of the integral (Eq. 3) the use of a Taylor series of second order.

In (Padula 2002), to avoid the arbitrariness of Robust Design formulation presented in (Drela 1998) and (Huyse 2001), an interesting methodology is proposed. The Mach number in the Eq. 1 is iteratively modified to calculate the integral Eq. 3 using the trapezoid rule.

A new and interesting approach is proposed in (Wu Li 2003) where the authors have demonstrated that Robust Design problem has to be solved using a Multi Objective Optimization Approach. By the definition of stability, the numerical formulation of the problem becomes:

$$\min_{D, \alpha(M)} (E(c_d), \sigma^2(c_d)) \tag{4}$$

subjected to

$$c_l(D, \alpha(M), M) = c_l^* \quad \text{for} \quad M \in \Omega \tag{5}$$

Mean and variance of C_d are defined as follows:

$$E(c_d) = \int_{M_{\min}}^{M_{\max}} c_d(D, \alpha, M) p(M) dM \tag{6}$$

$$\sigma^2(c_d) = \int_{M_{\min}}^{M_{\max}} (c_d(D, \alpha, M) - E(c_d))^2 p(M) dM \tag{7}$$

where $p(M)$ is the probability density function of Mach Number defined in the interval $M_{min} < M < M_{max}$. The solution of the above formulation normally is not easy, for these reasons the authors propone a Monte Carlo approach.

This Robust Design formulation (Eqs. 4-7) gives the possibility to determine two directions in the optimization: by the variance of drag coefficient, it is possible to minimize the off-design performance degradation (fig.2, design at the bottom); on the other hand, by the optimization of drag coefficient mean value, the performances will be privileged (fig.2, design at the top). Let us notice that the Robust Design formulation given by Eq.4 is based on a Multi Objective approach, so the final result will be a Pareto frontier, i.e. the set of solutions with the best compromise between the objective functions (fig.2, all the design of the Pareto frontier). The difficulty in this kind of approach could be find the number of high fidelity analysis needed to find the Pareto frontier; for this reason an alternative methodology is proposed using a descend direction that could reduce drag simultaneously and proportionally over the given range of Mach number.

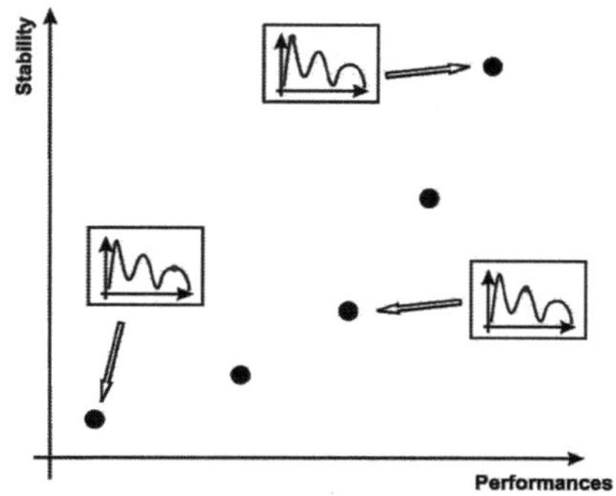

Fig. 2. Pareto Frontier obtained by Robust Design Optimization (Performances vs. Stability Degradation)

2.1 Why We Need a Multi Objective Approach

In this work a new method for Robust Design optimization is presented. The main idea is to use a multi objective approach to reach the best possible compromise between performance and stability of design. Referring to Fig. 3, the function has an absolute extreme and a relative one respectively corresponding to the coordinates *x1* and *x2*; in this case the uncertainties are represented by the tolerance Δ on the input parameter x (the same case of manufacturing tolerance). Obviously a standard optimization, without considering the fluctuations, would find out the point *x1* as the best solution that is the absolute maxima. But in *x1* the objective function has poor stability. In the case of Robust Design optimization (considering tolerance Δ) two different objectives have to be considered: mean performances and stability of solutions, according to the ideas presented in (Wu Li 2003). Considering the mean performance inside the tolerance Δ, the best configuration would be represented by the point *x1*, since the mean value of the function is the highest. But for the stability, which corresponds to an evaluation of the variance of function $f(x)$ inside the field Δ (Eq. 7), the best configuration is represented by the point *x2*, because the function is characterized by a lower variability inside the tolerance around to the point *x2*.

Consequently it is interesting to observe that when Robust Design optimization is performed, it is possible that the more stable region doesn't correspond to the more performing one. So to perform an optimization under fluctuations the best way is to define two different objectives for every function to optimize: its mean value and its variance. In mathematical term it is:

$$\begin{aligned} & f:\Re^n \Rightarrow \Re^m \\ & \max \quad E(f_i) = \int_q f_i(x,q)p(q)dq \\ & \min \quad \sigma^2(f_i) = \int_q [f_i(x,q) - E(f_i)]^2 p(q)dq \end{aligned} \tag{8}$$

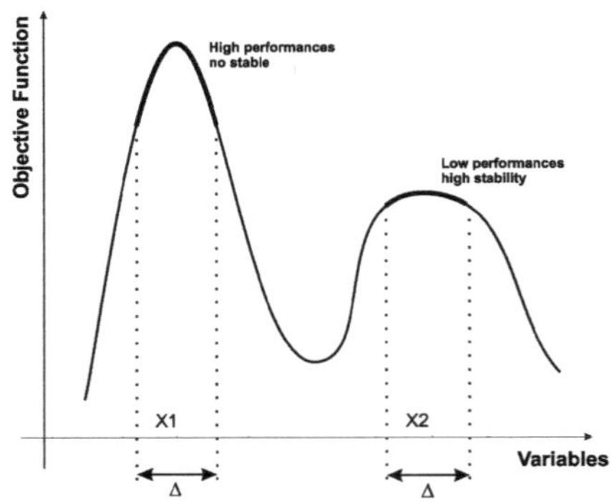

Fig. 3. Function with two different extremes: x1 absolute no stable extreme, x2 relative stable extreme

where f is the multi objective (in more general terms) function to be maximized and q are the uncertainties parameters, modeled by the probability density function p(q).

In this way the problem of an optimization under uncertainties becomes a Multi Objective Optimization problem where the objectives are the stability and the performance; to solve this problem we need to adopt the Game Theory (see chapter 2), which is the best methodology to solve a real multi objective problem without using a weighted function as:

$$\max \quad f_w = w_1 \bar{f} + w_2 \sigma_f \tag{9}$$

In fact it is tricky to assign a value to weights w_i, and this is the reason because it is better to refer to Game Theory approach. It is interesting to notice that after the optimization phase, using a Pareto Frontier approach, the designer does not get only one solution but a set of solutions (Pareto Frontier) which represents the best possible compromise between the objectives. An example of a Pareto Front for a Robust Design Optimization can be observed in Fig. 2; among the Pareto frontier it is possible to choose different compromises between performance and stability, with more flexibility than a standard optimization, where the solution is unique.

After finding the Pareto Frontier, the next phase it is the choice of the best design; the choice could be easy when there are not so many designs in the Pareto Frontier, in this case the designer could easily compare the different solutions and chose the best for his purpose. In this paper we present, completing our previous works (Pediroda, 2006), a Multi Criteria Decision Making methodology that helps the designer when the Pareto Frontier is complex, caused by many different objective functions or by many different designs. In this case, applying the algorithm, it will possible to realize an automatic ranking of the solutions, simplifying the choice of the designer.

It is important to underline that it is possible to face a wide range of problems with Robust Design approach (small manufacturing process errors, fluctuations in the operative conditions, unknown input parameters, etc.). The method is also extendible to more than one function to optimize, for example it is possible to improve the lift and drag of an airfoil with fluctuations in the flight speed, without the need of a weighted function to tie the two different performances.

3 Game Theory on Robust Design

Game Strategies, defined mathematically by J.Nash (Nash, 1951), have found their first applications in economics, in particular to solve the problems concerning the decisions that have some effects on different and often competitive fields.

These strategies may however be adopted also in the industrial design, and in particular they can be combined with evolutionary algorithms, in order to optimize a product according several criteria and contrasting objectives.

We shortly describe the basic formulation of two typologies of Game Strategies (co-operative and competitive), and then we will show how it is possible to implement practically these algorithms to solve multi-objective optimization cases.

In a problem of minimization of two functions $f_A(x,y)$ and $f_B(x,y)$, we define the variables space $(x,y)\in X \cup Y$ as the set of rational strategies. Thus, we decompose the variable space between two "players", called A and B, that are in charge respectively of the variable space X and Y; it follows that each pair $(x,y)\in X \cup Y$ represents a combination of the strategies played by the two players.

The Pareto front may be seen as the result of a co-operative game, in which the two players A and B try to minimize both the two functions; in other words, each strategy played by the players is evaluated by the fitness of the two functions.

Not a single solution is found, but instead a set of solutions, which is called Pareto front. This set is characterized by the fact that there does not exits a solution such that both the two functions have a better fitness of any point of the front. In mathematical terms:

$(x^*,y^*)\in X \cup Y$ belong to Pareto front if and only if:

$$(x',y')\in X\times Y \ : \begin{cases} f_A(x',y') \le f_A(x^*,y^*) \\ f_B(x',y') \le f_B(x^*,y^*) \end{cases} \tag{10}$$

These definitions can of course be generalized in the case of n functions f_i.

4 Exhaustive Example: Multi Objective Robust Design Optimization of an AIRFOIL

Using the Multi Objective Robust Design theory developed, we perform a more realistic optimization case consisting in the design of a non-symmetric airfoil based on RAE28222 geometry, using as flow solver the Navier-Stokes version of MUFLO and AIRFOIL codes (Haase 1983), which uses as turbulence model the Johnson-Coakley equations (fig. 4). The upper and lower side of profile are defined by two 10-degree Bèzier curves, and the co-ordinates of their control points are the variables of optimization (fig. 4). In total we have 18 design variables, which represent the position of the control points.

The uncertainties concern Mach number ($M=0.73\pm0.05$) and the angle of attack ($a=2°\pm0.5°$).

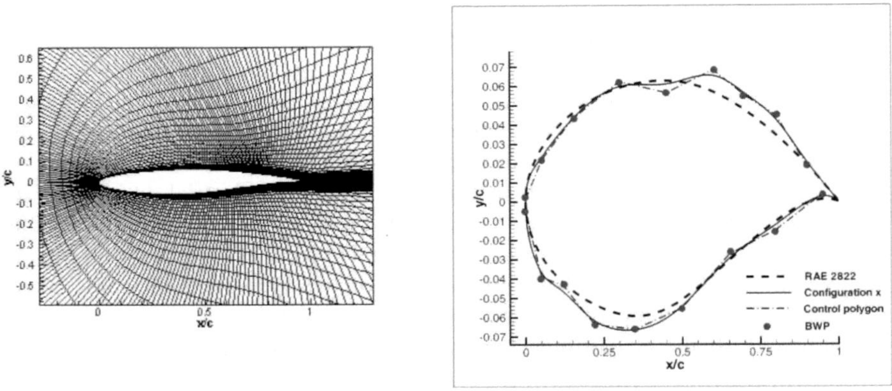

Fig. 4. Airfoil mesh with MUFLO (a) and airfoil parameterization using Bezier curves (b)

The optimization goal is to find out an airfoil geometry which yields better results respect to performances and stability, taking in account of the two uncertain parameters (angle of attack and free Mach number). From a mathematical point of view the optimization problem becomes:

$$
\begin{aligned}
&\min_{\Delta M,\Delta\alpha}\left(E(c_d),\sigma^2(c_l),\sigma^2(c_d)\right) \quad \max_{\Delta M,\Delta\alpha}\left(E(c_l)\right)\\
&\text{with}\\
&E(c_l)\geq E(c_l)^{RAE2822} \qquad \sigma^2(c_l)\leq\sigma^2(c_l)^{RAE2822}\\
&E(c_d)\leq E(c_d)^{RAE2822} \qquad \sigma^2(c_d)\leq\sigma^2(c_d)^{RAE2822}\\
&|E(c_m)|\leq\left|E(c_m)^{RAE2822}\right| \qquad \sigma^2(c_m)\leq\sigma^2(c_m)^{RAE2822}
\end{aligned}
\tag{11}
$$

We set seven constraints to optimization problem: the thickness of profile is fixed to be higher than 12% of the chord length, and the new configuration should present values better than or equal to the original RAE2822 airfoil corresponding to the mean and variance of drag, lift and pitching momentum coefficients.

We handle the constraints in optimization algorithm by mean penalty function approach with tolerance fixed to a value of 10 percent of the constrain value (Poloni et al., 2000).

For the objective function calculation needed in Eq. 11, we use an adaptive Monte Carlo methodology based on response surfaces (Poloni, 2003); on average for a design we need 10 high fidelity analyses.

5 Results

MOGA (Poloni, 1997) (Multi Objective Genetic Algorithm) has been used to solve the Multi Objective Robust Design Optimization of airfoils in transonic field and modeFrontier is the software used to implement MOGA. The problem has been set using 40 individuals per generation with 16 generations; for reducing the number of high fidelity analyses and ensure a good approximation of the Pareto Frontier, we implemented in the MOGA elitism and Multi Objective Directional Crossover (Yamamoto et al., 1995). In Figure 6 the trend of objective functions during the optimization process is shown: it is possible to notice that the desirable trends have been reached. In particular it is possible to underline that a remarkable improvement has been achieved regarding the standard deviation of drag coefficient. In fact, the peculiarity to face an optimization of airfoils in transonic field according to the principles of Robust Design is to be able to look for stable solutions but at the same time with as much performances as possible. The RAE2822 was designed to have the highest performances achievable corresponding to the operating condition considered but in this case it is evident that it has been possible to find more stable solutions especially concerning the drag coefficient value. The shock waves are presented in transonic field and their position change with the operative condition, so this result is directly linked to the high variability of shock wave position.

Having defined 4 objectives, according Pareto theory, the final solution is not unique but will be a set of solutions, which are the best compromises between the different objectives (Pareto frontier). Fig. 5 compares the configurations that belong to Pareto frontier (mean lift versus mean drag, and variance of lift versus variance of drag). It is possible to check that the optimization has been completed with success: in fact you can observe the position of the original design (RAE2822) compared with the others solutions. We obtain better results for stability and drag performance. But not all solutions belonging to the Pareto dominate the RAE design; for the mean lift coefficient it is possible to note that some solutions show a small worsening of the performance; this behavior is related to the tolerance for the constraint penalty function.

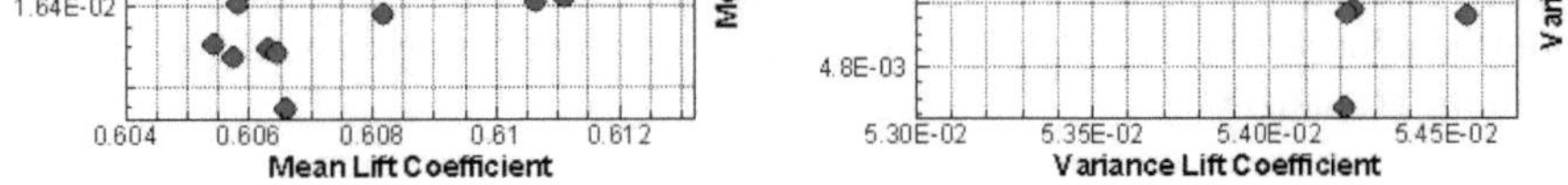

Fig. 5. Pareto Frontier representation in comparison with the original RAE2822 solution

6 Multi Criteria Decision Making

As in many other real-world problems, characterized by multiple objectives, attributes and different types of measures, which have to be satisfied simultaneously, the decision maker (DM) needs to articulate his preferences in terms of tradeoffs among objectives. Whether a multiple criteria decision problem appears naturally in life, in engineering design it has to be necessarily (Sen &Yang, 1998): i.e. only one design among all the alternatives could be putted on production.

Design selection problems are concerned with the evaluation or ranking of a set of available candidate designs in terms of multiple attributes and they form one important class of engineering decision problems. Due to the complexity and often huge amount of data, the analysis of an engineering decision problem usually requires the support of a computerized system and this usually requires the mathematical modeling of the decision problem (Yang et al. 1996).

In this work, in order to choose the more appropriate airfoil of Pareto frontier, a Multi Attribute Decision Making (MADM) method has been used: CODASID (Yang et al. 1996). This algorithm is based on an extended concordance analysis and a modified discordance analysis using raw data represented by a decision matrix and relative weights (with this method the DM may also assign veto threshold values to each attribute: this kind of information has not been used in this work). The new concordance and discordance analyses are used to generate three new indices, namely a preference concordance index, an evaluation concordance index and a discordance index. These three indices provide independent measures for evaluation of each alternative design and span a new space for ultimate ranking of alternative designs. A distance measure is defined in the new space to capture the similarities between a feasible design and given reference designs, which may, for example, be the best/least preferred (or ideal/nadir) designs. The basic idea of defining such a distance measure originates from the TOPSIS method (Hwang and Yoon 1981). The new distance measure, however, is more general and able to take into account a limited compensation.

In order to elicit and capture the DM's preferences and to calculate a priority vector, the Analytic Hierarchy Process (AHP) coupled with the eigenvector method have been employed (Saaty 1980, 2003) on the pairwise comparison matrix shown in Table 1. AHP is a widely used multi-criteria decision analysis method; unlike the conventional methods AHP uses pairwise comparisons, which allows verbal judgments and should enhance the precision of the result. Skipping the whole theory of AHP, it is worth to note that the comparison matrix, which is used to quantify how much more important a criterion is compared to another one by using a linear scale 1/9, 1/7,…, 1, 2, …, 9, should be consistent; $A=\{a_{ij}\}$ matrix will have complete consistency, if the following conditions are satisfied:

$$a_{ij} = \frac{1}{a_{ji}} \quad and \quad a_{ij}a_{ik} = a_{ik} \quad \forall i,j,k = 1,\ldots,n; i \neq j \tag{12}$$

In Table 1. it is possible to observe the relative importance of the different attributes: mainly the lift and drag performance are the most important objectives, followed by the stability (with a small privilege of the drag stability over the lift stability); the designer after the optimization decided to take into account the moment coefficient too, but with a minor relative importance.

In Table 2. the designer decided that any attribute don't have a very strong importance when it is compared to the others; the result is a weight vector that allows the algorithm to find the best compromise design from all points of view.

Table 1. Pairwise comparison matrix of the objective functions (performances privilege case)

	Mean C_L	**Mean C_D**	σ_{CL}	σ_{CD}	σ_{CM}	**Mean C_M**
Mean C_L	1					
Mean C_D	1	1				
σ_{CL}	1/5	1/5	1			
σ_{CD}	¼	¼	½	1		
σ_{CM}	1/8	1/8	1/8	1/8	1	
Mean C_M	1/3	1/3	2	2	2	1

Table 2. Pairwise comparison matrix of the objective functions (compromises privilege case)

	Mean C_L	**Mean C_D**	σ_{CL}	σ_{CD}	σ_{CM}	**Mean C_M**
Mean C_L	1					
Mean C_D	2	1				
σ_{CL}	½	1/3	1			
σ_{CD}	1	1	2	1		
σ_{CM}	1/5	1/5	1/5	1/5	1	
Mean C_M	1/3	1/4	1/2	1/2	2	1

How it is possible to observe in Fig. 6-7, the application of the different choices in the preference tables gives after the MCDM application different designs. The different airfoils have different shapes, especially regarding the suction side of the profile, where the interactions between geometry and shock waves are bigger. This is an

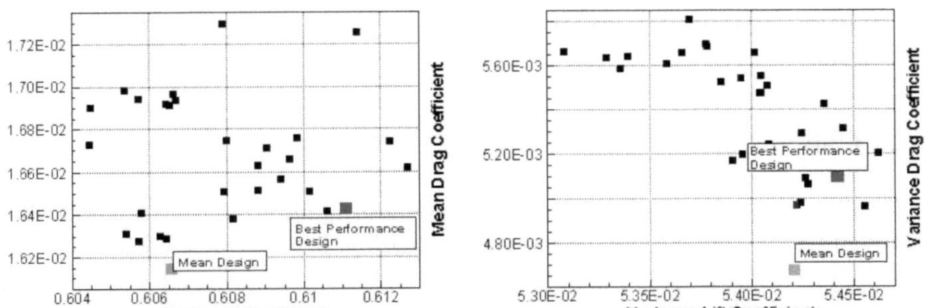

Fig. 6. Performance-Stability comparison inside the Pareto Frontier with the designs choose by MCDM

interesting consideration, when we remember that the performances stability of the airfoil is directly correlated whit the shock waves behavior in transonic case.

In Tab. 3 the numerical comparison between the original airfoil and the optimized (after MCDM choice) is presented. An important improvement is presented especially for the drag coefficient: stability (variance) and performance (mean) are both decreased. For the lift, the mean value is stable, and a small improvement is reached for the stability.

Regarding the designs chosen after MCDM, the most interesting difference is the lift mean value, where the design with performances privilege case gets higher value.

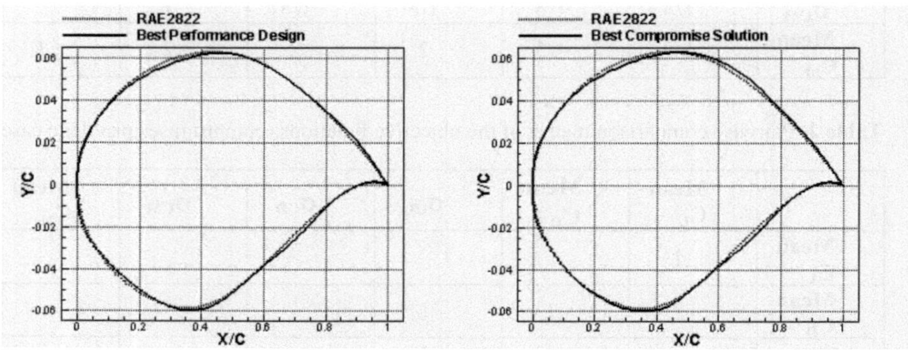

Fig. 7. Geometry profile comparison between the original airfoil (RAE2822) and the two designs choose by MCDM

Table 3. Performance-Stability comparison between the original airfoil (RAE2822) and designs choose by MCDM after Multi Objective Optimization

	Mean C_L	Mean C_D	Mean C_M	σ_{CL}	σ_{CD}	σ_{CM}
RAE2822	0.610	0.0172	-0.0889	0.0546	0.0061	0.0082
Best Performance Design	0.611	0.0164	-0.0867	0.0544	0.0051	0.0071
Best Compromise Design	0.606	0.0161	-0.0885	0.0542	0.0046	0.0070

7 Conclusion

In this paper a new approach for Robust Design Optimization is presented in order to design a transonic airfoil with uncertainties on free Mach number and angle of attack. The main idea is to use a multi objective approach to reach the best possible between performance and stability of design. When uncertainties are present, a Multi Objective methodology is requested, so a Game Theory approach has been used, in particular a co-operative Game Theory has been performed, according to Pareto. To optimize a MOGA (Multi Objective Genetic Algorithm) has been used and the final set of optimal solutions is the Pareto Frontier. To select the final optimal solution, the designer has used a CODASID algorithm where he decided a pairwise comparison matrix for selecting the weights of MCDM algorithm. The final solution is compared with a well known airfoil. The new design performs as the original one respect to every objective, especially concerning the lift and drag stability.

Acknowledgements

This work was supported by a research contract sponsored by ES.TEC.O srl (www.esteco.com). The integration platform used was modeFrontier 3.1 provided by ES.TEC.O to the authors.

References

10th AIAA/ISSMO Multidisciplinary Analysis and Optimization Conference, 30 August-1 September 2004, Albany, New York

1st AIAA Multidisciplinary Design Optimization Specialist Conference,18 - 21 Apr 2005 Hyatt Regency Austin, Austin, Texas

AIAA 1998, "Guide for verification and validation of computational fluid dynamic simulation", *AIAA Guide* G-077-1998, 1998

Drela M., Pros and Cons of Airfoil Optimization, 1998, *Frontier of Computational Fluid Dynamics*, ed. by D. A. Caughey and M. M. Hafez, World Scientific, , pp. 363-381.

Haase W., Wagner B., Jamenson A., 1983 Development of a Navier-Stokes method based on a finite volume technique for unsteady Euler equations, in *Proceeding of the 5th GAMM conference on Numerical Methods in Fluid Mechanics, Friedrich Viewer und Sohn, Braunschweigh* Wiesbaden,

Hicks R. M. and Vanderplaats G. N., 1977 Application of numerical optimization to the design of supercritical airfoils without drag-creep, SAE Paper No. 770440, *Business Aircraft Meeting,* Wichita

Huyse L., R. M. Lewis, 2001, Aerodynamic Shape Optimization of Two-dimensional Airfoils Under Uncertain Conditions, *NASA/CR-2001-210648 ICASE Report* No. 2001-1

Yang J. B., Sen P. and Meldrum P., "Multiple attribute evaluation in engineering decision support using limited compensation and reference designs", *Information and Systems Engineering*, Vol.2, No.3 & 4, pp.159-181, 1996, ISSN 0929-9610.

Nash, J.F., 1951 "Non-cooperative Games", *Annals of Mathematics*,

Padula S. L., W. Li, 2002 Options for Robust Airfoil Optimization under uncertainties, *9th AIAA/USAF/NASA/ISSMO Symposium on Multidisciplinary analysis and Optimization*, September 4-6, 2003 AIAA paper 2002-5602

Pediroda V., Poloni C., 2006, Robust Design, Approximation Methods and Self Organizing Maps Techniques for MDO problems, *Introduction to Optimization and Multidisciplinary Design, VKI Lecture Series 2006-03*, March 6-10, 2006

Poloni C., Pediroda V., 1997, GA coupled with computationally expensive simulations: tool to improve efficiency, in *Genetic Algoritms and Evolution Strategies in Engeneering and Computer Science*, edited by Wiley Sons.

Poloni C., Pediroda V., 2000, "Multi-criteria optimisation, constraint handling with Gas", *Genetic Algorithms for optimisation in aeronautics and turbomachinery*, von Karman Institute for Fuid Dynamics, Lecture Series 2000-07.

Poloni C., Pediroda V., Padovan L. (2003), Multi Objective Robust Design Optimization of airfoils in transonica field, *Eurogen 2003*, Barcellona, 15-17 September 2003

Saaty, T.L. (1980*). The Analytic Hierarchy Process*, McGraw-Hill, New York.

Saaty, T.L. (2003). "Decision-making with the AHP: Why is the principal eigenvector necessary". *European Journal of Operational Research* 145, 85–91

Sacks J., Welch W.J., Michell T.J., Wynn H. P. (1989), "*Design and Analysis of Computer Experiments*", Statistical Science, Vol 4 No.4, pp.409-453.

Sen P.,Yang J.B. (1998). *Multiple Criteria Decision Support in Engineering Design*. Springer.

Taghuci G. 1978, "Performance Analysis Design", *International Journal of Production Research* 16 (6), pp. 521-530

Trosset M.W., N. M. Alexandrov, L.T. Watson, 2005 "New Methods for Robust Design Using Computer Simulations", *Proceeding of the section on physical and engineering science*, America Statistical Association,

Hwang C.L., Yoon K. (1981), *Multiple Attribute Decision Making Methods and Applications*, Springer-Verlag, Berlin.

Wu Li, Sharon Padula, 2003 "Performance Trades Study for Robust Airfoil Shape Optimization", *21th Applied Aerodynamics Conference*, AIAA-2003-3790.

Yamamoto K., Inoue O, 1995, New Evolutionary direction operator for genetic algorithms, *AIAA Journal*, vol 33-10, pp. 1990-1993, 1995

MOGA-II for an Automotive Cooling Duct Optimization on Distributed Resources

Silvia Poles[1], Paolo Geremia[1], F. Campos[2], S. Weston[2], and M. Islam[3]

[1] ESTECO, Padriciano 99, 34012 Trieste, Italy
[2] ICON, Rofel House, Colet Gardens, London, W14 9DH, UK
[3] Audi AG, Wind-Tunnel Centre, 85045 Ingolstadt, Germany

Abstract. In this paper a procedure for the multi-objective optimization of an automotive cooling duct is described. The two objectives considered are the minimization of the pressure drop between the inlet and the outlet of the duct and the maximization of the outlet flow velocity. Since there is no a single optimum to be found, the MOGA-II was used as multi-objective genetic algorithm. The optimization of the duct was obtained employing a parametric model, performing flow analysis with an open source suite and using a multi-objective optimization product. The distributed optimization search exploited the parallelization capabilities of the MOGA-II algorithm which allowed the evaluation of several designs configurations by running concurrent threads of the flow analysis solver. The results obtained are very satisfactory, and the procedure described can be applied to even more complex problems.

1 Problem Description

The present work deals with designing the optimal shape of a duct in order to minimize the pressure drop between the inlet and the outlet of the duct and to maximize the outlet flow velocity (see Fig. 1). This process is normally complex, time-consuming and relies heavily on engineering experience. In order to reduce the product development time and satisfy the growing design requirements to stay competitive in the market, designers are giving more and more importance to the quality of their work complying with the principle of finding the best solution with the minimum effort. In this context, the designer can take full advantage of efficient optimization algorithms that allow concurrent designs evaluation on distributed computational resources.

The duct geometry in this example is located in the underbonnet of a vehicle, and is responsible for channelling air from the side-grill (see Fig. 2) towards the transmission for the purpose of cooling. Like most underbonnet regions in modern vehicles there is a very compact and crowded compartment with complex paths for air movement. Providing convective cooling with a side-duct is considered beneficial for the reduction in the transmission surface and oil temperatures.

The optimization procedure consisted of generating a parametric model of the duct, performing a *Computational Fluid Dynamics* (CFD) analysis and using

S. Obayashi et al. (Eds.): EMO 2007, LNCS 4403, pp. 633–644, 2007.

Fig. 1. Side grill inlet on a typical production car (Image courtesy of Audi AG)

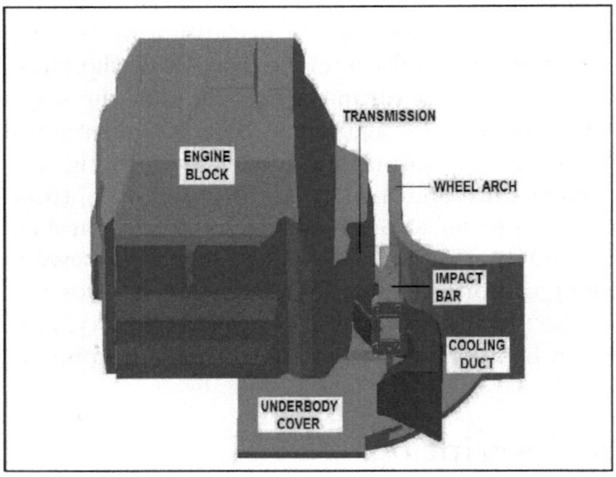

Fig. 2. Underbonnet components included in the duct optimization model (Image courtesy of Audi AG)

the multi-objective optimization software modeFRONTIER [4] to describe the optimization process and find the optimal design layout.

Steps to perform the cooling duct optimization can be enumerated as follows:

1. Retrieve data and generate the parametric model
2. Create automatically a polyhedral mesh of the flow domain
3. Setup of CFD analysis running on a multi-CPUs cluster
4. Setup an automatized optimization strategy using modeFRONTIER
5. Perform automatic multi-objective optimization
6. Multi-Criteria Decision Making (MCDM) tool usage
7. Robust Design analysis
8. Final selection of best solution

The first stage in the work was to retrieve all the available data for the full underbonnet simulation, which also included the baseline duct geometry. The final geometry is presented in Fig. 2.

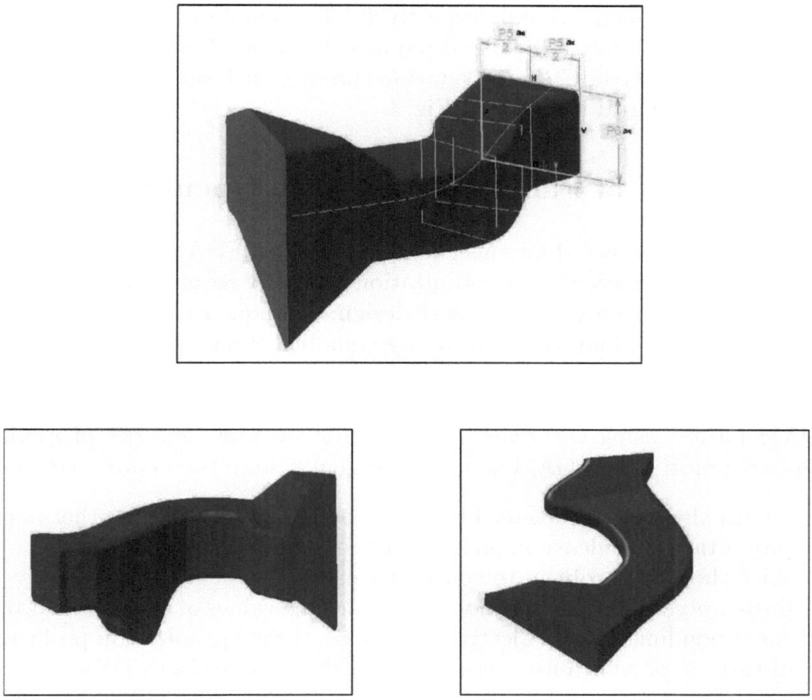

Fig. 3. Example of duct parameters and duct shapes resulting from extreme parameter combinations

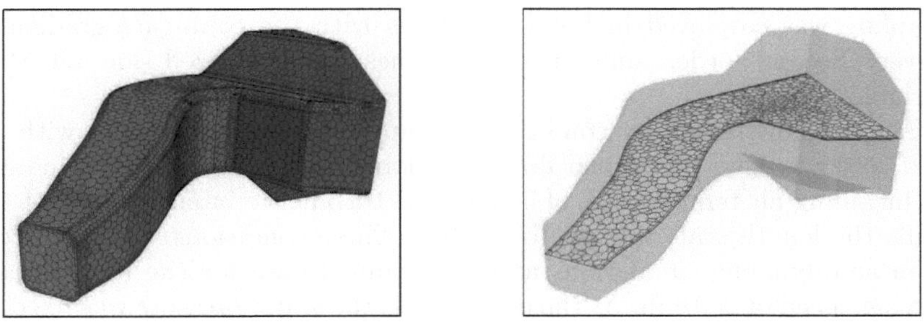

Fig. 4. Example of a polyhedral mesh in the duct geometry

The next step established the parametrization and the range of variation of the parameters controlling the geometry of the duct model. As a result, a total of 16 parameters were adopted; Fig. 3 illustrates the admissible variation of two of these parameters along with two configurations of the duct geometry for deliberately extreme values.

The cooling duct geometry was then imported into the chosen mesh generator (STAR–Design). The resulting grids were polyhedral with two wall extrusion layers of 0.6 mm and 1.1 mm respectively. The number of cells in the meshes totalled approximately 35,000 and required less than 3 minutes to be generated in a single Intel Xeon CPU workstation running on Linux. An instance of such a computational grid is shown in Fig. 4.

2 Summary of Flow Modeling with OpenFOAM

The CFD solver selected for this example was OpenFOAM [5], and open source CFD suite. Automatic design optimization generally requires many CFD computations to converge towards the best design; consequently, an open source solution offers the freedom to exploit a large number of hardware resources without the cost limitations imposed by commercial licensing.

The polyhedral mesh created in STAR–Design was first imported into OpenFOAM format using the *ccm24Foam* conversion tool. A series of input script files, written in OpenFOAM scripting language, were then employed to:

- set up the CFD problem, i.e.: apply boundary conditions, thermophysical properties, turbulence models, solver controls, etc.;
- solve the CFD problem to convergence;
- post-process the CFD solutions to extract the values of the flow field variables corresponding to the objectives functions of the optimization problem. Each of these actions is automatically handled by modeFRONTIER.

The air flow was computed as an ideal-compressible-subsonic turbulent gas. The standard high Reynolds $\kappa - \epsilon$ turbulence model with non-equilibrium wall function was applied. The y^+ values were verified at the walls on the model for selected geometries. The SIMPLE solution procedure for pressure-velocity coupling was employed in the calculations with the conjugate gradient linear solver. Second order differencing schemes were utilized for all the flow variables.

A mass flow boundary condition was applied at the duct inlet with a value of 0.506 kg/s. The prescribed flow direction was normal to the boundary with an incoming air temperature of 300 K. The turbulence intensity was set to 10%, while the length scale was set to 0.025 m. These conditions were approximated from aerodynamic simulations and wind-tunnel data for the full vehicle for a driving speed of 250 kph. At the duct outlet, the static pressure was fixed at 0 Pa (relative to the operating pressure of 101,325 Pa). The surfaces representing walls in the model were all defined to be adiabatic and no-slip stationary boundaries.

An upper limit of 500 iterations was specified in the event that a case would not converge. Post-processing of the simulation results was also automated so that modeFRONTIER could run the model analysis and extract/post-process the necessary information in batch mode. This information included the area average total pressure difference between the inlet and outlet as well as the area weighted average velocity on the outlet boundary.

3 Optimization Phase

The main effort in the optimization phase is dedicated to the work-flow definition, presented in Fig. 5 for the cooling duct application. The different components in the work-flow define each of the stages in the automated optimization process: the input variables, the objectives, the optimization loop settings, including the initial points and the optimization algorithm.

Figure 5 summarizes a complex modeFRONTIER's work-flow where Catia V5 [2] was used to perform the parametric shape variation running on a local Windows machine; STAR–Design was run in batch mode on a single CPU remote Linux machine by using a special purpose DOS remote shell; OpenFOAM was run on a remote 8 CPU Linux cluster. Since the concurrent run of threads of CFD solvers had to meet the license availability for both CatiaV5 and STAR–Design tools, python scripts were implemented in modeFRONTIER to perform the license polling as well as the concurrent execution of the CFD analyses for each design evaluation, thus maximizing the exploitation of the computational resources. *Grid computing* promises to deliver benefits by making use of all the available hardware resources. Grid technology [11] already showed its value in scientific research. More precisely, this technology allows reduction of CFD running time, by means of simultaneous computations on parallel hardware resources. As a result, large numbers of design configurations are computed in appreciably shorter time if compared with analogous tasks performed in traditional serial architectures.

Finally, given the multi-objective nature of the problem, a distributed and multi-objective genetic algorithm was selected in modeFRONTIER. Since several preliminary studies [6] showed no qualitative difference between the MOGA-II and other state-of-the-art methods in the forefront of multi-objective optimization (such as NSGA-II [1]), here we limited the study to MOGA-II. An in-depth description of MOGA-II is provided in the Sect. 4.

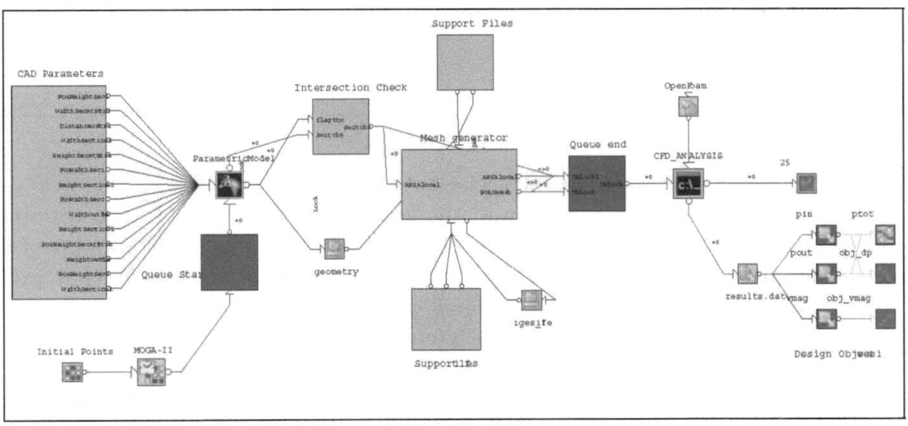

Fig. 5. modeFRONTIER work-flow for duct optimization

4 MOGA-II

MOGA-II is an improved version of MOGA (Multi-Objective Genetic Algorithm) of Poloni [8]. It uses a smart multi-search elitism for robustness and directional crossover for fast convergence. The efficiency of MOGA-II is controlled by its operators (classical crossover, directional crossover, mutation and selection) and by the use of *elitism*. The internal encoding of MOGA-II is implemented as in classical genetic algorithms [3]. Each variable is represented as a binary string where the length of the string depends on the base (the number of allowed values for the variable). Elitism plays a crucial role in multi-objective optimization because it helps preserving the individuals that are closest to the Pareto front and the ones that have the best dispersion.

MOGA-II uses four different operators for reproduction: one point crossover, directional crossover, mutation and selection. At each step of the reproduction process, one of the four operators is chosen (with regard to the predefined operator probabilities) and applied to the current individual. Algorithm 1 shows the reproduction in pseudo code.

Algorithm 1. Pseudo code of the reproduction used in MOGA-II

```
with (individual Ind_i ∈ generation G) do
   choose reproduction operator
   if (operator is one point crossover) then
      j ← TournamentSelection, where j ≠ i
      NewInd_i ← OnePointCrossover(Ind_i, Ind_j)
   else if (operator is directional crossover) then
      j ← RandomWalk(i)
      k ← RandomWalk(i), where k ≠ j ≠ i
      NewInd_i ← DirectionalCrossover(Ind_i, Ind_j, Ind_k)
   else if (operator is mutation) then
      NewInd_i ← Mutation(Ind_i)
   else if (operator is selection) then
      NewInd_i ← Ind_i
   end if
end with
```

One point crossover is the most classical operator for reproduction. Two parents are chosen and some portion of the genetic material (the design variables) is exchanged between the parent variables vectors (see Fig. 6). The point of the crossing site is randomly chosen and the binary strings are cut at that point. The two head pieces are then swapped and rejoined with the two tail pieces. From the resulting individuals, usually called children, one is randomly selected to be the new individual. In MOGA-II, one point crossover starts by taking the current individual Ind_i as the first parent. The second parent Ind_j is chosen by means of a multi-objective tournament selection on a randomly selected population subset: this operator returns the first non-dominated solution in the subset.

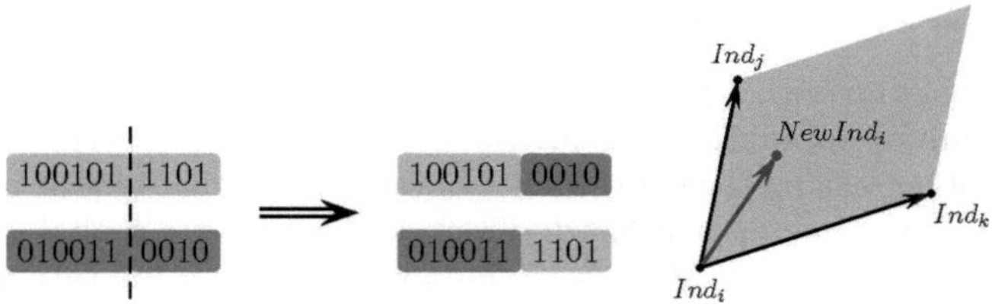

Fig. 6. One point crossover (left) and directional crossover between individuals Ind_i, Ind_j and Ind_k (right)

Fig. 7. Mutation example with DNA string mutation ratio set to 40%

Directional crossover is slightly different and assumes that a *direction of improvement* can be detected comparing the fitness values of two reference individuals. In [10] a novel operator called *evolutionary direction crossover* was introduced and it was shown that even in the case of a complex multi-modal function this operator outperforms classical crossover. The direction of improvement is evaluated by comparing the fitness of the individual Ind_i from generation t with the fitness of its parents belonging to generation $t-1$. The new individual is then created by moving in a randomly weighted direction that lies within the ones individuated by the given individual and his parents (see Fig. 6). A similar concept can be however applied on the basis of directions not necessarily linked to the evolution but detected by selecting two other individuals Ind_j and Ind_k in the same generation (like shown in Algorithm 1).

The *selection* of individuals Ind_j and Ind_k can be done using any available selection schema. In MOGA-II local tournament with random steps in a toroidal grid is used. First of all, the individual subject to reproduction is chosen as the starting point. Other individuals met in a random walk of assigned number of steps from that starting point are then marked as possible candidates for the first "parent" Ind_j. The list of all possible candidates for the second "parent" Ind_k is selected in the same way in a successive (and generally different) random walk from the same starting point. When the set of candidates is generated, the candidate with the best fitness is chosen. The number of steps N in the random walk remains fixed during the entire optimization run and is proportional to the population size.

Mutation is an operator that ensures *diversity* from one generation to the next. Using plain words we can say that mutation guarantees the algorithm robustness. In MOGA-II it is possible to define the value of the so-called *DNA String Mutation Ratio*. This value gives the percentage of the binary string that is perturbed by the mutation operator.

4.1 MOGA-II Results

MOGA-II was run with a population of 28 initial designs, evolving for 50 generations. The initial population was provided by the *Sobol*[9] method because of its capability of increasing the convergence of multi-objective genetic algorithms [7]. The following parameter were used: directional cross-over probability 50%, classical cross-over probability 35%, selection probability 5%, mutation 10%, elitism and steady evolution. When steady evolution is enabled, MOGA-II uses all the values as soon as they are available in a *first in – first out* way guaranteeing the complete parallelization of the optimization process. The total number of evaluated designs was 1,400 over a period of time of 72 hours. Each single complete evaluation took an average of 15 minutes, the process was parallelized on 8 CPUs.

Figure 8 shows the optimization results for the two objectives: total pressure drop *dp* and discharge velocity v_{mag}. The chart clearly illustrates that the algorithm generates a well-spread set of non-dominated points. Unfortunately, due to the complexity of the problem, nothing can guarantee that this represents the true Pareto front. Anyhow, the robustness demonstrated by MOGA-II on several numerical tests [6] gives us high hopes that, at least, these points represent a set of good solutions.

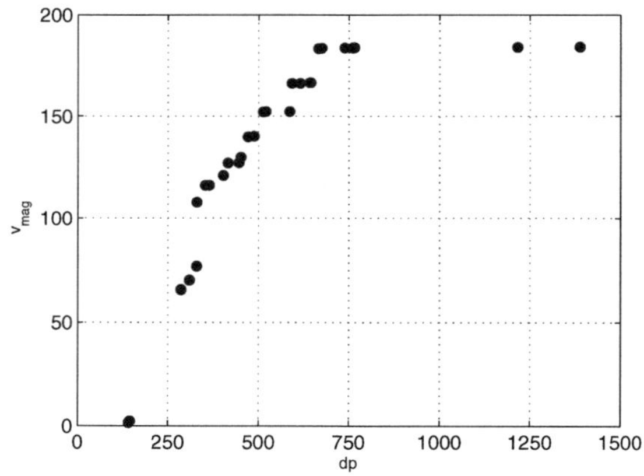

Fig. 8. Total pressure drop *dp* versus discharge velocity v_{mag}. The set of non-dominated designs coming out from MOGA-II optimization.

5 Multi-criteria Decision Making

Since there are more than one conflicting objective to be optimized simultaneously, there is no longer a single solution, but rather a whole set of possible solutions performing differently on the different objectives. Even though several solutions may exist only one has to be chosen and ranking between all the alternatives is a delicate issue. In order to facilitate the choice of a good

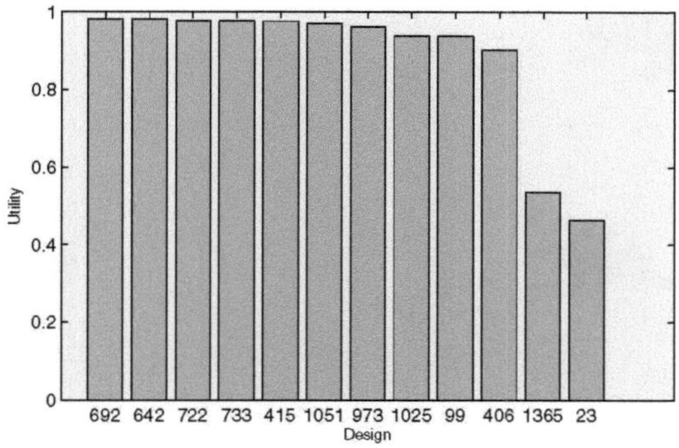

Fig. 9. Ranking of a subset of the best designs produced by the MCDM tool (right)

compromise between objectives from a Pareto front, the *Multi-Criteria Decision Making* (MCDM) tool featured in modeFRONTIER was employed.

In the case in study, the most important objective was to reduce the pressure drop *dp* while getting the maximum possible values for velocity magnitude v_{mag} at the outlet. To avoid excessive Mach numbers inside the duct, the acceptable velocity was set in the range $90 \leq v_{\mathrm{mag}} \leq 120\,\mathrm{m/s}$.

The MCDM tool was then employed in order to impose rational and transitive (i.e. coherent) relationships in terms of pair-wise comparisons only. Following this approach, all the designs falling under the range of the acceptable velocities were considered. The set of non-dominated desings was then scanned by taking into account the new range imposed on the velocity magnitude. Unfortunately, no non-dominated solutions were found in the range of $[90, 100]\,\mathrm{m/s}$. Consequently, for the latter range of velocity, the best designs in terms of pressure drop were added to the set of the existing non-dominated points, thus leading to a final set of candidate solutions composed of twelve points in total. The two objectives were then weighed within the MCDM tool, by assigning to both the objectives the same level of importance in the range of acceptable velocities. As a result, the existing non-dominated solutions were ranked, as shown in Fig. 9. Finally, the top three designs were extracted from the output list produced by the MCDM tool.

6 Robustness of Solutions

In order to select the final design, a robust analysis was performed on the first three candidate solutions achieved with the MCDM tool. All the deterministic input variables have to be re-defined as normal stochastic distributions with a specific value of standard deviation. In this way, for each design selected in the previous step, $N = 130$ sample designs are created according to a multivariate normal distribution with mean vector $\mu = (x_1, \ldots, x_{16})$ corresponding to the nominal value of each geometrical parameter. All the variables are considered as independent, hence the covariance matrix is a diagonal matrix and each standard deviation is equal to the 10% of the nominal value (i.e. $\sigma_i = 0.1 x_i$).

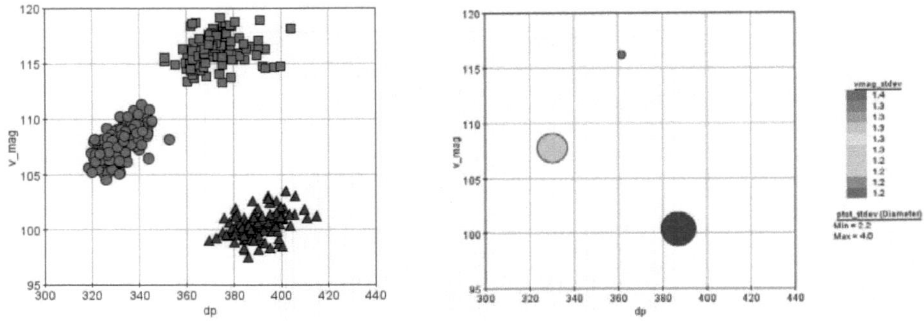

Fig. 10. Robustness analysis results: chart of the three different distribution samples (left) and bubble chart of the standard deviation with respect to the two objectives (right)

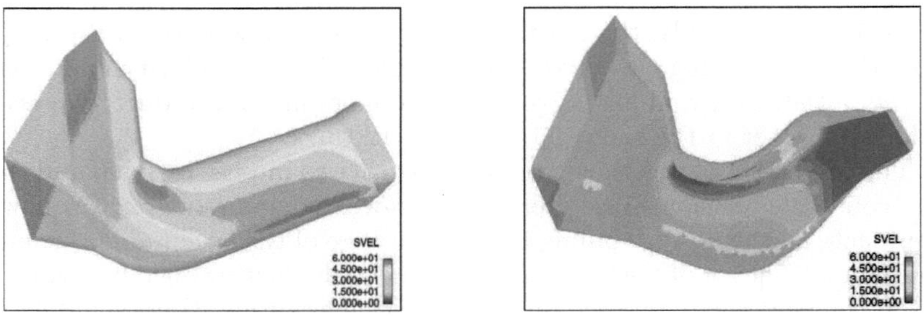

Fig. 11. Velocity magnitude at mid-span: baseline design (left) and optimized design (right)

Each of the three different distributions was created using *Latin Hypercube Sampling* (LHS) that is a particular Monte Carlo: more precisely it is a constrained Monte Carlo scheme. The constraint refers to the way each variable is sampled: the statistical distribution is split into $N = 130$ equally probable intervals, and then a random value is selected within each interval. In this way the N points are relatively uniformly distributed over the density function range.

The results from the sensitivity analysis allows the selection of the best robust design from the set of non-dominated solutions. The final configuration can now be chosen by considering the lowest standard deviation of the objectives as the main criterion, in addition to the standard optimization of the design objectives.

Figure 10 shows the response of the three candidate solutions based on the design objectives. The color of each bubble indicates the standard deviation for velocity magnitude, while the size of the bubble is proportional to the standard deviation value for pressure drop. From this, the candidate with the lowest pressure drop can be considered the best compromise in terms of velocity magnitude and, most importantly, standard deviation of the two objectives. Consequently, it can also be considered to be the best robust solution of our multi-objective optimization problem.

Results for velocity magnitude are compared in Fig. 11 for both the baseline design and the optimized configuration. Overall, the pressure drop was reduced from 357 to 330 Pa, while the discharge velocity was increased from 39.4 to 107.8 m/s.

7 Concluding Remarks

This paper aims to show the benefits of evolutionary multi-objective algorithms such as MOGA-II applied to design process, by using parametric models and distributed computational resources for flow analysis. The traditional design process based on the "trial-and-error" approach was replaced by an efficient methodology in order to find out and screen the optimal layout of a cooling duct. Table 1 summarizes the enhancement obtained from the original design in the described optimization procedure. The fast and easy integration of OpenFOAM in modeFRONTIER allowed improvement in the geometry of a cooling duct located in the underbonnet region of a vehicle. The duct geometry was modified through 16 parameters to minimize the pressure drop from inlet to outlet and to maximize the discharge velocity in order to improve the cooling performance of the vehicles transmission. The open source package OpenFOAM coupled with the MOGA-II algorithm allowed a concurrent designs evaluation in a very short time by an effective exploitation of all the computational resources available.

Table 1. Performance improvements

	dp	v_{mag}
baseline design	357 Pa	39.4 m/s
optimized design	330 Pa	107.8 m/s
total improvement	27 Pa	68.4 m/s
% of improvement	7.56%	173.60%

A further robustness analysis was then performed on the candidate solutions of the multi-objective problem extracted by using the MCDM decision support tool, thus finally yielding the best stable design configuration in terms of performance.

References

1. K. Deb, S. Agrawal, A. Pratab, and T. Meyarivan. "A Fast Elitist Non-Dominated Sorting Genetic Algorithm for Multi-Objective Optimization: NSGA-II." Proceedings of the Parallel Problem Solving from Nature VI Conference, Springer. Lecture Notes in Computer Science No. 1917, Paris, France, pp. 849–858, 2000
2. CATIA V5, See `http://www.ibm.com/catia`
3. D. E. Goldberg. "Genetic Algorithms in Search, Optimization and Machine Learning". Addison–Wesley, Reading Mass, USA, 1989.

4. modeFRONTIER version 3 Documentation. See http://www.esteco.com
5. OpenFOAM: The Open Source CFD Toolbox.
See http://www.opencfd.co.uk/openfoam
6. S. Poles, "Bench-marking MOGA-II". Technical report 2004-001, Esteco, Trieste, 2003.
7. S. Poles, Y. Fu, E. Rigoni "The Effect of Initial Population Sampling on the Convergence of Multi-Objective Genetic Algorithms". MOPGP June 2006 Loire Valley, France
8. C. Poloni and V. Pediroda. "GA coupled with computationally expensive simulations: tools to improve efficiency". In *Genetic Algorithms and Evolution Strategies in Engineering and Computer Science*, pages 267–288, John Wiley and Sons, England, 1997.
9. I. M. Sobol "On the Systematic Search in a Hypercube". SIAM Journal on Numerical Analysis, Vol. 16, No. 5 (Oct., 1979) , pp. 790-793.
10. K. Yamamoto and O. Inoue. "New evolutionary direction operator for genetic algorithms". *AIAA Journal*, volume 33, number 10, pages 1990–1993, 1995.
11. X. Yang and M. Hayes "Application of Grid techniques in the CFD field". Proceedings of Integrating CFD and Experiments in Aerodynamics, Glasgow, September 2003

Individual Evaluation Scheduling for Experiment-Based Evolutionary Multi-objective Optimization

Hirotaka Kaji[1] and Hajime Kita[2]

[1] Research and Development Operations, Yamaha Motor Co. Ltd.,
2500 Shingai, Iwata, Shizuoka, Japan
kajih@yamaha-motor.co.jp
[2] Academic Center for Computing and Media Studies, Kyoto Univercity,
Yoshida nihonmatsu-cho, Sakyo-ku, Kyoto, Japan
kita@media.kyoto-u.ac.jp

Abstract. Since the pioneer work of Evolution Strategies, experiment-based optimization is one of the promising applications of evolutionary computation. Recent progress in automatic control and instrumentation provides a smart environment called Hardware In the Loop Simulation (HILS) for such application. However, since optimization through experiment has severe condition of limited evaluation time and fluctuation of observation, we have to develop methodologies that overcome these problems. This paper discusses application of Multi-Objective Evolutionary Algorithms (MOEAs) to experiment-based optimization of control parameters of dynamical systems. In such applications, we have to apply various parameter candidates spreading near the Pareto frontier to the system, and it causes fluctuation of the observed criteria due to the transient response by parameter switching. For reduction of loss time caused by such transient response in evaluation of criteria, we propose techniques called Evaluation Order Scheduling and Evaluation Time Scheduling. Numerical experiments using a formal test problem and experiment in a HILS environment for real internal-combustion engines have demonstrated the effectiveness of the proposed methods.

1 Introduction

In these years, in automotive internal-combustion engines, their evaluation criteria such as environmental emissions (CO, HC, NO_x), fuel-consumption and engine torque, have to be balanced simultaneously at high level. To achieve such goals, many electronic control devices are mounted to the engines, and lots of parameters of the Engine Control Units (ECUs) have to be adjusted adequately. So far, this problem is solved by operator's manual calibration through experiment. However, to achieve higher engine performance on one hand and

S. Obayashi et al. (Eds.): EMO 2007, LNCS 4403, pp. 645–659, 2007.

to enhance productivity of design process on the other, automatic design based on multi-objective optimization is needed. One candidate is simulation-based optimization. However, to construct a physical model of an internal-combustion engine in detail requires a lot of effort, and it is not cost effective.

Since the pioneer work of Evolution Strategies [1,12,14], experiment-based optimization[1] is one of the promising applications of evolutionary computation. Recent progress in automatic control and instrumentation provides a smart environment called Hardware In the Loop Simulation (HILS) for such application. However, since optimization through experiment has to be completed under severe restriction of evaluation time and fluctuation of observation, we have to develop methodologies that overcome these problems.

In this paper, we discuss an experiment-based Evolutionary Multi-objective Optimization (EMO) to calibrate control parameters of an automotive internal-combustion engine using HILS. We have proposed a MOEA for noisy fitness functions and a crossover operator for periodic functions [8,9] to overcome these problems. However, because conventional MOEAs have been studied for simulation-based optimization, no MOEAs proposed so far can handle the adverse effect of the system dynamics appropriately. In this paper, we propose Individual Evaluation Scheduling (IES) which is composed of Evaluation Order Scheduling (EOS) and Evaluation Time Scheduling (ETS) for MOEAs to overcome this problem by reducing loss time for waiting for diminishing of transient response of engine control caused by parameter switching. The EOS is constructed based on a local search method for the traveling salesman problem so as to reduce the total magnitude of change in the parameters among population. Additionally, the ETS is defined based on the Euclidean distances between individuals so as to provide adequate estimate of length of loss time to wait.

This paper is organized as follows. In Section 2, our current studies are introduced. In Section 3, to handle the system dynamics appropriately, individual evaluation scheduling for MOEAs is introduced. The results of a test function with dynamics and a real engine experiment are shown in Sections 4 and 5 respectively. As a result, it is shown that the proposed method improves search ability of MOEAs in experiment-based optimization of dynamical systems.

2 Current Studies

2.1 Experiment-Based Optimization Under Hardware in the Loop Simulation Environment

In recent automotive development, we can use a smart environment called Hardware In the Loop Simulation (HILS) for the experiment-based optimization, thanks to the progress of automatic control and instrumentation. The HILS is a technique for simulating a whole system by synchronizing a simulator with a real machine to evaluate the performance of a large-scale system accurately in advance.

---

[1] In this paper, the term of "experiment-based optimization" is used in the meaning of "the system parameters of a real system are optimized directly by optimization techniques in real time through experiments".

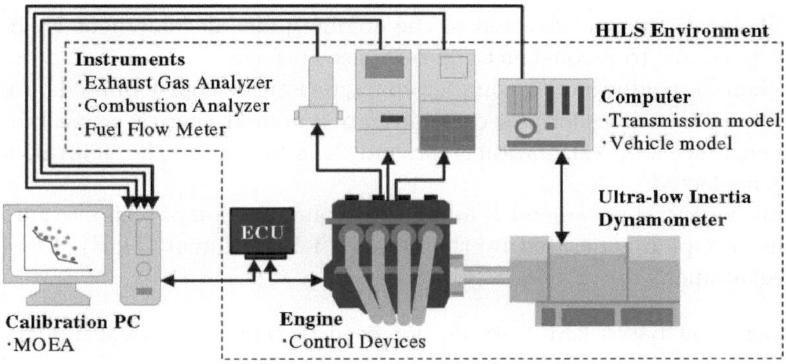

Fig. 1. Brock diagram of the engine HILS environment and the calibration PC implementing MOEAs

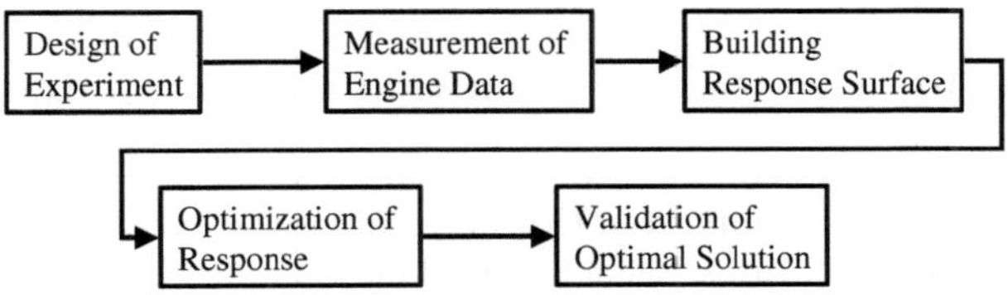

Fig. 2. The General optimization procedure using the RSM

Fig. 1 illustrates a engine HILS environment constructed by an engine test bed and a real internal-combustion engine. The engine test bed consists mainly of an ultra-low inertia dynamometer and a dynamo controlling computer having I/O interfaces. The ultra-low inertia dynamometer is connected with the crankshaft of the engine and the load is controlled in real time. Transmission and vehicle models are implemented in the computer to evaluate the engine on a condition almost equal to a real car. Moreover, an exhaust gas analyzer, a fuel flow meter, and a combustion analyzer, etc. are connected to measure the performance of engine. The ECU for engine control is connected with a calibration PC, and the control parameters of the ECU can be changed by the PC freely. In addition, the calibration PC monitors the outputs from the engine test bed, the instruments, and the ECU. Therefore, the MOEAs implemented in the calibration PC can handle this environment including a real machine in the similar way to a simulation.

In these days, the Response Surface Methodology (RSM) [10], which is a offline optimization technique using statistical models, is widely used for engine parameter calibration [3]. Fig. 2 shows general optimization procedure that uses the RSM. However, The HILS environment supports the automated measurement of engine data, but other processes have to be executed manually. Additionally, the RSM has following problems for the engine development:

- Whenever the specification of the engine under development is changed, it is necessary to reconstruct the response surfaces.
- When the engine having complex characteristics is approximated by a straightforward function such as second-order polynomial, there is a possibility for the occurrence large estimation error, and the influence on the calibration can not be neglected.
- If new control parameter is added, the configuration of response surfaces and the test plan generated by the Design of Experiment (DoE) [10] should be reexamined.

In experiment-based EMO, we do not need to build statistical models, and we are able to use the result directly. Therefore, some processes such as the DoE and the validation of optimal solution can be omitted, and the automated data measurement using the HILS lightens the burden of parameter adjustment. Since a certain number of the trial and error by MOEAs can be allowed, the experiment-based EMO using HILS environment can be a smarter calibration technique.

2.2 Multi-objective Genetic Algorithm for Noisy Fitness Functions

While MOEAs are studied intensively by many researchers (e.g. [2]), most of the work assume noise free evaluation. However, under the existence of noise, conventional MOEAs do not work satisfactorily. Hence, we have proposed the Memory-based Fitness Estimation and Distribution-based Selection GA (MFE-DSGA) [8] as a MOEA for noisy fitness functions considering its application to experiment-based optimization. The MFE-DSGA introduces three features: (1) a fitness estimation method, which is an enhancement of the MFEGA proposed by Sano et al. [13] for EMO, (2) a selection method that pays attention to distribution of individuals, and (3) the α-domination strategy proposed by Ikeda et al. [7] for removing individuals which are misconceived as non-dominated individuals by observation noises. We applied this method to the optimization of a controller for real engine in a HILS. Emissions of HC and NO_x were optimized as a two-objective optimization problem.The result of optimization are shown in Fig. 3.

Through the discussion with the experts of engine calibration, it was confirmed that the result was appropriate as a performance of engine used by this experiment, and the proposed method is useful for parameter calibration of ECUs because the trade-off is visualized. On the other hand, it was understood that the experts does not obtain the whole Pareto optimal set such as the EMO, but the time necessary to find similar knowledge is less than half needed by the EMO. To put this method into practical use, it is necessary to reduce the total time for optimization to make it shorter than the operation time by the experts at least. There are two approaches to shorten the optimization time: one is removing factors which decrease search accuracy to attain fast convergence, the other is reducing the number of evaluation of MOEAs. In this paper, the former approach is examined. we have already proposed a crossover operator for periodic functions which typically appear to the engine control and which is introduced in Section 2.3. Moreover, to reduce the influence of dynamics of

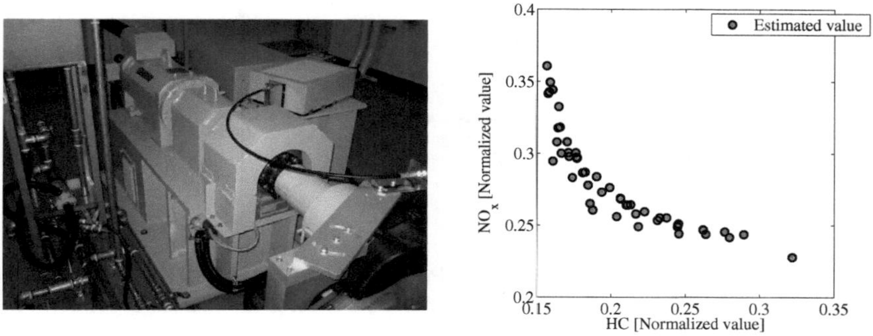

Fig. 3. Ultra-low inertia dynamometer used for the HILS (left) and population distribution on the objective function space in multi-objective optimization of a real engine (right)

the target system, we newly propose Individual Evaluation Scheduling which is explained in Section 3. After that, these methods are combined and evaluated in a real experiment in Section 5.

2.3 Crossover Operator for Periodic Functions

In engineering, a timing optimization for apparatus with cycling mechanism such as internal-combustion engine can be defined as an optimization problem of periodic function. For instance, in general four-stroke gasoline engines, the fuel injectors inject fuel into the cylinders once while the crank shaft makes two rotations. As a result, the fuel injection timing 0 and 720 degrees BTDC (Before Top Dead Centre) have same combustion effect, and outputs of engine become periodic functions for the fuel injection timing. However, naive application of a real-coded GA to such problem faces difficulties of sampling bias[2] [4,5] and evolutionary stagnation[3].

For these problems, we have proposed the UNDX-P (Unimodal Normal Distribution Crossover for Periodic function) [9] that enhances the UNDX proposed by Ono et al. [11], a crossover operator for real-coded GA. Conceptual diagram of UNDX-P is illustrated in Fig. 4, and the proposed algorithm is as follows:

1. Obtain a point $(x^{\mathrm{p}}, y^{\mathrm{p}}) = (\cos\theta^{\mathrm{p}}, \sin\theta^{\mathrm{p}})$ on a unit circle S^1 corresponding θ^{p} a variable of periodic function.
2. The UNDX is adopted for parents converted to points on a unit circle $S^1 = \{(x, y) \in \mathbf{R}^2 | x^2 + y^2 = 1\}$, and children $(x^{\mathrm{c}}, y^{\mathrm{c}})$ are generated as points on the two-dimensional space $\mathbf{R}^2$

[2] Since most crossover operators of real-coded GAs often generate offspring to the vicinity of the center of the search space, it is difficult to optimize functions whose optimal solution exists near the boundary.

[3] When plural powerful local optimal solutions exist far apart, offspring are generated in the areas which step over them, and the search stagnates.

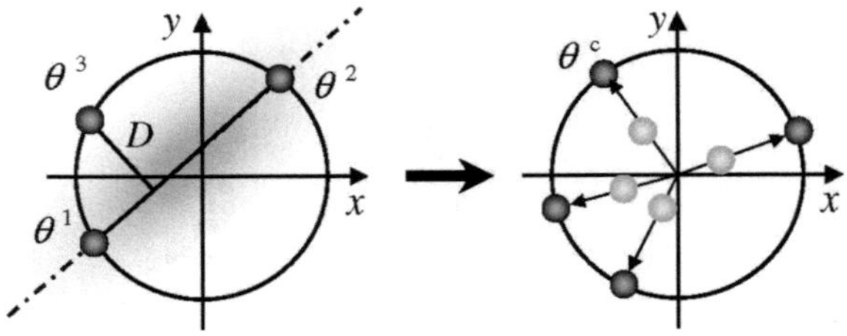

Fig. 4. Concept diagram of UNDX for Periodic function. The UNDX is adopted for three parents θ^1, θ^2 and θ^3 converted to the points on a unit circle S^1 (left). Angular variables θ_c of the generated children on the $\mathbf{R}^2$ are calculated by Eq. (1) (right).

3. For each of the generated children, angular variable θ^c is calculated by

$$\theta^c = \begin{cases} \tan^{-1}\frac{y^c}{x^c} & \text{if } x^c \geq \epsilon \\ \mathrm{sgn}(y^c)\cdot\left(\pi - \tan^{-1}\left|\frac{y^c}{x^c}\right|\right) & \text{if } x^c \leq -\epsilon \\ \mathrm{sgn}(y^c)\cdot\frac{\pi}{2} & \text{if } |x^c| < \epsilon,\ |y^c| \geq \epsilon \\ 0 & \text{if } |x^c| < \epsilon,\ |y^c| < \epsilon \end{cases} \quad (1)$$

where sgn(·) is the signum function, and ϵ is a sufficiently small positive number[4].

3 Individual Evaluation Scheduling for Dynamical Systems

Experiment-based optimization has to be carried out under uncertainty such as system and observation noise within the quite limited evaluation time which is restricted by operation time and durability of machine. If the target system to be optimized is a dynamical system, we have to wait until the transient response caused by switching of system parameters are diminished and optimized, and reduce its influence on observed performances. Since we have to apply many parameter candidates distributed widely as population, we have to manage such problems in evolutionary approach, especially in EMO.

As an example of a dynamical system, a time series of output torque where control parameters of an engine was switched at constant intervals is shown in Fig. 5. It is understood that the change of parameters cause the transient responses. Hence, when the MOEAs are applied for the optimization of system parameters of dynamical systems, the following dilemmas are caused:

- The performance of an individual should be measured after it settles enough to evade the influence of the uncertainty caused by dynamics.

[4] Generally, Eq. (1) is implemented in programming languages as a function called atan2(y, x).

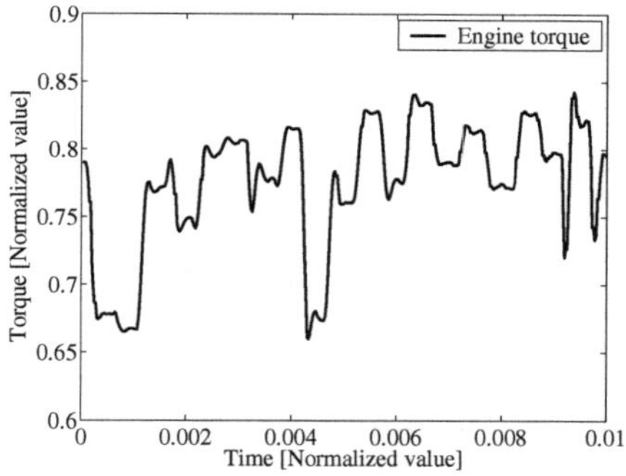

Fig. 5. Transient response of an engine torque

– The measurement time of individual should be shortened as much as possible since MOEAs require lots of evaluation times under the limited total evaluation time.

In this paper, we propose a method called Individual Evaluation Scheduling (IES) to improve performance of experiment-based optimization of system having dynamics. It consists of two ideas, i.e., Evaluation Order Scheduling (EOS) and Evaluation Time Scheduling (ETS). The former one is a technique of deciding evaluation order in the population to improve accuracy of the performance by reducing the total magnitude of parametric change. The latter one is to adjust waiting time for transient response.

3.1 Evaluation Order Scheduling

Consider a target to be optimized which is a stable dynamical system:

$$\dot{\boldsymbol{x}} = f(\boldsymbol{x}, \boldsymbol{a}), \tag{2}$$

where $\boldsymbol{x} = (x_1, x_2, \ldots, x_k)$ is a state variable vector, and $\boldsymbol{a} = (a_1, a_2, \ldots, a_n)$ is a system parameter vector represented by an individual of MOEAs and switched at prescribed intervals. The transient response according to individual change is caused by e.g., the shifting of the equilibrium point of Eq. (2). Fig. 6 shows concept diagram of the invalid evaluation time generated by the transient response due to the switching parameters.

It can be expected that the nearer parameter change makes the smaller equilibrium point changes excluding nonlinear phenomena such as bifurcation. As a result, the transient response can be expected to be suppressed, and if the system settles at its steady state faster, the accuracy of the estimated value can be improved. Hence, the problem is to find the order of evaluation of the population that achieves the minimal total magnitude of parametric change. It can be formulated as a similar problem of Traveling Salesman Problem (TSP). For a TSP, when a certain circuit shown in Fig. 7 is given, a circuit replaced

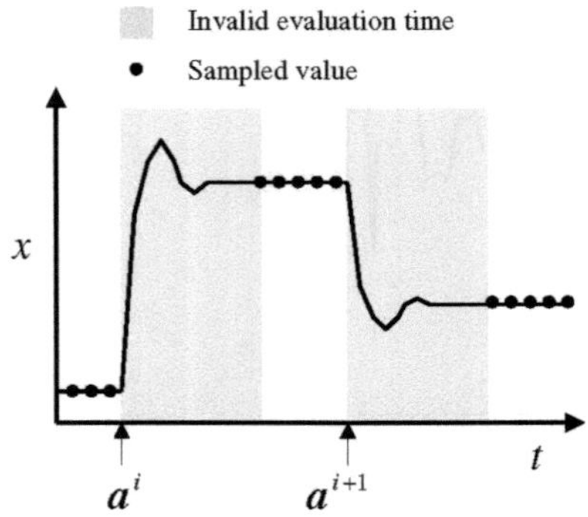

Fig. 6. Invalid evaluation time

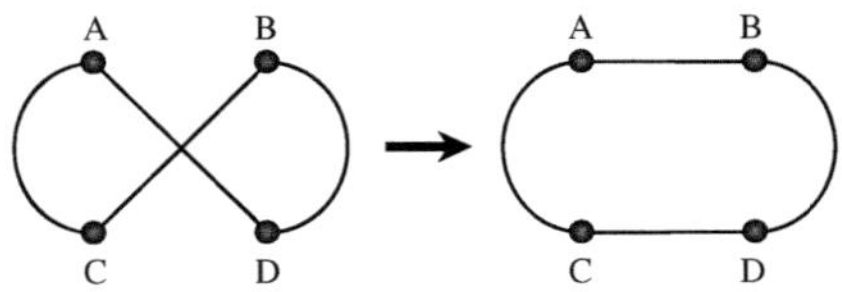

Fig. 7. 2-opt neighborhood

two arbitrary edges is called a 2-opt neighborhood. The local search using 2-opt neighborhood is known as a simple but effective heuristics of the TSP. Note that we have to find the shortest path with the starting point determined by the system parameter used in the current operation. The EOS algorithm based on the 2-opt method is described as follows:

1. A population whose order is optimized is defined as $A^0 = \{\boldsymbol{a}^0, A\}$, where $A = \{\boldsymbol{a}^1, \boldsymbol{a}^2, \ldots, \boldsymbol{a}^N\}$ is a population to be evaluated, N is the number of individuals, and $\boldsymbol{a}^0$ is an individual which was evaluated at last in the previous generation. Note that the order of $\boldsymbol{a}^0$ is fixed at the first one.
2. A permutation of A is defined as $Z = (z_1, z_2, \ldots, z_N)$, and Z is initialized.
3. The path length d_{total} is calculated by
$$d_{\text{total}} = \sum_{i=1}^{N} d_{z_{i-1}, z_i}, \tag{3}$$
where $z_0 = 0$, $d_{z_{i-1}, z_i} = \sqrt{\sum_{l=1}^{n} w_l (a_l^{z_{i-1}} - a_l^{z_i})^2}$, and w_l is weight parameter. It should be noted that we do not need to obtain a closed path, and therefore we exclude the length return to $\boldsymbol{a}^0$.
4. The path length of 2-opt neighborhoods of A^0 given by Z are examined.
5. If there exists a path in the 2-opt neighborhood whose d_{total} is shorter than that of the current path, it is employed as a new path, and then return to Step 4. Otherwise, Z is read out as a locally optimum permutation, that is, the evaluation order of A.

Fig. 8 shows the concept diagram of the EOS. In the EOS, adequate normalization of decision variables should be employed in advance, since we use distance among parameter sets.

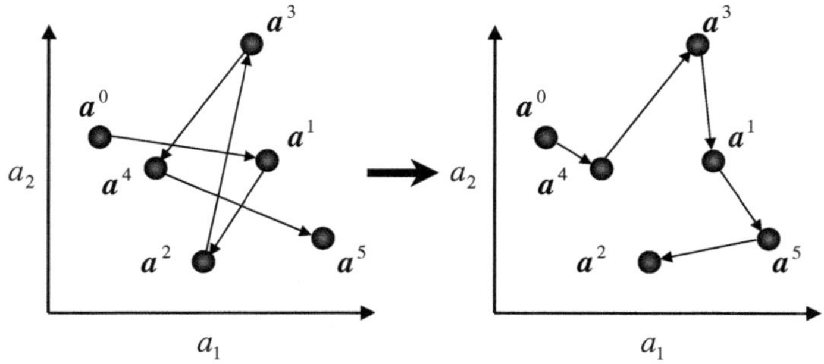

Fig. 8. Concept diagram of the Evaluation Order Scheduling

3.2 Evaluation Time Scheduling

Along with convergence of the population, shift of a equilibrium point of a dynamical system gradually diminishes. Consequently, the transient response generated by switching individuals gradually becomes small as well.

Thus, we attempt to change the invalid evaluation time. Assume that the EOS is adopted for the initial population $A(0)$ and the current population $A(t)$ as generation t. First, the mean edge length of initial population is calculated as $\bar{d}(0) = d_{\text{total}}(0)/(|A(0)|-1)$, where $d_{\text{total}}(0)$ is the path length of $A(0)$, and $|A(0)|$ is the size of $A(0)$. The invalid evaluation time of a individual $\boldsymbol{a}^i$ is adjusted as follows:

$$IET_i = IET_{\max} \cdot \frac{d_{i-1,i}}{\bar{d}(0)}, \; i = 1, 2, \ldots, N, \tag{4}$$

where $IET_{\max}$ is called the maximum invalid evaluation time, and it is a parameter to be set in advance.

4 Numerical Experiment

4.1 Experiment Settings and Measures

In this section, the performances of the proposed and conventional methods were compared through simulation of a dynamical system. Consider the following system consisting of four independent mass-damper-spring systems that share common adjustable parameters K_1 and K_2:

$$M\ddot{x}_1 + D\dot{x}_1 + K_1 x_1 = -Mg \tag{5}$$
$$M\ddot{x}_2 + D\dot{x}_2 + K_2 x_2 = -Mg \tag{6}$$
$$M\ddot{x}_3 + D\dot{x}_3 + 2K_1 x_3 = -Mg \tag{7}$$
$$M\ddot{x}_4 + D\dot{x}_4 + 2K_2 x_4 = -Mg, \tag{8}$$

where x_i is the position of the mass in which freedom length of spring is assumed to be zero, M is the mass, D is a damping coefficient, and g is the gravity

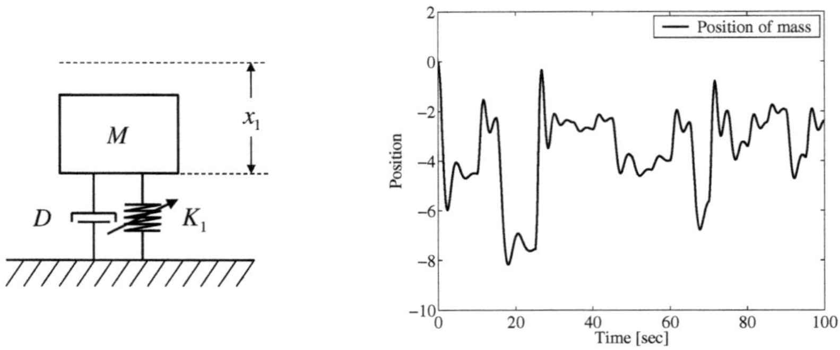

Fig. 9. Mass-damper-spring system (left) and its transient response (right)

acceleration, respectively. Consider a problem of putting positions x_1, x_2, x_3, x_4 to the desired position x_d by adjusting K_1, K_2. Since the gravity acceleration g affects vertical direction of each mass, the transient state is caused by the movement of equilibrium points when K_1, K_2 are switched. With this system, fitness functions are defined as:

$$f_1 = \sum_{i=1}^{2}(x_\mathrm{d} - \hat{x}_i)^2, \quad f_2 = \sum_{i=3}^{4}(x_\mathrm{d} - \hat{x}_i)^2, \tag{9}$$

where $\hat{x}$ is the estimated value of the steady state position. The mass-damper-spring system and the time series of Eq. (5) of which parameter K_1 was switched at random every five seconds are shown in Fig. 9. We can see that the transient response caused by switching K conform to the example in Fig. 5 very well.

In this system, because the equilibrium points are different between Eqs. (5)(6) and Eqs. (7)(8) for the same parameter set, there exists trade-off between f_1 and f_2. For $\hat{x}$, we used mean of sampled values taken in one second after the invalid evaluation time since an individual was switched. The sampling rate was 100ms on the simulation. As for parameter values, $M = 1$, $D = 1$, $K_1, K_2 \in [1, 5]$, $g = 9.81$, $x_\mathrm{d} = -3$ and $w_1 = w_2 = 1$ were used.

In this paper, the NSGA-II proposed by Deb et al. [2] was employed as a MOEA. Individual was coded as $\boldsymbol{a} = (K_1, K_2)$. The population size $|P| = 50$, and the offspring population size $|Q| = 50$ were used. The UNDX was used for crossover. Since the evaluation value includes uncertainty due to dynamic behavior of the system, the population was re-evaluated in this numerical experiment, i. e., $A(t) = P(t) \cup Q(t)$ was used as the evaluation population. The search was ended at the evaluation time of 5000 seconds on the simulation. The true fitness functions excluding the system dynamics are calculated theoretically as follows:

$$f_1^\mathrm{true} = \sum_{i=1}^{2}\left(x_\mathrm{d} - \frac{Mg}{K_i}\right)^2, \quad f_2^\mathrm{true} = \sum_{i=1}^{2}\left(x_\mathrm{d} - \frac{Mg}{2K_i}\right)^2. \tag{10}$$

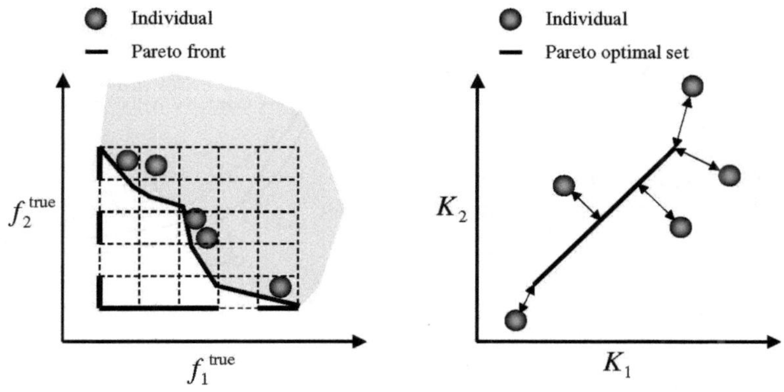

Fig. 10. Concept diagrams of coverage measure ($m = 2$, left) and mean absolute error measure (right)

In this experiment, 30 trials that used different initial populations were executed, and performances of the MOEAs were compared by the mean values of the trials for the two evaluation measures described below:

Coverage: This measure is proposed by Hiroyasu et al. [6], and it indicates the ratio of Pareto-front which is covered by the population. The coverage is defined as

$$C = \frac{1}{m} \sum_{i=1}^{m} \frac{c_i}{c_{\max}}, \tag{11}$$

where m is the number of the objective functions, $c_{\max}$ is the number of small areas where a hyper-plane composed of $m-1$ objective functions are evenly divided, and c_i is the number of areas including the true fitness of individuals projected to the hyper-plane. In our case, $m = 2$ and $c_{\max} = 25$ were used.

Mean Absolute Error: This measure indicates the error of the population for the true Pareto-optimal set. In this case, it is given by $K^* = \{K_1, K_2 \in [1.635, 3.27] | K_1 = K_2\}$. The mean absolute error is defined as the mean value of Euclidean distances from each individual to the nearest solution in K^*.

The concepts of the coverage measure and the mean absolute error measure are illustrated in Fig. 10. We compared the following cases:

Case 1: normal NSGA-II, $IET_{\max} = 1$ [sec].
Case 2: normal NSGA-II, $IET_{\max} = 4$ [sec].
Case 3: NSGA-II+EOS, $IET_{\max} = 1$ [sec].
Case 4: NSGA-II+EOS, $IET_{\max} = 4$ [sec].
Case 5: NSGA-II+EOS+ETS, $IET_{\max} = 4$ [sec].

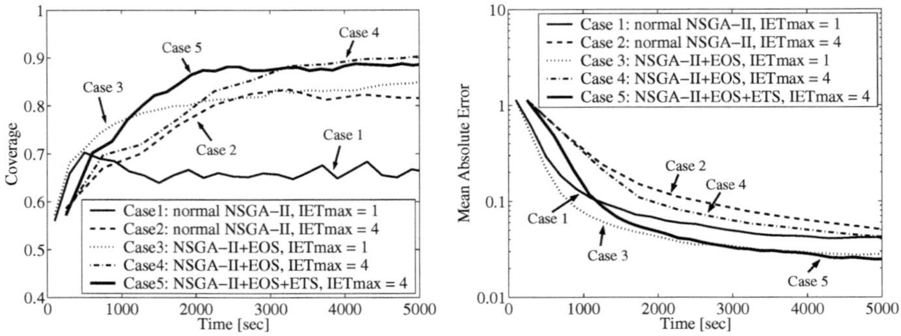

Fig. 11. Transition of the coverage (left) and the mean absolute error (right)

4.2 Discussion of Results

The results are shown in Fig. 11. The coverage and the mean absolute error were greatly improved by EOS in the comparison between Case 1 and 3, while the effect of the improvement was not too large in the comparison between Case 2 and 4. This is because the effect of the EOS became small since the transient response were settled by the enough invalid evaluation time on Case 2. Therefore, the EOS which suppresses the uncertainty caused by the system dynamics is effective for real environments, in which it is difficult to secure the long invalid evaluation time.

At the early stage of the search, the coverage of Case 3 was better than Case 4 because the number of generation could be increased in a same optimization time. However, the coverage of Case 4 overcame the Case 3 at about 2000 seconds and obtained the best result finally. On the other hand, for the mean absolute error, the convergence velocity of Case 3 was the fastest.

For the ETS, the coverage of Case 5 improved at the rising velocity comparably to that of Case 3, and the coverage of Case 5 that nearly equals to that of Case 4 was obtained in the end. Additionally, the mean absolute error of Case 5 was slightly better than that of Case 4 after 3000 seconds. Table 1 indicates the mean and standard deviation of the coverage and the mean absolute error, and the case showing the best performance is indicated in the bold-font. In conclusion, it is understood that the Pareto-optimal solutions having the high coverage and the small mean absolute error were obtained by the EOS and the ETS.

5 Real Engine Experiment

5.1 Experiment Settings

To examine the effectiveness of proposed method, two MOEAs, the normal NSGA-II and the NSGA-II+IES[5] were implemented in the calibration PC respectively, and the PC was connected to the ECU by serial communication.

[5] NSGA-II+IES means NSGA-II+EOS+ETS.

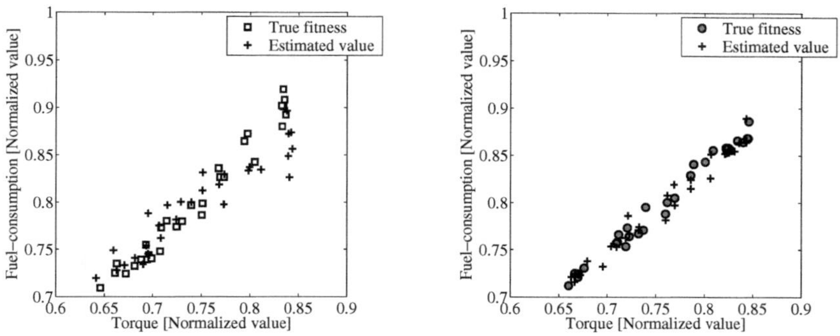

Fig. 12. Population distribution in the objective function space (left: NSGA-II, right: NSGA-II+IES)

Table 1. Performance comparison of the test problem for five cases

Case	Coverage		Mean Absolute Error	
	Mean	Std. Dev.	Mean	Std. Dev.
Case 1	0.66133	0.05251	0.04101	0.01050
Case 2	0.81533	0.02763	0.04765	0.00935
Case 3	0.85200	0.04888	0.02740	0.00746
Case 4	**0.90667**	0.03614	0.04106	0.00818
Case 5	0.88733	0.03503	**0.02501**	0.00615

Objective functions were to maximize engine torque and to minimize fuel-consumption. They were measured at a constant engine speed. Decision variables were fuel-injection timing, ignition timing, target air-fuel ratio[6] and valve control parameter. In general, there is a trade-off between the engine torque and the fuel-consumption, and this trade-off changes complicatedly by changing the above-mentioned engine control parameters.

The population size $|P| = 30$, and the offspring population size $|Q| = 30$ were used. Because the fuel-injection timing is a periodic function, the UNDX-P was adopted as a crossover method. The population P was re-evaluated as we did in the previous experiment. Search was terminated when the number of evaluations reached 1530. The optimization sequence is as follows:

1. The calibration PC converts an evaluated individual into control parameters, and then transmits it to the ECU.
2. The ECU operates the engine for a prescribed period with the received control parameters. Simultaneously, it transmits the sampled data of the engine torque and the fuel-consumption to the PC in sampling intervals.
3. The PC calculates the fitness of control parameters using the received data and executes the NSGA-II(+IES).

[6] It is ratio of air and gasoline mass.

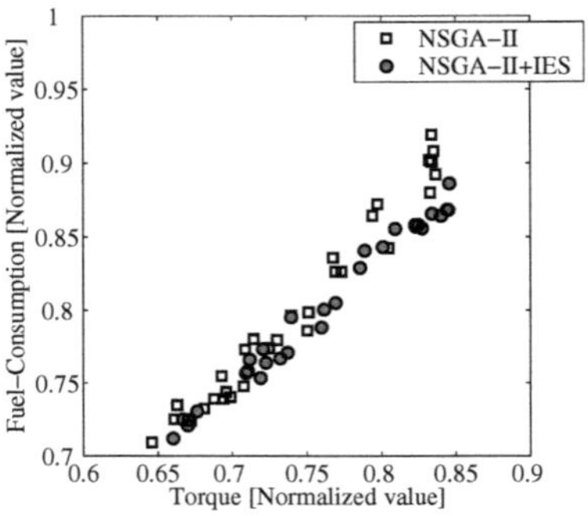

Fig. 13. Comparison of the Pareto frontier approximations by true fitness

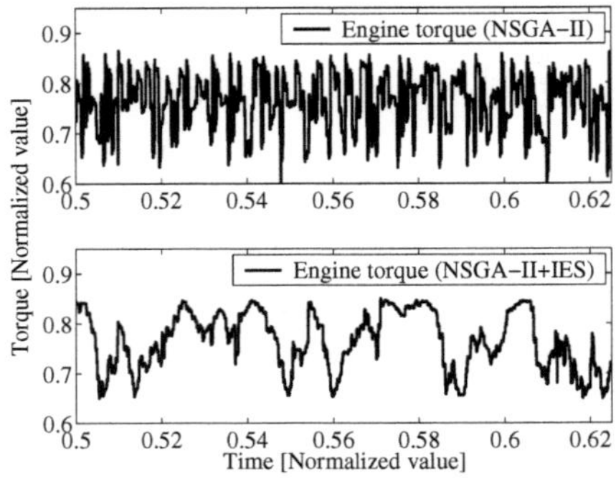

Fig. 14. Time series of engine torque (upper: NSGA-II, lower: NSGA-II+IES)

After optimization, the true fitness of the individuals in the final generation were verified by the measurement for a longer period.

5.2 Discussion of Results

Fig. 12 shows the population distribution of the estimated value and the true fitness in the objective function space. Each axis is shown in normalized scales. From this figure, we can see that the Pareto-optimal set shows straight line shape, and the accuracy of estimated value of the NSGA-II+IES is higher than that of the normal NSGA-II. We also show the comparison of the Pareto frontier approximations by true fitness in Fig. 13. This figure indicates that the NSGA-II+IES found a better convergence of Pareto-optimal set than the normal NSGA-II, especially in high torque area around 0.8.

Fig. 14 shows the time series of the engine torque. We can see that though the time series of the normal NSGA-II is largely vibrating, that of the NSGA-II+IES change more smoothly. These results indicate the EOS can suppress the transient response and improve the accuracy of the estimated values. Additionally, though the search accuracy was improved, the search time of the NSGA-II+IES was shortened about 13% to that of the normal NSGA-II by the ETS in the same number of evaluation.

Through discussion with experts of engine calibration, we confirmed that the characteristic of the Pareto-optimal set was appropriate enough as the performance of the engine, although the optimization time was not reached somewhat to the operation time of the experts. Therefore, it can be concluded that the IES is an effective technique for the experiment-based EMO of real engines.

6 Conclusions

In this paper, we proposed the Individual Evaluation Scheduling for the experiment-based evolutionary multi-objective optimization. Through numerical experiment using a formal test problem and experiment using a HILS environment for real engines, it was shown that the proposed method improved search accuracy and search time of MOEAs for dynamical systems simultaneously. Because the proposed method is independent from specific MOEA, the IES is applicable to various MOEAs. As future works, we will integrate this method and the MFE-DSGA, and apply it to emissions optimization of engine control systems.

References

1. T. Bäck: *Evolutionary Algorithms in Theory and Practice*, Oxford (1996)
2. K. Deb, A. Pratap, S. Agarwal, and T. Meyarivan: A Fast and Elitist Multiobjective Genetic Algorithm: NSGA-II, *IEEE Trans. on Evolutionary Computation*, Vol. 6, No. 2, pp. 182-197 (2002)
3. T. Dvorak, L. Malone, and R. Hoekstra: Statistical Process Control and Design of Experiment Process Improvement Methods for the Powertrain Laboratory, SAE paper 2003-01-3208 (2003)
4. L. Eshelman, and J. Schaffer: Real-Coded Genetic Algorithms and Interval-Schemata. *Foundations of Genetic Algorithms 2*, pp. 187-202 (1993)
5. L. Eshelman, and K. Mathias, J. Schaffer: Crossover Operator Biases : Exploiting the Population Distribution. *Proc. of ICGA97*, pp. 354-361 (1997)
6. T. Hiroyasu, M. Miki, and S. Watanabe: Divided Range Genetic Algorithms in Multiobjective Optimization Problems, *Proc. of IWES ´99*, pp. 57-65 (1999)
7. K. Ikeda, H. Kita, and S. Kobayashi: Failure of Pareto-based MOEAs, Does Non-dominated Really Mean Near to Optimal? *Proc. of CEC2001*, pp. 957-962 (2001)
8. H. Kaji: Proposal of a Multi-objective Genetic Algorithm for Noisy Fitness Functions (in Japanese), *Trans. of ISCIE*, Vol. 18, No. 12, pp. 423-432 (2005)
9. H. Kaji, and H. Kita: A Crossover Method of Genetic Algorithms for Periodic Function Optimization (in Japanese), *Proc. of the 32th SICE Symposium on Intelligent Systems*, pp. 157-162 (2005)
10. R. Myers, and D. Montgomery: *Response Surface Methodology 2nd edition*, John Wiley & Sons (2002)
11. I. Ono, and S. Kobayashi: A Real-coded Genetic Algorithm for Function Optimization Using Unimodal Normal Distribution Crossover, *Proc. of the 7th ICGA*, pp. 246-253 (1997)
12. I. Rechenberg: Evolution Strategy, *Computational Intelligence Imitating Life*, IEEE Press, pp. 147-159 (1995)
13. Y. Sano, and H. Kita: Optimization of Noisy Fitness Functions by Means of Genetic Algorithms Using History of Search with Test of Estimation, *Proc. of CEC2002*, pp. 360-365 (2002)
14. H. -P. Schwefel: *Evolution and Optimum Seeking*, John Wiley & Sons (1995)

A Multiobjectivization Approach for Vehicle Routing Problems

Shinya Watanabe and Kazutoshi Sakakibara

College of Information Science & Engineering, Ritsumeikan Univ.
1-1-1 Nojihigashi, Kusatsu
Shiga 525-8577, Japan
{sin,sakaki}@sys.ci.ritsumei.ac.jp
http://sys.ci.ritsumei.ac.jp/~sin/

Abstract. This paper presents a new approach for vehicle routing problems (VRPs), which are defined as problems of minimizing the total travel distance. The proposed approach treats VRPs as multi-objective problems using the concept of multiobjectivization. The multiobjectivization approach translates single-objective optimization problems into multi-objective optimization problems and then applies EMO to the translated problem. In the proposed approach, a newly defined objective related to assignment of customers is added, because the assignment has a more important influence on the search results than routing in VRPs. We investigated the characteristics and effectiveness of the proposed approaches by comparing the performance on conventional approaches and the proposed approaches.

Keywords: Vehicle Routing Problems(VRPs), Multiobjectivization.

1 Introduction

Vehicle Routing Problems (VRPs) are well known combinatorial optimization problems arising in many distribution and transportation systems, such as postal delivery, school bus routing, newspaper distribution, etc. VRPs have attracted a great deal of attention in recent years due to their wide applicability and economic importance.

VRPs can be described as the problem of minimizing the route cost from one depot to a set of geographically scattered customers (points, cities, stores, etc.). The routes must be designed in such a way that each customer is visited only once by exactly one vehicle, all routes start and end at the depot, and the total demands of any route must not exceed the capacity of the vehicle. In general, the objective of VRPs is to find the minimum number of routes or the minimum total travel distance [2]. As VRPs are good for exercising new heuristic search techniques and have a high degree of complexity, metaheuristics such as Local Search (LS) [1], Tabu Search (TS) and Genetic Algorithms (GA) [10] have been proposed over the last number of years [2].

S. Obayashi et al. (Eds.): EMO 2007, LNCS 4403, pp. 660–672, 2007.

On the other hand, there has been a great deal of progress in the study of evolutionary computation of multi-objective optimization (EMO) over the last decade [5]. In the field of EMO, there have been a few reports concerning the unique approach of "multiobjectivization" [7]. The multiobjectivization approach translates single-objective optimization problems into multi-objective optimization problems and then applies EMO to the translated problem. The most important feature of multiobjectivization is that it provides more freedom to explore and to reduce the likelihood of becoming trapped in local optima by adding additional objectives.

In this paper, we propose a new approach for VRPs, which treats VRPs as multi-objective problems using the concept of multiobjectivization. There have been many studies using EMO to optimize multi-objective VRPs [9], with objectives including the number of routes and total travel distance or number of routes and duration of routes, etc. In these studies, EMO treats the original objective of VRPs directly as multi-objective.

On the other hand, our approach deals not only with the original objective of VRPs but also with newly defined objectives related to assignment of customers. Generally, VRPs seem to have two different determinations: the assignment of customers and the order of the route. The assignment of customers is known to have a stronger influence on the search than the order of the route in many studies [6]. Therefore, we expect that the proposed approach will get better solutions in minimization of the total travel distance than the approach using only the total travel distance as a single objective.

Main concern of this multiobjectivization is whether additional objectives will work to accelerate the search in the original problem. We investigated the characteristics and effectiveness of the proposed approach by comparing the performance of the conventional approach and multiobjectivization approach. In numerical experiments, we used Taillard's test functions as a benchmark problem. In addition, we used NSGA-II reported by Deb et al. [3] in implementing our approach. Through numerical examples, we showed that the proposed multiobjectivization approach can obtain the solution with good quality and little variation in VRPs.

2 Vehicle Routing Problem

This paper deals with the most elementary version of VRPs, the capacitated VRPs (CVRPs), which can be described as follows [2]:

- All vehicles start from the depot and visit the assigned customer points, then return to the depot. Here, a route is formed by the sequence of the depot and the customer points visited by a vehicle. Therefore the number of vehicles is same as the number of route. Moreover, each customer is visited only once by exactly one vehicle.

- Each customer asks for a weight $w_i (i = 1, \ldots, N)$[1] of goods and a vehicle of capacity W is available to deliver the goods. In this paper, we used the same capacity W for all vehicles.
- A solution of the CVRP is a collection of routes where the total route demand is at most W.

VRPs have a number of objectives, such as minimization of the total travel distance, minimization of the number of routes, minimization of the duration of the routes, etc. In this paper, we used minimization of the total travel distance (F_{sum}) as the objective of the VRPs. The formula of the objective is as follows:

$$\text{minimize } F_{\text{sum}} = \sum_{m=1}^{M} c^m \tag{1}$$

where M is the total number of routes and c^m indicates mth route distance. The formula for c^m is as follows:

$$c^m = c^m_{0,u_1^m} + \sum_{i=1}^{n_m - 1} c^m_{u_i^m, u_{i+1}^m} + c^m_{u_{n_m}^m, 0} \tag{2}$$

where $c^m_{i,j}$ indicates the distance from customer i to customer j. u_i^m represents the ith customer to be routed in the mth route and "0" is the depot. n_m indicates the total number of customers in the mth route. Here, the total number of customers is $N = \sum_{m=1}^{M} n_m$.

VRPs have a constraint on the vehicle capacity W. In this paper, we used the same capacity W for all vehicles. The formula of the vehicle capacity W is as follows:

$$W \geq w^m = \sum_{i=1}^{n_m} w_{u_i^m}, \quad (m = 1, \ldots, M) \tag{3}$$

where w^m indicates the amount of customers' weight in the mth route and $w_{u_i^m}$ represents the weight of the goods for the ith customer to be routed in the mth route.

As noted above, in VRPs it is necessary to find a set of sequences of customers that will minimize the total travel distance. In addition, it is necessary to determine the following two points:

1) assignment of customers
2) routing (the order of coustomers)

3 The Multiobjectivization of Vehicle Routing Problem

In this paper, we propose a multiobjectivization approach, a newly defined objective related to assignment of customers is added for explicitly taking into account of the evaluation of customer assignments. In this section we describe the purpose of multiobjectivization and the evaluation method related to assignment of customers.

[1] N is the number of customers.

3.1 The Purpose of the Proposed Multiobjectivization Approach

As described above, two types of decision elements should be considered in VRPs. Among two decisions, the assignment of customers has a stronger influence on the search than the order of customers, because the order of customers can be determined under the fixed assignment of customers to the specific vehicle. If the assignment determination is not appropriate, good solutions cannot be obtained even if the best order is determined for all routes.

However, there have been no previous reports of VRPs explicitly taking into account evaluation of customer assignments. As a hierarchical search between the assignments and order determination, Bent et al. proposed a two-stage hybrid local search that first minimizes the number of vehicles using SA, and then minimizes the total travel distance using a large neighbourhood search [1]. In addition, Nanry et al. reported a hierarchical search using Reactive Tabu Search (RTS) to the customer assignments and the order determination [8]. However, these approaches do not evaluate the customer assignments directly, and evaluate only the original objective of VRPs: the total travel distance and the number of vehicles.

The proposed approach treats two types of decision elements independently. But this approach uses the same solution strategy for these decisions, not using individual solution strategy as a hierarchical search. Treating VRP as multi-objective problem, we can handle two different decisions concurrently and independently.

3.2 The Evaluation Method Related to Assignment of Customers

In the proposed approach, we capitalize on the objective functions using "Multi-objective clustering with automatic determination of the number of clusters (MOCK) [4]" to evaluate customer assignments.

Clusters and data points in clustering problems can be assumed as routes and customers in VRPs, respectively. Therefore, the objective functions used by MOCK can be diverted to evaluation of the customer assignments in VRPs.

MOCK adopts the following two functions, which reflect two fundamentally different aspects of good clustering solutions.

1) The global concept of the compactness of clusters.
2) The local concept of the connectedness of data points.

The first of these clustering objectives evaluates the overall density of clusters as the compactness, and the latter evaluates the degree to which neighbouring data points have been placed in the same cluster as the connectedness.

The overall density of clustering solutions, which reflects the overall intra-cluster spread of the data, is computed as:

$$\mathrm{Dev}(C) = \sum_{C_k \in C} \sum_{i \in C_k} \delta(i, \mu_k) \tag{4}$$

where C is the set of all clusters, μ_k is the centroid of cluster C_k and $\delta(.,.)$ represents the distance function (Euclidean distance).

The second objective function, connectivity, evaluates the degree to which neighbourhood data points have been placed in the same cluster. The second objective function is presented as the following formula:

$$\mathrm{Conn}(C) = \sum_{i=1}^{N} \left(\sum_{j=1}^{L} x_{i,nn_i(j)} \right), x_{i,nn_i(j)} = \begin{cases} \frac{1}{j} & \text{if } \nexists C_k : i, nn_i(j) \in C_k \\ 0 & \text{otherwise,} \end{cases} \quad (5)$$

where $nn_i(j)$ is the jth nearest neighbour of datum i, and L is a parameter determining the number of neighbours that contribute to the connectivity measure. $x_{i,nn_i(j)}$ represents the penalty value related to whether data i and jth nearest neighbour of data i are placed in the same cluster or not.

In Eq.(5), if data i and jth nearest neighbour of data i are not placed in the same cluster , $\frac{1}{j}$ is added as the penalty value $x_{i,nn_i(j)}$. Therefore, greater values of Eq.(5) indicate the tendency of a clustering solution in which neighbouring data are not placed in the same cluster.

In these different objectives, it is very important that the value of Dev(C) is decreased with increasing number of clusters, while Conn(C) is increased by increasing the number of clusters[2]. Therefore, Dev(C) and Conn(C) are in a trade-off relationship with the number of clusters.

Here, we examine the effectiveness of multiobjectivization to VRPs in which one or both of the above-mentioned objectives are added to the original objective.

4 Implementation of GA

In this section, we describe the implementation of GA to multi-objective VRPs based on multiobjectivization as described above.

4.1 Gene Expression (String Representation)

Various coding methods for VRPs have been proposed. In this numerical experiment, we used all routes directly as the genotype. In other words, the genotype in this numerical experiment is the same as the phenotype that represents the order of all routes. Therefore, there is no need for translation between genotype and phenotype.

4.2 Population Initialization

The average number of customers in one route can be calculated using the vehicle capacity W and a weight $w_i (i = 1, \ldots, N)$ for each customer. In this numerical experiment, the initial population was generated by random sampling that all routes for each individual must be equal to the average number of customers. The initialization process starts by inserting customers one by one into an empty route in random order until the number of customers in the route is equal to the average number of customers.

[2] Because it becomes difficult to place near-neighbour data in the same cluster.

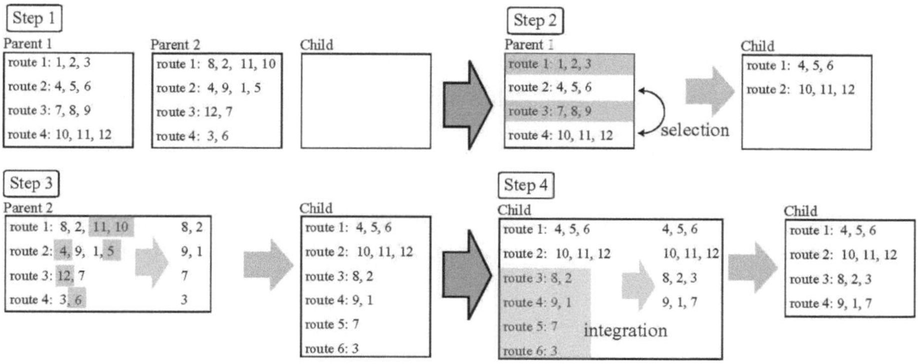

Fig. 1. The concept figure of PRIC

If an initial solution is not feasible, we used the repair method (stated in Section 4.6) to make it feasible. Therefore, all solutions of the initial population are feasible.

4.3 Crossover

As genotype has the same coding as phenotype, general crossover operators for VRPs, such as PMX, OX and EX, could not be adopted in this numerical experiment. Therefore, we implemented a new crossover operator, Partial Route Inheritance Crossover (PRIC), which aims to inherit as much as possible of the parents' route information.

In PRIC, the first child inherits half the routes of one parent directly, and then the remaining customers that are not in one parent are inherited using the route information of the other parent. The ratio of direct inheritance from parents to children should have strong effects on the search performance. Here, we designed the method to copy the half of routes in one parent directly.

Fig. 1 shows the procedure of PRIC, the details of that are described as follows:

Step 1: Selecting two parents (Parent1 and Parent2) randomly.

Step 2: Copying the half of routes in Parent1 to child.

Step 3: The remaining customers that are not in Parent1 are inherited by the routes of Parent2. If the number of routes in the child is complete, the simulation goes to Step 4. If not, the simulation is terminated.

Step 4: Until the number of routes in the child becomes equal to the number of routes in Parent1, routes copied in Step 3 are integrated using the following procedure:

- Step 4-1: The routes copied in Step 3 are sorted according to increasing number of customers included.
- Step 4-2: Each of the routes according to the sorted order is integrated into the nearest route. The distance between routes is calculated using Euclidean distance between the centred coordinates of the route (including

the depot). Therefore, the integration of routes is performed between the closest routes related to the centred coordinates. In this integration, customers are added to the bottom of the nearest route.

In Step 4 above, the routes with a small number of customers are integrated to decrease the total number of routes. Since the remaining customers that are not selected in Parent1 are picked out from Parent 2 routes, a lot of routes with small numbers of customers are produced (Step 3 in Fig. 1).

PRIC includes not only the effect of the inheritance of parents' routes but also the effects of the integration of routes and the re-shuffling of customers between routes.

4.4 Mutation

In this paper, we used six kinds of operators as mutation:

1) 2-opt*(asterisk) [2]
2) or-opt [2]
3) Relocate Operator [2]
4) Exchange Operator [2]
5) Integration of different routes into one route
6) Division of a route into two routes

2-opt*(asterisk) swaps sub-routes between different two routes, and or-opt replaces the sub-route with L customers in a random chosen route. Also, Relocate Operator simply moves a customer from one route to another and Exchange Operator swaps two customers in different routes.

We randomly selected one out of the 6 operators as mutation operator at each generation.

4.5 The Decision of Start and End Point in a Route

In this paper, the start and end customers in a route are decided in the evaluation phase. As the decision of start and end customers determines where to insert the depot in the sequence of customers, we used saving method [2] to insert the depot with the minimum total travel distance. Therefore, the optimal insertion point of the depot in the sequence of customers can be decided.

4.6 Treatment of a Solution with Constraint Violation

As VRPs have the constraint of the vehicle capacity, we should implement a repair method as a constraint handling technique. We used the repair method to divide an infeasible route into two routes. In this technique, a set of customer sequences satisfying the capacity constraint forms one route, and then the remaining customer sequences forms another route. Therefore, all solutions in this example are feasible.

Table 1. GA Parameters

population size	200
crossover rate	1.0
mutation rate	1/bit length
number of trials	30

Table 2. Problem Instance

Problem	N	W	$\bar{w}$	$C(w)$
tai75c($C = 0$)	75	1122	127	0.0
tai75c($C = 0.6$)	75	1122	163.2	0.6
tai75c(original)	75	1122	126.9	1.6
tai100d($C = 0$)	100	1297	136	0.0
tai100d($C = 0.8$)	100	1297	135.7	0.8
tai100d(original)	100	1297	135.7	1.6

5 Numerical Examples

In this study, we investigated the characteristics and effectiveness of the proposed approach by comparing the performance of both the conventional approaches and multiobjectivization approaches. To verify the effectiveness of multiobjectivization of VRPs, VRP instances provided by Taillard et al.[3] were used. In implementing our proposed approach, we used NSGA-II proposed by Deb et al. [3]. Table 1 shows the GA parameters.

5.1 VRPs Instances

We used six types of instances: tai75a, tai100d and newly defined four changing the variation coefficient (C) of the customer weight $w_i(i = 1, \ldots, N)$ in tai75a and tai100d[4]. The characteristics of the instance are described in Table 2. Table 2 represents the number of customers(N), the vehicle capacity(W), the average of the customer weight($\bar{w}$), standard deviation($\sigma(w)$) of the customer weight $w_i(i = 1, \ldots, N)$, and the variation coefficient ($C(w)$) of the customer weight.

In Table 2, tai75c($C = 0$) and tai75c($C = 0.6$) are the problems of changing the amount of the customer weight $w_i(i = 1, \ldots, N)$ in tai75c(original) so that the amounts of the variation coefficient of customer weights are about 0 and 0.6 respectively (the customer location and the vehicle capacity of these instances are the same as original instance). In the same way, tai100d($C = 0$) and tai100d($C = 0.8$) are modified so that the amounts of the variation coefficient of tai100d(original) are about 0 and 0.8 respectively.

The small value of the variation coefficient indicates that customer weights are homogenized. More homogenized customer weights make it easier to decide the assignment of customers, because simple heuristic approach, in which neighboring customers merge into the same cluster, can be worked effectively in this case. In contradiction to this, the higher value of the variation coefficient make it more difficult to assign customer. In this case, the performance related to assignment of customers seems to influence the quality of the obtained solutions more strongly.

[3] These test problems are available at `http://neo.lcc.uma.es/radi-aeb/WebVRP/`

[4] The variation coefficient (C) is the value that standard deviation was divided by the mean value. C is the index that represents the degree of variation.

Table 3. The four type experiments of NSGA-II

method	f_1	f_2	f_3
Conventional 1	$f^{Eq.1}$	$f^{Eq.1}$	—
Conventional 2	$f^{Eq.1}$	The number of routes	—
Proposed 1	$f^{Eq.1}$	$f^{Eq.4}$	—
Proposed 2	$f^{Eq.1}$	$f^{Eq.5}$	—
Proposed 3	$f^{Eq.1}$	$f^{Eq.4}$	$f^{Eq.5}$

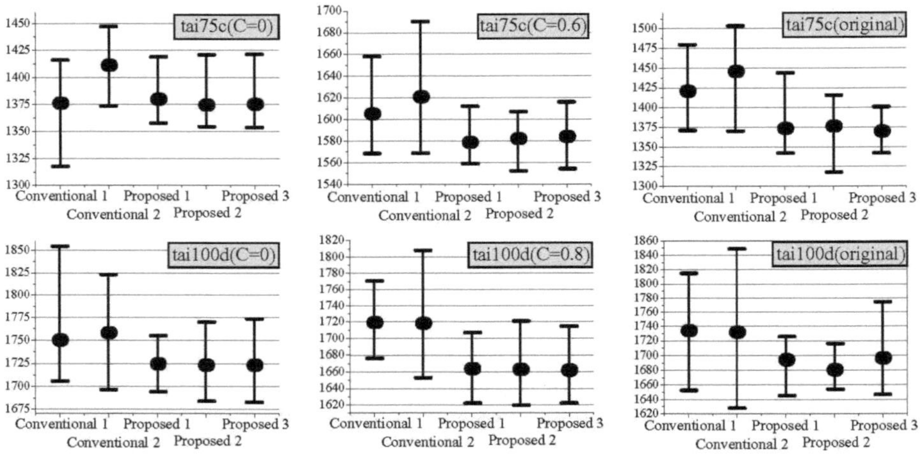

Fig. 2. The results of the total travel distance

5.2 Results and Analysis

In this study, we used five types of NSGA-II experiment based on the implementation of objectives (f_1 and f_2). Table 3 shows the 5 experiments. In Table 3, the first objective, which is common to all experiments, is the total travel distance (Eq.(1)).

In this experiment, termination conditions of the instances with 75 customers (tai75c($C = 0$), tai75c($C = 0.6$) and tai75c(original)) was 5000 generations, and those of the instances with 100 customers (tai100d($C = 0$), tai100d($C = 0.8$) and tai100d(original)) was set to 7500 generations.

We performed 30 trials and all results are shown as averages of 30 trials. The results of the 6 instances are shown in Fig. 2. Fig. 2 shows the minimum, maximum and average values in which the objective value represents the total travel distance. Also, Fig. 3 shows the standard deviation of the solutions so as to evaluate the degree of variation of the solutions. In multiobjectivization approaches, the solution with the minimum total travel distance is treated as the final best solution in each trial. In concrete terms, the solutions with the minimum f_1 value of each experiment are used as the final results.

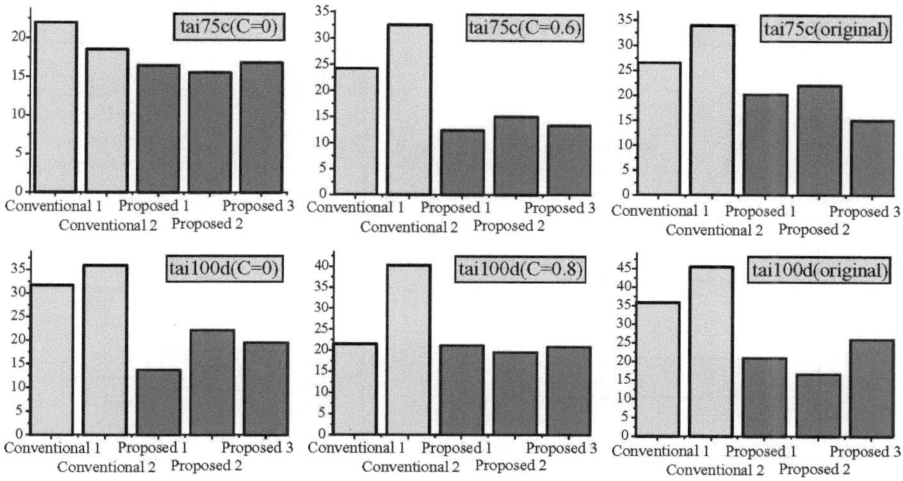

Fig. 3. The standard deviation of the solutions

As shown in Fig. 2, the solutions of multiobjectivization (three proposed approaches) were better than those of the conventional approaches. Especially, the difference in quality of the obtained solutions between the conventional and the proposed approaches increased with larger customer size and variation coefficient (C) value. From the width between minimum and maximum in Fig. 2 and the degree of variation in Fig. 3, it was also clear that the proposed approaches can obtain the solution with good quality and little variation at each trial. These results confirmed that proposed multiobjectivization approaches are more effective for VRPs than both the conventional approaches.

The increase in the number of objectives usually degrades the convergence of solutions to the Pareto front. But the multiobjectivization in this paper doesn't show this tendency, since additional objectives don't yield the trade-off relationship between original and additional objectives. In this paper, additional objectives can help to accelerate the search for original objective. The results of Fig. 2 enhance the legitimacy of this inference.

The transition of the objective values

Here, we describe the transition of the objective values in each approach. The transitions of the objective values in tai100d(original) are shown in Fig. 4. The four objective values in Fig. 4 are the total travel distance(Eq.(1)), the number of routes, compactness ($\mathrm{Dev}(C)$) and connectivity ($\mathrm{Conn}(C)$). As these objective values of Fig. 4, we used the objective value of the individuals with minimum value of the total travel distance in the generation[5]. In these figures, the horizontal axes indicate generation and the vertical axes indicate each objective value

[5] Therefore, all objective values except the total travel distance may make a change for the worse.

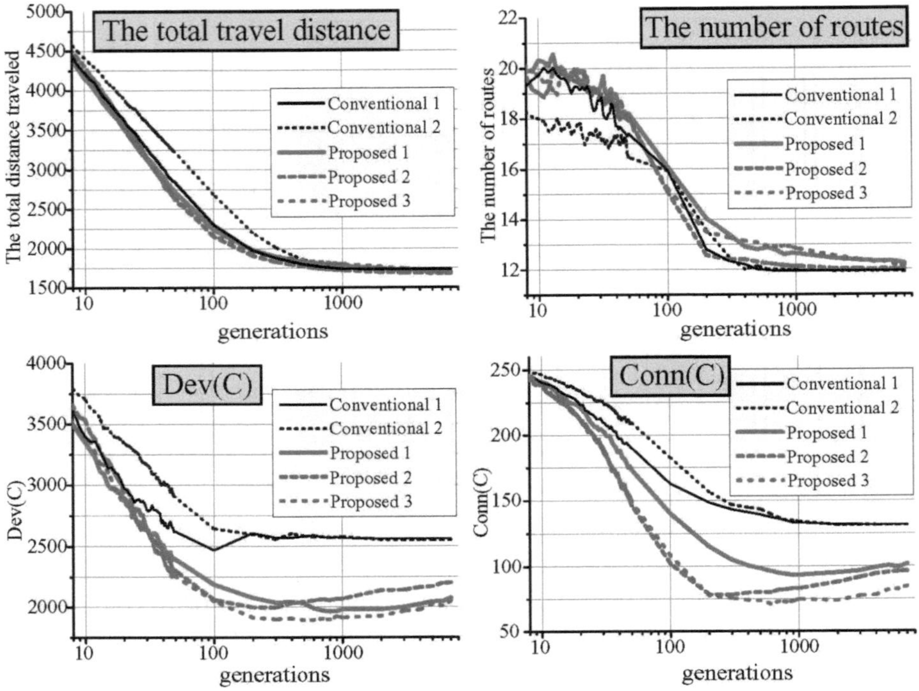

Fig. 4. The transition of the objective values

and the values of generation are described on a log scale (common logarithm). Also, all results are shown as averages of 30 trials.

From Fig. 4, there seem to be some sort of relationships between each objective values, because all objective values were decreased by a large generation number. But the values of compactness and connectivity in three proposed approaches were increased with more than 200 generations. Therefore the correlations of compactness and connectivity with near the minimum value of the total travel distance are guessed to be low or slight trade-off relation.

The transitions of the total travel distance had similar tendencies in all approaches. But the transition of Conventional2 with the total travel distance and the number of routes as objectives were a more slower slope for less than 150 generations as compared to the other approaches. On the other hand, when considering the transition of the number of route, Conventional2 got the smaller value in earlier generations as compared to the other approaches. This was because Conventional2 explicitly evaluate the number of route.

And in terms of the objective value of compactness and connectivity, three proposed approaches yield better results. It is interesting that Proposed3 with both objective functions of MOCK was better than Proposed1 and Proposed2 with one or other of two MOCK functions. Therefore it is apparent from Fig. 4 that Proposed3 is better in terms of assignment of customers than other multiobjectivization approaches with one or other of MOCK functions. On the other hand, the results concerning compactness and connectivity of Conventional2 with the minimization of the number of routes as second objective function and that of Conventional1 are not so different and worse than that of Proposed3. Therefore, the minimization of the number of routes doesn't have a strong effect on the assignment of customers.

6 Conclusions

In this paper, we proposed a new approach based on multiobjectivization for vehicle routing problems (VRPs) with a single objective. This approach treats VRPs as multi-objective problems in which a newly defined objective related to assignment of customers is added. As the objective related to assignment of customers, we used two objective functions used by MOCK, i.e., the compactness of clusters and the connectedness of data points. This multiobjectivization aims to accelerate the search for original objective by adding supplementary objective. We investigated the effectiveness of the proposed multiobjectivization approaches by comparison of its performance with that of the conventional approaches. Numerical experiments clarified the following points:

1) multiobjectivization approaches can obtain the solution with better quality and less variation at each trial. Therefore, Also, the experimental results indicate that multiobjectivization using both additional objectives is more effective than using either alone.
2) From the results of the transition of the objective values in the course of the search process, it was confirmed that the multiobjectivization approach using MOCK functions is very effective for the assignment of customers, while the minimization of the number of routes have little effect on the assignment of customers. Also, the three multiobjectivization approach using both objective functions of MOCK can derive better solutions than other approaches with only one MOCK function.

References

1. R. Bent and P. Van Hentenryck. A two stage hybrid local search for the vehicle routing problem with time windows. In *Transportation Science*, volume 38, pages 515–530, 2004.
2. O. Braysy and M. Gendreau. Vehicle routing problem with time windows, part i: Route construction and local search algorithms. *Transportation Science*, 39(1):104–118, 2005.
3. K. Deb, A. Pratap, S. Agarwal, and T. Meyarivan. A Fast and Elitist Multiobjective Genetic Algorithm: NSGA–II. *IEEE Transactions on Evolutionary Computation*, 6(2):182–197, April 2002.
4. J.Handl and J.Knowles. Exploiting the Trade-Off—The Benefits of Multiple Objectives in Data Clustering. In *Evolutionary Multi-Criterion Optimization. Third International Conference, EMO 2005*, pages 547–560, 2005.

5. K.Deb. *Multi-Objective Optimization using Evolutionary Algorithms.* Chichester, UK:Wiley, 2001.
6. K.F.Doerner, R.F. Hartl, and M. Lucka. A parallel version of the d-ant algorithm for the vehicle routing problem. In *Theory an Applications*, pages 109–118, 2005.
7. D. Knowles, A. Watson, and W. Corne. Reducing local optima in single-objective problems by multi-objectivization. In *Evolutionary Multi-Criterion Optimization. First International Conference, EMO 2001*, pages 268–282, 2001.
8. W. P. Nanry and J. W. Barnes. Solving the pickup and delivery problem with time windows using reactive tabu search. *Transportation Research Part B*, 34:107–121, 2000.
9. N.Jozefowiez, F.Semet, and E.Talbi. Parallel and Hybrid Models for Multi-objective Optimization: Application to the Vehicle Routing Problem. In *Parallel Problem Solving from Nature—PPSN VII*, pages 271–280, 2002.
10. J.Y. Potvin and S. Bengio. The vehicle routing problem with time windows - part ii: Genetic search. In *INFORMS Journal on Computing*, volume 8, pages 165–172, 1996.

Designing Traffic-Sensitive Controllers for Multi-Car Elevators Through Evolutionary Multi-objective Optimization

Kokolo Ikeda[1], Hiromichi Suzuki[2], Sandor Markon[2], and Hajime Kita[1]

[1] Kyoto University, Yoshida-Nihonmatsu, Sakyo, Kyoto 606-8501, Japan
kokolo@media.kyoto-u.ac.jp, kita@media.kyoto-u.ac.jp
http://www.ipe.media.kyoto-u.ac.jp/indexE.html
[2] Fujitec Co. Ltd, Big Wing, Hikone, Shiga 522-8588, Japan

Abstract. Multi-Car Elevator (MCE) that has several elevator cars in a single shaft attracts attention for improvement of transportation in high-rise buildings. However, because of lack of experience of such novel systems, design of controller for MCE is very difficult engineering problem. One of the promising approaches is application of evolutionary optimization to from-scratch optimization of the controller through discrete event simulation of the MCE system. In the present paper, the authors propose application of evolutionary multi-objective optimization to design of traffic-sensitive MCE controller. The controller for MCE is optimized for different traffic conditions in multi-objective way. By combining the multi-objective optimization with the exemplar-based policy (EBP) representation that has adequate flexibility and generalization ability as a controller, we can successfully design a controller that performs well both in the different traffic conditions and works adequately by generalization in the conditions not used in the optimization process.

1 Introduction

The elevator system is a critical component of high-rise buildings, and its design and control have been studied for many years. The control of cooperating elevator cars for efficient service of passengers is known as "the elevator group control problem". This problem is recognized as a difficult control task, involving stochastic, online scheduling with high combinatorial complexity and real-time response requirements. Since no effective analytical solution has been found to date, current commercial systems are controlled by using a combination of heuristic and artificial intelligence methods [Kim et al., 1998][Beielstein et al., 2003] [Zhou et al., 2005].

Recently, with increasing building heights and more complex usage patterns, multi-car elevators (MCEs) consisting of several cars in a single elevator shaft, usually driven by linear motors, are receiving increasing interest as high-performance transportation systems [Kita et al., 2002] [Sudo et al., 2002]. However, the accumulated knowledge for conventional elevators is not readily applicable to MCEs, which exhibit distinctly different behavior.

S. Obayashi et al. (Eds.): EMO 2007, LNCS 4403, pp. 673–686, 2007.

The most promising approach for MCE control appears to be simulation-based optimization, in which the policy of controller is represented by a function model and the parameters are optimized through a simulation. Sudo et al. have shown that the approach with genetic algorithms (GAs) is hopeful [Sudo et al., 2002] [Takahashi et al., 2003]. In those researches, MCE control is performed by assigning a hall-call to a certain car, with using the linear-sum weights $[\alpha_i]$. When a new call occurs, for all available cars, several feature values $[w_i^k]$ expressing the state of the car k are calculated. Then the car with the minimum linear-weighted sum $\sum \alpha_i w_i^k$ is assigned.

In [Ikeda et al., 2006], an exemplar-based policy representation (EBP) [Ikeda, 2005] is employed as a non-linear controller for MCE systems. An advantage of EBP to the controller of linear-sum type is the ability to control flexibly according to the current situation, and the result of numerical experiments has shown its superiority in MCE control.

In those simulation-based researches, the policy of control with adjustable parameters has been evaluated and optimized in a single traffic situation. So, there is no guarantee that the optimized policy works adequately in the other situations, such as in the other building or in the other traffic condition which changes largely depending on such as time-of-day. For practical use, considering the cost and difficulty of detecting and switching the policy depending on the situations, it is preferable that one policy works adequately in various situations as much as possible. In large part of conventional control methods, the current situation is detected by the set of rules such as fuzzy rules, and the corresponding control policy tuned separately is performed. Normally the rules are written out by experts and so very expensive.

In this paper, we employ the multi-objective optimization approach [Deb et al., 2000] [Obayashi and Sasaki 2004] in order to obtain the traffic-sensitive controller for MCE. In this approach, the policy with parameters is evaluated in multiple situations, the objective functions are defined respectively, and multi-objective optimization method is applied. The advantages of EBP for this approach are the ability to control flexibly according to the situation and the ability to generalize it.

This paper is organized as follows. In Section 2, a brief overview of the MCE system and its controller are shown. In Section 3, the simulation-based policy optimization of MCE controller, single-objective and multi-objective, are explained. In Section 4, experiments are done and the result is analyzed, and in Section 5, the paper is concluded.

2 MCE System and Controllers

2.1 Multi-Car Elevator Systems

The almost same MCE system described in [Takahashi et al., 2003] is considered in this study. The elements comprising the system are as follows (see Fig. 1).

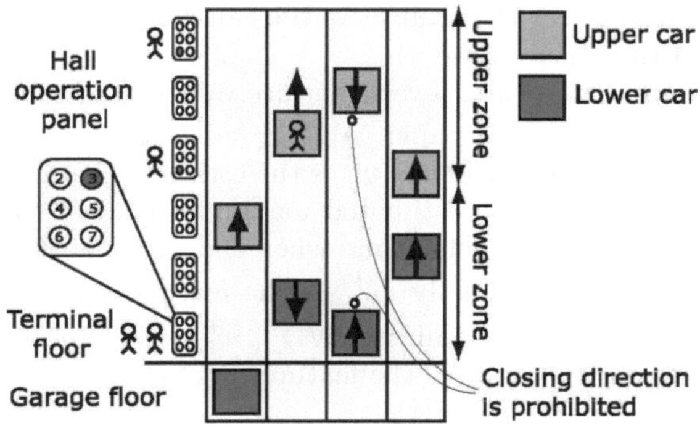

Fig. 1. MCE system

Floors. The lowest level of the building is assumed to be the sole point of entry (and exit) to the building, and thus experiences the highest traffic (10 times higher than other floors in this simulation). The lowest floor is called "the terminal floor". The other floors are assumed to be identical in terms of traffic demand, and are called "general floors".

Elevator Shafts. Shafts represent the space in which elevator cars (or cages) move. In the present simulation, the building has 4 shafts.

Elevator Cars. Each elevator shaft is considered to host two cars, which can only move vertically and cannot pass each other. Furthermore, to avoid collision and dead lock, cars in a single shaft are not allowed to approach each other simultaneously. These constraints make efficient control of MCEs difficult to achieve.

Registration of Destination Floor. It is assumed that the passengers register their destination floors not in the car but in the hall, and that passengers are guided to the car serving their need.

Zone Operation. For ease of operation, the floors are divided into upper and lower zones. The upper car in each shaft serves only the traffic demands whose origin or destination is in the upper zone. The lower car serves only the lower zone.

Garage Floor. To allow the upper car to serve the terminal floor, a garage floor at which the lower car stops is introduced below the terminal floor.

2.2 MCE Controller

In [Sudo et al., 2002], [Takahashi et al., 2003], [Ikeda et al., 2006] and this paper, MCE control is performed by assigning a hall call to a certain car. When a new call occurs, the call is assigned to a car by the following procedure:

1. For each shaft, the car that can serve the call is nominated, according to the definition of the service zone.
2. For all the nominated cars, several feature values expressing the state of the car are calculated. In this paper, $N_{\text{features}} = 4$ features $[w_1^k, w_2^k, w_3^k, w_4^{(k)}]$ are utilized where k is the car index. w_1^k is the estimated waiting time of the new call if assigned, w_2^k is the estimated maximum load of the car if assigned, and w_3^k is the estimated delay time when the car pass through the call and the next car serves it. Finally $w_4^{(k)}$ is the feature expressing the degree of current traffic, which is calculated by $\sum_k w_1^k$ and $\sum_k w_2^k$. The feature $w_4^{(k)}$ is common to all the cars. All the features are normalized so that almost all features are distributed in $[0, 1]$.
3. The preferences of cars are evaluated using the calculated feature values (or feature vectors), and the most preferable car is assigned to the call.

This decision of the car based on the feature values at the Step 3. in the above procedure is the central issue of design problem, and we have proposed the following two methods.

2.3 Linear-Sum Policy Controller

In this controller, given feature vectors are evaluated by a linear-weighted sum function, that has been used in [Sudo et al., 2002] [Takahashi et al., 2003]. Given weights α_i, the car with the minimum weighted sum $k^* = \arg\min_k \sum_{i=1}^{4} \alpha_i w_i^k$ is assigned to the given call. For example, if $(\alpha_1, \alpha_2, \alpha_3, \alpha_4) = (1, 0, 0, 0)$, the policy assigns the earliest car to a call. And if $(\alpha_1, \alpha_2, \alpha_3, \alpha_4) = (0, 1, 0, 0)$, the policy assigns the lightest car.

This approach, referred to as the linear-sum policy (LSP), is very simple and easy to implement. However, it is unable to make decisions flexibly depending on the state common to all the cars, such as whether the traffic is light or heavy. In other words, the weights of each feature value in the LSP approach are fixed, and do not vary according to the situation.

2.4 Exemplar-Based Policy Controller

In [Ikeda et al., 2006], an exemplar-based policy (EBP) representation was employed as a non-linear evaluator of candidate feature vectors.

An EBP consists of a set of exemplars, and an exemplar is defined as the pair of feature vectors $(v_j^1, v_j^2) \in \mathbb{R}^{N_{\text{features}}} \times \mathbb{R}^{N_{\text{features}}}$, meaning "to assign the call to the car with the feature vector v_j^1 is better than to assign it to the car with v_j^2". When a set of candidate feature vectors $C = \{w_c\}_c$ is given, a tournament is created, and $|C| - 1$ competitions based on the set of exemplars are conducted, finally the car corresponding to the winning vector is assigned to a given hall call. In this procedure, only the set of nearer exemplars to the feature vectors are referred (For detail, see Appendix A).

3 Simulation-Based Policy Optimization

It is assumed that the zone boundary is fixed, and that the calculation of feature vectors is also fixed. Then, optimization of the controller is performed in terms of the parameters of the policy selecting the most preferable vector from the candidates. The parameters are optimized through a simulation of the MCE and a Genetic Algorithm (GA).

3.1 Evaluation Using MCE Simulation

For this simulation, the same simulator used in [Takahashi et al., 2003] is employed. This simulator is based on the discrete event model called the Extended State Machine (ESM), which models the system using finite state machines with timers, among which messages are exchanged for synchronization [Kita et al., 2002][Mimaki et al., 1999]. In the ESM model, each of the elevator cars and the corresponding doors is represented by an ESM.

Table 1. Specifications of building and MCE

Item	Value
No. of Floors	30
Zone boundary : the lowest floor of the upper zone	16
No. of Elevator Shafts	4
No. of Cars/Shaft	2
Floor Height	4.34 m
dv^2/d^2t of Car	2.0 m/s^3
Max. Car Acceleration	1.1 m/s^2
Max. Load (persons/car)	20
Time Needed for	
Opening Doors	1.8 s
Closing Doors	2.4 s
Riding/Leaving	1.2 s/person
Passenges to serve (/hour)	750 to 2250
Traffic distribution (Terminal Floor $\leftrightarrow$ General Floor : General Floor $\leftrightarrow$ General Floor)	(10:1)

For evaluating and selecting in a GA, the fitness of a solution (policy) is defined by the averaged squared waiting time (*ASWT*) over the period of simulation (90 min in this case). To reduce the effect of transient stage of traffic, simulation result for a certain period (30 min) is excluded from evaluation. The specifications of the building considered in the simulation are listed in Table 1. Simulations were performed using a supercomputer Fujitsu HPC2500 of Kyoto University, using 32 CPUs among 128 CPUs in a node in a master-slave architecture for parallel computing.

3.2 Obtaining Traffic Sensitive Controller Through Single and Multi Objective Optimization

Now, the purpose of optimization is defined as obtaining the control policy of MCE, which performs adequately in the wide-range conditions, i.e. from the light traffic (1000 persons/hour) to the heavy traffic (2000 persons/hour).

An important ability expected for the policy representation is the condition-sensitive control. Unfortunately, it is unable for LSP to make decisions flexibly depending on the condition, whether the traffic is light or heavy. Another important ability is the generalization. Strictly, this purpose can be formalized as the 1001-objective optimization problem, $\min\{ASWT_i(x)\}_{1000 \leq i \leq 2000}$, where $ASWT_i(x)$ is the $ASWT$ of the policy x when i passengers occur per hour. However, evaluations of 1001-objective functions are very expensive and not necessarily required if generalization ability is expected for the policy representation.

For comparison study of combination of the controller representation and the type of GA, we employ two styles of controllers EBP and LSP, and compared five types of GAs to optimize their policies.

One GA, we call GA_{1000}, is carried out to attain the policy for light traffic situation (1000 persons/hour), only the $ASWT$ for the situation ($ASWT_{1000}$) is used for selection. In the same way, GA_{1500} is a GA in which only the $ASWT_{1500}$ is used for selection, and GA_{2000} is a GA in which only the $ASWT_{2000}$ is used. By applying such single-objective optimization, the optimized policy is expected to perform well at least in the considered situation.

In GA_{1000}, GA_{1500} and GA_{2000}, only one fitness function is considered and the others are ignored. In GA_{moop}, both of $ASWT_{1000}$ and $ASWT_{2000}$ are referred, and multi-objective optimization method is carried out. Generally, the purpose of multi-objective optimization is not to attain "the best solution" but to attain the set of non-dominated (Pareto) solutions. However, if the policy representation has enough condition-sensitive control ability, the optimized policy will perform well at both of light and heavy traffics.

In GA_{comb}, the combined single fitness $ASWT_{comb} = 2 \times ASWT_{1000} + ASWT_{2000}$ is used for selection. GA_{comb} is the GA to attain one of Pareto solutions using a fixed tradeoff rate. So, this is a single-objective optimization but two situations are considered as GA_{moop}.

The characteristics of GAs we employ are summarized in Table 2.

Table 2. Characteristics of five GAs we employ

Name	use $ASWT_{1000}$	use $ASWT_{1500}$	use $ASWT_{2000}$	optimization method
GA_{1000}	yes	no	no	single objective
GA_{1500}	no	yes	no	single objective
GA_{2000}	no	no	yes	single objective
GA_{comb}	yes	no	yes	single objective (combined)
GA_{moop}	yes	no	yes	multi objective

3.3 GA for Single Objective Optimization

The parameters to be optimized for LSP is the set of weights $\alpha_i \in \mathbb{R}^{N_{\text{features}}}$, and the parameters to be optimized for EBP is the set of exemplars that one of them is $(v_j^1, v_j^2) \in \mathbb{R}^{N_{\text{features}}} \times \mathbb{R}^{N_{\text{features}}}$. The common framework of GA [Ikeda and Kobayashi, 2002] is used for EBP and LSP single optimization as follows:

1. Parameters such as N_{pop} are fixed (see Table 3). In this research, they were selected by some exploratory experiments.
2. As the population, N_{pop} solutions are initialized. If LSP, each solution, N_{features} weights, $\alpha_i \in \mathbb{R}^{N_{\text{features}}}$ is randomly generated. If EBP, each solution E_i, set of $N_{\text{exemplars}}$ exemplars are randomly generated. An exemplar $e_{i,j} \in E_i = (v_{i,j}^1, v_{i,j}^2)$ is generated such that $v_{i,j}^1 + v_{i,j}^2 \in [0,2]^{N_{\text{features}}}$ and $v_{i,j}^2 - v_{i,j}^1 \in [-1,1]^{N_{\text{features}}}$.
3. N_{pop} solutions are randomly ordered, s_1, s_2, ..., $s_{N_{\text{pop}}}$. Then N_{pop} pairs (s_1, s_2), (s_2, s_3), ..., ($s_{N_{\text{pop}}}$, s_1) are passed to the following alternation procedure.
   (a) Parents (p_1, p_2) are given.
   (b) Children are reproduced by applying the crossover operator N_{children} times. For LSP, UNDX [Ono 1997] is used as crossover operator of real value vectors, and mixture of exemplars [Ikeda et al., 2006] is used for EBP (For detail, see Appendix B).
   (c) The evaluation value for each policy of the family (p_1 and children) is calculated by simulation of the MCE. To reduce the random fluctuation of evaluation values, N_{sims} simulation runs are performed independently and the average of the evaluation criterion is used. Such a GA is referred to as a N_{sims}-sample GA.
   (d) The policy p_* having the best evaluation value (lowest $ASWT_{1000}$, $ASWT_{1500}$, $ASWT_{2000}$ or $ASWT_{\text{comb}}$) in the family is selected, and p_1 in the population is replaced by p_*.
4. Step 3. is repeated $N_{\text{generations}}$ times, after which the final result, trained MCE control policy is obtained.

3.4 GA for Multi Objective Optimization

As GA_{moop}, a common framework of multi-objective optimization is used for both of EBP and LSP. Considering the noisy fitness evaluation and then uncertainty of ranking, NSGA-II [Deb et al., 2000] with some minor modification is employed as follows:

1. Parameters are fixed (see Table 3).
2. As the population, N_{pop} solutions are initialized by the same procedure to Section 3.3.
3. N_{children} solutions are reproduced by applying the crossover operator (used in Section 3.3). In this step, parents are randomly selected for each reproduction.

Table 3. Notation and parameter values used in optimization

Symbol	Explanation	Value
N_{pop}	Number of solutions (policies) in a population	30(single-objective), 60(multi-objective)
$N_{children}$	Number of children produced per reproduction step	6(single-objective), 150(multi-objective)
N_{sims}	Number of simulations for one evaluation	4(EBP), 8(LSP)
$N_{generations}$	Number of generations	80(EBP), 40(LSP)
$N_{exemplars}$	Number of exemplars in a EBP	900
k_{NN}	Localization parameter (the smaller, the localized)	30
E_i	The ith EBP, the set of exemplars of the ith policy	-
$e_{i,j}$	The jth exemplar of E_i	-

4. $N_{pop} + N_{children}$ solutions are evaluated, *i.e.* $ASWT_{1000}$ and $ASWT_{2000}$ are calculated by N_{sims} simulation runs each.
5. For each solution, the dominance-rank and the crowding-distance are calculated. As the crowding-distance, the Euclid distance to the nearest solutions with even-or-better rank is used.
6. The best N_{pop} solutions are selected to survive. The solution with the lower rank wins, and the solution with the smaller distance wins if draw in their ranks. Further, when draw in both ranks and distances, their distances to the second nearest solutions are compared.
7. Step 3. to Step 6. are repeated $N_{generations}$ times, after which the final result, trained MCE control policies with varieties are obtained.

Please note, though the different parameters are used, the total evaluation times are of the same numbers for EBP/LSP and for single-objective/multi-objective optimization.

4 Experiments

The specifications of the building considered in the experiments are listed in Table 1, and the parameters used for the GAs are shown in Table 3.

We employed two styles of controllers, EBP and LSP, and five types of GAs, GA_{1000}, GA_{1500}, GA_{2000}, GA_{comb} and GA_{moop}. To assess the performance of the optimization procedure, five independent GA trials with different random seeds were conducted for each series.

4.1 Evolution Process

Measured by a single criterion. Figure 2(left) shows the evolution processes of GA_{1000}, GA_{comb} and GA_{moop} for EBP. The average $ASWT_{1000}$ of a period is calculated for each trial, and their averages and standard deviations are shown.

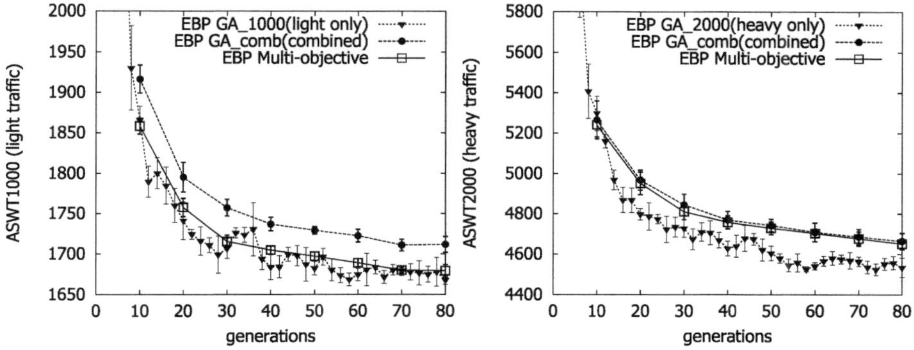

Fig. 2. Evolution process of EBP, $ASWT_{1000}$(left) and $ASWT_{2000}$(right)

Figure 2(right) shows the evolution processes of GA_{2000}, GA_{comb} and GA_{moop} for EBP. We can find that the GAs for one criterion, GA_{1000} and GA_{2000} are superior to others in their niches. But the averages of GA_{comb} and GA_{moop} were also soundly decreasing, this means that both $ASWT_{1000}$ and $ASWT_{2000}$ were simultaneously improved.

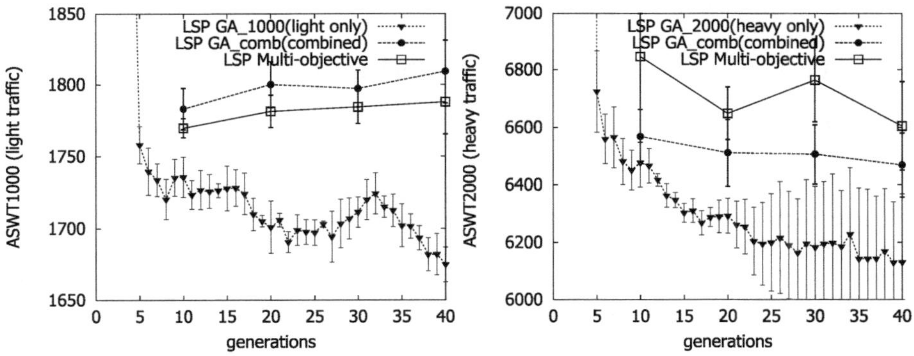

Fig. 3. Evolution processes of LSP, $ASWT_{1000}$(left) and $ASWT_{2000}$(right)

Figure 3 shows the evolution processes of LSP. Like as the case of EBP, that GA for one criterion, GA_{1000} and GA_{2000} are superior to others in their niches, and their $ASWT$ was decreasing over generations. However, especially in $ASWT_{1000}$ (left figure), the averaged performances of policies of GA_{comb} and GA_{moop} were getting worse. This suggests that performance of $ASWT_{1000}$ was sacrificed for the improvement of $ASWT_{2000}$. In other words, both $ASWT_{1000}$ and $ASWT_{2000}$ couldn't be simultaneously improved in LSP framework.

Measured by two criteria. In Figure 4, the sets of solutions in a period of a trial of GA_{moop}, LSP and EBP, are plotted. Triangles shows the sets of

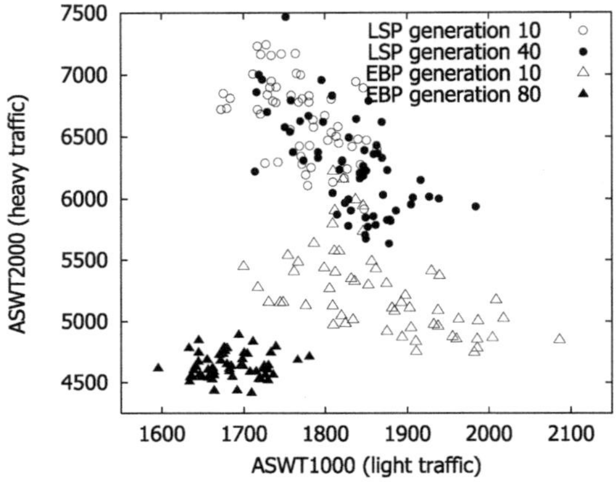

Fig. 4. Performance plot of solutions in a period of $GA_{\rm moop}$

performances $(x, y) = (ASWT_{1000}, ASWT_{2000})$ of EBP, generation 10 and generation 80. From the figure, both criterions are simultaneously improved. On the other hand, circles show the set of performances of LSP, generation 10 and generation 40. The improvement is little, and the slight slide to right($ASWT_{1000}$ worse) and down($ASWT_{2000}$ better) can be observed.

4.2 Performance Comparison of Policies Obtained

In this section, we focus on the performances of "elite" solutions of the evolved ones, instead of the averaged performances. About GA_{1000}, GA_{1500}, GA_{2000} and $GA_{\rm comb}$, all individuals are carefully (30 times for each) re-evaluated per 10 generations. Their temporal elites are re-evaluated (180 times for each) and finally the elite of the trial is selected. By this selection, totally 40 solutions are given (LSP/EBP, five trials, four GAs).

About $GA_{\rm moop}$, all individuals of the final generations are re-evaluated 30 times, and three elites, the solution with the best $ASWT_{1000}$, $ASWT_{2000}$, $ASWT_{\rm comb}$ are selected. By this selection, totally 30 solutions are given (LSP/EBP, five trials, three solutions each) with few duplications.

All elites are again re-evaluated 300 times for the comparison.

Comparison of Four GAs. Figure 5(left) shows the performances $(x, y) = (ASWT_{1000}, ASWT_{2000})$ of LSP. Pareto curve is very usual as multi-objective problems. In more detail, there observed three groups of LSP. One is such as $(\alpha_1, \alpha_2, \alpha_3, \alpha_4) = (1, 5, 0, 0)$, they prefer to assign the lighter car, and perform well at the heavy traffic (right bottom) . One other is such as $(\alpha_1, \alpha_2, \alpha_3, \alpha_4) = (1, 0, 1, 0)$, they avoid assigning the near-followed car in order to maintain an adequate distance between cars. The last is such as $(\alpha_1, \alpha_2, \alpha_3, \alpha_4) = (1, -1, 0, 0)$, they prefer to assign the heavier car in order to bias the loads to keep an adequate distance, and perform well at the light

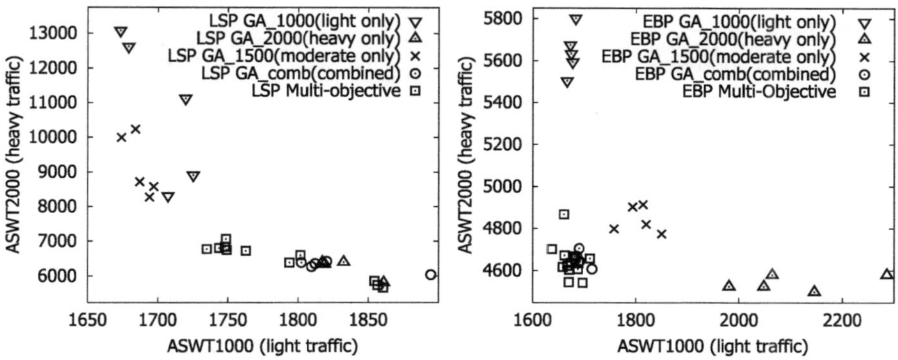

Fig. 5. Performance plot of elite solutions, LSP(left) and EBP(right)

traffic (left top). We can conclude that LSP has no condition-sensitive ability, and there is no versatile policy in LSP.

Figure 5(right) shows the performances $(ASWT_{1000}, ASWT_{2000})$ of EBP. In contrast to the case of LSP, the elites of GA_{comb} and GA_{moop} are better as GA_{1000} in $ASWT_{1000}$ and better as GA_{2000} in $ASWT_{2000}$. This fact suggests that such EBP can automatically detect the current situation (for example from the fourth feature) and make decision depending on it, by its localizing mechanism. In other words, EBP has the enough condition-sensitive control ability.

GA_{1500} works not so bad, but its performance is worse than GA_{comb} and GA_{moop}. This fact suggests that the training in the two conditions helps the generalization ability of EBP.

Performance in Wide-Range Traffic. To show the generalization ability of policies, seven delegates are selected from elites, LSP/EBP elites of GA_{1000}, GA_{1500} (EBP only), GA_{2000} and GA_{moop}. They are re-evaluated in several traffic situations, from 750 persons/hour to 2250 persons/hour. Their performances of a traffic are measured by the overrun ratio of $ASWT$ to the best of the seven delegates in the traffic.

Figure 6(left) shows the performances of LSP. In this case, as the prediction from Figure 5(left), the elite from GA_{moop} is not versatile but just intermediate performance.

Figure 6(right) shows the performances of EBP. The elite from GA_{1000} works well at traffic is light, but the performance is increasingly worse when the traffic is heavier. The elite from GA_{2000} has the opposite problem. The elite from GA_{1500} performs not so bad for all conditions, and the elite from GA_{moop} performs better than it in almost all conditions.

Through the experiments, the localization ability of EBP for condition-sensitive control, and the generalization ability for unknown conditions has been shown. By the multi-objective optimization with $ASWT_{1000}$ and $ASWT_{2000}$,

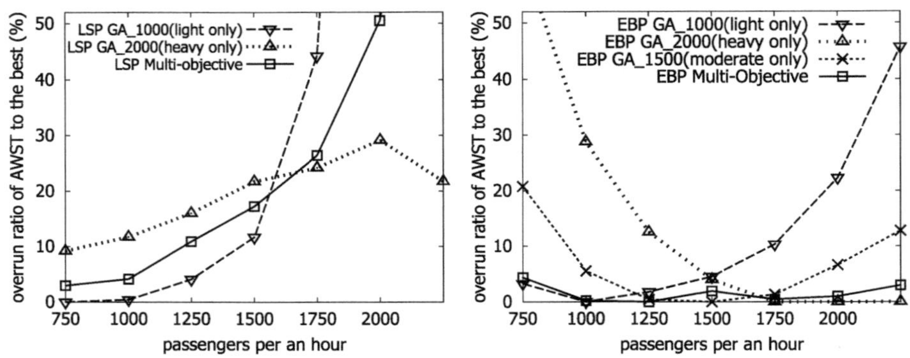

Fig. 6. Comparison of elites from GA_{1000}, GA_{1500}, GA_{2000} and GA_{moop}, LSP(left) and EBP(right), in wide-range traffics

the EBP is optimized as well as both of GA_{1000} and GA_{2000}, further, the EBP performs well also in the conditions that has not been experienced.

5 Conclusion

We presented a multi-objective optimization approach for learning condition-sensitive policy, and showed its effectiveness on the difficult problem of controlling multi-car elevators. The policy with parameters was evaluated in two traffic situations, and the objective functions were defined respectively, and a multi-objective optimization method was applied. We compared conventional linear-sum policy expression and exemplar-based policy (EBP) expression, and compared the multi-objective optimization approach and single-objective approach only for single situation. As the result, it was found that the EBP obtained by the multi-objective optimization worked adequately for wide-range situations. This fact suggests that EBP has the localization ability for condition-sensitive control, and the generalization ability for unknown conditions.

For practical use, the policy should be applicable to much wider situations, such as weekday and holiday, beginning of office hours, lunch hour and clock-out hours. For this demand, two subjects for future work exist. One is to improve the localization ability of EBP to detect the situation and the generalization ability to perform well in intermediate situations which are not tested. Another is to modify the multi-objective optimization method for such problem that has many objectives and the evaluation value is noisy.

Acknowledgments

The simulations were carried out on Fujitsu HPC2500, supercomputer systems in Kyoto University, and this work was supported in a part by a Grant-in-Aid for Young Scientists, Computing Service Group, ACCMS and IIMC, Kyoto University.

References

[Beielstein et al., 2003] T. Beielstein, C. P. Ewald, and S. Markon, "Optimal elevator group control by evolution strategies," *Genetic and Evolutionary Computation*, vol. 2724 of *LNCS*, 1963–1974, 2003.

[Deb et al., 2000] K. Deb, S. Arawal, A. Pratap and T. Meyarivan, "Fast Elitist Non-dominated Sorting Genetic Algorithm for Multi-objective Optimization : NSGA-II," *Parallel Problem Solving from Nature 6*, 849–858, 2000

[Ikeda, 2005] K. Ikeda, "Exemplar-Based Direct Policy Search with Evolutionary Optimization," *Congress on Evolutionary Computation*, pp. 2357–2364, 2005.

[Ikeda and Kobayashi, 2002] K. Ikeda and S. Kobayashi, "Deterministic Multi-Step Crossover Fusion: A Handy Crossover Composition for GAs," *Parallel Problem Solving from Nature*, 162–171, 2002

[Ikeda et al., 2006] K. Ikeda, H. Suzuki, S. Markon and H. Kita, "Evolutionary Optimization of a Controller for Multi-Car Elevators," *International Conference on Industrial Technology*, accepted, 2006

[Kim et al., 1998] C. B. Kim, K. A. Seong, H. Lee-Kwang and J. O. Kim, "Design and implementation of a fuzzy elevator group control system," *IEEE Transactions on Systems, Man, and Cybernetics, Part A*, 28(3): 277–287, 1998.

[Kita et al., 2002] H. Kita, S. Markon, T. Sudo and H. Suzuki, "A Study on Control of Multi-Car Elevators," *SICE Symposium on Autonomous and Decentralized System*, 63–66, 2002 (in Japanese).

[Mimaki et al., 1999] K. Mimaki, S. Markon, H. Kita, Y. Komoriya and Y. Nishikawa, "Modeling and Analysis of Complex Traffic in Buildings," *Proc. IEEE SMC'99*, IV, 589–594, 1999.

[Obayashi and Sasaki 2004] S. Obayashi and D. Sasaki, "Multi-Objective Optimization for Aerodynamic Designs by Using ARMOGAs," *7th Intr. Conf. on High Performance Computing and Grid in Asia Pacific Region*, 2004

[Ono 1997] I. Ono, "A Real-coded Genetic Algorithm for Function Optimization Using Unimodal Normal Distribution Crossover," *Proc. of 7th Intr. Conf. on Genetic Algorithms*, 246–253, 1997.

[Sudo et al., 2002] T. Sudo, H. Suzuki, S. Markon and H. Kita, "Effectiveness and control strategies of multi-car elevators for high-rise buildings," *TRANSLOG02*, 2002 (in Japanese).

[Takahashi et al., 2003] S. Takahashi, H. Kita, H. Suzuki, T. Sudo and S. Markon, "Simulation-based Optimization of a Controller for Multi-Car Elevators Using a Genetic Algorithm for Noisy Fitness Function," *Congress on Evolutionary Computation*, 1582–1587, 2003.

[Zhou et al., 2005] J. Zhou, T. Eguchi, K. Hirasawa, J. Hu, and S. Markon, "Elevator group supervisory control system using genetic network programming with reinforcement learning," *IEEE Congress on Evolutionary Computation*, vol 1, 336–342, 2005

A. Selection of the Best Feature Vector

When the feature vectors are given, an EBP selects the best one from them using exemplars as following procedure [Ikeda et al., 2006] (see Fig. 7).

1. Exemplars $E = \{(v_j^1, v_j^2)\}_j$ are given, where $v_j^1, v_j^2 \in \mathbb{R}^{N_{\text{features}}}$.
2. Feature vectors corresponding to possible cars, candidates, $C = \{w_c\}_c$ are given to be evaluated, where $w_c \in \mathbb{R}^{N_{\text{features}}}$.

3. An unbiased tournament for C is randomly created (the transitive law may not necessarily hold in this competition procedure).
4. A pair of competitors $w_{c^1} \in C$ and $w_{c^2} \in C$ are taken by following the tournament.
5. For each exemplar $(v_j^1, v_j^2) \in E$, the distance to the competitors $dist_j = |\frac{w_{c^1}+w_{c^2}}{2} - \frac{v_j^1+v_j^2}{2}|$ is calculated.
6. $E_{\text{local}} \in E$, the top k_{NN} exemplars nearest within $dist_j$ are selected (k_{NN} is the localization parameter).
7. For each exemplar $(v_j^1, v_j^2) \in E_{\text{local}}$, the direction $\overrightarrow{v_j^2 - v_j^1}$ and the inner product $IP_j = \overrightarrow{v_j^2 - v_j^1} \cdot \overrightarrow{w_{c^2} - w_{c^1}}$ are calculated. When $IP_j > 0$, the exemplar suggests that "w_{c^1} is better than w_{c^2}".
8. The number of exemplars in E_{local} for which $IP_j > 0$, *i.e.* $|\{(v_j^1, v_j^2) \in E_{\text{local}}, IP_j > 0\}|$ is counted. When the number is larger than $|E_{\text{local}}|/2$, w_{c^1} survives the competition (otherwise the opposite judgment is obtained).
9. After $|C|-1$ competitions have been completed, the winner is selected.

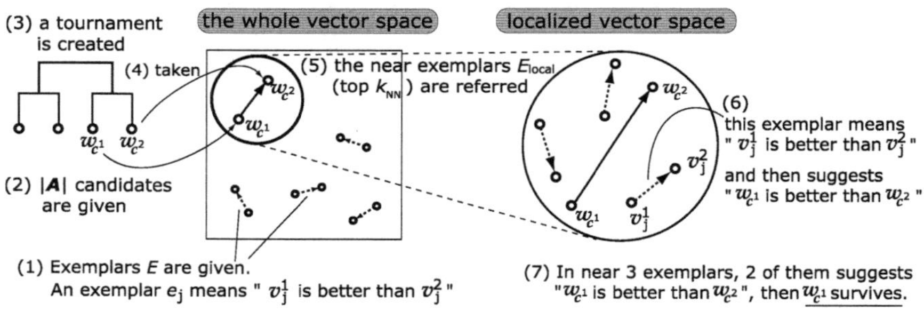

Fig. 7. Selection of the most preferable vector from candidates

B. Crossover Operator for EBP-GA

The crossover operator produces a new set of exemplars using parents. In this paper, $N_{\text{fusion}} = 36$ exemplars are newly created by a procedure called "fusion", and the rest exemplars are copied from parents. The crossover operation is performed as follows.

1. The parents E_{p1} and E_{p2} are given, and E_c is initialized as an empty set.
2. An exemplar $e_{p1,j} \in E_{p1}$ is selected randomly, and the exemplar $e_{p2,j*} \in E_{p2}$ nearest to $e_{p1,j}$ in E_{p2} is selected.
3. The rates $0 < \alpha < 1$ and $\beta = 1 - \alpha$ are fixed. For the exemplars $e_{p1,j} = (v_{p1,j}^1, v_{p1,j}^2)$ and $e_{p2,j*} = (v_{p2,j*}^1, v_{p2,j*}^2)$, an exemplar $e = (\alpha v_{p1,j}^1 + \beta v_{p2,j*}^1, \alpha v_{p1,j}^2 + \beta v_{p2,j*}^2)$ is newly created and added to E_c.
4. Steps 2. and 3., fusion procedure, are repeated N_{fusion} times.
5. An exemplar $e \in E_{p1} \cup E_{p2}$ is selected randomly. If $e \notin E_c$, e is added to E_c.
6. Step 5. is repeated until $|E_c| = N_{\text{exemplars}}$.

On the Interactive Resolution of Multi-objective Vehicle Routing Problems

Martin Josef Geiger and Wolf Wenger

Lehrstuhl für Industriebetriebslehre, Universität Hohenheim, 70593 Stuttgart, Germany
{mjgeiger,w-wenger}@uni-hohenheim.de
http://www.ibl.uni-hohenheim.de/

Abstract. The article presents a framework for the resolution of rich vehicle routing problems which are difficult to address with standard optimization techniques. We use local search on the basis on variable neighborhood search for the construction of the solutions, but embed the techniques in a flexible framework that allows the consideration of complex side constraints of the problem such as time windows, multiple depots, heterogeneous fleets, and, in particular, multiple optimization criteria. In order to identify a compromise alternative that meets the requirements of the decision maker, an interactive procedure is integrated in the resolution of the problem, allowing the modification of the preference information articulated by the decision maker. The framework is implemented in a computer system. Results of test runs on multiple depot multi-objective vehicle routing problems with time windows are reported.

Keywords: User-guided search, interactive optimization, multi-objective optimization, multi depot vehicle routing problem with time windows, variable neighborhood search.

1 Introduction

The vehicle routing problem (VRP) is one of the classical optimization problems known from operations research with numerous applications in real world logistics. In brief, a given set of customers has to be served with vehicles from a depot such that a particular criterion is optimized. The most comprehensive model therefore consists of a complete graph $G = (V, A)$, where $V = \{v_0, v_1, \ldots, v_n\}$ denotes a set of vertices and $A = \{(v_i, v_j) \mid v_i, v_j \in V, i \neq j\}$ denotes the connecting arcs. The depot is represented by v_0, and m vehicles are stationed at this location to service the customers $v_i, \ldots, v_n$. Each customer v_i demands a nonnegative quantity q_i of goods and service results in a nonnegative service time d_i. Traveling on a connecting arc (v_i, v_j) results in a cost c_{ij} or travel time t_{ij}. The most basic vehicle routing problem aims to identify a solution that serves all customers, not exceeding the maximum capacity of the vehicles Q_k and their maximum travel time T_k while minimizing the total distances/costs of the routes.

S. Obayashi et al. (Eds.): EMO 2007, LNCS 4403, pp. 687–699, 2007.

Various extensions have been proposed to this general problem type. Most of them introduce additional constraints to the problem domain such as time windows, defining for each customer v_i an interval $[e_i, l_i]$ of service. While arrival before e_i results in a waiting time, arrival after l_i is usually considered to be infeasible [1]. In other approaches, the time windows may be violated, leading to a tardy service at some customers. Violations of time windows are either integrated in the overall evaluation of solutions by means of penalty functions [2], or treated as separate objectives in multi-objective approaches [3].

Some problems introduce multiple depots as opposed to only a single depot in the classical case. Along with this sometimes comes the additional decision of open routes, where vehicles do not return to the place they depart from but to some other depot. Also, different types of vehicles may be considered, leading to a heterogeneous fleet in terms of the abilities of the vehicles.

Unfortunately, most problems of this domain are $\mathcal{NP}$-hard. As a result, heuristics and more recently metaheuristics have been developed with increasing success [4,5,6]. In order to improve known results, more and more refined techniques have been proposed that are able to solve, or at least approximate very closely, a large number of established benchmark instances [7]. It has to be mentioned however, that with the increasing specialization of techniques a decrease in generality of the resolution approaches follows. As a result, heuristic optimization frameworks such as HotFrame [8], EasyLocal++ [9] or ParadisEO [10] try to address this issue by providing generic libraries for the resolution of optimization problems.

While the optimality criterion of minimizing the total traveled distances is the most common, more recent approaches recognize the vehicle routing problem as a multi-objective optimization problem [11, 3, 12, 13, 14]. Important objectives besides the minimization of the total traveled distances are in particular the minimization of the number of vehicles in use, the minimization of the total tardiness of the orders, and the equal balancing of the routes. Following these objectives, it is desired to obtain solutions that provide a high quality of delivery service while minimizing the resulting costs. As many objectives are however of conflicting nature, not a single solution exists that optimizes all relevant criteria simultaneously. Instead, the overall problem lies in identifying the set of Pareto-optimal solutions P and selecting a most-preferred solution $x^* \in P$. In this context, three different general strategies of solving multi-objective optimization problems can be implemented:

1. *A priori* approaches reduce the multi-objective problem to a single-objective surrogate problem by formulating and maximizing a utility function. The advantage of this approach can be seen in its simplicity given the possibility to specify the precise utility function of the decision maker. The concept may however not be used if the decision maker is not able to state his/her preferences in the required way.
2. *A posteriori* approaches first identify the Pareto set P, and then allow the decision maker to select a most-preferred solution $x^* \in P$. The main advantage of this resolution principle is, that the computation of the optimal

solutions can be done offline without the immediate participation of the decision maker. A large number of elements of the Pareto set are on the other hand discarded later during the decision making procedure.
3. *Interactive* approaches allow the gradual articulation of preferences by the decision maker and compute a sequence of solutions based on his/her individual statements. Several advantages result from this concept. First, the computational effort is smaller in comparison to the identification of the entire Pareto set. Second, the gradual articulation of preferences allows the decision maker to reflect the chosen settings in the light of the obtained results and therefore adapt and react to the optimization procedure. A disadvantage of interactive multi-objective optimization procedures is however the need of the presence of a decision maker and the availability of an interactive software to present the results. Also, comparably little time for computations is allowed as the system should be able to react in (almost) real-time to inputs of the decision maker.

While it has been stressed already quite early, that combining computer programs with interactive planning procedures may be a beneficial way of tackling complex routing problems [15, 16, 17, 18], research in interactively solving multi-objective metaheuristics is a rather newly emerging field of research [19]. Given the increasing computing abilities of modern computers however, approaches can become increasingly interesting as they allow the resolution of complex problems under the consideration of interactive, individual guidance towards interesting solutions.

The article is organized as follows. In the following Section 2, a framework for interactive multi-objective vehicle routing is presented that aims to address two critical issues:

1. The necessary generality of resolution approaches when trying to solve a range of problems of different characteristics.
2. The integration of multiple objectives and the consideration of individually articulated preferences of the decision maker during the resolution procedure of the problem.

An implementation of the framework for multi-objective vehicle routing problems is presented in Section 3. The system is used to solve instances of multi-objective vehicle routing problems. Conclusions are presented in Section 4.

2 A Framework for Interactive Multi-objective Vehicle Routing

Independent from the precise characteristics of the particular VRP, two types of decisions have to be made when solving the problem.

1. Assignment of customers to vehicles (clustering).
2. Construction of a route for a given set of customers (sequencing).

It is well-known that both types of decisions influence each other to a considerable extent. While the clustering of customers to vehicles is an important input for the sequencing, the sequencing itself is of relevance when adding customers to routes as constraints of maximum distances have to be respected. The two types of decisions can be made either sequential (cluster first-route second vs. route first-cluster second) or in parallel.

Therefore, the framework presented here proposes the use of a set of elements to handle this issue with upmost generality. Figure 1 gives an overview about the elements used.

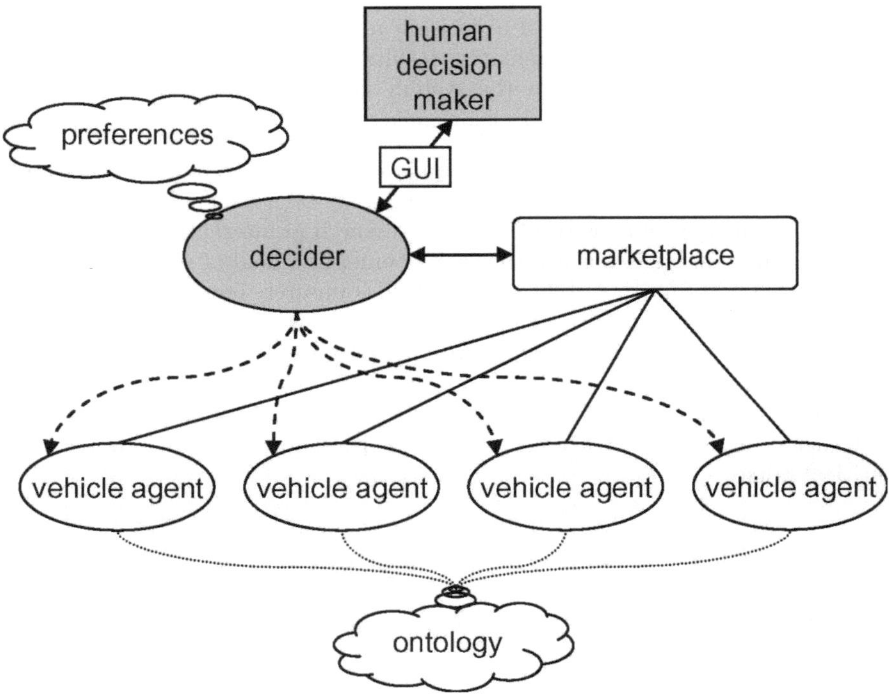

Fig. 1. Illustration of the framework for interactive multi-objective vehicle routing

- The *marketplace* represents the element where orders are offered for transportation. This elements is particularly necessary to allow an exchange of information gathered during the execution of the optimization procedure.
- *Vehicle agents* place bids for orders on the marketplace. These bids take into consideration the current routes of the vehicles and the potential change when integrating an additional order. Integrating additional orders into existing routes leads to an increase in terms of traveled routes and/or time window violations. This information is reported back to the marketplace.

- An *ontology* describes the precise properties of the vehicles such as their capacity, availability, current location, etc. This easily allows the consideration of different types of vehicles. It also helps to model open routes, where vehicle do not necessarily return to the depot where they depart from.
- A *decider* communicates with the human decision maker via a graphical user interface (GUI) and stores his/her individual preferences. In comparison to generic graphical user interfaces for multi objective optimization such as GUIMOO [20] we chose an approach that also visualizes the actual solution on a map, not only the evaluation of the currently considered solution.

  The decider also assigns orders to vehicles, taking into consideration the bids placed for the specific orders.

A solution is constructed by placing the orders on the marketplace, collecting bids from the vehicle agents, and assigning orders to vehicles while constantly updating the bids. Route construction by the vehicle agents is done in parallel using local search heuristics so that a route can be identified that maximizes the preferences of the decision maker. Reviewing possible ways of solving the clustering/sequencing problems, the presented approach follows the concept of combining both decisions in parallel.

In the proposed framework, the decision maker is allowed to change his/her preferences during the construction of the solution. If this happens, the *decider* updates the stored preference information and in consequence, the vehicles resequence their orders such that the updated preference information is met.

3 Implementation and Experimental Investigation

3.1 Configuration of the System

The framework has been implemented in a computer system. In the experiments that have been carried out, two objective functions are considered, the total traveled distances $DIST$ and the total tardiness $TARDY$ caused by vehicles arriving after the upper bound l_i of the time window. It should be noticed however, that neither the concept presented in Section 2 nor the actual implementation are restricted to two objective functions only. A sensible choice however had to be made in order to investigate the system in a quantitative way in a controllable experimental setting.

The preferences of the decision maker are represented introducing a weighted sum of both objective functions. Using the relative importance of the distances w_{DIST}, the overall utility UTILITY of a particular solution can be computed as given in Expression (1).

$$UTILITY = w_{DIST}\ DIST + (1 - w_{DIST})\ TARDY \tag{1}$$

The vehicle agents are able to modify the sequence of their orders using four different local search neighborhoods.

- Inverting the sequence of the orders between positions p_1 and p_2, $p_1 \neq p_2$. While this may be beneficial with respect to the distances, it may pose a problem for the time windows as usually orders are served in the sequence of their time windows.
- Exchanging the positions p_1 and p_2, $p_1 \neq p_2$ of two orders.
- Moving an order from position p_1 and reinserting it at position p_2, $p_1 < p_2$ (forward shift).
- Moving an order from position p_1 and reinserting it at position p_2, $p_1 > p_2$ (backward shift).

In each step of the local search procedure, a neighborhood is randomly picked from the set of neighborhoods and a move is computed and accepted given an improvement. We select each neighborhood with equal probability of $\frac{1}{4}$.

Bids for orders on the marketplace are generated by the vehicle agents, taking into consideration all possible insertion points in the current route. The sum of the weighted increase in distance DIST and tardiness TARDY gives the prize for the order. This price reflects the individual preferences articulated by the decision maker using the w_{DIST} parameter which expresses the tradeoff between distances and time window violations.

The decider assigns orders to vehicles such that the maximum regret when *not* assigning the order to a particular vehicle, and therefore having to assign it to some other vehicle, is minimized. It also analyzes the progress of the improvement procedures. Given no improvement for a certain number of iterations, the decider forces the vehicle agents to place back orders on the market such that they may be reallocated. In the current setting, the vehicle agents are allowed to compute 1000 neighboring solution without any further improvements before they are contacted by the decider to place back one order on the marketplace. The order to be placed back is the one of the current route that, when removing it from the route, leads to the biggest improvement with respect to the overall evaluation of the route.

3.2 Experiments

The optimization framework has been tested on ten benchmark instances taken from [21]. The instances range from 48 to 288 customers that have to be served from 4 to 6 depots, each of which possesses 2 to 7 vehicles vehicles. The precise description of the instances is given in [21] and therefore not repeated here. Download of the problem files is e. g. possible from `http://neo.lcc.uma.es/radi-aeb/WebVRP/`.

We simulated a decision maker changing the relative importance w_{DIST} during the optimization procedure. First, a decision maker starting with a $w_{DIST} = 1$ and successively decreasing it to 0, second a decision maker starting with a $w_{DIST} = 0$ and increasing it to 1, and third a decision maker starting with a $w_{DIST} = 0.5$, increasing it to 1 and decreasing it again to 0. Between adjusting the values of w_{DIST} in steps of 0.1, enough time for computations has been given to the system to allow a convergence to (at least) a local optimum. The system then has to follow the updated preference information, resequencing and reassigning the customers using the implemented local search metaheuristics.

The linear, additive model is one of the possibilities to describe a utility function, established in the literature. It is well-known that not any utility function follows the described approach, but we nevertheless introduce it as it has several advantages, one being the decision makers familiarity with graphical user interfaces where slider bars are used to modify the weight settings. Given the interaction of the decision maker with the system by means of a slider bar, it may equally be possible that the decision maker changes the weight settings in rather large steps instead of the small linear steps of size 0.1. In this case, we expect that the experimental setup does not lead to different results. Given appropriate computer hardware, the system should simply change the solutions quickly from one solution to another without necessarily showing intermediate alternatives.

Figure 2 to 6 plot the results obtained during the test runs.

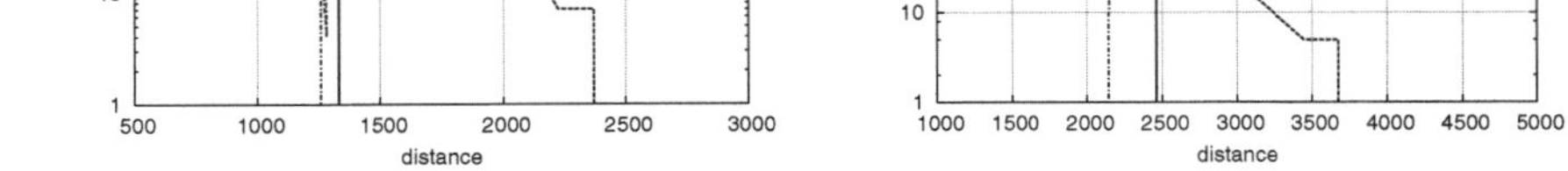

Fig. 2. Results of the test runs on instance 1a and 2a

It can be seen, that the results are significantly different depending on the initial chosen value of w_{DIST}. For initial values of $w_{DIST} = 0.5$, the framework follows more closely the Pareto front compared to other initial parameter settings.

To illustrate this behavior more closely, we are going to discuss the results for instance '1a' more closely and verbally. The first decision maker starts with $DIST = 975$, $TARDY = 6246$ and moves to $DIST = 1412$, $TARDY = 0$ while the second starts with $DIST = 2953$, $TARDY = 0$ and moves to $DIST = 1326$, $TARDY = 3654$. Clearly, the first strategy outperforms the second. While an initial value of $w_{DIST} = 0$ allows the identification of a solution with zero tardiness, it tends to construct routes that, when decreasing the relative importance

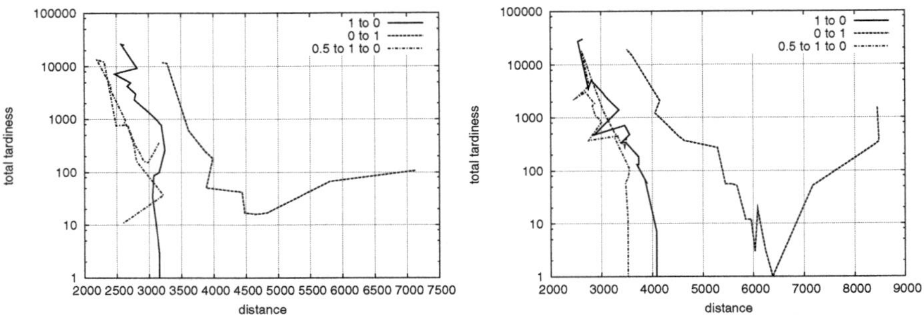

Fig. 3. Results of the test runs on instance 3a and 4a

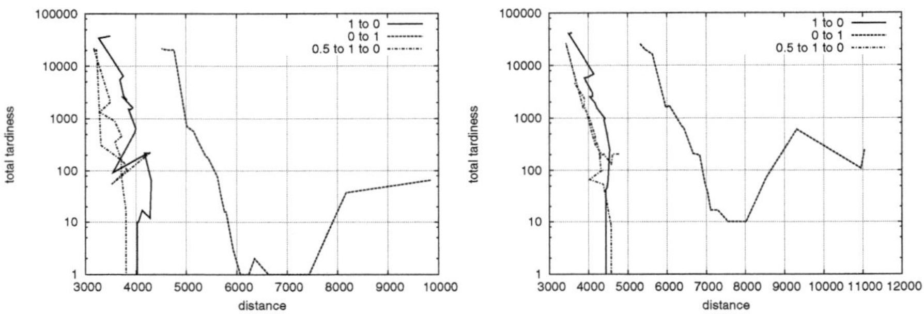

Fig. 4. Results of the test runs on instance 5a and 6a

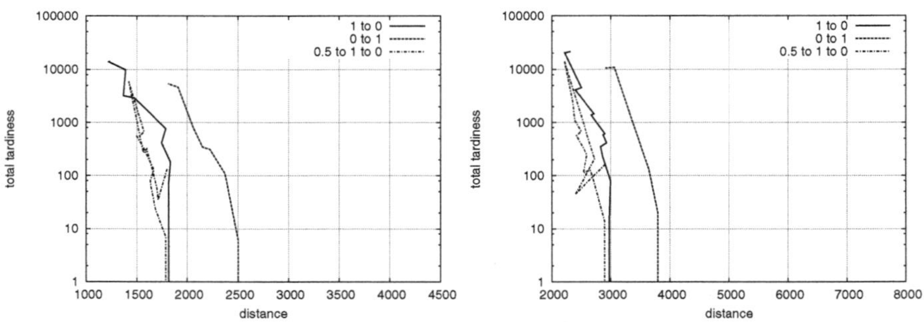

Fig. 5. Results of the test runs on instance 7a and 8a

of the tardiness, turn out to be hard to adapt. In comparison to the strategy starting with a $w_{DIST} = 1$, the clustering of customers turns out the be prohibitive for a later improvement.

When comparing the third strategy of starting with a $w_{DIST} = 0.5$, it becomes obvious that this outperforms both other ways of interacting with the system. Here, the solutions start with $DIST = 1245$, $TARDY = 63$, go to $DIST = 946$,

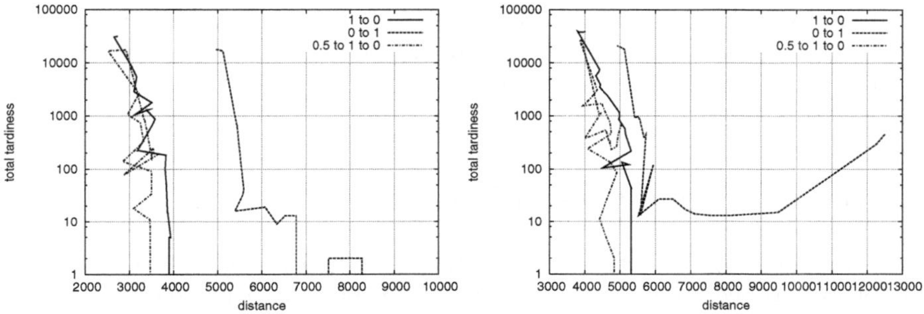

Fig. 6. Results of the test runs on instance 9a and 10a

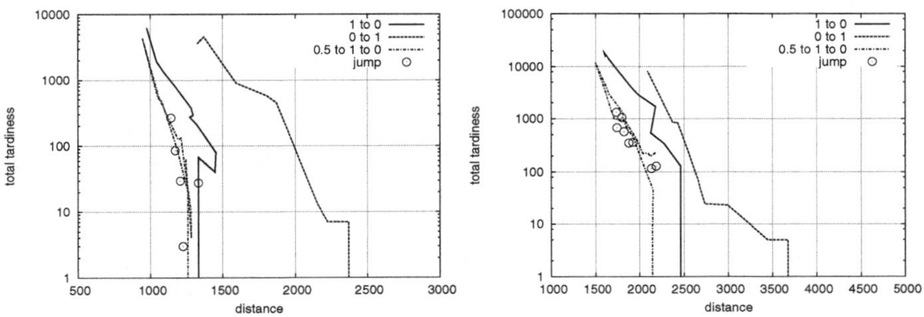

Fig. 7. Results of the test runs on instance 1a and 2a, compared to discrete jumps around $w_{DIST} = 0.7$

$TARDY = 4342$, and finally to $DIST = 1335$, $TARDY = 0$. Apparently, starting with a compromise solution is beneficial even for both extreme values of $DIST$ and $TARDY$.

To further investigate cases where the decision maker changes the weight setting in bigger discrete steps, we simulated a decision maker finally approaching a weight value of $w_{DIST} = 0.7$, starting from values around 0.7 with decreasing distance. The precise sequence of weights in the experiment is $w_{DIST} = \{0.4, 0.9, 0.6, 0.8, 0.65, 0.75, 0.7\}$, and the distances to the final, desired weight setting of the decision maker are $-0.3, +0.2, -0.1, +0.1, -0.05, +0.05, 0$. The setting of the experiment is based on the assumption that starting from an initial weight setting, the decision maker approaches the actually desired one by an alternating process of over- and underestimating the true value of w_{DIST}. The distances to the desired w_{DIST} decrease in this process as the decision maker reflects upon the chosen weight combination and the solution presented by the system.

Figure 7 plots the obtained results for the instances 1a and 2a. We chose to omit the plots for the other instances as the following interpretation and the resulting conclusions are identical.

It can be seen, that the discrete changes of the weight values lead to results that closely follow the curve obtained for the third decision maker. As suspected above, the results are comparable to the strategy of changing w_{DIST} in steps of 0.1.

4 Summary and Conclusions

A framework for the interactive resolution of multi-objective vehicle routing problems has been presented. The concept has been implemented in a computer system. Results on a benchmark instance have been reported, compared, and analyzed.

First investigations indicate that the concept may successfully solve vehicle routing problems under multiple objectives and complex side constraints. In this context, an interaction with the system is provided by a graphical user interface. The relative importance of the objective functions can be modified by means of a slider bar, resulting in different solutions which are computed in real time by the system, therefore providing an immediate feedback to the user. Figure 8 shows two extreme solutions that have been interactively obtained by the system.

As a result of the experiments, it becomes clear that for the investigated case, a compromise value of $w_{DIST} = 0.5$ should be chosen for the computation of a first solution before starting an interaction with the system. The so constructed alternative can be modified towards the minimization of the traveled distances as well as towards the minimization of the total tardiness.

Besides this theoretically gained insight, the contribution of the framework can also be seen in describing a general concept for the resolution of complex vehicle routing problems. As practical problems often vary in terms of their characteristics, this may turn out to be beneficial when problems with different side constraints have to be addressed using a single optimization procedure. An additional use can be found for dynamic vehicle routing problems. The market mechanism provides a platform for the matching of offers to vehicles without the immediate need of accepting them, yet still obtaining feasible solutions and gathering a prize for acceptance of offers which may be reported back to the customer.

Future developments are manifold. First, other ways of representing preferences than a weighted sum approach may be beneficial to investigate. While the comparable easy interaction with the GUI by means of a slider bar enables the user to directly change the relative importance of the objective functions, it prohibits the definition of more complex preference information, e. g. involving aspiration levels.

Second, different and improved ways of implementing the market mechanism have to be investigated. First results indicate that the quality of the solutions is biased with respect to the initial setting of the relative importance of the optimality criteria. It appears as if more complex reallocations of orders between vehicles are needed to address this issue.

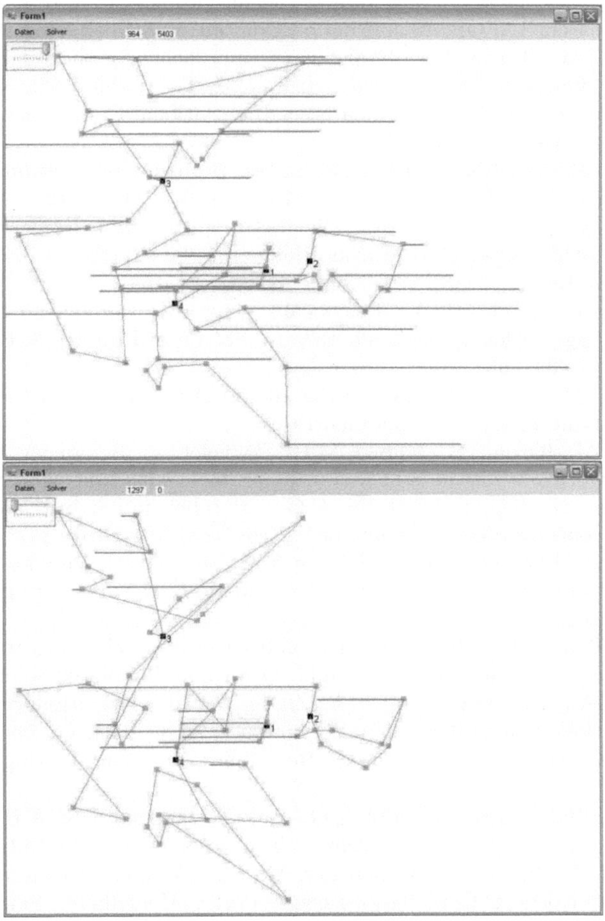

Fig. 8. Two screenshots of the graphical user interface. On the top, a short solution with high tardiness, on the bottom, a solution with low tardiness but long traveling distances.

References

1. Solomon, M.M., Desrosiers, J.: Time window constrained routing and scheduling problems. Transportation Science **22**(1) (February 1988) 1–13
2. Éric Taillard, Badeau, P., Gendreau, M., Guertin, F., Potvin, J.Y.: A tabu search heuristic for the vehicle routing problem with soft time windows. Transportation Science **31**(2) (May 1997) 170–186
3. Geiger, M.J.: Genetic algorithms for multiple objective vehicle routing. [22] 349–353
4. Potvin, J.Y., Kervahut, T., Garcia, B.L., Rousseau, J.M.: The vehicle routing problem with time windows part i: Tabu search. INFORMS Journal on Computing **8**(2) (Spring 1996) 158–164

5. Potvin, J.Y., Bengio, S.: The vehicle routing problem with time windows part ii: Genetic search. INFORMS Journal on Computing **8**(2) (Spring 1996) 165–172
6. Gendreau, M., Bräysy, O.: Metaheuristic approaches for the vehicle routing problem with time windows: A survey. In Ibaraki, T., Yagiura, M., Nonobe, K., eds.: Proceedings of the Fifth Metaheuristics International Conference - MIC2003, Kyoto, Japan (August 2003) 1–10
7. Beasley, J.E.: OR-library: Distributing test problems by electronic mail. Journal of the Operational Research Society **41**(11) (1990) 1069–1072
8. Fink, A., Voß, S.: HotFrame: A heuristic optimization framework. In Voß, S., Woodruff, D.L., eds.: Optimization Software Class Libraries. Kluwer Academic Publishers, Boston, Dordrecht, London (2002) 81–154
9. Di Gaspero, L., Schaerf, A.: Easylocal++: an object-oriented framework for the flexible design of local-search algorithms. Software: Practice & Experience **33**(8) (July 2003) 733–765
10. Cahon, S., Melab, N., Talbi, E.G.: ParadisEO: A framework for the reusable design of parallel and distributed metaheuristics
11. Rahoual, M., Kitoun, B., Mabed, M.H., Bachelet, V., Benameur, F.: Multicriteria genetic algorithms for the vehicle routing problem with time windows. [22] 527–532
12. Jozefowiez, N., Semet, F., Talbi, E.G.: Parallel and hybrid models for multi-objective optimization: Application to the vehicle routing problem. In et al., J.M.G., ed.: Parallel Problem Solving from Nature VII. Volume 2439 of Lecture Notes in Computer Science., Berlin, Heidelberg, New York, Springer Verlag (2002) 271–280
13. Murata, T., Itai, R.: Multi-objective vehicle routing problems using two-fold EMO algorithms to enhance solution similarity on non-dominated solutions. In Coello Coello, C.A., Hernández Aguirre, A., Zitzler, E., eds.: Evolutionary Multi-Criterion Optimization, Third International Conference, EMO 2005. Volume 3410 of Lecture Notes in Computer Science., Berlin, Heidelberg, New York, Springer Verlag (2005) 885–896
14. Jozefowiez, N., Semet, F., Talbi, E.G.: Enhancements of NSGA II and its application to the vehicle routing problem with route balancing. In Talbi, E.G., Liardet, P., Collet, P., Lutton, E., Schoenauer, M., eds.: Artificial Evolution. Volume 3871 of Lecture Notes in Computer Science., Berlin, Heidelberg, New York, Springer Verlag (2006) 131–142
15. Stacey, P.J.: Practical vehicle routeing using computer programs. Journal of the Operational Research Society **34**(10) (1983) 975–981
16. Waters, C.D.J.: Interactice vehicle routeing. Journal of the Operational Research Society **35**(9) (1984) 821–826
17. Park, Y.B., Koelling, C.P.: A solution of vehicle routing problems in a multiple objective environment. Engineering Costs and Production Economics (10) (1986) 121–132
18. Park, Y.B., Koelling, C.P.: An interactive computerized algorithm for multicriteria vehicle routing problems. Computers & Industrial Engineering **16**(4) (1989) 477–490
19. Phelps, S.P., Köksalan, M.: An interactive evolutionary metaheuristic for multi-objective combinatorial optimization. Management Science **49** (2003) 1726–1738

20. Cahon, S., Simarik, T., Vironda, T.: A graphical user interface for multi objective optimization guimoo http://guimoo.gforge.inria.fr/
21. Cordeau, J.F., Laporte, G., Mercier, A.: A unified tabu search heuristic for vehicle routing problems with time windows. Journal of the Operational Research Society (52) (2001) 928–936
22. de Sousa, J.P., ed.: Proceedings of the Metaheuristics International Conference MIC 2001. In de Sousa, J.P., ed.: Proceedings of the Metaheuristics International Conference MIC 2001, Porto, Portugal (July 2001)

Radar Waveform Optimisation as a Many-Objective Application Benchmark

Evan J. Hughes

Department of Aerospace, Power and Sensors,
Cranfield University, Shrivenham, Swindon,
Wiltshire, England. SN6 8LA
e.j.hughes@cranfield.ac.uk

Abstract. This paper introduces a real, unmodified *Many-Objective* optimisation problem for use in optimisation algorithm benchmarking. The radar waveform design problem has 9 objectives and an integer decision space that can be scaled from 4 to 12 decision variables. Proprietary radar waveform design software has been encapsulated in a fast and portable form to facilitate research groups in studying high-order optimisation of real engineering problems.

1 Introduction

Real engineering problems are often characterised by many objectives. Pareto ranking has been exploited in recent years to develop a large number of excellent multi-objective optimisation algorithms which can solve bi-objective optimisation problems effectively and reliably, for example, NSGA-II [2]. However, it is known that Pareto ranking alone does not scale well to problems with large numbers of objectives (4+ typically cause problems) [9,5]. Currently, there are few algorithms that are designed specifically to tackle many-objective problems.

This paper describes a real, unmodified engineering problem and the software is provided to allow optimisation with many-objectives to be studied, and hopefully efficient optimisation algorithms developed.

The problem has 9 objectives, and from 4 to 12 integer decision variables, each in the range [500,1500] inclusive, giving 1001 alleles per decision variable. It is known that some of the objectives are not totally independent, and it is suspected that the Pareto set is concave in places, with regions of low density.

The problem is the design of a waveform for a *Pulsed Doppler Radar*, typical of many airborne fighter radar systems. The radar system is required to measure both range and velocity of targets. Unfortunately, with the very long ranges (100 nautical miles typical) and very high velocities (Mach 5 possible), with a simple waveform it is only possible to measure either: range unambiguously but ambiguous velocity; velocity unambiguously but with the range ambiguous; or with both range and velocity ambiguous. For example, if velocity is measured, then target range may only be known modulo by say 5 kilometres, i.e. a target at 108km would appear at 3km.

To allow full unambiguous measurements, a set of simple waveforms is transmitted, each subtly different from the last. The results of the multiple waveforms are then combined in order to resolve the ambiguities. The problem is how to choose the set of simple

S. Obayashi et al. (Eds.): EMO 2007, LNCS 4403, pp. 700–714, 2007.

waveforms. Previous work in this area has led to the development of an evolutionary algorithm capable of designing practical waveforms [1].

This radar waveform design problem is interesting in that in a practical radar system, an entire set of non-dominated solutions would need to be created prior to each mission. While the radar is active, it will choose a general location on the non-dominated surface, based on current radar operating conditions, then select a waveform randomly which is local to this chosen location. The random choice helps prevent 3rd-party interception of the waveform as it is changing constantly, yet the waveform is biased towards an optimal radar configuration. Thus the radar chooses its operating point on the non-dominated surface dynamically on-line, from a non-dominated set that is likely to remain fixed for each mission.

An initial analysis of the properties of the objective surface has been performed and a demonstration of the typical behaviour of two different optimisation algorithms, NSGA-II and MSOPS, on the function is presented.

Section 2 details the radar design problem and section 3 describes the format of the objective function software. Section 4 introduces initial results from analysing the non-dominated surface and section 5 describes the results of comparing the performance of two example optimisation algorithms. Finally section 6 concludes.

2 Radar Waveform Design

2.1 Introduction

Radar systems are categorised by the rate at which they transmit pulses of energy toward the target, called the Pulse Repetition Frequency, or *PRF* [10]. There are three broad categories: Low PRF with few pulses (20 typical) and big gaps between them (1 millisecond typical); High PRF with many pulses (thousands) and short gaps (few micro seconds); and Medium PRF where there are a moderate number of pulses (64 typical) and moderate gaps (100 μS typical).

Low PRF radar systems can measure range exactly, but velocity measurements are ambiguous for any velocities greater than the maximum unambiguous veloctiy V_{mu}, given in (1) where F_{prf} is the pulse repetition frequency in Hertz and λ is the wavelength of the transmitted pulses.

$$V_{\text{mu}} = \frac{F_{\text{prf}}\lambda}{2} \tag{1}$$

The maximum unambiguous range of the radar is given by (2), where $c \approx 3\times10^8\,\text{ms}^{-1}$ is the speed of propagation of the pulse.

$$R_{\text{mu}} = \frac{c}{2F_{\text{prf}}} \tag{2}$$

A typical Low PRF radar may have a PRF of 1kHz, yielding a maximum unambiguous range of $R_{\text{mu}} = 150\text{km}$ and a maximum unambiguous velocity of $V_{\text{mu}} = 15\text{ms}^{-1}$, assuming a 10GHz transmission frequency ($\lambda = c/F_{\text{TX}}$, therefore $\lambda = 0.03\text{m}$).

The main advantage of low-PRF radar is the ability to measure target range directly using simple pulse delay ranging. However, low-PRF radar suffers from a lack of velocity visibility, since unwanted ground returns and undesired slow moving targets get folded over and over into the small velocity window. Low-PRF radar is best suited to operation in the absence of ground clutter returns, for example where a radar is looking up at high-flying targets, rather than looking down and the radar beam is striking the ground.

A typical High PRF radar may have a PRF of 100kHz, yielding a maximum unambiguous range of $R_{\mathrm{mu}} = 1.5$km and a maximum unambiguous velocity of $V_{\mathrm{mu}} = 1500\mathrm{ms}^{-1}$, assuming a 10GHz transmission frequency.

The principle advantage of high-PRF radar, is the ability to detect high closing-rate targets, in what is essentially a noise-limited environment. However, detection performance is poor in tail aspect (low closing-rate) engagements, where targets compete directly with the velocity spectrum of the sidelobe clutter, where transmissions out of the side of the antenna beam strike the ground and provide echoes back. Furthermore, the highly ambiguous range response causes the sidelobe clutter to fold within the ambiguous range interval. Consequently, sidelobe clutter can only be discarded by resolving in velocity. High PRF radar is very good where small relative velocities are not often seen, or when exceptionally good antennas are available that have very little spurious radiation out of the side of the beam.

Medium-PRF radar is a compromise solution designed to overcome some of the limitations of both low and high-PRF radar. By operating above the low-PRF region, the ambiguous repetitions of the ground clutter spectrum may be sufficiently separated without incurring unreasonable range ambiguities. Consequently, the radar is better able to reject mainbeam clutter when in a look-down scenario through velocity filtering without rejecting too many targets. By operating below the high-PRF region, the radar's ability to contend with sidelobe clutter in tail-chase engagements is improved. Targets may now be extracted from sidelobe clutter using a combination of velocity filtering and range gating.

For example, the mainbeam clutter may be $20\mathrm{ms}^{-1}$ wide, so can be 'notched' out as long as $V_{\mathrm{mu}} > 20\mathrm{ms}^{-1}$. However, if targets are folded in to the notched region, they cannot be detected and the region is said to be *blind*. When the pulse is transmitted, the receiver is turned off to protect it and the ranges at multiples of Rmu are now *eclipsed* and no targets may be detected here either. A second eclipsing region may also be applied to help reject the effects of the sidelobe clutter from the ground. The region will extend from a range that is just shorter than the aircrafts altitude (i.e. the first range at which an echo from the ground could occur) to often a few kilometres ahead of the altitude return.

A typical Medium PRF radar may have a PRF of 10kHz, yielding a maximum unambiguous range of $R_{\mathrm{mu}} = 15$km and a maximum unambiguous velocity of $V_{\mathrm{mu}} = 150\mathrm{ms}^{-1}$, assuming a 10GHz transmission frequency. However, in many applications, both range and velocity will now be ambiguous.

Medium PRF radars possess excellent clutter rejection characteristics which render them an attractive proposition for airborne intercept (AI), fire control systems, ground based air surveillance, weapon locating radar and a variety of other applications.

2.2 PRF Selection

Each PRF is characterised by regions of blind velocities and ranges associated with the velocity filtering of mainbeam clutter and time gating of sidelobe clutter and associated eclipsing losses. These blind zones are depicted in black on a blind zone map, as in figures 1 & 2.

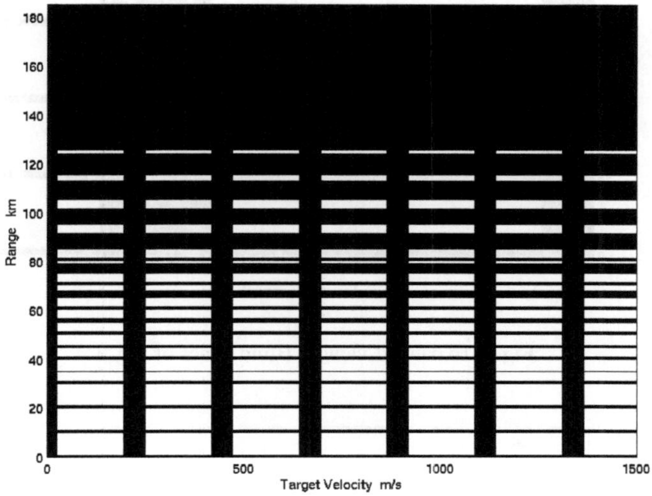

Fig. 1. Blind zones for a single, clutter limited, medium PRF waveform with PRF 14.9kHz

Multiple bursts of pulses are required in order to perform target detection and to resolve range and Doppler ambiguities. This is achieved by transmitting burst of pulses at a number of PRFs within the dwell time on target and sequentially measuring and comparing the ambiguous information received from every PRF. For example, eight different PRFs may be used but must all be able to be transmitted sequentially within the dwell time on the target, with each PRF burst having 64 processed pulses (64-point Fast Fourier Transform (FFT)) and a short period of time in which to change over PRFs. In practice, the change-over time is to allow the first pulse to reach, and return, from the furthest possible target of interest. Thus extra pulses are transmitted in a process termed *Space Charging*. For example, if the maximum range was 185km, and $R_{\text{mu}} = 15km$, 13 extra pulses would be sent giving a total of 77, but only 64 would be processed, making sure that 64 returns from both the closest and furthest targets were contained within the processing window.

The positions of blind zones vary with PRF, therefore, by applying suitable PRFs in a multiple-PRF detection scheme, not only may range and Doppler ambiguities be resolved, but also the blind zones may be staggered to improve target visibility. Ground clutter returns received through the antenna sidelobes may be strong enough to overwhelm weak target signals, consequently blind ranges tend to worsen with increasing range, as shown in figure 3.

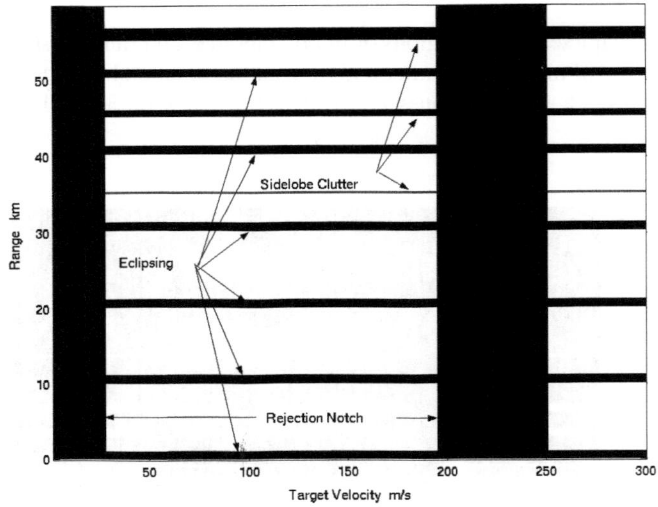

Fig. 2. Expanded view of Blind zones of Fig. 1

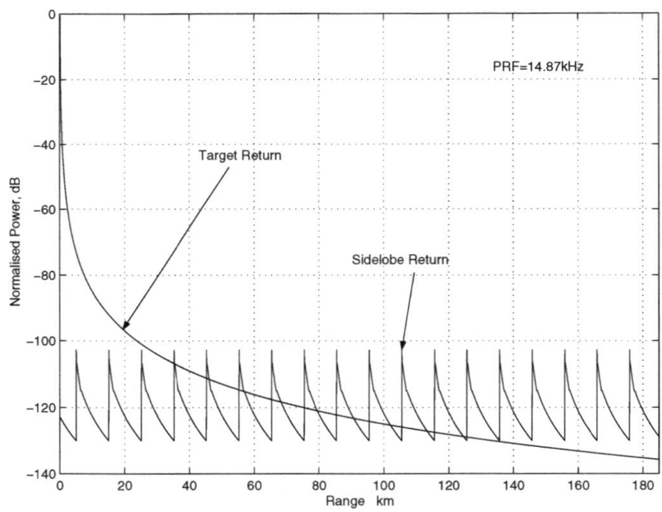

Fig. 3. Comparison of target return and sidelobe clutter for a single, noise limited, medium PRF waveform with PRF 14.9kHz

Conventionally, three PRFs are required to be clear in range and Doppler in order to resolve range and Doppler ambiguities and to declare a target detection.

The number of PRFs within a schedule must be selected carefully; too few and the ability to overcome range-velocity blind zones will be hindered. With too many PRFs, then, depending on the average PRF, there may be insufficient time to transmit the entire PRF schedule within the dwell time on target. Typically, eight PRFs are employed spanning about an octave.

If significant harmonic relationships exist between any of the PRFs chosen, then it may not be possible to resolve all of the ambiguities and the schedule is not *decodable*. In reality, targets have a physical size too and extend outside of individual range or velocity cells. It is desirable to make sure that a schedule is not only technically decodable, but also decodable in the presence of range or velocity extended targets. If the decodability criteria is broken by a large target, then false targets or *ghosts* will be observed. Unfortunately there is no means of distinguishing a ghost target from a real target and so all must be processed as true detections, leading to false alarms.

Because of the relatively wide size of the rejection notches, the possibility remains for a PRF schedule to be decodable and still have some rejection notch overlap; this is found to be a particular problem at the first repetitions of the ambiguous velocity intervals. The consequences of such occurrences are bands of velocities in which the radar is blind, or nearly blind (three PRFs clear only), at *all* ranges, thereby allowing a target to approach at a particular velocity with minimum risk of detection. Nothing can be done about the rejection notches, centred on zero Hz, which blind the radar to crossing targets.

After the pulses have been received by the radar, they are decoded using the coincidence algorithm [8]. The coincidence algorithm operates by taking the target returns and for each range bin, performing an FFT across all pulses in the PRF. Thus a map of range-velocity is produced. The regions of heavy clutter are then notched out and a detection algorithm is then used to identify potential targets within this 'folded' (ambiguous) range-velocity map. The process is repeated for each of the PRFs.

The next stage is to decode the targets and resolve the ambiguities. This is performed by taking each range-velocity map and repeating them until the maximum range-velocity extent of interest has been covered. For a single PRF, this will give many ranges and velocities at which a target may be present. The process is repeated for all the PRFs and the results overlaid. If a true target is present, it will appear in the same position in all PRIs (yet may not be detected, or may have been eclipsed or notched out). Any region of the range-velocity map that has 3 or more coincident detections is declared as being a true target. The process works well but issues can arise where very fast targets have moved between range cells between the first and last PRF being transmitted and do not necessarily align in the coincidence process. The problem is known as *range-walk* and is accounted for in the software.

In the radar problem encapsulated for this paper, the length of the transmitted pulse is directly proportional to the delay before the next pulse. This keeps the duty cycle of the transmitter constant. The radar is also frequency-hopping in that the transmission frequency changes for each PRF. The result is that the wavelength will also change and so the order of transmission for the PRFs is important.

The selection of PRFs in a medium PRF set is therefore based on the following constraints:

1. A spread of values which enable the resolution of range and velocity ambiguities to ensure basic decodability,

2. Removal of totally blind ranges and velocities,
3. The total time required for transmission of the waveform must be within the target dwell time.

The objectives then become:

1. Maximise the size of the target in range that can be decoded without ghosting,
2. Maximise the size of the target in velocity that can be decoded without ghosting,
3. Maximise the size of the clutter patch in range that can be tolerated before blind ranges occur,
4. Maximise the size of the clutter patch in velocity that can be tolerated before blind velocities occur,
5. Minimise the total time required for transmission of the waveform.

Objectives 1 to 4 may be calculated using the process outlined in [7] by calculating what is the target extent at any range/ velocity that can be tolerated before problems arise. If any of the objectives are zero or negative, i.e. a negative target size is the maximum that can be tolerated, then it implies that one of the constraints has been violated.

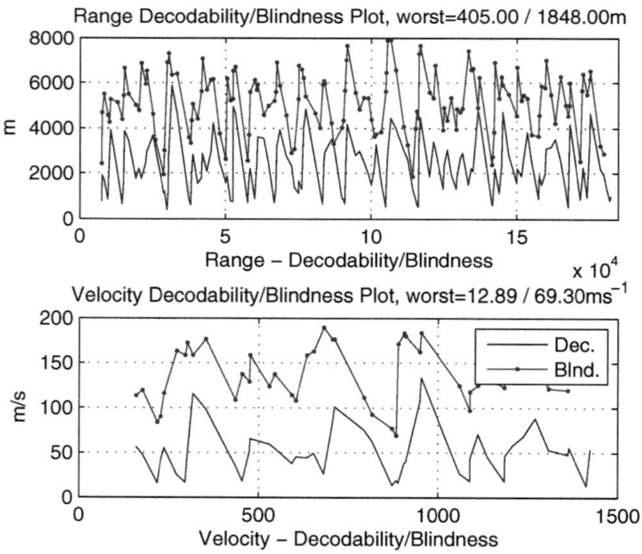

Fig. 4. Minimum target sizes for PRI set $[50.8\ 50.0\ 55.6\ 64.0\ 86.2\ 70.1\ 67.4\ 96.8]\mu s$

For the practical radar design in section 2.3, as each range and velocity has an associated minimum target size as demonstrated in figure 4, the worst case and median performance are actually of interest. The result is that for objectives 1 to 4, both the median overall performance and minimum overall performance need to be maximised, yielding 9 objectives in total. The constraints can be applied easily in objective space after the optimisation process. For the PRI set $[50.8 50.0 55.6 64.0 86.2 70.1 67.4 96.8]\mu s$ results shown figure 4, the corresponding objectives are:

1. Median range decodability: 2205.0 metres,
2. Median velocity decodability: 47.9 ms^{-1},
3. Median range blindness: 5310.0 metres,
4. Median velocity blindness: 135.5ms^{-1},
5. Minimum range decodability: 405.0 metres,
6. Minimum velocity decodability: 12.9ms^{-1},
7. Minimum range blindness: 1848.0 metres,
8. Minimum velocity blindness: 69.3ms^{-1},
9. Dwell time: 44.8ms.

As objectives 1 to 8 are positive, and objective 9 is less than 50ms, the PRI set shown will form a viable radar waveform.

2.3 The Radar Model

A radar model based on an airborne fire control type application was derived to trial the fitness of PRF sets. The model assumes approximately 10GHz operation, 64-point FFT processing, 10% fixed duty ratio of pulse length to Pulse-Repetition Interval (PRI), linear FM pulse compression achieving a variable compression ratio with the PRF and that platform motion compensation is applied (i.e. the location of the ground is shifted in velocity back to zero, rather than being left at the platforms forward speed). The maximum target velocity with respect to the ground was taken as 1500 m/s and the maximum range was taken to be 185 km (100 nmi). These and other operational characteristics are summarised in Table 1. It is intended that the model should be representative of the types currently in service or about to enter service.

Table 1. Summary of the radar model's characteristics

Parameter	Value
Carrier frequency	9.97 GHz for PRF_1, each following PRF -30MHz
Minimum PRI	50 μs
Maximum PRI	150 μs
Compressed pulsewidth	0.5 μs
Receiver recovery time	1.0 μs
Range resolution	75m
FFT size	64 bins
Duty cycle	10% fixed
Maximum target dwell time	50ms
Maximum target Velocity	$\pm 1500ms^{-1}$
Maximum detection range	185.2 km (100 nmi)
Number of PRFs/PRIs	4 to 12
Number of PRFs for coincidence	3

3 Software Structure

The software for the radar design problem is available for download from [3]. As the exact design analysis algorithms are proprietary, the software source is not provided, rather a compiled but portable binary file. In the interest of maximising portability between platform types, and simultaneously protecting the proprietary algorithms, Matlab P-code format has been used. Matlab P-code is a platform independent pre-parsed binary format used by the Matlab engine, helping to reduce the options for de-compilation that 'C', Fortran or Java would present.

The function `testpris.p` takes a $1 \times N$ vector as an input, where $N \in [4, 12]$ is the number of decision variables, and outputs a 1×9 vector of metrics. Each of the decision variables is an integer in the range [500,1500] inclusive and represents the set of Pulse Repetition Interval values between $[50.0\mu s, 150.0\mu s]$ in steps of $0.1\mu s$.

The 9 metrics that are output represent:

1. Median range extent of target before schedule is not decodable (in metres),
2. Median velocity extent of target before schedule is not decodable (in ms^{-1}),
3. Median range extent of target before schedule has blind regions (in metres),
4. Median velocity extent of target before schedule has blind regions (in ms^{-1}),
5. Minimum range extent of target before schedule is not decodable (in metres),
6. Minimum velocity extent of target before schedule is not decodable (in ms^{-1}),
7. Minimum range extent of target before schedule has blind regions (in metres),
8. Minimum velocity extent of target before schedule has blind regions (in ms^{-1}),
9. Time required to transmit total waveform (in milliseconds, to be **minimised**)

The metrics 1 to 8 are to be maximised, while metric 9 is to be minimised. There are 9 corresponding constraints: the first eight metrics must all be greater than zero, and the 9th metric must be less than 50 ms. In order to simplify the conversion of the objectives all to minimisation, and to simplify the constraint process, a wrapper function has been provided `objpri.m` that will allow a $P \times N$ matrix to be provided, and a $P \times 9$ matrix is returned with all of the metrics arranged for minimisation, and aligned so that if any are greater than zero (the maximisation is converted to a minimisation by negating), then the solution can be considered not feasible as a practical waveform. The function also allows an entire population (size P in the example above) to be passed as one matrix.

The current version of the MSOPS [4] optimisation algorithm code used to generate the results found later in this paper is also provided as an example of how the objective function may be implemented.

The run-time of the objective function is quite short, considering it is an un-modified engineering application. Under Matlab and on a 1.8GHz Pentium 4 processor, Microsoft Windows XP, table 2 indicates the observed processing times for 10000 evaluations, and therefore times for single objective vector calculations.

4 Initial Objective Surface Analysis

The objective surface consists of 9 objectives and from 4 to 12 decision variables. Some relationships are known to exist between pairs of objectives, and also between the chromosomes and objectives.

Table 2. Example processing times for objective vector calculation

N	Time 10000 eval (sec)	Time 1 eval (ms)
4	21.6	2.16ms
8	33.6	3.36ms
12	46.4	4.64ms

The first main relationship is that if the number of decision variables is less than 9, then the objective region must be a projection of the lower-dimensional decision space manifold into the 9-objective space: thus not every possible objective vector is defined. With greater than 9 decision variables, the converse is true and there is likely to exist extensive many-to-one mappings between decision space and objective space. At present, it is not clear if one particular choice of decision space dimensionality provides the entire Pareto surface of the problem. It is hypothesised however that this is not the case and that the full Pareto surface will be comprised of sections where the decision space dimensionality changes. From a radar design perspective, the number of PRFs used and therefore the decision space dimensionality are not critical, as long as the schedule is valid and performs well. Short schedules are often attractive as they tend to require less processing time, although this processing aspect is a design preference rather than an objective, and is useful for refining the choice of PRF schedule from the full Pareto set.

The second relationship is that the first 4 objectives are median values, and objectives 5-to-8 are the minimum values, of the same 4 data sets. Thus objective 1 will always be better than (or equal to) objective 5 etc. Objectives 1 & 3, 5 & 7 are metrics associated with the performance in range and tend to have a degree of correlation, i.e. they may not conflict strongly. Similarly, Objectives 2 & 4, 6 & 7 are metrics associated with the performance in velocity and again do not tend to conflict strongly with each other either, however they do conflict with the objectives associated with range.

The third relationship is between the decision variables and the dwell time (objective 9). The objective is calculated from equation 3, where $F_9(\mathbf{x})$ is objective 9, $\mathbf{x}$ is the decision vector (which is integers in units of $0.1\mu s$) and $\lceil\cdot\rceil$ is a rounding-up operation. The first part of the sum accounts for the time to transmit 64 pulses for the FFT. The second part of the sum calculates how many pulses are required to space-charge to the maximum range of interest $R_{\max}$. The objective has a nearly linear relationship, apart from the quantised space-charge offset. Interestingly, the order of the decision variables also has no influence on the objective, implying that it is therefore approximately unimodal (i.e. no local optima), but multi-global (i.e. more than one global optima exist). At first sight, the minimum value appears to occur when all of the decision variables are at their minimum, i.e. for an 8 PRI system, the minimum total dwell time would be 35.6ms, giving an objective value of -14.4. However, if unambiguous range of the minimum possible PRI is not an integer fraction of the maximum range, then the constant space-charge term may be rounded up. It may therefore be possible that under some conditions, the global optima may not quite occur when all the decision variables are at their minimum values.

$$F_9(\mathbf{x}) = 1000 \sum_{i=1}^{N} \left(64 \frac{x_i}{10^7} + \left\lceil \frac{2 \times 10^7 R_{\max}}{x_i c} \right\rceil \frac{x_i}{10^7} \right) - 50 \qquad (3)$$

Similarly a fourth observation can be made that the velocity-related objectives **are** dependent on the order of the decision variables (i.e. if the elements of a decision vector are re-ordered, the objective values may change). This coupling is due to the frequency-hopping radar design: the target velocity produces a *Doppler shift* of the carrier frequency; the shift amount is dependent on the carrier frequency itself. As the first PRI described by the first decision variable is transmitted at 9.97GHz, the second at 9.94Ghz etc., the order of the decision variables will change the effective transmission frequency, and therefore the velocity performance of the waveform. The objectives that are associated with range however are only very weakly correlated to the decision variable ordering. The modification of the objective value that occurs with a re-order is due to the effects of target range-walk. Given that a relatively large range resolution of 75 metres is used in this design, in a dwell of 50 milliseconds, a target must be travelling at $1500ms^{-1}$ or faster in order to move range cells during the dwell. The effect is thus only very small in this example as only targets at the limit of the velocity of interest will be affected. Objectives 1 & 3 are the most likely to undergo any change as these are calculated based on the medians. Objectives 5 & 7 are calculated using minimum and although possible, it is unlikely that any order-dependence will be observed.

The number of decision variables influences both performance, and also ultimately which constraints are most difficult to satisfy.

5 Algorithm Comparison

An initial examination of the ability of multi-objective optimisers to explore the objective surface was performed. Two primary optimisers: NSGA-II and MSOPS were used, along with a 3rd which is a steady-state derivative of the MSOPS algorithm and is currently under development (unpublished prototype which is run in a 'Pareto ranking mode' to aid confirmation of NSGA-II results). The experiments were to generate non-dominated surfaces for the application problem using 8 decision variables. Although NSGA-II is known to be less suitable for many-objective problems when compared to bi-objective problems, it has been included as a useful reference algorithm.

Each optimiser was run 30 times for 20,000 function evaluations. In each run of each optimiser, all 20,000 points that were generated were collected and the non-dominated surface of these points established, rather than relying on the contents of the final population alone e.g. as is common in NSGA-II.

The non-dominated surfaces of the 90 independent experiments were collated into a single group consisting of 775,140 points. Initially, an attempt was made to create the composite non-dominated surface to establish the contribution from each algorithm. However after over 24 hours of processing, the non-dominated set was still incomplete, but it was clear that all of the runs of all of the algorithms made a significant contribution to the surface – in the high dimensionality of the objective space, the proportion of non-dominated solutions is very large. The group of 775,140 points was used to establish a lower-bound reference point, and a range for each of the objectives for scaling purposes.

The lower reference used was: [-7035.0 -81.3 -27150.0 -296.9 -2700.0 -22.3 -7660.5 -100.5 -14.4]. The range of each objective was calculated as: [7110.0 84.7 23130.0 249.5 2775.0 27.0 10511.0 100.3 78.0]. Figure 5 shows a plot of 5000 example non-dominated points, and figure 6 shows the same points, but normalised. It is clear that the relationship between the objectives is non-trivial.

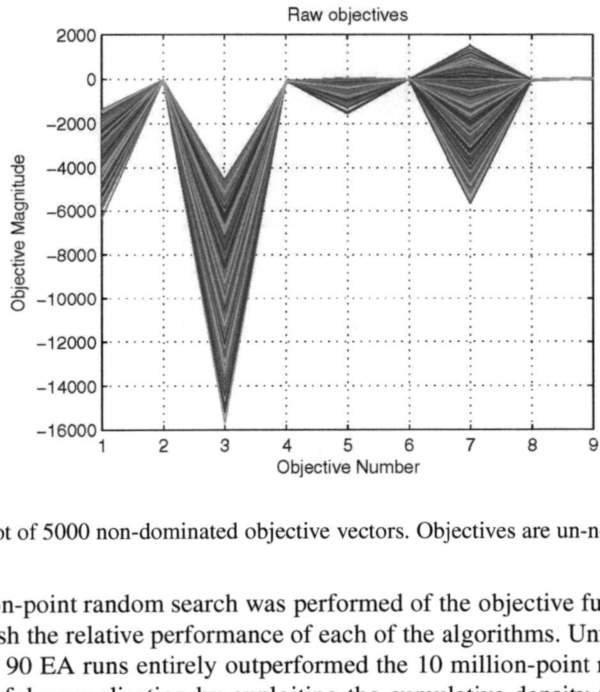

Fig. 5. Plot of 5000 non-dominated objective vectors. Objectives are un-normalised.

A 10 million-point random search was performed of the objective function in order to help establish the relative performance of each of the algorithms. Unfortunately, every one of the 90 EA runs entirely outperformed the 10 million-point random search, preventing useful normalisation by exploiting the cumulative density function of the aggregated objectives [6].

Figure 7 shows an approximation of the distribution of the median attainment surface of the 3 algorithms over the 30 runs. Each of the 3 lines of the graph is calculated by:

1. Generate 200 approximately uniformly distributed unit length target vectors over the entire objective space.
2. For each of the 200 target vectors in turn, calculate the weighted min-max result for all the points in each of the 30 sets of repeated experiments. The minimum value in each of the 30 sets is taken, yielding a 30 by 200 matrix for each of the 3 algorithms.
3. The median of each column of this matrix is taken, creating a 200 element vector for each algorithm tested: the vector is a sampled approximation of the median attainment surface.
4. Each of the 200 element vectors is sorted to build a cumulative distribution, and then plotted.

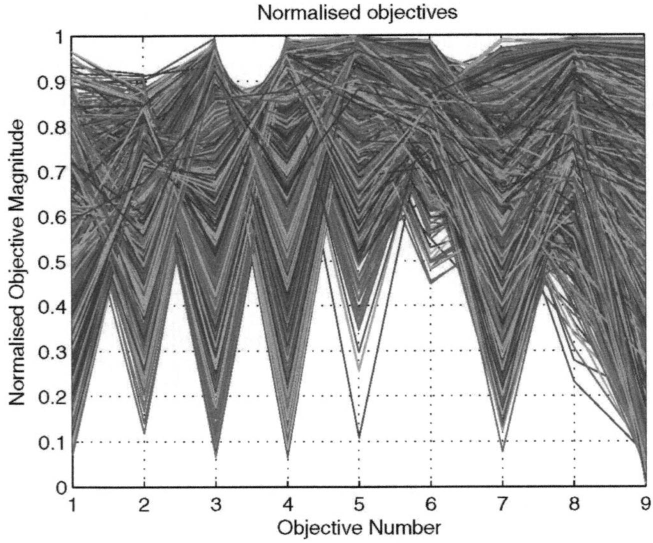

Fig. 6. Plot of 5000 non-dominated objective vectors. Objectives are normalised so that full 775,140 point vector set lies in range [0,1].

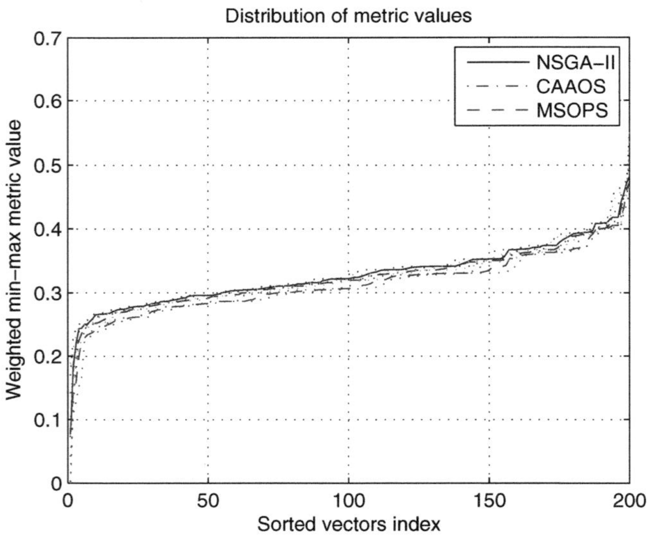

Fig. 7. Sampled median attainment surface distributions for NSGA-II (solid), MSOPS (dashed) and prototype algorithm (labelled CAAOS, dash-dot). Lower values are superior. Dotted lines show 95% confidence intervals obtained through 10 independent choices of the 200 target vectors.

As a 200 point sample of a 9-dimensional objective space is very sparse, the process was repeated 10 times, each with a different set of 200 points. The results were used to create 95% confidence interval bounds and plotted as dotted lines on figure 7, with the median of the 10 results as the thick-line. It is clear that despite 200 points being a sparse sample, the results are reassuringly accurate.

The performance of the three algorithms is very similar, with MSOPS leading slightly everywhere, as anticipated from previous studies [5]. As all 3 algorithms have produced similar results, despite entirely independent trials and algorithms, it is hypothesised that the obtained non-dominated sets are quite close to the Pareto optimal solution, but as the difference between MSOPS and NSGA-II shows, the set can only be considered non-dominated rather than Pareto. Additionally, as this is a 9-objective problem and NSGA-II has performed reasonably well with 20,000 evaluations, it is suspected that the overall density over the majority of the Pareto surface is high, allowing the problem to be approximated very well within 20,000 evaluations. However some of the results from the MSOPS trials suggest that there are regions of low density (the weighted min-max metric can converge well, however the Vector-Angle Distance Scaling metric is poor, suggesting a low density of points). Thus it is anticipated that the objective may be viable for study where only very few function evaluations are available (typical for on-line optimisation within a radar system).

It has been observed that there are concavities in certain dimensions (e.g. between objectives 1 & 2), and it may be the edges of the set that are sparse. Early indications from analysis using MSOPS also suggests that there may be regions of discontinuities and possibly sections of disconnected objective space.

6 Conclusions

This paper has presented and described a real engineering application of many-objective optimisation, and also provided access to software that allows the application to be studied by other researchers in the field.

The objectives can be calculated quickly enough to allow for practical optimisation algorithm development, and the surface appears complex enough to test the performance of visualisation and surface analysis tools.

At times, industrial acceptance of multi and many-objective optimisation algorithms has been slow. It is hoped that by providing an un-simplified real-world problem to use as an empirical benchmark, others can be encouraged to do the same for other problems, allowing more credibility to be attached to optimisation algorithm performance.

I also hope that as researchers in the field develop better algorithms for many-objective optimisation, the results can be collated and the true Pareto set for this optimisation problem approached. This collected set would be made available to extend the non-dominated data already provided from the production of this paper.

References

1. C. M. Alabaster, E. J. Hughes, and J. H. Matthew. Medium PRF radar PRF selection using evolutionary algorithms. In *IEEE Transactions on Aerospace and Electronic Systems*, volume 39, pages 990–1001, July 2003.

2. K. Deb, S. Agrawal, A. Pratap, and T. Meyarivan. A fast elitist non-dominated sorting genetic algorithm for multi-objective optimization: NSGA-II. In M. Schoenauer, K. Deb, G. Rudolph, X. Yao, E. Lutton, J. J. Merelo, and H.-P. Schwefel, editors, *Parallel Problem Solving from Nature – PPSN VI*, pages 849–858, Berlin, 2000. Springer.
3. E. J. Hughes. Many-objective radar design software. http://code.evanhughes.org.
4. E. J. Hughes. Multiple single objective pareto sampling. In *Congress on Evolutionary Computation 2003*, pages 2678–2684, Canberra, Australia, 8-12 December 2003. IEEE.
5. E. J. Hughes. Evolutionary many-objective optimisation: Many once or one many? In *IEEE Congress on Evolutionary Computation, 2005*, volume 1, pages 222–227, Sept. 2005.
6. E. J. Hughes. Multi-objective equivalent random search. In *Parallel Problem Solving From Nature, PPSN IX*, Reykjavik, Iceland, September 2006. Springer LNCS-4193.
7. A. Kinghorn and N. Williams. The decodability of multiple-prf radar waveforms. In *Radar 97 (Conf. Publ. No. 449)*, pages 544 – 547, Oct. 1997.
8. W. H. Long and K. A. Harringer. Medium PRF for the AN/APG-66 radar. *Proceedings of the IEEE*, 73(2):301–311, Feb. 1985.
9. R. C. Purshouse. Evolutionary many-objective optimisation: An exploratory analysis. In *The 2003 Congress on Evolutionary Computation (CEC 2003)*, volume 3, pages 2066–2073, Canberra, Australia, 8–12 December 2003. IEEE.
10. M. I. Skolnik, editor. *Radar Handbook*. McGraw-Hill, 2nd edition, 1990. ISBN 0-07-057913-X.

Robust Multi-Objective Optimization in High Dimensional Spaces

André Sülflow, Nicole Drechsler, and Rolf Drechsler

Institute of Computer Science
University of Bremen
28359 Bremen, Germany
{suelflow,nd,drechsle}@informatik.uni-bremen.de

Abstract. In most real world optimization problems several optimization goals have to be considered in parallel. For this reason, there has been a growing interest in Multi-Objective Optimization (MOO) in the past years. Several alternative approaches have been proposed to cope with the occurring problems, e.g. how to compare and rank the different elements. The available techniques produce very good results, but they have mainly been studied for problems of "low dimension", i.e. with less than 10 optimization objectives.

In this paper we study MOO for high dimensional spaces. We first review existing techniques and discuss them in our context. The pros and cons are pointed out. A new relation called *ϵ-Preferred* is presented that extends existing approaches and clearly outperforms these for high dimensions. Experimental results are presented for a very complex industrial scheduling problem, i.e. a utilization planning problem for a hospital. This problem is also well known as *nurse rostering*, and in our application has more than 20 optimization targets. It is solved using an evolutionary approach. The new algorithms based on relation *ϵ-Preferred* do not only yield better results regarding quality, but also enhances the robustness significantly.

1 Introduction

To solve complex optimization problems today, it is often not sufficient to only consider a single optimization criteria. In contrast, many real world problems have several – often contradicting – optimization goals. Thus, in the recent past several techniques for *Multi-Objective Optimization* (MOO) have been proposed.

One of the first approaches in this direction was the use of *Pareto-optimal elements*. This has been discussed in the context of *Evolutionary Algorithms* (EAs) in [1]. The goal is to determine elements from the *Pareto set*. To guide this search, there exist several alternative methods (see e.g. [2,3]) where the core is a relation that allows to compare different elements. E.g. the relation *Dominate* proposed in [1] can be applied. These methods are well known and have been studied intensively. But so far these studies mainly consider problems with a small number of optimization criteria, e.g. in [2] comparisons for dimensions two or three are given.

S. Obayashi et al. (Eds.): EMO 2007, LNCS 4403, pp. 715–726, 2007.

For higher dimensional spaces there only exist a few studies (see e.g. [4,5,6,7]). As testcases scalable test functions proposed in [4] are considered. For example, in [3] it is reported that the number of individuals in the Pareto set, i.e. the non-dominated solutions, increase with the number of optimization objectives. Experiments have shown that for 20 objectives the percentage of solutions that cannot be distinguished using relation *Dominates* in random populations is nearly 100%. For this reason, new measures and relations have to be defined that help to automate and guide the optimization process.

In general for higher dimensions *weighted sums* or *aggregation* have been proposed, since they are easy to describe and, on a first sight, scale well. But for high dimensions these techniques reach their limits, since it is hard (or even impossible) to determine good weights or the fitness of the optimal solution is not known in advance, respectively.

In [8,9] an alternative relation called *Preferred* (originally introduced as *Favour*) has been proposed and applied for five dimensions. Experiments have shown that *Preferred* clearly outperformed relation *Dominates* and an approach based on weighted sums. But in all cases described above the dimensions considered are rather small, i.e. less than 10. However, for complex optimization problems, where especially EAs are frequently used, often a higher number of dimensions occur. Of course, the standard algorithms can also be applied in the case of higher dimensions, but it will be shown in this paper by a detailed discussion and also by experimental studies for an industrial application that other techniques should be applied.

In this paper we first discuss the existing techniques and point out their main properties. Then, an experimental study shows the weaknesses of the above techniques for higher dimensions. For the experiments an industrial application where a very complex scheduling problem with many constraints occurs is considered. I.e. the *nurse rostering problem* [10], a well-known problem in mixed integer optimization, where a highly constraint schedule for employees in a hospital is generated. In this problem, 25 optimization goals are considered in parallel. As an additional difficulty, there are different types of constraints. Some can be seen as "hard constraints" that are enforced by state laws, while others are "soft constraints" that should be fulfilled as good as possible. It is demonstrated by experiments that for high dimensions the approach using relation *Preferred* outperforms traditional methods based on non-dominated sorting (relation *Dominates*). But relation *Preferred* is not robust for high dimensions and has to be extended accordingly. Therefore, we propose an extension of *Preferred* that also takes the relative difference over all dimensions into account. It considers environments of radius ϵ, where elements outside this region are "punished". The new relation is called ϵ-*Preferred*. Experimental results show that the new approach results in higher quality and, additionally, gives very robust optimization results.

The paper is structured as follows: In Section 2 previous work is reviewed and properties of the different relations are discussed. An experimental study for a complex scheduling problem is presented in Section 3. This study clearly shows the weaknesses of the existing techniques. Our new approach including experimental evaluations is introduced in Section 4. Finally, in Section 5 the results are summarized and directions for future research are pointed out.

2 Preliminaries

To make the paper self-contained, a brief review of proposed relations for comparing MO solutions is given. In the second part the MOO methods used for the experiments are described.

2.1 Relations

A multi-objective optimization problem is defined as follows: Given a search space Ω, an evaluation function $F : \Omega \rightarrow \mathbb{R}^m$ is defined to calculate the fitness vector of size m: $F(A) : \forall A \in \Omega$. Then the optimization goal is to minimize (or maximize) the elements of $F(A)$. In the following we assume, without loss of generality, that F has to be minimized for all objectives. According to [1] it holds:

Definition 1. *Let $A, B \in \Omega$. $A \prec_{dominates} B :\Leftrightarrow$
$\exists j : F_j(A) < F_j(B) \wedge \forall i \neq j : F_i(A) \leqslant F_i(B), 1 \leqslant i \leqslant m$.*

Based on this, we can describe a *Pareto set* (*non-dominated set*) as χ: $\forall p \in \chi : \nexists q \in \chi : q \prec_{dominates} p$.

As can be seen from the definition above, if two elements $A, B \in \Omega$ are compared with relation *Dominates*, then A dominates B only if it is less or equal to B in all objectives and if it is better in at least one objective. Using relation *Dominates*, a set of elements can be classified into several levels of non-dominated solutions. Thus, first the non-dominated set is computed. Then, disregarding the non-dominated set, the next level of non-dominated elements is found. This is repeated, until all elements have been considered. The resulting procedure is called non-dominated sorting [3].

In comparison relation *Preferred* [8] respects the number of objectives in which A differs from B:

Definition 2. *Let $A, B \in \Omega$. $A \prec_{preferred} B :\Leftrightarrow$
$|\{i : F_i(A) < F_i(B), 1 \leqslant i \leqslant m\}| > |\{j : F_j(B) < F_j(A), 1 \leqslant j \leqslant m\}|$.*

A is then said to be *preferred* to B if A is better than B in a larger number of objectives. The Relation *Preferred* is not transitive. This means it is possible to have cycles in the relation graph of the elements of Ω.

Analogously to non-dominated sorting a set of elements, e.g. a population, can be grouped into several levels by using relation *Preferred* [8].

2.2 Methods

For our analysis three different methods are used. Based on relation *Dominates* are the methods `Dominates` [11] and `NSGA II` [12]. Based on relation *Preferred* the proposed strongly-connected components building algorithm `Preferred` is used [8].

The method `Dominates` is a part of the `Evolving Object Library` [11] used as backbone for our study. This method counts for each individual the number of individuals which are dominated. Thus, if the number is zero, then this individual is in the Pareto-front. The best rating is given to the individuals without any dominators. Then the elements with one dominator follow and so on. Thus, in contrast to non-dominated sorting [1], only the first Pareto-front is built and considered using the method of [11]. Note, that the "distribution" of the elements in the solution space is not taken into account. By this, it might happen that all elements from the same region are favoured, while other regions are not considered.

To avoid this concentration on a small part of the search space, the `NSGA II` algorithm has been proposed [12]. The idea of `NSGA II` is as follows: The individuals in a population are classified by non-dominated sorting [1]. Then the algorithm for computing the crowding distance is used to ensure that the Pareto-front is widely spread. This also helps to preserve a diversity in the set of possible solutions, e.g. in the population in the case of an EA.

The algorithm `Preferred` from [8] builds all strongly-connected components in the relation graph that result from the pairwise comparison of all individuals of the population. All individuals in the same component get the same fitness (ranking value). Then all components are hierarchically ordered, followed by an assignment of ranking values. For more details see [8].

3 Application of Models

While previous methods and algorithms have been successfully applied in many fields ranging from graph problems to circuit design, all the studies have in common that MOO problems of low dimension have been considered, i.e. typically only three or four. The situation changes, if higher dimensions are the main focus. In this section we first introduce a very complex industrial scheduling problem, i.e. a utilization planning problem for a hospital. The problem is taken from a real-world environment of a hospital in Austria. An experimental study follows, using the techniques introduced in Section 2.2.

3.1 Utilization Planing Problem

The problem of utilization planing, i.e. the *nurse rostering problem* [10], is very complex and cannot be described here in all details. We briefly highlight the main aspects to give an idea of the underlying optimization problem.

The problem is to determine a schedule for the employees at a hospital. In the experiments schedules for ten persons for a planning period of 30 days have

Name/Day	1	2	3	4	5	6	7	8	9	...	30
C. Meyer	S	S	D	D	V	V	V	-	N	...	N
J. Smith	D	D	-	-	N	N	N	-	-	...	V
J. Doe	-	-	D	D	D	D	-	-	-	...	D
J. Blow	F	F	-	L	L	L	-	-	F	...	-
J. Bloggs	S	V	V	-	-	N	N	N	-	...	-
...											

Fig. 1. Example nurse rostering schedule

to be computed. The computation of the fitness can be roughly categorized in three main areas:

1. Rules resulting from ergonomics, e.g. having regular shifts
2. Restrictions by law, e.g. maximal hours of work per day or maximal working days per month
3. Rules of the nurse station, e.g. sufficient nurses per shift

Some of these constraints are "hard" in the sense that they have to be fulfilled, while others are "soft", i.e. they improve the fitness, but a violation does not invalidate the schedule. Altogether 25 optimization objectives are influencing the fitness function. Each one might have a different influence, e.g. some are linear while others are exponential regarding the influence.

Example 1. To give a better understanding of the algorithm a sketch of a schedule is given in Figure 1. Depending on the grade of training the optimization algorithm assigns exactly one shift to a person per day. In this example all given shifts are marked with a letter. These letters have the following meaning: Day shift (D), Night shift (N), Late shift (L), Vacation (V), Free shift (F), Stand-by shift (S), No shift (-).

For more details see [10].

3.2 Implementation

The core of the optimization is directly based on a schedule similar to Figure 1. In the optimization algorithm one mutation and two recombination operators are used. They are applied with a probability of 50% for the mutation operator and 25% for each recombination operator. The mutation operator sets a legal shift for a randomly chosen person and day. Both recombination operators exchange a block of shifts, specified by a random choice of employees and days. Since the focus of this paper is not on the optimization technique, i.e. the EA approach,

Table 1. Fitness for generation 3000

Algorithm	$AVG_{\Psi,3000}$	$AVG_{\Psi,3000}$ in percent	σ_{3000}	$\sigma_{0..3000}$
Dominates	429805	100%	11%	7%
NSGA II	421840	98%	13%	7%
Preferred	196842	46%	67%	88%

the details are left out. In contrast, it is emphasized that the approach presented in this paper is applicable for other MOO techniques as well.

As metric to compare the results of the 25 dimensional optimization, a weighted sum approach, which reduces the fitness vector to one dimension, is used. The weighted sum metric is in general defined as follows: $\Psi : \mathbb{R}^m \to \mathbb{R}$ with $\Psi(A) = \sum_{i=1}^{m} w_i \cdot F_i(A)$. The justification of the weights results from the experience of an expert, several months of development in the area of nurse rostering and the given constraints. Note, that a lot of time and experience was necessary to adapt these weights.

To measure the influence of random seeds on the results, the random number generator has been initialized with 10 different values. But they were chosen as constants in each run. The results presented in Section 3.3 and 4.3 give the average value AVG_Ψ for these 10 runs. With the best weighted sum of generation g for random seed i denoted as $\Psi_{i,g}$, the average value $AVG_{\Psi,g} : \mathbb{R} \to \mathbb{R}$ of the ten runs $\Psi_{i,g} : 1 \leqslant i \leqslant 10$ is calculated as follows:

$$AVG_{\Psi,g} = \frac{\sum_{i=1}^{10} \Psi_{i,g}}{10}$$

Additionally to the average value, the standard deviation σ_g has been calculated in percent from $AVG_{\Psi,g}$ as follows:

$$\sigma_g = \sqrt{\frac{1}{10-1} \sum_{i=1}^{10} (\frac{\Psi_{i,g}}{AVG_{\Psi,g}} - 1)^2}$$

3.3 Experimental Evaluation

The experiments are based on the standard setting of the EA as applied in the industrial setting (see Section 3.2) . No fine-tuning has been done for any of the algorithms.

The results of all experiments are given for a population size of 10 and a run of 3000 generations. The final average fitness values are shown in Table 1 in column $AVG_{\Psi,3000}$ for the methods `Dominates`, `NSGA II` and `Preferred`, respectively. For comparison, in the next column $AVG_{\Psi,3000}$ is given as percentage normalized for `Dominates`. As can be seen, `Dominates` and `NSGA II` perform almost identical, while `Preferred` gives a reduction of more than 50%.

The next column gives the standard deviation. Here `Dominates` and `NSGA II` have values from 10-15%, while `Preferred` has an “unstable” value of over 60%.

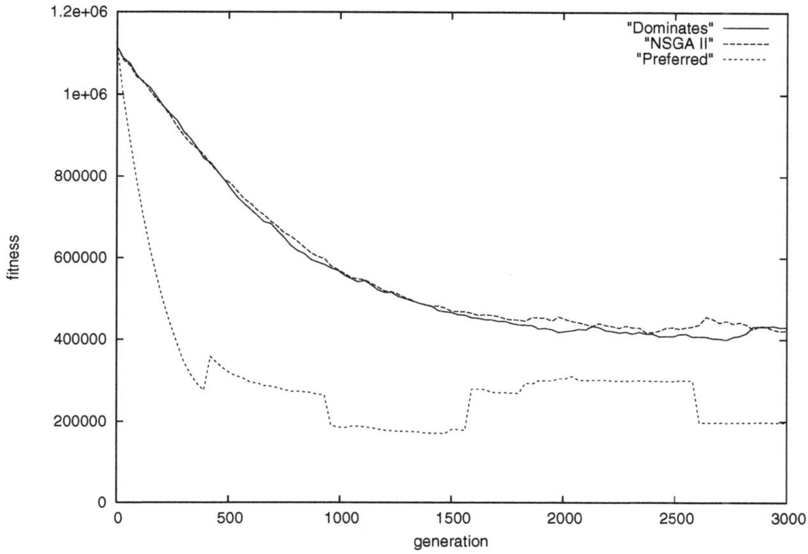

Fig. 2. Fitness for `Dominate`, `NSGA II` and `Preferred`

The result gets even worse if the average value of the standard deviation from the first to the last generation is considered (see last column). The results of the complete run are shown in Figure 2.

In summary the experimental study showed:

- The performance of `Preferred` was significantly better than that of `Dominate` and `NSGA II` for high dimensional MOO.
- Measuring the standard deviation showed that `Preferred` is not very robust, i.e. the algorithms should generate good solutions for each random seed.

3.4 Discussion

In this section the experiments are discussed to explain the observed behavior.

A more detailed analysis of the method `Preferred` showed the reason for the behavior: Relation *Preferred* favours a solution A, if it is better in more dimensions than a solution B. But a "problem" of relation *Preferred* is that the other dimensions, i.e. those where B is better than A, are not considered at all. It might happen that the negative effect of these dimensions is very strong and, as a result "jumps" occur in the weighted sum metric. This effect can, for example, be observed in Figure 2 between generation 450 and 500 or 1600 and 2600.

The main reasons for the weak performance regarding the fitness of relation *Dominates* can be explained as follows: Due to the high number of dimensions it rarely happens that an element is better in all dimensions. In fact, in the EA run 93% of the solutions were not comparable to each other using `Dominates`. Only solutions which are better or equal in all dimensions were distinguishable from other solutions. This also applies to `NSGA II` and is a reason, why the relation *Dominates* based methods `Dominates` and `NSGA II` could not guide to good solutions.

In [3] this problem has been described for up to 20 objectives. There it has been suggested to use a larger population size or to use modified fitness assigning techniques. Both approaches will not work in our application. Increasing the population size means higher run times. Furthermore, this might reduce the number of non-comparable elements. To modify the fitness assignment might help to preserve a good Pareto-front, but not to improve the path to an optimum.

So approaches based on *Preferred* should be used for guiding the search in high dimensional spaces, because it is possible to compare solutions, which are not comparable with relation *Dominates. Preferred* uses the number of better objectives as a criterion for the comparison. But as a result the technique suffers from unstable behavior as explained above.

4 Robust MOO

To improve the robustness of the approach based on *Preferred*, an extension called ϵ-*Preferred* is introduced in the following. Before we give a formal definition, the main idea is briefly sketched.

4.1 Overall Idea

One principle in multi-objective optimization is to model the criteria of human decision making. In our application it has been observed that a human planner rejects solutions, if specific limits of the objectives quality are not satisfied. Hence, the idea is to define fitness limits for each dimension. The resulting relation is called ϵ-*Preferred*, where an ϵ-value is defined for each optimization objective. A solution is rejected if it exceeds one or more ϵ-limits.

For a motivation of our idea look at the following example:

Example 2. Consider solutions A and B and a fitness function F for a minimization problem. Let $F(A) = (1, 1, 100)$ and $F(B) = (5, 5, 5)$. Then relation *Preferred* would hold: $A \prec_{preferred} B$

But, dependent on the application considered, solution A is not a satisfying solution, because the third component does not fulfill the planners expectations.

To overcome this problem, a maximum environment ϵ_i is set for each optimization objective $1 \leqslant i \leqslant m$.

As can be seen in Example 2, if we set $\epsilon_3 = 50$, solution A becomes worse than B, because the third component of solution A does not satisfy the given quality limits.

4.2 Relation ϵ-*Preferred*

In this section an extension of *Preferred*, denoted as ϵ-*Preferred*, is formally introduced.

Table 2. Fitness for generation 3000

Algorithm	$AVG_{\Psi,3000}$	$AVG_{\Psi,3000}$ in percent	σ_{3000}	$\sigma_{0..3000}$
Dominates	429805	100%	11%	7%
Preferred	196842	46%	67%	88%
ϵ-Preferred	116594	27%	10%	6,6%

Definition 3. *Let $A, B \in \Omega$ and ϵ_i, $1 \leqslant i \leqslant m$*

$$A \prec_{\epsilon-exceed} B \Leftrightarrow$$
$$|\{i : F_i(A) < F_i(B) \wedge |F_i(A) - F_i(B)| > \epsilon_i\}|$$
$$> |\{j : F_j(A) > F_j(B) \wedge |F_j(A) - F_j(B)| > \epsilon_j\}|.$$

The relation ϵ-*exceed* counts how often a solution exceeds the given limits ϵ_i. Then solution A is better than solution B with respect to the limits ϵ_i, if A has less exceedings than B.

Using ϵ-*exceed* the extension ϵ-*Preferred* is defined as follows:

Definition 4. *Given two solutions $A, B \in \Omega$,*
$A \prec_{\epsilon-preferred} B \Leftrightarrow A \prec_{\epsilon-exceed} B \vee (B \not\prec_{\epsilon-exceed} A \wedge A \prec_{preferred} B)$

First it is counted how often a solution exceeds the ϵ-limits and the better solution is determined. If both solutions are in the given range, *Preferred* is used for comparison.

By building the relation graph with the newly proposed relation ϵ-*Preferred*, it is possible to create cycles, as relation *Preferred* does, too. For this reason we use the same strongly-connected components building algorithm as suggested in [8].

4.3 Experimental Evaluation

The experimental setting is the same as described in Section 3.3, i.e. we studied the run for 3000 generations. The underlying EA was identical for all approaches and only the MOO relation was changed.

In the experiments for all dimensions the parameter ϵ is set to 10000. This choice is very conservative and only a weak restriction for the algorithm. But, it can be avoided that the algorithm explores regions of the search space that are not of the planners interest.

The same information as above is now given for ϵ-`Preferred` in Table 2 and Figure 3. Since `Dominates` and `NSGA II` behave almost the same, only `Dominates` is shown in the table. Compared to `Preferred`, the results are further improved by over 30% by using relation ϵ-*Preferred*. But the even more remarkable observation is the robustness of the technique. While it is significantly better than `Preferred`, it is even better than `Dominates`. This can be seen very well in Figure 3, where the fitness over 3000 generations is shown. There are no "jumps" any more.

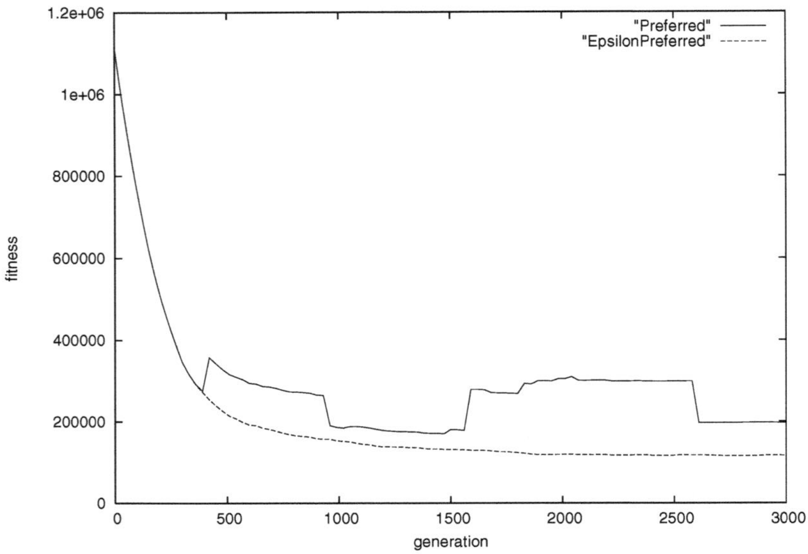

Fig. 3. Fitness comparison for *Preferred* and ϵ-*Preferred*

Table 3. Fitness cut-out of the 18 most important dimensions

Algorithm	$AVG_{\Psi,3000}$	$AVG_{\Psi,3000}$ in percent	σ_{3000}	$\sigma_{0..3000}$
ϵ-Preferred	45188	100%	16%	11%
ϵ-Preferred-2500	38998	86%	15%	13%
ϵ-Preferred-1000	33018	73%	8%	10%

Influence of Epsilon Values. For our experiments above large epsilon values have been used. Only in cases where a solution was "very bad" in one or more dimension it was rejected. In this section we briefly discuss the influence of alternative choices. The experiments are summarized in Table 3. By this, directions for future work are pointed out (see also next section).

In a first run, denoted as ϵ-*Preferred-2500*, the epsilon values of 18 out of the 25 optimization objectives were reduced from 10000 to 2500. The 18 dimensions were the ones that the human expert considered "most important". In a second run (ϵ-*Preferred-1000*) we additionally reduced two out of the 18 dimensions to the value 1000 to consider those as "very important" criterias. As can be seen even by these first experiments, the quality could be further improved, while obtaining the same robustness.

If the value of ϵ was too low (e.g. close to zero) for one dimension, then each solution with small deterioration in this direction was immediately rejected without respecting possible improvements in other dimensions. The same problem occurred in the case of relation *Dominates*.

In summary, based on relation ϵ-*Preferred* the quality measured by the fitness value could be significantly improved, while robustness was obtained at the same time.

5 Conclusions and Future Work

With more complex applications, MOO is becoming a more important topic. To compare two solutions relations have to be defined. These have mainly been evaluated on problems of low dimension, i.e. with up to 10 optimization goals.

In this paper a complex industrial scheduling problem has been investigated. The problem has more than 20 optimization goals with different "levels" of importance. Previously proposed relations have been evaluated and discussed in the context of high dimensional MOO. It turned out that the methods either suffer from low quality or low robustness.

A new relation called ϵ-*Preferred* has been suggested and experimentally studied. It was shown that very high quality results could be obtained, i.e. an improvement of more than 30% so far, and in addition the robustness could be improved.

It is focus of current work to develop automatic techniques for determining the epsilon values automatically. In this context also the dynamic reduction during the optimization run will be investigated. First experiments showed that there is still significant potential for improvement.

Acknowledgment

The authors like to thank Peter Bäume from S2-Engineering, Bremen, for support and many helpful discussions.

References

1. D.E. Goldberg. *Genetic Algorithms in Search, Optimization & Machine Learning.* Addision-Wesley Publisher Company, Inc., 1989.
2. E. Zitzler and L. Thiele. Multiobjective evolutionary algorithms: A comparative case study and the strength pareto approach. *IEEE Trans. on Evolutionary Comp.*, 3(4):257–271, 1999.
3. K. Deb. *Multi-objective Optimization using Evolutionary Algorithms.* John Wiley and Sons, New York, 2001.
4. K. Deb, L. Thiele, M. Laumanns, and E. Zitzler. Scalable test problems for evolutionary multi-objective optimization. Technical Report 112, Computer Engineering and Networks Laboratory (TIK), Swiss Federal Institute of Technology (ETH), Zurich, Switzerland, 2001.
5. V. R. Khare, X. Yao, and K. Deb. Performance scaling of multi-objective evolutionary algorithms. In *EMO 2003*, volume 2632 of *Lecture Notes in Computer Science*, pages 376–390, 2003.

6. K. Deb and S. Sundar. Reference point based multi-objective optimization using evolutionary algorithms. In *GECCO '06: Proceedings of the 8th Annual Conference on Genetic and Evolutionary Computation*, pages 635–642, 2006.
7. K. Deb, S. Chaudhuri, and K. Miettinen. Towards estimating nadir objective vector using evolutionary approaches. In *GECCO '06: Proceedings of the 8th Annual Conference on Genetic and Evolutionary Computation*, pages 643–650, 2006.
8. N. Drechsler, R. Drechsler, and B. Becker. Multi-objective optimisation based on relation favour. In *Int'l Conference on Evolutionary Multi-Criterion Optimization*, pages 154–166, 2001.
9. F. Schmiedle, N. Drechsler, D. Große, and R. Drechsler. Priorities in multi-objective optimization for genetic programming. In *GECCO '01: Proceedings of the 6th Annual Conference on Genetic and Evolutionary Computation*, pages 129–136, 2001.
10. E.K. Burke, P. De Causmaecker, G.V. Berghe, and H.V. Landeghem. The state of the art of nurse rostering. *Journal of Scheduling*, 7:441–499, 2004.
11. M. Keijzer, J.J. Merelo, G. Romero, and M. Schoenhauer. Evolving objects: a general purpose evolutionary computation library. *Artificial Evolution*, pages 321–244, 2001.
12. K. Deb, A. Pratap, S. Agarwal, and T. Meyarivan. A fast and elitist multiobjective genetic algorithm: NSGA-II. *IEEE Trans. on Evolutionary Comp.*, 6:182–197, 2000.

Substitute Distance Assignments in NSGA-II for Handling Many-Objective Optimization Problems

Mario Köppen and Kaori Yoshida

Kyushu Institute of Technology, Dept. Artificial Intelligence,
680-4, Kawazu, Iizuka, Fukuoka 820-8502 Japan
{mkoeppen, kaori}@pluto.ai.kyutech.ac.jp

Abstract. Many-objective optimization refers to optimization problems with a number of objectives considerably larger than two or three. In this paper, a study on the performance of the Fast Elitist Non-dominated Sorting Genetic Algorithm (NSGA-II) for handling such many-objective optimization problems is presented. In its basic form, the algorithm is not well suited for the handling of a larger number of objectives. The main reason for this is the decreasing probability of having Pareto-dominated solutions in the initial external population. To overcome this problem, substitute distance assignment schemes are proposed that can replace the crowding distance assignment, which is normally used in NSGA-II. These distances are based on measurement procedures for the highest degree, to which a solution is nearly Pareto-dominated by any other solution: like the number of smaller objectives, the magnitude of all smaller or larger objectives, or a multi-criterion derived from the former ones. For a number of many-objective test problems, all proposed substitute distance assignments resulted into a strongly improved performance of the NSGA-II.

1 Introduction

Recently, there has been increasing awareness for the specific application of evolutionary multi-objective optimization algorithms to problems with a number of objectives considerably larger than two or three. Fleming et al. [8] note the common appearance of such problems in design optimization, and suggested the use of the term *many-objective optimization.* Most evolutionary multi-objective optimization algorithms (EMOs) show a rather decreasing performance, or rapidly increasing search effort for an increasing number of objectives. Other problems with the handling of many objectives are related to the missing means for performance assessment, to difficulties in visualizing results, and to the low number of existing, well-studied test problems. The DTLZ suite of test problems [6,7] defines most of their problems for an arbitrary number of objectives. Results here have been reported for up to 8 objectives [10]. The Pareto-Box problem [12] was also defined for an arbitrary number of objectives, and results were given for up to 15 objectives.

S. Obayashi et al. (Eds.): EMO 2007, LNCS 4403, pp. 727–741, 2007.

The reason for the decreasing algorithm performance is strongly related to the (often even exponentially) growing problem complexity. This growing complexity can be measured by several means. One example for this is, if considering a randomly initialized population, the rapidly decreasing probability of having a pair of solutions, where one solution Pareto-dominates the other. Within the unit hypercube, the expectation value for the number of non-dominated solutions among m randomly selected solutions can be computed by [12]:

$$e_m(n) = m - \sum_{k=1}^{m} \frac{(-1)^{k+1}}{k^{n-1}} \binom{m}{k} \tag{1}$$

where m stands for the number of individuals, and n for the number of objectives. For example, for 15 objectives and 10 individuals, the expectation value for the number of dominated solutions is already as low as 0.0027.

Among the most successful and most often applied EMOs we find the Fast Elitist Non-dominated Sorting Genetic Algorithm (NSGA-II) [3,5]. But the poor performance of the NSGA-II algorithm for a large number of objectives has already been reported as well, see e.g. [10,9]. This can be considered a kind of misfortune, as otherwise, the NSGA-II is one of the most attractive EMOs today, due to its simple structure, its availability, the elaborated design of its operations [1], the existence of experience in practical applications, and its excellent performance on the majority of test problems.

This paper attempts to overcome this drawback by analyzing the reasons for NSGA-II's failure in the many-objective optimization domain, and by providing corresponding countermeasures. The main approach, as will be more detailed in section 2, is to replace the crowding distance assignment that is used for secondary ranking among individuals of the same rank. Four methods will be considered here, which all suit better to a larger number of objectives. Section 3 will present results for the convergence metric and Pareto front coverage for a number of many-objective test problems, and section 4 will render conclusions from these results.

2 Substitute Distance Assignments in NSGA-II

2.1 Structure of NSGA-II Algorithm

The outline of the NSGA-II algorithm can be seen in the following listing. Here, we are focussing on a multi-objective minimization problems.

NSGA-II:

$R_t = P_t \cup Q_t$	combine parent and children population
$F = \texttt{fast\_nondominated\_sort}(R_t)$	$F = (F_1, F_2, \cdots)$ all non-dominated fronts of R_t
$P_{t+1} = \emptyset$, $i = 1$	init next parent population
while $\vert P_{t+1}\vert < N$ do	until the parent population is filled

`secondary_ranking_assignment`(F_i)	calculate ranking values in F_i
$P_{t+1} = P_{t+1} \cup F_i$, $i = i+1$	include i-th non-dominated front in the
end	parent population
Sort$(P_{t+1}, \geq_n)$	sort in descending order using $\geq_n$
$P_{t+1} = P_{t+1}[0:N]$	choose the first N elements of P_{t+1}
$Q_{t+1} =$ `make_new_population`(P_{t+1})	use selection, crossover and mutation
$t = t+1$	to create a new population Q_{t+1}

For each generation t, the algorithm maintains an external population of N parent individuals P_t and creates a child population Q_t from the parents. Both populations, fused together, are lexicographically sorted by two different global ranking measures. The first is the non-dominated sorting, as result of the procedure `fast_nondominated_sort`. For details of its implementation, see [3]. The main outcome of this procedure is the assignment of a *rank* to each solution in the set R_t. Two solutions of the same rank do not dominate each other, but for each solution of rank $r > 1$, there exists at least one dominating individual of lower rank. The rank 1 is assigned to all solutions that are in the Pareto set. Thus, the rank value implies a total ordering of the set of solutions in the algorithm for each generation.

To yield a more competitive ordering, NSGA-II also assigns a secondary ranking measure to each solution. So far, only the `crowding-distance-assignment` has been considered, as given in the following listing:

CROW-DIST: `crowding-distance-assignment`(I)

$l = \lvert I \rvert$	number of solutions in I
for each i, set $I[i].dist = 0$	initialize distance
for each objective m do	
$I = \text{sort}(I, m)$	sort using each objective value
$I[1].dist = I[l].dist = \infty$	so that boundary points are always selected
for $i = 2$ to $(l-1)$ do	for all other points
$I[i].dist = I[i].dist + (I[i+1].m - I[i-1].m)$	
end	
end	larger *dist* count better

This distance measure is well suited for a later stage of the algorithms' application, where the population is already close to the true Pareto front of the problem (hopefully). It forces the solutions to keep distance to their neighboring solutions in objective space. Using this distance in addition to the ranking, the comparison of two solutions is based on the ordering relation $\geq_n$:

$$i \geq_n j \text{ if } (i_{rank} < j_{rank}) \text{ or } ((i_{rank} = j_{rank}) \text{ and } (i_{dist} > j_{dist})) \quad (2)$$

2.2 NSGA-II and Many Objectives

In the case of a larger number of objectives, the performance of NSGA-II is notably dropping, down to a level, where its behavior resembles more or less a random search [12]. Considering the plot in fig. 1, some insight into this phenomenon can be yielded. For an initial parent population of 100 individuals, ranks have been computed. The plot shows the average number of rank 1 solutions over 100 such random initializations and with increasing number of objectives for the DTLZ2 and DTLZ6 problems. For more than 2 or 3 objectives, the amount of rank 1 solutions sharply increases. For the (also known to be more complex) DTLZ6 problem, the rank 1 rapidly accounts for more than 90 percent of the population. For the partial ordering used in the NSGA-II, this means that most of the ranking now is delegated to the secondary ranking assignment, i.e. the crowding distance comparison. However, measuring crowding in an initial population is randomized as well, and as a result, the algorithm gets stuck right from the beginning.

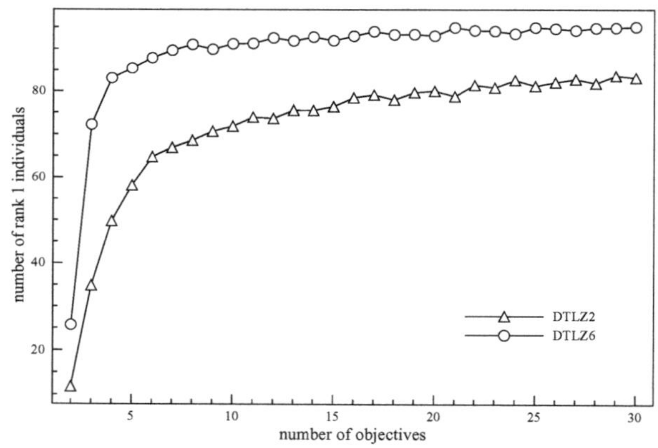

Fig. 1. Average number of rank 1 individuals in an initial random parent population of 100 individuals for the DTLZ2 and DTLZ6 problems with 2 to 30 objectives

It seems suitable to consider a different way for secondary ranking assignment in the first (explorative) generations of the algorithm, in order to avoid the algorithm getting stuck. This will be discussed in the next subsection.

2.3 Secondary Ranking Assignment by Pareto Dominance Degrees

As it was already pointed out in the introduction, with increasing number of objectives the appearance of Pareto-dominance among the solutions becomes more and more unlikely. However, two solutions can be close to the situation where one solution Pareto-dominates the other. As a basic idea, we are going to measure this kind of closeness and use such measurements instead of the crowding distance for the secondary ranking in the NSGA-II algorithm.

The degree, by which a solution A is nearly-dominated by a solution B, can be related to more than one criterion. Basically, the following independent cases can be considered:

– the number of smaller or larger objectives;
– the magnitude of all smaller or larger objectives; or
– a multi-criterion based on the former ones.

In the following, we are going to consider measurements for all these cases. The measurements take advantage of the fact that in NSGA-II, due to the non-dominated sorting, the secondary ranking is only applied to solution sets I where no solution Pareto-dominates any other solution of the same set: $pareto\_set(I) = I$.

Subvector dominance (SV-DOM): given two solutions A and B, the procedure $svd(A, B)$ directly counts the number of objectives of B that are smaller than the corresponding objectives in A. For each solution $I[i]$ in a set I of solutions, the largest such value among all other solutions is assigned as distance value to $I[i]$. The smaller this value, the smaller is the number of lower objectives that appear among all other members of the set I. Such a solution is more close to being not Pareto-dominated by any other solution. For a strongly Pareto-dominated solution, its distance equals the number of objectives. In [2], such a measure was used for the so-called *efficiency of order k*-selection among Pareto optimal solutions. The pseudo-code for computing SV-DOM is as follows:

SV-DOM: `subvector-dominance-assignment`(I)

def $svd(i, j)$	comparing solution i with j
$cnt = 0$	initialize counter
for each objective m do	
$cnt = cnt + 1$ if $I[j].m < I[i].m$	count number of smaller objectives
return cnt	
end	
for each $i = 1, \cdots, \|I\|$ do	for all solutions $I[i]$
set $I[i].dist = 0$	initialize distance
for each $j \neq i$ do	among all other solutions j
$v = svd(i, j)$	find the one with the largest
if $I[i].dist < v$ then $I[i].dist = v$	number of smaller objectives;
end	this j gives the distance value for i
end	smaller $dist$ count better

-eps-dominance ($-\epsilon$-DOM): for two solutions A and B of the solution set, the procedure $mepsd(A, B)$ considers all objectives of B that are larger than the corresponding objectives of A (i.e. worse). It computes the smallest value ϵ, which, if subtracted from all objectives of B, makes B Pareto-dominating A. This corresponds to the concept of additive ϵ-dominance. For each solution $I[i]$ in a set I of solutions, the smallest such value among all other solutions is assigned as distance value to $I[i]$. The larger this distance for a solution, the higher the "effort" that would be needed to make the other solutions Pareto-dominating the former. For a Pareto-dominated solution, the distance is 0. The ϵ-DOM distance can also be computed as follows:

-eps-DOM: `meps-dominance-assignment`(I)

```
def mepsd(i, j)                              comparing solution i with j
  max = 0                                    initialize maximum variable
  for each objective m do
    if I[j].m > I[i].m then                  for all larger objectives
      max = max [I[j].m − I[i].m, max]       get largest differing objective
  end
  return max
end
for each i = 1, ··· , |I| do                 for all solutions I[i]
  set I[i].dist = ∞                          initialize distance
  for each j ≠ i do                          among all other solutions j
    v = mepsd(i, j)                          find the one with the smallest
    if I[i].dist > v then I[i].dist = v      maximal differing larger objective;
  end                                        this j gives the distance value for i
end                                          larger dist count better
```

Fuzzy Pareto dominance (FPD): Given two solutions A and B, this procedure accounts for all objectives of B that are also larger than the corresponding objectives of A (i.e. worse). Instead of seeking the maximum difference (as we did for ϵ-DOM), and thus basing the comparison onto a single objective only, we are going to fuse all the magnitudes of larger objectives into a single value. The procedure equals the Fuzzy-Pareto-Dominance relation as presented in [11]. It uses the notion of bounded division of two reals x and y from $[0, 1]$:

$$\left[\frac{x}{y}\right] = \begin{cases} 1, & \text{if } y \le x \\ x/y, & \text{if } x < y \end{cases} \tag{3}$$

All bounded quotients of corresponding objectives in A and B are multiplied. For a smaller objective in B, this gives a factor of 1. Thus, if A is Pareto-dominated by B, the measure becomes 1. For each solution $I[i]$ in a set I of solutions, the largest product value from all other solutions is assigned as distance value to $I[i]$. The smaller this value, the lower the degree by which a solution is dominated by any other solution in I. The pseudo-code for FPD distance measure is as follows:

FPD: `fuzzy-pareto-dominance-assignment`(I)

def $fpd(i,j)$	comparing solution i with j
$cv = 1$	initialize comparison value
for each objective m do	
$cv = cv \cdot [I[i].m/I[j].m]$	multiply bounded quotient
end	
return cv	
end	
for each $i = 1, \cdots, \lvert I\rvert$ do	for all solutions $I[i]$
set $I[i].dist = 0$	initialize distance
for each $j \neq i$ do	among all other solutions j
$v = fpd(i,j)$	find the one with the largest
if $I[i].dist < v$ then $I[i].dist = v$	comparison value to i;
end	this j gives the distance value for i
end	smaller *dist* count better

Sub-objective dominance count (SOD-CNT): None of the methods introduced so far regards for *all* aspects of the ranking relation between two solutions. If the comparison is based on all larger objectives, the information about smaller objectives is neglected, and vice versa. If the number of larger objectives is considered, nothing is known about the difference in the magnitudes of these objectives. Thus, we are also considering a multi-criterion here, and provide a distance assignment procedure for such a multi-criterion ranking.

Taking any solution A of a (non-dominated) solution set I, we derive a set S_A of all pairs of two single-criterion distance measures to all other solutions B of the set. In this study, we take the pair $M - svd(A, B)$ (M is the number of objectives) from SV-DOM distance and $mepsd(A, B)$ from $-\epsilon$-DOM distance. This set S_A has a Pareto set, which is composed of all solutions that "perform well" against A. Each solution in I gets the number of occurrences in all the possible Pareto sets PSO_A assigned. The higher this number, the more often the corresponding solution "performs well" against some other solution in I. In pseudo-code:

SOD-CNT: `subobjective-dominance-count-assignment`(I)

for each i, set $I[i].dist = 0$	initialize distance
for each $i = 1, \cdots, \lvert I\rvert$ do	for all solutions $I[i]$
$S_i = \{(M - svd(i,j), mepsd(i,j)) \mid j \neq i\}$	
	all pairs of subvector dominance
	and -eps-dominance distances
$POS_i = pareto\_set(S_i)$	get the Pareto set of S_i
for each $j \in PSO_i$ do	for each solution j in that Pareto set
$I[j].dist = I[j].dist + 1$	increment counter of solution j;
end	larger *dist* count better

For better understanding, we will provide an example for the computation of SOD-CNT. Consider the four vectors $A = (2,1,3), B = (1,2,4), C = (3,3,1)$ and $D = (1,4,1)$. The values for $M - svd(row, column)$ and $meps(row, column)$ are given in the following tables:

	(2,1,3)	(1,2,4)	(3,3,1)	(1,4,1)
(2,1,3)	-	2	2	1
(1,2,4)	1	-	2	2
(3,3,1)	1	1	-	2
(1,4,1)	2	2	2	-

$M - svd(row, column)$

	(2,1,3)	(1,2,4)	(3,3,1)	(1,4,1)
(2,1,3)	-	1	2	3
(1,2,4)	1	-	2	2
(3,3,1)	2	3	-	1
(1,4,1)	2	3	2	-

$mepsd(row, column)$

For example, the entry "2" in the second column of the first row in the left-hand table indicates that the solution $(1,2,4)$ has one smaller objective than $(2,1,3)$. Thus, the entry is $3 - 1 = 2$. The corresponding entry in the right-hand table indicates that at least 1 has to be subtracted from all objectives in $(1,2,4)$ to make it Pareto-dominating $(2,1,3)$.

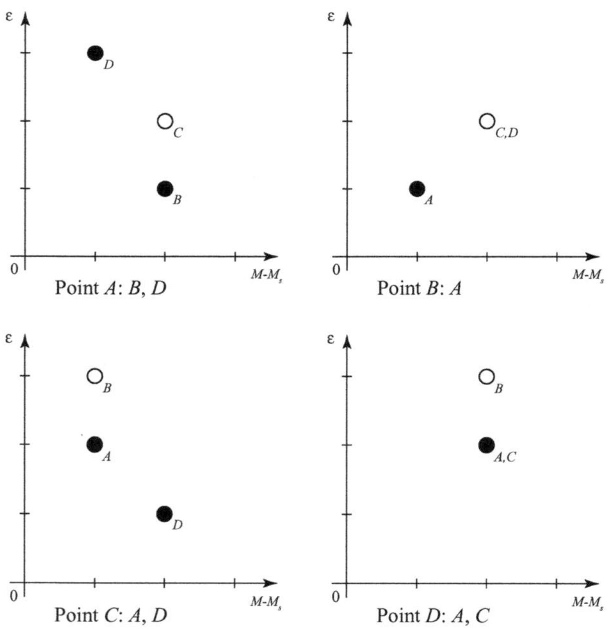

Fig. 2. Example for the computation of the sub-objective dominance count (SOD-CNT) secondary ranking measure

Figure 2 shows the following evaluation for each solution. For example, for solution A, we read the three pairs $(2,1)$ for comparing with B, $(2,2)$ for comparing with C, and $(1,3)$ for comparing with D from the tables above. The Pareto set of these three pairs is the set $\{(2,1),(1,3)\}$, which refers to the solutions B and D (black circles in the figure). Doing this for all four solutions, A appears three times in such a Pareto set, B one time, C one time and D two times. This equals the distance assignment to the four solutions, and gives A to be of higher value for the secondary ranking, followed by D, and B, C having lowest ranking value.

2.4 Using the Substitute Distance Assignments

In the NSGA-II algorithm, the four proposed distance assignment procedures are used the same way as the crowding distance, as given by eq. (2). For SV-DOM and FPD, the ">" has to be replaced by "<", as for these procedures, smaller values count better.

3 Results

3.1 Convergence Metric

In this subsection, we present some results that were obtained using the newly introduced substitute distance assignments. As test problems, DTLZ2, DTLZ3 and DTLZ6 for 2, 8 and 15 objectives have been used. Since these test problems are well covered in literature, and also for limited space reason, we are not going to provide details of the definitions of these test problems here. For details, the reader is kindly referenced to the literature [6,7]. Also, the genuine many-objective Pareto-Box test problem, as introduced in [12], was studied. Here, to any point $x \in [0,1]^M$, the M objectives $|x_i - 0.5|$ were assigned.

As performance measure, the convergence metric [4] was used. In case of DTLZ2, DTLZ3 and DTLZ6, this measure simplifies to $|I|-1$, where I is a solution (vector of objectives). In case of the Pareto-Box problem, the convergence metric can be also simply computed by $|I|$, as the task here is to come close to the mid-point of the unit hypercube.

The settings for the NSGA-II algorithm were the same as used in [10]: crossover probability 0.7, distribution index for SBX 15, mutation probability $1/M$, and distribution index for polynomial mutation 20. However, due to a better fit to the many-objective optimization domain, search effort was kept small. In all cases, a population of 20 individuals was used, and each experiment went over 300 generations. This is equal to the smallest settings that were used in [10].

The results listed in table 1 were achieved by averaging the minimal convergence metric of a population over the 300 generations for 30 runs each. The reason that no archive was used is as follows: for a larger number of objectives, any solution tends to be included in the archive, as Pareto dominance is becoming more unlikely. Thus, the procedure for reducing archive size equals more or less the procedure of elitist selection in the population itself, and it is easily observed that the values of the convergence metric for both sets do not differentiate much. The presented substitute distance assignments could be considered for adding to an archive in a many-objective optimization problem as well. This

Table 1. Results of the application of the substitute distance assignments to common test problems with increasing complexity

Obj.	CROW-DIST	SV-DOM	-ϵ-DOM	FPD	SOD-CNT
	Convergence Metric for Pareto-Box				
2	$(3 \pm 3) \cdot 10^{-5}$	$(5 \pm 7) \cdot 10^{-5}$	$(3 \pm 4) \cdot 10^{-5}$	$(7 \pm 9) \cdot 10^{-5}$	$(8 \pm 10) \cdot 10^{-5}$
8	0.49 ± 0.04	$(16 \pm 18) \cdot 10^{-4}$	0.028 ± 0.006	$(6 \pm 15) \cdot 10^{-4}$	$(8 \pm 5) \cdot 10^{-4}$
15	0.98 ± 0.07	0.02 ± 0.01	0.066 ± 0.008	0.09 ± 0.08	0.005 ± 0.001
	Convergence Metric for DTLZ2				
2	$(9 \pm 2) \cdot 10^{-4}$	$(3 \pm 2) \cdot 10^{-4}$	$(8 \pm 2) \cdot 10^{-4}$	$(5 \pm 2) \cdot 10^{-4}$	$(7 \pm 4) \cdot 10^{-5}$
8	0.80 ± 0.07	$(3 \pm 2) \cdot 10^{-4}$	0.029 ± 0.007	0.15 ± 0.06	$(11 \pm 10) \cdot 10^{-5}$
15	0.81 ± 0.06	0.002 ± 0.002	0.06 ± 0.01	0.3 ± 0.1	$(14 \pm 17) \cdot 10^{-5}$
	Convergence Metric for DTLZ3				
2	22 ± 9	20 ± 10	15 ± 9	16 ± 10	16 ± 6
8	890 ± 60	50 ± 20	40 ± 20	230 ± 50	30 ± 10
15	990 ± 80	80 ± 30	60 ± 20	400 ± 100	30 ± 10
	Convergence Metric for DTLZ6				
2	0.7 ± 0.2	0.7 ± 0.2	0.51 ± 0.06	0.8 ± 0.3	0.6 ± 0.1
8	9.05 ± 0.09	7.4 ± 0.6	6.6 ± 0.9	8.7 ± 0.3	3.8 ± 0.9
15	9.05 ± 0.08	8.1 ± 0.4	8.2 ± 0.5	8.9 ± 0.1	4.8 ± 0.8

is the topic of an on-going study of the authors. The $\pm$-values in table 1 refer to the corresponding standard deviation from the sample average.

The results clearly demonstrate the better performance of *all* substitute distance assignments, even in the case of two objectives. Between the substitutes, SOD-CNT achieves the best results, and FPD the worst (but generally still better than CROW-DIST). Table 2 gives a comparison to results from literature. This shows the modified NSGA-II also to be highly competitive, especially regarding the comparable low effort that was needed to achieve the given convergence metric values.

3.2 Pareto Front Coverage

Having found substantially better convergence metric values in the former subsection, the question about Pareto front coverage has to be considered as well. However, the quantitative assessment of the coverage is not simple. So far, several ad hoc approaches for certain test problems, with a focus on visualization have been presented (as e.g. in [10]). For doing similarly for many objectives, we propose a class of test problems that allows for easy visualization and evaluation of Pareto front coverage, which is referred to as P^* *problem* for indicating the variable number of points from which the objectives are derived.

Given is a set P of m points P_i in the Euclidian plane (the case of two dimensional Euclidian space is completely sufficient for the present analysis). The feature space F equals the Euclidian plane, where the points P_i are located. The objective space O is an m-dimensional vector space. For a given point x in

Table 2. Comparison of results with the results reported in [10]. Column 2 also lists average convergence metric values for 1000 randomly initialized vectors. The entries in the columns entitled Effort are population size times number of generations that were used to achieve the reported results.

Obj.	Random	PESA[10]	NSGA-II[10]	Effort	SOD-CNT	Effort
Convergence Metric for DTLZ2						
2	0.83	0.00008	0.00180	$20 \cdot 300$	0.00007	$20 \cdot 300$
8	0.84	0.00689	2.30766	$600 \cdot 600$	0.00011	$20 \cdot 300$
15	0.83	-	-	-	0.00014	$20 \cdot 300$
Convergence Metric for DTLZ3						
2	1077.7	22.52023	21.32032	$20 \cdot 500$	16.0	$20 \cdot 300$
8	1082.2	7.23062	1753.41364	$600 \cdot 1000$	30.0	$20 \cdot 300$
15	1079.3	-	-	-	30.0	$20 \cdot 300$
Convergence Metric for DTLZ6						
2	9.10	0.79397	0.63697	$20 \cdot 500$	0.6	$20 \cdot 300$
8	9.08	6.32247	10.27306	$600 \cdot 1000$	3.8	$20 \cdot 300$
15	9.08	-	-	-	4.8	$20 \cdot 300$

the feature space, its objective vector $o(x)$ is the vector with the components $o_i = d(x, P_i)$ for $i = 1$ to m, where $d(a, b)$ is the Euclidian distance of two points $a, b \in F$. Thus, the objectives to minimize are the distances to a given collection of points, where the distance to any of these point is treated as an independent objective.

The Pareto set of this problem, i.e. the set of feature vectors giving objective vectors that are not dominated by any other feasible objective vector (in other words, are closest to all points P_i), equals the convex closure of the points P_i. Here, convex closure means the union of the volume enclosed by the convex hull and the convex hull itself.

To see this, consider the left subfigure of fig. 3. Consider any point X in the plane that does not belong to the convex closure of the points P_i. Assume that the convex hull of the points P_i is being established by the poly-line $\overline{A_1 A_2 \dots A_n A_{n+1=1}}$, where each $A_i \in P$. As X is outside the enclosed area, there must be a connection $\overline{A_i A_{i+1}}$ such that all points of P are either on the connecting line, or on the opposite side of the connecting line than X. Dropping a perpendicular from X to the connecting line, one can see that any point Y between X and the line is more close to any point on the opposite side of the connecting line, and more close to any point on the connecting line as well. Thus, for any point X outside the convex closure there is a point that is more close to all points of P, and X does not belong to the Pareto set.

To see that none of the points of the convex closure dominates any other, consider the right subfigure of fig. 3. By connecting any two points U and V of the convex closure and drawing the perpendiculars to this line trough U and through V, the convex closure is segmented into three parts. There is at least one point of the point set A (and thus of P) located to the l.h.s. of the perpendicular

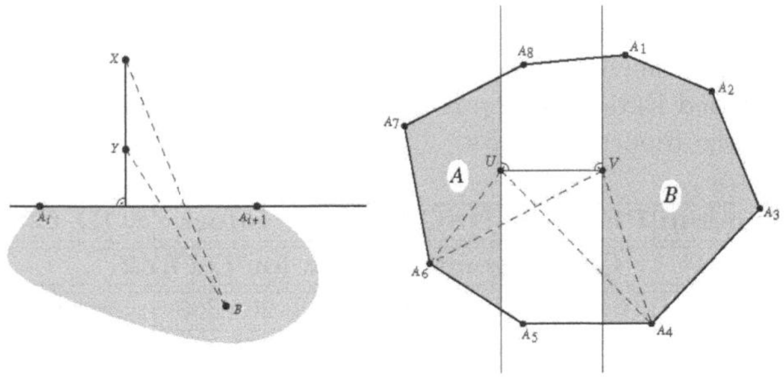

Fig. 3. Illustrating the proof that the Pareto set of the P* problem for the points A_i is the convex closure of these points

through U (indicated by encircled A in the figure), or located on this line, and there is at least one point of A belonging to the r.h.s. of the perpendicular through V or on it (indicated by encircled B). Otherwise, the shape would not be convex. Now, point U is more close to any point of A than V, and point V is more close to any point of B than U. Neither U nor V can dominate the other.

Having thus a rather simple solution structure in the feature space (not objective space, which is high-dimensional), the problem is worth a study for a heuristic algorithm for several reasons:

- the number of objectives can be easily scaled
- by reducing the area enclosed by the convex closure, the effort for random search (the "Monte-Carlo Barrier") can be easily increased
- typical performance measures (as average distance to Pareto front, number of individuals belonging to the Pareto front) can be directly computed
- as the feature space is two-dimensional, the results can be directly visualized; however, the extension to higher-dimensional spaces is straightforward
- the search space is not bounded
- the problem is a continuous optimization problem
- boundary conditions can be directly included
- crowding in objective space directly corresponds to crowding in feature space
- modeling of algorithm behavior seems feasible
- by using the distance to the center of gravity of the points instead, a comparison to the single-objective case becomes possible

We have studied the performance of the modified NSGA-II algorithms on such a 15-objective P* problem.

Figure 4 gives a result that demonstrates how differently the considered methods are behaving. Note that this figure shows the Pareto front of the algorithms and the test problem in feature space, and not in the (15-dimensional) objective space. The best coverage of the polygon is achieved with the FPD method, closely followed by $-\epsilon$-DOM. These are the methods that employed the magnitudes of larger objectives of solutions directly. SVD shows a rather small coverage

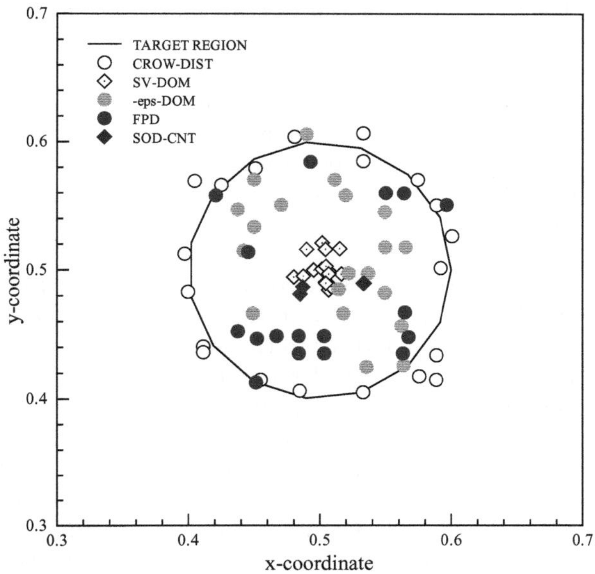

Fig. 4. Final populations using the considered secondary ranking methods on a 15-objective P* problem after 100 generations. Population size is 20.

of the Pareto front, and SOD-CNT, having by far the best convergence metric values, nearly collapses into a single point. Notable also the distribution of the "default" crowding distance measure: as the crowding distance, by construction, keeps extreme individuals in the objective space, it favors individuals that are near to the corners of the polygon. For the P* test problem, this feature of the crowding distance is obviously a drawback.

4 Conclusions

We have studied a number of modifications of the NSGA-II algorithm, to make this algorithm better capable of solving many-objective optimization problems. The modifications were substitutes for the crowding distance assignment, based on closeness of solutions to the case that one solution Pareto-dominates the other. The results and experiences, also taking non-measurable aspects into account, can be summarized as follows:

SV-DOM. This distance measure is very easy to implement, and needs the lowest computation time. It showed the second best results for the convergence metric, but failed in the Pareto front coverage study.

$-\epsilon$-DOM. This distance measure can also be easily implemented, and needs a little more computational effort than SV-DOM. Convergence metric performance was average among the considered modifications, but is accompanied by a good Pareto front coverage. Altogether, this makes this method a good trade-off among all the studied modifications.

FPD. This method has a rather high computational effort, and it also has some formal weaknesses regarding issues of division by 0 etc. Among the studied modifications, the convergence metric performance was worst (however, still better than the crowding distance), but it demonstrated the best Pareto set coverage for the P* problem.

SOD-CNT. The convergence metric of this method is excellent. This good result is weakened by a very poor Pareto front coverage, and also higher computational effort. Moreover, in contrary to the other methods, the processing time is not predictable, as the size of the Pareto sets for the single solutions may vary. During the experiments, we faced an unexpected very long processing time for larger populations and a small number of objectives. Also, if the non-dominated sets gets smaller, the differentiation among the individuals by this measure becomes low. In a few cases, we could even observe convergence of the algorithm, despite of the use of a mutation operation. Some issues regarding this approach, which showed a very good convergence metric, still need further investigation.
To summarize, the results of this study indicate two promising strategies for the application of NSGA-II to many-objective optimization problems: first is to replace the crowding distance completely by $-\epsilon$-dominance distance, second is to use the sub-objective dominance count distance SOD-CNT for the first generations of the algorithm, as long as most of the individuals get rank 1 assigned, and switch to the crowding distance, once the SOD-CNT values tends to be equalized over the whole population.

Acknowledgment

A researcher involved in this study has been supported by a JSPS grant.

References

1. Meghna Babbar, Ashvin Lakshmikantha, and David E. Goldberg. A Modified NSGA-II to Solve Noisy Multiobjective Problems. In James Foster, editor, *2003 Genetic and Evolutionary Computation Conference. Late-Breaking Papers*, pages 21–27, Chicago, Illinois, USA, July 2003. AAAI.
2. I. Das. A preference ordering among various pareto optimal alternatives. *Structural and Multidisciplinary Optimization*, 18(1):30–35, 1999.
3. Kalyanmoy Deb, Samir Agrawal, Amrit Pratab, and T. Meyarivan. A Fast Elitist Non-Dominated Sorting Genetic Algorithm for Multi-Objective Optimization: NSGA-II. In Marc Schoenauer, Kalyanmoy Deb, Günter Rudolph, Xin Yao, Evelyne Lutton, Juan Julian Merelo, and Hans-Paul Schwefel, editors, *Proceedings of the Parallel Problem Solving from Nature VI Conference*, pages 849–858, Paris, France, 2000. Springer. Lecture Notes in Computer Science No. 1917.

4. Kalyanmoy Deb and Sachin Jain. Running Performance Metrics for Evolutionary Multi-Objective Optimization. In Lipo Wang, Kay Chen Tan, Takeshi Furuhashi, Jong-Hwan Kim, and Xin Yao, editors, *Proceedings of the 4th Asia-Pacific Conference on Simulated Evolution and Learning (SEAL'02)*, volume 1, pages 13–20, Orchid Country Club, Singapore, November 2002. Nanyang Technical University.
5. Kalyanmoy Deb, Amrit Pratap, Sameer Agarwal, and T. Meyarivan. A Fast and Elitist Multiobjective Genetic Algorithm: NSGA–II. *IEEE Transactions on Evolutionary Computation*, 6(2):182–197, April 2002.
6. Kalyanmoy Deb, Lothar Thiele, Marco Laumanns, and Eckart Zitzler. Scalable Test Problems for Evolutionary Multi-Objective Optimization. Technical Report 112, Computer Engineering and Networks Laboratory (TIK), Swiss Federal Institute of Technology (ETH), Zurich, Switzerland, 2001.
7. Kalyanmoy Deb, Lothar Thiele, Marco Laumanns, and Eckart Zitzler. Scalable Multi-Objective Optimization Test Problems. In *Congress on Evolutionary Computation (CEC'2002)*, volume 1, pages 825–830, Piscataway, New Jersey, May 2002. IEEE Service Center.
8. Peter Fleming, Robin C. Purshouse, and Robert J. Lygoe. Many-Objective Optimization: An Engineering Design Perspective. In Carlos A. Coello Coello, Arturo Hernández Aguirre, and Eckart Zitzler, editors, *Evolutionary Multi-Criterion Optimization. Third International Conference, EMO 2005*, pages 14–32, Guanajuato, México, March 2005. Springer. Lecture Notes in Computer Science Vol. 3410.
9. Crina Grosan. Multiobjective adaptive representation evolutionary algorithm (MAREA) - a new evolutionary algorithm for multiobjective optimization. In A. Abraham, Bernard De Baets, Mario Köppen, and Bertram Nickolay, editors, *Applied Soft Computing Technologies: The Challenge of Complexity*, Advances in Soft Computing, pages 113–121. Springer Verlag, Germany, 2006.
10. V. Khare, X. Yao, and K. Deb. Performance Scaling of Multi-objective Evolutionary Algorithms. In Carlos M. Fonseca, Peter J. Fleming, Eckart Zitzler, Kalyanmoy Deb, and Lothar Thiele, editors, *Evolutionary Multi-Criterion Optimization. Second International Conference, EMO 2003*, pages 376–390, Faro, Portugal, April 2003. Springer. Lecture Notes in Computer Science. Volume 2632.
11. Mario Köppen and Raul Vicente Garcia. A fuzzy scheme for the ranking of multivariate data and its application. In *Proceedings of the 2004 Annual Meeting of the NAFIPS (CD-ROM)*, pages 140–145, Banff, Alberta, Canada, 2004.
12. Mario Köppen, Raul Vicente-Garcia, and Betram Nickolay. Fuzzy-Pareto-Dominance and Its Application in Evolutionary Multi-objective Optimization. In Carlos A. Coello Coello, Arturo Hernández Aguirre, and Eckart Zitzler, editors, *Evolutionary Multi-Criterion Optimization. Third International Conference, EMO 2005*, pages 399–412, Guanajuato, México, March 2005. Springer. Lecture Notes in Computer Science Vol. 3410.

Pareto-, Aggregation-, and Indicator-Based Methods in Many-Objective Optimization

Tobias Wagner[1], Nicola Beume[2], and Boris Naujoks[2]

[1]Institut für Spanende Fertigung (ISF)
[2]Chair of Algorithm Engineering
University of Dortmund, 44221 Dortmund, Germany
wagner@isf.de,{nicola.beume,boris.naujoks}@uni-dortmund.de
http://ls11-www.cs.uni-dortmund.de, http://www.isf.de

Abstract. Research within the area of Evolutionary Multi-objective Optimization (EMO) focused on two- and three-dimensional objective functions, so far. Most algorithms have been developed for and tested on this limited application area. To broaden the insight in the behavior of EMO algorithms (EMOA) in higher dimensional objective spaces, a comprehensive benchmarking is presented, featuring several state-of-the-art EMOA, as well as an aggregative approach and a restart strategy on established scalable test problems with three to six objectives. It is demonstrated why the performance of well-established EMOA (NSGA-II, SPEA2) rapidly degradates with increasing dimension. Newer EMOA like ε-MOEA, MSOPS, IBEA and SMS-EMOA cope very well with high-dimensional objective spaces. Their specific advantages and drawbacks are illustrated, thus giving valuable hints for practitioners which EMOA to choose depending on the optimization scenario. Additionally, a new method for the generation of weight vectors usable in aggregation methods is presented.

1 Introduction

In the field of evolutionary multi-objective optimization, a lot of test problems and applications with two or three objectives have been studied. Problems with more than three objectives, which have been termed *many*-objective problems by Farina and Amato [1], have been tackled only rarely. Many techniques that work well for only a few objectives are anticipated to have difficulties in high-dimensional objective spaces. Thus, many-objective optimization is significantly more challenging than scenarios usually being analyzed.

Within multi-objective optimization, we consider d-dimensional vectors of objective values for a problem of d objective functions $\mathbf{f} = (f_1, \ldots, f_d)$. Among these vectors, a partial order holds concerning the considered minimization problems. For details on often used terms and definitions like Pareto dominance, Pareto set and front, books on EMOA by Deb [2] or Coello Coello et al. [3] are suggested.

The selection module of an EMO algorithm (EMOA) requires a mapping of an objective vector to a ranking criterion to establish a complete order among

S. Obayashi et al. (Eds.): EMO 2007, LNCS 4403, pp. 742–756, 2007.

individuals. Popular EMOA usually consist of two selection operators. The primary selection operator is based on Pareto dominance and favors non-dominated solutions over dominated ones. The secondary operator is constituted diversity preserving and rates solutions incomparable concerning the primary operator.

This concept of selection already documents the insight that Pareto dominance may not be sufficient as a sole selection operator, due to the large amount of possibly incomparable solutions. More precisely, a d-dimensional objective vector is only comparable with a fraction of $1/2^{d-1}$ of an (infinite) objective space (cf. Farina and Amato [1]). The importance of the secondary selection operator grows with increasing dimension of the objective space since the incomparability concerning the Pareto-based operator becomes the typical case.

Few previous studies on many-objective optimization by Purshouse and Fleming [4] and Hughes [5] focus to demonstrate the bad performance of NSGA-II by Deb et al. [6]. Hughes observed a simple single-objective restart strategy outperforming NSGA-II on a six-objective function in a two-dimensional decision space. Upon this, he implied a generalization to all Pareto-based techniques.

In contradiction, the work at hand includes positive results by demonstrating that some modern EMOA using Pareto-concepts cope very well with high-dimensional objective spaces. We ascribe the good performance of ε-MOEA, IBEA, SMS-EMOA, and MSOPS to new concepts of aggregation and indicator functions and explain how and why these EMOA work successfully. A comprehensive benchmark is presented on the established test functions of the DTLZ function family, which feature a high dimensional decision and a scalable objective space. Moreover, a slight modification to NSGA-II is suggested, which causes a better performance. Our motivation is not to modify NSGA-II but to demonstrate which aspects of classic EMOA are responsible for the problems within many-objective optimization.

The aggregation method MSOPS by Hughes [5] is studied more detailedly. The problems using aggregation are described and solution concepts are presented with a focus on suitable sets of weight vectors.

The considered test functions, performance measures and basic settings of the EMOA are described in the following section. Section 3 deals with the behavior of Pareto-based EMOA, Section 4 with aggregation methods, and Section 5 with methods utilizing indicator functions for selection. In these sections, algorithms are presented and their performances are described with help of the quality measures. Section 6 summarizes the findings and gives an outlook on how to further deepen insight in many-objective optimization.

2 Benchmark Settings

All algorithms, except otherwise mentioned, have been implemented within the PISA framework[1] [7] since an integrative framework simplifies comparisons. The same variation operators are used with exactly the same parameterization, which

[1] PISA - Platform and Programming Language Independent Interface for Search Algorithms, ETH Zürich (www.tik.ee.ethz.ch/pisa/)

is chosen according to the studies of Deb et al. [8]. Simulated binary crossover (SBX) and polynomial mutation (PM) as described by Deb [2] are applied with mutation probability $p_m = 1/n$ per decision variable and recombination probability $p_c = 1$ per individual. The distribution indices $\eta_c = 15$ and $\eta_m = 20$ are used. If not otherwise stated, a $(\mu + \mu)$ strategy and a binary tournament for mating selection are applied. A number of 30.000 function evaluations is accomplished and the population size $\mu = 100$ is chosen. For each EMOA, besides SMS-EMOA, on each test function, 20 runs are performed. Due to the exponential runtime and the small standard deviation in the observed runs, SMS-EMOA is only repeated 5 times.

2.1 Test Functions

To benchmark the performance of the considered EMOA, the functions DTLZ1 and DTLZ2 of the DTLZ test function family [9] are invoked. These functions are scalable in the number of objectives and thus allow for a many-objective study. The decision vector is divided into two subvectors. The first one of length $d-1$ contains the parameters defining the position on the given surface while the second of length ν specifies the distance to the Pareto front. This results in dimension $d + \nu - 1$ of the decision space. According to Deb et al. [9], $\nu = 5$ is used in DTLZ1 and $\nu = 10$ is used in DTLZ2 respectively.

The Pareto front of DTLZ1 is a linear hyperplane. DTLZ2 features a Pareto front that corresponds to the positive part of the unit hypersphere ($|\mathbf{f}(\mathbf{x})| = 1$). Here, the interaction between objectives is nonlinear. The domain of all decision variables is $[0, 1]$. Due to different scaling constants in the distance function, the codomain of objective values for DTLZ1 is $[0, 1 + 225\nu]$ and $[0, 1 + 0.25\nu]$ for DTLZ2, respectively. The Pareto set of both test functions corresponds to $x_d, \ldots, x_n = 0.5$ with arbitrary values for $x_1, \ldots, x_{d-1}$.

2.2 Performance Assessment

For performance assessment, $\mathcal{S}$-metric by Zitzler and Thiele [10] and convergence measure [8] are considered. The $\mathcal{S}$-metric determines the size of the dominated hypervolume in objective space bounded by a reference point $\mathbf{r}$. In EMO research it is of outstanding importance due to its theoretical properties. The values depend on proximity to the Pareto front as well as on distribution of points. The maximal $\mathcal{S}$-metric value is reached by the Pareto front. The reference points $\mathbf{r} = 0.7^d$ for DTLZ1 and $\mathbf{r} = 1.1^d$ for DTLZ2 were used in previous studies [8,11] and are close to the Pareto front in order to emphasize on the distribution of optimal solutions. Points that do not dominate the reference point are discarded for metric calculation. The metric values are normalized by calculating the fraction of the analytical optimal value. Note that exactly 100% are unreachable with a finite number of points.

The convergence measure describes the average distance of the approximation to the Pareto front in objective space. In contrast to the study of Deb et al. [8], the euclidean distance to the nearest optimal solution is determined analytically without using a reference set. This is possible due to the special structure of the employed Pareto fronts.

3 Pareto-based EMOA

As *Pareto-based* EMOA, we classify EMOA with selection criteria that are mainly based on the qualitative information of Pareto-dominance, Pareto-based ranking, or counting. Thus, NSGA-II, SPEA2, and ε-MOEA are considered here.

NSGA-II. The *Elitist Non-dominated Sorting Genetic Algorithm* (NSGA-II) by Deb et al. [6] applies the rank assigned to each solution by non-dominated sorting as primary selection criterion. Non-dominated individuals are assigned rank one and the set of individuals with equal rank is called a front. Those individuals that are non-dominated if the first front was removed are assigned rank two. The third front is decided within the population discarding the first and the second front and so on. Individuals with equal ranks are evaluated using a secondary selection criterion called crowding distance. This subsumes the distances to the next higher and lower values in each dimension, respectively. Currently, the NSGA-II is supposed to be the best known and most frequently applied EMOA. Jensen [12] improves the non-dominated sorting algorithm, determining the overall runtime of NSGA-II, to run in $O(\mu \log^{d-1} \mu)$ per generation.

SPEA-2. The *Strength Pareto Evolutionary Algorithm* (SPEA2) by Zitzler et al. [13] uses two ranking criteria as well. It is an elitist algorithm with an archive of constant size, which is chosen to be the population size μ in the experiments at hand. As primary selection criterion, a strength value that gives the number of individuals in the population dominated by the current individual is assigned. Based on these values a raw fitness is computed as the sum of the strength values of every individual that dominates it. Thus, every non-dominated individual's raw fitness equals zero. In a second step, a density estimation is performed based on the euclidean distances between all individuals. The primary fitness value is the raw fitness plus the reciprocal of the sum of the distance to the k-nearest neighbor [14][2]. To fill the archive for the next generation, the individuals with the best fitness are copied. In case of individuals with equal fitness, the distance to the k-nearest neighbor for increasing k is used as further criterion. Given $d \geq 3$, these methods require a runtime in $O(d\mu^2)$ per generation [12].

ε-MOEA. Laumanns et al. [15] proposed the ε-MOEA to combine the convergence properties of an elitist MOEA like suggested by Rudolph and Agapie [16] with the need to preserve a diverse set of solutions. The objective space is divided into a grid of boxes, whose size can be adjusted by the choice of ε. Dominance is checked according to the boxes where the solutions are positioned. The

[2] In PISA k is chosen as 1.

archive **E** holds one solution for each non-dominated box. If the box of a new solution dominates other boxes in the archive, the associated archive members are rejected. In case of two solutions belonging to the same box, Laumanns et al. decline the new solution except it dominates the old one. Later, Deb et al. [8] propose to select the solution, which is closer to the best corner of the box. They also administrate a co-evaluated population **P** of constant size. If a new solution is not dominated by any member of the population, it replaces a randomly chosen member favoring dominated solutions. They also suggest a steady-state approach, where the offspring is generated by a parent from **P** and a parent from **E**. A binary tournament regarding the dominance relation is performed to choose the member of **P** for mating. The parent from **E** is chosen equiprobable. Because no further diversity measures are computed, the runtime of a generation of ε-MOEA is $O(d|\mathbf{E}|)$.

3.1 Experimental Results

NSGA-II and SPEA2 rapidly decrease in quality with increasing dimension of objective space. If more than four objectives are considered, these algorithms do not converge to the Pareto set as indicated by the high distance values (cf. Tab. 1). With dimension greater than four, no relative hypervolume is measured because no point dominating the reference point is achieved (cf. Tab. 2).

Further studies with these algorithms have been performed to exhibit if any convergence occurs with a higher number of function evaluations. As shown in Fig. 1, both algorithms increase the distance to the Pareto front in the first generations because the diversity based selection criteria favor higher distances between solutions. Special emphasis is given to extremal solutions with values near zero in one or more objectives. These solutions remain non-dominated and the distance cannot be decreased thereafter.

Table 1. The convergence measure for the pareto dominance based algorithms

		DTLZ1			DTLZ2		
obj.	algorithm	mean	std.dev	median	mean	std.dev	median
3	ε-MOEA	0.00614	0.00413	0.00484	0.00102	0.00022	0.00105
	NSGA-II	0.06333	0.15581	0.01002	0.01049	0.00162	0.01027
	SPEA2	0.06783	0.16435	0.00792	0.00801	0.00112	0.00806
4	ε-MOEA	0.15990	0.34073	0.01990	0.00129	0.00024	0.00126
	NSGA-II	1.70260	1.95260	0.69515	0.08522	0.02580	0.08060
	SPEA2	3.47990	4.78910	1.66910	0.08164	0.01676	0.08901
5	ε-MOEA	0.22348	0.41685	0.01941	0.02681	0.00120	0.02670
	NSGA-II	300.416	37.2461	317.506	1.06780	0.14504	1.07770
	SPEA2	358.818	25.0853	366.236	1.30970	0.15758	1.27760
6	ε-MOEA	0.97014	1.39920	0.27217	0.00272	0.00067	0.00266
	NSGA-II	393.674	17.6076	388.689	2.15610	0.09584	2.16910
	SPEA2	482.742	13.6757	479.577	2.32000	0.09617	2.36070

Table 2. The relative hypervolume for the pareto dominance based algorithms

		DTLZ1, $\mathbf{r} = 0.7^d$			DTLZ2, $\mathbf{r} = 1.1^d$		
obj.	algorithm	mean	std.dev	median	mean	std.dev	median
3	ε-MOEA	0.94560	0.01005	0.94662	0.92858	0.00118	0.92836
	NSGA-II	0.94333	0.11423	0.96923	0.86913	0.00803	0.86918
	SPEA2	0.98010	0.00152	0.98068	0.90760	0.00350	0.90782
4	ε-MOEA	0.85493	0.18655	0.92697	0.87722	0.00186	0.87766
	NSGA-II	0.45730	0.40600	0.46204	0.71644	0.01971	0.71733
	SPEA2	0.62316	0.34319	0.72224	0.78461	0.01258	0.78202
5	ε-MOEA	0.82261	0.16668	0.86933	0.83847	0.00308	0.83809
	NSGA-II	0	0	0	0.11570	0.06842	0.11734
	SPEA2	0	0	0	0.12528	0.06942	0.12864
6	ε-MOEA	0.64563	0.38344	0.81552	0.85332	0.01434	0.85497
	NSGA-II	0	0	0	0	0	0
	SPEA2	0	0	0	0	0	0

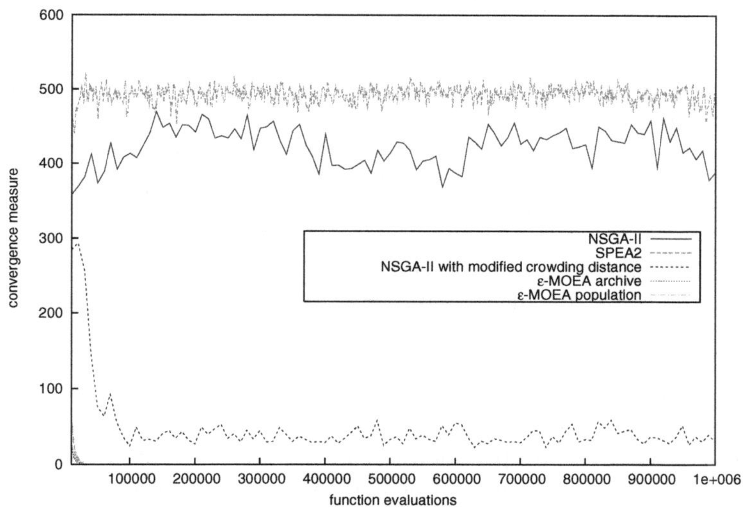

Fig. 1. Convergence measure during the optimization run performing the median result on six-objective DTLZ1

To confirm this assumptions and improve NSGA-II, a slight modification of crowding distance is studied. Originally, an individual without a neighbor regarding one dimension of the objective space is assigned an infinite crowding distance. Instead of that, a value of zero is used, causing that non-dominated solutions with extremal values are rejected. Although this variant is not able to converge to the Pareto front, an improvement of the average distance within the first 100,000 function evaluations is obvious (Fig. 1). Then, most of the decision variables have reached their optimal value. Only one or two of them remain in a local optimum. This experiment shows that a diversity measure with emphasis on a spread of the population can misguide the MOEA to deterioration and the loss of promising non-dominated solutions.

The performance of ε-MOEA highly depends on the choice of ε. We choose it such that **E** finally contains about 100 solutions.[3] The ε-MOEA is able to produce optimal solutions within the allowed number of function evaluations for all considered numbers of objectives. This is shown in the lower left part of figure 1. The active dominance-preserving function of the archive, combined with an utopia point distance criterion for non-dominated individuals in the same hyperbox avoids the effects of deterioration and thus ensures convergence even for the co-evolving set **P**. Though, the hypergrid guarantees an uniform distribution of individuals, the obtained hypervolume values are only for DTLZ2 competitive with the best considered algorithms. This is due to the trend of the hyperbox method to avoid extremal solutions, as described by Deb et al. [8].

4 Aggregation-Based EMOA

Basic aggregation methods are single-objective optimizers, which multiply the objective values with weights and accumulate them to a scalar value. The EMOA considered here, enhance aggregation concepts in order to produce a set of solutions. In contrast to the other EMOA considered, aggregation-based approaches require the a priori definition of relations between objective functions. This results in a certain focus during the optimization.

MSOPS. *Multiple Single Objective Pareto Sampling* (MSOPS) does not feature Pareto methods, but handles all objectives in parallel. The decision maker has to choose T vectors of weights for every objective function to enable an aggregation. Hughes [17] recommends weighted *min-max* (MSOPS 1) and a combination of this approach with *Vector-Angle-Distance-Scaling (VADS)* called *dual optimisation* (MSOPS 2). Depending on the aggregation strategy, one receives a set of T or $2T$ aggregated scores per solution. The scores are held in a score matrix S, where each row belongs to a solution and each column represents an aggregated score. Each column of the matrix S is ranked, giving the best performing population member rank one. The rank values are stored in a matrix R. Each row of R is sorted ascending, resulting in a lexicographical order of the individuals. The runtime is in $O(\mu T d)$ for the computation of the aggregated scores, and in $O(\mu T \log T)$ and $O(T \mu \log \mu)$ respectively to perform the sort. Thus, the runtime of MSOPS is $O(\mu T(d + \log T + \log \mu)$ per generation.

Obviously, the choice of weight vectors determines the distribution properties of MSOPS. Each weight vector $\mathbf{w} = (w_1, \ldots, w_d)$ corresponds to a direction,

[3] d=3, DTLZ1: $\varepsilon = (0.03, 0.03, 0.03)$, DTLZ2: $\varepsilon = (0.058, 0.058, 0.058)$.
d=4, DTLZ1: $\varepsilon = (0.047, 0.047, 0.047, 0.047)$, DTLZ2: ε=(0.125, 0.125, 0.125, 0.125).
d=5, DTLZ1: $\varepsilon = (0.057, \ldots, 0.057)$, DTLZ2, $\varepsilon = (0.18, \ldots, 0.18)$.
d=6, DTLZ1: $\varepsilon = (0.066, \ldots, 0.066)$, DTLZ2, $\varepsilon = (0.232, \ldots, 0.232)$.

given analytically by a target vector starting in the origin. The aim of the aggregation methods is to reach the point on the corresponding direction vector which is as close as possible to the origin. To this end, weighted *min-max* focuses on the distance to the origin, while $VADS$ favors solutions whose position vector has a small intersecting angle with the target vector.

In this study, the optimization shall not have a special focus, but an approximation of the whole Pareto front is desired and the weight vectors have to be chosen appropriately. In Hughes [5] benchmarking '*50 target vectors spread uniformly across the search space*' are used. The target vectors $\mathbf{t} = (t_1, \ldots, t_d)$ are created by calculating an initial number of steps $s = \lfloor \sqrt[d]{T} \rfloor$ and constructing each possible vector containing multiples of $1/s$ between 0 and 1. Afterward, these target vectors are normalized and doubles are removed. If the number of targets is lower as desired, s is incremented and the procedure is repeated. At the end, a next neighbor technique is used to prune the set of target vectors to the desired size. Because the PISA implementation of MSOPS uses weight vectors, a transformation of the target vectors into weights is necessary. The authors recommend – deviant from Hughes [5] – the following procedure for transformation, that can also be used to transform a set of utopia or reference points into weights and avoids numerically unstable calculations in many cases.

From the aggregation methods can be referred that a weight vector for a specified target fulfills the following $d-1$ conditions:

$$w_1 \cdot t_1 = w_2 \cdot t_2, \qquad w_2 \cdot t_2 = w_3 \cdot t_3, \qquad \ldots \qquad w_{d-1} \cdot t_{d-1} = w_d \cdots t_d$$

The normalizing condition $w_1 + \ldots + w_d = 1$ is added in order to obtain a completely defined system of equations. Thus, the components of the corresponding weight vector can be computed as follows:

$$w_i = \frac{\prod_{j \neq i} t_j}{\sum_{k=1}^{d} \prod_{j \neq k} t_j} (i = 1, \ldots, d) \tag{1}$$

To extremal solutions with value 0 in $d-1$ objectives, a small ε needs to be added to allow the above calculation. Hughes [17] generally recommends to use a number of target vectors that is lower than the population size. Besides, he states that the number of target vectors has to be increased for more objectives. To cover both needs, three different sets of target vectors are used. The first contains 50 vectors, the second 100 vectors, and the third 200 vectors.

RSO. A restart strategy of a conventional single-objective evolutionary optimizer is applied as well and abbreviated *RSO (Repeated Single Objective)* according to Hughes [5]. Here, a single-objective run is performed for each of the 100 weight vectors. Thus, the number of function evaluations has to be divided among them, resulting in only 300 evaluations per run.

The derandomized mutation operator by Ostermeier et al. [18] is applied in a $(1,10)$-evolution strategy. This operator was a first step towards the popular Covariance Matrix Adaptation (CMA) operator by Hansen and Ostermeier [19], which is known to produce good results within limited function evaluations. To handle multiple objectives in a single-objective EA, the weighted *min-max* approach was chosen like in MSOPS.

4.1 Experimental Results

The methods using aggregation show an obvious convergence in all scenarios considered because they benefit from the property of the *min-max* method to minimize all objectives at once. While MSOPS obtains very promising results, RSO does not succeed in reaching the Pareto front. This is due to a too small number of function evaluations per run and the loss of information with every restart. Confirming the observations of Hughes [5], RSO outperforms NSGA-II and SPEA2 in case of five and six objectives.

Table 3. The convergence measure for the aggregation algorithms

		DTLZ1			DTLZ2		
obj.	algorithm	mean	std.dev	median	mean	std.dev	median
3	MSOPS 1 50	0.00276	0.00235	0.00185	0.00013	0.00014	$9.0{\cdot}10^{-5}$
	MSOPS 1 100	0.00278	0.00241	0.00244	0.00015	0.00010	0.00015
	MSOPS 1 200	0.00234	0.00156	0.00210	0.00080	0.00020	0.00076
	MSOPS 2 50	0.00214	0.00221	0.00161	$9.0{\cdot}10^{-5}$	$5.9{\cdot}10^{-5}$	$8.4{\cdot}10^{-5}$
	MSOPS 2 100	0.00222	0.00172	0.00191	0.00037	0.00013	0.00035
	MSOPS 2 200	0.00128	0.00074	0.00116	0.00168	0.00034	0.00168
	RSO	62.9990	15.2960	59.7140	0.26753	0.04901	0.26776
4	MSOPS 1 50	0.00392	0.00451	0.00269	0.00023	0.00023	0.00012
	MSOPS 1 100	0.00292	0.00252	0.00231	0.00024	0.00039	0.00013
	MSOPS 1 200	0.00365	0.00319	0.00264	0.00072	0.00028	0.00067
	MSOPS 2 50	0.00246	0.00216	0.00182	0.00016	0.00010	0.00012
	MSOPS 2 100	0.00849	0.02369	0.00282	0.00074	0.00024	0.00072
	MSOPS 2 200	0.00439	0.00378	0.00260	0,00203	0.00047	0.00195
	RSO	118.260	33.4420	121.190	0.56473	0.07953	0.57386
5	MSOPS 1 50	0.08016	0.31475	0.00814	0.00059	0.00027	0.00060
	MSOPS 1 100	0.05667	0.23459	0.00337	0.00017	0.00023	$7.1{\cdot}10^{-5}$
	MSOPS 1 200	0.00779	0.00556	0.00651	0.00096	0.00033	0.00092
	MSOPS 2 50	0.13676	0.26271	0.01882	0.00113	0.00038	0.00097
	MSOPS 2 100	0.03308	0.11179	0.00614	0.00138	0.00065	0.00119
	MSOPS 2 200	0.00870	0.01079	0.00535	0,00231	0.00059	0.00233
	RSO	111.960	35.1240	112.140	0.73556	0.15491	0.72211
6	MSOPS 1 50	0.02207	0.06509	0.00604	0.00044	0.00030	0.00044
	MSOPS 1 100	0.00936	0.01579	0.00406	0.00012	$8.7{\cdot}10^{-5}$	$9.7{\cdot}10^{-5}$
	MSOPS 1 200	0.00734	0.00420	0.00712	0.00048	0.00028	0.00039
	MSOPS 2 50	0.27890	0.63926	0.02603	0.00091	0.00058	0.00069
	MSOPS 2 100	0.18106	0.32499	0.02496	0.00190	0.00097	0.00180
	MSOPS 2 200	0.01344	0.01134	0.01026	0.00118	0.00056	0.00116
	RSO	110.910	42.7920	113.600	0.67628	0.13970	0.69903

Almost all variants of MSOPS attain very low average distances indicating that only optimal solutions have been found. Only for five or six objectives, variants using a lower number of target vectors fail to converge to the Pareto front in some of the runs. In the table, this behavior can be inferred from a high standard deviation and high differences between the mean and the median value. From the obtained hypervolume can be concluded that the distribution properties can be slightly improved by the supporting use of VADS. Hughes assumption that the number of target vectors should be increased if more objectives are concerned is confirmed. For three objectives, the variants of MSOPS using 50 target vectors obtain the maximal hypervolume among the aggregation methods. With increasing objectives, the best values can be obtained with a higher number of target vectors. In general, the results show that the method used to design the target vectors is able to generate well distributed Pareto front approximations. Even for three objectives, NSGA-II and ε-MOEA (DTLZ1), respectively

Table 4. The relative hypervolume of the aggregation algorithms

		DTLZ1, $\mathbf{r} = 0.7^d$			DTLZ2, $\mathbf{r} = 1.1^d$		
obj.	algorithm	mean	std.dev	median	mean	std.dev	median
3	MSOPS 1 50	0.97142	0.00127	0.97184	0.89663	0.00717	0.89817
	MSOPS 1 100	0.96484	0.00171	0.96537	0.88344	0.00208	0.88341
	MSOPS 1 200	0.96180	0.00955	0.96625	0.88752	0.02681	0.88490
	MSOPS 2 50	0.97278	0.00111	0.97317	0.89822	0.00054	0.89799
	MSOPS 2 100	0.96719	0.00623	0.96776	0.91774	0.01203	0.92105
	MSOPS 2 200	0.95744	0.00965	0.96020	0.91117	0.00775	0.91253
	RSO	0	0	0	0.67735	0.03730	0.68188
4	MSOPS 1 50	0.96590	0.00107	0.96623	0.84765	0.01438	0.85238
	MSOPS 1 100	0.94724	0.00573	0.94887	0.72575	0.03761	0.73177
	MSOPS 1 200	0.94764	0.01187	0.94968	0.81489	0.03289	0.82292
	MSOPS 2 50	0.96726	0.00062	0.96730	0.85284	0.00049	0.85273
	MSOPS 2 100	0.96908	0.00258	0.96955	0.86206	0.00609	0.86445
	MSOPS 2 200	0.95605	0.00561	0.95742	0.85938	0.01289	0.86395
	RSO	0	0	0	0.39649	0.02363	0.39435
5	MSOPS 1 50	0.97740	0.00614	0.97956	0.78971	0.05479	0.80668
	MSOPS 1 100	0.96312	0.01848	0.97160	0.48432	0.32422	0.72034
	MSOPS 1 200	0.97749	0.00584	0.97694	0.82177	0.01404	0.82490
	MSOPS 2 50	0.93235	0.16743	0.98387	0.81037	0.00915	0.80863
	MSOPS 2 100	0.98743	0.00119	0.98762	0.86497	0.00606	0.86565
	MSOPS 2 200	0.97966	0.00296	0.97987	0.84002	0.01467	0.84609
	RSO	0	0	0	0.04960	0.03184	0.05873
6	MSOPS 1 50	0.98688	0.00469	0.98770	0.70669	0.18905	0.76654
	MSOPS 1 100	0.95343	0.02840	0.96312	0.63285	0.13323	0.68515
	MSOPS 1 200	0.99046	0.00169	0.99056	0.81435	0.03071	0.81964
	MSOPS 2 50	0.92549	0.18116	0.99355	0.84659	0.00215	0.84627
	MSOPS 2 100	0.96533	0.06398	0.98592	0.79881	0.01918	0.79436
	MSOPS 2 200	0.99122	0.00160	0.99154	0.81208	0.11049	0.83925
	RSO	0	0	0	0.16333	0.03440	0.15121

NSGA-II and SPEA2 (DTLZ2) can be outperformed regarding the $\mathcal{S}$-metric. Note that the given method to generate the target vectors only performs well on continuous Pareto fronts. As observed by Hughes [17], a refinement of the targets is necessary for more complicated problems.

5 Indicator-Based EMOA

The term *indicator-based EA (IBEA)* was introduced by Zitzler and Künzli [20] for EMOA guided by a general preference information. The EMOA's selection operator uses a preference function (indicator) as a single-objective substitute for the d-dimensional objective function. In contrast to the aggregation methods, this preference information describes a general aim. No specification of weights or targets is needed. As already stated in Sec. 1, classic EMOA use two ranking criterions: one regarding the dominance relation and the other for distribution aspects. Here, a single indicator is used to optimize a desired property of the approximation set.

IBEA. In Zitzler's and Künzli's [20] IBEA framework, binary performance metrics that map an ordered pair of individuals to a scalar value are suggested as indicator functions. Each individual is compared with all others, thus $O(\mu^2)$ indicator values must be calculated. A suitable indicator has to be *dominance preserving* [20], which sloppily means that the indicator must not evaluate a vector better than another that dominates it. Two efficiently computable indicators have been suggested in [20]. The additive ϵ-indicator subsumes the translations in each dimension of objective space that are necessary to create a weakly dominated solution. The hypervolume indicator measures the dominated hypervolume that is only dominated by one vector and not by the other. Both indicators can be computed in linear time regarding the dimension of the objective space. This results in a runtime $O(\mu^2 d)$ per generation. For both indicators, negative values mean that the first individual of the argument pair dominates the other. For each individual, its indicator values are charged in a sum of an exponential function to get a fitness value. A positive scaling constant is invoked, which is chosen as $\kappa = 0.05$ as recommended in [20] for the applied adaptive variant of IBEA. For dominance preserving indicators holds that the fitness value of a vector is worse than the fitness value of a vector that dominates it.

SMS-EMOA. The *$\mathcal{S}$-metric Selection-EMOA* (SMS-EMOA) by Emmerich et al. [21,11] aims at maximizing the $\mathcal{S}$-metric value of the population. This optimization aim rewards progression toward the Pareto front as well as a good distribution of individuals. The maximal $\mathcal{S}$-metric value is reached by the Pareto front. Thus, optimizing the $\mathcal{S}$-metric value is a very general purpose. Contrary to most other EMOA, a steady-state selection scheme and an equiprobable mating selection are applied. SMS-EMOA invokes the non-dominated sorting procedure as primary selection criterion and the selection occurs among the members of the worst ranked front. The secondary criterion applied to the last front is the hypervolume contribution, which is defined as the exclusively dominated hypervolume of an objective vector. The individual with the lowest hypervolume contribution is discarded. The non-dominated sorting can alternatively be omitted, which hardly influences the algorithms performance. The runtime of a generation of SMS-EMOA is $O(\mu^{d/2+1})$ as described by Beume and Rudolph [22].

5.1 Experimental Results

As can be inferred from the convergence measure, both IBEA variants reach the Pareto front of DTLZ2. On DTLZ1, only IBEA$_{\epsilon+}$ converges towards the Pareto front for all dimensions. IBEA$_{HD}$ reaches a very good distance value on DTLZ1 with three dimensions but fails in case of more objectives. This is due to the normalization of objective values to $[0, 1]$, tending the hypervolume indicator to favor extremal solutions, which hinder the progression.

Surprisingly, the IBEA$_{\epsilon+}$ using the additive ϵ-indicator reaches better $\mathcal{S}$-metric values than the IBEA$_{HD}$ invoking the hypervolume indicator. The consideration of translation lengths in the additive ϵ-indicator causes a good distribution of solutions. Contrary, the approximation of the hypervolume contribution through the binary hypervolume indicator tends to spiral downward with increasing dimension of objective space. Both adaptive IBEA fail to produce a good distribution on DTLZ1, which we ascribe to the high-scaled co-domain and the resulting difficulties in the scaling of the fitness values.

SMS-EMOA reaches the best $\mathcal{S}$-metric values of all considered algorithms. The distance values are very good as well and all runs except one reached the Pareto front. This run on six-objective DTLZ1 stagnated since one decision variable –which defines the distance– remains static at a non-optimal value due to an unusual loss of diversity in decision space in the beginning of the optimization process. Since the selector modules in PISA only decide regarding the objective values, this effect cannot be blamed to the selection properties of SMS-EMOA. Figure 2 exemplarily pictures the distribution of an usual six-objective result set

Table 5. The convergence measure of the indicator-based EMOA

		DTLZ1			DTLZ2		
obj.	algorithm	mean	std.dev.	median	mean	std.dev.	median
3	IBEA$_{\epsilon+}$	0.04399	0.17481	0.00057	0.00015	$5.0 \cdot 10^{-5}$	0.00014
	IBEA$_{HD}$	0.00137	0.00337	0.00029	$1.3 \cdot 10^{-5}$	$5.3 \cdot 10^{-6}$	$1.2 \cdot 10^{-5}$
	SMS-EMOA	0.00110	0.00148	0.00039	$3.4 \cdot 10^{-6}$	$1.2 \cdot 10^{-6}$	$2.8 \cdot 10^{-6}$
4	IBEA$_{\epsilon+}$	0.01790	0.02940	0.00096	0.00071	0.00012	0.00069
	IBEA$_{HD}$	76.1230	119.550	0.00136	$4.5 \cdot 10^{-5}$	$1.3 \cdot 10^{-5}$	$4.2 \cdot 10^{-5}$
	SMS-EMOA	0.00193	0.00176	0.00100	$1.4 \cdot 10^{-5}$	$5.0 \cdot 10^{-6}$	$1.2 \cdot 10^{-5}$
5	IBEA$_{\epsilon+}$	0.02056	0.06678	0.00129	0.00115	0.00019	0.00112
	IBEA$_{HD}$	151.310	131.820	215.000	0.00013	0.00014	0.00010
	SMS-EMOA	0.00333	0.00215	0.00351	$3.7 \cdot 10^{-5}$	$9.2 \cdot 10^{-6}$	$3.8 \cdot 10^{-5}$
6	IBEA$_{\epsilon+}$	0.00467	0.00450	0.00256	0.00187	0.00031	0.00184
	IBEA$_{HD}$	82.1580	116.410	0.00182	0.00015	$5.6 \cdot 10^{-5}$	0.00014
	SMS-EMOA	0.10278	0.22310	0.00444	$5.4 \cdot 10^{-5}$	$1.1 \cdot 10^{-5}$	$5.2 \cdot 10^{-5}$

Table 6. The relative hypervolume of the indicator-based algorithms

		DTLZ1, $\mathbf{r} = 0.7^d$			DTLZ2, $\mathbf{r} = 1.1^d$		
obj.	algorithm	mean	std.dev.	median	mean	std.dev	median
3	$\text{IBEA}_{\epsilon+}$	0.77693	0.03182	0.78033	0.92991	0.00075	0.93002
	IBEA_{HD}	0.73929	0.03144	0.74208	0.92023	0.00071	0.92008
	SMS-EMOA	0.98352	0.00071	0.98387	0.93870	$6.3 \cdot 10^{-5}$	0.93873
4	$\text{IBEA}_{\epsilon+}$	0.82920	0.02445	0.83425	0.89477	0.00059	0.89484
	IBEA_{HD}	0.51417	0.35620	0.70647	0.88633	0.00090	0.88619
	SMS-EMOA	0.97612	0.00034	0.97627	0.90370	$6.4 \cdot 10^{-5}$	0.90368
5	$\text{IBEA}_{\epsilon+}$	0.87018	0.02777	0.86961	0.88571	0.00097	0.88584
	IBEA_{HD}	0.26292	0.33673	0	0.88250	0.00122	0.88259
	SMS-EMOA	0.99182	0.00019	0.99182	0.89619	$9.5 \cdot 10^{-5}$	0.89624
6	$\text{IBEA}_{\epsilon+}$	0.89146	0.03569	0.90029	0.89283	0.00130	0.89322
	IBEA_{HD}	0.40153	0.30853	0.53634	0.88431	0.02231	0.89124
	SMS-EMOA	0.96688	0.06741	0.99698	0.90483	0.00014	0.90481

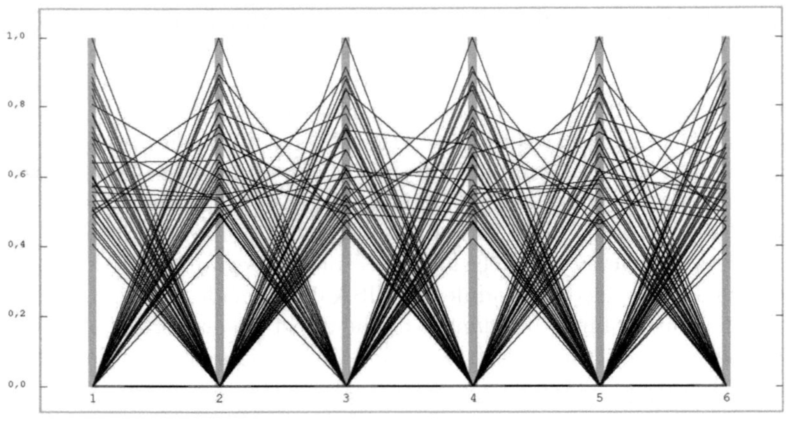

Fig. 2. Results of one run of SMS-EMOA on six-objective DTLZ2. In the parallel plot, each column corresponds to one objective.

of SMS-EMOA in a parallel plot. Every objective is covered and the structure of the set is almost symmetric, indicating a uniformly spread distribution of solutions over the whole Pareto front.

6 Summary and Outlook

The bad performance of early Pareto-based methods like NSGA-II and SPEA2 observed by Hughes [5] and Purshouse and Fleming [4] is confirmed. They show a rapid degradation with increasing number of objectives. Some additional studies show that they do not converge to the Pareto front at all and stagnate far away from it. The performance of ε-MOEA refutes the hypothesis of Hughes that a Pareto-based approach cannot succeed on many-objective problem instances. Instead, favoring extremal solutions has been shown to hinder the progression in many-objective spaces, which is also obviously for IBEA.

It is shown that more recent EMOA using indicators, which feature more than just distribution aspects, perform very well in many-objective optimization. Especially, SMS-EMOA, which optimizes the population's dominated hypervolume, outperforms the other algorithms on all considered test functions. Moreover, an aggregation-based EMOA, namely MSOPS, performs well with respect to convergence aspects. A sophisticated scheme for the generation of weight vectors is introduced and also produces well distributed solution sets. In comparison to the simple restart strategy RSO, MSOPS benefits from structural equalities of good solutions by optimizing all weight vectors in parallel.

Future research will deepen the insights in the behavior of indicator-based algorithms in particular. Theoretical statements are aspired for the convergence of the MOEA showing promising results in this study. Statistically guided parameter studies should be performed to obtain suitable parametrizations for many-objective problems. Especially, the size of the population and the offspring are to be studied. Furthermore, relations between the Pareto front and the Pareto set are studied all together resulting in new optimization techniques. These feature good convergence and distribution properties in objective space as well as in decision space.

Acknowledgements

This work was supported by the *Deutsche Forschungsgemeinschaft (DFG)* as part of the *Collaborative Research Center 'Computational Intelligence' (SFB 531)* and the *Collaborative Research Center SFB/TR TRR 30* as well as the research project *SCHW 361/15-1 "Ein Verfahren zur Optimierung von aus Mehrkomponenten bestehenden Schiffsantrieben"*.

References

1. Farina, M., Amato, P.: On the optimal solution definition for many-criteria optimization problems. In Keller, J., Nasraoui, O., eds.: Proc. of the NAFIPS-FLINT Int'l Conf. 2002, IEEE Press, Piscataway NJ (2002) 233–238
2. Deb, K.: Multi-Objective Optimization using Evolutionary Algorithms. Wiley, Chichester, UK (2001)
3. Coello Coello, C.A., Van Veldhuizen, D.A., Lamont, G.B.: Evolutionary Algorithms for Solving Multi-Objective Problems. Kluwer, New York (2002)
4. Purshouse, R.C., Fleming, P.J.: Evolutionary Multi-Objective Optimisation: An Exploratory Analysis. In: Proc. of the 2003 Congress on Evolutionary Computation (CEC'2003). Volume 3., Canberra, Australia, IEEE Press (2003) 2066–2073
5. Hughes, E.J.: Evolutionary Many-Objective Optimisation: Many Once or One Many? In: Evolutionary Computation 'Congress (CEC'05), Edinburgh, UK. Volume 1., IEEE Press, Piscataway NJ (2005) 222–227

6. Deb, K., Pratap, A., Agarwal, S., Meyarivan, T.: A Fast and Elitist Multiobjective Genetic Algorithm: NSGA–II. IEEE Transactions on Evolutionary Computation **6**(2) (2002) 182–197
7. Bleuler, S., Laumanns, M., Thiele, L., Zitzler, E.: PISA — a platform and programming language independent interface for search algorithms. In Fonseca, C.M., et al., eds.: Evolutionary Multi-Criterion Optimization, 2nd Int'l Conf. (EMO 2003). LNCS 2632, Springer, Berlin (2003) 494 – 508
8. Deb, K., Mohan, M., Mishra, S.: A Fast Multi-objective Evolutionary Algorithm for Finding Well-Spread Pareto-Optimal Solutions. KanGAL report 2003002, Indian Institute of Technology, Kanpur, India (2003)
9. Deb, K., Thiele, L., Laumanns, M., Zitzler, E.: Scalable Multi-objective Optimization Test Problems. In: Proc. of the 2002 Congress on Evolutionary Computation (CEC 2002). Volume 1., IEEE Press, Piscataway NJ (2002) 825–830
10. Zitzler, E., Thiele, L.: Multiobjective Optimization Using Evolutionary Algorithms—A Comparative Case Study. In Eiben, A.E., ed.: Parallel Problem Solving from Nature V. LNCS 1498, Springer, Berlin (1998) 292–301
11. Naujoks, B., Beume, N., Emmerich, M.: Multi-objective optimisation using S-metric selection: Application to three-dimensional solution spaces. In: Evolutionary Computation Congress (CEC'05), Edinburgh, UK. Volume 2., Piscataway NJ, IEEE Press (2005) 1282–1289
12. Jensen, M.T.: Reducing the run-time complexity of multiobjective EAs: The NSGA-II and other algorithms. IEEE Transactions On Evolutionary Computation **7**(5) (2003) 503–515
13. Zitzler, E., Laumanns, M., Thiele, L.: SPEA2: Improving the Strength Pareto Evolutionary Algorithm. Technical Report 103, Computer Engineering and Networks Laboratory (TIK), Swiss Federal Institute of Technology (ETH) Zürich, Switzerland (2001)
14. Silverman, B.W.: Density estimation for statistics and data analysis. Chapman and Hall, London (1986)
15. Laumanns, M., Thiele, L., Deb, K., Zitzler, E.: Combining convergence and diversity in evolutionary multi-objective optimization. Evolutionary Computation **10**(3) (2002) 263–282
16. Rudolph, G., Agapie, A.: Convergence properties of some multi-objective evolutionary algorithms. In Zalzala, A., Eberhart, R., eds.: Congress on Evolutionary Computation (CEC2000). Volume 2., Piscataway NJ, IEEE Press (2000) 1010–1016
17. Hughes, E.J.: Multiple Single Objective Pareto Sampling. In: Congress on Evolutionary Computation (CEC'03), IEEE Press, Piscataway NJ (2003)
18. Ostermeier, A., Gawelczyk, A., Hansen, N.: Step-size adaptation based on non-local use of selection information. In Davidor, Y., et al., eds.: Parallel Problem Solving from Nature (PPSN 1994). LNCS 866, Springer, Berlin (1994) 189–198
19. Hansen, N., Ostermeier, A.: Completely Derandomized Self-Adaptation in Evolution Strategies. IEEE Computational Intelligence Magazine **9**(2) (2001) 159–195
20. Zitzler, E., Künzli, S.: Indicator-based selection in multiobjective search. In Yao, X., et al., eds.: 8th Int'l Conf. on Parallel Problem Solving from Nature (PPSN VIII), UK, Springer (2004) 832–842
21. Emmerich, M., Beume, N., Naujoks, B.: An EMO algorithm using the hypervolume measure as selection criterion. In Coello, C.A.C., et al., eds.: Evolutionary Multi-Criterion Optimization: 3rd Int'l Conf. (EMO 2005), Springer, Berlin (2005) 62–76
22. Beume, N., Rudolph, G.: Faster S-Metric Calculation by Considering Dominated Hypervolume as Klee's Measure Problem. In: International Conference on Computational Intelligence (CI 2006). (2006) (in print).

Quantifying the Effects of Objective Space Dimension in Evolutionary Multiobjective Optimization

Joshua Knowles[1] and David Corne[2]

[1] School of Computer Science, Kilburn Building, University of Manchester, Manchester M13 9PL, UK
j.knowles@manchester.ac.uk
[2] School of Mathematical and Computer Sciences (MACS), Heriot-Watt University, UK

Abstract. The scalability of EMO algorithms is an issue of significant concern for both algorithm developers and users. A key aspect of the issue is scalability to objective space dimension, other things being equal. Here, we make some observations about the efficiency of search in discrete spaces as a function of the number of objectives, considering both uncorrelated and correlated objective values. Efficiency is expressed in terms of a cardinality-based (scaling-independent) performance indicator. Considering random sampling of the search space, we measure, empirically, the fraction of the true PF covered after p iterations, as the number of objectives grows, and for different correlations. A general analytical expression for the expected performance of random search is derived, and is shown to agree with the empirical results. We postulate that for even moderately large numbers of objectives, random search will be competitive with an EMO algorithm and show that this is the case empirically: on a function where each objective is relatively easy for an EA to optimize (an NK-landscape with K=2), random search compares favourably to a well-known EMO algorithm when objective space dimension is ten, for a range of inter-objective correlation values. The analytical methods presented here may be useful for benchmarking of other EMO algorithms.

Keywords: multiobjective optimization, nondominated sorting, nondominated ranking, random search, coverage indicator, inter-objective correlation, many objectives.

1 Introduction

The past two decades have seen the development of more and more effective and efficient evolutionary multiobjective optimization (EMO) algorithms [4,6]. These methods are often run with the goal of approximating the whole Pareto front and most EMO algorithms are designed to do this on problems of arbitrary parameter space and objective space dimension. Yet, the scalability of these methods, in practice, remains an issue of concern for the field.

Empirical testing of EMO algorithms relies on both test functions (see [11] for a review) and performance assessment methods [10,17,21,24]. Today, some test

S. Obayashi et al. (Eds.): EMO 2007, LNCS 4403, pp. 757–771, 2007.

functions are scalable in both parameter and objective dimension [11]; and some performance indicators are also suitable for many objective problems (notably those based on counting, i.e. cardinality-based indicators). These advances have made it possible to compare performance of EMO algorithms when the number of objectives is scaled up beyond the typical two or three. Thus, recently, researchers have shown empirically that some EMO algorithms (especially those based on dominance ranking for selection) perform poorly when the number of objectives d is greater than three [7,8,12,13,18], some suggesting alternative approaches.

However, it would be useful to know to what extent EMO algorithms are really performing poorly, relative to some absolute level of performance. In other words, it would be good to get some idea of the intrinsic difficulty of search as a function of objective space dimension. More precisely, we would like to know how particular performance statistics change as a function of objective space dimension, for a baseline method on a baseline/generic problem.

In this paper, we consider the performance of random search as an informative baseline, which is in line with a suggestion in [14]. Further, it is possible to be independent of the specifics of an objective function: if a one-to-one mapping from parameter to objective space is assumed, and sampling is random, points can be chosen from the objective space rather than the parameter space, without affecting the outcome, and hence a parameter space is not needed at all. Thus, discrete data sets consisting of objective vectors only are used, and two parameters are varied: the objective dimension, and a covariance term which influences the degree of inter-objective correlation. Since we are considering only cardinal, scaling-independent performance indicators, our conclusions are also independent of the data distribution in each objective (Gaussians are used for convenience). We also develop an analytic equation for predicting the expected performance of random search, and show that on these data it works.

The rest of the paper is organized as follows. Section 2 sets out some definitions and methods used in the remainder of the paper: it recalls the dominance relation; the performance indicator, *coverage*; ranking methods used in EMO fitness-assignment; and the data generation method and data sets used here. Section 3 presents empirical distributions for the number of nondominated points in our data sets and the distribution of coverage values obtained from runs of random search. In Section 4, we derive analytical expressions for the expected coverage indicator value for random sampling and show this agrees with the empirical results. Section 5 presents a case study where PESA-II is compared with random search on NK landscapes of varying dimension and inter-objective correlation. PESA-II compares unfavourably with the analytic performance of random search for 10 objectives, and this is confirmed empirically. Section 6 discusses the findings and suggests directions for further investigation.

2 Definitions and Methods

The standard definition of Pareto dominance in the objective space is used. Assuming mininimization, without loss of generality, x dominates y is written as $x \prec y$ and has the following meaning:

$$x \prec y \iff \forall i \in 1..d, x_i \leq y_i \wedge \exists j \in 1..d, x_j < y_j, \tag{1}$$

where x and y are d-dimensional objective vectors.

The Pareto front (PF) of an objective space $Y \subset \mathbb{R}^d$ is then the set,

$$\{y \in Y \mid \neg \exists x \in Y, x \prec y\}. \tag{2}$$

In iterative search, the size of the Pareto front may be an important determining factor of how successful a given search will be, e.g. if 'coverage' (see below) of this set is the metric of performance. Trivially, an iterative searcher will need at least p iterations to cover a Pareto front of cardinality p.

2.1 Quality Indicators for Performance Assessment

To assess performance as a function of objective space dimension, we consider a cardinality-based quality indicator from the literature. This is motivated by two factors: (i) cardinality-based indicators are scaling independent, whereas distance based measures usually are not; and (ii) cardinality-based measures are computationally less expensive to compute than distance based measures in a high-dimensional space.

Coverage Indicator [23,24]. The coverage $C(A, B)$ is a nonsymmetric indicator assessing the fraction of points (in the objective space) in B that are 'covered' by those in A, where 'covered' means dominated by or equal to:

$$C(A, B) = \frac{|\{y \in B \mid \exists z \in A, z \preceq y\}|}{|B|}. \tag{3}$$

If B is the true Pareto front, then a coverage of 1.0 indicates perfectly solving the Pareto optimization problem. In this paper, 'coverage' is always meant in this sense, i.e. of $C(A, PF)$. Note that coverage is thus a unary indicator [24]. Comparing two sets A and B, if $C(A, PF) > C(B, PF)$ then this implies that B cannot be *better* than A.[1] However, given two sets A and B, where A is *better* than B, coverage may not be able to detect this, e.g. if neither set contains any Pareto front point. Thus, coverage is compatible with $\ntriangleleft$ but not complete with respect to $\triangleleft$ [24].

2.2 Ranking

Many EMO algorithms are based on some form of explicit or implicit dominance ranking over a population or sample set of points. The ranking serves to assign fitness (or reproductive opportunity) to the solutions and is thought to be one of the most important factors governing MOEA performance.

Some algorithms, such as PAES [16] and PESA-II [5], are based on a binary ranking of solutions, i.e., a (truncated) nondominated set of solutions only is

[1] *A better* than B, written $A \triangleleft B$, means that every vector in B is covered (or weakly dominated) by at least one vector in A, and A is not equal to B.

maintained, whereas dominated solutions are always discarded and play no part in generating new solutions. Other algorithms use a more fine-grained ranking of the population, based on dominance. We consider two of the more popular of these. The first, nondominated sorting was suggested by Goldberg and later implemented by Srinivas and Deb [20]. In this, the rank of a solution is the nondominated 'layer' in which it lies within the set of points being compared. The second is nondominated ranking, proposed by Fonseca and Fleming [9], in which the rank of a point is one plus the number of points dominating it.

2.3 Data Suite

As alluded to above, our data sets consist of distributions of points in objective space only, rather than a function from a parameter space to an objective space. Since we are interested only in dominance and scaling-independent indicators of performance, we do not need to be concerned about the distribution in each objective, of these points, as this will have no effect. However, what will affect results, is any correlations between objectives. Thus, for convenience, we choose to use a multi-variate Gaussian generator, which is able to generate Gaussians approximating a specified covariance matrix, provided it is positive semi-definite. For this, we use the 'R' statistical software function, 'mvrnorm()' and generate sets of data with the off-diagonal elements of the covariance matrix all set to the same constant positive value. We generated a suite of data consisting of 10 sets of 1000 points each, for each level of dimension and covariance. We used dimensions $2, 5, 10, 20$, and covariances $0, 0.25, 0.5$, giving 120 data sets altogether.

3 Empirical Distributions

Figure 1 plots the number of internally nondominated points as a function of data set size n, objective dimension d, and covariance value used in generation of the data. These plots are similar to those shown in [22] for lattice data (for uncorrelated objectives only), and also those in [19].

Analytical expressions for the (expected) number of maximal vectors in a set of size n and dimension d, where the components of each vector are independent and continuous, have been given in several works [3,2,1,22]. These show that the number of vectors is $O(\ln^{d-1}(n))$ for large n and constant d, and the precise expected value can be computed using recurrence relations, as those given in [22].

From our plots, it seems possible that, for correlated data, the number of vectors does not follow $O(\ln^{d-1}(n))$, since we were not able to fit such a curve. However, it could also be that it is possible, but we need to use larger values of n. This remains an open question.

3.1 Nondominated Ranking Distributions

Figure 2 shows the nondominated sorting and nondominated ranking ranks for the search space as a whole, for each objective dimension and degree of correlation.

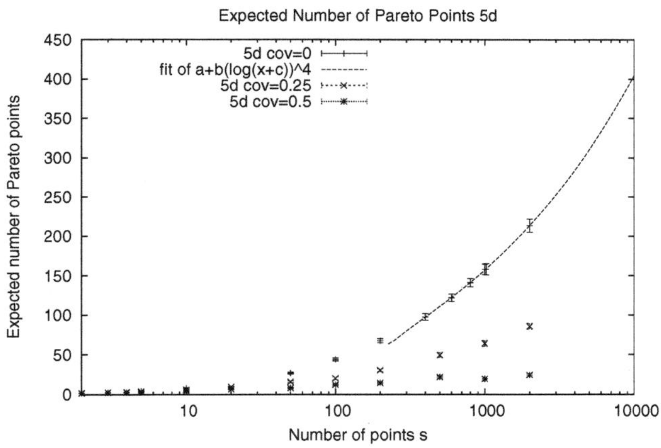

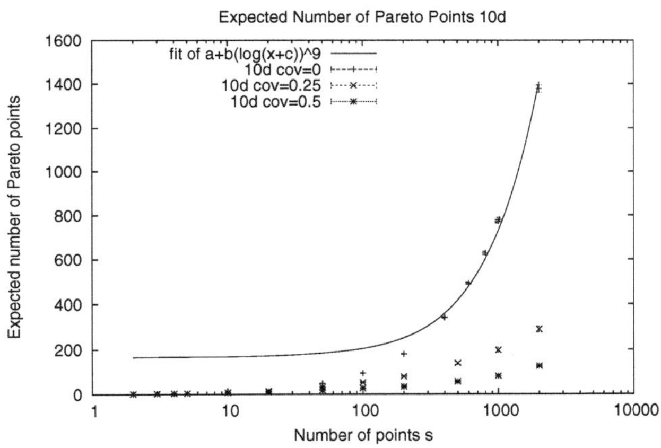

Fig. 1. Empirical distributions of the number of internally nondominated points in a sample of s points for 5 and 10 objectives and three correlations arising from the use of different covariance matrices. For a correlation of 0.0, a curve of $O(\ln^{d-1}(s))$ has been fitted through the largest five values, using least squares estimation. Specifically a curve of the form $a + b(\ln(s + c)^{d-1})$ was used; for the 5d curve $a = 63.3$, $b = 0.0479$, $c = -226$; for the 10d curve $a = 133$, $b = 1.37 \cdot 10^{-5}$, $c = 164$.

As objective dimension increases, nondominated ranking is found to maintain a better resolution (more differences in rank) for longer, although for large dimension and no correlation, almost all vectors become nondominated in samples of this size. It is notable that for nondominated sorting, the most frequent rank shifts away from 1, when objective dimension increases, particularly for correlated data. This could lead, in an MOEA based on nondominated sorting, to over-promotion of the middle ranks if, say, tournament selection is used.

More generally, these plots illustrate that, with more than ten objectives, the principal basis of selection pressure is lost (whether for simple discarding

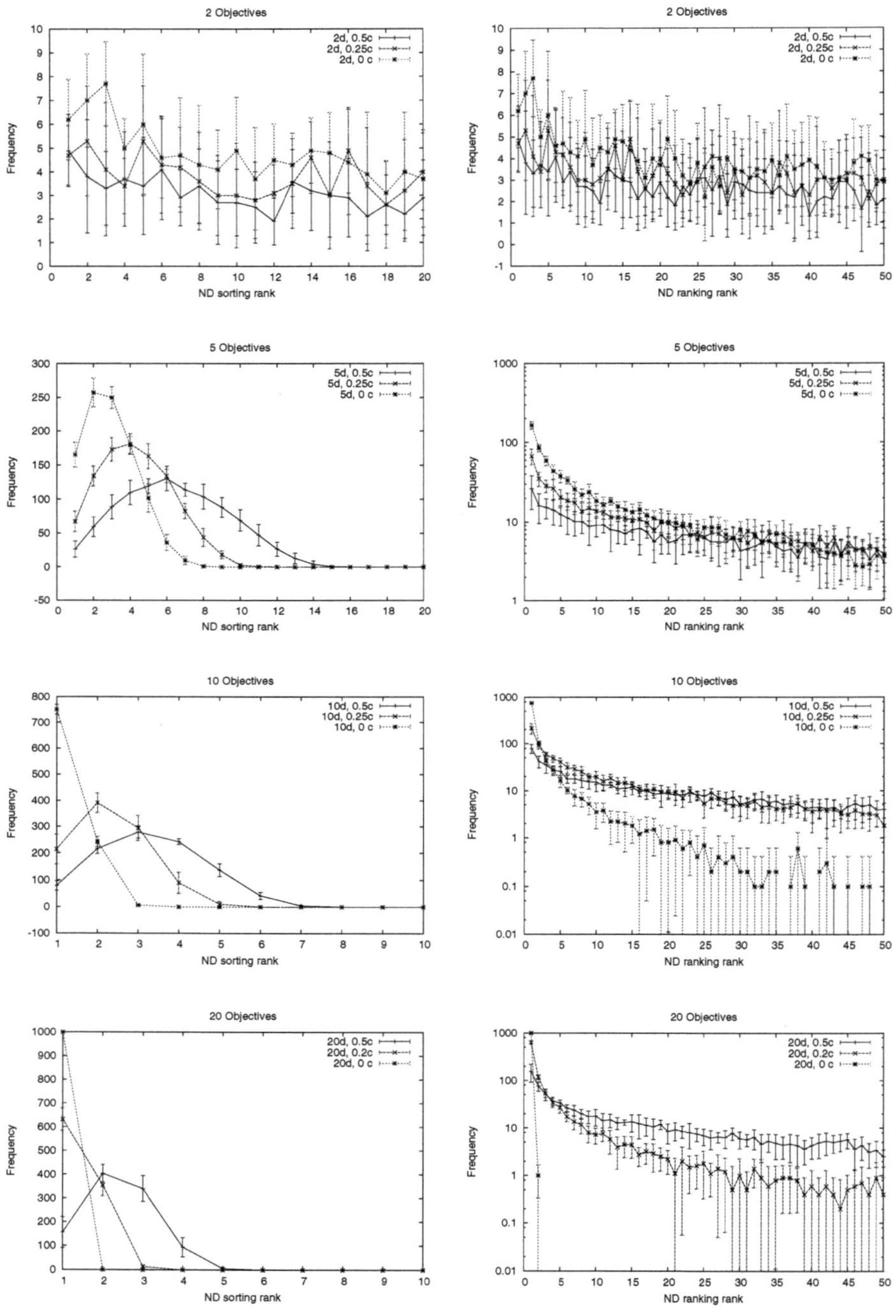

Fig. 2. The effects of dimension and correlation on the distribution of ranks seen under nondominated sorting (Goldberg) and nondominated ranking (Fonseca and Fleming)

of dominated points, nondominated sorting or nondominated ranking), because nearly all points share the same rank. Further, this reiterates that the pursuit of the whole PF is not appropriate in most high dimensional objective spaces. However, even if a representation of the PF only, is sought, ranking will cease to play a helpful role in selection of individuals for d above 20, and perhaps even d above 10. These empirical results confirm long-known limitations of dominance ranking in high-dimensional spaces [9].

3.2 Empirical Distributions of Coverage of the PF

Distributions of the coverage indicator for different random search sample sizes are given in Figure 3, comparing 2d with 20d objective spaces and comparing covariances of 0.0 and 0.5. Observations about the trends seen in these plots are made in the figure caption.

4 Analytical Methods and Results

For random search, the expected value of the coverage indicator is related closely to the probability $\text{gpd}(k, p, w, n)$ of picking precisely k distinct winners in a lottery in which p picks from a hat containing n distinct numbers are made (with replacement), where $w \leq n$ of the numbers are 'winners'.
Notice that:

$$\text{gpd}(0, p, w, n) = \left(\frac{n - w}{n}\right)^p \tag{4}$$

and that:

$$\text{gpd}(k, k, w, n) = \frac{w}{n}.\frac{w-1}{n} \cdots \frac{w-(l-1)}{n} = \frac{w!}{n^k (w-k)!} \tag{5}$$

Meanwhile we can also express $\text{gpd}(k, p, w, n)$, where $p > k$, as follows:

$$\text{gpd}(k, p, w, n) = \frac{(n-w)+k}{n}.\text{gpd}(k, p-1) + \frac{w-(k-1)}{n}.\text{gpd}(k-1, p-1). \tag{6}$$

The two terms represent the two distinct ways that k distinct winners from p picks can arise in the pth pick; in the first term we have the chance that all k distinct winners are already present, in which case the pth pick must be a non-winner or an already-picked winner; in the second term we have the chance that we have distinct winners so far, and so the next pick must be one of the as-yet-unpicked winners.

These three expressions for $\text{gpd}(k, p, w, n)$ enable us to calculate it for any valid set of values, using a dynamic programming procedure.

4.1 Analytical Expression for Expected Value of Coverage

To compute an expected value for the coverage indicator, we note that the degree to which a random sample of p points from n covers the Pareto optima is the

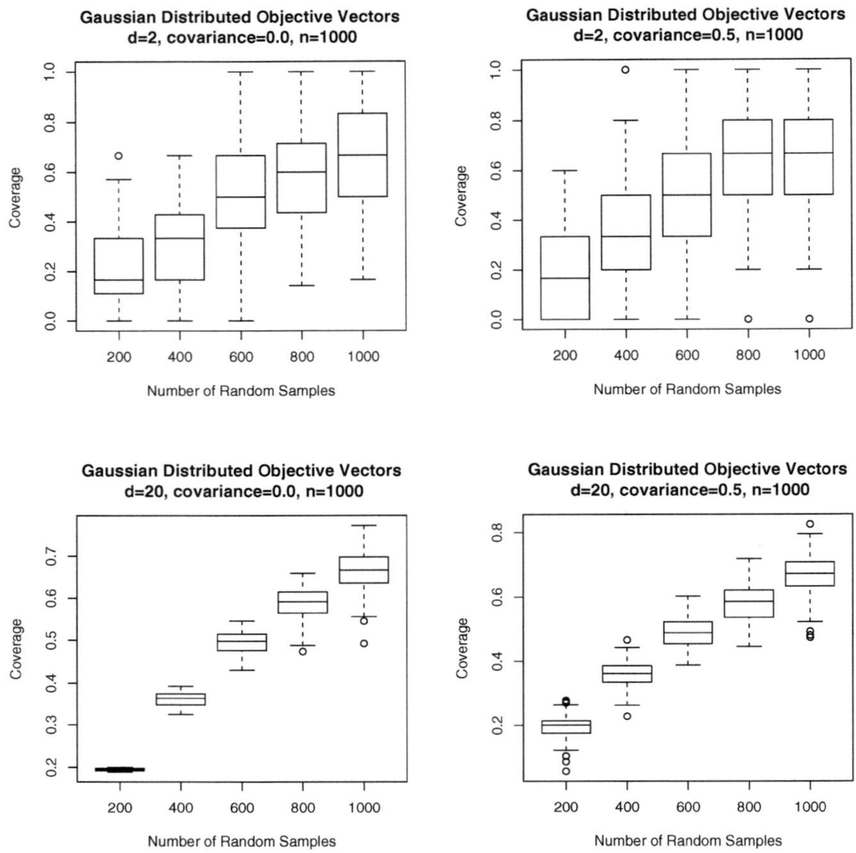

Fig. 3. Box and whisker plots of coverage indicator values, C(sample,PF), for 10 independent runs each of random search for each number of samples, and on ten different data sets for each combination of dimension and covariance. With high objective space dimension, the variance in the coverage is low because nearly every point is nondominated, especially when there is no correlation between objectives. For low dimensions, random search can do very poorly or very well, yielding a high variance in coverage, since only a few points dominate all the rest. Inter-objective correlation reduces the number of nondominated points in the search space, which also leads to a larger variance in coverage. As more samples are taken, there is also an increasing variance in coverage because there is an increasing opportunity for re-sampling the same points. Overall, note that the median value of coverage in these plots is relatively stable for a given number of samples, independently of the objective space dimension and correlations.

same as the degree to which that sample covers the search space as a whole. That is, if a sample covers a proportion q of the search space, it will cover the same proportion q of the Pareto optima of that search space. Hence, for random search, the expected value of the coverage indicator will be independent of w. It will thus also be independent of the objective space dimension and the correlation between objectives for random search.

We calculate it by using a three-argument version of gpd, where $\text{gpd}(k,p,n)$ gives the probability of obtaining k distinct points from a random sample with replacement of size p from a space of size n. The recurrence equations in this case, are as follows:[2]

$$\text{gpd}(1,p,n) = \left(\frac{1}{n}\right)^{p-1} \tag{7}$$

$$\text{gpd}(k,k,n) = \frac{n}{n}.\frac{n-1}{n}\cdots\frac{n-(k-1)}{n} = \frac{n!}{n^k(n-k)!} \tag{8}$$

and for $k > 1$:

$$\text{gpd}(k,p,n) = \frac{k}{n}.\text{gpd}(k,p-1,n) + \frac{n-(k-1)}{n}.\text{gpd}(k-1,p-1,n). \tag{9}$$

(noting that $\text{gpd}(k,0,n) = 0$ for any k).

The expected coverage of a sample of size p can then be expressed as follows:

$$E(C(\text{sample}, PF)) = \sum_{k=1}^{k=p} \text{gpd}(k,p,n).k \tag{10}$$

For example, we computed the expected coverage after p iterations of random search, using the above formula, and compared it with the empirical distribution for 100 trials of random search, for search spaces each of 2d and 20d. The result is plotted in Figure 4.

5 Case Study Using NK Landscapes

Here, we test the predictions of the above analytical expression for coverage by applying the techniques to a multiobjective *function* (i.e. an actual mapping from a decision space to an objective space). We use, for each objective, the well known NK landscape [15] and set $K = 2$, giving a relatively low degree of ruggedness, making it suitable for an EA search.

We study ten instances (in all) each of size $N = 10$, giving a search space of $n = 1024$. As before, we wish to study the effect of correlations between objectives. NK landscapes are based upon tables of uniformly random variates in $[0, 1]$; to generate a target correlation of $c = 0.5$ between the first and any other objective i, we simply copy a fraction $c = 0.5$ of the values used in the table for objective 1 (selected at random), to the corresponding entries in the table for objective i, using randomly drawn variates for the remaining entries. We

[2] Note that Equations (8) and (9) are equivalent to (5) and (6), respectively, except with $w = n$, but equation (4) is replaced by (7), since the boundary nonzero probability cases are now those of finding 1 distinct point in $p > 1$ picks, rather than 0 distinct winners.

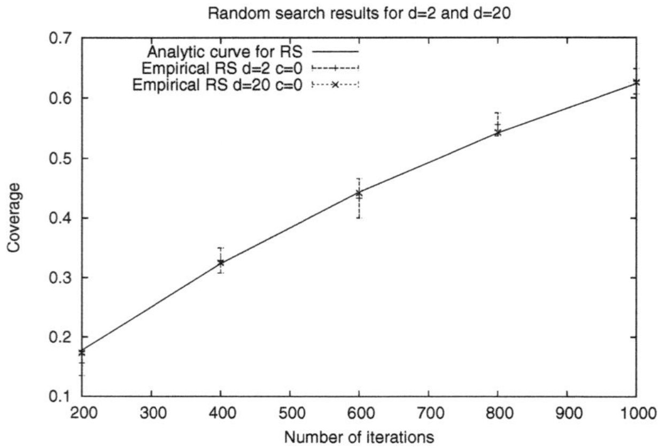

Fig. 4. Predicting the expected coverage indicator value for random search on instances of differing dimension, 2d and 20d. The mean and standard error are indicated by the error bars for the empirical data.

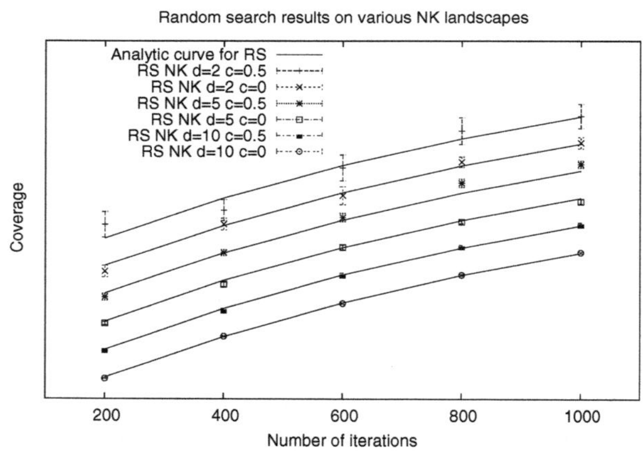

Fig. 5. Predicted expected coverage value for random search, and the mean and standard error of coverage for random search, for various NK instances with different d and c values. The curves have been shifted to aid viewing: the mean performance of random search is approximately the same, independently of dimension and correlation, as predicted by equation 10.

verified that the Spearman rank correlation between objective 1 and objective i was approximately c, as a result of this procedure.

Nine of the instances are created by the combinations of objective dimension $d \in \{2, 5, 10\}$ and correlation $c \in \{0, 0.25, 0.5\}$. Additionally, there is one instance of a ten dimensional NK landscape, where the correlation $c = -0.8$ between the first and all other objectives; to generate this, a procedure analogous to the one

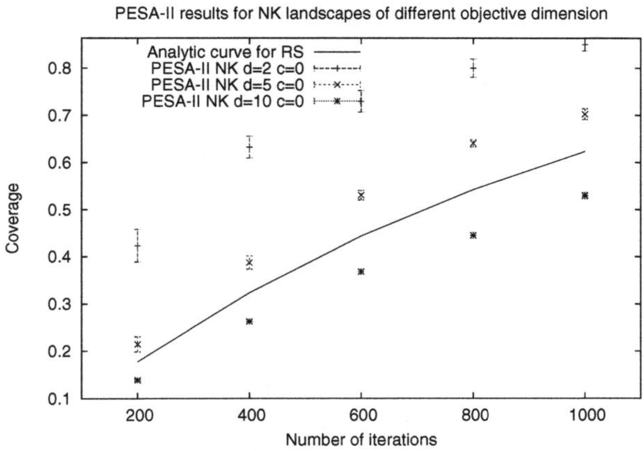

Fig. 6. Predicted expected coverage value for random search, and the mean and standard error of coverage for PESA-II, for various NK instances with different d values and no (or little) inter-objective correlation

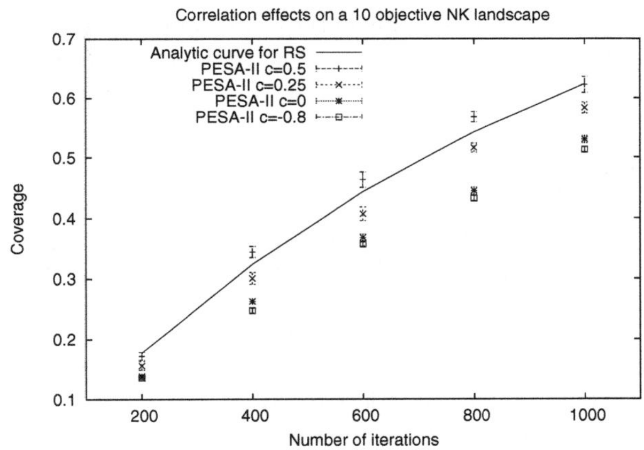

Fig. 7. Predicted expected coverage value for random search, and the mean and standard error of coverage for PESA-II, for NK instances with $d = 10$ and various correlations $c \in \{0.5, 0.25, 0.0, -0.8\}$ between the first and each of the other objectives

for positive correlations was employed, except that values v copied from objective one, were substituted by $1 - v$ in the tables for the other objectives.

We compare the performance of random search with that of PESA-II [5], across the ten instances. PESA-II was used with an internal population size of 2, external population size unlimited (by setting it to 1024), uniform crossover rate of 0.7, per-bit mutation rate of 0.1, and number of objective space boxes of $10^2 = 100$, $3^5 = 243$ and $= 2^{10} = 1024$ respectively for the 2, 5, and 10 objective instances. Twenty independent runs were performed of each algorithm on each of the ten instances. The mean and standard error of the coverage are shown in Figures 5–7.

Figure 5 indicates that the analytical expression for the expected random search performance, in terms of coverage, predicts the empirical results accurately for a range of NK instances with differing d and c values. Thus, the analytical expression can be trusted to predict performance of random search, even on a single problem instance about which nothing is known. Thus, a baseline performance comparison between an EMO algorithm and random search is rendered possible without the necessity of actually making numerous runs of random search.

Figure 6 indicates the mean performance of PESA-II on the three NK problems with independent objectives ($c = 0$), and $d = 2, 5$ and 10. The suitability of NK landscapes for an EA is shown by the performance of PESA-II on the $d = 2$ instance, where it performs relatively well, compared to random search. But, ten objectives is more than sufficient for the performance of PESA-II to deteriorate to significantly below that of random search.

Figure 7 shows the effect of correlation on the performance of PESA-II for the four ten-dimensional NK instances. A positive correlation of 0.5 between the first and each other objective reduces the size of the Pareto front sufficiently for PESA-II to work slightly more efficiently than random search. For lesser values of correlation, or negative correlation, PESA-II optimizes less efficiently than random search.

6 Conclusion

Assessing the scalability of EMO algorithms to objective space dimension remains an issue of concern to the EMO field. To get a proper handle on this, it is good to have a firm understanding of how this parameter affects the number of nondominated points, and the effect of this, in turn, on certain performance indicators. We used random sampling to determine these effects, and we considered also the influence of inter-objective correlation. The results presented are orthogonal to the influence of parameter space 'topology' and any other (e.g. algorithmic) influences.

We derived a general analytic equation for computing the expected coverage of the Pareto front using p iterations of random search. We observe that this is independent of the relative size of the Pareto front, and hence also of the objective space dimension and correlations between objectives. Thus, expected random search performance, in terms of coverage, can be predicted without any knowledge of a problem instance.[3] We confirmed this empirically on Gaussian distributed data and on instances of a multiobjective NK landscape problem.

A general observation from the study comes from quantifying the performance of PESA-II, relative to the baseline of random search, on the NK landscapes. It is apparent that provided the number of target (Pareto optimal) solutions is

[3] Given that we assume a one-to-one mapping of parameter to objective space, as stated earlier.

less than approximately one quarter of the search space, PESA-II outperforms random search, in terms of coverage. This suggests that MOEAs can still be efficient optimizers for problems of 10 objectives and more, provided that not all of the Pareto front is sought. Thus, our findings reiterate that for many-objective problems, the use of preference information to limit the number of target solutions is advisable. However, if preference information is unavailable, then random search may be a sensible alternative to an evolutionary algorithm.

A number of effects were not investigated. Perhaps the most important one is degeneracy (i.e. several distinct parameter space points mapping to the same objective vector). This *would* change the analytical expression for the performance of random search, the effect being greater the more non-uniformly the degeneracy occurs. In problem instances where non-uniform degeneracy occurs to any great degree, the methods presented here would not be appropriate. We have also only considered coverage of the Pareto front, whereas MOEAs are usually used with the intention of building only a *representation* of the Pareto front; this is true, but doesn't change the fact that assessing performance of an MOEA with respect to random search, using coverage, does still give an indication of its relative efficiency, provided the MOEA is not operated in a mode where it actively discards points, e.g. by archive truncation. Still, the consideration of other performance indicators is of course an important avenue for future work.

Acknowledgments. JK is supported by a David Phillips fellowship from the Biotechnology and Biological Sciences Research Council (BBSRC), UK.

References

1. Z. Bai, C. Chao, H. Hwang, and W. Liang. On the Variance of the Number of Maxima in Random Vectors and Its Applications. *The Annals of Applied Probability*, 8(3):886–895, 1998.
2. Z. Bai, L. Devroye, H. Hwang, and T. Tsai. Maxima in hypercubes. *Random Structures Algorithms*, 27:290–309, 2005.
3. J. Bentley, H. Kung, M. Schkolnick, and C. Thompson. On the Average Number of Maxima in a Set of Vectors and Applications. *Journal of the ACM (JACM)*, 25(4):536–543, 1978.
4. C. A. Coello Coello. An Updated Survey of GA-Based Multiobjective Optimization Techniques. *ACM Computing Surveys*, 32(2):109–143, June 2000.
5. D. W. Corne, N. R. Jerram, J. D. Knowles, and M. J. Oates. PESA-II: Region-based Selection in Evolutionary Multiobjective Optimization. In L. Spector, E. D. Goodman, A. Wu, W. Langdon, H.-M. Voigt, M. Gen, S. Sen, M. Dorigo, S. Pezeshk, M. H. Garzon, and E. Burke, editors, *Proceedings of the Genetic and Evolutionary Computation Conference (GECCO'2001)*, pages 283–290, San Francisco, California, 2001. Morgan Kaufmann Publishers.
6. K. Deb. *Multi-Objective Optimization using Evolutionary Algorithms.* John Wiley & Sons, Chichester, UK, 2001. ISBN 0-471-87339-X.

7. N. Drechsler, R. Drechsler, and B. Becker. Multi-objective Optimisation Based on Relation *favour*. In E. Zitzler, K. Deb, L. Thiele, C. A. C. Coello, and D. Corne, editors, *First International Conference on Evolutionary Multi-Criterion Optimization*, pages 154–166. Springer-Verlag. Lecture Notes in Computer Science No. 1993, 2001.
8. M. Farina. A Neural Network Based Generalized Response Surface Multiobjective Evolutionary Algorithm. In *Congress on Evolutionary Computation (CEC'2002)*, volume 1, pages 956–961, Piscataway, New Jersey, May 2002. IEEE Service Center.
9. C. M. Fonseca and P. J. Fleming. An Overview of Evolutionary Algorithms in Multiobjective Optimization. Technical report, Department of Automatic Control and Systems Engineering, University of Sheffield, Sheffield, U. K., 1994.
10. C. M. Fonseca and P. J. Fleming. On the Performance Assessment and Comparison of Stochastic Multiobjective Optimizers. In H.-M. Voigt, W. Ebeling, I. Rechenberg, and H.-P. Schwefel, editors, *Parallel Problem Solving from Nature—PPSN IV*, Lecture Notes in Computer Science, pages 584–593, Berlin, Germany, September 1996. Springer-Verlag.
11. S. Huband, P. Hingston, L. Barone, and L. While. A Review of Multiobjective Test Problems and a Scalable Test Problem Toolkit. *IEEE Transactions on Evolutionary Computation*, 10(5):477–506, 2006.
12. E. J. Hughes. Multiple Single Objective Pareto Sampling. In *Proceedings of the 2003 Congress on Evolutionary Computation (CEC'2003)*, volume 4, pages 2678–2684, Canberra, Australia, December 2003. IEEE Press.
13. E. J. Hughes. Evolutionary many-objective optimisation: Many once or once many? In *IEEE Congress on Evolutionary Computation (CEC 2005)*, pages 222–227. IEEE, 2005.
14. E. J. Hughes. Assessing robustness of optimisation performance for problems with expensive evaluation functions. In *IEEE Congress on Evolutionary Computation — CEC 2006*, pages 9825–9830. IEEE Press, 2006.
15. S. A. Kauffman. Adaptation on rugged fitness landscapes. In D. Stein, editor, *SFI Studies in the Sciences of Complexity, Lecture Volume 1*, pages 527–618. Addison Wesley, 1989.
16. J. D. Knowles and D. W. Corne. Approximating the Nondominated Front Using the Pareto Archived Evolution Strategy. *Evolutionary Computation*, 8(2):149–172, 2000.
17. J. D. Knowles, L. Thiele, and E. Zitzler. A tutorial on the performance assessment of stochastive multiobjective optimizers. Technical Report TIK-Report No. 214b, Computer Engineering and Networks Laboratory, ETH Zurich, 2006.
18. K. Maneeratana, K. Boonlong, and N. Chaiyaratana. Compressed-objective genetic algorithm. In *Parallel Problem Solving from Nature - PPSN IX*, volume 4193 of *LNCS*, pages 473–482. Springer, 2006.
19. R. C. Purshouse. *On the Evolutionary Optimisation of Many Objectives.* PhD thesis, Department of Automatic Control and Systems Engineering, The University of Sheffield, Sheffield, UK, September 2003.
20. N. Srinivas and K. Deb. Multiobjective optimization using nondominated sorting in genetic algorithms. Technical report, Department of Mechanical Engineering, Indian Institute of Technology, Kanpur, India, 1993.
21. D. A. V. Veldhuizen and G. B. Lamont. Evolutionary Computation and Convergence to a Pareto Front. In J. R. Koza, editor, *Late Breaking Papers at the Genetic Programming 1998 Conference*, pages 221–228, Stanford University, California, July 1998. Stanford University Bookstore.

22. M. Yukish. *Algorithms to Identify Pareto Points in Multi-Dimensional Data Sets.* PhD thesis, Pennsylvania State University, PENN, USA, 2004.
23. E. Zitzler. *Evolutionary Algorithms for Multiobjective Optimization: Methods and Applications.* PhD thesis, Swiss Federal Institute of Technology (ETH), Zurich, Switzerland, November 1999.
24. E. Zitzler, L. Thiele, M. Laumanns, C. M. Fonseca, and V. G. da Fonseca. Performance Assessment of Multiobjective Optimizers: An Analysis and Review. *IEEE Transactions on Evolutionary Computation*, 7(2):117–132, April 2003.

Non-linear Dimensionality Reduction Procedures for Certain Large-Dimensional Multi-objective Optimization Problems: Employing Correntropy and a Novel Maximum Variance Unfolding

Dhish Kumar Saxena and Kalyanmoy Deb

Kanpur Genetic Algorithms Laboratory (KanGAL)
Department of Mechanical Engineering
Indian Institute of Technology, PIN 208016, India
{dksaxena,deb}@iitk.ac.in
http://www.iitk.ac.in/kangal

Abstract. In our recent publication [1], we began with an understanding that many real-world applications of multi-objective optimization involve a large number (10 or more) of objectives but then, existing evolutionary multi-objective optimization (EMO) methods have primarily been applied to problems having smaller number of objectives (5 or less). After highlighting the major impediments in handling large number of objectives, we proposed a principal component analysis (PCA) based EMO procedure, for dimensionality reduction, whose efficacy was demonstrated by solving upto 50-objective optimization problems. Here, we are addressing the fact that, when the data points live on a non-linear manifold or that the data structure is non-gaussian, PCA which yields a smaller dimensional 'linear' subspace may be ineffective in revealing the underlying dimensionality. To overcome this, we propose two new non-linear dimensionality reduction algorithms for evolutionary multi-objective optimization, namely C-PCA-NSGA-II and MVU-PCA-NSGA-II. While the former is based on the newly introduced correntropy PCA [2], the later implements maximum variance unfolding principle [3,4,5], in a novel way. We also establish the superiority of these new EMO procedures over the earlier PCA-based procedure, both in terms of accuracy and computational time, by solving upto 50-objective optimization problems.

1 Introduction

In formulating a multi-objective optimization problem, designers and decision-makers prefer to put every performance index related to the problem as an objective, rather than as a constraint, thereby totalling a large number of objectives. However, evolutionary multi-objective optimization (EMO) methods which find a representative set of solutions in the Pareto-optimal front [6], are, in general, found to be vulnerable to large-objective optimization problems. In [1], we had illustrated this 'curse of dimensionality' on the elitist non-dominated sorting GA or NSGA-II [7]. While solving DTLZ2(10), we had shown that approximately only 4% solutions could come to the Pareto-optimal front. When large number of objectives exist, the probability of having any two arbitrary

S. Obayashi et al. (Eds.): EMO 2007, LNCS 4403, pp. 772–787, 2007.

solutions to be non-dominated to each other increases, as there are many objectives in which a trade-off (one is better in one objective but worse in any other objective) can occur. While dealing with a finite-sized population-based approach, the proportion of non-dominated solutions in the population increases. Since EMO algorithms provide more emphasis to the non-dominated solutions, a large proportion of the old population gets emphasized, thereby not leaving much room for new solutions to be included in the population. This, in effect, reduces the selection pressure for better solutions in the population and results in poor convergence. Over and above this difficulty at algorithmic level, handling large-objectives is not only computationally expensive, it is also a challenge for proper decision-making, as visualizing a Pareto-optimal frontier which is more than three-dimensional, is extremely difficult. Amidst all these, a natural question arises, if it is even worth applying EMO methods, for large-objective problems. In [1], we had highlighted that there may exist large-objective problems, which have redundant objectives, that is, although the problem may have, say **M** objectives but the Pareto-optimal front involves a much lower-dimensional interaction. There, we addressed solving such problems by suggesting a principal component analysis (PCA) based NSGA-II procedure which progresses iteratively from within the search space towards Pareto-optimal region by adaptively finding most anti-correlated lower-dimensional interactions. While PCA yields a smaller dimensional linear subspace that best represents the full data according to a minimum square-error criterion, it may be ineffective in revealing the underlying dimensionality when the data points live on a non-linear manifold (manifolds are spaces that are locally linear but unlike Euclidean subspaces, they can be globally non-linear) or that the data structure is nongaussian. The strength of our earlier proposal of PCA-NSGA-II algorithm emerged from the fact that we could relate most important directions in the data set (in terms of variance) to the importance of objectives, given a multi-objective optimization problem. Now if the determination of important directions in data set is erroneous, the inferences drawn about importance of objectives and hence the determination of redundant objectives will be meaningless. Hence, it would be worthwhile to assess situations in which PCA is likely to extract erroneous directions. Such situations can be best examined under the question: "Does the data live in a low-dimensional subspace" or "Does the data live on a low-dimensional sub manifold", which we examine in the following section.

2 Difficulties with PCA

To highlight difficulties with standard PCA, let us begin with a concrete example of DTLZ5(2,5) [1]. Our earlier proposed PCA-NSGA-II , when tested for this problem, brought out f_3 and f_5 as critical objectives and the rest as redundant. Let us investigate these results in light of two facts. (i) Fact one relates to the property of DTLZ5(2,M) problems where the Pareto-optimal front corresponds to the last and any one of the rest, objective. In this context, declaration of f_3 and f_5 as critical is right. (ii) Fact two relates to the criteria of judging an objective set as critical. PCA-NSGA-II is expected to declare those objectives as critical which apart from being in conflict with each other, also account for variances larger than those declared redundant. From Figure 1, f_4 and f_5 can be seen to account for largest variance, amongst the set of five objectives. This

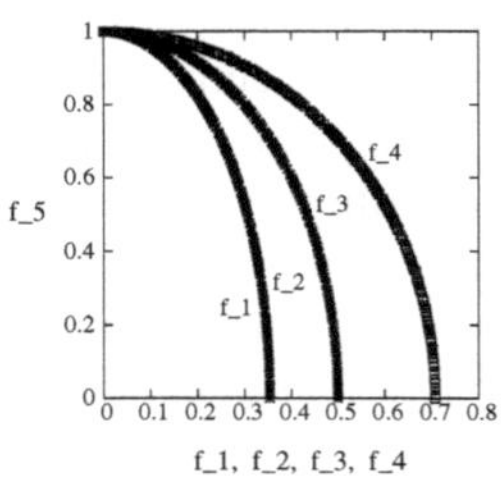

Fig. 1. Relative Variance measures in DTLZ5(2,5)

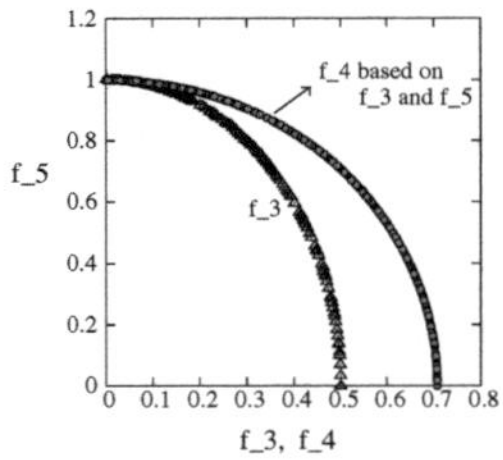

Fig. 2. Reconstruction of actually critical objective, declared redundant

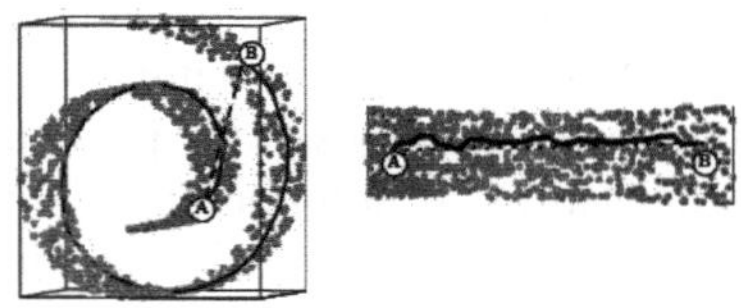

Fig. 3. Submanifold: a problem for standard PCA (taken from [8])

infact, is generalizable to all DTLZ5(2,M) problems, where objectives with indices M and $(M - 1)$ will collectively account for largest variance in the data set of all given objectives. Hence, the last two objectives must come up as the critical objectives, given any DTLZ5(2,M) problem. This poses a question on declaration of f_3 as critical (over f_4) along with f_5. As far as this test problem is considered, this still is not an inaccuracy in result. Figure 2, shows the plot of f_4 computed from the variable set obtained from an NSGA-II run with two objectives declared critical, namely f_3 and f_5. This plot of reconstructed f_4 can be seen to identically match with that obtained from a separate NSGA-II run, with objectives f_4 and f_5, as is shown in Figure 1. Hence, for this problem, selection of f_3 and f_5 is at worst a misrepresentation of results and not a case of inaccuracy, since the variable set achieved with either f_3 with f_5 or f_4 with f_5, is the same. However, what cannot be negated is the fact that PCA could actually not conduct what it was expected to, in terms of capturing required objectives with larger variances. Well then, the hint to what went wrong lies in Figure 3. Two very small regions are marked as 'A', 'B' respectively on both - the manifold (on left) and the unfolded representation (on right). By euclidean measures, points 'A' and 'B' not being amongst the farthest, will not be found to account for the largest variance, by PCA. However, as is evident from the right segment of the figure, largest variance lies trapped between 'A' and 'B'. This highlights the limitation of PCA in general (also discussed in [8]) and implies the need to distinguish between subspaces and submanifolds. While in the case of subspaces, PCA would be effective, it is likely to mislead in case when data lives on a submanifold. Given this problem, the natural strategy could be to embed the data into a space where the patterns can be discovered as linear relations, which is where our focus lies in the following sections.

3 Non-linear Dimensionality Reduction

While, we have already seen the application (PCA [1]) of one of the linear dimensionality reduction methods (Independent Component Analysis (ICA), Singular Value Decomposition (SVD), Factor Analysis, Metric Multidimensional Scaling (MDS) being others), in the context of large dimensional multi-objective optimization, the need is to resort to non-linear dimensionality reduction methods. The later can be broadly categorized in two groups: those which are based on non-linear mappings and those that are based on proximity matrices (distance measurements that just give a visualization). The principal method amongst those that provide a mapping from the high dimensional space to the embedded space is the *kernel PCA*. This method provides a non-linear PCA through the use of kernel functions. However, for many kernel functions, kernel PCA actually increases the dimension of the data. On the other hand, a method based on proximity matrices is one where the data is presented to the algorithm in the form of a similarity matrix or a distance matrix. These methods, prime of which are Isomap [9], graph Laplacian Eigenmaps [10], locally linear embeddings (LLE) [11] (all these, also interpreted as instances of kernel PCA in [12]) and maximum variance unfolding, all fall under the broader class of multidimensional scaling [13], the variations emanating based on the differences in how the proximity data is computed. We consider and employ below, one approach from each category (mapping and proximity matrices) for non-linear dimensionality reduction in context of large dimensional multi-objective optimization.

4 Methods Based on Non-linear Mappings

4.1 Main Ingredients of Kernel Methods

We now highlight the four key aspects of kernel approach, which are as follows: (i) Data items are embedded (through non-linear mapping) into a vector space called the feature space. (ii) Linear relations are sought among the images of the data items in the feature space. (iii) The algorithms are implemented in such a way that the co-ordinates of the embedded points are not needed, only their pairwise inner products. (iv) The pairwise inner products can be computed efficiently directly from the original data items using a kernel function, which can be defined as a function κ that for all $\mathbf{x}, \mathbf{z} \in \mathbf{X}$ satisfies $\kappa(\mathbf{x}, \mathbf{z}) = \langle \phi(\mathbf{x}), \phi(\mathbf{z}) \rangle$, where ϕ is a mapping from $\mathbf{X}$ to an (inner product) feature space $\mathbf{F}$ ($\phi : \mathbf{x} \longmapsto \phi(\mathbf{x}) \in \mathbf{F}$). To help clarify some key concepts, let us consider a two-dimensional input space $\mathbf{X} \subseteq \Re^2$ together with the feature map $\phi : \mathbf{x} = (x_1, x_2) \longmapsto \phi(\mathbf{x}) = (x_1^2, x_2^2, \sqrt{2}x_1x_2) \in \mathbf{F} = \Re^3$. The comparisons of the feature map with the inner product in the feature space can be evaluated as follows:

$$\langle \phi(\mathbf{x}), \phi(\mathbf{z}) \rangle = (x_1z_1 + x_2z_2)^2 = \langle \mathbf{x}, \mathbf{z} \rangle^2$$

Hence, the function $\kappa(\mathbf{x}, \mathbf{z}) = \langle \mathbf{x}, \mathbf{z} \rangle^2$ is a kernel function with $\mathbf{F}$ its corresponding feature space. While, the above is one way of knowing whether a given function is a kernel or not, another, which is of more practical utility, is by way of what is referred as *Characterization of kernels* [14], according to which, a function $\kappa : (\mathbf{X} \times \mathbf{X}) \longmapsto \Re$, which is either continuous or has a finite domain, can be decomposed into a feature map ϕ in a Hilbert space F applied to both its arguments, followed by evaluation of the inner product in F *if and only if* it satisfies positive-semidefinite property. Some widely used kernel functions are the linear, polynomial and Gaussian kernels, and are given, respectively by:

$$\kappa(\mathbf{x}, \mathbf{z}) = \mathbf{x} \cdot \mathbf{z}; \qquad (1 + \mathbf{x} \cdot \mathbf{z})^p; \qquad e^{\frac{-|\mathbf{x}-\mathbf{z}|^2}{2\sigma^2}} \tag{1}$$

With this basic understanding of kernels, let us highlight two key features of kernels which will be in later sections used, to answer some critical questions, for example, the justification for the usage of different kernel forms for the same problem or the usage of the same kernel to different problems. They are: (i) a feature space is not uniquely determined by the kernel function. It can be seen that the kernel $\kappa(\mathbf{x},\mathbf{z}) = \langle \mathbf{x},\mathbf{z}\rangle^2$also computes the inner product (hence, defines) corresponding to the four-dimensional feature map $\phi : \mathbf{x} = (x_1, x_2) \longmapsto \phi(\mathbf{x}) = (x_1^2, x_2^2, x_1x_2, x_2x_1) \in \mathbf{F} = \Re^4$ (ii) the second feature relates to defining new kernels, given a set of kernels. The fact that kernel functions satisfy a series of closure properties, open up the possibility of defining kernels by successive adjustments, either performing successive embeddings or manipulating the given kernel function [14]. In this case, the embedding corresponding to overall kernel can be composed through successive embeddings, i.e., $\phi(\mathbf{x}) = \psi_1(\mathbf{x}) \circ \psi_2(\mathbf{x}) \circ \ldots \circ \psi_n(\mathbf{x})$ where, $\psi_i(\mathbf{x})$ is an intermediate embedding. We leave this section, reiterating the main advanage of Kernel-based methods, in that the input data is mapped to a feature space by a non-linear mapping, where the inner products in the feature space can be computed directly by a kernel function without knowing the non-linear mapping explicitly.

4.2 From PCA to Kernel-PCA (K-PCA): Difficulties in Generalization

In PCA, given a data set $\mathbf{X}$ of size $M \times N$, where M denotes the number of 'measurement types' (in current context–objectives) and N denotes the number of time samples (in current context–population members), the correlation matrix $\mathbf{R}$ to be eigen-decomposed and given by $\frac{1}{N}\mathbf{X}\mathbf{X}^T$ turns to be of dimension $M \times M$. Scholkopf et al. [15,16] non-linearly mapped the input space (time sample by sample), reformulated the PCA by using the 'kernel trick', in that substituting a kernel function for the inner product, to propose a non-linear form of PCA, and referred as Kernel PCA. The new eigenvalue problem (emerging from diagonalization of covariance matrix in feature space) shapes up as $\mathbf{K}\alpha = (N\lambda)\alpha$, where $(N\lambda)$ denotes the eigenvalue, α–the eigenvector and $\mathbf{K}$ is referred as Kernel Gram Matrix. Composed by $K_{ij} = \kappa(\mathbf{x}_i, \mathbf{x}_j)$, $\forall\ i, j = 1..N$ ($\mathbf{x}_i$ representing columns of $\mathbf{X}$), $\mathbf{K}$ turns to be of dimension $N \times N$, making it incompatible (since, in context of evolutionary multi-objective optimization algorithms, $M \ll N$) with our earlier proposed dimensionality reduction scheme [1]. Hence employment of kernel PCA procedure would not be possible, given the current context.

4.3 Correntropy PCA (C-PCA)

The correntropy PCA, introduced in [2], is a non-linear PCA technique based on the generalized correlation function, referred by its authors as correntropy. Correntropy function is defined as $\mathbf{V}(\mathbf{x},\mathbf{y}) = E\left[\kappa(\mathbf{x},\mathbf{y})\right]$ for two random variables $\mathbf{x}$ and $\mathbf{y}$. As a positive-definite kernel function is imposed on the argument inside the expectation, the correntropy function becomes positive-definite. Given this, by Moore-Aronszajn theorem, there must exist a unique RKHS (Reproducing Kernel Hilbert Space) associated with correntropy function, say $\langle \Pi(\mathbf{x}), \Pi(\mathbf{y})\rangle = \mathbf{V}(\mathbf{x},\mathbf{y})$. This forms the basis of employing correntropy function for non-linear principal component analysis. Unlike in kernel PCA which transforms data into a feature space sample by sample, here the data is mapped component wise into a feature space (associated with the correntropy function). Let $\mathbf{x}_j$, $j = 1, ..., N$, $\mathbf{x}_j \in \Re^M$, $\sum_{j=1}^{N} \mathbf{x}_j = \mathbf{0}$, be a set of zero mean vector observations and Π be a function defined as,

$$\Pi : \Re^M \longmapsto F; \qquad \mathbf{x} \longmapsto [\Pi(x_1), \Pi(x_2), ..., \Pi(x_M)]$$

where x_i denotes the ith component of the original input data sample $\mathbf{x}$. As this non-linear mapping which transforms data component wise into a high dimensional RKHS, is associated with the correntropy function, we have

$$\langle \Pi(x_i), \Pi(x_j) \rangle = \mathbf{V}(x_i, x_j) = E\left[\kappa(\mathbf{x}_i, \mathbf{x}_j)\right] = \frac{1}{N} \sum_{k=1}^{N} \kappa(x_{ik}, x_{jk}) \forall i, j = 1..M \quad (2)$$

The general equation which also accounts for centering of data is as follows:

$$\mathbf{V}_{ij} = E\left[\kappa(\mathbf{x}_i - \mathbf{x}_j)\right] - \frac{1}{N^2} \sum_{k=1}^{N} \sum_{m=1}^{N} \kappa(x_{ik} - x_{jm}) \quad (3)$$

As can be found in [2], the new eigenvalue problem (based on covariance matrix of the transformed data in feature space) shapes up as $\mathbf{V}\alpha = (M\lambda)\alpha$, where $\mathbf{V}$ is the Correntropy matrix, given by Equation 3, and of dimension $M \times M$. This eigen-decomposition is not only dimensionally compatible with that in PCA-NSGA-II [1] but would also be computationally efficient as it promises a non-linear dimensionality reduction, with the underlying matrix, dimensionally remaining the same, as in the linear PCA case. Hence, we will employ this approach to refer the resulting algorithm as C-PCA-NSGA-II.

5 Methods Based on Proximity Matrices

It can be realized that the success and efficacy of methods discussed above, would primarily depend on the choice of the kernel, as different kernels may extract or conceal different types of low-dimensional structures. Hence, given an unknown problem, the learning system or the feature selection scheme actually would have two tasks to address, that of choosing a kernel from a family of kernels (even zeroing upon a family of kernels would reflect our prior expectation about the functions we may be expected to learn) and subsequently of selecting features in the feature space of the chosen kernel. It is now our endeavour to customize the choice of the kernel, specifically for a given problem at hand and hence infuse credibility to the determined 'underlying submanifold' in general or to the set of objectives determined as critical (or otherwise - redundant) for large-dimensional multi-objective optimization problems in particular. For this, we are adopting a method based on proximity matrices, namely maximum variance unfolding with details discussed below.

5.1 Maximum Variance Unfolding: The Concept

Weinberger et al [3,4,5], in their pursuit of unsupervised learning of manifolds, have proposed a way of learning the kernel matrix, as an instance of semidefinite programming, where the kernel matrix is constructed by maximizing the variance in feature space subject to local constraints that preserve the angles and distances between nearest neighbours. Their proposal of maximum variance unfolding can conceptually be justified by the observation that any slack in a piece of string serves to decrease the Euclidean distance between its two ends, just as any furling of the flag brings its corners together. Hence in general, while any "fold" between two points on a manifold decreases the Euclidean distance between the points, they equated the task of "unfolding" the manifold (a task due for the implicit mapping in the kernel matrix to be learned) to maximizing the sum total of the pairwise distance (evident in Figure 4) between the inputs $\mathbf{x}_i$ such that the distances and angles between an input point and its neighbours remains unchanged, giving way to the outputs from the final state of this transformation. The effect of this transformation can be easily visualized by imagining the beads on a necklace (coiled up in three dimensions) as inputs. By pulling the necklace taut, the beads would then be alligned in a line, resulting in a case of non-linear dimensionality reduction from $\Re^3$ to $\Re^1$.

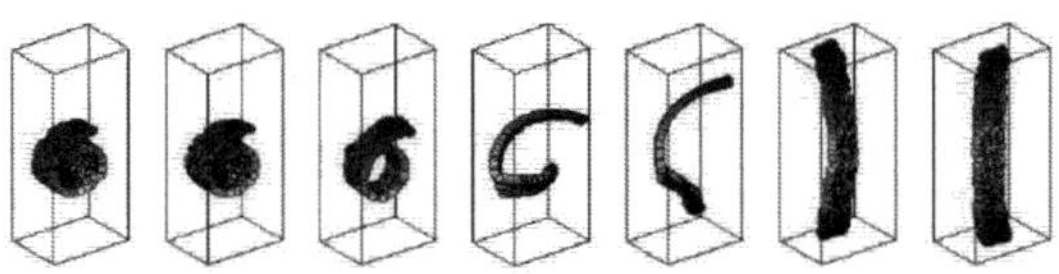

Fig. 4. Maximum Variance Unfolding (taken from [8])

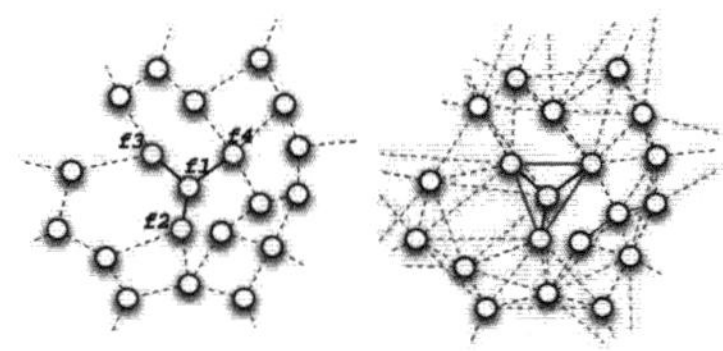

Fig. 5. Preservation of local isometry (taken from [3] and edited)

5.2 Maximum Variance Unfolding: A Novel Implementation

While the proposed algorithm by Weinberger et al. can be found in [3,4,5], let us highlight that they treat a time sample (a vector of dimension equal to number of measurement types) as an input. In the context of multi-objective optimization, it implies that each of the N solutions–a population member with dimension equaling number of objectives (M), would be an input. Consequently, the learned matrix would be of size $N \times N$, leading to dimensional incompatibility (similar to the case of kernel PCA, section 4.2) with our earlier proposed scheme of correlating importance of objectives to directions of large variances (obtained from eigen-decomposition of an appropriate matrix). To beat this curse of dimensional incompatibility, we suggest to treat the vector of each 'objective function' (of dimension equaling the number of population members) as an input. Since, we would still be dealing with vectors, the underlying philosophy of maximum variance unfolding would still hold. This would allow the 'to be learned' kernel matrix of dimension $M \times M$ and hence, also the application of our earlier proposed PCA based scheme to the 'unfolded' manifold. With this background we present the original–maximum variance unfolding algorithm, now dimensionally customized to the generalization needs of PCA-NSGA-II [1] algorithm. Let us begin with a one-to-one correspondence relation between the set of inputs: objective functions $\mathbf{x}_1, \mathbf{x}_2, \ldots, \mathbf{x}_M$ and features: $\phi(\mathbf{x}_1), \phi(\mathbf{x}_2), \ldots, \phi(\mathbf{x}_M)$. Then, define an $M \times M$ neighborhood matrix η whose each element $\eta_{ij} \in \{0, 1\}$, being one only when $\mathbf{x}_j$ happens to be amongst *'k' nearest* neighbour of $\mathbf{x}_i$ (implications of *'k'* discussed in Section 7.2). Further after defining the inner-product matrix $K_{ij} = \langle \phi(\mathbf{x}_i), \phi(\mathbf{x}_j) \rangle$ and the Gram matrix of the inputs $G_{ij} = \mathbf{x}_i \cdot \mathbf{x}_j$, the optimization problem is posed as a *semidefinite program* (SDP) as shown below:

Maximize trace(**K**) = $\frac{1}{2M} \sum_{ij}(K_{ii} - 2K_{ij} + K_{jj})$ subject to:

$$\begin{array}{ll} (1) & K_{ii} - 2K_{ij} + K_{jj} = G_{ii} - 2G_{ij} + G_{jj} \quad \text{for all (i,j) with } \eta_{ij} = 1 \\ (2) & \sum_{ij} K_{ij} = 0 \\ (3) & K \succeq 0 \end{array} \tag{4}$$

While, the first constraint ensures that distances between nearby inputs match distances between nearby outputs, the second is for the centering the objective functions in the feature space which will ensure a unique solution (up to rotation). The third constraint ensures positive-semidefiniteness of **K**, a condition required to interpret the kernel matrix as storing the inner products of vectors in a Hilbert space. Unlike the original quadratic program, this SDP is convex and can be solved efficiently in polynomial time through any off-the shelf solvers available in public domain. We have used the SeDuMi toolbox (Sturm [1999]) in MATLAB, for the same. The **R** matrix employed in PCA-NSGA-II is now replaced by **K** matrix, giving way to what we will be referring as MVU-PCA-NSGA-II algorithm.

6 Proposal: C-PCA-NSGA-II or MVU-PCA-NSGA-II

6.1 Algorithmic Details

The entire structure of these two algorithms would remain the same as that of our earlier PCA-NSGA-II [1]. They would differ, just in the matrix with which they begin i.e., employment of **V** for former and **K** for the later. Furthermore, to make the scheme more robust, we are incorporating certain changes in the manner we interpret the first principal component. For the sake of completeness, they are as follows.

(1) Eigenvalue Analysis for Dimensionality Reduction. We compute the eigenvalues (and corresponding eigenvectors) of the **V** or **K**, which are then ranked in the decreasing order of their magnitudes. The first principal component (eigenvector corresponding to the largest eigenvalue) is designated as 'PCA1'. The first component of this vector denotes the contribution of first objective function towards this vector, the second element denotes the contribution of the second objective, and so on. For a three-objective problem, the three contributions could be treated as the direction cosines defining a directed-ray in the objective space. A positive value denotes an increase in objective value moving along this principal component (axes) and a negative value denotes a decrease. Thus, if we consider the objectives corresponding to the most-positive and most-negative elements of this vector, they denote the objectives which have the maximum contribution for an increase or a decrease in the principal component. Thus, by picking the most-negative and the most-positive elements from a principal component, the two most conflicting (hence important) objectives along that direction can be identified.

(2) Effect of Multiple Principal Components. By above argument, each principal component is analyzed for the two main objectives causing a conflict and the information about the overall conflicting objectives is gathered. We suggest a procedure which starts with analyzing the the first principal component and then proceed to analyze the second principal component and so on, till all the significant components are considered. For this purpose, we pre-define a threshold cut (TC) and when the cumulative contribution of all previously principal components exceeds TC, we do not analyze any more principal components. Based on our experience with several test problems we suggest TC equal to 95%. **V** or **K** being a positive-definite matrix, usage of matrix $\mathbf{VV^T}$ instead of **V** or $\mathbf{KK^T}$ instead of **K**, would only square up the eigenvalues while the eigenvector would remain unaltered. As the variance contribution of a principal component relates to the ratio of corresponding eigenvalue of the total, now, lesser number of principal components would have to be considered to meet the predefined threshold (TC). As, usage of $\mathbf{VV^T}$ or $\mathbf{KK^T}$ would logically make the analysis more compact, it is these which we have employed for the proposed scheme. To make the dimensionality-reduction procedure effective and applicable to various scenarios, we suggest the following additional procedure. As the first principal component captures the significant portion of the total variance in data set, we would want to capture any signal of a conflicting objective. Hence, for the first principal component, along with the objective corresponding to the most-positive element, we consider as important, any/all objectives which correspond to a negative component, howsoever small. If in some case, all the elements along PCA-1 are positive, we pick up the objectives corresponding to the first two most positive elements. For subsequent principal components, we first check if the corresponding eigenvalue is greater than 0.1 or not. If not, we only choose the objective corresponding to the highest absolute element in the eigenvector. If yes and also if the cumulative contribution of eigenvalues is less than TC, we consider various cases. If all elements of the eigenvector are positive, we only choose the objective corresponding to the highest element. If all elements of the eigenvector are negative, we choose all objectives. Otherwise, if the value of the highest positive element (p) is less than the absolute value of the most-negative element (n), we consider two different scenarios. If $p \geq 0.9|n|$, we choose two objectives corresponding to p and n. On the other hand [1], if $p < 0.9|n|$, we only choose the objective corresponding to n. Similarly, if $p > |n|$, then we consider two other scenarios. If $p \geq 0.8|n|$, we choose both objectives corresponding to p and n. On the other hand, if $p < 0.8|n|$, we only choose the objective corresponding to p. So what we have now, is a set of critical or non-redundant objectives, based on the criterion of variance.

(3) Final Reduction Using the Correlation Matrix. Hopefully, the above procedure identifies most of the redundant objectives dictated by the data set. To consider if more reduction in the number of objectives is possible, we then return to a *reduced* correlation matrix (only columns and rows corresponding to non-redundant objectives) and investigate if there still exists a set of objectives having identical positive or negative correlation coefficients (with respect to their signs) with other objectives and having a positive correlation among themselves. This will suggest that any one member from such group would be enough to establish the conflicting relationships with the remaining objectives. In such a case, we retain the one which was chosen the earliest (corresponding to the larger eigenvalue) by the PCA analysis. However, if members of such a group come from the same PCA, then the one having larger absolute value is retained. Still if there magnitudes are equal (less likely, though) then the one having more significant contribution (larger absolute value) along next PCA, is picked. Other objectives from the set are not considered further. It should be mentioned, that once NSGA-II is run for sufficiently large number of generations, the correlation matrix **R** stabilizes and correlation patterns turn invariant over number of generations. Hence, in a broad sense, we have two different criterions of dimensionality reduction embedded in our algorithm. While the first is based on variance (eigenvalue decomposition and analysis), the second exploits the situation when two objectives are identically correlated with all the rest (those belonging to 'the then' important objective set). It is very important, here, to highlight the usage of the term correlation matrix **R** as against correntropy matrix **V** or **K**, in this final reduction stage. While **R** has embedded in it, the original inter-relationships among objectives, given an objective set; **V** or **K** on the other hand, is representative of the feature space, where the original relationships between objectives are not retained (the mapping or unfolding of data, only retains the isometry - locally). Hence, while we utilize **V** or **K** for the first stage of variance based reduction, we resort to the use of **R** to exploit the original inter-relationships between objectives.

6.2 Overall C-PCA-NSGA-II or MVU-PCA-NSGA-II Procedure

Step 1: Set an iteration counter $t = 0$ and initial set of objectives $\mathbb{I}_0 = \{1, 2, \ldots, M\}$.

Step 2: Initialize a random population for all objectives in the set $\mathbb{I}_t$, run an EMO, and obtain a population P_t.

Step 3: Make the final (that corresponding to P_t) data 'centered', by deducting objective wise, the mean.

Step 4: Perform C-PCA-NSGA-II or MVU-PCA-NSGA-II analysis on P_t using $\mathbb{I}_t$ to choose a reduced set of objectives $\mathbb{I}_{t+1}$ using the predefined TC. Substeps include:

1. Computation of the correlation matrix **R** (as defined in Section 4.2).
2. Selection of a valid kernel and computation of the correntropy matrix **V** using Equation 3 or Selection of the free parameter *'k'* and construction of kernel matrix **K** by solving SDP, discussed in Section 5.2.
3. Computation of eigenvalues and eigenvectors of **V** or **K** and picking non-redundant objectives using the procedure discussed in Sections 6.1-(1) and (2)
4. Reduce the number of objectives further, if possible, by interpreting **R** for the non-redundant objectives found in item 2 above, using the guidelines discussed in Section 6.1-(3).

Step 5: If $\mathbb{I}_{t+1} = \mathbb{I}_t$, stop and declare the obtained front. Else set $t = t + 1$ and go to Step 2.

7 Simulation Results

To test the efficacy of **C-PCA-NSGA-II** and **MVU-PCA-NSGA-II** procedures for non-linear dimensionality reduction in context of MOPs, we chose the DTLZ5(I,M) [1] problems (minor variants of the well-known DTLZ5 problem [17] having M objectives) and DTLZ2 problems [1]. To provide a reasonable computational effort, we have used a population size and number of generations, respectively as, 200 and 1,000 (for 3-objective), 400 and 2,000 (for 5-objective) and 800 and 10,000 for rest of the problems.

7.1 C-PCA-NSGA-II

To evaluate the correntropy matrix, we employed polynomial and gaussian kernels with degree three, two, one and half. We begin by illustrating the simulation results corresponding to polynomial kernel with degree three. In this case, all DTLZ5(2,M) problems, could be solved accurately, upto 50-objectives (M=50), in just one iteration (as against multiple iterations when PCA-NSGA-II was employed [1], Table 3 illustrates this fact for the DTLZ5(2,10)) and that too, in most cases required only the first principal component (as it was found to account for variance $\geq 95\%$). The fact that the last two objectives (as they were theoretically argued for in Section 2) could be extracted as critical ones in each case, suggests that the earlier problem of misreading important variance directions, when data lived on submanifold, has been duly rectified. As the trend of correlation matrix was observed to be identical, for brevity, we have shown aside, the correlation matrix only for one case of DTLZ5(2,M) and DTLZ2(M), each. Given the nature of DTLZ5(2,M) problems, the Pareto-optimal front being sought–corresponding to last two objectives, is identical for all cases. For sample illustration, Figure 6 and Figure 7 show repectively, the algorithmically obtained and theoretically constructed Pareto-optimal fronts for DTLZ5(2,10). The next set of experiment

Table 1. DTLZ5(2,3)

Iter. 1: PCA-1 (0.989)	f_1 f_3
Iter. 1: Post Eig. Analysis	f_1 f_3
Iter. 1: Post Corr. Analysis	f_1 f_3

	f_1	f_3
f_1	+	-
f_3	-	+

Table 2. DTLZ5(2,10)

Iter. 1: PCA-1 (0.982)	f_9 f_{10}
Iter. 1: Post Eig. Analysis	f_9 f_{10}
Iter. 1: Post Corr. Analysis	f_9 f_{10}

Table 3. DTLZ5(2,10) : using PCA

Iter. 1 (3 Prin. Comp.)	f_1 f_5 f_9 f_{10}
Iter. 2 (2 Prin. Comp.)	f_5 f_9 f_{10}
Iter. 3 (1 Prin. Comp.)	f_9 f_{10}
Iter. 4 (1 Prin. Comp.)	f_9 f_{10}

Table 4. DTLZ5(2,20)

Iter. 1: PCA-1 (0.952)	f_{19} f_{20}
Iter. 1: Post Eig. Analysis	f_{19} f_{20}
Iter. 1: Post Corr. Analysis	f_{19} f_{20}

Table 5. DTLZ5(2,30)

Iter. 1: PCA-1 (0.967)	f_{29} f_{30}
Iter. 1: Post Eig. Analysis	f_{29} f_{30}
Iter. 1: Post Corr. Analysis	f_{29} f_{30}

Table 6. DTLZ5(2,50)

Iter. 1: PCA-1 (0.941)	f_{49} f_{50}
PCA-2 (0.055)	f_{49}
Iter. 1: Post Eig. Analysis	f_{49} f_{50}
Iter. 1: Post Corr. Analysis	f_{49} f_{50}

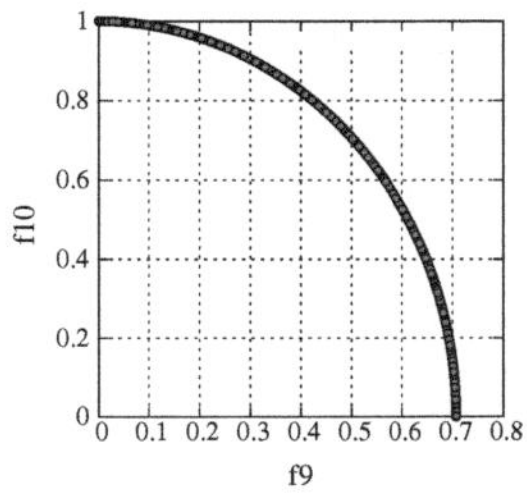

Fig. 6. Pop. obtained after the second iteration of C-PCA-NSGAII on DTLZ5(2,10)

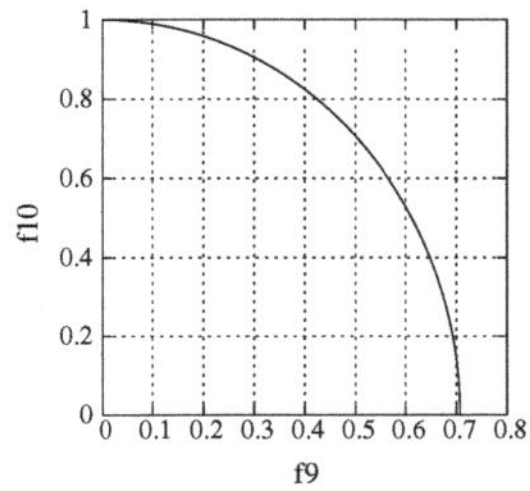

Fig. 7. Theoretical POF in f10-f9 plane for DTLZ5(2,10)

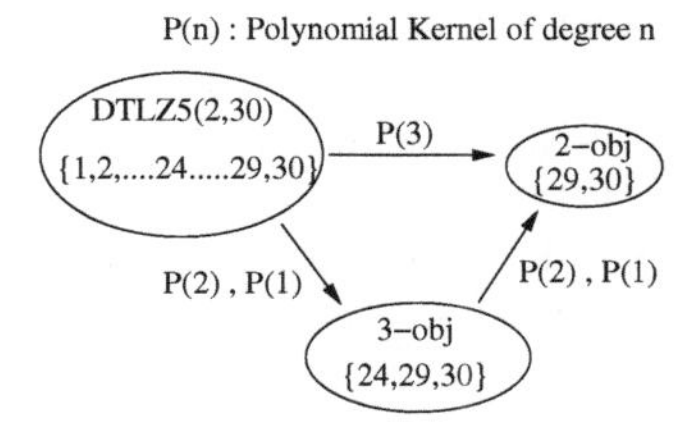

Fig. 8. Defining Kernels with successive adjustments

Table 7. DTLZ2(3)

		f_1	f_2	f_3
Iter. 1:	PCA-1 (0.548)	f_1		f_3
	PCA-2 (0.445)		f_2	
Iter. 1:	Post Eig. Analysis	f_1	f_2	f_3
Iter. 1:	Post Corr. Analysis	f_1	f_2	f_3

	f_1	f_2	f_3
f_1	+	-	-
f_2	-	+	-
f_3	-	-	+

Table 8. DTLZ2(5)

		f_1	f_2	f_3	f_4	f_5
Iter. 1:	PCA-1 (0.328)	f_1	f_2	f_3	f_4	
	PCA-2 (0.287)		f_2		f_4	
	PCA-3 (0.205)	f_1				
	PCA-4 (0.168)					f_5
Iter. 1:	Post Eig. Analysis	f_1	f_2	f_3	f_4	f_5
Iter. 1:	Post Corr. Analysis	f_1	f_2	f_3	f_4	f_5

involved polynomial kernels of degree two and one, respectively. Interestingly, not only were both found to yield accurate results but also in most cases, the objective set obtained as important between two successive iterations (wherever applicable), was found to match. However, the manner in which these matching objective sets emerged from different principal components or subsequent consideration of reduced correlation matrix, differed. Table 9 and Table 10 highlight this observation for for first iteration of DTLZ5(2,10) problem, solved with polynomial kernel with degrees two and one, respectively. At the end of first iteration, both these cases projected f_7, f_9 and f_{10} as important and in the second iteration f_9 and f_{10}. The intent here, is to highlight the difference in the manner in which f_7, f_9 and f_{10} were adjudged important at first iteration. Figure 8 shows that the emdedding required to unfold the data set in DTLZ5(2,30) is either achieved by polynomial kernel of degree three or equivalently by two successive embeddings of degree two or one. The first embedding corresponding to degree two and the second with degree one and vice versa, could also have led to the desired overall embedding. Experiments were also performed with Gaussian kernels with degree three, two, one and half, respectively and in each case accurate results were observed, i.e., in each case of DTLZ5(2,M), last two objectives emerged as critical ones. Also, all objectives were assessed as critical in case of DTLZ2(3) and DTLZ2(5) problems. What differed amongst these experiments with different degrees of kernel, were the principal components which contributed to these objectives as critical, as has been shown above. Failing cases were observed in polynomial kernel type with degree as half. In this case, only DTLZ2(3), DTLZ2(5) and DTLZ5(2,3) could be solved successfully. In rest all

Table 9. Iteration-1 : DTLZ5(2,10) with Poly. kernel: degree:2

PCA-1 (0.930)	f_7 f_9 f_{10}
PCA-2 (0.031)	f_9 f_{10}
Post Eig. Analysis	f_7 f_9 f_{10}
Post Corr. Analysis	f_7 f_9 f_{10}

	f_7	f_9	f_{10}
f_7	+	-	-
f_9	-	+	-
f_{10}	-	-	+

Table 10. Iteration-1 : DTLZ5(2,10) with Poly.kernel: degree:1

PCA-1 (0.449)	$f_3 \cdots f_7$ f_8 f_9 f_{10}
PCA-2 (0.336)	f_9 f_{10}
PCA-3 (0.211)	f_8
Post Eig. Analysis	f_3 f_7 f_8 f_9 f_{10}
Post Corr. Analysis	f_7 f_9 f_{10}

cases, it ran into numerical errors. The reason is that while handling centered data, the dot product of two vectors in consideration may have a value less than one and square root of a negative number, leads nowhere. Hence as cited in [18], a fractional power polynomial does not necessarily define a kernel function as it might not even define a positive-semidefinite Gram matrix.

7.2 MVU-PCA-NSGA-II

The only free parameter involved in the maximum variance unfolding method is *'k'*, which for a particular input point, physically signifies (as evident in Figure 5) the number of neighbours with whom its distances and included angles have to be retained as invariant during the 'unfolding' process. While too high a value of *'k'* would overconstrain and hence delay the 'unfolding' process, too low a value is riskier as it may lead to distortion of local topology and hence may lead to 'erroneous unfolding', in that

Table 11. Implication of *'k'* on 'unfolding'

'k' parameter	D.O.F = No. of (unknowns - constraints)
$k{=}M-1$	$\frac{M(M+1)}{2} - \frac{M(M-1)}{2} - 2 = M - 2$
$k{=}\bar{k}$	Min$\equiv[\eta_{ij} = 1 \implies \eta_{ji} \neq 1]$: $\frac{M(M+1)}{2} - min(\binom{M}{2}, M\bar{k}) - 2$

Table 12. DTLZ5(2,10), *'k'* = 9

Iter. 1: PCA-1 (0.999)	f_1 f_2 f_7 f_8 f_{10}
Iter. 2: 2 Prin. Comp.	f_1 f_2 f_7 f_8

Table 13. DTLZ5(2,20), *'k'* = 19

Iter. 1: 13 Prin. Comp.	f_1 f_5 f_{17}

Table 14. DTLZ5(2,30), *'k'* = 29

Iter. 1: 25 Prin. Comp.	f_6 f_7 f_{18} f_{24} f_{27} f_{29}

Table 15. DTLZ5(2,50), *'k'* = 49

Iter. 1: 44 Prin. Comp.	f_{36} f_{44} f_{46} f_{47} f_{48} f_{49}

in fulfilling the objective of maximizing the sum of all pairwise distances between inputs, even the neighbours may be stretched. Hence, proper selection of *'k'* is crucial. Given an M-objective problem, for which size of *'to be learned'* kernel matrix would be $M \times M$, while the number of independent (accounting symmetry) unknown matrix

Table 16. Final Results with *'k'*=$\lceil\sqrt{M}\rceil$

Problem	*'k'*=$\lceil\sqrt{M}\rceil$	Number of iterations required	Final objective set
DTLZ5(2,5)	*'k'*=3	one	f_4 and f_5
DTLZ5(2,10)	*'k'*=4	two	f_9 and f_{10}
DTLZ5(2,20)	*'k'*=5	two	f_{19} and f_{20}
DTLZ5(2,30)	*'k'*=6	two	f_{29} and f_{30}
DTLZ5(2,50)	*'k'*=8	two	f_{49} and f_{50}
DTLZ2(5)	*'k'*=3	one	f_1 to f_5

elements will be fixed as $\frac{M(M+1)}{2}$, the number of constraints given by Equation 4-1 (those in Equation 4-2, 4-3, together contributing 2 constraints), would vary with *'k'*. Defining, degree of freedom (D.O.F) as number of points in input space (here, objectives) which are free for unfolding (stretching, to enhance the pairwise distances) w.r.t. their neighbours, we discuss the effect of *'k'* on D.O.F in Table 11. Its noteworthy that even in the safest case of preservation of isometry, wherein each point is considered to be a neighbour of all the rest, enough D.O.F is available for attainment of objective (in Equation 4). We experimented with MVU-PCA-NSGA-II, for varying values of *'k'*. While accurate results (last two objectives for DTLZ5(2,M) upto $M = 50$ and all objectives for DTLZ2(M) upto $M = 5$) were obtained for *'k'*=$M - 1$ (case of all neighbours considered as *'k'*-nearest), inaccuracy creeped in the most unconstrained case of *'k'*=1 (case of only the first nearest neighbour considered amongst *'k'*-nearest). In the later case, though enough D.O.F was available for attainment of objective, it came at the cost of distortion of local-isometry, wherein even the actual neighbours get stretched. This is evident in Tables 12, 13, 14, 15, where one of the last two objectives were found to be lost in the first iteration itself, defying the property of the problem, leaving no motivation to continue. While, in literature, for most of the studies *'k'*=4 is taken, to strike an intermediate value which depends on number of input points, we tested for *'k'*=$\lceil\sqrt{M}\rceil$ and it found accurate results for all test problems, as shown in Table 16. We hence prescribe *'k'*=$\lceil\sqrt{M}\rceil$ to be a suitable standard, for problems considered here and even in general.

8 Conclusions

We started by highlighting the limitations of standard PCA and consequently the need for non-linear dimensionality reduction scheme, in general. We have substantiated the above arguments by evaluating a five-objective optimization problem in terms of *'what is'* (accounting for underlying subspaces) and *'what ought to be'* (need to account for underlying submanifolds). Reiterating the basic issues in the concept and utility of kernel functions in non-linear dimensionality reduction, we have investigated and explained the non applicability of kernel PCA. The fact that recently published correntropy PCA, fitted well with our earlier scheme and that it does not suffer like kernel-PCA, from curse of computational overhead, we have adopted it to propose a non-linear dimensionality reduction scheme for multi-objective optimization problems, to be referred as C-PCA-NSGA-II. This however suffered from the difficulty of picking up a suitable kernel function for a given problem at hand. To do away with this problem, we adopted a way to learn the kernel matrix, by way of maximum variance unfolding, to eventually propose MVU-PCA-NSGA-II. The efficacy of the former has been tested for varying kernel functions like Gaussian and polynomial, and with varying degrees. These experiments have shown that such variations did not have a significant effect on their predictive performance for the problems tested here, as most led to accurate results (though with varying efficacy). In case of the later, experiments for varying *'k'*– the only free parameter involved therein, have been utilized to suggest its appropriately safe value. To summarize, this paper has overcome the limitation of our earlier proposed PCA-NSGA-II algorithm under the broader context of non-linear dimensionality reduction.

Acknowledgements

The authors appreciate Arnab Sinha, research scholar at Department of Electrical Engineering, Indian Institute of Technology Kanpur, for fruitful discussions with the first author and for his valuable suggestions. Authors also acknowledge the support provided by STMicroelectronics, Italy, Singapore and India for performing this study.

References

1. Deb, K., Saxena, D.: Searching for pareto-optimal solutions through dimensionality reduction for certain large-dimensional multi-objective optimization problems. In: IEEE Congress on Evolutionary Computation. (2006) 3353–3360
2. Wu, J.X., Pokharel, P., Paiva, A., Principe, J.C.: Nonlinear component analysis based on correntropy. In: IEEE Congress on Evolutionary Computation. (2006) 3517–3521
3. Weinberger, K.Q., Saul, L.K.: Unsupervised learning of image manifolds by semidefinite programming. International Journal of Computer Vision **70** (2006) 77–90
4. Weinberger, K.Q., Saul, L.K.: An introduction to nonlinear dimensionality reduction by maximum variance unfolding. In: In Proceedings of the National Conference on Artificial Intelligence (AAAI), Nectar paper. (2006)
5. Weinberger, K.Q., Sha, F., Saul, L.K.: Learning a kernel matrix for nonlinear dimensionality reduction. In: Proceedings of the Twenty First International Conference on Machine Learning (ICML). (2004) 839–846
6. Deb, K.: Multi-objective optimization using evolutionary algorithms. Wiley, Chichester, UK: (2001)
7. Deb, K., Agrawal, S., Pratap, A., Meyarivan, T.: A fast and elitist multi-objective genetic algorithm: NSGA-II. IEEE Transactions on Evolutionary Computation **6** (2002) 182–197
8. Saul, L.K., Weinberger, K.Q., Ham, J.H., Sha, F., Lee, D.D.: Spectral methods for dimensionality reduction. In: Semisupervised Learning (to appear) (smdr_ssl06.pdf at http://www.cs.ucsd.edu/ saul/abstracts.html). MIT Press, Cambridge, MA (2006)
9. Tenenbaum, J.B., de Silva, V., Langford, J.C.: A global geometric framework for nonlinear dimensionality reduction. Science **290** (2000) 2319–2323
10. Belkin, M., Niyogi, P.: Laplacian eigenmaps for dimensionality reduction and data representation. Neural Computation **15** (2003) 1373–1396

11. Roweis, S., Saul, L.: Nonlinear dimensionality reduction by locally linear embedding. Science **290** (2000) 2323–2326
12. Ham, J.H., Lee, D.D., Mika, S.: A kernel view of the dimensionality reduction of manifolds. Technical Reports TR-110, Max Planck Institute for Biological Cybernatics, Tubingen, Germany (2003)
13. Cox, T.F., Cox, M.A.A.: Multidimensional Scaling. Chapman and Hall, London (1994)
14. Taylor, J.S., Cristianini, N.: Kernel methods for pattern analysis. Cambridge University Press (2004)
15. Scholkopf, B., Smola, A.J.: Learning With Kernels. MIT Press, Cambridge (2002)
16. Scholkopf, B., Smola, A., Muller, K.R.: Nonlinear component analysis as a kernel eigenvalue problem. Neural Computation **10** (1998) 1299–1319
17. Deb, K., Thiele, L., Laumanns, M., Zitzler, E.: Scalable test problems for evolutionary multiobjective optimization. In: Congress on Evolutionary Computation (CEC). Volume 1. (2002) 825–830
18. Liu, C.: Capitalize on dimensionality increasing techniques for improving face recognition grand challenge performance. IEEE Transactions on Pattern Analysis and Machine Intelligence **28** (2006) 725–737

I-MODE: An Interactive Multi-objective Optimization and Decision-Making Using Evolutionary Methods

Kalyanmoy Deb and Shamik Chaudhuri

Kanpur Genetic Algorithms Laboratory (KanGAL)
Indian Institute of Technology, Kanpur
Kanpur, PIN 208016, India
{deb,shamik}@iitk.ac.in

Abstract. With the popularity of efficient multi-objective evolutionary optimization (EMO) techniques and the need for such problem-solving activities in practice, EMO methodologies and EMO research and application have received a great deal of attention in the recent past. The first decade of research in EMO area has been spent on developing efficient algorithms for finding a well-converged and well-distributed set of Pareto-optimal solutions, although EMO researchers were always aware of the importance of procedures which would help choose one particular solution from the Pareto-optimal set for implementation. In this paper, we address this long-standing issue and suggest an interactive EMO procedure by collating most salient research in EMO and putting together a step-by-step EMO and decision-making procedure. The idea is implemented in a GUI-based, user-friendly software which allows a user to supply the problem mathematically or by using user-defined macros and enables the user to evaluate solutions directly or by calling an executable software, such as popularly-used MATLAB software for a local search or ANSYS software for finite element analysis, etc. Starting with standard EMO applications, continuing to finding robust, partial, and user-defined preferred frontiers through standard MCDM procedures, the well-coordinated software allows the user to first have an idea of the complete trade-off frontier, then systematically focus in preferred regions, and finally choose a single solution for implementation.

1 Introduction

In the past decade of research and application activities of evolutionary multi-criterion optimization (EMO), major focus has been made in finding a set of trade-off solutions, representing the entire Pareto-optimal front. Although these efforts were the first steps in evaluating the potential of EMO methodologies as a true multi-objective optimizer, it is now time to address an equally important matter of choosing a single solution from the Pareto-optimal front for implementation. Such a task should involve a decision-making activity in which higher-level information must be provided by the decision-maker. It is obvious to

S. Obayashi et al. (Eds.): EMO 2007, LNCS 4403, pp. 788–802, 2007.

realize that such a decision-making activity is subjective and must depend on the problem being solved. Thus, any effort in this direction must be spent on devising a procedure which will help a decision-maker (DM) to arrive at a solution of his/her choice, rather than one which will recommend a solution automatically. The multi-criterion decision-making (MCDM) approaches address a similar issue and some MCDM ideas can be borrowed to address the decision-making issue in an EMO study. Besides the higher-level decision-making approaches, there are some other more direct decisions which most decision-makers may like to follow. Some such decision-making ideas may include (i) preference of a robust frontier, instead of a Pareto-optimal frontier, (ii) preference of locally-optimal solutions obtained from EMO solutions, instead of simply choosing the EMO solutions, preference of knee solutions and preference of some specific regions detected by various means, instead of the entire trade-off Pareto-optimal frontier.

In this paper, we give shape to an earlier proposal by the authors [6] in combining EMO procedures with a number of direct (less subjective) decision-making tools and a number of higher-level (subjective) decision-making tools with a procedure which can go back and forth between many such tools and an EMO procedure. The main motivation behind such a repetitive procedure is that often the choice of a higher-level decision-making tool or fixation of parameter values associated with such a tool cannot be done a priori. When an idea of the entire trade-off frontier is obtained, a decision-making tool with all its associated parameters can be chosen adequately. The decision-making task is subjective to the DM and the final outcome of such a task will be dependent on the desires of the DM. To make the task of decision-making easier and possible, we also develop a GUI-based software (currently developed for a linux operating system) with visualization tools. Starting with a set of trade-off solutions, the developed I-MODE software will allow a decision-maker to finally choose a single preferred solution by performing a number of decision-making tasks. Currently, the procedure can be used for any number of objectives, but the software is restricted to a maximum of three objectives due to lack of suitable efficient visualization procedures. The working of the procedure is demonstrated on a welded-beam design problem having two objectives. The proposed methodology is one of many possible implementations of hybrid EMO and decision-making tools.

2 Existing Methodologies for Hybrid Multi-objective Optimization and Decision-Making

There exist different interactive multi-objective optimization methods in the literature based on the classical optimization methods. Some popular methodologies, as described in [12] are as follows: Interactive Surrogate Worth Trade-off (ISWT) method [2], Reference point method [15], NIMBUS approach [12] etc. Each method is different from each other, but uses a single solution in each iteration. A guess solution is usually modified to another solution iteratively and by gathering some information from a DM. Since a single solution is used in an iteration, the DM only can find local information (such as a local trade-off or search direction) and cannot make a decision using a more global picture of the true Pareto-optimal front. However, in the context of an EMO, there do not exist many interactive studies. Tan and his students developed a GUI-based MOEA toolbox for multi-objective optimization [14]. The toolbox was designed with some classical decision-making aides, such as goal and priority settings. But a clear procedure of arriving at a single preferred solution was not present in the toolbox. Fonseca and Fleming [11] devised a GUI-based procedure which allowed some target values to be set for each objective and the trade-off objective information of different solutions found using an EMO procedure was demonstrated. However, the procedure lacked any quantitative statistical analysis of the solutions and also clearly did not provide any indication of the location of chosen solutions vis-a-vis the Pareto-optimal front. Another interactive GUI-based EMO software was *Guimoo*, developed by INRIA, but it lacked any decision-making facility.

3 Interactive Multi-objective Optimization and Decision-Making Using Evolutionary Methods (I-MODE)

In the proposed interactive EMO procedure, we attempt to put together some recent salient research results of EMO (described below) along with salient decision-making principles to constitute, a hybrid interactive multi-criterion decision-making procedure. The existing EMO procedures used in I-MODE are as follows:

1. An EMO which is capable of finding the entire or a partial Pareto-optimal set, as desired [3,1].
2. An EMO which capable of finding a preferred region of interest on the Pareto-optimal frontier using the reference point approach [10].
3. An EMO which is capable of finding a *robust* frontier [8], instead of Pareto-optimal frontier.
4. An EMO with a local search procedure which provides a better convergence properties [9,5].
5. An EMO which is capable of handling multiple disconnected objective regions and constitute a parallel search.

I-MODE also uses the following single-objective optimization procedures, mostly for the purpose of verifying the multi-objective trade-off frontier obtained by an EMO:

1. A procedure for finding individual optimal solution(s) corresponding to each objective function subject to satisfaction of all supplied constraints [12].
2. The ϵ-constraint method of finding a single Pareto-optimal solution [12].
3. The multi-objective version of the ϵ-constraint method in which any number of original objectives can be kept as objectives and remaining original objectives can be constrained to some ϵ values. This procedure is expected to find a lower-dimensional Pareto-optimal front which would be a subset of the high-dimensional Pareto-optimal front.

Finally, for the decision-making purpose, we have borrowed a number of MCDM methodologies:

1. Tchebycheff methods with different "norms",
2. Reference point method [15],
3. Utility function method including weighted-sum approach and pseudo-weight method,
4. Surrogate worth trade-off method [12].

Using above procedures, we have designed an interactive procedure which allow a systematic procedure of performing any of the above tasks alone, in combination with each other or in sequence to each other in a manner which provides adequate flexibility to a decision-maker. We present the procedure in the step-by-step format. The parameters which are expected to be supplied by the decision-maker (DM) are mentioned in parenthesis.

Step 1: Obtain an approximate non-dominated front with following options:
- 1.1 Compute the complete front (DM: no parameter)
- 1.2 Compute a partial front (DM: limiting trade-off values)
- 1.3 Compute Pareto-optimal solutions near the reference points only (DM: reference point and limiting spread parameter)
- 1.4 Compute the *robust* Pareto-frontier (DM: robustness parameters)

Outcome: An approximate trade-off frontier

Step 2: Improve the obtained non-dominated front using other optimization methods:
- 2.1 Single-objective local searches from selected solutions:
  - 2.1.1 Automated selection: Clustering (DM: number of desired solutions)
  - 2.1.2 User-defined selection: (i) Weighted-sum approach (DM: weight vectors), (ii) Utility function based approach (DM: utility functions), (iii) Tchebycheff function approach (DM: ideal points and L_p norm), (iv) Using trade-off information between objectives (DM: Trade-off values).
- 2.2 Obtain a better trade-off frontier with specific solutions obtained using ϵ-constraint method (DM: ϵ values)

Outcome: A near-optimal and well-distributed trade-off frontier

Step 3: Verify obtained front with other optimization tasks:
- 3.1 ϵ-constraint method (single or multi-objective) (DM: ϵ-vector)
- 3.2 Optimization of individual objectives

Outcome: A verified (and confident) trade-off frontier

Step 4: Make decisions and choose regions of interest using one or more of the following methods:
- 4.1 Weighted-sum approach (DM: weight vectors)
- 4.2 Utility function based approach (DM: utility functions)
- 4.3 Tchebycheff function approach (DM: ideal points and L_p norm)
- 4.4 Using trade-off information between objectives (DM: Trade-off values)
- 4.5 Checking robustness of solutions (DM: robustness parameters)

Outcome: One or more regions of interest identified

Step 5: Until satisfied, go to Step 1 and focus further study in the above regions of interest, else declare the chosen solution(s)

3.1 Description of the I-MODE Procedure

The main difference between our proposed approach and the existing classical interactive methods described in [12] is that in our approach, we first attempt to find and show the DM the extent and shape of the Pareto-optimal front using a few representative solutions. This procedure, in addition to providing estimated ideal and nadir points of the problem, will also paint a good picture in the mind of the DM about the shape of the Pareto-optimal frontier which will help the DM later to concentrate on a particular region on the front. However, if the DM, for some reason, is interested in focusing on a particular region on the frontier, such information regarding his/her preference can be provided.

Thus, in the very first step of the I-MODE, in most situations, the DM applies an EMO (NSGA-II, an efficient multi-objective optimizer [4], is used here) on the problem to obtain a non-dominated front (Step 1). The EMO algorithm can start with two types of initial population. If some problem information is available then a biased population honoring the problem information can be generated, otherwise a completely random set of solutions can be chosen. Without any preference to any particular region on the trade-off frontier, the DM can find a representative set of solutions on the entire Pareto-optimal frontier. If, however, the DM is interested in a portion of the entire frontier, a number of options are available. A useful procedure would be to suggest a surrogate worth trade-off information (such as a 100% sacrifice in one objective must bring in at least a ζ% gain in another objective and so on) and find a partial frontier using the guided-domination based EMO [1]. Another useful way to find a biased set of trade-off optimal solutions is to use a number of reference points (or aspiration points) and use I-MODE to find optimal solutions close to these reference points [10]. This way, the DM gets to know Pareto-optimal solutions which are near his/her chosen reference points and are not on the entire Pareto-optimal frontier. In practice, solutions are only possible to be implemented with a finite precision. If this uncertainty in decision variables cause the objective and constraint function values to change by a large amount, the solution is declared as a non-robust solution. As another alternative, right in the beginning, the DM can opt for finding *robust* solutions which are less sensitive to parameter perturbations using a robust EMO procedure [8]. To find a robust frontier (which would be, in general, different from the original Pareto-optimal frontier), the user needs to define a range of likely perturbation in decision variables or parameters and an allowable change in functions (called as the robustness parameter) for defining a robust solution. The DM chooses one of the two robust optimization procedures described in [8] and can find the corresponding robust frontier.

Once the preliminary front is established through Step 1, the next step is to improve the obtained frontier by means of other optimization concepts. This step is necessary simply because evolutionary algorithms do not have a mathematical convergence proof for any arbitrary problem and certain portion of the frontier may not converge close to the true optimal frontier. One of the ways to have confidence about the optimality of obtained solutions is to try various optimization concepts and check to see if no further improvement on the obtained solutions are possible by various optimization runs. We attempt to improve the solutions using a local search procedure. For this task, a few solutions are picked from the non-dominated set and an individual local search is initiated from each of these solutions. For each solution, a combined single objective is constructed by computing a normalized pseudo-weight vector based on the location of the solution in the Pareto-optimal frontier [3,7]. After several local searches are performed, the new non-dominated frontier is constructed. The solutions chosen for a local search can be randomly picked from the obtained set in Step 1 or some preferred solutions picked using some decision-making tools. In some occasions, there may exist wide gaps in the frontier obtained after the local searches and new single-objective ϵ-constraint procedures can be initiated with ϵ vectors chosen inside such gaps to find a number of representative solutions there. At the end of Step 2, the DM expects to come up with a well-converged and well-distributed set of trade-off solutions.

The next step (Step 3) is to verify the obtained frontier by a number of single-objective optimizations. The extreme solutions of the non-dominated front can be verified by using a single-objective genetic algorithm on each objective independently. The intermediate trade-off solutions can be verified by using the ϵ-constraint method in which only j ($\in [1, M]$) objectives can be kept as objectives and the remaining $(M - j)$ objectives can be converted into constraints [2]. If $j = 1$ is chosen, a single-objective optimization and if $j > 1$ is chosen, a multi-objective optimization procedure can be used to find one or more optimal solutions. These solutions should theoretically fall on the trade-off frontier obtained by I-MODE.

After obtaining the non-dominated frontier in Step 1, improving it through local searches in Step 2 and verifying the frontier through several other optimizations in Step 3, the DM is confident enough on the near-optimality of the obtained frontier and is ready to perform some decision-making tasks by getting an idea of the range of trade-off objective values. In Step 4, the DM can use a number of decision-making tools to concentrate one or more regions of preference by analyzing different regions of the trade-off frontier. For this purpose, pseudo-weight selection, Tchebycheff metric method with different norms, reference point method, surrogate worth trade-off method, etc. can be used depending on the appropriateness of the procedure to a particular problem. The DM can simply decide to choose the robust solutions from the frontier. In the case of a systematic evaluation of the frontier, multiple regions of interest can be selected simultaneously for further investigation. A completely new multi-objective optimization run with or without considering all the above-discussed steps (robust optimization, local searches, decision-making etc.) can be repeated to find more trade-off solutions in the chosen regions of interest and to help choose one or more subregions for further investigation. This procedure can be continued till the DM is satisfied (in Step 5) with a preferred solution.

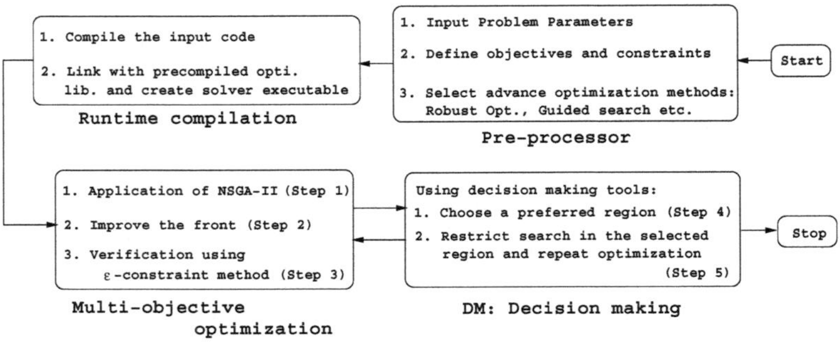

Fig. 1. Structure of I-MODE software

3.2 I-MODE Software Implementation

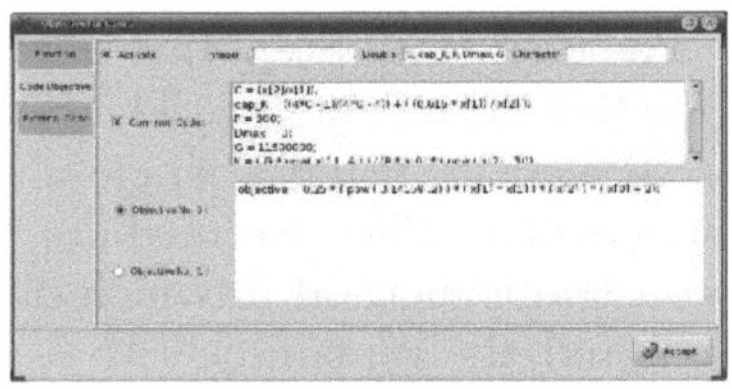

Fig. 2. Window for coding objective functions

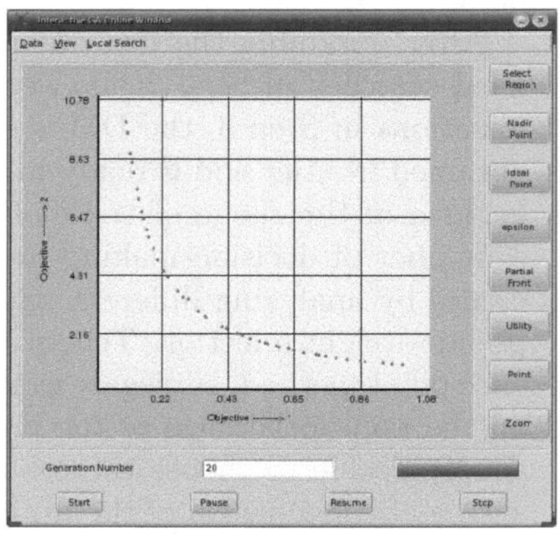

Fig. 3. Plot window for two-objective problem

I-MODE procedure is extensively user-dependent, where the DM has to interact with the software frequently for an effective run. To carry out such a rigorous interactive activity, we need a software with a powerful GUI, through which the DM can specify his/her preferences. At the time of development of the above features, we kept the GUI simple but effective for the decision-maker. The whole software is developed using C-language on the Linux platform and GUI is developed using GTK toolkit. This provides a robust structure of the code which can handle a large optimization problem where the memory requirement may be high. The I-MODE software has three broad modules, namely the pre-processor module, the optimization module, and the decision-making module (Figure 1).

In the pre-processor module, the DM specifies the optimization problem by setting the number of objectives, variables, and constraints. The DM also codes the objective function in a GUI window (Figure 2) once for any subsequent operations or can supply the program through a C-code or can be linked with an external evaluation software, such as a finite element software, MATLAB or others. Different GA parameters are also specified in the preprocessor module. The next module is the optimization module in which the DM can execute various optimization runs (Steps 1, 2 and 3 of the I-MODE procedure). Finally in the decision-making module, the DM uses different decision-making tasks (Step 4) to choose preferred solutions or regions. Figure 3 shows the online visualization window where the DM can observe the real-time animation of the optimization run. On this window, several `Menu buttons` are available, such as `Point Menu`, `Utility Menu`, `ϵ-constraint Menu`, `Ideal and Nadir point Menu`, and `Select region Menu`. The DM can choose one or more such menus and proceed with the software.

4 Case Study: A Welded Beam Design Problem

In this problem, a beam is welded on another beam and carry a certain load. This design problem with one objective is a particularly well-studied [13] one, but here we modify the problem to include a second objective:

$$\begin{array}{ll} \text{Minimize} & f_1(\mathbf{x}) = 1.10471h^2\ell + 0.04811tb(14.0+\ell), \\ \text{Minimize} & f_2(\mathbf{x}) = \delta(\mathbf{x}) = \frac{2.1952}{t^3 b}, \\ \text{Subject to} & g_1(\mathbf{x}) \equiv 13,600 - \tau(\mathbf{x}) \geq 0, \quad g_2(\mathbf{x}) \equiv 30,000 - \sigma(\mathbf{x}) \geq 0, \\ & g_3(\mathbf{x}) \equiv b - h \geq 0, \quad g_4(\mathbf{x}) \equiv P_c(\mathbf{x}) - 6,000 \geq 0, \\ & 0.125 \leq h, b \leq 5.0, \quad 0.1 \leq \ell, t \leq 10.0. \end{array} \tag{1}$$

The first objective is the cost of fabrication and second objective is the end deflection, both of which are to be minimized. Four non-linear constraints are related to limitations on normal stress $\sigma(\mathbf{x})$, shear stress $\tau(\mathbf{x})$, buckling load $P_c(\mathbf{x})$ and a dimensional practicality. There are four design variables: thickness of the beam b, width of the beam t, length of weld ℓ, and weld thickness h, each bounded between lower and upper bounds. The non-linear terms for stress and buckling are given elsewhere [13]. The problem is coded in the pre-processor phase of I-MODE and following systematic procedure is used to obtain a single solution from a two-objective consideration.

4.1 Step 1: Find an Approximate Front

First, we find an idea of the Pareto-optimal front using I-MODE. We set following parameter values: Population size=100; maximum generation=100; crossover probability=0.9; mutation probability=0.1; distribution indices for SBX recombination and polynomial mutation are 10 and 20, respectively. Figure 4 shows the obtained front (solutions marked from A to B) using NSGA-II. To validate the obtained front, we independently find the estimated ideal point by individual minimization of each objective and the estimated nadir point by using the nadir point estimation procedure of I-MODE. Estimated ideal and nadir points are joined to show the range of obtained trade-off optimal solutions. It is clear from the figure that the non-dominated points obtained through NSGA-II do not cover the entire range defined by the nadir and ideal objective vectors. Thus, we can conclude that the NSGA-II procedure is unable to find the entire optimal front and we need to improve this frontier.

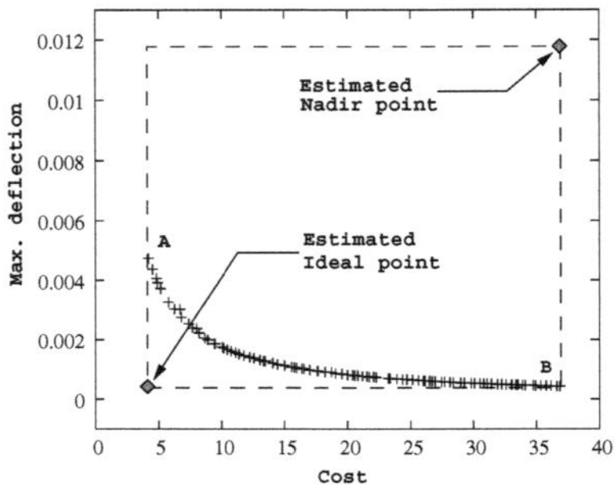

Fig. 4. Pareto-optimal front after original NSGA-II run

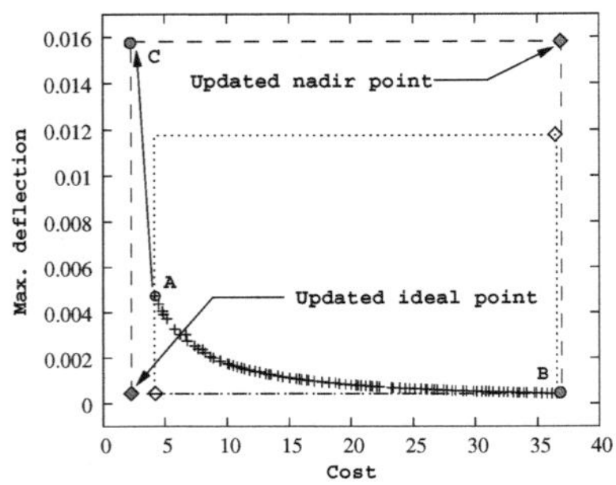

Fig. 5. Updated extreme solutions of the non-dominated front

4.2 Step 2: Improve the Trade-Off Frontier

To obtain a better front, we use the local search procedure on the end points. For this purpose, we use the `fmincon` optimization procedure of MATLAB procedure, which is a classical SQP method. We link MATLAB with the I-MODE software through the local search option: `User defined`. Figure 5 shows the solutions after the local searches. The minimum-cost solution A obtained through NSGA-II gets largely improved to solution C, but the minimum-deflection solution B gets improved slightly. The updated ideal and nadir points are found to be different from those obtained earlier, due to the difficulty in obtaining the minimum-cost solution in this problem. Now from these updated results, we observe that there is a gap between solutions A and C. So we try to find the missing part of the Pareto-optimal front by using the ϵ-constraint method by minimizing f_1 and constraining f_2 to several ϵ_2 values. The obtained solutions are then passed through a non-domination check with the original NSGA-II solutions and the front is modified with new solutions. Figure 6 shows the updated front consisting of previous NSGA-II run and ϵ-constraint single-objective solutions.

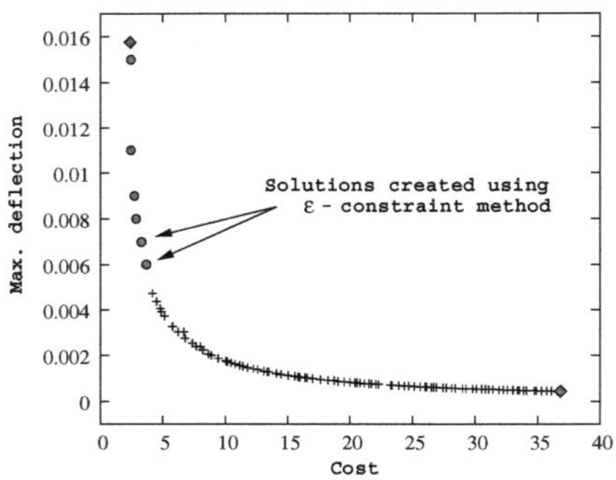

Fig. 6. Updated Pareto-optimal front of welded beam problem

4.3 Step 3: Verify Obtained Front

Since the new ϵ-constraint optimization runs find solutions which are well-matched with the NSGA-II frontier, we skip the verification process at this first iteration of the proposed I-MODE procedure.

4.4 Step 4: Make Decisions and Choose Regions of Interest

The above steps helped us get an idea of the range of Pareto-optimal solutions. The next step is to find one or more regions of interest based on a higher-level consideration. Here, we use two criteria. First we are interested in concentrating in a region which is robust (less sensitive to the variable perturbation). We have already seen in subsection 4.2 that minimum-cost region is sensitive to parameter values and difficult to optimize. To perform the robustness study, we assume that the beam dimensions t and b are expected to vary with $\pm 2\%$ from their chosen values and weld dimensions h and l vary with $\pm 4\%$. These values are kept this way to take into account the fact that parameters t and b are obtained by a machining operation and are expected to have a better control on dimensional tolerance compared to the weld dimensions. We use the robustness of type II [8] procedure and obtain the robust frontier with robustness parameter $\eta = 0.01$, meaning that a maximum of 1% difference in average perturbation in objective values from their original values due to uncertainty is allowed. Figure 7 shows the robust frontier. It is interesting to note that minimum-cost solutions are sensitive. Since the minimum-cost solution corresponds to minimal use of materials, the solution tends to make most constraints active. With an expected fluctuation in design variables, such solutions can easily become infeasible and cannot qualify to be robust. Thus, the robust consideration as a direct decision-making tool enables us to keep away from choosing a solution close to the minimum-cost solution. However, still in this problem we observe a wide variety of solutions which qualify as robust solutions.

To reduce our focus further, we now use a subjective decision-making procedure of surrogate worth trade-off. Of the robust solutions, we are interested in solutions for which a 100% sacrifice in the cost value, at-least 150% improvement in deflection occurs. That is, from a solution if we double the cost value, we are interested in solutions which reduces the deflection 2.5 times. Simultaneously, we would also like to ensure that a saving of at-least 25% cost for a 100% sacrifice in deflection. To find such solutions from the remaining portion of the trade-off

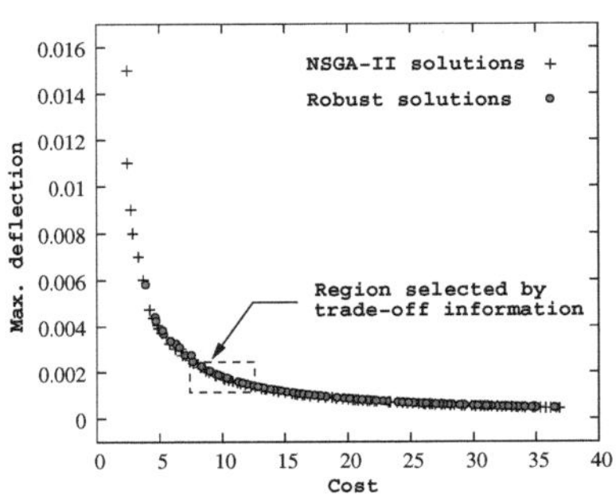

Fig. 7. Robust solutions obtained using robust NSGA-II

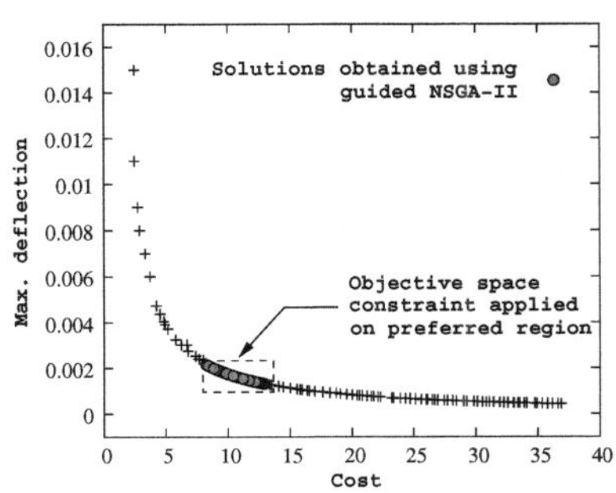

Fig. 8. Solutions obtained using guided NSGA-II

frontier, we specify the following matrix and obtain a partial frontier by the I-EMO software (Figure 8):

$$\texttt{Tradeoff matrix} = \begin{bmatrix} 1.0 & 1.5 \\ 0.25 & 1.0 \end{bmatrix} \quad (2)$$

It is interesting to note that only a small portion in the intermediate portion of the robust frontier becomes the preferred region of solutions corresponding to above trade-off information.

4.5 Step 5: Termination Criterion

This completes one iteration of the I-EMO procedure. Since we have not converged to a single solution yet, we move to Step 1 for another round of I-EMO but concentrate only in the trade-off region obtained at the end of Step 4.

4.6 Step 1: Find More Solutions in Preferred Region by NSGA-II

We run NSGA-II with the guided-domination concept and obtained more solutions in the preferred region of interest. Figure 9 shows 100 solutions obtained with a rerun of guided NSGA-II.

4.7 Step 2: Improve the Front

We ignore this step due to a robustness study planned in subsection 4.9.

4.8 Step 3: Verify Obtained Front

Here, we perform five ϵ-constraint single-objective minimizations of f_2 by constraining f_1 into different cost values in the current range $[8.2, 13.0]$. The obtained solutions are shown in Figure 9 with circles, which suggests that the NSGA-II front and these ϵ-constraint solutions more or less agree, thereby gaining confidence on the obtained NSGA-II solutions.

We perform another verification process here. Since in the previous iteration we expected certain trade-off (given in equation 2), we compute the pseudo-weight vector of five widely-separated solutions (in diamonds). Table 1 shows the objective function values and weight vectors for these selected points. Recall that for an identical proportion of loss in either objective a more stringent gain in deflection objective was set by the matrix. From the above table, we observe that the second objective is given more importance than the first objective in the selected region. These calculations give us confidence in our approach and we now proceed to make further decisions to choose a single preferred solution.

Table 1. Pseudo-weight for selected solutions

Solution	Cost	Max. deflection	Pseudo-weight	
	f_1	f_2	w_1	w_2
A	8.262	0.002153	0.48	0.52
B	9.541	0.001838	0.47	0.53
C	11.087	0.001558	0.45	0.55
D	12.914	0.001321	0.42	0.58

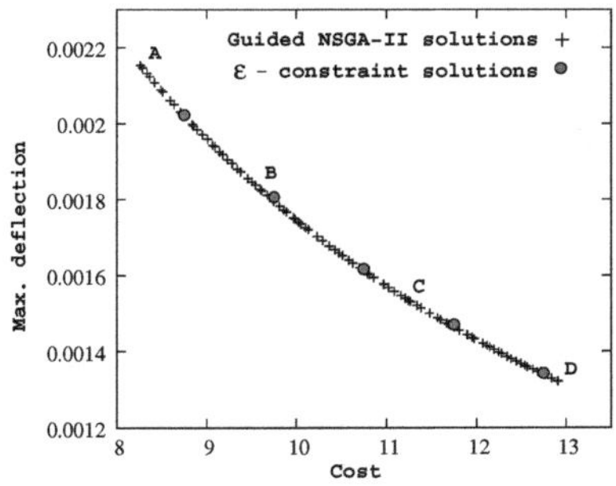

Fig. 9. Guided NSGA-II run is confirmed by ϵ-constraint method

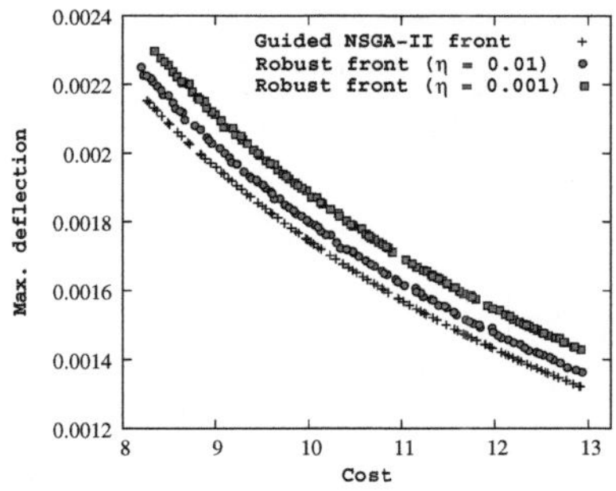

Fig. 10. Robust optimal fronts

4.9 Step 4: Make Decisions and Choose Subregions of Interest

To choose a subregion of interest, we first investigate the robustness of the current trade-off frontier using two different robustness parameter values of 0.01 and 0.001. Figure 10 shows that with a stricter requirement in fluctuation in function values due to perturbations in design variables, the robust frontier gets worse. To investigate how the solutions (design variables) change with a more strict requirement on objective fluctuations, we plot two design variables (t and b) versus the cost objective in Figure 11. Other two variables are found to have a similar 'almost constant' behavior as that of t. The robust solutions are different from the original solutions and it is interesting to note that the variable b of robust and Pareto-optimal solutions follow a relationship with the cost objective: the beam thickness variable must be increased linearly with large-cost solutions. Interestingly, the change in beam thickness does not seem to depend much on the chosen robustness parameter, but demand a significant change in t. These informations are interesting and useful and are found as a by-product of I-MODE

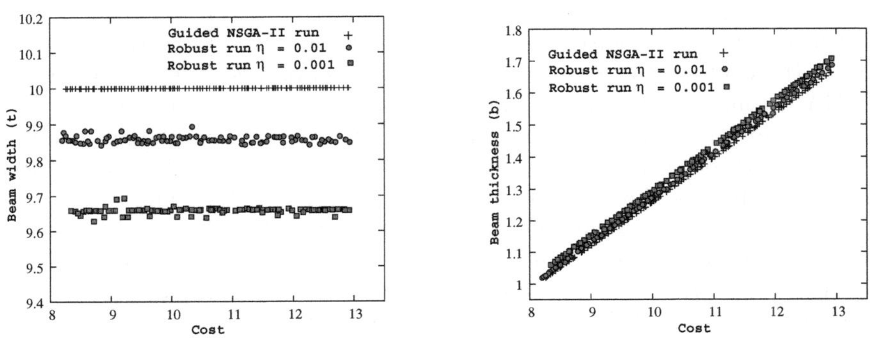

Fig. 11. Variable sensitivity of the welded beam problem

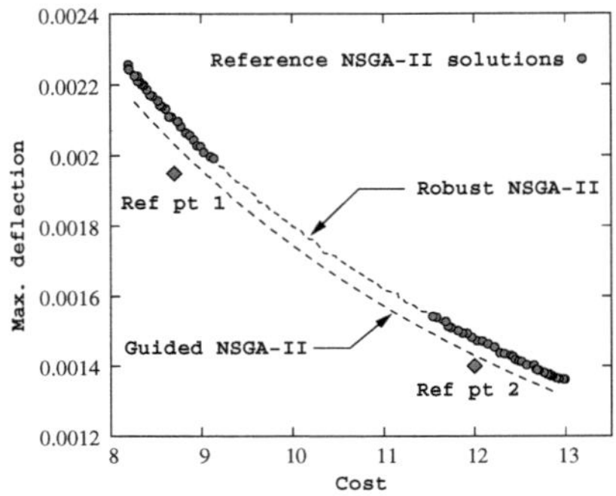

Fig. 12. Robust Pareto-optimal solutions based on the target values

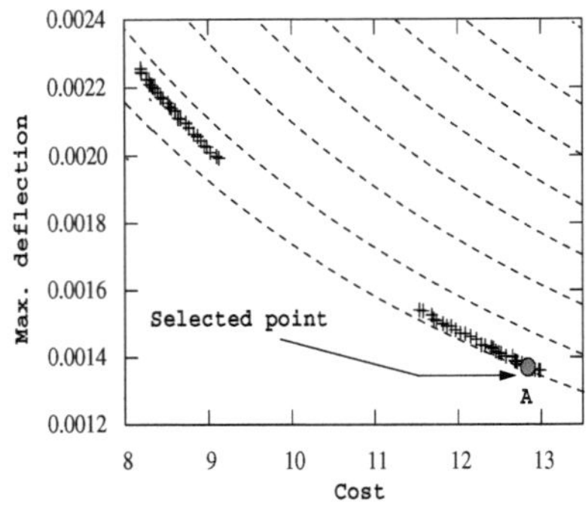

Fig. 13. Final solution selected based on utility function

procedure. Based on these plots, we decide to fix the robustness parameter to $\eta = 0.01$ and proceed with the rest of the study.

To narrow down the preferred region, next we consider a subjective decision-making tool with reference points. Say, we are interested in solutions towards two extreme regions of the remaining trade-off front and specify following two reference (aspiration) points: $(8.7, 0.00195)^T$ and $(12.0, 0.0014)^T$. To get a reasonable spread of solutions, we choose (by trial-and-error here) a spread parameter of $\epsilon = 0.001$. Figure 12 shows the final solutions obtained by the reference NSGA-II run on both reference points simultaneously. Reference points are also shown in the figure.

Finally, we decide to use another subjective decision-making tool based on the utility function approach. We decide to use the following utility function: Minimize $U(f_1, f_2) = f_1 \times f_2$. Since cost and deflection are conflicting to each other, a product of the two objective values in the regions of our interest may be thought as a combined utility measure, minimizing which may result a solution having small values of both objectives. Figure 13 shows the contour plot of the above utility function and reference point based NSGA-II solutions. The utility function is tangential with the reference NSGA-II solutions at point A, thereby meaning that the solution A is the most preferred solution with respect to the chosen utility. The decision variables and objective function values of this solution are shown in Table 2. It is important to note here that the above task of finding the best solution based on a utility function is performed on a chosen *range* of *robust* solutions and on solutions exhibiting certain *trade-off* information and is not to be confused with finding the best solution for a fixed utility function on the entire Pareto-optimal solution.

Table 2. Most preferred solution of the welded-beam problem with supplied decision-making aides

Design Variables (in)				Objective Values	
h	l	t	b	Cost	Deflection
0.917	1.009	9.856	1.672	12.838	0.00137

4.10 Step 5: Select the Most Preferred Solution

Since the outcome is a single solution, we terminate the I-MODE procedure and declare solution A is the preferred outcome of the complete multi-objective optimization and decision-making procedure.

Here, we have followed a sequence of steps with some subjective decision-making tools to come up with a preferred solution, which is *robust*, *near-optimal*, having desired *trade-off* in objectives, close to preferred *aspiration* points, and possessing *optimal desired utility*. It is obvious that the outcome of the study would change if any major change in the sequence of operation of steps is chosen or a different decision-making tool or different parameter values are chosen. There are certainly other procedures possible which will result in a different solution. This is the unique feature of multi-objective optimization. But, what we have demonstrated here is a systematic procedure of using such mixed optimization-cum-decision-making strategies for arriving at a preferred solution.

5 Conclusions

In this paper, we, have proposed an interactive optimization and decision-making procedure for solving two and three-objective optimization problems. In order to arrive at the procedure with a GUI-based software, we have used salient research results from classical and evolutionary multi-objective optimization literatures in a synergistic manner. The procedure not only finds near Pareto-optimal fronts and then helps the DM to choose a particular solution, the procedure provides options for doing *checks and balances* at various stages so that the DM is more confident in arriving at a particular solution. Till now, no such combined (classical and EMO) software is available for this task. With the ground-breaking research and application studies using EMO so far, it is now time for researchers to think and develop such interactive hybrid methodologies which will give multi-objective optimization studies a real practical flavor which they rightfully deserve.

Acknowledgement

Authors acknowledge the financial support provided by STMicroelectronics, Italy, Singapore and India during the course of this study.

References

1. J. Branke, T. Kauβler, and H. Schmeck. Guidance in Evolutionary Multi-objective Optimization. *Advances in Engineering Software*, 32:499–507, 2001.
2. V. Chankong and Y. Y. Haimes. *Multiobjective Decision Making Theory and Methodology*. New York: North-Holland, 1983.
3. K. Deb. *Multi-objective optimization using evolutionary algorithms*. Chichester, UK: Wiley, 2001.

4. K. Deb, S. Agrawal, A. Pratap, and T. Meyarivan. A fast and elitist multi-objective genetic algorithm: NSGA-II. *IEEE Transactions on Evolutionary Computation*, 6(2):182–197, 2002.
5. K. Deb and S. Chaudhuri. Automated discovery of innovative designs of mechanical components using evolutionary multi-objective algorithms. In N. Nedjah and L. de Macedo M, editors, *Evolutionary Machine Design: Methodology and Applications*, chapter 6, pages 143–168. Nova Science Publishers, Inc, New York, 2005.
6. K. Deb and S. Chaudhuri. I-EMO: An interactive evolutionary multi-objective optimization tool. In *Pattern Recognition and Machine Intelligence: First International Conference (PReMI-2005)*, pages 690–695. Springer, 2005.
7. K. Deb and T. Goel. A hybrid multi-objective evolutionary approach to engineering shape design. In *Proceedings of the First International Conference on Evolutionary Multi-Criterion Optimization (EMO-01)*, pages 385–399, 2001.
8. K. Deb and H. Gupta. Searching for robust Pareto-optimal solutions in multi-objective optimization. In *Proceedings of the Third Evolutionary Multi-Criteria Optimization (EMO-05) Conference (Also Lecture Notes on Computer Science 3410)*, pages 150–164, 2005.
9. K. Deb and S. Jain. Running performance metrics for evolutionary multi-objective optimization. In *Proceedings of the Fourth Asia-Pacific Conference on Simulated Evolution and Learning (SEAL-02)*, pages 13–20, 2002.
10. K. Deb, J. Sundar, U. B. Rao N., and S. Chaudhuri. Reference point based multi-objective optimization using evolutionary algorithms. *International Journal of Computational Intelligence Research*, 2(3):273–286, 2006.
11. C. M. Fonseca and P. J. Fleming. Multiobjective optimization and multiple constraint handling with evolutionary algorithms–Part II: Application example. *IEEE Transactions on Systems, Man, and Cybernetics: Part A: Systems and Humans*, 28(1):38–47, 1998.
12. K. Miettinen. *Nonlinear Multiobjective Optimization.* Kluwer, Boston, 1999.
13. G. V. Reklaitis, A. Ravindran, and K. M. Ragsdell. *Engineering Optimization Methods and Applications.* New York : Wiley, 1983.
14. K. C. Tan, T. H. Lee, D. Khoo, and E. F. Khor. A multiobjective evolutionay algorithm toolbox for computer-aided multiobjective optimization. *IEEE Transactions on Systems,Man, and Cybernetics - Part B: Cybernetics*, 31(4):537–556, 2001.
15. A. P. Wierzbicki. The use of reference objectives in multiobjective optimization. In G. Fandel and T. Gal, editors, *Multiple Criteria Decision Making Theory and Applications*, pages 468–486. Berlin: Springer-Verlag, 1980.

Dynamic Multi-objective Optimization and Decision-Making Using Modified NSGA-II: A Case Study on Hydro-thermal Power Scheduling

Kalyanmoy Deb, Udaya Bhaskara Rao N., and S. Karthik

Kanpur Genetic Algorithms Laboratory (KanGAL)
Indian Institute of Technology Kanpur, PIN 208016, India
deb@iitk.ac.in, uday.iitk@gmail.com, ksindhya@iitk.ac.in

Abstract. Most real-world optimization problems involve objectives, constraints, and parameters which constantly change with time. Treating such problems as a stationary optimization problem demand the knowledge of the pattern of change a priori and even then the procedure can be computationally expensive. Although dynamic consideration using evolutionary algorithms has been made for single-objective optimization problems, there has been a lukewarm interest in formulating and solving dynamic multi-objective optimization problems. In this paper, we modify the commonly-used NSGA-II procedure in tracking a new Pareto-optimal front, as soon as there is a change in the problem. Introduction of a few random solutions or a few mutated solutions are investigated in detail. The approaches are tested and compared on a test problem and a real-world optimization of a hydro-thermal power scheduling problem. This systematic study is able to find a minimum frequency of change allowed in a problem for two dynamic EMO procedures to adequately track Pareto-optimal frontiers on-line. Based on these results, this paper also suggests an automatic decision-making procedure for arriving at a dynamic single optimal solution on-line.

1 Introduction

A dynamic optimization problem involves objective functions, constraint functions and problem parameters which can change with time. Such problems often arise in real-world problem solving, particularly in optimal control problems or problems requiring an on-line optimization. There are two computational procedures usually followed. In one approach, optimal control laws or rules are evolved by solving an off-line optimization problem formed by evaluating a solution on a number of real scenarios of the dynamic problem [11]. This approach is useful in problems which are computationally too expensive for any optimization algorithm to be applied on-line. The other approach is a direct optimization procedure on-line. In such a case, the problem is considered stationary for some time

S. Obayashi et al. (Eds.): EMO 2007, LNCS 4403, pp. 803–817, 2007.

period and an optimization algorithm be allowed to find optimal or near-optimal solution(s) within the time span in which the problem remains stationary. Thereafter, a new problem is constructed based on the current problem scenario and a new optimization is performed for the new time period. Although this procedure is approximate due to the static consideration of the problem during the time for optimization, efforts are made to develop efficient optimization algorithms which can track the optimal solution(s) within a small number of iterations so that the required time period for fixing the problem is small and the approximation error is reduced. In this paper, we consider solving dynamic optimization problems having more than one objective function using the direct on-line optimization procedure described above.

Although single-objective dynamic optimization has received some attention in the past [2], the dynamic multi-objective optimization is yet to receive a significant attention. When a multi-objective optimization problem changes with time in stepped manner, the task of an dynamic EMO procedure is to find or track the Pareto-optimal front as and when there is a change. After the idea has been put forward earlier [6], there has been a lukewarm interest on this topic [8,7]. In this paper, we suggest two variations of NSGA-II for tracking dynamic Pareto-optimal frontiers. The effect of frequency of change in a problem and the proportion of added random or mutated solutions are parameters which are systematically studied to evaluate the developed procedures for their tracking efficiency. The proposed NSGA-II procedures are applied to a complex hydro-thermal power scheduling problem involving two conflicting objectives. The change in problem appears due to a change in demand in power with time. The efficacy of modified NSGA-II procedures is illustrated by finding the smallest frequency of change which can be allowed before the EMO procedures can track the optimal front with a significant confidence. Finally, a decision-making aid is coupled with the dynamic NSGA-II procedures to help identify one solution from the obtained front automatically (on-line). Interesting conclusions about the particular problem and about dynamic multi-objective optimization problem, in general, are made from this study.

2 Dynamic Problems as On-Line Optimization Problems

Many search and optimization problems in practice change with time and therefore must be treated as an on-line optimization problems. The change in the problem with time t can be either in its objective functions or in its constraint functions or in its variable boundaries or in any combination of above. Such an optimization problem ideally must be solved at every time instant t or whenever there is a change in any of the above functions with t. In such optimization problems, the time parameter can be mapped with the iteration counter τ of the optimization algorithm. One difficulty which may arise in solving the above on-line optimization task is that the underlying optimization algorithm may not get too many iterations to find the optimal solutions before there is a change in the problem. If the change is too frequent, the best hope of an optimization task is to *track* the optimal solutions as closely as possible within the time span allowed to iterate. However, for steady changes in a problem (which is usually the case in practice), there lies an interesting trade-off which we discuss next. Let us assume that the change in the optimization problem is gradual in t. Let us also assume that each optimization iteration requires a finite time G to execute and that τ_T iterations are needed (or allowed) to track the optimal frontier. Here, we assume that problem does not change (or assumed to be constant) within a time interval t_T, and $G\tau_T < t_T$. Here, initial $G\tau_T$ time is taken up by the optimization algorithm to track the new trade-off frontier and to make a decision for implementing a particular solution from the frontier. Here, we choose $\alpha = G\tau_T/t_T$ to be a small value (say 0.25), such that after the optimal frontier is tracked, $(1-\alpha)t_T$ time is spent on using the outcome for the time period. Figure 1 illustrates this dynamic procedure. Thus, if we allow a large value of t_T (allowing a proportionately large number of optimization iterations τ_T), a large change in the problem is expected, but the change occurs only after a large number of iterations of the optimization algorithm. Thus, despite the large change in the problem, the optimization algorithm may have enough iterations to track the trade-off optimal solutions. On the other hand, if we choose a small τ_T, the change in the problem is frequent (which approximates the real scenario more closely), but a lesser number of iterations are allowed to track new optimal solutions for a problem which has also undergone a small change. Obviously, there lies a lower limit to τ_T below which, albeit a small change in the problem, the number of iterations are not enough for an algorithm to track the new optimal solutions adequately. Such a limiting τ_T will depend on the nature of the dynamic problem and the chosen algorithm, but importantly allows the best scenario (and closest approximation to the original problem) which an algorithm can achieve. Here, we investigate this aspect in the context of dynamic multi-objective optimization problem and find such a limiting τ_T for two variants of NSGA-II algorithm. The procedure adopted in this study is generic and can be applied to other dynamic optimization problems as well, including $t_T = \tau_T = 1$ case.

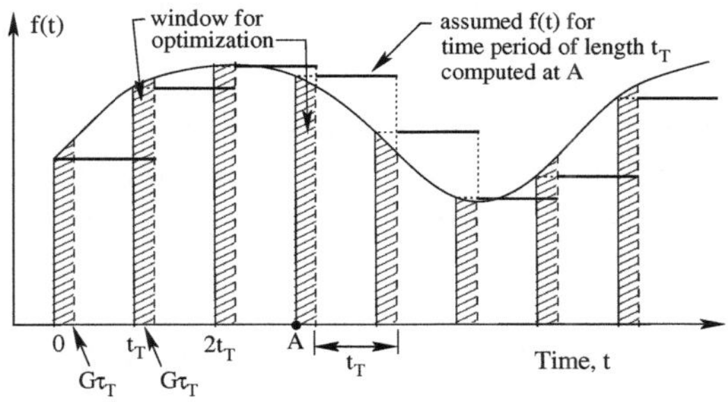

Fig. 1. The on-line optimization procedure adopted in this study. For simplicity, only one objective is shown.

3 Proposed Modifications to NSGA-II

We make some changes to the original NSGA-II procedure to handle dynamic optimization problems. First, we introduce a test to identify whether there is a change in the problem at every generation. For this purpose, we randomly pick a few solutions from the parent population (10% population members used here, but in a deterministic problem one member is enough) and re-evaluate them. If there is a change in any of the objectives and constraint functions, we establish that there is a change in the problem. In the event of a change, all parent solutions are re-evaluated before merging parent and child population into a bigger pool. This process allows both offspring and parent solutions to be evaluated using the changed objectives and constraints.

In the first version (DNSGA-II-A) of the proposed dynamic NSGA-II, we introduce new random solutions whenever there is a change in the problem. A ζ% of the new population is replaced with randomly created solutions. This helps to introduce new (random) solutions whenever there is a change in the problem. This method may perform better in problems undergoing a large change in the objectives and constraints. In the second version (DNSGA-II-B), instead of introducing random solutions, ζ% of the population is replaced with mutated solutions of existing solutions (chosen randomly), similar in principle to hypermutation based GAs for single-objective optimization [3]. This way, the new solutions introduced in the population are related to the existing population. This method may work well in problems undergoing a small change in the problem.

4 Simulation Results on a Test Problem

Farina, Deb and Amato [6] proposed five dynamic test problems. FDA2 is a Type-II unconstrained problem, in which the Pareto-optimal front changes from convex to non-convex shapes in the objective space with time and a part of the decision variables ($\mathbf{x}_{III}$) also changes with time. Here is a modified version of FDA2:

$$\begin{array}{ll}\text{Minimize } f_1(\mathbf{x}_I) = x_1, & \\ \text{Minimize } f_2(\mathbf{x}) = g \times h, & \\ \text{where} & g(\mathbf{x}_{II}) = 1 + \sum_{x_i \in X_{II}} x_i^2, \quad h(\mathbf{x}_{III}, f_1, g) = 1 - \left(\frac{f_1}{g}\right)^{2\left(H(t) + \sum_{x_i \in X_{III}} (x_i - H(t)/4)^2\right)}, \\ & H(t) = 2\sin(0.5\pi(t-1)), \quad t = 2\lfloor \frac{\tau}{\tau_T} \rfloor \frac{\tau_T}{\tau^{\max} - \tau_T}, \\ & \mathbf{x}_I = x_1 \in [0,1], \quad \mathbf{x}_{II}, \mathbf{x}_{II} \in [-1,1]. \end{array} \tag{1}$$

There are five variables in $\mathbf{x}_{II}$ and seven variables in $\mathbf{x}_{III}$, thereby making a total of 13 variables. Here we use a maximum generation of $\tau^{\max} = 200$. We consider that the problem remains fixed for τ_T generations and thereafter the parameter t changes by an amount $2\tau_T/(\tau^{\max} - \tau_T)$ (thereby making $\alpha = 1$). Thus, the above problem simulates the following scenario. The time parameter t changes within $[0,2]$, independent to the value of τ_T. If τ_T is large, the problem changes less frequently but the amount of change is large. Since a large number of iterations are allowed, an optimization procedure may not have difficulties in tracking the new optimal front. On the other hand, if τ_T is small, the problem changes frequently, but the amount of change is small. It would then be interesting to find a critical τ_t below which an algorithm will not perform well due to the availability of too few generations in tracking the new frontier.

First, we study the effect of frequency of change (τ_t = 50, 25, 20, 10, and 5) on problem FDA2. We fix $\zeta = 0.2$. At a particular τ_t value, the performance will degrade so much that the optimization procedure will not be able to track the Pareto-optimal frontier. NSGA-II parameters used in this study are as follows: Population size is 100, SBX crossover probability is 0.9, polynomial mutation probability is $1/n$ (where n is the number of variables), and distribution indices for crossover and mutation are 10 and 20, respectively. To illustrate the deterioration of fronts for two cases of $\tau_T = 20$ and 10, we have plotted all 200/20 or 10 and 200/10 or 20 fronts obtained using DNSGA-II-A (shown with circles) against the true Pareto-optimal fronts (shown in solid lines) in Figures 2 and 3, respectively. It is somewhat clear from these two figures that the fronts close to the middle of the time period (when there is a comparatively larger shift in the front), a change in every 10 generations is not adequate.

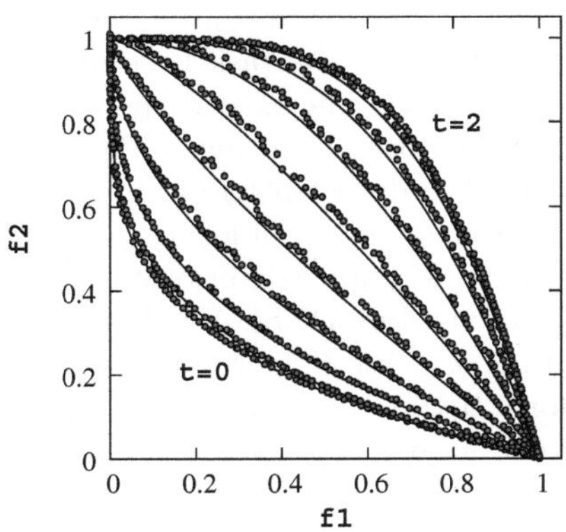

Fig. 2. Obtained fronts against theoretical fronts with $\tau_T = 20$ in FDA2

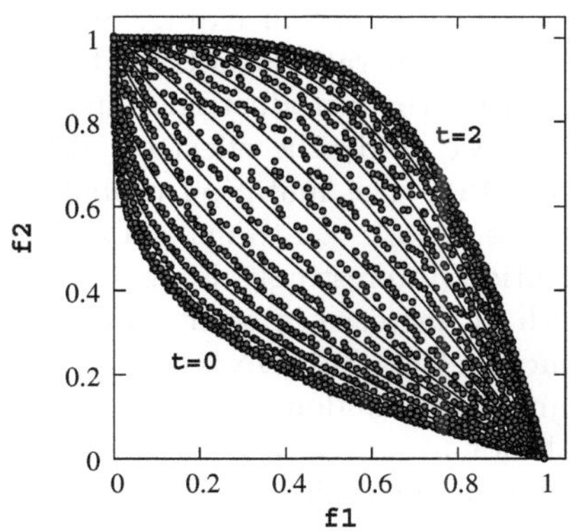

Fig. 3. Obtained fronts against theoretical fronts with $\tau_T = 10$ in FDA2

To perform this study, we consider the performance index to be the ratio of hypervolumes of achieved and true trade-off fronts (obtained mathematically) with respect to fixed reference points. Figure 4 shows the average ratio of hypervolumes of different τ_T values with time (generation). It is observed that with a more frequent change in the problem, the performance deteriorates. If a hypervolume ratio smaller than 94% (say) is considered to be a threshold for indicating a poor performance, then a change more frequent than $\tau_t = 20$ is considered to produce poor performance by the DNSGA-II-A procedure with a 20% change in population by random solutions.

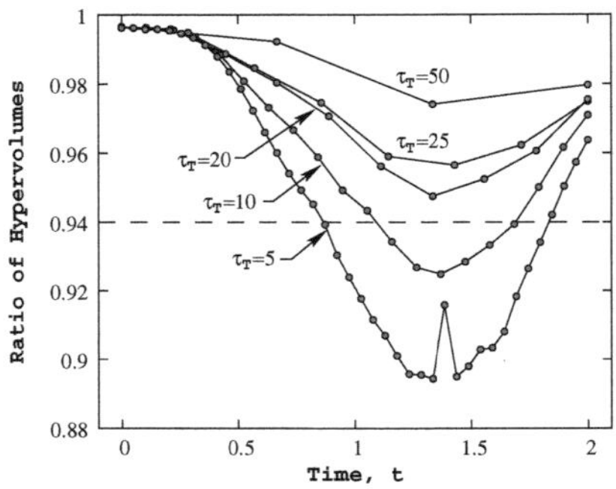

Fig. 4. DNSGA-II-A results on FDA2 ($\zeta = 0.2$)

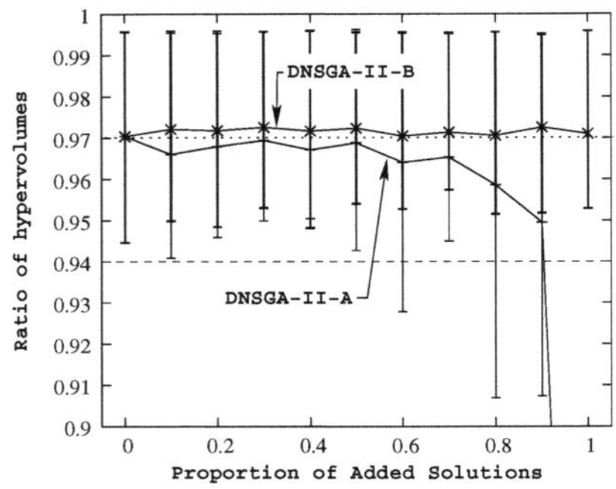

Fig. 5. Effect of varying ζ in DNSGA-II-A and DNSGA-II-B in FDA2 ($\tau_T = 20$)

Next, we perform a parametric study of varying ζ on the FDA2 problem with a variation of the problem after every $\tau_T = 20$ generations. Figure 5 shows the variation of the ratio of obtained hypervolume to the exact hypervolume (and best and worst values through errorbars) with ζ using both DNSGA-II-A and DNSGA-II-B. The figure shows that with an introduction of more random solutions, the performance of DNSGA-II-A (random solution addition) deteriorates. With 20 generations to track the new optimal frontier, the task becomes difficult with the introduction of more random solutions in the existing population. Next, we study the effect of adding mutated solutions by using DNSGA-II-B on FDA2. The mutation probability is doubled and the distribution index is reduced to $\eta_m = 4$ to make a significant change in some variables in an existing solution. Interestingly, Figure 5 shows that the performance deteriorates slightly with an increase in addition of mutated solutions, but DNSGA-II-B performs much better than DNSGA-II-A. The addition of a limited proportion of new mutated or random solutions seems to perform better than not adding any new solution at all. With DNSGA-II-B procedure, almost any proportion of addition of mutated solution produce better performance of the algorithm, whereas with DNSGA-II-A, about 20-40% addition of random solutions is better on multiple runs. With this background, we are now ready to apply dynamic NSGA-II procedures to hydro-thermal power scheduling problems.

5 A Case Study: Hydro-thermal Power Scheduling

In a hydro-thermal power generation systems, both the hydroelectric and thermal generating units are utilized to meet the total power demand. The optimum power scheduling problem involves the allocation of power to all concerned units, so that the total fuel cost of thermal generation and emission properties are minimized, while satisfying all constraints in the hydraulic and power system networks [15]. To solve the hydro-thermal scheduling problem, many different conventional such as Newton's method [17], Lagrange multiplier method [12], dynamic programming [16] and soft computing methodologies such as genetic algorithms [10], evolutionary programming [13], simulated annealing [14] etc. have been tried to solve the single-objective optimization problem. The problem is dynamic due to the changing nature of power demand with time. Thus, ideally the optimal power scheduling problem is truly a on-line dynamic optimization problem in which solutions must be found as and when there is a change in the power demand. In such situations, what can be expected of an optimization algorithm is that it tracks the new optimal solutions as quickly as possible, whenever there is a change.

To understand the insights about the complexity of the problem, at first, we formulate and solve the *stationary* problem using NSGA-II by converting it as an off-line optimization problem. This also facilitates us to compare NSGA-II with a simulated annealing based procedure exist in the literature on the same stationary problem [1]. Gaining the confidence on NSGA-II's ability to solve the constrained problem, we then consider a dynamic version of the problem and solve using the proposed dynamic NSGA-II procedures.

5.1 Optimization Problem Formulation

The original formulation of the problem was given in Basu [1]. The hydro-thermal power generation system is optimized for a total scheduling period of T. However, the system is assumed to remain fixed for a period of t_T so that there are a total of $M = T/t_T$ changes in the problem during the total scheduling period. In this off-line optimization problem, we assume that the demand in all M time intervals are known a priori and an optimization needs to be made to find the overall schedule before starting the operation. In Section 6, we shall consider the problem as a dynamic optimization problem.

Let us also assume that the system consists of N_h number of hydroelectric (P_{ht}) and N_s number of thermal (P_{st}) generating units sharing the total power demand, such that $\mathbf{x} = (P_{ht}, P_{st})$. The bi-objective optimization problem is given as follows:

$$\begin{array}{ll}
\text{Minimize } f_1(\mathbf{x}) = \sum_{t=1}^{M}\sum_{s=1}^{N_s} t_T[a_s + b_s P_{st} + c_s P_{st}^2 + |d_s \sin\{e_s(P_s^{min} - P_{st})\}|], \\
\text{Minimize } f_2(\mathbf{x}) = \sum_{t=1}^{M}\sum_{s=1}^{N_s} t_T[\alpha_s + \beta_s P_{st} + \gamma_s P_{st}^2 + \eta_s \exp(\delta_s P_{st})], \\
\text{subject to } \sum_{s=1}^{N_s} P_{st} + \sum_{h=1}^{N_h} P_{ht} - P_{Dt} - P_{Lt} = 0, \quad t = 1, 2 \ldots, M, \\
\qquad \sum_{t=1}^{M} t_T(a_{0h} + a_{1h}P_{ht} + a_{2h}P_{ht}^2) - W_h = 0, \quad h = 1, 2, \ldots, N_h, \\
\qquad P_s^{min} \le P_{st} \le P_s^{max}, \quad s = 1, 2, \ldots, N_s, t = 1, 2, \ldots, M, \\
\qquad P_h^{min} \le P_{ht} \le P_h^{max}, \quad h = 1, 2, \ldots, N_h, t = 1, 2, \ldots, M.
\end{array} \tag{2}$$

The transmission loss P_{Lt} term at the t-th interval is given as follows:

$$P_{Lt} = \sum_{i=1}^{N_h+N_s} \sum_{j=1}^{N_h+N_s} P_{it} B_{ij} P_{jt}. \tag{3}$$

This constraint involves both thermal and hydroelectric power generation units. Four power demand values of 900, 1,100, 1,000 and 1,300 MW are considered for the four time periods, respectively. All parameters mentioned in the above formulation are presented in the appendix. The water availability constraint (second set of constraints) requires hydroelectric unit values from different time intervals and makes a dynamic optimization task difficult. We shall discuss about this difficulty more in Section 6. In the present context of solving the problem as an off-line optimization problem, such a dependency is not a matter.

Thus, the bi-objective problem involves $(M(N_s + N_h))$ variables, two objectives, $(M + N_h)$ quadratic equality constraints and $(2M(N_s + N_h))$ variable bounds. The specific stationary case considered here involves only four $(M = 4)$ changes in demand over $T = 48$ hours having a time window of statis of $t_T = 12$ hours. The corresponding problem has six (two hydroelectric $(N_h = 2)$ and four thermal $(N_s = 4)$) power units. For the above data, the optimization problem has 24 variables, two objectives, six equality constraints, and 48 variable bounds.

Handling quadratic equality constraints: First, we consider the water availability constraints. Each equality constraint (for a hydroelectric unit h) can be used to replace one of the M power generation values $(P_{h\mu})$ by finding the roots of the quadratic equation and by fixing other P_{ht} as they are in the GA solution:

$$P_{h\mu}^2 + \frac{a_{1h}}{a_{2h}} P_{h\mu} + \frac{1}{t_\mu a_{2h}} \left(-W_h + a_{0h}T + \sum_{\substack{t=1\\ m\neq\mu}}^{M} t_T a_{1h} P_{ht} + \sum_{\substack{t=1\\ m\neq\mu}}^{M} t_T a_{2h} P_{ht}^2 \right) = 0. \quad (4)$$

Since $\frac{a_{1h}}{a_{2h}}$ is always positive, only one root can be positive and we accept this root as $P_{h\mu}$. To maintain the structure of the solution, we maintain the ratio of M different P_{ht} values, as they are in a NSGA-II solution. That is, if the original value of μ-th hydroelectric unit was $P_{h\bar{\mu}}$, other units are replaced as follows: $P_{ht} \leftarrow (P_{h\mu}/P_{h\bar{\mu}})P_{ht}$ for $t = 1, 2, \ldots, M$ and $t \neq \mu$. If the above repair mechanism for all N_h hydroelectric units is not successful, we declare the GA solution as infeasible and no further consideration of power balance constraints nor the computation of objective functions are performed. Recall that NSGA-II employs a constraint handling which does not require objective values for infeasible solutions, thereby suiting the above procedure.

We follow a similar procedure as above for handling the power balance constraint and repair a particular thermal unit $P_{\psi t}$ of four thermal units for each time slot. The quadratic equation for this variable can be written as follows:

$$B_{\psi\psi} P_{\psi m}^2 + (2\sum_{j=1}^{n-1} B_{\psi j} P_{jt} - 1) P_{\psi m} + (P_{Dt} + \sum_{i=1}^{n-1}\sum_{j=1}^{n-1} P_{it} B_{ij} P_{jt} - \sum_{i=1}^{n-1} B_{\psi i} P_{it}) = 0, \quad (5)$$

where $n = N_h + N_s$. Since the hydroelectric power units (P_{ht}) are available, the above equation can be solved for $P_{\psi t}$. If this particular value comes within the variable bounds, then the variable is accepted and we go for next constraint involving P_{st} of the next time period. Otherwise, another root-finding equation is tried for the next thermal unit. If for a time period, none of the N_s thermal units resulted in a successful replacement, a penalty is computed and the solution is declared infeasible.

5.2 Simulation Results on the Stationary Problem

NSGA-II is combined with the above-discussed constraint handling method for solving the hydro-thermal scheduling problem. Here, we only consider four changes in the problem in the entire period of 48 hours. Thus, the off-line optimization problem has two objectives, 4(2 + 4) or 24 variables, and six constraints. NSGA-II parameters used in this study are as follows: Population size = 240, Number of generations = 2,000, Crossover probability = 0.9, Mutation probability = 0.04, Distribution indices for crossover and mutation = 10 and 20, respectively. To validate the obtained NSGA-II front, we employ a

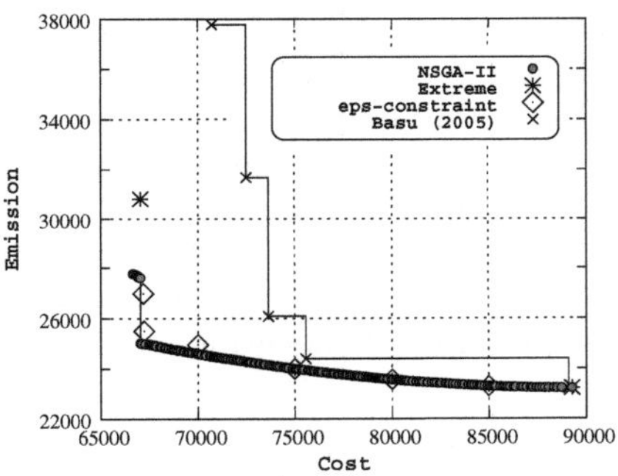

Fig. 6. Pareto-optimal front obtained by NSGA-II, verified by single-objective methods, and by a previous study

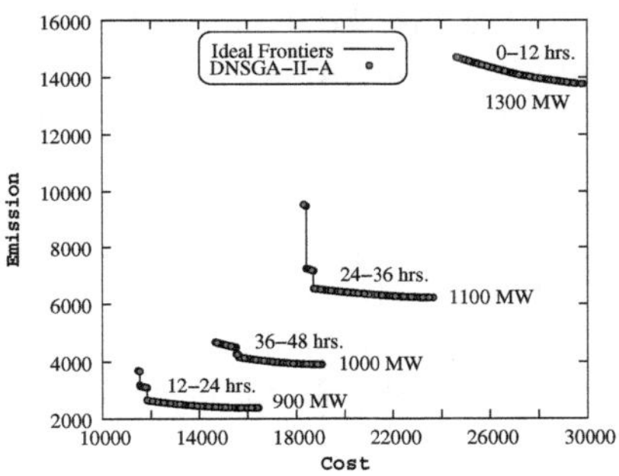

Fig. 7. Four fronts, each change in demand, obtained using DNSGA-II-A with $\zeta = 0.2$

single-objective GA and solve several ϵ-constraint problems [9] by fixing f_1 value at different levels. These points are shown in Figure 6 and it is observed that all these points more or less match with those obtained by NSGA-II. Each objective is also optimized independently by a GA and two solutions obtained are plotted in the same figure. One of the extreme points (minimum f_1) is dominated by a NSGA-II solution and the minimum emission solution is matched by a NSGA-II solution. These multiple optimization procedures give us confidence about the optimality of the obtained NSGA-II frontier.

Basu [1] used a simulated annealing (SA) procedure to solve the same problem. That study used a naive penalty function approach in which if any SA solution if found infeasible, it is simply penalized. For different weight vectors scalarizing both objectives, the study presented a set of optimized solutions. A comparison of these results with our NSGA-II approach (in Figure 6) reveals that the front obtained by NSGA-II *dominates* that obtained in the previous study. One of reasons for a better performance of our approach is the use of a better constraint handling strategy. These results give us confidence in our approach of handling constraints and using NSGA-II for the bi-objective hydro-thermal power dispatch problem. Now, we apply the two proposed dynamic NSGA-II methodologies to the dynamic version of the problem.

6 Dynamic Hydro-thermal Power Scheduling Problem

The dynamic version of the problem involves more frequent changes in the demand P_{Dt}. To make the demand varying in a continuous manner, we make a piece-wise linear interpolation of power demand values with the following (t, P_{dm}) values: (0, 1,300), (12, 900), (24, 1,100), (36, 1,000) and (48, 1,300) in (Hrs, MW). We keep the overall time window of $T = 48$ hours, but increase the frequency of changes (that is, increase M from four to 192, so that the time window t_T for each demand level varies from 12 hours to 48/192 hours or 15 minutes. It will then be an interesting task to find the smallest time window of statis which a specific multi-objective optimization algorithm can solve successfully. We run the dynamic NSGA-II procedures for $960/M$ (M is the number of changes in the problem) generations for each change in the problem.

Equation 2 requires hydroelectric power generation units from different time intervals to be used together to satisfy the equation. In an dynamic optimization problem, this is a difficulty, as this means that an information about all hydroelectric units are needed right in the first generation. This constraint equates the total required water head to be identical to the available value for each hydroelectric system. In this study, we use a simple principle of allocating an identical water head W_h/M for each time interval.

6.1 Simulation Results

We apply the two dynamic NSGA-II procedures (DNSGA-II-A and DNSGA-II-B) discussed above to solve the dynamic optimization problem. The parameters used are the same as in the off-line optimization case presented before. To compare the dynamic NSGA-II procedures, we first treat each problem as a static optimization problem and apply the original NSGA-II procedure [4] for a large number (500) of generations so that no further improvement is likely. We call these fronts as ideal fronts and compute the hypervolume measure using a reference point which is the nadir point of the ideal front. Thereafter, we apply each dynamic NSGA-II and find an optimized non-dominated front. Then for each front, we compute the hypervolume using the same reference point and then compute the ratio of this hypervolume value with that of the ideal front. This way, the maximum value of the ratio of hypervolume for an algorithm is one and as the ratio becomes smaller than one, the performance of the algorithm gets poorer. First, we consider the problem in which we consider a change after every 12 hours ($M = 4$). Figure 7 shows the four Pareto-optimal fronts obtained using DNSGA-II-A with 20% addition of random solutions every time there is a change in the problem. The DNSGA-II-A procedure is able to find a set of solutions very close to the ideal frontiers in all four time periods. The figure makes one aspect clear. As the demand is more, the power production demands larger cost and emission values.

Increasing number of changes in the problem: Figures 8 to 11 show the hypervolume ratio for different number of changes ($\tau_T = 4$ to 192) in the problem

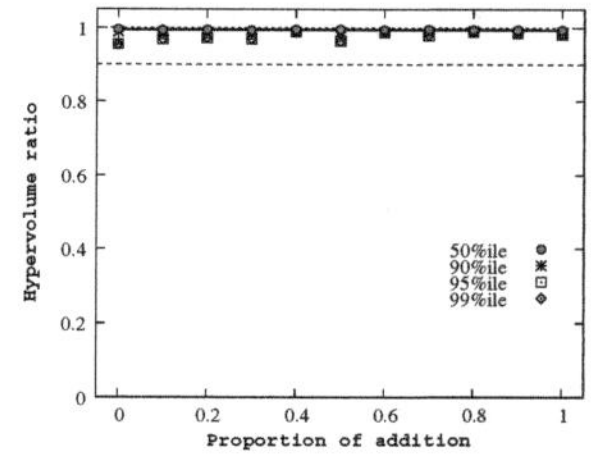

Fig. 8. 3-hourly ($M = 16$) change with DNSGA-II-A

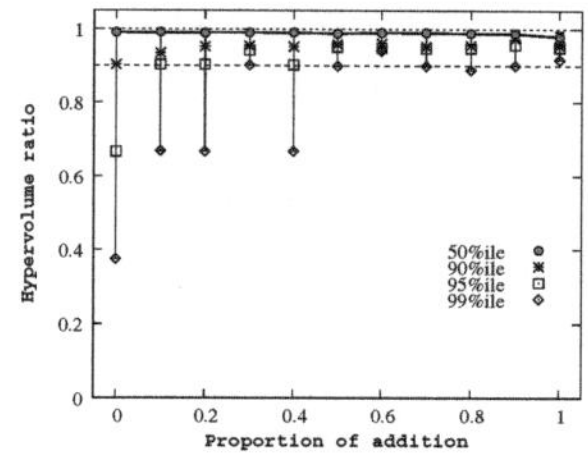

Fig. 9. 1-hourly ($M = 48$) change with DNSGA-II-A

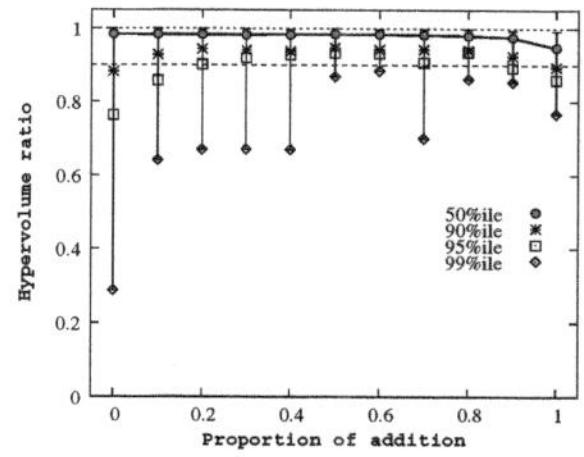

Fig. 10. 30-min. ($M = 96$) change with DNSGA-II-A

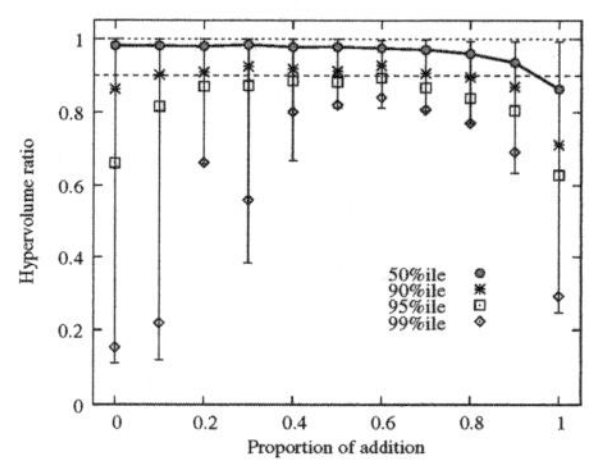

Fig. 11. 15-min.($M = 192$) change with DNSGA-II-A

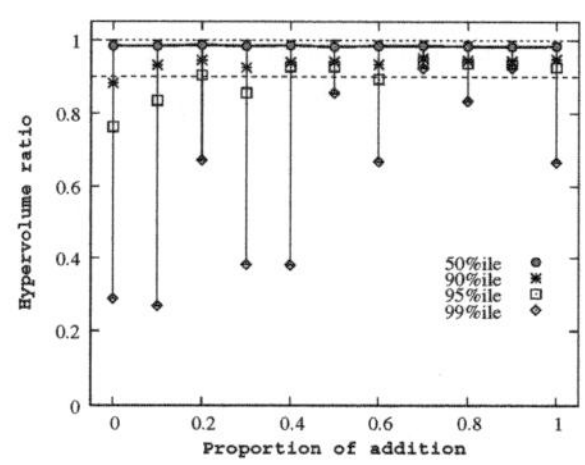

Fig. 12. 30-min. ($M = 96$) change with DNSGA-II-B

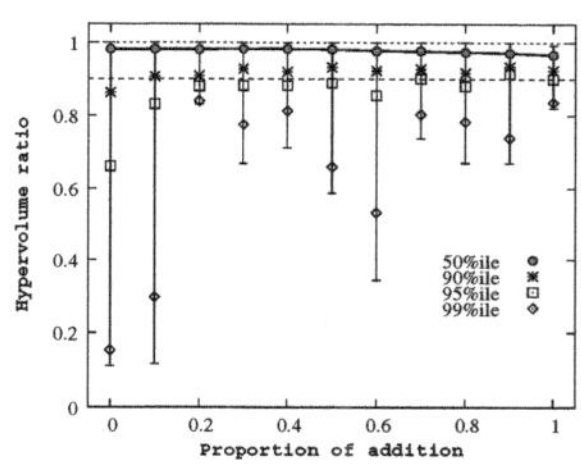

Fig. 13. 15-min. ($M = 192$) change with DNSGA-II-B

with different proportion of addition of random solutions, ζ, using DNSGA-II-A. The figures also mark the 50th, 90th, 95th and 99th percentile of hypervolume ratio, meaning the cut-off hypervolume ratio which is obtained by the best 50, 90, 95, and 99 percent of M frontiers in a problem with M changes. Figures reveal that as M increases, the performance of the algorithm gets poorer due to the fact that a smaller number of generations ($960/M$) was allowed to meet the time constraint. If a 90% hypervolume ratio is assumed to be the minimum required hypervolume ratio for a reasonable performance of an algorithm and if we consider 95 percentile performance is adequate, the figures show that we can allow a maximum of 96 changes (with a 30-min. change) in the problem. For this case, about 20 to 70% random solutions can be added whenever there is a change in the problem to start the next optimization. Too low addition does not introduce much diversity to start the new problem and too large addition of random solutions destroys the population structure which would have helped for the new problem. The wide range of addition for a successful run suggests the robustness of the DNSGA-II procedure for this problem. Next, we consider DNSGA-II-B procedure in which mutated solutions are added instead of random solutions. Mutations are performed with double the mutation probability and with a $\eta_m = 2$. Figures 12 to 13 show the performance plots for two M values. Here, the effect is somewhat different. In general, with an increase in addition of mutated solutions, the performance is better, as mutations perturb existing solutions locally, thereby helping to introduce adequate diversity needed for the next problem. Once again, 96 changes in the problem in 48 hours seem to be the largest number of changes allowed for the algorithm to perform reasonably well. However, addition of mutated solutions over $\zeta = 40\%$ of the population seems to perform well. Once again, DNSGA-II-B procedure is also found to work well with a wide variety of ζ values.

7 Decision Making in Dynamic EMO

One of the issues which is not discussed enough in the EMO literature is the decision-making aspect after a set of trade-off solutions are found. Some studies in this direction for stationary problems have been just begun [5] and more such studies are called for. In dynamic multi-objective optimization problem, there is an additional problem with the decision-making task. A solution is to be chosen and implemented as quickly as the trade-off frontier is found, and in most situations before the next change in the problem has taken place. This definitely calls for an automatic procedure for decision-making with some pre-specified utility function or some other procedure. In this paper, we choose a utility measure which is related to the relative importance given to both cost and emission objectives. First, we consider a case in which equal importance to both cost and emission are given. As soon as a frontier is found for the forthcoming time period, we compute the pseudo-weight w_1 (for cost objective) for every solution $\mathbf{x}$ using the following term:

$$w_1(\mathbf{x}) = \frac{(f_1^{\max} - f_1(\mathbf{x}))/(f_1^{\max} - f_1^{\min})}{(f_1^{\max} - f_1(\mathbf{x}))/(f_1^{\max} - f_1^{\min}) + (f_2^{\max} - f_2(\mathbf{x}))/(f_2^{\max} - f_2^{\min})}. \quad (6)$$

Thereafter, we choose the solution with $w_1(\mathbf{x})$ closest to 0.5. A little thought will reveal that this task is different from performing a weighted-sum approach with equal weights for each objective. The task here is to choose the middle point in the trade-off frontier providing a solution equi-distant from individual optimal solutions (irrespective of whether the frontier is convex or non-convex). Since the Pareto-optimal frontier is not known a priori, getting the frontier first and then choosing the desired solution is the only viable approach for achieving the task.

To demonstrate the utility of this dynamic decision-making procedure, we consider the hydro-thermal problem with 48 time periods (meaning an hourly change in the problem). Figure 14 shows the obtained frontiers in solid lines and the corresponding preferred (operating) solution with a circle. It can be observed that due to the preferred importance of 50-50% to cost and emission, the solution comes nearly in the middle of each frontier. To meet the water availability constraint, the hydroelectric units of $T_{h1} = 219.76$ MW and $T_{h2} = 398.11$ MW are computed and kept constant over time. However, four thermal power units must produce power to meet the remaining demand and these values for all 48 time periods are shown in Figure 15. The changing pattern in overall computation of thermal power varies similar to that in the remaining demand in power.

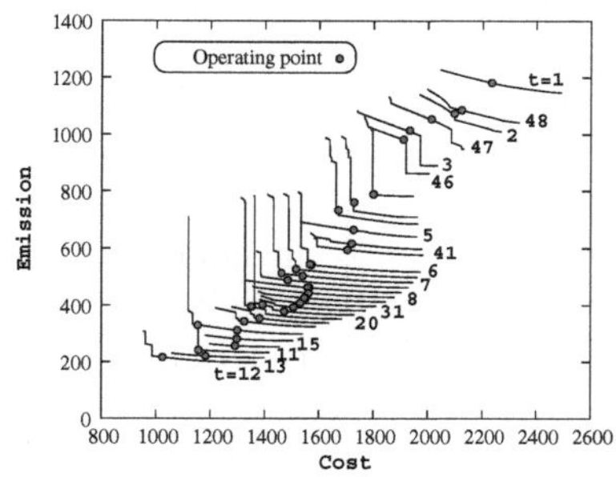

Fig. 14. Operating solution for 50-50% cost-emission case

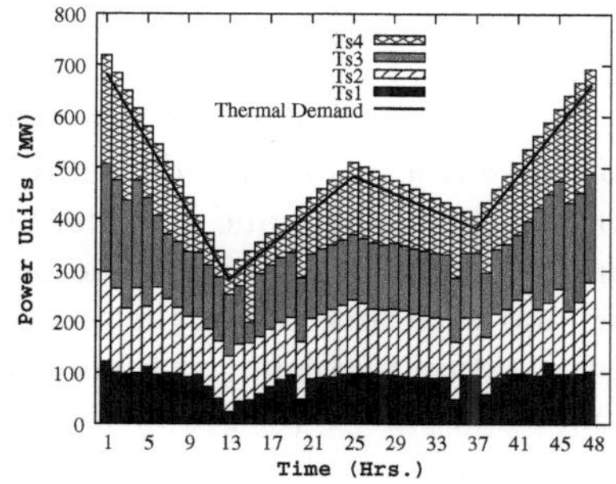

Fig. 15. Variation of thermal power production for 50-50% cost-emission case

The figure also shows a slight over-generation of power to meet the loss term P_{Lt} given in equation 3.

Next, we compare the above operating schedule of power generation with two other extreme cases: (i) 100-0% importance to cost and emission and (ii) 0-100% importance to cost and emission. Figure 16 shows the variation of cost for all the three cases. First, the optimal cost values fluctuate the way the power demand varies. Second, the case with 100% importance to cost requires minimum cost, but causes large emission values and the case with 100% importance to emission causes minimum emission values, but with large costs. A comparison of overall cost and emission values for the entire 48-hour operation for these three cases is summarized in the above inset table which demonstrates this fact.

Case	Cost	Emission
50-50%	74239.07	25314.44
100-0%	69354.73	27689.08
0-100%	87196.50	23916.09

8 Conclusions

In this paper, we have suggested and demonstrated the solution of a dynamic multi-objective optimization task in a systematic manner. Although the procedure can be used on-line, the current implementation assumes that the problem remains unchanged for a time period (statis) and the optimization algorithm is run for an initial fraction of the statis and the outcome is used for the remaining period. To restart the EMO procedure (NSGA-II has been used here) for the changed problem, two different strategies are suggested: Introduction

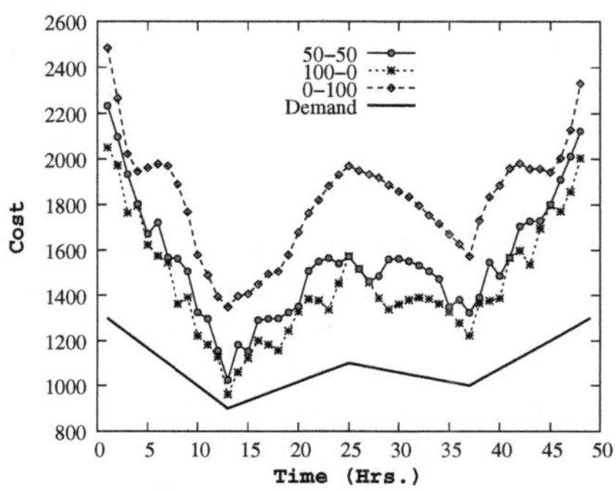

Fig. 16. Variation of operating cost with time for $M = 48$ (1-hourly change)

of random solutions (DNSGA-II-A) and introduction of mutated solutions (DNSGA-II-B). The number of added solutions relative to the population size (ζ) are kept as a parameter for the study. The procedure is tested on a two-objective test problem and to a hydro-thermal power dispatch problem involving both hydro-electric and thermal power generation units with coupled and non-linear equality constraints. The problem is dynamic due to the change of power demand with time. First, the problem has been solved considering it as an off-line optimization problem (with known power demand) and a better Pareto-optimal front than that reported in an earlier study has been found here. Thereafter, the dynamic problem is solved and the effect of discretization (length of statis) on the performance of both dynamic NSGA-II procedures has been elaborated. NSGA-II with addition of random solutions works the best with about 20-70% addition of new solutions, whereas NSGA-II with addition of mutated solutions works the best for 40-100% addition of new solutions. For the 48-hour overall time range of operation, this systematic study has found that allowing at least an every 30-minute change in the problem is better solved by both proposed dynamic NSGA-II procedures. We are currently investigating a true on-line optimization procedure in which the problem is assumed to remain unchanged only during one generation of the dynamic NSGA-II procedure. A mixed addition of random and mutated solutions can also be tried. Nevertheless, this study proposes and demonstrates the working of two viable dynamic EMO procedures for on-line optimization problems and further studies are imminent to test and fine-tune the procedures for their practical use.

Acknowledgements

Authors acknowledge the financial support from STMicroelectronics Pvt. Ltd., Italy, Singapore and India during the course of this study. Authors would also like to thank Ms. Barnali Kar for introducing them to the hydro-thermal power dispatch optimization problem.

References

1. M. Basu. A simulated annealing-based goal-attainment method for economic emission load disptch of fixed head hydrothermal power systems. *Electric Power and Energy Systems*, 27(2):147–153, 2005.
2. Jurgen Branke. *Evolutionary Optimization in Dynamic Environments.* Kluwer Academic Publishers, Dordrecht, 2002.
3. H. G. Cobb. An investigation into the use of hypermutation as an adaptive operator in gas having continuous, time-dependent nonstationary environments. Technical Report AIC-90-001, Naval Research Laboratory, Wash, DC: USA, 1990.
4. K. Deb, S. Agrawal, A. Pratap, and T. Meyarivan. A fast and elitist multi-objective genetic algorithm: NSGA-II. *IEEE Transactions on Evolutionary Computation*, 6(2):182–197, 2002.
5. K. Deb and S. Chaudhuri. Towards estimating nadir objective vector using evolutionary approaches. In *Proceedings of the Genetic and Evolutionary Computation Conference (GECCO-2006)*, pages 643–650. New York: The Association of Computing Machinery (ACM), 2006.

6. M. Farina, K. Deb, and P. Amato. Dynamic multiobjective optimization problems: Test cases, approximations, and applications. *IEEE Transactions on Evolutionary Computation*, 8(5):425–442, 2000.
7. I. Hatzakis and D. Wallace. Dynamic multi-objective optimization with evolutionary algorithms: A forward-looking approach. In *Proceedings of the 8th Annual Conference on Genetic and Evolutionary Computation*, pages 1201–1208, 2006.
8. Y. Jin and B. Sendoff. Constructing dynamic optimization test problems using the multiobjective optimization concept. In *Lecture Notes in Computer Science, 3005*, pages 525–536, 2004.
9. K. Miettinen. *Nonlinear Multiobjective Optimization*. Kluwer, Boston, 1999.
10. S. O. Orero and M.R. Irving. A genetic algorithm modeling framework and solution technique for short term optimal hydrothermal scheduling. *IEEE Transactions on Power Systems*, 13(2), 1998.
11. D. Pratihar, K. Deb, and A. Ghosh. Fuzzy-genetic algorithms and time-optimal obstacle-free path generation for mobile robots. *Engineering Optimization*, 32:117–142, 1999.
12. A. H. A. Rashid and K. M. Nor. An efficient method for optimal scheduling of fixed head hydro and thermal plants. *IEEE Transactions PWRS*, 6(2), 1991.
13. N. Sinha, R. Chakraborty, and P.K. Chattopadhyay. Fast evolutionary programming techniques for short-term hydrothermal scheduling. *Electric Power Systems Research*, 66:97–103, 2003.
14. K. P. Wong and Y. W. Wong. Short-term hydrothermal scheduling part i: simulated annealing approach. *IEE Proc-C Gener., Trans., Distrib*, 141(5), 1994.
15. A. J. Wood and B. F. Woolenberg. *Power Generation, Operation and Control*. John-Wiley & Sons, 1986.
16. J. S. Yang and N. Chen. Short-term hydrothermal co-ordination using multipass dynamic programming. *IEEE Transactions PWRS*, 4(3), 1989.
17. M. F. Zaghlool and F. C. Trutt. Efficient methods for optimal scheduling of fixed head hydrothermal power systems. *IEEE Transactions PWRS*, 3(1), 1988.

A Parameters for Hydro-thermal Problem

The following parameters are taken from a previous study [1].

Hydroelectric system data						
Unit	a_{0h}	a_{1h}	a_{2h}	W_h	P_h^{min}	P_h^{max}
1	260	8.5	0.00986	125000	0	250
2	250	9.8	0.01140	286000	0	500

,

Cost related thermal system data							
Unit	a_s	b_s	c_s	d_s	e_s	P_s^{min}	P_s^{max}
3	60.0	1.8	0.0030	140	0.040	20	125
4	100.0	2.1	0.0012	160	0.038	30	175
5	120.0	2.1	0.0010	180	0.037	40	250
6	40.0	1.8	0.0015	200	0.035	50	300

,

Emission related thermal system data					
Unit	α_s	β_s	γ_s	η_s	δ_s
3	50	-0.555	0.0150	0.5773	0.02446
4	60	-1.355	0.0105	0.4968	0.02270
5	45	-0.600	0.0080	0.4860	0.01948
6	30	-0.555	0.0120	0.5035	0.02075

$$,B = \begin{bmatrix} 49 & 14 & 15 & 15 & 20 & 17 \\ 14 & 45 & 16 & 20 & 18 & 15 \\ 15 & 16 & 39 & 10 & 12 & 12 \\ 15 & 20 & 10 & 40 & 14 & 10 \\ 20 & 18 & 12 & 14 & 35 & 11 \\ 17 & 15 & 12 & 10 & 11 & 36 \end{bmatrix} \times 10^{-6}.$$

Acceleration of Experiment-Based Evolutionary Multi-objective Optimization Using Fitness Estimation

Hirotaka Kaji[1] and Hajime Kita[2]

[1] Research and Development Operations, Yamaha Motor Co. Ltd.,
2500 Shingai, Iwata, Shizuoka, Japan
kajih@yamaha-motor.co.jp
[2] Academic Center for Computing and Media Studies, Kyoto Univercity,
Yoshida nihonmatsu-cho, Sakyo-ku, Kyoto, Japan
kita@media.kyoto-u.ac.jp

Abstract. Evolutionary Multi-objective Optimization (EMO) is expected to be a powerful optimization framework for real world problems such as engineering design. Recent progress in automatic control and instrumentation provides a smart environment called Hardware In the Loop Simulation (HILS). It is available for our target application, that is, the experiment-based optimization. However, since Multi-objective Evolutionary Algorithms (MOEAs) require a large number of evaluations, it is difficult to apply it to real world problems of costly evaluation. To make experiment-based EMO using the HILS environment feasible, the most important pre-requisite is to reduce the number of necessary fitness evaluations. In the experiment-based EMO, the performance analysis of the evaluation reduction under the uncertainty such as observation noise is highly important, although the previous works assume noise-free environments. In this paper, we propose an evaluation reduction to overcome the above-mentioned problem by selecting the solution candidates by means of the estimated fitness before applying them to the real experiment in MOEAs. We call this technique Pre-selection. For the estimation of fitness, we adopt locally weighted regression. The effectiveness of the proposed method is examined by numerical experiments.

1 Introduction

In recent automotive engine development, a number of parameters of Engine Control Units (ECUs) mounted to engines have to be adjusted adequately to achieve higher engine performance. Because plural criteria such as environmental emissions (CO, HC, NO_x), fuel-consumption and engine torque need to be balanced at higher level, this operation called calibration becomes time-consuming and complex process year after year.

To attain such demanding goals, automatic design based on multi-objective optimization is needed as the new methodology that takes the place of conventional operator's manual calibration. Recent progress in automatic control and instrumentation provides a smart environment called Hardware In the Loop

S. Obayashi et al. (Eds.): EMO 2007, LNCS 4403, pp. 818–831, 2007.

Simulation (HILS) for the calibration of control parameters of engines. The HILS environment is composed of a real engine and an engine test bed which can simulate vehicle running conditions using an ultra-low inertia dynamometer controlled by a computer. Applying Evolutionary Multi-objective Optimization (EMO) to the HILS environment is a promising field of application.

However, since the evaluation of the real engine experiment is costly, the requirement for many evaluation Multi-objective Evolutionary Algorithms (MOEAs) causes a serious problem, that is, tremendous optimization time. Parallelization is one of the solutions for real world problems which require enormous evaluation costs, but the parallelization of the HILS environment is not realistic choice in view of the installation cost and space. As another approach for the time consuming problems in the airfoil design using Computational Fluid Dynamics (CFD) [12] for instance, evaluation reduction methods for EAs have been studied actively[1]. In the past decade, main current of the evaluation reduction is to use statistical approximation model which is constructed by search histories of individuals evaluated in the real environment in the past . Generally speaking, because an evaluation cost of the approximation model is smaller enough than the real environment, it is possible to decrease the total evaluation time substantially.

To make experiment-based EMO using the HILS environment feasible, reduction of fitness evaluation is the most important requirement. However, the researches of the evaluation reduction method for MOEAs are still few compared with that for single objective EAs. Additionally, in our target application, that is, the experiment-based EMO for the real engine, the performance analysis of the evaluation reduction under uncertainty environments is highly important, although previous works [6,10,14,15] assume noise-free environment.

In this paper, we discuss a Pre-selection as an evaluation reduction method for the experiment-based EMO. To apply to the uncertainty environment such as HILS, we propose a general Pre-selection algorithm which can employ approximation modeling techniques having a robustness for observation noise. The algorithm is constructed in the manner that the offspring generated in the area having sparse distribution of non-dominated individuals in the archived population is preferentially selected. This paper is organized as follows. In Section 2, we introduce Locally Weighted Regression (LWR) [1] as an approximation modeling technique suitable for noise environment. In Section 3, we propose a general Pre-selection algorithm which does not depend on features of modeling techniques for MOEAs. Moreover, we examine effectiveness of the proposed method under noise-free and observation noise environments through numerical experiments in Section 4. Section 5 is the conclusion of this paper.

2 Background of Locally Weighted Regression

In conventional evaluation reduction for EAs, many researchers have employed many kinds of approximation modeling technique for fitness function. For

[1] See the detailed survey by Jin et al. [11].

example, polynomials, artificial neural networks, radial basis functions, and Kriging (DACE) [17] are adopted. Among the previous works, we paid special attention to the researches by Branke et al. [2,3]. They have proposed a Pre-selection that used Locally Weighted Regression (LWR) to estimate fitness values of candidates as an evaluation reduction for single objective EAs, and have shown its effectiveness by detailed examinations [2]. They also have demonstrated that the fitness estimation by the LWR was effective for the problem which has the uncertainty by observation noise [3]. Therefore, because our goal is to develop the evaluation reduction for the experiment-based EMO which includes uncertainty such as observation noise, we employ the LWR as an approximation modeling technique.

An introduction of the LWR is described below. Consider a set of search history $H = \{(\boldsymbol{h}^1, \boldsymbol{f}(\boldsymbol{h}^1)), (\boldsymbol{h}^2, \boldsymbol{f}(\boldsymbol{h}^2)), \ldots, (\boldsymbol{h}^l, \boldsymbol{f}(\boldsymbol{h}^l))\}$ which stores information of the search process of MOEAs, where $\boldsymbol{h} = [h_1 \ \cdots \ h_n]^{\mathrm{T}}$ is n-dimensional decision variable vector which was evaluated as an individual in a real environment in past. $\boldsymbol{f}(\boldsymbol{h}) = [f_1(\boldsymbol{h}) \ \cdots \ f_m(\boldsymbol{h})]^{\mathrm{T}}$ is m-dimensional objective value vector, l is the number of individuals stored in H.

The LWR is a method of constructing approximation model from data set H. The weighted polynomial regression is applied for a neighborhood of an individual (query) which should obtain the estimated value of its fitness vector. In this paper, second-order model without interaction terms is employed as a local model. When a neighborhood set Ω of an individual $\boldsymbol{x} = [x_1 \ \cdots \ x_n]^{\mathrm{T}}$ is generated from H by k-Nearest Neighbors (k-NN) method based on the Euclidean distance $d_{\mathrm{E}}(\boldsymbol{h}, \boldsymbol{x}) = \sqrt{(\boldsymbol{h} - \boldsymbol{x})^{\mathrm{T}}(\boldsymbol{h} - \boldsymbol{x})}$, its estimated value $\hat{\boldsymbol{f}}(\boldsymbol{x}) = \left[\hat{f}_1(\boldsymbol{x}) \ \cdots \hat{f}_m(\boldsymbol{x})\right]^{\mathrm{T}}$ is calculated by the following equations:

$$\hat{f}_i(\boldsymbol{x}) = \begin{bmatrix}1 & x_1 & \cdots & x_n & x_1^2 & \cdots & x_n^2\end{bmatrix} \boldsymbol{b}_i \tag{1}$$

$$\boldsymbol{b}_i = \left(\boldsymbol{X}^{\mathrm{T}} \boldsymbol{W} \boldsymbol{X}\right)^{-1} \boldsymbol{X}^{\mathrm{T}} \boldsymbol{W} \boldsymbol{y}_i \tag{2}$$

$$\boldsymbol{X} = \begin{bmatrix} 1 & h_{11} & \cdots & h_{1n} & h_{11}^2 & \cdots & h_{1n}^2 \\ \vdots & \vdots & \vdots & \vdots & \vdots & \vdots & \vdots \\ 1 & h_{k1} & \cdots & h_{kn} & h_{k1}^2 & \cdots & h_{kn}^2 \end{bmatrix} \tag{3}$$

$$\boldsymbol{y}_i = \left[f_i(\boldsymbol{h}^1) \ f_i(\boldsymbol{h}^2) \ \cdots \ f_i(\boldsymbol{h}^k)\right]^{\mathrm{T}}, \tag{4}$$

where h_{ij} is jth element of $\boldsymbol{h}^i$ which is an individual near to ith from $\boldsymbol{x}$ in Ω, and $\boldsymbol{W}$ is called weighted matrix which is a diagonal matrix with diagonal elements

$$w_i = \sqrt{K(d_{\mathrm{E}}(\boldsymbol{h}_i, \boldsymbol{x}))}, \ i = 1, 2, \ldots, k. \tag{5}$$

$K(\cdot)$ is called weighted function or kernel function and is used to calculate weight of search histories. In this paper, Gaussian kernel

$$K(d) = \exp\left(-\frac{d^2}{u}\right) \tag{6}$$

is used, where u is smoothing parameter and is set to be the distance to the kth nearest search history in this paper.

In MOEAs, we can assume the distribution of the population gradually converges to the Pareto optimal set through the search. Hence, a lot of individuals near the Pareto optimal set will be preserved in the search history set H. If convergence is taken place, strong correlation between decision variables may appear when the effective dimension of Pareto optimal set degenerates in the decision variable space[2]. Then, multicollinearity should be considered. That is estimation accuracy of regression coefficients vector $\boldsymbol{b}$ deteriorates, because the matrix $(\boldsymbol{X}^{\mathrm{T}}\boldsymbol{W}\boldsymbol{X})$ becomes singular and calculation of the inverse matrix $(\boldsymbol{X}^{\mathrm{T}}\boldsymbol{W}\boldsymbol{X})^{-1}$ is unstable numerically.

To avoid the influence of multicollinearity, we employed the ridge regression[8]. Instead of Eq. (2), The ridge regression uses the following equation:

$$\boldsymbol{b}_i^{\mathrm{R}} = \left(\boldsymbol{X}^{\mathrm{T}}\boldsymbol{W}\boldsymbol{X} + \lambda\boldsymbol{I}\right)^{-1}\boldsymbol{X}^{\mathrm{T}}\boldsymbol{W}\boldsymbol{y}_i, \tag{7}$$

and can avoid that the matrix $\boldsymbol{X}^{\mathrm{T}}\boldsymbol{W}\boldsymbol{X}$ to become singular by adding constant λ to the diagonal elements.

3 Pre-selection for MOEAs

Since EAs are stochastic optimization method, offspring may be generated at a position far from an area containing the optimal solution. It is undesirable to evaluate such non-promising offspring on the real environment of costly evaluation. If the offspring can be evaluated by the approximation model of the fitness, non-promising offspring can be excluded beforehand to make optimization efficient. Such a technique called Pre-selection or Pre-screening winnows promising offspring based on the estimated fitness value obtained by the approximation model. The Pre-selection has the feature which does not lose the advantage of direct search and updates the approximation model every generation.

As a Pre-selection for MOEAs, Emmerich et al. have proposed a method which used Kriging [17] for fitness estimation [6]. They paid attention to the fact that the Kriging can predict the mean square error of the estimated fitness value. To select individuals that have high possibility of improvement, they used values considering the mean square error for the ranking. However, because the Kriging makes such approximation model as the one passing all the search histories, it is not suitable for the experiment-based EMO involving the uncertainty such as the observation noise. Hence, for employing the LWR mentioned above, we have to develop more general Pre-selection algorithm which does not depend on features of modeling techniques.

[2] For instance, SCH [5], typical test function, has the Pareto optimal set which is straight line shape in the n-dimensional space.

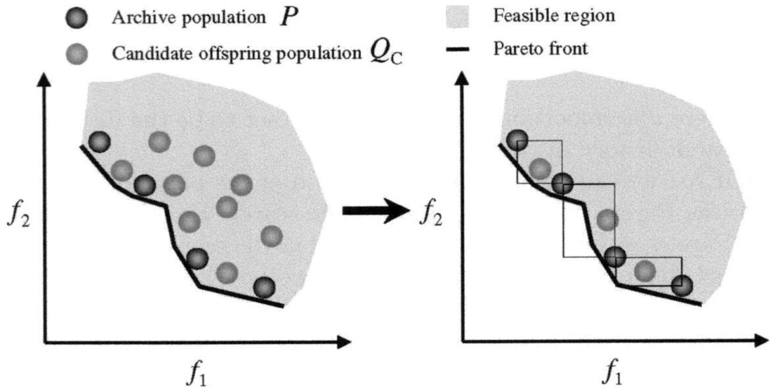

Fig. 1. Concept diagram of the Pre-selection algorithm for MOEAs. Left figure shows the set $R_C = P \cup Q_C$, and right figure depicts crowding distance calculation of candidate offspring which became non-dominated individuals.

An important point of Pre-selection for MOEAs is how to select the offspring which should be evaluated in the real environment when lots of promising candidate offspring of equal rank exist. Then, instead of the mean square error prediction of the Kriging, we use a sparsity criterion of each promising candidate offspring in the archived population for useful and effective search. The algorithm is constructed in the manner that the offspring generated in the area having sparse distribution of non-dominated individuals in the archived population is preferentially selected. In this paper, The crowding distance proposed by Deb et al. [4] is used as a sparsity criterion. Fig. 1 shows the concept chart of the proposed method. A detailed algorithm will be described later.

In experiment-based optimization, due to observation noise or estimation error, there is a possibility that non-Pareto optimal individual may survive as a apparently non-dominated one. Fig. 2 shows a situation in which a weak Pareto individual is treated as a non-dominated individual. Assume that the estimation error is caused to the true fitness $\boldsymbol{f}(\boldsymbol{x})$ of the weak Pareto individual $\boldsymbol{x}$ within the range shown in gray circle. If observation noise or estimation error make the value of objective function f_1 apparently good, $\boldsymbol{x}$ is treated as the non-dominated individual and is stored in the archived population as an elite. Additionally, since large crowding distance is allocated easily for $\boldsymbol{x}$ when a problem has wide weak Pareto frontier, there will be high probability for selecting such individual.

To solve this problem, we use the α-domination strategy proposed by Ikeda et al. [9] for the ranking applied to the estimated fitness values. The α-domination strategy is defined as an expansion of superiority comparing as follows:

Definition 1. (**α-domination**) Consider a m-objective minimization problem

$$\min_{\boldsymbol{x}}(f_1(\boldsymbol{x}), \ldots, f_m(\boldsymbol{x})) \quad \boldsymbol{x} \in X \subset \mathbf{R}^n.$$

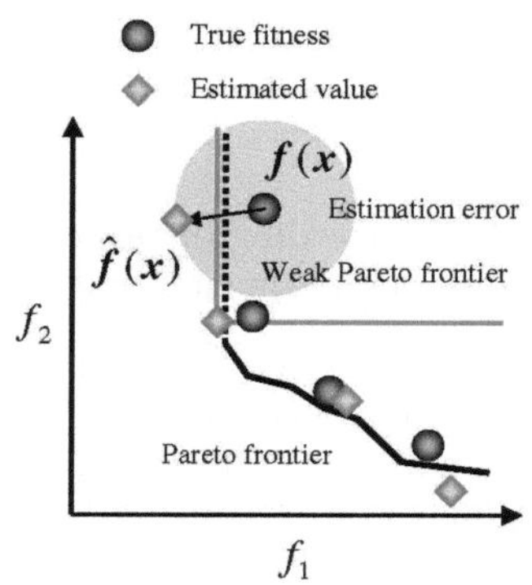

Fig. 2. Concept diagram of the fake non-dominated individual

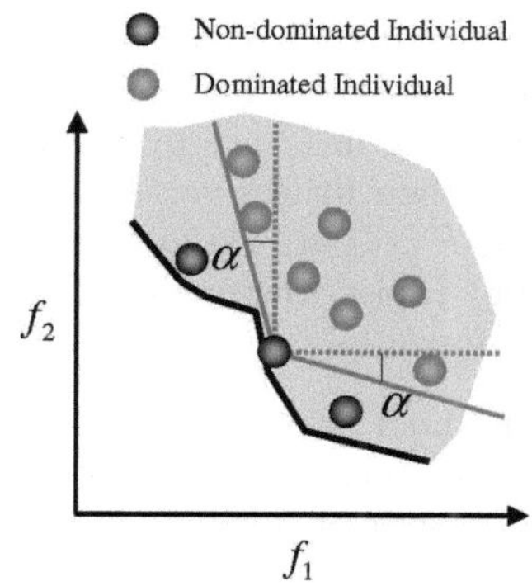

Fig. 3. Concept diagram of the α-domination strategy

A solution $\boldsymbol{x}$ α-dominates a solution $\boldsymbol{y} \in X$

$$\Leftrightarrow \forall\, i\ g_i(\boldsymbol{x}, \boldsymbol{y}) \leq 0 \ \wedge\ \exists\, i\ g_i(\boldsymbol{x}, \boldsymbol{y}) < 0,$$

where

$$g_i(\boldsymbol{x}, \boldsymbol{y}) = f_i(\boldsymbol{x}) - f_i(\boldsymbol{y}) + \sum_{j \neq i}^{n} \alpha_{ij}(f_j(\boldsymbol{x}) - f_j(\boldsymbol{y})).$$

The concept diagram of the α-domination strategy is shown in Fig. 3. When an objective function f_i is compared by the α-domination strategy, another objective function f_j is considered in a ratio of α_{ij}. For example, an individual that need to greatly corrupt f_j to improve f_i a little is dominated easily. As a result, fake non-dominated individuals are excluded efficiently.

Based on the aforesaid discussion, we propose a general Pre-selection algorithm which can be combined with any approximation modeling technique for MOEAs. The prototype algorithm of the proposed method is shown below:

1. All fitness vector $\boldsymbol{f}(\boldsymbol{x})$ of initial population $P(0)$ are evaluated in a real environment, and individuals and their fitness are preserved in search history set H.
2. Candidate offspring population Q_{C} is generated from the archived population P by applying selection, crossover, and mutation operators.
3. Estimated values $\hat{\boldsymbol{f}}(\boldsymbol{x})$ of $R_{\mathrm{C}} = P \cup Q_{\mathrm{C}}$ are calculated using the approximation model constructed by use of H.
4. The ranking of R_{C} is done by using the α-domination strategy [9] based on $\hat{\boldsymbol{f}}(\boldsymbol{x})$.
5. A candidate offspring which becomes a non-dominated individual is added to P, and its crowding distance is calculated. This operation is adopted for all the non-dominated candidate offspring.

6. The evaluated offspring population Q is selected from the non-dominated candidate offspring assigned with good crowding distance.
7. Offspring in Q are evaluated in the real environment, and they are stored with their fitness vector in search history.
8. $R = P \cup Q$ are assigned with the rank and the crowding distance based on $\boldsymbol{f}(\boldsymbol{x})$ (noise-free environment) or $\hat{\boldsymbol{f}}(\boldsymbol{x})$ (noise environment), and return to 2. after applying the generation alternation.

If the number of individuals which was stored in the search history set H is not enough to construct the approximation model, individuals are evaluated in the real environment until it reaches a necessary number.

4 Numerical Experiments

4.1 Problems, Parameter Settings and Measures

From the test problem proposed by Deb et al. [4,5], we employ two objective optimization problems SCH, FON, and ZDT1 and three objective optimization problem DTLZ2 for numerical experiment.

- SCH ($n = 10$)

$$f_1(\boldsymbol{x}) = \frac{1}{n}\sum_{i=1}^{n} x_i^2$$
$$f_2(\boldsymbol{x}) = \frac{1}{n}\sum_{i=1}^{n} (x_1 - 2)^2$$
$$x_i \in [-4, 4], \quad i = 1, 2, \ldots, n$$

- FON ($n = 10$)

$$f_1(\boldsymbol{x}) = 1 - \exp\left(-\sum_{i-1}^{n}\left(x_i - \frac{1}{\sqrt{n}}\right)^2\right)$$
$$f_2(\boldsymbol{x}) = 1 - \exp\left(-\sum_{i-1}^{n}\left(x_i + \frac{1}{\sqrt{n}}\right)^2\right)$$
$$x_i \in [-2, 2], \quad i = 1, 2, \ldots, n$$

- ZDT1 ($n = 10$)

$$f_1(\boldsymbol{x}) = x_1$$
$$f_2(\boldsymbol{x}) = g(\boldsymbol{x})(1 - \sqrt{f_1/g(\boldsymbol{x})})$$
$$g(\boldsymbol{x}) = 1 + 9 \cdot \sum_{i=2}^{n} \frac{x_i}{n-1}$$
$$x_i \in [0, 1], \quad i = 1, 2, \ldots, n$$

– DTLZ2 ($n = 10$)

$$\begin{aligned} f_1(\boldsymbol{x}) &= (1 + g(\boldsymbol{x}))\cos(x_1\pi/2)\cos(x_2\pi/2) \\ f_2(\boldsymbol{x}) &= (1 + g(\boldsymbol{x}))\cos(x_1\pi/2)\sin(x_2\pi/2) \\ f_3(\boldsymbol{x}) &= (1 + g(\boldsymbol{x}))\sin(x_1\pi/2) \\ g(\boldsymbol{x}) &= \sum_{x=3}^{n}(x_i - 0.5)^2 \\ x_i &\in [0, 1], \quad i = 1, 2, \ldots, n \end{aligned}$$

In this paper, The NSGA-II [4] is employed as an MOEA. The archived population size $|P| = 30$, the candidate offspring population size $|Q_{\mathrm{C}}| = 100$ and evaluated offspring population size $|Q| = 4$ were used for the Pre-selection NSGA-II. For crossover, the Unimodal Normal Distribution Crossover (UNDX) proposed by Ono et. al. was used [16]. Since it is known that the UNDX shows good performance without mutation, we did not use mutation operation. The neighborhood size k of k-NN method for the LWR was 5% of the number of individuals that stored in the search history set H. However, minimum value is $k = 30$. The parameter of α-domination strategy employed $\alpha = 0.05$, and the ridge parameter was $\lambda = 0.0001$. The normal NSGA-II ($|P| = 30$, $|Q| = 30$) was used for comparison. In all these experiments, the number of evaluation was 2030 ($|Q| = 4$) and 2040 times ($|Q| = 30$). Because the times of function evaluation is the most restrictive factor for experiment-based optimization, comparison was carried out with the same number of function evaluation, although computation time is another candidate condition. Thirty trials with different initial populations were conducted.

Performances of two methods were compared by the mean values of the trials for the two evaluation measures described below:

Coverage: This measure is proposed by Hiroyasu et al. [7], and it indicates the ratio of Pareto frontier which is covered by the population. The coverage is defined as

$$C = \frac{1}{m}\sum_{i=1}^{m}\frac{c_i}{c_{\max}}, \tag{8}$$

where m is the number of the objective functions, $c_{\max}$ is the number of small areas where a hyper-plane composed of $m - 1$ objective functions are evenly divided, and c_i is the number of areas including the true fitness of individuals projected to the hyper-plane. In our case, $c_{\max} = 15$ ($m = 2$) and $c_{\max} = 6$ ($m = 3$) were used.

Mean Absolute Error: This measure indicates the error of the population for the true Pareto optimal set. In this case, it is given by

$$\begin{aligned} X^*_{\mathrm{SCH}} &= \{x_1, \ldots, x_n \in [0, 2] \mid x_1 = \ldots = x_n\}, \\ X^*_{\mathrm{FON}} &= \{x_1, \ldots, x_n \in [-1/\sqrt{n}, 1\sqrt{n}] \mid x_1 = \ldots = x_n\}, \\ X^*_{\mathrm{ZDT1}} &= \{x_1 \in [0, 1],\ x_2 = \ldots = x_n = 0\}, \\ X^*_{\mathrm{DTLZ2}} &= \{x_1, x_2 \in [0, 1],\ x_3 = \ldots = x_n = 0.5\}. \end{aligned}$$

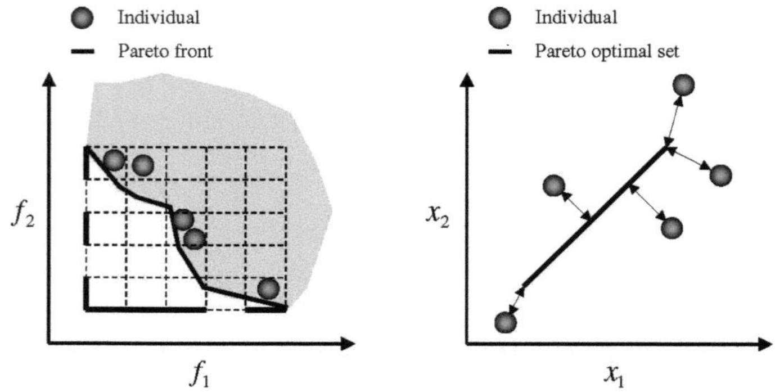

Fig. 4. Coverage measure ($m = 2$, left) and mean absolute error measure (right)

The mean absolute error is defined as the mean value of Euclidean distances from each individual to the nearest solution in X^*.

The concepts of the coverage measure and the mean absolute error measure are illustrated in Fig. 4.

4.2 Performance Analysis Under Noise-Free Environments

The comparison result of the proposed and conventional method in noise-free environment is shown in Table 1. The best performance in each problem is indicated in the bold font. From Table 1, it is understood that the proposed method outperforms the NSGA-II. As an example, transitions of coverage and mean absolute error for FON are shown in Fig. 5. It indicates that the proposed method reduces the number of evaluation to about one fifth, because the proposed method achieved the coverage and mean absolute error that had been finally obtained by the conventional technique by about 400 evaluation. This tendency was similar on the other three test problems, too.

Next, the influences of the Pre-selection parameters were examined on the following condition[3]:

- Candidate offspring population size $|Q_{\mathrm{C}}| = 30,\ 100,\ 300$
- Evaluated offspring population size $|Q| = 1,\ 4,\ 10$
- Parameters of the UNDX[4]
  $(\sigma_\xi, \sigma_\eta) = (0.25, 0.175/\sqrt{n}), (0.5, 0.35/\sqrt{n}), (1.0, 0.7/\sqrt{n})$

These parameters were compared with the basic setting of the algorithm $|Q_{\mathrm{C}}| = 100$, $|Q| = 4$, $(\sigma_\xi, \sigma_\eta) = (0.5, 0.35/\sqrt{n})$. The FON ($n = 10$) was used as a test function. Table 2 shows the experiment result. The following is understood from Table 2.

[3] Neither k nor α were analyzed because they are the setting parameters for estimation.
[4] $\sigma_\xi = 0.5$ and $\sigma_\eta = 0.35/\sqrt{n}$ are recommended parameters in [16].

Table 1. Performance comparison of the NSGA-II with pre-selection and the normal NSGA-II for four test functions in noise-free environment

Test Functions	Method	Coverage		Mean Absolute Error	
		Mean	Std. Dev.	Mean	Std. Dev.
SCH	Pre-selection	**0.981111**	0.018944	**0.112788**	0.018500
	NSGA-II	0.914444	0.034667	0.392743	0.0793348
FON	Pre-selection	**0.988889**	0.018222	**0.034481**	0.005383
	NSGA-II	0.965556	0.030929	0.140479	0.023860
ZDT1	Pre-selection	**0.983333**	0.020991	**0.005772**	0.004124
	NSGA-II	0.420000	0.189696	0.076696	0.022411
DTLZ2	Pre-selection	**0.370679**	0.052575	**0.100439**	0.016921
	NSGA-II	0.248765	0.043343	0.153413	0.022653

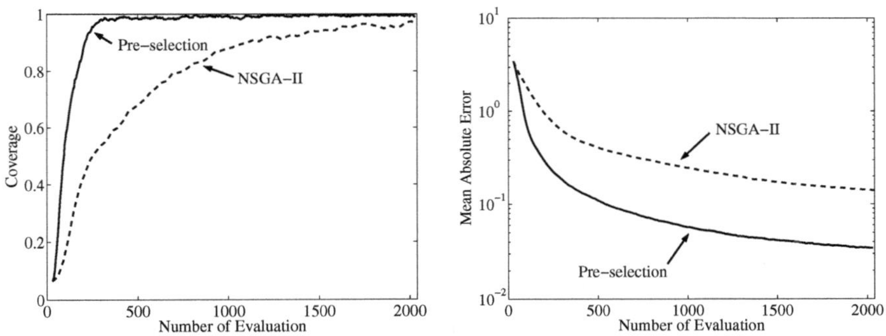

Fig. 5. Transitions of the coverage and the mean absolute error of FON in the noise-free environment

- The search performances slightly improved whenever $|Q_C|$ was increased from 30 to 300. However, since the difference between the performance of $|Q_C| = 100$ and $|Q_C| = 300$ was little, it is not a good approach to simply increase the size of Q_C in view of calculation cost.
- The performances of $|Q| = 1$ and $|Q| = 4$ were better than $|Q| = 10$. This result indicates that the promising offspring were surely added to the archived population P when the size of evaluated offspring population was small. However, the number of candidate offspring per evaluated offspring increases if $|Q|$ is set small. Hence, it should be adjusted according to the calculation cost of the approximation model.
- For σ_ξ and σ_η, the recommended values got the best result. When the parameters are small, that is, the offspring are generated in small region, extrapolation is hardly expected. Therefore, mean absolute error was extremely deteriorated. On the other hand, when they are generated in large region, there is only a little difference at the coverage for the recommended values

Table 2. Performance comparison of design parameters of the pre-selection algorithm and the UNDX for FON

Parameters	Value	Coverage		Mean Absolute Error	
		Mean	Std. Dev.	Mean	Std. Dev.
Basic setting	–	0.988889	0.018222	0.034481	0.005383
$\lvert Q_C \rvert$	30	0.991111	0.014993	0.049873	0.007122
	300	0.985556	0.020869	0.032054	0.004573
$\lvert Q \rvert$	1	1.000000	0.000000	0.025436	0.004441
	10	0.986667	0.018775	0.049873	0.007662
$(\sigma_\xi, \sigma_\eta)$	$(0.25, 0.175/\sqrt{n})$	0.980000	0.034575	0.121922	0.041719
	$(1.0, 0.7/\sqrt{n})$	0.995556	0.011525	0.060645	0.006968

Table 3. Performance comparison of the NSGA-II with pre-selection and the normal NSGA-II for four test functions in observation noise environment

Test Functions	Method	Coverage		Mean Absolute Error	
		Mean	Std. Dev.	Mean	Std. Dev.
SCH	Pre-selection	**0.534444**	0.089906	0.874871	0.127909
	PS ($\lambda = 0.01$)	0.513333	0.100421	0.920226	0.151003
	PS ($\lambda = 0.1$)	0.528889	0.134402	0.929779	0.110923
	PS ($\lambda = 1$)	0.238889	0.164488	**0.662967**	0.213107
	NSGA-II	0.538889	0.125054	1.399185	0.253036
FON	Pre-selection	0.586667	0.308947	0.248030	0.070666
	PS ($\lambda = 0.01$)	0.813333	0.206522	0.233200	0.037559
	PS ($\lambda = 0.1$)	**0.914444**	0.037837	0.214436	0.029421
	PS ($\lambda = 1$)	0.886667	0.088668	**0.167076**	0.016192
	NSGA-II	0.591111	0.175716	0.465918	0.118304
ZDT1	Pre-selection	0.701111	0.150728	**0.022994**	0.005718
	PS ($\lambda = 0.01$)	0.854444	0.108095	0.023906	0.006079
	PS ($\lambda = 0.1$)	**0.867778**	0.076053	0.024900	0.005042
	PS ($\lambda = 1$)	0.443333	0.166816	0.027209	0.004265
	NSGA-II	0.271111	0.133257	0.164172	0.024604
DTLZ2	Pre-selection	**0.370062**	0.040286	**0.091117**	0.010480
	PS ($\lambda = 0.01$)	0.330556	0.054975	0.109558	0.016236
	PS ($\lambda = 0.1$)	0.288889	0.040426	0.138259	0.020927
	PS ($\lambda = 1$)	0.139197	0.047368	0.212676	0.020506
	NSGA-II	0.178704	0.054814	0.162716	0.021747

while the mean absolute error is decreased. Thus, if the parameters are too much, the density of the candidate offspring generated in the vicinity of the Pareto optimal set becomes small, and convergence of the population slows.

The difference of each performance excluding small value of the UNDX parameters was relatively small for the performance of the normal NSGA-II. Hence, the proposed method has robustness for the design parameter selection.

4.3 Performance Analysis Under Observation Noise Environments

The performance under the observation noise was examined because it is the objective of this paper to apply evaluation reduction to experiment-based EMO. A simple expression of noisy test function is defined as follows:

$$F_i(\boldsymbol{x}) = f_i(\boldsymbol{x}) + \delta_i, \quad \delta_i \sim N(0, \sigma_i^2), \quad i = 1, 2, \ldots, m \tag{9}$$

where, $N(0, \sigma_i^2)$ is the normal distribution random number, and the standard deviation σ_i is 10% of the i th range of the Pareto front.

In the LWR, the observation noise can be filtered by appropriately adjusting parameters such as parameters of weighted function $K(\cdot)$, smoothing parameter u, ridge parameter λ, and choice of local model structure [1]. In this experiment, the ridge parameter was employed for filtering the observation noise. The performances of $\lambda = 0.01,\ 0.1,\ 1$ were examined with the basic setting $\lambda = 0.0001$

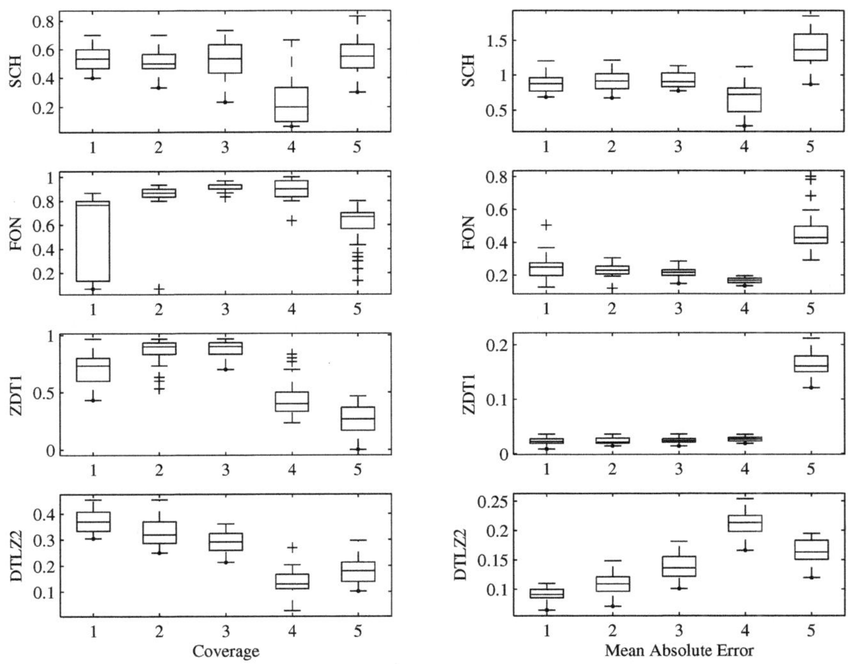

Fig. 6. This figure shows Table 3 by box plot. Each number on x axis corresponds to the following method respectively. 1: Pre-selection, 2: Pre-selection ($\lambda = 0.01$), 3: Pre-selection ($\lambda = 0.1$), 4: Pre-selection ($\lambda = 1$), 5: normal NSGA-II.

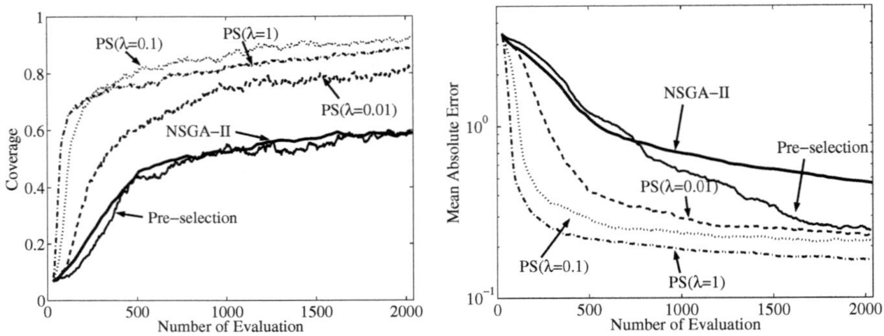

Fig. 7. Transitions of the coverage and the mean absolute error of FON in the observation noise environment

which was used at noise-free environment to avoid the influence of the multi-collinearity.

Table 3 and Fig. 6 show the results of experiment. From these results, it is understood that the effect of the ridge parameter adjustment is obvious, but it is dependent on the test problems. Parameter value $\lambda = 0.01$ indicates more stable performance than the other values. The ridge parameter should be determined by using the cross validation etc., but its calculation cost can not be neglected. Thus, an indicator of the selection of a more appropriate ridge parameter is one of the future tasks. However, while it is derived from small experience, we propose to use $\lambda = 0.01$ or less under the noise environment, and set to small value in such extent that the inverse matrix calculation does not become unstable under the noise free environment as primal choice.

Transitions of coverage and mean absolute error of FON are shown in Fig. 7. Fig. 7 indicates that the performance was improved by adjusting the ridge parameter. The result in the noise-free environment was improved as well.

From the above-mentioned discussion, it was shown that the proposed method had the better performance than normal NSGA-II under the observation noise environment.

5 Conclusion

In this paper, we proposed a Pre-selection algorithm for experiment-based evolutionary multi-objective optimization and examined the performance of proposed method through numerical experiments. As a result, we confirmed that the Pre-selection was able to reduce the number of evaluations greatly for four test functions regardless of the presence of the observation noise. Moreover, the search performance did not change two much when the setting parameters of Pre-selection and UNDX was changed. These facts indicate that the proposed method has robustness for the setting parameters.

As future works, we will examine the performance of our Pre-selection for more complicated test functions, and apply to the real problems such as the control parameter calibration of a real engine [13].

References

1. C. G. Atkeson, A. W. Moore, and S. Schaal: Locally Weighted Learning, *Artificial Intelligence Review*, Vol. 11, pp. 11-73 (1997)
2. J. Branke, and C. Schmidt: Faster Convergence by means of fitness Estimation, *Soft Computing*, Vol. 9, No. 1, pp. 13-20 (2005)
3. J. Branke, C. Schmidt, and H. Schmeck: Efficient Fitness Estimation in Noisy Environments, *Proc. of GECCO2001*, pp. 243-250 (2001)
4. K. Deb, A. Pratap, S. Agarwal, and T. Meyarivan: A Fast and Elitist Multiobjective Genetic Algorithm: NSGA-II, *IEEE Trans. on Evolutionary Computation*, Vol. 6, No. 2, pp. 182-197 (2002)
5. K. Deb, L. Thiele, M. Laumanns, and E. Zitzlar: Scalable Test Problems for Evolutionary Multi-Objective Optimization, TIK-Technical Report No. 112 (2001)
6. M. Emmerich, and B. Naujoks: Metamodel Assisted Multiobjective Optimisation Strategies and their Application in Airfoil Design, *Proc. of ACDM VI*, pp. 249-260 (2004)
7. T. Hiroyasu, M. Miki, and S. Watanabe: Divided Range Genetic Algorithms in Multiobjective Optimization Problems, *Proc. of IWES 99*, pp. 57-65 (1999)
8. A. Hoerl, and R. Kennard: Ridge Regression: Biased Estimation for Nonorthogonal Problems, *Technometrics*, Vol. 12, No. 1, pp. 55-67 (1970)
9. K. Ikeda, H. Kita, and S. Kobayashi: Failure of Pareto-based MOEAs, Does Non-dominated Really Mean Near to Optimal?, *Proc. of CEC2001*, pp. 957-962 (2001)
10. S. Jeong, K. Yamamoto, and S. Obayashi: Kriging-Based Probabilistic Method for Constrained Multi-Objective Optimization Problem, AIAA Paper 2004-6437 (2004)
11. Y. Jin, and J. Branke: Evolutionary Optimization in Uncertain Environments – A Survey, *IEEE Trans. on Evolutionary Computation*, Vol. 9, No. 3, pp. 303-317 (2005)
12. Y. Jin, M. Olhofer, and B. Sendhoff: Managing Approximate Models in Evolutionary Aerodynamic Design Optimization, *Proc. of CEC2001*, pp. 592-599 (2001)
13. H. Kaji: Proposal of a Multi-objective Genetic Algorithm for Noisy Fitness Functions (in Japanese), *Trans. of ISCIE*, Vol. 18, No. 12, pp. 423-432 (2005)
14. J. Knowles: ParEGO: A Hybrid Algorithm with On-line Landscape Approximation for Expensive Multiobjective Optimization Problems, *IEEE Trans. on Evolutionary Computation*, Vol. 10, No. 1, pp. 50-66 (2006)
15. P. Nain and K. Deb: A Multi-Objective Optimization Procedure with Successive Approximate Models, KanGAL Report Number 2005002 (2005)
16. I. Ono, and S. Kobayashi: A Real-coded Genetic Algorithm for Function Optimization Using Unimodal Normal Distribution Crossover, *Proc. of the 7th ICGA*, pp. 246-253 (1997)
17. J. Sacks, W. Welch, T. Michell, and H. Wynn: Design and analysis of computer experiments, *Statistical Science*, vol. 4, No. 4, pp. 409-435 (1989)

Prediction-Based Population Re-initialization for Evolutionary Dynamic Multi-objective Optimization

Aimin Zhou[1], Yaochu Jin[2], Qingfu Zhang[1], Bernhard Sendhoff[2], and Edward Tsang[1]

[1] Department of Computer Science, University of Essex, Colchester, CO4 3SQ, U.K.
[2] Honda Research Institute Europe, Carl-Legien-Str. 30, 63073, Offenbach, Germany

Abstract. Optimization in changing environment is a challenging task, especially when multiple objectives are to be optimized simultaneously. The basic idea to address dynamic optimization problems is to utilize history information to guide future search. In this paper, two strategies for population re-initialization are introduced when a change in the environment is detected. The first strategy is to predict the new location of individuals from the location changes that have occurred in the history. The current population is then partially or completely replaced by the new individuals generated based on prediction. The second strategy is to perturb the current population with a Gaussian noise whose variance is estimated according to previous changes. The prediction based population re-initialization strategies, together with the random re-initialization method, are then compared on two bi-objective test problems. Conclusions on the different re-initialization strategies are drawn based on the preliminary empirical results.

1 Introduction

In this paper, we consider the following continuous dynamic multi-objective optimization problems (DMOP):

$$\begin{aligned} &\text{minimize } \boldsymbol{F}(x,t) = (f_1(x,t), f_2(x,t), \ldots, f_m(x,t))^T, \\ &\text{subject to } \quad x \in X, \end{aligned} \tag{1}$$

where $t = 0, 1, 2, \cdots$ represents time, $x = (x_1, \ldots, x_n)^T \in R^n$ is the decision variable vector and $X \subset R^n$ is the decision space. R^m is the objective space. $\boldsymbol{F} : (X, t) \rightarrow R^m$ consists of m real-valued objective functions $f_i(x, t)$ $(i = 1, 2, \ldots, m)$, each of which is continuous with respect to x over X. The Pareto front (PF) in the objective space and the Pareto set (PS) in the decision space change over time. The task of a dynamic multi-objective optimization algorithm is to trace the movement of the PF and PS with reasonable computational costs.

Inspired by the success of evolutionary algorithms on dynamic scalar optimization problems [1,2,3], research work on evolutionary dynamic multi-objective optimization (EDMO) has very recently been conducted by several researchers. In the following, we briefly review the current work on EDMO:

i **Test Problems:** Benchmarks are important for developing and testing algorithms for solving DMOPs. In [4], Jin and Sendhoff proposed a method for constructing

S. Obayashi et al. (Eds.): EMO 2007, LNCS 4403, pp. 832–846, 2007.

dynamic multi-objective test problems by aggregating different objectives of existing stationary multi-objective problems and changing the weights dynamically. Test problems in [5] and [6] are created by adding time-varying terms to the objectives in stationary MOP test problems.

ii **Algorithms:** Several attempts for solving DMOPs by evolutionary algorithms have been reported recently. Stationary multi-objective evolutionary algorithms such as NSGA-II [7], SPEA2 [8], MSOPS [9] and OMOEA-II [10] have been directly applied to DMOPs [6,11]. A few evolutionary algorithms for solving dynamic single objective optimization problems have also been extended to the case of multi-objective problems [12]. Several strategies have been proposed by extending stationary multi-objective evolutionary algorithms for tracking the movement of the PS [13,14,15] or uncertain objectives [16,17].

iii **Performance Indicators:** It is hard to measure the performance of algorithms for DMOPs for the following reasons. Firstly, the measure must be able to evaluate the quality of approximation of a solution set, which itself is not trivial. Secondly, the PS is changing over time. It is natural to draw the PF for stationary multi-objective optimization, but it is no longer practical to plot the changing PFs in dynamic environment. In [12], two convergence performance measures have been suggested. In [6], the generational distance with time was plotted to show the convergence. In addition, a distribution indicator, known as the PL-metric, has also been introduced [6].

Arguably, diversity maintenance is essential in dynamic scalar objective evolutionary optimization algorithms. It is, however, interesting to note that in multi-objective evolutionary algorithms, the diversity of population is inherently maintained due to the multi-objective nature. Thus, it is probably of greater importance to ensure that the population is able to follow the moving PF more quickly. To this end, a good guess of the new location of the changed PS is of great interest.

In this paper, we study how to generate an initial population close to a changed PF when a change is detected in a dynamic environment. Inspired by the prediction strategy in [14,15], we build prediction models to predict the location of the new PS based on the information collected from the previous search. Different to the method in [14,15] where only the new locations of two anchor points and the Closest-To-Ideal point are predicted, we predict the new locations of a number of Pareto solutions in the decision space once a change is detected. Individuals in the initial population for the changed problem are generated around these predicted points. In such a way, the changed PS and PF can be found more effectively by the algorithm.

Four methods for re-initialization have been studied and compared in this paper. They are 1) Random re-initialization method in which the initial populations are randomly generated in the search space; 2) Variation method in which the individuals in the current population are perturbed using a Gaussian noise whose variance is determined by changes in the history; 3) Prediction method in which the new trial solutions are generated around predicted locations; and 4) A naive hybrid method, in which half of population is generated by strategy 2 and half is created by strategy 3.

The remainder of the paper is organized as follows. In Section 2, the four re-initialization methods are described in detail. Section 3 presents the two test functions and performance indicators used in this paper. The empirical results are shown in Section 4. The paper is concluded with Section 5.

2 Re-initialization Strategies for Dynamic Multi-objective Optimization

2.1 The Algorithm Framework

The main steps of the dynamic multi-objective evolutionary algorithm with predicted re-initialization (DMEA/PRI) are described as follows.

DMEA/PRI

Step 0. Set generation index $\tau := 1$ and time window $t := 1$. Initialize population P_τ.
Step 1. If a change is detected,
- **1.1.** Store P_τ in memory: Set $Q_t = P_\tau$.
- **1.2.** Re-initialization: generate an initial population $P_{\tau+1}$ based on information from Q_k, $k = t, t-1, \cdots$.
- **1.3** $t := t + 1$.

Step 2. If no change is detected, create an offspring population and do selection:
- **2.1.** Create an offspring population P from P_τ using an offspring generator for stationary optimization.
- **2.2.** Select $P_{\tau+1}$ from $P \cup P_\tau$.

Step 3. If the stop criterion is met, stop; else set $\tau := \tau + 1$ and go to **Step 1**.

In this paper, we concentrate on population re-initialization when a change in the environment is detected. Other genetic operators are introduced briefly as follows.

To detect the environment change, a naive strategy is to recalculate the function values of some individuals at the beginning of each generation. If their objective values change, environmental change has occurred.

In [18,19,20,21], we have proposed model-based algorithms to tackle stationary multi-objective optimization problems. By taking into account the regularity property of MOPs, these methods can approximate the PF efficiently. In this paper, the method proposed in [20] is used in **Step 2.1** of the above framework to generate offspring. The basic idea is to use the local PCA [22] algorithm to build a $(m-1)$-D (where m is objective number) piecewise continuous manifold in decision space, and then sample new trial solutions from the model thus built.

A modified selection based on non-dominated sorting of NSGA-II [7], which is called NDS-Selection [20] is used in **Step 2.2**.

We assume that inside a time window, there is no change in the environment and thus the dynamic optimization problem can be considered as a stationary problem. In the above framework, the algorithm works as an algorithm for solving stationary multi-objective optimization without **Step 1**. Details of **Step 1** will be discussed in the following subsection.

2.2 Prediction-Based Population Re-initialization

Diversity is maintained inherently in multi-objective optimization. Thus, we concentrate ourselves on faster convergence to the new PF when a change is detected in the environment by predicting the new locations of the Pareto optimal solutions using historical information. We assume that the recorded solutions in the previous time windows when a change is detected, i.e., $Q_t, \cdots, Q_1$, can provide information for predicting the new location of the PS PS_{t+1} at time window $t+1$. We further assume that the location of PS_{t+1} is a function of the locations $Q_t, \cdots, Q_1$:

$$Q_{t+1} = F(Q_t, \cdots, Q_1, t),$$

where Q_{t+1} denotes the new location of the PS for time window $t+1$.

The problem now becomes how to use the historical information $(Q_t, \cdots, Q_1)$ to generate new individuals as the initial population for time window $t+1$. In practice, function $F(\cdot)$ is not known and must be estimated using a certain technique. In the following, we discuss how to generate initial solutions for time window $t+1$.

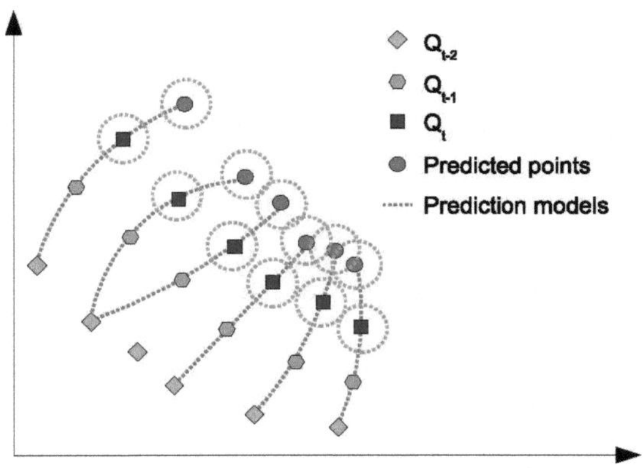

Fig. 1. Illustration of creating an initial population at the beginning of a time window (in decision space)

Prediction Model. Suppose that $x_t, x_{t-1}, \ldots, x_1$, $x_i \in Q_i, i = t, \cdots, 1$ are a series of points in the decision space that describes the movements of the PS, a generic model to predict the location of the initial individuals for the $(t+1)$-th time window can be formulated as follows:

$$x_{t+1} = F(x_t, x_{t-1}, \cdots, x_{t-K+1}, t), \tag{2}$$

where K represents the number of the previous time windows that x_{t+1} is dependent on in the prediction model. An example with $K = 3$ is illustrated in Fig. 1.

Any time series models [23] can be used for modeling F in (2). The major problem in making a prediction is that it is very difficult to identify the relationship between the stored solutions in $Q_t, \cdots, Q_1$ to build a time series. In this paper, we adopted a heuristic approach to identifying such time series. For a point $x_t \in Q_t$, its parent location in the previous time window can be defined as the nearest point in Q_{t-1}, i.e.,

$$x_{t-1} = \arg\min_{x \in Q_{t-1}} ||x - x_t||_2.$$

Once a time series is identified for each individual in the population, any linear or nonlinear prediction model can be used to predict the location of the individual for the next time window. In this paper, the following simple linear model is adopted:

$$x_{t+1} = F(x_t, x_{t-1}) = x_t + (x_t - x_{t-1}). \tag{3}$$

Variation with a "Predicted" Noise. The assumption that the movement of the PS can be described by a time series might be too strict. To improve the chance of the initial population to cover the PS in the new time window, a "predicted" Gaussian noise can be added to the current population and/or predicted locations. The standard deviation of the noise is estimated by looking at the changes occurred before:

$$\varepsilon \sim N(0, \delta I), \tag{4}$$

where I is an identity matrix and δ is the standard deviation, which is defined by

$$\delta^2 = \frac{1}{4n} ||x_t - x_{t-1}||_2^2,$$

where n is the number of decision vector. See the example in Fig. 1.

Re-initialization Methods. In the following, we describe the four methods for re-initializing population that we will empirically study in this paper.

i **Random (RND) Method.** All new solutions are randomly initialized in the search space:

$$x_{t+1} = \text{rand}(x^l, x^u),$$

where $\text{rand}(x^l, x^u)$ returns a random vector within the lower boundary x^l and upper boundary x^u of the search space.
In this restart method, no historical information is used.

ii **Variation (VAR) Method.** All new solutions are created by varying the solution in the last time window with a "predicted" Gaussian noise:

$$x_{t+1} = x_t + \varepsilon,$$

where ε is defined in (4).

In this method, only the information in the last time window is used and no models are built. We believe the new solutions should be close to the solutions in the last time window. Hopefully, the new trial solutions can cover the PS of the new time window.

iii **Prediction (PRE) Method.** All new solutions are sampled around the predicted locations:

$$x_{t+1} = F(x_t, x_{t-1}) + \varepsilon,$$

where F is defined in (3) and ε is defined in (4).

In this method, the last two time windows are used to predict new trial locations. By considering historical information, we hope the prediction model can capture the moving trend.

iv **Variation and Prediction (V&P) Method.** In this strategy, half of initial population for the $(t+1)$-th time window is sampled around the predicted locations and half is created by varying the points in the last time window (the current population when a change is detected). This method can be formulated as:

$$x_{t+1} = \begin{cases} F(x_t, x_{t-1}) + \varepsilon, & \text{if rand()} < 0.5; \\ x_t + \varepsilon & \text{otherwise.} \end{cases}$$

In the above equation, rand() returns a uniform random number in $[0, 1]$, F is defined in (3) and ε is defined in (4).

By hybridizing VAR method and PRE method, both historical location information and prediction models are used in re-initializing population.

3 Experimental Setup

3.1 Benchmark Problems

Two test problems are used in our simulation studies. The first one is the FDA1 [12], which is defined as follows:

$$\begin{cases} f_1(x,t) = x_1 \\ f_2(x,t) = 1 - \sqrt{f_1/g} \\ g(x,t) = 1 + \sum_{i=2}^{n} (x_i - G(t))^2 \\ G(t) = \sin(0.5\pi t) \\ x \in [0,1] \times [-1,1]^{n-1}, t = \frac{1}{n_T} \lfloor \frac{\tau}{\tau_T} \rfloor \end{cases}, \tag{5}$$

where τ is the generation counter, τ_T is the number of generations in each time window, and n_T controls the distance between two consecutive PSs. In fact, τ_T and n_T represent the *frequency of change* and *severity of change* respectively.

As shown in Fig. 2, the PSs of FDA1 are line segments parallel to coordinates, and the PFs are convex and remains unchanged. To study the performance of the proposed

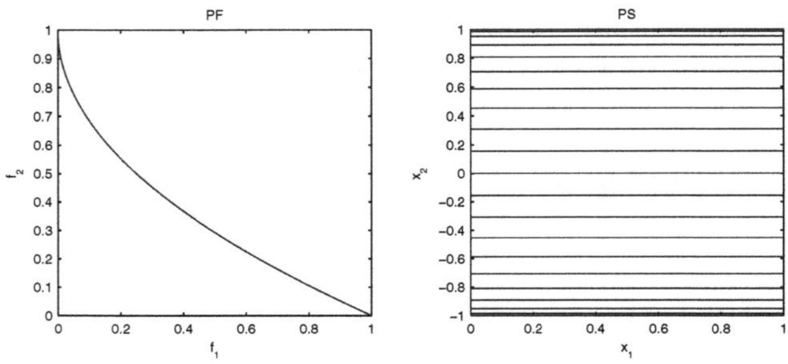

Fig. 2. Illustration of Pareto front and Pareto sets of FDA1 with $n_T = 10$

algorithm on problems with variable linkages, we modify the above FDA1 by using the method proposed in [24]. The modified test problem, which is named ZJZ, is as follows:

$$\begin{cases} f_1(x,t) = x_1 \\ f_2(x,t) = 1 - (f_1/g)^{H(t)} \\ g(x,t) = 1 + \sum_{i=2}^{n} (x_i + G(t) - x_1^{H(t)})^2 \\ H(t) = 1.5 + G(t) \\ G(t) = \sin(0.5\pi t) \\ x \in [0,1] \times [-1,2]^{n-1}, t = \frac{1}{n_T} \lfloor \frac{\tau}{\tau_T} \rfloor \end{cases} . \tag{6}$$

In ZJZ, both the PF and the PS are changing and there are nonlinear linkages between the decision variables. The PFs and PSs are illustrated in Fig. 3.

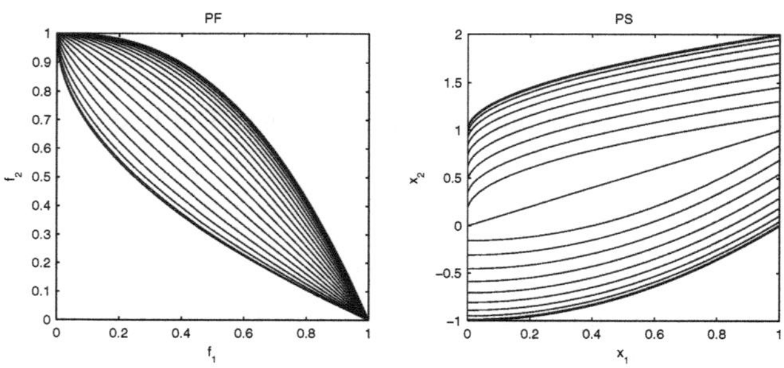

Fig. 3. Illustration of Pareto fronts and Pareto sets of ZJZ with $n_T = 10$

3.2 Performance Indicators

It is not trivial to assess the performance of evolutionary algorithms for solving dynamic multi-objective optimization problems. Let $M(P_\tau)$ measure the performance (the smaller, the better) of population P_τ at generation τ, it is natural to plot performance indicator M against time (see for example in Figs. 4 and 5 in Section 4.2). From the plot, we can observe the performance at any given time or the trend within a longer time period. To compare two algorithms, we may need a scalar value to indicate the quality of an algorithm on a given problem.

Suppose an algorithm is run N times on a given problem, and P_τ^i is the population at generation τ in the i-th run. Inspired by the idea of the offline error metric [1], we use

$$\text{Ave}(M(P_\tau)) = \frac{1}{N}\sum_{i=1}^{N} M(P_\tau^i),$$

and

$$\text{Std}(M(P_\tau)) = \sqrt{\frac{1}{N-1}\sum_{i=1}^{N}[M(P_\tau^i) - \text{Ave}(M(P_\tau))]^2}$$

to denote the mean and standard deviation of the performance indicator M at generation τ.

To assess the performance of an algorithm fairly, we record the following averages of the means and the deviations over t in our experiments:

$$\text{Ave}(M) = \frac{1}{T}\sum_{\tau=1}^{T} \text{Ave}(M(P_\tau)), \tag{7}$$

$$\text{Std}(M) = \frac{1}{T}\sum_{\tau=1}^{T} \text{Std}(M(P_\tau)). \tag{8}$$

In this paper, a distance-based performance indicator $D(P)$ suggested in [20] and the hypervolume difference ($I_H^-(P)$) proposed in [25] are used,

$$D(P) = \frac{1}{|P*|}\sum_{x\in P*} ||x - y(x)||_2,$$

where P is an obtained nondominated set, P^* is a reference PF, and $y(x) = \arg\min_{y\in P} ||x - y||_2$.

$$I_H^-(P) = I_H(P^*) - I_H(P),$$

where $I_H(P)$ is the hypervolume [26] of set P.

Both $D(P)$ and $I_H^-(P)$ can measure the approximation quality in convergence and diversity.

4 Experimental Results

4.1 Parameter Settings

In the experiments, 5 randomly selected individuals are re-calculated to detect environmental changes at the beginning of each generation. The parameter of model based offspring generator in **Step 2** of DMEA/PRI framework, i.e., the number of cluster in the local PCA algorithm is set to 5.

The other parameters are as follows: the population size is 100, and the number of decision variables is 10 for both FDA1 and ZJZ. In all experiments, the algorithm will stop after 60 environmental changes have been detected. The severity of change, n_T is set to be 5 and 10. The frequency of change, τ_T is set to be 5, 10, 15, 20, 30 or 40 generations, thus the time window size will be 500, 1000, 1500, 2000, 3000 and 4000 in terms of fitness evaluations. The results are based on 20 independent runs.

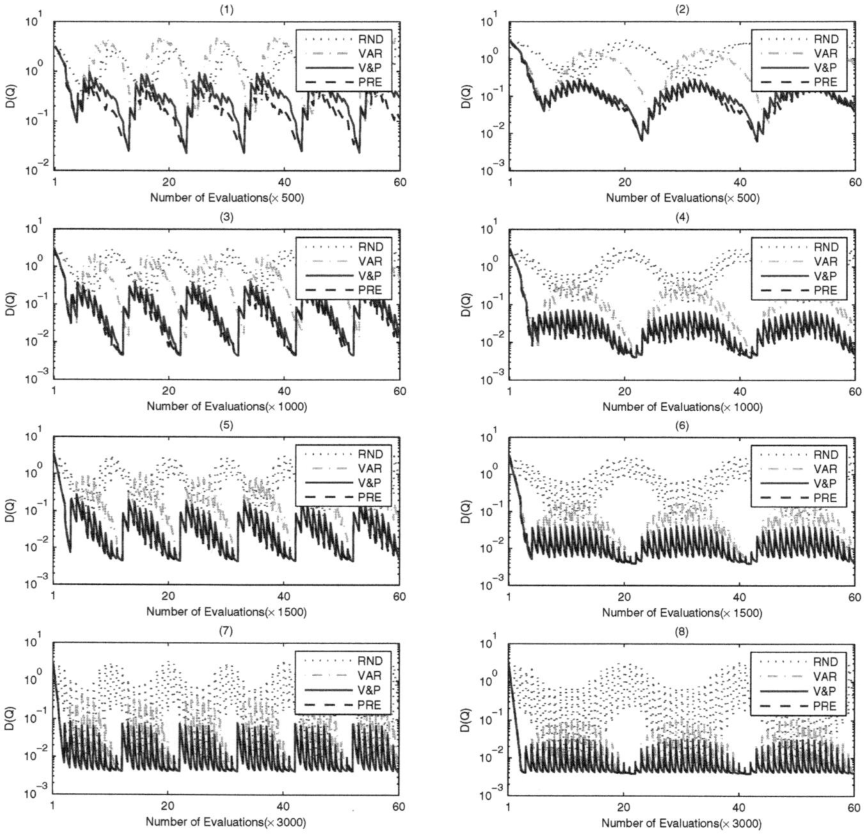

Fig. 4. The evolution of the average $D(Q_t)$ over time among 20 independent runs with 60 time changes for four re-initialization strategies on FDA1, the left column is with $n_T = 5$ and right column with $n_T = 10$

4.2 Results and Discussions

The statistical results on FDA1 and ZJZ with D and I_H^- indicators are shown in Figs. 6 and 7. Due to space limit, only curve of D over time on FDA1 and curve of I_H^- over time on ZJZ are drawn in Figs. 4 and 5. The results on both D and I_H^- are consistent.

We firstly consider the following three factors which influence the performance of the four re-initialization strategies.

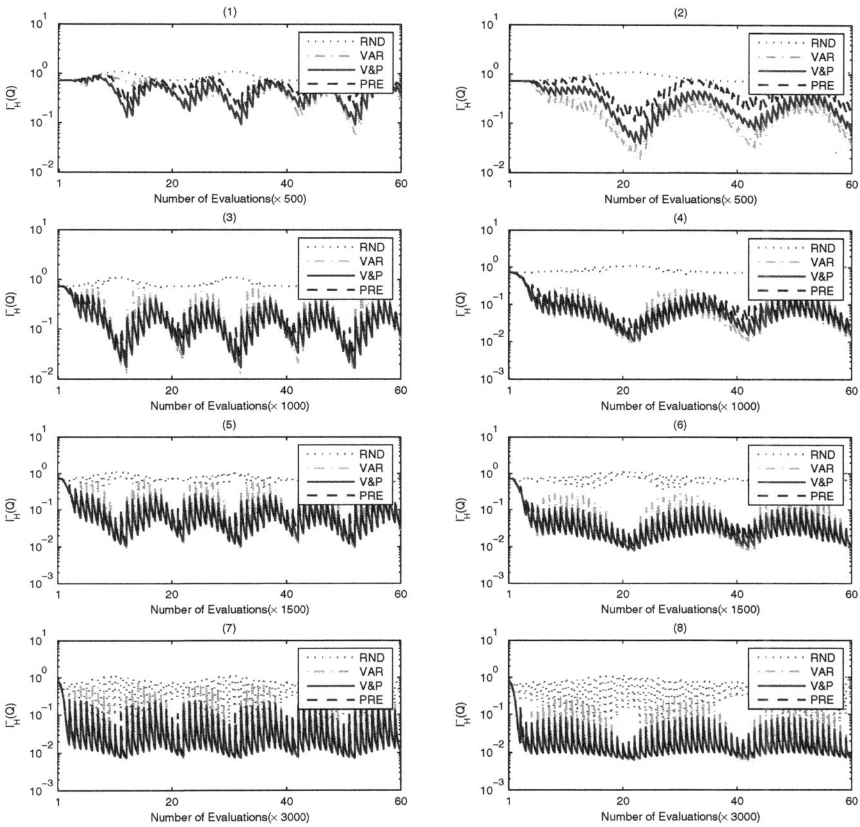

Fig. 5. The evolution of the average $I_H^-(Q_t)$ over time among 20 independent runs with 60 time changes for four re-initialization strategies on ZJZ, the left column is with $n_T = 5$ and right column with $n_T = 10$

Frequency of change. The parameter τ_T represents the frequency of change, i.e., the width of time windows. It is evident that the more evaluations used in each time window, the better the performance. From Figs. 6 and 7, we can see the performance of four strategies becomes better as the time window increases. For FDA1 and ZJZ, when $\tau_T \geq 1000$, the performance of PRE and V&P become very similar. The reason is that the difference on the initial populations is covered by the long run in the time window.

Severity of change. The parameter n_T controls severity of change, i.e., the closeness between two consecutive PSs. When n_T changes from 5 to 10, both the mean and variance of performance indicators become small in Figs. 6 and 7.

Problem characteristic. FDA1 has linear linkages between decision variables while ZJZ has nonlinear linkages between decision variables which can be seen from Figs. 2 and 3. From Fig. 6, we can see that PRE strategy works better than the others, it is because in this case, PRE can successfully predict the future locations. While from

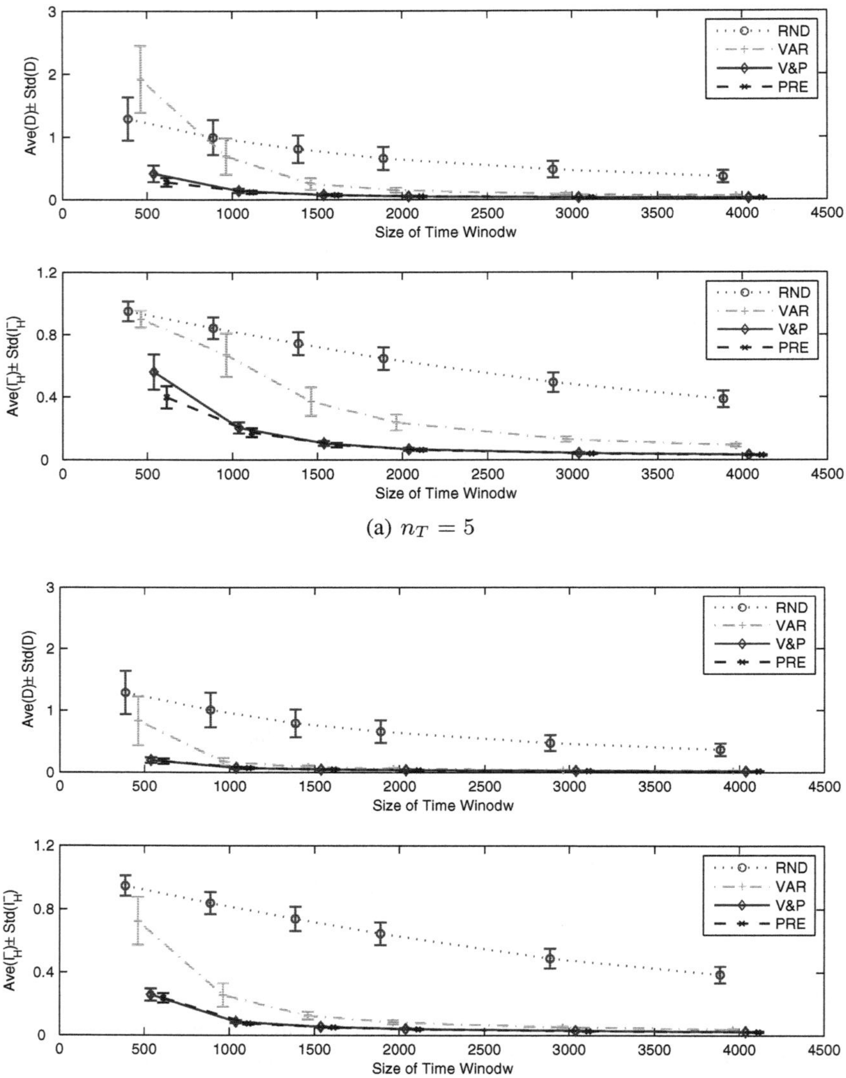

(a) $n_T = 5$

(b) $n_T = 10$

Fig. 6. The statistical results of Ave$(D) \pm$ Std(D) and Ave$(I_H^-) \pm$ Std(I_H^-) from time among 20 independent runs for four re-initialization strategies on FDA1 (To improve the clarity of the figures, the error bars are shifted a little along the horizontal coordinate)

Fig. 7, the VAR and V&P strategies work better than PRE strategy, the reason might be that the linear model in PRE strategy fails to capture the trend of moving PS. On the other hand, FDA1 and ZJZ are problems with cycles, and the results in Figs. 4 and 5 show such cycles.

We further analyze the performance of the four re-initialization strategies.

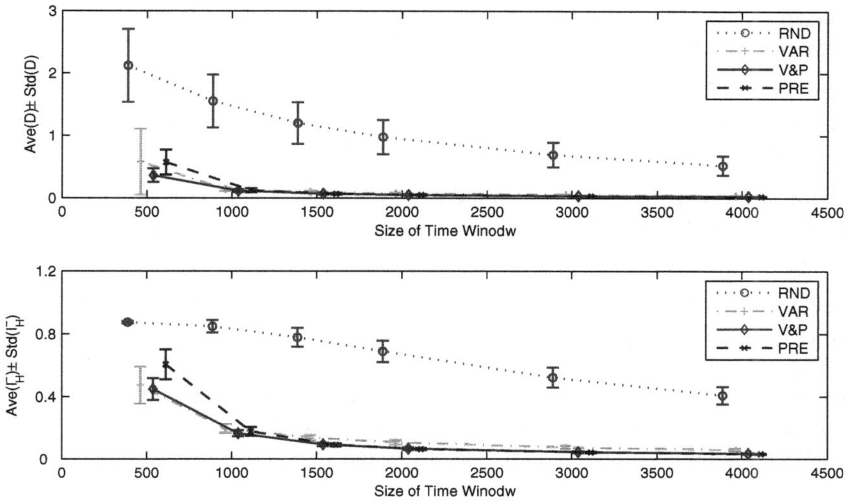

(a) $n_T = 5$

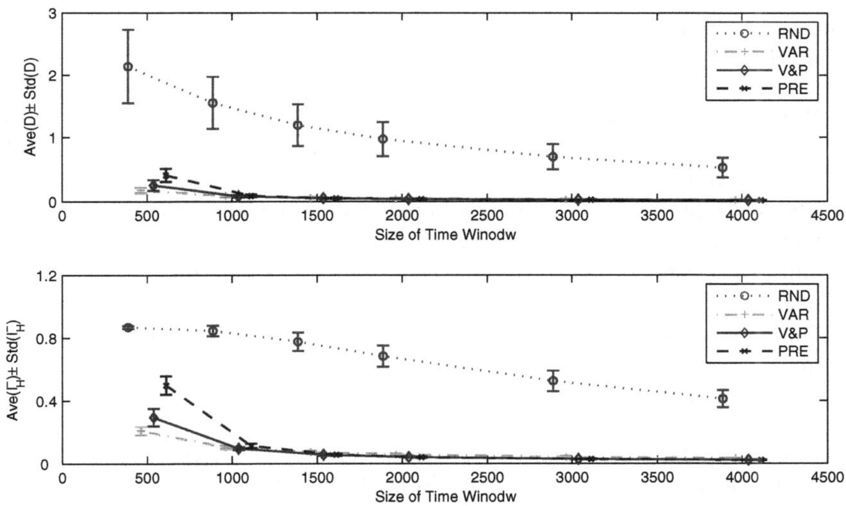

(b) $n_T = 10$

Fig. 7. The statistical results of Ave$(D) \pm$ Std(D) and Ave$(I_H^-) \pm$ Std(I_H^-) from time among 20 independent runs for four re-initialization strategies on ZJZ (To improve the clarity of the figures, the error bars are shifted a little along the horizontal coordinate)

RND. RND strategy does not work as well as the other three strategies. Besides, the results in Figs. 4 and 5 of the RND method are inconsistent with those of the other strategies. The reason might be as follows. First, the RND method does not consider the previous results in generating the initial population for the next time window. Second, it is easier to tackle a stationary problem whose PS locates at the 'center' of search space than a problem whose PS lies near the boundary.

VAR. For ZJZ problem, VAR strategy performs better than others because there are nonlinear linkages between decision variables, see Figs. 3 and 7. In these cases, it might be hard to build efficient prediction models which may mislead the search.

PRE. PRE strategy is able to catch the trend and thus predict new trial locations when there are linear linkages between decision variables and outperforms other strategies, see Figs. 2 and 6.

V&P V&P hybridizes VAR strategy and PRE strategy. For PDA1 problem, its performance is near PRE and for ZJZ problem, its performance is near VAR.

Overall, among the four strategies, the RND method is of no practical interest. The PRE method and the VAR method show some advantages over the other two strategies, depending on the characteristics of test problems and the width of the time windows. The V&P method is more likely to perform better in practice.

5 Conclusions

In this paper, four strategies for re-initializing populations are empirically studied on two test problems. The experimental results indicate that:

- When a change occurs, strategies which utilize historical information can accelerate the search process. The method that re-initializes a population purely randomly does not work.
- The width of the time window has significant influence on the performance. The wider the time window is, the better the performance. Other characteristics of problems, such as the distance between the neighboring Pareto sets, will also affect the results.
- Different strategies should be applied in different situations. In general, a hybrid method, i.e., the V&P method, might be more recommendable when little information about the problems is known.

To verify the proposed algorithms, a test problem, called ZJZ, has also been introduced. The location of both the Pareto front and the Pareto set of ZJZ changes over time. In addition, there are nonlinear linkages between decision variables. To assess the performance of the proposed algorithms, a performance indicator is suggested as well by combining the offline error measure in dynamic single objective optimization and the performance indicators in multi-objective optimization.

The research on dynamic multi-objective optimization is still in its very infancy and our work presented in this paper is also rather preliminary. Much work remains to be done in the future, for example, not only detecting environmental changes but also estimating the severity of changes, analyzing the problem structure of DMOPs, designing dedicated offspring generators and selection strategies by taking into account the problem structure, testing the suggested methods on more benchmarks, and comparing them with other methods.

References

1. Jürgen Branke. *Evolutionary Optimization in Dynamic Environments*, volume 3 of *Genetic Algorithms and Evolutionary Computation*. Kluwer Academic Publishers, 2002.
2. Yaochu Jin and Jürgen Branke. Evolutionary optimization in uncertain environments-a survey. *IEEE Transactions on Evolutionary Computation*, 9(3):303–317, 2005.
3. Shengxiang Yang and Xin Yao. Experimental study on population-based incremental learning algorithms for dynamic optimization problems. *Soft Computing*, 9(11):815–834, 2005.
4. Yaochu Jin and Bernhard Sendhoff. Constructing dynamic test problems using the multi-objective optimization concept. In *Applications of Evolutionary Computing*, volume 3005 of *Lecture Notes in Computer Science*, pages 525–536, Coimbra, Portugal, April 2004. Springer.
5. Marco Farina, Kalyanmoy Deb, and Paolo Amato. Dynamic multiobjective optimization problems: test cases, approximations, and applications. *IEEE Transactions on Evolutionary Computation*, 8(5):425–442, Oct. 2004.
6. Jörn Mehnen, Tobias Wagner, and Günter Rudolph. Evolutionary optimization of dynamic multiobjective functions. Technical Report CI-204/06, University of Dortmund, 2006.
7. Kalyanmoy Deb. A fast and elitist multiobjective genetic algorithm: NSGA-II. *IEEE Transactions on Evolutionary Computation*, 6(2):182–197, 2002.
8. Eckart Zitzler, Marco Laumanns, and Lothar Thiele. SPEA2: improving the strength Pareto evolutionary algorithm for multiobjective optimization. In *Evolutionary Methods for Design, Optimisation and Control*, pages 95–100, Barcelona, Spain, 2002. CIMNE.
9. Evan J. Hughes. Multiple single objective Pareto sampling. In *Proceedings of the Congress on Evolutionary Computation (CEC 2003)*, pages 2678–2684, Canberra, 8-12 December 2003. IEEE Press.
10. Sanyou Zeng, Shuzhen Yao, Lishan Kang, and Yong Liu. An efficient multi-objective evolutionary algorithm: OMOEA-II. In *Third International Conference on Evolutionary Multi-Criterion Optimization (EMO 2005)*, volume 3410 of *Lecture Notes in Computer Science*, pages 108–119, Guanajuato, Mexico, March 2005. Springer.
11. Sanyou Zeng, Guang Chen, Liangfeng Zhang, Hui Shi, Hugo de Garis, Lixin Ding, and Lishan Kang. A dynamic multi-objective evolutionary algorithm based on an orthogonal design. In *Proceedings of the Congress on Evolutionary Computation (CEC 2006)*, pages 2588–2595, Vancouver, BC, Canada, July 2006. IEEE Press.
12. M Farina and P Amato. Linked interpolation-optimization strategies for multicriteria optimization problems. *Soft Computing*, 9(1):54–65, 2005.
13. P. Amato and M. Farina. An ALife-inspired evolutionary algorithm for dynamic multiobjective optimization problems. In *World Conference on Soft Computing*, 2003.
14. Iason Hatzakis and David Wallace. Dynamic multi-objective optimization with evolutionary algorithms: A forward-looking approach. In *Proceedings of Genetic and Evolutionary Computation Conference (GECCO 2006)*, pages 1201–1208, Seattle, Washington, USA, July 2006.
15. Iason Hatzakis and David Wallace. Topology of anticipatory populations for evolutionary dynamic multi-objective optimization. In *11th AIAA/ISSMO Multidisciplinary Analysis and Optimization Conference*, Portsmouth, Virginia, USA, September 2006.
16. Jürgen Teich. Pareto-front exploration with uncertain objectives. In *First International Conference on Evolutionary Multi-Criterion Optimization (EMO 2001)*, volume 1993 of *Lecture Notes in Computer Science*, pages 314–328, Zurich, Switzerland, March 2001. Springer.
17. Evan J. Hughes. Evolutionary multi-objective ranking with uncertainty and noise. In *First International Conference on Evolutionary Multi-Criterion Optimization (EMO 2001)*, volume 1993 of *Lecture Notes in Computer Science*, pages 329–343, Zurich, Switzerland, March 2001. Springer.

18. Aimin Zhou, Qingfu Zhang, Yaochu Jin, Edward Tsang, and Tatsuya Okabe. A model-based evolutionary algorithm for bi-objective optimization. In *Proceedings of the Congress on Evolutionary Computation (CEC 2005)*, pages 2568–2575, Edinburgh, U.K, September 2005. IEEE Press.
19. Aimin Zhou, Yaochu Jin, Qingfu Zhang, Bernhard Sendhoff, and Edward Tsang. Combining model-based and genetics-based offspring generation for multi-objective optimization using a convergence criterion. In *Proceedings of the Congress on Evolutionary Computation (CEC 2006)*, pages 3234–3241, Vancouver, BC, Canada, July 2006. IEEE Press.
20. Qingfu Zhang, Aimin Zhou, and Yaochu Jin. Modelling the regularity in estimation of distribution algorithm for continuous multi-objective evolutionary optimization with variable linkages. *IEEE Transactions on Evolutionary Computation*, 2006. Submitted.
21. Aimin Zhou, Qingfu Zhang, Yaochu Jin, Bernhard Sendhoff, and Edward Tsang. Modelling the population distribution in multi-objective optimization by generative topographic mapping. In *Parallel Problem Solving From Nature - PPSN IX*, pages 443–452, 2006.
22. Nandakishore Kambhatla and Todd K. Leen. Dimension reduction by local principal component analysis. *Neural Computation*, 9(7):1493–1516, October 1997.
23. Christopher Chatfield. *The Analysis of Time Series: An Introduction.* CRC Press, 2004.
24. Hui Li and Qingfu Zhang. A multiobjective differential evolution based on decomposition for multiobjective optimization with variable linkages. In *International Conference on Parallel Problem Solving from Nature (PPSN IX)*, pages 583–592, 2006.
25. Joshua D. Knowles, Lothar Thiele, and Eckart Zitzler. A tutorial on the performance assessment of stochastic multiobjective optimizers. Technical Report 214, Computer Engineering and Networks Laboratory, ETH Zurich, Gloriastrasse 35, 8092 Zurich, Switzerland, 2006.
26. Eckart Zitzler and Lothar Thiele. Multiobjective optimization using evolutionary algorithms - a comparative case study. In *Parallel Problem Solving from Nature - PPSN V*, volume 1498 of *Lecture Notes in Computer Science*, pages 292–304, Amsterdam, The Netherlands, September 1998. Springer.

multi-Multi-Objective Optimization Problem and Its Solution by a MOEA

Gideon Avigad

School of mechanical engineering, the Iby and Aladar Faculty of Engineering
Tel-Aviv Universty, Israel
gideon@eng.tau.ac.il

Abstract. In this paper, a new type of Multi-Objective Problems (MOPs) is introduced and formulated. The new type is an outcome of a motivation to find optimal solutions for different MOPs, which are coupled through communal components. Therefore, in such cases a multi-Multi-Objective Optimization Problem (m-MOOP) has to be considered. The solution to the m-MOOP is defined and an approach to search for it by applying an EMO algorithm sequentially is presented. This method, although not always resulting in the individual MOPs' Pareto fronts, nevertheless gives solutions to the m-MOOP problem in hand. Several measures that allow the assessment of the introduced approach are offered. To demonstrate the approach and its applicability, academic examples as well as a "real-life," engineering example, are given.

Keywords: Communality, Family of designs, Engineering design.

1 Introduction

Sharing components among products is an effective way to cut costs. Expenses are decreased through the reduction in components design time as well as through savings in manufacturing costs and inventory (see e.g., [1]). Robertson and Ulrich, [2], point out that "By sharing components and production processes among products, companies can develop differentiated products efficiently, increase the flexibility and responsiveness of their manufacturing processes, and take market share away from competitors that develop only one product at a time." An example of the importance of sharing is Black & Decker's universal electric motor. According to Lehnerd [3], in the 1970s Black & Decker developed a family of universal motors for their power tools in response to a new 'double insulation' safety regulation. Prior to that, they used different motors in each of their 122 basic tools with hundreds of variations. By paying attention to standardization and exploiting platform scaling around the motor stack length, material costs dropped from \$0.77 to \$0.42 per motor while labor costs fell from \$0.248 to \$0.045 per motor, yielding an annual savings of \$1.82M per year.

Most of the attempts to share components between products are associated with the design of a product family. A product family is a group of related products that share common components and/or subsystems – yet satisfy a variety of market niches (e.g., [4]).

S. Obayashi et al. (Eds.): EMO 2007, LNCS 4403, pp. 847–861, 2007.

A design for components' commonality, is not restricted to hardware components as done e.g., in [1], but is also associated with software components. According to [5], software product families, aim at decreasing the costs and time required to produce a customer specific product. Oftentimes, commonality of software components is referred to as Component Based Software Development (CBSD). According to [6], there are two main benefits specific to CBSD. First, it gives structure to system design and system development, thus making system verification and maintenance more tractable. Second, it allows reuse of development effort by allowing components to be re-used across products and in the longer term by paving the way for a market for software components. Studies and approaches for CBSD may be found in many citations (e.g. [7],).

The focus of this paper is on the commonality of hardware components rather than on software components. According to [1], two types of component sharing can be used when selecting a product platform. The first is *component sharing*, in which one or more components are common to several products. The second is the sharing of "scaled" versions of components. Mathematically this can be described as *variable sharing*.

Successful engineering design of products generally requires the resolution of various conflicting design objectives ([8]). In case of contradicting objectives within a MOP, there is no universally accepted definition of an 'optimum' as in a single-objective optimization (see [9]). In such a case, there is no single global solution and it is often useful to determine a set of solutions that fits a predetermined definition for an optimum and let a Decision Maker (DM) choose between them. The predominant concept in defining such a set point is that of Pareto optimality ([10]). By definition, Pareto solutions, which belong to the Pareto optimality set, are considered optimal because there are no other designs that are superior in all objectives (e.g., [11]). The current focus is on different products whose designs require the solution of a MOP for each, and share common components.

The search for optimal solutions for a MOP is commonly termed Multi Objective Optimization (MOO). A comprehensive survey and comparison between most multi-objective optimization techniques and algorithms can be found in [12].

Searching a multi-objective design space, for optimal solutions, by Evolutionary Computation (EC) approaches (such as genetic algorithms) is commonly referred to as Evolutionary Multi-objective Optimization, (EMOO/EMO). Multi-Objective Evolutionary Algorithm (MOEA) is an EMOO algorithm, which searches for a solution in a multi-criteria space using some inspiration from evolutionary theories. Most MOEAs use genetic algorithms for the evolutionary search.

Research on MOEA has grown considerably in the last few years (see: Coello's web site http://www.lania.mx/~ccoello/EMOO/EMOObib.html). A number of algorithms, such as the Multiple Objective Genetic Algorithm (MOGA) of [13], are known to advance the use of EMO to solve MOPs. These algorithms utilize the non-dominancy notation ([14]), to direct the search towards a Pareto front. They use 'sharing' to allow the spreading of solutions along the front. According to Coello, ([15]), the later generation of Pareto-based algorithms, such as NSGA-II, ([16]), involves three major elements. The first element concerns the creation of a search pressure towards the Pareto set. This is commonly achieved by one of the known Pareto-based fitness assignment (dominance-based) techniques. The second element is set to avoid convergence to a single solution, and preserve diversity. The third element is elitism, which helps to prevent losing non-dominated solutions, which are diversified. Detailed descriptions of multi-objective evolutionary techniques could be found in [17]. The approximation and diversity of the numerically obtained set are the main issues to be considered with respect to EMOO algorithm performances. Laumanns *et al*, ([18]), discussed the important issue of convergence versus diversity of the solutions as attained by an EMOO algorithm, and introduced the epsilon measure to improve MOEAs with respect to the above two goals. To analyze and compare MOEAs with respect to these goals, performance metrics and measures have been also suggested by others (e.g., [19]).

When considering the solution of MOPs for product families design, it must be realized that commonality often causes a penalty with respect to individual product performance ([20]). Therefore the design challenge is how to maximize commonality and optimize the family products while satisfying individual constraints and minimizing performance losses. Several EMOO approaches have been developed to help designers resolve this tradeoff and determine the best design variable settings for the product platform and individual products within the corresponding family. The approaches to evolve the families are either sequential or simultaneous. In both approaches a Pareto front is developed. For example, Rai and Allada, ([21]), introduced a sequential approach to tackle the modular product family design problem. The first step performs a multi-objective optimization using a multi-agent framework to determine the Pareto-design solutions for a given product. The second step performs post-optimization analysis to determine the optimal platform level for a related set of product families and their variants. This is commonly done by choosing solutions from the MOPs' Pareto fronts (e.g., [1]). An example for the second approach is a MOEA approach, which has been taken in [22]. In [22], NSGA-II is utilized for a one – stage optimization algorithm that search simultaneously for the products optimizing the commonality. The objectives of such commonality related MOPs are commonly associated with the variation in design variables and a deviation function from a given engineering goal (e.g., [22]). In such a case, the representation in the objective space is not associated directly with the objectives but rather with the utility of the objectives (one axis) and with a measure for commonality (the other axis).

The main difference between the hereby introduced approach and former works lies in the consideration of different MOPs, which are coupled by common components. Here, the products do not share the same objective space and therefore neither utility of objectives nor same objective space setting is applicable. To further explain the difference between this paper approach and others, it is noted that the same cockpit for different aircrafts is not an example to the current approach. Here a search for a robotic arm to move an object from one place to the other, coupled with a search for a conveyor to move another object, is an appropriate example. The objectives associated with the search for the robotic arm might be the minimization of the integral square of the end- effector's error and the minimization of its deflection. The search for a conveyor may be associated with the maximization of the object transfer speed and the minimization of the overall control force. The coupling between the MOPs is dictated by the need to use the same motor for both designs.

2 Methodology

2.1 Problem Definition

In the classical multi-objective search problem, such as dealt with in [15], the set of Pareto optimal solutions is sought from the set of all possible particular solutions. Any particular solution is characterized by specific values of the problem decision variables representing a point in the problem decision space. The set of Pareto optimal solutions is found by comparing the performances of all particular solutions in the objective space for non-dominancy. The representation, in the objective space, of the set of non-dominated solutions is known as the Pareto front. The classical MOP is commonly formalized as follows:

$$\min F(x) \qquad (1)$$

$$\text{s. t.} \qquad x \in X \subseteq S \subseteq R^n$$

where x is the vector of decision variables. In general, x might be subjected to equality and/or inequality constraints, which commonly include some bounds on the decision variables. A solution $x \in X \subseteq S \subseteq R^n$, which satisfies all the constraints, is called a feasible solution. The set X, of all feasible solutions, is called the feasible region in the search space S. The MOP deals with minimizing y=F(x), the vector of K objective functions where,

$$F(x) = [f_1(x), f_2(x), \ldots, f_K(x)]^T \qquad K \geq 2 \qquad (2)$$

It can be shown that problems involving maximization, or a mixture of both min and max with respect to different objectives, may easily be transformed to the above problem. Furthermore, it should be noted that usually, due to contradicting objectives, there is no single solution to the above problem. The interest, in the classical MOP, is therefore on the trade-offs with respect to the objectives. The well-known concept of Pareto dominance supports exploring such trade-offs. The development of an optimality-based Pareto front in the objective space is based on a comparison between solutions using the idea of vector domination ([14]).

The problem of the m-MOOPs is formulized as:

$$\min F(x, y)$$

$$\text{s.t} \quad \begin{cases} x \in X \subseteq S \subseteq R^n \\ y = y_m;\ m = 1, \ldots, n_{mop};\ y_m \in Y_m \subseteq U_m \subseteq R^m \end{cases} \qquad (3)$$

where: $$F(x, y) = \{F^m(x, y_m), \text{for } m = 1, \ldots, n_{MOP}\} \qquad (4)$$

$$F^m(x, y_m) = [f_1^m(x, y_m), f_2^m(x, y_m), \ldots, f_K^m(x, y_m)]^T \qquad K \geq 2$$

The m-MOOPs is a problem, which involves n_{mop} objective spaces and associated design spaces. The design spaces possess communality through the communal search space X. The relations between the design spaces and the objective spaces may be elucidated by considering a 2-MOOPs problem and by observing figure 1. The figure depicts two design spaces (left side of the figure). One space is constructed out of two sub-spaces X and Y_1 (designated as XY_1). The other design space has also two sub-spaces, X and Y_2. A solution (e.g., xy_1) to a MOP is constructed out of parameters values taken from one of the unshared sub-spaces (e.g., y_1 from Y_1) and from the shared sub-space (x from X). Each such solution has its performances in its related objective space (e.g., Z_1). Each set of objective functions $F^m(x, y_m)$ maps a particular solutions xy_m belonging to the space XY_m to an objective space point z_m. Such a mapping is depicted in figure 1 designated by the arrows.

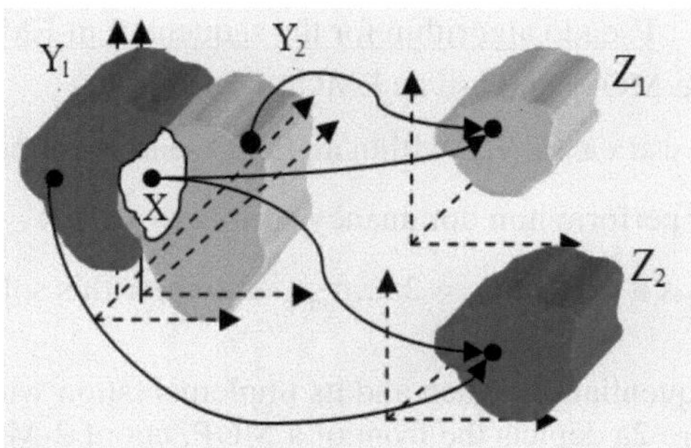

Fig. 1. Solutions in a bi-design space and corresponding MOPs' objective spaces

The solution to the m-th MOP belongs to the m-MOOPs optimal set, $MP^* = \bigcup_{m=1}^{n_{mop}} MP_m^*$, where MP_m^* is defined as follows:

$$MP_m^* := \{(x, y_m)^* \in XY \,|\, \forall i = 1,.., n_{mop} \neg\exists\, x' \in X : F(x', y_i) \prec F(x, y_m)^*\} \quad (5)$$

where $XY = \bigcup_{m=1}^{n_{mop}} XY_m$ m=1,...n_{mop} and the associated fronts are:

$$MPF_m^* := \{z_m^* \in Z \,|\, z_m^* = F(x, y_m)^* : (x, y_m)^* \in MP_m^*\} \quad m=1,\dots, n_{mop} \quad (6)$$

where $Z = \bigcup_{\forall x \in X \wedge y_m \in Y_m \; m=1,..,n_{mop}} F(xy_m)$ is the combined objective space

The solution to the m-MOOPs problem, as defend in equations 5, 6, is a set of solutions which possess communal components, for which, no change of a communal component will result in an improvement of performances in one MOP without loss of performances in other MOPs.

2.2 A Sequential EMO Approach

It is suggested that in the current study the search for solutions to the m-MOOP will be done by a sequential approach. This approach is introduced in the following

The sequential approach is designed to solve the MOOPs by a lixographic method. In such a case, one of the MOOPs is solved, gaining its front. The optimal set for that MOOP contain x^*. The search for the set x^* is then relaxed and the remaining MOPs are optimized using its values as constants. This means that the problem in equation 3 is decomposed to:

$$\begin{array}{ll} \min(f(x,y)) & \text{if } m=1 \\ \min\limits_{x^*=con}(f(y)) & \text{if } m\neq 1 \end{array} \quad (7)$$

The EMO implementation is associated with solving sequentially n_{mop} MOPs by using a MOEA (e.g., NSGA-II) as detailed in the following pseudo algorithm.

Pseudo algorithm for the sequential m-EMO

a. Choose one of the MOPs and use an EMO to find $(x,y_1)^*$

b. For $i=2,...,n_{mop}$ use $x\in(x,y_1)^*$ within an EMO search to find $(x,y_i)^*$ for all $i\neq 1$

c. For $i=2,...,n_{mop}$ perform non dominancy sorting to $(x,y_i)^*$

d. If $\exists x\in(x,y_1)^* \wedge x\notin(x,y_i)^*$ $y_i=2,...,n_{mop}$ eliminate this solution in MOP_1

To elucidate the sequential approach and its implementation within an m-MOOP refer to figure 2a-c. Figure 2a depicts the front of a MOP, out of 2-MOOPs, which is solved first (step 'a' above). Each solution performances, is designated with a different symbol, emphasizing that each may be associated with different values for the communal parameters (the x parameters). Using each of the x values found by the solution of the first MOP, may result in a front when the second MOP is solved (step 'b' above). This means that for each point in the objective space of the first MOP there may be a set of solutions' performances in the second objective space. These are designated by corresponding symbols in figure 2b.

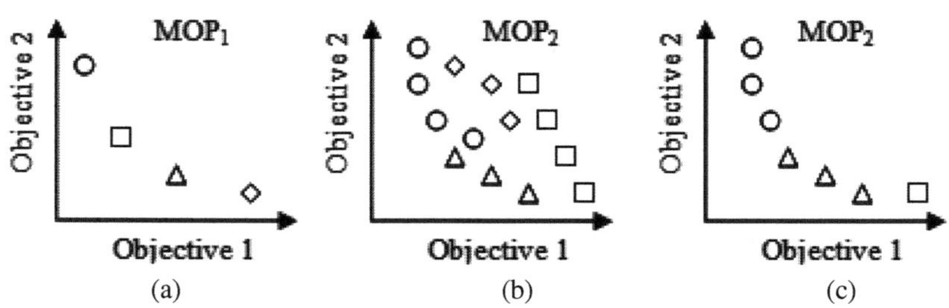

Fig. 2. Obtaining the m-MOOP front by a sequential approach

Following the demand for optimality of solutions to all MOPs, the fronts depicted in figure 2b are sorted for non-dominancy to result in the front (step 'c' above), which is depicted in figure 2c. As a last step (step 'd' above), the solution which is associated with the diamond in figure 2a is removed (as it is not associated with optimal solutions in both MOPs). The solution to the m-MOOP is the two fronts, which are depicted in both figures 2a and 2c after eliminating the diamond solution in figure 2a.

While searching for solutions to m-MOOPs sequentially, concern should be given to three possible cases. The three cases are explained using figure 3, which is related to an m-MOOP with two coupled MOPs. Figure 3a depicts the front of a MOP, MOP_1 which is found first. A gray boundary designates the feasible region for both MOPs.

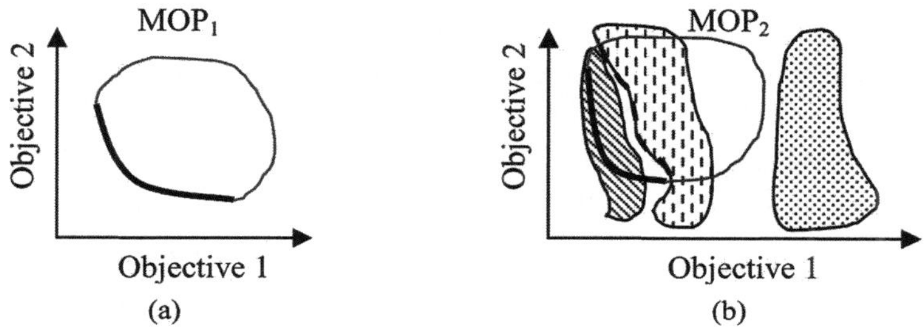

Fig. 3. Front of one MOP and possible m-MOOP fronts in the second MOP

Using x^*found by solving MOP_1 for the second MOP, MOP_2, may result in solutions, fundamentally located at three sub-spaces within the objective space of MOP_2. These three sub-spaces, depicted in figure 3b, are associated with the three cases declared upon. In the first case, the solutions are within the sub-space, which is designated by dots. In such a case, no feasible solutions are present for MOP_2 and the m-MOOP is unsolved. The second case is associated with the sub-space, which is designated by short-vertical lines. In this case a back-located front may be found (see dashed line in figure 3b). In the last case, the solutions are within the sub-space, which is designated by tilted lines in figure 3b. In this case, the Pareto fronts or part of the Pareto fronts of all MOPs, are associated with the solution to the m-MOOP. It is noted that the second case is associated with a loss of optimality as further discussed in section 2. 3.

2.3 Assessment Measures

In this section, measures to assess the successes of the algorithm to evolve the problem's fronts are presented. These measures allow comparing between different initializations of the algorithm. The measures introduced include a Waste of Resources Measure (WRM), Loss of Optimality Measure (LOM) and the computational time measure. These measures are utilized in section 3, where examples are used to demonstrate the use and the applicability of the proposed methodology.

Efficient use of computational resources, WRM: As briefly noted in section 2.2, the sequential approach may utilize the resources of available individuals inefficiently (compare figures 2a and 2b with relation to the diamond solution). This may happen wherever the optimal communal variables (x^*) do not span the Pareto fronts of <u>all MOPs</u>. If this is the case, the sequential algorithm "wastes" some of its resources (of available individuals) to find solutions that are not solutions to the m-MOOPs. This waste of resources may be quantified by counting the unshared solutions to obtain WRM as follows:

$$\forall x_i^* \in (x, y_1)^* \wedge x_i^* \notin x^*;\ \text{WRM} = \sum_{i=1}^{n_p} x_i^* \tag{8}$$

Loss of optimality measure, LOM: To assess the success of an algorithm to find a solution to the m-MOOPs the proximity indicator, ([19]), is utilized. For each MOP, a representing set OS_m, m=1,., n_{mop} is found such that it is spread diversely on the MOP's Pareto front. It serves as the optimal approximation set. If the set may not be found analytically the representing set is a set found by individually running each MOP to find OS_m^*, m=1,., n_{mop}. The minimal Euclidian distance between the j-th evolved solution within the m-th MOP with a representative of a set, AS_m, which is the front achieved in the last generation (of the m-MOOPs algorithm), $D_{j,OS \to AS}^m$ is found. Due to the inherent uncertainty in the preference of objectives within MOPs, any solution may be chosen by the designers. Therefore, the worst case should be considered for the comparison. Therefore a measure termed Loss of Optimality Measure, LOM, is computed: LOM= max(LOM_m) where:

$$\text{LOM}_m = \frac{\max_j D_{j,AS \to OS}^m}{B^m} \qquad m = 1, \ldots, n_{mop} \tag{9}$$

where B^m is the maximal Euclidean distance between solution performances within the m-th MOP. This serves as a scaling factor. A lower LOM means less loss of optimality and is viewed as an advantage. The LOM is utilized in section 3.

Computational time: The computational complexity that influences the search run-time, using a traditional EMO, depends primarily on the number of generations, and the size of the population (e.g., [20]. In such a traditional case, it is assumed that there is no substantial difference in the computational time of the performances from individual to individual. In the sequential m-EMO the run-time is influenced by the difference in times to solve the computational models of the different MOPs and by the order at which the MOPs are solved. To elucidate this declaration, let the time to compute the model of MOP_1 be 't_1' while to compute the model of MOP_2 takes 't_2' and $t_1 > t_2$. Further suppose that each MOP is solved with a population of 'p' individuals evolved for 'g' generations. If MOP_2 is solved first and then the results are utilized to evolve MOP_1, the computational time T, which is associated with the computations of the model, would be $T = gp(t_1 + zt_2)$. This time is less than if the order is reversed (in that case $T = gp(t_2 + zt_1)$). This difference is demonstrated in section 3.3.

3 Test-Cases

In this section bi-MOOP examples are given to demonstrate the issues that were introduced in the methodology (see section 2). Specifically cases 2 and 3 (see section

2.2) are demonstrated. Case 1 is not demonstrated as it is a trivial one. In all of the following examples NSGA-II with 20 individuals, 50% cross-over and 5% mutation is used over 300 generations for the sequential algorithm. An 8 bit code is used for all design parameters. The simulations were done by using MATLAB™.

3.1 Academic Example – 1

In the first example the sequential EMO is utilized to solve a bi-MOOP, which is associated with the third case (see section 2.2). The MOPs involved are:

$$\text{MOP}_1\text{: } \min(f_1^1, f_2^1) \text{ where, } \quad \begin{aligned} f_1^1 &= x \\ f_2^1 &= 1 + {y_1}^2 - x + 0.2\sin(\pi x) \end{aligned}$$

$$\text{and: } \quad -2.0 \le x \le 2.0,\ 0 \le y_1 \le 2.0$$

$$\text{MOP}_2\text{: } \min(f_1^2, f_2^2) \text{ where, } \quad \begin{aligned} f_1^2 &= -10(e^{-0.2(x^2+{y_2}^2)^{0.5}} + 1) \\ f_2^2 &= (x-0.5)^2 + (y_2-0.5)^2 \end{aligned}$$

$$\text{and: } \quad -2.0 \le x \le 2.0,\ -2.0 \le y_1 \le 2.0$$

In the above equations x serve as the communal variable representing the communal component shared by the two different design problems.

The sequential approach (see section 2.2) begins by solving MOP_1 by using NSGA-II. The front of this problem is depicted in figure 4, designated by squares.

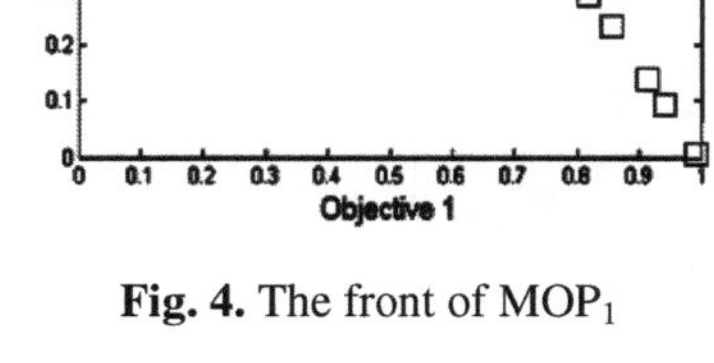

Fig. 4. The front of MOP_1

The next step within this approach is to use the x^* values as constants within the second MOP. Each such value is associated with a front within the objective space of MOP_2, as depicted (for some of the solutions) in figures 5a, 5b. The numbers depicted in those figures are the values for x, found by solving the first MOP. It is noted that the results are shown in two different figures for the sake of clarity and to highlight the waste of resources (see section 2.3) as clearly depicted in figure 5a. The m-MOOP solution is obtained by sorting for non-dominancy the fronts of figures 5a and 5b. The resulting front (of MOP_2) is depicted in figure 6a. Also depicted in that figure is MOP_2 front designated by dots. It is noted that following the sorting, all solutions shown if figure 5a are not part of the front and figure 6a shows that the resulting front is associated with the communal solutions shown in figure 5b. Figure 6b depicts the fronts of the m-MOOP on the same graph showing the communal related solutions only. By comparing figures 4 and 6b it may be observed that the m-MOOP front associated with MOP_1 is just a part of the overall front of MOP_2. This means that just some of the solutions obtained by the solution of MOP_1 may be used as communal components when optimality is desired.

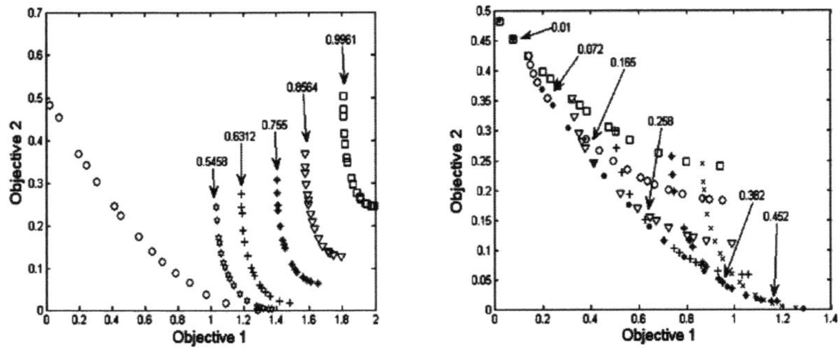

(a) (b)

Fig. 5. Solving MOP_2 for each of the solutions found by solving MOP_1

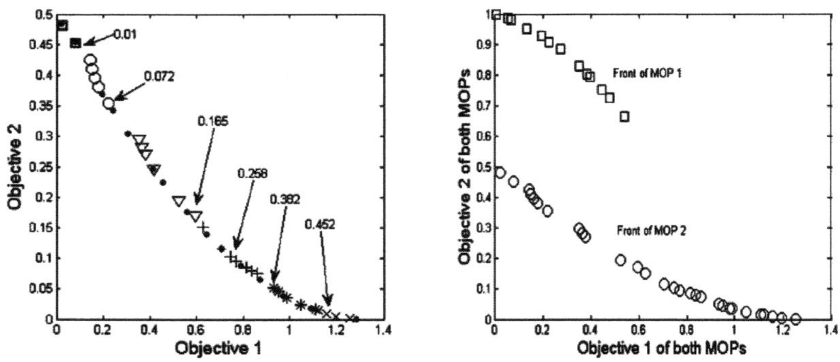

(a) (b)

Fig. 6. (a) Result of the non dominancy sorting (b) The m-MOOP fronts

3.2 Academic Example - 2

This example demonstrates an m-MOOP, which is associated with the second case (see section 2.2). The m-MOOP is associated with the following MOPs, with x is the communal component.

MOP_1: $\min(f_1^1, f_2^1)$ where, $f_1^1 = (x-3)^2 + (y_1 - 3)\text{^}2$

$f_2^1 = 1.2 + {y_1}^2 - x + 0.1\sin(3\pi x)$

and: $0.01 \le x \le 4,\ 1.0 \le y_1 \le 2.0$

MOP_2: $\min(f_1^2, f_2^2)$ where, $f_1^2 = x$

$f_2^2 = 1 + {y_2}^2 - x - 0.2\sin(\pi x)$

and: $0.01 \le x \le 4,\ -1.0 \le y_2 \le 2.0$

Using the sequential algorithm presented in section 2.2 to solve the m-MOOP, once by starting with MOP_2 and once starting with MOP_1 results in the fronts, which are depicted in figures 7a and 7b respectively. In these figures the front, which is found first is designated by blank symbols while the second front to be found is designated by filled symbols. Also depicted in each figure, is the front of the second MOP, which is obtained separately. These are designated by dots.

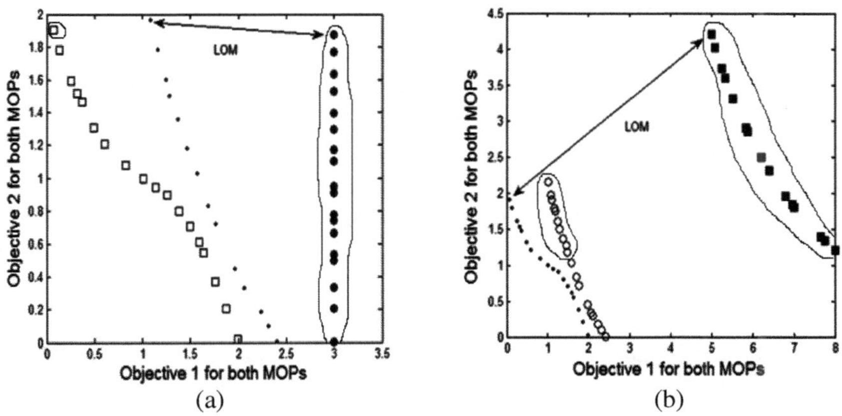

Fig. 7. (a) Starting with MOP_2 (b) Starting with MOP_1

After eliminating non optimal solutions (see sequential m-MOOP procedure), the solution to the m-MOOP when starting with MOP_2, are the solutions that are associated with the encircled performances in figure 7a. In the same manner the solution to the m-MOOP when starting with MOP_1, are the solutions that are associated with the encircled performances in figure 7b. The values of the measures, introduced in section 2.3, for this example are summarized in table 1 (where 't' is used to reduce non-important numerical data).

Table 1. Measures' values

Measure / Start	WRM	LOM	T
MOP_2	18	0.53	1.1t
MOP_1	10	1.95	t

Observing the results, it is depicted that if the procedure starts with MOP_1 the waste of resources is high. In fact 18 solutions out of 19 on the front of MOP_1 do not have representatives on the front of MOP_2. This means that there exists just one optimal related communal component. Changing the order at which the MOPs are solved results in a better use of the resources (a waste of 10 instead of 18). Nonetheless the loss of optimality is much higher. The difference in computational time is minor. By analyzing the results it is concluded that the solution to the m-MOOP by the sequential approach is sensitive to the order at which the MOPs are solved. It is further noted that in general the sequential approach might be associated with all cases presented in section 2.2. Therefore it is possible that by starting with one MOP a solution to the m-MOOP is found (cases 2, 3) but starting from another may lead to not finding any solution (case 1). It is hereby noted that if there exists a hierarchy of MOPs, that is one MOP which is more important than the others, it should be solved first. This is due to the fact that the first solved MOP is not associated with a loss of optimality.

3.3 Real World Example

This bi-MOPs is associated with the following MOPs. The First MOP is a mechatronic example where the structure and the control of a one-arm manipulator are to be designed. The arm is considered rigid, is made of aluminum (E=70000Mpa, $\rho = 2700 \text{kg}/\text{m}^3$) and has a 10 by 10 mm cross section. The arm is depicted in figure 8a (side view). A load of M_L=0.8kg is to be raised in a distance of x=1.5m. An arm of length L, and weight m (depends on length) is used for the lifting.

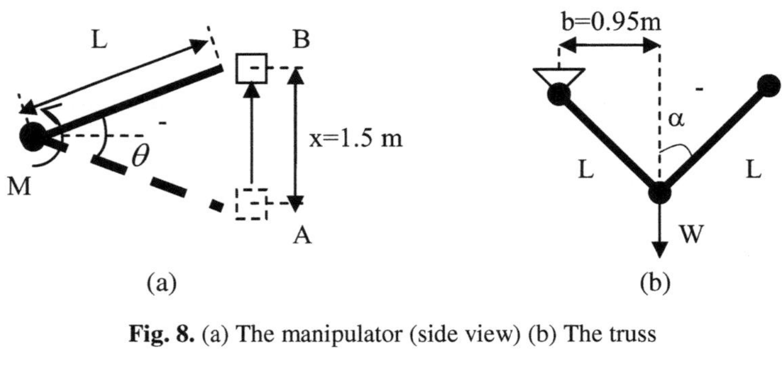

Fig. 8. (a) The manipulator (side view) (b) The truss

The arm is manipulated by a torque M at its base, determined by a controller. The bi-objective problem involves the minimization of: $f_1^1 = \text{ISE}, f_2^1 = \Delta f$ where Δf, is the deflection of the end affecter, $\text{ISE} \equiv \int_0^{t_{final}} \text{error} \cdot dt$. The model for the first MOP is detailed in [23]. The mechatronic design includes decisions on the length of the arm, L, that may change between 1m to 6m (dictated by inventory), as well as on the PID controller parameters (see [23] for details). The front for this MOP is depicted in figure 9a designated by blank squares. The front includes designs with arm lengths that vary between 1m and 2.2m. The shorter arms are associated with less deflection but higher ISE. The higher ISE is a result of the bigger change of angle, which has a non-linear expression in the model. Therefore the linear controller is less successful in controlling it. The second MOP is associated with a design of a symmetric truss, which is loaded with a load W as depicted in figure 8b. The objectives of the second MOP are to minimize $f_2^1 = T = 0.5W(\cos\alpha)^{-1}, f_2^2 = \Delta f = \sqrt{2}\frac{WL}{AE}\sqrt{1+\sin 2\alpha}$, which are the tension in a bar and the deflection of the junction. The front for the second MOP is depicted in figure 9b designated by blank circles. The longer bars are associated with less tension (smaller angle) and with higher deflection. The m-MOOP is to optimize the solutions for both MOPs with a demand for commonality, such that, the truss bars should be the same part as the manipulator arm. Using the sequential approach and starting with MOP_2, then with MOP_1 results in the fronts, which are designated by filled symbols in figures 9a and 9b respectively. The numerical results of the measures for both cases are summarized in table 2.

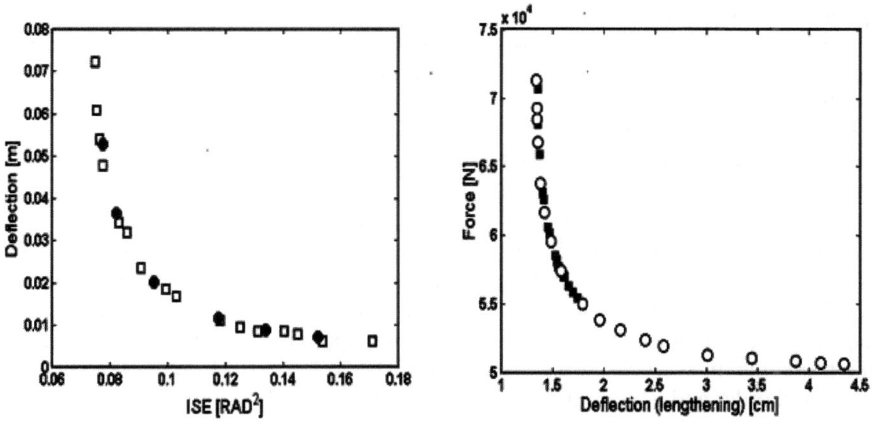

(a) (b)

Fig. 9. Starting with MOP_2 (b) Starting with MOP_1

Table 2. Measures' values

Measure / Start	WRM	LOM	T
MOP_2	16	0.008	52t
MOP_1	15	0.009	t

From the results it is depicted that the loss of optimality is minimal. This is due to the fact that the communal components that were found, guarantee optimal performances with respect to both the m-MOOP and with respect to each of the the MOPs. The waste of resources do exists, nevertheless, some 5 different communal solutions are found in each case. A profound difference is depicted in the computational time. It takes an average of 52 times more time to find the fronts of the m-MOOP when MOP_2 is the first to be solved.

4 Summary Conclusions and Future Work

In this paper, a new type of MOPs is introduced. This involves the search for optimal solutions for different MOPs, which although distinct, possess communal components. The solution to the posed m-MOOP problem is also described and the design of a search for such a solution by means of a sequential EMO is proposed. The result of the proposed search is a set of fronts, which are not necessarily the individual MOPs' Pareto fronts, but are the solution to the m-MOOP. Several measures, which can be used to assess the performances of the proposed algorithm, are presented. These measures can not only serve for comparison between different orders at which the sequential approach is implemented, but can also serve as a base for a future comparison with other approaches. Academic examples as well as an engineering design example have been used to demonstrate the introduced algorithm and its applicability to "real-life" problems.

It has been demonstrated in this paper that the sequential approach is applicable in solving the m-MOOP problem. It is appropriate for solving an m-MOOP which involves a high preference to one MOP, which should be solved first. Moreover it finds communal solutions with no loss of optimality in case 3 related problems. Nonetheless, it seems to possess several drawbacks including: a) not all solutions on the front of one MOP have optimal corresponding solutions in other MOPs. This means that there is an apparent waste of computational resources. b) The results are dependent on the sequential order in which the MOPs are selected and solved. This means that different fronts and therefore different solutions will be obtained for different sequential orders. c) The overall computation time is highly dependent on the order at which the sequential approach is implemented. d) Loss of optimality can be detected in case-2-like problems. In such cases, the loss of optimality is dependent on the selection of the sequential order and is present in all MOPs, but not in the first one.

As a future work, the above drawbacks should be resolved mainly by taking a new approach, which solves the problem simultaneously. Such a simultaneous algorithm is currently under investigation. The so-far obtained results are promising, showing an improvement with respect to all of this paper introduced measures.

Bibliography

1. Fellini, R., Kokkolaras, M., Panos, Papalambros Y., and Perez-Duarte, A. Platform Selection Under Performance Loss Constraints in Optimal Design of Product Families, Proceedings of DETC.02 ASME 2002 Design Engineering Technical Conferences and Computer and Information in Engineering Conference Montreal, Canada, September 29-October 2, 2002.
2. Robertson, D. and Ulrich, K. Planning Product Platforms, Sloan Management Review, 39(4): 19–31, 1998.
3. Lehnerd, A. P., 1987, "Revitalizing the Manufacture and Design of Mature Global Products," *Technology and Global Industry: Companies and Nations in the World Economy,* B. R. Guile and H. Brooks, eds., National Academy Press, Washington, D.C., pp. 49-64.

4. Simpson , T.W., and D'Souza, B.S. Assessing Variable Levels of Platform Commonality Within a Product Family Using a Multiobjective Genetic Algorithm., *Journal of Concurrent Engineering: Research and Applications*, 12(2): 119-129, 2004.
5. Project IST-2001-34820 Advanced Real-Time Systems, Component-based Design and Integration Platforms W1.A2.N1.Y1, 2003, http://www.artist-embedded.org/
6. Weyuker, E.J., *Testing Component-Based Software: A Cautionary Tale.* IEEE Software, 15(5):p. 54-59, 1998.
7. C. Dellarocas. The synthesis environment for component based software development. In *Proceedings of 8th International Workshop on Software Technology and Engineering Practice*, London, UK, July 1997.
8. Mattson, C. A., and Messac, A. Pareto Frontier Based Concept Selection under Uncertainty, with Visualization, *Optimization and Engineering*, 6: 85–115, 2005.
9. Van Veldhuizen, D.A., and Lamont, G.B. Multiobjective Evolutionary Algorithms: Analyzing the State-of-the-Art, *Evolutionary Computation*, 8(2): 125-147, 2000.
10. Pareto, V. *Cours D'Economic Politique*, Volume 1 and 2 Frouge, Lausanne, 1896.
11. Steuer, R. E. Multiple Criteria Optimization, Theory Computations and Applications, John Wiley & Sons, Inc.: New York, 1986.
12. Marler, R.T., and Arora, J.S. Review of Multi-Objective Optimization Concepts and Methods for Engineering. *Technical Report* Number ODL-01.01, University of Iowa, Optimal Design Laboratory, Iowa City, IA, 2003.
13. Fonseca, C.M, Fleming P.J. Genetic algorithms for multiobjective optimization: formulation, discussion and generalization. *Proceedings of the Fifth International Conference on Genetic Algorithms*; 416–23, 1993.
14. Goldberg, D.E. *Genetic algorithms in search, optimization and machine learning*, Reading, MA: Addison-Wesley, 1989.
15. Coello, C.A.C. Recent Trends in Evolutionary Multiobjective Optimization, in Ajith Abraham, Lakhmi Jain and Robert Goldberg (editors), *Evolutionary Multiobjective Optimization: Theoretical Advances And Applications*, pp. 7-32, Springer-Verlag, London, 2005.
16. Deb, K., Pratap, A., Agarwal, S., and Meyarivan, T., 'A Fast and Elitist Multiobjective Genetic Algorithm: NSGA–II,' *IEEE Transactions on Evolutionary Computation*, 6(2):182–197, April 2002.
17. Deb, K., *Multi-Objective Optimization using Evolutionary Algorithms.* J. Wiley & Sons, Ltd., 2001.
18. Laumanns, M., Thiele, L., Deb, K., and Zitzler, E. Combining convergence and diversity in evolutionary multi-objective optimization. *Evolutionary Computation, 10*(3): 263.282, 2002.
19. Bosman, P.A.N. and Thierend, D. The balance between Proximity and diversity in multiobjective evolutionary algorithms, *IEEE Transactions on Evolutionary Computation*, 7 (2), 2003.
20. Nelson, S.A., Parkinson, M.B., and Papalambros, P.Y. "Multicriteria Optimization in Product Platform Design," *Journal of Mechanical Design*, 123(2):199-204, 2001.
21. Rai, R., and Allada, V. Modular product family design: agent-based Pareto-optimization and quality loss function-based post-optimal analysis, *International Journal of Production Research*, 41(17), 2003.
22. Simpson , T.W., and D'Souza, B.S. Assessing Variable Levels of Platform Commonality Within a Product Family Using a Multiobjective Genetic Algorithm., *Journal of Concurrent Engineering: Research and Applications*, 12(2): 119-129, 2004.
23. G. Avigad, A. Moshaiov and N. Brauner, "Towards a general tool for mechatronic design," in the *Proc. of the 2003 IEEE CSS Conference on Control Applications, CCA 2003,* June 23-25, 2003, Istanbul, Turkey.

The Hypervolume Indicator Revisited: On the Design of Pareto-compliant Indicators Via Weighted Integration

Eckart Zitzler, Dimo Brockhoff, and Lothar Thiele

Computer Engineering (TIK), ETH Zurich
{zitzler,brockho,thiele}@tik.ee.ethz.ch
http://www.tik.ee.ethz.ch/sop/

Abstract. The design of quality measures for approximations of the Pareto-optimal set is of high importance not only for the performance assessment, but also for the construction of multiobjective optimizers. Various measures have been proposed in the literature with the intention to capture different preferences of the decision maker. A quality measure that possesses a highly desirable feature is the hypervolume measure: whenever one approximation completely dominates another approximation, the hypervolume of the former will be greater than the hypervolume of the latter. Unfortunately, this measure—as any measure inducing a total order on the search space—is biased, in particular towards convex, inner portions of the objective space. Thus, an open question in this context is whether it can be modified such that other preferences such as a bias towards extreme solutions can be obtained. This paper proposes a methodology for quality measure design based on the hypervolume measure and demonstrates its usefulness for three types of preferences.

1 Motivation

Using the hypervolume of the dominated portion of the objective space as a measure for the quality of Pareto set approximations has received more and more attention in recent years. The reason is that this measure has two important advantages over other set measures:

1. It is sensitive to any type of improvements, i.e., whenever an approximation set A dominates another approximation set B, then the measure yields a strictly better quality value for the former than for the latter set [23].
2. As a result from the first property, the hypervolume measure guarantees that any approximation set A that achieves the maximally possible quality value for a particular problem contains all Pareto-optimal objective vectors [5].

So far, this is the only measure known in the literature on evolutionary multi-criterion optimization that possesses these properties.

The hypervolume measure—or *hypervolume indicator* [23]—was first proposed and employed in [21,22] where it was denoted as 'size of the space covered'; later,

S. Obayashi et al. (Eds.): EMO 2007, LNCS 4403, pp. 862–876, 2007.

also other terms such as 'hyperarea metric' [14], 'S-metric' [18], and 'Lebesgue measure' [11,5] were used. On the one hand, the hypervolume indicator is meanwhile among the most popular measures for the performance assessment of multiobjective optimizers and in this context it has been subject to several theoretical investigations [8,5,23,15]. On the other hand, there are some studies that discuss the usage of this measure for multiobjective search [10,20,4] and in particular the issue of fast hypervolume calculation has been a focus of research recently [16,17,6,1].

Despite the aforementioned advantages of the hypervolume indicator, it inevitably has its biases. There is some freedom with respect to the choice of the reference point, but nevertheless it represents only one particular class of preference information that may not be appropriate in specific situations. This discussion directly leads to the question of whether it is possible to design quality measures that (i) share the two above properties of the hypervolume indicator, while (ii) standing for a different type of preferences of the decision maker. The fact that besides the hypervolume no other measures with these properties are known indicates that the formalization of *arbitrary* preferences in terms of a quality measure may be difficult. However, not being aware of such measures does not imply that such indicators do not exist.

This paper presents a first step to tackle this issue: it demonstrates that novel quality measures with the aforementioned properties can be designed and proposes a general design methodology on the basis of the hypervolume indicator. In detail, the key contributions are:

- A generalized definition of the hypervolume indicator using attainment functions [2] that can be used for any type of dominance relation;
- A weighted-integration approach to directly manipulate and control the influence of certain regions in the objective space for the hypervolume indicator;
- Three new example set measures for biobjective problems that provide the same sensitivity as the hypervolume indicator, but represent different types of preference information: (i) the preference of extreme solutions, (ii) the preference of predefined reference points, and (iii) bias towards one of the objectives.

The usefulness of the methodology and the three proposed measures is demonstrated on selected test problems.

2 Mathematical Framework

2.1 Preliminaries

Without loss of generalization, we consider a maximization problem with n objective functions $f_i : X \to (0,1)^n$, $1 \leq i \leq n$. Requiring the objective values to lay between 0 and 1 instead of using $\mathbb{R}^n$ as objective space simplifies the following discussions, but does not represent a serious limitation as there exists a bijective mapping from $\mathbb{R}$ into the open interval $(0,1) \subset \mathbb{R}$. The objective functions map a solution $\boldsymbol{x} \in X$ in the *decision space* to an objective vector $f(\boldsymbol{x}) = (f_1(\boldsymbol{x}), \ldots, f_n(\boldsymbol{x})) \in (0,1)^n$ in the *objective space* $Z = (0,1)^n$.

In the following, the (weak) Pareto-dominance relation $\succeq$ is used as a preference relation on the search space X indicating that a solution $\boldsymbol{x}$ is at least as good as a solution $\boldsymbol{y}$ ($\boldsymbol{x} \succeq \boldsymbol{y}$) if and only if $\forall 1 \leq i \leq n : f_i(\boldsymbol{x}) \geq f_i(\boldsymbol{y})$.[1] This relation can be canonically extended to sets of solutions where a set $A \subseteq X$ weakly dominates a set $B \subseteq X$ ($A \succeq B$) iff $\forall \boldsymbol{y} \in B\ \exists \boldsymbol{x} \in A : \boldsymbol{x} \succeq \boldsymbol{y}$ [23]. Note that any other type of dominance relation, e.g., based on arbitrary convex cones [13], could be used as well, and the considerations in this paper apply to any dominance relation.

Given the preference relation $\succeq$, we consider the optimization goal to identify a set of solutions that approximates the set of Pareto-optimal solutions and ideally this set is not strictly dominated by any other approximation set. For reasons of simplicity though, we assume that the outcome of a multiobjective optimizer is a set of objective vectors, also denoted as *approximation set*, and the set of all possible objective vector sets is denoted as $\Omega := 2^Z$. Therefore, we will also use the symbol $\succeq$ for objective vectors and objective vector sets, although it is originally defined on X. In practice, one always obtains a set of decision vectors instead of objective vectors, but most often only the objective vectors are considered to evaluate the quality of a solution set.

Since the generalized weak Pareto dominance relation $\succeq$ defines only a partial order on Ω, there may be incomparable sets in Ω which may cause difficulties with respect to search and performance assessment. These difficulties become more serious as the number of objectives in the problem formulation increases, see [3] for details. One way to circumvent this problem is to define a total order on Ω which guarantees that any two objective vector sets are mutually comparable. To this end, quality indicators have been introduced that assign, in the simplest case, each approximation set a real number, i.e., a (unary) indicator I is a function $I : \Omega \to \mathbb{R}$, cf. [23]. One important feature an indicator should have is *Pareto compliance* [9], i.e., it must not contradict the order induced by the Pareto dominance relation. In detail, this means that whenever $A \succeq B \wedge B \not\succeq A$, then the indicator value of A must not be worse than the indicator value of B. A stricter version of compliance would be to require that $A \succeq B \wedge B \not\succeq A$ implies that the indicator value of A is strictly better than the indicator value of B (if better means a higher indicator value):

$$A \succeq B \ \wedge \ B \not\succeq A \Rightarrow I(A) > I(B)$$

So far, the hypervolume indicator has been the only known indicator with this property.

[1] If $\boldsymbol{x} \succeq \boldsymbol{y}$, we say $\boldsymbol{x}$ *weakly dominates* $\boldsymbol{y}$. Two solutions $\boldsymbol{x}$ and $\boldsymbol{y}$ are called *incomparable* if neither weakly dominates the other one. If for two solutions $\boldsymbol{x}$ and $\boldsymbol{y}$ both $\boldsymbol{x} \succeq \boldsymbol{y}$ and $\boldsymbol{y} \not\succeq \boldsymbol{x}$ holds, we say that $\boldsymbol{x}$ is strictly better than $\boldsymbol{y}$ or $\boldsymbol{x}$ *strictly dominates* $\boldsymbol{y}$ ($\boldsymbol{x} \succ \boldsymbol{y}$). A solution $\boldsymbol{x}^* \in X$ is called *Pareto optimal* if there is no other $\boldsymbol{x} \in X$ that strictly dominates $\boldsymbol{x}^*$.

2.2 The Hypervolume Indicator

The classical definitions of the hypervolume indicator are based on volumes of polytopes [22] or hypercubes [5] and assume that Pareto dominance is the underlying preference relation. Here, we give a generalized definition based on attainment functions that allows to consider arbitrary dominance relations.

The attainment function [2] gives, roughly speaking, for each objective vector in Z the probability that it is weakly dominated by the outcome of a particular multiobjective optimizer. As only single sets are considered here, we can take a slightly simplified definition of the attainment function:

Definition 1 (Attainment function for an objective vector set). *Given a set $A \in \Omega$, the attainment function $\alpha_A : [0,1]^n \rightarrow \{0,1\}$ for A is defined as*

$$\alpha_A(\boldsymbol{z}) := \begin{cases} 1 & \textit{if } A \succeq \{\boldsymbol{z}\} \\ 0 & \textit{else} \end{cases}$$

for $\boldsymbol{z} \in Z$.

This definition is illustrated for weak Pareto dominance in Fig. 1 and applies to any type of dominance relation.

The concept of attainment functions can now be used to give a formal definition of the well known hypervolume indicator. It is simply defined as the volume of the objective space enclosed by the attainment function and the axes.

Definition 2 (Hypervolume indicator). *The hypervolume indicator I_H^* with reference point $(0,\ldots,0)$ can be formulated via the attainment function as*

$$I_H^*(A) := \int_{(0,\ldots,0)}^{(1,\ldots,1)} \alpha_A(\boldsymbol{z}) d\boldsymbol{z}$$

where A is any objective vector set in Ω.

In the following section, we will give a rough overview about the new concepts that are introduced in the paper and illustrate how $I_H^*(A)$ can be modified to incorporate preference information without violating compliance to Pareto dominance.

3 Introductory Example and Outline of the Proposed Approach

The attainment function, the integration over which gives the hypervolume for a given set A, is a binary function: all weakly dominated objective vectors are assigned 1, while the remaining objective vectors are assigned 0. That means all weakly dominated objective vectors have the same weight and contribute equally to the overall indicator value.

The main idea behind the approach proposed in this paper is to give different weights to different regions in the objective space. This can be achieved by

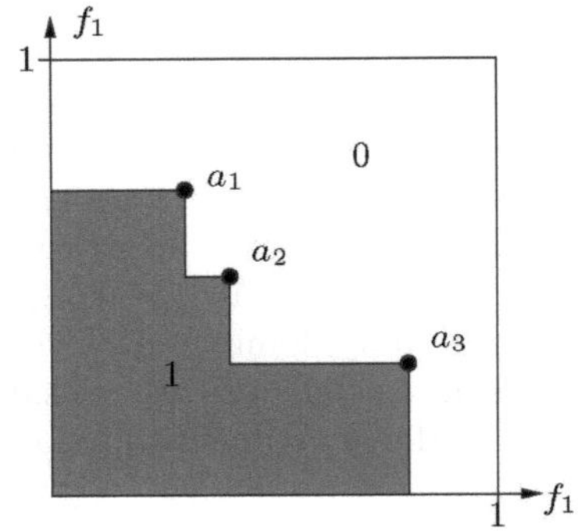

Fig. 1. Illustration of the attainment function α_A for $A = \{\boldsymbol{a}_1, \boldsymbol{a}_2, \boldsymbol{a}_3\}$ in the two-dimensional case

defining a weight distribution over the objective space such that the value that a particular weakly dominated objective vector contributes to the overall indicator value can be any real value greater than 0—provided the integral over the resulting function still exists. To this end, we introduce a weight distribution function $w : Z \to \mathbb{R}^+$, and the hypervolume is calculated as the integral over the product of the weight distribution function and the attainment function:

$$I_H^w(A) := \int_{(0,\ldots,0)}^{(1,\ldots,1)} w(\boldsymbol{z}) \cdot \alpha_A(\boldsymbol{z}) d\boldsymbol{z}$$

As will be shown later, thereby the basic hypervolume indicator can be modified such that (a) the compliance to Pareto dominance is preserved and (b) preference information can be flexibly introduced.

To see what the effect of different weight distribution functions is on the behavior of the corresponding modified hypervolume indicator I_H^w, it is helpful to consider equi-indicator surfaces. An equi-indicator surface $S(I, K)$ for a given indicator function I and an indicator value K is defined as the set of points $\boldsymbol{z} \in Z$ that each has an indicator value K, i.e.:

$$S(I, K) = \{\boldsymbol{z} \in Z \, : \, I(\{z\}) = K\}$$

In other words, the equi-indicator surface represents the indicator field for approximation sets with a single element.

If we consider a uniform weight distribution function with $w(\boldsymbol{z}) = 1$ for $\boldsymbol{z} \in Z$, we obtain the standard hypervolume indicator I_H^*. In this case, the equi-indicator surfaces looks for $n = 2$ as depicted in Fig. 2a; from the representation of these curves, one can conclude that the standard hypervolume indicator has convex equi-indicator surfaces and therefore implicitly introduces a preference towards solutions close the the diagonal. For instance, consider the point $(0.5, 0.5)$ located on the diagonal. To obtain the same indicator value for a point not on the diagonal, the degradation in one objective needs to be compensated by a larger

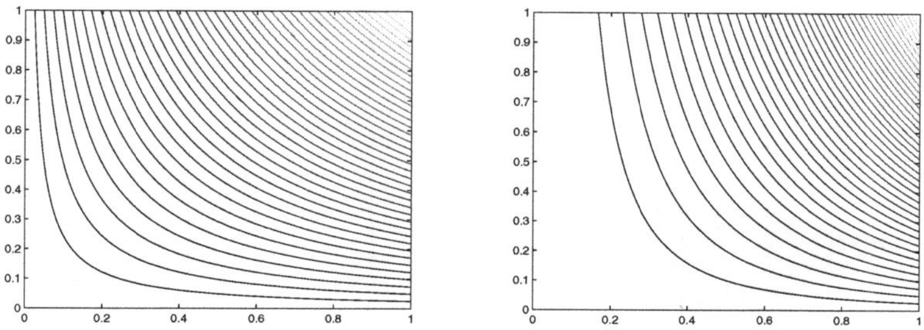

(a) The hypervolume indicator

(b) A biased indicator

Fig. 2. Equi-indicator surfaces for simple indicators in the biobjective case. The abscissa in these two-dimensional examples denotes f_1 and the ordinate f_2. Figure (a) shows (a sample of) surfaces for the hypervolume indicator I_H^* (weight distribution function $w((z_1, z_2)) = 1$), Figure (b) illustrates a biased, modified indicator with weight distribution function $w((z_1, z_2)) = z_1$. Points on one equipotential curve share the same indicator value.

improvement in the other objective, e.g., $(0.25, 1)$ instead of $(0.25, 0.75)$ where degradation and improvement would be both 0.25.

If we now change the weight distribution function to $w(\boldsymbol{z}) = z_1$ with $\boldsymbol{z} = (z_1, z_2, \ldots, z_n)$, then in the biobjective case the equi-indicator surfaces shown in Fig. 2b are obtained. Obviously, solutions with objective vectors that have large components in the direction of z_1 are preferred.

Another possibility is to impose special emphasis on the border of the objective space, see Fig. 3a. The objective vectors in the 'center' of the objective space have weight 1, while the objective vectors on the axes are assigned a substantially larger weight.[2] The corresponding equi-indicator surfaces are shown in Fig. 3b. Here, the bias of the original hypervolume indicator for a single solution towards the diagonal is removed by putting more emphasis on the areas close to the coordinate axes.

The above two examples illustrate how weight distribution functions on the objective space can be used to change the bias of the hypervolume indicator. Based on these informal observations, we will describe the underlying methodology next.

4 Methodology: The Weighted-Integration Approach

The main concept of the approach proposed in this paper is to extend the basic hypervolume indicator by a weight distribution function $w : [0, 1]^n \rightarrow \mathbb{R}^+$ which serves to emphasize certain regions of the objective space:

[2] Since the borders have zero width, they will actually not influence the integral; therefore, dirac-type functions need to be used to make the border weights effective.

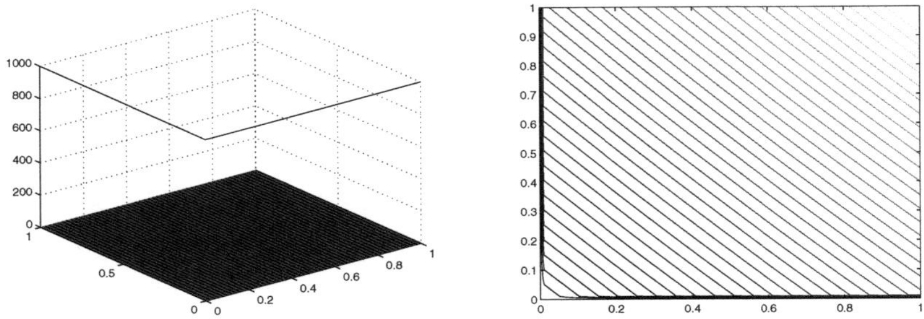

(a) Weight distribution function with emphasis on the coordinate axes

(b) (Sample of the) equi-indicator surfaces for the corresponding indicator

Fig. 3. Weight distribution function (left) and corresponding indicator (right) when stressing on coordinate axes

Definition 3 (Generalized Hypervolume Indicator). *The generalized hypervolume indicator I_H^w with weight distribution function $w : [0,1]^n \rightarrow \mathbb{R}^+$ is defined as the weighted integral*

$$I_H^w(A) := \int_{(0,\ldots,0)}^{(1,\ldots,1)} w(\boldsymbol{z}) \cdot \alpha_A(\boldsymbol{z}) d\boldsymbol{z}$$

where A is an approximation set in Ω.

If using this indicator as the basis for optimization algorithms or performance assessment tools, it would be important to know whether it is compliant with the concept of Pareto-dominance. This property will be shown next.

Theorem 1. *Let w be a weight distribution function $w : [0,1]^n \rightarrow \mathbb{R}^+$ such that the corresponding generalized hypervolume indicator I_H^w is well-defined for all $A \in \Omega$. Then for any two arbitrary approximation sets $A \in \Omega$ and $B \in \Omega$, it holds*

$$A \succeq B \ \wedge \ B \not\succeq A \Rightarrow I_H^w(A) > I_H^w(B).$$

Proof. If we have $A \succeq B \ \wedge \ B \not\succeq A$, then the following two conditions hold: $\forall \boldsymbol{y} \in B \ \exists \boldsymbol{x} \in A : \boldsymbol{x} \succeq \boldsymbol{y}$ and $\exists \boldsymbol{x} \in A \ \ \nexists \boldsymbol{y} \in B : \boldsymbol{y} \succeq \boldsymbol{x}$. Now we can easily see that the attainment functions of A and B satisfy $(\alpha_A(\boldsymbol{z}) = 1) \Rightarrow (\alpha_B(\boldsymbol{z}) = 1)$ as $A \succeq B$. Every point in the objective space that is weakly dominated by some element in B is also weakly dominated by some element in A. In addition, as $B \not\succeq A$ there are some points in the objective space that are weakly dominated by points in A but not weakly dominated by points in B. Therefore, there exists a region $R \subset Z$ with $(\alpha_A(\boldsymbol{z}) = 1) \wedge (\alpha_B(\boldsymbol{z}) = 0)$ for $\boldsymbol{z} \in R$; in particular:

$$\int_{(0,\ldots,0)}^{(1,\ldots,1)} (\alpha_A(\boldsymbol{z}) - \alpha_B(\boldsymbol{z})) d\boldsymbol{z} > 0$$

Using the definition of the generalized hypervolume indicator and noting that $w(\boldsymbol{z}) > 0$, we find $I_H^w(A) > I_H^w(B)$. □

In order to simplify the definition of weight distribution functions and to avoid the use of dirac-type functions, we use a slightly different representation of the generalized hypervolume indicator where line segments can be used to establish emphasis on zero-width regions such as axes. Every line segment l_i is specified by a start point $\boldsymbol{s}_i \in Z$, an end point $\boldsymbol{e}_i \in Z$, and a corresponding weight distribution function $\overline{w}_i : [0,1] \to \mathbb{R}_0^+$. Using these notation, we can rewrite the generalized hypervolume indicator according to Def. 3 as follows

$$I_H^{w,\overline{w}_1,\overline{w}_2,\ldots,\overline{w}_L}(A) := \int_{(0,\ldots,0)}^{(1,\ldots,1)} w(\boldsymbol{z}) \cdot \alpha_A(\boldsymbol{z}) \cdot d\boldsymbol{z} +$$

$$\sum_{i \in \{1,2,\ldots,L\}} \int_0^1 \overline{w}_i(\boldsymbol{z}) \cdot \alpha_A(\boldsymbol{s}_i + t \cdot (\boldsymbol{e}_i - \boldsymbol{s}_i)) \cdot dt$$

Assuming that the weight distribution functions are chosen such that all integrals are well-defined, it is easy to see that the property proven in Theorem 1 is preserved.

In the following, we will discuss three examples of useful weight distribution functions that will also be used for experimental results.

1. The first weight distribution function is the sum of two exponential functions in direction of the axes:

   $$w^{ext}(\boldsymbol{z}) = (e^{20 \cdot z_1} + e^{20 \cdot z_2})/(2 \cdot e^{20})$$

   with $L = 0$. The effect is an indicator with preference of extremal solutions. Because of the weight distribution function's steep slope near the two axes, a Pareto front approximation with solutions crowded near the axes yield a larger indicator value than a population with solutions in the interior region of the objective space where the weight distribution function contribute less to the indicator value.
2. The second weight distribution function focuses on the second objective by using an exponential function in f_2-direction:

   $$w^{asym}(\boldsymbol{z}) = e^{20 \cdot z_2}/e^{20}$$

   In addition, the following line segment with a constant weight distribution function on the f_1-axis is used:

   $$\overline{w}_1^{asym}(\boldsymbol{z}) = 400, \quad \boldsymbol{s}_1 = (0,0), \quad \boldsymbol{e}_1 = (1,0)$$

   where $L = 1$. This combination results in an indicator preferring solutions with extreme f_2 values and an additional solution near the f_1 axis. The additional line segment along the f_1 axis used here instead of an additional exponential function in f_1 direction yields only a single additional solution lying near the f_1 axis instead of many solutions with large f_1 value as with the weight distribution function defined above.

3. Often, a decision maker has some idea which points in the search space are the most important ones. With the third weight distribution function, we can integrate such information into a Pareto-compliant indicator. A point (a, b) of interest, also called reference point, can be chosen in advance. The weight distribution function defined below will direct the search of indicator based algorithms towards this point. Multiple reference points can be considered simultaneously by adding up the corresponding distinct weight distribution functions.

   The following weight distribution function is based on a ridge-like function through the reference point (a, b), parallel to the diagonal:

$$w^{ref}(\boldsymbol{z}) = \begin{cases} c + \frac{(2-((2(x-a))^2+(2(y-b))^2))}{(0.01+(2(x-a)-2(y-b))^2)} & \text{if } |z_1 - a| < 0.5 \wedge |z_2 - b| < 0.5 \\ c & \text{else} \end{cases}$$

   with $L = 0$. The constant c should be chosen small in comparison to the values of the ridge; here, we use $c = 10^{-5}$.

The computation of the generalized hypervolume indicator is based on the representation described above. It first partitions the whole unit hypercube $[0, 1]^n$ into smaller hyperrectangles based on the objective vectors contained in the set A, and then the weighted volumes of these hyperrectangles are added. To this end, the above weight distribution functions have been symbolically integrated using a commercial symbolic mathematics tool.

5 Proof-of-Principle Results

5.1 Simple Indicator-Based Optimization Algorithm

For the experimental validation of the weighted-integration approach, a simple indicator-based evolutionary algorithm (SIBEA) is considered that uses similar concepts as proposed in [10,20,4,7]. As the purpose of this section is to show the influence of preference information which has been incorporated into the generalized hypervolume indicator and not to compare different optimization algorithms, methods to improve the convergence rate such as fitness-based mating selection are not taken into account.

SIBEA (Simple Indicator-Based Evolutionary Algorithm)

Input: population size μ; number of generations N; indicator function I;

Output: approximation of Pareto-optimal set A;

Step 1 (Initialization): Generate an initial set of decision vectors P of size μ; set the generation counter $m := 0$.

Step 2 (Environmental Selection): Iterate the following three steps until the size of the population does no longer exceed μ:

1. Rank the population using Pareto dominance and determine the set of individuals $P' \subseteq P$ with the worst rank.

2. For each solution $\boldsymbol{x} \in P'$ determine the loss $d(\boldsymbol{x})$ w.r.t. the indicator I if it is removed from P', i.e., $d(\boldsymbol{x}) := I(P') - I(P' \setminus \{x\})$.
3. Remove the solution with the smallest loss $d(\boldsymbol{x})$ from the population P (ties are broken randomly).

Step 3 (Termination): If $m \geq N$ then set $A := P$ and stop; otherwise set $m := m + 1$.

Step 4 (Mating): Randomly select elements from P to form a temporary mating pool Q of size μ. Apply variation operators such as recombination and mutation to the mating pool Q which yields Q'. Set $P := P + Q'$ (multi-set union) and continue with Step 2.

As to the environmental selection step, an issue are dominated individuals in the population: they never lead to a change in the indicator value which is entirely determined by the nondominated front of the population. Therefore, the population is first partitioned into fronts (Step 2.1) using the dominance rank (number of dominating individuals)[3], and only individuals located in the worst front are investigated for deletion.

Furthermore, we consider two scaling variants to obtain the maximum effect of the weighted integral: online and offline scaling. In the online variant, the objective function values are scaled to the interval $[0, 1]$ within each generation; to guarantee that boundary solutions contribute a non-zero hypervolume to the overall indicator value, for each axis a line segment with a constant weight distribution function is added. The offline variant does not scale the objective function values but the weighting distribution function. In detail, the weighted integral is only computed over and scaled to the region of the Pareto front, which needs to be known in advance. Since any approximation set outside this region would yield an indicator value of zero, the standard hypervolume indicator value, down-scaled such that is does not interfere with the weighted integral, is added.

5.2 Experiments

We now show how the three weight distribution functions defined above influence the search process of SIBEA for three biobjective test problems. For each weight distribution function, we derive two indicators, one for the online scaling method and one for offline scaling, resulting in six different indicators overall. We name the corresponding indicators I_H^{ext}, I_H^{asym}, and I_H^{ref} respectively, and distinguish between the online and the offline version. The same holds for the usual hypervolume indicator I_H, where we also distinguish between the two scaling methods.

The test functions ZDT1, ZDT3, and ZDT6, cf. [19], are optimized by a SIBEA run with population size 20 for 1000 generations.[4] Note, that the ZDT

[3] A nondominating sorting could be used as well.

[4] The individuals are coded as real vectors with 30 (ZDT1 and ZDT3) and 10 (ZDT6 decision variables, where the SBX-20 operator is used for recombination and a polynomial distribution for mutation. The recombination and mutation probabilities were set to 1.0, according to [3].

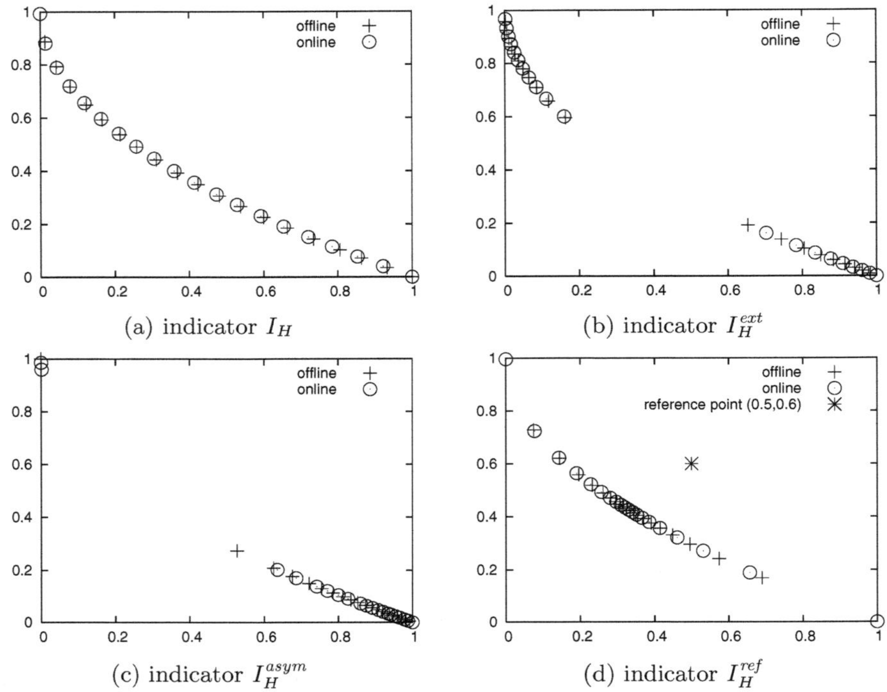

Fig. 4. Pareto front approximations for the three different indicators based on weight distribution functions on the function ZDT1. For reference, the generated Pareto front approximation using the usual hypervolume indicator I_H is given in (a). The two scaling methods are plotted for comparison.

functions are to be minimized. Thus, an internal transformation is performed, independent whether the online or offline scaling is enabled.

The figures Fig 4, Fig. 5, and Fig. 6 show the computed Pareto front approximations after 1000 generations for the three ZDT functions and the three indicators I_H^{ext}, I_H^{asym}, and I_H^{ref} with both scaling methods. The reference point for I_H^{ref} is chosen as $(0.5, 0.6)$ for ZDT1 and ZDT6 and as $(0.5, 1.2)$ for ZDT3[5]. The approximation derived with the established hypervolume indicator I_H is also shown as golden reference.

The experiments show two main aspects. Firstly, the behavior of the evolutionary algorithm is similar for all three problems if always the same indicator is used—independent of the front shape and the scaling method used. With the indicator I_H^{ext} the solutions accumulate near the extremal points. When using the indicator I_H^{asym}, mainly the f_2 values are minimized. Due to the additional weight on the line segment, at least one solution with large f_1 value is also kept in the population if I_H^{asym} is used. With the indicator I_H^{ref}, the population moves towards

[5] The reference point is changed for ZDT3 due to the larger Pareto-optimal front of the ZDT3 problem.

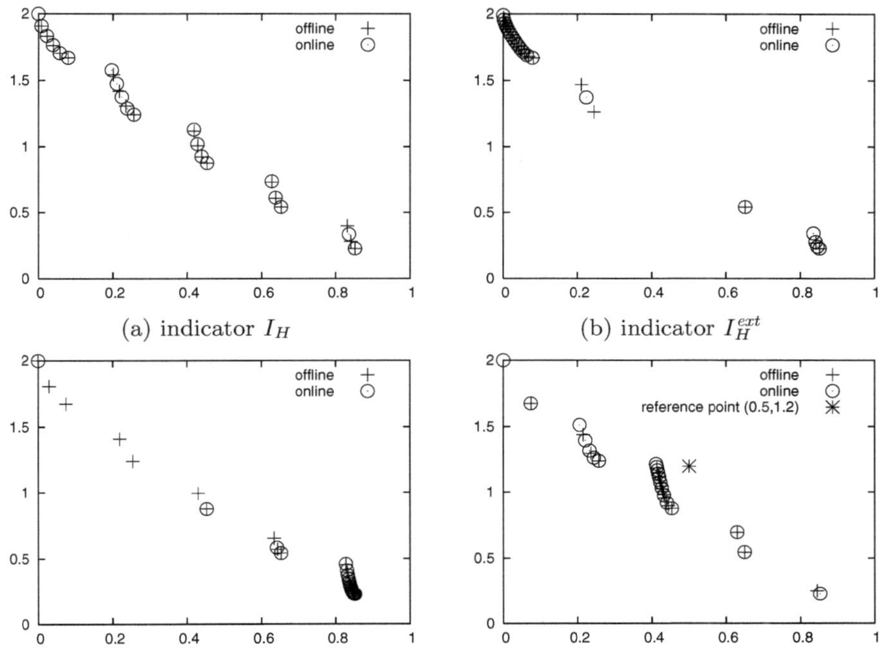

(a) indicator I_H

(b) indicator I_H^{ext}

(c) indicator I_H^{asym}

(d) indicator I_H^{ref}

Fig. 5. Pareto front approximations for the three different indicators and the two scaling methods on the function ZDT3. For reference, the generated Pareto front approximation using the usual hypervolume indicator I_H is given in (a). Due to the larger Pareto-optimal front, the reference point for I_H^{ref} is chosen as $(0.5, 1.2)$.

the predefined reference point $(0.5, 0.6)$, and $(0.5, 1.2)$ respectively. Secondly, the weighted-integration approach seems to be feasible for designing new Pareto-compliant indicators according to specific preferences. The simple indicator-based algorithm was indeed attracted to those regions in the objective space that were particularly emphasized by means of large weight values.

When comparing the two scaling variants, online and offline, only slight differences can be observed with the test cases studied in this paper. Online scaling has the advantage that the preferences are always adapted according to the current shape of the Pareto front approximation. However, thereby the actual global indicator changes during the run and potentially cycles can occur during the optimization process—a phenomenon, cf. [12], that emerges with most algorithms. Cycling is not necessarily a problem in the biobjective case, but as the number of objectives increases, it is likely that this behavior causes difficulties. The alternative is offline scaling. Here, the indicator remains fixed and can be used for comparing the outcomes of different methods. The drawback of this approach is the requirement that domain knowledge is available: either about the location of the Pareto front or about regions of interest. This problem holds basically for all types of indicators.

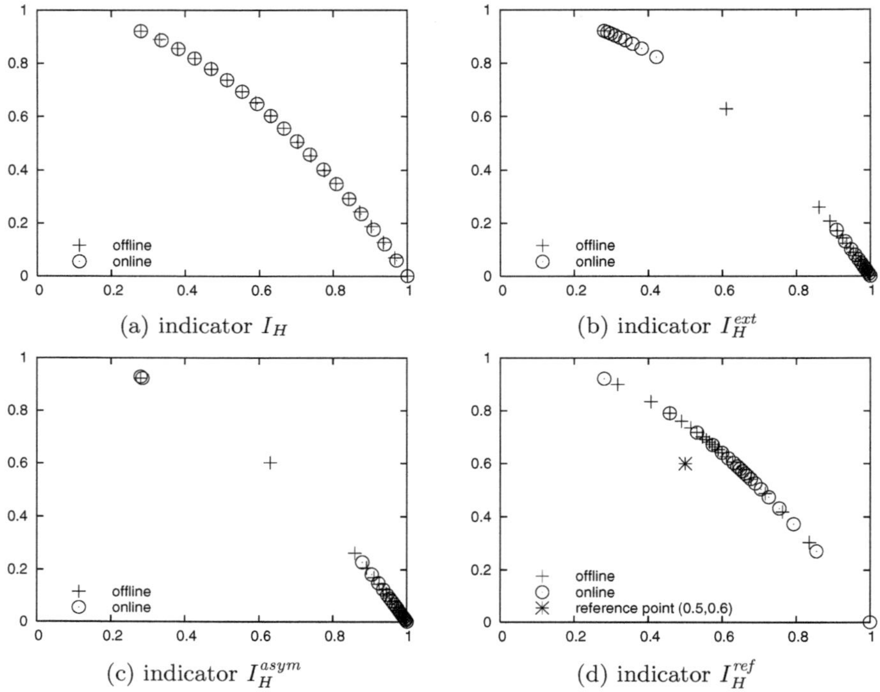

Fig. 6. Pareto front approximations for the three different indicators based on weight distribution functions on the function ZDT6. Plot (a) shows the generated Pareto front approximation using the usual hypervolume indicator I_H for comparison. The reference point for I_H^{ref} is chosen as $(0.5, 0.6)$.

6 Discussion

This paper has introduced a novel methodology to design Pareto-compliant indicators on the basis of the hypervolume indicator. Different preferences can be integrated, while an important property of the hypervolume indicator, sensitivity to dominance, is preserved. This is insofar an important result as up to now the pure hypervolume indicator has been the only one with this property. The possibility to design dominance-sensitive and Pareto-compliant indicators that can guide the search towards extreme solutions or reference points offers therefore more flexibility to tune the search with respect to the decision maker's preferences. We have demonstrated how this approach works and can be used for three example indicators in biobjective scenarios. As expected, the outcomes reflected the encoded preferences.

The presented methodology offers new ways for multiobjective optimization and performance assessment. However, this paper is just a first step in this direction, and the capabilities as well as the limitations of the weighted-integration approach need to be explored and require more research. In particular, the following considerations may point to interesting future research topics:

- The presented new indicators are designed for biobjective problems, but clearly one is interested in general indicators for n objectives; the first two indicators can be easily extended to higher dimensions, but for the ridge-based indicators this extension is not straight forward. The definition of general indicator classes for an arbitrary number of objectives will be one of the next steps to take.
- The efficient computation of the generalized hypervolume indicator based on weight distribution functions is especially an issue, if it is hard to obtain a function for the integral in closed form; here, numerical approximation might be a solution, although it is unclear how such an approach could work in practice.
- Whether novel indicators require new algorithms is an open issue; this holds in particular when other dominance relations based an arbitrary convex cones are used.

Acknowledgements

The authors would like to thank Stefan Bleuler, Tim Hohm, and Simon Künzli for valuable discussions. Dimo Brockhoff has been supported by the Swiss National Science Foundation (SNF) under grant 112079.

References

1. N. Beume and G. Rudolph. Faster S-Metric Calculation by Considering Dominated Hypervolume as Klee's Measure Problem. Technical Report CI-216/06, Sonderforschungsbereich 531 Computational Intelligence, Universität Dortmund, July 2006. shorter version published at IASTED International Conference on Computational Intelligence (CI 2006).
2. V. G. da Fonseca, C. M. Fonseca, and A. O. Hall. Inferential performance assessment of stochastic optimisers and the attainment function. In E. Zitzler, K. Deb, L. Thiele, C. A. C. Coello, and D. Corne, editors, *Evolutionary Multi-Criterion Optimization, First International Conference, EMO 2001*, pages 213–225. Springer-Verlag, Berlin, 2001.
3. K. Deb, L. Thiele, M. Laumanns, and E. Zitzler. Scalable test problems for evolutionary multi-objective optimization. In A. Abraham, R. Jain, and R. Goldberg, editors, *Evolutionary Multiobjective Optimization: Theoretical Advances and Applications*, chapter 6, pages 105–145. Springer, 2005.
4. M. Emmerich, N. Beume, and B. Naujoks. An EMO Algorithm Using the Hypervolume Measure as Selection Criterion. In *Evolutionary Multi-Criterion Optimization (EMO). Third International Conference*, pages 62–76. Springer-Verlag Berlin, 2005.
5. M. Fleischer. The measure of Pareto optima. Applications to multi-objective metaheuristics. In C. M. Fonseca, P. J. Fleming, E. Zitzler, K. Deb, and L. Thiele, editors, *Evolutionary Multi-Criterion Optimization. Second International Conference, EMO 2003*, pages 519–533, Faro, Portugal, April 2003. Springer. Lecture Notes in Computer Science. Volume 2632.
6. C. M. Fonseca, L. Paquete, and M. López-Ibáñez. An Improved Dimension-Sweep Algorithm for the Hypervolume Indicator. In *IEEE Congress on Evolutionary Computation (CEC 2006)*, pages 1157–1163, Sheraton Vancouver Wall Centre Hotel, Vancouver, BC Canada, July 2006. IEEE Press.

7. C. Igel, N. Hansen, and S. Roth. The multi-objective variable metric evolution strategy, part i. Technical report, Institut für Neuroinformatik, Ruhr-Universität Bochum, Germany, 2005. IRINI-2005-04.
8. J. Knowles and D. Corne. On metrics for comparing non-dominated sets. In *Congress on Evolutionary Computation (CEC 2002)*, pages 711–716, Piscataway, NJ, 2002. IEEE Press.
9. J. Knowles, L. Thiele, and E. Zitzler. A Tutorial on the Performance Assessment of Stochastic Multiobjective Optimizers. TIK Report 214, Computer Engineering and Networks Laboratory (TIK), ETH Zurich, Feb. 2006.
10. J. D. Knowles, D. W. Corne, and M. Fleischer. Bounded Archiving using the Lebesgue Measure. In *Proceedings of the 2003 Congress on Evolutionary Computation (CEC'2003*, pages 2490–2497, Canberra, Australia, Dec. 2006. IEEE Press.
11. M. Laumanns, G. Rudolph, and H.-P. Schwefel. Approximating the Pareto Set: Concepts, Diversity Issues, and Performance Assessment. Technical Report CI-72/99, 1999.
12. M. Laumanns, L. Thiele, K. Deb, and E. Zitzler. Combining convergence and diversity in evolutionary multiobjective optimization. *Evolutionary Computation*, 10(3):263–282, 2002.
13. K. Miettinen. *Nonlinear Multiobjective Optimization.* Kluwer, Boston, 1999.
14. D. A. V. Veldhuizen. *Multiobjective Evolutionary Algorithms: Classifications, Analyses, and New Innovations.* PhD thesis, Graduate School of Engineering, Air Force Institute of Technology, Air University, June 1999.
15. L. While. A New Analysis of the LebMeasure Algorithm for Calculating Hypervolume. In *Evolutionary Multi-Criterion Optimization. Third International Conference, EMO 2005*, pages 326–340, Guanajuato, México, Mar. 2005. Springer.
16. L. While, L. Bradstreet, L. Barone, and P. Hingston. Heuristics for Optimising the Calculation of Hypervolume for Multi-objective Optimisation Problems. In *IEEE Congress on Evolutionary Computation (CEC'2005)*, pages 2225–2232, IEEE Service Center, Edinburgh, Scotland, Sept. 2005. IEEE Press.
17. L. While, P. Hingston, L. Barone, and S. Huband. A Faster Algorithm for Calculating Hypervolume. *IEEE Transactions on Evolutionary Computation*, 10(1):29–38, Feb. 2006.
18. E. Zitzler. *Evolutionary Algorithms for Multiobjective Optimization: Methods and Applications.* PhD thesis, Swiss Federal Institute of Technology (ETH) Zürich, Switzerland, 1999.
19. E. Zitzler, K. Deb, and L. Thiele. Comparison of multiobjective evolutionary algorithms: Empirical results. *Evolutionary Computation*, 8(2):173–195, 2000.
20. E. Zitzler and S. Künzli. Indicator-based selection in multiobjective search. In *Proc. 8th International Conference on Parallel Problem Solving from Nature (PPSN VIII)*, Birmingham, UK, September 2004.
21. E. Zitzler and L. Thiele. Multiobjective Optimization Using Evolutionary Algorithms - A Comparative case study. In *PPSN-V*, pages 292–301, Amsterdam, Sept. 1998.
22. E. Zitzler and L. Thiele. Multiobjective Evolutionary Algorithms: A Comparative Case Study and the Strength Pareto Approach. *IEEE Transactions on Evolutionary Computation*, 3(4):257–271, 1999.
23. E. Zitzler, L. Thiele, M. Laumanns, C. M. Foneseca, and V. Grunert da Fonseca. Performance assessment of multiobjective optimizers: An analysis and review. *IEEE Transactions on Evolutionary Computation*, 7(2):117–132, 2003.

The Multiple Multi Objective Problem – Definition, Solution and Evaluation

Wolfgang Ponweiser and Markus Vincze

Vienna University of Technology
Automation and Control Institute
Gusshausstr. 27-29/E376, A-1040 Vienna, Austria
{ponweiser,vincze}@acin.tuwien.ac.at
http://www.acin.tuwien.ac.at*

Abstract. Considering external parameters during any evaluation leads to an optimization problem which has to handle several concurrent multi objective problems at once. This novel challenge, the Multiple Multi Objective Problem M-MOP, is defined and analyzed. Guidelines and metrics for the development of M-MOP optimizers are generated and exemplary demonstrated at an extended version of Deb's NSGA-II algorithm. The relationship to the classical MOPs is highlighted and the usage of performance metrics for the M-MOP is discussed. Due to the increased number of dimensions the M-MOP represents a complex optimization task that should be settled in the optimization community.

Keywords: Multiple Multi Optimization Problem M-MOP, Performance Evaluation, Genetic Optimization.

1 Introduction

Since most practical problems are characterized by contradicting targets the extension to optimize several objectives has got more and more attention. The purpose of this type of problem, the *multi objective problem MOP*, is to find the so called Pareto Set, containing all non-dominated solutions [1].

This paper presents a novel extenion to MOP, the so called *multiple multi objective problem M-MOP*. The challenge is to find the optimal solutions for several similar MOP problems at once. Since all the single problems are itself MOPs, optimality is still Pareto optimality. The similarity of the problems is, that they share exactly the same input space, the decision space.

A domain that naturally leads to a M-MOP is any kind of evaluation. A MOP results from the evaluation of an algorithm by a set of configuration parameters. Whereas these parameters represent the input (the decision space) the evaluated performance represents the output (the objective space). Due to the fact that the final preferences between the several performance metrics are usually not known

* This work is supported by projects S9101, S9103 and S9106 of the Austrian Science Foundation.

S. Obayashi et al. (Eds.): EMO 2007, LNCS 4403, pp. 877–892, 2007.

at evaluation time, this results in a classical MOP. According to Coello [2] this is called 'a posteriori' preference articulation.

In fact the resulting performance further depends on some external parameters which are properties of the test site. These properties are further called *context*. To enable the user of the evaluation data to adjust the evaluation result to fit its own application the explicit consideration of this context is required. In cases where these context parameters can be configured concurrently on the test site a single evaluation run delivers several results for each of the realized contexts (see Fig. 1). Consider that a single evaluation run has just a single input vector. Every relationship between this input and the output of a single context is the task of a single MOP. In contrary the relation between the single input and the multiple outputs is the task of a multiple MOP, the M-MOP.

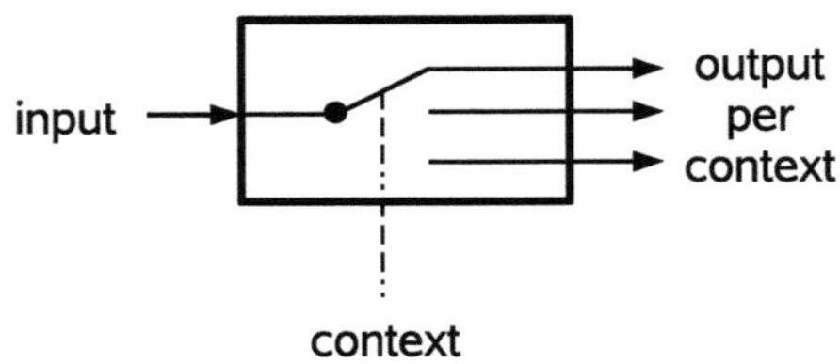

Fig. 1. The incorporation of external parameters (the context) as generator of *multiple* multi objective output

A typical example is the evaluation of object detection algorithms in the computer vision domain [3]. Such algorithms search for pre-specified objects in an image. There exist a lot of different algorithms [4,5,6] and evaluation endeavors [7,8] for this type of algorithms. Obviously their performance depends not only on the configuration of the algorithms itself, but additionally on the properties of the images and the objects used in the test databases. These properties build the context of the evaluation.

For the integration of an object detection algorithm it is important that the following requirements are specified:

- The performance characteristic that needs to be met. Typical examples for object detection algorithms are values for success rate, processing time and spatial accuracy.
- The expected context of the application. For the object detection task typical examples are the appearance and expected size of the objects to detect, lighting conditions and image noise.

In the same way as the requirements are specified the evaluation has to represent its results. Consequently a contex sensitive evaluation is required.

At first glance the context can be integrated as additional input. Due to the mechanisms of optimization methods bad context conditions will never be analyzed. Another approach is to simply expand the output by the context values. This leads to an optimization of the context itself, which is not the intention of the entire challenge. A third approach is the exhaustive calculation of the MOP for every single context separately, which leads to an enormous computational effort. A detailed description of these dependencies is presented in Section 2. For that reason it becomes reasonable to efficiently use the concurrency potential by creating a new problem type, the M-MOP.

To the best knowledge of the authors there is no scientifically work published dealing with M-MOP. Preliminary results by the authors have been presented in [9]. The complete approach and demonstration results are given here. Since this challenge is an extension of the classical MOP all further treatments during the next sections include the state of the art related to them.

Considering the M-MOP as a new challenge all classical MOP tools and approaches need to be reconsidered. Especially the optimization techniques for solving MOP have to be adapted. Section 2 introduces a detailed definition of the M-MOP problem. Section 3 presents an optimization algorithm by adapting Deb's NAGA-II [10]. Further more the performance metrics used to evaluate the optimization techniques have to be adapted. Section 4 details this phenomena and presents a newly developed performance metric that takes the specific of the M-MOP into account. Finally Section 5 summarises the paper and presents an outlook for further research.

2 Definition of the M-MOP

The mathematical background of the M-MOP is still the same as the background of a single MOP. The single MOP maps an input vector $\boldsymbol{x}$ of the m dimensional input (decision space) $\boldsymbol{x} \in X^m$ to an output vector $\boldsymbol{f}(\boldsymbol{x})$ of the n dimensional output (objective space) $\boldsymbol{f}(\boldsymbol{x}) \in F^n$ as can be seen in Figure 2. Using the domination relation defined in the objective space the Pareto front, the Pareto set and the Pareto rank of optimal solutions for a single MOP are defined [2] (see Fig. 2).

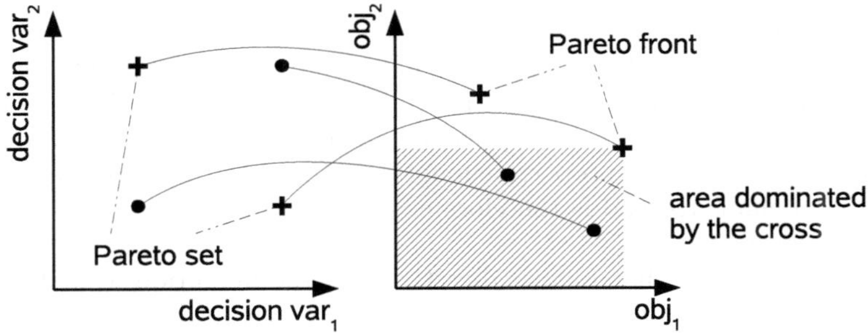

Fig. 2. The mapping of a classical Multi Objective Problem (MOP) with dominated solutions, the Pareto front and the Pareto set

In cases where this mapping is further influenced by some external conditions there are several options to integrate these context parameters $\boldsymbol{c}$ of the context space $\boldsymbol{c} \in C^o$ into a classical MOP, which are discussed below (see Fig. 3):

- Extending the input by the context (see Fig. 3(a)): The resulting mapping will be $\boldsymbol{f}(\boldsymbol{x}, \boldsymbol{c})$. A common optimizer for a single MOP tries to find all optimal solutions, the Pareto set, of the entire decision space. Since the real input space and the context space are unified, just the optimal solutions over this unified decision space will be found. Considering a difficult context that generally degrades the output still provides optimal solutions regarding to only this context. However these solutions will not appear as output of the optimizer since other easier contexts would produce better outputs, which will dominate the solutions of the bad context. For that reason such an integration of the context is not preferred.
- Extending the output by the context (see Fig. 3(b)): The resulting mapping will be $\boldsymbol{fc}(\boldsymbol{x})$ with $\boldsymbol{fc} \in FC^{n+o}$. This approach is mathematically not appropriate, since the context dimensions do not need to have an order. The different values in the set of a specific context dimension can be distinguished but they do not need to have an ordering relationship. An example is the topology of an object, which can be 'convex', 'concave' or 'with holes'. Although these properties influence the appearance of the object in an image and will therefore generate different detection rates there is no inherent order between these instances. As a result if there is no ordering relation, the dominance relation is not defined and hence no Pareto front and Pareto set can be calculated. Consequently the entire optimization definition gets lost.
- Treating every context as a single MOP (see Fig. 3(c)): The resulting mapping will be $\forall \boldsymbol{c} \in C^o\ \boldsymbol{f}_{\boldsymbol{c}}(\boldsymbol{x}_{\boldsymbol{c}})$. This corresponds to an exhaustive search for all possible context values. It is obvious that such an approach is not efficient and therefore not appropriate.
- *The M-MOP mapping* (see Fig. 3(d)): The characteristic that constitutes the extension to a M-MOP is that all different MOPs share the same input space, where the outputs are generated concurrently. Accordingly the suggested mapping is $\forall \boldsymbol{c} \in C^o\ \boldsymbol{f}_{\boldsymbol{c}}(\boldsymbol{x})$.

Out of this definition some properties and values can be extracted that explicit deal with M-MOP and support the design of optimization algorithms and metrics. One of the core property that characerises the M-MOP is the fact of a shared decision space, respectively the independence of the decision space from the context ($x \perp c$). This property should be efficiently used when dealing with this type of challenge and is therefore essential for the design of M-MOP optimization algorithms.

Another property that follows directly from the definition is that for a single input there are multiple outputs, one for every context. After collecting a set of solutions the Pareto rank for every solution in every context can be calculated. To compress this information to a single value just the best Pareto rank value (the lowest) is selected and further defined as *best Pareto rank* $bPr = min\{Pr_{c_1}, Pr_{c_2}, ..., Pr_{c_o}\}$. This value packs the quality of a solution into a single value. As a result it is of great interest for optimization algorithms, because they require a metric enabling the comparison of different solutions. Due to the underlying Pareto ranking of a single context a normalization takes place between

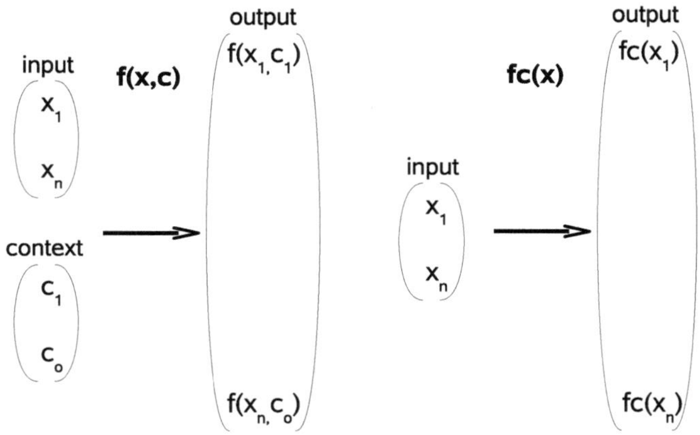

(a) Extending the input by the context.

(b) Extending the output by the context.

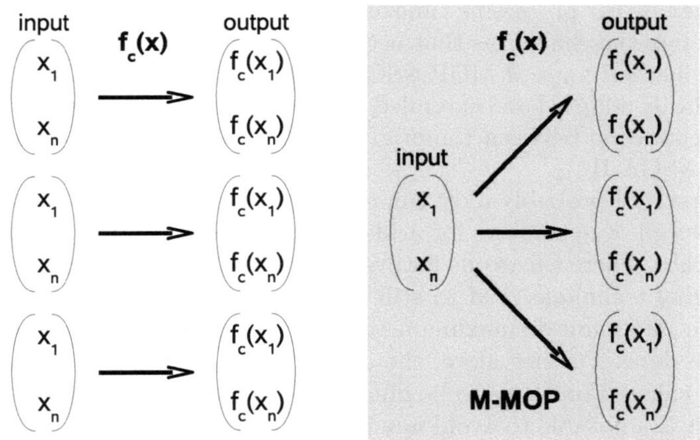

(c) Treating every context as a single MOP.

(d) *The M-MOP mapping.*

Fig. 3. Extending the classical MOP mapping by introducing context. Only the M-MOP mapping generates optima for all contexts, is mathematically feasible and efficient.

the different contexts. Thus a suppression of bad contexts like in the 'context-into-input' mapping version is avoided. Furthermore the minimum operator of the best Pareto rank guarantees the focussing on optimal solutions.

Using the concept of the best Pareto rank and the according normalization between different contexts the *optimality* can be redefined as *a solution with a best Pareto rank of 1.* Having optimality defined directly leads to the next definition, the so called *best Pareto Rank set.* This is the set of solution out of all possible solutions with a best Pareto rank of 1. It contains all solutions that are Pareto optimal in at least one single context and represents therefore the goal of an optimization of a M-MOP.

Analysing these concepts the relation between M-MOP as an extension of MOP gets obvious. The best Pareto rank replaces the usual Pareto rank as relationship value between different solutions, and the best Pareto rank set replaces the Pareto set as the collection of optimal solutions. Finally, note the fact that the dominance relation between solutions is still only defined for one single context.

3 Solving a M-MOP

There is a large set of different approaches for optimizing classical MOPs. An excellent survey can be found in [2]. By far the most scientific effort is spent on genetic algorithms for finding the optimal solutions of a MOP. Some of the most prominent algorithms which are reflecting the development during the last two decades are VEGA [11], MOGA [12], PAES [13], NSGA-II [10] and SPEA2 [14]. The main contradicting properties these methods have to deal with are the convergence to the Pareto front and the diversity of the solutions found as well as the efficiency represented by the processing time or the number of evaluations required.

To express the challenges that need to be considered when designing an optimization method for a M-MOP NSGA-II [10] is used as a base. Its methodology is exemplarily adapted and extended to handle the M-MOP properties. To enable a clear separation between the original and the extended versions, it is further called E-NSGA-II.

The first and probably most important property that needs to be considered for the design of optimizers for a M-MOP are the multiple instances of the objective space. There exists one for every context. Although there can be any kind of averaging technique used to still operate in this space (e.g. mean, standard deviation, minimum or maximum value, ...), a careful selection of these values has to be done. Further more, the computational effort increased enormously, since the calculations have to be done for every context separately. Out of these insights it is advisable to avoid any operation in the objective space.

In classical MOP optimization methods the objective space is intuitively used to calculate the density of solutions. Examples are SPEA2 [14] and NSGA-II [10]. A solution for the M-MOP is to relocate the density calculations from the objective space to the decision space. The placement of these measures to either the decision space or the objective space was subject to a long scientific discussion [1] where the practical implementations nowadays tend to use the objective space. Consequently such a relocation is conventional.

For the E-NSGA-II this simply results in the relocation of the diversity value calculation from the objective space to the decision space. Especially in cases where the decision space consists of discrete dimensions like in the example of the visual object detection task, this enables an easy normalization of the different dimensions, by simply counting the number of digits between solutions.

Another effect that needs to be considered is the extension of the Pareto rank to the best Pareto rank. Especially in NSGA-II the Pareto rank is the main selection criteria. Accordingly the best Pareto rank is used in the E-NSGA-II. A drawback of the best Pareto rank is its computational effort. The complexity of a single Pareto rank calculation is multiplied by the number of contexts. Still the normalization and optimization character of the best Pareto rank intends its use.

The practical implementation of the E-NSGA-II optimization algorithm for the visual object detection M-MOP had to deal with some further task specific challenges. An important property is the enormous evaluation time. Typical processing times for an object detection algorithm are in the order of seconds. Since a single evaluation run needs to process several hundreds of images to achieve statistically plausible evaluation results a single solution evaluation can take several minutes. According to this fact it is the main goal of the E-NSGA-II implementation to keep the number of evaluations as low as possible.

Another property of this task is the high number of dimensions combined with a very low quantification at each of these dimensions:

- The decision space depends on the vision method evaluated. It has 3 to 4 dimensions with a resolution of 3 to 9 values each.
- The objective space consists of 8 dimensions consisting of discrete and continuous metrics.
- The context space consists of 8 dimensions with a resolution of 3 to 4 values each.

Especially the low quantification of the decision space and the context space will be increased in the near future. Nevertheless the resulting Pareto front is expected to be cliffy.

The last property is more related to the use of the evaluation. Since perfect performing computer vision algorithms for object detection are just available for very restricted setups[1], the final user who selects an object detection algorithm based on the evaluation cannot expect an optimal performing algorithm. As a result it is not absolutely necessary to deliver the global optimal algorithm configuration for the specified context. An algorithm that performs close to the optimum will be sufficient.

To deal with this difficult task a dynamic population size in relation to the number of Pareto optimal solutions and the number of overall evaluated individuals is integrated. Further more the mutation probability is set by an inverse relation to the number of new generated Pareto optimal solutions. These two mechanisms enable an reactive behavior of the algorithm behavior to achieve diversity when necessary.

Generally speaking the optimization of an M-MOP is a parallel problem solving strategy. Due to the use of the concurrent evaluation of several contexts a lot of computational effort can be saved. Enabling a measurement of this fact as well as the general optimization behavior is the content of the following section.

4 Evaluating M-MOP Optimizers

This section presents the development of metrics measuring the performance of optimization algorithms for M-MOPs. Like for the classical MOP there are

[1] Computer vision is successfully deployed in industrial application where environmental conditions like light, viewing direction or occlusion can be controlled.

at least three different properties that have to be expressed by performance metrics [15,16,17]:

- Convergence expresses the displacement between the real Pareto front and the Pareto front build by the solutions found so far.
- Diversity expresses the distribution of the found solutions over the entire objective space.
- Efficiency expresses the computational effort required.

The efficiency property is the easiest one to measure. It is usually specified by the processing time or the number of evaluations required [17]. The same paper presents a detailed insight in the behavior of a lot of state of the art performance metrics. Typical examples are the generational distance [18] and $D1_R$ [19] for convergence, the spread [20] and maximum spread [21] measure as well as the 'diversity' measure by Deb&Jain [22] for diversity and the hypervolume indicator [23,24] also called S metric and Lebesgue measure as a combined metric.

Of course most of these metrics are completely defined in the objective space. As already mentioned in the previous section this contradicts with the guidelines established for dealing with a M-MOP. However in contrary to the optimization methods the metrices completely relay on the objective space. A relocation to the decision space as done for the diversity at the E-NSGA-II development would lead to a complete modification of the meaning of the metric. Therefore an relocation approach for the reuse of these metrics for M-MOP is not appropriate.

The only reasonable alternative is to calculate the metrics for every single objective space and using some averaging technique to conclude to a final value. As already discussed in the previous section averaging over all contexts has to be done carefully. For example most of the performance metrics of a single MOP are itself already the outcome of some averaging. A typical example is the generational distance [18]. Let F_i and F^{GT} be a set of non-dominated solutions and the set of all Pareto-optimal solutions. The generational distance is the average distance from each solution in F_i to its nearest Pareto-optimal solution F^{GT}. Since in the M-MOP case there exist several F_i and F^{GT} sets the averaging can be done in two steps. The first one is averaging by the number of solutions in F_i in every context separately. In the second step the average for all contexts is calculated from all the single values of step one. Another approach is to calculate the average from all distances of all solutions in all F_i of every context in one single step. To keep the meaning of the metric as similar as possible to the MOP metric the two step approach is preferred. Still keep in mind that some contexts which for example contain very few Pareto points can influence the metric enormously. A first attempt to reduce the intensive use of the objective space is presented in the next subsection.

4.1 Average Pareto Rank Difference

The key idea for this novel performance metric is to avoid some objective specific normalization by the reuse of the Pareto rank concept. In contrary to the generational distance or the $D1_R$ metrics that utilize the euclidian distance between the non-dominated solutions and the real optimal solutions, the difference in terms of the Pareto rank is used for the new metric.

With P_i as the population under investigation, F_i the non-dominated individuals of P_i related to P_i, the entire set of possible individuals as 'entire ground truth' P^{GT} and F^{GT} as the non-dominated individuals of P^{GT}, the Average Pareto Rank Difference $APRD$ is defined as:

$$APRD(F_i, P^{GT}) = \frac{\sum_{\boldsymbol{u} \in F^{GT}} RD(F_i, \boldsymbol{u})}{card\{F^{GT}\}} \tag{1}$$

With RD as the ranking difference function:

$$RD(F_i, \boldsymbol{u}) = \begin{cases} \max_{\boldsymbol{v} \in F_i \,|\, \boldsymbol{u} \succeq \boldsymbol{v}} PR(\boldsymbol{v} \,|\, P^{GT}) \\ \qquad if\; card\{\forall \boldsymbol{v} \in F_i \,|\, \boldsymbol{u} \succeq \boldsymbol{v}\} > 0 \\ \max\{2, \max_{\boldsymbol{v} \in F_i \,|\, c_{min} * \boldsymbol{u} \succeq \boldsymbol{v}} PR(\boldsymbol{v} \,|\, P^{GT})\} \\ \qquad if\; card\{\forall \boldsymbol{v} \in F_i \,|\, \boldsymbol{u} \succeq \boldsymbol{v}\} = 0 \end{cases} \tag{2}$$

$$c_{min} = \min\{\forall c \in \mathbb{R} \,|\, card\{\forall \boldsymbol{v} \in F_i \,|\, c\boldsymbol{u} \succeq \boldsymbol{v}\} > 0\} \tag{3}$$

Where $PR(\boldsymbol{v} \,|\, P^{GT})$ is the Pareto rank of the solution $\boldsymbol{v}$ in the entire ground truth population P^{GT}, and $card\{X\}$ is the number of elements of the set X.

To understand the APRD it is important to differentiate between the following two sets (see Fig. 4(a)): the target set (F^{GT}), as the Pareto set of optimal solutions over all possible solutions (P^{GT}), and the status set (F_i), as the Pareto optimal solutions of the population under investigation (P_i). Obviously all status points are non-dominated in the set of the population under investigation (P_i) and have therefore a Pareto rank of 1 regarding the population under investigation (P_i). The key is, that additionally all status points (F_i) can be attributed with a Pareto rank according to *all* possible solutions (P^{GT}) that can be different from 1.

The idea of the APRD is to search for Pareto optimal solutions (F_i) in the population under investigation, that are probably selected instead of the real optimal solutions (F^{GT}). As expressed by Equation 2, there are two different cases: The first and propably usual one is that of a target point that dominates at least one status point (see Fig. 4(b)). For this case the worst Pareto rank of all these dominated status points counts for the APRD. In the second case the target point dominates no status point. To still find a status point that is probably selected instead, the target point is linear extended in the objective space, until it dominates a status point (see Fig. 5).

Finally, the average of all Pareto rank values using the number of target points is calculated. The range of the APRD is ≥ 1. The optimal value is 1. In this optimal case every target point is found and hence equals the status points, which accordingly have a Pareto rank of 1 regarding all possible solutions.

While this metric is still defined regarding a MOP the extension to the M-MOP follows the suggestion given before by averaging over all context specific APRD values. Consider, that due to the use of the Pareto rank normalization also sparse occupied objective spaces contribute no big outliers and therefore cannot distort the final APRD value.

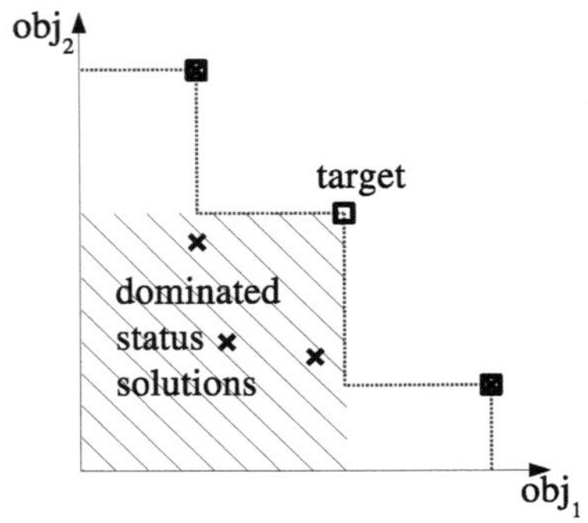

(a) The potential status points, that could be selected instead of the missed target point (maximization assumed).

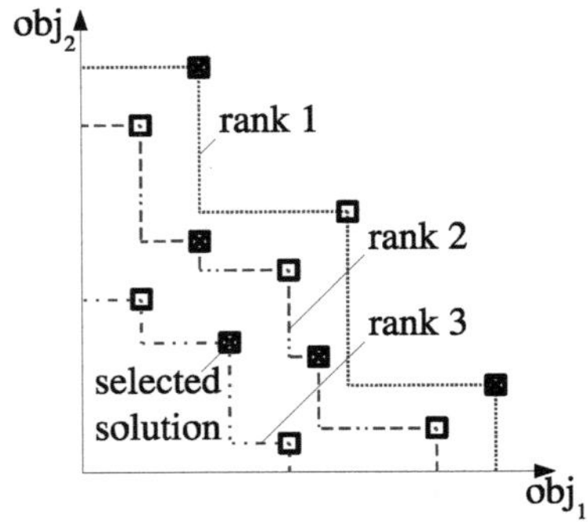

(b) The Pareto rank assignment of the status points related to the ground truth, and the selected local worst case of rank 3 (maximization assumed).

Fig. 4. The selection and ranking for the 'Average Pareto Rank Difference' APRD performance metric

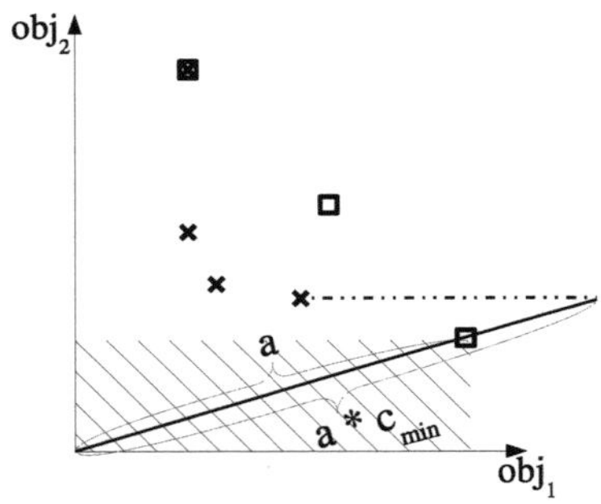

Fig. 5. The target point dominates no status solutions. Hence it is linear extended using a factor c_{min} to find the next status point (maximization assumed).

Even so the handling of the target solutions that originally do not dominate status points is done in the objective space. The normalization mechanism used there is the linear extension of the solution itself.

The consequent search of the APRD for solutions that are selected instead of the real optimal ones is further motivated by a user point of view. The benefit of the entire optimization mechanism is to avoid a practically not possible exhaustive search for the best solution. Due to the shortcomings associated with optimization techniques the real optimal solutions cannot be guaranteed. As a result the main interest of the final user is the loss in performance generated by the miss of optimal solutions. To enable a single metric to represent this performance loss a normalization has to be done. For the APRD the Pareto rank as a preference free normalization mechanism is used.

Even though the APRD mainly measures the convergence of a population, the counting and averaging according to all optimal ground truth solutions provides some diversity properties. An important assumption for the APRD is that the decision space is discrete. Together with the related requirement of the Pareto ranks for all possible solutions this is the main disadvantage of the APRD.

Finally, it has to be stated that the APRD is a conservative metric. For all available solutions that are dominated by the missed ground truth point the one with the worst Pareto rank counts. Accordingly the APRD represents a lower bound for the performance loss. As a result a real selected solutions can perform better than expressed by the APRD metric.

4.2 Performance of the E-NSGA-II for M-MOP

The evaluation is done using the data provided by two different object detection algorithms. The first one is based on color histograms, further called GCH [25], where the second one uses a support vector machine SVM [26] to discriminate different image patches. The dynamic population size is handled using two relationship parameters. The 'Optimal-Pop-Size-Multiplier' sets the maximum of individuals of a population regarding the number of optimal individuals found so far and is set to 3. The 'Individual-Pop-Size-Multiplier' sets the maximum of individuals of a population regarding the number of individuals generated and is set to 0.8. As long as not explicitely mentioned, the standard values for the following experiments for the initial population size is 10, for the number of generations is 25, for the crossover propability is 0.8 and for the starting mutation rate is 0.12. To explain the dependency to random values all experiments are repeated 10 times with different seed values.

The first evaluation compares the performance of the E-NSGA-II algorithm to a purely random search. This is done regarding the number of evaluations, and the so called success rate. This value measures the rate of Pareto optimal solutions found. Table 1 points out the gain of using the genetic optimization algorithm.

Table 1. The mean of the evaluated success rate for the GCH method of the E-NSGA-II algorithm compared to a random search at different percentages of evaluations done regarding the number of possible evaluations

number of evaluations	E-NSGA-II	random
10%	0.366	0.092198
20%	0.713	0.168085
30%	0.954	0.234752
40%	0.993	0.296454

Even though these values are not encouraging. For finding 95% of the real optimal solutions of this dataset 30% of all possible solutions have to be evaluated. However the real benefit of the genetic algorithm is clarified by comparing not the pure success rate but using the APRD value of Section 4.1. This metric considers

Table 2. The mean of evaluated 'Average Pareto Rank Difference' APRD for the GCH method of the E-NSGA-II algorithm compared to a random search at different percentages of evaluations done regarding the number of possible evaluations

number of evaluations	E-NSGA-II	random
10%	2.432	2.73125
20%	1.908	2.58669
30%	1.572	2.27775
40%	1.309	2.1864

that a not found Pareto optimal point can be replaced by a relatively well performing solution. Table 2 presents the results for the E-NSGA-II algorithm and a random search.

A more detailed analysis for only the E-NSGA-II algorithm is presented in Table 3. The convergence behavior of especially the worst case (the maximum value) shows the robustness of the approach. After 10 generations a selection of at least the third best solution is provided.

Table 3. The evolution of the E-NSGA-II algorithm for the GCH object detector in terms of the APRD

generations	mean value	min value	max value
5	3.3722	2.30341	4.98024
10	2.40762	1.6826	2.86687
15	2.02137	1.38715	2.74748
20	1.79079	1.31474	2.44171
25	1.75323	1.30659	2.43833

To express the difficulty related to the low quantification of the underlying computer vision task the same object detection method is evaluated with an even worse parameter resolution (see Table 4). The mean and best (minimum) values promise a well performing optimization. Even so the real reason for these apparently good values is the low resolution of possible solutions. Due to this fact the maximum number of solutions theoretically dominated is in the order of 5. This behavior can be extracted by the slow convergence of the worst case (the maximum value).

Table 4. The evolution of the E-NSGA-II algorithm for the GCH object detector with reduced decision space resolution in terms of the APRD

generations	mean value	min value	max value
5	2.39391	1.66234	3.59369
10	2.14373	1.495	3.07269
15	1.89555	1.0	3.01923
20	1.41196	1.0	3.00433
25	1.30955	1.0	3.00433

Table 5. The evolution of the E-NSGA-II algorithm for the SVM object detector with reduced decision space resolution in terms of the APRD

generations	mean value	min value	max value
5	2.71488	2.29564	3.29721
10	2.06048	1.67064	2.48775
15	1.89104	1.43546	2.22319
20	1.83273	1.34223	2.22319
25	1.76618	1.15343	2.18197

Table 6. A comparison of different mutation rate values for the evolution of the E-NSGA-II algorithm for the GCH object detector in terms of the mean APRD values

mutation rate generations	0.10	0.12	0.14
5	3.36471	3.3722	2.86929
10	2.66545	2.40762	2.25021
15	2.0521	2.02137	1.88017
20	1.75965	1.79079	1.73598
25	1.74585	1.75323	1.71969

As already stated previously the main interest of the final user is not finding the global optimum. They require good solutions but in a fast and especially robust way. To indicate the reliability of the E-NSGA-II algorithm Table 5 presents the performance for the second object detection algorithm, the SVM.

Finally, Table 6 presents the dependency of the performance of the E-NSGA-II algorithm to the mutation rate. For this dataset performing more than 20 generations delivers robustness against minor variations of the mutation rate.

5 Conclusion

This paper contributed a novel challenge for the optimization community, the multiple multi objective problem M-MOP. It is an extension of the multi objective optimization (MOP) problem by the fact that the evaluation of a single solution provides a set of outputs where each output itself consists of multiple objective values.

As an example from the computer vision domain, the context sensitive evaluation and selection of object detection algorithms is used. The concurrent appearance of different objects with different contextual properties like topology or reflectance in one single image permit the evaluation of these algorithms using a single evaluation run. Consequently several evaluation outputs are obtained concurrently for a single object detection algorithm configuration.

Several straight forward approaches to handle this type of challenge are presented which turned out to fail either mathematically or at least by their computational effort. The main idea for the solution is to intensively use the concurrency property these problems provide. This leads to guidelines and metrics that support the development of optimization algorithms and supports the handling of the challenge. One of these guidelines is that computation in the objective space should be avoided. The most important metric defined is the one of the best Pareto rank. It enables the compression of the performance of a solution over all contexts to a single value.

A first optimization technique for finding the Pareto optimal solutions of a M-MOP is presented. It is based on the Deb's NSGA-II approach for the classical MOP. The core adaptations done according to the presented proposals are the use of the best Pareto rank value as the main selection criteria and the relocation of the diversity measure from the objective space to the decision space.

Further more the properties of the M-MOP are analyzed according to the use of performance metrics of classical MOPs. Especially the averaging of these metrics over the different contexts has to be done carefully. A new performance metric the Average Pareto Rank Difference APRD is introduced which partly relaxes these problems. Additionally this metric is perfectly dedicated for the users of optimizers since it represents the loss of performance between using the optimizer result instead of the theoretical optimum.

A detailed evaluation of the developed E-NSGA-II algorithm is performed. The results justify the guidelines presented in Section 3 for the development of optimization algorithms for M-MOP. Finally the motivation of the new developed Average Pareto Rank Difference APRD performance metric is confirmed.

Whereas this paper defines the M-MOP challenge and presents some first methods to handle it there is still a lot of space for improvements. First of all the adaptation of several common MOP optimization algorithms to the task of a M-MOP needs to be done. On top of that the generation of a completely new optimization algorithm for M-MOP could provide higher performance.

Another important topic is to experiment with several classical performance metrics and to evaluate their behavior at the M-MOP. Related is the extension and creation of test databases of this specific task. Especially the availability of data from several different domains as well as artificial data is important to enhance the reliability of evaluation results and to increase the motivation for this new optimization problem. Finally, a detailed analyse of the diversity of the best Pareto rank sets could provide fundamental theoretical insight enabling more advanced distribution techniques.

References

1. Goldberg, D.: Genetic algorithms in search, optimization, and machine learning. Addison-Wesle (1989)
2. Coello, C., Veldhuizen, D., Lamont, G.: Evolutionary algorithms for solving multi-objective problems. Kluwer Academic/Plenum, New York, USA (2002)
3. Fisher, R., Dawson-Howe, K., Fitzgibbon, A., Robertson, C., Trucco, E.: Dictionary of computer vision and image processing. Wiley, Chichester (2005)
4. Draper, B.: From knowledge bases to markov models to pca. In: nvited talk for the Workshop on Computer Vision System Control Architectures (held in conjunction with ICVS), Graz, Austria (2003)

5. Leonardis, A., Bischof, H.: Robust recognition using eigenimages. Computer Vision and Image Understanding: CVIU **78**(1) (2000) 99–118
6. Obdrzálek, S., Matas, J.: Sub-linear indexing for large scale object recognition. In: Proceedings of the British Machine Vision Conference. Volume 1. (2005) 1–10
7. Appenzeller, G., Crowley, J.: Automatic parameter control for experimental evaluation of visual systems. In: Proceedings of the Asian Computer Vision Conference, Singapour (1995)
8. Forstner, W.: Pros and cons against performance characterization of vision algorithms (1996)
9. Ponweiser, W., Vincze, M.: Robust handling of multiple multi-objective optimisations. In: accepted for the International Workshop on Biologically-Inspired Optimisation Methods for Parallel and Distributed Architectures: Algorithms, Systems and Applications. (to appear 2006)
10. Deb, K., Agrawal, S., Pratab, A., Meyarivan, T.: A Fast Elitist Non-Dominated Sorting Genetic Algorithm for Multi-Objective Optimization: NSGA-II. In Schoenauer, M., Deb, K., Rudolph, G., Yao, X., Lutton, E., Merelo, J.J., Schwefel, H.P., eds.: Proceedings of the Parallel Problem Solving from Nature VI Conference, Paris, France, Springer. Lecture Notes in Computer Science No. 1917 (2000) 849–858
11. Schaffer, J.D.: Multiple objective optimization with vector evaluated genetic algorithms. In Grefenstette, J.J., ed.: ICGA, Lawrence Erlbaum Associates (1985) 93–100
12. Fonseca, C.M., Fleming, P.J.: Genetic algorithms for multiobjective optimization: Formulation, discussion and generalization. In Forrest, S., ed.: ICGA, Morgan Kaufmann (1993) 416–423
13. Knowles, J., Corne, D.: The pareto archived evolution strategy: A new baseline algorithm for pareto multiobjective optimisation. In Angeline, P.J., Michalewicz, Z., Schoenauer, M., Yao, X., Zalzala, A., eds.: Proceedings of the Congress on Evolutionary Computation. Volume 1., Mayflower Hotel, Washington D.C., USA, IEEE Press (1999) 98–105
14. Zitzler, E., Laumanns, M., Thiele, L.: SPEA2: Improving the Strength Pareto Evolutionary Algorithm. Technical Report 103, Gloriastrasse 35, CH-8092 Zurich, Switzerland (2001)
15. Knowles, J., Corne, D.: On metrics for comparing non-dominated sets (2002) In Congress on Evolutionary Computation (CEC 2002).
16. Okabe, T., Jin, Y., Sendhoff, B.: A critical survey of performance indices for multi-objective optimization. In: Proceedings of the IEEE Congress on Evolutionary Computation. Volume 2. (2003) 878–885
17. Zitzler, E., Thiele, L., Laumanns, M., Fonseca, C., Fonseca, V.: Performance assessment of multiobjective optimizers: an analysis and review (2002)
18. Van Veldhuizen, D.A.: Multiobjective Evolutionary Algorithms: Classifications, Analyses, and New Innovations. PhD thesis, Wright-Patterson AFB, OH (1999)
19. Czyzak, P., Jaszkiewicz, A.: Pareto-simulated annealing – a metaheuristic technique for multi-objective combinatorial optimization. Journal of Multi-Criteria Decision Analysis **7**(1) (1998) 34–47
20. Ishibuchi, H., Narukawa, K.: Recombination of similar parents in emo algorithms. [27] 265–279
21. Zitzler, E.: Evolutionary Algorithms for Multiobjective Optimization: Methods and Applications. PhD thesis, Shaker Verlag, Aachen (1999)
22. Deb, K., Jain, S.: Running performance metrics for evolutionary multi-objective optimization. Technical Report 2002, KanGAL Report (2002)

23. Emmerich, M., Beume, N., Naujoks, B.: An emo algorithm using the hypervolume measure as selection criterion. [27] 62–76
24. Zitzler, E., Thiele, L.: Multiobjective optimization using evolutionary algorithms — A comparative case study. Lecture Notes in Computer Science **1498** (1998) 292–301
25. Swain M.J., B.D.: Indexing via color histograms. In: Proceedings, Third International Conference on Computer Vision. (1990) 390–393
26. Roobaert, D., Zillich, M., Eklundh, J.O.: A pure learning approach to background-invariant object recognition using pedagogical support vector learning. In: CVPR (2), IEEE Computer Society (2001) 351–357
27. Coello, C.A.C., Aguirre, A.H., Zitzler, E., eds.: Evolutionary Multi-Criterion Optimization, Third International Conference, EMO 2005, Guanajuato, Mexico, March 9-11, 2005, Proceedings. In Coello, C.A.C., Aguirre, A.H., Zitzler, E., eds.: EMO. Volume 3410 of Lecture Notes in Computer Science., Springer (2005)

Adequacy of Empirical Performance Assessment for Multiobjective Evolutionary Optimizer

Swee Chiang Chiam[1], Chi Keong Goh, and Kay Chen Tan

[1] Department of Electrical and Computer Engineering
National University of Singapore
4 Engineering Drive 3
Singapore 117576
g0500055@nus.edu.sg

Abstract. Recent studies show that evolutionary optimizers are effective tools in solving real-world problem with complex and competing specifications. As more advanced multiobjective evolutionary optimizers (MOEO) are being designed and proposed, the issue of performance assessment has become increasingly important. While performance assessment could be done via theoretical and empirical approach, the latter is more practical and effective and has been adopted as the de facto approach in the evolutionary multiobjective optimization community. However, researches pertinent to empirical study have focused mainly on its individual components like test metrics and functions, there are limited discussions on the overall adequacy of empirical test in substantiating their statements made on the performance and behavior of the evaluated algorithm. As such, this paper aims to provide a holistic perspective towards the empirical investigation of MOEO and present a conceptual framework, which researchers could consider in the design and implementation of MOEO experimental study. This framework comprises of a structural algorithmic development plan and a general theory of adequacy in the context of evolutionary multiobjective optimization.

Keywords: Multiobjective Optimization, Evolutionary Computation, Adequacy, Performance assessment.

1 Introduction

Many real-world applications involve complex optimization problem with various competing specifications and constraints that are often difficult, if not impossible, to be solved without the aid of powerful and efficient optimization algorithms. Over the years, many multi-objective evolutionary optimizers (MOEO), a class of stochastic search technique, have been developed for this purpose, ranging from evolutionary algorithm, evolutionary strategy, genetic programming, to newly proposed algorithmic models, like ant colony optimization, particle swarm optimization, estimation of distribution algorithm and etc. and they have been demonstrated to be very powerful and applicable for solving such problems.

S. Obayashi et al. (Eds.): EMO 2007, LNCS 4403, pp. 893–907, 2007.

The algorithmic development of MOEO involves an iterative process of designer intuition and validation, where performance assessment is carried out continuously to examine and improve algorithmic design. In addition, performance assessment plays a crucial role in improving our understanding of MOEO and the interplay between its different components. The knowledge gained will greatly aid in the future development of better MOEO. Therefore, as more advanced MOEO are being designed and proposed, the issue of performance assessment has become increasingly important. However, the assessment of MOEO capability is not a trivial task. Due to its stochastic nature, the capability of MOEO cannot be precisely determined before its actual application. Furthermore, performance assessment is complicated in the context of evolutionary multiobjective optimization (EMOO), where the various conflicting goals of EMOO have a profound impact on the performance assessment of MOEO. Interestingly, Bosman and Thierens [1] noted that state-of-the-art MOEO have similar or incomparable performances due to these conflicting optimization goals.

The most practical and effective means for assessing the performance of MOEO is via an empirical study, where the evaluated algorithm will be applied to a set of test functions and the evolved solutions will be taken as an indication of the algorithmic performance. Although performance assessment can also be done via theoretical study [2], this approach often lacks the flexibility and practicality of empirical investigation. Furthermore, due to the stochastic nature of MOEO and its complex relationship with the optimization problem, it is difficult, if not impossible, to establish any formal mathematical treatment of algorithmic performance. Consequently, researchers will either get lost in the mire of complexity or resort to substantial simplifications before any analysis can be done. Due to the limitations of theoretical studies, performance assessment via the empirical approach has been adopted as the *de facto* approach in the EMOO community.

Research pertinent to empirical study has been focused on the development of test functions and performance metrics, resulting in great strides in these areas. Initial empirical studies are usually based on simple extension of single objective optimization problems, which reveals little or no characteristics of the algorithm under investigation. To this end, benchmark test suites have been formalized [3], [4] to challenge the MOEO in various aspects of problem difficulty [5]. In order to quantify the evolved tradeoffs, different metrics have been proposed over the years to measure the various goals in EMOO i.e. proximity, diversity and distribution. The fact that the interpretation of experimental results is largely dependent on the accuracy of performance indicators has initiated much research in this aspect also [6], [7], [8].

Although much work has been done to improve the reliability of empirical studies, there are little or no discussions at all on how it should be conducted with adequate substantiality on their statements made on the performance and behavior of the evaluated algorithm. As such, in contrast to existing works, this paper provides a holistic perspective towards the empirical investigation of MOEO and presents a conceptual framework for the design and implementation of MOEO empirical study and its definition of adequacy in the context of EMOO. For this purpose, the various aspects of MOEO empirical study will be considered, which includes the delineation of its essential components and the description and discussion of related design and implementation issues.

2 Algorithm Development of MOEO

The development process of MOEO comprises of two main stages, namely the algorithm formulation and the empirical study, as illustrated in figure 1. Usually, the process initiates from the algorithm formulation, where the motivation of his work is identified and the MOEO is crafted out accordingly to address the associated problem. Subsequently, the MOEO will be validated empirically. Through the empirical study, a better understanding of the MOEO will be achieved and ways to further improve the MOEO could be identified and implemented. After which, the development cycle will be repeated.

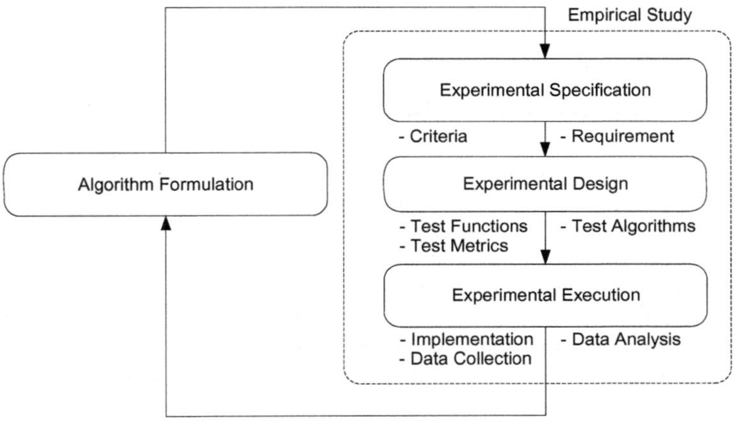

Fig. 1. MOEO Development

The experimental study can further be decomposed into three distinct stages:

- *Experimental Specification*: The criteria to evaluate the MOEO will be determined, for example, assess its algorithmic capability in the identified problem or explore for further improvements. This will dictate the requirements for the experimental study.
- *Experimental Design*: This stage is concerned with the overall structure of the entire empirical study, where appropriate test functions, test metrics and test algorithms will be selected based on the experimental requirements.
- *Experimental Execution:* The execution of the empirical study is carried in this stage. Issues include the fairness of the implementation and the manner in which the experimental data should be collected and analyzed, so that the algorithmic performance can be properly interpreted.

The various stages of the empirical study will be described in detail subsequently in this section. Algorithm formulation will be briefly discussed also, as it has a direct influence on the type of empirical study being conducted.

2.1 Algorithm Formulation

Algorithm development usually begins with algorithm formulation, where the entire MOEO is conceptualized and designed based on some underlying motivations to resolve certain problems or issues, for instance the improvement of existing MOEO or the investigation of unexplored problem domains. In general, these newly developed MOEO can be broadly classified according to the level of improvements or the scope of the problems explored, as illustrated in table 1.

Table 1. Classification of newly developed algorithms

Algorithmic formulation	
Algorithm level	Problem level
- Improvement of existing operators	- Similar problem
- Introduction of new operators	- Extended version of problem
- Completely new algorithm framework	- Unexplored domain

The improvement in the algorithm level can be in the form of operator improvement, for instance, replacing the traditional single point crossover operator by uniform crossover operator, or the introduction of additional operators to complement the existing evolutionary optimization framework. At the extreme case, it can be a new algorithmic framework, like PSO and ACO. As for the problem level, the MOEO could be designed for similar problems tackled previously by other optimizers, like the typical benchmark problems in EMOO. Alternatively, it could refer to extended version of problems, like the consideration of more cities and/or constraints in the traveling salesman problem. Lastly, the problem has not been solved by any MOEO before i.e. a single objective optimization problem being extended into the multiobjective domain.

2.2 Experimental Specification

The objective of empirical study can range from the validation of one's design intuition, to the understanding of some theory, to the comparison of MOEO performances. Before the design and implementation of the empirical study, a set of criteria that highlights the extent and depth of the empirical test should be defined, as they will dictate the requirements of the empirical study.

Empirical assessment can be broadly categorized into individual and comparative assessment and each category can be further divided into several types, according to their level of analysis. Table 2 summarizes the different types of empirical assessment, highlighting the different types of criteria and their corresponding experimental requirements.

Individual assessment will solely validate the MOEO formulated, while comparative assessment will evaluate its significance with respect to the current state-of-the-art. Type I assessment is the most preliminary level of study and evaluates the capability of the MOEO in satisfying the goals of proximity, diversity and distribution

Table 2. Different category of empirical assessment

Individual Assessment		
Type	Criteria	Experimental Requirements
I	- Assess algorithmic capability	- Test algorithm capability to converge And maintain diversity
II	- Evaluate robustness of algorithmic performance	- Test and analyze effects of parameter variation on algorithmic performance
III	- Verify correctness of algorithm	- Characterize population or individual dynamics
Comparative Assessment		
Type	Criteria	Experimental Requirements
I	- Assess the significance of algorithmic capability	- Compare algorithm capability to converge And maintain diversity
II	- Evaluate the relative robustness of the algorithmic performance	- Compare effects of parameter variation on algorithmic performance
III	- Validate theory	- Compare population or individual dynamics

in EMOO, while Type II assessment is a parameter sensitivity test to analyze its various components and their relationship. Lastly, Type III will analyze the dynamics of the evolutionary process and unveil insights to the algorithmic behavior and characteristics. Naturally, different criteria demands varying experimental requirements due to the type of statements that could be made on the algorithmic performance and behavior.

It is recommended that an empirical study should address all these criteria for a reasonable degree of comprehensiveness. However, this is not widely witnessed in practice, where most studies bypass individual assessment and proceed directly to comparative assessment. Furthermore, it is apparent that there is a clear lack of studies involving type III assessment, as MOEO is commonly being assessed as a black box where only its external behavior is being investigated, without giving any hint or analysis on its underlying dynamics and operations.

2.3 Experimental Design

The empirical study should account for all the experimental criteria outlined in the prior stage. This ensures the adequacy of the empirical observations in substantiating any statements made about the algorithmic behavior or performance of the evaluated MOEO. The main issues considered in this stage include test functions, test metrics and test algorithms.

1) *Test functions:* Test functions are used to gauge the effectiveness and efficiency of MOEO in dealing with real-world problems. These test functions should be simple in implementation to facilitate the extraction of the algorithmic behavior yet, at the same time, complex enough to allow conjectures to the real-world. In practice, test suites usually comprised of theoretical benchmark problems or simplified model of real-world problems.

Many guidelines for the construction of test functions and test suites have been suggested in the literature. In general, the EMOO community has mostly agreed with Deb *et al.* [5] on the type of problem difficulties to be utilized in experimental studies and at present, more than 50 multiobjective test functions with different features that can pose difficulties in converging to the Pareto front and maintaining a diverse solution set have been applied in the MOEO literature [4]. As these test functions are used to emulate real-world optimization problems, related work should report clearly on how the selected suite of test functions serve this purpose.

The choice of test functions ultimately depends on the problem and issues considered during the algorithm formulation. However, it is often hard to make a selection given the wide array of choices available. One possible approach is to consider the MOEO functionalities challenged by the test functions. Basically, the different problem features can be classified into two broad categories of primary and secondary according to the functionalities challenged:

- Primary: Bias, Non-convexity, Discontinuities, Deception, Parameter interaction, Isolated Optimal, Multi-modality, High-Dimensionality.
- Secondary: Robustness, Dynamic, Noise, Goal/Preferences, Constraints.

The latter category provides challenges beyond the difficulties posed by the former category to the basic ability of MOEO in discovering a near-optimal and diverse Pareto-front. These secondary test functions should be considered only if it is relevant to the underlying motivation in the algorithmic formulation. Nevertheless, regardless of the secondary MOEO features considered, primary test functions must be included to test the basic capability of MOEO.

2) *Test Metrics:* Performance metrics or indicators play an important role in reflecting the quality of the solution set found. These metrics can be either unary or binary. The former reflects a certain quality aspect of the approximated solution set, while the latter provides an indication of performance disparity between pairs of approximated solution set. Examples of unary metrics include generational distance and maximum spread, while binary metrics include coverage and binary hyper-volume.

There has been increasing concerns on the accuracy of performance metrics. To this end, Knowles and Corne [6] and Zitzler *et al.* [8] have discussed at length, the suitability and limitations of various performance metrics. In particular, Zitzler *et al.* [8] highlighted the limitations of unary indicators due to their non-conformance to the dominance criterion and suggested the use of binary measures. Nevertheless, unary indicators are still important in providing specific information pertinent to the optimization goals of proximity, diversity and distribution.

Furthermore, it is instructive to consider the characteristics of the different metrics and how they complement each other. Basically, the different metrics can be classified into

a) Specific indicators: quantifies the approximate solution set in one aspect of EMOO i.e. proximity, diversity and distribution.
b) General indicator: provide a general indication of solution set quality in the various EMOO goals
c) Pareto indicators: quantifies the solution set based on dominance criterion

An example to illustrate the need of complementary indicators from the various categories is as such: The use of multiple general indicators, like hyper-volume ratio and inverse generational distance, should be complemented with Pareto indicators as they cannot provide specific information about dominance performance of the MOEO.

3) *Test algorithms:* Apart from the selection of test problems and test metrics, it is also necessary to have an appropriate set of test algorithms. The purpose of the test algorithms is to assess the significance of the evaluated MOEO i.e. whether its performance is comparable to other existing algorithms. The type of test algorithm can range from classical optimization techniques, stochastic meta-heuristics to related MOEO. At the minimal, the selected set of test algorithms should be truly reflective of the issues considered and represents the state-of-the-art. For example, common test algorithms for evolutionary algorithms for general optimization purposes will include NSGAII, SPEA2 and PAES.

2.4 Experimental Execution

In this final stage, the empirical study is conducted and the experimental results obtained will be analyzed. Practicality and fairness is of significant concern here.

1) *Implementation:* Actual realization of the test algorithms, test functions and test metrics selected during the experimental design stage involved both hardware and software considerations. Practical hardware constraints include processor speed, single or distributed processing, memory requirements and etc. Different computational platforms have varying memory, system resources allocation system, computing time in loading and running program executable. On the other hand, examples of relevant software consideration are the operating system and programming language used to implement the algorithm.

One particular issue in this stage is the implementation of test algorithms. This depends mainly on their availability for download and the designer's programming skill. While it may be more practical to use the existing codes, it is well-known that different implementations, with respect to compiler platform or programming style, can have severe impact on algorithmic performance. The other alternative thus is to recode the test algorithms and run all of them on a common computational platform, but this will be subjected to the programming capability of the designer. Nevertheless, implementation details should be comprehensively furnished to facilitate future related works.

2) *Data Collection:* Appropriate experimental data should be collected during the experimentation process to facilitate analysis in the later stage. The type of data gathered is largely dependent on the type of analysis warranted. In the most general case, the coordinates of the final set of Pareto optimum solutions found in the objective and search space will suffice for the plotting of the Pareto front approximated and the calculation of performance metrics. It is highly likely that these data may require further processing, before they could be interpreted and analyzed. Lastly, due to high computational cost, careful considerations must be made to avoid repeating the experiments again.

3) *Data Analysis:* In this stage, the experimental data obtained earlier will be analyzed and evaluated with respect to the algorithmic motivation and the experimental specifications.

For typical EMOO empirical studies, statistical information of the various test metrics will be reported, i.e. the mean, median, standard deviation, maximum and minimum, to quantify the reliability and validity of the algorithmic performance. These statistics can either be summarized in tabular form or displayed graphically via box plots. While the former displays the precise value obtained for the various metrics, the use of the latter can highlight the relative differences between test algorithms, which are useful in comparative assessment. Nevertheless, a mere difference in the average of the metrics cannot be blindly regarded as performance difference between the various MOEO. As such, statistical techniques like ANOVA, t-test and etc, should be included to quantitate the significance of the differences amongst the optimizers.

Besides the statistical information, the Pareto fronts obtained are usually illustrated to reflect clearly the multiobjective nature of the optimization problem under investigation. Graphical visualization of the Pareto front allows a quick overview on the performance of the MOEO and cross validation between the statistical results earlier. Generally, authors will present the Pareto front of a randomly chosen run or select the run that could best complement the statistical information displayed earlier.

3 Ideality of Empirical Assessment

The ideal empirical study should be able to provide the basis of complete assurance in any statements made on the algorithmic performance and behavior. In other words, the ideal empirical study should be both reliable and valid. Reliability requires that the results from the empirical test should be consistent i.e. if an empirical test has proven the superiority of a particular MOEO in a particular set of problems, then that MOEO should always perform better on other similar problems under identical circumstances. On the other hand, validity requires that the test must always produce a meaningful result i.e. if the MOEO is incapable of solving a particular type of problem, the empirical test should be capable of revealing this defect and conversely, if the MOEO is indeed capable, the test should not show the otherwise. This requirement is similar to the completeness of test data in software development [9].

This naturally leads to the question of what constitutes a reliable and valid empirical test for MOEO. Unfortunately, these theoretically simple concepts are practically hard to attain in implementation. This is further complicated by the availability of the wide array of choices for test functions, test metrics indicators and test algorithms. Naturally, this leads to the question of whether exhaustive testing, defined in terms of all possible combinations of assessment tools is another viable option. The answer is an emphatic no for two reasons.

- It is an impractical attempt to achieve a reliable test
- The validity of the concept of problem difficulty hinges on our perceived notion of how a real-world problem behaves.

The crux of the problem hence lies in the definition of an appropriately sized empirical study that is adequate enough to substantiate the statements it made on the algorithmic performance and behavior.

4 Adequacy of Empirical Assessment

Empirical study forms one of the most integral components in EMOO. Besides understanding the algorithmic development of MOEO, it is important to determine whether the empirical study conducted is adequate to derive any conclusive statement about its performance. Despite its importance, there is alarmingly little research in this area. Due to the lack of proper theoretical foundation, rules of thumb serve as the guidelines for researchers in designing their evaluation study. As such, there is a pressing need to establish standard guidelines in assessing the adequacy of the empirical study.

The concept of (experimental) adequacy is first coined by Goodenough and Gerhart [9] in the field of software engineering in order to formalize a framework for software testing. Likewise, this section aims to motivate the use of similar concepts to guide our investigation of algorithmic capability and behavior. For this purpose, a set of criterion to evaluate the adequacy of an empirical study will be formally defined. Also, a general axiomatic theory of adequacy will be developed to make explicit some underlying assumptions in general empirical study.

4.1 Adequacy Criterion

The adequacy criterion can be regarded as a predicate that defines what must be exercised to constitute a comprehensive test i.e. one which is able to substantiate the conclusive remark of the corresponding analysis. The adequacy criterion *C* could be formulated as a function that take in the algorithm, *A* of a multiobjective optimizer and its corresponding set of specifications, *S* and empirical test, *T*.

Definition 1. (Adequacy Criterion): *The adequacy criterion C is the function:*

$$C : A \times S \times T \rightarrow \{True, False\} \tag{1}$$

where $C(A,S,T) = True$ *means that the evaluation test T is adequate for testing algorithm A against specification S, otherwise T is inadequate.*

Of course, the adequacy criterion need not necessarily be restricted to a discrete form of measurement. It is also possible to fuzzify the above relationship and use real numbers to quantify the degree of adequacy.

Definition 2. (Adequacy Criterion-Fuzzy): *The evaluation adequacy criterion C is the function:*

$$C : A \times S \times T \rightarrow [0,1] \tag{2}$$

where $C(A,S,T) = r$ *means that the evaluation test T is adequate to a degree of r for testing algorithm A against specification S, otherwise T is inadequate. A larger value of r will correspond to a higher degree of the test adequacy.*

Designing a set of adequacy criterion that can meet the dual requirement of reliability and flexibility is quite practically infeasible. Hence, instead of seeking such ideal criterion, this requirements will be relaxed and inadequacy criterion will be defined instead i.e. criteria that define the inadequacy of the evaluation suite. Furthermore, since the empirical study can be segregated into the various components discussed earlier, the adequacy criterion could consider them individually also.

Definition 3. (Inadequacy Criterion): *The evaluation adequacy criterions C is the function:*

$$C: A\times S\times T_1\times T_2\times T_3\times T_4\times T_5\times T_6 \rightarrow \{True, False\} \tag{3}$$

where $C(A,S,T_1,T_2,T_3,T_4,T_5,T_6) = False$ *if any of the components of the test (Test function,* T_1*; Test metrics,* T_2*; Test algorithms,* T_3*; Implementation,* T_4*; Data collection,* T_5*;Data Analysis,* T_6*) is not adequate for testing algorithm A against specification S, otherwise T is adequate.*

This definition provides a simple approach to evaluate the adequacy of MOEO empirical test in general, where each of the different components of the empirical test could be evaluated for their adequacy based on the experimental specifications. If each of the components cannot be proven to be inadequate with respect to the experimental specification, this will imply that the empirical test is adequate to substantiate conclusive statement on the algorithmic performance and behavior of the MOEO. A detailed discussion on the adequacy of the different components will be provided.

4.2 Axiomatization of Empirical Analysis Adequacy

Weyuker [10] proposed several fundamental properties to evaluate the adequacy of test data in software engineering. Inspired by this study, a set of axioms to make explicit the underlying expectations of any empirical study will be presented in this section. These axioms will aid in the designing of empirical study and serve as a guideline to determine its adequacy.

The first and most important property of the adequacy criterion is the existence of one. Without this axiom, there will not be any basis for empirical studies, as no matter how comprehensive and extensive the studies were, it will still be inadequate to derive any conclusion.

Axiom I (Existence): *For every algorithm, A and a set of specification S, there exists an adequate empirical test T.*

Of course, the existence of an adequate evaluation does not guarantee its feasibility. A possible refinement to this axiom will be to require the existence of a finite empirical study for every algorithm and is adequate in all means as well. However, this will depend on the experimental specifications, as they could be finite or infinite in nature. For example, an all-encompassing MOEO designed to solve all problems in general obviously has an infinite specification and the corresponding adequate empirical study will be infinite. Conversely, for a MOEO specially designed for the traveling salesman problem, a test suite of such problems is more than adequate. As such, if one requires a finite evaluation test, it will require that the experimental specifications are finite in the first place. Failure to comply with this simple logic will result in an inadequate evaluation suite. This relationship can be formalized as such

Axiom II (Non-exhaustive Existence): *There exists A and T such that A is adequately tested by T, and T is not exhaustive if and only if the specification S is finite.*

Essentially, the adequacy criterion should denote a minimum degree of evaluation adequacy. Surely, if an existing evaluation is adequate, including more benchmark function or measuring the performance with more performance metrics, which is redundant also, should not make it inadequate.

Axiom III (Monotonicity): *If T is adequate for A and* $T \subseteq T'$ *then T' is adequate for A.*

Another fundamental property of adequacy criterion is that if nothing has been done, the test should not be deemed adequate at all. Even though the algorithm might achieve its desired goals, adequacy criterion is ultimately related to the evaluation study instead of its actual performance. Otherwise, it might be reasonably argued that the empty set is reasonable set for such algorithms.

Axiom IV (Empty Test): *The empty set is not adequate for any algorithms.*

The first four axioms that were introduced are general in nature and viewed the algorithm as a whole. However, all MOEO could be decomposed into its different operators O_1, O_2,...O_N. For example the canonical genetic algorithm is essentially made up of the initial population generator, selection operator, crossover and mutation operator and the fitness evaluator. Thus, the next three axioms will decompose the algorithm into their various operators and relate them to the adequacy criterion of empirical analysis. The following three axioms are based on reasonable intuitive grounds and known empirical phenomenon of algorithmic behavior.

Suppose an improvement to the existing was proposed by modifying the crossover and mutation operator and plugging them individually into a baseline genetic algorithm yield superior result than conventional operators. This does not mean that this is adequate to conclude that they are collectively better, as these two operators might be conflicting in nature. Essentially, this is depicted in the following axiom.

Axiom V (Anti Decomposition): *There is an algorithm A with operators* O_1, O_2... *such that the individual evaluation of* O_1, O_2... *is not adequate for the evaluation of A.*

Similarly, in the same scenario, if the operators were evaluated collectively instead, superior result obtained is not conclusive of the efficiency of the individual operators, as the improvements could be due to only one operator. As such, the following axiom is postulated.

Axiom VI (Anti Generalization): *There is an algorithm A with operators* O_1, O_2... *such that the evaluation of the A is not adequate for the evaluation of* O_1, O_2...

Lastly, suppose an improvement to the initial population generator was proposed for two MOEO, say EA and PSO. Individual evaluation of this operator in EA could not be extended to PSO.

Axiom VII (Anti Extensionality): *There is an algorithm A with operators* $C \in C_1$... *and algorithm B with operators* $D \in D_1$... *where* $D \cap C_1 \neq NULL$ *such that the individual evaluation of* C_1 *in A is not adequate for the individual evaluation of* C_1 *in B.*

These seven axioms make explicit the underlying assumptions in any general empirical assessment of MOEO. While the first four axioms form the basis of empirical study, the remaining three describe the analyzing of algorithm in operator levels.

4.3 Discussion of Adequacy Test Criterion

The adequacy of an empirical assessment test depends heavily on the type of experimental specification and the corresponding experimental requirements. Amongst the six components discussed earlier, data analysis is closely related to the ultimate conclusion derived from the empirical test, while the rest of the components are rather influenced by the type of analysis required. As such, the adequacy test criterion for each component will be discussed via a bottom up approach.

Data Analysis: The experimental analysis should be sufficiently adequate to quantify the correctness and viability of the MOEO, test and verify any hypothesis and assumption made during the algorithm design and uncover pertinent parameter relationship and/or dynamic behavior within the algorithm. Naturally, the analysis should be synchronized to the underlying motivation during the algorithm formulation. Of course, the depth of the analysis will depend on the level of specification desired.

Individual assessment (type I) mainly assesses the algorithmic capability of a particular MOEO. To ensure minimal adequacy, the MOEO should be tested with benchmark problem in related domain and the analysis should reflect its capability in satisfying the three EMOO goals of proximity, diversity and distribution. Extending to type II assessment, the effects of varying the relevant parameters on the algorithmic performance of the MOEO will be investigated. For instance, if new operators were proposed, the relevant parameters will be those introduced by the additional operators, while for a new algorithm, all the significant parameters should be investigated. Furthermore, there should be sufficient variation to establish a clear relationship between the parameters and the algorithmic performance. For this purpose, it is recommended that there should be at least several different configurations for each parameter under investigation. As for type III assessment, the type of analysis will depend on the nature of the optimizers i.e. its underlying theoretical basis. For example, a variation operator was proposed to improve the algorithmic performance of MOEO via emphasized exploration and exploitation efforts in the different stages of the evolutionary progress [11]. The population dynamics of the MOEO in the objective and search space along the evolutionary progress was analyzed to verify the theoretical hypothesis.

As for comparative assessment (type I), the objective is to measure the significance of the MOEO with existing ones. Thus, similar analysis in individual assessment (type I) will be now applied to several algorithms simultaneously to compare their algorithmic capability. Also, statistical tools should be employed to add creditability to the analysis. To reflect the relative robustness of the MOEO for type II assessment, the parameter variation effects for the various test algorithms should be investigated also. The experimental results should show the superiority of the evaluated MOEO over the range of parameters investigated or its robustness, which is measured by its insensitiveness to the parameter variations. Lastly, type III assessment will justify why the evaluated MOEO did perform better/worse in the earlier analysis i.e. provide the explanation for the performance difference in type I and II assessment. Sadly, this type of assessment is lacking in modern literature. The conventional approach is a black box style of assessment, where only the external behavior is being investigated, without giving any hint or analysis for why it is working. Of course, this is mainly due to the difficulty involved with this type of analysis.

Data Collection: The type of experimental data required is highly dependent on the analysis. For type I assessment, recording the coordinates of the final evolved population in the objective and search space will be sufficient for the calculation of performance metrics and the plotting of Pareto fronts. Also, the computational time and fitness evaluation could be recorded and used to evaluate the algorithmic efficiency. For type II assessment, similar data should be obtained under different parameter configurations. To unearth key dynamics of the evaluated MOEO for type III assessment, it might be necessary to obtain coordinate traces of the evolving population in both objective and search space. Special attention should be taken for comparative assessment to ensure that the data are collected in similar manner from all the test algorithms i.e. during the same stage of the evolutionary process.

Implementation: Since implementation merely execute the empirical study crafted during the experimental design stage, the issue of adequacy in implementation centered on whether the empirical test is being conducted fairly, specifically for comparative assessment. As such, steps must be taken to ensure fairness of all experiments conducted is not compromised.

Firstly, all of the algorithms evaluated should be run under identical computational platforms, in both hardware and software aspects. This is to ensure fairness in the comparison as briefly mentioned earlier. As such, the performances of the coded test algorithms should be at least comparable, if not better than reported in the original work. This is especially important if computational efficiency is concerned.

Also, as far as possible, justifications should be made for the algorithmic parameter settings such as crossover, mutation rates etc. Furthermore, it should be ensured that the same initial population is applied for each test algorithm at every run, since the algorithmic performance of MOEO is extremely sensitive to the nature of the population. Lastly, the stochastic nature of MOEO demands that multiple runs should be conducted, so as to obtain the necessary statistical information about its algorithmic performance and behavior.

Test functions: The test function suite is independent of the type of assessment. Rather it depends on the problem scope of the MOEO, whether it is designed for general optimization purpose or for specific problems. For the former, it is important that the test suite selected should contain all the primary test functions and if applicable, secondary test functions should be included to supplement the analysis. On the other hand, if the MOEO is designed for specific application in certain nature or domain instead, an adequate test suite should include a set of such related problems. Nevertheless, the test functions should possess a good balance of features that challenge the MOEO in satisfying the optimization goals of convergence, diversity and distribution [5]. Furthermore, in light of recent investigations [12], [13], test suites should contain problems with high-dimensional objective space or non-linear and non-separable in nature [14].

Test Metrics: The choice of performance metrics is highly dependent on the experimental specification. For type I performance assessment, specific indicators to measure convergence, diversity and distribution respectively should be considered.

Furthermore for comparative assessment, Pareto indicators should be considered to assess the dominance relationships between the various test algorithms. Extending to type II analysis, which is concerned with parameter sensitivity analysis, similar metrics could still be employed to measure changes over the various algorithmic configurations. Lastly, the adequacy of performance metrics for type III analysis will depend on the nature of the MOEO and the algorithmic motivation. For example, to assess the capability of a particular noise-handling feature, the issue of robustness should be considered as another performance metrics. In some cases, specific metrics should be defined to supplement the existing ones.

Test algorithms: For individual assessment, there is obviously no need for any test algorithms. Nevertheless, it might be insightful to include a baseline algorithm to serve as a basis of gauge rather than comparison. For instance, if some evolutionary operators are proposed, they will have to be tested under different parameter values for type II analysis. In such a case, test algorithm in the form of baseline algorithm containing the operators individually and collectively could be included, so that their individual and synergetic effects could be adequately assessed.

The nature of comparative assessment makes test algorithms an essential component. These algorithms should be truly reflective of the state-of-the arts and of the issues considered. Otherwise, the experimental results that form the basis of comparative analysis will not be reliable. Also, the type of algorithms depends largely on its algorithmic motivation. If the MOEO represents a novel approach, the test algorithms should include a set of baseline algorithms from the field of evolutionary computation like evolutionary algorithm, particle swarm optimization, simulated annealing and etc. However, if the optimizer is the improved model of an existing field, using a set of related MOEO that represents the state of the art in that particular field will be necessary. Lastly, if improvements were done in the operator level, it will be necessary to apply the proposed operators and conventional operators into state-of-the-art MOEO to evaluate their performance difference.

4.4 Summary

From the above discussion, a systematic procedure to assess the adequacy of an empirical study has been formalized. First, it should be ensured that the experimental specifications defined are finite to ensure the existence of a finite empirical study (Axiom II). Also, Axiom V to VII should not be violated if operator level analysis was involved. The empirical study will then be segregated into its various components and their adequacy will be individually assessed with respect to the experimental specification. Finally, using the definition of inadequacy criterion (3), the overall adequacy of the empirical study will be satisfied, if each and every component is not deemed inadequate under the experimental specifications as discussed in section 4.3.

Ultimately, the adequacy of the empirical study is bounded by the experimental specifications; this means that the extent of statement that can be made on the algorithmic performance and behavior are restricted by the type of experimental criteria. For instance, a type I individual assessment can only validate the capability of the MOEO in satisfying the various optimization goals. Any statements made on its significance and behavior based on this empirical study will be rendered void.

5 Conclusion

This paper addresses the issue of adequacy for the empirical performance assessment of MOEO. For this purpose, a set of adequacy criterion is presented, which evaluates the general adequacy of an empirical study based on its various components. Details of the various components on how they relate to the overall algorithmic development process and their adequacy under different experimental specifications are provided. Furthermore, a set of axioms that made explicit the underlying assumptions of general empirical study was formulated. Future works include considering case studies of previous empirical study in related literature and evaluating their adequacy. Hopefully, this work can motivate discussion in this area from different perspectives to further improve empirical study techniques in general.

References

1. Bosman, P.A.N., Thierens, D.: The balance between proximity and diversity in multiobjective evolutionary algorithms. IEEE Transactions on Evolutionary Computation, Vol. 7, No. 2 (2003) 174-188
2. He, J., Yao, X.: Towards an analytic framework for analyzing the computation time of evolutionary algorithms. Artificial Intelligence, Vol. 145 (2003) 59-97
3. Veldhuizen, D.A.V., Lamont, G.B.: Multiobjective Evolutionary Algorithm Test Suites. ACM Symposium on Applied Computing (1999) 351-357
4. Okabe, T.: Evolutionary Multi-Objective Optimization-On the Distribution of Offspring in Parameter and Fitness Space. PhD thesis, Technische Fakultaet der Universitaet Bielefeld, Shaker Verlag (2004)
5. Deb, K., Thiele, L., Laumanns, M., Zitzler E.: Scalable MultiObjective Optimization Test Problems. Congress on Evolutionary Computation, Vol. 1 (2002) 825-830
6. Knowles, J.D., Corne, D.W.: On metrics for comparing non-dominated sets. Congress on Evolutionary Computation Conference (2002) 711-716
7. Hansen, M. P., Jaszkiewicz, A.: Evaluating the quality of approximations to the non-dominated set. Technical Report IMM-REP1998 -7, Technical University of Denmark (1998)
8. Zitzler, E., Thiele, L., Laumanns, M., Fonseca, C. M., Fonseca V.G.: Performance assessment of multiobjective optimizers: an analysis and review. IEEE Transactions of Evolutionary Computation, Vol. 7, No. 2 (2003) 117-132
9. Goodenough, J.B., Gerhart, S. L.: Towards a theory of test data selection. IEEE Transactions on Software Engineering, Vol. 1, No. 2 (1975) 156-173
10. Weyuker, E.J.: Axiomatizing software test data adequacy. IEEE Transactions on Software Engineering, Vol. 12, No. 12 (1986) 1128-1138
11. Teoh, E. J., Chiam, S. C., Goh, C. K., Tan, K. C.: Adapting evolutionary dynamics of variation for multi-objective optimization. IEEE Congress on Evolutionary Computation, Vol. 2 (2005) 1290-1297
12. Khare, V., Yao, X., Deb, K.: Performance scaling of multi-objective evolutionary algorithms. Evolutionary Multi-Criterion Optimization Conference (2003) 376-390
13. Hughes, E.J.: Evolutionary many-objective optimization: many once or one many? IEEE Congress on Evolutionary Computation, Vol. 1 (2005) 222-227
14. Huband, S., Barone, L., While, R.L., Hingston, P.: A Scalable Multi-objective Test Problem Toolkit. Evolutionary Multi-Criterion Optimization Conference (2005) 280-295

A Comparative Study of Progressive Preference Articulation Techniques for Multiobjective Optimisation

Salem F. Adra, Ian Griffin, and Peter J. Fleming

Department of Automatic Control and Systems Engineering
University of Sheffield, Mappin Street
Sheffield S1 3JD, UK
{s.adra, i.griffin, p.fleming}@sheffield.ac.uk

Abstract. Multiobjective optimisation has traditionally focused on problems consisting of 2 or 3 objectives. Real-world problems often require the optimisation of a larger number of objectives. Research has shown that conclusions drawn from experimentations carried out on 2 or 3 objectives cannot be generalized for a higher number of objectives. The curse of dimensionality is a problem that faces decision makers when confronted with many objectives. Preference articulation techniques, and especially progressive preference articulation (PPA) techniques are effective methods for supporting the decision maker. In this paper, some of the most recent and most established PPA techniques are examined, and their utility for tackling *many*-objective optimisation problems is discussed and compared from the viewpoint of the decision maker.

Keywords: Progressive preference articulation, Multiobjective optimisation.

1 Introduction

Real-world problems commonly require the simultaneous consideration of multiple performance measures. Solving such problems is therefore concerned with finding an ideal solution that satisfies the decision maker's (DM) preferences and meets the goal values for the problem objectives without violating certain constraints. Conventionally, evolutionary multiobjective optimisation (EMO) has focused on dealing with optimisation problems comprising 2 or 3 objectives, mainly for the convenience of graphical demonstration and illustration. Conclusions drawn from such low-dimensional multiobjective frameworks used to be generalized for the multiobjective branch of evolutionary optimisation problems. Recently, research [1, 2] has shown that the case of high-dimensional optimisation problems (more than 3 objectives) also termed as "Many Objective Optimisation Problems" is a special case of evolutionary multiobjective problems that requires further investigation. Indeed a different set of difficulties and challenges can be associated the Evolutionary *Many*-Objective Optimisation sub-category, most importantly the unambiguous conflict of solutions convergence and diversity in such scenarios. Convergence and diversity are two of the primary requirements of a multiobjective optimiser. Other evolutionary *many*-objective optimisation difficulties involve the obvious dimensionality increase of the Pareto front, and the difficulty of visualizing such scenarios.

S. Obayashi et al. (Eds.): EMO 2007, LNCS 4403, pp. 908–921, 2007.

Reducing the dimensionality and therefore the complexity of a problem is a straightforward way for dealing with the high-dimensional problems. Early approaches such as the weighted sum or the Tchebyshev method [3] consist of scaling techniques to convert multiobjective problems into a single objective counterpart. Such approaches presented several shortcomings, mainly the absence of the desired parallel search capacity. More recent techniques of dimensionality reduction for dealing with multiobjective optimisation problems consist of techniques to identify objectives redundancy and to eliminate it. Principal Component Analysis [4] ,[1] is an example of such a technique. Its aim is to identify redundant objectives whose absence has no substantial effect on the optimisation process, thereby simplifying the complexity of certain high dimensional problems and reducing the hyperspaces of solutions. While dimensionality reduction is a remedial measure to tackle multiobjective optimisation problems, it can only be deployed in reducible scenarios where redundancy or objective relationships such as independence or harmony are existent and detectable.

In scenarios, where insufficient redundancy can be detected in a high-dimensional problem, progressive preference articulation (PPA) is a proven useful alternative remedial measure. The incorporation of DM preference into evolutionary multiobjective optimisation algorithms is very useful for guiding the search into pertinent regions of interest (ROI), which are relevant to the decision maker. Coello [5] has produced a comprehensive survey about handling preferences in EMO. It can also provide advantages over the use of pure Pareto-optimality, which is unfettered in its search and is liable to produce solutions outside the ROI as well as within it. Until recently, most EMO research has focused on bi-objective problems where the need for incorporating the decision maker's preferences is less apparent. The aim of this paper is to encourage and promote the research of incorporating progressive preference articulation techniques into evolutionary multiobjective optimisation. In this paper, some of the most recent preference articulation techniques are discussed and upgraded to their progressive versions for incorporation into evolutionary multiobjective optimisation processes. Their major strengths and weaknesses for tackling *many*-objective optimisation problems are discussed and illustrated on a straightforward bi-objective scenario for simplicity. The preference articulation techniques investigated in this work include Branke's "Guided Dominance Scheme" [5], Deb's "Biased Crowding Technique" [6], the manipulation of the ε-dominance concept within the framework of Deb's steady state ε –MOEA [7] and Fonseca and Fleming's preferability operator (FF-PPA) [8], which we believe was the first truly PPA technique for EMO.

In section 2 of this paper, the requirements of a multiobjective optimiser are presented. In section 3, a brief description of the preference articulation techniques inspected in this paper is given. In section 4, the usefulness and practicality of the studied progressive preference articulation techniques are visually illustrated on simple bi-objective scenarios. (In some cases, we have introduced a progressive capability into existing preference articulation techniques.) In section 5, the strengths, weaknesses, user-friendliness and efficiency of these PPA techniques, in a many objective optimisation context, are discussed from the viewpoint of the decision maker. Lastly, some concluding remarks are presented.

2 Requirements of Multiobjective Optimisers

It is not possible to find a single Utopian solution when tackling a multiobjective optimisation problem with many conflicting objectives, instead a set of optimal solutions, also called a Pareto front, is anticipated. The Pareto front is the set of solutions whose members cannot be improved further in terms of a certain objective without introducing some deterioration in terms of one or more of the other competing objectives. Solving a multiobjective optimisation problem is therefore best approached using a population-based technique, such as evolutionary algorithms.

The set of solutions achieved by an optimiser, also called the approximation set, is required to be the closest possible to the true Pareto front. Because of the non-existence of an ideal single solution, the set of optimized solutions is also required to be well spread and covering wide areas of the Pareto front, presenting the decision maker with a well distributed set of solutions to choose from, based on certain preferences such as objective priorities or regions of interest. Furthermore, the approximation set has to be achieved within an acceptable amount of time. Therefore, convergence, pertinence to the DM (ROI), speed of convergence and diversity are all desired and essential requirements of multi-objective optimisers and constitute their assessment basis. In Figure 1 the ideal solution for a multiobjective optimisation problem is illustrated for a discontinuous Pareto front.

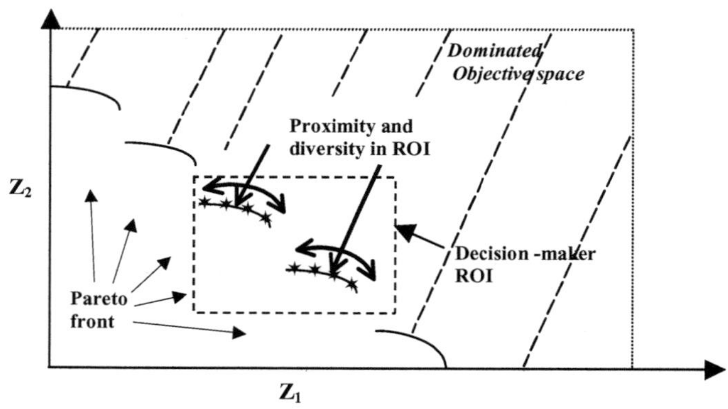

Fig. 1. The Ideal Solution to a Multi-Objective Optimisation Problem

3 The Investigated Preference Articulation Techniques

In this section a brief description of some of the most recent preference articulation techniques is provided:

3.1 Guided Dominance for Evolutionary Multi-objective Optimisation

Branke et al. introduced the guided dominance principle within the context of a novel optimiser, termed the Guided Multi-Objective Evolutionary Algorithm (G-MOEA) [9]. The principle of guided dominance manifested the DM's preferences through a modification of the definition of dominance. The user has to determine all maximally acceptable tradeoffs between all pairs of objectives.

To illustrate this concept, consider an optimisation problem consisting of two competing objectives. In order to use the guided dominance scheme, the DM has to decide *a priori* the maximum acceptable amount of degradation in terms of objective 2 which can be deemed worthy to be recompensed by a single unit of improvement in terms of objective 1, and vice versa. The standard dominance concept,

$$x \succ y \Leftrightarrow \forall i \in \{1,2\}, f_i(x) \leq f_i(y)) \wedge \exists i \in \{1,2\} : f_i(x) < f_i(y)) \tag{1}$$

where a solution x is said to dominate another solution y if x is less or equal than y in terms of all the objective values, with at least one strict inequality, becomes

$$x \succ y \Leftrightarrow (f_1(x) + m_{12} f_2(x) \leq f_1(y) + m_{12} f_2(y)) \wedge (m_{21} f_1(x) + f_2(x) \leq m_{21} f_1(y) + f_2(y)) \tag{2}$$

with an inequality in at least one case. In equation 2, m_{12} and m_{21} denote correspondingly the maximum acceptable amount of degradation in terms of objective 1 and 2 which are compensated by a single unit of improvement in terms of objective 2 and 1 respectively.

The guided dominance scheme corresponds to a simple transformation of the objective space, which makes its incorporation into dominance-based evolutionary algorithms straightforward and practical. Using the guided dominance approach, it will be possible to emphasize any part of a convex Pareto front by carefully setting suitable trade-off values.

3.2 Biased Crowding Distance

The biased crowding distance is one of the state-of-the-art preference articulation techniques that allows the user to efficiently focus on certain regions of interest on a convex or concave Pareto front. This technique has its roots in the biased fitness sharing approach [10] which was employed in the Non Dominated Sorting Genetic Algorithm (NSGA) [11]. The biased crowding measure is defined for any solution k on any particular front all along the optimisation process as follows:

$$D_k = d_k \left(\frac{d_k^{'}}{d_k} \right)^{\alpha} \tag{3}$$

where d_k is the original crowding measure [12] for the solution k based on its neighbouring solutions, and d_k' is the crowding measure of the projected solutions on the plane whose direction is specified by the user to express a certain region of interest on the Pareto front. α is the parameter responsible for controlling the bias intensity.

As a result, solutions located on the region of the front which is tangent to the DM's devised projection plane, which reflects a certain preference of a ROI, will be biased and favourite to be maintained because the ratio d_k'/ d_k will be close to unity,

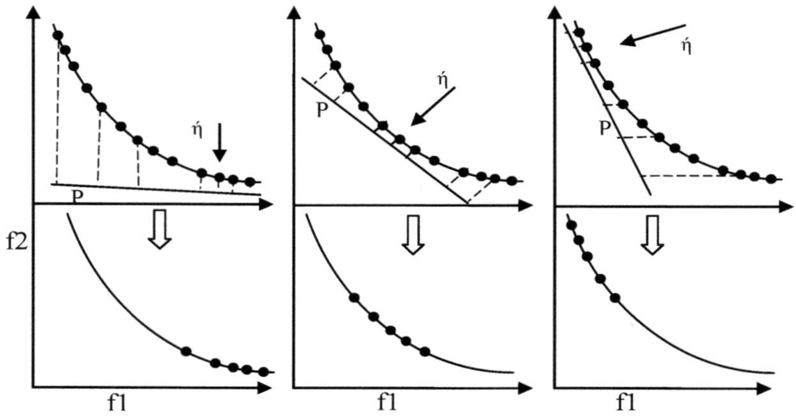

Fig. 2. Biased Crowding Distance (P = Projection Plane, ή = Projection Direction)

and therefore D_k will be approximately the same measure as the original crowding value for such solutions. Figure 2 illustrates the concept of the biased crowding distance for 3 different ROI on a convex front.

3.3 ε –MOEA: Manipulating the ε-Dominance

ε-MOEA [13] is one of the state-of-the-art multiobjective evolutionary optimisers. It is a steady-state algorithm composed of two populations of solutions, which co-evolve simultaneously, but independently. The ε-dominance concept illustrated in Figure 3 is adopted for archive inclusion. A solution can only be included in the archive, which eventually should contain a representative bounded set of solutions which form the Pareto front, if it is not ε-dominated by any of the other members of the archive. ε-MOEA uses a grid-like strategy similar to PAES [14], but more sophisticated, to divide the objective space into hyperboxes and promote solutions diversity without setting an upper limit on the archive size prior to the approximation. Instead the strategy used in ε-MOEA ensures that the archive will eventually get bounded with a well-distributed and limited number of solutions, which represent the Pareto front. Despite the sophistication and usefulness of ε-MOEA, the deployed ε-dominance concept is the reason behind choosing this optimisation technique as a preference articulation technique to be investigated along with the other techniques used in this work. By setting a vector of progressively articulated ε-values, instead of a single fixed value, to form the basis for solutions selection and inclusion in the archive, ε-MOEA is upgraded to a PPA technique which enhances its overall performance, at least from a decision maker's point of view within a *many*-objective optimisation context. Each single ε value will correspond to the accuracy or tolerance in terms of a certain specific dimension or objective. The motivation behind this upgrade is to investigate and exploit the efficacy of the ε-dominance concept as a PPA technique.

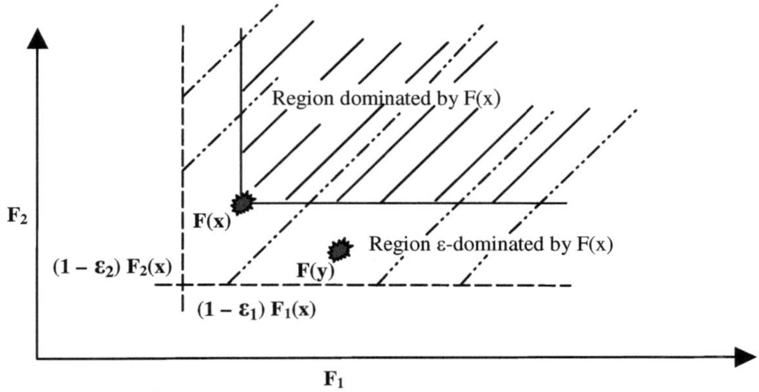

Fig. 3. ε-Pareto dominance for 2 objectives

3.4 FF-PPA Technique

Formulated and implemented in 1998, Fonseca and Fleming's PPA technique [9] remains an important approach to progressive preference articulation. The underlying layer of this technique is based on a combination of concepts such as Pareto optimality, constraint optimisation and satisfaction, the lexicographic method and goal programming. The core of this PPA technique is based on the *Preferability operator*, which is a transitive relational operator that incorporates goal and priority information about the objectives and which consequently modifies the dominance definition.

Using FF-PPA, two alternative solutions A and B are first compared in terms of their objectives with the highest priority while disregarding the objectives of this priority class that meets their goal values. In the case where the objectives, belonging to the same priority class, of solutions A and B meet all their goal values or contrarily violate some or all of their goal values in an exact similar way, the next priority class will be considered. This process continues until reaching the lowest priority class, where solutions are compared based on the pure Pareto optimality concept. Through a user-friendly interface, the DM can set goal values for the objectives being optimised and can change the priorities of the objectives in a progressive fashion at any time during the optimisation process; the dominance concept gets updated accordingly. In other words, using this technique the DM has full control of the optimisation process and can efficiently focus on any region of interest at any time and upon request. Figure 4 illustrates the user interface of this PPA technique which includes the parallel coordinates graph [15], an efficient, FF-PPA independent, visualization technique for any problem dimension. Here, each line in the graph connects the performance objectives achieved by an individual member of the population and represents a potential solution to the design problem. This is in contrast to the usual Cartesian method of representation and has the advantage of being able to handle representations where the number of objectives exceeds 3.

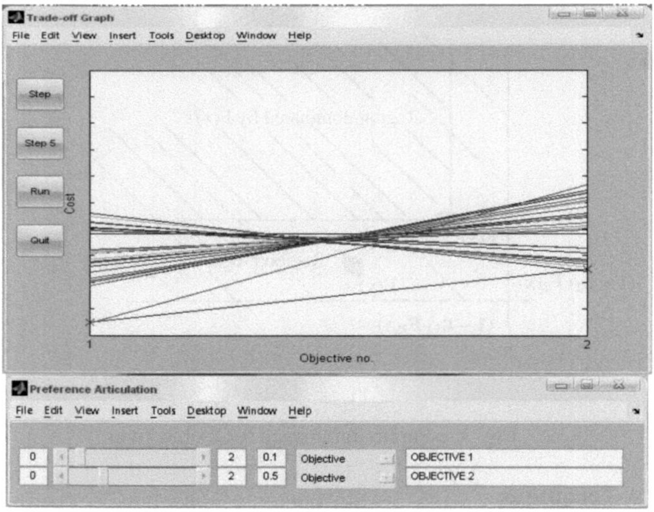

Fig. 4. The user interface of Fonseca and Fleming's PPA technique

In the next section, a graphical demonstration of the above mentioned PPA techniques will be presented illustrating the accuracy, efficiency and usefulness of these techniques.

4 PPA Techniques in Practice

In this section, the performance of the guided dominance concept, the biased crowding measure, the use of ε-dominance and FF-PPA technique will be assessed. The efficiency and practicality of these techniques will be examined from the decision maker's point of view, assuming that their expertise in evolutionary computation might be very limited or nil. The goal is to highlight the utility of these PPA techniques to a DM, mainly in terms of reducing the search space, and focusing on ROI, which is a remedial measure for use in high-dimensional problems. Several bi-objective scenarios, convenient for graphical illustrations, will be deployed to highlight the strengths and weaknesses of these techniques, and will permit the inference of well-based conclusions for the high dimensional cases. Note that except for the ε – MOEA, NSGA-II was chosen to be the underlying optimiser for hybridizing the biased crowding technique, the guided dominance and the FF-PPA.

4.1 Demonstration of ε –Dominance as a PPA Technique in the ε –MOEA Context

In Figure 5 and 6, the convex test function ZDT1 [16], was deployed to investigate the utility of manipulating the ε –dominance concept in terms of each dimension separately as an attempt to simulate a preference articulation scheme.

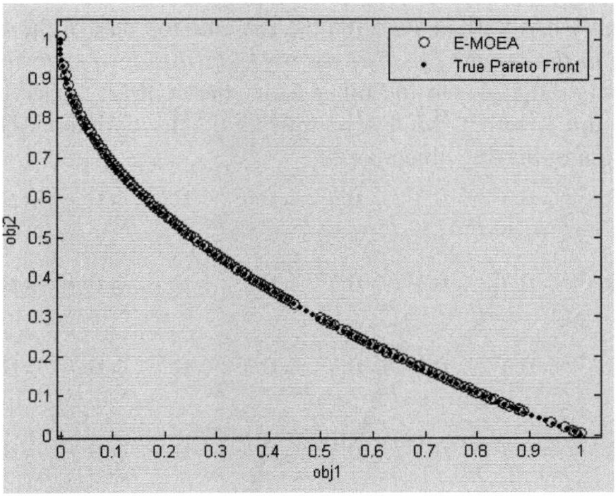

Fig. 5. ε –MOEA running on ZDT1 with ε1 = ε2 = 0.05

In Figure 5, ε –MOEA was executed on ZDT1 using the same configuration used in [7]. Starting with ε1 = ε2 = 0.05, it was clear that the end results were conforming to the desired preference which states no priorities between the 2 objectives.

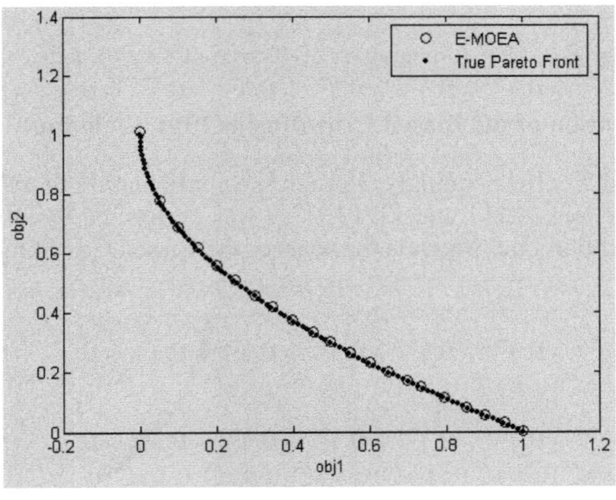

Fig. 6. ε –MOEA running on ZDT1 with ε1 = 0.05 and ε2 = 0.005

In Figure 6, the value of ε2 was decreased to 0.005 to denote a preference in terms of objective 2 while ε1 retained its previous value. In other words, a solution **"x"** favouring objective 2 will dominate a larger number of solutions favouring objective 1 and which used to be considered as non-dominated solutions alongside **"x"** when the epsilon values were equal. In Figure 6, and after running the algorithm for a reasonable amount of time, it was visible that the results reflected a minor bias in terms of objective 2, but good solutions in terms of objective 1 were still present. In other words, the density of solutions lying on the extreme top part of the true Pareto front - *corresponding to the ranges [0 - 0.2] for obj1 and [0.75 - 1] for obj2* - of the ZDT1 test function was reduced. On the other hand, the contrary observations can be deduced from Figure 7, with ε –MOEA optimising the discontinuous ZDT3 test function with the intention of biasing objective 1.

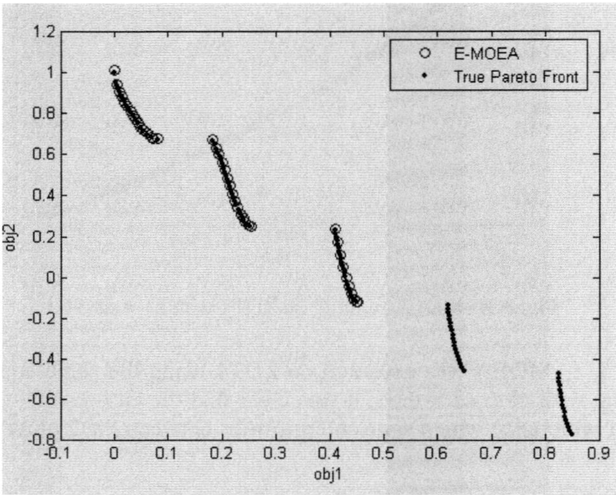

Fig. 7. ε –MOEA running on ZDT3 with ε1 = 0.005 and ε2 = 0.05

4.2 Demonstration of the Biased Crowding as PPA Technique

In Figures 8 and 9 correspondingly, a progressive articulation of a ROI was expressed for the test functions ZDT1 and ZDT3. This was performed by modifying the direction of the projection line, which is the basis of the Biased Crowding technique.

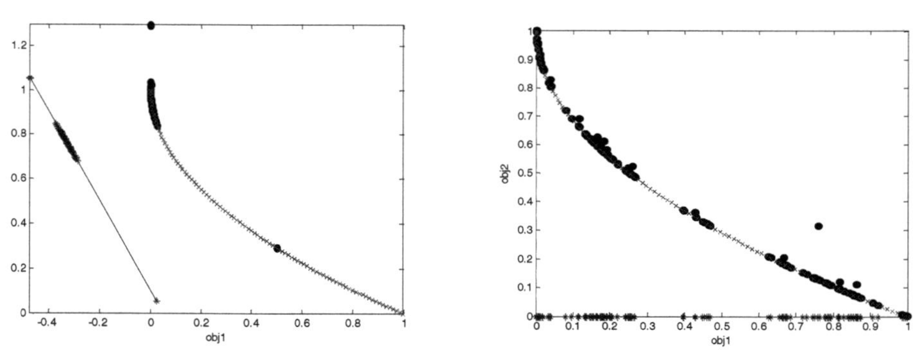

Fig. 8. NSGA-II/Biased Crowding running on ZDT1

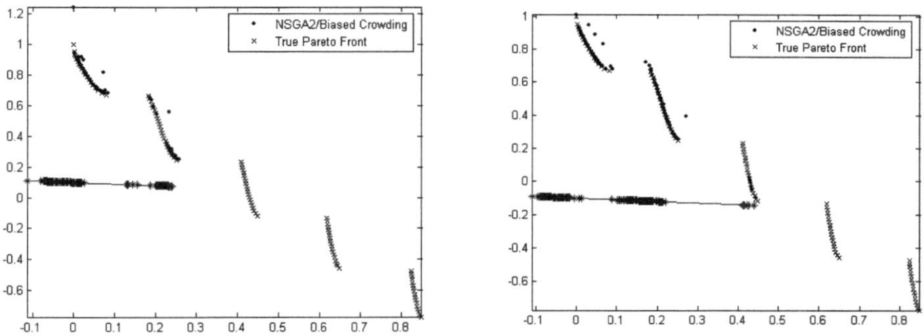

Fig. 9. NSGA-II/Biased Crowding running on ZDT2

For these continuous well-shaped Pareto fronts, the biased crowding seems to perform well, although the obtained results may fall a little outside the ROI, as well as within it. The alpha parameter in the biased crowding measure is a remedial measure to control results accuracy, though setting its value together with the right plane direction is not a straightforward decision, or attractive to a DM especially as the dimensionality of the problem increases.

4.3 Demonstration of the Guided Dominance Principle as a PPA Technique

Figures 10 and 11 illustrate, respectively, the results achieved by G-MOEA for the convex and discontinuous test functions ZDT1 and ZDT3 for 2 different consecutive preferences (a and b). In Figure 10a, there were no preferences among the 2 objectives, but the ROI was decided vaguely by choosing an equal maximal amount of acceptable degradation for the 2 objectives when the other objective improves by a single unit. The bounds of the decision maker's ROI cannot be simply expressed; instead there is a need for an intermediate translation of the DM preferences into line slopes that delimit the desired ROI.

In Figure 10b, the amount of degradation in terms of objective 1 that merits a unit improvement in terms of objective 2 was increased, therefore favouring objective 2. The bias was observed, although it was still not a straightforward method from the DM's point of view to execute a specific detailed optimisation and search scenario. The guided-dominance was then seen to be inconsistent in terms of biasing one of the 2 objectives when applied to ZDT3 (Figure 11). By just switching the preferences, it was clear that the results were not balanced, as the results were too numerous and varied when objective 1 was the prioritised objective (Figure 11b) as opposed to the opposite case of the same continuous scenario (Figure 11a).

4.4 Demonstration of FF-PPA Technique

Figures 12 and 13 illustrate an optimisation scenario solving accordingly ZDT3 and ZDT1. The PPA of the DM was expressed using Fonseca and Fleming's preferability operator. In Figure 12 (a-b), the 2 objectives had the same priority but

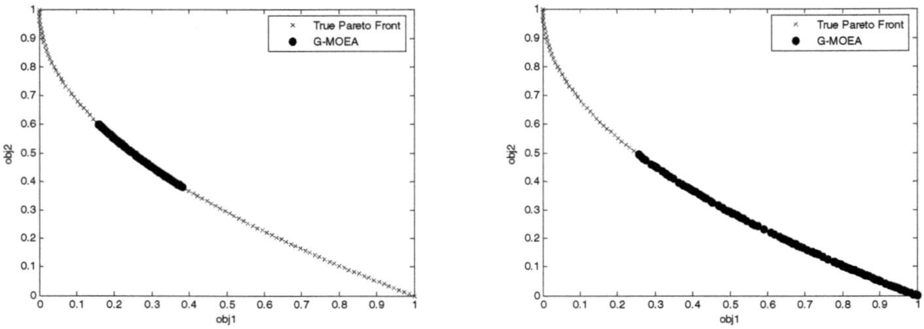

a b

Fig. 10. G-MOEA running on ZDT1

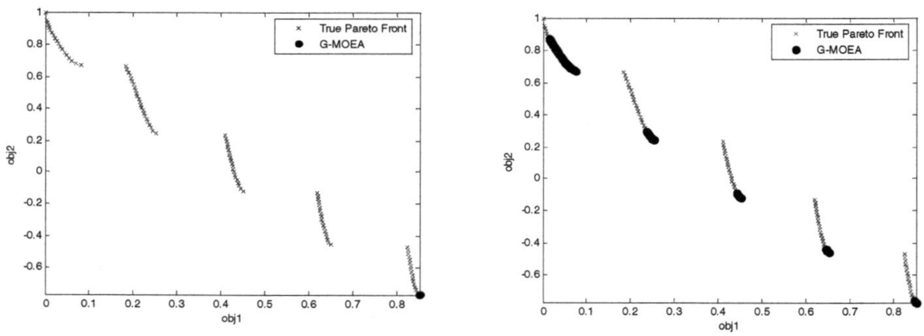

a b

Fig. 11. G-MOEA running on ZDT3

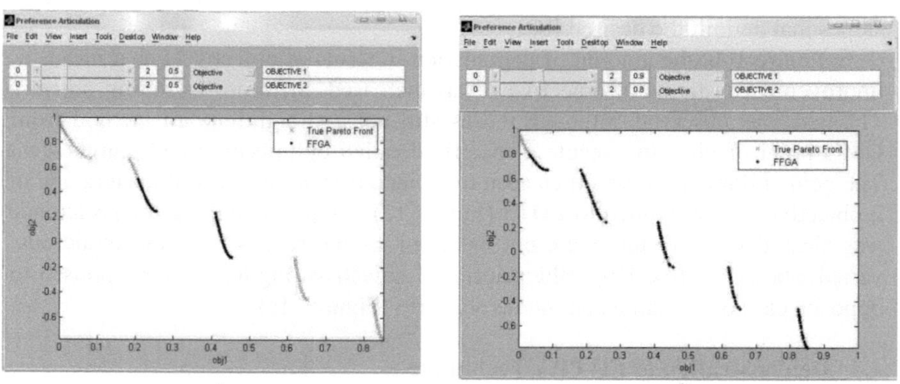

a b

Fig. 12. Fonseca and Fleming's PPA technique running on ZDT3

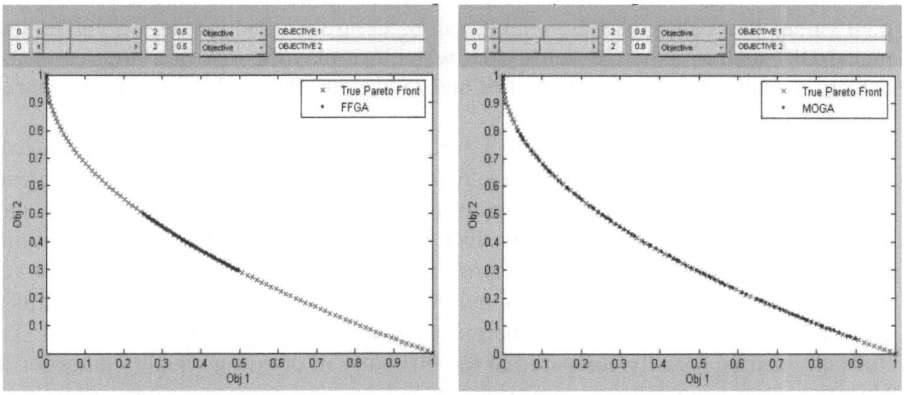

c d

Fig. 13. Fonseca and Fleming's PPA technique running on ZDT1

Different desired goals in order to emphasize the facility of reducing the search space andfocusing on regions of interest. The desired goal values for objective 1 and objective 2 were as follows respectively: Figure 12(a) →(0.5,0.5), Figure 12(b) →(0.9,0.8).

From Figure 12, it is very obvious that FF-PPA technique is a precise and DM-oriented facility. The numerical goal value in each dimension is the only information required from a DM to reduce the search space and focus on certain parts of a Pareto front. The goal values progressively articulated by the DM are input to the optimiser to modify the concept of dominance and steer the search and selection process in the desired search region.

5 Discussion and Concluding Remarks

Progressive preference articulation is a useful approach for reducing high-dimensional spaces and tackling evolutionary many-objective optimisation problems. It has benefits when compared with its a priori preference articulation technique counterpart, which requires the DM to know his/her preferences in advance, and which makes no use of the information that becomes available during the search process. In the previous section, experiments were carried out using some of the most recent and most established PPA techniques. Although the deployed scenarios consisted of 2 dimensional scenarios only, for visualization purposes, the strengths, weaknesses, and therefore the efficiency and suitability of these PPA techniques for the many-objective optimisation were apparent. The FF-PPA technique clearly still stands as an efficient, truly "progressive" articulation technique. It is a user-friendly and direct technique. The accuracy and pertinence of the results achieved by the FF-PPA technique are realised with modest computational effort, and easily scales to any number of objectives. In evolutionary many-objective optimisation scenarios, supervised by an application-expert DM, the FF-PPA technique is a suitable optimisation technique. The technique's traditionally criticised negative influence on the underlying search process which can be caused by setting easily achievable, or contrarily very optimistic goal values that can hinder the search, can be controlled and monitored via an automated DM such as expert systems [17] which can play the role of a progress sensor detecting such optimisation anomalies and modifying the goal values as appropriate with or without DM intervention.

The biased crowding concept is a well-established preference articulation technique that can be used in a progressive manner to focus on a certain ROI. It is mostly useful and practical from the DM's point of view when used with convex or concave optimisation problems with no more than 3 objectives. When dealing with multimodal, ill-behaved or high-dimensional problems, using the biased crowding is not very efficient, especially in high-dimensional problems, because it can be very confusing for a DM to devise a plane or hyperplane of interest for solutions projection. This difficulty can be broadly compared to the difficulty of devising weight values for the objectives in a weighted sum approach. Indeed, even in scenarios where a projection hyperplane might be suggested, this technique will require a considerable appreciation of high dimensional geometry and calculation, which can be very complex to operate or automate. Even in the best scenarios, it was actually noted that the resolution of the ROI achieved by the biased crowding based PPA technique is not as precisely aligned with the DM preferences when compared with the FF-PPA technique, although this fuzziness can be suitable for addressing vague user preferences.

On the other hand, setting epsilon values for each objective and ensuring the facility of progressively modifying the epsilon values can establish the ε –dominance concept within the ε-MOEA as another PPA technique. From the process operator or DM's point of view, it will remain, however, a complicated approach, involving the manipulation of ε-values rather than numerical goal values for the objectives. The accuracy and pertinence of the results achieved by such a PPA approach can be quite imprecise with a fuzzy response to certain preferences. On the other hand, the diversity promotion mechanism employed in ε-MOEA is a state-of-the-art multiobjective optimisation technique for limiting archive size and promoting diversity, and is highly commendable. Although it seems so far that the use of ε –dominance is better reserved for defining results precision and the magnitude of computational requirements - which can be used to reduce the effect of dominance resistance in high dimensional problems- future research into using ε –dominance as a method to articulate preferences is definitely desired.

Lastly, despite its simplicity and practicality for certain optimisation problems, the use of the guided dominance scheme as a PPA technique suffers from several weaknesses. Because the modification of the dominance scheme implicitly assumes linear utility functions, it can be quite complicated to handle multimodal and non-convex optimisation problems [9], [6]. In addition, when tackling high-dimensional problems this technique can be computationally expensive and demanding [6], especially from a DM point of view, as the number of required pair-wise tradeoff values for this technique becomes very high.

PPA methods are a very important area of application for *many*-objective optimisation problems and new approaches are welcomed. Meanwhile, it is planned to perform controlled studies of relatively novice DMs solving selected *many*-objective optimisation tasks using the four methods described above.

References

1. Purshouse, R.C., *On the Evolutionary Optimisation of Many Objectives*. 2003, Department of Automatic Control and Systems Engineering, The University of Sheffield: Sheffield, UK.
2. Khare, V., X. Yao, and K. Deb. *Performance Scaling of Multi-Objective Evolutionary Algorithms*. in *Proceedings of the Second International Conference on Evolutionary Multi-Criterion Optimization (EMO 2003)*. 2003. Berlin: Springer.
3. Miettinen, K., *Nonlinear Multiobjective Optimization*. 1999, Boston: Kluwer.
4. Deb, K. and D.K. Saxena, *On Finding Pareto-Optimal Solutions Through Dimensionality Reduction for Certain Large-Dimensional Multi-Objective Optimization Problems*. 2005, Kanpur Genetic Algorithms Laboratory (KanGAL) Indian Institute of Technology Kanpur: Kanpur.
5. Branke, J.u., T. Kau\ssler, and H. Schmeck, *Guidance in Evolutionary Multi-Objective Optimization*. Advances in Engineering Software, 2001. **32**: p. 499-507.
6. Branke, J.u. and K. Deb, *Integrating User Preferences into Evolutionary Multi-Objective Optimization*, in *Knowledge Incorporation in Evolutionary Computation*, Y. Jin, Editor. 2005, Springer: Berlin Heidelberg. p. 461-477.
7. Deb, K., M. Mohan, and S. Mishra, *Evaluating the ϵ-Domination Based Multi-Objective Evolutionary Algorithm for a Quick Computation of Pareto-Optimal Solutions*. Evolutionary Computation, 2005. **13**(4): p. 501-525.
8. Fonseca, C.M. and P.J. Fleming, *Multiobjective Optimization and Multiple Constraint Handling with Evolutionary Algorithms -- Part I: A Unified Formulation*. IEEE Transactions on Systems, Man, and Cybernetics, Part A: Systems and Humans, 1998. **28**(1): p. 26-37.
9. Branke, J., T. KauBler, and H. Schmeck, *Guidance in Evolutionary Multi-Objective Optimization*. Advances in Engineering Software, 2001. **32**: p. 499-507.
10. Deb, K., *Multi-objective Evolutionary Algorithms: Introducing Bias Among Pareto-optimal Solutions*, in *Advances in Evolutionary Computing. Theory and Applications*, A. Ghosh and S. Tsutsui, Editors. 2003, Springer: Berlin. p. 263-292.
11. Srinivas, N. and K. Deb, *Multiobjective Optimization Using Nondominated Sorting in Genetic Algorithms*. Evolutionary Computation, 1994. **2**(3): p. 221-248.
12. Deb, K., et al. *A Fast Elitist Non-Dominated Sorting Genetic Algorithm for Multi-Objective Optimization: NSGA-II*. in *Proceedings of the Parallel Problem Solving from Nature VI Conference*. 2000. Paris, France: Springer. Lecture Notes in Computer Science No. 1917.
13. Laumanns, M., et al., *Combining Convergence and Diversity in Evolutionary Multi-Objective Optimization*. Evolutionary Computation, 2002. **10**(3): p. 263-282.
14. Knowles, J.D. and D.W. Corne, *Approximating the Nondominated Front Using the Pareto Archived Evolution Strategy*. Evolutionary Computation, 2000. **8**(2): p. 149-172.
15. 15. Inselberg, A., *The Plane with Parallel Coordinates*. The Visual Computer, 1985. **1**: p. 69-91.
16. Zitzler, E., K. Deb, and L. Thiele, *Comparison of Multiobjective Evolutionary Algorithms: Empirical Results*. Evolutionary Computation, 2000. **8**(2): p. 173-195.
17. Fonseca, C.M. and P.J. Fleming. *Genetic Algorithms for Multiobjective Optimization: Formulation, Discussion and Generalization*. in *Proc. of the Fifth Int. Conf. on Genetic Algorithms*. 1993. San Mateo, CA: Morgan Kaufmann.

Test Problems Based on Lamé Superspheres

Michael T.M. Emmerich and André H. Deutz

Leiden Institute of Advanced Computer Science, Leiden University,
2333 CA Leiden, The Netherlands
{deutz,emmerich}@liacs.nl
http://www.liacs.nl

Abstract. Pareto optimization methods are usually expected to find well-distributed approximations of Pareto fronts with basic geometry, such as smooth, convex and concave surfaces. In this contribution, test-problems are proposed for which the Pareto front is the intersection of a Lamé supersphere with the positive $\mathbb{R}^n$-orthant. Besides scalability in the number of objectives and decision variables, the proposed test problems are also scalable in a characteristic we introduce as *resolvability of conflict*, which is closely related to convexity/concavity, curvature and the position of knee-points of the Pareto fronts.

As a very basic bi-objective problem we propose a generalization of Schaffer's problem. We derive closed-form expressions for the efficient sets and the Pareto fronts, which are arcs of Lamé supercircles. Adopting the bottom-up approach of test problem construction, as used for the DTLZ test-problem suite, we derive test problems of higher dimension that result in Pareto fronts of superspherical geometry.

Geometrical properties of these test-problems, such as concavity and convexity and the position of knee-points are studied. Our focus is on geometrical properties that are useful for performance assessment, such as the dominated hypervolume measure of the Pareto fronts. The use of these test problems is exemplified with a case-study using the SMS-EMOA, for which we study the distribution of solution points on different 3-D Pareto fronts.

1 Introduction

Next to introducing a manageable mathematical foundation for meta-heuristic approaches in multiobjective optimization, constructing a repository of scalable and multimodal test problems is of vital importance [1, 3, 6, 8, 10]. In analyzing a test problem family with well-defined properties, we make a contribution to this ongoing effort of the multiobjective optimization community.

One reason for obtaining the complete Pareto front (PF) of a problem instead of a single non-dominated solution, is that the shape of the PF provides the decision maker with useful extra information about the nature of the conflict. A qualitative approach to this problem is to distinguish between concave, convex and linear (parts of) PFs, as it is well known that on convex PFs it is easier to

S. Obayashi et al. (Eds.): EMO 2007, LNCS 4403, pp. 922–936, 2007.

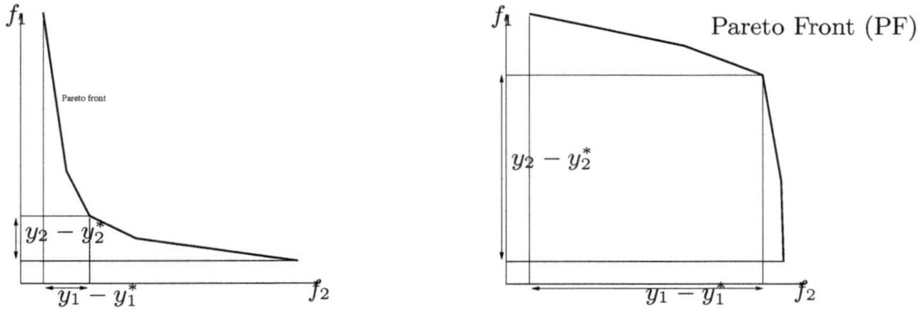

Fig. 1. Visualization of a measure for conflict resolvability. The left figure displays a scenario in which a good compromise exists, while in the scenario displayed on the right hand side they do not exists. The conflict resolvability is computed as the maximum of $y_1 - y_1^*$ and $y_2 - y_2^*$ at the position which minimizes this value, where y_1^* and y_2^* are the coordinates of the ideal point.

find good compromises than on concave ones, and also the behavior of algorithms is often different on both types of geometry.

Taking this into account, we aim for test problems that capture all three types of PFs (concave, convex, and linear). However, the test-case we propose allows also to scale quantitatively the resolvability of conflict. As a measure of the *resolvability of conflict* one may consider:

$$\mathrm{RoC}(PF) = 1 - \frac{\min_{\mathbf{y} \in PF} \max_{i \in \{1,\ldots,m\}} |y_i - y_i^*|}{\max_{\mathbf{y} \in PF} \max_{i \in \{1,\ldots,m\}} |y_i - y_i^*|} \tag{1}$$

Here, $\mathbf{y}^*$ denotes the ideal solution, m is the number of objective functions, and *PF* denotes the PF. Note, that in case of the denominator being zero, we define RoC(*PF*) to be one. Ideally, this value should be close to 1, meaning that all objectives are complementary.

In Figure 1 two situations are depicted. In the left figure a convex PF for a problem is depicted, for which good compromise solutions exists. In the right figure, a concave PF is depicted for which there exists no good *compromise*. We construct a highly symmetrical class of functions for which the geometry of the PF can be varied gradually from convex shapes with high resolvability of conflicts, to linear shapes, and concave shapes with low resolvability of conflicts (cf. figure 1). The problem family we propose is highly symmetrical and only introduces the difficulty of obtaining well-spread solutions on the different shapes of PFs. We consider these problems as interesting, as they can be used for analyzing metaheuristics in a controlled way, i. e. by isolating difficulties. However, we note that the complexity of the test problems can be gradually increased by adding difficulties in a managed way.

The problems can be considered as generalizations of models with spherical symmetries, that are frequently used as elementary test problems in single

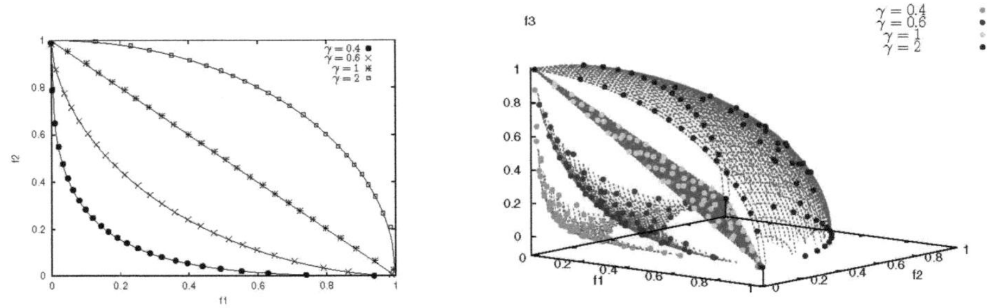

Fig. 2. PFs of different shapes as obtained with the test problem generator. The parameter γ controls the curvature and also the convexity, linearity and concavity of the PF. The points exemplify an approximation sets achieved with the SMS-EMOA.

objective optimization. Moreover we provide an analysis of the geometry of the PFs.

For the 2-D objective space, we generalize Schaffer's well known test problem $f_1(x) = x^2, f_2(x) = (1-x)^2$ to higher dimensional search spaces. The generalized Schaffer problem [15], which has been used in some applied studies, reads: $f_1(x) = \frac{1}{n^\alpha}(\sum_{i=1}^n x_i^2)^\alpha \rightarrow \min$ and $f_2(x) = \frac{1}{n^\alpha}(\sum_{i=1}^n (1-x_i)^2)^\alpha \rightarrow \min$ for $x_i \in \mathbb{R}_+$, where $i = 1, ..., n$. The parameter $\alpha \in \mathbb{R}_+$ controls the convexity/concavity and the resolvability of the PF, as it will be obtained in our analysis. For these generalized Schaffer functions explicit descriptions of the PF ($f_2 = (1-f_1^\gamma)^{\frac{1}{\gamma}}$, $\gamma = \frac{1}{2\alpha}$) and efficient set are derived . More importantly the concavity changes for different choices of γ gradually from concave ($0 < \gamma < 1$), to linear ($\gamma = 1$) and convex ($1 < \gamma$). An interesting observation is that the shapes of the different Pareto curves are arcs of Lamé supercircles.

Adopting the bottom-up-approach of test-problem generation [3], the concept can be generalized to arbitrary numbers of objective functions. Again we obtain problems with super-spherical geometry of the PF for which the convexity/concavity and resolvability can be controlled by means of a single parameter. For higher dimensions the geometrical properties of such PFs are not obvious to see. Thus, we provide a detailed analysis of the geometry, focussing on properties like convexity/concavity and the size of dominated hypervolume. Based on the geometrical analysis, we provide explicit formulas for computing standard performance metrics that measure the quality of finite set approximations to the PFs, e.g., the average distance to the PF and percentage of dominated hypervolume. This will help the practitioner, who wants to assess the performance of meta-heuristics on these test problems.

The structure of this article is as follows: After the preliminaries (Section 2), Section 3 focusses on problems with two criteria. We derive an explicit formula for the PF of the generalized Schaffer problem. In Section 4 we study a general class of m-dimensional PFs for which the solution of the Schaffer problem is the

2-dimensional instance. In particular we will focus on the influence of the control parameter γ on the convexity/concavity of the PF. Adopting the bottom-up approach of multiobjective test problem construction, we provide in Section 5 a family of test problems with scalable geometrical properties. Section 6 deals with performance metrics. Finally, Section 7 illustrates the use of the test problems by means of a case study. Concluding discussions are in section 8.

2 Mathematical Preliminaries

Next, we are going to outline notions and definitions of Pareto optimality and non-dominance as they are used throughout this article. The notation is mainly borrowed from Ehrgott [4].

Definition 1. *Given two vectors* $\mathbf{y} \in \mathbb{R}^m$, $\mathbf{y}' \in \mathbb{R}^m$ *we say* $\mathbf{y}$ *dominates* $\mathbf{y}'$ *(in symbols:* $\mathbf{y} \prec \mathbf{y}'$ *, iff* $\forall i = 1, \ldots, m : y_i \leq y_i'$ *and* $\exists i \in \{1, \ldots, m\} : y_i < y_i'$. *Moreover, we define* $\mathbf{y} \preceq \mathbf{y}' \Leftrightarrow \mathbf{y} \prec \mathbf{y}' \vee \mathbf{y} = \mathbf{y}'$.

Definition 2. *Given a set of points* $\mathcal{Y}$, *a point* $\mathbf{y}$ *is said to be non-dominated with respect to* $\mathcal{Y}$, *iff there does not exist* $\mathbf{y}' \in \mathcal{Y} : \mathbf{y}' \prec \mathbf{y}$. *Moreover, the subset of non-dominated points* $\mathbf{y}$ *in* $\mathcal{Y}$ *with respect to* $\mathcal{Y}$ *is called the non-dominated set of* $\mathcal{Y}$. *Also this set is referred to as the Pareto front (PF) of* $\mathcal{Y}$.

In the context of optimization problems $f_i(x) \to \min, i = 1, \ldots, m,\ x \in \mathcal{X}$ the concept of dominance is also defined on the search space.

Definition 3. *We say* $x \prec x' :\Leftrightarrow (f_1(x), \ldots, f_m(x)) \prec (f_1(x'), \ldots, f_m(x'))$. *Also, we define* $x \preceq x' :\Leftrightarrow (f_1(x), \ldots, f_m(x)) \preceq (f_1(x'), \ldots, f_m(x'))$.

Definition 4. *Given a set of points* $\mathcal{X}$ *the non-dominated subset of the set* $\{\mathbf{y} \mid y_1 = f_1(x), \ldots, y_m = f_m(x), x \in \mathcal{X}\}$ *is called the Pareto front (PF) with respect to the optimization problem. Moreover, the inverse image of this set in* $\mathcal{X}$ *is called the efficient set in* $\mathcal{X}$, *see[4]. The elements of this set are called efficient points.*

In the sequel $\mathbb{R}_+^m$ denotes the set $\{(y_1, \ldots, y_m) \in \mathbb{R}^m \mid y_i \geqq 0, i = 1, \ldots, m\}$ and by $\mathbf{1}$ we denote $(1, \ldots, 1) \in \mathbb{R}^m$ where m is clear from the context.

3 Efficient Set and Pareto Front for the Generalized Schaffer Problem

In this section we derive a closed form expression for the solution of the problem

$$f_1(\mathbf{x}) = \frac{1}{n^\alpha}(\sum_{i=1}^n x_i^2)^\alpha \to \min,\ f_2(\mathbf{x}) = \frac{1}{n^\alpha}(\sum_{i=1}^n (x_i - 1)^2)^\alpha \to \min,\ \mathbf{x} \in \mathbb{R}_+^n \quad (2)$$

Moreover, we will show that the efficient set of this problem will be the line segment $\mathcal{L}_n$:

$$\mathcal{L}_n = \{\mathbf{x} \in \mathbb{R}^n | \mathbf{x} = \lambda\mathbf{1}, \lambda \in [0, 1]\} \quad (3)$$

Next, a number of lemmata for $\alpha = 1$ will be derived that provide building blocks for proving the interesting result for general $\alpha > 0$.

Lemma 1. *Decision vectors on the line segment $\lambda\mathbf{1}, \lambda \in [0,1]$ are mutually non dominated with respect to problem (2) and $\alpha = 1$.*

Proof. For any $(\lambda, \lambda') \in [0,1]^2$ with $\lambda' > \lambda$: $f_1(\lambda\mathbf{1}) = n\lambda^2 < n(\lambda')^2 = f_1(\lambda'\mathbf{1})$ and $f_2(\lambda\mathbf{1}) = n(1-\lambda)^2 > n(1-\lambda')^2 = f_2(\lambda'\mathbf{1})$ □

Lemma 2. *Let $\mathbf{x} \in \mathbb{R}^n_+ - \mathcal{L}_n$. Then $\lambda\mathbf{1} \prec \mathbf{x}$ for some $\lambda \in (0,1]$.*

Proof. Let $\mathbf{x} \in \mathbb{R}^n_+ - \mathcal{L}_n$. We consider the number $\lambda = \sqrt{\frac{1}{n}\sum_{i=1}^n x_i^2}$. We first show that for this λ, $\lambda\mathbf{1} \prec \mathbf{x}$. Note that $\lambda > 0$ and possibly $\lambda > 1$. In case $\lambda \leq 1$ we are done. In case $\lambda > 1$ we easily see that $\mathbf{1} \prec \mathbf{x}$ (since $\mathbf{1} \prec \lambda\mathbf{1} \prec \mathbf{x}$). So in both cases the lemma obtains. In the remainder we will show that for the above chosen λ, $\lambda\mathbf{1} \prec \mathbf{x}$ holds:

It is clear that $f_1(\lambda\mathbf{1}) = f_1(\mathbf{x})$ holds, since $f_1(\lambda\mathbf{1}) = \frac{1}{n}\sum_{i=1}^n \lambda^2 = \lambda^2 = \frac{1}{n}\sum_{i=1}^n x_i^2 = f_1(\mathbf{x})$. Moreover, we can show that under the given assumptions $f_2(\lambda\mathbf{1}) < f_2(\mathbf{x})$ holds: $f_2(\lambda\mathbf{1}) < f_2(\mathbf{x}) \Leftrightarrow \frac{1}{n}\sum_{i=1}^n (1-\lambda)^2 < \frac{1}{n}\sum_{i=1}^n (1-x_i)^2 \Leftrightarrow 1 - 2\lambda + \lambda^2 < 1 - 2\frac{1}{n}\sum_{i=1}^n x_i + \frac{1}{n}\sum_{i=1}^n x_i^2 \Leftrightarrow \lambda^2 > \frac{1}{n^2}(\sum_{i=1}^n x_i)^2 \Leftrightarrow n\sum_{i=1}^n x_i^2 > (\sum_{i=1}^n x_i)^2$. The latter inequality holds in general for positive x_i, if for some pair $(i,j) \in \{1,\ldots,n\}^2$ the inequality $x_i \neq x_j$ holds. Since for at least one pair $(i,j) \in \{1,\ldots,n\}^2$ the inequality $x_i \neq x_j$ holds, we can show the relevant inequality as follows. To make the structure of the sum on the right hand side more visible let us format the expression in a matrix form:

$$(\sum_{i=1}^n x_i)^2 = \begin{array}{l} x_1x_1 + \cdots + x_1x_j + \cdots + x_1x_n + \\ \quad\vdots \qquad\qquad \vdots \\ x_ix_1 + \cdots + x_ix_j + \cdots + x_ix_n + \\ \quad\vdots \qquad\qquad \vdots \\ x_nx_1 + \cdots + x_1x_j + \cdots + x_nx_n \end{array} \tag{4}$$

Now, with $x_ix_j + x_jx_i \leq x_i^2 + x_j^2$ and in particular $x_ix_j + x_jx_i < x_i^2 + x_j^2$ for those pairs $(i,j) \in \{1,\ldots,n\}^2$ with $x_i \neq x_j$ the result can be simply obtained by overestimating all of the $(n-1)/n$ expressions $x_ix_j + x_jx_i$ by $x_i^2 + x_j^2$ for distinct i and j and then adding the diagonal x_i^2 which, in summary, results in an (strict) overestimator $n\sum_{i=1}^n x_i^2$. In other words we have shown that for the chosen λ, $\lambda\mathbf{1} \prec \mathbf{x}$ holds. □

Now, the lemmata can be assembled to prove the following central lemma:

Lemma 3. *The efficient set for problem (2) with $\alpha = 1$ is given by $\mathcal{L}_n$.*

Proof. Firstly we will show that $\mathcal{L}_n$ is a subset of the efficient set (ES) of problem (2). Secondly we will show that ES of problem (2) is a subset of $\mathcal{L}_n$. Let $\mathbf{x} \in \mathcal{L}_\mathbf{n}$. We want to show that $\mathbf{x}$ belongs to ES. Suppose the contrary. That is, $\exists \mathbf{x}' \in \mathbb{R}^\mathbf{n}_+$ such that $\mathbf{x}' \prec \mathbf{x}$. We distinguish two cases. Case I: $\mathbf{x}' \in \mathcal{L}_\mathbf{n}$. This leads to a contradiction because of Lemma 1. Next consider Case II: $\mathbf{x}' \in \mathbb{R}^{\mathbf{n}}_{+} - \mathcal{L}_{\mathbf{n}}$. This also leads to a contradiction because of Lemma 2 and Lemma 1.

Secondly we will show that ES is a subset of $\mathcal{L}_n$. Or equivalently: We will show that the assumption that there exists an $\mathbf{x}$ in ES such that $\mathbf{x} \notin \mathcal{L}_{\mathbf{n}}$ leads to a contradiction. Assume the existence of such an $\mathbf{x}$. Since $\mathbf{x} \notin \mathcal{L}_{\mathbf{n}}$ we know – according to Lemma 2 that there exists $\mathbf{x}' \in \mathcal{L}_{\mathbf{n}}$ such that $\mathbf{x}' \prec \mathbf{x}$. A contradiction. □

Now, by assuming general $\lambda > 0$ the proof can be extended to the central theorem of this section as follows:

Theorem 1. *The efficient set for* $f_1(\mathbf{x}) = (\frac{1}{n}\sum_{i=1}^{n} x_i^2)^\alpha \rightarrow \min$, $f_2(\mathbf{x}) = (\frac{1}{n}\sum_{i=1}^{n}(1 - x_i)^2)^\alpha \rightarrow \min$ *and* $x \in \mathbb{R}^n_+$ *is given as* $\mathcal{L}_n = \{\lambda\mathbf{1} | \lambda \in [0, 1]\}$. *Moreover the PF of this problem is* $y_2 = (1 - y_1^{1/2\alpha})^{2\alpha}$, $y_1 \in [0, 1]$.

Proof. The generalization to $\alpha > 0$ follows from the fact, that f_1 and f_2 are transformed by the same strictly monotonous function $y \mapsto y^\theta$, $\theta > 0$, such that for any two points $\mathbf{x}$ and $\mathbf{x}'$: $f_1(\mathbf{x}) > f_1(\mathbf{x}') \Leftrightarrow f_1(\mathbf{x})^\theta > f_1(\mathbf{x}')^\theta$, $f_1(\mathbf{x}) \geq f_1(\mathbf{x}') \Leftrightarrow f_1(\mathbf{x})^\theta \geq f_1(\mathbf{x}')^\theta$ and the same for f_2. Hence, also the pre-order defined on the decision space remains equal for the problem with $\alpha = 1$ and any other $\alpha > 0$. The expression for the PF can be derived as follows: Let $\mathbf{x}$ denote an arbitrary vector $(\lambda, \ldots, \lambda) \in \mathcal{L}_n$, $\lambda \in [0, 1]$. Then $f_1(\mathbf{x}) = (\lambda^2)^\alpha$ and $f_2(\mathbf{x}) = ((1 - \lambda)^2)^\alpha$. From the first equation we get $\lambda = f_1^{1/2\alpha}$, which is then to be substituted in f_2, resulting in $f_2 = (1 - f_1^{1/2\alpha})^{2\alpha}$ □

For $\alpha = 1$ an alternative geometric proof of Theorem 1 can be given.

One thing that is apparent is that the parameter α plays an important role for the shape of the PF. It is easily seen that for $\alpha = 0.5$ the PF is linear, for $\alpha > 0.5$ it gets convex and for $\alpha < 0.5$ it gets concave (see also Theorem 2 and Fig. 2). Moreover the PF is symmetric w.r.t the main bisector line between the f_1 and f_2 coordinate axes and takes its extremal values (extremal solutions) in the points $\mathbf{y}_1^* = (0, 1)^T$ and $\mathbf{y}_2^* = (1, 0)^T$. The Nadir point is $\mathbf{y}^N = (1, 1)^T$. Note, that for the more general problem $1/n^\alpha(\sum_{i=1}^{n} |x_i|^q)^\alpha \rightarrow \min$ and $1/n^\alpha(\sum_{i=1}^{n} |1 - x_i|^q)^\alpha \rightarrow \min$ similar expressions for the non-dominated front can be found. However, for $0 \leq q \leq 1$ the efficient set is no longer a line segment. Emmerich [6] derived that the efficient set for $q = 1$ is the hypercube of dimension m. The problem was used in [5]. An open question for future research would be how to extend further Schaffer's problem for more than two objective functions, where each objective function is a distance function to a fixed point in $\mathbb{R}^n$. For linearly independent points, we conjecture that the convex hull of the points is the efficient set.

A crucial observation is, that the PFs of this class of problems are Lamé supercircles [12], i. e. zero sets of $|y_1|^\gamma + |y_2|^\gamma$, intersected with $\mathbb{R}^2_+$. The connection between γ and α is established as $\gamma = \frac{1}{2\alpha}$. These curves have interesting geometric properties that can be exploited to compute performance metrics (cf. section 6). A generalization of Lamé curves are the m-dimensional superspheres discussed in the next section.

4 N-Dimensional Pareto Fronts with Superspherical Geometry

In this section we will define and look at superspheres of arbitrary dimension. More specifically we will consider parts of superspheres which consist of mutually non-dominated points, of which the solutions of the aforementioned generalized Schaffer problems are special cases. It is easy to see that they arise as zero sets of strictly concave or strictly convex functions. Alternatively they can be viewed as graphs of strictly convex or strictly concave functions. Since the sets we are considering consist of mutually non dominated points, we can view them as PFs arising from multiobjective optimization problems.

4.1 Convexity and Concavity of Superspheres

In the sequel we will use notions on concavity and convexity such as convexity of sets, convexity/concavity of functions, strict convexity/concavity of functions etcetera without defining them. Instead we refer the reader to the standard literature (for instance, Convex Analysis by R. Tyrrel Rockafeller [14]).

The following definition introduces the main building blocks of the test problems we study: postive parts of superspheres and hyperspheres.

Definition 5. *Consider the set*

$$\{(y_1, \ldots, y_m) \in \mathbb{R}^m \mid |y_1|^\gamma + \cdots + |y_m|^\gamma - 1 = 0\}, \tag{5}$$

where $\gamma \in \mathbb{R}_+$ *is arbitrary and fixed. We will call such zero-sets* γ–superspheres *or more precisely the* $m-1$-dimensional γ–supersphere *(notation:* S_γ^{m-1}*). The supersphere which arises for* $\gamma = 2$ *is usually called the* $m-1$-dimensional hypersphere *(notation:* S^{m-1}*).*

In the sequel we will only consider the "positive" parts of the γ-superspheres, i.e., we consider sets of the form:

$$\{(y_1, \ldots, y_m) \in \mathbb{R}^m_+ \mid y_1^\gamma + \cdots + y_m^\gamma = 1\}, \tag{6}$$

where $\gamma \in \mathbb{R}_+$ is arbitrary but fixed. We denote these "positive" parts of hyperspheres ($\gamma = 2$) by $S^{m-1,+}$ and those of superspheres by $S_\gamma^{m-1,+}$

Theorem 2 shows that we can view the (positive parts) of the γ-superspheres as graphs of concave ($\gamma > 1$) or convex ($0 < \gamma < 1$) functions.

Theorem 2. *Let* $\gamma \in \mathbb{R}_+$ *and let* $\mathcal{X}_\gamma = \{(y_1, \ldots, y_{m-1}) \in \mathbb{R}^{m-1}_+ \mid y_1^\gamma + \cdots + y_{m-1}^\gamma \leq 1\}$*. For each positive* γ*, define a function* $h_\gamma : \mathcal{X}_\gamma \longrightarrow \mathbb{R}$ *by* $h(\mathbf{y}) = (1 - (y_1^\gamma + \cdots + y_{m-1}^\gamma))^{\frac{1}{\gamma}}$*, where* $\mathbf{y} \in \mathbb{R}^{m-1}$*. Then* h_γ *is strictly concave for* $\gamma > 1$ *and strictly convex for* $0 < \gamma < 1$*. For* $\gamma = 1$*,* h_γ *is convex and concave (but neither is strict).* □

Alternatively, the next Theorem shows that the positive parts of γ-superspheres can also be viewed as zero-sets of strictly concave or strictly convex functions – we again omit the proof of this theorem:

Theorem 3. *Let* $\gamma \in \mathbb{R}_+$ *and let* $f_\gamma : \mathbb{R}^m_+ \longrightarrow \mathbb{R}$ *be the function defined by* $f_\gamma(\mathbf{y}) := y_1^\gamma + \cdots + y_m^\gamma - 1$.

1. *for* $\gamma > 1$, *the function* f_γ *is strictly convex and the below set* $B = \{\mathbf{y} \in \mathbb{R}^m_+ \mid f(\mathbf{y}) \leq 0\}$ *is convex and* $S^{m-1,+}_\gamma = \{\mathbf{y} \in \mathbb{R}^m_+ \mid f(\mathbf{y}) = 0\}$ *is a subset of the boundary of* B. *(The remaining boundary points lie in the coordinate hyperplanes and the set of these remaining points is described by* $\bigcup_{i=1}^m \{(y_1, \ldots, y_{i-1}, 0, y_{i+1}, \ldots, y_m) \in \mathbb{R}^m_+ \mid y_1^\gamma + \cdots + x_{i-1}^\gamma + 0 + x_{i+1}^\gamma + \cdots y_m^\gamma - 1 \leq 0\}$. *Or equivalently by* $\bigcup_{i=1}^n B \cap \{(y_1, \ldots, y_m) \in \mathbb{R}^m_+ \mid y_i = 0\}$*.)*
2. *for* $0 < \gamma < 1$, *the function* f_γ *is strictly concave and the above set* $A = \{\mathbf{y} \in \mathbb{R}^m_+ \mid f(\mathbf{y}) \geq 0\}$ *is convex and* $S^{m-1,+}_\gamma = \{\mathbf{y} \in \mathbb{R}^m_+ \mid f(\mathbf{y}) = 0\}$ *is a subset of the boundary of* A. *(The remaining boundary points of* A *lie in the coordinate hyperplanes and the set of these remaining boundary points is described by* $\bigcup_{i=1}^m \{(y_1, \ldots, y_{i-1}, 0, y_{i+1}, \ldots, y_m) \in \mathbb{R}^m_+ \mid y_1^\gamma + \cdots + x_{i-1}^\gamma + 0 + x_{i+1}^\gamma + \cdots y_m^\gamma - 1 \geq 0\}$. *Or equivalently by* $\bigcup_{i=1}^m A \cap \{(y_1, \ldots, y_m) \in \mathbb{R}^m_+ \mid y_i = 0\}$*.)*

Moreover the part of the boundary which is equal to $S^{m-1,+}_\gamma$ *in each of the above cases is equal to the graph of the function* h_γ *defined in Theorem 2.* □

For a proof of these theorems the reader is referred to [7].

4.2 Resolvability/Intractability of Conflict Versus γ

Next, we discuss how the measure of resolvability of conflict is related to γ. It would be desirable for the user, to provide only the desired value for the RoC and then compute a corresponding value for γ. In order to come up with an expression for γ, we exploit that the Minkowski distance to the ideal point gets minimized in points with $y_1 = \cdots = y_m$. Moreover, we use $y_1^\gamma + \ldots y_m^\gamma = 1$. Combining these expressions yields the desired equation:

$$\gamma = \frac{\log(1/m)}{\log(1 - \mathrm{RoC}_m)} \tag{7}$$

For a visualization and some explicitly computed values, see Figure 3.

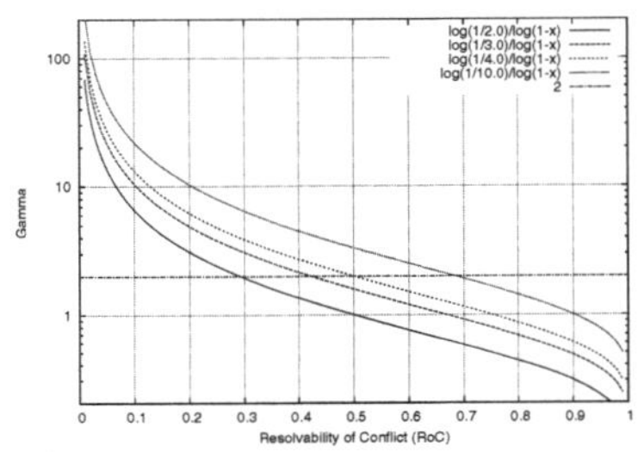

m	γ	$\mathrm{RoC}_m(\gamma)$
2	$\frac{1}{2}$	$\frac{3}{4}$
2	1	$\frac{1}{2}$
2	2	$1 - \frac{\sqrt{2}}{2}$
3	$\frac{1}{2}$	$\frac{8}{9}$
3	1	$\frac{2}{3}$
3	2	$1 - \frac{\sqrt{3}}{3}$

n	γ	$\mathrm{RoC}_m(\gamma)$
4	$\frac{1}{2}$	$\frac{15}{16}$
4	1	$\frac{3}{4}$
4	2	$\frac{1}{2}$
10	$\frac{1}{2}$	$\frac{99}{100}$
10	1	$\frac{9}{10}$
10	2	$1 - \frac{\sqrt{10}}{10}$

Fig. 3. RoC versus γ (logscale)

5 Construction of Test Problems

Now we are ready to use the superspheres as PFs of test problems. First we introduce parametrizations for superspheres. After that we can apply the methods introduced by Deb et al. [3] to generate test problems of which the superspheres are the PFs. We are afforded scalability in the number of objectives and also in the number of decision variables and a complete control over the extent of convexity/concavity and therefore resolvability of conflict.

5.1 Parametrizations of Hyperspheres and Superspheres

Consider the following representation of an $m-1$-dimensional γ-super-sphere (notation: S_γ^{m-1}) as a subset of $\mathbb{R}^m$

$$S_\gamma^{m-1} = \{(y_1, y_2, ..., y_m) \in \mathbb{R}^m \mid |y_1|^\gamma + |y_2|^\gamma + \ldots |y_m|^\gamma = 1\} \tag{8}$$

where $\gamma \in \mathbb{R}_+$ is fixed. Since super-spheres with $\gamma = 2$ – they are usually called hyperspheres – admit parametrizations, we easily get parametrizations for any γ-super-sphere. For consider an (m-1)-dimensional hypersphere S^{m-1}:

$$S^{m-1} = \{(y_1, y_2, ..., y_m) \in \mathbb{R}^m \mid y_1^2 + y_2^2 + \ldots y_m^2 = 1\} \tag{9}$$

Such a hypersphere admits parametrizations (for example:

$$\left\{ \begin{array}{rcl} y_1 &=& \cos(\theta_1) \\ y_2 &=& \sin(\theta_1)\cos(\theta_2) \\ y_3 &=& \sin(\theta_1)\sin(\theta_2)\cos(\theta_3) \\ \ldots &=& \ldots \\ y_{m-1} &=& \sin(\theta_1)\sin(\theta_2)\ldots\sin(\theta_{m-2})\cos(\theta_{m-1}) \\ y_m &=& \sin(\theta_1)\sin(\theta_2)\ldots\sin(\theta_{m-2})\sin(\theta_{m-1}), \end{array} \right. \tag{10}$$

where $\theta_1 \ldots \theta_{m-1}$ are in $\mathbb{R}$ (or if need be in bounded intervals of $\mathbb{R}$)). For each parametrization of the hypersphere S^{m-1}, we obtain a parametrization of an $(m-1)-\gamma$-super-sphere, for a fixed $\gamma \in \mathbb{R}_+$. For let

$$y_i = p_i(\theta_1, \ldots, \theta_{m-1}), \tag{11}$$

where $i = 1, \ldots, m$ be a parametrization of S^{m-1}, then

$$y_i = \pm(p_i(\theta_1, \ldots, \theta_{m-1}))^{\frac{2}{\gamma}}, \tag{12}$$

where $i = 1, \ldots, m$ is a parametrization of S_γ^{m-1} (if we are willing to tread carefully with raising to the power $\frac{2}{\gamma}$ in (12), i.e., first raise to the power 2 and subsequently to the power $\frac{1}{\gamma}$).

5.2 Test Problems

Next we proceed to use parametrizations for the *positive part* of γ-superspheres in the construction of test problems. We will single out one family of parametrizations for the γ-superspheres which arise from the parametrization of the hypersphere as in (10). In more detail we define the positive parts of γ-superspheres as follows. Let $\gamma > 0$ be fixed. We let $\tilde{p}_1(\theta_1, \ldots, \theta_{m-1}) = (\cos(\theta_1))^{\frac{2}{\gamma}}$, $\tilde{p}_i(\theta_1, \ldots, \theta_{m-1}) = (\sin(\theta_1) \cdot \sin(\theta_2) \cdot \ldots \cdot \sin(\theta_{i-1}) \cdot \cos(\theta_i))^{\frac{2}{\gamma}}$. where $1 < i < m$, and finally $\tilde{p}_m(\theta_1, \ldots, \theta_{m-1}) = (\sin(\theta_1) \cdot \sin(\theta_2) \cdot \ldots \cdot \sin(\theta_{m-1}))^{\frac{2}{\gamma}}$ with $0 \leq \theta_j \leq \frac{\Pi}{2}$ and $j = 1, \ldots, m-1$ be a parametrization of $S_\gamma^{m-1,+}$.

Adopting the approach of Deb et al. [3] for spheres and linear surfaces, the parametrization for the γ-superspheres can be used to construct test problems as follows. We let

$$f_i(\theta, r) = (1 + g(r))\tilde{p}_i(\theta) \tag{13}$$

for $i = 1, \ldots, m$ and where $0 \leq \theta_j \leq \frac{\Pi}{2}$ for $j = 1, \ldots, m-1$ and $g : \mathbb{R}_+ \longrightarrow \mathbb{R}_+$ and the test problem consists of the minimization of the f_i for $i = 1, \ldots, m$ under the constraints $0 \leq \theta_j \leq \frac{\pi}{2}$, for $j = 1, \ldots, m-1$.

The variables θ_i with $i = 1, \ldots, m-1$ are viewed as meta-variables and one can, for instance, map the decision variables to θ_i as follows: $\theta_i = \frac{\pi}{2} x_i$ for $i = 1, \ldots, m-1$ with the restriction $0 \leq x_i \leq 1$. But one can choose other mappings as well. Also given the function g we can view the body we obtain as the image of $(f_i)_{i=1}^m$ as a layered 'onion' where each layer corresponds to a function value of g. Each of the layers can be described as follows: $\{(x_1, \ldots, x_m) \in \mathbb{R}_+^m \mid x_1^\gamma + \cdots + x_m^\gamma = (1 + g(r))^\gamma\}$ if we fix r or if we fix a function value $g(r)$. Also r can be considered as a meta-variable. The PF occurs for the minimum of the function g.

From the above it is clear that it is straightforward to apply the methods developed in Deb et al. [3] to the case of superspheres.

5.3 Uni- and Multimodal Test Problems and Their Mirror Problems

As an *unimodal test problem* with $n \geq m$ decision variables we propose the problem ED1: $g(r) = r$, $r = \sqrt{(x_m^2 + \cdots + x_n^2)}$.

For a given γ with convex (concave) PF, we can obtain a problem (we will term it the *mirror problem*) with *congruent* concave (convex) PF by setting: $f_i(\mathbf{x}) = 1/(g(\mathbf{x}) + 1)\tilde{p}_i(\theta_1, \ldots, \theta_{m-1}), i = 1, \ldots, m$ For $\gamma = 2$ the PF of the mirror problem is given in Figure 4 (left).

In order to make the problem multimodal, we propose to choose $g(r) = F_{\text{natmin}}(r)$, for a function F_{natmin} that we define next (cf. Figure 4 (right)):

$$F_{\text{natmin}}(r) = b + (r - a) + 0.5 + 0.5\cos(2\pi(r-a) + \pi), a \approx 0.051373, b \approx 0.0253235 \tag{14}$$

The function F_{natmin} has the 'nice' property that it takes its minima at the natural numbers. The optimal function values are the natural numbers. Therefore, it can be checked which of the local PFs has been reached. The constants

a and b have been computed numerically by setting the derivative of F_{natmin} to zero and at the same time setting the function F_{natmin} to zero and solving (numerically) for a and b.

The resulting optimization problem, we will name it ED2, is a multimodal test-problem with equidistant local Pareto-fronts with respect to the radii $||\mathbf{x}||_\gamma$. Many-to-one mappings can be introduced by replacing this function by the function $\sin^2(x/\pi)$ in a similar manner as discussed for the F_{natmin} function.

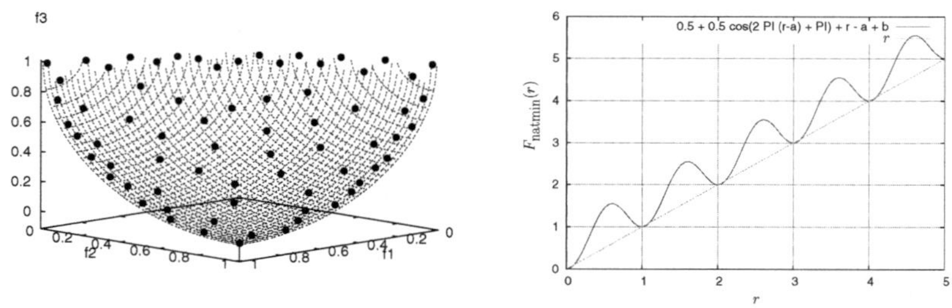

Fig. 4. (l) Convex PF of the mirror problem for $\gamma = 2$. (r) The function $F_{\text{natmin}} : \mathbb{R}_+ \longrightarrow \mathbb{R}_+$ that takes its minima at the natural numbers is depicted.

In conclusion, we note that in the bicriteria case and $\gamma = 2$ the new test problems result in the same PF as Schaffer's 1D problem [15]. For $m > 2$ and $\gamma = 1$ and $\gamma = 2$ the PFs we propose are equivalent to those of DTLZ problems [3]. However, we extended this benchmark by introducing scalability w.r.t. the RoC. Huband et al. [8] compiled a list of recommendations and desirable features. Regarding this list we note that we focus on the difficulties of scalability in dimensions, number of objective functions, and curvature (e.g. concavity/convexity, RoC). Moreover, we discussed multimodal versions of the test problems. Many extra difficulties, such as non-separability, plateaus etc., can be introduced in a managed way by designing adequate functions for g and using coordinate transformations.

6 Implementation of Performance Metrics

The measure of dominated hypervolume and the average distance of points to the PF are two measures that are frequently used to measure the quality of a pareto set approximation [2, 5].

6.1 Dominated Hypervolume of Pareto Fronts

Next, we provide expression for the measure of the dominated hypervolume [16] for PFs of the problem. Let $\mathcal{S}(P)$ denote the measure of the dominated hypervolume of a PF P. For different γ and m it suffices to provide formulas for a reference point of $\mathbf{r} = \mathbf{1} \in \mathbb{R}^m$ as it just dominates all points of the PF. Note, that in case of reference points that are dominated by $\mathbf{r}$ the computation of the remaining part reduces to computing a sum of hyperbox volumes. The Lebesgue measure of the dominated hypervolume for $\mathbf{r} = \mathbf{1}$ and different γ and m will be denoted with $\mathcal{S}_{\gamma,m}$.

For the objective space with dimension $m = 2$ (and $m = 3$) the equations are obtained as complements of quarters of areas of super-circles (super-spheres) [12, 11]:

$$1 - \mathcal{S}_{\gamma,2} = \frac{4^{1-\frac{1}{\gamma}}\sqrt{\pi}\Gamma(1+\frac{1}{\gamma})}{\Gamma(\frac{1}{2}+\frac{1}{\gamma})}/4, \quad 1 - \mathcal{S}_{\gamma,3} = \frac{8[\Gamma(1+1/\gamma)]^3}{\Gamma(1+3/\gamma)}/8 \qquad (15)$$

Note, that we considered only the first quadrant and a constant radius of $r = 1$. For higher dimension, we know no general formula for the hypervolume. However, for $\gamma = 2$ the volume of the m-dimensional unit hypersphere S^m can be used to compute the $\mathcal{S}_{2,m}$.

$$1 - \mathcal{S}_{2,m} = \frac{\pi^{m/2}}{\Gamma(0.5m)}/(2^m) \qquad (16)$$

Expressions for integer m allow for a simplified formulas (cf. [13]). Finally, for the linear case we get the simple expression $\mathcal{S}_{1,m} = 1 - 1/m!$. Some special cases are reported in the table below[1]:

$\mathcal{S}_{\gamma,m}$	$\gamma = 2$	$\gamma = 1$	$\gamma = 0.5$	$\gamma > 0$
$m = 2$	$1 - \pi/4$	$\frac{1}{2}$	$1/6$	$\mathcal{S}_{\gamma,2}$
$m = 3$	$1 - \frac{\pi}{6}$	1- $\frac{1}{6}$	$1 - 8/6!$	$\mathcal{S}_{\gamma,3}$
$m > 3$	$\frac{\pi^{m/2}}{\Gamma(\frac{m}{2}+1)}$	$1 - \frac{1}{m!}$	?	?

6.2 Distance to the Pareto Front

The average distance of points to the PF is another measure that can be used to measure the quality of PF approximations. Usually, it is combined with other measures that take the coverage of the PF into account. This measure also turns out to be useful as an indicator for local convergence to a local PF, as it may easily occur for the multimodal test function proposed in Section 5.3.

A straightforward approach to compute the distance of a point to the Pareto-front that also fits with the detection of local convergence is to compute directly the supersphere radius r_γ of a point with respect to γ, i.e.: $d(\mathbf{x}) = -1 + r_\gamma(\mathbf{x})$. This measure becomes zero, if and only if the point $\mathbf{y}$ lies on the PF. Otherwise it is a value that decreases with the Minkowski distance of that point to the PF. Moreover, the value can be used to detect the local PF to which a run has converged.

[1] Question marks indicate results unknown to us.

6.3 A Note on Knee-Points

It has been often stated that, in case of convex parts of the PF it is desirable to obtain knee-points in Pareto optimization, as they are considered to be good compromise solutions. In case of *our* test problems knee-points for convex PFs are exactly given by $(1 - \mathrm{RoC}_m(\gamma))\mathbf{1} \in \mathbb{R}^m$. Hence, the distance of a solution to a knee-point can be easily checked.

7 Case Study

In a case study, illustrating the use of the test problem, we looked at the results of the SMS-EMOA [5, 9]. This steady state EMO algorithm aims at finding well-spread solution sets by maximizing their dominated hypervolume. It is an interesting question how exactly this algorithm distributes points on PFs of different shape. Test runs for the 2-D case are reported in Figure 2 (left). A population size of 15 was used and 15000 objective function evaluations. The generalized Schaffer problem with $\mathbf{x} \in [-2, 2]^5$ has been solved for different values of γ $(= \frac{1}{2\alpha})$. The results indicate well-distributed point sets and the almost all solutions are on the PF. For high RoC the density of points grows near the knee point, while for the linear case points are distributed uniformly. This corresponds to the optimal distribution w.r.t. S-metric maximization [7].

In addition we computed PFs for the following 3-D problem: Setting $g = (\sum_{i=m}^{n} x_i^2)^{0.5}$ and

$$f_1 = ((\cos(x_1))^2)^{\frac{1}{\gamma}}(1 + g(\mathbf{x})) \to \min \tag{17}$$

$$f_2 = ((\sin(x_1)\cos(x_2))^2)^{\frac{1}{\gamma}}(1 + g(\mathbf{x})) \to \min \tag{18}$$

$$f_3 = ((\sin(x_1)\sin(x_1))^2)^{\frac{1}{\gamma}}(1 + g(\mathbf{x})) \to \min \tag{19}$$

$$\mathbf{x} \in [0, \pi/2]^2 \times [0, 1]^{n-2}, n = 7 \tag{20}$$

The population size was changed to 70, while the other settings remained the same. for different γ are displayed in Figure 5 and, for the convex mirror problem of $\gamma = 2$, in 4 (right). From another viewpoint the same results are shown in Figure 2 (right). Again, in the linear case points are distributed evenly across the PF. In case of convex and concave fronts, compromise regions as well as regions at the boundary are sampled with higher density. The results for $\gamma = 1$ and $\gamma = 2$ conform to the SMS-EMOA results on the DTLZ test-suite reported in [9]. A more detailed study of the SMS-EMOA (and similar algorithms) on the new test problems, though interesting, is beyond the scope of this paper .

8 Conclusions

We proposed and studied test problems with PFs being parts of Lamé superspheres, i.e. zero sets of $y_1^\gamma + \cdots + y_m^\gamma$, for $\gamma \in \mathbb{R}^+$. They can be scaled from concave problems with low conflict resolvability, to linear problems, and finally convex problems with high conflict resolvability.

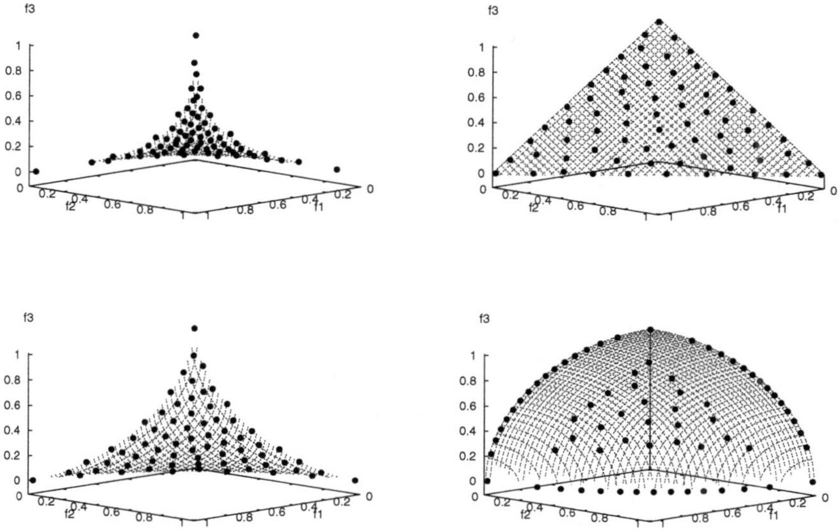

Fig. 5. Results of the SMS-EMOA for different settings of γ on the 3-D test-problems

As a first class of such functions we introduced generalized versions of Schaffer's test-problem $f_1(\mathbf{x}) = \frac{1}{m^\alpha}(\sum_{i=1}^m |x_i|^q)^\alpha \rightarrow \min$, $f_2(\mathbf{x}) = \frac{1}{m^\alpha}(\sum_{i=1}^m |1 - x_i|^q)^\alpha \rightarrow \min$ and provided closed form expressions for their efficient set and PFs, which turned out to be arcs of Lamé supercircles with $\gamma = \frac{1}{2\alpha}$. Then, adopting the construction methods proposed by Deb et al. [3], we constructed test problems that result in PFs that are parts of m-dimensional super-spheres. For all test problems we provided means to compute performance measures, such as the dominated hypervolume measure, the (average) distance to the PF and distance of solutions to the knee point. A case study with the SMS-EMOA illustrated the use of these test problems.

Our primary goal was to introduce and study a set of test problems with basic geometry, rather than benchmark problems that include all kinds of difficulties in an intertwined way. This way, properties of the algorithm and its solution sets can be studied in a isolated manner on well-understood geometries. However, these elementary problems can be used as building blocks for more complex test problems in benchmark suites.

In future, it would be interesting to extend the test problems, in order to capture problems, for which subsets of the objectives lead to highly resolvable problems and for which other subsets do not. A promising approach is to use individual γ values for different dimensions, i.e. look at zero sets of the form $y_1^{\gamma_1} + \cdots + y_m^{\gamma_m} = 1$. However, such families of PFs deserve a thorough study that would extend the scope of this paper.

Supporting material (`C++`-implementation of test problems, related technical reports) is provided under `www.liacs.nl/~emmerich/superspheres.html`.

Acknowledgments. The authors gratefully acknowledge support from the Netherlands Organization for Scientific Research.

References

[1] K. Deb: Multi-objective genetic algorithms: Problem difficulties and construction of test problems. ECJ 7 (1999) pp. 205-230
[2] K. Deb, S. Agrawal, A. Pratab, T. Meyarivan: *A Fast Elitist Non-Dominated Sorting Genetic Algorithm for Multi-Objective Optimization: NSGA-II*, PPSN VI Conference, Springer. Lecture Notes in Computer Science No. 1917, Paris, France, Marc Schoenauer et al. (Edt.), pp. 849–858, 2000
[3] K. Deb, L. Thiele, M. Laumanns, and E. Zitzler: *Scalable Test Problems for Evolutionary Multiobjective Optimization*, TIK-Technical Report 112, Institut für Technische Informatik und Kommunikationsnetze, ETH Zürich, Switzerland, 2001
[4] M. Ehrgott: *Multicriteria Optimization*, Springer, Berlin, 2005
[5] M. Emmerich, N. Beume, and B. Naujoks: *An EMO Algorithm Using the Hypervolume Measure as Selection Criterion*, C. A. Coello Coello, A. Hernández Aguirre and E. Zitzler, Proc. Evolutionary Multi-Criterion Optimization: Third Int'l Conference (EMO 2005), Springer, Berlin, 2005, pp. 62–76, Lecture Notes in Computer Science, 3410, 2005
[6] M. Emmerich: *A Rigorous Analysis of Two Bi-Criteria Problem Families with Scalable Curvature of the Pareto Fronts*, LIACS-TR 2005-05, Leiden Univ., 2005
[7] M. Emmerich and André Deutz: Multiobjective Test Problems with Superspherical Pareto Fronts, LIACS TR 2006-05, Leiden Univ., 2006
[8] S. Huband, Ph. Hingston, L. Barone, L. Whyle: *A Review of Multiobjective Test Problems and a Scalable Test Problem Kit*, IEEE Transactions on Evolutionary Computation, Vol. 10, No5, pp 477-506, 2006
[9] B. Naujoks, N. Beume, and M. Emmerich: Multi-objective optimisation using S-metric selection: Application to three-dimensional solution spaces: Proc. Conf. IEEE-CEC'05, IEEE-Press, Piscataway, NJ, 2005, pp. 1282-1289
[10] V.A. Van Veldhuizen and G.B. Lamont: Multiobjective Evolutionary Algorithm Test Suites, 1999 ACM Symp. on Applied Computing, ACM (1999), pp. 351-357
[11] E. W. Weissstein: *Superellipsoid.* From MathWorld - A Wolfram Web Resource. `http://mathworld.wolfram.com/Superellipsoid.html,August2004`
[12] E. W. Weissstein: *Superellipse.* From MathWorld - A Wolfram Web Resource. `http://mathworld.wolfram.com/Superellipse.html,September2005`
[13] E. W. Weissstein: *Hypersphere.* From MathWorld - A Wolfram Web Resource. `http://mathworld.wolfram.com/Hypersphere.html`, April 2006
[14] R. Tyrrel Rockafeller: *Convex Analysis*, Princeton University Press, 1996
[15] J. D. Schaffer: *Multiple objective optimization with vector evaluated genetic algorithms.* In: Proceedings of the first International Conference on Genetic Algorithms. Lawrence Erlbaum Ass. (1985),
[16] E. Zitzler. Evolutionary Algorithms for Multiobjective Optimization: Methods and Applications. PhD thesis, Swiss Federal Institute of Technology (ETH), Zurich, Switzerland, November 1999.

Overview of Artificial Immune Systems for Multi-objective Optimization

Felipe Campelo[1], Frederico G. Guimarães[2], and Hajime Igarashi[1]

[1] Hokkaido University, Laboratory of Hybrid Systems, Graduate School of Information Science and Technology, Sapporo, Japan
pinto@em-si.eng.hokudai.ac.jp, igarashi@ssi.ist.hokudai.ac.jp
[2] Universidade Federal de Minas Gerais, Departamento de Engenharia Elétrica, Belo Horizonte, MG, Brazil
fgg@ufmg.br

Abstract. Evolutionary algorithms have become a very popular approach for multiobjective optimization in many fields of engineering. Due to the outstanding performance of such techniques, new approaches are constantly been developed and tested to improve convergence, tackle new problems, and reduce computational cost. Recently, a new class of algorithms, based on ideas from the immune system, have begun to emerge as problem solvers in the evolutionary multiobjective optimization field. Although all these immune algorithms present unique, individual characteristics, there are some trends and common characteristics that, if explored, can lead to a better understanding of the mechanisms governing the behavior of these techniques. In this paper we propose a common framework for the description and analysis of multiobjective immune algorithms.

1 Introduction

Multiobjective problems arise in many engineering and scientific applications, where many conflicting goals have to be achieved simultaneously. In this class of problems, evolutionary algorithms in general have been demonstrated to be an effective and efficient tool for finding the set of trade-off solutions that characterize the Pareto-optimal set. For a good overview of the current state-of-art in multiobjective evolutionary techniques, we refer to some of the main books in the field[21,8,37] and also to the Online EMO Repository [11].

During the last decade [16], a new paradigm based on principles of the immune system has been employed for developing interesting algorithms for both mono and multiobjective optimization (MOO). Artificial immune systems (AIS) [20] have found applications in many fields such as pattern recognition, computer defense, optimization, and others. Since then, many multiobjective AIS algorithms have appeared in a variety of conference proceedings and technical journals, some of them not specialized in evolutionary computation. The main objective of this paper is to present a broad overview of the current MO-AIS techniques available in literature. Performance comparisons, however, are outside of the scope of this work, due to space constraints. This paper proposes a

S. Obayashi et al. (Eds.): EMO 2007, LNCS 4403, pp. 937–951, 2007.

common framework for MO-AIS algorithms, presenting a canonical MO-AIS algorithm from which all other MO-AIS algorithms reviewed can be instantiated. This common framework can simplify the comparative analysis of the algorithms, as well as the introduction of new characteristics and the study of their effects. Finally, we discuss the employment of other AIS principles in the improvement of multiobjective techniques, and present a brief overview of other immunological principles that could be employed for the development of new algorithms.

2 Multiobjective Optimization and the Immune System

In multiobjective optimization, we consider the following general problem:

$$\begin{aligned} \mathcal{X}^* &= \arg\min f(x) \\ \text{subject to: } & x \in \mathcal{F} \subseteq \mathcal{X} \end{aligned} \tag{1}$$

in which $x \in \mathcal{X}$ represents the optimization variables. The objective functions are $f : \mathcal{X} \mapsto \mathbb{R}^m$, that is, they map the optimization variables into real values. The set $\mathcal{F}$ represents the feasible set, mathematically defined as:

$$\mathcal{F} = \{x \in \mathcal{X} : g(x) \leq 0\} \tag{2}$$

where $g : \mathcal{X} \mapsto \mathbb{R}^p$ are the constraint functions. If the problem is unconstrained, $\mathcal{F}$ and $\mathcal{X}$ are equivalent.

In the multiobjective context, there is not only one solution, but a set of trade-off or Pareto-optimal solutions defined as:

$$\mathcal{X}^* \triangleq \{x \in \mathcal{F} : \nexists z \in \mathcal{F} | f(z) \leq f(x), f(z) \neq f(x)\} \tag{3}$$

Since $f(z)$ and $f(x)$ are vectors in $\mathbb{R}^m$, we need to define the relational operators $\leq$ and $\neq$:

$$f(z) \leq f(x) \Leftrightarrow f_i(z) \leq f_i(x), \forall i = 1, \ldots, m \tag{4}$$

$$f(z) \neq f(x) \Leftrightarrow \exists i = \{1, \ldots, m\} : f_i(z) \neq f_i(x) \tag{5}$$

The evolutionary multiobjective techniques are designed to find a set of non-dominated solutions that best represents the Pareto-optimal set. The search is performed through the consecutive application of stochastic and heuristic operators, balancing global and local search capabilities, over a population of candidate solutions. For a good overview of evolutionary multiobjective algorithms, see References [21,8].

Figure 1 shows the outline of a general population-based algorithm. This algorithm presents the fundamental ingredients for designing an evolutionary multiobjective technique, with the implementation details of each operator (e.g., whether the initialization procedure is random or deterministic, or the way to implement the Selection) varying from one algorithm to another.

1. Define the search space $\mathcal{X}$, population size N, objective $f(\cdot)$ and constraint $g(\cdot)$ functions;
2. $A(t=0) \leftarrow$ Initialize offline population;
3. $B(t=0) = \left\{b^{(1)}, \ldots, b^{(N)}\right\} \leftarrow$ Initialize online population;
4. **While** ($\neg$ stop criterion) **do**:
   (a) Evaluate population using $f(\cdot)$ and $g(\cdot)$;
   (b) $\Psi(t) = \left\{\psi(b^{(1)}), \ldots, \psi(b^{(N)})\right\} \leftarrow$ Evaluate scalar quality $(B(t))$;
   (c) $C(t) \leftarrow$ Selection $(A(t), B(t), \Psi(t))$;
   (d) $D(t) \leftarrow$ Variation $(C(t))$;
   (e) $A(t+1) \leftarrow \text{Update}(A(t), B(t))$;
   (f) $B(t+1) \leftarrow D(t)$;
   (g) $t \leftarrow t+1$;

Fig. 1. Outline of a population-based algorithm

The offline population, also termed memory or archive population, will store the "best solutions" achieved by the algorithm, i.e., the representation of the Pareto-optimal set. The update method for the offline population should consider dominance relations and also a good representation of the Pareto front. Actually, the adoption of an offline population $A(t)$ in multiobjective evolutionary algorithms is nowadays considered as an essential characteristic, being also used as the dividing line between the first and second generation of evolutionary multiobjective techniques [10]. Thus, all algorithms that represent the current state-of-the-art in evolutionary multiobjective optimization employ such explicit elitism method, see References [12,22,41] amongst others. Another important characteristic highlighted in Figure 1 is the calculation of a scalar quantity that measures the quality of the solution in a multiobjective context. Finally, the selection and variation steps represent the heuristic search of the algorithm, responsible for generating the next population based on the current online population $B(t)$.

In general, any population-based technique can be adequately represented by the baseline algorithm above. The fundamental differences in all these algorithms in literature reside basically in: (i) the scalar quality calculation; (ii) the update of the offline population and the mechanism for preserving diversity in $A(t)$; (iii) the selection mechanism; (iv) the variation mechanism, that is, how the next population is generated based on the current one.

Recently, another class of evolutionary techniques have been developed for multiobjective optimization: the multiobjective artificial immune systems (MO-AIS) algorithms, motivated by principles and models of the immunology. These algorithms demonstrate that AIS can be effectively employed for improving existing evolutionary techniques or designing new methods under different principles. The next section gives a broad overview of the multiobjective AIS techniques, but first, we discuss the analogies between the immune system and the multiobjective optimization problem in (1).

2.1 The Immune System

In a broad sense, the natural immune system (NIS) can be considered as the sum of the defenses of a given organism against foreign or endogenous threats, such as microorganisms, toxic substances, cancer cells, etc.. These defenses can be in the form of mechanical barriers (e.g., skin), biochemical barriers (body fluids containing destructive enzymes), immune cells (*leukocytes*) or molecular responses (interferons and other cytokines). Together, these defense lines are responsible for most of the body resistance to invasions and malfunctions that would otherwise weaken, damage or kill it.

The immune system is composed by the innate and the adaptive parts. As the name suggests, the innate immune system is born together with the organism, and represents a first-line defense against unknown pathogens. The cells of the innate immune system (*granulocytes* and *macrophages*, two kinds of *leukocytes*) are immediately available to defend the body against a large number of *antigens*, without requiring previous exposure to them and/or infection-specific adaptation. This part of the system plays an essential role on the early immune response against a given intruder, since the evolutionary process of the adaptive immune system has slower dynamics and may take a number of days before starting to function effectively.

In the adaptive immune system, the most important cells are a class of *leukocytes* called *lymphocytes*, which possess the ability to undergo a process of evolutionary adaptation when activated by a strong-binding external *antigens* or another *lymphocytes*' paratope [20]. This ability makes the adaptive NIS far more versatile than the innate immune system.

Each *naïve lymphocyte* (one that has not been involved in an immune response) carries specific receptors on its surface. Since there are millions of lymphocytes circulating through the body at any time, we have a large repertoire of molecular patterns that can be recognized with varying intensities. Once the *lymphocyte* binds to an antigen with a binding strength over a certain threshold, it starts proliferating and producing clones that undergo a process called *affinity maturation*, which is responsible for small variations in the shape of the receptors. Eventually, one of these slightly different clones will present a stronger binding to the *antigen*, and will start dominating the immune response. So, it can be said that the lymphocytes of the adaptive immune system undergo a process in all aspects similar to Darwinian evolution by natural selection, within a given individual.

This evolutionary process, called *Clonal Selection Principle* (CSP), is the basis for most of the multiobjective optimization algorithms based on the immune approach. Another popular theory amongst algorithm designers is the *Idiotypic Network Theory*, sometimes also called *Immune Network Theory*. This theory models the dynamic behavior of the immune system as a network of interacting elements, with the antigens recognizing (and suppressing) not only antigens, but also other antibodies, in a self-regulatory process. Both the CSP and the immune network are described in detail elsewhere [20], and will therefore not be discussed in depth here.

There are a number of other models and theories that are sometimes used in the AIS field. A notable example is the MOIA algorithm, explained later in this paper, which models a large number of biological processes in its iterative process and coding of the candidate solutions. The discussion of these theories is also out of the scope of this paper. Comprehensive reviews on immunology can be found in the main textbooks of this field [1].

2.2 Terminology in AIS Optimization

The antigen recognition by the immune system can be seen as a searching problem since it needs to find the antibody that best binds to a given antigen. In this sense, the problem (1) can be seen as the antigen or, in the case of many objectives and constraints, problem (1) can be seen as a polyvalent antigen.

The candidate solutions in the algorithm are named antibodies. The binding intensity between one antigen and one antibody is called antigen-antibody affinity. The binding intensity between two antibodies is called antibody-antibody affinity. Finally, the affinity degree between a given antibody and a polyvalent antigen is called avidity. In multiobjective optimization, affinity values are represented by the objective and constraint values. The avidity value is a scalar value giving the overall binding intensity between the antigen, represented by problem (1), and the antibody, i.e., the solution $x \in \mathcal{X}$. Hence, the avidity value measures the quality of the solution, and its definition varies among different algorithms. The antibody-antibody affinity can be associated to the similarity degree between the solutions. The calculation of this similarity degree depends of the representation system, for example, either binary or real.

3 Multiobjective AIS Algorithms

3.1 Yoo and Hajela's Algorithm

The first multiobjective technique that employed AIS ideas was Yoo and Hajela's algorithm [39]. They used a genetic algorithm, with normal selection, crossover, and mutation operators, but employing immune-based ideas for modifying the fitness values. In their algorithm, the memory population $A(t)$ containing the nondominated solutions is called antigen population. The online population $B(t)$ is called antibody population. One antigen is randomly selected from $A(t)$ and S antibodies are randomly selected from $B(t)$. The affinity (similarity) between antigen and antibodies is calculated and the one with the highest affinity has his fitness value increased. This process is repeated a given number of times. This approach was tested on a number of structural design problems, including two truss design problems and a I-beam problem. Although Yoo and Hajela's algorithm can not be considered a true MO-AIS, it is pioneer in using AIS ideas in multiobjective optimization.

3.2 I-PAES

Another hybrid approach, the Immune Pareto Archived Evolution Strategy (I-PAES) proposed by Cutello et.al. [13] is based on the PAES algorithm [29], with a local search phase based on the clonal selection principle. The original PAES is a multiobjective (1+1) local search evolution strategy, that proposes a grid-based approach for maintaining diversity in the offline population. I-PAES modifies the variation mechanism in the original PAES(1+1) by using immune inspired operators, specifically cloning and hypermutation [13]. The authors demonstrate the application of I-PAES in protein structure prediction problems.

3.3 Luh and Chueh's MOIA

Luh and Chueh's multiobjective Immune Algorithm (MOIA) was first proposed in 2003 [32], and then adapted to deal with constrained multiobjective problems in 2004 [33]. It is a complex algorithm with a strong biological motivation, based on the *Clonal Selection* theory, DNA library building, distinction between heavy and light protein chains in the antibodies, interleukin interactions, and a number of other immunological models that fall out of the scope of this paper (see Reference [6] for further details).

MOIA uses binary representation of the search space, with each variable of the candidate solutions represented by a *heavy chain* part (the most significant bits) and a *light chain* (less significant bits). This distinction is used at a certain stage of the algorithm to implement a local search procedure around the most promising solutions of the population.

After generating a random initial online population, MOIA enters the iterative cycle by evaluating this population over all objectives and constraints, and using these values to calculate the rank of the antibodies. It must be noted that in MOIA the constrained problems are transformed in an unconstrained one by means of a special kind of penalization of the objective functions by the constraint violations [33]. After calculating the rank of each antibody, the non-dominated ones are selected for a local search procedure, implemented through cloning and hypermutation of the *light chains* in the bitstring. The best solutions found by this process are copied to both the offline and online populations. The offline population is then cleaned from dominated or unfeasible solutions, with a few dominated feasible solutions being stored for insertion in the *Germ-line DNA library* in a later step of the algorithm.

After the local search procedure, the *avidity* value is calculated for all antibodies. In MOIA, the *avidity* accounts both for the performance of the solutions and for the similarity between them. The algorithm then assembles the *Germ-line DNA library*, by combining a few antibodies expelled from the offline population (as mentioned above) with ones selected by Tournament from the current population. This *Germ-line DNA library* is then used to generate the next-generation population, by combining fragments of the *donor* solutions into new points in the search space, a procedure similar to the *Crossover* operator used in genetic algorithms.

A last step in this algorithm is to apply a number of diversity generation operators over the next-generation population, according to user-defined probabilities. These operators [33] are responsible for keeping the exploration feature of the algorithm, avoiding the premature convergence to a local Pareto Front. The iterative cycle is repeated until a given number of generations are completed.

MOIA was used to solve some analytical benchmarck problems [32], and the results obtained were superior to those obtained by SPEA, MOGA, NPGA and NSGA. Also, the constrained version of MOIA [33] was used in the design of truss structures.

3.4 MISA

The Multi-objective Immune System Algorithm (MISA) [7,9,14] is an immune algorithm based on the *Clonal Selection Principle* with elitism. The algorithm uses a grid-based random generation of the initial population in the search space, and presents an interesting selection strategy for choosing the antibodies to be cloned, based both on dominance and feasibility. Also, the number of clones each selected antibody receives is regulated by a niching procedure in the objective space, in order to drive the evolution towards a fair sampling of the Pareto front.

The performance of the MISA is tested on various analytical benchmark problems [7], and compared to other state-of-the-art multiobjective optimization algorithms [9], where it has shown a competitive performance when compared to NSGA-2, micro-GA, and PAES.

3.5 MOCSA

The Multi-Objective Clonal Selection Algorithm (MOCSA) [5] combines ideas from CLONALG [18] and opt-AINet [19] in a MO-AIS algorithm for real-valued optimization. In MOCSA, the quality values are calculated using nondominated sorting. The population is sorted based on these values, and the first N_c best solutions are selected for cloning. MOCSA uses real-coding and Gaussian mutation similarly to opt-AINet. The number of clones in the proliferation phase depends on the ranking of the individual in the nondominated fronts.

The mutated clones are evaluated and combined with the original ones. They are sorted again using nondominated sorting and the N_c best solutions are preserved. MOCSA also employs a diversity generation mechanism in such a way that the worst individuals, i.e., those not selected for cloning, are eliminated and substituted by randomly generated individuals.

MOCSA was used to solve an analytical benchmark problem, as well as the problem of designing an electrostatic micromotor [5]. The analytical results were compared to results obtained by MISA, NSGA-2, PAES and micro-GA, with MOCSA producing competitive results according to two standard metrics (*generational distance* and *spacing* [9]). In 2006, an improved version of the MOCSA was employed for the solution of a 3-objective design of a superconducting magnetic energy storage (SMES) system [26].

3.6 VAIS

The Vector Artificial Immune System (VAIS) [23,25] is a multiobjective version of the opt-AINet algorithm [19]. The immune network theory states that antibodies can recognize other antibodies and this chain of recognition either stimulates or suppresses their proliferation. In the original opt-AINet, the memory population stores the sub-optimal solutions in a single objective optimization problem. A suppression operator is applied to the memory population in order to eliminate redundancy.

In VAIS, the author adapts these ideas for developing a multiobjective algorithm whose memory population now stores the nondominated solutions. VAIS employs real representation of the variables and quality-proportional Gaussian mutation, as in the opt-AINet. In order to evaluate the quality values, VAIS utilizes strength values likewise in SPEA2 [41], but without the use of density values, since the suppression mechanism in the memory population already deals with dense regions. The suppression mechanism is also modified in VAIS. It considers similarity in the objective space, not in the parameter space as in opt-AINet.

VAIS was tested in a number of analytical problems[25], in which VAIS showed similar or better results when compared to NSGA-2. A modified version of the VAIS, called VIS, was presented in 2006 [24], along with several examples of application on analytical constrained and unconstrained benchmark problems.

3.7 IDCMA

The Immune Dominance Clonal Multi-objective Algorithm (IDCMA) [27,34] introduces a new similarity measure between antibodies, based on distances in the objective space: the immune differential degree. Again, this similarity measure is used to reduce the size of the offline population in the update step.

The algorithm also presents a different selection mechanism for cloning. In this mechanism, one antibody is randomly selected from the offline population in the beginning of each iteration. The quality value of each individual in the online population is computed based on the antibody-antibody affinity, that is, similarity in the representation of the solutions. The population is sorted based on these affinity values, and the first N_c ones are selected for cloning. Finally, the solutions in the clone population undergo recombination and mutation to generate the next population.

IDCMA was used to solve a 0/1 Knapsack problem, and the results are compared against those obtained by a number of first-generation algorithms, including NSGA, NPGA and VEGA [34]. It was observed that the solutions found by IDCMA dominated the ones obtained by the other algorithms, and could therefore be qualified as better ones.

3.8 IFMOA

From the same group that proposed IDCMA, the Immune Forgetting Multiobjective Optimization Algorithm (IFMOA) [31,40] is also based on the clonal selection principle for the variation step. The scalar quality values for the solutions are calculated in the same way as in SPEA2 [41]. The selection for cloning is deterministic and the same number of clones is used for each solution. The immune forgetting operator, proposed in IFMOA, consists of substituting a given number of solutions from the online population, randomly selected, by individuals from the offline population, also randomly selected. However, the authors do not clarify the benefit of this operator to the algorithm.

In order to show the applicability of the IFMOA, a number of comparisons against the SPEA2 and MOGA were performed, for diverse analytical benchmarck problems [31]. It was shown that the Pareto fronts obtained by IFMOA were significantly better than the ones from the other algorithms. IFMOA has been also used for solving problems related to unsupervised feature selection [40].

3.9 ACSAMO

Another multiobjective algorithm based on the Clonal Selection Principle is the Adaptive Clonal Selection Algorithm for Multiobjective Optimization (ACSAMO) [38], proposed in 2006 by Wang and Mahfouf. ACSAMO generates a fixed number of clones for all antibodies and presents an quality-proportional mutation, like in the VAIS. The antibody-antigen affinities are calculated by using a dynamically adjusted weighted approach, in which an evolutionary pressure in the direction of the "best so far" and "best this generation" solutions is applied over the online population. At each generation, the two "best" solutions are found according to a random-weight linear aggregation of objectives, as in (6):

$$\mathrm{WF}_i = \sum_{j=1}^{m} w_j f_j \left(Ab_i\right); \quad \sum_{j=1}^{m} w_j = 1 \tag{6}$$

The "best" solutions are chosen based on the random variation of the weights w_j. The affinity is then calculated as:

$$\mathrm{af}_i = dist\left(Ab_i, Ab_c\right) + dist\left(Ab_i, Ab_g\right) \tag{7}$$

where Ab_c and Ab_g are the "best so far" and the "best this generation" solutions according to the aggregation function (6), respectively.

The selection operator for the offline population is based on the Pareto dominance (nondominated solutions are copied to the offline population); if the maximum size of the offline population is exceeded, a crowding procedure is used for eliminating solutions from the most crowded regions of the Pareto front.

ACSAMO was tested on some analytical benchmark problems, in which it outperformed both SPEA and NSGA-2 [38].

4 A Common Framework for MO-AIS Algorithms

As we can see from the previous section, we have many different ways of implementing a MO-AIS algorithm. Nonetheless, these algorithms share some common characteristics. Except from the Yoo and Hajela's algorithm, all of them employ the clonal selection principle to a certain extent, including MOIA, that employs clonal proliferation and mutation as a local search procedure for the offline population. This principle is extensively used for designing AIS-based optimization algorithms and it forms a fundamental ingredient in defining the variation step, see Fig. 1, within a MO-AIS technique. In this section, we propose the outline of a canonical MO-AIS, which is shown in Fig. 2.

1. Define the search space $\mathcal{X}$, population size N, objective $f(\cdot)$ and constraint $g(\cdot)$ functions;
2. $A(t=0) \leftarrow$ Initialize offline population;
3. $B(t=0) = \left\{b^{(1)}, \ldots, b^{(N)}\right\} \leftarrow$ Initialize online population;
4. **While** ($\neg$ stop criterion) **do**:
   (a) Evaluate antibody-antigen affinities using $f(\cdot)$ and $g(\cdot)$;
   (b) $\Psi(t) = \left\{\psi(b^{(1)}), \ldots, \psi(b^{(N)})\right\} \leftarrow$ Evaluate avidity $(B(t))$;
   (c) $C(t) = \left\{c^{(1)}, \ldots, c^{(N_c)}\right\} \leftarrow$ Selection for Cloning $(A(t), B(t), \Psi(t))$;
   (d) $D(t) = \left\{d^{(1)}, \ldots, d^{(N_c)}\right\} \leftarrow$ Proliferation and Mutation $(C(t))$;
   (e) $E(t) = \left\{e^{(1)}, \ldots, e^{(N_d)}\right\} \leftarrow$ Diversification;
   (f) $A(t+1) \leftarrow$ Update$(A(t), B(t))$;
   (g) $B(t+1) \leftarrow D(t) \cup E(t)$;
   (h) $t \leftarrow t+1$;

Fig. 2. Outline of the canonical MO-AIS

This canonical algorithm presents the fundamental ingredients for designing a given MO-AIS. The first important ingredient is the avidity evaluation, i.e., the computation of the scalar quality of the individuals. This can be made using procedures already known for multi-objective evolutionary algorithms and their variations. AIS introduces the additional possibility of using antibody-antibody recognition to define quality, as used in the Yoo and Hajela's algorithm and also in ACSAMO. With these values, we can proceed to the selection for cloning, which can be either deterministic or stochastic. In a deterministic selection the best N_c solutions based on the avidity values are selected. A stochastic selection can use, for instance, a tournament selection based on antibody-antibody affinity, where one antibody from the offline population is randomly selected as reference antibody for the tournament.

Proliferation and hypermutation are important elements in the variation mechanism of MO-AIS. The affinity-proportional mutation rate (likewise in opt-AINet) and the affinity-proportional number of clones (likewise in CLONALG) can be combined together, introducing a very interesting balance between local and global search in the searching process. This is done in MOCSA, for example. The number of clones can be also defined based on the idea of generating more clones in the best regions, and/or more clones in the less crowded regions of the Pareto front estimative, as proposed in MISA.

The diversification step is another important element in MO-AIS, and it is more related to the global search capability of the algorithm. Not all MO-AIS in literature present such an explicit diversity mechanism. In general, the diversity generation is performed by the introduction of new random solutions, but other implementations are also possible. MOIA's operators for generating the next population from the DNA germinal library can all be seen as a sophisticated diversity mechanism.

Therefore, with the adequate adaptations, all algorithms reviewed in the previous section can be seen as particular instances of this canonical algorithm. In some of them the steps 4.(b) to 4.(e) are not so evident and easily distinguishable. This canonical MO-AIS is useful to identify similarities and differences among the algorithms and make the comparison easier. By delineating these basic ingredients, designers can identify these steps in each algorithm and examine different approaches for implementing the same step in a given MO-AIS algorithm. Moreover, the canonical MO-AIS makes evident the differences between AIS-based algorithms and other multi-objective evolutionary algorithms, mainly in the variation mechanism, by highlighting the specific operators of a truly MO-AIS technique.

5 Other Immune Principles

From the algorithms reviewed, it is clear that the most common idea used in the development of MO-AIS algorithms is by far the CSP, with some other principles (e.g., immune network, interleukin interactions) also being employed. There are, however, a number of immunological principles that have shown great promise in other areas of engineering, and could theoretically be used either in the development of new tools for multiobjective optimization or in the improvement of the existing algorithms. In this section we introduce two of such principles, along with some general ideas on how they could be applied for MOO.

5.1 Negative Selection

In the natural immune system, negative selection (NS) is responsible for the inhibition/death of a given lymphocyte upon being activated. This principle is used basically to model the elimination of antibodies that react against self-antigens, which could eventually cause auto-immune diseases. In engineering applications, NS-based algorithms have been used for intrusions detection [28], anomaly and fault detection [36,17], among other areas related to pattern recognition. To the knowledge of the authors, however, there has been no use so far of NS-based systems for optimization, either mono or multiobjective.

In general, MO-AIS algorithms present a explicit diversity generation mechanism at some point of the iterative process, which usually involve the insertion of newly generated random individuals in the population. While this unsupervised generation of new individuals has, as intended, the potential to explore new regions of the search space, it may also generate solutions in regions of the space that have already been explored in previous iterations of the algorithm. This potential waste of valuable function evaluations is particularly aggravated in the later generations of the algorithms, when larger portions of the search space have been covered.

This kind of problem could in theory be reduced by having a NS routine embedded in the diversity-generation operators: during the optimization process, the previously explored regions of the search space would be stored as a "self" set, and new solutions generated by the operator would be created in a way similar to the one used for creating new detectors in NS-based fault detection algorithms [17]. With this, the exploration of new regions by the diversity generation algorithm would be guaranteed, therefore improving the overall algorithm performace.

5.2 Danger Theory

While mainstream immunology supports a view of the NIS classification abilities in terms of *self-nonself* discrimination, there are a number of phenomena that do not fit in this model. For instance, the fact that the immune system does not react against the bacterial flora in the gut, but is triggered by chemicals emmited by stressed *self* cells indicates that some extra mechanism for the recognition of potential threats is present. The Danger Theory (DT) [35] proposes an explanation to these behaviors of the NIS, based not on a *self-nonself* distinction but instead on a measure of the level of threat represented by a given antigen.

In AIS, a number of aplications of the DT have been proposed [2], including intrusion detection systems for computers [3] and anomaly detection. A possible application of this principle in MO-AIS would be to have under-explored regions of the Pareto front sending "danger" signals to the immune algorithm, indicating the need of an "immune response" (i.e., a better exploration of the space) in that direction. Another possibility is the use of DT principles for online decision-making in multiobjective optimization, with some regions of the Pareto emmiting "danger" signals in order to guide the evolution towards specific trade-offs between the various objectives.

6 Discussion

This paper has presented an overview of current MO-AIS in literature and suggested a common framework for MO-AIS algorithms. The CSP is largely employed in the design of optimization algorithms, especially for defining their variation mechanism. Nevertheless, other principles and theories from AIS have been used in the quality assignment of the population, in the promotion of diversity in the online population, and in the update of the offline population. Moreover, some works have employed AIS ideas for improving constraint handling in evolutionary techniques [15,4] for the multiobjective case.

On the other hand, some AIS theories are not well explored in the field of optimization, such as Negative Selection and Danger Theory. These immune models have been employed successfully in other fields of engineering, and an investigation of their potential as tools for the improvement of MO-AIS may be an interesting area of research.

From this overview, however, it is apparent that there is probably little need for more new MO-AIS algorithms, but instead an extensive comparison of the available methods and the available implementations of the fundamental steps outlined in Fig.2 should be pursued. The definition of what a MO-AIS algoritmh is and what it must have to be considered as such can help the design of meaningful comparison experiments. This work tries to fill this gap before proceeding to the comparison of methods, which is the logical next step on this research.

References

1. Abbas, A.K., Lichtman, A.H.: Cellular and Molecular Immunology. (paperback) 5th Edition, Saunders, March 2005.
2. Aickelin, U., Cayzer, S.: The danger theory and its application to artificial immune systems. 1st International Conference on Artificial Immune Systems, University of Kent at Canterbury, England, September, (2002) 141–148
3. Aickelin, U., Bentley, P., Cayzer, S., Kim, J., McLeod, J.: Danger Theory: The Link between AIS and IDS? 2nd International Conference on Artificial Immune Systems, Edinburgh, UK, September 1-3 (2003) 147–155
4. Balicki, J.: Multi-criterion evolutionary algorithm with model of the immune system to handle constraints for task assignments. 7th International Conference on Artificial Intelligence and Soft Computing, Zakopane, Poland, June, Lecture Notes in Computer Science **3070** (2004) 394–399
5. Campelo, F., Guimarães, F.G., Saldanha, R.R., Igarashi, H., Noguchi, S., Lowther, D.A., Ramirez, J.A.: A novel multiobjective immune algorithm using nondominated sorting. 11th International IGTE Symposium on Numerical Field Calculation in Electrical Engineering, Seggauberg, Austria, (2004)
6. Chueh, C-H., An Immune Algorithm for Engineering Optimization, PhD thesis, Department of Mechanical Engineering, Tatung University, Taipei, Taiwan, July (2004)
7. Coello, C.A.C., Cruz-Cortés, N.: An approach to solve multiobjective optimization problems based on an artificial immune system. 1st International Conference on Artificial Immune Systems, University of Kent at Canterbury, England, September, (2002) 212–221
8. Coello, C.A.C., Van Veldhuizen, D.A., Lamont, G.B.: Evolutionary Algorithms for Solving Multi-Objective Problems. Kluwer Academic Publishers, New York, March 2002.
9. Coello, C.A.C., Cruz-Cortés, N.: Solving multiobjective optimization problems using an artificial immune system. Genetic Programming and Evolvable Machines **6** 2 (2005) 163–190
10. Coello, C.A.C.: Evolutionary multi-objective optimization: a historical view of the field. IEEE Computational Intelligence Magazine, vol. 1, no. 1, p. 28-36, February 2006.
11. Coello, C.A.C.: EMO repository. At: `http://delta.cs.cinvestav.mx/~ccoello/EMOO`

12. Corne, D.W., Knowles, J.D., Oates, M.J.: The pareto envelope-based selection algorithm for multiobjective optimization. Parallel Problem Solving from Nature VI Conference, 2000, Paris, France, Lecture Notes in Computer Science **1917** (2000) 839–848
13. Cutello, V., Narzisi, G., Nicosia, G.: A class of pareto archived evolution strategy algorithms using immune inspired operators for ab-initio protein structure prediction. Applications of Evolutionary Computing, Evoworkshops 2005, Lausanne, Switzerland, March 30-April 1, Lecture Notes in Computer Science **3449** (2005) 54–63
14. Cruz-Cortés, N., Coello, C.A.C.: Multiobjective Optimization Using Ideas from the Clonal Selection Principle. Genetic and Evolutionary Computation Conference, GECCO 2003, Lecture Notes in Computer Science **2723** (2003) 158–170
15. Cruz-Cortés, N., Trejo-Pérez, D., Coello, C.A.C.: Handling constraints in global optimization using an artificial immune system. 4th International Conference on Artificial Immune Systems, Banff, Alberta, Canada, August 14-17, Lecture Notes in Computer Science **3627** (2005) 234–247
16. Dasgupta, D., Ji, Z., Gonzalez, F.: Artificial immune system (AIS) research in the last five years. Proceedings of CEC 2003: IEEE Congress on Evolutionary Computation, **1** (2003) 123–130
17. Dasgupta, D., KrishnaKumar, K., Barry, M.: Negative Selection Algorithm for Aircraft Fault Detection The 3rd International Conference on Artificial Immune Systems, Catania, Italy, September 13-16, Lecture Notes in Computer Science **3239** (2004) 1–13
18. de Castro, L.N., von Zuben, F.J.: Learning and optimization using the clonal selection principle. IEEE Transactions on Evolutionary Computation, **6** 3 (2002) 239–251.
19. de Castro, L.N., Timmis, J.: An artificial immune network for multimodal function optimization. Proceedings of CEC 2002: IEEE Congress on Evolutionary Computation, **1** (2002) 699–674.
20. de Castro, L.N., Timmis, J.: Artificial immune systems: a new computational intelligence approach. Springer-Verlag, 2002.
21. Deb, K.: Multi-Objective Optimization Using Evolutionary Algorithms. John Wiley & Sons, Chichester, UK, 2001.
22. Deb, K., Pratap, A., Agarwal, S., Meyarivan, T.: A fast and elitist multiobjective genetic algorithm: NSGA-II. IEEE Transactions on Evolutionary Computation, **6** 2 (2002) 182–197.
23. Freschi, F.: Multi-Objective Artificial Immune System for Optimization in Electrical Engineering, PhD thesis, Department of Electrical Engineering, Politecnico di Torino, Torino, Italy, (2006)
24. Freschi, F., Repetto, M.: VIS: an artificial immune network for multi-objective optimization, Engineering Optimization, **38** 8 (2006) 975–996.
25. Freschi, F., Repetto, M.: Multiobjective optimization by a modified artificial immune system algorithm. 4th International Conference on Artificial Immune Systems, Banff, Alberta, Canada, August 14-17, Lecture Notes in Computer Science **3627** (2005) 248–261
26. Guimarães, F.G., Campelo, F., Saldanha, R.R., Igarashi, H., Takahashi, R.H.C., Ramirez, J.A.: A multiobjective proposal for the TEAM benchmark problem 22. IEEE Transactions on Magnetics **42** 4 (2006) 1471–1474

27. Jiao, L., Gong, M., Shang, R., Du, H., Lu, B.: Clonal selection with immune dominance and energy based multiobjective optimization. 3rd International Conference on Evolutionary Multi-Criterion Optimization, Guanajuato, Mexico, March 9-11, Lecture Notes in Computer Science **3410** (2005) 474–489
28. Kim, J., Bentley, P.: Negative selection and niching by an artificial immune system for network intrusion detection. Late Breaking Papers at the 1999 Genetic and Evolutionary Computation Conference, *in* Scott Brave and Annie S. Wu, Orlando, Florida (1999) 149–158
29. Knowles, J.D., Corne, D.: Approximating the Nondominated Front Using the Pareto Archived Evolution Strategy. Evolutionary Computation **8** 2 (2000) 149–172
30. Kurapati, A., Azarm, S.: Immune network simulation with multiobjective genetic algorithms for multidisciplinary design optimization. Engineering Optimization **33** 2 (2000) 245–260
31. Lu, B., Jiao, L., Du, H., Gong, M.: IFMOA: immune forgetting multiobjective optimization algorithm. 1st International Conference on Advances in Natural Computation, Changsha, China, August 27-29, Part III Lecture Notes in Computer Science **3612** (2005), 399–408
32. Luh, G-C., Chueh, C-H., Liu, W-W.: MOIA: multi-objective immune algorithm. Engineering Optimization **35** 2 (2003) 143–164
33. Luh, G-C., Chueh, C-H., Liu, W-W.: Multi-objective optimal design of truss structure with immune algorithm. Computers and Structures **82** (2004) 829–844
34. Ma, W., Jiao, L., Gong, M., Liu, F.: A novel artificial immune systems multiobjective optimization algorithm for 0/1 knapsack problems. International Conference on Computational Intelligence and Security, Part I, Xi'an, China, December 15-19, Lecture Notes in Artificial Intelligence **3801** (2005) 793–798
35. Matzinger, P.: Tolerance Danger and the Extended Family. Annual reviews of Immunology **12** (1994), 991–1045.
36. Singh S.: Anomaly detection using negative selection based on the r-contiguous matching rule. 1st International Conference on Artificial Immune Systems, University of Kent at Canterbury, England, September, (2002) 99–106
37. Tan, K.C., Khor, E.F., Lee T.H.: Multiobjective Evolutionary Algorithms and Applications (Advanced Information and Knowledge Processing). (Hardcover) Springer; 1st edition, May 2005.
38. Wang, X.L., Mahfouf, M.: ACSAMO: an adaptive multiobjective optimization algorithm using the clonal selection principle. 2nd European Symposium on Nature-inspired Smart Information Systems, Puerto de la Cruz, Tenerife, Spain, November 29 - December 1 (2006)
39. Yoo, J., Hajela, P.: Immune network simulations in multicriterion design. Structural Optimization **18** 2-3 (1999) 85–94
40. Zhang, X., Lu, B., Gou, S., Jiao, L.: Immune multiobjective optimization algorithm for unsupervised feature selection. Applications of Evolutionary Computing, EvoWorkshops 2006, Budapest, Hungary, April 10-12, Lecture Notes in Computer Science **3907** (2006) 484–494
41. Zitzler, E., Laumanns, M., Thiele, L.: SPEA2: improving the strength pareto evolutionary algorithm. Technical report; Gloriastrasse 35, CH-8092 Zurich, Switzerland, 2001

Author Index

Lecture Notes in Computer Science

For information about Vols. 1–4300

please contact your bookseller or Springer

Vol. 4429: C. Ullrich, J.H. Siekmann, R. Lu (Eds.), Cognitive Systems. X, 162 pages. 2007. (Sublibrary LNAI).

Vol. 4405: L. Padgham, F. Zambonelli (Eds.), Agent-Oriented Software Engineering VII. XII, 225 pages. 2007.

Vol. 4403: S. Obayashi, K. Deb, C. Poloni, T. Hiroyasu, T. Murata (Eds.), Evolutionary Multi-Criterion Optimization. XIX, 954 pages. 2007.

Vol. 4397: C. Stephanidis, M. Pieper (Eds.), Universal Access in Ambient Intelligence Environments. XV, 467 pages. 2007.

Vol. 4396: J. García-Vidal, L. Cerdà-Alabern (Eds.), Wireless Systems and Mobility in Next Generation Internet. IX, 271 pages. 2007.

Vol. 4394: A. Gelbukh (Ed.), Computational Linguistics and Intelligent Text Processing. XVI, 648 pages. 2007.

Vol. 4393: W. Thomas, P. Weil (Eds.), STACS 2007. XVIII, 708 pages. 2007.

Vol. 4392: S.P. Vadhan (Ed.), Theory of Cryptography. XI, 595 pages. 2007.

Vol. 4390: S.O. Kuznetsov, S. Schmidt (Eds.), Formal Concept Analysis. X, 329 pages. 2007. (Sublibrary LNAI).

Vol. 4385: K. Coninx, K. Luyten, K.A. Schneider (Eds.), Task Models and Diagrams for Users Interface Design. XI, 355 pages. 2007.

Vol. 4384: T. Washio, K. Satoh, H. Takeda, A. Inokuchi (Eds.), New Frontiers in Artificial Intelligence. IX, 401 pages. 2007. (Sublibrary LNAI).

Vol. 4383: E. Bin, A. Ziv, S. Ur (Eds.), Hardware and Software, Verification and Testing. XII, 235 pages. 2007.

Vol. 4381: J. Akiyama, W.Y.C. Chen, M. Kano, X. Li, Q. Yu (Eds.), Discrete Geometry, Combinatorics and Graph Theory. XI, 289 pages. 2007.

Vol. 4380: S. Spaccapietra, P. Atzeni, F. Fages, M.-S. Hacid, M. Kifer, J. Mylopoulos, B. Pernici, P. Shvaiko, J. Trujillo, I. Zaihrayeu (Eds.), Journal on Data Semantics VIII. XV, 219 pages. 2007.

Vol. 4378: I. Virbitskaite, A. Voronkov (Eds.), Perspectives of Systems Informatics. XIV, 496 pages. 2007.

Vol. 4377: M. Abe (Ed.), Topics in Cryptology – CT-RSA 2007. XI, 403 pages. 2006.

Vol. 4376: E. Frachtenberg, U. Schwiegelshohn (Eds.), Job Scheduling Strategies for Parallel Processing. VII, 257 pages. 2007.

Vol. 4373: K. Langendoen, T. Voigt (Eds.), Wireless Sensor Networks. XIII, 358 pages. 2007.

Vol. 4372: M. Kaufmann, D. Wagner (Eds.), Graph Drawing. XIV, 454 pages. 2007.

Vol. 4371: K. Inoue, K. Satoh, F. Toni (Eds.), Computational Logic in Multi-Agent Systems. X, 315 pages. 2007. (Sublibrary LNAI).

Vol. 4370: P.P Lévy, B. Le Grand, F. Poulet, M. Soto, L. Darago, L. Toubiana, J.-F. Vibert (Eds.), Pixelization Paradigm. XV, 279 pages. 2007.

Vol. 4369: M. Umeda, A. Wolf, O. Bartenstein, U. Geske, D. Seipel, O. Takata (Eds.), Declarative Programming for Knowledge Management. X, 229 pages. 2006. (Sublibrary LNAI).

Vol. 4368: T. Erlebach, C. Kaklamanis (Eds.), Approximation and Online Algorithms. X, 345 pages. 2007.

Vol. 4367: K. De Bosschere, D. Kaeli, P. Stenström, D. Whalley, T. Ungerer (Eds.), High Performance Embedded Architectures and Compilers. XI, 307 pages. 2007.

Vol. 4366: K. Tuyls, R. Westra, Y. Saeys, A. Nowé (Eds.), Knowledge Discovery and Emergent Complexity in Bioinformatics. IX, 183 pages. 2007. (Sublibrary LNBI).

Vol. 4364: T. Kühne (Ed.), Models in Software Engineering. XI, 332 pages. 2007.

Vol. 4362: J. van Leeuwen, G.F. Italiano, W. van der Hoek, C. Meinel, H. Sack, F. Plášil (Eds.), SOFSEM 2007: Theory and Practice of Computer Science. XXI, 937 pages. 2007.

Vol. 4361: H.J. Hoogeboom, G. Păun, G. Rozenberg, A. Salomaa (Eds.), Membrane Computing. IX, 555 pages. 2006.

Vol. 4360: W. Dubitzky, A. Schuster, P.M.A. Sloot, M. Schroeder, M. Romberg (Eds.), Distributed, High-Performance and Grid Computing in Computational Biology. X, 192 pages. 2007. (Sublibrary LNBI).

Vol. 4358: R. Vidal, A. Heyden, Y. Ma (Eds.), Dynamical Vision. IX, 329 pages. 2007.

Vol. 4357: L. Buttyán, V. Gligor, D. Westhoff (Eds.), Security and Privacy in Ad-Hoc and Sensor Networks. X, 193 pages. 2006.

Vol. 4355: J. Julliand, O. Kouchnarenko (Eds.), B 2007: Formal Specification and Development in B. XIII, 293 pages. 2006.

Vol. 4354: M. Hanus (Ed.), Practical Aspects of Declarative Languages. X, 335 pages. 2006.

Vol. 4353: T. Schwentick, D. Suciu (Eds.), Database Theory – ICDT 2007. XI, 419 pages. 2006.

Vol. 4352: T.-J. Cham, J. Cai, C. Dorai, D. Rajan, T.-S. Chua, L.-T. Chia (Eds.), Advances in Multimedia Modeling, Part II. XVIII, 743 pages. 2006.

Vol. 4351: T.-J. Cham, J. Cai, C. Dorai, D. Rajan, T.-S. Chua, L.-T. Chia (Eds.), Advances in Multimedia Modeling, Part I. XIX, 797 pages. 2006.

Vol. 4349: B. Cook, A. Podelski (Eds.), Verification, Model Checking, and Abstract Interpretation. XI, 395 pages. 2007.

Vol. 4348: S.T. Taft, R.A. Duff, R.L. Brukardt, E. Ploedereder, P. Leroy (Eds.), Ada 2005 Reference Manual. XXII, 765 pages. 2006.

Vol. 4347: J. Lopez (Ed.), Critical Information Infrastructures Security. X, 286 pages. 2006.

Vol. 4346: L. Brim, B. Haverkort, M. Leucker, J. van de Pol (Eds.), Formal Methods: Applications and Technology. X, 363 pages. 2007.

Vol. 4345: N. Maglaveras, I. Chouvarda, V. Koutkias, R. Brause (Eds.), Biological and Medical Data Analysis. XIII, 496 pages. 2006. (Sublibrary LNBI).

Vol. 4344: V. Gruhn, F. Oquendo (Eds.), Software Architecture. X, 245 pages. 2006.

Vol. 4342: H. de Swart, E. Orłowska, G. Schmidt, M. Roubens (Eds.), Theory and Applications of Relational Structures as Knowledge Instruments II. X, 373 pages. 2006. (Sublibrary LNAI).

Vol. 4341: P.Q. Nguyen (Ed.), Progress in Cryptology - VIETCRYPT 2006. XI, 385 pages. 2006.

Vol. 4340: R. Prodan, T. Fahringer, Grid Computing. XXIII, 317 pages. 2007.

Vol. 4339: E. Ayguadé, G. Baumgartner, J. Ramanujam, P. Sadayappan (Eds.), Languages and Compilers for Parallel Computing. XI, 476 pages. 2006.

Vol. 4338: P. Kalra, S. Peleg (Eds.), Computer Vision, Graphics and Image Processing. XV, 965 pages. 2006.

Vol. 4337: S. Arun-Kumar, N. Garg (Eds.), FSTTCS 2006: Foundations of Software Technology and Theoretical Computer Science. XIII, 430 pages. 2006.

Vol. 4335: S.A. Brueckner, S. Hassas, M. Jelasity, D. Yamins (Eds.), Engineering Self-Organising Systems. XII, 212 pages. 2007. (Sublibrary LNAI).

Vol. 4334: B. Beckert, R. Hähnle, P.H. Schmitt (Eds.), Verification of Object-Oriented Software. XXIX, 658 pages. 2007. (Sublibrary LNAI).

Vol. 4333: U. Reimer, D. Karagiannis (Eds.), Practical Aspects of Knowledge Management. XII, 338 pages. 2006. (Sublibrary LNAI).

Vol. 4332: A. Bagchi, V. Atluri (Eds.), Information Systems Security. XV, 382 pages. 2006.

Vol. 4331: G. Min, B. Di Martino, L.T. Yang, M. Guo, G. Ruenger (Eds.), Frontiers of High Performance Computing and Networking – ISPA 2006 Workshops. XXXVII, 1141 pages. 2006.

Vol. 4330: M. Guo, L.T. Yang, B. Di Martino, H.P. Zima, J. Dongarra, F. Tang (Eds.), Parallel and Distributed Processing and Applications. XVIII, 953 pages. 2006.

Vol. 4329: R. Barua, T. Lange (Eds.), Progress in Cryptology - INDOCRYPT 2006. X, 454 pages. 2006.

Vol. 4328: D. Penkler, M. Reitenspiess, F. Tam (Eds.), Service Availability. X, 289 pages. 2006.

Vol. 4327: M. Baldoni, U. Endriss (Eds.), Declarative Agent Languages and Technologies IV. VIII, 257 pages. 2006. (Sublibrary LNAI).

Vol. 4326: S. Göbel, R. Malkewitz, I. Iurgel (Eds.), Technologies for Interactive Digital Storytelling and Entertainment. X, 384 pages. 2006.

Vol. 4325: J. Cao, I. Stojmenovic, X. Jia, S.K. Das (Eds.), Mobile Ad-hoc and Sensor Networks. XIX, 887 pages. 2006.

Vol. 4323: G. Doherty, A. Blandford (Eds.), Interactive Systems. XI, 269 pages. 2007.

Vol. 4320: R. Gotzhein, R. Reed (Eds.), System Analysis and Modeling: Language Profiles. X, 229 pages. 2006.

Vol. 4319: L.-W. Chang, W.-N. Lie (Eds.), Advances in Image and Video Technology. XXVI, 1347 pages. 2006.

Vol. 4318: H. Lipmaa, M. Yung, D. Lin (Eds.), Information Security and Cryptology. XI, 305 pages. 2006.

Vol. 4317: S.K. Madria, K.T. Claypool, R. Kannan, P. Uppuluri, M.M. Gore (Eds.), Distributed Computing and Internet Technology. XIX, 466 pages. 2006.

Vol. 4316: M.M. Dalkilic, S. Kim, J. Yang (Eds.), Data Mining and Bioinformatics. VIII, 197 pages. 2006. (Sublibrary LNBI).

Vol. 4314: C. Freksa, M. Kohlhase, K. Schill (Eds.), KI 2006: Advances in Artificial Intelligence. XII, 458 pages. 2007. (Sublibrary LNAI).

Vol. 4313: T. Margaria, B. Steffen (Eds.), Leveraging Applications of Formal Methods. IX, 197 pages. 2006.

Vol. 4312: S. Sugimoto, J. Hunter, A. Rauber, A. Morishima (Eds.), Digital Libraries: Achievements, Challenges and Opportunities. XVIII, 571 pages. 2006.

Vol. 4311: K. Cho, P. Jacquet (Eds.), Technologies for Advanced Heterogeneous Networks II. XI, 253 pages. 2006.

Vol. 4310: T. Boyanov, S. Dimova, K. Georgiev, G. Nikolov (Eds.), Numerical Methods and Applications. XIII, 715 pages. 2007.

Vol. 4309: P. Inverardi, M. Jazayeri (Eds.), Software Engineering Education in the Modern Age. VIII, 207 pages. 2006.

Vol. 4308: S. Chaudhuri, S.R. Das, H.S. Paul, S. Tirthapura (Eds.), Distributed Computing and Networking. XIX, 608 pages. 2006.

Vol. 4307: P. Ning, S. Qing, N. Li (Eds.), Information and Communications Security. XIV, 558 pages. 2006.

Vol. 4306: Y. Avrithis, Y. Kompatsiaris, S. Staab, N.E. O'Connor (Eds.), Semantic Multimedia. XII, 241 pages. 2006.

Vol. 4305: A.A. Shvartsman (Ed.), Principles of Distributed Systems. XIII, 441 pages. 2006.

Vol. 4304: A. Sattar, B.-H. Kang (Eds.), AI 2006: Advances in Artificial Intelligence. XXVII, 1303 pages. 2006. (Sublibrary LNAI).

Vol. 4303: A. Hoffmann, B.-H. Kang, D. Richards, S. Tsumoto (Eds.), Advances in Knowledge Acquisition and Management. XI, 259 pages. 2006. (Sublibrary LNAI).

Vol. 4302: J. Domingo-Ferrer, L. Franconi (Eds.), Privacy in Statistical Databases. XI, 383 pages. 2006.

Vol. 4301: D. Pointcheval, Y. Mu, K. Chen (Eds.), Cryptology and Network Security. XIII, 381 pages. 2006.

MIX
Papier aus verantwortungsvollen Quellen
Paper from responsible sources
FSC® C105338

If you have any concerns about our products,
you can contact us on
ProductSafety@springernature.com

In case Publisher is established outside the EU,
the EU authorized representative is:
Springer Nature Customer Service Center GmbH
Europaplatz 3, 69115 Heidelberg, Germany

Printed by Libri Plureos GmbH
in Hamburg, Germany